INSTRUCTOR'S SOLUTIONS MANUAL

IRVIN A. MILLER

PHYSICS

PRINCIPLES WITH APPLICATIONS

Fifth Edition

GIANCOLI

PRENTICE HALL, Upper Saddle River, NJ 07458

Assistant Editor: ***Wendy A. Rivers***
Production Editor: ***Mindy DePalma***
Special Projects Manager: ***Barbara A. Murray***
Supplement Cover Manager: ***Paul Gourhan***
Production Coordinator: ***Ben Smith***

A Pearson Education Company
Upper Saddle River, NJ 07458

Printed with corrections August 1999

Printed in the United States of America

10 9 8 7

ISBN 0-13-627985-6

Prentice-Hall International (UK) Limited, *London*
Prentice-Hall of Australia Pty. Limited, *Sydney*
Prentice-Hall Canada, Inc., *Toronto*
Prentice-Hall Hispanoamericana, S.A., *Mexico*
Prentice-Hall of India Private Limited, *New Delhi*
Prentice-Hall of Japan, Inc., *Tokyo*
Simon & Schuster Asia Pte. Ltd., *Singapore*
Editora Prentice-Hall do Brasil, Ltda., *Rio de Janeiro*

PREFACE

This Solution Manual contains worked-out solutions to all the problems in *Physics: Principles with Applications*, 5th Edition, by Douglas C. Giancoli. At the end of the solutions is a list of answers to the even-numbered problems, except for any required figures, as well as grids which relate the problems from the 4th edition with the problems of the 5th edition.

I have tried to aim the presentation for the new instructor or graduate teaching assistant, which assumes some understanding of the physics and some mathematical sophistication. The emphasis is on presenting the application of physical concepts that are appropriate to arrive at a solution of a particular problem. The best pedagogical approach has been given primary consideration. Details of mathematical manipulations that are necessary to obtain a numerical answer, such as solving two equations simultaneously, are generally not included. Occasionally it is necessary or convenient to solve an equation numerically. In such cases, the use of a spreadsheet and/or computer graphing capability is helpful. Comments have been added to some problems to indicate extensions or variations.

Many people have been involved with the production of this manual. I want to thank Doug Giancoli for his helpful conversations, via both human voice and electronic voice. Among the staff at Prentice Hall, Wendy Rivers has been extremely helpful in guiding me through the process of turning thoughts into a publication.

Much effort has been put into having accurate answers for the problems. I want to acknowledge the benefit of having solutions from the previous edition, prepared by Bo Lou. His approach to many of the problems has been used or adapted here and he also provided answer checking for this edition. David Curott also provided answer checking and our back and forth e-mail discussions of some of the problems were much appreciated.

Even with all the assistance of others, the final responsibility for the content of this manual is mine. I have tried to ensure that the physics and the answers are correct.

I hope that you will find the presentation of solutions useful and, in at least some cases, thought provoking.

Irv Miller
Philadelphia, PA

CONTENTS

CHAPTER 1

1. (*a*) Assuming one significant figure, we have
   10 billion yr = 10×10^9 yr = $\boxed{1 \times 10^{10}\text{ yr.}}$
   (*b*) $(1 \times 10^{10}\text{ yr})(3 \times 10^7\text{ s/yr}) = \boxed{3 \times 10^{17}\text{ s.}}$

2. (*a*) $8.69 \times 10^4 = \boxed{86{,}900.}$
   (*b*) $7.1 \times 10^3 = \boxed{7{,}100.}$
   (*c*) $6.6 \times 10^{-1} = \boxed{0.66.}$
   (*d*) $8.76 \times 10^2 = \boxed{876.}$
   (*e*) $8.62 \times 10^{-5} = \boxed{0.000\,086\,2.}$

3. (*a*) Assuming the zeros are not significant, we have $1{,}156{,}000 = \boxed{1.156 \times 10^6.}$
   (*b*) $218 = \boxed{2.18 \times 10^2.}$
   (*c*) $0.0068 = \boxed{6.8 \times 10^{-3}.}$
   (*d*) $27.635 = \boxed{2.7635 \times 10^1.}$
   (*e*) $0.21 = \boxed{2.1 \times 10^{-1}.}$
   (*f*) $22 = \boxed{2.2 \times 10^1.}$

4. (*a*) $\boxed{\text{3 significant figures.}}$
   (*b*) Because the zero is not needed for placement, we have $\boxed{\text{4 significant figures.}}$
   (*c*) $\boxed{\text{3 significant figures.}}$
   (*d*) Because the zeros are for placement only, we have $\boxed{\text{1 significant figure.}}$
   (*e*) Because the zeros are for placement only, we have $\boxed{\text{2 significant figures.}}$
   (*f*) $\boxed{\text{4 significant figures.}}$
   (*g*) $\boxed{\text{2, 3, or 4 significant figures,}}$ depending on the significance of the zeros.

5. % uncertainty = [(0.25 m)/(2.26 m)] 100 = $\boxed{11\%.}$
   Because the uncertainty has 2 significant figures, the % uncertainty has 2 significant figures.

6. We assume an uncertainty of 1 in the last place, i. e., 0.01, so we have
   % uncertainty = [(0.01)/(1.67)] 100 = $\boxed{0.6\%.}$
   Because the uncertainty has 1 significant figure, the % uncertainty has 1 significant figure.

7. We assume an uncertainty of 0.5 s.
   (*a*) % uncertainty = [(0.5 s)/(5 s)] 100 = $\boxed{1 \times 10^1\%.}$
   Because the uncertainty has 1 significant figure, the % uncertainty has 1 significant figure.
   (*b*) % uncertainty = [(0.5 s)/(50 s)] 100 = $\boxed{1\%.}$
   Because the uncertainty has 1 significant figure, the % uncertainty has 1 significant figure.
   (*c*) % uncertainty = [(0.5 s)/(5 min)(60 s/min)] 100 = $\boxed{0.2\%.}$
   Because the uncertainty has 1 significant figure, the % uncertainty has 1 significant figure.

8. For multiplication, the number of significant figures in the result is the least number from the multipliers; in this case 2 from the second value.
   $(2.079 \times 10^2\text{ m})(0.072 \times 10^{-1}) = 0.15 \times 10^1\text{ m} = \boxed{1.5\text{ m.}}$

9. To add, we make all of the exponents the same:

$$7.2\times10^3\text{ s} + 8.3\times10^4\text{ s} + 0.09\times10^6\text{ s} = 0.72\times10^4\text{ s} + 8.3\times10^4\text{ s} + 9\times10^4\text{ s}$$
$$= 18.02\times10^4\text{ s} = \boxed{1.8\times10^5\text{ s.}}$$

Because we are adding, the location of the uncertain figure for the result is the one furthest to the left. In this case, it is fixed by the third value.

10. We assume an uncertainty of 0.1×10^4 cm. We compare the area for the specified radius to the area for the extreme radius.

$$A_1 = \pi R_1^2 = \pi(2.8\times10^4\text{ cm})^2 = 2.46\times10^9\text{ cm}^2\,; \quad A_2 = \pi R_2^2 = \pi[(2.8 + 0.1)\times10^4\text{ cm}]^2 = 2.64\times10^9\text{ cm}^2,$$

so the uncertainty in the area is $\Delta A = A_2 - A_1 = 0.18\times10^9\text{ cm}^2 = 0.2\times10^9\text{ cm}^2$.

We write the area as $\boxed{A = (2.5 \pm 0.2)\times10^9\text{ cm}^2.}$

11. We compare the volume with the specified radius to the volume for the extreme radius.

$$V_1 = \tfrac{4}{3}\pi R_1^3 = \tfrac{4}{3}\pi(3.86\text{ m})^3 = 241\text{ m}^3; \quad V_2 = \tfrac{4}{3}\pi R_2^3 = \tfrac{4}{3}\pi(3.86\text{ m} + 0.08\text{ m})^3 = 256\text{ m}^3,$$

so the uncertainty in the volume is $\Delta V = V_2 - V_1 = 15\text{ m}^3$; and the % uncertainty is

$$\%\text{ uncertainty} = [(15\text{ m}^3)/(241\text{ m}^3)]\,100 = \boxed{6\%.}$$

12. (*a*) 10^6 volts = 1 megavolt = $\boxed{\text{1 Mvolt.}}$
(*b*) 10^{-6} meters = 1 micrometer = $\boxed{1\ \mu\text{m.}}$
(*c*) 5×10^3 days = 5 kilodays = $\boxed{\text{5 kdays.}}$
(*d*) 8×10^2 bucks = 8 hectobucks = $\boxed{\text{0.8 kbucks.}}$
(*e*) 8×10^{-9} pieces = 8 nanopieces = $\boxed{\text{8 npieces.}}$

13. (*a*) 86.6 mm = 86.6×10^{-3} m = $\boxed{\text{0.086 6 m.}}$
(*b*) 35 μV = 35×10^{-6} V = $\boxed{\text{0.000 035 V.}}$
(*c*) 860 mg = 860×10^{-3} g = $\boxed{\text{0.860 g.}}$ This assumes that the last zero is significant.
(*d*) 600 picoseconds = 600×10^{-12} s = $\boxed{\text{0.000 000 000 600 s.}}$ This assumes that both zeros are significant.
(*e*) 12.5 femtometers = 12.5×10^{-15} m = $\boxed{\text{0.000 000 000 000 012 5 m.}}$
(*f*) 250 gigavolts = 250×10^9 volts = $\boxed{\text{250,000,000,000 volts.}}$

14. 50 hectokisses = 50×10^2 kisses = $\boxed{\text{5,000 kisses.}}$
1 megabuck/yr = 1×10^6 bucks/yr = 1,000,000 bucks/yr $\boxed{\text{(millionaire).}}$

15. If we assume a height of 5 ft 10 in, we have

height = 5 ft 10 in = 70 in = (70 in)[(1 m)/(39.37 in)] = $\boxed{\text{1.8 m.}}$

16. (*a*) 93 million mi = 93×10^6 mi = $(93\times10^6$ mi)[(1610 m)/(1 mi)] = $\boxed{1.5\times10^{11}\text{ m.}}$
(*b*) 1.5×10^{11} m = 150×10^9 m = $\boxed{\text{150Gm.}}$

17. (*a*) 1.0×10^{-10} m = $(1.0\times10^{-10}$ m)[(1 in)/(2.54 cm)][(100 cm)/(1 m)] = $\boxed{3.9\times10^{-9}\text{ in.}}$
(*b*) We let the units lead us to the answer:

(1.0 cm)[(1 m)/(100 cm)][(1 atom)/(1.0×10^{-10} m)] = $\boxed{1.0\times10^8\text{ atoms.}}$

18. To add, we make all of the units the same:

$$1.00\text{ m} + 142.5\text{ cm} + 1.24\times10^5\ \mu\text{m} = 1.00\text{ m} + 1.425\text{ m} + 0.124\text{ m}$$
$$= 2.549\text{ m} = \boxed{2.55\text{ m.}}$$

Because we are adding, the location of the uncertain figure for the result is the one furthest to the left. In this case, it is fixed by the first value.

19. (*a*) $1\ \text{km/h} = (1\ \text{km/h})[(0.621\ \text{mi})/(1\ \text{km})] =$ $\boxed{0.621\ \text{mi/h.}}$
(*b*) $1\ \text{m/s} = (1\ \text{m/s})[(1\ \text{ft})/(0.305\ \text{m})] =$ $\boxed{3.28\ \text{ft/s.}}$
(*c*) $1\ \text{km/h} = (1\ \text{km/h})[(1000\ \text{m})/(1\ \text{km})][(1\ \text{h})/(3600\ \text{s})] =$ $\boxed{0.278\ \text{m/s.}}$
A useful alternative is
$$1\ \text{km/h} = (1\ \text{km/h})[(1\ \text{h})/(3.600\ \text{ks})] = 0.278\ \text{m/s.}$$

20. For the length of a one-mile race in m, we have
$$1\ \text{mi} = (1\ \text{mi})[(1610\ \text{m})/(1\ \text{mi})] = 1610\ \text{m};$$ so the difference is 110 m.
The % difference is
$$\%\ \text{difference} = [(110\ \text{m})/(1500\ \text{m})]\ 100 = \boxed{7.3\%.}$$

21. (*a*) $1.00\ \text{ly} = (2.998 \times 10^8\ \text{m/s})(1.00\ \text{yr})[(365.25\ \text{days})/(1\ \text{yr})][(24\ \text{h})/(1\ \text{day})][(3600\ \text{s})/(1\ \text{h})]$
$= \boxed{9.46 \times 10^{15}\ \text{m.}}$
(*b*) $1.00\ \text{ly} = (9.46 \times 10^{15}\ \text{m})[(1\ \text{AU})/(1.50 \times 10^8\ \text{km})][(1\ \text{km})/(1000\ \text{m})] =$ $\boxed{6.31 \times 10^4\ \text{AU.}}$
(*c*) speed of light $= (2.998 \times 10^8\ \text{m/s})[(1\ \text{AU})/(1.50 \times 10^8\ \text{km})][(1\ \text{km})/(1000\ \text{m})][(3600\ \text{s})/(1\ \text{h})]$
$= \boxed{7.20\ \text{AU/h.}}$

22. For the surface area of a sphere, we have
$$A_{\text{moon}} = 4\pi r_{\text{moon}}^2 = 4\pi[\tfrac{1}{2}(3.48 \times 10^6\ \text{m})]^2 = \boxed{3.80 \times 10^{13}\ \text{m}^2.}$$
We compare this to the area of the earth by finding the ratio:
$$A_{\text{moon}}/A_{\text{Earth}} = 4\pi r_{\text{moon}}^2 / 4\pi r_{\text{Earth}}^2 = (r_{\text{moon}}/r_{\text{Earth}})^2 = [(1.74 \times 10^3\ \text{km})/(6.38 \times 10^3\ \text{km})]^2 = 7.44 \times 10^{-2}.$$
Thus we have
$$\boxed{A_{\text{moon}} = 7.44 \times 10^{-2}\ A_{\text{Earth}}.}$$

23. (*a*) $7800 = 7.8 \times 10^3 \approx 10 \times 10^3 =$ $\boxed{10^4.}$
(*b*) $9.630 \times 10^2 \approx 10 \times 10^2 =$ $\boxed{10^3.}$
(*c*) $0.00076 = 0.76 \times 10^{-3} \approx$ $\boxed{10^{-3}.}$
(*d*) $150 \times 10^8 = 1.50 \times 10^{10} \approx$ $\boxed{10^{10}.}$

24. We assume that a good runner can run 6 mi/h (equivalent to a 10-min mile) for 5 h/day. Using 3000 mi for the distance across the U. S., we have
$$\text{time} = (3000\ \text{mi})/(6\ \text{mi/h})(5\ \text{h/day}) \approx \boxed{100\ \text{days.}}$$

25. We assume a rectangular house 40 ft × 30 ft, 8 ft high; so the total wall area is
$$A_{\text{total}} = [2(40\ \text{ft}) + 2(30\ \text{ft})](8\ \text{ft}) \approx 1000\ \text{ft}^2.$$
If we assume there are 12 windows with dimensions 3 ft × 5 ft, the window area is
$$A_{\text{window}} = 12(3\ \text{ft})(5\ \text{ft}) \approx 200\ \text{ft}^2.$$
Thus we have
$$\%\ \text{window area} = [A_{\text{window}}/A_{\text{total}}](100) = [(200\ \text{ft}^2)/(1000\ \text{ft}^2)](100) \approx \boxed{20\%.}$$

26. If we take an average lifetime to be 70 years and the average pulse to be 60 beats/min, we have
$$N = (60\ \text{beats/min})(70\ \text{yr})(365\ \text{day/yr})(24\ \text{h/day})(60\ \text{min/h}) \approx \boxed{2 \times 10^9\ \text{beats.}}$$

27. If we approximate the body as a box with dimensions 6 ft, 1 ft, and 8 in, we have
$$\text{volume} = (72\ \text{in})(12\ \text{in})(8\ \text{in})(2.54\ \text{cm/in})^3 \approx \boxed{1 \times 10^5\ \text{cm}^3.}$$

28. We assume the distance from Beijing to Paris is 10,000 mi.
   (*a*) If we assume that today's race car can travel for an extended period at an average speed of 40 mi/h, we have
   time = [(10,000 mi)/(40 mi/h)](1 day/24 h) ≈ **10 days.**
   (*b*) If we assume that in 1906 a race car could travel for an extended period at an average speed of 5 mi/h, we have
   time = [(10,000 mi)/(5 mi/h)](1 day/24 h) ≈ **80 days.**

29. We assume that 12 patients visit a dentist each day. If a dentist works 8 h/day, 5 days/wk for 48 wk/yr, the total number of visits per year for a dentist is
   n_{dentist} = (12 visits/day)(5 days/wk)(48 wk/yr) ≈ 3000 visits/yr.
   We assume that each person visits the dentist 2 times/yr.
   (*a*) We assume that the population of San Francisco is 700,000. We let the units lead us to the answer:
   N = (700,000 persons)(2 visits/yr)/(3000 visits/yr·person) ≈ **500 dentists.**
   (*b*) Left to the reader to estimate the population.

30. If we assume that the person can mow at a speed of 1 m/s and the width of the mower cut is 0.5 m, the rate at which the field is mown is (1 m/s)(0.5 m) = 0.5 m^2/s.
   If we take the dimensions of the field to be 110 m by 50 m, we have
   time = [(110 m)(50 m)/(0.5 m^2/s)]/(3600 s/h) **≈ 3 h.**

31. We assume an average time of 3 yr for the tire to wear d = 1 cm and the tire has a radius of r = 30 cm and a width of w = 10 cm. Thus the volume of rubber lost by a tire in 3 yr is $V = wd2\pi r$. If we assume there are 100 million vehicles, each with 4 tires, we have
   m = (100 × 10^6 vehicles)(4 tires/vehicle)(0.1 m)(0.01 m)2π(0.3 m)(1200 kg/m^3)/(3 yr)
   ≈ 3 × 10^8 kg/yr.

32. (*a*) 1.0 Å = (1.0 × 10^{-10} m)/(10^{-9} m/nm) = **0.10 nm.**
   (*b*) 1.0 Å = (10^{-10} m)/(10^{-15} m/fm) = **1.0 × 10^5 fm.**
   (*c*) 1.0 m = (1.0 m)/(10^{-10} m/Å) = **1.0 × 10^{10} Å.**
   (*d*) From the result for Problem 21, we have
   1.0 ly = (9.5 × 10^{15} m)/(10^{-10} m/Å) = **9.5 × 10^{25} Å.**

33. (*a*) 1.00 yr = (365.25 days)(24 h/day)(3600 s/h) = **3.16 × 10^7 s.**
   (*b*) 1.00 yr = (3.16 × 10^7 s)/(10^{-9} s/ns) = **3.16 × 10^{16} ns.**
   (*c*) 1.00 s = (1.00 s)/(3.16 × 10^7 s/yr) = **3.17 × 10^{-8} yr.**

34. (*a*) The maximum number of buses is needed during rush hour. If we assume that at any time there are 40,000 persons commuting by bus and each bus has 30 passengers, we have
   N = (40,000 commuters)/(30 passengers/bus) ≈ 1000 buses ≈ **1,000 drivers.**
   (*b*) Left to the reader.

35. If we ignore any loss of material from the slicing, we find the number of wafers from
   (30 cm)(10 mm/cm)/(0.60 mm/wafer) = 500 wafers.
   For the maximum number of chips, we have
   (500 wafers)(100 chips/wafer) = **50,000 chips.**

36. If we assume there is 1 automobile for 2 persons and a U. S. population of 250 million, we have 125 million automobiles. We estimate that each automobile travels 15,000 miles in a year and averages 20 mi/gal. Thus we have
$N = (125 \times 10^6 \text{ automobiles})(15{,}000 \text{ mi/yr})/(20 \text{ mi/gal}) \approx$ $\boxed{1 \times 10^{11} \text{ gal/yr.}}$

37. We let D represent the diameter of a gumball. Because there are air gaps around the gumballs, we estimate the volume occupied by a gumball as a cube with volume D^3. The machine has a square cross-section with sides equivalent to 10 gumballs and is about 14 gumballs high, so we have
$N = \text{volume of machine/volume of gumball} = (14D)(10D)^2/D^3 \approx$ $\boxed{1.4 \times 10^3 \text{ gumballs.}}$

38. The volume used in one year is
$V = [(40{,}000 \text{ persons})/(4 \text{ persons/family})](1200 \text{ L/family}\cdot\text{day})(365 \text{ days/yr})(10^{-3} \text{ m}^3/\text{L})$
$\approx 4 \times 10^6 \text{ m}^3.$
If we let d represent the loss in depth, we have
$d = V/\text{area} = (4 \times 10^6 \text{ m}^3)/(50 \text{ km}^2)(10^3 \text{ m/km})^2 \approx 0.09 \text{ m} \approx$ $\boxed{9 \text{ cm.}}$

39. For the volume of a 1-ton rock, we have
$V = (2000 \text{ lb})/(3)(62 \text{ lb/ft}^3) \approx 11 \text{ ft}^3.$
If we assume the rock is a sphere, we find the radius from
$11 \text{ ft}^3 = \frac{4}{3}\pi r^3$, which gives $r \approx 1.4$ ft, so the diameter would be $\boxed{\approx 3 \text{ ft} \approx 1 \text{ m.}}$

40. We find the amount of water from its volume:
$m = (5 \text{ km})(8 \text{ km})(1.0 \text{ cm})(10^5 \text{ cm/km})^2(10^{-3} \text{ kg/cm}^3)/(10^3 \text{ kg/t}) = 40 \times 10^4 \text{ t} \approx$ $\boxed{4 \times 10^5 \text{ t.}}$

41. We will use a pencil with a diameter of 5 mm and assume that it is held 0.5 m from the eye. Because the triangles AOD and BOC are similar, we can equate the ratio of distances:
BC/AD = OQ/OP;
$\text{BC}/(0.005 \text{ m}) = (3.8 \times 10^5 \text{ km})/(0.5 \text{ m})$,
which gives
BC $\boxed{\approx 4 \times 10^3 \text{ km.}}$

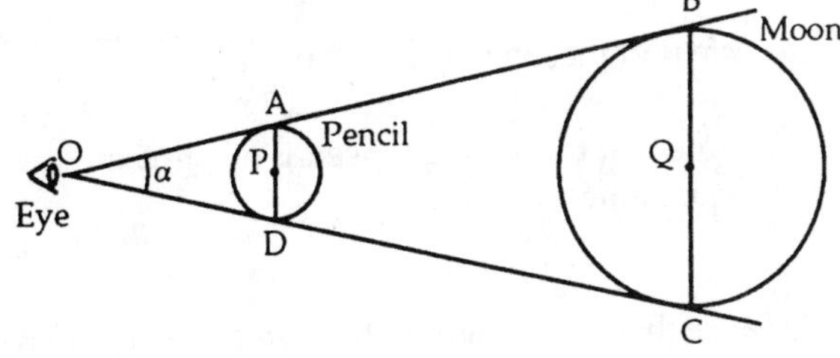

42. (*a*) $V = (1{,}000 \text{ m}^3)(10^2 \text{ cm/m})^3 =$ $\boxed{1.00 \times 10^9 \text{ cm}^3.}$
(*b*) $V = (1{,}000 \text{ m}^3)(3.28 \text{ ft/m})^3 =$ $\boxed{3.53 \times 10^4 \text{ ft}^3.}$
(*c*) $V = (1{,}000 \text{ m}^3)(10^2 \text{ cm/m})^3/(2.54 \text{ cm/in})^3 =$ $\boxed{6.10 \times 10^7 \text{ in}^3.}$

43. We assume that we can walk an average of 15 miles a day. If we ignore the impossibility of walking on water and travel around the equator, the time required is
$\text{time} = 2\pi R_{\text{Earth}}/\text{speed} = 2\pi(6 \times 10^3 \text{ km})(0.621 \text{ mi/km})/(15 \text{ mi/day})(365 \text{ days/yr})$ $\boxed{\approx 4 \text{ yr.}}$

44. If we use 0.5 m for the cubit, for the dimensions we have
$\boxed{\text{150 m long, 25 m wide, and 15 m high.}}$

CHAPTER 2

1. We find the average speed from
   average speed = d/t = (230 km)/(3.25 h) = **70.8 km/h.**

2. We find the time from
   average speed = d/t;
   25 km/h = (15 km)/t , which gives **t = 0.60 h (36 min).**

3. We find the distance traveled from
   average speed = d/t;
   (110 km/h)/(3600 s/h) = d/(2.0 s), which gives **d= 6.1 ×10^{-2} km = 61 m.**

4. (*a*) 65 mi/h = (65 mi/h)(1.61 km/1 mi)= **105 km/h.**
   (*b*) 65 mi/h = (65 mi/h)(1610 m/1 mi)/(3600 s/1 h) = **29 m/s.**
   (*c*) 65 mi/h = (65 mi/h)(5280 ft/1 mi)/(3600 s/1 h) = **95 ft/s.**

5. (*a*) We find the elapsed time before the speed change from
   speed = d_1/t_1 ;
   65 mi/h = (130 mi)/t_1 , which gives t_1 = 2.0 h.
   Thus the time at the lower speed is
   $t_2 = T - t_1$ = 3.33 h – 2.0 h = 1.33 h.
   We find the distance traveled at the lower speed from
   speed = d_2/t_2 ;
   55 mi/h = d_2/(1.33 h), which gives d_2 = 73 mi.
   The total distance traveled is
   $D = d_1 + d_2$ = 130 mi + 73 mi = **203 mi.**
   (*b*) We find the average speed from
   average speed = d/t = (203 mi)/(3.33 h) = **61 mi/h.**
   Note that the average speed is <u>not</u> $\frac{1}{2}$(65 mi/h + 55 mi/h). The two speeds were not maintained for equal times.

6. Because there is no elapsed time when the light arrives, the sound travels one mile in 5 seconds. We find the speed of sound from
   speed = d/t = (1 mi)(1610 m/1 mi)/(5 s) **≈ 300 m/s.**

7. (*a*) We find the average speed from
   average speed = d/t = 8(0.25 mi)(1610 m/1 mi)/(12.5 min)(60 s/min) = **4.29 m/s.**
   (*b*) Because the person finishes at the starting point, there is no displacement; thus the average velocity is
   $\bar{v} = \Delta x/\Delta t$ = **0.**

8. (*a*) We find the average speed from
   average speed = d/t = (130 m + 65 m)/(14.0 s + 4.8 s) = **10.4 m/s.**
   (*b*) The displacement away from the trainer is 130 m – 65 m = 65 m; thus the average velocity is
   $\bar{v} = \Delta x/\Delta t$ = (65 m)/(14.0 s + 4.8 s) = **+ 3.5 m/s.**

9. Because the two locomotives are traveling with equal speeds in opposite directions, each locomotive will travel half the distance, 4.25 km. We find the elapsed time from

speed = d_1/t_1 ;
(95 km/h)/(60 min/h) = (4.25 km)/t, which gives **t = 2.7 min.**

10. We find the total time for the trip by adding the times for each leg:

$T = t_1 + t_2 = (d_1/v_1) + (d_2/v_2)$
= [(2100 km)/(800 km/h)] + [(1800 km)/(1000 km/h)] = **4.43 h.**

We find the average speed from

average speed = $(d_1 + d_2)/T$ = (2100 km + 1800 km)/(4.43 h) = **881 km/h.**

Note that the average speed is not $\frac{1}{2}$(800 km/h + 1000 km/h). The two speeds were not maintained for equal times.

11. We find the time for the outgoing 200 km from

$t_1 = d_1/v_1$ = (200 km)/(90 km/h) = 2.22 h.

We find the time for the return 200 km from

$t_2 = d_2/v_2$ = (200 km)/(50 km/h) = 4.00 h.

We find the average speed from

average speed = $(d_1 + d_2)/(t_1 + t_{lunch} + t_2)$
= (200 km + 200 km)/(2.22 h + 1.00 h + 4.00 h) = **55 km/h.**

Because the trip finishes at the starting point, there is no displacement; thus the average velocity is

$\bar{v} = \Delta x/\Delta t$ = **0.**

12. We find the time for the sound to travel the length of the lane from

$t_{sound} = d/v_{sound}$ = (16.5 m)/(340 m/s) = 0.0485 s.

We find the speed of the bowling ball from

$v = d/(T - t_{sound})$
= (16.5 m)/(2.50 s – 0.0485 s) = **6.73 m/s.**

13. We find the average acceleration from

$\bar{a} = \Delta v/\Delta t$
= [(95 km/h)(1 h/3.6 ks) – 0]/(6.2 s) = **4.3 m/s^2.**

14. We find the time from

$\bar{a} = \Delta v/\Delta t$;
1.6 m/s^2 = (110 km/h – 80 km/h)(1 h/3.6 ks)/Δt, which gives **Δt = 5.2 s**

15. (*a*) We find the acceleration from

$\bar{a} = \Delta v/\Delta t$
= (10 m/s – 0)/(1.35 s) = **7.41 m/s^2.**

(*b*) $\bar{a}$ = (7.41 m/s^2)(1 km/1000m)(3600 s/1h)^2 = **9.60 × 10^4 km/h^2.**

16. We find the acceleration (assumed to be constant) from

$v^2 = v_0^2 + 2a(x_2 - x_1)$;
0 = [(90 km/h)/(3.6 ks/h)]^2 + 2a(50 m), which gives **a = – 6.3 m/s^2.**

The number of g's is

$N = |a|/g$ = (6.3 m/s^2)/(9.80 m/s^2) = **0.64.**

17. (*a*) We take the average velocity during a time interval as the instantaneous velocity at the midpoint of the time interval:

$v_{\text{midpoint}} = \bar{v} = \Delta x/\Delta t.$

Thus for the first interval we have

$v_{0.125\text{ s}} = (0.11\text{ m} - 0)/(0.25\text{ s} - 0) = 0.44\text{ m/s}.$

(*b*) We take the average acceleration during a time interval as the instantaneous acceleration at the midpoint of the time interval:

$a_{\text{midpoint}} = \bar{a} = \Delta v/\Delta t.$

Thus for the first interval in the velocity column we have

$a_{0.25\text{ s}} = (1.4\text{ m/s} - 0.44\text{ m/s})/(0.375\text{ s} - 0.125\text{ s}) = 3.8\text{ m/s}^2.$

The results are presented in the following table and graph.

t(s)	x(m)
0.0	0.0
0.25	0.11
0.50	0.46
0.75	1.06
1.00	1.94
1.50	4.62
2.00	8.55
2.50	13.79
3.00	20.36
3.50	28.31
4.00	37.65
4.50	48.37
5.00	60.30
5.50	73.26
6.00	87.16

t(s)	v(m/s)
0.0	0.0
0.125	0.44
0.375	1.4
0.625	2.4
0.875	3.5
1.25	5.36
1.75	7.85
2.25	10.5
2.75	13.1
3.25	15.9
3.75	18.7
4.25	21.4
4.75	23.9
5.25	25.9
5.75	27.8

t(s)	a(m/s²)
0.25	3.8
0.50	4.0
0.75	4.5
1.06	4.9
1.50	5.0
2.00	5.2
2.50	5.3
3.00	5.5
3.50	5.6
4.00	5.5
4.50	4.8
5.00	4.1
5.50	3.8

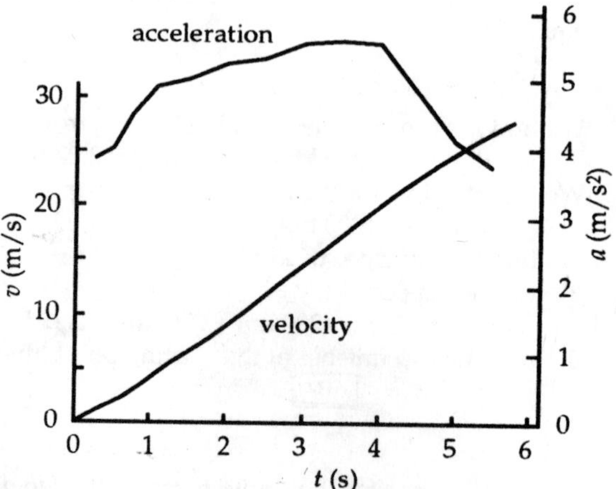

Note that we do not know the acceleration at $t = 0$.

18. When $x_0 = 0$ and $v_0 = 0$, we see that

$v = v_0 + at$ becomes $v = at$;

$x = x_0 + v_0t + \frac{1}{2}at^2$ becomes $x = \frac{1}{2}at^2$;

$v^2 = v_0^2 + 2a(x - x_0)$ becomes $v^2 = 2ax$; and

$\bar{v} = \frac{1}{2}(v + v_0)$ becomes $\bar{v} = \frac{1}{2}v$.

19. We find the acceleration from

$v = v_0 + a(t - t_0);$

$25\text{ m/s} = 12\text{ m/s} + a(6.0\text{ s})$, which gives $a = 2.2\text{ m/s}^2.$

We find the distance traveled from

$x = \frac{1}{2}(v + v_0)t$

$= \frac{1}{2}(25\text{ m/s} + 12\text{ m/s})(6.0\text{ s}) = 1.1 \times 10^2\text{ m}.$

20. We find the acceleration (assumed constant) from

$v^2 = v_0^2 + 2a(x_2 - x_1);$

$0 = (20\text{ m/s})^2 + 2a(85\text{ m})$, which gives $a = -2.4\text{ m/s}^2.$

21. We find the length of the runway from

$v^2 = v_0^2 + 2aL;$

$(30\text{ m/s})^2 = 0 + 2(3.0\text{ m/s}^2)L$, which gives $L = 1.5 \times 10^2\text{ m}.$

22. We find the average acceleration from

$v^2 = v_0^2 + 2\bar{a}(x_2 - x_1)$;

$(11.5\text{ m/s})^2 = 0 + 2\bar{a}(15.0\text{ m})$, which gives $\boxed{\bar{a} = 4.41\text{ m/s}^2.}$

We find the time required from

$x = \frac{1}{2}(v + v_0)t$;

$15.0\text{ m} = \frac{1}{2}(11.5\text{ m/s} + 0)t$, which gives $\boxed{t = 2.61\text{ s.}}$

23. For an assumed constant acceleration the average speed is $\frac{1}{2}(v + v_0)$, thus

$x = \frac{1}{2}(v + v_0)t$:

$= \frac{1}{2}(0 + 25.0\text{ m/s})(5.00\text{ s}) = \boxed{62.5\text{ m.}}$

24. We find the speed of the car from

$v^2 = v_0^2 + 2a(x_1 - x_0)$;

$0 = v_0^2 + 2(-7.00\text{ m/s}^2)(80\text{ m})$, which gives $\boxed{v_0 = 33\text{ m/s.}}$

25. We convert the units for the speed: $(45\text{ km/h})/(3.6\text{ ks/h}) = 12.5\text{ m/s}$.

(*a*) We find the distance the car travels before stopping from

$v^2 = v_0^2 + 2a(x_1 - x_0)$;

$0 = (12.5\text{ m/s})^2 + 2(-0.50\text{ m/s}^2)(x_1 - x_0)$, which gives $\boxed{x_1 - x_0 = 1.6\times10^2\text{ m.}}$

(*b*) We find the time it takes to stop the car from

$v = v_0 + at$;

$0 = 12.5\text{ m/s} + (-0.50\text{ m/s}^2)t$, which gives $\boxed{t = 25\text{ s.}}$

(*c*) With the origin at the beginning of the coast, we find the position at a time *t* from

$x = v_0t + \frac{1}{2}at^2$. Thus we find

$x_1 = (12.5\text{ m/s})(1.0\text{ s}) + \frac{1}{2}(-0.50\text{ m/s}^2)(1.0\text{ s})^2 = 12\text{ m}$;

$x_4 = (12.5\text{ m/s})(4.0\text{ s}) + \frac{1}{2}(-0.50\text{ m/s}^2)(4.0\text{ s})^2 = 46\text{ m}$;

$x_5 = (12.5\text{ m/s})(5.0\text{ s}) + \frac{1}{2}(-0.50\text{ m/s}^2)(5.0\text{ s})^2 = 56\text{ m}$.

During the first second the car travels $12\text{ m} - 0 = \boxed{12\text{ m.}}$

During the fifth second the car travels $56\text{ m} - 46\text{ m} = \boxed{10\text{ m.}}$

26. We find the average acceleration from

$v^2 = v_0^2 + 2\bar{a}(x_2 - x_1)$;

$0 = [(90\text{ km/h})/(3.6\text{ ks/h})]^2 + 2\bar{a}(0.80\text{ m})$, which gives $\boxed{\bar{a} = -3.9\times10^2\text{ m/s}^2.}$

The number of *g*'s is

$|\bar{a}| = (3.9\times10^2\text{ m/s}^2)/[(9.80\text{ m/s}^2)/g] = \boxed{40g.}$

27. We convert the units for the speed: $(90\text{ km/h})/(3.6\text{ ks/h}) = 25\text{ m/s}$.

With the origin at the beginning of the reaction, the location when the brakes are applied is

$x_0 = v_0t = (25\text{ m/s})(1.0\text{ s}) = 25\text{ m}$.

(*a*) We find the location of the car after the brakes are applied from

$v^2 = v_0^2 + 2a_1(x_1 - x_0)$;

$0 = (25\text{ m/s})^2 + 2(-4.0\text{ m/s}^2)(x_1 - 25\text{ m})$, which gives $\boxed{x_1 = 103\text{ m.}}$

(*b*) We repeat the calculation for the new acceleration:

$v^2 = v_0^2 + 2a_2(x_2 - x_0)$;

$0 = (25\text{ m/s})^2 + 2(-8.0\text{ m/s}^2)(x_2 - 25\text{ m})$, which gives $\boxed{x_2 = 64\text{ m.}}$

28. With the origin at the beginning of the reaction, the location when the brakes are applied is
    $d_0 = v_0 t_R$.
    We find the location of the car after the brakes are applied, which is the total stopping distance, from
    $v^2 = 0 = v_0^2 + 2a(d_S - d_0)$, which gives $d_S = v_0 t_R - v_0^2/(2a)$.
    Note that a is negative.

29. We convert the units:
    (120 km/h)/(3.6 ks/h) = 33.3 m/s.
    (10.0 km/h/s)/(3.6 ks/h) = 2.78 m/s^2.
    We use a coordinate system with the origin where the motorist passes the police officer, as shown in the diagram. 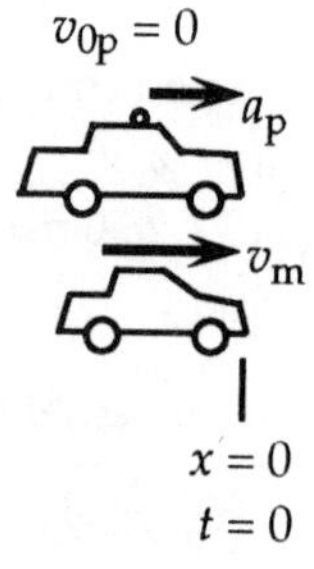
    The location of the speeding motorist is given by
    $x_m = x_0 + v_m t = 0 + (33.3 \text{ m/s})t$.
    The location of the police officer is given by
    $x_p = x_0 + v_{0p} t + \frac{1}{2}a_p t^2 = 0 + 0 + \frac{1}{2}(2.78 \text{ m/s}^2)t^2$.
    The officer will reach the speeder when these locations coincide, so we have
    $x_p = x_m$;
    $(33.3 \text{ m/s})t = \frac{1}{2}(2.78 \text{ m/s}^2)t^2$, which gives $t = 0$ (the original passing) and **$t = 24.0$ s.**
    We find the speed of the officer from
    $v_p = v_{0p} + a_p t$;
    $= 0 + (2.78 \text{ m/s}^2)(24.0 \text{ s}) = 66.7 \text{ m/s} =$ **240 km/h** (about 150 mi/h!).

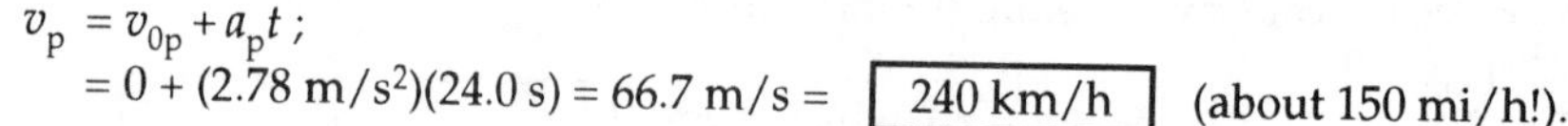

30. We use a coordinate system with the origin where the initial action takes place, as shown in the diagram. 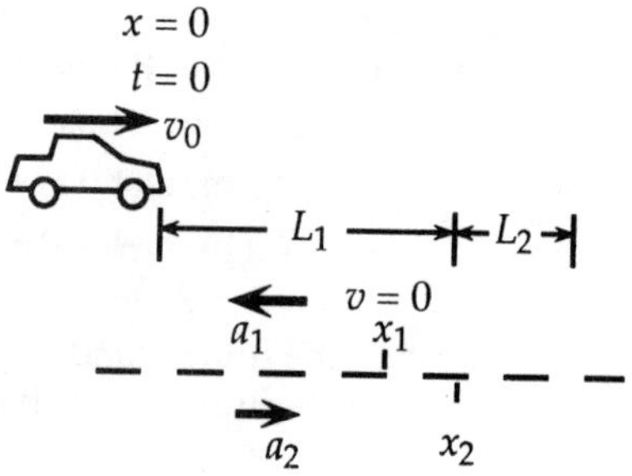
    The initial speed is (50 km/h)/(3.6 ks/h) = 13.9 m/s.
    If she decides to stop, we find the minimum stopping distance from
    $v_1^2 = v_0^2 + 2a_1(x_1 - x_0)$;
    $0 = (13.9 \text{ m/s})^2 + 2(-6.0 \text{ m/s}^2)x_1$, which gives $x_1 = 16$ m.
    Because this is less than L_1, the distance to the intersection, she can safely stop in time.
    If she decides to increase her speed, we find the acceleration from the time to go from 50 km/h to 70 km/h (19.4 m/s):
    $v = v_0 + a_2 t$;
    $19.4 \text{ m/s} = 13.9 \text{ m/s} + a_2(6.0 \text{ s})$, which gives $a_2 = 0.917 \text{ m/s}^2$.
    We find her location when the light turns red from
    $x_2 = x_0 + v_0 t_2 + \frac{1}{2}a_2 t_2^2 = 0 + (13.9 \text{ m/s})(2.0 \text{ s}) + \frac{1}{2}(0.917 \text{ m/s}^2)(2.0 \text{ s})^2 = 30$ m.
    Because this is L_1, she is at the beginning of the intersection, but moving at high speed.
    She should decide to stop!

31. We find the assumed constant speed for the first 27.0 min from
    $v_0 = \Delta x/\Delta t = (10{,}000 \text{ m} - 1100 \text{ m})/(27.0 \text{ min})(60 \text{ s/min}) = 5.49 \text{ m/s}$.
    The runner must cover the last 1100 m in 3.0 min (180 s). If the runner accelerates for t s, the new speed will be
    $v = v_0 + at = 5.49 \text{ m/s} + (0.20 \text{ m/s}^2)t$;
    and the distance covered during the acceleration will be
    $x_1 = v_0 t + \frac{1}{2}at^2 = (5.49 \text{ m/s})t + \frac{1}{2}(0.20 \text{ m/s}^2)t^2$.
    The remaining distance will be run at the new speed, so we have
    $1100 \text{ m} - x_1 = v(180 \text{ s} - t)$; or
    $1100 \text{ m} - (5.49 \text{ m/s})t - \frac{1}{2}(0.20 \text{ m/s}^2)t^2 = [5.49 \text{ m/s} + (0.20 \text{ m/s}^2)t](180 \text{ s} - t)$.
    This is a quadratic equation:
    $0.10\,t^2 - 36\,t + 111.8 = 0$, with the solutions $t = -363$ s, +3.1 s.
    Because $t = 0$ when the acceleration begins, the positive answer is the physical answer: **$t = 3.1$ s.**

32. $(280\ \text{m/s}^2)(1\ g/9.80\ \text{m/s}^2) =$ **28.6 g.**

33. We use a coordinate system with the origin at the top of the cliff and down positive.
To find the time for the object to acquire the velocity, we have
$v = v_0 + at$;
$(90\ \text{km/h})/(3.6\ \text{ks/h}) = 0 + (9.80\ \text{m/s}^2)t$, which gives **$t = 2.6$ s.**

34. We use a coordinate system with the origin at the top of the cliff and down positive.
To find the height of the cliff, we have
$y = y_0 + v_0t + \frac{1}{2}at^2$
$= 0 + 0 + \frac{1}{2}(9.80\ \text{m/s}^2)(3.50\ \text{s})^2 =$ **60.0 m.**

35. We use a coordinate system with the origin at the top of the building and down positive.
(*a*) To find the time of fall, we have
$y = y_0 + v_0t + \frac{1}{2}at^2$;
$380\ \text{m} = 0 + 0 + \frac{1}{2}(9.80\ \text{m/s}^2)t^2$, which gives **$t = 8.81$ s.**
(*b*) We find the velocity just before landing from
$v = v_0 + at$
$= 0 + (9.80\ \text{m/s}^2)(8.81\ \text{s}) =$ **86.3 m/s (down).**

36. We use a coordinate system with the origin at the ground and up positive.
(*a*) At the top of the motion the velocity is zero, so we find the height h from
$v^2 = v_0^2 + 2ah$;
$0 = (25\ \text{m/s})^2 + 2(-9.80\ \text{m/s}^2)h$, which gives **$h = 32$ m.**
(*b*) When the ball returns to the ground, its displacement is zero, so we have
$y = y_0 + v_0t + \frac{1}{2}at^2$
$0 = 0 + (25\ \text{m/s})t + \frac{1}{2}(-9.80\ \text{m/s}^2)t^2$,
which gives $t = 0$ (when the ball starts up), and **$t = 5.1$ s.**

37. We use a coordinate system with the origin at the ground and up positive.
We can find the initial velocity from the maximum height (where the velocity is zero):
$v^2 = v_0^2 + 2ah$;
$0 = v_0^2 + 2(-9.80\ \text{m/s}^2)(2.7\ \text{m})$, which gives $v_0 = 7.27\ \text{m/s}$.
When the kangaroo returns to the ground, its displacement is zero. For the entire jump we have
$y = y_0 + v_0t + \frac{1}{2}at^2$;
$0 = 0 + (7.27\ \text{m/s})t + \frac{1}{2}(-9.80\ \text{m/s}^2)t^2$,
which gives $t = 0$ (when the kangaroo jumps), and **$t = 1.5$ s.**

38. We use a coordinate system with the origin at the ground and up positive.
When the ball returns to the ground, its displacement is zero, so we have
$y = y_0 + v_0t + \frac{1}{2}at^2$;
$0 = 0 + v_0(3.3\ \text{s}) + \frac{1}{2}(-9.80\ \text{m/s}^2)(3.3\ \text{s})^2$, which gives $v_0 = 16\ \text{m/s}$.
At the top of the motion the velocity is zero, so we find the height h from
$v^2 = v_0^2 + 2ah$;
$0 = (16\ \text{m/s})^2 + 2(-9.80\ \text{m/s}^2)h$, which gives **$h = 13$ m.**

39. With the origin at the release point and the initial condition of $v_0 = 0$, we have
   (*a*) $v = v_0 + at = 0 + gt = (9.80\ \text{m/s}^2)t$;
   (*b*) $y = y_0 + v_0t + \frac{1}{2}at^2 = 0 + 0 + \frac{1}{2}gt^2 = (4.90\ \text{m/s}^2)t^2$.

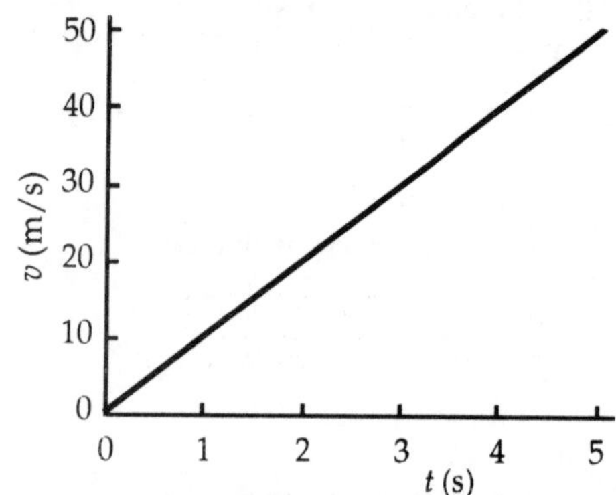

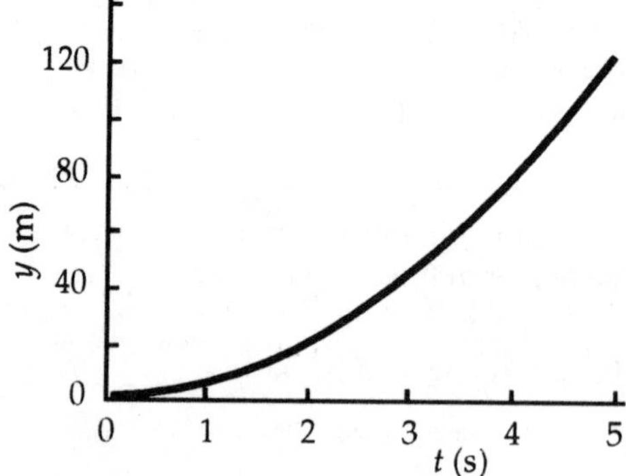

40. We use a coordinate system with the origin at the ground and up positive.
   (*a*) We can find the initial velocity from the maximum height (where the velocity is zero):
   $v^2 = v_0^2 + 2ah$;
   $0 = v_0^2 + 2(-9.80\ \text{m/s}^2)(1.20\ \text{m})$, which gives $\boxed{v_0 = 4.85\ \text{m/s.}}$
   (*b*) When the player returns to the ground, the displacement is zero. For the entire jump we have
   $y = y_0 + v_0t + \frac{1}{2}at^2$;
   $0 = 0 + (4.85\ \text{m/s})t + \frac{1}{2}(-9.80\ \text{m/s}^2)t^2$,
   which gives $t = 0$ (when the player jumps), and $\boxed{t = 0.990\ \text{s.}}$

41. We use a coordinate system with the origin at the ground and up positive. When the package returns to the ground, its displacement is zero, so we have
   $y = y_0 + v_0t + \frac{1}{2}at^2$;
   $0 = 105\ \text{m} + (5.50\ \text{m/s})t + \frac{1}{2}(-9.80\ \text{m/s}^2)t^2$.
   The solutions of this quadratic equation are $t = -4.10$ s, and $t = 5.22$ s. Because the package is released at $t = 0$, the positive answer is the physical answer: $\boxed{5.22\ \text{s.}}$

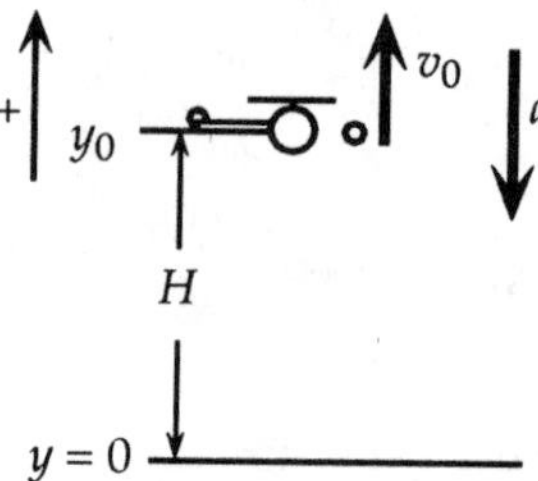

42. We use a coordinate system with the origin at the release point and down positive. Because the object starts from rest, $v_0 = 0$. The position of the object is given by
   $y = y_0 + v_0t + \frac{1}{2}at^2 = 0 + 0 + \frac{1}{2}gt^2$.
   The positions at one-second intervals are
   $y_0 = 0$;
   $y_1 = \frac{1}{2}g(1\ \text{s})^2 = \frac{1}{2}g$;
   $y_2 = \frac{1}{2}g(2\ \text{s})^2 = (4)\frac{1}{2}g$;
   $y_3 = \frac{1}{2}g(3\ \text{s})^2 = (9)\frac{1}{2}g$; … .
   The distances traveled during each second are
   $d_1 = y_1 - y_0 = \frac{1}{2}g$;
   $d_2 = y_2 - y_1 = (4 - 1)\frac{1}{2}g = 3(\frac{1}{2}g)$;
   $d_3 = y_3 - y_2 = (9 - 4)\frac{1}{2}g = 5(\frac{1}{2}g)$; … .

43. We use a coordinate system with the origin at the ground and up positive. Without air resistance, the acceleration is constant, so we have

$v^2 = v_0^2 + 2a(y - y_0)$;

$v^2 = v_0^2 + 2(-9.80 \text{ m/s}^2)(0 - 0) = v_0^2$, which gives $v = \pm v_0$.

The two signs represent the two directions of the velocity at the ground. The magnitudes, and thus the speeds, are the same.

44. We use a coordinate system with the origin at the ground and up positive.

(*a*) We find the velocity from

$v^2 = v_0^2 + 2a(y - y_0)$;

$v^2 = (20.0 \text{ m/s})^2 + 2(-9.80 \text{ m/s}^2)(12.0 \text{ m} - 0)$, which gives $v = \boxed{\pm 12.8 \text{ m/s.}}$

The stone reaches this height on the way up (the positive sign) and on the way down (the negative sign).

(*b*) We find the time to reach the height from

$v = v_0 + at$;

$\pm 12.8 \text{ m/s} = 20.0 \text{ m/s} + (-9.80 \text{ m/s}^2)t$, which gives $\boxed{t = 0.735 \text{ s}, 3.35 \text{ s.}}$

(*c*) There are two answers because the stone reaches this height on the way up (t = 0.735 s) and on the way down (t = 3.35 s).

45. We use a coordinate system with the origin at the release point and down positive. On paper the apple measures 7 mm, which we will call 7 mmp. If its true diameter is 10 cm, the conversion is 0.10 m/7 mmp. The images of the apple immediately after release overlap. We will use the first clear image which is 8 mmp below the release point. The final image is 61 mmp below the release point, and there are 7 intervals between these two images.

The position of the apple is given by

$y = y_0 + v_0t + \frac{1}{2}at^2 = 0 + 0 + \frac{1}{2}gt^2$.

For the two selected images we have

$y_1 = \frac{1}{2}gt_1^2$; (8 mmp)(0.10 m/7 mmp) $= \frac{1}{2}(9.80 \text{ m/s}^2)t_1^2$, which gives $t_1 = 0.153$ s;

$y_2 = \frac{1}{2}gt_2^2$; (61 mmp)(0.10 m/7 mmp) $= \frac{1}{2}(9.80 \text{ m/s}^2)t_2^2$, which gives $t_2 = 0.422$ s.

Thus the time interval between successive images is

$\Delta t = (t_2 - t_1)/7 = (0.422 \text{ s} - 0.153 \text{ s})/7 = \boxed{0.038 \text{ s.}}$

46. We use a coordinate system with the origin at the top of the window and down positive. We can find the velocity at the top of the window from the motion past the window:

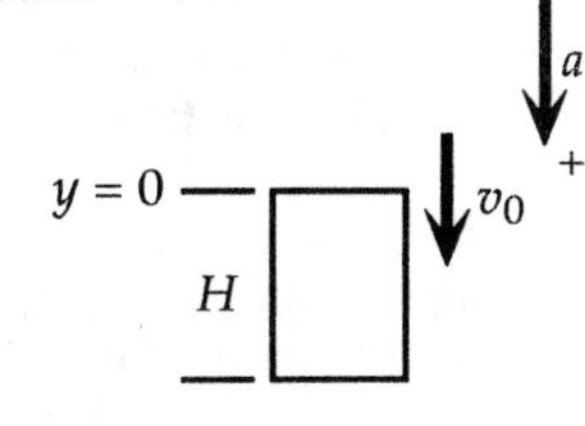

$y = y_0 + v_0t + \frac{1}{2}at_A^2$;

$2.2 \text{ m} = 0 + v_0(0.30 \text{ s}) + \frac{1}{2}(9.80 \text{ m/s}^2)(0.30 \text{ s})^2$, which gives $v_0 = 5.86$ m/s.

For the motion from the release point to the top of the window, we have

$v_0^2 = v_{\text{release}}^2 + 2g(y_0 - y_{\text{release}})$;

$(5.86 \text{ m/s})^2 = 0 + 2(9.80 \text{ m/s}^2)(0 - y_{\text{release}})$, which gives $y_{\text{release}} = -1.8$ m.

The stone was released $\boxed{1.8 \text{ m}}$ above the top of the window.

47. If the height of the cliff is H, the time for the sound to travel from the ocean to the top is

$t_{\text{sound}} = H/v_{\text{sound}}$.

The time of fall for the rock is $T - t_{\text{sound}}$. We use a coordinate system with the origin at the top of the cliff and down positive. For the falling motion we have

$y = y_0 + v_0t + \frac{1}{2}at^2$;

$H = 0 + 0 + \frac{1}{2}a(T - t_{\text{sound}})^2 = \frac{1}{2}(9.80 \text{ m/s}^2)[3.4 \text{ s} - H/(340 \text{ m/s})]^2$.

This is a quadratic equation for H:

$4.24 \times 10^{-5} H^2 - 1.098H + 56.64 = 0$, with H in m; which has the solutions $H = 52$ m, 2.58×10^4 m.

The larger result corresponds to t_{sound} greater than 3.4 s, so the height of the cliff is $\boxed{52 \text{ m.}}$

48. We use a coordinate system with the origin at the nozzle and up positive.
For the motion of the water from the nozzle to the ground, we have

$y = y_0 + v_0 t + \frac{1}{2}at^2;$

$-1.5\text{ m} = 0 + v_0(2.0\text{ s}) + \frac{1}{2}(-9.80\text{ m/s}^2)(2.0\text{ s})^2$, which gives $\boxed{v_0 = 9.1\text{ m/s.}}$

49. We use a coordinate system with the origin at the top of the cliff and up positive.

(*a*) For the motion of the stone from the top of the cliff to the ground, we have

$y = y_0 + v_0 t + \frac{1}{2}at^2;$

$-75.0\text{ m} = 0 + (12.0\text{ m/s})t + \frac{1}{2}(-9.80\text{ m/s}^2)t^2.$

This is a quadratic equation for t, which has the solutions $t = -2.88$ s, 5.33 s.
Because the stone starts at $t = 0$, the time is $\boxed{5.33\text{ s.}}$

(*b*) We find the speed from

$v = v_0 + at$

$= 12.0\text{ m/s} + (-9.80\text{ m/s}^2)(5.33\text{ s}) = -40.2\text{ m/s}.$

The negative sign indicates the downward direction, so the speed is $\boxed{40.2\text{ m/s.}}$

(*c*) The total distance includes the distance up to the maximum height, down to the top of the cliff, and down to the bottom. We find the maximum height from

$v^2 = v_0^2 + 2ah;$

$0 = (12.0\text{ m/s})^2 + 2(-9.80\text{ m/s}^2)h$, which gives $h = 7.35$ m.

The total distance traveled is

$d = 7.35\text{ m} + 7.35\text{ m} + 75.0\text{ m} = \boxed{89.7\text{ m.}}$

50. We use a coordinate system with the origin at the ground and up positive.

(*a*) We find the initial speed from the motion to the window:

$v_1^2 = v_0^2 + 2a(y_1 - y_0);$

$(12\text{ m/s})^2 = v_0^2 + 2(-9.80\text{ m/s}^2)(25\text{ m} - 0)$, which gives $\boxed{v_0 = 25\text{ m/s.}}$

(*b*) We find the maximum altitude from

$v_2^2 = v_0^2 + 2a(y_2 - y_0);$

$0 = (25\text{ m/s})^2 + 2(-9.80\text{ m/s}^2)(y_2 - 0)$, which gives $y_2 = \boxed{32\text{ m.}}$

(*c*) We find the time from the motion to the window:

$v_1 = v_0 + at_1$

$12\text{ m/s} = 25\text{ m/s} + (-9.80\text{ m/s}^2)t_1$, which gives $t_1 = \boxed{1.3\text{ s.}}$

(*d*) We find the time to reach the street from

$y = y_0 + v_0 t + \frac{1}{2}at^2;$

$0 = 0 + (25\text{ m/s})t + \frac{1}{2}(-9.80\text{ m/s}^2)t^2.$

This is a quadratic equation for t, which has the solutions $t = 0$ (the initial throw), 5.1 s.
Thus the time after the baseball passed the window is $5.1\text{ s} - 1.3\text{ s} = \boxed{3.8\text{ s.}}$

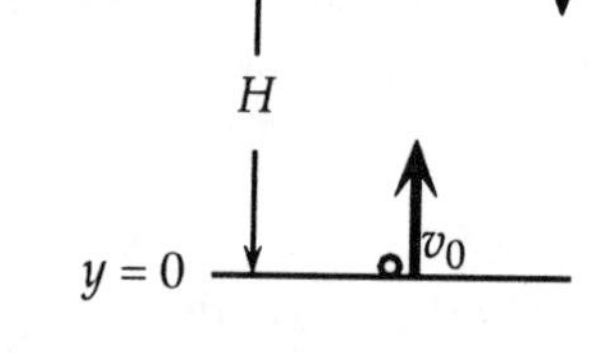

51. (*a*) We find the instantaneous velocity from the slope of the straight line from $t = 0$ to $t = 10$ s:

$v_{10} = \Delta x/\Delta t = (2.8\text{ m} - 0)/(10.0\text{ s} - 0) = \boxed{0.28\text{ m/s.}}$

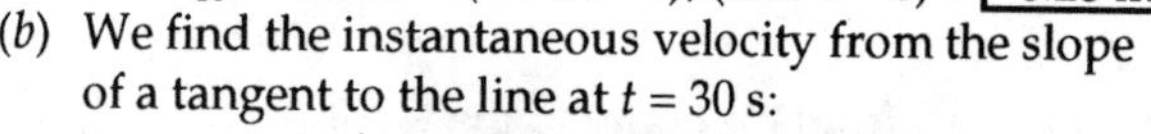

(*b*) We find the instantaneous velocity from the slope of a tangent to the line at $t = 30$ s:

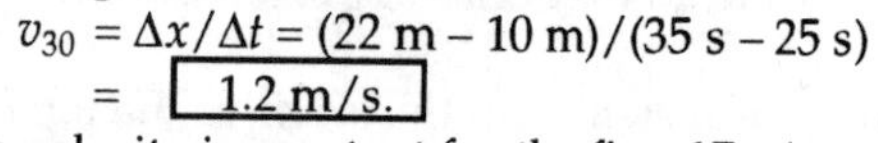

$v_{30} = \Delta x/\Delta t = (22\text{ m} - 10\text{ m})/(35\text{ s} - 25\text{ s})$

$= \boxed{1.2\text{ m/s.}}$

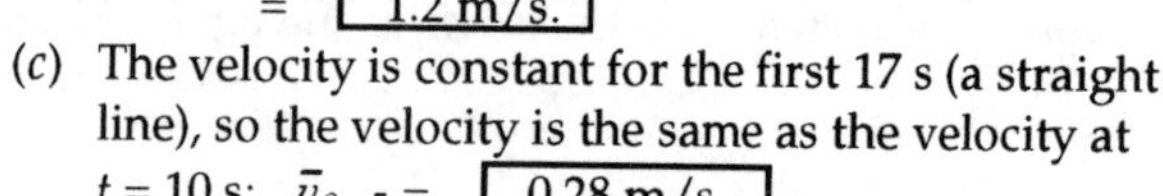

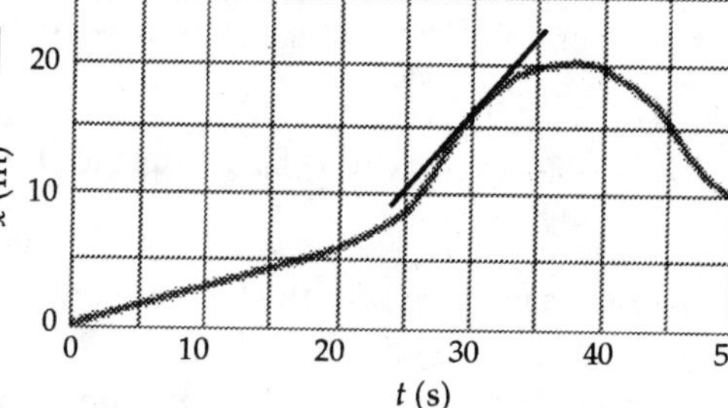

(*c*) The velocity is constant for the first 17 s (a straight line), so the velocity is the same as the velocity at $t = 10$ s: $\bar{v}_{0\to5} = \boxed{0.28\text{ m/s.}}$

(*d*) For the average velocity we have

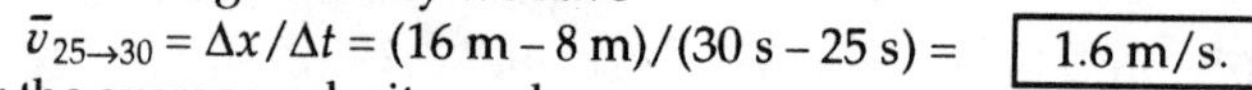

$\bar{v}_{25\to30} = \Delta x/\Delta t = (16\text{ m} - 8\text{ m})/(30\text{ s} - 25\text{ s}) = \boxed{1.6\text{ m/s.}}$

(*e*) For the average velocity we have

$\bar{v}_{40\to50} = \Delta x/\Delta t = (10\text{ m} - 20\text{ m})/(50\text{ s} - 40\text{ s}) = \boxed{-1.0\text{ m/s.}}$

52. (*a*) Constant velocity is indicated by a straight line, which occurs from $\boxed{t = 0 \text{ to } 17 \text{ s.}}$
(*b*) The maximum velocity is when the slope is greatest: $\boxed{t = 28 \text{ s.}}$
(*c*) Zero velocity is indicated by a zero slope. The tangent is horizontal at $\boxed{t = 38 \text{ s.}}$
(*d*) Because the curve has both positive and negative slopes, the motion is in $\boxed{\text{both directions.}}$

53. (*a*) The maximum velocity is indicated by the highest point, which occurs at $\boxed{t = 50 \text{ s.}}$
(*b*) Constant velocity is indicated by a horizontal slope, which occurs from $\boxed{t = 90 \text{ s to } 107 \text{ s.}}$
(*c*) Constant acceleration is indicated by a straight line, which occurs from
$\boxed{t = 0 \text{ to } 20 \text{ s, and } t = 90 \text{ s to } 107 \text{ s.}}$
(*d*) The maximum acceleration is when the slope is greatest: $\boxed{t = 75 \text{ s.}}$

54. (*a*) For the average acceleration we have

$\bar{a}_2 = \Delta v/\Delta t$
$= (24 \text{ m/s} - 14 \text{ m/s})/(8 \text{ s} - 4 \text{ s}) = \boxed{2.5 \text{ m/s}^2;}$

$\bar{a}_4 = \Delta v/\Delta t$
$= (44 \text{ m/s} - 37 \text{ m/s})/(27 \text{ s} - 16 \text{ s}) = \boxed{0.6 \text{ m/s}^2.}$

(*b*) The distance traveled is represented by the area under the curve, which we approximate as a rectangle with a height equal to the mean velocity:

$d = v_{\text{mean}}\Delta t$

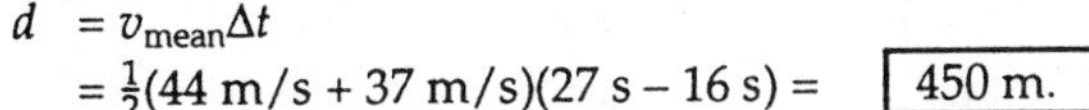

$= \frac{1}{2}(44 \text{ m/s} + 37 \text{ m/s})(27 \text{ s} - 16 \text{ s}) = \boxed{450 \text{ m.}}$

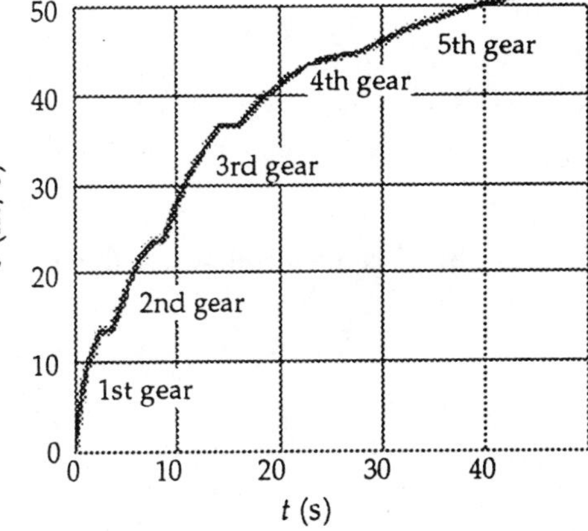

55. (*a*) For the average acceleration we have

$\bar{a}_1 = \Delta v/\Delta t = (14 \text{ m/s} - 0)/(3 \text{ s} - 0) = \boxed{4.7 \text{ m/s}^2.}$

(*b*) For the average acceleration we have

$\bar{a}_3 = \Delta v/\Delta t = (37 \text{ m/s} - 24 \text{ m/s})/(14 \text{ s} - 8 \text{ s}) = \boxed{2.2 \text{ m/s}^2.}$

(*c*) For the average acceleration we have

$\bar{a}_5 = \Delta v/\Delta t = (52 \text{ m/s} - 44 \text{ m/s})/(50 \text{ s} - 27 \text{ s}) = \boxed{0.3 \text{ m/s}^2.}$

(*d*) For the average acceleration we have

$\bar{a}_{1\to4} = \Delta v/\Delta t = (44 \text{ m/s} - 0)/(27 \text{ s} - 0) = \boxed{1.6 \text{ m/s}^2.}$

Note that we cannot add the four average accelerations and divide by 4.

56. The distance is represented by the area under the curve. We will estimate it by counting the number of blocks, each of which represents (10 m/s)(10 s) = 100 m.

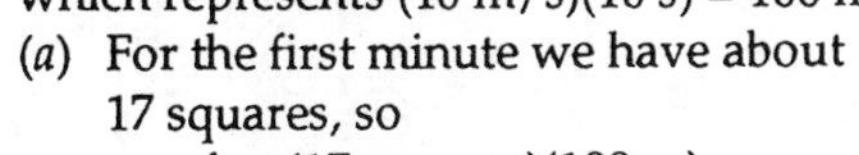

(*a*) For the first minute we have about 17 squares, so

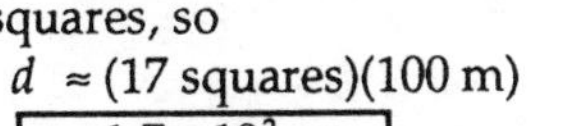

$d \approx (17 \text{ squares})(100 \text{ m})$

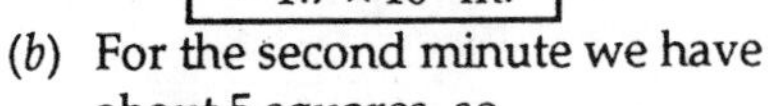

$\boxed{\approx 1.7 \times 10^3 \text{ m.}}$

(*b*) For the second minute we have about 5 squares, so

$d \approx (5 \text{ squares})(100 \text{ m})$ $\boxed{\approx 5 \times 10^2 \text{ m.}}$

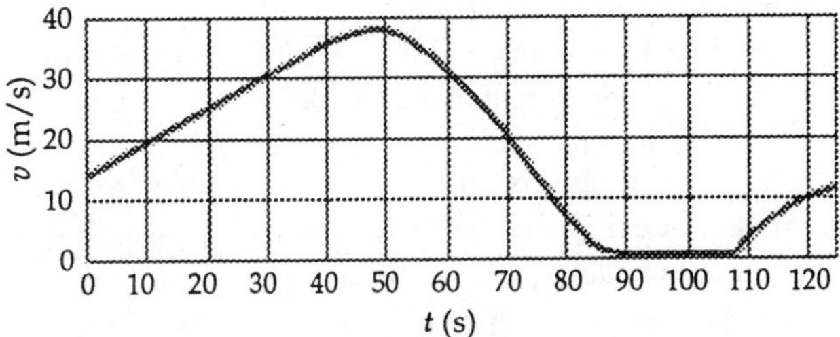

57. The instantaneous velocity is the slope of the x vs. t graph:

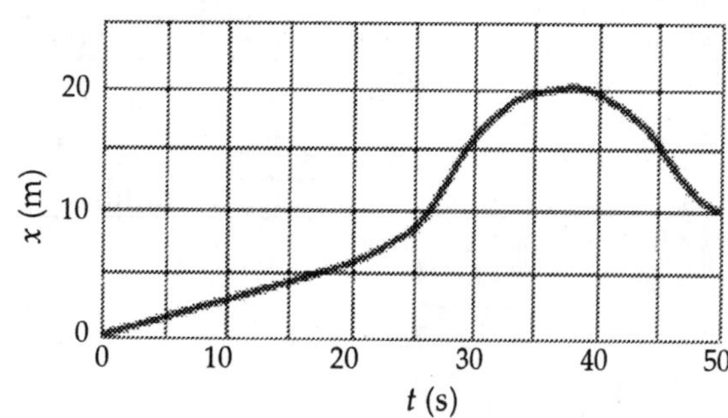

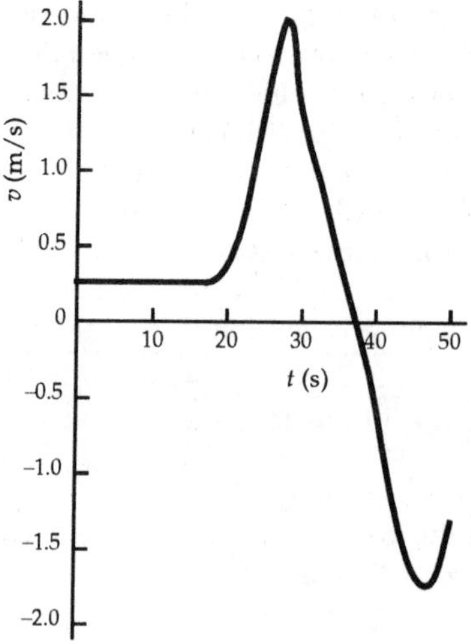

58. The displacement is the area under the v vs. t graph:

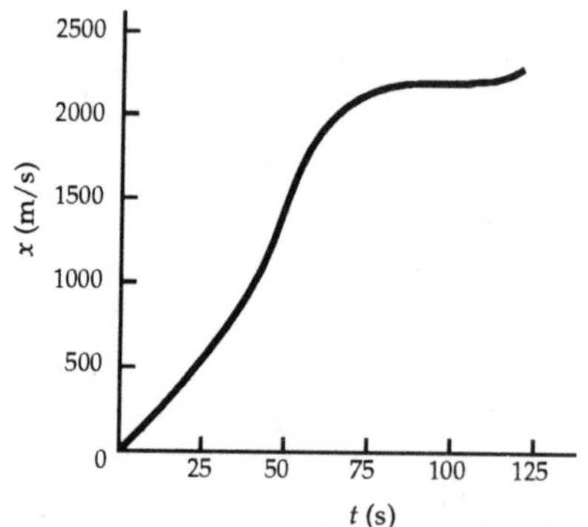

59. For the motion from A to B,
   (*a*) The object is moving in the **negative** direction. The slope (the instantaneous velocity) is negative; the x-value is decreasing.
   (*b*) Because the slope is becoming more negative (greater magnitude of the velocity), the object is **speeding up.**
   (*c*) Because the velocity is becoming more negative, the acceleration is **negative.**

   For the motion from D to E,
   (*d*) The object is moving in the **positive** direction. The slope (the instantaneous velocity) is positive; the x-value is increasing.
   (*e*) Because the slope is becoming more positive (greater magnitude of the velocity), the object is **speeding up.**
   (*f*) Because the velocity is becoming more positive, the acceleration is **positive.**
   (*g*) The position is constant, so the object is **not moving,** the velocity and the acceleration are **zero.**

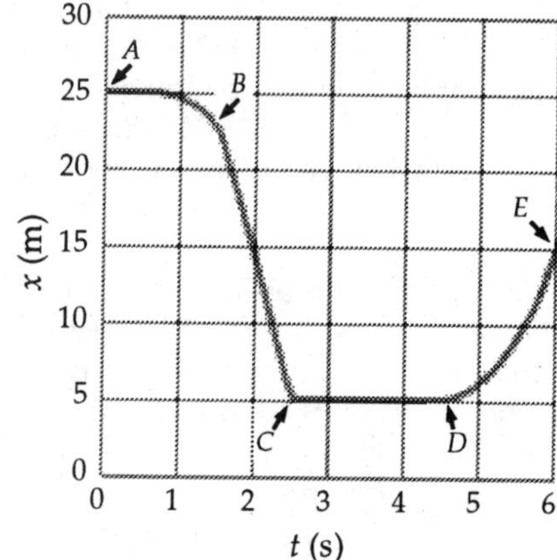

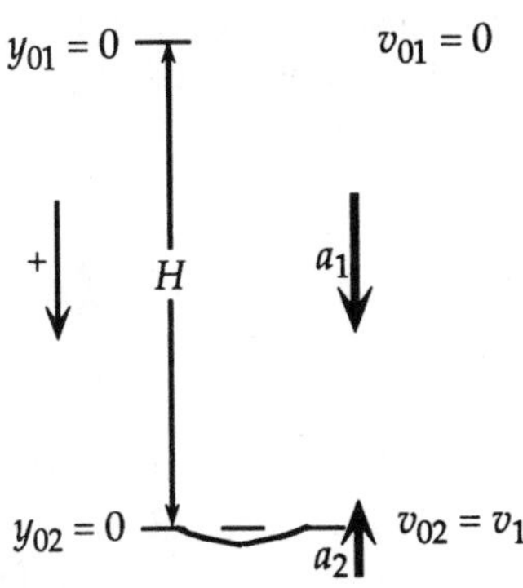

60. For the falling motion, we use a coordinate system with the origin at the fourth-story window and down positive. For the stopping motion in the net, we use a coordinate system with the origin at the original position of the net and down positive.
   (*a*) We find the velocity of the person at the unstretched net (which is the initial velocity for the stretching of the net) from the free fall:
   $v_{02}^2 = v_{01}^2 + 2a_1(y_1 - y_{01}) = 0 + 2(9.80 \text{ m/s}^2)(15.0 \text{ m} - 0),$
   which gives $v_{02} = 17.1$ m/s.
   We find the acceleration during the stretching of the net from
   $v_2^2 = v_{02}^2 + 2a_2(y_2 - y_{02});$
   $0 = (17.1 \text{ m/s})^2 + 2a_2(1.0 \text{ m} - 0),$
   which gives $a_2 =$ $\boxed{-1.5 \times 10^2 \text{ m/s}^2.}$
   (*b*) To produce the same velocity change with a smaller acceleration requires a greater displacement. Thus **the net should be loosened.**

61. The height reached is determined by the initial velocity. We assume the same initial velocity of the object on the moon and Earth. With a vertical velocity of 0 at the highest point, we have
   $v^2 = v_0^2 + 2ah;$
   $0 = v_0^2 + 2(-g)h$, so we get
   $v_0^2 = 2g_{\text{Earth}}h_{\text{Earth}} = 2g_{\text{moon}}h_{\text{moon}}$, or $h_{\text{moon}} = (g_{\text{Earth}}/g_{\text{moon}})h_{\text{Earth}} =$ $\boxed{6h_{\text{Earth}}\,.}$

62. We assume that the seat belt keeps the occupant fixed with respect to the car. The distance the occupant moves with respect to the front end is the distance the front end collapses, so we have
   $v^2 = v_0^2 + 2a(x - x_0);$
   $0 = [(100 \text{ km/h})/(3.6 \text{ ks/h})]^2 + 2(-30)(9.80 \text{ m/s}^2)(x - 0)$, which gives $x =$ $\boxed{1.3 \text{ m.}}$

63. If the lap distance is D, the time for the first 9 laps is
   $t_1 = 9D/(199 \text{ km/h})$, the time for the last lap is
   $t_2 = D/\bar{v}$, and the time for the entire trial is
   $T = 10D/(200 \text{ km/h})$.
   Thus we have
   $T = t_1 + t_2\,;$
   $10D/(200 \text{ km/h}) = 9D/(199 \text{ km/h}) + D/\bar{v}$, which gives $\boxed{\bar{v} = 209.5 \text{ km/h.}}$

64. We use a coordinate system with the origin at the release point and down positive.
   (*a*) The speed at the end of the fall is found from
   $v^2 = v_0^2 + 2a(x - x_0)$
   $= 0 + 2g(H - 0)$, which gives $v = (2gH)^{1/2}$.
   (*b*) To achieve a speed of 50 km/h, we have
   $v = (2gH)^{1/2}$; $(50 \text{ km/h})/(3.6 \text{ ks/h}) = [2(9.80 \text{ m/s}^2)H_{50}]^{1/2}$, which gives $\boxed{H_{50} = 9.8 \text{ m.}}$
   (*c*) To achieve a speed of 100 km/h, we have
   $v = (2gH)^{1/2}$; $(100 \text{ km/h})/(3.6 \text{ ks/h}) = [2(9.80 \text{ m/s}^2)H_{100}]^{1/2}$, which gives $\boxed{H_{100} = 39 \text{ m.}}$

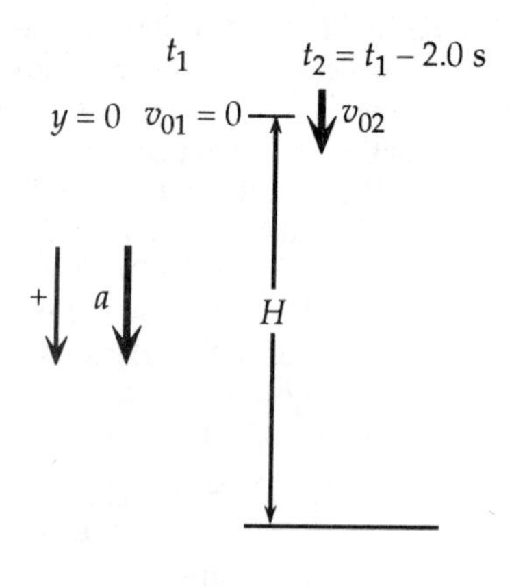

65. We use a coordinate system with the origin at the roof of the building and down positive, and call the height of the building H.
  (*a*) For the first stone, we have
  $$y_1 = y_{01} + v_{01}t_1 + \tfrac{1}{2}at_1^2;$$
  $$H = 0 + 0 + \tfrac{1}{2}(g)t_1^2, \text{ or } H = \tfrac{1}{2}gt_1^2.$$
  For the second stone, we have
  $$y_2 = y_{02} + v_{02}t_2 + \tfrac{1}{2}at_2^2;$$
  $$H = 0 + (30.0 \text{ m/s})t_2 + \tfrac{1}{2}(g)t_2^2$$
  $$= (30.0 \text{ m/s})(t_1 - 2.00 \text{ s}) + \tfrac{1}{2}(g)(t_1 - 2.00 \text{ s})^2$$
  $$= (30.0 \text{ m/s})t_1 - 60.0 \text{ m} + \tfrac{1}{2}gt_1^2 - (2.00 \text{ s})gt_1 + (2.00 \text{ s}^2)g.$$
  When we eliminate H from the two equations, we get
  $$0 = (30.0 \text{ m/s})t_1 - 60.0 \text{ m} - (2.00 \text{ s})gt_1 + (2.00 \text{ s}^2)g, \text{ which gives}$$
  $$\boxed{t_1 = 3.88 \text{ s.}}$$
  (*b*) We use the motion of the first stone to find the height of the building:
  $$H = \tfrac{1}{2}gt_1^2 = \tfrac{1}{2}(9.80 \text{ m/s}^2)(3.88 \text{ s})^2 = \boxed{73.9 \text{ m.}}$$
  (*c*) We find the speeds from
  $$v_1 = v_{01} + at_1 = 0 + (9.80 \text{ m/s}^2)(3.88 \text{ s}) = \boxed{38.0 \text{ m/s;}}$$
  $$v_2 = v_{02} + at_2 = 30.0 \text{ m/s} + (9.80 \text{ m/s}^2)(3.88 \text{ s} - 2.00 \text{ s}) = \boxed{48.4 \text{ m/s.}}$$

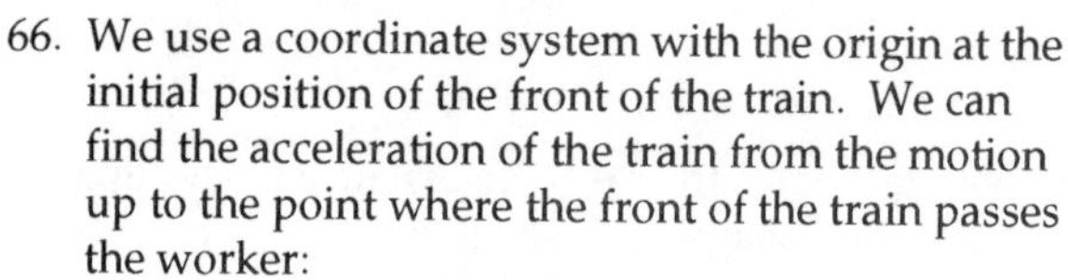

66. We use a coordinate system with the origin at the initial position of the front of the train. We can find the acceleration of the train from the motion up to the point where the front of the train passes the worker:
  $$v_1^2 = v_0^2 + 2a(D - 0);$$
  $$(20 \text{ m/s})^2 = 0 + 2a(180 \text{ m} - 0), \text{ which gives } a = 1.11 \text{ m/s}^2.$$
  Now we consider the motion of the last car, which starts at $-L$, to the point where it passes the worker:
  $$v_2^2 = v_0^2 + 2a[D - (-L)]$$
  $$= 0 + 2(1.11 \text{ m/s}^2)(180 \text{ m} + 90 \text{ m}), \text{ which gives } v_2 = \boxed{24 \text{ m/s.}}$$

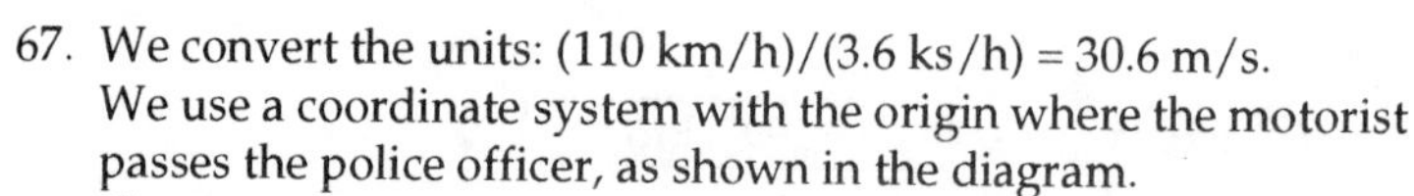

67. We convert the units: (110 km/h)/(3.6 ks/h) = 30.6 m/s.
  We use a coordinate system with the origin where the motorist passes the police officer, as shown in the diagram.
  (*a*)

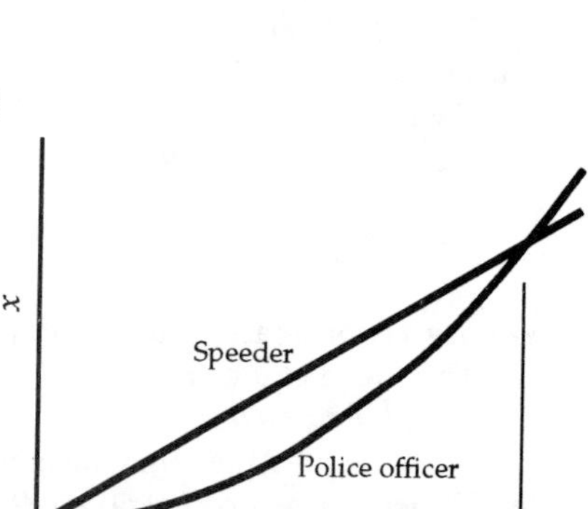

  (*b*) The location of the speeding motorist is given by
  $$x_m = x_0 + v_m t,$$
  which we use to find the time required:
  $$700 \text{ m} = (30.6 \text{ m/s})t, \text{ which gives } \boxed{t = 22.9 \text{ s.}}$$
  (*c*) The location of the police car is given by
  $$x_p = x_0 + v_{0p}t + \tfrac{1}{2}a_p t^2 = 0 + 0 + \tfrac{1}{2}a_p t^2,$$
  which we use to find the acceleration:
  $$700 \text{ m} = \tfrac{1}{2}a_p(22.9 \text{ s})^2, \text{ which gives } \boxed{a_p = 2.67 \text{ m/s}^2.}$$

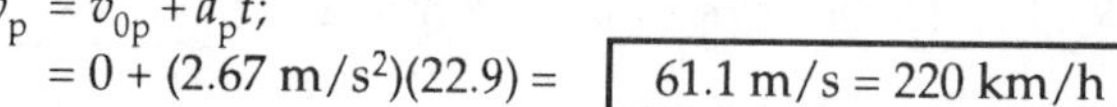

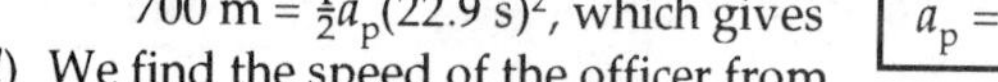

  (*d*) We find the speed of the officer from
  $$v_p = v_{0p} + a_p t;$$
  $$= 0 + (2.67 \text{ m/s}^2)(22.9) = \boxed{61.1 \text{ m/s} = 220 \text{ km/h}} \text{ (about 135 mi/h!).}$$

68. We convert the maximum speed units: $v_{max} = (90 \text{ km/h})/(3.6 \text{ ks/h}) = 25 \text{ m/s}$.

(*a*) There are $(36 \text{ km})/0.80 \text{ km}) = 45$ trip segments, which means 46 stations (with 44 intermediate stations). In each segment there are three motions.

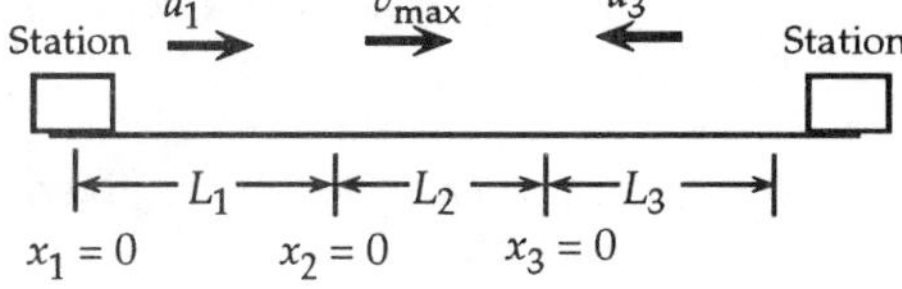

Motion 1 is the acceleration to v_{max}.

We find the time for this motion from

$v_{max} = v_{01} + a_1 t_1;$

$25 \text{ m/s} = 0 + (1.1 \text{ m/s}^2)t_1$, which gives $t_1 = 22.7 \text{ s}$.

We find the distance for this motion from

$x_1 = x_{01} + v_{01}t + \frac{1}{2}a_1 t_1^2; \quad L_1 = 0 + 0 + \frac{1}{2}(1.1 \text{ m/s}^2)(22.7 \text{ s})^2 = 284 \text{ m}.$

Motion 2 is the constant speed of v_{max}, for which we have

$x_2 = x_{02} + v_{max}t_2; \quad L_2 = 0 + v_{max}t_2.$

Motion 3 is the acceleration from v_{max} to 0.

We find the time for this motion from

$0 = v_{max} + a_3 t_3 = 25 \text{ m/s} + (-2.0 \text{ m/s}^2)t_3$, which gives $t_3 = 12.5 \text{ s}$.

We find the distance for this motion from

$x_3 = x_{03} + v_{max}t + \frac{1}{2}a_3 t_3^2; \quad L_3 = 0 + (25 \text{ m/s})(12.5 \text{ s}) + \frac{1}{2}(-2.0 \text{ m/s}^2)(12.5 \text{ s})^2 = 156 \text{ m}.$

The distance for Motion 2 is

$L_2 = 800 \text{ m} - L_1 - L_3 = 800 \text{ m} - 284 \text{ m} - 156 \text{ m} = 360 \text{ m}$, so the time for Motion 2 is

$t_2 = L_2/v_{max} = (360 \text{ m})/(25 \text{ m/s}) = 14.4 \text{ s}.$

Thus the total time for the 45 segments and 44 stops is

$T = 45(t_1 + t_2 + t_3) + 44(20 \text{ s}) = 45(22.7 \text{ s} + 14.4 \text{ s} + 12.5 \text{ s}) + 44(20 \text{ s}) = 3112 \text{ s} =$ **52 min.**

(*b*) There are $(36 \text{ km})/3.0 \text{ km}) = 12$ trip segments, which means 13 stations (with 11 intermediate stations.)

The results for Motion 1 and Motion 3 are the same:

$t_1 = 22.7 \text{ s}, \; L_1 = 284 \text{ m}, \; t_3 = 12.5 \text{ s}, \; L_3 = 156 \text{ m}.$

The distance for Motion 2 is

$L_2 = 3000 \text{ m} - L_1 - L_3 = 3000 \text{ m} - 284 \text{ m} - 156 \text{ m} = 2560 \text{ m}$, so the time for Motion 2 is

$t_2 = L_2/v_{max} = (2560 \text{ m})/(25 \text{ m/s}) = 102 \text{ s}.$

Thus the total time for the 12 segments and 11 stops is

$T = 12(t_1 + t_2 + t_3) + 11(20 \text{ s}) = 12(22.7 \text{ s} + 102 \text{ s} + 12.5 \text{ s}) + 11(20 \text{ s}) = 1870 \text{ s} =$ **31 min.**

This means there is a higher average speed for stations farther apart.

69. We use a coordinate system with the origin at the start of the pelican's dive and down positive. We find the time for the pelican to reach the water from

$y_1 = y_0 + v_0 t + \frac{1}{2}at_1^2;$

$16.0 \text{ m} = 0 + 0 + \frac{1}{2}(9.80 \text{ m/s}^2)t^2$, which gives $t_1 = 1.81 \text{ s}$.

This means that the fish must spot the pelican $1.81 \text{ s} - 0.20 \text{ s} = 1.61 \text{ s}$ after the pelican starts its dive. We find the distance the pelican has fallen at this time from

$y_2 = y_0 + v_0 t + \frac{1}{2}at_2^2;$

$= 0 + 0 + (9.80 \text{ m/s}^2)(1.61 \text{ s})^2 = 12.7 \text{ m}.$

Therefore the fish must spot the pelican at a height of $16.0 \text{ m} - 12.7 \text{ m} =$ **3.3 m.**

70. In each case we use a coordinate system with the origin at the beginning of the putt and the positive direction in the direction of the putt. The limits on the putting distance are $6.0 \text{ m} < x < 8.0 \text{ m}$.

For the downhill putt we have:

$v^2 = v_{0down}^2 + 2a_{down}(x - x_0); \quad 0 = v_{0down}^2 + 2(-2.0 \text{ m/s}^2)x.$

When we use the limits for x, we get $4.9 \text{ m/s} < v_{0down} < 5.7 \text{ m/s}$, or **$\Delta v_{0down} = 0.8 \text{ m/s}$.**

For the uphill putt we have:

$v^2 = v_{0up}^2 + 2a_{up}(x - x_0); \quad 0 = v_{0up}^2 + 2(-3.0 \text{ m/s}^2)x.$

When we use the limits for x, we get $6.0 \text{ m/s} < v_{0up} < 6.9 \text{ m/s}$, or **$\Delta v_{0up} = 0.9 \text{ m/s}$.**

The smaller spread in allowable initial velocities makes the downhill putt more difficult.

71. We use a coordinate system with the origin at the initial position of the car.
The passing car's position is given by
$x_1 = x_{01} + v_0t + \frac{1}{2}a_1t^2 = 0 + v_0t + \frac{1}{2}a_1t^2.$
The truck's position is given by
$x_{\text{truck}} = x_{0\text{truck}} + v_0t = D + v_0t.$
The oncoming car's position is given by
$x_2 = x_{02} - v_0t = L - v_0t.$
For the car to be safely past the truck, we must have
$x_1 - x_{\text{truck}} = 10\text{ m};$
$v_0t + \frac{1}{2}a_1t^2 - (D + v_0t) = \frac{1}{2}a_1t^2 - D = 10\text{ m},$
which allows us to find the time required for passing:
$\frac{1}{2}(1.0\text{ m/s}^2)t^2 - 30\text{ m} = 10\text{ m}$, which gives $t = 8.94\text{ s}$.
At this time the car's location will be
$x_1 = v_0t + \frac{1}{2}a_1t^2 = (25\text{ m/s})(8.94\text{ s}) + \frac{1}{2}(1.0\text{ m/s}^2)(8.94\text{ s})^2 = 264\text{ m}$ from the origin.
At this time the oncoming car's location will be
$x_2 = L - v_0t = 400\text{ m} - (25\text{ m/s})(8.94\text{ s}) = 176\text{ m}$ from the origin.
Because this is closer to the origin, the two cars will have collided,
so **the passing attempt should not be made.**

72. We use a coordinate system with the origin at the roof of the building and down positive.
We find the time of fall for the second stone from
$v_2 = v_{02} + at_2;$
$12.0\text{ m/s} = 0 + (9.80\text{ m/s}^2)t_2$, which gives $t_2 = 1.22\text{ s}$.
During this time, the second stone fell
$y_2 = y_{02} + v_{02}t_2 + \frac{1}{2}at_2^2 = 0 + 0 + \frac{1}{2}(9.80\text{ m/s}^2)(1.22\text{ s})^2 = 7.29\text{ m}.$
The time of fall for the first stone is
$t_1 = t_2 + 1.50\text{ s} = 1.22\text{ s} + 1.50\text{ s} = 2.72\text{ s}.$
During this time, the first stone fell
$y_1 = y_{01} + v_{01}t_1 + \frac{1}{2}at_1^2 = 0 + 0 + \frac{1}{2}(9.80\text{ m/s}^2)(2.72\text{ s})^2 = 36.3\text{ m}.$
Thus the distance between the two stones is
$y_1 - y_2 = 36.3\text{ m} - 7.29\text{ m} =$ **29.0 m.**

73. For the vertical motion of James Bond we use a coordinate system with the origin at the ground and up positive. We can find the time for his fall to the level of the truck bed from

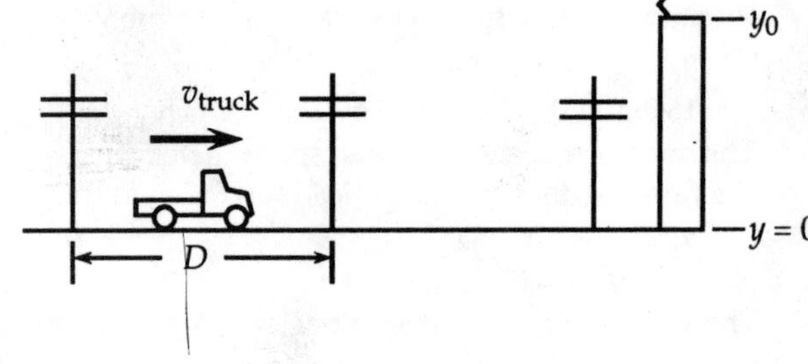

$y = y_0 + v_0t + \frac{1}{2}at^2;$
$1.5\text{ m} = 10\text{ m} + 0 + \frac{1}{2}(-9.80\text{ m/s}^2)t^2,$
which gives $t = 1.32\text{ s}$.
During this time the distance the truck will travel is
$x = x_0 + v_{\text{truck}}t = 0 + (30\text{ m/s})(1.32\text{ s}) = 39.6\text{ m}.$
Because the poles are 20 m apart, he should jump when the truck is **2 poles** away, assuming that there is a pole at the bridge.

CHAPTER 3

1. We choose the west and south coordinate system shown.
For the components of the resultant we have
$R_W = D_1 + D_2 \cos 45°$
$= (125 \text{ km}) + (65 \text{ km}) \cos 45° = 171 \text{ km};$
$R_S = 0 + D_2$
$= 0 + (65 \text{ km}) \sin 45° = 46 \text{ km}.$
We find the resultant displacement from
$R = (R_W^2 + R_S^2)^{1/2} = [(171 \text{ km})^2 + (46 \text{ km})^2]^{1/2} =$ $\boxed{177 \text{ km};}$
$\tan \theta = R_S/R_W = (46 \text{ km})/(171 \text{ km}) = 0.269$, which gives $\boxed{\theta = 15° \text{ S of W.}}$

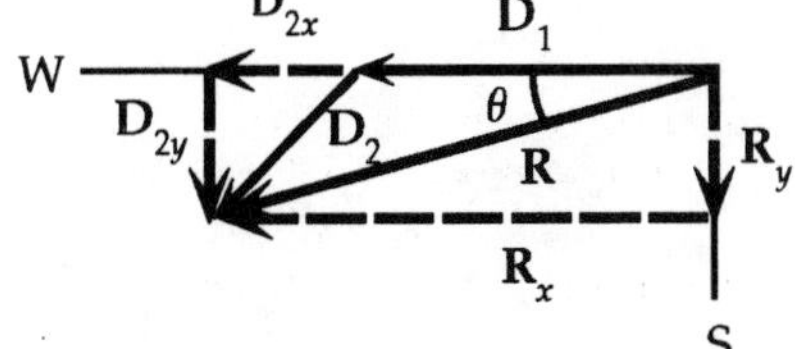

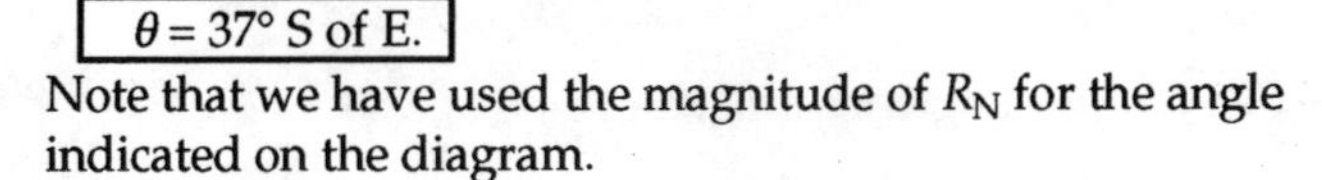

2. We choose the north and east coordinate system shown.
For the components of the resultant we have
$R_E = D_2 = 16 \text{ blocks};$
$R_N = D_1 - D_3 = 14 \text{ blocks} - 26 \text{ blocks} = -12 \text{ blocks}.$
We find the resultant displacement from
$R = (R_E^2 + R_N^2)^{1/2} = [(16 \text{ blocks})^2 + (-12 \text{ blocks})^2]^{1/2}$
$= \boxed{20 \text{ blocks;}}$
$\tan \theta = R_N/R_E = (12 \text{ blocks})/(16 \text{ blocks}) = 0.75$, which gives
$\boxed{\theta = 37° \text{ S of E.}}$
Note that we have used the magnitude of R_N for the angle indicated on the diagram.

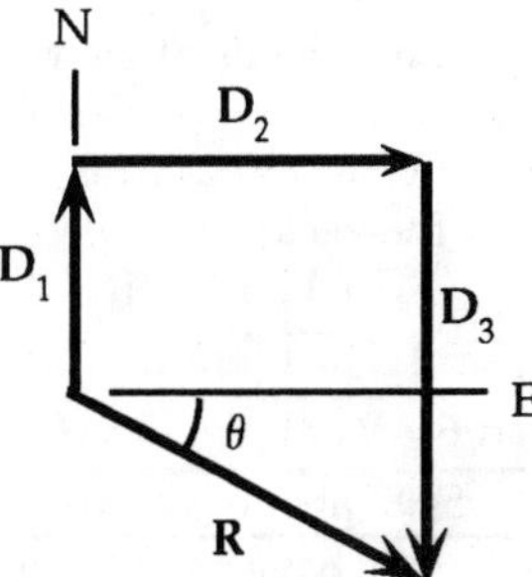

3. We show the six head-to-tail additions:

$\mathbf{V}_1 + \mathbf{V}_2 + \mathbf{V}_3$

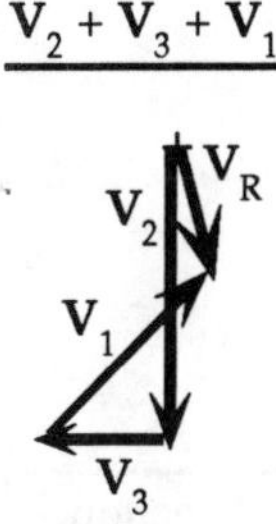

$\mathbf{V}_2 + \mathbf{V}_3 + \mathbf{V}_1$

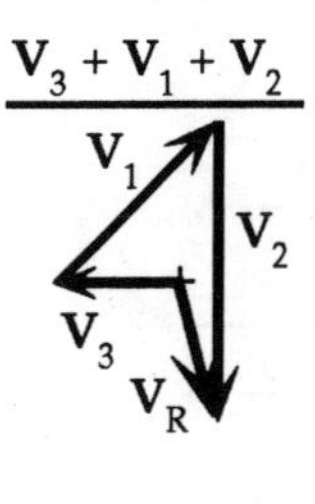

$\mathbf{V}_3 + \mathbf{V}_1 + \mathbf{V}_2$

V1
V2
V3
VR

$\mathbf{V}_1 + \mathbf{V}_3 + \mathbf{V}_2$

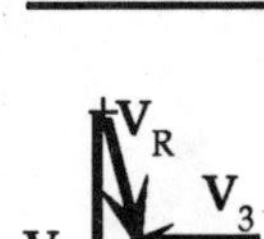

$\mathbf{V}_2 + \mathbf{V}_1 + \mathbf{V}_3$

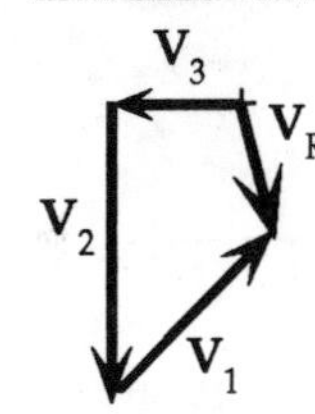

$\mathbf{V}_3 + \mathbf{V}_2 + \mathbf{V}_1$

V3
VR
V2
V1

4. We find the vector from
$V = (V_x^2 + V_y^2)^{1/2} = [(18.80)^2 + (-16.40)^2]^{1/2} =$ $\boxed{24.95;}$
$\tan \theta = V_y/V_x = (-16.40)/(18.80) = -0.8723$, which gives
$\boxed{\theta = 41.10° \text{ below the } x\text{-axis.}}$

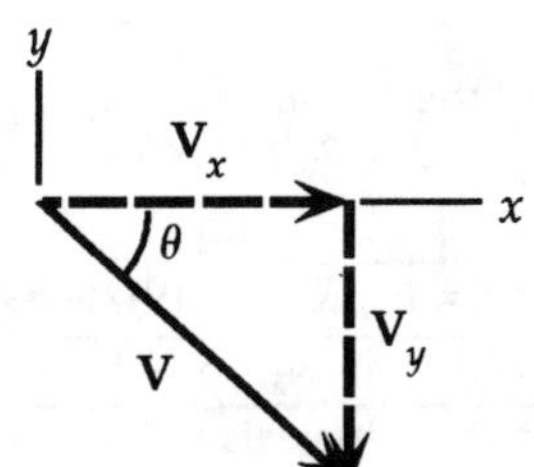

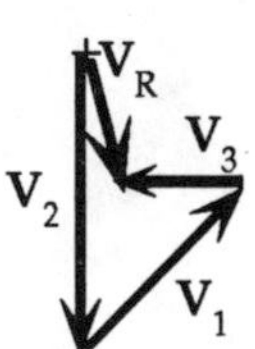

5. The resultant is $\boxed{31 \text{ m}, 44° \text{ N of E.}}$

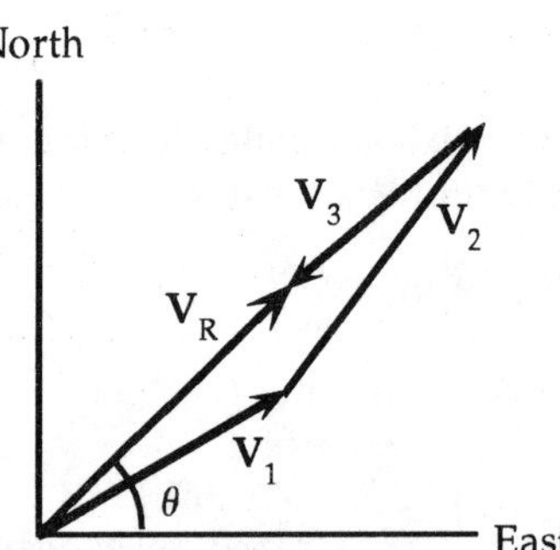

6. (*a*)

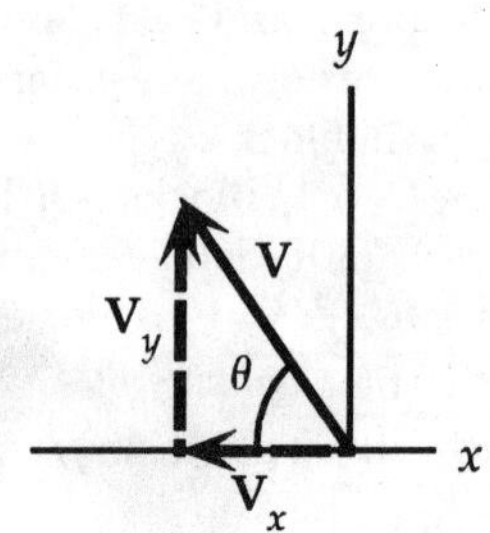

(*b*) For the components of the vector we have

$V_x = -V\cos\theta = -24.3 \cos 54.8° = \boxed{-14.0;}$

$V_y = V\sin\theta = 24.3 \sin 54.8° = \boxed{19.9.}$

(*c*) We find the vector from

$V = (V_x^2 + V_y^2)^{1/2} = [(-14.0)^2 + (19.9)^2]^{1/2}$

$= \boxed{24.3;}$

$\tan\theta = Vy/V_x = (19.9)/(14.0) = 1.42$, which gives

$\boxed{\theta = 54.8° \text{ above } -x\text{-axis.}}$

Note that we have used the magnitude of V_x for the angle indicated on the diagram.

7. Because the vectors are parallel, the direction can be indicated by the sign.

(*a*) $\mathbf{C} = \mathbf{A} + \mathbf{B} = 8.31 + (-5.55)$

$= \boxed{2.76 \text{ in the } +x\text{-direction.}}$

(*b*) $\mathbf{C} = \mathbf{A} - \mathbf{B} = 8.31 - (-5.55)$

$= \boxed{13.86 \text{ in the } +x\text{-direction.}}$

(*c*) $\mathbf{C} = \mathbf{B} - \mathbf{A} = -5.55 - (8.31)$

$= -13.86$ in the $+x$-direction or $\boxed{13.86 \text{ in the } -x\text{-direction.}}$

(*a*) $\mathbf{C} = \mathbf{A} + \mathbf{B}$

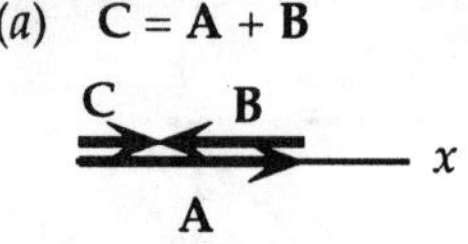

(*b*) $\mathbf{C} = \mathbf{A} - \mathbf{B}$

C x A −B

(*c*) $\mathbf{C} = \mathbf{B} - \mathbf{A}$

C x −A B

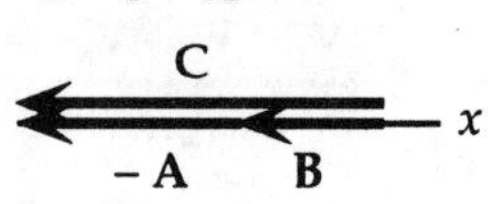

8. (*a*) $\boxed{\begin{array}{l} V_{1x} = -8.08, \quad V_{1y} = 0; \\ V_{2x} = V_2 \cos 45° = 4.51 \cos 45° = 3.19, \\ V_{2y} = V_2 \sin 45° = 4.51 \sin 45° = 3.19. \end{array}}$

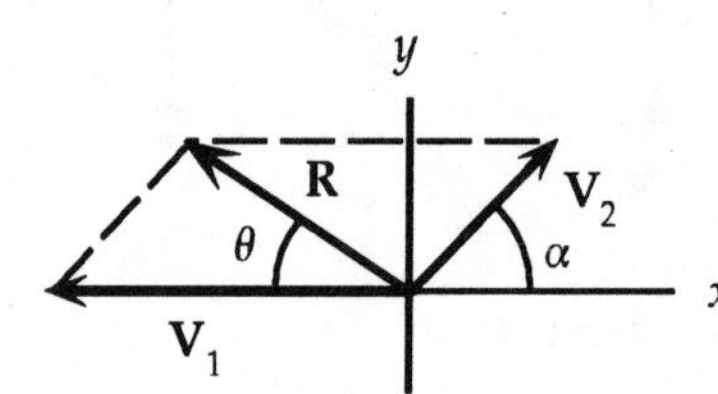

(*b*) For the components of the resultant we have

$R_x = V_{1x} + V_{2x} = -8.08 + 3.19 = -4.89;$

$R_y = V_{1y} + V_{2y} = 0 + 3.19 = 3.19.$

We find the resultant from

$R = (R_x^2 + R_y^2)^{1/2} = [(-4.89)^2 + (3.19)^2]^{1/2}$

$= \boxed{5.84;}$

$\tan\theta = R_y/R_x = (3.19)/(4.89) = 0.652$, which gives

$\boxed{\theta = 33.1° \text{ above } -x\text{-axis.}}$

Note that we have used the magnitude of R_x for the angle indicated on the diagram.

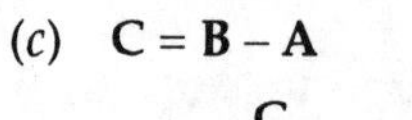

9. (*a*) Using the given angle, we find the components from

$V_N = V \cos 38.5° = (785 \text{ km/h}) \cos 38.5° = \boxed{614 \text{ km/h};}$

$V_W = V \sin 38.5° = (785 \text{ km/h}) \sin 38.5° = \boxed{489 \text{ km/h}.}$

(*b*) We use the velocity components to find the displacement components:

$d_N = V_N t = (614 \text{ km/h})(3.00 \text{ h}) = \boxed{1.84 \times 10^3 \text{ km};}$

$d_W = V_W t = (489 \text{ km/h})(3.00 \text{ h}) = \boxed{1.47 \times 10^3 \text{ km}.}$

10. We find the components of the sum by adding the components:

$\mathbf{V} = \mathbf{V}_1 + \mathbf{V}_2 = (3.0, 2.7, 0.0) + (2.9, -4.1, -1.4) = \boxed{(5.9, -1.4, -1.4).}$

By extending the Pythagorean theorem, we find the magnitude from

$V = (V_x^2 + V_y^2 + V_z^2)^{1/2} = [(5.9)^2 + (-1.4)^2 + (-1.4)^2]^{1/2} = \boxed{6.2.}$

11. (*a*) For the components we have

$R_x = A_x + B_x + C_x$

$= 66.0 \cos 28.0° - 40.0 \cos 56.0° + 0 = \boxed{35.9;}$

$R_y = A_y + B_y + C_y$

$= 66.0 \sin 28.0° + 40.0 \sin 56.0° - 46.8 = \boxed{17.3.}$

(*b*) We find the resultant from

$R = (R_x^2 + R_y^2)^{1/2} = [(35.9)^2 + (17.3)^2]^{1/2} = \boxed{39.9;}$

$\tan \theta = R_y/R_x = (17.3)/(35.9) = 0.483$, which gives

$\boxed{\theta = 25.8° \text{ above } + x\text{-axis.}}$

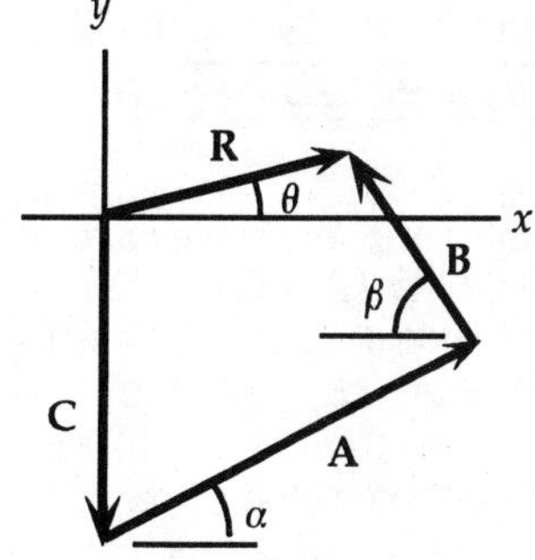

12. For the components we have

$R_x = A_x - C_x$

$= 66.0 \cos 28.0° - 0 = 58.3;$

$R_y = A_y - C_y$

$= 66.0 \sin 28.0° - (-46.8) = 77.8.$

We find the resultant from

$R = (R_x^2 + R_y^2)^{1/2} = [(58.3)^2 + (77.8)^2]^{1/2} = \boxed{97.2;}$

$\tan \theta = R_y/R_x = (77.8)/(58.3) = 1.33$, which gives

$\boxed{\theta = 53.1° \text{ above } + x\text{-axis.}}$

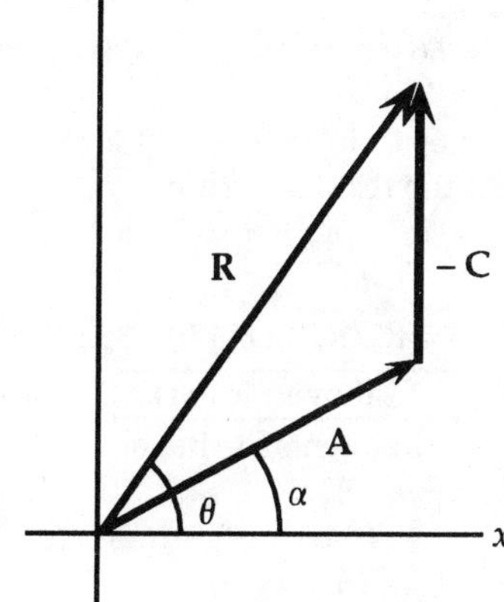

13\. (*a*) For the components we have

$R_x = B_x - A_x$
$= -40.0 \cos 56.0° - 66.0 \cos 28.0° = -80.7;$

$R_y = B_y - A_y$
$= 40.0 \sin 56.0° - 66.0 \sin 28.0° = 2.2.$

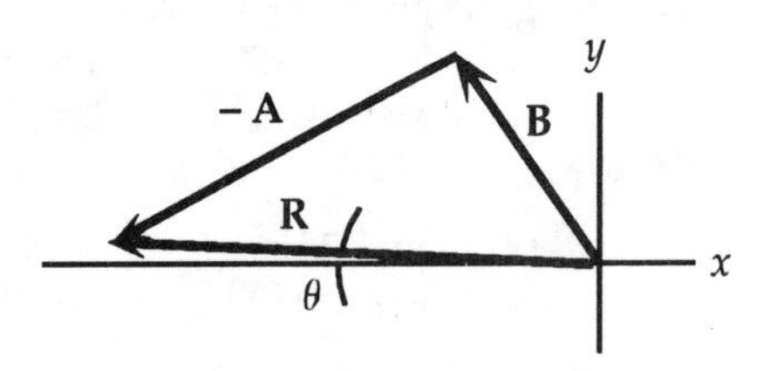

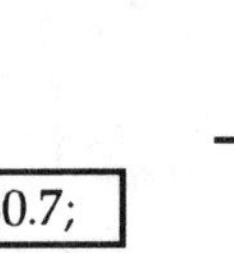

We find the resultant from

$R = (R_x^2 + R_y^2)^{1/2} = [(-80.7)^2 + (2.2)^2]^{1/2} =$ **80.7;**

$\tan \theta = R_y/R_x = (2.2)/(80.7) = 0.0273$, which gives

$\theta = 1.56°$ above $-x$-axis.

Note that we have used the magnitude of R_x for the angle indicated on the diagram.

(*b*) For the components we have

$R_x = A_x - B_x$
$= 66.0 \cos 28.0° - (-40.0 \cos 56.0°) = 80.7;$

$R_y = A_y - B_y$
$= 66.0 \sin 28.0° - 40.0 \sin 56.0° = -2.2.$

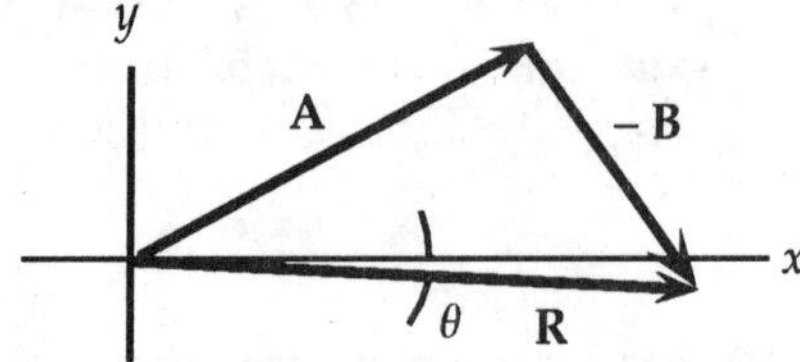

We find the resultant from

$R = (R_x^2 + R_y^2)^{1/2} = [(80.7)^2 + (-2.2)^2]^{1/2} =$ **80.7;**

$\tan \theta = R_y/R_x = (2.2)/(80.7) = 0.0273$, which gives

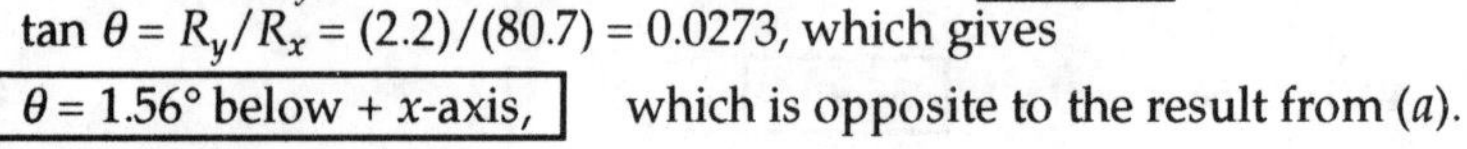

$\theta = 1.56°$ below $+x$-axis, which is opposite to the result from (*a*).

14\. (*a*) For the components we have

$R_x = A_x - B_x + C_x$
$= 66.0 \cos 28.0° - (-40.0 \cos 56.0°) + 0 = 80.7;$

$R_y = A_y - B_y + C_y$
$= 66.0 \sin 28.0° - 40.0 \sin 56.0° - 46.8 = -49.0.$

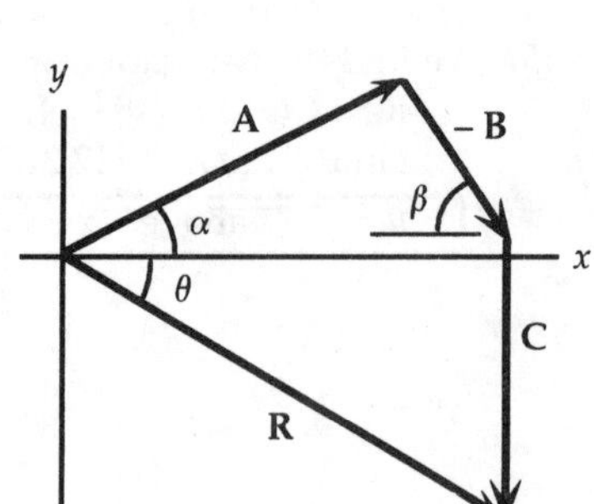

We find the resultant from

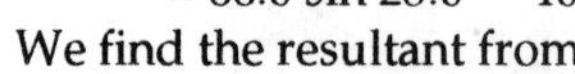

$R = (R_x^2 + R_y^2)^{1/2} = [(80.7)^2 + (-49.0)^2]^{1/2}$

$=$ **94.4;**

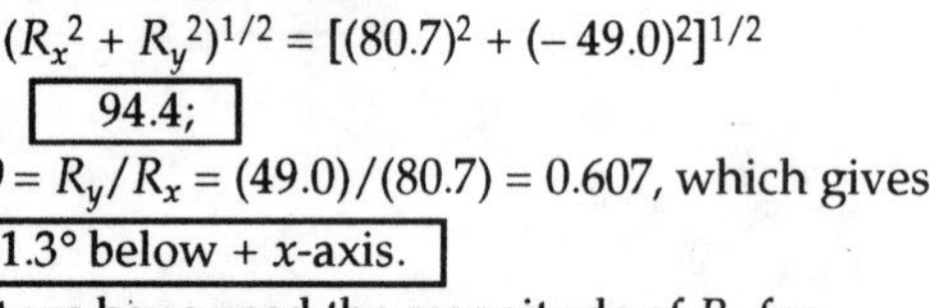

$\tan \theta = R_y/R_x = (49.0)/(80.7) = 0.607$, which gives

$\theta = 31.3°$ below $+x$-axis.

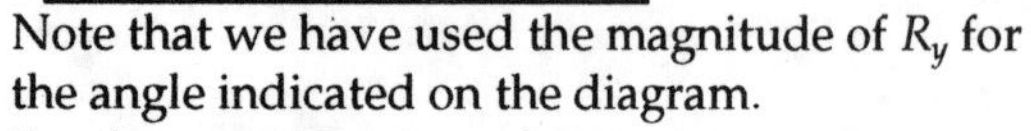

Note that we have used the magnitude of R_y for the angle indicated on the diagram.

(*b*) For the components we have

$R_x = A_x + B_x - C_x$
$= 66.0 \cos 28.0° + (-40.0 \cos 56.0°) - 0 = 35.9;$

$R_y = A_y + B_y - C_y$
$= 66.0 \sin 28.0° + 40.0 \sin 56.0° - (-46.8) = 111.0.$

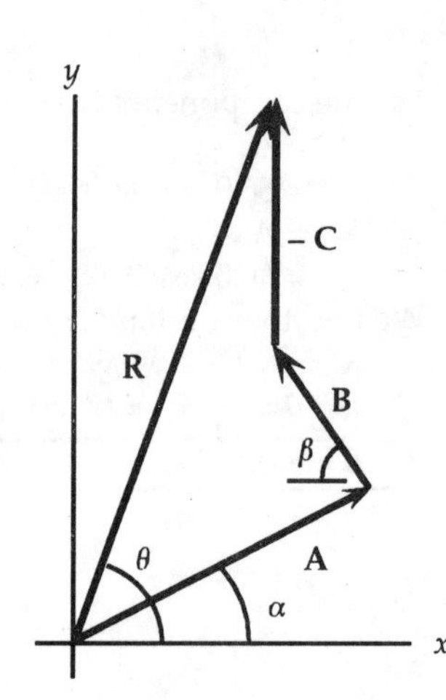

We find the resultant from

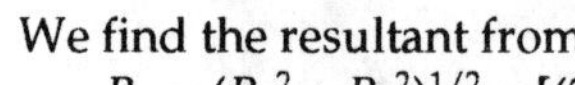

$R = (R_x^2 + R_y^2)^{1/2} = [(35.9)^2 + (111.0)^2]^{1/2}$

$=$ **117;**

$\tan \theta = R_y/R_x = (111.0)/(35.9) = 3.09$, which gives

$\theta = 72.1°$ above $+x$-axis.

(*c*) For the components we have

$R_x = B_x - 2A_x$
$= -40.0 \cos 56.0° - 2(66.0 \cos 28.0°) = -139.0;$

$R_y = B_y - 2A_y$
$= 40.0 \sin 56.0° - 2(66.0 \sin 28.0°) = -28.8.$

We find the resultant from

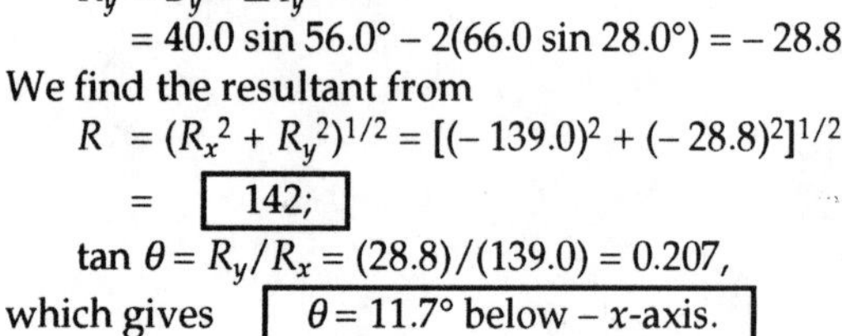

$R = (R_x^2 + R_y^2)^{1/2} = [(-139.0)^2 + (-28.8)^2]^{1/2}$

$=$ **142;**

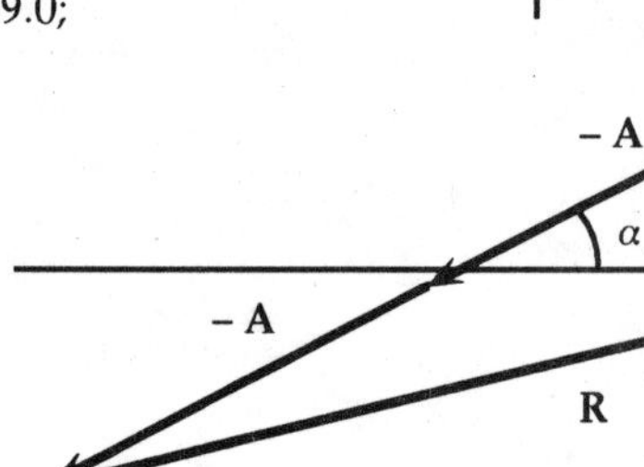

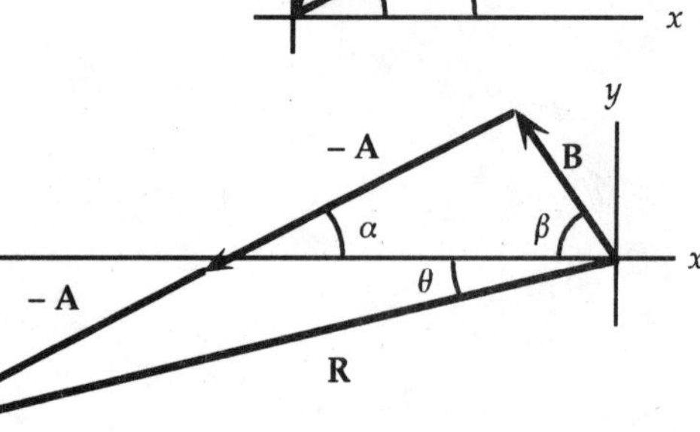

$\tan \theta = R_y/R_x = (28.8)/(139.0) = 0.207$,

which gives **$\theta = 11.7°$ below $-x$-axis.**

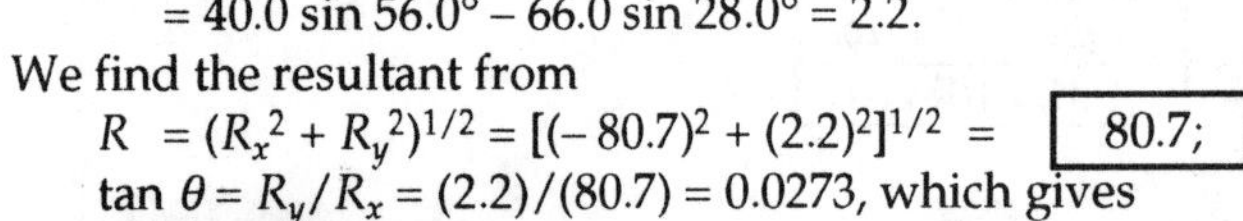

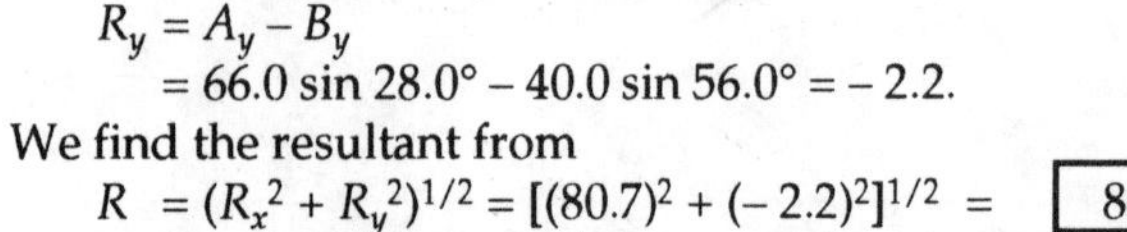

15. (*a*) For the components we have

$R_x = C_x - A_x - B_x$
$= 0 - 66.0 \cos 28.0° - (-40.0 \cos 56.0°)$
$= -35.9;$

$R_y = C_y - A_y - B_y$
$= -46.8 - 66.0 \sin 28.0° - 40.0 \sin 56.0°$
$= -111.0.$

We find the resultant from

$R = (R_x^2 + R_y^2)^{1/2} = [(-35.9)^2 + (-111.0)^2]^{1/2}$
$= \boxed{117;}$

$\tan \theta = R_y/R_x = (111.0)/(35.9) = 3.09,$

which gives

$\boxed{\theta = 72.1° \text{ below } -x\text{-axis.}}$

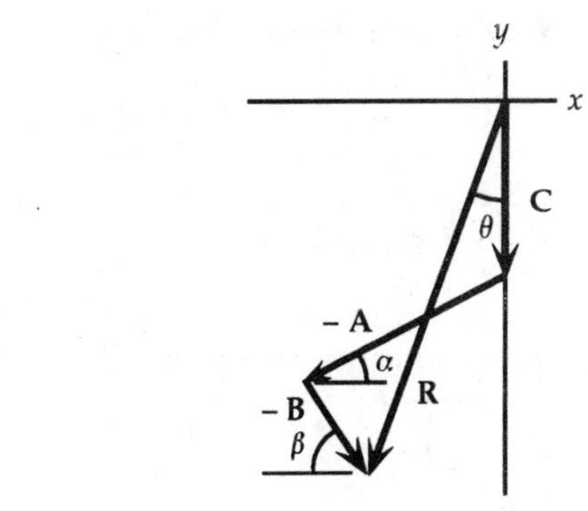

(*b*) For the components we have

$R_x = 2A_x - 3B_x + 2C_x$
$= 2(66.0 \cos 28.0°) - 3(-40.0 \cos 56.0°) + 2(0)$
$= 183.8;$

$R_y = 2A_y - 3B_y + 2C_y$
$= 2(66.0 \sin 28.0°) - 3(40.0 \sin 56.0°) + 2(-46.8)$
$= -131.2.$

We find the resultant from

$R = (R_x^2 + R_y^2)^{1/2} = [(183.8)^2 + (-131.2)^2]^{1/2}$
$= \boxed{226;}$

$\tan \theta = R_y/R_x = (131.2)/(183.8) = 0.714,$

which gives

$\boxed{\theta = 35.5° \text{ below } +x\text{-axis.}}$

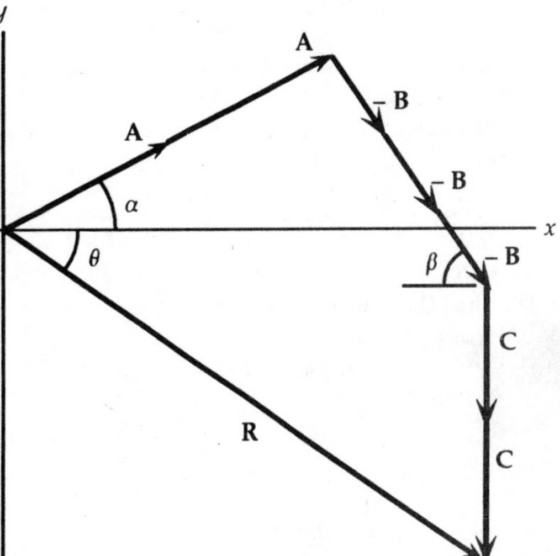

16. (*a*) For the vertical component we have

$a_V = (3.80 \text{ m/s}^2) \sin 30.0° = \boxed{1.90 \text{ m/s}^2 \text{ down.}}$

(*b*) Because the elevation change is the vertical displacement, we find the time from the vertical motion, taking down as the positive direction:

$y = v_{0y}t + \frac{1}{2}a_V t^2;$

$335 \text{ m} = 0 + \frac{1}{2}(1.90 \text{ m/s}^2)t^2$, which gives $\boxed{t = 18.8 \text{ s.}}$

17. For the components we have

$D_x = H_x = -4580 \sin 32.4° = \boxed{-2454 \text{ m;}}$

$D_y = H_y = +4580 \cos 32.4° = \boxed{+3867 \text{ m;}}$

$D_z = V = \boxed{+2085 \text{ m.}}$

By extending the Pythagorean theorem, we find the length from

$D = (D_x^2 + D_y^2 + D_z^2)^{1/2}$
$= [(-2454 \text{ m})^2 + (3867 \text{ m})^2 + (2085 \text{ m})^2]^{1/2}$
$= \boxed{5032 \text{ m.}}$

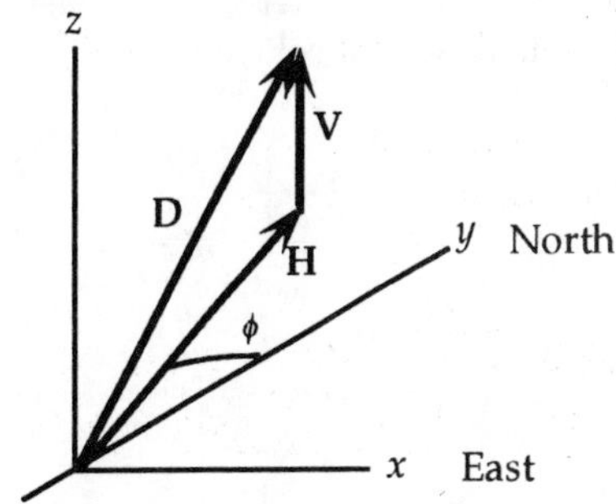

18. (*a*) We find the x-component from

$A^2 = A_x^2 + A_y^2$;

$(90.0)^2 = A_x^2 + (-55.0)^2$; which gives $\boxed{A_x = \pm 71.2.}$

(*b*) If we call the new vector **B**, we have

$R_x = A_x + B_x$;

$-80.0 = +71.2 + B_x$, which gives $\boxed{B_x = -151.2;}$

$R_y = A_y + B_y$;

$0 = -55.0 + B_y$, which gives $\boxed{B_y = +55.0.}$

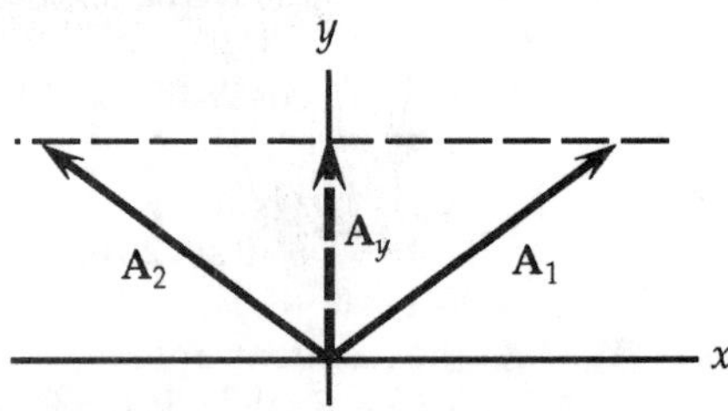

We find the resultant from

$B = (B_x^2 + B_y^2)^{1/2} = [(-151.2)^2 + (+55.0)^2]^{1/2} = \boxed{161;}$

$\tan\theta = B_y/B_x = (55.0)/(151.2) = 0.364$, which gives $\boxed{\theta = 20° \text{ above } -x\text{-axis.}}$

19. We choose a coordinate system with the origin at the takeoff point, with x horizontal and y vertical, with the positive direction down. We find the time for the tiger to reach the ground from its vertical motion:

$y = y_0 + v_{0y}t + \tfrac{1}{2}a_y t^2$;

$7.5 \text{ m} = 0 + 0 + \tfrac{1}{2}(9.80 \text{ m/s}^2)t^2$, which gives $t = 1.24$ s.

The horizontal motion will have constant velocity.

We find the distance from the base of the rock from

$x = x_0 + v_{0x}t$;

$x = 0 + (4.5 \text{ m/s})(1.24 \text{ s}) = \boxed{5.6 \text{ m.}}$

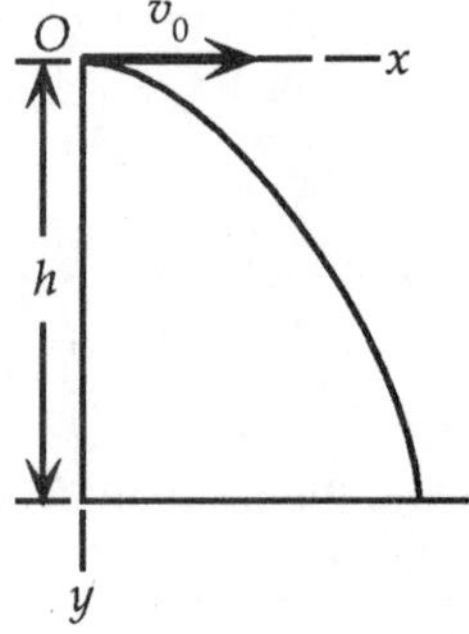

20. We choose a coordinate system with the origin at the takeoff point, with x horizontal and y vertical, with the positive direction down. We find the height of the cliff from the vertical displacement:

$y = y_0 + v_{0y}t + \tfrac{1}{2}a_y t^2$;

$y = 0 + 0 + \tfrac{1}{2}(9.80 \text{ m/s}^2)(3.0 \text{ s})^2 = \boxed{44 \text{ m.}}$

The horizontal motion will have constant velocity.

We find the distance from the base of the cliff from

$x = x_0 + v_{0x}t$;

$x = 0 + (1.6 \text{ m/s})(3.0 \text{ s}) = \boxed{4.8 \text{ m.}}$

21. Because the water returns to the same level, we can use the expression for the horizontal range:

$R = v_0^2 \sin(2\theta)/g$;

$2.0 \text{ m} = (6.5 \text{ m/s})^2 \sin(2\theta)/(9.80 \text{ m/s}^2)$, which gives

$\sin(2\theta) = 0.463$, or $2\theta = 28°$ and $152°$,

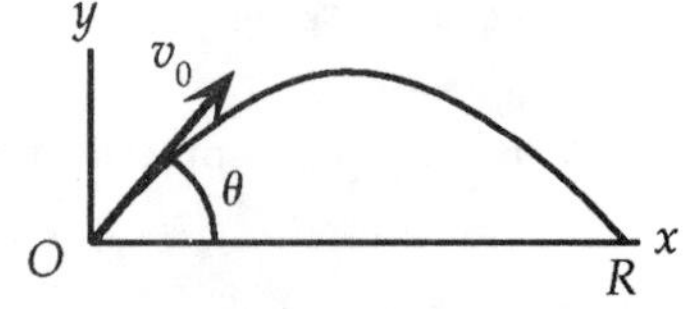

so the angles are $\boxed{14° \text{ and } 76°.}$

At the larger angle the water has a smaller horizontal velocity but spends more time in the air, because of the larger initial vertical velocity. Thus the horizontal displacement is the same for the two angles.

22. The horizontal velocity is constant, and the vertical velocity will be zero when the pebbles hit the window. Using the coordinate system shown, we find the vertical component of the initial velocity from

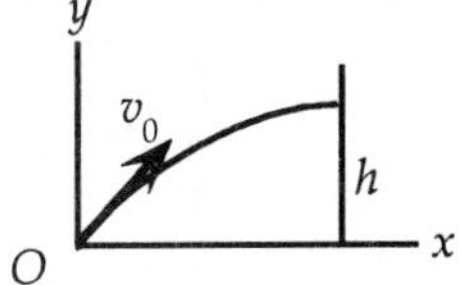

$v_y^2 = v_{0y}^2 + 2a_y(h - y_0)$;
$0 = v_{0y}^2 + 2(-9.80\ \text{m/s}^2)(8.0\ \text{m} - 0)$, which gives $v_{0y} = 12.5\ \text{m/s}$.
(We choose the positive square root because we know that the pebbles are thrown upward.)
We find the time for the pebbles to hit the window from the vertical motion:
$v_y = v_{0y} + a_yt$;
$0 = 12.5\ \text{m/s} + (9.80\ \text{m/s}^2)t$, which gives $t = 1.28\ \text{s}$.
For the horizontal motion we have
$x = x_0 + v_{0x}t$;
$9.0\ \text{m} = 0 + v_{0x}(1.28\ \text{s})$, which gives $\boxed{v_{0x} = 7.0\ \text{m/s}.}$
Because the pebbles are traveling horizontally when they hit the window, this is their speed.

23. The ball passes the goal posts when it has traveled the horizontal distance of 36.0 m. From this we can find the time when it passes the goal posts:
$x = v_{0x}t$; $36.0\ \text{m} = (20.0\ \text{m/s}) \cos 37.0°\ t$, which gives $t = 2.25\ \text{s}$.
To see if the kick is successful, we must find the height of the ball at this time:
$y = y_0 + v_{0y}t + \frac{1}{2}a_yt^2 = 0 + (20.0\ \text{m/s}) \sin 37.0°\ (2.25\ \text{s}) + \frac{1}{2}(-9.80\ \text{m/s}^2)(2.25\ \text{s})^2 = 2.24\ \text{m}$.
Thus the kick is $\boxed{\text{unsuccessful}}$ because it passes 0.76 m below the bar.
To have a successful kick, the ball must pass the goal posts with an elevation of at least 3.00 m. We find the time when the ball has this height from
$y = y_0 + v_{0y}t + \frac{1}{2}a_yt^2$;
$3.00\ \text{m} = 0 + (20.0\ \text{m/s}) \sin 37.0°\ t + \frac{1}{2}(-9.80\ \text{m/s}^2)t^2$.
The two solutions of this quadratic equation are $t = 0.28\ \text{s}, 2.17\ \text{s}$.
The horizontal distance traveled by the ball is found from
$x = v_{0x}t = (20.0\ \text{m/s}) \cos 37.0°\ t$.
For the two times, we get $x = 4.5\ \text{m}, 34.7\ \text{m}$.
Thus the kick must be made no farther than $\boxed{34.7\ \text{m}}$ from the goal posts (and no nearer than 4.5 m).
If the vertical velocity is found at these two times from
$v_y = v_{0y} + a_yt = (20.0\ \text{m/s}) \sin 37.0° + (-9.80\ \text{m/s}^2)t = +9.3\ \text{m/s}, -9.3\ \text{m/s}$,
we see that the ball is falling at the goal posts for a kick from 34.7 m and rising at the goal posts for a kick from 4.5 m.

24. We choose a coordinate system with the origin at the release point, with x horizontal and y vertical, with the positive direction down. We find the time of fall from the vertical displacement:

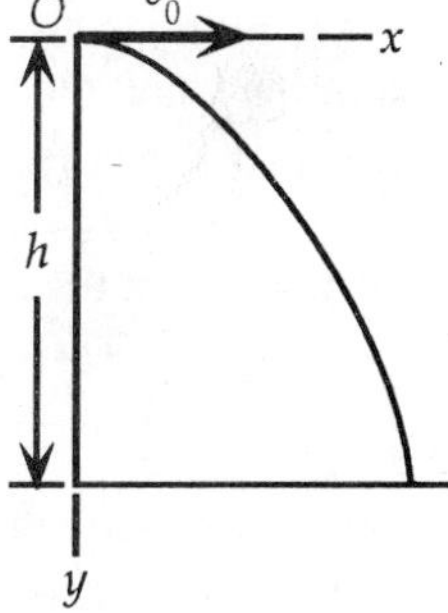

$y = y_0 + v_{0y}t + \frac{1}{2}a_yt^2$;
$56\ \text{m} = 0 + 0 + \frac{1}{2}(9.80\ \text{m/s}^2)t^2$, which gives $t = 3.38\ \text{s}$.
The horizontal motion will have constant velocity.
We find the initial speed from
$x = x_0 + v_{0x}t$;
$45\ \text{m} = 0 + v_0(3.38\ \text{s})$, which gives $\boxed{v_0 = 13\ \text{m/s}.}$

25. The horizontal motion will have constant velocity, v_{0x}.
Because the projectile lands at the same elevation, we find the vertical velocity on impact from
$v_y^2 = v_{0y}^2 + 2a_y(y - y_0) = v_{0y}^2 + 0 = v_{0y}^2$, so $v_y = -v_{0y}$.
(We choose the negative square root because we know that the projectile is coming down.)
The speed at impact is
$v = (v_x^2 + v_y^2)^{1/2} = [v_{0x}^2 + (-v_{0y})^2]^{1/2} = (v_{0x}^2 + v_{0y}^2)^{1/2}$, which is the initial speed.

26. We find the time of flight from the vertical displacement:

$y = y_0 + v_{0y}t + \frac{1}{2}a_y t^2$;

$0 = 0 + (20.0 \text{ m/s})(\sin 37.0°)t + \frac{1}{2}(-9.80 \text{ m/s}^2)t^2$, which gives $t = 0, 2.46$ s.

The ball is kicked at $t = 0$, so the football hits the ground **2.46 s** later.

27. We choose a coordinate system with the origin at the release point, with x horizontal and y vertical, with the positive direction down.

The horizontal motion will have constant velocity. We find the time required for the fall from

$x = x_0 + v_{0x}t$;

$36.0 \text{ m} = 0 + (22.2 \text{ m/s})t$, which gives $t = 1.62$ s.

We find the height from the vertical motion:

$y = y_0 + v_{0y}t + \frac{1}{2}a_y t^2$;

$h = 0 + 0 + \frac{1}{2}(9.80 \text{ m/s}^2)(1.62 \text{ s})^2 =$ **12.9 m.**

28. We choose a coordinate system with the origin at the release point, with x horizontal and y vertical, with the positive direction up.

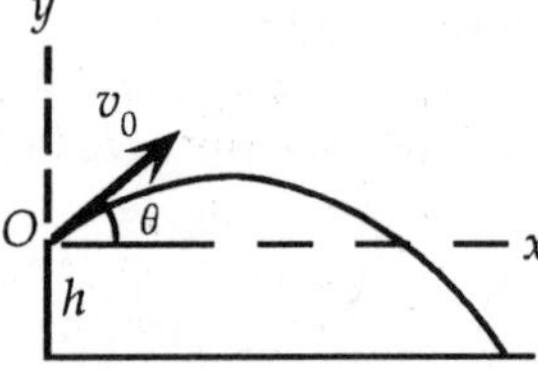

We find the time required for the fall from the vertical motion:

$y = y_0 + v_{0y}t + \frac{1}{2}a_y t^2$;

$-2.2 \text{ m} = 0 + (14 \text{ m/s})(\sin 40°)t + \frac{1}{2}(-9.80 \text{ m/s}^2)t^2$.

The solutions of this quadratic equation are $t = -0.22$ s, 2.06 s.

Because the shot is released at $t = 0$, the physical answer is 2.06 s.

We find the horizontal distance from

$x = x_0 + v_{0x}t$;

$x = 0 + (14 \text{ m/s})(\cos 40°)(2.06 \text{ s}) =$ **22 m.**

29. Because the initial and final locations are at the same level, we can use the expression for the horizontal range. The horizontal range on Earth is given by

$R = v_0^2 \sin(2\theta_0)/g$, whereas on the moon it is

$R_{moon} = v_0^2 \sin(2\theta_0)/g_{moon}$.

Because we have the same v_0 and θ_0, when we divide the two equations, we get

$R_{moon}/R = g/g_{moon}$, or

$R_{moon} = (g/g_{moon})R = [g/(g/6)]R = 6R$, so a person could jump **six times as far.**

30. (*a*) Because the athlete lands at the same level, we can use the expression for the horizontal range:

$R = v_0^2 \sin(2\theta_0)/g$;

$7.80 \text{ m} = v_0^2 \sin[2(30°)]/(9.80 \text{ m/s}^2)$, which gives **$v_0 = 9.39$ m/s.**

(*b*) For an increase of 5%, the initial speed becomes $v_0' = (1 + 0.05)v_0 = (1.05)v_0$, and the new range is

$R' = v_0'^2 \sin(2\theta_0)/g = (1.05)^2 v_0^2 \sin(2\theta_0)/g = 1.10R$.

Thus the increase in the length of the jump is

$R' - R = (1.10 - 1)R = 0.10(7.80 \text{ m}) =$ **0.80 m.**

31. We choose a coordinate system with the origin at the release point, with x horizontal and y vertical, with the positive direction down.

Because the horizontal velocity of the package is constant at the horizontal velocity of the airplane, the package is always directly under the airplane. The time interval is the time required for the fall, which we find from the vertical motion:

$y = y_0 + v_{0y}t + \frac{1}{2}a_y t^2$;

$160 \text{ m} = 0 + 0 + \frac{1}{2}(9.80 \text{ m/s}^2)t^2$, which gives **$t = 5.71$ s.**

32. (*a*) We choose a coordinate system with the origin at the release point, with x horizontal and y vertical, with the positive direction down. We find the time of flight from the horizontal motion:

$x = x_0 + v_{0x}t;$

$120 \text{ m} = 0 + (250 \text{ m/s})t$, which gives $t = 0.480$ s.

We find the distance the bullet falls from

$y = y_0 + v_{0y}t + \frac{1}{2}a_y t^2;$

$y = 0 + 0 + \frac{1}{2}(9.80 \text{ m/s}^2)(0.480 \text{ s})^2 =$ **1.13 m.**

(*b*) The bullet will hit the target at the same elevation, so we can use the expression for the horizontal range:

$R = v_0^2 \sin(2\theta_0)/g;$

$120 \text{ m} = (250 \text{ m/s})^2 \sin(2\theta_0)/(9.80 \text{ m/s}^2)$, which gives $\sin(2\theta_0) = 0.0188$, or $2\theta_0 = 1.08°$, $\boxed{\theta_0 = 0.54°.}$

The larger angle of 89.5° is not realistic.

33. We choose a coordinate system with the origin at the release point, with y vertical and the positive direction up. At the highest point the vertical velocity will be zero, so we find the time to reach the highest point from

$v_y = v_{0y} + a_y t_{up};$

$0 = v_{0y} + (-g)t_{up}$, which gives $t_{up} = v_{0y}/g$.

We find the elevation h at the highest point from

$v_y^2 = v_{0y}^2 + 2a_y(y - y_0);$

$0 = v_{0y}^2 + 2(-g)h$, which gives $h = v_{0y}^2/2g$.

We find the time to fall from the highest point from

$y = y_0 + v_{0y}t + \frac{1}{2}a_y t_{down}^2;$

$0 = h + 0 + \frac{1}{2}(-g)t_{down}^2$, which gives

$t_{down} = (2h/g)^{1/2} = [2(v_{0y}^2/2g)/g]^{1/2} = v_{0y}/g$, which is the same as t_{up}.

34. To plot the trajectory, we need a relationship between x and y, which can be obtained by eliminating t from the equations for the two components of the motion:

$x = v_{0x}t = v_0 (\cos \theta)t;$

$y = y_0 + v_{0y}t + \frac{1}{2}a_y t^2 = 0 + v_0 (\sin \theta)t + \frac{1}{2}(-g)t^2.$

The relationship is

$y = (\tan \theta)x - \frac{1}{2}g(x/v_0 \cos \theta)^2.$

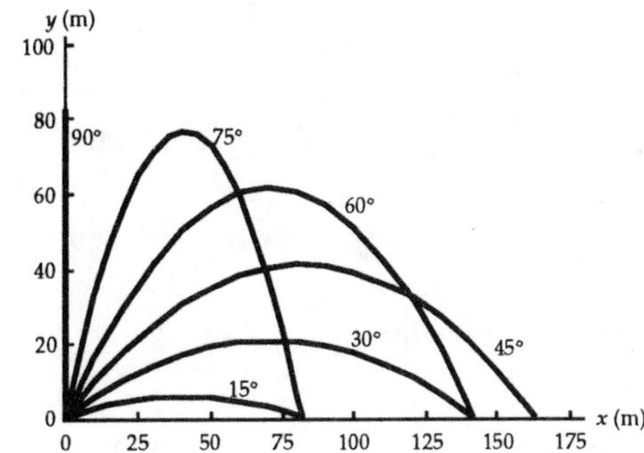

35. (*a*) At the highest point, the vertical velocity $v_y = 0$. We find the maximum height h from

$v_y^2 = v_{0y}^2 + 2a_y(y - y_0);$

$0 = [(75.2 \text{ m/s}) \sin 34.5°]^2 + 2(-9.80 \text{ m/s}^2)(h - 0)$, which gives $\boxed{h = 92.6 \text{ m.}}$

(*b*) Because the projectile returns to the same elevation, we have

$y = y_0 + v_{0y}t + \frac{1}{2}a_y t^2;$

$0 = 0 + (75.2 \text{ m/s})(\sin 34.5°)t + \frac{1}{2}(-9.80 \text{ m/s}^2)t^2$, which gives $t = 0$, and 8.69 s.

Because $t = 0$ was the launch time, the total time in the air was **8.69 s.**

(*c*) We find the horizontal distance from

$x = v_{0x}t = (75.2 \text{ m/s})(\cos 34.5°)(8.69 \text{ s}) =$ **539 m.**

(*d*) The horizontal velocity will be constant: $v_x = v_{0x} = (75.2 \text{ m/s}) \cos 34.5° = 62.0 \text{ m/s}$.

We find the vertical velocity from

$v_y = v_{0y} + a_y t = (75.2 \text{ m/s}) \sin 34.5° + (-9.80 \text{ m/s}^2)(1.50 \text{ s}) = 27.9 \text{ m/s}.$

The magnitude of the velocity is

$v = (v_x^2 + v_y^2)^{1/2} = [(62.0 \text{ m/s})^2 + (27.9 \text{ m/s})^2]^{1/2} =$ **68.0 m/s.**

We find the angle from

$\tan \theta = v_y/v_x = (27.9 \text{ m/s})/(62.0 \text{ m/s}) = 0.450$, which gives $\boxed{\theta = 24.2° \text{ above the horizontal.}}$

36. (*a*) We choose a coordinate system with the origin at the base of the cliff, with x horizontal and y vertical, with the positive direction up.

We find the time required for the fall from the vertical motion:

$y = y_0 + v_{0y}t + \frac{1}{2}a_yt^2;$

$0 = 125\text{ m} + (105\text{ m/s})(\sin 37.0°)t + \frac{1}{2}(-9.80\text{ m/s}^2)t^2,$

which gives $t = -1.74,\ 14.6$ s.

Because the projectile starts at $t = 0$, we have $\boxed{t = 14.6\text{ s.}}$

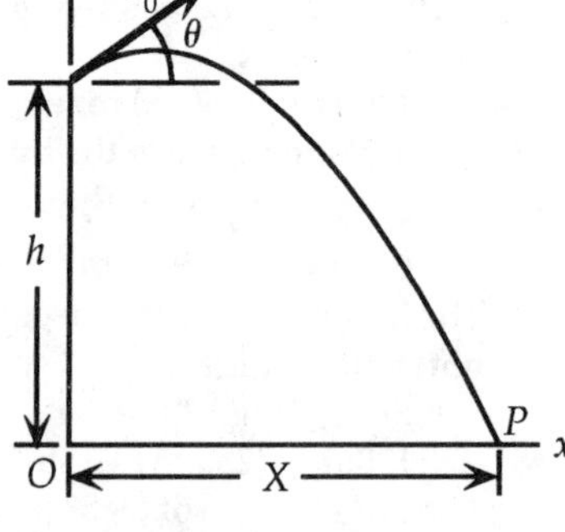

(*b*) We find the range from the horizontal motion:

$X = v_{0x}t = (105\text{ m/s})(\cos 37.0°)(14.6\text{ s})$

$= 1.22 \times 10^3\text{ m} = \boxed{1.22\text{ km.}}$

(*c*) For the velocity components, we have

$v_x = v_{0x} = (105\text{ m/s})\cos 37.0° = \boxed{83.9\text{ m/s.}}$

$v_y = v_{0y} + a_yt = (105\text{ m/s})\sin 37.0° + (-9.80\text{ m/s}^2)(14.6\text{ s}) = \boxed{-79.9\text{ m/s.}}$

(*d*) When we combine these components, we get

$v = (v_x^2 + v_y^2)^{1/2} = [(83.9\text{ m/s})^2 + (-79.9\text{ m/s})^2]^{1/2} = \boxed{116\text{ m/s.}}$

(*e*) We find the angle from

$\tan\theta = v_y/v_x = (79.9\text{ m/s})/(83.9\text{ m/s}) = 0.952$, which gives $\boxed{\theta = 43.6°\text{ below the horizontal.}}$

37. We use the coordinate system shown in the diagram.

To see if the water balloon hits the boy, we will find the location of the water balloon and the boy when the water balloon passes the vertical line that the boy follows.

We find the time at which this occurs from the horizontal motion of the water balloon:

$d = v_{0x}t = (v_0\cos\theta_0)t$, which gives $t = d/(v_0\cos\theta_0)$.

At this time the location of the boy is

$y_{\text{boy}} = y_{0\text{boy}} + v_{0y\text{boy}}t + \frac{1}{2}a_yt^2$

$= h + 0 + \frac{1}{2}(-g)[d/(v_0\cos\theta_0)]^2 = h - \frac{1}{2}g[d/(v_0\cos\theta_0)]^2.$

The vertical position of the water balloon will be

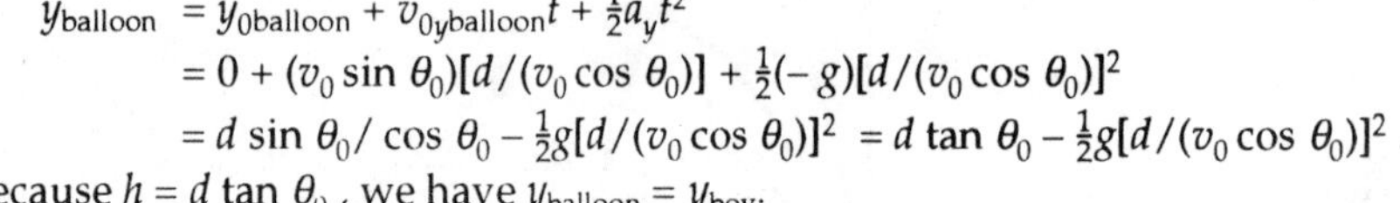

$y_{\text{balloon}} = y_{0\text{balloon}} + v_{0y\text{balloon}}t + \frac{1}{2}a_yt^2$

$= 0 + (v_0\sin\theta_0)[d/(v_0\cos\theta_0)] + \frac{1}{2}(-g)[d/(v_0\cos\theta_0)]^2$

$= d\sin\theta_0/\cos\theta_0 - \frac{1}{2}g[d/(v_0\cos\theta_0)]^2 = d\tan\theta_0 - \frac{1}{2}g[d/(v_0\cos\theta_0)]^2.$

Because $h = d\tan\theta_0$, we have $y_{\text{balloon}} = y_{\text{boy}}$.

38. (*a*) The time to fall is found from $y = y_0 + v_{0y}t + \frac{1}{2}a_yt^2;$

$0 = 235\text{ m} + 0 + \frac{1}{2}(-9.80\text{ m/s}^2)t_1^2$, which gives $t_1 = 6.93$ s.

Because the goods have the same horizontal speed as the airplane, they will always be directly below the airplane and will have a horizontal displacement of

$x_1 = v_{0x}t_1 = (69.4\text{ m/s})(6.93\text{ s}) = \boxed{480\text{ m,}}$

which is the distance from the climbers that the goods must be dropped.

(*b*) If the goods are given a vertical velocity, they will still have the horizontal velocity of the airplane. We find the time to fall from

$x_2 = v_{0x}t_2;\quad 425\text{ m} = (69.4\text{ m/s})t_2$, which gives $t_2 = 6.12$ s.

We find the necessary vertical velocity from

$y = y_0 + v_{0y}t_2 + \frac{1}{2}a_yt_2^2;$

$0 = 235\text{ m} + v_{0y}(6.12\text{ s}) + \frac{1}{2}(-9.80\text{ m/s}^2)(6.12\text{ s})^2$, which gives $\boxed{v_{0y} = -8.41\text{ m/s (down).}}$

(*c*) For the velocity components after the fall, we have

$v_x = v_{0x} = 69.4\text{ m/s}.$

$v_y = v_{0y} + a_yt_2 = -8.41\text{ m/s} + (-9.80\text{ m/s}^2)(6.12\text{ s}) = -68.4\text{ m/s}.$

When we combine these components, we get

$v = (v_x^2 + v_y^2)^{1/2} = [(69.4\text{ m/s})^2 + (-68.4\text{ m/s})^2]^{1/2} = \boxed{97.5\text{ m/s.}}$

39. We will take down as the positive direction. The direction of motion is the direction of the velocity. For the velocity components, we have

$v_x = v_{0x} = v_0$.

$v_y = v_{0y} + a_y t = 0 + gt = gt.$

We find the angle that the velocity vector makes with the horizontal from

$\tan\theta = v_y/v_x = gt/v_0$, or $\boxed{\theta = \tan^{-1}(gt/v_0) \text{ below the horizontal.}}$

40. If $\mathbf{v}_{sw}$ is the velocity of the ship with respect to the water,

$\mathbf{v}_{js}$ the velocity of the jogger with respect to the ship, and

$\mathbf{v}_{jw}$ the velocity of the jogger with respect to the water, then

$\mathbf{v}_{jw} = \mathbf{v}_{js} + \mathbf{v}_{sw}$.

We choose the direction of the ship as the positive direction. Because all vectors are parallel, in each case the motion is one-dimensional.

When the jogger is moving toward the bow, we have

$v_{jw} = v_{js} + v_{sw} = 2.0 \text{ m/s} + 8.5 \text{ m/s} = \boxed{10.5 \text{ m/s in the direction of the ship's motion.}}$

When the jogger is moving toward the stern, we have

$v_{jw} = -v_{js} + v_{sw} = -2.0 \text{ m/s} + 8.5 \text{ m/s} = \boxed{6.5 \text{ m/s in the direction of the ship's motion.}}$

41. If $\mathbf{v}_{Hr}$ is the velocity of Huck with respect to the raft,

$\mathbf{v}_{Hb}$ the velocity of Huck with respect to the bank, and

$\mathbf{v}_{rb}$ the velocity of the raft with respect to the bank, then

$\mathbf{v}_{Hb} = \mathbf{v}_{Hr} + \mathbf{v}_{rb}$, as shown in the diagram.

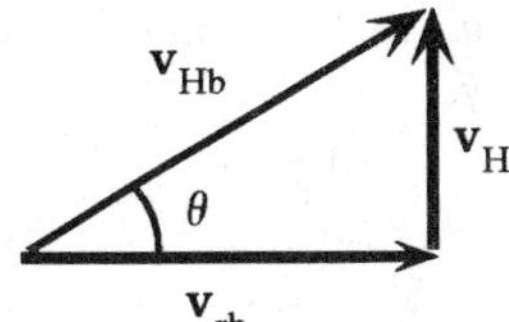

From the diagram we get

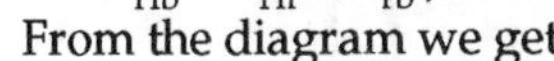

$v_{Hb} = (v_{Hr}^2 + v_{rb}^2)^{1/2} = [(1.0 \text{ m/s})^2 + (2.7 \text{ m/s})^2]^{1/2} = \boxed{2.9 \text{ m/s.}}$

We find the angle from

$\tan\theta = v_{Hr}/v_{rb} = (1.0 \text{ m/s})/(2.7 \text{ m/s}) = 0.37$, which gives $\boxed{\theta = 20° \text{ from the river bank.}}$

42. If $\mathbf{v}_{SG}$ is the velocity of the snow with respect to the ground,

$\mathbf{v}_{CG}$ the velocity of the car with respect to the ground, and

$\mathbf{v}_{SC}$ the velocity of the snow with respect to the car, then

$\mathbf{v}_{SC} = \mathbf{v}_{SG} + \mathbf{v}_{GC} = \mathbf{v}_{SG} - \mathbf{v}_{CG}$, as shown in the diagram.

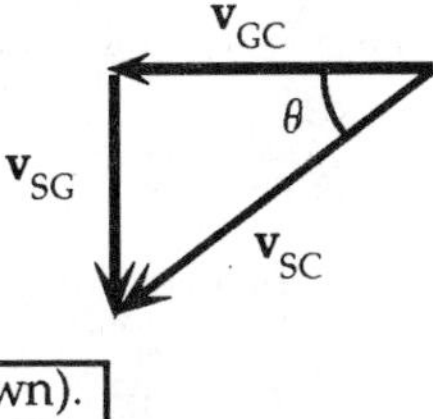

From the diagram we get

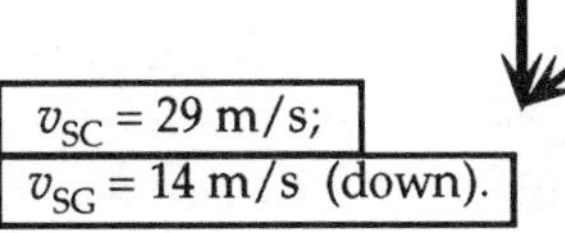

$\cos\theta = v_{GC}/v_{SC}$, or $\cos 30° = (25 \text{ m/s})/v_{SC}$, which gives $\boxed{v_{SC} = 29 \text{ m/s;}}$

$\tan\theta = v_{SG}/v_{GC}$, or $\tan 30° = v_{SG}/(25 \text{ m/s})$, which gives $\boxed{v_{SG} = 14 \text{ m/s (down).}}$

43. (*a*) If $\mathbf{v}_{bs}$ is the velocity of the boat with respect to the shore,

$\mathbf{v}_{bw}$ the velocity of the boat with respect to the water, and

$\mathbf{v}_{ws}$ the velocity of the water with respect to the shore, then

$\mathbf{v}_{bs} = \mathbf{v}_{bw} + \mathbf{v}_{ws}$, as shown in the diagram.

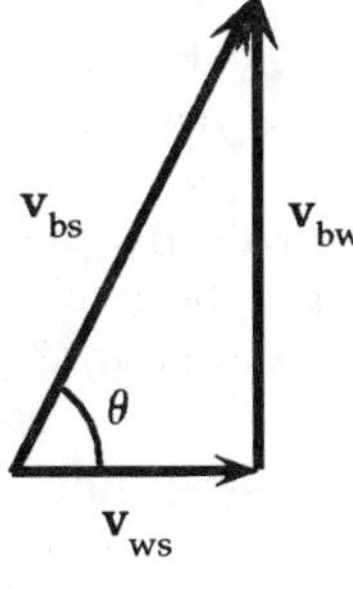

From the diagram we get

$v_{bs} = (v_{bw}^2 + v_{ws}^2)^{1/2} = [(2.30 \text{ m/s})^2 + (1.20 \text{ m/s})^2]^{1/2} = \boxed{2.59 \text{ m/s.}}$

We find the angle from

$\tan\theta = v_{bw}/v_{wr} = (2.30 \text{ m/s})/(1.20 \text{ m/s}) = 1.92$, which gives

$\boxed{\theta = 62.4° \text{ from the shore.}}$

(*b*) Because the boat will move with constant velocity, the displacement will be

$d = v_{bs}t = (2.59 \text{ m/s})(3.00 \text{ s}) = \boxed{7.77 \text{ m at } 62.4° \text{ to the shore.}}$

44. Because the planes are approaching along the same line, for the relative velocity we have

$v = v_1 - v_2 = 835 \text{ km/h} - (-835 \text{ km/h}) = 1670 \text{ km/h}.$

We find the time before they would meet from

$t = d/v = (10.0 \text{ km})/(1670 \text{ km/h}) = \boxed{0.00600 \text{ h} = 21.6 \text{ s.}}$

45. If $\mathbf{v}_{pa}$ is the velocity of the airplane with respect to the air, $\mathbf{v}_{pg}$ the velocity of the airplane with respect to the ground, and $\mathbf{v}_{ag}$ the velocity of the air(wind) with respect to the ground, then $\mathbf{v}_{pg} = \mathbf{v}_{pa} + \mathbf{v}_{ag}$, as shown in the diagram.

(*a*) From the diagram we find the two components of $\mathbf{v}_{pg}$:

$v_{pgE} = v_{ag} \cos 45° = (100 \text{ km/h}) \cos 45° = 70.7 \text{ km/h};$

$v_{pgS} = v_{pa} - v_{ag} \sin 45° = 500 \text{ km/h} - (100 \text{ km/h}) \cos 45° = 429 \text{ km/h}.$

For the magnitude we have

$v_{pg} = (v_{pgE}^2 + v_{pgS}^2)^{1/2} = [(70.7 \text{ km/h})^2 + (429 \text{ km/h})^2]^{1/2} =$ **435 km/h.**

We find the angle from

$\tan \theta = v_{pgE}/v_{pgS} = (70.7 \text{ km/h})/(429 \text{ km/h}) = 0.165$, which gives

$\theta = 9.36°$ east of south.

(*b*) Because the pilot is expecting to move south, we find the easterly distance from this line from

$d = v_{pgE}t = (70.7 \text{ km/h})(10 \text{ min})/(60 \text{ min/h}) =$ **12 km.**

Of course the airplane will also not be as far south as it would be without the wind.

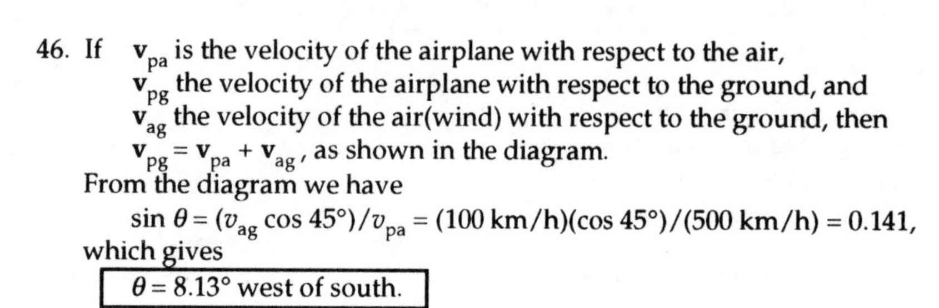

46. If $\mathbf{v}_{pa}$ is the velocity of the airplane with respect to the air, $\mathbf{v}_{pg}$ the velocity of the airplane with respect to the ground, and $\mathbf{v}_{ag}$ the velocity of the air(wind) with respect to the ground, then $\mathbf{v}_{pg} = \mathbf{v}_{pa} + \mathbf{v}_{ag}$, as shown in the diagram.

From the diagram we have

$\sin \theta = (v_{ag} \cos 45°)/v_{pa} = (100 \text{ km/h})(\cos 45°)/(500 \text{ km/h}) = 0.141,$

which gives

$\theta = 8.13°$ west of south.

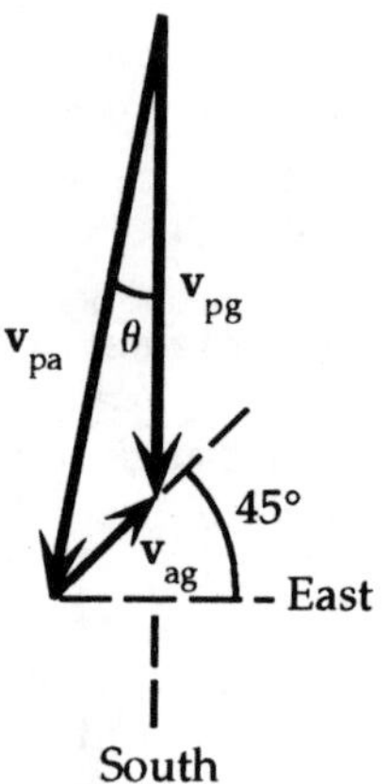

47. From the vector diagram of Example 3–10, we have

$v_{BW}^2 = v_{BS}^2 + v_{WS}^2;$

$(1.85 \text{ m/s})^2 = v_{BS}^2 + (1.20 \text{ m/s})^2$, which gives **$v_{BS} = 1.41$ m/s.**

48. If $\mathbf{v}_{pw}$ is the velocity of the passenger with respect to the water, $\mathbf{v}_{pb}$ the velocity of the passenger with respect to the boat, and $\mathbf{v}_{bw}$ the velocity of the boat with respect to the water, then $\mathbf{v}_{pw} = \mathbf{v}_{pb} + \mathbf{v}_{bw}$, as shown in the diagram.

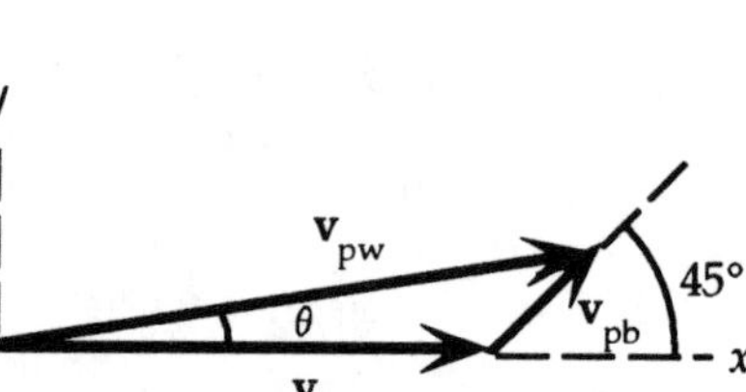

From the diagram we find the two components of $\mathbf{v}_{pw}$:

$v_{pwx} = v_{pb} \cos 45° + v_{bw}$

$= (0.50 \text{ m/s}) \cos 45° + 1.50 \text{ m/s} = 1.85 \text{ m/s};$

$v_{pwy} = v_{pb} \sin 45° = (0.50 \text{ m/s}) \sin 45° = 0.354 \text{ m/s}.$

For the magnitude we have

$v_{pw} = (v_{pwx}^2 + v_{pwy}^2)^{1/2} = [(1.85 \text{ m/s})^2 + (0.354 \text{ m/s})^2]^{1/2} =$ **1.9 m/s.**

We find the angle from

$\tan \theta = v_{pwy}/v_{pwx} = (0.354 \text{ m/s})/(1.85 \text{ m/s}) = 0.191$, which gives

$\theta = 11°$ above the water.

49. If $\mathbf{v}_{bs}$ is the velocity of the boat with respect to the shore, $\mathbf{v}_{bw}$ the velocity of the boat with respect to the water, and $\mathbf{v}_{ws}$ the velocity of the water with respect to the shore, then $\mathbf{v}_{bs} = \mathbf{v}_{bw} + \mathbf{v}_{ws}$, as shown in the diagram.

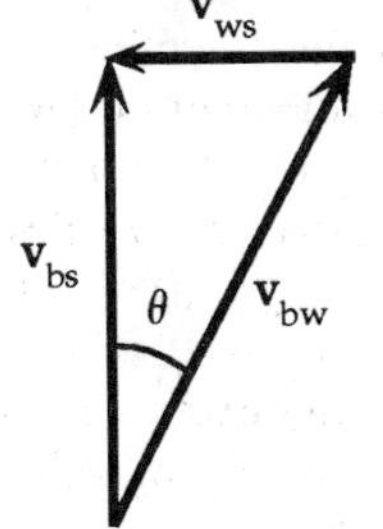

(a) From the diagram we have

$v_{ws} = v_{bw} \sin \theta = (3.60 \text{ m/s}) \sin 27.5° = \boxed{1.66 \text{ m/s.}}$

(b) $v_{bs} = v_{bw} \cos \theta = (3.60 \text{ m/s}) \cos 27.5° = \boxed{3.19 \text{ m/s.}}$

50. If $\mathbf{v}_{bs}$ is the velocity of the boat with respect to the shore, $\mathbf{v}_{bw}$ the velocity of the boat with respect to the water, and $\mathbf{v}_{ws}$ the velocity of the water with respect to the shore, then $\mathbf{v}_{bs} = \mathbf{v}_{bw} + \mathbf{v}_{ws}$, as shown in the diagram.

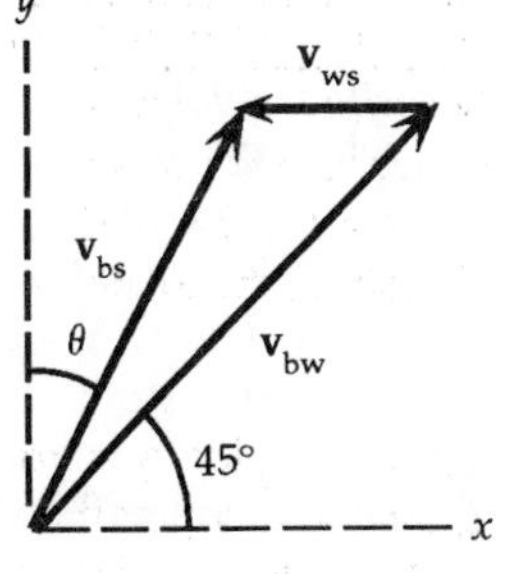

We find the angle of the boat's motion from the distances:

$\tan \theta = d_{shore}/d_{river} = (110 \text{ m})/(260 \text{ m}) = 0.423$, which gives $\theta = 22.9°$.

The y-component of $\mathbf{v}_{bs}$ is also the y-component of $\mathbf{v}_{bw}$:

$v_{bsy} = v_{bwy} = (2.20 \text{ m/s}) \sin 45° = 1.56 \text{ m/s}.$

We find the x-component from

$v_{bsx} = v_{bsy} \tan \theta = (1.56 \text{ m/s}) \tan 22.9° = 0.657 \text{ m/s}.$

For the x-component of the relative velocity, we use the diagram to get

$|v_{ws}| = v_{bwx} - v_{bsx} = (2.20 \text{ m/s}) \cos 45° - 0.657 \text{ m/s} = \boxed{0.90 \text{ m/s.}}$

51. If $\mathbf{v}_{sb}$ is the velocity of the swimmer with respect to the bank, $\mathbf{v}_{sw}$ the velocity of the swimmer with respect to the water, and $\mathbf{v}_{wb}$ the velocity of the water with respect to the bank, then $\mathbf{v}_{sb} = \mathbf{v}_{sw} + \mathbf{v}_{wb}$, as shown in the diagram.

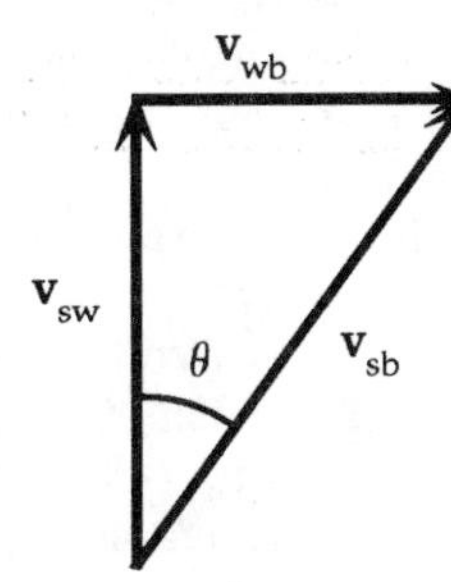

(a) We find the angle from

$\tan \theta = v_{wb}/v_{sw} = (0.80 \text{ m/s})/(1.00 \text{ m/s}) = 0.80$, which gives $\theta = 38.7°$.

Because the swimmer travels in a straight line, we have

$\tan \theta = d_{shore}/d_{river}$; $0.80 = d_{shore}/(150 \text{ m})$, which gives $d_{shore} = \boxed{120 \text{ m.}}$

(b) We can find how long it takes by using the components across the river:

$t = d_{river}/v_{sw} = (150 \text{ m})/(1.00 \text{ m/s}) = \boxed{150 \text{ s } (2.5 \text{ min}).}$

52. We have a new diagram, as shown. From the diagram, we have

$\sin \theta = v_{wb}/v_{sw} = (0.80 \text{ m/s})/(1.00 \text{ m/s}) = 0.80,$

which gives $\boxed{\theta = 53°.}$

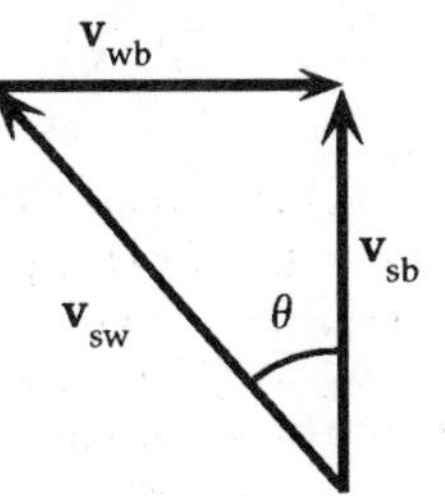

53. If $\mathbf{v}_{pw}$ is the velocity of the airplane with respect to the wind, $\mathbf{v}_{pg}$ the velocity of the airplane with respect to the ground, and $\mathbf{v}_{wg}$ the velocity of the wind with respect to the ground, then $\mathbf{v}_{pg} = \mathbf{v}_{pw} + \mathbf{v}_{wg}$, as shown in the diagram.

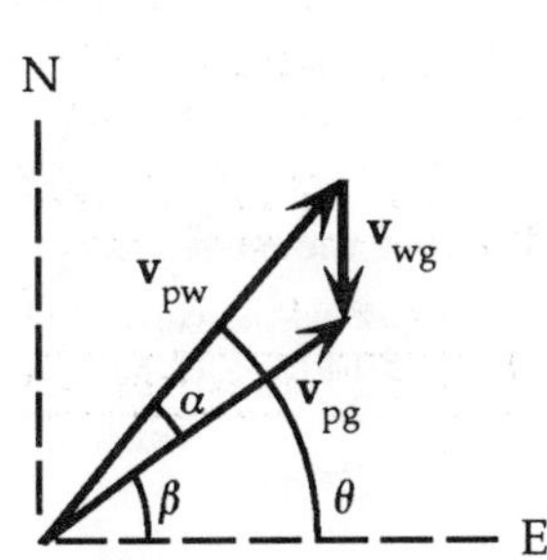

We have two unknowns: $\mathbf{v}_{pg}$ and θ (or α).

If we use the law of sines for the vector triangle, we have

$v_{pw}/\sin (90° + \beta) = v_{wg}/\sin \alpha$, or

$\sin \alpha = (v_{wg}/v_{pw})\sin (90° + \beta)$

$= [(100 \text{ km/h})/(600 \text{ km/h})] \sin 125.0° = 0.137$, or $\alpha = 7.85°$.

Thus we have $\theta = \alpha + \beta = 7.85° + 35.0° = \boxed{42.9° \text{ N of E.}}$

54. If $\mathbf{v}_{cg}$ is the velocity of the car with respect to the ground,
$\mathbf{v}_{mg}$ the velocity of the motorcycle with respect to the ground, and
$\mathbf{v}_{mc}$ the velocity of the motorcycle with respect to the car, then
$\mathbf{v}_{mc} = \mathbf{v}_{mg} - \mathbf{v}_{cg}$.
Because the motion is in one dimension, for the initial relative velocity we have
$v_{mc} = v_{mg} - v_{cg} = (90.0 \text{ km/h} - 75.0 \text{ km/h})/(3.6 \text{ ks/h}) = 4.17 \text{ m/s}.$
For the linear motion, in the reference frame of the car we have
$x = x_0 + v_0 t + \frac{1}{2}at^2;$
$60.0 \text{ m} = 0 + (4.17 \text{ m/s})(10.0 \text{ s}) + \frac{1}{2}a(10.0 \text{ s})^2$, which gives $\boxed{a = 0.366 \text{ m/s}^2.}$
Note that this is also the acceleration in the reference frame of the ground.

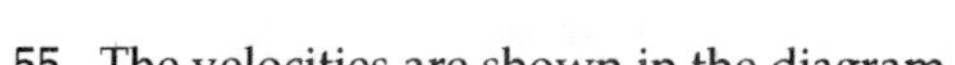

55. The velocities are shown in the diagram.
For the relative velocity of car 1 with respect to car 2, we have
$\mathbf{v}_{12} = \mathbf{v}_{1g} - \mathbf{v}_{2g}$.
For the magnitude we have
$v_{12} = (v_{1g}^2 + v_{2g}^2)^{1/2} = [(30 \text{ km/h})^2 + (50 \text{ km/h})^2]^{1/2} = \boxed{58 \text{ km/h}.}$
We find the angle from
$\tan\theta = v_{1g}/v_{2g} = (30 \text{ km/h})/(50 \text{ km/h}) = 0.60$, which gives $\boxed{\theta = 31°.}$
For the relative velocity of car 2 with respect to car 1, we have
$\mathbf{v}_{21} = \mathbf{v}_{2g} - \mathbf{v}_{1g} = -(\mathbf{v}_{1g} - \mathbf{v}_{2g}) = -\mathbf{v}_{12} = \boxed{58 \text{ km/h opposite to } \mathbf{v}_{12}.}$

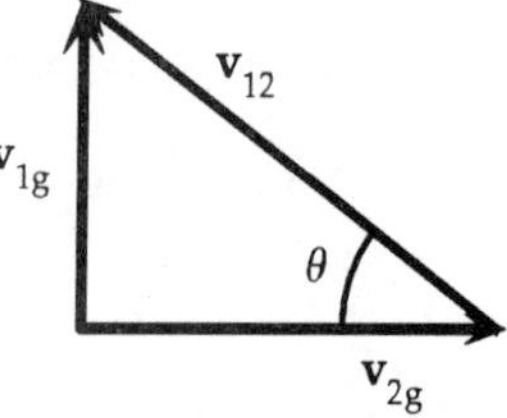

56. If $\mathbf{v}_{sg}$ is the velocity of the speeder with respect to the ground,
$\mathbf{v}_{pg}$ the velocity of the police car with respect to the ground, and
$\mathbf{v}_{ps}$ the velocity of the police car with respect to the speeder, then
$\mathbf{v}_{ps} = \mathbf{v}_{pg} - \mathbf{v}_{sg}$.
Because the motion is in one dimension, for the initial relative velocity we have
$v_{0ps} = v_{0pg} - v_{sg} = (90 \text{ km/h} - 140 \text{ km/h})/(3.6 \text{ ks/h}) = -13.9 \text{ m/s}.$
If we use a coordinate system in the reference frame of the speeder with the origin at the speeder, the distance that the police car travels before its acceleration is
$x_0 = v_{0ps}t_1 = (-13.9 \text{ m/s})(1.00 \text{ s}) = -13.9 \text{ m}.$
The negative sign indicates that the police car is falling behind.
For the linear motion of the police car after accelerating, in the reference frame of the car we have
$x = x_0 + v_{0ps}t_2 + \frac{1}{2}at_2^2;$
$0 = -13.9 \text{ m} + (-13.9 \text{ m/s})t_2 + \frac{1}{2}(2.00 \text{ m/s}^2)t_2^2$, which gives $t_2 = -0.94 \text{ s}, 14.8 \text{ s}.$
Because the police car starts accelerating at $t_2 = 0$, the physical answer is 14.8 s, so the total time is
$t = t_1 + t_2 = 1.00 \text{ s} + 14.8 \text{ s} = \boxed{15.8 \text{ s}.}$

57. With an initial relative velocity of v_{0ps} in m/s, if we use a coordinate system with the origin at the speeder, the distance that the police car travels before its acceleration is
$x_0 = v_{0ps}t_1 = v_{0ps}(1.00 \text{ s}).$
For the 6.00 s of the linear motion of the police car after accelerating, in the reference frame of the car we have
$x = x_0 + v_0 t_2 + \frac{1}{2}at_2^2;$
$0 = v_{ps}(1.00 \text{ s}) + v_{0ps}(6.00 \text{ s}) + \frac{1}{2}(2.00 \text{ m/s}^2)(6.00 \text{ s})^2$, which gives $v_{0ps} = -5.14 \text{ m/s}.$
Because the motion is in one dimension, for the initial relative velocity we have
$v_{0ps} = v_{0pg} - v_{sg};$
$(-5.14 \text{ m/s})(3.6 \text{ ks/h}) = (90 \text{ km/h}) - v_{sg}$, which gives $\boxed{v_{sg} = 109 \text{ km/h}.}$

58. The arrow will hit the apple at the same elevation, so we can use the expression for the horizontal range:

$R = v_0^2 \sin(2\theta_0)/g;$

$27 \text{ m} = (35 \text{ m/s})^2 \sin(2\theta_0)/(9.80 \text{ m/s}^2)$, which gives

$\sin(2\theta_0) = 0.216$, or $2\theta_0 = 12.4°$, $\boxed{\theta_0 = 6.2°.}$

59. (*a*) For the magnitude of the resultant to be equal to the sum of the magnitudes, the two vectors must be $\boxed{\text{parallel.}}$

(*b*) The expression is the one we use when we find the magnitude of a vector from its rectangular components. Thus the two vectors must be $\boxed{\text{perpendicular.}}$

(*c*) The only way to have the sum and difference of two magnitudes be equal is for $V_2 = -V_2$, or $V_2 = 0$. Only a zero vector has zero magnitude: $\boxed{\mathbf{V}_2 = 0.}$

60. The displacement is shown in the diagram.

For the components, we have

$\boxed{D_x = 50 \text{ m}, D_y = -25 \text{ m}, D_z = -10 \text{ m}.}$

To find the magnitude we extend the process for two dimensions:

$D = (D_x^2 + D_y^2 + D_z^2)^{1/2}$

$= [(50 \text{ m})^2 + (-25 \text{ m})^2 + (-10 \text{ m})^2]^{1/2} = \boxed{57 \text{ m.}}$

The direction is specified by the two angles shown, which we find from

$\tan \theta_h = D_y/D_x = (25 \text{ m})/(50 \text{ m}) = 0.5,$

which gives $\boxed{\theta_h = 27° \text{ from the } x\text{-axis toward the } -y\text{-axis;}}$

$\tan \theta_v = D_z/(D_x^2 + D_y^2)^{1/2}$

$= (10 \text{ m})/[(50 \text{ m})^2 + (-25 \text{ m})^2]^{1/2} = 0.178$, which gives $\boxed{\theta_v = 10° \text{ below the horizontal.}}$

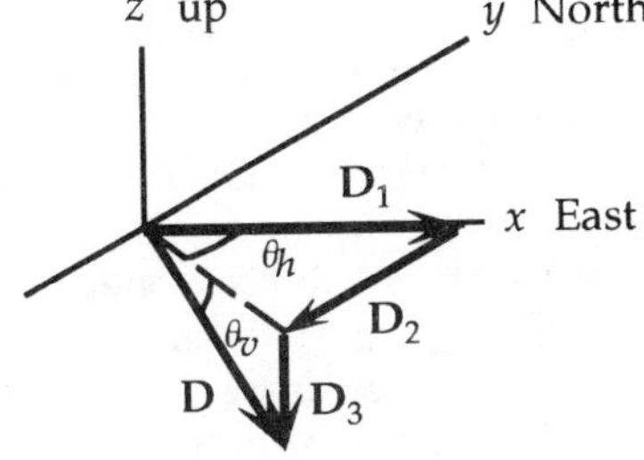

61. If we assume constant acceleration along the slope, we have

$v = v_0 + at;$

$0 = [(120 \text{ km/h})/(3.6 \text{ ks/h})] + a(12 \text{ s})$, which gives $a = -2.78 \text{ m/s}^2$ along the slope.

For the components we have

$a_{\text{horizontal}} = a \cos 30° = (-2.78 \text{ m/s}^2) \cos 30° = \boxed{-2.4 \text{ m/s}^2 \text{ (opposite to the truck's motion).}}$

$a_{\text{vertical}} = a \sin 30° = (-2.78 \text{ m/s}^2) \sin 30° = \boxed{-1.4 \text{ m/s}^2 \text{ (down).}}$

62. We see from the diagram that

$\cos \theta = A_x/A = (75.4)/(88.5) = 0.852$, or $\boxed{\theta = \pm 31.6°.}$

For the y-component, we have

$A_y = A \sin \theta = (88.5) \sin (\pm 31.6°) = \boxed{\pm 46.3.}$

The second vector (below the x-axis) is not shown in the diagram.

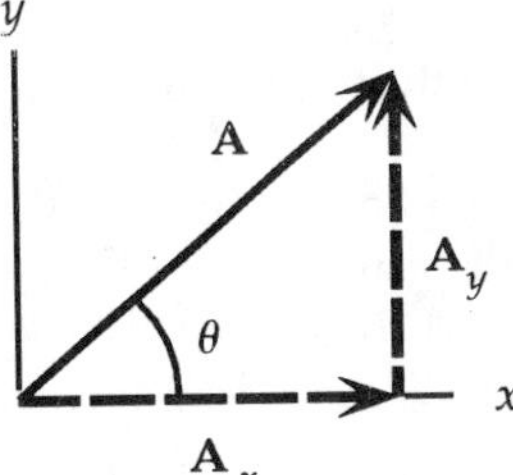

63. The velocity of the rain with respect to the train is $\mathbf{v}_{rT} = \mathbf{v}_r - \mathbf{v}_T$, where

$\mathbf{v}_r$ is the velocity of the rain with respect to the ground and

$\mathbf{v}_T$ is the velocity of the train with respect to the ground.

From the diagram we have

$\tan \theta = v_T/v_r$, or $\boxed{v_r = v_T/\tan \theta.}$

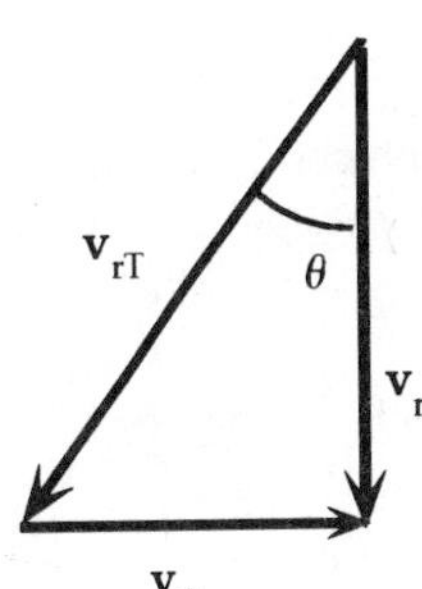

64. If $\mathbf{v}_{pw}$ is the velocity of the airplane with respect to the wind,
$\mathbf{v}_{pg}$ the velocity of the airplane with respect to the ground, and
$\mathbf{v}_{wg}$ the velocity of the wind with respect to the ground, then
$\mathbf{v}_{pg} = \mathbf{v}_{pw} + \mathbf{v}_{wg}$, as shown in the diagram.

Because the plane has covered 125 km in 1.00 hour, $v_{pg} = 125$ km/h.
We use the diagram to write the component equations:

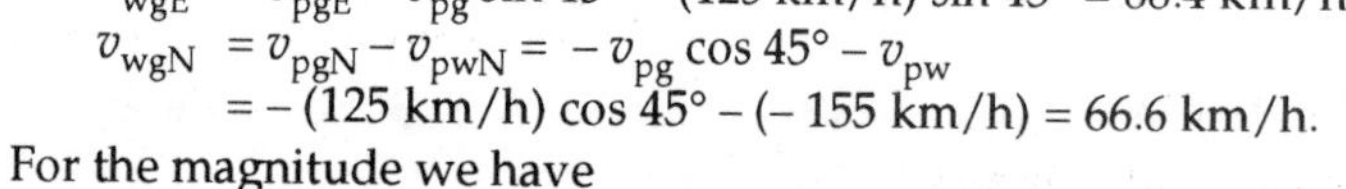

$v_{wgE} = v_{pgE} = v_{pg} \sin 45° = (125 \text{ km/h}) \sin 45° = 88.4 \text{ km/h};$

$v_{wgN} = v_{pgN} - v_{pwN} = -v_{pg} \cos 45° - v_{pw}$
$= -(125 \text{ km/h}) \cos 45° - (-155 \text{ km/h}) = 66.6 \text{ km/h}.$

For the magnitude we have

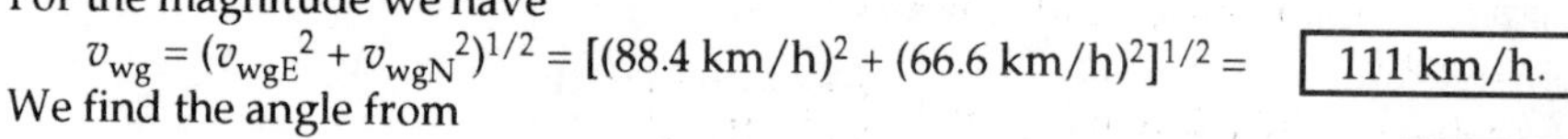

$v_{wg} = (v_{wgE}^2 + v_{wgN}^2)^{1/2} = [(88.4 \text{ km/h})^2 + (66.6 \text{ km/h})^2]^{1/2} =$ **111 km/h.**

We find the angle from

$\tan \theta = v_{wgN}/v_{wgE} = (66.6 \text{ km/h})/(88.4 \text{ km/h}) = 0.754$, which gives 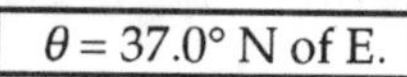 $\theta = 37.0°$ N of E.

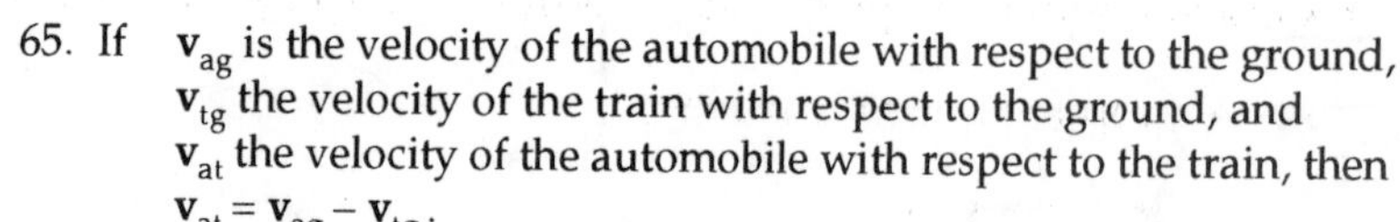

65. If $\mathbf{v}_{ag}$ is the velocity of the automobile with respect to the ground,
$\mathbf{v}_{tg}$ the velocity of the train with respect to the ground, and
$\mathbf{v}_{at}$ the velocity of the automobile with respect to the train, then
$\mathbf{v}_{at} = \mathbf{v}_{ag} - \mathbf{v}_{tg}$.

Because the motion is in one dimension, for the initial relative velocity we have

$v_{at} = v_{ag} - v_{tg} = (95 \text{ km/h} - 75 \text{ km/h})/(3.6 \text{ ks/h}) = 5.56 \text{ m/s}.$

If we use a coordinate system in the reference frame of the train with the origin at the back of the train, we find the time to pass from

$x_1 = v_{at} t_1;$

$1000 \text{ m} = (5.56 \text{ m/s}) t_1$, which gives $t_1 = 1.8 \times 10^2$ s.

With respect to the ground, the automobile will have traveled

$x_{1g} = v_{ag} t_1 = [(95 \text{ km/h})/(3.6 \text{ ks/h})](1.8 \times 10^2 \text{ s}) = 4.8 \times 10^3 \text{ m} =$ **4.8 km.**

If they move in opposite directions, we have

$v_{at} = v_{ag} - v_{tg} = [95 \text{ km/h} - (-75 \text{ km/h})]/(3.6 \text{ ks/h}) = 47.2 \text{ m/s}.$

We find the time to pass from

$x_2 = v_{at} t_2;$

$1000 \text{ m} = (47.2 \text{ m/s}) t_2$, which gives $t_2 = 21$ s.

With respect to the ground, the automobile will have traveled

$x_{2g} = v_{ag} t_2 = [(95 \text{ km/h})/(3.6 \text{ ks/h})](21 \text{ s}) = 5.6 \times 10^2 \text{ m} =$ **0.56 km.**

66. We can find the time for the jump from the horizontal motion:

$x = v_x t;$

$t = x/v_x = (8.0 \text{ m})/(9.1 \text{ m/s}) =$ **0.88 s.**

It takes half of this time for the jumper to reach the maximum height or to fall from the maximum height. If we consider the latter, we have

$y = y_0 + v_{0y} t + \frac{1}{2} a_y t^2,$

$0 = h_{max} + 0 + \frac{1}{2}(-9.80 \text{ m/s}^2)(0.44 \text{ s})^2$, which gives $h_{max} = 0.95$ m.

67. Because the golf ball returns to the same elevation, we can use the expression for the horizontal range. The horizontal range on earth is given by $R = v_0^2 \sin(2\theta_0)/g$, whereas on the moon it is

$R_{moon} = v_0^2 \sin(2\theta_0)/g_{moon}.$

Because we assume the same v_0 and θ_0, when we divide the two equations, we get

$R_{moon}/R = g/g_{moon}$, or

$g_{moon} = (R/R_{moon}) g = [(30 \text{ m})/(180 \text{ m})](9.80 \text{ m/s}^2) =$ **1.6 m/s^2.**

68. We choose a coordinate system with the origin at home plate, x horizontal and y up, as shown in the diagram.
The minimum speed of the ball is that which will have the ball just clear the fence. The horizontal motion is

$x = v_{0x}t;$

$95 \text{ m} = v_0 \cos 40° \ t$, which gives $v_0t = 124 \text{ m}$.

The vertical motion is

$y = y_0 + v_{0y}t + \frac{1}{2}a_yt^2;$

$12 \text{ m} = 1.0 \text{ m} + v_0 \sin 40° \ t + \frac{1}{2}(-9.80 \text{ m/s}^2)t^2.$

We can use the first equation to eliminate v_0t from the second and solve for t, which gives $t = 3.74$ s.
When this value is used in the first equation, we get $v_0 =$ **33 m/s.**

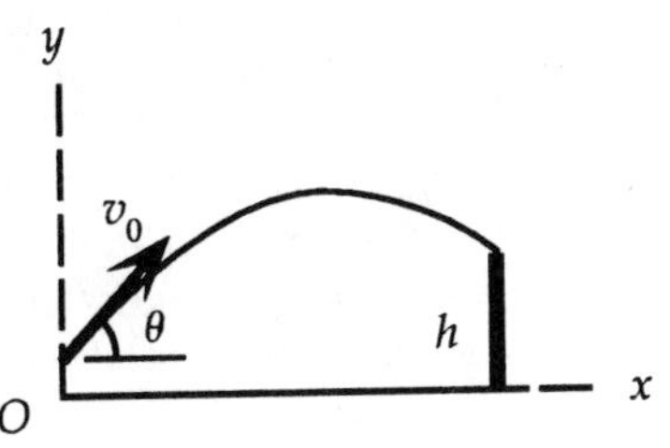

69. We choose a coordinate system with the origin at the takeoff point, with x horizontal and y vertical, with the positive direction down.
We find the time for the diver to reach the water from the vertical motion:

$y = y_0 + v_{0y}t + \frac{1}{2}a_yt^2;$

$35 \text{ m} = 0 + 0 + \frac{1}{2}(9.80 \text{ m/s}^2)t^2$, which gives **$t = 2.7$ s.**

The horizontal motion will have constant velocity.
We find the minimum horizontal initial velocity needed to land beyond the rocky outcrop from

$x = x_0 + v_{0x}t;$

$5.0 \text{ m} = 0 + v_0(2.7 \text{ s})$, which gives **$v_0 = 1.9$ m/s.**

70. We use the coordinate system shown in the diagram.
We find the time for the ball to reach the net from the vertical motion:

$y = y_0 + v_{0y}t + \frac{1}{2}a_yt^2;$

$0.90 \text{ m} = 2.50 \text{ m} + 0 + \frac{1}{2}(-9.80 \text{ m/s}^2)t^2$, which gives $t = 0.571$ s.

We find the initial velocity from the horizontal motion:

$x = v_{0x}t;$

$15.0 \text{ m} = v_0 \ (0.571 \text{ s})$, which gives **$v_0 = 26.3$ m/s.**

We find the time for the ball to reach the ground from the vertical motion:

$y = y_0 + v_{0y}t_2 + \frac{1}{2}a_yt_2^2;$

$0 = 2.50 \text{ m} + 0 + \frac{1}{2}(-9.80 \text{ m/s}^2)t_2^2$, which gives **$t_2 = 0.714$ s.**

We find where it lands from the horizontal motion:

$x_2 = v_0t_2 = (26.3 \text{ m/s})(0.714 \text{ s}) = 18.8 \text{ m}.$

Because this is $18.8 \text{ m} - 15.0 \text{ m} =$ **3.8 m beyond the net,** which is less than 7.0 m, the serve is **good.**

71. If $\mathbf{v}_{ag}$ is the velocity of the automobile with respect to the ground, $\mathbf{v}_{hg}$ the velocity of the helicopter with respect to the ground, and $\mathbf{v}_{ha}$ the velocity of the helicopter with respect to the automobile, then

$\mathbf{v}_{ha} = \mathbf{v}_{hg} - \mathbf{v}_{ag}$.

For the horizontal relative velocity we have

$v_{ha} = v_{hg} - v_{ag} = (185\ \text{km/h} - 145\ \text{km/h})/(3.6\ \text{ks/h}) = 11.1\ \text{m/s}$.

This is the initial (horizontal) velocity of the explosive, so we can find the time of fall from

$y = y_0 + v_{0y}t + \frac{1}{2}a_y t^2$;

$88.0\ \text{m} = 0 + 0 + \frac{1}{2}(+\ 9.80\ \text{m/s}^2)t^2$, which gives $t = 4.24$ s.

During this time, we find the horizontal distance the explosive travels with respect to the car from

$x = v_{ha}t = (11.1\ \text{m/s})(4.24\ \text{s}) = 47.1\ \text{m}$.

Because the helicopter is always directly above the explosive, this is how far behind the automobile the helicopter must be when the explosive is dropped. Thus we find the angle from

$\tan\theta = y/x = (88.0\ \text{m})/(47.1\ \text{m}) = 1.87$, which gives $\boxed{\theta = 61.8° \text{ below the horizontal.}}$

72. (*a*) When the boat moves upstream the speed with respect to the bank is $v - u$. When the boat moves downstream the speed with respect to the bank is $v + u$. For each leg the distance traveled is $\frac{1}{2}D$, so the total time is

$t = [\frac{1}{2}D/(v - u)] + [\frac{1}{2}D/(v + u)] = \frac{1}{2}D(v + u + v - u)/(v - u)(v + u) =$ 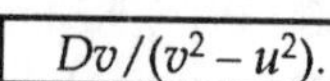 $\boxed{Dv/(v^2 - u^2).}$

(*b*) To move directly across the river the boat must head at an angle θ, as shown in the diagram. From the diagram we see that the speed with respect to the shore is

$v_{bs} = (v^2 - u^2)^{1/2}$,

during both legs of the trip. Thus the total time is

$t = [\frac{1}{2}D/(v^2 - u^2)^{1/2}] + [\frac{1}{2}D/(v^2 - u^2)^{1/2}] =$ $\boxed{D/(v^2 - u^2)^{1/2}.}$

We must assume that $u < v$, otherwise the boat will be swept downstream and never make it across the river. This appears in our answers as a negative time in (*a*) and the square root of a negative number in (*b*).

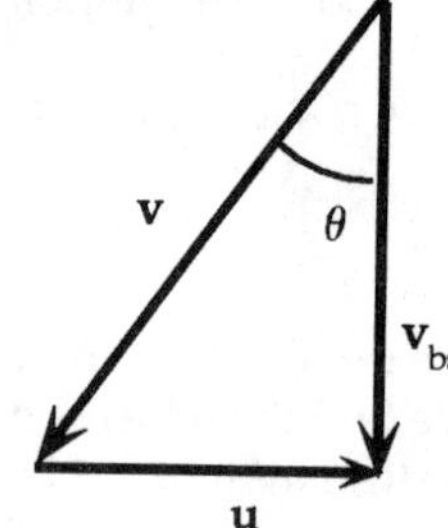

73. We use the coordinate system shown in the diagram, with up positive. For the horizontal motion, we have

$x = v_{0x}t$;

$L = (v_0 \cos\theta)t$;

$195\ \text{m} = (v_0 \cos\theta)(7.6\ \text{s})$, which gives $v_0 \cos\theta = 25.7\ \text{m/s}$.

For the vertical motion, we have

$y = y_0 + v_{0y}t + \frac{1}{2}a_y t^2$;

$H = 0 + (v_0 \sin\theta)t + \frac{1}{2}(-g)t^2$;

$155\ \text{m} = (v_0 \sin\theta)(7.6\ \text{s}) + \frac{1}{2}(-\ 9.80\ \text{m/s}^2)(7.6\ \text{s})^2$,

which gives $v_0 \sin\theta = 57.6\ \text{m/s}$.

We can find the initial angle θ by dividing the two results:

$\tan\theta = (v_0 \sin\theta)/(v_0 \cos\theta) = (57.6\ \text{m/s})/(25.7\ \text{m/s}) = 2.24$, which gives $\theta = 66.0°$.

Now we can use one of the previous results to find the initial velocity:

$v_0 = (25.7\ \text{m/s})/\cos\theta = (25.7\ \text{m/s})/\cos 66.0° = 63\ \text{m/s}$.

Thus the initial velocity is $\boxed{63\ \text{m/s}, 66° \text{ above the horizontal.}}$

CHAPTER 4

1. If we select the sled and child as the object, we apply Newton's second law to find the force:
   $\Sigma F = ma;$
   $F = (60.0 \text{ kg})(1.15 \text{ m/s}^2) = \boxed{69.0 \text{ N.}}$

2. If we select the bike and rider as the object, we apply Newton's second law to find the mass:
   $\Sigma F = ma;$
   $255 \text{ N} = m(2.20 \text{ m/s}^2)$, which gives $\boxed{m = 116 \text{ kg.}}$

3. We apply Newton's second law to the object:
   $\Sigma F = ma;$
   $F = (9.0 \times 10^{-3} \text{ kg})(10{,}000)(9.80 \text{ m/s}^2) = \boxed{8.8 \times 10^2 \text{ N.}}$

4. Without friction, the only horizontal force is the tension. We apply Newton's second law to the car:
   $\Sigma F = ma;$
   $F_T = (1050 \text{ kg})(1.20 \text{ m/s}^2) = \boxed{1.26 \times 10^3 \text{ N.}}$

5. We find the weight from the value of g.
   (*a*) Earth: $F_G = mg = (66 \text{ kg})(9.80 \text{ m/s}^2) = \boxed{6.5 \times 10^2 \text{ N.}}$
   (*b*) Moon: $F_G = mg = (66 \text{ kg})(1.7 \text{ m/s}^2) = \boxed{1.1 \times 10^2 \text{ N.}}$
   (*c*) Mars: $F_G = mg = (66 \text{ kg})(3.7 \text{ m/s}^2) = \boxed{2.4 \times 10^2 \text{ N.}}$
   (*d*) Space: $F_G = mg = (66 \text{ kg})(0 \text{ m/s}^2) = \boxed{0.}$

6. (*a*) The weight of the box depends on the value of g:
   $F_G = m_2g = (20.0 \text{ kg})(9.80 \text{ m/s}^2) = \boxed{196 \text{ N.}}$
   We find the normal force from
   $\Sigma F_y = ma_y;$
   $F_N - m_2g = 0$, which gives $F_N = m_2g = \boxed{196 \text{ N.}}$
   (*b*) We select both blocks as the object and apply Newton's second law:
   $\Sigma F_y = ma_y;$
   $F_{N2} - m_1g - m_2g = 0$, which gives
   $F_{N2} = (m_1 + m_2)g = (10.0 \text{ kg} + 20.0 \text{ kg})(9.80 \text{ m/s}^2) = \boxed{294 \text{ N.}}$
   If we select the top block as the object, we have
   $\Sigma F_y = ma_y;$
   $F_{N1} - m_1g = 0$, which gives
   $F_{N1} = m_1g = (10.0 \text{ kg})(9.80 \text{ m/s}^2) = \boxed{98.0 \text{ N.}}$

(*a*)

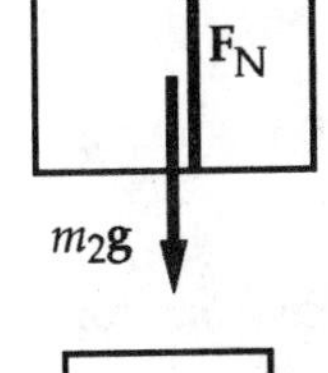

(*b*)

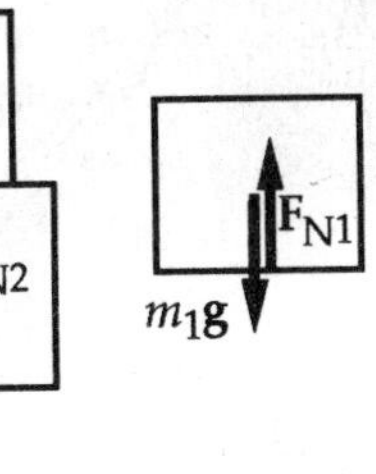

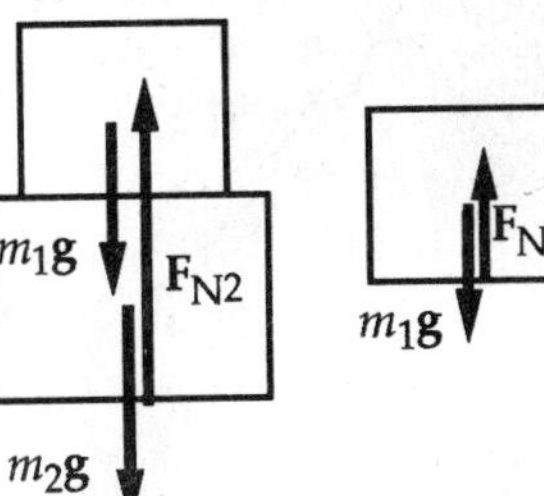

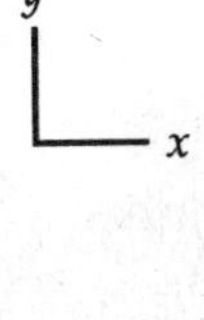

7. The acceleration can be found from the car's one-dimensional motion:
   $v = v_0 + at;$
   $0 = [(90 \text{ km/h})/(3.6 \text{ ks/h})] + a(8.0 \text{ s})$, which gives $a = -3.13 \text{ m/s}^2$.
   We apply Newton's second law to find the required average force
   $\Sigma F = ma;$
   $F = (1100 \text{ kg})(-3.13 \text{ m/s}^2) = \boxed{-3.4 \times 10^3 \text{ N.}}$
   The negative sign indicates that the force is opposite to the velocity.

8. The required average acceleration can be found from the one-dimensional motion:

   $v^2 = v_0^2 + 2a(x - x_0)$;

   $(175 \text{ m/s})^2 = 0 + 2a(0.700 \text{ m} - 0)$, which gives $a = 2.19 \times 10^4 \text{ m/s}^2$.

   We apply Newton's second law to find the required average force

   $\Sigma F = ma$;

   $F = (7.00 \times 10^{-3} \text{ kg})(2.19 \times 10^4 \text{ m/s}^2) =$ **153 N.**

9. Because the line snapped, the tension $F_T > 22$ N.

   We write $\Sigma \mathbf{F} = m\mathbf{a}$ from the force diagram for the fish:

   y-component: $F_T - mg = ma$, or $F_T = m(a + g)$.

   We find the minimum mass from the minimum tension:

   $22 \text{ N} = m_{min}(4.5 \text{ m/s}^2 + 9.80 \text{ m/s}^2)$, which gives $m_{min} = 1.5$ kg.

   Thus we can say **$m > 1.5$ kg.**

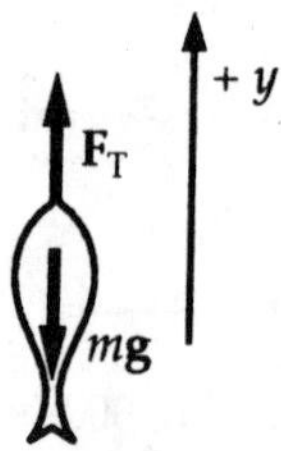

10. The average acceleration of the ball can be found from the one-dimensional motion:

    $v^2 = v_0^2 + 2a(x - x_0)$;

    $0 = (45.0 \text{ m/s})^2 + 2a(0.110 \text{ m} - 0)$, which gives $a = -9.20 \times 10^3 \text{ m/s}^2$.

    We apply Newton's second law to find the required average force applied to the ball:

    $\Sigma F = ma$;

    $F = (0.140 \text{ kg})(-9.20 \times 10^3 \text{ m/s}^2) = -1.29 \times 10^3$ N.

    The force on the glove has the same magnitude but the opposite direction:

    1.29×10^3 N in the direction of the ball's motion.

11. The average acceleration of the shot can be found from the one-dimensional motion:

    $v^2 = v_0^2 + 2a(x - x_0)$;

    $(13 \text{ m/s})^2 = 0 + 2a(2.8 \text{ m} - 0)$, which gives $a = 30.2 \text{ m/s}^2$.

    We apply Newton's second law to find the required average force applied to the shot:

    $\Sigma F = ma$;

    $F = (7.0 \text{ kg})(30.2 \text{ m/s}^2) =$ **2.1×10^2 N.**

12. We write $\Sigma \mathbf{F} = m\mathbf{a}$ from the force diagram for the car:

    y-component: $F_T - mg = ma$, or

    $F_T = m(a + g) = (1200 \text{ kg})(0.80 \text{ m/s}^2 + 9.80 \text{ m/s}^2) =$ **1.3×10^4 N.**

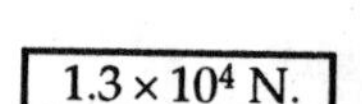

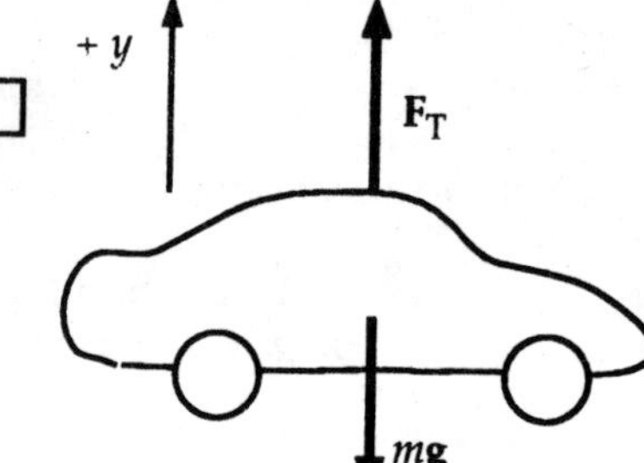

13. We write $\Sigma \mathbf{F} = m\mathbf{a}$ from the force diagram for the bucket:

    y-component: $F_T - mg = ma$;

    $63 \text{ N} - (10 \text{ kg})(9.80 \text{ m/s}^2) = (10 \text{ kg})a$,

    which gives **$a = -3.5 \text{ m/s}^2$ (down).**

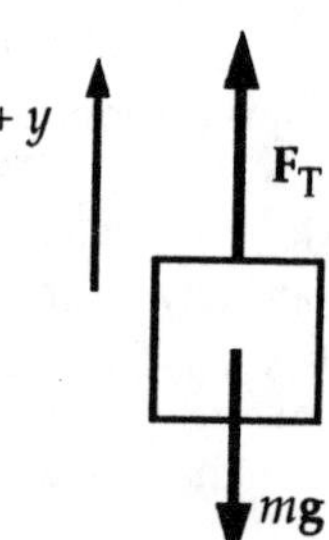

14. The maximum tension will be exerted by the motor when the elevator is accelerating upward.
We write $\Sigma\mathbf{F} = m\mathbf{a}$ from the force diagram for the elevator:
y-component: $F_{Tmax} - mg = ma$, or
$F_{Tmax} = m(a + g) = (4850 \text{ kg})(0.0600 + 1)(9.80 \text{ m/s}^2) =$ $\boxed{5.04 \times 10^4 \text{ N.}}$
The minimum tension will be exerted by the motor when the elevator is accelerating downward. We write $\Sigma\mathbf{F} = m\mathbf{a}$ from the force diagram for the car:
y-component: $F_{Tmin} - mg = ma$, or
$F_{Tmin} = m(a + g) = (4850 \text{ kg})(-0.0600 + 1)(9.80 \text{ m/s}^2) =$ $\boxed{4.47 \times 10^4 \text{ N.}}$

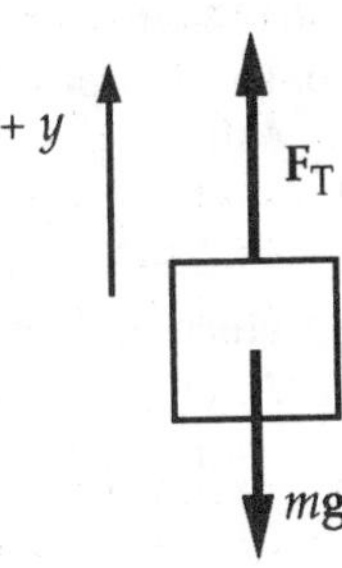

15. To have the tension less than the weight, the thief must have a downward acceleration so that the tension $F_T \leq m_{effective}g$.
We write $\Sigma\mathbf{F} = m\mathbf{a}$ from the force diagram for the thief:
y-component: $mg - F_T = ma$.
We find the minimum acceleration from the minimum tension:
$mg - m_{effective}g = ma_{min}$;
$(75 \text{ kg} - 58 \text{ kg})(9.80 \text{ m/s}^2) = (75 \text{ kg})a_{min}$, which gives $a_{min} = 2.2 \text{ m/s}^2$.
Thus we can say $\boxed{a \text{ (downward)} \geq 2.2 \text{ m/s}^2.}$

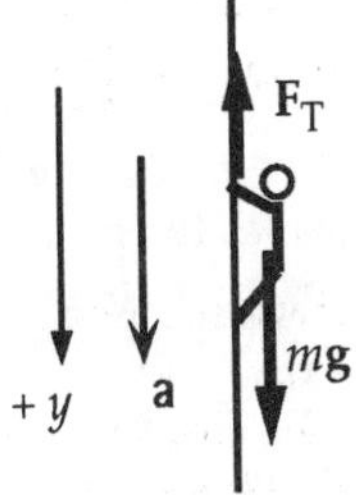

16. The scale reads the force the person exerts on the scale. From Newton's third law, this is also the magnitude of the normal force acting on the person.
We write $\Sigma\mathbf{F} = m\mathbf{a}$ from the force diagram for the person:
y-component: $F_N - mg = ma$, or
$0.75mg - mg = ma$, which gives $a = (0.75 - 1)(9.80 \text{ m/s}^2) =$ $\boxed{-2.5 \text{ m/s}^2 \text{ (down).}}$

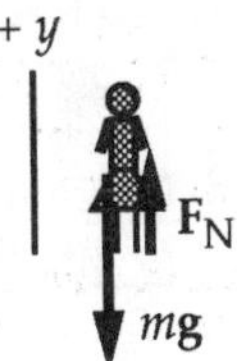

17. The maximum tension will be exerted by the motor when the elevator has the maximum acceleration.
We write $\Sigma\mathbf{F} = m\mathbf{a}$ from the force diagram for the elevator:
y-component: $F_{Tmax} - mg = ma_{max}$;
$21{,}750 \text{ N} - (2100 \text{ kg})(9.80 \text{ m/s}^2) = (2100 \text{ kg})a_{max}$,
which gives $\boxed{a_{max} = 0.557 \text{ m/s}^2.}$

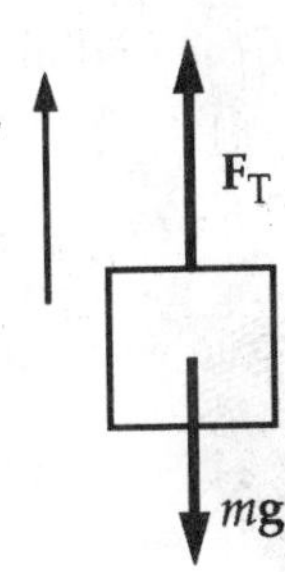

18. With down positive, we write $\Sigma\mathbf{F} = m\mathbf{a}$ from the force diagram for the skydivers:
$mg - F_R = ma$;
(a) Before the parachute opens, we have
$mg - \frac{1}{4}mg = ma$, which gives $\boxed{a = \frac{3}{4}g = 7.4 \text{ m/s}^2 \text{ (down).}}$
(b) Falling at constant speed means the acceleration is zero, so we have
$mg - F_R = ma = 0$, which gives
$F_R = mg = (120.0 \text{ kg})(9.80 \text{ m/s}^2) =$ $\boxed{1176 \text{ N.}}$

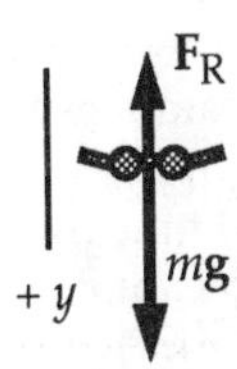

19. From Newton's third law, the gases will exert a force on the rocket that is equal and opposite to the force the rocket exerts on the gases.

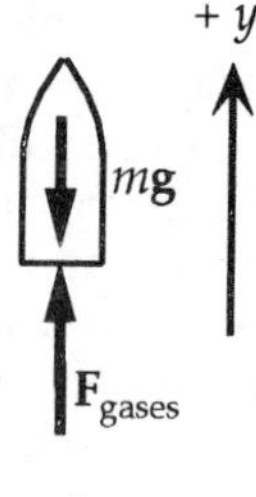

(*a*) With up positive, we write $\Sigma\mathbf{F} = m\mathbf{a}$ from the force diagram for the rocket:

$F_{\text{gases}} - mg = ma;$

$33 \times 10^6 \text{ N} - (2.75 \times 10^6 \text{ kg})(9.80 \text{ m/s}^2) = (2.75 \times 10^6 \text{ kg})a,$

which gives $\boxed{a = 2.2 \text{ m/s}^2.}$

(*b*) If we ignore the mass of the gas expelled and any change in g, we can assume a constant acceleration. We find the velocity from

$v = v_0 + at = 0 + (2.2 \text{ m/s}^2)(8.0 \text{ s}) = \boxed{18 \text{ m/s.}}$

(*c*) We find the time to achieve the height from

$y = y_0 + v_0t + \frac{1}{2}at^2;$

$9500 \text{ m} = 0 + 0 + \frac{1}{2}(2.2 \text{ m/s}^2)t^2$, which gives $\boxed{t = 93 \text{ s.}}$

20. We find the velocity necessary for the jump from the motion when the person leaves the ground to the highest point, where the velocity is zero:

+ y

F_N mg

$v^2 = v_{\text{jump}}^2 + 2(-g)h;$

$0 = v_{\text{jump}}^2 + 2(-9.80 \text{ m/s}^2)(0.80 \text{ m})$, which gives $v_{\text{jump}} = 3.96 \text{ m/s}.$

We can find the acceleration required to achieve this velocity during the crouch from

$v_{\text{jump}}^2 = v_0^2 + 2a(y - y_0);$

$(3.96 \text{ m/s})^2 = 0 + 2a(0.20 \text{ m} - 0)$, which gives $a = 39.2 \text{ m/s}^2.$

Using the force diagram for the person during the crouch, we can write $\Sigma\mathbf{F} = m\mathbf{a}$:

$F_N - mg = ma;$

$F_N - (66 \text{ kg})(9.80 \text{ m/s}^2) = (66 \text{ kg})(39.2 \text{ m/s}^2)$, which gives $F_N = 3.2 \times 10^3 \text{ N}.$

From Newton's third law, the person will exert an equal and opposite force on the ground:
$\boxed{3.2 \times 10^3 \text{ N downward.}}$

21. (*a*) We find the velocity just before striking the ground from

mg F_{legs} + y

$v_1^2 = v_0^2 + 2(-g)h;$

$v_1^2 = 0 + 2(9.80 \text{ m/s}^2)(4.5 \text{ m})$, which gives $v_1 = \boxed{9.4 \text{ m/s.}}$

(*b*) We can find the average acceleration required to bring the person to rest from

$v^2 = v_1^2 + 2a(y - y_0);$

$0 = (9.4 \text{ m/s})^2 + 2a(0.70 \text{ m} - 0)$, which gives $a = -63 \text{ m/s}^2.$

Using the force diagram for the person during the crouch, we can write $\Sigma\mathbf{F} = m\mathbf{a}$:

$mg - F_{\text{legs}} = ma;$

$(45 \text{ kg})(9.80 \text{ m/s}^2) - F_{\text{legs}} = (45 \text{ kg})(-63 \text{ m/s}^2)$, which gives $\boxed{F_{\text{legs}} = 3.3 \times 10^3 \text{ N up.}}$

22. (*a*) If we assume that he accelerates for a time t_1 over the first 50 m and reaches a top speed of v, we have

$x_1 = \frac{1}{2}(v_0 + v)t_1 = \frac{1}{2}vt_1$, or $t_1 = 2x_1/v = 2(50 \text{ m})/v = (100 \text{ m})/v.$

Because he maintains this top speed for the last 50 m, we have

$t_2 = (50 \text{ m})/v.$

Thus the total time is $T = t_1 + t_2 = (100 \text{ m})/v + (50 \text{ m})/v = 10.0 \text{ s}.$

When we solve for v, we get $v = 15.0 \text{ m/s}$; so the acceleration time is

$t_1 = (100 \text{ m})/(15.0 \text{ m/s}) = 6.67 \text{ s}.$

We find the constant acceleration for the first 50 m from

$a = \Delta v/\Delta t = (15.0 \text{ m/s} - 0)/(6.67 \text{ s}) = 2.25 \text{ m/s}^2.$

We find the horizontal force component that will produce this acceleration from

$F = ma = (65 \text{ kg})(2.25 \text{ m/s}^2) = \boxed{1.5 \times 10^2 \text{ N.}}$

(*b*) As we found in part (*a*): $\boxed{v = 15.0 \text{ m/s.}}$

23. If the boxes do not move, the acceleration is zero.
Using the force diagrams, we write $\Sigma \mathbf{F} = m\mathbf{a}$.
For the hanging box, we have
$F_T - F_{G2} = 0$, or $F_T = F_{G2}$.
For the box on the floor, we have
$F_T - F_{G1} + F_N = 0$, or $F_N = F_{G1} - F_T = F_{G1} - F_{G2}$.
(*a*) $F_N = 70\text{ N} - 30\text{ N} =$ **40 N.**
(*b*) $F_N = 70\text{ N} - 60\text{ N} =$ **10 N.**
(*c*) $F_N = 70\text{ N} - 90\text{ N} = -20\text{ N}$.
Because the floor can only push up on the box, it is not possible to have a negative normal force. Thus the normal force is **0;** the box will leave the floor and accelerate upward.

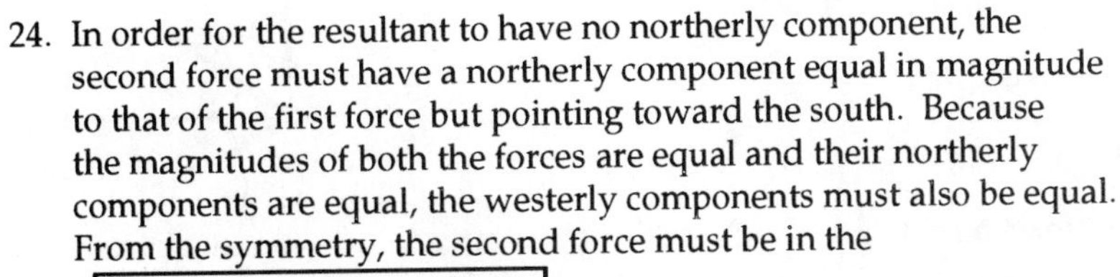

24. In order for the resultant to have no northerly component, the second force must have a northerly component equal in magnitude to that of the first force but pointing toward the south. Because the magnitudes of both the forces are equal and their northerly components are equal, the westerly components must also be equal. From the symmetry, the second force must be in the **southwesterly direction.**

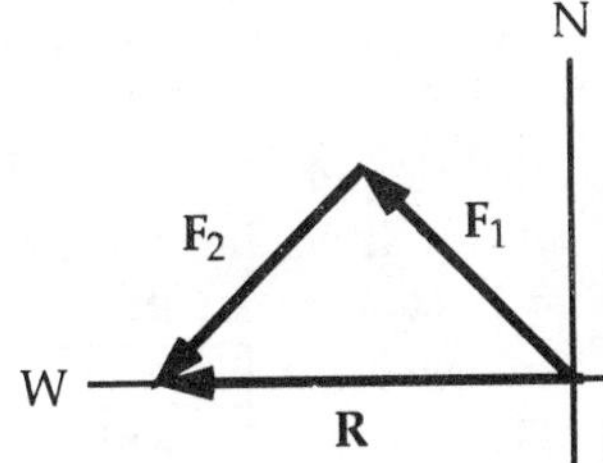

25. (*a*)

(*b*)

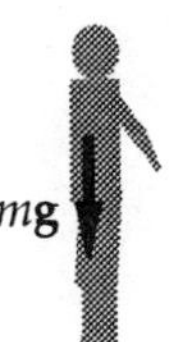

26. (*a*)

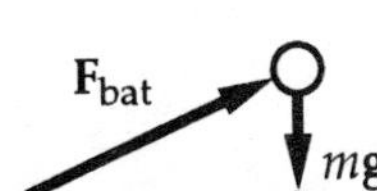

(*b*)

Note that we ignore air resistance.

27. (*a*) For the components of the net force we have

$F_{ax} = -F_1 = -10.2$ N;

$F_{ay} = -F_2 = -16.0$ N.

We find the magnitude from

$F_a^2 = F_{ax}^2 + F_{ay}^2 = (-10.2\text{ N})^2 + (-16.0\text{ N})^2$,

which gives $\boxed{F_a = 19.0\text{ N.}}$

We find the direction from

$\tan\alpha = |F_{ay}|/|F_{ax}| = (16.0\text{ N})/(10.2\text{ N}) = 1.57$,

which gives $\boxed{\alpha = 57.5° \text{ below } -x\text{-axis.}}$

The acceleration will be in the direction of the net force:

$a_a = F_a/m = (19.0\text{ N})/(27.0\text{ kg}) = \boxed{0.702\text{ m/s}^2, 57.5° \text{ below } -x\text{-axis.}}$

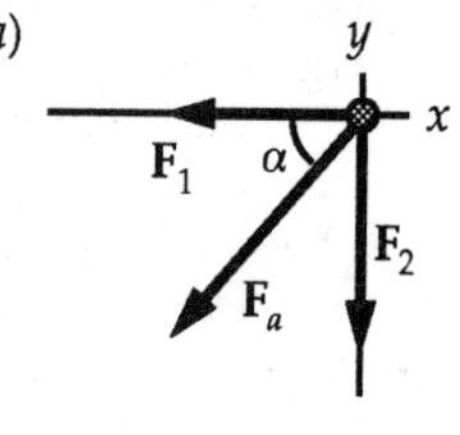

(*b*) For the components of the net force we have

$F_{bx} = F_1\cos\theta = (10.2\text{ N})\cos 30° = 8.83$ N;

$F_{by} = F_2 - F_1\sin\theta = 16.0\text{ N} - (10.2\text{ N})\sin 30° = 10.9$ N.

We find the magnitude from

$F_b^2 = F_{bx}^2 + F_{by}^2 = (8.83\text{ N})^2 + (10.9\text{ N})^2$,

which gives $\boxed{F_b = 14.0\text{ N.}}$

We find the direction from

$\tan\beta = |F_{by}|/|F_{bx}| = (10.9\text{ N})/(8.83\text{ N}) = 1.23$,

which gives $\boxed{\beta = 51.0° \text{ above } +x\text{-axis.}}$

The acceleration will be in the direction of the net force:

$a_b = F_b/m = (14.0\text{ N})/(27.0\text{ kg}) = \boxed{0.519\text{ m/s}^2, 51.0° \text{ above } +x\text{-axis.}}$

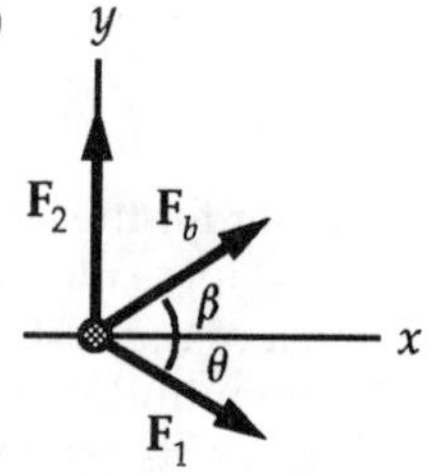

28.

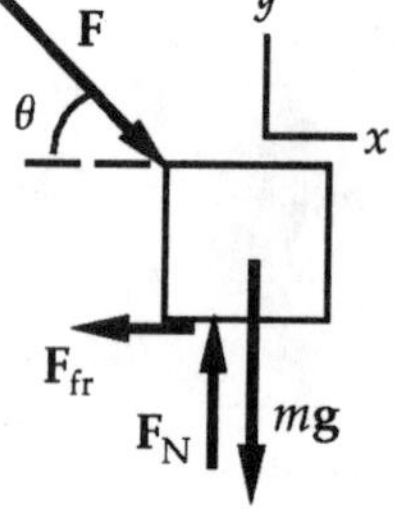

(*b*) Because the velocity is constant, the acceleration is zero.

We write $\Sigma\mathbf{F} = m\mathbf{a}$ from the force diagram for the mower:

x-component: $F\cos\theta - F_{fr} = ma = 0$, which gives

$F_{fr} = (88.0\text{ N})\cos 45° = \boxed{62.2\text{ N.}}$

(*c*) *y*-component: $F_N - mg - F\sin\theta = ma = 0$, which gives

$F_{fr} = (14.5\text{ kg})(9.80\text{ m/s}^2) + (88.0\text{ N})\sin 45° = \boxed{204\text{ N.}}$

(*d*) We can find the acceleration from the motion of the mower:

$a = \Delta v/\Delta t = (1.5\text{ m/s} - 0)/(2.5\text{ s}) = 0.60\text{ m/s}^2$.

For the *x*-component of $\Sigma\mathbf{F} = m\mathbf{a}$ we now have

$F\cos\theta - F_{fr} = ma$;

$F\cos 45° - 62.2\text{ N} = (14.5\text{ kg})(0.60\text{ m/s}^2)$, which gives $F = \boxed{100\text{ N.}}$

29. (*a*) We find the horizontal acceleration from the horizontal component of the force exerted on the sprinter, which is the reaction to the force the sprinter exerts on the block:

$F\cos\theta = ma$;

$(800\text{ N})\cos 22° = (65\text{ kg})a$, which gives $a = \boxed{11\text{ m/s}^2.}$

(*b*) For the motion of the sprinter we can write

$v = v_0 + at = 0 + (11.4\text{ m/s}^2)(0.38\text{ s}) = \boxed{4.3\text{ m/s.}}$

30. (*a*) Because the buckets are at rest, the acceleration is zero.
We write $\Sigma \mathbf{F} = m\mathbf{a}$ from the force diagram for each bucket:
lower bucket: $F_{T2} - m_2 g = m_2 a = 0$, which gives
$F_{T2} = m_2 g = (3.0 \text{ kg})(9.80 \text{ m/s}^2) = \boxed{29 \text{ N.}}$
upper bucket: $F_{T1} - F_{T2} - m_1 g = m_1 a = 0$, which gives
$F_{T1} = F_{T2} + m_1 g = 29 \text{ N} + (3.0 \text{ kg})(9.80 \text{ m/s}^2) = \boxed{58 \text{ N.}}$

(*b*) The two buckets must have the same acceleration.
We write $\Sigma \mathbf{F} = m\mathbf{a}$ from the force diagram for each bucket:
lower bucket: $F_{T2} - m_2 g = m_2 a$, which gives
$F_{T2} = m_2(g + a)$
$= (3.0 \text{ kg})(9.80 \text{ m/s}^2 + 1.60 \text{ m/s}^2) = \boxed{34 \text{ N.}}$
upper bucket: $F_{T1} - F_{T2} - m_1 g = m_1 a$, which gives
$F_{T1} = F_{T2} + m_1(g + a)$
$= 34 \text{ N} + (3.0 \text{ kg})(9.80 \text{ m/s}^2 + 1.60 \text{ m/s}^2) = \boxed{68 \text{ N.}}$

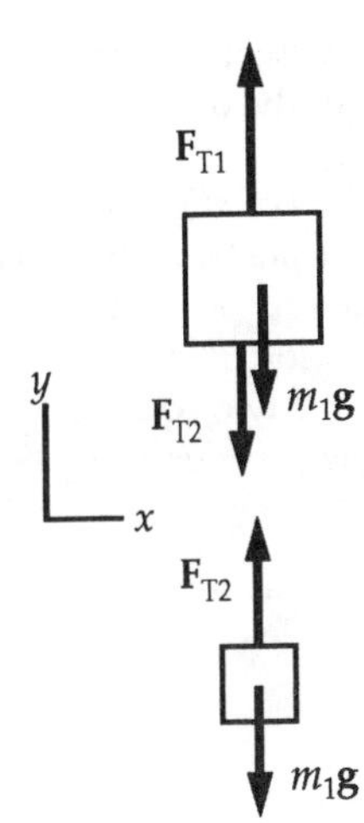

31. (*a*) We select the helicopter and the car as the system.
We write $\Sigma \mathbf{F} = m\mathbf{a}$ from the force diagram:
$F_{air} - m_h g - m_c g = (m_h + m_c)a$, which gives
$F_{air} = (m_h + m_c)(a + g)$
$= (6500 \text{ kg} + 1200 \text{ kg})(0.60 \text{ m/s}^2 + 9.80 \text{ m/s}^2)$
$= \boxed{8.01 \times 10^4 \text{ N.}}$

(*b*) We select the car as the system.
We write $\Sigma \mathbf{F} = m\mathbf{a}$ from the force diagram:
$F_T - m_c g = m_c a$, which gives
$F_T = m_c(a + g)$
$= (1200 \text{ kg})(0.60 \text{ m/s}^2 + 9.80 \text{ m/s}^2)$
$= \boxed{1.25 \times 10^4 \text{ N.}}$

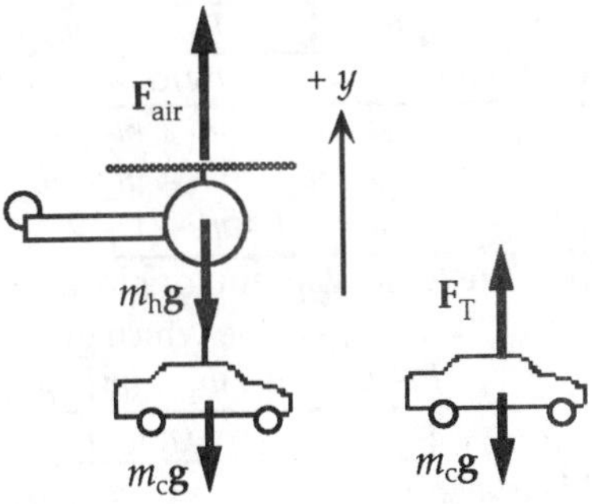

32. (*a*) Because the speed is constant, the acceleration is zero.
We write $\Sigma \mathbf{F} = m\mathbf{a}$ from the force diagram:
$F_T + F_T - mg = ma = 0$, which gives
$F_T = \frac{1}{2}mg = \frac{1}{2}(65 \text{ kg})(9.80 \text{ m/s}^2) = \boxed{3.2 \times 10^2 \text{ N.}}$

(*b*) Now we have:
$F_T' + F_T' - mg = ma;$
$2(1.10)(\frac{1}{2}mg) - mg = ma$, which gives
$a = 0.10g = 0.10(9.80 \text{ m/s}^2) = \boxed{0.98 \text{ m/s}^2.}$

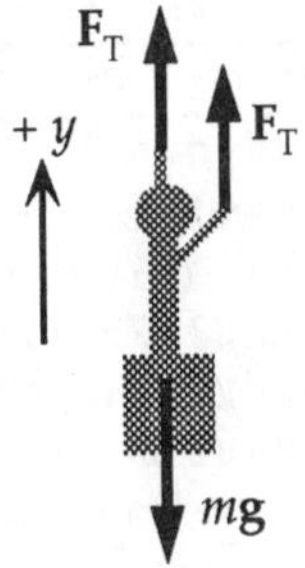

33. If we select the first and second cars as the system, the only horizontal force is the tension in the coupling between the locomotive and the first car. From the force diagram, we have
$\Sigma F_x = (m_1 + m_2)a_x$, or $F_1 = (m + m)a = 2ma$.
If we select the second car as the system, the only horizontal force is the tension in the coupling between the first car and the second car. From the force diagram, we have
$\Sigma F_x = m_2 a_x$, or $F_2 = ma$.
Thus we have $F_1/F_2 = 2ma/ma = 2$, for any nonzero acceleration.

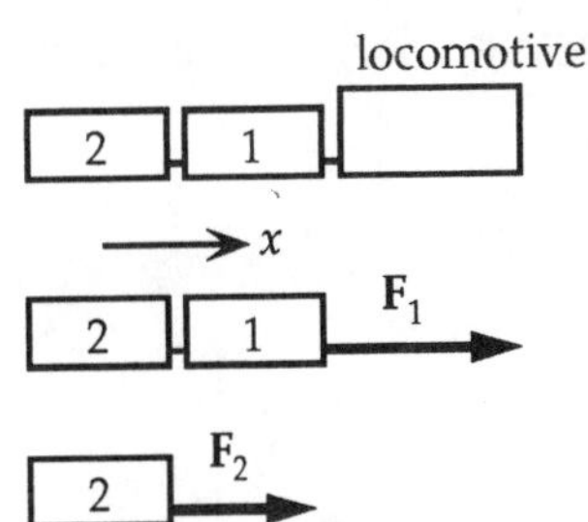

34. From the motion of the car we can find its acceleration, which is the acceleration of the dice:

$v = v_0 + at;$

$(20\ \text{m/s}) = 0 + a(5.0\ \text{s})$, which gives $a = 4.0\ \text{m/s}^2$.

We write $\Sigma\mathbf{F} = m\mathbf{a}$ from the force diagram for the dice:

x-component: $F_T \sin\theta = ma;$

y-component: $F_T \cos\theta - mg = 0.$

If we divide the two equations, we get

$\tan\theta = a/g = (4.0\ \text{m/s}^2)/(9.80\ \text{m/s}^2) = 0.408$, which gives $\boxed{\theta = 22°.}$

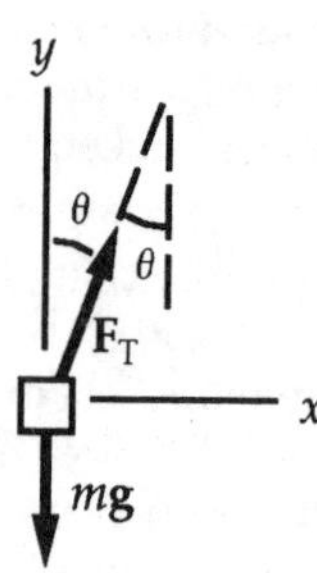
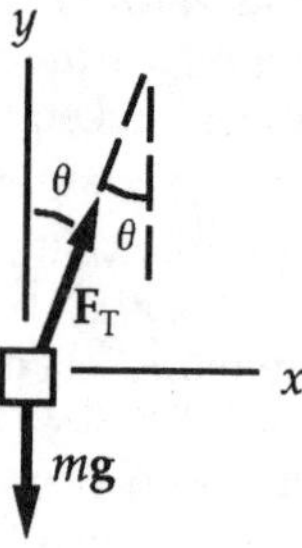

35.

(*a*)

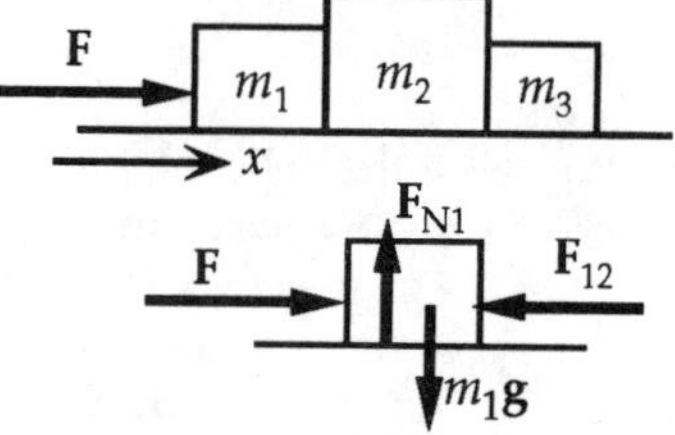

(*b*) If we select all three blocks as the system, we have

$\Sigma F_x = ma_x$: $F = (m_1 + m_2 + m_3)a$,

which gives $\boxed{a = F/(m_1 + m_2 + m_3).}$

(*c*) For the three blocks individually, for $\Sigma F_x = ma_x$ we have

$$\boxed{\begin{aligned}F_{\text{net1}} &= m_1a = m_1F/(m_1 + m_2 + m_3);\\ F_{\text{net2}} &= m_2a = m_2F/(m_1 + m_2 + m_3);\\ F_{\text{net3}} &= m_3a = m_3F/(m_1 + m_2 + m_3).\end{aligned}}$$

(*d*) From the force diagram for block 1 we have

$F_{\text{net1}} = F - F_{12} = m_1a$, which gives

$F_{12} = F - m_1a = F - m_1F/(m_1 + m_2 + m_3)]$

$= \boxed{F(m_2 + m_3)/(m_1 + m_2 + m_3).}$

This is also F_{21} (Newton's third law).

From the force diagram for block 2 we have

$F_{\text{net2}} = F_{21} - F_{23} = m_2a$, which gives

$F_{23} = F_{21} - m_2a = F - m_1a - m_2a$

$= F - (m_1 + m_2)F/(m_1 + m_2 + m_3)]$

$= \boxed{Fm_3/(m_1 + m_2 + m_3).}$

This is also F_{32} (Newton's third law).

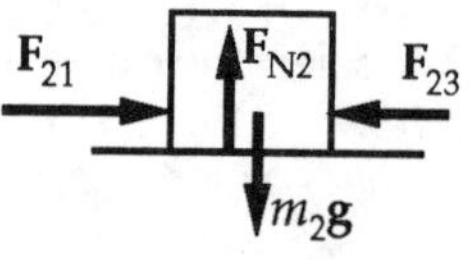

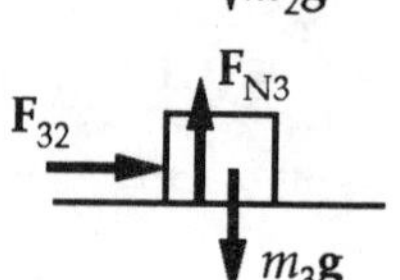

(*e*) When we use the given values, we get

$a = F/(m_1 + m_2 + m_3) = (96.0\ \text{N})/(12.0\ \text{kg} + 12.0\ \text{kg} + 12.0\ \text{kg}) = \boxed{2.67\ \text{m/s}^2.}$

$F_{\text{net1}} = m_1a = (12.0\ \text{kg})(2.67\ \text{m/s}^2) = 32\ \text{N}.$

$F_{\text{net2}} = m_2a = (12.0\ \text{kg})(2.67\ \text{m/s}^2) = 32\ \text{N}.$

$F_{\text{net3}} = m_3a = (12.0\ \text{kg})(2.67\ \text{m/s}^2) = 32\ \text{N}.$

Because the blocks have the same mass and the same acceleration, we expect

$\boxed{F_{\text{net1}} = F_{\text{net2}} = F_{\text{net3}} = 32\ \text{N}.}$

For the forces between the blocks we have

$F_{21} = F_{12} = F - m_1a = 96.0\ \text{N} - (12.0\ \text{kg})(2.67\ \text{m/s}^2) = \boxed{64\ \text{N}.}$

$F_{32} = F_{23} = F - m_1a - m_2a = 96.0\ \text{N} - (12.0\ \text{kg})(2.67\ \text{m/s}^2) - (12.0\ \text{kg})(2.67\ \text{m/s}^2) = \boxed{32\ \text{N}.}$

36. Forces are drawn for each of the blocks. Because the string doesn't stretch, the tension is the same at each end of the string, and the accelerations of the blocks have the same magnitude. Note that we take the positive direction in the direction of the acceleration for each block.

We write $\Sigma \mathbf{F} = m\mathbf{a}$ from the force diagram for each block:

y-component (block 1): $F_T - m_1 g = m_1 a$;

y-component (block 2): $m_2 g - F_T = m_2 a$.

By adding the equations, we find the acceleration:

$a = (m_2 - m_1)g/(m_1 + m_2)$

$= (3.2 \text{ kg} - 2.2 \text{ kg})(9.80 \text{ m/s}^2)/(3.2 \text{ kg} + 2.2 \text{ kg})$

$= 1.81 \text{ m/s}^2$ for both blocks.

For the motion of block 1 we take the origin at the ground and up positive. Until block 2 hits the ground, we have

$v_1^2 = v_{01}^2 + 2a(y_1 - y_{01})$

$= 0 + 2(1.81 \text{ m/s}^2)(3.60 \text{ m} - 1.80 \text{ m})$, which gives

$v_1 = 2.56 \text{ m/s}$.

Once block 2 hits the floor, $F_T \to 0$ and block 1 will have the downward acceleration of g. For this motion of block 1 up to the highest point reached, we have

$v^2 = v_1^2 + 2a(h - y_1)$

$0 = (2.56 \text{ m/s})^2 + 2(-9.80 \text{ m/s}^2)(h - 3.60 \text{ m})$, which gives $\boxed{h = 3.93 \text{ m.}}$

37.

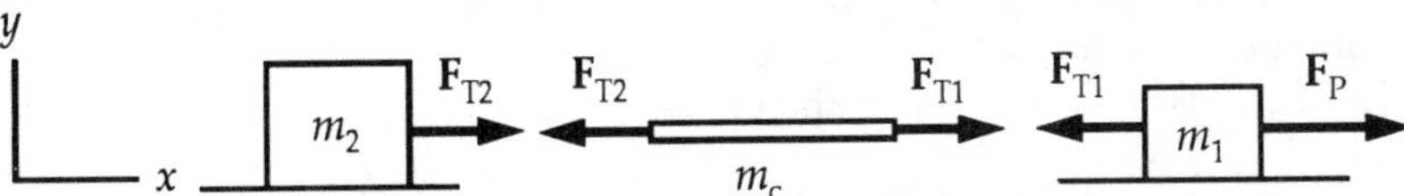

The blocks and the cord will have the same acceleration. If we select the two blocks and cord as the system, we have

$\Sigma F_x = ma_x$: $F_P = (m_1 + m_2 + m_c)a$, which gives

$a = F_P/(m_1 + m_2 + m_c) = (40.0 \text{ N})/(10.0 \text{ kg} + 1.0 \text{ kg} + 12.0 \text{ kg}) = \boxed{1.74 \text{ m/s}^2.}$

For block 1 we have $\Sigma F_x = ma_x$:

$F_P - F_{T1} = m_1 a$;

$40.0 \text{ N} - F_{T1} = (10.0 \text{ kg})(1.74 \text{ m/s}^2)$, which gives $\boxed{F_{T1} = 22.6 \text{ N.}}$

For block 2 we have $\Sigma F_x = ma_x$:

$F_{T2} = m_2 a$;

$F_{T2} = (12.0 \text{ kg})(1.74 \text{ m/s}^2) = \boxed{20.9 \text{ N.}}$

Note that we can see if these agree with the analysis of $\Sigma F_x = ma_x$ for block 3:

$F_{T1} - F_{T2} = m_3 a$:

$22.6 \text{ N} - 20.9 \text{ N} = (1.0 \text{ kg})a$, which gives $a = 1.7 \text{ m/s}^2$.

38. The friction is kinetic, so $F_{fr} = \mu_k F_N$. With constant velocity, the acceleration is zero.

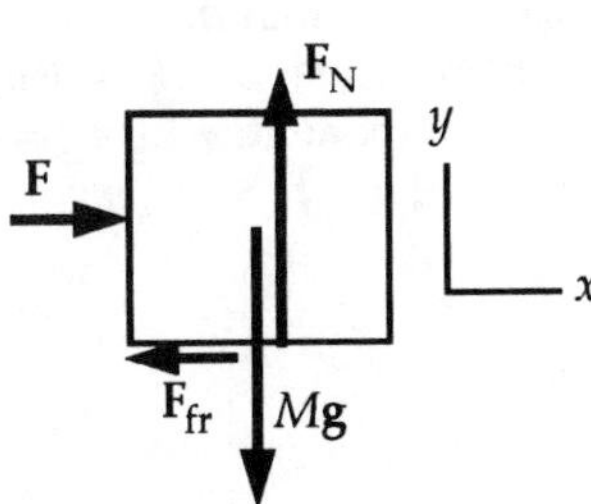

Using the force diagram for the crate, we can write $\Sigma \mathbf{F} = m\mathbf{a}$:

x-component: $F - \mu_k F_N = 0$;

y-component: $F_N - Mg = 0$.

Thus $F_N = Mg$, and

$F = \mu_k F_N = \mu_k Mg = (0.30)(35 \text{ kg})(9.80 \text{ m/s}^2) = \boxed{1.0 \times 10^2 \text{ N.}}$

If $\mu_k = 0$, there is $\boxed{\text{no force}}$ required to maintain constant speed.

39. (*a*) In general, static friction is given by $F_{sfr} \le \mu_s F_N$. Immediately before the box starts to move, the static friction force reaches its maximum value: $F_{sfr,max} = \mu_s F_N$. For the instant before the box starts to move, the acceleration is zero.
Using the force diagram for the box, we can write $\Sigma\mathbf{F} = m\mathbf{a}$:
    x-component: $F - \mu_s F_N = 0$;
    y-component: $F_N - Mg = 0$.
Thus $F_N = Mg$, and
    $F = \mu_s F_N = \mu_s Mg$;
    $40.0\ \text{N} = \mu_s(5.0\ \text{kg})(9.80\ \text{m/s}^2)$, which gives 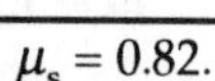 $\boxed{\mu_s = 0.82.}$

(*b*) When the box accelerates and the friction changes to kinetic, we have
    $F - \mu_k F_N = Ma$;
    $40.0\ \text{N} - \mu_k(5.0\ \text{kg})(9.80\ \text{m/s}^2) = (5.0\ \text{kg})(0.70\ \text{m/s}^2)$, which gives $\boxed{\mu_k = 0.74.}$

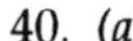
40. (*a*) (*b*) (*c*)

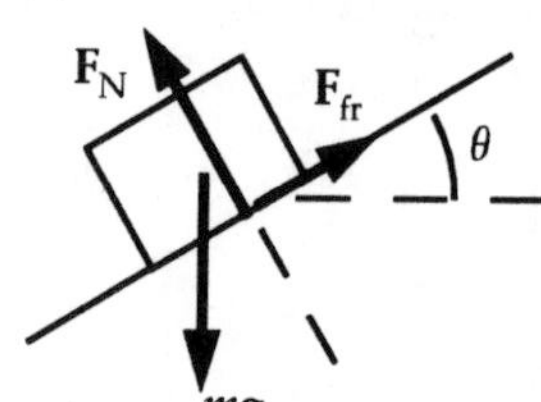

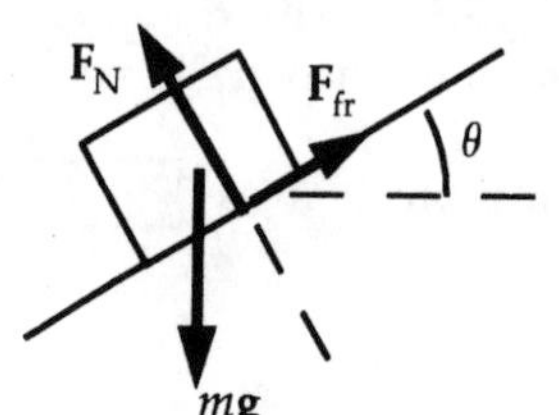

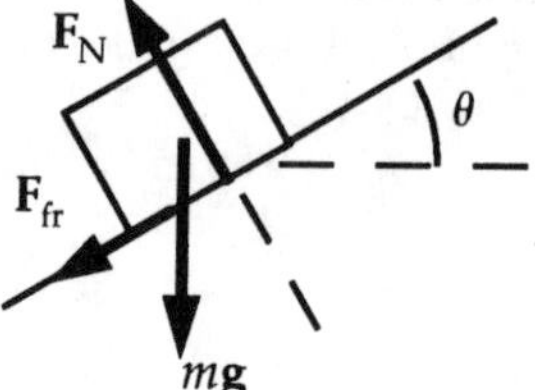

In (*a*) the friction is static and opposes the impending motion down the plane.
In (*b*) the friction is kinetic and opposes the motion down the plane.
In (*c*) the friction is kinetic and opposes the motion up the plane.

41. The drawer will suddenly open when the resisting static friction force reaches its maximum value: $F_{sfr,max} = \mu_s F_N$. Frequently drawers are stuck from pressure on the sides and top of the drawer. Here we assume that the friction force is produced only by the normal force on the bottom of the drawer.
For $\Sigma\mathbf{F} = m\mathbf{a}$ we have
    x-component: $F - \mu_s F_N = 0$;
    y-component: $F_N - Mg = 0$.
Thus $F_N = Mg$, and
    $F = \mu_s F_N = \mu_s Mg$;
    $8.0\ \text{N} = \mu_s(2.0\ \text{kg})(9.80\ \text{m/s}^2)$, which gives $\boxed{\mu_s = 0.41.}$

42. We can find the required acceleration, assumed constant, from
    $x = v_0 t + \frac{1}{2}at^2$;
    $(0.250\ \text{mi})(1610\ \text{m/mi}) = 0 + \frac{1}{2}a\,(6.0\ \text{s})^2$, which gives $a = 22.4\ \text{m/s}^2$.
If we assume that the tires are just on the verge of slipping, $F_{sfr,max} = \mu_s F_N$, so we have
    x-component: $\mu_s F_N = ma$;
    y-component: $F_N - mg = 0$.
Thus we have $\mu_s = ma/mg = a/g = (22.4\ \text{m/s}^2)/(9.80\ \text{m/s}^2) = \boxed{2.3.}$

43. If motion is just about to begin, the static friction force will be maximum: $F_{sfr,max} = \mu_s F_N$, and the acceleration will be zero. We write $\Sigma\mathbf{F} = m\mathbf{a}$ from the force diagram for each block:

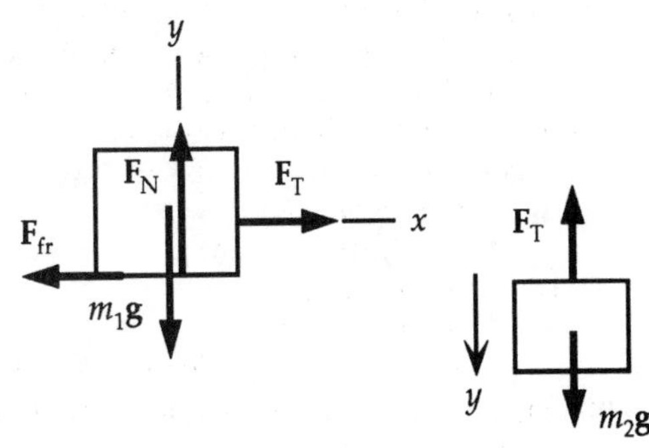

x-component (block 1): $F_T - \mu_s F_N = 0$;
y-component (block 1): $F_N - m_1 g = 0$;
y-component (block 2): $m_2 g - F_T = 0$.

Thus we see that

$F_T = \mu_s m_1 g = m_2 g$.

This allows us to find the minimum mass of body 1:

$(0.30)m_1 = 2.0$ kg, which gives $\boxed{m_1 = 6.7 \text{ kg.}}$

44. The kinetic friction force provides the acceleration. For $\Sigma\mathbf{F} = m\mathbf{a}$ we have

x-component: $-\mu_k F_N = ma$;
y-component: $F_N - mg = 0$.

Thus we see that

$a = -\mu_k g = -(0.20)(9.80 \text{ m/s}^2) = -1.96 \text{ m/s}^2$.

We can find the distance from the motion data:

$v^2 = v_0^2 + 2a(x - x_0)$;
$0 = (4.0 \text{ m/s})^2 + 2(-1.96 \text{ m/s}^2)(x - 0)$, which gives $x = \boxed{4.1 \text{ m.}}$

45. (a) The two crates must have the same acceleration.

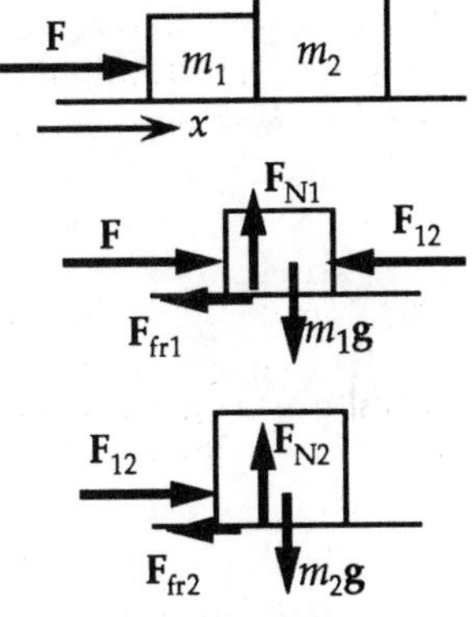

From the force diagram for crate 1 we have

x-component: $F - F_{12} - \mu_k F_{N1} = m_1 a$;
y-component: $F_{N1} - m_1 g = 0$, or $F_{N1} = m_1 g$.

From the force diagram for crate 2 we have

x-component: $F_{12} - \mu_k F_{N2} = m_2 a$;
y-component: $F_{N2} - m_2 g = 0$, or $F_{N2} = m_2 g$.

If we add the two x-equations, we get

$F - \mu_k m_1 g - \mu_k m_2 g = m_1 a + m_2 a$;
$730 \text{ N} - (0.15)(75 \text{ kg})(9.80 \text{ m/s}^2) - (0.15)(110 \text{ kg})(9.80 \text{ m/s}^2) =$
$(75 \text{ kg} + 110 \text{ kg})a$, which gives $a = \boxed{2.5 \text{ m/s}^2.}$

(b) We can find the force between the crates from the x-equation for crate 2:

$F_{12} - \mu_k m_2 g = m_2 a$;
$F_{12} - (0.15)(110 \text{ kg})(9.80 \text{ m/s}^2) = (110 \text{ kg})(2.5 \text{ m/s}^2)$, which gives
$F_{12} = \boxed{4.4 \times 10^2 \text{ N.}}$

46. (a) If the automobile does not skid, the friction is static, with $F_{sfr} \leq \mu_s F_N$. On a level road, the normal force is $F_N = mg$. The static friction force is the only force slowing the automobile and will be maximum in order to produce the minimum stopping distance. We find the acceleration from the horizontal component of $\Sigma\mathbf{F} = m\mathbf{a}$:

$-\mu_s mg = ma$, which gives $a = -\mu_s g$.

For the motion until the automobile stops, we have

$v_{final}^2 = v_0^2 + 2a(x - x_0)$;
$0 = v^2 + 2(-\mu_s g)(x_{min})$, which gives $x_{min} = v^2/2\mu_s g$.

(b) For the given data we have

$x_{min} = [(95 \text{ km/h})/(3.6 \text{ ks/h})]^2/2(0.75)(9.80 \text{ m/s}^2) = \boxed{47 \text{ m.}}$

(c) The only change is in the value of g:

$x_{min} = [(95 \text{ km/h})/(3.6 \text{ ks/h})]^2/2(0.75)(1.63 \text{ m/s}^2) = \boxed{2.8 \times 10^2 \text{ m.}}$

47. If the crate does not slide, it must have the same acceleration as the truck. The friction is static, with $F_{sfr} \le \mu_s F_N$. On a level road, the normal force is $F_N = mg$. If we consider the crate as the system, the static friction force will be opposite to the direction of motion (to oppose the impending motion of the crate toward the front of the truck), is the only force providing the acceleration, and will be maximum in order to produce the maximum acceleration. We find the acceleration from $\Sigma F_x = ma_x$:

    $\mu_s mg = ma$, which gives

    $a = \mu_s g = -(0.75)(9.80\ \text{m/s}^2) =$ $\boxed{-7.4\ \text{m/s}^2.}$

48. We simplify the forces to the three shown in the diagram. If the car does not skid, the friction is static, with $F_{sfr} \le \mu_s F_N$. The static friction force will be maximum just before the car slips. We write $\Sigma \mathbf{F} = m\mathbf{a}$ from the force diagram:

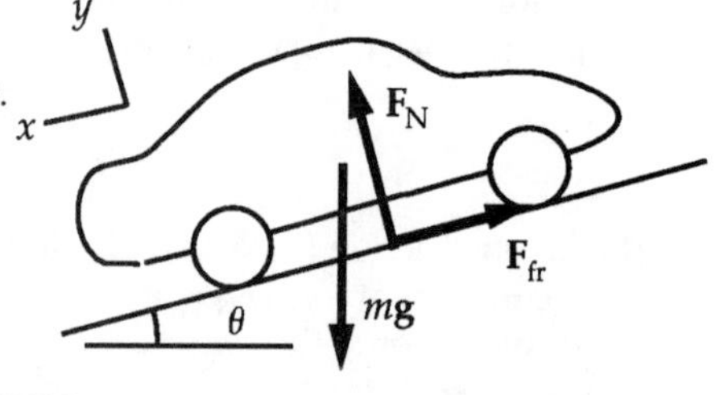

    x-component: $mg \sin \theta_{max} - \mu_s F_N = 0$;

    y-component: $F_N - mg \cos \theta_{max} = 0$.

    When we combine these, we get

    $\tan \theta_{max} = \mu_s = 0.15$, or $\theta_{max} = 8.5°$.

    Thus a car will slip on any driveway with an incline greater than 8.5°. The only driveway safe to park in is 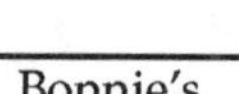 $\boxed{\text{Bonnie's.}}$

49. The kinetic friction force will be up the slide to oppose the motion. We choose the positive direction in the direction of the acceleration. From the force diagram for the child, we have $\Sigma \mathbf{F} = m\mathbf{a}$:

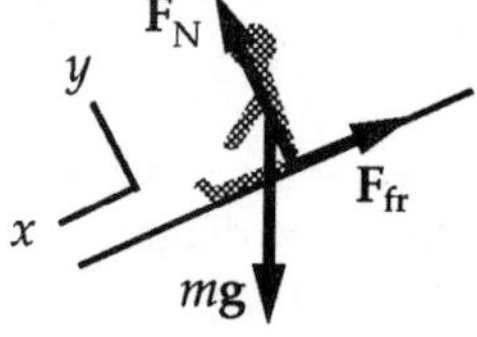

    x-component: $mg \sin \theta - F_{fr} = ma$;

    y-component: $F_N - mg \cos \theta = 0$.

    When we combine these, we get

    $a = g \sin \theta - \mu_k g \cos \theta = g(\sin \theta - \mu_k \cos \theta)$.

    For the motion of the child, we have

    $v^2 = v_0^2 + 2a(x - x_0) = 0 + 2ad$, where d is the distance along the slide.

    If we form the ratio for the two slides, we get

    $(v_{friction}/v_{none})^2 = a_{friction}/a_{none} = (\sin \theta - \mu_k \cos \theta)/\sin \theta$;

    $(\frac{1}{2})^2 = (\sin 28° - \mu_k \cos 28°)/\sin 28°$, which gives $\boxed{\mu_k = 0.40.}$

50. We find the maximum permissible deceleration from the motion until the automobile stops:

    $v = v_0 + at$;

    $0 = [(40\ \text{km/h})/(3.6\ \text{ks/h})] + a_{max}(3.5\ \text{s})$, which gives $a_{max} = -3.17\ \text{m/s}^2$.

    The minimum time for deceleration without the cup sliding means that the static friction force, which is the force producing the deceleration of the cup, is maximum. On a level road, the normal force is $F_N = mg$. The maximum static friction force is $F_{sfr,max} = \mu_s F_N$. For the horizontal component of $\Sigma \mathbf{F} = m\mathbf{a}$, we have

    $-\mu_s mg = ma_{max}$, which gives

    $\mu_s = -a_{max}/g = -(-3.17\ \text{m/s}^2)/(9.80\ \text{m/s}^2) =$ $\boxed{0.32.}$

51. From the force diagram for the soap, we have $\Sigma \mathbf{F} = m\mathbf{a}$:

    x-component: $mg \sin \theta = ma$;

    y-component: $F_N - mg \cos \theta = 0$.

    From the x-equation we find the acceleration:

    $a = g \sin \theta = (9.80\ \text{m/s}^2) \sin 7.3° = 1.25\ \text{m/s}^2$.

    For the motion of the soap, we find the time from

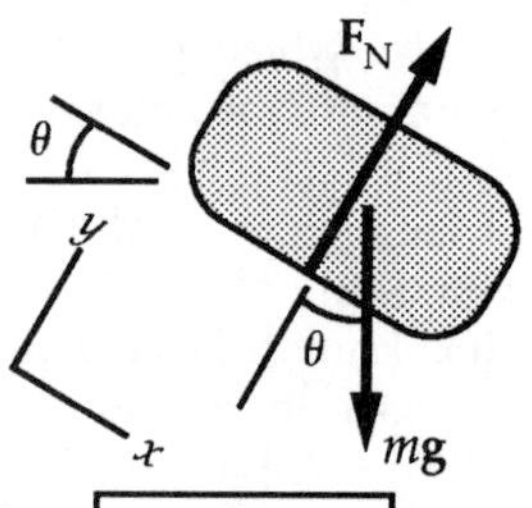

    $x = x_0 + v_0 t + \frac{1}{2}at^2$;

    $2.0\ \text{m} = 0 + 0 + \frac{1}{2}(1.25\ \text{m/s}^2)t^2$, which gives $\boxed{t = 1.8\ \text{s.}}$

    Because the acceleration does not depend on the mass, there will be $\boxed{\text{no change.}}$

52. (*a*) From the force diagram for the block, we have $\Sigma\mathbf{F} = m\mathbf{a}$:

    x-component: $mg \sin\theta = ma$;

    y-component: $F_N - mg\cos\theta = 0$.

    From the x-equation we find the acceleration:

    $a = g\sin\theta = (9.80 \text{ m/s}^2)\sin 22.0° = \boxed{3.67 \text{ m/s}^2.}$

    (*b*) For the motion of the block, we find the speed from

    $v^2 = v_0^2 + 2a(x - x_0)$;

    $v^2 = 0 + 2(3.67 \text{ m/s}^2)(9.10 \text{ m} - 0)$, which gives $v = \boxed{8.17 \text{ m/s}.}$

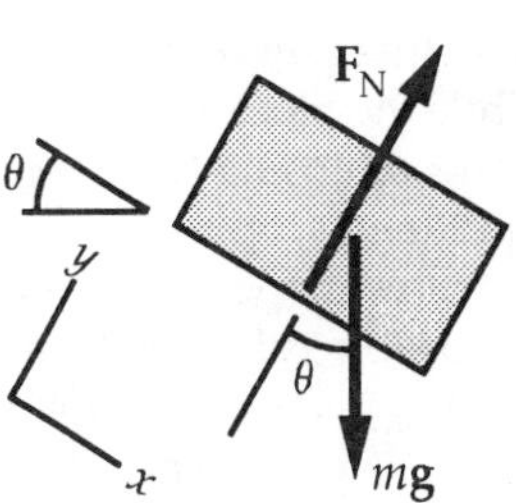

53. We choose the origin for x at the bottom of the plane.
    Note that down the plane (the direction of the acceleration) is positive.

    (*a*) From the force diagram for the block, we have $\Sigma\mathbf{F} = m\mathbf{a}$:

    x-component: $mg\sin\theta = ma$;

    y-component: $F_N - mg\cos\theta = 0$.

    From the x-equation we find the acceleration:

    $a = g\sin\theta = (9.80 \text{ m/s}^2)\sin 22.0° = 3.67 \text{ m/s}^2$.

    For the motion of the block, we find the distance from

    $v^2 = v_0^2 + 2a(x - x_0)$;

    $0 = (-3.0 \text{ m/s})^2 + 2(3.67 \text{ m/s}^2)(x - 0)$, which gives $x = -1.2$ m.

    Thus the block travels $\boxed{1.2 \text{ m up the plane.}}$

    (*b*) We find the time to return to the bottom from

    $x = x_0 + v_0t + \frac{1}{2}at^2$;

    $0 = 0 + (-3.0 \text{ m/s})t + \frac{1}{2}(3.67 \text{ m/s}^2)t^2$, which gives $t = 0$ (the start), and $\boxed{t = 1.6 \text{ s.}}$

54. Note that down is the positive direction in both problems. While the block is sliding down, friction will be up the plane, opposing the motion. The x-component of the force equation becomes

x-component: $mg \sin \theta - F_{fr} = ma_{down}$.

Thus the acceleration is now

$$a_{down} = g \sin \theta - \mu_k g \cos \theta = g(\sin \theta - \mu_k \cos \theta)$$
$$= (9.80 \text{ m/s}^2)[\sin 22.0° - (0.20) \cos 22.0°] = 1.85 \text{ m/s}^2.$$

While the block is sliding up, friction will be down the plane, opposing the motion. The x-component of the force equation becomes

x-component: $mg \sin \theta + F_{fr} = ma_{up}$.

Thus the acceleration is now

$$a_{up} = g \sin \theta + \mu_k g \cos \theta = g(\sin \theta + \mu_k \cos \theta)$$
$$= (9.80 \text{ m/s}^2)[\sin 22.0° + (0.20) \cos 22.0°] = 5.49 \text{ m/s}^2.$$

(*a*) For the motion of the block down the plane, the acceleration will be **1.85 m/s².**

We find the speed from

$v^2 = v_0^2 + 2a_{down}(x - x_0)$;

$v^2 = 0 + 2(1.85 \text{ m/s}^2)(9.10 \text{ m} - 0)$, which gives $v =$ **5.81 m/s.**

(*b*) For the motion of the block up the plane, we find the distance from

$v^2 = v_0^2 + 2a_{up}(x - x_0)$;

$0 = (-3.0 \text{ m/s})^2 + 2(5.49 \text{ m/s}^2)(x - 0)$, which gives $x = -0.82$ m.

Thus the block travels **0.82 m up the plane.**

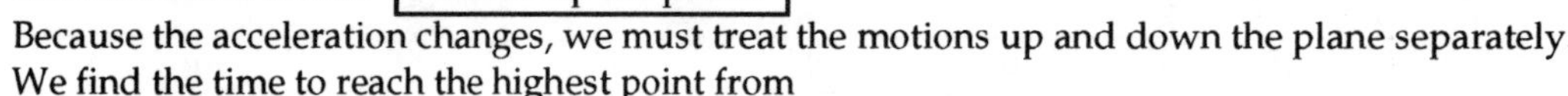

Because the acceleration changes, we must treat the motions up and down the plane separately.

We find the time to reach the highest point from

$v = v_0 + a_{up}t_{up}$;

$0 = (-3.0 \text{ m/s}) + (5.49 \text{ m/s}^2)t_{up}$, which gives $t_{up} = 0.55$ s.

We find the time to return to the bottom from

$x = x_0 + v_0 t_{down} + \frac{1}{2}a_{down}t_{down}^2$;

$0 = -0.82 \text{ m} + 0 + \frac{1}{2}(1.85 \text{ m/s}^2)t^2$, which gives $t_{down} = 0.94$ s.

Thus the total time is

$T = t_{up} + t_{down} = 0.55 \text{ s} + 0.94 \text{ s} =$ **1.49 s.**

55. While the roller coaster is sliding down, friction will be up the plane, opposing the motion. From the force diagram for the roller coaster, we have $\Sigma\mathbf{F} = m\mathbf{a}$:

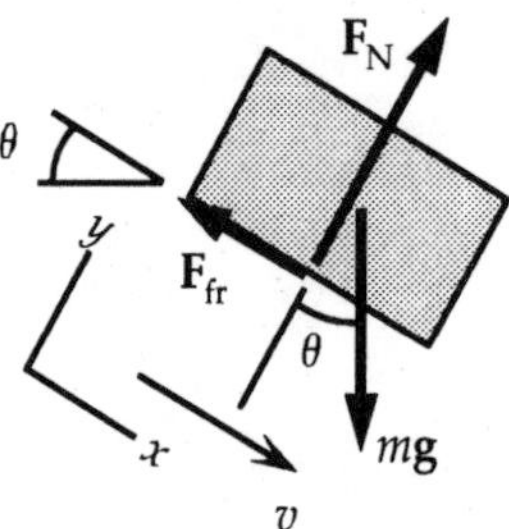

x-component: $mg \sin \theta - F_{fr} = mg \sin \theta - \mu_k F_N = ma$.

y-component: $F_N - mg \cos \theta = 0$.

We combine these equations to find the acceleration:

$$a = g \sin \theta - \mu_k g \cos \theta = g(\sin \theta - \mu_k \cos \theta)$$
$$= (9.80 \text{ m/s}^2)[\sin 45° - (0.12) \cos 45°] = 6.10 \text{ m/s}^2.$$

For the motion of the roller coaster, we find the speed from

$v^2 = v_0^2 + 2a(x - x_0)$;

$v^2 = [(6.0 \text{ km/h})/(3.6 \text{ ks/h})]^2 + 2(6.10 \text{ m/s}^2)(45.0 \text{ m} - 0)$, which gives $v =$

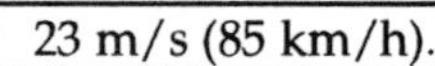

23 m/s (85 km/h).

56. While the box is sliding down, friction will be up the plane, opposing the motion. From the force diagram for the box, we have $\Sigma\mathbf{F} = m\mathbf{a}$:

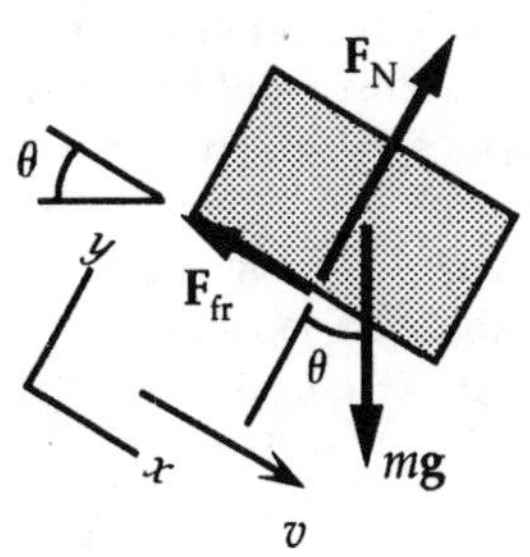

x-component: $mg \sin\theta - F_{fr} = ma$.
y-component: $F_N - mg\cos\theta = 0$.

From the x-equation, we have

$F_{fr} = mg\sin\theta - ma = m(g\sin\theta - a)$
$= (18.0\text{ kg})[(9.80\text{ m/s}^2)\sin 37.0° - (0.270\text{ m/s}^2)]$
$=$ **101 N.**

Because the friction is kinetic, we have

$F_{fr} = \mu_k F_N = \mu_k mg\cos\theta$;
$101.3\text{ N} = \mu_k(18.0\text{ kg})(9.80\text{ m/s}^2)\cos 37.0°$, which gives **$\mu_k = 0.719$.**

57. When the cart is pushed up the ramp at constant speed, friction will be down, opposing the motion. From the force diagram for the cart, we have $\Sigma\mathbf{F} = m\mathbf{a}$:

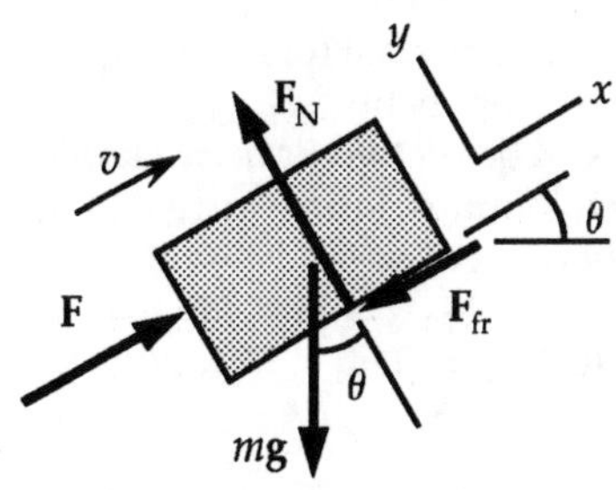

x-component: $F - mg\sin\theta - F_{fr} = ma = 0$;
y-component: $F_N - mg\cos\theta = 0$;

with $F_{fr} = \mu_k F_N$.

For a 5° slope we have

$F = mg(\sin\theta + \mu_k \sin\theta)$
$= (30\text{ kg})[(9.80\text{ m/s}^2)[\sin 5° + (0.10)\cos 5°]$
$= 55\text{ N}$.

Because this is greater than 50 N, a 5° slope is **too steep.**

58. For each object we take the direction of the acceleration as the positive direction. The kinetic friction from the table will oppose the motion of the bowl.

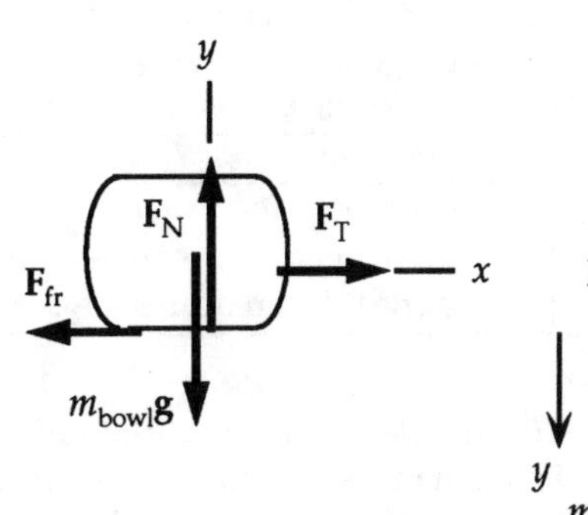

(*a*) From the force diagrams, we have $\Sigma\mathbf{F} = m\mathbf{a}$:

x-component(bowl): $F_T - F_{fr} = m_{bowl}a$;
y-component(bowl): $F_N - m_{bowl}g = 0$;
y-component(cat): $m_{cat}g - F_T = m_{cat}a$.

With $F_{fr} = \mu_k F_N$, we have

$F_T = F_{fr} = m_{bowl}a + \mu_k m_{bowl}g$.

When we eliminate F_T, we get

$(m_{cat} - \mu_k m_{bowl})g = (m_{cat} + m_{bowl})a$;
$[(5.0\text{ kg}) - (0.44)(11\text{ kg})](9.80\text{ m/s}^2) = (5.0\text{ kg} + 11\text{ kg})a$,

which gives **$a = 0.098\text{ m/s}^2$.**

(*b*) We find the time for the bowl to reach the edge of the table from

$x = x_0 + v_0t + \tfrac{1}{2}at^2$;
$0.9\text{ m} = 0 + 0 + \tfrac{1}{2}(0.098\text{ m/s}^2)t^2$, which gives **$t = 4.3\text{ s}$.**

59. Before the mass slides, the friction is static, with $F_{sfr} \le \mu_s F_N$. The static friction force will be maximum just before the mass slides. We write $\Sigma\mathbf{F} = m\mathbf{a}$ from the force diagram:
    x-component: $mg \sin \theta_{max} - \mu_s F_N = 0$;
    y-component: $F_N - mg \cos \theta_{max} = 0$.
    When we combine these, we get
    $\tan \theta_{max} = \mu_s = 0.60$, or $\boxed{\theta_{max} = 31°.}$

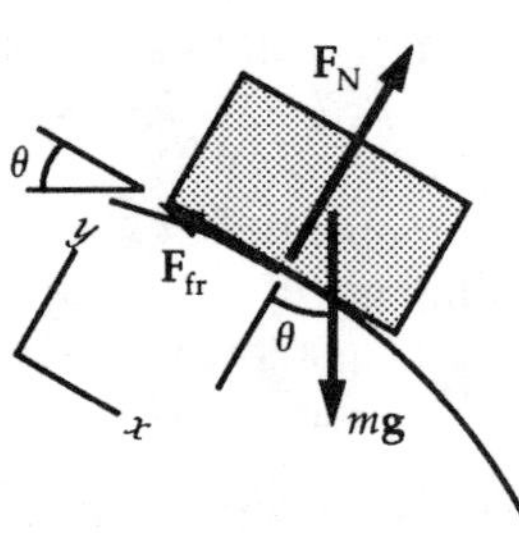

60. We assume there is no tension in the rope and simplify the forces to those shown. From the force diagram, we have $\Sigma\mathbf{F} = m\mathbf{a}$:
    x-component: $F_{Nshoes} - F_{Nwall} = 0$,
    so the two normal forces are equal: $F_{Nshoes} = F_{Nwall} = F_N$;
    y-component: $F_{frshoes} + F_{frwall} - mg = 0$.
    For a static friction force, we know that $F_{sfr} \le \mu_s F_N$.
    The minimum normal force will be exerted when the static friction forces are at the limit:
    $\mu_{sshoes} F_{Nshoes} + \mu_{swall} F_{Nwall} = mg$;
    $(0.80 + 0.60)F_N = (70 \text{ kg})(9.80 \text{ m/s}^2)$, which gives $\boxed{F_N = 4.9 \times 10^2 \text{ N.}}$

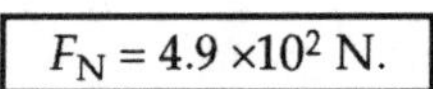

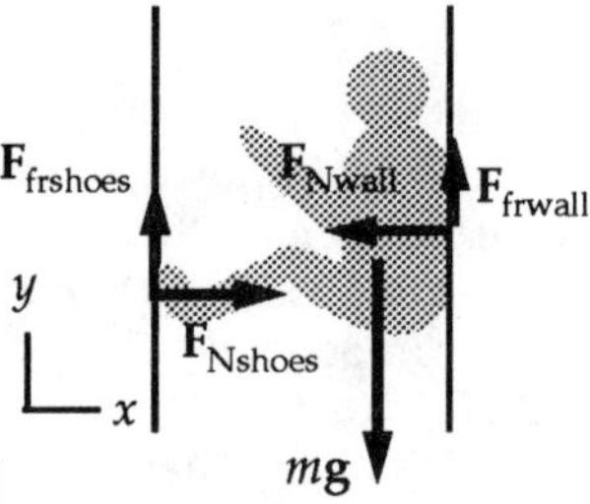

61. (*a*) The forces and coordinate systems are shown in the diagram. From the force diagram, with the block m_2 as the system, we can write $\Sigma\mathbf{F} = M\mathbf{a}$:
    y-component: $m_2 g - F_T = m_2 a$.
    From the force diagram, with the block m_1 as the system, we can write $\Sigma\mathbf{F} = m\mathbf{a}$:
    x-component: $F_T - m_1 g \sin \theta = m_1 a$.
    When we eliminate F_T between these two equations, we get
    $\boxed{a = (m_2 - m_1 \sin \theta)g/(m_1 + m_2).}$
    (*b*) Because up the plane is our positive direction, we have
    a down (negative) requires $m_2 < m_1 \sin \theta$.
    a up (positive) requires $m_2 > m_1 \sin \theta$.

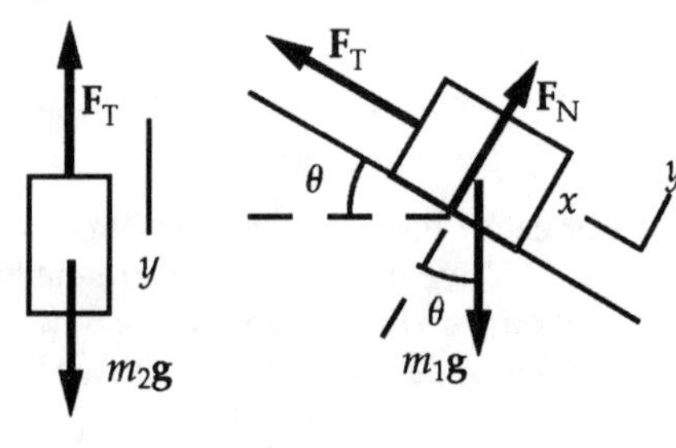

62. The direction of the kinetic friction force is determined by the direction of the velocity, not the direction of the acceleration. The block on the plane is moving up, so the friction force is down. The forces and coordinate systems are shown in the diagram. From the force diagram, with the block m_2 as the system, we can write $\Sigma\mathbf{F} = M\mathbf{a}$:
    y-component: $m_2 g - F_T = m_2 a$.
    From the force diagram, with the block m_1 as the system, we can write $\Sigma\mathbf{F} = m\mathbf{a}$:
    x-component: $F_T - F_{fr} - m_1 g \sin \theta = m_1 a$;
    y-component: $F_N - m_1 g \cos \theta = 0$; with $F_{fr} = \mu_k F_N$.
    When we eliminate F_T between these two equations, we get
    $a = (m_2 - m_1 \sin \theta - \mu_k m_1 \cos \theta)g/(m_1 + m_2)$
    $= [2.7 \text{ kg} - (2.7 \text{ kg}) \sin 25° - (0.15)(2.7 \text{ kg}) \cos 25°](9.80 \text{ m/s}^2)/(2.7 \text{ kg} + 2.7 \text{ kg})$
    $= \boxed{2.2 \text{ m/s}^2 \text{ up the plane.}}$
    The acceleration is up the plane because the answer is positive.

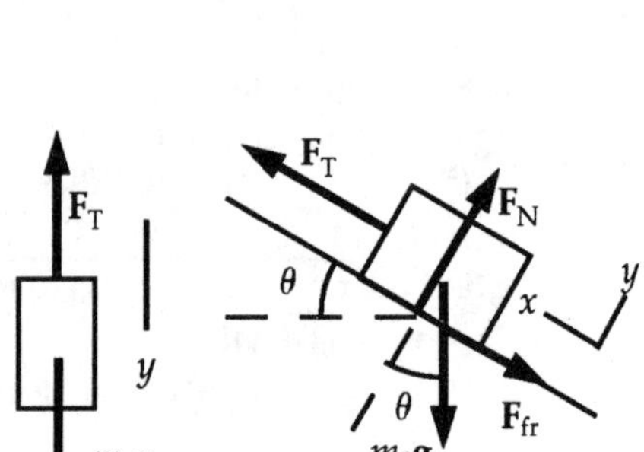

63. With the block moving up the plane, we find the coefficient of kinetic friction required for zero acceleration from

$a = (m_2 - m_1 \sin\theta - \mu_k m_1 \cos\theta)g/(m_1 + m_2);$

$0 = [2.7\text{ kg} - (2.7\text{ kg})\sin 25° - \mu_k(2.7\text{ kg})\cos 25°](9.80\text{ m/s}^2)/(2.7\text{ kg} + 2.7\text{ kg}).$

which gives $\boxed{\mu_k = 0.64.}$

64. Both motions have constant velocity, so the acceleration is zero. From the force diagram for the motion coasting down the hill, we can write $\Sigma\mathbf{F} = m\mathbf{a}$:

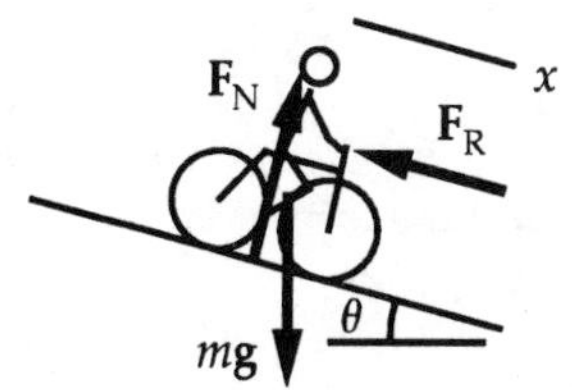

x-component: $mg \sin\theta - F_R = 0$, or $F_R = mg \sin\theta$.

From the force diagram for the motion climbing up the hill, we can write $\Sigma\mathbf{F} = m\mathbf{a}$:

x-component: $F - mg \sin\theta - F_R = 0$, so

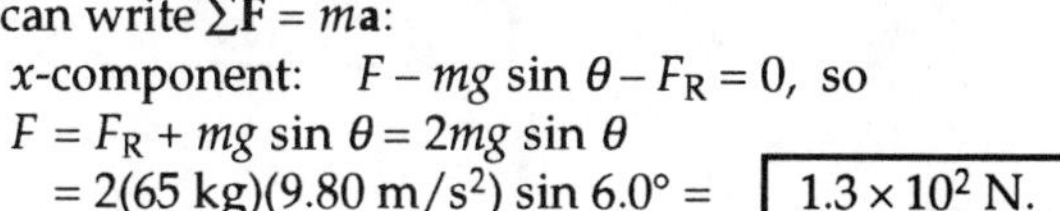

$F = F_R + mg \sin\theta = 2mg \sin\theta$

$= 2(65\text{ kg})(9.80\text{ m/s}^2)\sin 6.0° = \boxed{1.3 \times 10^2\text{ N.}}$

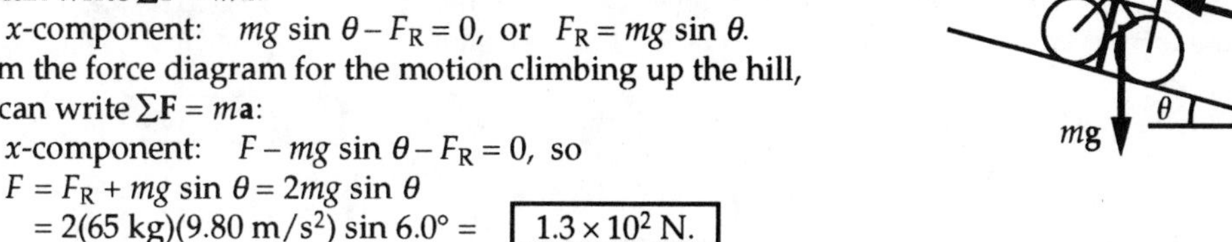

65. The acceleration can be found from the blood's one-dimensional motion:

$v = v_0 + at;$

$0.35\text{ m/s} = (0.25\text{ m/s}) + a(0.10\text{ s})$, which gives $a = 1.00\text{ m/s}^2$.

We apply Newton's second law to find the required force

$\Sigma F = ma;$

$F = (20 \times 10^{-3}\text{ kg})(1.00\text{ m/s}^2) = \boxed{2.0 \times 10^{-2}\text{ N.}}$

66. We apply Newton's second law to the person:

$\Sigma F = ma;$

$F = (70\text{ kg})(-30)(9.80\text{ m/s}^2) = \boxed{-2.1 \times 10^4\text{ N}}$ (opposite to motion).

We find the distance from

$v^2 = v_0^2 + 2a(x - x_0)$

$0 = [(90\text{ km/h})/(3.6\text{ ks/h})]^2 + 2(-30)(9.80\text{ m/s}^2)(x - x_0)$, which gives $x - x_0 = \boxed{1.1\text{ m.}}$

67. (*a*) For the object to move with the ground, the static friction force must provide the same acceleration. With the standard coordinate system, for $\Sigma\mathbf{F} = m\mathbf{a}$ we have

x-component: $F_{sfr} = ma;$

y-component: $F_N - mg = 0.$

For static friction, $F_{sfr} \le \mu_s F_N$, or $ma \le \mu_s mg$; thus $\mu_s \ge a/g$.

(*b*) For the greatest acceleration, the minimum required coefficient is

$\mu_s = a/g = (4.0\text{ m/s}^2)/(9.80\text{ m/s}^2) = 0.41.$

Because this is greater than 0.25, the chair $\boxed{\text{will slide.}}$

68. From the force diagram for the car we have

x-component: $F - F_T = m_{car}a$;

y-component : $F_{N,car} - m_{car}g = 0$.

From the force diagram for the trailer we have

x-component: $F_T - \mu_k F_{N,trailer} = m_{trailer}a$;

y-component: $F_{N,trailer} - m_{trailer}g = 0$, or

$F_{N,trailer} = m_{trailer}g$.

If we add the two x-equations, we get

$F - \mu_k m_{trailer}g = m_{car}a + m_{trailer}a$;

3.5×10^3 N – (0.15)(450 kg)(9.80 m/s²) = (1000 kg + 450 kg)a, which gives a = 1.96 m/s².

We can find the force on the trailer from the x-equation for the trailer:

$F_T - \mu_k m_{trailer}g = m_{trailer}a$;

F_T – (0.15)(450 kg)(9.80 m/s²) = (450 kg)(1.96 m/s²), which gives F_T = **1.5×10^3 N.**

69. On a level road, the normal force is $F_N = mg$. The kinetic friction force is the only force slowing the automobile. We find the acceleration from the horizontal component of $\Sigma\mathbf{F} = m\mathbf{a}$:

$-\mu_k mg = ma$, which gives

$a = -\mu_k g = -(0.80)(9.80\ \text{m/s}^2) = -7.84\ \text{m/s}^2$.

For the motion until the automobile stops, we have

$v^2 = v_0^2 + 2a(x - x_0)$;

$0 = v_0^2 + 2(-7.84\ \text{m/s}^2)(80\ \text{m})$, which gives v_0 = **35 m/s (130 km/h).**

70. We find the angle of the hill from

$\sin\theta = 1/4$, which gives $\theta = 14.5°$.

The kinetic friction force will be up the hill to oppose the motion. From the force diagram for the car, we have $\Sigma\mathbf{F} = m\mathbf{a}$:

x-component: $mg\sin\theta - F_{fr} = ma$;

y-component: $F_N - mg\cos\theta = 0$.

When we combine these, we get

$a = g\sin\theta - \mu_k g\cos\theta = g(\sin\theta - \mu_k\cos\theta)$.

(*a*) If there is no friction, we have

$a = g\sin\theta = (9.80\ \text{m/s}^2)\sin 14.5° = 2.45\ \text{m/s}^2$.

We find the car's speed from

$v^2 = v_0^2 + 2a(x - x_0)$;

$v^2 = 0 + 2(2.45\ \text{m/s}^2)(50\ \text{m})$, which gives v = **16 m/s.**

(*b*) If there is friction, we have

$a = g(\sin\theta - \mu_k\cos\theta) = (9.80\ \text{m/s}^2)[\sin 14.5° - (0.10)\cos 14.5°] = 1.50\ \text{m/s}^2$.

We find the car's speed from

$v^2 = v_0^2 + 2a(x - x_0)$;

$v^2 = 0 + 2(1.50\ \text{m/s}^2)(50\ \text{m})$, which gives v = **12 m/s.**

71. We write $\Sigma\mathbf{F} = m\mathbf{a}$ from the force diagram for the fish:

y-component: $F_T - mg = ma$, or $F_T = m(a + g)$.

(*a*) At constant speed, $a = 0$, so we have

$F_{Tmax} = m_{max}g$;

45 N = m_{max}(9.80 m/s²), which gives **m_{max} = 4.6 kg.**

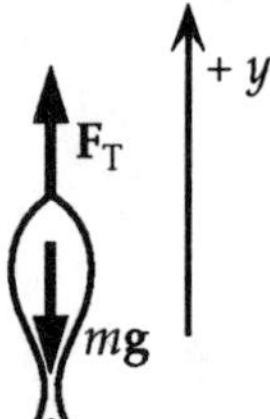

(*b*) For an upward acceleration, we have

$F_{Tmax} = m_{max}(a + g)$;

45 N = m_{max}(2.0 m/s² + 9.80 m/s²), which gives **m_{max} = 3.8 kg.**

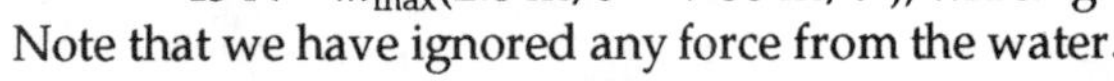

Note that we have ignored any force from the water.

72. For the motion until the elevator stops, we have

$v^2 = v_0^2 + 2a(x - x_0)$;

$0 = (3.5 \text{ m/s})^2 + 2a\,(3.0 \text{ m})$, which gives $a = -2.04 \text{ m/s}^2$.

We write $\Sigma \mathbf{F} = m\mathbf{a}$ from the force diagram for the elevator:

y-component: $mg - F_T = ma$; or

$(1300 \text{ kg})(9.80 \text{ m/s}^2) - F_T = (1300 \text{ kg})(-2.04 \text{ m/s}^2)$,

which gives

$F_T =$ $\boxed{1.5 \times 10^4 \text{ N.}}$

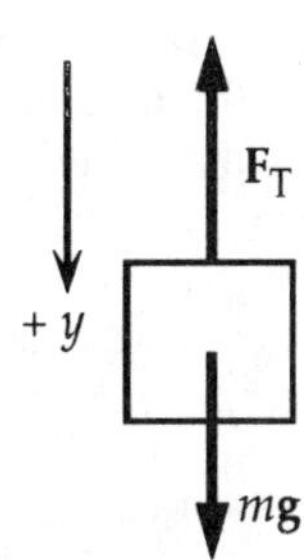

73. (*a*) We assume that the block will slide down; friction will be up the plane, opposing the motion. From the force diagram for box 1, without the tension, we have $\Sigma \mathbf{F} = m\mathbf{a}$:

x-component: $m_1 g \sin\theta - \mu_{k1} F_{N1} = m_1 a_1$;

y-component: $F_{N1} - m_1 g \cos\theta = 0$.

Combining these, we get the acceleration:

$a_1 = (\sin\theta - \mu_{k1}\cos\theta)g$

$= [\sin 30° - (0.10)\cos 30°](9.80 \text{ m/s}^2)$

$=$ $\boxed{4.1 \text{ m/s}^2.}$

From the force diagram for box 2, without the tension, we have $\Sigma \mathbf{F} = m\mathbf{a}$:

x-component: $m_2 g \sin\theta - \mu_{k2} F_{N2} = m_2 a_2$;

y-component: $F_{N2} - m_2 g \cos\theta = 0$.

Combining these, we get the acceleration:

$a_2 = (\sin\theta - \mu_{k2}\cos\theta)g$

$= [\sin 30° - (0.20)\cos 30°](9.80 \text{ m/s}^2) =$ $\boxed{3.2 \text{ m/s}^2.}$

(*b*) We see from part (*a*) that the upper box will have a greater acceleration. Any tension in the string would increase this effect. Thus the upper box will slide faster than the lower box and the string will become lax, with no tension. Until the upper box reaches the lower box, when a normal force is created, the accelerations will be the same as in part (*a*):

$\boxed{a_1 = 4.1 \text{ m/s}^2, \text{ and } a_2 = 3.2 \text{ m/s}^2.}$

(*c*) In this configuration, we see that the lower box will try to pull away from the upper one, creating a tension in the string. Because the two boxes are tied together, they will have the same acceleration. The x-components of the force equation will be

box 1 (lower): $m_1 g \sin\theta - F_T - \mu_{k1} m_1 g \cos\theta = m_1 a$;

box 2 (upper): $m_2 g \sin\theta + F_T - \mu_{k2} m_2 g \cos\theta = m_2 a$.

We add these equations to eliminate F_T, and get the acceleration:

$a = \{\sin\theta - [(\mu_{k1} m_1 + \mu_{k2} m_2)/(m_1 + m_2)]\cos\theta\}g$

$= (\sin 30° - \{(0.10)(1.0 \text{ kg}) + (0.20)(2.0 \text{ kg})]/(1.0 \text{ kg} + 2.0 \text{ kg})\}\cos 30°)(9.80 \text{ m/s}^2) =$ $\boxed{3.5 \text{ m/s}^2.}$

74. We take the positive direction upward.
The scale reads the force the person exerts on the scale. From Newton's third law, this is also the magnitude of the normal force acting on the person. The effective mass on the scale is

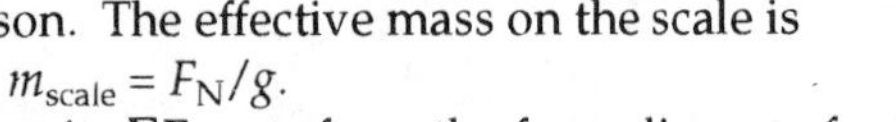

$m_{scale} = F_N/g.$

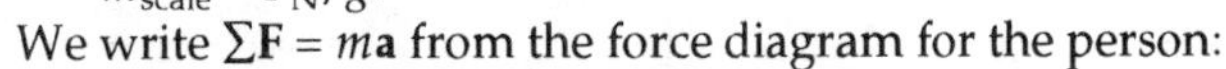

We write $\Sigma\mathbf{F} = m\mathbf{a}$ from the force diagram for the person:

y-component: $F_N - mg = ma$, or

$m_{scale} = F_N/g = m(a + g)/g.$

(*a*) When the elevator is at rest, $a = 0$:

$m_{scale} = m(0 + g)/g = m =$ **75.0 kg.**

(*b*) When the elevator is climbing at constant speed, $a = 0$:

$m_{scale} = m(0 + g)/g = m =$ **75.0 kg.**

(*c*) When the elevator is falling at constant speed, $a = 0$:

$m_{scale} = m(0 + g)/g = m =$ **75.0 kg.**

(*d*) When the elevator is accelerating upward, a is positive:

$m_{scale} = m(a + g)/g = (75.0\text{ kg})(3.0\text{ m/s}^2 + 9.80\text{ m/s}^2)/(9.80\text{ m/s}^2) =$ **98.0 kg.**

(*e*) When the elevator is accelerating downward, a is negative:

$m_{scale} = m(a + g)/g = (75.0\text{ kg})(-3.0\text{ m/s}^2 + 9.80\text{ m/s}^2)/(9.80\text{ m/s}^2) =$ **52.0 kg.**

75. (*a*) We select the origin at the bottom of the ramp, with up positive. We find the acceleration from the motion up the ramp:

$v^2 = v_0^2 + 2a(x - x_0);$

$0 = v_0^2 + 2a(d - 0)$, which gives $a = -v_0^2/2d$.

When the block slides up the ramp, kinetic friction will be down, opposing the motion. From the force diagram for the block, we have $\Sigma\mathbf{F} = m\mathbf{a}$:

x-component: $-mg\sin\theta - \mu_k F_N = ma;$

y-component: $F_N - mg\cos\theta = 0.$

When we eliminate F_N from the two equations and use the result for a, we get

$-mg\sin\theta - \mu_k mg\cos\theta = m(-v_0^2/2d)$, which gives

$\mu_k = (v_0^2/2gd\cos\theta) - \tan\theta.$

(*b*) Once the block stops, the friction becomes static and will be up the plane, to oppose the impending motion down. If the block remains at rest, the acceleration is zero. The static friction force must be $\leq \mu_s F_N$ and we have

x-component: $-mg\sin\theta + F_{frs} = 0$, or $F_{frs} = mg\sin\theta \leq \mu_s mg\cos\theta.$

Thus we know that **$\mu_s \geq \tan\theta.$**

76. On a level road, the normal force is $F_N = mg$. The kinetic friction force is the only force slowing the motorcycle. We find the acceleration from the horizontal component of $\Sigma\mathbf{F} = m\mathbf{a}$:

$-\mu_k mg = ma$, which gives

$a = -\mu_k g = -(0.80)(9.80\text{ m/s}^2) = -7.84\text{ m/s}^2.$

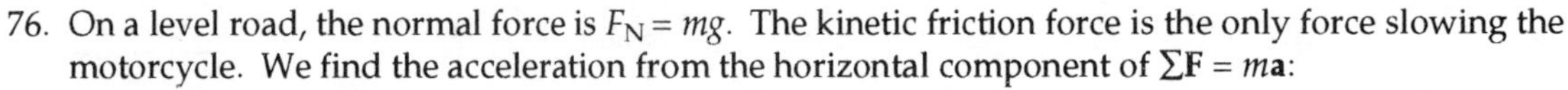

For the motion through the sandy stretch, we have

$v^2 = v_0^2 + 2a(x - x_0);$

$v^2 = (17\text{ m/s})^2 + 2(-7.84\text{ m/s}^2)(15\text{ m})$, which gives $v = \pm 7.3\text{ m/s}$.

The negative sign corresponds to the motorcycle going beyond the sandy stretch and returning, assuming the same negative acceleration after the motorcycle comes to rest. This will not occur, so the motorcycle

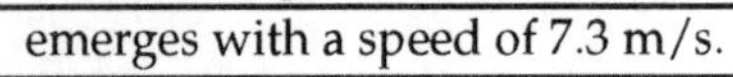
emerges with a speed of 7.3 m/s.

If the motorcycle did not emerge, we would get a negative value for v^2, indicating that there is no real value for v.

77. We find the acceleration of the car on the level from

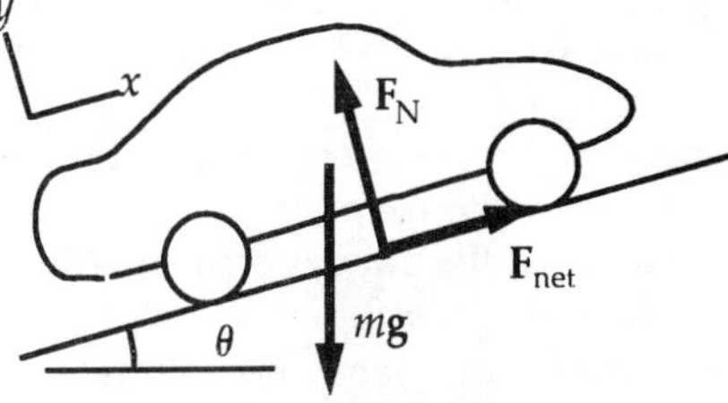

$v = v_0 + at;$

$(21 \text{ m/s}) = 0 + a(14.0 \text{ s})$, which gives $a = 1.5 \text{ m/s}^2$.

This acceleration is produced by the net force:

$F_{net} = ma = (1100 \text{ kg})(1.5 \text{ m/s}^2) = 1650 \text{ N}.$

If we assume the same net force on the hill, with no acceleration on the steepest hill, from the force diagram we have

x-component: $F_{net} - mg \sin \theta = 0;$

$1650 \text{ N} - (1100 \text{ kg})(9.80 \text{ m/s}^2) \sin \theta = 0$, which gives $\sin \theta = 0.153$, or $\boxed{\theta = 8.8°.}$

78. The velocity is constant, so the acceleration is zero.

(*a*) From the force diagram for the bicycle, we can write $\Sigma \mathbf{F} = m\mathbf{a}$:

x-component: $mg \sin \theta - F_R = 0$, or

$mg \sin \theta = cv;$

$(80 \text{ kg})(9.80 \text{ m/s}^2) \sin 5.0° = c(6.0 \text{ km/h})/(3.6 \text{ m/s})$, which gives

$\boxed{c = 41 \text{ kg/s.}}$

(*b*) We have an additional force in $\Sigma \mathbf{F} = m\mathbf{a}$:

x-component: $F + mg \sin \theta - F_R = 0$, so

$F = cv - mg \sin \theta$

$= [(41 \text{ kg/s})(20 \text{ km/h})/(3.6 \text{ m/s})] - (80 \text{ kg})(9.80 \text{ m/s}^2) \sin 5.0° = \boxed{1.6 \times 10^2 \text{ N.}}$

79. From the force diagram for the watch, we have $\Sigma \mathbf{F} = m\mathbf{a}$:

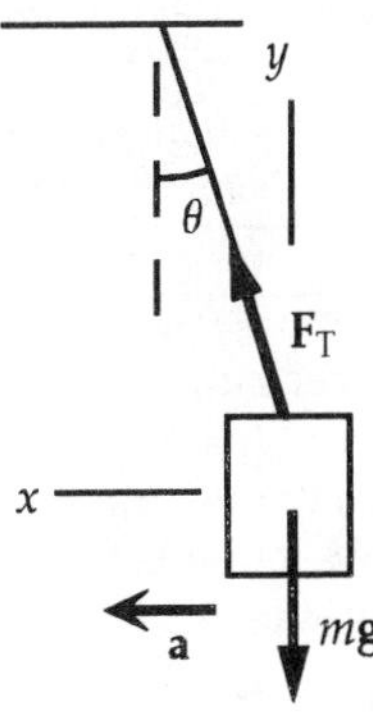

x-component: $F_T \sin \theta = ma;$

y-component: $F_T \cos \theta - mg = 0$, or $F_T \cos \theta = mg$.

If we divide the two equations, we find the acceleration:

$a = g \tan \theta = (9.80 \text{ m/s}^2) \tan 25° = 4.56 \text{ m/s}^2.$

For the motion of the aircraft, we find the takeoff speed from

$v = v_0 + at = 0 - (4.56 \text{ m/s}^2)(18 \text{ s}) = \boxed{82 \text{ m/s } (300 \text{ km/h}).}$

80. There are only three tensions. The tension in the rope that goes around both pulleys is constant:

    $F_{T1} = F_{T2} = F.$

    We choose up positive and assume that the piano is lifted with no acceleration.
    If the masses of the pulleys are negligible, we can write $\Sigma F_y = ma_y$.

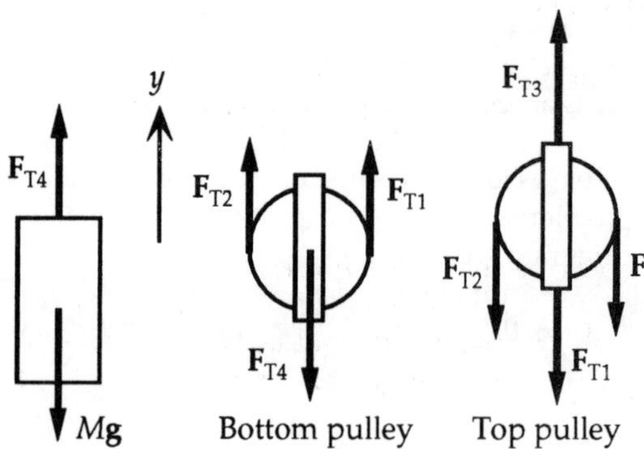

(*a*) If we select the piano and bottom pulley as the system, we have

$F_{T2} + F_{T1} - Mg = 0$, which gives

$2F_{T1} = Mg$, or $F_{T1} = \boxed{F = \tfrac{1}{2}Mg.}$

(*b*) For the individual elements, we have

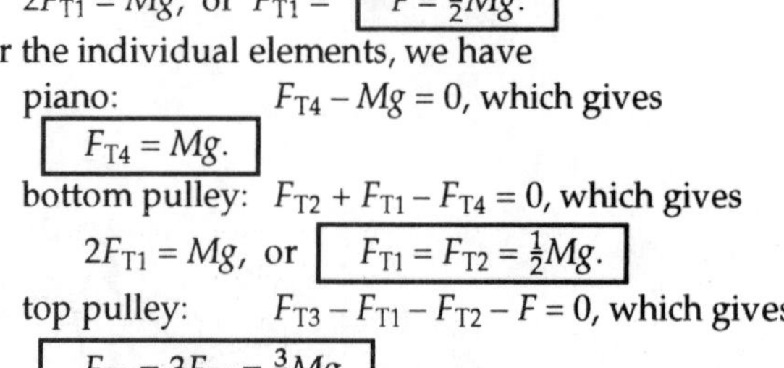

piano: $F_{T4} - Mg = 0$, which gives

$\boxed{F_{T4} = Mg.}$

bottom pulley: $F_{T2} + F_{T1} - F_{T4} = 0$, which gives

$2F_{T1} = Mg$, or $\boxed{F_{T1} = F_{T2} = \tfrac{1}{2}Mg.}$

top pulley: $F_{T3} - F_{T1} - F_{T2} - F = 0$, which gives

$\boxed{F_{T3} = 3F_{T1} = \tfrac{3}{2}Mg.}$

81. (*a*) If motion is just about to begin, the static friction force on the block will be maximum: $F_{sfr,max} = \mu_s F_N$, and the acceleration will be zero. We write $\Sigma \mathbf{F} = m\mathbf{a}$ from the force diagram for each object:

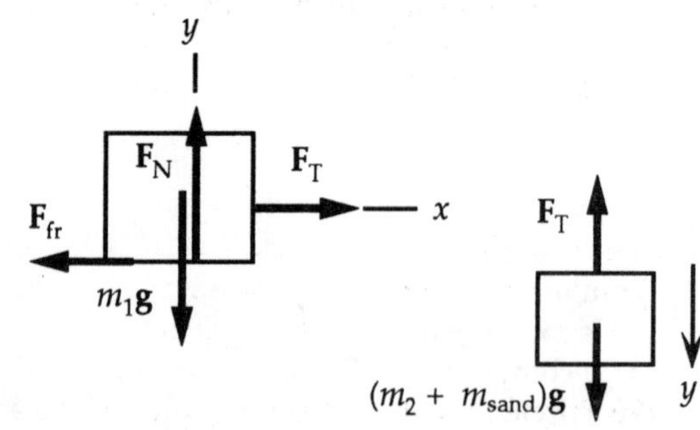

x-component (block): $F_T - \mu_s F_N = 0;$

y-component (block): $F_N - m_1 g = 0;$

y-component (bucket): $(m_2 + m_{sand})g - F_T = 0.$

When we combine these equations, we get

$(m_2 + m_{sand})g = \mu_s m_1 g$, which gives

$m_{sand} = \mu_s m_1 + m_2 = (0.450)(28.0\text{ kg}) - 1.00\text{ kg}$

$= \boxed{11.6\text{ kg}.}$

(*b*) When the system starts moving, the friction becomes kinetic, and the tension changes. The force equations become

x-component (block): $F_{T2} - \mu_k F_N = m_1 a;$

y-component (block): $F_N - m_1 g = 0;$

y-component (bucket): $(m_2 + m_{sand})g - F_{T2} = m_2 a.$

When we combine these equations, we get

$(m_2 + m_{sand} - \mu_k m_1)g = (m_1 + m_2 + m_{sand})a$, which gives

$a = [1.00\text{ kg} + 11.6\text{ kg} - (0.320)(28.0\text{ kg})](9.80\text{ m/s}^2)/(28.0\text{ kg} + 1.00\text{ kg} + 11.6\text{ kg}) = \boxed{0.879\text{ m/s}^2.}$

CHAPTER 5

1. The centripetal acceleration is
$a_R = v^2/r = (500 \text{ m/s})^2/(6.00 \times 10^3 \text{ m})(9.80 \text{ m/s}^2/g) =$ $\boxed{4.25g \text{ up.}}$

2. (*a*) The centripetal acceleration is
$a_R = v^2/r = (1.35 \text{ m/s})^2/(1.20 \text{ m}) =$ $\boxed{1.52 \text{ m/s}^2 \text{ toward the center.}}$
(*b*) The net horizontal force that produces this acceleration is
$F_{net} = ma_R = (25.0 \text{ kg})(1.52 \text{ m/s}^2) =$ $\boxed{38.0 \text{ N toward the center.}}$

3. The centripetal acceleration of the Earth is
$a_R = v^2/r = (2\pi r/T)^2/r = 4\pi^2 r/T^2$
$= 4\pi^2(1.50 \times 10^{11} \text{ m})/(3.16 \times 10^7 \text{ s})^2 =$ $\boxed{5.93 \times 10^{-3} \text{ m/s}^2 \text{ toward the Sun.}}$
The net force that produces this acceleration is
$F_{net} = m_E a_R = (5.98 \times 10^{24} \text{ kg})(5.93 \times 10^{-3} \text{ m/s}^2) =$ $\boxed{3.55 \times 10^{22} \text{ N toward the Sun.}}$
This force is the gravitational attraction from the $\boxed{\text{Sun.}}$

4. The force on the discus produces the centripetal acceleration:
$F = ma_R = mv^2/r;$
$280 \text{ N} = (2.0 \text{ kg})v^2/(1.00 \text{ m})$, which gives $v =$ $\boxed{12 \text{ m/s.}}$

5. We write $\Sigma\mathbf{F} = m\mathbf{a}$ from the force diagram for the stationary hanging mass, with down positive:
$mg - F_T = ma = 0;$ which gives
$F_T = mg.$
For the rotating puck, the tension provides the centripetal acceleration, $\Sigma F_R = Ma_R$:
$F_T = Mv^2/R.$
When we combine the two equations, we have
$Mv^2/R = mg$, which gives $v = (mgR/M)^{1/2}$.

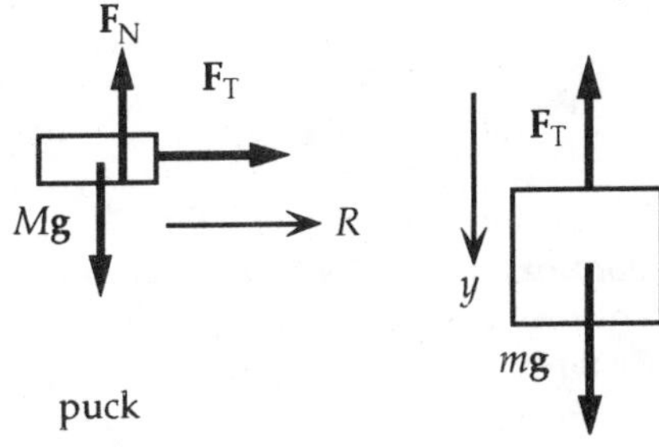

6. For the rotating ball, the tension provides the centripetal acceleration, $\Sigma F_R = Ma_R$:
$F_T = Mv^2/R.$
We see that the tension increases if the speed increases, so the maximum tension determines the maximum speed:
$F_{Tmax} = Mv_{max}^2/R;$
$60 \text{ N} = (0.40 \text{ kg})v_{max}^2/(1.3 \text{ m})$, which gives $\boxed{v_{max} = 14 \text{ m/s.}}$
If there were friction, it would be kinetic opposing the motion of the ball around the circle. Because this is perpendicular to the radius and the tension, it would have $\boxed{\text{no effect.}}$

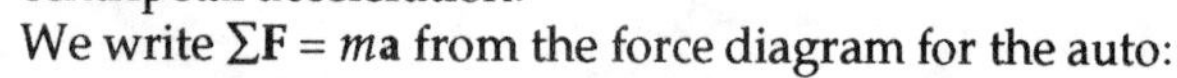

7. If the car does not skid, the friction is static, with $F_{fr} \le \mu_s F_N$.
This friction force provides the centripetal acceleration. We take a coordinate system with the x-axis in the direction of the centripetal acceleration.
We write $\Sigma \mathbf{F} = m\mathbf{a}$ from the force diagram for the auto:
   x-component: $F_{fr} = ma = mv^2/R$;
   y-component: $F_N - mg = 0$.
The speed is maximum when $F_{fr} = F_{fr,max} = \mu_s F_N$.
When we combine the equations, the mass cancels, and we get
   $\mu_s g = v_{max}^2/R$;
   $(0.80)(9.80\ \text{m/s}^2) = v_{max}^2/(70\ \text{m})$, which gives $\boxed{v_{max} = 23\ \text{m/s}.}$
The mass canceled, so the result is 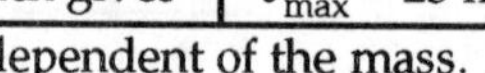
independent of the mass.

8. At each position we take the positive direction in the direction of the acceleration.
   (*a*) At the top of the path, the tension and the weight are downward.
   We write $\Sigma \mathbf{F} = m\mathbf{a}$ from the force diagram for the ball:
      $F_{T1} + mg = mv^2/R$;
      $F_{T1} + (0.300\ \text{kg})(9.80\ \text{m/s}^2) = (0.300\ \text{kg})(4.15\ \text{m/s})^2/(0.850\ \text{m})$,
   which gives $\boxed{F_{T1} = 3.14\ \text{N}.}$
   (*b*) At the bottom of the path, the tension is upward and the weight is downward. We write $\Sigma \mathbf{F} = m\mathbf{a}$ from the force diagram for the ball:
      $F_{T2} - mg = mv^2/R$;
      $F_{T2} - (0.300\ \text{kg})(9.80\ \text{m/s}^2) = (0.300\ \text{kg})(4.15\ \text{m/s})^2/(0.850\ \text{m})$,
   which gives $\boxed{F_{T2} = 9.02\ \text{N}.}$

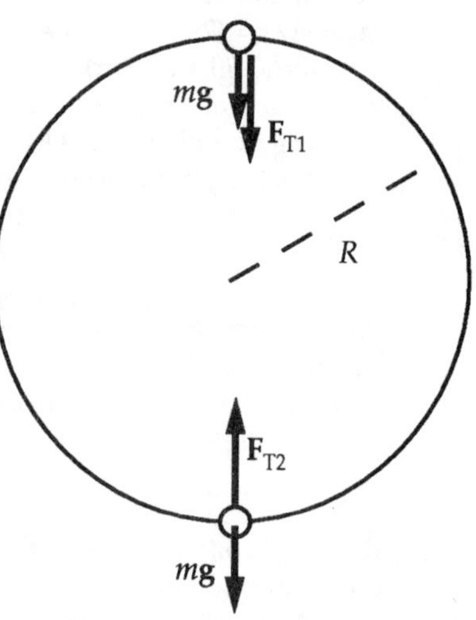

9. The friction force provides the centripetal acceleration. We take a coordinate system with the x-axis in the direction of the centripetal acceleration.
We write $\Sigma \mathbf{F} = m\mathbf{a}$ from the force diagram for the auto:
   x-component: $F_{fr} = ma = mv^2/R$;
   y-component: $F_N - mg = 0$.
If the car does not skid, the friction is static, with $F_{fr} \le \mu_s F_N$.
Thus we have
   $mv^2/R \le \mu_s mg$, or
   $\mu_s \ge v^2/gR = [(95\ \text{km/h})/(3.6\ \text{ks/h})]^2/(9.80\ \text{m/s}^2)(85\ \text{m})$. Thus $\boxed{\mu_s \ge 0.84.}$

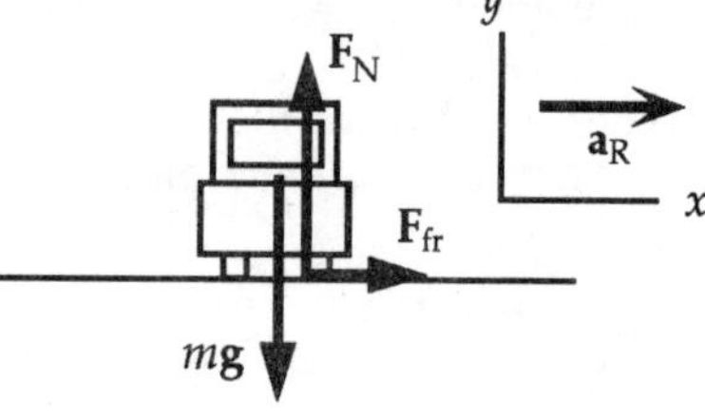

10. The horizontal force on the astronaut produces the centripetal acceleration:
   $F = ma_R = mv^2/r$;
   $(7.75)(2.0\ \text{kg})(9.80\ \text{m/s}^2) = (2.0\ \text{kg})v^2/(10.0\ \text{m})$, which gives $v = \boxed{27.6\ \text{m/s}.}$
The rotation rate is
   Rate $= v/2\pi r = (27.6\ \text{m/s})/2\pi(10.0\ \text{m}) = \boxed{0.439\ \text{rev/s}.}$
Note that the results are independent of mass, and thus are the same for all astronauts.

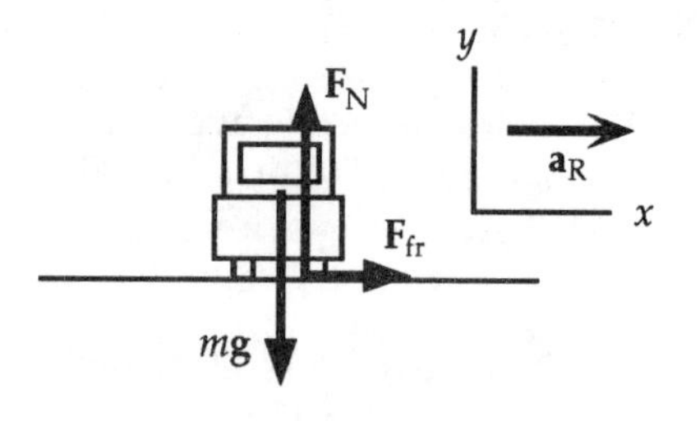

11. The static friction force provides the centripetal acceleration.
We write $\Sigma\mathbf{F} = m\mathbf{a}$ from the force diagram for the coin:

x-component: $F_{fr} = mv^2/R$;

y-component: $F_N - mg = 0$.

The highest speed without sliding requires $F_{fr,max} = \mu_s F_N$.
The maximum speed before sliding is

$$v_{max} = 2\pi R/T_{min} = 2\pi R f_{max}$$
$$= 2\pi(0.110\text{ m})(36/\text{min})/(60\text{ min/s}) = 0.415\text{ m/s}.$$

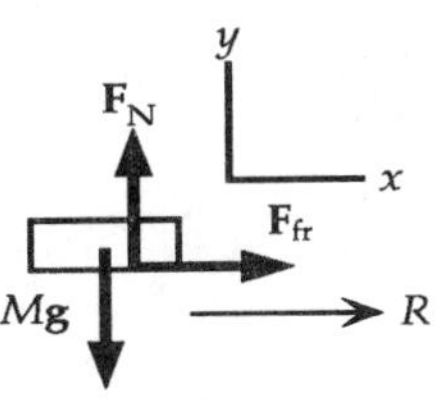

Thus we have

$$\mu_s mg = mv_{max}^2/R$$

$\mu_s(9.80\text{ m/s}^2) = (0.415\text{ m/s})^2/(0.110\text{ m})$, which gives $\mu_s =$ $\boxed{0.16.}$

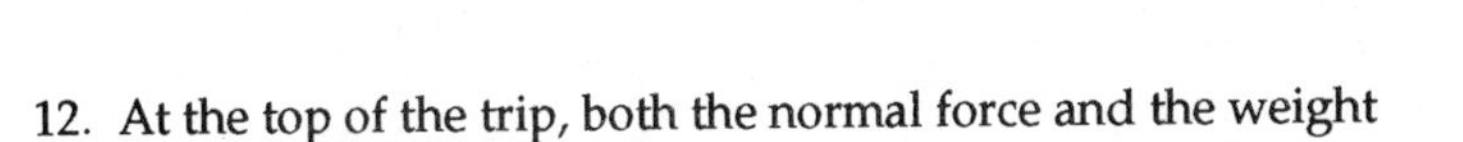

12. At the top of the trip, both the normal force and the weight are downward.
We write $\Sigma\mathbf{F} = m\mathbf{a}$ from the force diagram for the passenger:

y-component: $F_N + mg = mv^2/R$.

The speed v will be minimum when the normal force is minimum. The normal force can only push away from the seat, that is, with our coordinate system it must be positive, so $F_{N\min} = 0$.
Thus we have $v_{min}^2 = gR$, or

$$v_{min} = (gR)^{1/2}$$
$$= [(9.80\text{ m/s}^2)(8.6\text{ m})]^{1/2} = \boxed{9.2\text{ m/s.}}$$

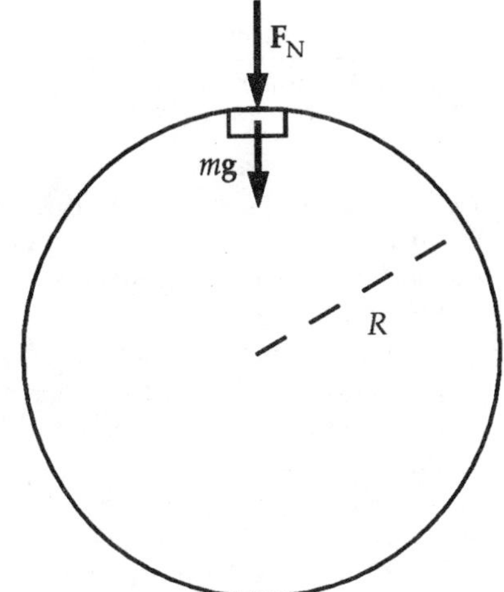

13. At the top of the hill, the normal force is upward and the weight is downward, which we select as the positive direction.

(*a*) We write $\Sigma\mathbf{F} = m\mathbf{a}$ from the force diagram for the car:

$$m_{car}g - F_{Ncar} = mv^2/R;$$
$$(1000\text{ kg})(9.80\text{ m/s}^2) - F_{Ncar} = (1000\text{ kg})(20\text{ m/s})^2/(100\text{ m}),$$

which gives $\boxed{F_{Ncar} = 5.8\times10^3\text{ N.}}$

(*b*) When we apply a similar analysis to the driver, we have

$$(70\text{ kg})(9.80\text{ m/s}^2) - F_{Npass} = (70\text{ kg})(20\text{ m/s})^2/(100\text{ m}),$$

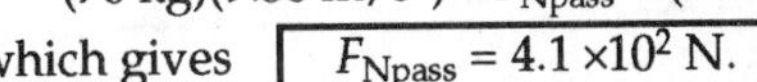

which gives $\boxed{F_{Npass} = 4.1\times10^2\text{ N.}}$

(*c*) For the normal force to be equal to zero, we have

$$(1000\text{ kg})(9.80\text{ m/s}^2) - 0 = (1000\text{ kg})v^2/(100\text{ m}),$$

which gives $\boxed{v = 31\text{ m/s}}$ (110 km/h or 70 mi/h).

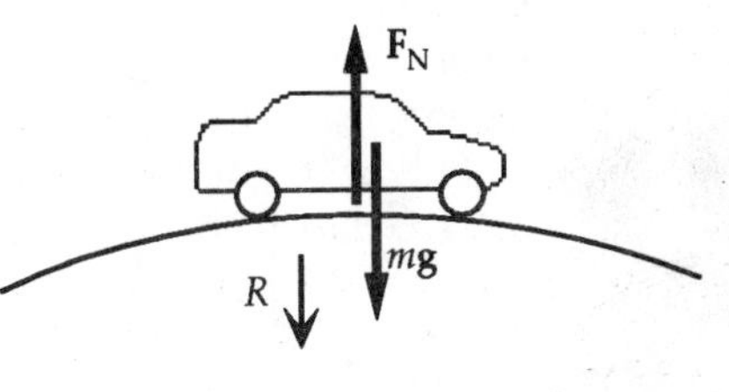

14. To feel "weightless" the normal force will be zero and the only force acting on a passenger will be that from gravity, which provides the centripetal acceleration:

$mg = mv^2/R$, or $v^2 = gR$;

$v^2 = (9.80\text{ m/s}^2)(7.5\text{ m})$, which gives $v = 8.57\text{ m/s}$.

We find the rotation rate from

Rate $= v/2\pi R = [(8.57\text{ m/s})/2\pi(7.5\text{ m})](60\text{ s/min}) =$ $\boxed{11\text{ rev/min.}}$

15. We check the form of $a_R = v^2/r$ by using the dimensions of each variable:
    $[a_R] = [v/t] = [d/t^2] = [L/T^2]$;
    $[v^2] = [(d/t)^2] = [(L/T)^2] = [L^2/T^2]$;
    $[r] = [d] = [L]$.
    Thus we have $[v^2/r] = [L^2/T^2]/[L] = [L/T^2]$, which are the dimensions of a_R.

16. The masses will have different velocities:
    $v_1 = 2\pi r_1/T = 2\pi r_1 f; \quad v_2 = 2\pi r_2/T = 2\pi r_2 f.$
    We choose the positive direction toward the center of the circle.
    For each mass we write $\Sigma F_r = ma_r$:
    m_1: $F_{T1} - F_{T2} = m_1 v_1^2/r_1 = 4\pi^2 m_1 r_1 f^2$;
    m_2: $F_{T2} = m_2 v_2^2/r_2 = 4\pi^2 m_2 r_2 f^2$.
    When we use this in the first equation, we get
    $F_{T1} = F_{T2} + 4\pi^2 m_1 r_1 f^2$; thus
    $\boxed{F_{T1} = 4\pi^2 f^2(m_1 r_1 + m_2 r_2); \quad F_{T2} = 4\pi^2 f^2 m_2 r_2.}$

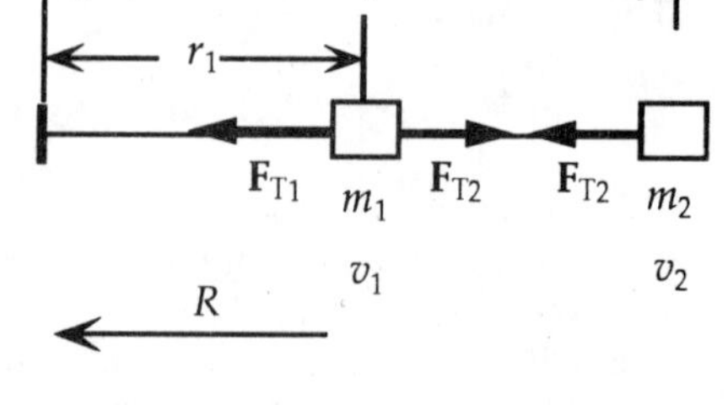

17. We convert the speed: (90 km/h)/(3.6 ks/h) = 25 m/s.
    We take the x-axis in the direction of the centripetal acceleration. We find the speed when there is no need for a friction force.
    We write $\Sigma \mathbf{F} = m\mathbf{a}$ from the force diagram for the car:
    x-component: $F_{N1} \sin\theta = ma_1 = mv_1^2/R$;
    y-component: $F_{N1} \cos\theta - mg = 0$.
    Combining these, we get
    $v_1^2 = gR \tan\theta = (9.80 \text{ m/s}^2)(70 \text{ m}) \tan 12°$,
    which gives $v_1 = 12.1$ m/s. Because the speed is greater than this, a friction force is required. Because the car will tend to slide up the slope, the friction force will be down the slope. We write $\Sigma \mathbf{F} = m\mathbf{a}$ from the force diagram for the car:
    x-component: $F_{N2} \sin\theta + F_{fr} \cos\theta = ma_2 = mv_2^2/R$;
    y-component: $F_{N2} \cos\theta - F_{fr} \sin\theta - mg = 0$.

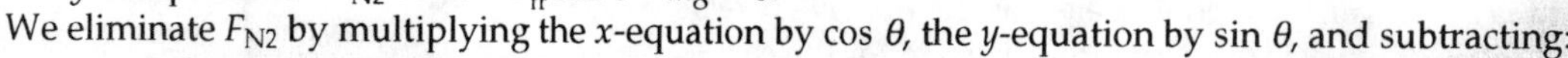

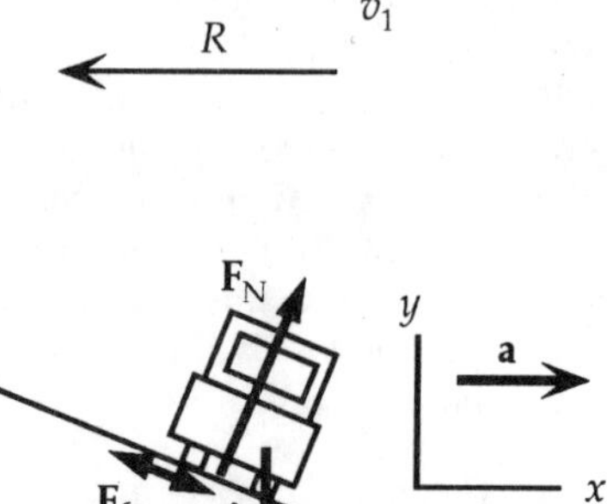

We eliminate F_{N2} by multiplying the x-equation by $\cos\theta$, the y-equation by $\sin\theta$, and subtracting:
    $F_{fr} = m\{[(v_2^2/R)\cos\theta] - g\sin\theta\}$
    $= (1200 \text{ kg})(\{[(25 \text{ m/s})^2/(70 \text{ m})] \cos 12°\} - (9.80 \text{ m/s}^2) \sin 12°) = \boxed{8.0 \times 10^3 \text{ N down the slope.}}$

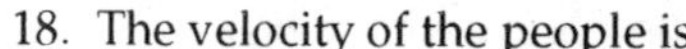

18. The velocity of the people is
    $v = 2\pi R/T = 2\pi Rf = 2\pi(5.0 \text{ m/rev})(0.50 \text{ rev/s}) = 15.7$ m/s.
    The force that prevents slipping is an upward friction force.
    The normal force provides the centripetal acceleration.
    We write $\Sigma \mathbf{F} = m\mathbf{a}$ from the force diagram for the person:
    x-component: $F_N = mv^2/R$;
    y-component: $F_{fr} - mg = 0$.
    Because the friction is static, we have
    $F_{fr} \leq \mu_s F_N$, or $mg \leq \mu_s mv^2/R$.
    Thus we have
    $\mu_s \geq gR/v^2 = (9.80 \text{ m/s}^2)(5.0 \text{ m})/(15.7 \text{ m/s})^2 = \boxed{0.20.}$
    There is no force pressing the people against the wall. They feel the normal force and thus are applying the reaction to this, which is an outward force <u>on the wall</u>. There is no horizontal force on the people except the normal force.

19. The mass moves in a circle of radius r and has a centripetal acceleration.
We write $\Sigma\mathbf{F} = m\mathbf{a}$ from the force diagram for the mass:

x-component: $F_T \cos\theta = mv^2/r$;

y-component: $F_T \sin\theta - mg = 0$.

Combining these, we get

$rg = v^2 \tan\theta$;

$(0.600\ \text{m})(9.80\ \text{m/s}^2) = (7.54\ \text{m/s})^2 \tan\theta$, which gives

$\tan\theta = 0.103$, or $\boxed{\theta = 5.91°.}$

We find the tension from

$F_T = mg/\sin\theta = (0.150\ \text{kg})(9.80\ \text{m/s}^2)/\sin 5.91° = \boxed{14.3\ \text{N.}}$

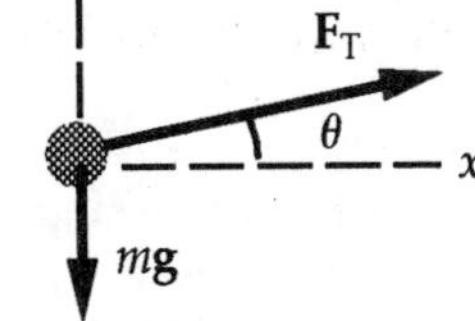

20. We convert the speeds:

$(70\ \text{km/h})/(3.6\ \text{ks/h}) = 19.4\ \text{m/s}$;

$(90\ \text{km/h})/(3.6\ \text{ks/h}) = 25.0\ \text{m/s}$.

At the speed for which the curve is banked perfectly, there is no need for a friction force. We take the x-axis in the direction of the centripetal acceleration.
We write $\Sigma\mathbf{F} = m\mathbf{a}$ from the force diagram for the car:

x-component: $F_{N1} \sin\theta = ma_1 = mv_1^2/R$;

y-component: $F_{N1} \cos\theta - mg = 0$.

Combining these, we get $v_1^2 = gR \tan\theta$.

$(19.4\ \text{m/s})^2 = (80\ \text{m})(9.80\ \text{m/s}^2)\tan\theta$, which gives

$\tan\theta = 0.482$, or $\theta = 25.7°$.

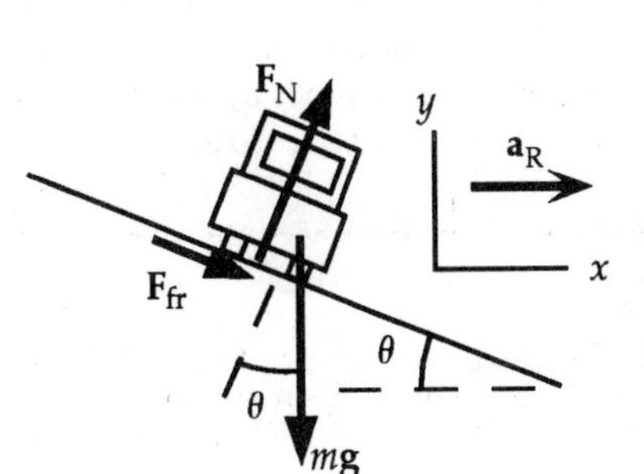

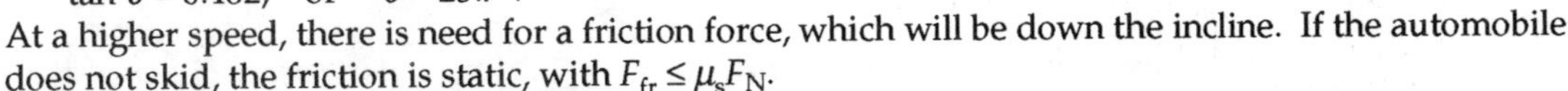
At a higher speed, there is need for a friction force, which will be down the incline. If the automobile does not skid, the friction is static, with $F_{fr} \le \mu_s F_N$.
We write $\Sigma\mathbf{F} = m\mathbf{a}$ from the force diagram for the car:

x-component: $F_{N2} \sin\theta + F_{fr} \cos\theta = ma_2 = mv_2^2/R$;

y-component: $F_{N2} \cos\theta - F_{fr} \sin\theta - mg = 0$.

We eliminate F_{fr} by multiplying the x-equation by $\sin\theta$, the y-equation by $\cos\theta$, and adding:

$F_{N2} = m\{[(v_2^2/R)\sin\theta] + g\cos\theta\}$.

By reversing the trig multipliers and subtracting, we eliminate F_{N2} to get

$F_{fr} = m\{[(v_2^2/R)\cos\theta] - g\sin\theta\}$.

If the automobile does not skid, the friction is static, with $F_{fr} \le \mu_s F_N$:

$m\{[(v_2^2/R)\cos\theta] - g\sin\theta\} \le \mu_s m\{[(v_2^2/R)\sin\theta] + g\cos\theta\}$, or

$\mu_s \ge \{[(v_2^2/R)\cos\theta] - g\sin\theta\}/\{[(v_2^2/R)\sin\theta] + g\cos\theta\} =$

$[(v_2^2/gR)] - \tan\theta]/\{[(v_2^2/gR)\tan\theta] + 1\}$.

When we express $\tan\theta$ in terms of the design speed, we get

$\mu_s \ge [(v_2^2/gR) - (v_1^2/gR)]/\{[(v_2^2/gR)(v_1^2/gR)] + 1\} = (1/gR)(v_2^2 - v_1^2)/[(v_1v_2/gR)^2 + 1]$

$= [1/(9.80\ \text{m/s}^2)(80\ \text{m})][(25.0)^2 - (19.4\ \text{m/s})^2]/\{[(19.4\ \text{m/s})(25.0\ \text{m/s})/(9.80\ \text{m/s}^2)(80\ \text{m})]^2 + 1\}$

$= \boxed{0.23.}$

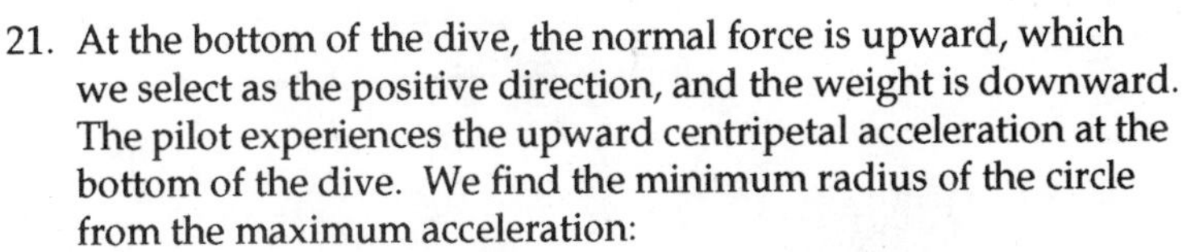

21. At the bottom of the dive, the normal force is upward, which we select as the positive direction, and the weight is downward. The pilot experiences the upward centripetal acceleration at the bottom of the dive. We find the minimum radius of the circle from the maximum acceleration:

$a_{max} = v^2/R_{min}$;

$(9.0)(9.80\ \text{m/s}^2) = (310\ \text{m/s})^2/R_{min}$, which gives $R_{min} = 1.1 \times 10^3\ \text{m}$.

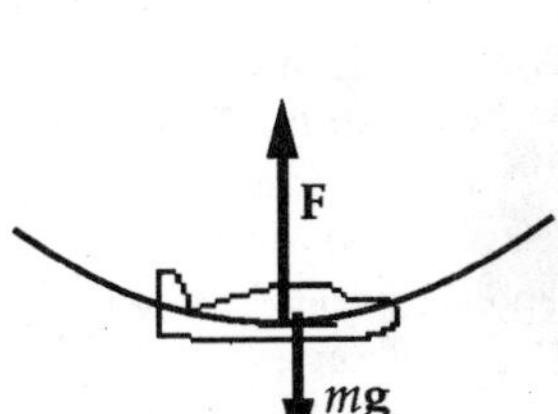

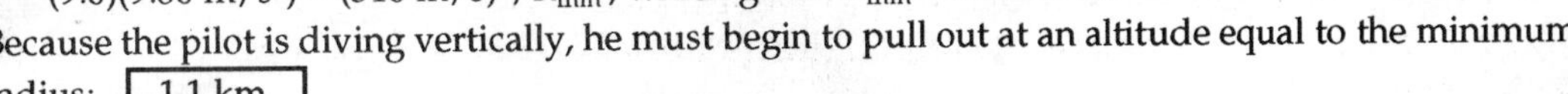
Because the pilot is diving vertically, he must begin to pull out at an altitude equal to the minimum radius: $\boxed{1.1\ \text{km.}}$

22. For the components of the net force we have
$\Sigma F_{\text{tan}} = ma_{\text{tan}} = (1000 \text{ kg})(3.2 \text{ m/s}^2) =$ $\boxed{3.2 \times 10^3 \text{ N};}$
$\Sigma F_{\text{R}} = ma_{\text{R}} = (1000 \text{ kg})(1.8 \text{ m/s}^2) =$ $\boxed{1.8 \times 10^3 \text{ N.}}$

23. We find the constant tangential acceleration from the motion around the turn:
$v_{\text{tan}}^2 = v_0^2 + 2a_{\text{tan}}(x_{\text{tan}} - x_0)$
$[(320 \text{ km/h})/(3.6 \text{ ks/h})]^2 = 0 + 2a_{\text{tan}}[\pi(200 \text{ m}) - 0]$, which gives $\boxed{a_{\text{tan}} = 6.29 \text{ m/s}^2.}$
The centripetal acceleration depends on the speed, so it will increase around the turn. We find the speed at the halfway point from
$v_1^2 = v_0^2 + 2a_{\text{tan}}(x_1 - x_0)$
$= 0 + 2(6.29 \text{ m/s}^2)[\pi(100 \text{ m}) - 0]$, which gives $v_1 = 62.8 \text{ m/s}$.
The radial acceleration is
$a_{\text{R}} = v_1^2/R = (62.8 \text{ m/s})^2/(200 \text{ m}) =$ $\boxed{19.7 \text{ m/s}^2.}$
The magnitude of the acceleration is
$a = (a_{\text{tan}}^2 + a_{\text{R}}^2)^{1/2} = [(6.29 \text{ m/s}^2)^2 + (19.7 \text{ m/s}^2)^2]^{1/2} = 20.7 \text{ m/s}^2.$
On a flat surface, $F_{\text{N}} = Mg$; and the friction force must provide the acceleration: $F_{\text{fr}} = ma$.
With no slipping the friction is static, so we have
$F_{\text{fr}} \leq \mu_s F_{\text{N}}$, or $Ma \leq \mu_s Mg$.
Thus we have
$\mu_s \geq a/g = (20.7 \text{ m/s}^2)/(9.80 \text{ m/s}^2) =$ $\boxed{2.11.}$

24. (*a*) We find the speed from the radial component of the acceleration:
$a_{\text{R}} = a \sin\theta = v_1^2/R$;
$(1.05 \text{ m/s}^2) \cos 32.0° = v_1^2/(2.70 \text{ m})$, which gives $v_1 =$ $\boxed{1.23 \text{ m/s.}}$
(*b*) Assuming constant tangential acceleration, we find the speed from
$v_2 = v_1 + a_{\text{tan}}t = (1.23 \text{ m/s}) + (1.05 \text{ m/s}^2)(\cos 32.0°)(2.00 \text{ s}) =$ $\boxed{3.01 \text{ m/s.}}$

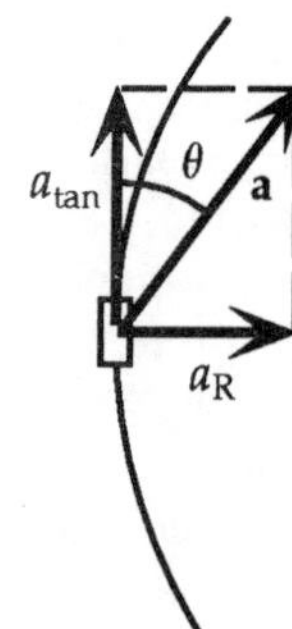

25. Because the spacecraft is 2 Earth radii above the surface, it is 3 Earth radii from the center. The gravitational force on the spacecraft is
$F = GMm/r^2$
$= (6.67 \times 10^{-11} \text{ N}\cdot\text{m}^2/\text{kg}^2)(5.98 \times 10^{24} \text{ kg})(1400 \text{ kg})/[3(6.38 \times 10^6 \text{ m})]^2 =$ $\boxed{1.52 \times 10^3 \text{ N.}}$

26. The acceleration due to gravity on the surface of a planet is
$g = F/m = GM/R^2$.
For the Moon we have
$g_{\text{Moon}} = (6.67 \times 10^{-11} \text{ N}\cdot\text{m}^2/\text{kg}^2)(7.35 \times 10^{22} \text{ kg})/(1.74 \times 10^6 \text{ m})^2 =$ $\boxed{1.62 \text{ m/s}^2.}$

27. The acceleration due to gravity on the surface of a planet is
$g = F/m = GM/R^2$.
If we form the ratio of the two accelerations, we have
$g_{\text{planet}}/g_{\text{Earth}} = (M_{\text{planet}}/M_{\text{Earth}})/(R_{\text{planet}}/R_{\text{Earth}})^2$, or
$g_{\text{planet}} = g_{\text{Earth}}(M_{\text{planet}}/M_{\text{Earth}})/(R_{\text{planet}}/R_{\text{Earth}})^2 = (9.80 \text{ m/s}^2)(1)/(2.5)^2 =$ $\boxed{1.6 \text{ m/s}^2.}$

28. The acceleration due to gravity on the surface of a planet is
$g = F/m = GM/R^2$.
If we form the ratio of the two accelerations, we have
$g_{planet}/g_{Earth} = (M_{planet}/M_{Earth})/(R_{planet}/R_{Earth})^2$, or
$g_{planet} = g_{Earth}(M_{planet}/M_{Earth})/(R_{planet}/R_{Earth})^2 = (9.80 \text{ m/s}^2)(2.5)/(1)^2 =$ $\boxed{24.5 \text{ m/s}^2.}$

29. (*a*) The mass does not depend on the gravitational force, so it is $\boxed{\text{2.10 kg on both.}}$
(*b*) For the weights we have
$w_{Earth} = mg_{Earth} = (2.10 \text{ kg})(9.80 \text{ m/s}^2) =$ $\boxed{20.6 \text{ N};}$
$w_{planet} = mg_{planet} = (2.10 \text{ kg})(12.0 \text{ m/s}^2) =$ $\boxed{25.2 \text{ N.}}$

30. The acceleration due to gravity at a distance r from the center of the Earth is
$g = F/m = GM_{Earth}/r^2$.
If we form the ratio of the two accelerations for the different distances, we have
$g_h/g_{surface} = [(R_{Earth})/(R_{Earth} + h)]^2 = [(6400 \text{ km})/(6400 \text{ km} + 300 \text{ km})]^2$
which gives $\boxed{g_h = 0.91 g_{surface}.}$

31. The acceleration due to gravity on the surface of the neutron star is
$g = F/m = GM/R^2 = (6.67 \times 10^{-11} \text{ N}\cdot\text{m}^2/\text{kg}^2)(5)(2.0 \times 10^{30} \text{ kg})/(10 \times 10^3 \text{ m})^2 =$ $\boxed{6.7 \times 10^{12} \text{ m/s}^2.}$

32. The acceleration due to gravity at a distance r from the center of the Earth is
$g = F/m = GM_{Earth}/r^2$.
If we form the ratio of the two accelerations for the different distances, we have
$g/g_{surface} = (R_{Earth}/R)^2$;
$1/10 = [(6400 \text{ km})/R]^2$, which gives $R =$ $\boxed{2.0 \times 10^7 \text{ m.}}$

33. The acceleration due to gravity on the surface of the white dwarf star is
$g = F/m = GM/R^2 = (6.67 \times 10^{-11} \text{ N}\cdot\text{m}^2/\text{kg}^2)(2.0 \times 10^{30} \text{ kg})/(1.74 \times 10^6 \text{ m})^2 =$ $\boxed{4.4 \times 10^7 \text{ m/s}^2.}$

34. The acceleration due to gravity at a distance r from the center of the Earth is
$g = F/m = GM_{Earth}/r^2$.
If we form the ratio of the two accelerations for the different distances, we have
$g/g_{surface} = [(R_{Earth})/(R_{Earth} + h)]^2$;
(*a*) $g = (9.80 \text{ m/s}^2)[(6400 \text{ km})/(6400 \text{ km} + 3.2 \text{ km})]^2 =$ $\boxed{9.8 \text{ m/s}^2.}$
(*b*) $g = (9.80 \text{ m/s}^2)[(6400 \text{ km})/(6400 \text{ km} + 3200 \text{ km})]^2 =$ $\boxed{4.3 \text{ m/s}^2.}$

35. We choose the coordinate system shown in the figure and find the force on the mass in the lower left corner.
Because the masses are equal, for the magnitudes of the forces from the other corners we have

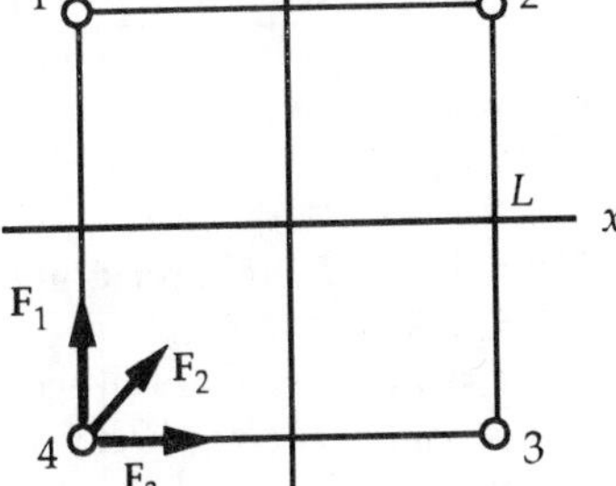

$F_1 = F_3 = Gmm/r_1^2$
$= (6.67 \times 10^{-11} \text{ N}\cdot\text{m}^2/\text{kg}^2)(7.5 \text{ kg})(7.5 \text{ kg})/(0.60 \text{ m})^2$
$= 1.04 \times 10^{-8} \text{ N}$;
$F_2 = Gmm/r_2^2$
$= (6.67 \times 10^{-11} \text{ N}\cdot\text{m}^2/\text{kg}^2)(7.5 \text{ kg})(7.5 \text{ kg})/[(0.60 \text{ m})/\cos 45°]^2$
$= 5.21 \times 10^{-9} \text{ N}$.
From the symmetry of the forces we see that the resultant will be along the diagonal. The resultant force is
$F = 2F_1 \cos 45° + F_2$
$= 2(1.04 \times 10^{-8} \text{ N}) \cos 45° + 5.21 \times 10^{-9} \text{ N} =$ $\boxed{2.0 \times 10^{-8} \text{ N toward center of the square.}}$

36. For the magnitude of each attractive gravitational force, we have

Sun Venus Earth Jupiter Saturn

$F_V = GM_EM_V/r_V^2 = Gf_VM_E^2/r_V^2$
$= (6.67 \times 10^{-11}\ \text{N}\cdot\text{m}^2/\text{kg}^2)(0.815)(5.98 \times 10^{24}\ \text{kg})^2/[(108 - 150) \times 10^9\ \text{m}]^2$
$= 1.10 \times 10^{18}\ \text{N};$
$F_J = GM_EM_J/r_J^2 = Gf_JM_E^2/r_J^2$
$= (6.67 \times 10^{-11}\ \text{N}\cdot\text{m}^2/\text{kg}^2)(318)(5.98 \times 10^{24}\ \text{kg})^2/[(778 - 150) \times 10^9\ \text{m}]^2$
$= 1.92 \times 10^{18}\ \text{N};$
$F_S = GM_EM_S/r_V^2 = Gf_SM_E^2/r_S^2$
$= (6.67 \times 10^{-11}\ \text{N}\cdot\text{m}^2/\text{kg}^2)(95.1)(5.98 \times 10^{24}\ \text{kg})^2/[(1430 - 150) \times 10^9\ \text{m}]^2$
$= 1.38 \times 10^{17}\ \text{N}.$

The force from Venus is toward the Sun; the forces from Jupiter and Saturn are away from the Sun. For the net force we have

$F_{net} = F_J + F_S - F_V = 1.92 \times 10^{18}\ \text{N} + 1.38 \times 10^{17}\ \text{N} - 1.10 \times 10^{18}\ \text{N} =$ $\boxed{9.6 \times 10^{17}\ \text{N away from the Sun.}}$

37. The acceleration due to gravity on the surface of a planet is

$g = F/m = GM/R^2.$

If we form the ratio of the two accelerations, we have

$g_{Mars}/g_{Earth} = (M_{Mars}/M_{Earth})/(R_{Mars}/R_{Earth})^2;$
$0.38 = [M_{Mars}/(6.0 \times 10^{24}\ \text{kg})]/(3400\ \text{km}/6400\ \text{km})^2$, which gives $M_{Mars} =$ $\boxed{6.4 \times 10^{23}\ \text{kg.}}$

38. We relate the speed of the Earth to the period of its orbit from

$v = 2\pi R/T.$

The gravitational attraction from the Sun must provide the centripetal acceleration for the circular orbit:

$GM_EM_S/R^2 = M_Ev^2/R = M_E(2\pi R/T)^2/R = M_E4\pi^2R/T^2$, so we have
$GM_S = 4\pi^2R^3/T^2;$
$(6.67 \times 10^{-11}\ \text{N}\cdot\text{m}^2/\text{kg}^2)M_S = 4\pi^2(1.50 \times 10^{11}\ \text{m})^3/(3.16 \times 10^7\ \text{s})^2$, which gives $M_S =$ $\boxed{2.0 \times 10^{30}\ \text{kg.}}$

This is the same as found in Example 5–17.

39. The gravitational attraction must provide the centripetal acceleration for the circular orbit:

$GM_Em/R^2 = mv^2/R$, or
$v^2 = GM_E/(R_E + h)$
$= (6.67 \times 10^{-11}\ \text{N}\cdot\text{m}^2/\text{kg}^2)(5.98 \times 10^{24}\ \text{kg})/(6.38 \times 10^6\ \text{m} + 3.6 \times 10^6\ \text{m}),$

which gives $v =$ $\boxed{6.3 \times 10^3\ \text{m/s.}}$

40. The greater tension will occur when the elevator is accelerating upward, which we take as the positive direction. We write $\Sigma\mathbf{F} = m\mathbf{a}$ from the force diagram for the monkey:

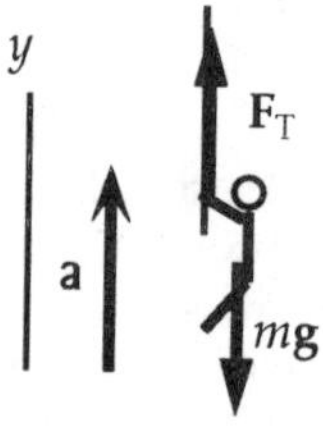

$F_T \cos\theta - mg = ma;$
$220\ \text{N} - (17.0\ \text{kg})(9.80\ \text{m/s}^2) = (17.0\ \text{kg})a$, which gives
$a =$ $\boxed{3.14\ \text{m/s}^2\ \text{upward.}}$

Because the rope broke, the tension was at least 220 N, so this is the minimum acceleration.

41. We relate the speed of rotation to the period of rotation from

$v = 2\pi R/T.$

For the required centripetal acceleration, we have

$a_R = v^2/R = (2\pi R/T)^2/R = 4\pi^2R/T^2;$
$\frac{1}{2}(9.80\ \text{m/s}^2) = 4\pi^2(16\ \text{m})/T^2$, which gives $T =$ $\boxed{11\ \text{s.}}$

42. We relate the speed to the period of revolution from
$v = 2\pi R/T$.
For the required centripetal acceleration, we have
$a_R = v^2/R = (2\pi R/T)^2/R = 4\pi^2 R/T^2$;
$9.80\ \text{m/s}^2 = 4\pi^2(6.38 \times 10^6\ \text{m})/T^2$, which gives $T =$ $\boxed{5.07 \times 10^3\ \text{s}\ (1.41\ \text{h}).}$
The result is $\boxed{\text{independent of the mass}}$ of the satellite.

43. We relate the speed to the period of revolution from
$v = 2\pi R/T$.
The required centripetal acceleration is provided by the gravitational attraction:
$GM_M m/R^2 = mv^2/R = m(2\pi R/T)^2/R = m4\pi^2 R/T^2$, so we have
$GM_M = 4\pi^2(R_M + h)^3/T^2$;
$(6.67 \times 10^{-11}\ \text{N}\cdot\text{m}^2/\text{kg}^2)(7.4 \times 10^{22}\ \text{kg}) = 4\pi^2(1.74 \times 10^6\ \text{m} + 1.00 \times 10^5\ \text{m})^3/T^2$,
which gives $T = 7.06 \times 10^3\ \text{s} =$ $\boxed{2.0\ \text{h.}}$

44. We take the positive direction upward. The spring scale reads the normal force expressed as an effective mass: F_N/g.
We write $\Sigma \mathbf{F} = m\mathbf{a}$ from the force diagram:
$F_N - mg = ma$, or $m_{\text{effective}} = F_N/g = m(1 + a/g)$.

(*a*) For a constant speed, there is no acceleration, so we have
$m_{\text{effective}} = m(1 + a/g) = m =$ $\boxed{58\ \text{kg.}}$
(*b*) For a constant speed, there is no acceleration, so we have
$m_{\text{effective}} = m(1 + a/g) = m =$ $\boxed{58\ \text{kg.}}$
(*c*) For the upward (positive) acceleration, we have
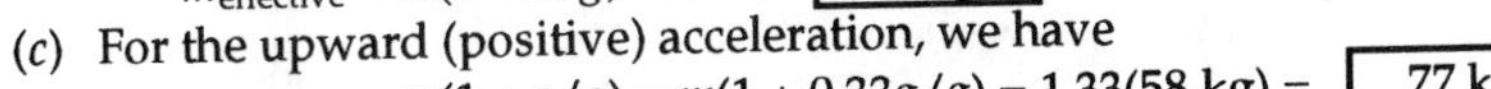
$m_{\text{effective}} = m(1 + a/g) = m(1 + 0.33g/g) = 1.33(58\ \text{kg}) =$ $\boxed{77\ \text{kg.}}$
(*d*) For the downward (negative) acceleration, we have
$m_{\text{effective}} = m(1 + a/g) = m(1 - 0.33g/g) = 0.67(58\ \text{kg}) =$ $\boxed{39\ \text{kg.}}$
(*e*) In free fall the acceleration is $-g$, so we have
$m_{\text{effective}} = m(1 + a/g) = m(1 - g/g) =$ $\boxed{0.}$

45. We relate the orbital speed to the period of revolution from
$v = 2\pi R/T$.
The required centripetal acceleration is provided by the gravitational attraction:
$GM_S m/R^2 = mv^2/R = m(2\pi R/T)^2/R = m4\pi^2 R/T^2$, so we have
$GM_S = 4\pi^2 R^3/T^2$.
For the two extreme orbits we have
$(6.67 \times 10^{-11}\ \text{N}\cdot\text{m}^2/\text{kg}^2)(5.69 \times 10^{26}\ \text{kg}) = 4\pi^2(7.3 \times 10^7\ \text{m})^3/T_{\text{inner}}^2$,
which gives $T_{\text{inner}} = 2.01 \times 10^4\ \text{s} =$ $\boxed{5\ \text{h}\ 35\ \text{min;}}$
$(6.67 \times 10^{-11}\ \text{N}\cdot\text{m}^2/\text{kg}^2)(5.69 \times 10^{26}\ \text{kg}) = 4\pi^2(17 \times 10^7\ \text{m})^3/T_{\text{outer}}^2$,
which gives $T_{\text{outer}} = 7.15 \times 10^4\ \text{s} =$ $\boxed{19\ \text{h}\ 50\ \text{min.}}$
Because the mean rotation period of Saturn is between the two results, with respect to a point on the surface of Saturn, the edges of the rings are moving in opposite directions.

46. The centripetal acceleration has a magnitude of
$a_R = v^2/R = (2\pi R/T)^2/R = 4\pi^2 R/T^2$
$= 4\pi^2(12.0\text{ m})/(12.5\text{ s})^2 = 3.03\text{ m/s}^2.$
At each position we take the positive direction in the direction of the acceleration. Because the seat swings, the normal force from the seat is upward and the weight is downward. The apparent weight is measured by the normal force.

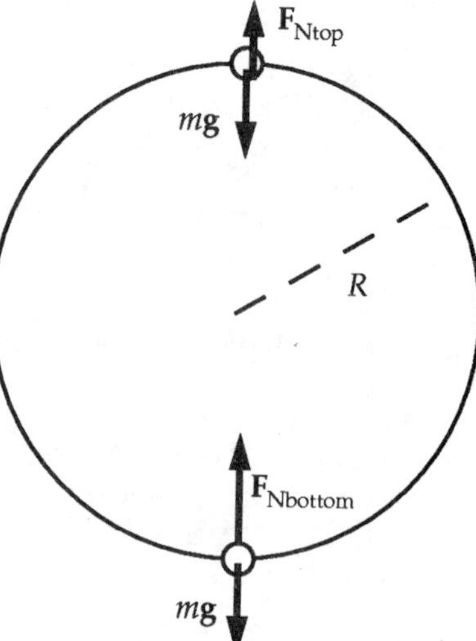

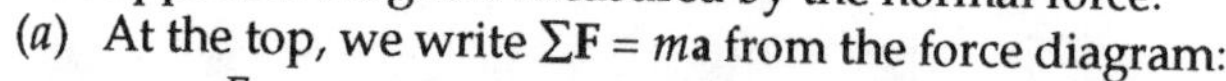

(*a*) At the top, we write $\Sigma F = ma$ from the force diagram:
$-F_{Ttop} + mg = ma_R$, or $F_{Ttop} = mg(1 - a_R/g)$.
For the fractional change we have
Fractional change $= (F_{Ttop} - mg)/mg = -a_R/g$
$= -(3.03\text{ m/s}^2)/(9.80\text{ m/s}^2) =$ **$-0.309\ (-30.9\%)$.**

(*b*) At the bottom, we write $\Sigma F = ma$ from the force diagram:
$F_{Tbottom} - mg = ma_R$, or $F_{Tbottom} = mg(1 + a_R/g)$.
For the fractional change we have
Fractional change $= (F_{Tbottom} - mg)/mg = +a_R/g$
$= +(3.03\text{ m/s}^2)/(9.80\text{ m/s}^2) =$ **$+0.309\ (+30.9\%)$.**

47. The acceleration due to gravity is
$g = F_{grav}/m = GM/R^2$
$= (6.67\times10^{-11}\text{ N}\cdot\text{m}^2/\text{kg}^2)(7.4\times10^{22}\text{ kg})/(4.2\times10^6\text{ m})^2 = 0.28\text{ m/s}^2.$
We take the positive direction toward the Moon. The apparent weight is measured by the normal force. We write $\Sigma F = ma$ from the force diagram:
$-F_N + mg = ma,$

(*a*) For a constant velocity, there is no acceleration, so we have
$-F_N + mg = 0$, or
$F_N = mg = (70\text{ kg})(0.28\text{ m/s}^2) =$ **20 N (toward the Moon).**

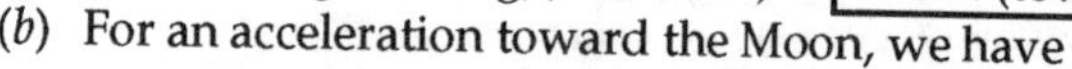

(*b*) For an acceleration toward the Moon, we have
$-F_N + mg = ma$, or
$F_N = m(g - a) = (70\text{ kg})(0.28\text{ m/s}^2 - 2.9\text{ m/s}^2) =$ **-1.8×10^2 N (away from the Moon).**

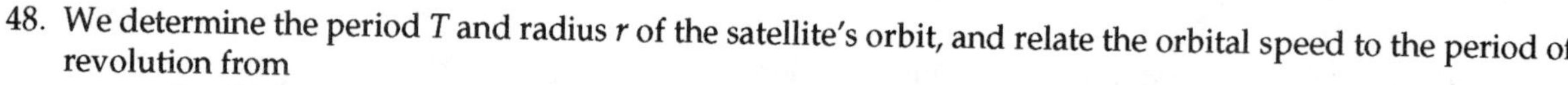

48. We determine the period T and radius r of the satellite's orbit, and relate the orbital speed to the period of revolution from
$v = 2\pi r/T.$
We know that the gravitational attraction provides the centripetal acceleration:
$GM_{planet}m/r^2 = mv^2/r = m(2\pi r/T)^2/r = m4\pi^2 r/T^2$, so we have
$M_{planet} = 4\pi^2 r^3/GT^2$.

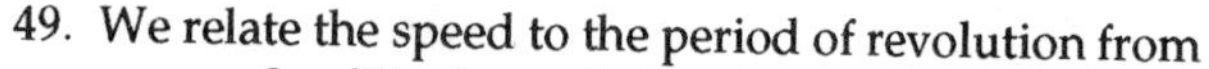

49. We relate the speed to the period of revolution from
$v = 2\pi r/T$, where r is the distance to the midpoint.
The gravitational attraction provides the centripetal acceleration:
$Gmm/(2r)^2 = mv^2/r = m(2\pi r/T)^2/r = m4\pi^2 r/T^2$, so we have
$m = 16\pi^2 r^3/GT^2$
$= 16\pi^2(180\times10^9\text{ m})^3/(6.67\times10^{-11}\text{ N}\cdot\text{m}^2/\text{kg}^2)[(5.0\text{ yr})(3.16\times10^7\text{ s/yr})]^2 =$ **5.5×10^{29} kg.**

50. (*a*) We relate the speed of rotation to the period of revolution from

$v = 2\pi R/T.$

We know that the gravitational attraction provides the centripetal acceleration:

$GM_{\text{planet}}m/R^2 = mv^2/R = m(2\pi R/T)^2/R = m4\pi^2R/T^2$, so we have

$M_{\text{planet}} = 4\pi^2R^3/GT^2.$

Thus the density is

$\rho = M_{\text{planet}}/V = [4\pi^2R^3/GT^2]/\frac{4}{3}\pi R^3 = 3\pi/GT^2.$

(*b*) For the Earth we have

$\rho = 3\pi/GT^2 = 3\pi/(6.67\times10^{-11}\ \text{N}\cdot\text{m}^2/\text{kg}^2)[(90\ \text{min})(60\ \text{s/min})]^2 =$ $\boxed{4.8\times10^3\ \text{kg/m}^3.}$

Note that the density of iron is $7.8\times10^3\ \text{kg/m}^3$.

51. From Kepler's third law, $T^2 = 4\pi^2R^3/GM_E$, we can relate the periods of the satellite and the Moon:

$(T/T_{\text{Moon}})^2 = (R/R_{\text{Moon}})^3;$

$(T/27.4\ \text{d})^2 = [(6.38\times10^6\ \text{m})/(3.84\times10^8\ \text{m})]^3$, which gives $T =$ $\boxed{0.0587\ \text{days}\ (1.41\ \text{h}).}$

52. From Kepler's third law, $T^2 = 4\pi^2R^3/GM_S$, we can relate the periods of Icarus and the Earth:

$(T_{\text{Icarus}}/T_{\text{Earth}})^2 = (R_{\text{Icarus}}/R_{\text{Earth}})^3;$

$(410\ \text{d}/365\ \text{d})^2 = [R_{\text{Icarus}}/(1.50\times10^{11}\ \text{m})]^3$, which gives $\boxed{R_{\text{Icarus}} = 1.62\times10^{11}\ \text{m.}}$

53. From Kepler's third law, $T^2 = 4\pi^2R^3/GM_S$, we can relate the periods of the Earth and Neptune:

$(T_{\text{Neptune}}/T_{\text{Earth}})^2 = (R_{\text{Neptune}}/R_{\text{Earth}})^3;$

$(T_{\text{Neptune}}/1\ \text{yr})^2 = [(4.5\times10^{12}\ \text{m})/(1.50\times10^{11}\ \text{m})]^3$, which gives $T_{\text{Neptune}} =$ $\boxed{1.6\times10^2\ \text{yr.}}$

54. We use Kepler's third law, $T^2 = 4\pi^2R^3/GM_E$, for the motion of the Moon around the Earth:

$T^2 = 4\pi^2R^3/GM_E;$

$[(27.4\ \text{d})(86{,}400\ \text{s/d})]^2 = 4\pi^2(3.84\times10^8\ \text{m})^3/(6.67\times10^{-11}\ \text{N}\cdot\text{m}^2/\text{kg}^2)M_E,$

which gives $M_E =$ $\boxed{5.97\times10^{24}\ \text{kg.}}$

55. From Kepler's third law, $T^2 = 4\pi^2R^3/GM_S$, we can relate the periods of Halley's comet and the Earth to find the mean distance of the comet from the Sun:

$(T_{\text{Halley}}/T_{\text{Earth}})^2 = (R_{\text{Halley}}/R_{\text{Earth}})^3;$

$(76\ \text{yr}/1\ \text{yr})^2 = [R_{\text{Halley}}/(1.50\times10^{11}\ \text{m})]^3$, which gives $R_{\text{Halley}} = 2.68\times10^{12}\ \text{m}.$

If we take the nearest distance to the Sun as zero, the farthest distance is

$d = 2R_{\text{Halley}} = 2(2.68\times10^{12}\ \text{m}) =$ $\boxed{5.4\times10^{12}\ \text{m.}}$

It is still orbiting the Sun and thus is $\boxed{\text{in the Solar System.}}$ The planet nearest it is $\boxed{\text{Pluto.}}$

56. We relate the speed to the period of revolution from

$v = 2\pi r/T$, where r is the distance to the center of the Milky Way.

The gravitational attraction provides the centripetal acceleration:

$GM_{\text{galaxy}}M_S/r^2 = M_Sv^2/r = M_S(2\pi r/T)^2/r = M_S4\pi^2r/T^2$, so we have

$$M_{\text{galaxy}} = 4\pi^2r^3/GT^2$$
$$= 4\pi^2[(30{,}000\ \text{ly})(9.5\times10^{15}\ \text{m/ly})]^3/(6.67\times10^{-11}\ \text{N}\cdot\text{m}^2/\text{kg}^2)[(200\times10^6\ \text{yr})(3.16\times10^7\ \text{s/yr})]^2 = \boxed{3.4\times10^{41}\ \text{kg.}}$$

The number of stars ("Suns") is

$(3.4\times10^{41}\ \text{kg})/(2.0\times10^{30}\ \text{kg}) =$ $\boxed{1.7\times10^{11}.}$

57. From Kepler's third law, $T^2 = 4\pi^2R^3/GM_{\text{Jupiter}}$, we have
$M_{\text{Jupiter}} = 4\pi^2R^3/GT^2$.
(*a*) $M_{\text{Jupiter}} = 4\pi^2R_{\text{Io}}{}^3/GT_{\text{Io}}{}^2$
$= 4\pi^2(422 \times 10^6\ \text{m})^3/(6.67 \times 10^{-11}\ \text{N}\cdot\text{m}^2/\text{kg}^2)[(1.77\ \text{d})(86{,}400\ \text{s/d})]^2 = \boxed{1.90 \times 10^{27}\ \text{kg.}}$
(*b*) $M_{\text{Jupiter}} = 4\pi^2R_{\text{Europa}}{}^3/GT_{\text{Europa}}{}^2$
$= 4\pi^2(671 \times 10^6\ \text{m})^3/(6.67 \times 10^{-11}\ \text{N}\cdot\text{m}^2/\text{kg}^2)[(3.55\ \text{d})(86{,}400\ \text{s/d})]^2 = \boxed{1.90 \times 10^{27}\ \text{kg;}}$
$M_{\text{Jupiter}} = 4\pi^2R_{\text{Ganymede}}{}^3/GT_{\text{Ganymede}}{}^2$
$= 4\pi^2(1070 \times 10^6\ \text{m})^3/(6.67 \times 10^{-11}\ \text{N}\cdot\text{m}^2/\text{kg}^2)[(7.16\ \text{d})(86{,}400\ \text{s/d})]^2 = \boxed{1.89 \times 10^{27}\ \text{kg;}}$
$M_{\text{Jupiter}} = 4\pi^2R_{\text{Callisto}}{}^3/GT_{\text{Callisto}}{}^2$
$= 4\pi^2(1883 \times 10^6\ \text{m})^3/(6.67 \times 10^{-11}\ \text{N}\cdot\text{m}^2/\text{kg}^2)[(16.7\ \text{d})(86{,}400\ \text{s/d})]^2 = \boxed{1.89 \times 10^{27}\ \text{kg.}}$
The results are consistent.

58. From Kepler's third law, $T^2 = 4\pi^2R^3/GM_{\text{Jupiter}}$, we can relate the distances of the moons:
$(R/R_{\text{Io}})^3 = (T/T_{\text{Io}})^2$.
Thus we have
$(R_{\text{Europa}}/422 \times 10^3\ \text{km})^3 = (3.55\ \text{d}/1.77\ \text{d})^2$, which gives $\boxed{R_{\text{Europa}} = 6.71 \times 10^5\ \text{km.}}$
$(R_{\text{Ganymede}}/422 \times 10^3\ \text{km})^3 = (7.16\ \text{d}/1.77\ \text{d})^2$, which gives $\boxed{R_{\text{Ganymede}} = 1.07 \times 10^6\ \text{km.}}$
$(R_{\text{Callisto}}/422 \times 10^3\ \text{km})^3 = (16.7\ \text{d}/1.77\ \text{d})^2$, which gives $\boxed{R_{\text{Callisto}} = 1.88 \times 10^6\ \text{km.}}$
All values agree with the table.

59. (*a*) From Kepler's third law, $T^2 = 4\pi^2R^3/GM_S$, we can relate the periods of the assumed planet and the Earth:
$(T_{\text{planet}}/T_{\text{Earth}})^2 = (R_{\text{planet}}/R_{\text{Earth}})^3$;
$(T_{\text{planet}}/1\ \text{yr})^2 = (3)^3$, which gives $\boxed{T_{\text{planet}} = 5.2\ \text{yr.}}$
(*b*) $\boxed{\text{No,}}$ the radius and period are independent of the mass of the orbiting body.

60. (*a*) In a short time interval *t*, the planet will travel a distance *vt* along its orbit. This distance has been exaggerated on the diagram. Kepler's second law states that the area swept out by a line from the Sun to the planet in time *t* is the same anywhere on the orbit. If we take the areas swept out at the nearest and farthest points, as shown on the diagram, and approximate the areas as triangles (which is a good approximation for very small *t*), we have

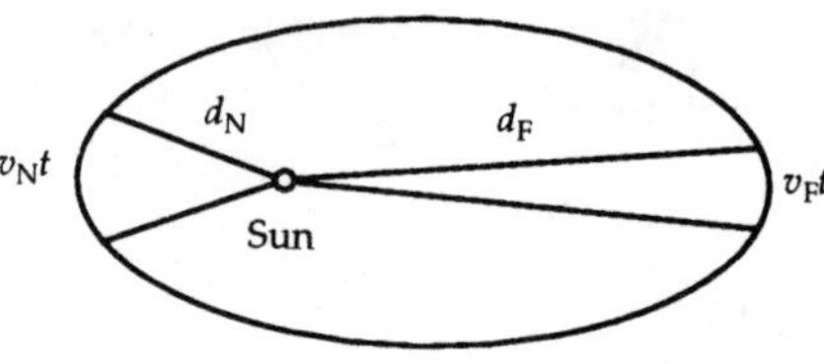

$\frac{1}{2}d_N(v_Nt) = \frac{1}{2}d_F(v_Ft)$, which gives $v_N/v_F = d_F/d_N$.
(*b*) For the average velocity we have
$\bar{v} = 2\pi[\frac{1}{2}(d_N + d_F)]/T = \pi(1.47 \times 10^{11}\ \text{m} + 1.52 \times 10^{11}\ \text{m})/(3.16 \times 10^7\ \text{s}) = 2.97 \times 10^4\ \text{m/s}$.
From the result for part (*a*), we have
$v_N/v_F = d_F/d_N = 1.52/1.47 = 1.034$, or v_N is 3.4% greater than v_F.
For this small change, we can take each of the extreme velocities to be ± 1.7% from the average. Thus we have
$v_N = 1.017(2.97 \times 10^4\ \text{m/s}) = \boxed{3.02 \times 10^4\ \text{m/s;}}$
$v_F = 0.983(2.97 \times 10^4\ \text{m/s}) = \boxed{2.92 \times 10^4\ \text{m/s.}}$

61. An apparent gravity of one g means that the normal force from the band is mg, where

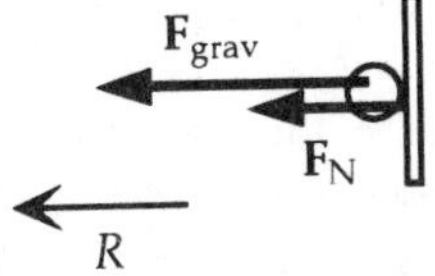

$g = GM_E/R_E^2$.

The normal force and the gravitational attraction from the Sun provide the centripetal acceleration:

$GM_Sm/R_{SE}^2 + mGM_E/R_E^2 = mv^2/R_{SE}$, or

$v^2 = G[(M_S/R_{SE}) + (M_ER_{SE}/R_E^2)]$

$= (6.67 \times 10^{-11}\ \mathrm{N \cdot m^2/kg^2})[(1.99 \times 10^{30}\ \mathrm{kg})/(1.50 \times 10^{11}\ \mathrm{m}) + (5.98 \times 10^{24}\ \mathrm{kg})(1.50 \times 10^{11}\ \mathrm{m})/(6.38 \times 10^6\ \mathrm{m})^2]$, which gives $v = 1.2 \times 10^6\ \mathrm{m/s}$.

For the period of revolution we have

$T = 2\pi R_{SE}/v = [2\pi(1.50 \times 10^{11}\ \mathrm{m})/(1.2 \times 10^6\ \mathrm{m/s})]/(86{,}400\ \mathrm{s/day}) =$ 9.0 Earth-days.

62. The acceleration due to gravity at a distance r from the center of the Earth is

$g = F/m = GM_{Earth}/r^2$.

If we form the ratio of the two accelerations for the different distances, we have

$g/g_{surface} = [R_{Earth}/(R_{Earth} + h)]^2$;

$1/2 = [(6.38 \times 10^3\ \mathrm{km})/(6.38 \times 10^3\ \mathrm{km} + h)]^2$, which gives $h =$ 2.6×10^3 km.

63. The net force on Tarzan will provide his centripetal acceleration, which we take as the positive direction.

We write $\Sigma\mathbf{F} = m\mathbf{a}$ from the force diagram for Tarzan:

$F_T - mg = ma = mv^2/R$.

The maximum speed will require the maximum tension that Tarzan can create:

$1400\ \mathrm{N} - (80\ \mathrm{kg})(9.80\ \mathrm{m/s^2}) = (80\ \mathrm{kg})v^2/(4.8\ \mathrm{m})$, which gives $v =$ 6.1 m/s.

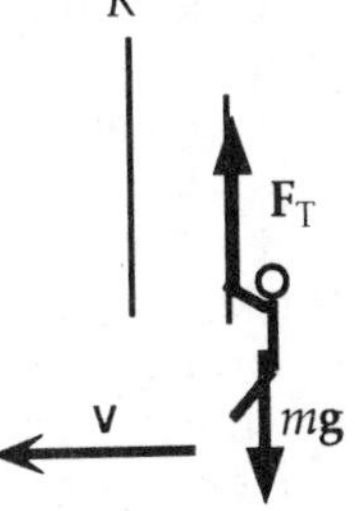

64. Yes. If the bucket is traveling fast enough at the top of the circle, in addition to the weight of the water a force from the bucket, similar to a normal force, is required to provide the necessary centripetal acceleration to make the water go in the circle. From the force diagram, we write

$F_N + mg = ma = mv_{top}^2/R$.

The minimum speed is that for which the normal force is zero:

$0 + mg = mv_{top,min}^2/R$, or $v_{top,min} = (gR)^{1/2}$.

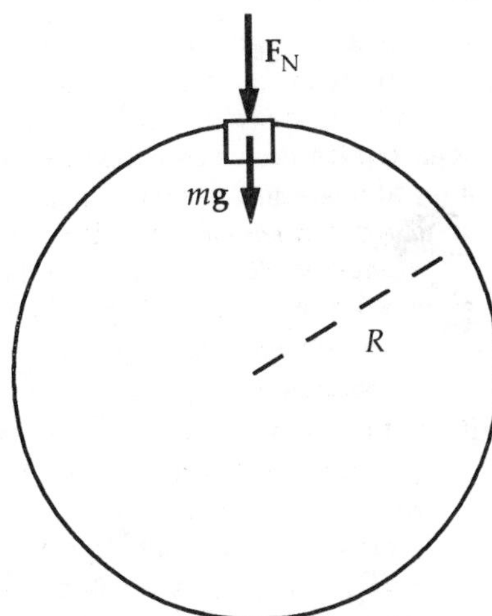

65. We find the speed of the skaters from the period of rotation:

$v = 2\pi r/T = 2\pi(0.80\ \mathrm{m})/(3.0\ \mathrm{s}) = 1.68\ \mathrm{m/s}$.

The pull or tension in their arms provides the centripetal acceleration:

$F_T = mv^2/R$;

$= (60.0\ \mathrm{kg})(1.68\ \mathrm{m/s})^2/(0.80\ \mathrm{m}) =$ 2.1×10^2 N.

66. If we consider a person standing on a scale, the apparent weight is measured by the normal force. The person is moving with the rotational speed of the surface of the Earth:
$v = 2\pi R_E/T = 2\pi(6.38 \times 10^6 \text{ m})/(86{,}400 \text{ s}) = 464 \text{ m/s}.$
We take down as positive and write $\Sigma F = ma$:
$-F_N + mg = ma = mv^2/R_E$, or $F_N = mg_{effective} = mg - mv^2/R_E$.
Thus $g_{effective} - g = -v^2/R_E = -(464 \text{ m/s})^2/(6.38 \times 10^6 \text{ m}) =$ $\boxed{-0.0337 \text{ m/s}^2 \text{ (0.343\% of } g).}$

67. Because the gravitational force is always attractive, the two forces will be in opposite directions. If we call the distance from the Earth to the Moon D and let x be the distance from the Earth where the magnitudes of the forces are equal, we have
$GM_Mm/(D-x)^2 = GM_Em/x^2$, which becomes $M_Mx^2 = M_E(D-x)^2$.
$(7.35 \times 10^{22} \text{ kg})x^2 = (5.98 \times 10^{24} \text{ kg})[(3.84 \times 10^8 \text{ m}) - x]^2$, which gives
$x = \boxed{3.46 \times 10^8 \text{ m from Earth's center.}}$

68. (*a*) People will be able to walk on the $\boxed{\text{inside surface farthest from the center,}}$ so the normal force can provide the centripetal acceleration.
(*b*) The centripetal acceleration must equal g:
$g = v^2/R = (2\pi R/T)^2/R = 4\pi^2R/T^2;$
$9.80 \text{ m/s}^2 = 4\pi^2(0.55 \times 10^3 \text{ m})/T^2$, which gives $T = 47.1$ s.
For the rotation speed we have
revolutions/day = (86,400 s/day)/(47.1 s/rev) = $\boxed{1.8 \times 10^3 \text{ rev/day.}}$

69. We take the positive direction upward. The spring scale reads the normal force expressed as an effective mass: F_N/g.
We write $\Sigma F = ma$ from the force diagram:
$F_N - mg = ma$, or $m_{effective} = F_N/g = m(1 + a/g)$;
$80 \text{ kg} = (60 \text{ kg})[1 + a/(9.80 \text{ m/s}^2)]$,
which gives $a = \boxed{3.3 \text{ m/s}^2 \text{ upward.}}$
The direction is given by the sign of the result.

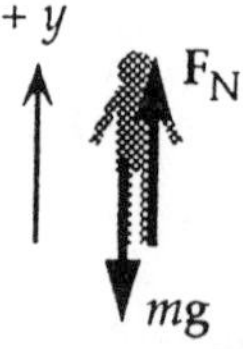

70. At each position we take the positive direction in the direction of the acceleration.
(*a*) The centripetal acceleration is
$a_R = v^2/R$, so we see that the radius is minimum for a maximum centripetal acceleration:
$(6.0)(9.80 \text{ m/s}^2) = [(1500 \text{ km/h})/(3.6 \text{ ks/h})]^2/R_{min}$,
which gives $\boxed{R_{min} = 3.0 \times 10^3 \text{ m} = 3.0 \text{ km.}}$
(*b*) At the bottom of the circle, the normal force is upward and the weight is downward. We write $\Sigma F = ma$ from the force diagram for the ball:
$F_{Nbottom} - mg = mv^2/R = m(6.0g);$
$F_{Nbottom} - (80 \text{ kg})(9.80 \text{ m/s}^2) = (80 \text{ kg})(6.0)(9.80 \text{ m/s}^2)$,
which gives $\boxed{F_{Nbottom} = 5.5 \times 10^3 \text{ N.}}$
(*c*) At the top of the path, both the normal force and the weight are downward. We write $\Sigma F = ma$ from the force diagram for the ball:
$F_{Ntop} + mg = mv^2/R;$
$F_{Ntop} + (80 \text{ kg})(9.80 \text{ m/s}^2) = (80 \text{ kg})(6.0)(9.80 \text{ m/s}^2)$,
which gives $\boxed{F_{Ntop} = 3.9 \times 10^3 \text{ N.}}$

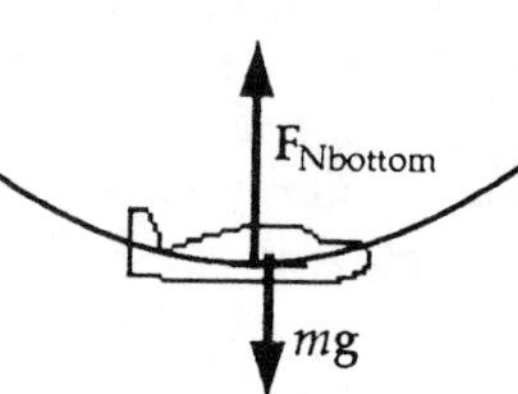

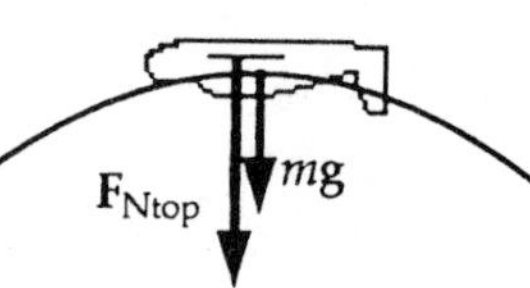

71. The acceleration due to gravity on the surface of a planet is
$g_P = F_{grav}/m = GM_P/r^2$, so we have
$\boxed{M_P = g_P r^2/G.}$

72. (*a*) The attractive gravitational force on the plumb bob is
$F_M = GmM_M/D_M^2$.
Because $\Sigma F = 0$, we see from the force diagram:
$\tan \theta = F_M/mg = (GmM_M/D_M^2)/(mGM_E/R_E^2)$,
where we have used GM_E/R_E^2 for g.
Thus we have
$\boxed{\theta = \tan^{-1} (M_M R_E^2/M_E D_M^2).}$

(*b*) For the mass of a cone with apex half-angle α, we have
$M_M = \rho V = \rho \frac{1}{3}\pi h^3 \tan^2 \alpha$
$= (3 \times 10^3\ \text{kg/m}^3)\frac{1}{3}\pi(4 \times 10^3\ \text{m})^3 \tan^2 30° = \boxed{7 \times 10^{13}\ \text{kg.}}$

(*c*) Using the result from part (*a*) for the angle, we have
$\tan \theta = M_M R_E^2/M_E D_M^2$
$= (7 \times 10^{13}\ \text{kg})(6.4 \times 10^6\ \text{m})^2/(5.97 \times 10^{24}\ \text{kg})(5 \times 10^3\ \text{m})^2 = 2 \times 10^{-5}$,
which gives $\boxed{\theta \approx (1 \times 10^{-3})°.}$

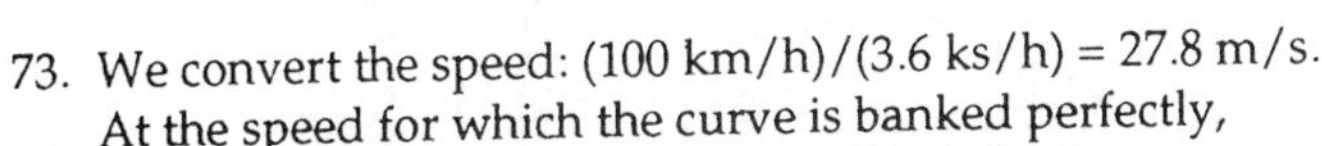

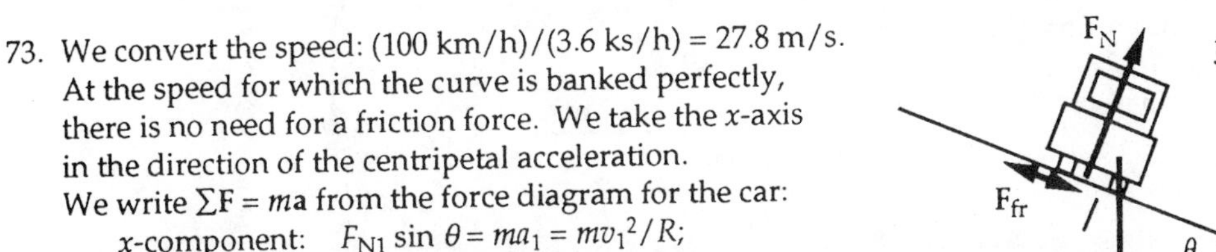

73. We convert the speed: (100 km/h)/(3.6 ks/h) = 27.8 m/s.
At the speed for which the curve is banked perfectly, there is no need for a friction force. We take the x-axis in the direction of the centripetal acceleration.
We write $\Sigma F = ma$ from the force diagram for the car:
x-component: $F_{N1} \sin \theta = ma_1 = mv_1^2/R$;
y-component: $F_{N1} \cos \theta - mg = 0$.
Combining these, we get $v_1^2 = gR \tan \theta$;
$(27.8\ \text{m/s})^2 = (60\ \text{m})(9.80\ \text{m/s}^2) \tan \theta$, which gives
$\tan \theta = 1.31$, or $\theta = 52.7°$.
At a higher speed, there is need for a friction force, which will be down the incline to help provide the greater centripetal acceleration. If the automobile does not skid, the friction is static, with $F_{fr} \leq \mu_s F_N$.
At the maximum speed, $F_{fr} = \mu_s F_N$. We write $\Sigma F = ma$ from the force diagram for the car:
x-component: $F_{N2} \sin \theta + \mu_s F_{N2} \cos \theta = ma_2 = mv_{max}^2/R$;
y-component: $F_{N2} \cos \theta - \mu_s F_{N2} \sin \theta - mg = 0$, or $F_{N2}(\cos \theta - \mu_s \sin \theta) = mg$.
When we eliminate F_{N2} by dividing the equations, we get
$v_{max}^2 = gR[(\sin \theta + \mu_s \cos \theta)/(\sin \theta - \mu_s \cos \theta)]$
$= (9.80\ \text{m/s}^2)(60\ \text{m})[(\sin 52.7° + 0.30 \cos 52.7°)/(\sin 52.7° - 0.30 \cos 52.7°)]$,
which gives $v_{max} = 39.5\ \text{m/s} = 140\ \text{km/h}$.
At a lower speed, there is need for a friction force, which will be up the incline to prevent the car from sliding down the incline. If the automobile does not skid, the friction is static, with $F_{fr} \leq \mu_s F_N$.
At the minimum speed, $F_{fr} = \mu_s F_N$. The reversal of the direction of F_{fr} can be incorporated in the above equations by changing the sign of μ_s, so we have
$v_{min}^2 = gR[(\sin \theta - \mu_s \cos \theta)/(\sin \theta + \mu_s \cos \theta)]$
$= (9.80\ \text{m/s}^2)(60\ \text{m})[(\sin 52.7° - 0.30 \cos 52.7°)/(\sin 52.7° + 0.30 \cos 52.7°)]$,
which gives $v_{min} = 20.7\ \text{m/s} = 74\ \text{km/h}$.
Thus the range of permissible speeds is $\boxed{74\ \text{km/h} < v < 140\ \text{km/h}.}$

74. We relate the speed to the period from
$v = 2\pi R/T.$
To be apparently weightless, the acceleration of gravity must be the required centripetal acceleration, so we have
$a_R = g = v^2/R = (2\pi R/T)^2/R = 4\pi^2 R/T^2;$
$9.80 \text{ m/s}^2 = 4\pi^2(6.38 \times 10^6 \text{ m})/T^2$, which gives $T =$ $\boxed{5.07 \times 10^3 \text{ s } (1.41 \text{ h}).}$

75. (*a*) The attractive gravitational force between the stars is providing the required centripetal acceleration for the circular motion.
(*b*) We relate the orbital speed to the period of revolution from
$v = 2\pi r/T$, where r is the distance to the midpoint.
The gravitational attraction provides the centripetal acceleration:
$Gmm/(2r)^2 = mv^2/r = m(2\pi r/T)^2/r = m4\pi^2 r/T^2$, so we have
$m = 16\pi^2 r^3/GT^2$
$= 16\pi^2(4.0 \times 10^{10} \text{ m})^3/(6.67 \times 10^{-11} \text{ N}\cdot\text{m}^2/\text{kg}^2)[(12.6 \text{ yr})(3.16 \times 10^7 \text{ s/yr})]^2 =$ $\boxed{9.6 \times 10^{26} \text{ kg.}}$

76. The chandelier swings out until the tension in the suspension provides the centripetal acceleration, which is the centripetal acceleration of the train. The forces are shown in the diagram.
We write $\Sigma \mathbf{F} = m\mathbf{a}$ from the force diagram for the chandelier:
x-component: $F_T \sin\theta = mv^2/r;$
y-component: $F_T \cos\theta - mg = 0.$
When these equations are combined, we get
$\tan\theta = v^2/rg;$
$\tan 17.5° = v^2/(275 \text{ m})(9.80 \text{ m/s}^2)$, which gives $v =$ $\boxed{29.2 \text{ m/s.}}$

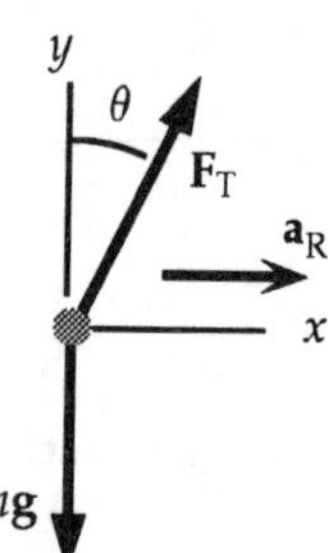

77. The acceleration due to gravity on the surface of a planet is
$g_P = F_{grav}/m = GM_P/R^2.$
If we form the ratio of the expressions for Jupiter and the Earth, we have
$g_{Jupiter}/g_{Earth} = (M_{Jupiter}/M_{Earth})(R_{Earth}/R_{Jupiter})^2;$
$g_{Jupiter}/g_{Earth} = [(1.9 \times 10^{27} \text{ kg})/(6.0 \times 10^{24} \text{ kg})][(6.38 \times 10^6 \text{ m})/(7.1 \times 10^7 \text{ m})]^2,$
which gives $g_{Jupiter} = 2.56 g_{Earth}$.
This has not taken into account the centripetal acceleration. We ignore the small contribution on Earth.
The centripetal acceleration on the equator of Jupiter is
$a_R = v^2/R = (2\pi R/T)^2/R = 4\pi^2 R/T^2$
$= 4\pi^2(7.1 \times 10^7 \text{ m})/[(595 \text{ min})(60 \text{ s/min})]^2 = 2.2 \text{ m/s}^2 = 0.22 g_{Earth}$.
The centripetal acceleration reduces the effective value of g:
$g'_{Jupiter} = g_{Jupiter} - a_R = 2.56 g_{Earth} - 0.22 g_{Earth} =$ $\boxed{2.3 g_{Earth}.}$

78. The gravitational attraction from the core must provide the centripetal acceleration for the orbiting stars:
$GM_{star}M_{core}/R^2 = M_{star}v^2/R$, so we have
$M_{core} = v^2 R/G$
$= (780 \text{ m/s})^2(5.7 \times 10^{17} \text{ m})/(6.67 \times 10^{-11} \text{ N}\cdot\text{m}^2/\text{kg}^2) =$ $\boxed{5.2 \times 10^{33} \text{ kg.}}$
If we compare this to our Sun, we get
$M_{core}/M_{Sun} = (5.2 \times 10^{33} \text{ kg})/(2.0 \times 10^{30} \text{ kg}) =$ $\boxed{2.6 \times 10^3 \times.}$

CHAPTER 6

1. Because there is no acceleration, the contact force must have the same magnitude as the weight. The displacement in the direction of this force is the vertical displacement. Thus,
   $W = F\,\Delta y = (mg)\,\Delta y = (75.0\text{ kg})(9.80\text{ m/s}^2)(10.0\text{ m}) = \boxed{7.35 \times 10^3\text{ J.}}$

2. (*a*) Because there is no acceleration, the horizontal applied force must have the same magnitude as the friction force. Thus,
   $W = F\,\Delta x = (180\text{ N})(6.0\text{ m}) = \boxed{1.1 \times 10^3\text{ J.}}$
   (*b*) Because there is no acceleration, the vertical applied force must have the same magnitude as the weight. Thus,
   $W = F\,\Delta y = mg\,\Delta y = (900\text{ N})(6.0\text{ m}) = \boxed{5.4 \times 10^3\text{ J.}}$

3. Because there is no acceleration, from the force diagram we see that
   $F_N = mg$, and $F = F_{fr} = \mu_k mg$.
   Thus,
   $W = Fx\cos 0° = \mu_k mg\,x\cos 0°$
   $= (0.70)(150\text{ kg})(9.80\text{ m/s}^2)(12.3\text{ m})(1) = \boxed{1.3 \times 10^4\text{ J.}}$

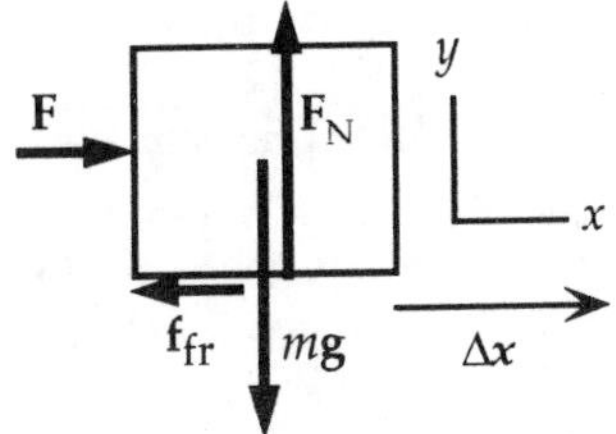

4. Because there is no acceleration, the net work is zero, that is, the (positive) work of the car and the (negative) work done by the average retarding force must add to zero. Thus,
   $W_{net} = W_{car} + F_{fr}\,\Delta x\cos 180° = 0$, or
   $F_{fr} = -\,W_{car}/\Delta x\cos 180° = -\,(7.0 \times 10^4\text{ J})/(2.8 \times 10^3\text{ m})(-1) = \boxed{25\text{ N.}}$

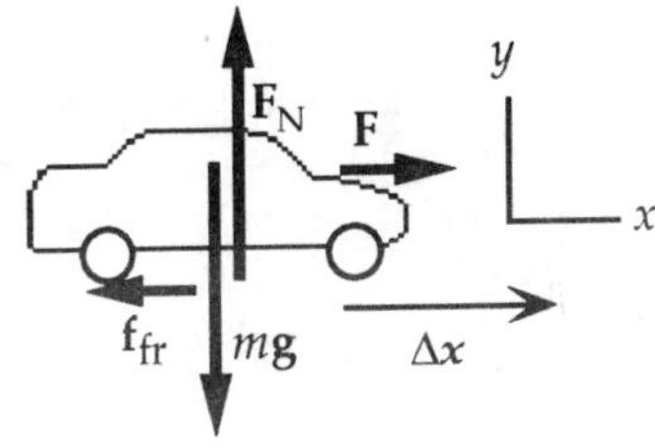

4, 6, 15

5. Because the speed is zero before the throw and when the rock reaches the highest point, the positive work of the throw and the (negative) work done by the (downward) weight must add to zero. Thus,
   $W_{net} = W_{throw} + mgh\cos 180° = 0$, or
   $h = -\,W_{throw}/mg\cos 180° = -\,(115\text{ J})/(0.325\text{ kg})(9.80\text{ m/s}^2)(-1) = \boxed{36.1\text{ m.}}$

6. The maximum amount of work that the hammer can do is the work that was done by the weight as the hammer fell:
   $W_{max} = mgh\cos 0° = (2.0\text{ kg})(9.80\text{ m/s}^2)(0.40\text{ m})(1) = \boxed{7.8\text{ J.}}$
   People add their own force to the hammer as it falls in order that additional work is done before the hammer hits the nail, and thus more work can be done on the nail.

7. The minimum work is needed when there is no acceleration.

   (*a*) From the force diagram, we write $\Sigma\mathbf{F} = m\mathbf{a}$:

   y-component: $F_N - mg\cos\theta = 0$;

   x-component: $F_{min} - mg\sin\theta = 0$.

   For a distance d along the incline, we have

   $$W_{min} = F_{min}d\cos 0° = mgd\sin\theta(1)$$
   $$= (1000\ \text{kg})(9.80\ \text{m/s}^2)(300\ \text{m})\sin 17.5°$$
   $$= \boxed{8.8\times10^5\ \text{J.}}$$

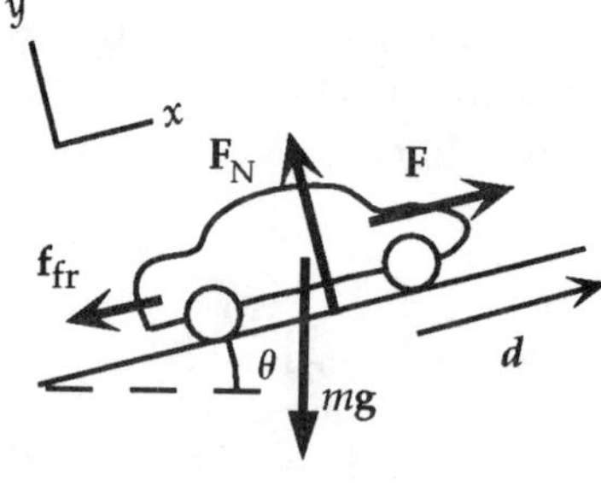

   (*b*) When there is friction, we have

   x-component: $F_{min} - mg\sin\theta - \mu_k F_N = 0$, or

   $F_{min} = mg\sin\theta + \mu_k mg\cos\theta = 0$,

   For a distance d along the incline, we have

   $$W_{min} = F_{min}d\cos 0° = mgd(\sin\theta + \mu_k\cos\theta)(1)$$
   $$= (1000\ \text{kg})(9.80\ \text{m/s}^2)(300\ \text{m})(\sin 17.5° + 0.25\cos 17.5°) = \boxed{1.6\times10^6\ \text{J.}}$$

8. Because the motion is in the x-direction, we see that the weight and normal forces do no work:

   $W_{FN} = W_{mg} = 0$.

   From the force diagram, we write $\Sigma\mathbf{F} = m\mathbf{a}$:

   x-component: $F\cos\theta - F_{fr} = 0$, or $F_{fr} = F\cos\theta$.

   For the work by these two forces, we have

   $W_F = F\,\Delta x\,\cos\theta = (12\ \text{N})(15\ \text{m})\cos 20° = 1.7\times10^2\ \text{J}$.

   $W_{fr} = F\cos\theta\,\Delta x\,\cos 180° = (12\ \text{N})\cos 20°(15\ \text{m})(-1) = -1.7\times10^2\ \text{J}$.

   As expected, the total work is zero: $\boxed{W_F = -W_{fr} = 1.7\times10^2\ \text{J.}}$

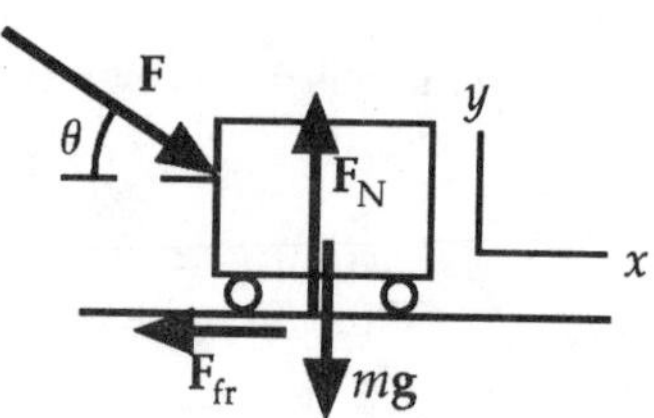

9. Because the net work must be zero, the work to stack the books will have the same magnitude as the work done by gravity. For each book the work is mg times the distance the center is raised (zero for the first book, one book-height for the second book, etc.).

   $W_1 = 0, W_2 = mgh, W_3 = mg2h; \dots$.

   Thus for eight books, we have

   $$W = W_1 + W_2 + W_3 + \dots + W_8 = mgh(0 + 1 + 2 + \dots + 7) = (1.8\ \text{kg})(9.80\ \text{m/s}^2)(0.046\ \text{m})(28) = \boxed{23\ \text{J.}}$$

10. (*a*) From the force diagram, we write $\Sigma\mathbf{F} = m\mathbf{a}$:

   y-component: $F_N - mg\cos\theta = 0$;

   x-component: $-F - \mu_k F_N + mg\sin\theta = 0$.

   Thus we have

   $$F = -\mu_k F_N + mg\sin\theta = -\mu_k mg\cos\theta + mg\sin\theta$$
   $$= mg(\sin\theta - \mu_k\cos\theta)$$

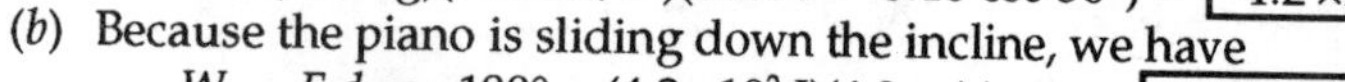

   $$= (280\ \text{kg})(9.80\ \text{m/s}^2)(\sin 30° - 0.40\cos 30°) = \boxed{4.2\times10^2\ \text{N.}}$$

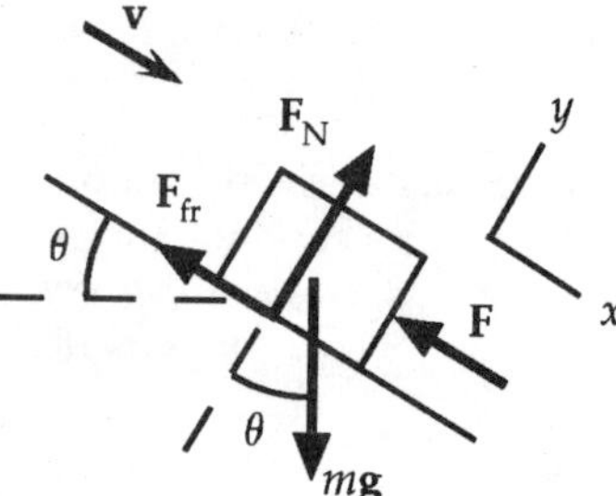

   (*b*) Because the piano is sliding down the incline, we have

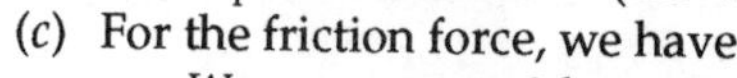

   $$W_F = F\,d\cos 180° = (4.2\times10^2\ \text{J})(4.3\ \text{m})(-1) = \boxed{-1.8\times10^3\ \text{J.}}$$

   (*c*) For the friction force, we have

   $$W_{fr} = \mu_k mg\cos\theta d\cos 180°$$

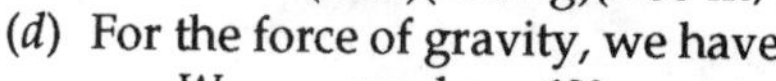

   $$= (0.40)(280\ \text{kg})(9.80\ \text{m/s}^2)(\cos 30°)(4.3\ \text{m})(-1) = \boxed{-4.1\times10^3\ \text{J.}}$$

   (*d*) For the force of gravity, we have

   $$W_{grav} = mg\,d\cos 60°$$
   $$= (280\ \text{kg})(9.80\ \text{m/s}^2)(4.3\ \text{m})(\cos 60°) = \boxed{5.9\times10^3\ \text{J.}}$$

   (*e*) Because the normal force does no work, we have

   $$W_{net} = W_{grav} + W_F + W_{fr} + W_N$$
   $$= 5.9\times10^3\ \text{J} - 1.8\times10^3\ \text{J} - 4.1\times10^3\ \text{J} + 0 = \boxed{0.}$$

11. (*a*) To find the required force, we use the force diagram to write $\Sigma F_y = ma_y$:
$F - Mg = Ma$, so we have
$F = M(a + g) = M(0.10g + g) =$ $\boxed{1.10Mg.}$

(*b*) For the work, we have
$W_F = Fh \cos 0° =$ $\boxed{1.10Mgh.}$

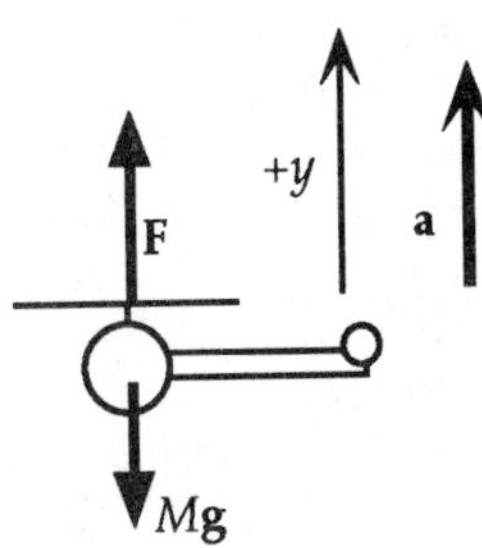

12. From the graph we obtain the forces at the two ends:
at $d_A = 10.0$ m, $(F \cos \theta)_A = 150$ N; at $d_B = 35.0$ m, $(F \cos \theta)_B = 250$ N.
The work done in moving the object is the area under the $F \cos \theta$ vs. x graph. If we assume the graph is a straight line, we have
$W \approx \frac{1}{2}[(F \cos \theta)_A + (F \cos \theta)_B](d_B - d_A) = \frac{1}{2}(150 \text{ N} + 250 \text{ N})(35.0 \text{ m} - 10.0 \text{ m}) =$ $\boxed{5.0 \times 10^3 \text{ J.}}$

13. The work done in moving the object is the area under the $F \cos \theta$ vs. x graph.

(*a*) For the motion from 0.0 m to 10.0 m, we find the area of two triangles and one rectangle:
$W = \frac{1}{2}(400 \text{ N})(3.0 \text{ m} - 0.0 \text{ m})$
$+ (400 \text{ N})(7.0 \text{ m} - 3.0 \text{ m}) +$
$\frac{1}{2}(400 \text{ N})(10.0 \text{ m} - 7.0 \text{ m})$
$=$ $\boxed{2.8 \times 10^3 \text{ J.}}$

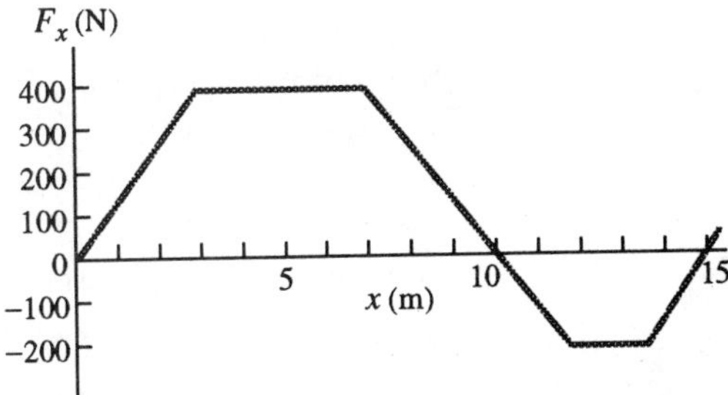

(*b*) For the motion from 0.0 m to 15.0 m, we add the negative area of two triangles and one rectangle:
$W = 2.8 \times 10^3 \text{ J} - \frac{1}{2}(200 \text{ N})(12.0 \text{ m} - 10.0 \text{ m}) -$
$(200 \text{ N})(13.5 \text{ m} - 12.0 \text{ m}) - \frac{1}{2}(200 \text{ N})(15.0 \text{ m} - 13.5 \text{ m}) =$ $\boxed{2.1 \times 10^3 \text{ J.}}$

14. We obtain the forces at the beginning and end of the motion:
at $x_1 = 0.038$ m, $F_1 = kx_1 = (88 \text{ N/m})(0.038 \text{ m}) = 3.34$ N;
at $x_2 = 0.058$ m, $F_2 = kx_2 = (88 \text{ N/m})(0.058 \text{ m}) = 5.10$ N.
From the graph the work done in stretching the object is the area under the F vs. x graph:
$W = \frac{1}{2}[F_1 + F_2](x_2 - x_1)$
$= \frac{1}{2}(3.34 \text{ N} + 5.10 \text{ N})(0.058 \text{ m} - 0.038 \text{ m}) =$ $\boxed{0.084 \text{ J.}}$

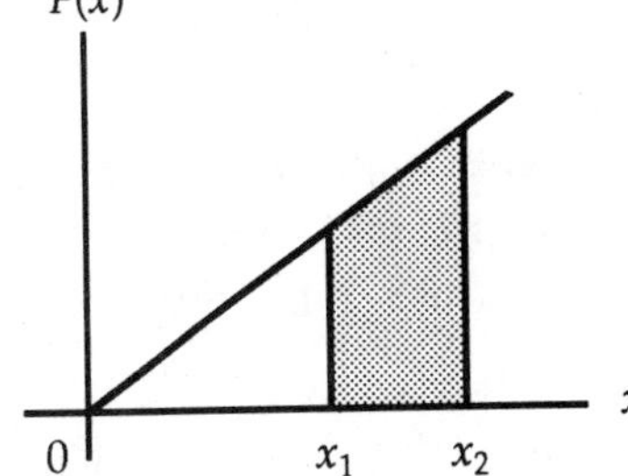

15. The work done in moving the object is the area under the F_x vs. x graph. For the motion from 0.0 m to 11.0 m, we find the area of two triangles and one rectangle:
$W = \frac{1}{2}(24.0 \text{ N})(3.0 \text{ m} - 0.0 \text{ m}) +$
$(24.0 \text{ N})(8.0 \text{ m} - 3.0 \text{ m}) +$
$\frac{1}{2}(24.0 \text{ N})(11.0 \text{ m} - 8.0 \text{ m}) =$ $\boxed{1.9 \times 10^2 \text{ J.}}$

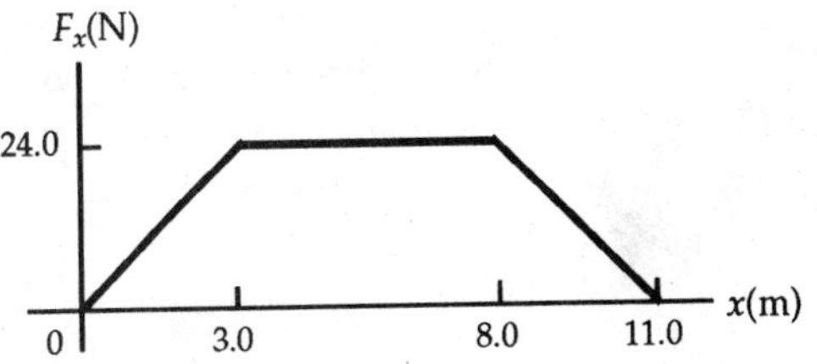

16. We obtain the forces at the beginning and end of the motion:
at $r_E + h = 6.38 \times 10^6$ m + 2.5×10^6 m = 8.88×10^6 m,
$F_2 = GM_E m/r^2$
$= (6.67 \times 10^{-11}\ \text{N}\cdot\text{m}^2/\text{kg}^2)(5.98 \times 10^{24}\ \text{kg})(1300\ \text{kg})/[(8.88 \times 10^6\ \text{m})]^2 = 6.58 \times 10^3\ \text{N}.$
at $r_E = 6.38 \times 10^6$ m,
$F_1 = GM_E m/r_E^2$
$= (6.67 \times 10^{-11}\ \text{N}\cdot\text{m}^2/\text{kg}^2)(5.98 \times 10^{24}\ \text{kg})(1300\ \text{kg})/[(6.38 \times 10^6\ \text{m})]^2 = 1.27 \times 10^4\ \text{N}.$

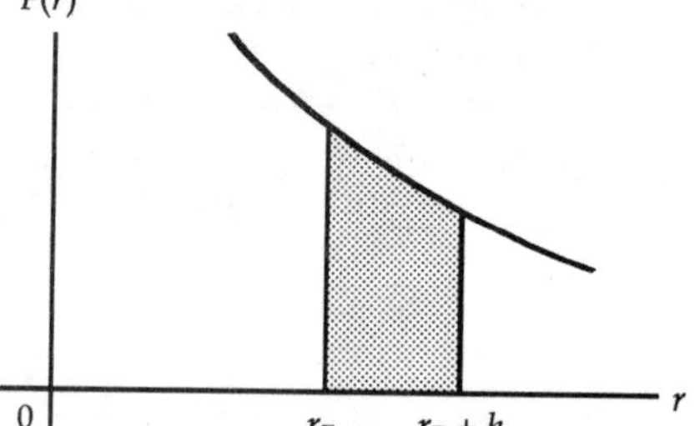

From the graph the work done in stretching the object is the area under the F vs. r graph, which we approximate as a straight line:
$W = \frac{1}{2}[F_1 + F_2]h$
$= \frac{1}{2}(6.58 \times 10^3\ \text{N} + 1.27 \times 10^4\ \text{N})(2.5 \times 10^6\ \text{m}) =$ $\boxed{2.4 \times 10^{10}\ \text{J}.}$
This will be a slight overestimate.

17. We find the speed from
$\text{KE} = \frac{1}{2}mv^2;$
$6.21 \times 10^{-21}\ \text{J} = \frac{1}{2}(5.31 \times 10^{-26}\ \text{kg})v^2$, which gives $v =$ $\boxed{484\ \text{m/s}.}$

18. (*a*) $\text{KE}_2 = \frac{1}{2}mv_2^2 = 2\text{KE}_1 = 2(\frac{1}{2}mv_1^2)$, which gives $v_2 = v_1\sqrt{2}$, so the speed increases by a factor of $\boxed{\sqrt{2}.}$
(*b*) $\text{KE}_2 = \frac{1}{2}mv_2^2 = \frac{1}{2}m(2v_1)^2 = 4(\frac{1}{2}mv_1^2) = 4\text{KE}_1$, so the kinetic energy increases by a factor of $\boxed{4.}$

19. The work done on the electron decreases its kinetic energy:
$W = \Delta\text{KE} = \frac{1}{2}mv^2 - \frac{1}{2}mv_0^2 = 0 - \frac{1}{2}(9.11 \times 10^{-31}\ \text{kg})(1.90 \times 10^6\ \text{m/s})^2 =$ $\boxed{-1.64 \times 10^{-18}\ \text{J}.}$

20. The work done on the car decreases its kinetic energy:
$W = \Delta\text{KE} = \frac{1}{2}mv^2 - \frac{1}{2}mv_0^2 = 0 - \frac{1}{2}(1000\ \text{kg})[(110\ \text{km/h})/(3.6\ \text{ks/h})]^2 =$ $\boxed{-4.67 \times 10^5\ \text{J}.}$

21. The percent increase in the kinetic energy is
% increase $= [(\frac{1}{2}mv_2^2 - \frac{1}{2}mv_1^2)/\frac{1}{2}mv_1^2](100\%) = (v_2^2 - v_1^2)(100\%)/v_1^2$
$= [(100\ \text{km/h})^2 - (90\ \text{km/h})^2](100\%)/(90\ \text{km/h})^2 =$ $\boxed{23\%.}$

22. The work done on the arrow increases its kinetic energy:
$W = Fd = \Delta\text{KE} = \frac{1}{2}mv^2 - \frac{1}{2}mv_0^2;$
$(95\ \text{N})(0.80\ \text{m}) = \frac{1}{2}(0.080\ \text{kg})v^2 - 0$, which gives $v =$ $\boxed{44\ \text{m/s}.}$

23. The work done by the force of the glove decreases the kinetic energy of the ball:
$W = Fd = \Delta\text{KE} = \frac{1}{2}mv^2 - \frac{1}{2}mv_0^2;$
$F(0.25\ \text{m}) = 0 - \frac{1}{2}(0.140\ \text{kg})(35\ \text{m/s})^2$, which gives $F = -3.4 \times 10^2$ N.
The force by the ball on the glove is the reaction to this force:
$\boxed{3.4 \times 10^2\ \text{N in the direction of the motion of the ball.}}$

24. The work done by the braking force decreases the kinetic energy of the car:

$W = \Delta\text{KE};$

$-Fd = \frac{1}{2}mv^2 - \frac{1}{2}mv_0^2 = 0 - \frac{1}{2}mv_0^2.$

Assuming the same braking force, we form the ratio:

$d_2/d_1 = (v_{02}/v_{01})^2 = (1.5)^2 =$ **2.25.**

25. On a level road, the normal force is mg, so the kinetic friction force is $\mu_k mg$. Because it is the (negative) work of the friction force that stops the car, we have

$W = \Delta\text{KE};$

$-\mu_k mg\, d = \frac{1}{2}mv^2 - \frac{1}{2}mv_0^2;$

$-(0.42)m(9.80\ \text{m/s}^2)(88\ \text{m}) = -\frac{1}{2}mv_0^2$, which gives $v_0 =$ **27 m/s (97 km/h or 60 mi/h).**

Because every term contains the mass, **it cancels.**

26. The work done by the air resistance decreases the kinetic energy of the ball:

$W = F_{air}d = \Delta\text{KE} = \frac{1}{2}mv^2 - \frac{1}{2}mv_0^2 = \frac{1}{2}m(0.90v_0)^2 - \frac{1}{2}mv_0^2 = \frac{1}{2}mv_0^2[(0.90)^2 - 1];$

$F_{air}(15\ \text{m}) = \frac{1}{2}(0.25\ \text{kg})[(95\ \text{km/h})/(3.6\ \text{ks/h})]^2[(0.90)^2 - 1]$, which gives $F_{air} =$ **– 1.1 N.**

27. With $m_1 = 2m_2$, for the initial condition we have

$\text{KE}_1 = \frac{1}{2}\text{KE}_2;$

$\frac{1}{2}m_1v_1^2 = \frac{1}{2}(\frac{1}{2}m_2v_2^2)$, or $2m_2 v_1^2 = \frac{1}{2}m_2v_2^2$, which gives $v_1 = \frac{1}{2}v_2$.

After a speed increase of Δv, we have

$\text{KE}_1' = \text{KE}_2';$

$\frac{1}{2}m_1(v_1 + \Delta v)^2 = \frac{1}{2}m_2(v_2 + \Delta v)^2;$

$2m_2(\frac{1}{2}v_2 + 5.0\ \text{m/s})^2 = m_2(v_2 + 5.0\ \text{m/s})^2.$

When we take the square root of both sides, we get

$\sqrt{2}(\frac{1}{2}v_2 + 5.0\ \text{m/s}) = \pm(v_2 + 5.0\ \text{m/s})$, which gives a positive result of $v_2 =$ **7.1 m/s.**

For the other speed we have $v_1 = \frac{1}{2}v_2 =$ **3.5 m/s.**

28. (*a*) From the force diagram we write $\Sigma F_y = ma_y$:

$F_T - mg = ma;$

$F_T - (220\ \text{kg})(9.80\ \text{m/s}^2) = (220\ \text{kg})(0.150)(9.80\ \text{m/s}^2),$

which gives $F_T =$ **2.48×10^3 N.**

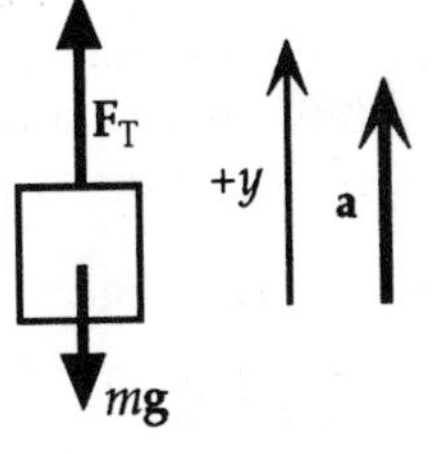

(*b*) The net work is done by the net force:

$W_{net} = F_{net}h = (F_T - mg)h$

$= [2.48 \times 10^3\ \text{N} - (220\ \text{kg})(9.80\ \text{m/s}^2)](21.0\ \text{m}) =$ **6.79×10^3 J.**

(*c*) The work is done by the cable is

$W_{cable} = F_T h$

$= (2.48 \times 10^3\ \text{N})(21.0\ \text{m}) =$ **5.21×10^4 J.**

(*d*) The work is done by gravity is

$W_{grav} = -mgh$

$= -(220\ \text{kg})(9.80\ \text{m/s}^2)(21.0\ \text{m}) =$ **-4.53×10^4 J.**

Note that $W_{net} = W_{cable} + W_{grav}$.

(*e*) The net work done on the load increases its kinetic energy:

$W_{net} = \Delta\text{KE} = \frac{1}{2}mv^2 - \frac{1}{2}mv_0^2;$

$6.79 \times 10^3\ \text{J} = \frac{1}{2}(220\ \text{kg})v^2 - 0$, which gives $v =$ **7.86 m/s.**

F_T $+y$ a mg

29. The potential energy of the spring is zero when the spring is not stretched ($x = 0$). For the stored potential energy, we have

$$\text{PE} = \tfrac{1}{2}kx_f^2 - 0;$$
$$25\text{ J} = \tfrac{1}{2}(440\text{ N/m})x_f^2 - 0, \text{ which gives } x_f = \boxed{0.34\text{ m.}}$$

30. For the potential energy change we have

$$\Delta\text{PE} = mg\,\Delta y = (6.0\text{ kg})(9.80\text{ m/s}^2)(1.2\text{ m}) = \boxed{71\text{ J.}}$$

31. For the potential energy change we have

$$\Delta\text{PE} = mg\,\Delta y = (64\text{ kg})(9.80\text{ m/s}^2)(4.0\text{ m}) = \boxed{2.5\times10^3\text{ J.}}$$

32. (*a*) With the reference level at the ground, for the potential energy we have

$$\text{PE}_a = mgy_a = (2.10\text{ kg})(9.80\text{ m/s}^2)(2.20\text{ m}) = \boxed{45.3\text{ J.}}$$

(*b*) With the reference level at the top of the head, for the potential energy we have

$$\text{PE}_b = mg(y_b - h) = (2.10\text{ kg})(9.80\text{ m/s}^2)(2.20\text{ m} - 1.60\text{ m}) = \boxed{12.3\text{ J.}}$$

(*c*) Because the person lifted the book from the reference level in part (*a*), the potential energy is equal to the work done: $\boxed{45.3\text{ J.}}$ In part (*b*) the initial potential energy was negative, so the final potential energy is not the work done.

33. (*a*) With the reference level at the ground, for the potential energy change we have

$$\Delta\text{PE} = mg\,\Delta y = (55\text{ kg})(9.80\text{ m/s}^2)(3100\text{ m} - 1600\text{ m}) = \boxed{8.1\times10^5\text{ J.}}$$

(*b*) The minimum work would be equal to the change in potential energy:

$$W_{\min} = \Delta\text{PE} = \boxed{8.1\times10^5\text{ J.}}$$

(*c*) $\boxed{\text{Yes,}}$ the actual work will be more than this. There will be additional work required for any kinetic energy change, and to overcome retarding forces, such as air resistance and ground deformation.

34. The potential energy of the spring is zero when the spring is not stretched ($x = 0$).

(*a*) For the change in potential energy, we have

$$\Delta\text{PE} = \tfrac{1}{2}kx^2 - \tfrac{1}{2}kx_0^2 = \boxed{\tfrac{1}{2}k(x^2 - x_0^2).}$$

(*b*) If we call compressing positive, we have

$$\Delta\text{PE}_{\text{compression}} = \tfrac{1}{2}k(+x_0)^2 - 0 = \tfrac{1}{2}kx_0^2;$$
$$\Delta\text{PE}_{\text{stretching}} = \tfrac{1}{2}k(-x_0)^2 - 0 = \boxed{\tfrac{1}{2}kx_0^2.}$$

The change in potential energy is the $\boxed{\text{same.}}$

35. We choose the potential energy to be zero at the ground ($y = 0$). Because the tension in the vine does no work, energy is conserved, so we have

$$E = \text{KE}_i + \text{PE}_i = \text{KE}_f + \text{PE}_f;$$
$$\tfrac{1}{2}mv_i^2 + mgy_i = \tfrac{1}{2}mv_f^2 + mgy_f;$$
$$\tfrac{1}{2}m(5.6\text{ m/s})^2 + m(9.80\text{ m/s}^2)(0) = \tfrac{1}{2}m(0)^2 + m(9.80\text{ m/s}^2)h,$$

which gives $\boxed{h = 1.6\text{ m.}}$

$\boxed{\text{No,}}$ the length of the vine does not affect the height; it affects the angle.

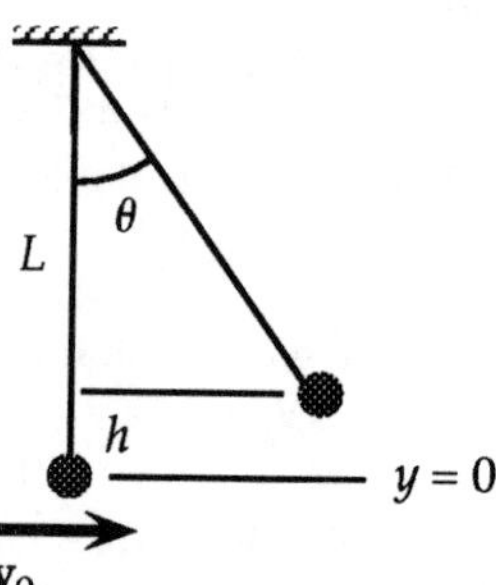

36. We choose the potential energy to be zero at the bottom ($y = 0$). Because there is no friction and the normal force does no work, energy is conserved, so we have

$$E = KE_i + PE_i = KE_f + PE_f;$$

$$\tfrac{1}{2}mv_i^2 + mgy_i = \tfrac{1}{2}mv_f^2 + mgy_f;$$

$\tfrac{1}{2}m(0)^2 + m(9.80\text{ m/s}^2)(125\text{ m}) = \tfrac{1}{2}mv_f^2 + m(9.80\text{ m/s}^2)(0)$, which gives $v_f =$ $\boxed{49.5\text{ m/s.}}$

This is 180 km/h! It is a good thing there is friction on the ski slopes.

37. We choose the potential energy to be zero at the bottom ($y = 0$). Because there is no friction and the normal force does no work, energy is conserved, so we have

$$E = KE_i + PE_i = KE_f + PE_f;$$

$$\tfrac{1}{2}mv_i^2 + mgy_i = \tfrac{1}{2}mv_f^2 + mgy_f;$$

$\tfrac{1}{2}mv_i^2 + m(9.80\text{ m/s}^2)(0) = \tfrac{1}{2}m(0)^2 + m(9.80\text{ m/s}^2)(1.35\text{ m})$, which gives $v_i =$ $\boxed{5.14\text{ m/s.}}$

38. We choose the potential energy to be zero at the ground ($y = 0$). We find the minimum speed by ignoring any frictional forces. Energy is conserved, so we have

$$E = KE_i + PE_i = KE_f + PE_f;$$

$$\tfrac{1}{2}mv_i^2 + mgy_i = \tfrac{1}{2}mv_f^2 + mgy_f;$$

$\tfrac{1}{2}mv_i^2 + m(9.80\text{ m/s}^2)(0) = \tfrac{1}{2}m(0.70\text{ m/s})^2 + m(9.80\text{ m/s}^2)(2.10\text{ m})$, which gives $v_i =$ $\boxed{6.5\text{ m/s.}}$

Note that the initial velocity will not be horizontal, but will have a horizontal component of 0.70 m/s.

39. We choose $y = 0$ at the level of the trampoline.

(*a*) We apply conservation of energy for the jump from the top of the platform to the trampoline:

$$E = KE_i + PE_i = KE_f + PE_f;$$

$$\tfrac{1}{2}mv_0^2 + mgH = \tfrac{1}{2}mv_1^2 + 0;$$

$$\tfrac{1}{2}m(5.0\text{ m/s})^2 + m(9.80\text{ m/s}^2)(3.0\text{ m}) = \tfrac{1}{2}mv_1^2,$$

which gives $v_1 =$ $\boxed{9.2\text{ m/s.}}$

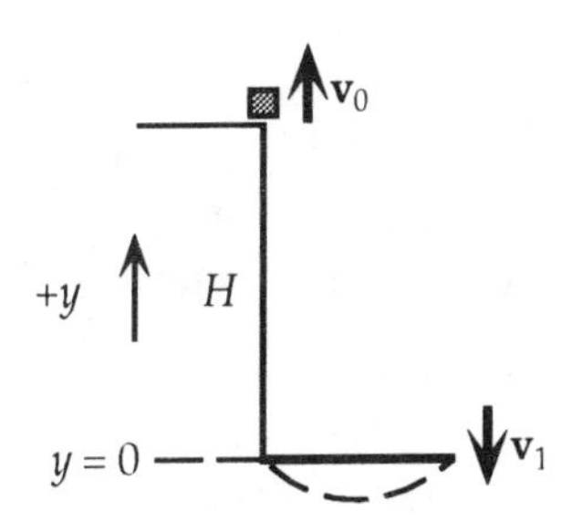

(*b*) We apply conservation of energy from the landing on the trampoline to the maximum depression of the trampoline. If we ignore the small change in gravitational potential energy, we have

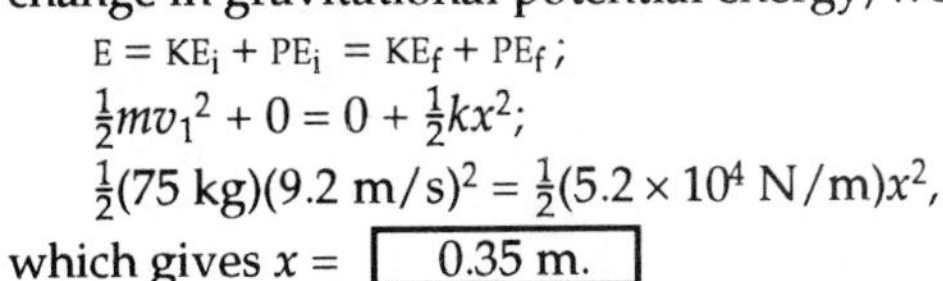

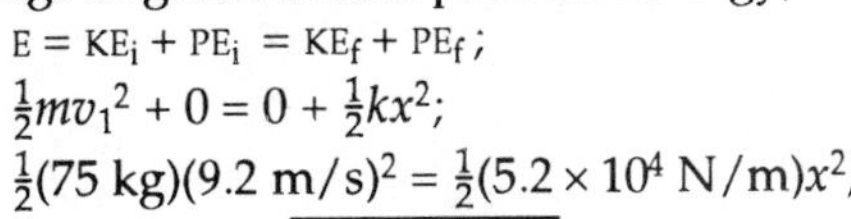

$$E = KE_i + PE_i = KE_f + PE_f;$$

$$\tfrac{1}{2}mv_1^2 + 0 = 0 + \tfrac{1}{2}kx^2;$$

$$\tfrac{1}{2}(75\text{ kg})(9.2\text{ m/s})^2 = \tfrac{1}{2}(5.2 \times 10^4\text{ N/m})x^2,$$

which gives $x =$ $\boxed{0.35\text{ m.}}$

This will increase slightly if the gravitational potential energy is taken into account.

40. We choose $y = 0$ at point B. With no friction, energy is conserved. The initial (and constant) energy is

$$E = E_A = mgy_A + \tfrac{1}{2}mv_A^2$$
$$= m(9.8\text{ m/s}^2)(30\text{ m}) + 0 = (294\text{ J/kg})m\,.$$

At point B we have

$$E = mgy_B + \tfrac{1}{2}mv_B^2;$$
$$(294\text{ J/kg})m = m(9.8\text{ m/s}^2)(0) + \tfrac{1}{2}mv_B^2,$$

which gives $\boxed{v_B = 24\text{ m/s.}}$

At point C we have

$$E = mgy_C + \tfrac{1}{2}mv_C^2;$$

$(294\text{ J/kg})m = m(9.8\text{ m/s}^2)(25\text{ m}) + \tfrac{1}{2}mv_C^2$, which gives $\boxed{v_C = 9.9\text{ m/s.}}$

At point D we have

$$E = mgy_D + \tfrac{1}{2}mv_D^2;$$

$(294\text{ J/kg})m = m(9.8\text{ m/s}^2)(12\text{ m}) + \tfrac{1}{2}mv_D^2$, which gives $\boxed{v_D = 19\text{ m/s.}}$

41. We choose the potential energy to be zero at the ground ($y = 0$). Energy is conserved, so we have

$$E = KE_i + PE_i = KE_f + PE_f;$$
$$\tfrac{1}{2}mv_i^2 + mgy_i = \tfrac{1}{2}mv_f^2 + mgy_f;$$

$\tfrac{1}{2}m(185\text{ m/s})^2 + m(9.80\text{ m/s}^2)(265\text{ m}) = \tfrac{1}{2}mv_f^2 + m(9.80\text{ m/s}^2)(0)$, which gives $v_f = \boxed{199\text{ m/s.}}$

Note that we have not found the direction of the velocity.

42. (*a*) For the motion from the bridge to the lowest point, we use energy conservation:

$$KE_i + PE_{gravi} + PE_{cordi} = KE_f + PE_{gravf} + PE_{cordf};$$
$$0 + 0 + 0 = 0 + mg(-h) + \tfrac{1}{2}k(h - L_0)^2;$$
$$0 = -(60\text{ kg})(9.80\text{ m/s}^2)(31\text{ m}) + \tfrac{1}{2}k(31\text{ m} - 12\text{ m})^2,$$

which gives $k = \boxed{1.0 \times 10^2\text{ N/m.}}$

(*b*) The maximum acceleration will occur at the lowest point, where the upward restoring force in the cord is maximum:

$$kx_{max} - mg = ma_{max};$$
$$(1.0 \times 10^2\text{ N/m})(31\text{ m} - 12\text{ m}) - (60\text{ kg})(9.80\text{ m/s}^2) = (60\text{ kg})a_{max},$$

which gives $a_{max} = \boxed{22\text{ m/s}^2.}$

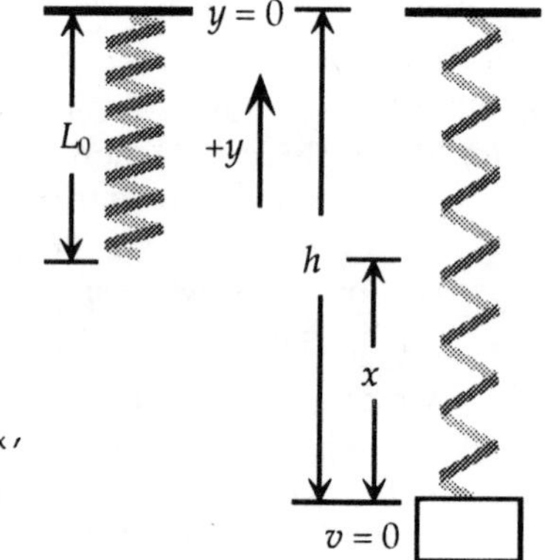

43. We choose the potential energy to be zero at the compressed position ($y = 0$).

(*a*) For the motion from the release point to where the ball leaves the spring, we use energy conservation:

$$KE_i + PE_{gravi} + PE_{springi} = KE_f + PE_{gravf} + PE_{springf};$$
$$0 + 0 + \tfrac{1}{2}kx^2 = \tfrac{1}{2}mv^2 + mgx + 0;$$
$$\tfrac{1}{2}(900\text{ N/m})(0.150\text{ m})^2 =$$
$$\tfrac{1}{2}(0.300\text{ kg})v^2 + (0.300\text{ kg})(9.80\text{ m/s}^2)(0.150\text{ m}),$$

which gives $v = \boxed{8.03\text{ m/s.}}$

(*b*) For the motion from the release point to the highest point, we use energy conservation:

$$KE_i + PE_{gravi} + PE_{springi} = KE_f + PE_{gravf} + PE_{springf};$$
$$0 + 0 + \tfrac{1}{2}kx^2 = 0 + mgh + 0;$$
$$0 + 0 + \tfrac{1}{2}(900\text{ N/m})(0.150\text{ m})^2 = (0.300\text{ kg})(9.80\text{ m/s}^2)h,$$

which gives $h = \boxed{3.44\text{ m.}}$

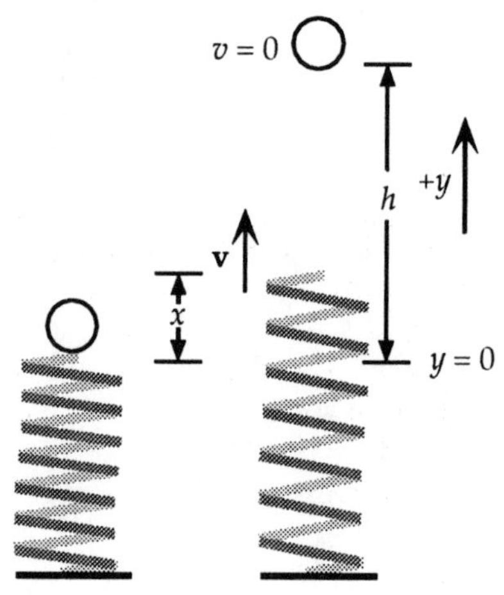

44. With $y = 0$ at the bottom of the circle, we call the start point A, the bottom of the circle B, and the top of the circle C. At the top of the circle we have the forces mg and F_N, both downward, that provide the centripetal acceleration:

$$mg + F_N = mv_C^2/r.$$

The minimum value of F_N is zero, so the minimum speed at C is found from

$$v_{Cmin}^2 = gr.$$

From energy conservation for the motion from A to C we have

$$KE_A + PE_A = KE_C + PE_C;$$

$$0 + mgh = \tfrac{1}{2}mv_C^2 + mg(2r),$$

thus the minimum height is found from

$$gh = \tfrac{1}{2}v_{Cmin}^2 + 2gr = \tfrac{1}{2}gr + 2gr, \text{ which gives } \boxed{h = 2.5r.}$$

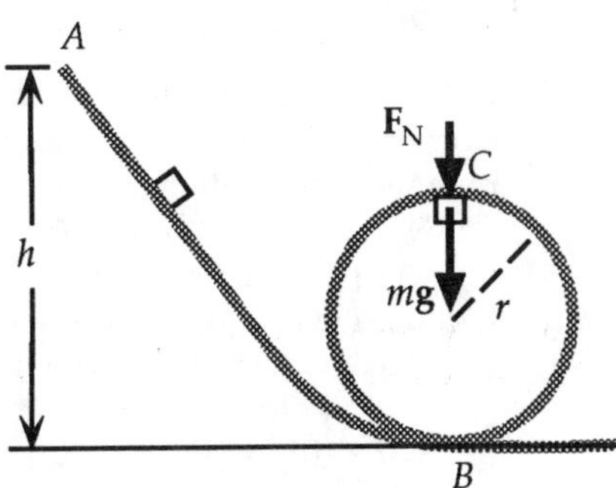

45. The potential energy is zero at $x = 0$. For the motion from the release point, we use energy conservation:

$$E = KE_i + PE_i = KE_f + PE_f;$$

$$E = 0 + \tfrac{1}{2}kx_0^2 = \tfrac{1}{2}mv^2 + \tfrac{1}{2}kx^2, \text{ which gives } \boxed{E = \tfrac{1}{2}kx_0^2 = \tfrac{1}{2}mv^2 + \tfrac{1}{2}kx^2.}$$

46. The maximum acceleration will occur at the lowest point, where the upward restoring force in the spring is maximum:

$$kx_{max} - Mg = Ma_{max} = M(5.0g), \text{ which gives } x_{max} = 6.0Mg/k.$$

With $y = 0$ at the initial position of the top of the spring, for the motion from the break point to the maximum compression of the spring, we use energy conservation:

$$KE_i + PE_{gravi} + PE_{springi} = KE_f + PE_{gravf} + PE_{springf};$$

$$0 + Mgh + 0 = 0 + Mg(-x_{max}) + \tfrac{1}{2}kx_{max}^2.$$

When we use the previous result, we get

$$Mgh = -[6.0(Mg)^2/k] + \tfrac{1}{2}k(6.0Mg/k)^2, \text{ which gives } \boxed{k = 12Mg/h.}$$

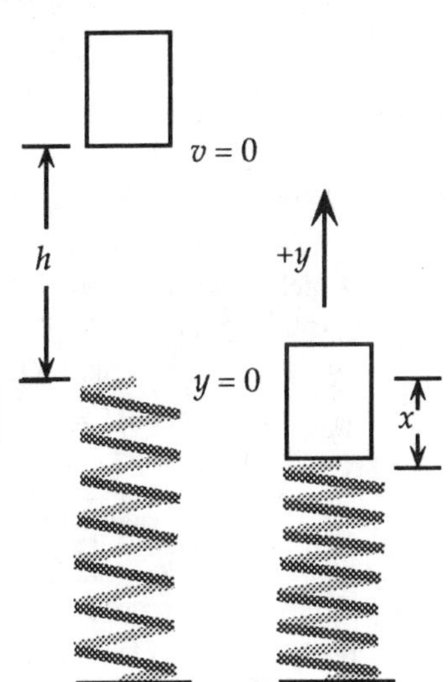

47. The maximum acceleration will occur at the maximum compression of the spring:

$$kx_{max} = Ma_{max} = M(5.0g), \text{ which gives } x_{max} = 5.0Mg/k.$$

For the motion from reaching the spring to the maximum compression of the spring, we use energy conservation:

$$KE_i + PE_{springi} = KE_f + PE_{springf};$$

$$\tfrac{1}{2}Mv^2 + 0 = 0 + \tfrac{1}{2}kx_{max}^2.$$

When we use the previous result, we get

$$\tfrac{1}{2}Mv^2 = \tfrac{1}{2}k(5.0Mg/k)^2, \text{ which gives}$$

$$k = 25Mg^2/v^2 = 25(1200\text{ kg})(9.80\text{ m/s}^2)^2/[(100\text{ km/h})/(3.6\text{ ks/h})]^2 = \boxed{3.7\times10^3\text{ N/m.}}$$

48. (*a*) The work done against gravity is the increase in the potential energy:

$W = mgh = (75.0 \text{ kg})(9.80 \text{ m/s}^2)(120 \text{ m}) = \boxed{8.82 \times 10^4 \text{ J.}}$

(*b*) If this work is done by the force on the pedals, we need to find the distance that the force acts over one revolution of the pedals and the number of revolutions to climb the hill. We find the number of revolutions from the distance along the incline:

$N = (h/\sin\theta)/(5.10 \text{ m/revolution})$

$= [(120 \text{ m})/\sin 7.50°]/(5.10 \text{ m/revolution}) = 180$ revolutions.

Because the force is always tangent to the circular path, in each revolution the force acts over a distance equal to the circumference of the path: πD. Thus we have

$W = NF\pi D$;

$8.82 \times 10^4 \text{ J} = (180 \text{ revolutions})F\pi(0.360 \text{ m})$, which gives $F = \boxed{433 \text{ N.}}$

49. The thermal energy is equal to the loss in kinetic energy:

$E_{\text{thermal}} = -\Delta KE = \frac{1}{2}mv_i^2 - \frac{1}{2}mv_f^2 = \frac{1}{2}(2)(6500 \text{ kg})[(95 \text{ km/h})/(3.6 \text{ ks/h})]^2 - 0 = \boxed{4.5 \times 10^6 \text{ J.}}$

50. We choose the bottom of the slide for the gravitational potential energy reference level. The thermal energy is the negative of the change in kinetic and potential energy:

$E_{\text{thermal}} = -(\Delta KE + \Delta PE) = \frac{1}{2}mv_i^2 - \frac{1}{2}mv_f^2 + mg(h_i - h_f)$

$= 0 - \frac{1}{2}(17 \text{ kg})(2.5 \text{ m/s})^2 + (17 \text{ kg})(9.80 \text{ m/s}^2)(3.5 \text{ m} - 0) = \boxed{5.3 \times 10^2 \text{ J.}}$

51. (*a*) We find the normal force from the force diagram for the ski:

y-component: $F_{N1} = mg\cos\theta$;

which gives the friction force: $F_{fr1} = \mu_k mg\cos\theta$.

For the work-energy principle, we have

$W_{NC} = \Delta KE + \Delta PE = (\frac{1}{2}mv_f^2 - \frac{1}{2}mv_i^2) + mg(h_f - h_i)$;

$-\mu_k mg\cos\theta\, L = (\frac{1}{2}mv_f^2 - 0) + mg(0 - L\sin\theta)$;

$-(0.090)(9.80 \text{ m/s}^2)\cos 20° (100 \text{ m}) =$

$\frac{1}{2}v_f^2 - (9.80 \text{ m/s}^2)(100 \text{ m})\sin 20°$,

which gives $v_f = \boxed{22 \text{ m/s.}}$

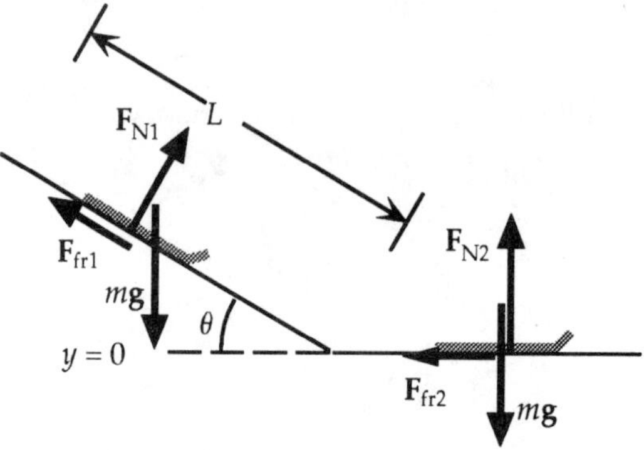

(*b*) On the level the normal force is $F_{N2} = mg$, so the friction force is $F_{fr2} = \mu_k mg$.

For the work-energy principle, we have

$W_{NC} = \Delta KE + \Delta PE = (\frac{1}{2}mv_f^2 - \frac{1}{2}mv_i^2) + mg(h_f - h_i)$;

$-\mu_k mg\, D = (0 - \frac{1}{2}mv_i^2) + mg(0 - 0)$;

$-(0.090)(9.80 \text{ m/s}^2)D = -\frac{1}{2}(22 \text{ m/s})^2$,

which gives $D = \boxed{2.9 \times 10^2 \text{ m.}}$

52. On the level the normal force is $F_N = mg$, so the friction force is $F_{fr} = \mu_k mg$.

For the work-energy principle, we have

$W_{NC} = \Delta KE + \Delta PE = (\frac{1}{2}mv_f^2 - \frac{1}{2}mv_i^2) + mg(h_f - h_i)$;

$F(L_1 + L_2) - \mu_k mg\, L_2 = (\frac{1}{2}mv_f^2 - 0) + mg(0 - 0)$;

$(350 \text{ N})(15 \text{ m} + 15 \text{ m}) - (0.30)(90 \text{ kg})(9.80 \text{ m/s}^2)(15 \text{ m}) = \frac{1}{2}(90 \text{ kg})v_f^2$,

which gives $v_f = \boxed{12 \text{ m/s.}}$

53. We choose $y = 0$ at point B. For the work-energy principle applied to the motion from A to B, we have

$$W_{NC} = \Delta KE + \Delta PE = (\tfrac{1}{2}mv_B^2 - \tfrac{1}{2}mv_A^2) + mg(h_B - h_A);$$
$$-0.2mgL = (\tfrac{1}{2}mv_B^2 - \tfrac{1}{2}mv_A^2) + mg(0 - h_A);$$
$$-0.2(9.80\text{ m/s}^2)(45.0\text{ m}) = \tfrac{1}{2}v_B^2 - \tfrac{1}{2}(1.70\text{ m/s})^2 - (9.80\text{ m/s}^2)(30\text{ m}),$$

which gives $v_B =$ **20 m/s.**

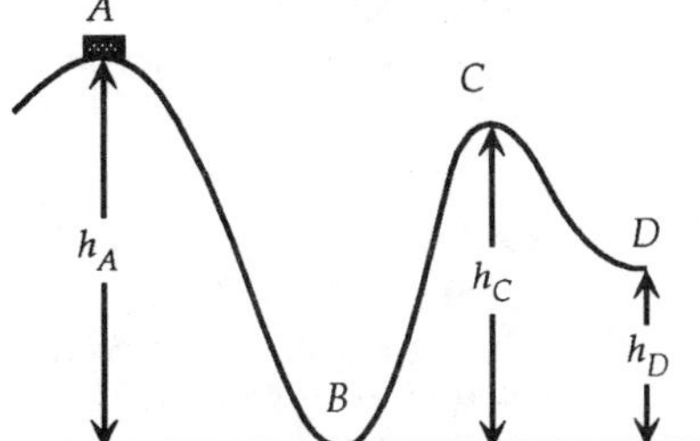

54. We find the normal force from the force diagram for the skier:

y-component: $F_N = mg\cos\theta$;

which gives the friction force: $F_{fr} = \mu_k mg\cos\theta$.

For the work-energy principle for the motion up the incline, we have

$$W_{NC} = \Delta KE + \Delta PE = (\tfrac{1}{2}mv_f^2 - \tfrac{1}{2}mv_i^2) + mg(h_f - h_i);$$
$$-\mu_k mg\cos\theta\, L = (0 - \tfrac{1}{2}mv_i^2) + mg(L\sin\theta - 0);$$
$$-\mu_k(9.80\text{ m/s}^2)\cos 18^\circ\,(12.2\text{ m}) = -\tfrac{1}{2}(12.0\text{ m/s})^2 + (9.80\text{ m/s}^2)(12.2\text{ m})\sin 18^\circ,$$

which gives $\mu_k =$ **0.31.**

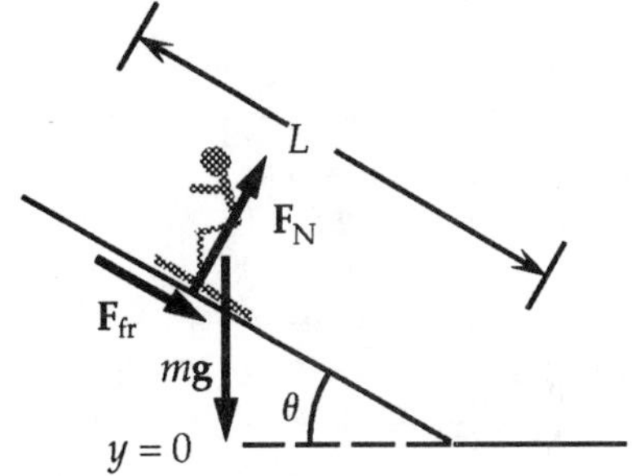

55. On the level the normal force is $F_N = mg$, so the friction force is $F_{fr} = \mu_k mg$.
The block is at rest at the release point and where it momentarily stops before turning back.
For the work-energy principle, we have

$$W_{NC} = \Delta KE + \Delta PE = (\tfrac{1}{2}mv_f^2 - \tfrac{1}{2}mv_i^2) + (\tfrac{1}{2}kx_f^2 - \tfrac{1}{2}kx_i^2);$$
$$-\mu_k mg\, L = (0 - 0) + \tfrac{1}{2}k(x_f^2 - x_i^2);$$
$$-\mu_k(0.520\text{ kg})(9.80\text{ m/s}^2)(0.050\text{ m} + 0.023\text{ m}) = \tfrac{1}{2}(180\text{ N/m})[(0.023\text{ m})^2 - (-0.050\text{ m})^2],$$

which gives $\mu_k =$ **0.48.**

56. We find the spring constant from the force required to compress the spring:

$$k = F/x_i = (-20\text{ N})/(-0.18\text{ m}) = 111\text{ N/m}.$$

On the level the normal force is $F_N = mg$, so the friction force is $F_{fr} = \mu_k mg$.
The block is at rest at the release point and where it momentarily stops before turning back.
For the work-energy principle, we have

$$W_{NC} = \Delta KE + \Delta PE = (\tfrac{1}{2}mv_f^2 - \tfrac{1}{2}mv_i^2) + (\tfrac{1}{2}kx_f^2 - \tfrac{1}{2}kx_i^2);$$
$$-\mu_k mg\, L = (0 - 0) + \tfrac{1}{2}k(x_f^2 - x_i^2);$$
$$-(0.30)(0.180\text{ kg})(9.80\text{ m/s}^2)(0.18\text{ m} + x_f) = \tfrac{1}{2}(111\text{ N/m})[x_f^2 - (-0.18\text{ m})^2].$$

This reduces to the quadratic equation

$$55.5x_f^2 + 0.529x_f - 1.70 = 0,$$ which has the solutions $x_f = 0.17$ m, -0.18 m.

The negative solution corresponds to no motion, so the physical result is $x_f =$ **0.17 m.**

57. We choose the potential energy to be zero at the ground ($y = 0$).
We convert the speeds: (500 km/h)/(3.6 ks/h) = 139 m/s; (200 km/h)/(3.6 ks/h) = 55.6 m/s.
(*a*) If there were no air resistance, energy would be conserved:
$$0 = \Delta\text{KE} + \Delta\text{PE} = (\tfrac{1}{2}mv_f^2 - \tfrac{1}{2}mv_i^2) + mg(h_f - h_i);$$
$$0 = \tfrac{1}{2}(1000\text{ kg})[(v_f^2 - (139\text{ m/s})^2] + (1000\text{ kg})(9.80\text{ m/s}^2)(0 - 3500\text{ m}),$$
which gives v_f = 297 m/s = **1.07 × 10³ km/h.**
(*b*) With air resistance we have
$$W_{NC} = \Delta\text{KE} + \Delta\text{PE} = (\tfrac{1}{2}mv_f^2 - \tfrac{1}{2}mv_i^2) + mg(h_f - h_i);$$
$$-F(h_i / \sin\theta) = \tfrac{1}{2}m(v_f^2 - v_i^2) + mg(0 - h_i);$$
$$-F(3500\text{ m})/\sin 10° = \tfrac{1}{2}(1000\text{ kg})[(55.6\text{ m/s})^2 - (139\text{ m/s})^2] + (1000\text{ kg})(9.80\text{ m/s}^2)(-3500\text{ m})$$
which gives F = **2.1 × 10³ N.**

58. The amount of work required is the increase in potential energy: $W = mg\,\Delta y$.
We find the time from
$$P = W/t = mg\,\Delta y/t;$$
1750 W = (285 kg)(9.80 m/s²)(16.0 m)/t, which gives t = **25.5 s.**

59. We find the equivalent force exerted by the engine from
$$P = Fv;$$
(18 hp)(746 W/hp) = F(90 km/h)/(3.6 ks/h), which gives F = 5.4 × 10² N.
At constant speed, this force is balanced by the average retarding force, which must be **5.4 × 10² N.**

60. (*a*) 1 hp = (550 ft·lb/s)(4.45 N/lb)/(3.281 ft/m) = 746 W.
(*b*) P = (100 W)/(746 W/hp) = **0.134 hp.**

61. (*a*) 1 kWh = (1000 Wh)(3600 s/h) = 3.6 × 10⁶ J.
(*b*) $W = Pt$ = (500 W)(1 kW/1000W)(1 mo)(30 day/mo)(24 h/day) = **360 kWh.**
(*c*) W = (360 kWh)(3.6 × 10⁶ J/kWh) = **1.3 × 10⁹ J.**
(*d*) Cost = W × rate = (360 kWh)($0.12/kWh) = **$43.20.**
The charge is for the amount of energy used, and thus is **independent of rate.**

62. We find the average resistance force from the acceleration:
$$R = ma = m\,\Delta v/\Delta t = (1000\text{ kg})(70\text{ km/h} - 90\text{ km/h})/(3.6\text{ ks/h})(6.0\text{ s}) = -933\text{ N}.$$
If we assume that this is the resistance force at 80 km/h, the engine must provide an equal and opposite force to maintain a constant speed. We find the power required from
$P = Fv$ = (933 N)(80 km/h)/(3.6 ks/h) = **2.1 × 10⁴ W** = (2.1 × 10⁴ W)/(746 W/hp) = **28 hp.**

63. We find the work from
$W = Pt$ = (3.0 hp)(746 W/hp)(1 h)(3600 s/h) = **8.1 × 10⁶ J.**

64. The work done by the shot-putter increases the kinetic energy of the shot. We find the power from
$$P = W/t = \Delta\text{KE}/t = (\tfrac{1}{2}mv_f^2 - \tfrac{1}{2}mv_i^2)/t$$
$= \tfrac{1}{2}$(7.3 kg)[(14 m/s)² – 0]/(2.0 s) = **3.6 × 10² W** (about 0.5 hp).

65. The work done by the pump increases the potential energy of the water. We find the power from
$$P = W/t = \Delta\text{PE}/t = mg(h_f - h_i)/t = (m/t)g(h_f - h_i)$$
= [(8.00 kg/min)/(60 s/min)](9.80 m/s²)(3.50 m – 0) = **4.57 W.**

66. The work done increases the potential energy of the player. We find the power from

$P = W/t = \Delta\text{PE}/t = mg(h_f - h_i)/t$

$= (105 \text{ kg})(9.80 \text{ m/s}^2)[(140 \text{ m}) \sin 30° - 0]/(61 \text{ s}) =$ $\boxed{1.2 \times 10^3 \text{ W}}$ (about 1.6 hp).

67. The work done increases the potential energy of the player. We find the speed from

$P = W/t = \Delta\text{PE}/t = mg(h_f - h_i)/t = mg(L \sin \theta - 0)/t = mgv \sin \theta$

$(0.25 \text{ hp})(746 \text{ W/hp}) = (70 \text{ kg})(9.80 \text{ m/s}^2)v \sin 6.0°$, which gives $v =$ $\boxed{2.6 \text{ m/s.}}$

68. From the force diagram for the car, we have:

x-component: $F - F_{fr} = mg \sin \theta.$

Because the power output is $P = Fv$, we have

$(P/v) - F_{fr} = mg \sin \theta.$

The maximum power determines the maximum angle:

$(P_{max}/v) - F_{fr} = mg \sin \theta_{max};$

$(120 \text{ hp})(746 \text{ W/hp})/[(70 \text{ km/h})/(3/6 \text{ ks/h})] - 600 \text{ N} =$

$(1000 \text{ kg})(9.80 \text{ m/s}^2) \sin \theta_{max},$

which gives $\sin \theta_{max} = 0.409$, or $\boxed{\theta_{max} = 24°.}$

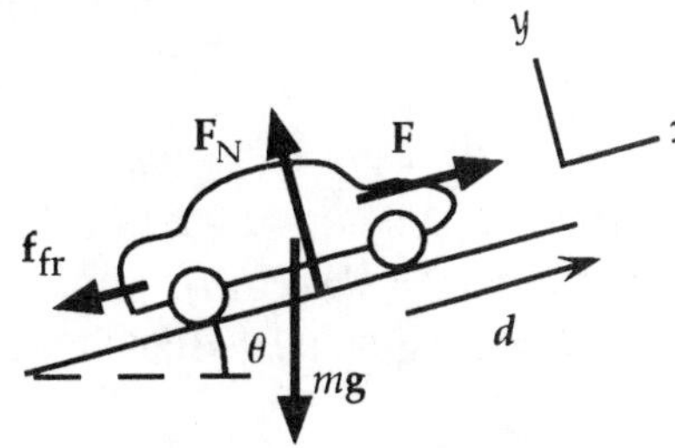

69. The work done by the lifts increases the potential energy of the people. We assume an average mass of 70 kg and find the power from

$P = W/t = \Delta\text{PE}/t = mg(h_f - h_i)/t = (m/t)g(h_f - h_i)$

$= [(47{,}000 \text{ people/h})(70 \text{ kg/person})/(3600 \text{ s/h})](9.80 \text{ m/s}^2)(200 \text{ m} - 0)$

$=$ $\boxed{1.8 \times 10^6 \text{ W}}$ (about 2.4×10^3 hp).

70. For the work-energy principle applied to coasting down the hill a distance L, we have

$W_{NC} = \Delta\text{KE} + \Delta\text{PE} = (\frac{1}{2}mv_f^2 - \frac{1}{2}mv_i^2) + mg(h_f - h_i);$

$-F_{fr}L = (\frac{1}{2}mv^2 - \frac{1}{2}mv^2) + mg(0 - L \sin \theta)$, which gives $F_{fr} = mg \sin \theta.$

Because the climb is at the same speed, we assume the resisting force is the same.

For the work-energy principle applied to climbing the hill a distance L, we have

$W_{NC} = \Delta\text{KE} + \Delta\text{PE} = (\frac{1}{2}mv_f^2 - \frac{1}{2}mv_i^2) + mg(h_f - h_i);$

$FL - F_{fr}L = (\frac{1}{2}mv^2 - \frac{1}{2}mv^2) + mg(0 - L \sin \theta);$

$(P/v) - mg \sin \theta = mg \sin \theta$, which gives

$P = 2mgv \sin \theta = 2(75 \text{ kg})(9.80 \text{ m/s}^2)(5.0 \text{ m/s}) \sin 7.0° =$ $\boxed{9.0 \times 10^2 \text{ W}}$ (about 1.2 hp).

71. (*a*) If we ignore the small change in potential energy when the snow brings the paratrooper to rest, the work done decreases the kinetic energy:

$W = \Delta\text{KE} = \frac{1}{2}mv_f^2 - \frac{1}{2}mv_i^2$

$= \frac{1}{2}(80 \text{ kg})[0 - (30 \text{ m/s})^2] =$ $\boxed{-3.6 \times 10^4 \text{ J.}}$

(*b*) We find the average force from

$F = W/d = (-3.6 \times 10^4 \text{ J})/(1.1 \text{ m}) =$ $\boxed{-3.3 \times 10^3 \text{ N.}}$

(*c*) With air resistance during the fall we have

$W_{NC} = \Delta\text{KE} + \Delta\text{PE} = (\frac{1}{2}mv_f^2 - \frac{1}{2}mv_i^2) + mg(h_f - h_i)$

$= \frac{1}{2}(80 \text{ kg})[(30 \text{ m/s})^2 - 0] + (80 \text{ kg})(9.80 \text{ m/s}^2)(0 - 370 \text{ m}) =$ $\boxed{-2.5 \times 10^5 \text{ J.}}$

72. For the motion during the impact until the car comes momentarily to rest, we use energy conservation:

$KE_i + PE_{springi} = KE_f + PE_{springf}$;

$\frac{1}{2}mv_i^2 + 0 = 0 + \frac{1}{2}kx^2$;

$(1400 \text{ kg kg})[(8 \text{ km/h})/(3/6 \text{ ks/h})]^2 = k(0.015 \text{ m})^2$, which gives $k = \boxed{3 \times 10^7 \text{ N/m.}}$

73. We let N represent the number of books of mass m that can be placed on a shelf. For each book the work increases the potential energy and thus is mg times the distance the center is raised. From the diagram we see that the work required to fill the nth shelf is

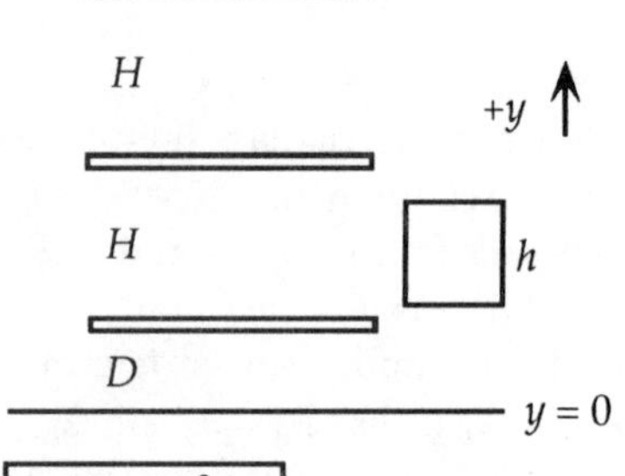

$W_n = Nmg[D + \frac{1}{2}h + (n - 1)H]$.

Thus for the five shelves, we have

$W = W_1 + W_2 + W_3 + W_4 + W_5$

$= Nmg[5(D + \frac{1}{2}h) + H + 2H + 3H + 4H]$

$= Nmg[5(D + \frac{1}{2}h) + 10H]$

$= (25)(1.5 \text{ kg})(9.80 \text{ m/s}^2)\{5[0.100 \text{ m} + \frac{1}{2}(0.20 \text{ m})] + 10(0.300 \text{ m})\} = \boxed{1.5 \times 10^3 \text{ J.}}$

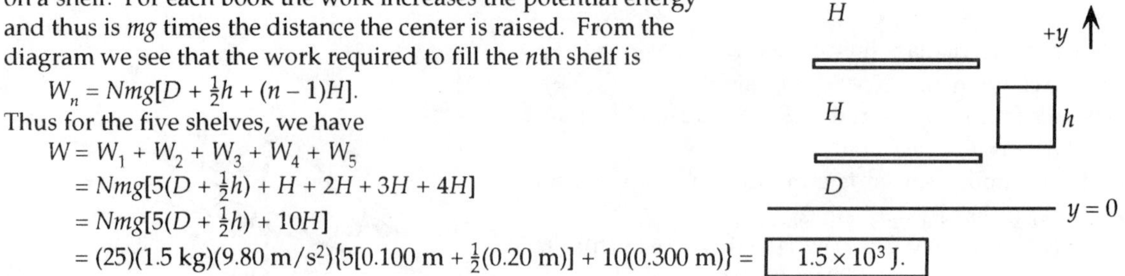

74. We choose the potential energy to be zero at the ground ($y = 0$). We find the minimum speed by ignoring any frictional forces. Energy is conserved, so we have

$E = KE_i + PE_i = KE_f + PE_f$;

$\frac{1}{2}mv_i^2 + mgy_i = \frac{1}{2}mv_f^2 + mgy_f$;

$\frac{1}{2}mv_i^2 + m(9.80 \text{ m/s}^2)(0) = \frac{1}{2}m(6.5 \text{ m/s})^2 + m(9.80 \text{ m/s}^2)(1.1 \text{ m})$, which gives $v_i = \boxed{8.0 \text{ m/s.}}$

Note that the initial velocity will not be horizontal, but will have a horizontal component of 6.5 m/s.

75. We choose the reference level for the gravitational potential energy at the ground.

(*a*) With no air resistance during the fall we have

$0 = \Delta KE + \Delta PE = (\frac{1}{2}mv_f^2 - \frac{1}{2}mv_i^2) + mg(h_f - h_i)$, or

$\frac{1}{2}(v_f^2 - 0) = -(9.80 \text{ m/s}^2)(0 - 18 \text{ m})$, which gives $v_f = \boxed{19 \text{ m/s.}}$

(*b*) With air resistance during the fall we have

$W_{NC} = \Delta KE + \Delta PE = (\frac{1}{2}mv_f^2 - \frac{1}{2}mv_i^2) + mg(h_f - h_i)$;

$F_{air}(18 \text{ m}) = \frac{1}{2}(0.20 \text{ kg})[(10.0 \text{ m/s})^2 - 0] + (0.20 \text{ kg})(9.80 \text{ m/s}^2)(0 - 18 \text{ m})$,

which gives $F_{air} = \boxed{-1.4 \text{ N.}}$

76. We choose the reference level for the gravitational potential energy at the lowest point. The tension in the cord is always perpendicular to the displacement and thus does no work.

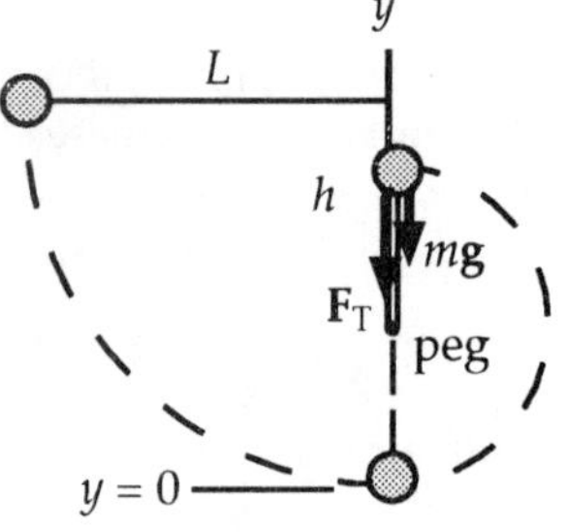

(*a*) With no air resistance during the fall, we have

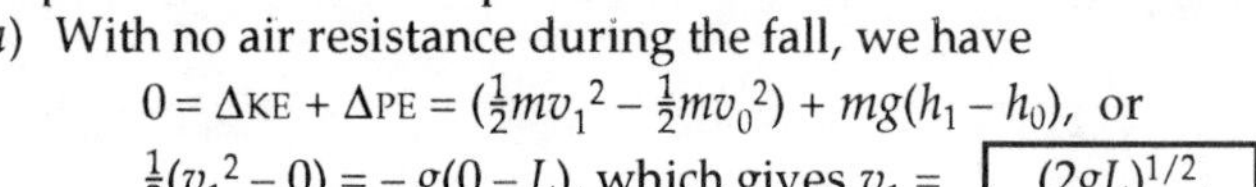

$0 = \Delta KE + \Delta PE = (\frac{1}{2}mv_1^2 - \frac{1}{2}mv_0^2) + mg(h_1 - h_0)$, or

$\frac{1}{2}(v_1^2 - 0) = -g(0 - L)$, which gives $v_1 = \boxed{(2gL)^{1/2}.}$

(*b*) For the motion from release to the rise around the peg, we have

$0 = \Delta KE + \Delta PE = (\frac{1}{2}mv_2^2 - \frac{1}{2}mv_0^2) + mg(h_2 - h_0)$, or

$\frac{1}{2}(v_2^2 - 0) = -g[2(L - h) - L] = g(2h - L) = 0.60gL$,

which gives $v_2 = \boxed{(1.2gL)^{1/2}.}$

77. (*a*) The work done against gravity is the increase in the potential energy:

$W = \Delta \text{PE} = mg(h_f - h_i) = (65\text{ kg})(9.80\text{ m/s}^2)(3900\text{ m} - 2200\text{ m}) =$ $\boxed{1.1 \times 10^6\text{ J.}}$

(*b*) We find the power from

$P = W/t = (1.1 \times 10^6\text{ J})/(5.0\text{ h})(3600\text{ s/h}) =$ $\boxed{60\text{ W} = 0.081\text{ hp.}}$

(*c*) We find the power input from

$P_{\text{input}} = P/\text{efficiency} = (60\text{ W})/(0.15) =$ $\boxed{4.0 \times 10^2\text{ W} = 0.54\text{ hp.}}$

78. The potential energy is zero at $x = 0$.

(*a*) Because energy is conserved, the maximum speed occurs at the minimum potential energy:

$\text{KE}_i + \text{PE}_i = \text{KE}_f + \text{PE}_f;$

$\frac{1}{2}mv_0^2 + \frac{1}{2}kx_0^2 = \frac{1}{2}mv_{\max}^2 + 0$, which gives $\boxed{v_{\max} = [v_0^2 + (kx_0^2/m)]^{1/2}.}$

(*b*) The maximum stretch occurs at the minimum kinetic energy:

$\text{KE}_i + \text{PE}_i = \text{KE}_f + \text{PE}_f;$

$\frac{1}{2}mv_0^2 + \frac{1}{2}kx_0^2 = 0 + \frac{1}{2}kx_{\max}^2$, which gives $\boxed{x_{\max} = [x_0^2 + (mv_0^2/k)]^{1/2}.}$

79. (*a*) With $y = 0$ at the bottom of the circle, we call the start point A, the bottom of the circle B, and the top of the circle C. At the top of the circle we have the forces mg and F_{Ntop}, both downward, that provide the centripetal acceleration:

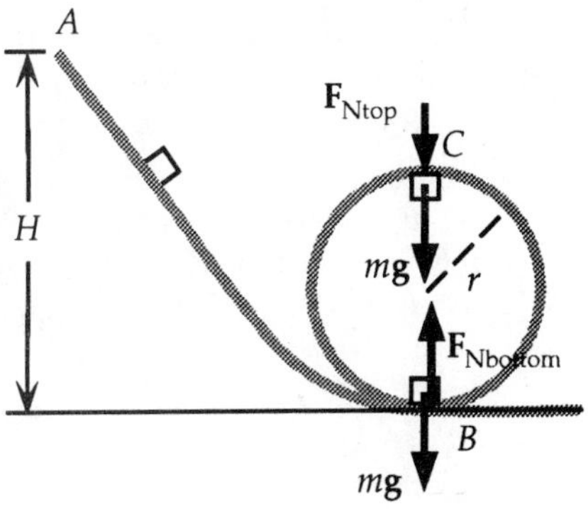

$mg + F_{\text{Ntop}} = mv_C^2/r.$

The minimum value of F_{Ntop} is zero, so the minimum speed at C is found from

$v_{C\min}^2 = gr.$

From energy conservation for the motion from A to C we have

$\text{KE}_A + \text{PE}_A = \text{KE}_C + \text{PE}_C;$

$0 + mgh = \frac{1}{2}mv_C^2 + mg(2r),$

thus the minimum height is found from

$gh = \frac{1}{2}v_{C\min}^2 + 2gr = \frac{1}{2}gr + 2gr$, which gives $\boxed{h = 2.5r.}$

(*b*) From energy conservation for the motion from A to B we have

$\text{KE}_A + \text{PE}_A = \text{KE}_B + \text{PE}_B;$

$0 + mg2h = 5mgr = \frac{1}{2}mv_B^2 + 0$, which gives $v_B^2 = 10gr$.

At the bottom of the circle we have the forces mg down and F_{Nbottom} up that provide the centripetal acceleration:

$-mg + F_{\text{Nbottom}} = mv_B^2/r.$

If we use the previous result, we get

$F_{\text{Nbottom}} = mv_B^2/r + mg =$ $\boxed{11mg.}$

(*c*) From energy conservation for the motion from A to C we have

$\text{KE}_A + \text{PE}_A = \text{KE}_C + \text{PE}_C;$

$0 + mg2h = 5mgr = \frac{1}{2}mv_C^2 + mg(2r)$, which gives $v_C^2 = 6gr$.

At the top of the circle we have the forces mg and F_{Ntop}, both down, that provide the centripetal acceleration:

$mg + F_{\text{Ntop}} = mv_C^2/r.$

If we use the previous result, we get

$F_{\text{Ntop}} = mv_C^2/r - mg =$ $\boxed{5mg.}$

(*d*) On the horizontal section we have $F_N =$ $\boxed{mg.}$

80. (*a*) The work done by gravity is the decrease in the potential energy:

$W_{grav} = -\Delta\text{PE} = -mg(h_f - h_i) = -(900\text{ kg})(9.80\text{ m/s}^2)(0 - 30\text{ m})$

$= \boxed{2.6\times10^5\text{ J.}}$

(*b*) The work done by gravity increases the kinetic energy:

$W_{grav} = \Delta\text{KE};$

$2.6\times10^5\text{ J} = \frac{1}{2}(900\text{ kg})v^2 - 0$, which gives $\boxed{v = 24\text{ m/s.}}$

(*c*) For the motion from the break point to the maximum compression of the spring, we use energy conservation:

$\text{KE}_i + \text{PE}_{gravi} + \text{PE}_{springi} = \text{KE}_f + \text{PE}_{gravf} + \text{PE}_{springf};$

$0 + mgh + 0 = 0 + mg(-x_{max}) + \frac{1}{2}kx_{max}^2;$

$(900\text{ kg})(9.80\text{ m/s}^2)(30\text{ m}) =$

$-(900\text{ kg})(9.80\text{ m/s}^2)x_{max} + \frac{1}{2}(4.0\times10^5\text{ N/m})x_{max}^2.$

This is a quadratic equation for x_{max}, which has the solutions

$x_{max} = -1.13\text{ m}, 1.17\text{ m}.$

Because x_{max} must be positive, the spring compresses $\boxed{1.2\text{ m.}}$

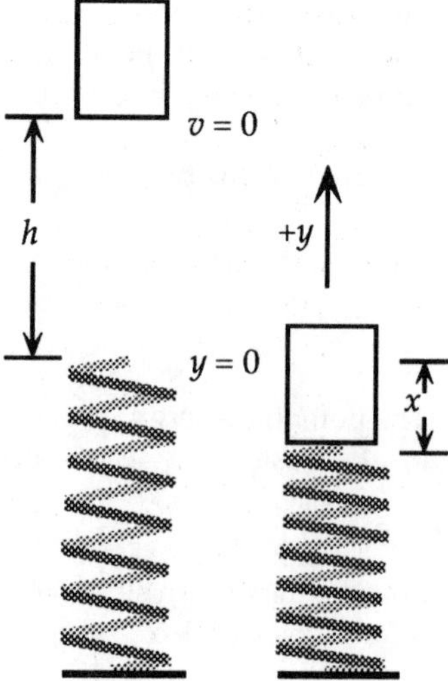

81. We choose the reference level for the gravitational potential energy at the lowest point.

(*a*) With no air resistance during the fall, we have

$0 = \Delta\text{KE} + \Delta\text{PE} = (\frac{1}{2}mv^2 - \frac{1}{2}mv_0^2) + mg(h - h_0)$, or

$\frac{1}{2}(v^2 - 0) = -g(0 - H)$, which gives

$v_1 = (2gH)^{1/2} = [2(9.80\text{ m/s}^2)(80\text{ m})] = \boxed{40\text{ m/s.}}$

(*b*) If 60% of the kinetic energy of the water is transferred, we have

$P = (0.60)\frac{1}{2}mv^2/t = (0.60)\frac{1}{2}(m/t)v^2$

$= (0.60)\frac{1}{2}(550\text{ kg/s})(40\text{ m/s})^2) = \boxed{2.6\times10^5\text{ W}}$ (about 3.5×10^2 hp).

82. We convert the speeds: (10 km/h)/(3.6 ks/h) = 2.78 m/s; (30 km/h)/(3.6 ks/h) = 8.33 m/s.

We use the work-energy principle applied to coasting down the hill a distance L to find b:

$W_{NC} = \Delta\text{KE} + \Delta\text{PE} = (\frac{1}{2}mv_f^2 - \frac{1}{2}mv_i^2) + mg(h_f - h_i);$

$-bv_1^2L = (\frac{1}{2}mv_1^2 - \frac{1}{2}mv_1^2) + mg(0 - L\sin\theta),$

which gives $b = (mg/v^2)\sin\theta = [(75\text{ kg})(9.80\text{ m/s}^2)/(2.78\text{ m/s})^2]\sin 4.0° = 6.63\text{ kg/m}.$

For the work-energy principle applied to speeding down the hill a distance L, the cyclist must provide a force so we have

$W_{NC} = \Delta\text{KE} + \Delta\text{PE} = (\frac{1}{2}mv_f^2 - \frac{1}{2}mv_i^2) + mg(h_f - h_i);$

$F_2L - bv_2^2L = (\frac{1}{2}mv_2^2 - \frac{1}{2}mv_2^2) + mg(0 - L\sin\theta)$, which gives $F_2 = bv_2^2 - mg\sin\theta.$

The power supplied by the cyclist is

$P = F_2v_2 = [(6.63\text{ kg/m})(8.33\text{ m/s})^2 - (75\text{ kg})(9.80\text{ m/s}^2)\sin 4.0°](8.33\text{ m/s}) = 3.41\times10^3\text{ W}.$

For the work-energy principle applied to climbing the hill a distance L, the cyclist will provide a force $F_3 = P/v_3$, so we have

$W_{NC} = \Delta\text{KE} + \Delta\text{PE} = (\frac{1}{2}mv_f^2 - \frac{1}{2}mv_i^2) + mg(h_f - h_i);$

$(P/v_3)L - bv_3^2L = (\frac{1}{2}mv_3^2 - \frac{1}{2}mv_3^2) + mg(L\sin\theta - 0),$

which gives $[(3.41\times10^3\text{ W})/v_3] - (6.63\text{ kg/m})v_3^2 = (75\text{ kg})(9.80\text{ m/s}^2)\sin 4.0°.$

This is a cubic equation for v_3, which has one real solution: $v_3 = 5.54\text{ m/s}.$

The speed is (5.54 m/s)(3.6 ks/h) = $\boxed{20\text{ km/h.}}$

83. We choose the reference level for the gravitational potential energy at the bottom. From energy conservation for the motion from top to bottom, we have

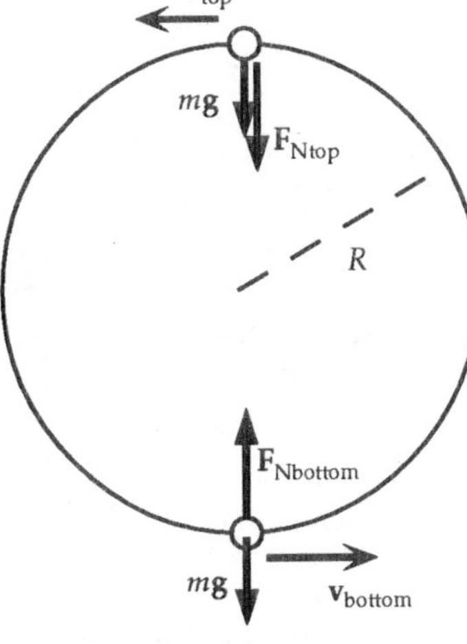

$KE_{top} + PE_{top} = KE_{bottom} + PE_{bottom}$;

$\frac{1}{2}mv_{top}^2 + mg2r = \frac{1}{2}mv_{bottom}^2 + 0$, which gives

$v_{bottom}^2 = v_{top}^2 + 4gr$.

At the bottom of the circle we have the forces mg down and $F_{Nbottom}$ up that provide the centripetal acceleration:

$-mg + F_{Nbottom} = mv_{bottom}^2/r$, which gives

$F_{Nbottom} = (mv_{bottom}^2/r) + mg$.

At the top of the circle we have the forces mg and F_{Ntop}, both down, that provide the centripetal acceleration:

$mg + F_{Ntop} = mv_{top}^2/r$, which gives

$F_{Ntop} = (mv_{top}^2/r) - mg$.

If we subtract the two equations, we get

$F_{Nbottom} - F_{Ntop} = (mv_{bottom}^2/r) + mg - [(mv_{top}^2/r) - mg]$

$= (m/r)(v_{bottom}^2 - v_{top}^2) + 2mg = 4mg + 2mg = 6mg$.

The speed must be above the minimum at the top so the roller coaster does not leave the track. From Problem 44, we know that we must have $h > 2.5r$.

The result we found does not depend on the radius or speed.

84. We choose $y = 0$ at the scale. We find the spring constant from the force (your weight) required to compress the spring:

$k = F_1/x_1 = (-700\text{ N})/(-0.50\times10^{-3}\text{ m}) = 1.4\times10^6\text{ N/m}$.

We apply conservation of energy for the jump to the scale. If we ignore the small change in gravitational potential energy when the scale compresses, we have

$KE_i + PE_i = KE_f + PE_f$;

$0 + mgH = 0 + \frac{1}{2}kx_2^2$;

$(700\text{ N})(1.0\text{ m}) = \frac{1}{2}(1.4\times10^6\text{ N/m})x_2^2$, which gives $x_2 = 0.032$ m.

The reading of the scale is

$F_2 = kx_2 = (1.40\times10^6\text{ N/m})(0.032\text{ m}) =$ **4.4×10^4 N.**

85. We choose the potential energy to be zero at the lowest point ($y = 0$).

(*a*) Because the tension in the vine does no work, energy is conserved, so we have

$KE_i + PE_i = KE_f + PE_f$;

$\frac{1}{2}mv_0^2 + 0 = 0 + mgh = mg(L - L\cos\theta) = mgL(1 - \cos\theta)$;

$\frac{1}{2}m(5.0\text{ m/s})^2 = m(9.80\text{ m/s}^2)(10.0\text{ m})(1 - \cos\theta)$

which gives $\cos\theta = 0.872$, or $\theta =$ **29°.**

(*b*) The velocity is zero just before he releases, so there is no centripetal acceleration. There is a tangential acceleration which has been decreasing his tangential velocity. For the radial direction we have

$F_T - mg\cos\theta = 0$; or

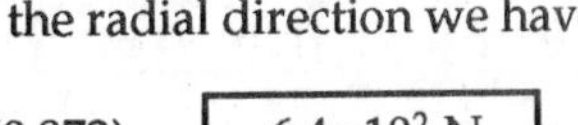
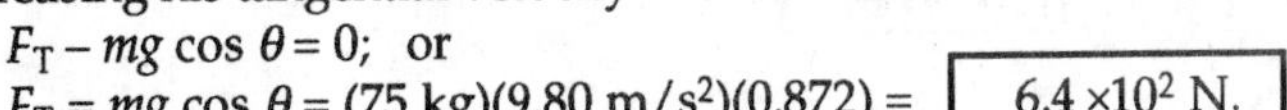

$F_T = mg\cos\theta = (75\text{ kg})(9.80\text{ m/s}^2)(0.872) =$ **6.4×10^2 N.**

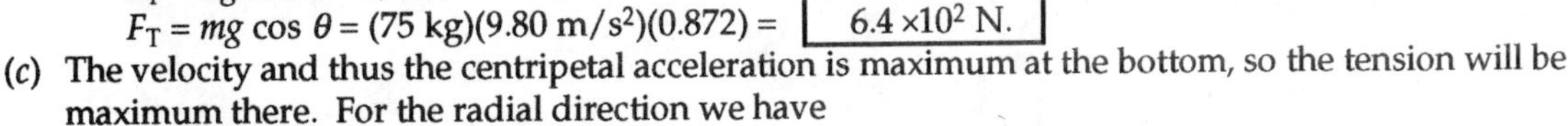

(*c*) The velocity and thus the centripetal acceleration is maximum at the bottom, so the tension will be maximum there. For the radial direction we have

$F_T - mg = mv_0^2/L$, or

$F_T = mg + mv_0^2/L = (75\text{ kg})[(9.80\text{ m/s}^2) + (5.0\text{ m/s})^2/(10.0\text{ m})] =$ **9.2×10^2 N.**

86. We choose the potential energy to be zero at the floor. The work done increases the potential energy of the athlete. We find the power from

$P = W/t = \Delta PE/t = mg(h_f - h_i)/t$

$= (70\text{ kg})(9.80\text{ m/s}^2)(5.0\text{ m} - 0)/(9.0\text{ s}) =$ **3.8×10^2 W** (about 0.5 hp).

CHAPTER 7

1. $p = mv = (0.022 \text{ kg})(8.1 \text{ m/s}) = \boxed{0.18 \text{ kg}\cdot\text{m/s.}}$

2. During the throwing we use momentum conservation for the one-dimensional motion:
   $0 = (m_{\text{boat}} + m_{\text{child}})v_{\text{boat}} + m_{\text{package}}v_{\text{package}}$;
   $0 = (55.0 \text{ kg} + 20.0 \text{ kg})v_{\text{boat}} + (5.40 \text{ kg})(10.0 \text{ m/s})$, which gives
   $v_{\text{boat}} = \boxed{-0.720 \text{ m/s (opposite to the direction of the package).}}$

3. We find the force on the expelled gases from
   $F = \Delta p/\Delta t = (\Delta m/\Delta t)v = (1300 \text{ kg/s})(40{,}000 \text{ m/s}) = 5.2 \times 10^7 \text{ N}.$
   An equal, but opposite, force will be exerted on the rocket: $\boxed{5.2 \times 10^7 \text{ N, up.}}$

4. For this one-dimensional motion, we take the direction of the halfback for the positive direction. For this perfectly inelastic collision, we use momentum conservation:
   $M_1v_1 + M_2v_2 = (M_1 + M_2)V$;
   $(95 \text{ kg})(4.1 \text{ m/s}) + (85 \text{ kg})(5.5 \text{ m/s}) = (95 \text{ kg} + 85 \text{ kg})V$, which gives $\boxed{V = 4.8 \text{ m/s.}}$

5. For the horizontal motion, we take the direction of the car for the positive direction. The load initially has no horizontal velocity. For this perfectly inelastic collision, we use momentum conservation:
   $M_1v_1 + M_2v_2 = (M_1 + M_2)V$;
   $(12{,}500 \text{ kg})(18.0 \text{ m/s}) + 0 = (12{,}500 \text{ kg} + 5750 \text{ kg})V$, which gives $\boxed{V = 12.3 \text{ m/s.}}$

6. For the one-dimensional motion, we take the direction of the first car for the positive direction. For this perfectly inelastic collision, we use momentum conservation:
   $M_1v_1 + M_2v_2 = (M_1 + M_2)V$;
   $(9500 \text{ kg})(16 \text{ m/s}) + 0 = (9500 \text{ kg} + M_2)(6.0 \text{ m/s})$, which gives $\boxed{M_2 = 1.6 \times 10^4 \text{ kg.}}$

7. We let V be the speed of the block and bullet immediately after the embedding and before the two start to rise.
   For this perfectly inelastic collision, we use momentum conservation:
   $mv + 0 = (M + m)V$;
   $(0.021 \text{ kg})(210 \text{ m/s}) = (0.021 \text{ kg} + 1.40 \text{ kg})V$, which gives $V = 3.10 \text{ m/s}$.
   For the rising motion we use energy conservation, with the potential reference level at the ground:
   $\text{KE}_i + \text{PE}_i = \text{KE}_f + \text{PE}_f$;
   $\frac{1}{2}(M + m)V^2 + 0 = 0 + (m + M)gh$, or
   $h = V^2/2g = (3.10 \text{ m/s})^2/2(9.80 \text{ m/s}^2) = \boxed{0.491 \text{ m.}}$

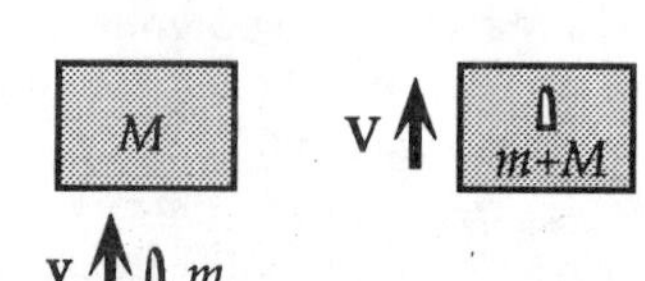

8. On the horizontal surface after the collision, the normal force is $F_N = (m + M)g$.
We find the common speed of the block and bullet immediately after the embedding by using the work-energy principle for the sliding motion:

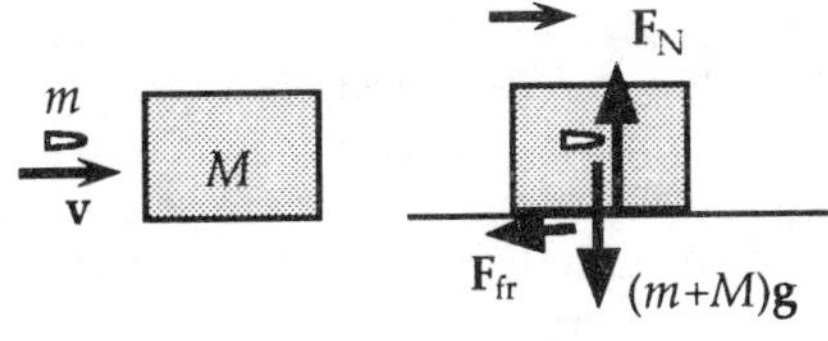

$W_{fr} = \Delta KE$;

$-\mu_k(m + M)gd = 0 - \frac{1}{2}(M + m)V^2$;

$0.25(9.80\ m/s^2)(9.5\ m) = \frac{1}{2}V^2$, which gives $V = 6.82\ m/s$.

For the collision, we use momentum conservation:

$mv + 0 = (M + m)V$;

$(0.015\ kg)v = (0.015\ kg + 1.10\ kg)(6.82\ m/s)$, which gives $v = 5.1 \times 10^2\ m/s.$

9. The new nucleus and the alpha particle will recoil in opposite directions.
Momentum conservation gives us

$0 = MV - m_\alpha v_\alpha$,

$0 = (57m_\alpha)V - m_\alpha(3.8 \times 10^5\ m/s)$, which gives $V = 6.7 \times 10^3\ m/s.$

10. Because mass is conserved, the mass of the new nucleus is $M_2 = 222\ u - 4.0\ u = 218\ u$.
Momentum conservation gives us

$M_1V_1 = M_2V_2 + m_\alpha v_\alpha$,

$(222\ u)(420\ m/s) = (218\ u)(350\ m/s) + (4.0\ u)v_\alpha$, which gives $v_\alpha = 4.2 \times 10^3\ m/s.$

11. Momentum conservation gives us

$mv_1 + Mv_2 = mv_1' + Mv_2'$,

$(0.013\ kg)(230\ m/s) + 0 = (0.013\ kg)(170\ m/s) + (2.0\ kg)v_2'$,

which gives $v_2' = 0.39\ m/s.$

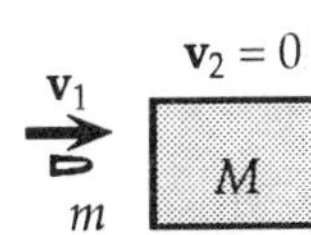

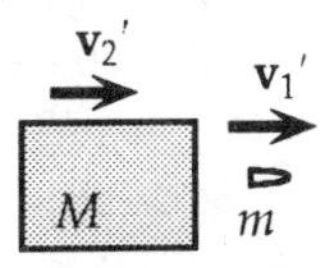

12. (*a*) With respect to the Earth after the explosion, one section will have a speed v_1' and the other will have a speed $v_2' = v_1' + v_{relative}$. Momentum conservation gives us

$mv = \frac{1}{2}mv_1' + \frac{1}{2}mv_2'$, or

$v = \frac{1}{2}v_1' + \frac{1}{2}(v_1' + v_{relative}) = v_1' + \frac{1}{2}v_{relative}$;

$5.80 \times 10^3\ m/s = v_1' + \frac{1}{2}(2.20 \times 10^3\ m/s)$, which gives $v_1' = 4.70 \times 10^3\ m/s.$

The other section will have

$v_2' = v_1' + v_{relative} = 4.70 \times 10^3\ m/s + 2.20 \times 10^3\ m/s = 6.90 \times 10^3\ m/s.$

(*b*) The energy supplied by the explosion increases the kinetic energy:

$E = \Delta KE = [\frac{1}{2}(\frac{1}{2}m)v_1'^2 + \frac{1}{2}(\frac{1}{2}m)v_2'^2] - \frac{1}{2}mv^2$

$= [\frac{1}{2}(\frac{1}{2}975\ kg)(4.70 \times 10^3\ m/s)^2 + \frac{1}{2}(\frac{1}{2}975\ kg)(6.90 \times 10^3\ m/s)^2] - \frac{1}{2}(975\ kg)(5.80 \times 10^3\ m/s)^2$

$= 5.90 \times 10^8\ J.$

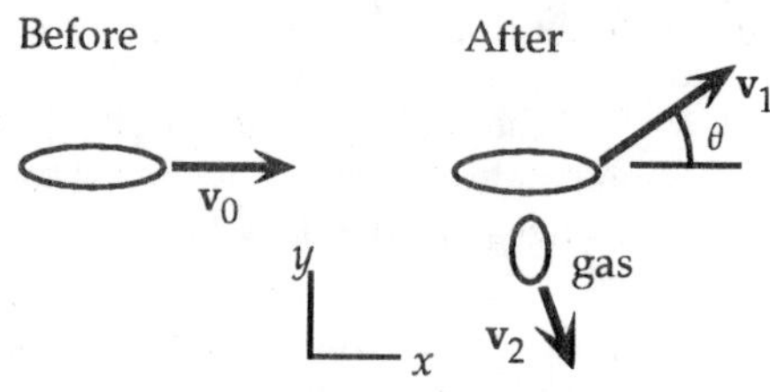

13. If M is the initial mass of the rocket and m_2 is the mass of the expelled gases, the final mass of the rocket is $m_1 = M - m_2$. Because the gas is expelled perpendicular to the rocket in the rocket's frame, it will still have the initial forward velocity, so the velocity of the rocket in the original direction will not change. We find the perpendicular component of the rocket's velocity after firing from

$v_{1\perp} = v_0 \tan \theta = (115 \text{ m/s}) \tan 35° = 80.5 \text{ m/s}.$

Using the coordinate system shown, for momentum conservation in the perpendicular direction we have

$0 + 0 = m_1 v_{1\perp} - m_2 v_{2\perp}$, or

$(M - m_2) v_{1\perp} = m_2 v_{2\perp};$

$(3180 \text{ kg} - m_2)(80.5 \text{ m/s}) = m_2(1750 \text{ m/s})$, which gives $m_2 =$ **140 kg.**

14. We find the average force on the ball from

$F = \Delta p/\Delta t = (\Delta mv/\Delta t) = [(0.0600 \text{ kg/s})(65.0 \text{ m/s}) - 0]/(0.0300 \text{ s}) =$ **130 N.**

Because the weight of a 60-kg person is ≈ 600 N, this force is **not large enough.**

15. We find the average force on the ball from

$F = \Delta p/\Delta t = m\,\Delta v/\Delta t = (0.145 \text{ kg})[(52.0 \text{ m/s}) - (-39.0 \text{ m/s})]/(1.00 \times 10^{-3} \text{ s}) =$ **1.32×10^4 N.**

16. (*a*) We find the impulse on the ball from

$\text{Impulse} = \Delta p = m\,\Delta v = (0.045 \text{ kg})(45 \text{ m/s} - 0) =$ **2.0 N·s.**

(*b*) The average force is

$F = \text{Impulse}/\Delta t = (2.0 \text{ N·s})/(5.0 \times 10^{-3} \text{ s}) =$ **4.0×10^2 N.**

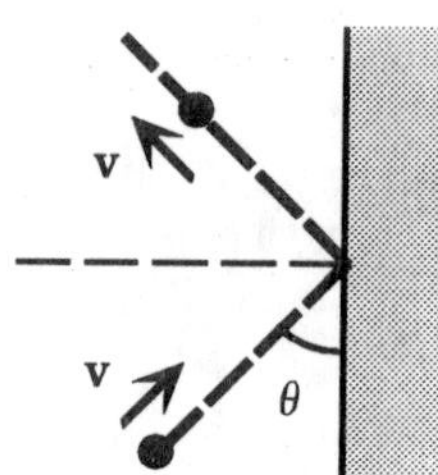

17. The momentum parallel to the wall does not change, therefore the impulse will be perpendicular to the wall. With the positive direction toward the wall, we find the impulse on the ball from

$\text{Impulse} = \Delta p_\perp = m\,\Delta v_\perp = m[(-v \sin \theta) - (v \sin \theta)]$

$= -2mv \sin \theta = 2(0.060 \text{ kg})(25 \text{ m/s}) \sin 45° = -2.1 \text{ N·s}.$

The impulse on the wall is in the opposite direction: **2.1 N·s.**

18. (*a*) With the positive direction in the direction of the fullback (East), the momentum is

$p = m_{\text{fullback}} v_{\text{fullback}} = (115 \text{ kg})(4.0 \text{ m/s}) =$ **4.6×10^2 kg·m/s (East).**

(*b*) We find the impulse on the fullback from

$\text{Impulse}_{\text{fullback}} = \Delta p_{\text{fullback}}$

$= 0 - 4.6 \times 10^2 \text{ kg·m/s} =$ **-4.6×10^2 kg·m/s (West).**

(*c*) We find the impulse on the tackler from

$\text{Impulse}_{\text{tackler}} = -\text{Impulse}_{\text{fullback}} =$ **$+4.6 \times 10^2$ kg·m/s (East).**

(*d*) We find the average force on the tackler from

$F_{\text{tackler}} = \text{Impulse}_{\text{tackler}}/\Delta t = (+4.6 \times 10^2 \text{ kg·m/s})/(0.75 \text{ s}) =$ **6.1×10^2 N (East).**

19. (*a*) The impulse is the area under the F vs. t curve.
The value of each block on the graph is
$1 \text{ block} = (50 \text{ N})(0.01 \text{ s}) = 0.50 \text{ N}\cdot\text{s}.$
We estimate there are 10 blocks under the curve, so the impulse is
$\text{Impulse} = (10 \text{ blocks})(0.50 \text{ N}\cdot\text{s/block})$ $\boxed{\approx 5.0 \text{ N}\cdot\text{s.}}$

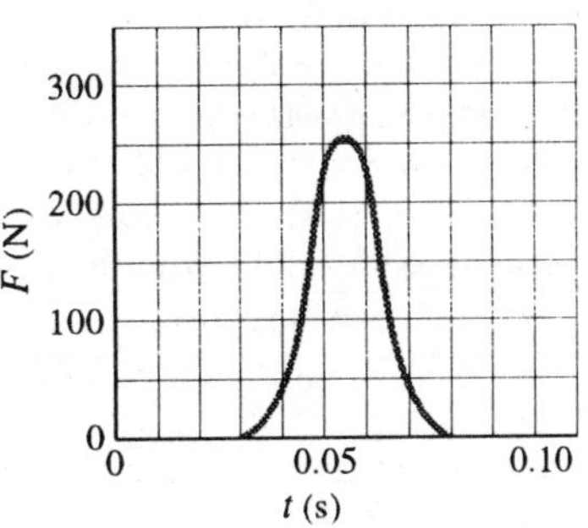

(*b*) We find the final velocity of the ball from
$\text{Impulse} = \Delta p = m\,\Delta v;$
$5.0 \text{ N}\cdot\text{s} = (0.060 \text{ kg})(v - 0)$, which gives $v = \boxed{83 \text{ m/s.}}$

20. The maximum force that each leg can exert without breaking is
$(170 \times 10^6 \text{ N/m}^2)(2.5 \times 10^{-4} \text{ m}^2) = 4.25 \times 10^4 \text{ N},$
so, if there is an even landing with both feet, the maximum force allowed on the body is 8.50×10^4 N.
We use the work-energy principle for the fall to find the landing speed:
$0 = \Delta\text{KE} + \Delta\text{PE};$
$0 = \frac{1}{2}mv_{\text{land}}^2 - 0 + (0 - mgh_{\max})$, or $v_{\text{land}}^2 = 2gh_{\max}$.
The impulse from the maximum force changes the momentum on landing. If we take down as the positive direction and assume the landing lasts for a time t, we have
$-F_{\max}t = m\,\Delta v = m(0 - v_{\text{land}})$, or $t = mv_{\text{land}}/F_{\max}$.
We have assumed a constant force, so the acceleration will be constant. For the landing we have
$y = v_{\text{land}}t + \frac{1}{2}at^2 = v_{\text{land}}(mv_{\text{land}}/F_{\max}) + \frac{1}{2}(-F_{\max}/m)(mv_{\text{land}}/F_{\max})^2 = \frac{1}{2}mv_{\text{land}}^2/F_{\max} = mgh_{\max}/F_{\max};$
$0.60 \text{ m} = (75 \text{ kg})(9.80 \text{ m/s}^2)h_{\max}/(8.50 \times 10^4 \text{ N})$, which gives $h_{\max} = \boxed{69 \text{ m.}}$

21. For the elastic collision of the two balls, we use momentum conservation:
$m_1v_1 + m_2v_2 = m_1v_1' + m_2v_2';$
$(0.440 \text{ kg})(3.70 \text{ m/s}) + (0.220 \text{ kg})(0) = (0.440 \text{ kg})v_1' + (0.220 \text{ kg})v_2'.$
Because the collision is elastic, the relative speed does not change:
$v_1 - v_2 = -(v_1' - v_2')$, or $3.70 \text{ m/s} - 0 = v_2' - v_1'$.
Combining these two equations, we get
$\boxed{v_1' = 1.23 \text{ m/s,}}$ and $\boxed{v_2' = 4.93 \text{ m/s.}}$

22. For the elastic collision of the two pucks, we use momentum conservation:
$m_1v_1 + m_2v_2 = m_1v_1' + m_2v_2';$
$(0.450 \text{ kg})(3.00 \text{ m/s}) + (0.900 \text{ kg})(0) = (0.450 \text{ kg})v_1' + (0.900 \text{ kg})v_2'.$
Because the collision is elastic, the relative speed does not change:
$v_1 - v_2 = -(v_1' - v_2')$, or $3.00 \text{ m/s} - 0 = v_2' - v_1'$.
Combining these two equations, we get
$\boxed{v_1' = -1.00 \text{ m/s (rebound),}}$ and $\boxed{v_2' = 2.00 \text{ m/s.}}$

23. For the elastic collision of the two billiard balls, we use momentum conservation:
$m_1v_1 + m_2v_2 = m_1v_1' + m_2v_2';$
$m(2.00 \text{ m/s}) + m(-3.00 \text{ m/s}) = mv_1' + mv_2'.$
Because the collision is elastic, the relative speed does not change:
$v_1 - v_2 = -(v_1' - v_2')$, or $2.00 \text{ m/s} - (-3.00 \text{ m/s}) = v_2' - v_1'$.
Combining these two equations, we get
$\boxed{v_1' = -3.00 \text{ m/s (rebound),}}$ and $\boxed{v_2' = 2.00 \text{ m/s.}}$
Note that the two billiard balls exchange velocities.

24. For the elastic collision of the two balls, we use momentum conservation:

$m_1v_1 + m_2v_2 = m_1v_1' + m_2v_2'$;

$(0.060 \text{ kg})(2.50 \text{ m/s}) + (0.090 \text{ kg})(1.00 \text{ m/s}) = (0.060 \text{ kg})v_1' + (0.090 \text{ kg})v_2'$.

Because the collision is elastic, the relative speed does not change:

$v_1 - v_2 = -(v_1' - v_2')$, or $2.50 \text{ m/s} - 1.00 \text{ m/s} = v_2' - v_1'$.

Combining these two equations, we get

$\boxed{v_1' = 0.70 \text{ m/s},}$ and $\boxed{v_2' = 2.20 \text{ m/s}.}$

25. (*a*) For the elastic collision of the two balls, we use momentum conservation:

$m_1v_1 + m_2v_2 = m_1v_1' + m_2v_2'$;

$(0.220 \text{ kg})(5.5 \text{ m/s}) + m_2(0) = (0.220 \text{ kg})(-3.7 \text{ m/s}) + m_2v_2'$.

Because the collision is elastic, the relative speed does not change:

$v_1 - v_2 = -(v_1' - v_2')$, or $5.5 \text{ m/s} - 0 = v_2' - (-3.7 \text{ m/s})$, which gives $\boxed{v_2' = 1.8 \text{ m/s}.}$

(*b*) Using the result for v_2' in the momentum equation, we get $\boxed{m_2 = 1.1 \text{ kg}.}$

26. (*a*) For the elastic collision of the two bumper cars, we use momentum conservation:

$m_1v_1 + m_2v_2 = m_1v_1' + m_2v_2'$;

$(450 \text{ kg})(4.50 \text{ m/s}) + (550 \text{ kg})(3.70 \text{ m/s}) = (450 \text{ kg})v_1' + (550 \text{ kg})v_2'$.

Because the collision is elastic, the relative speed does not change:

$v_1 - v_2 = -(v_1' - v_2')$, or $4.5 \text{ m/s} - 3.7 \text{ m/s} = v_2' - v_1'$.

Combining these two equations, we get

$\boxed{v_1' = 3.62 \text{ m/s},}$ and $\boxed{v_2' = 4.42 \text{ m/s}.}$

(*b*) For the change in momentum of each we have

$\Delta p_1 = m_1(v_1' - v_1) = (450 \text{ kg})(3.62 \text{ m/s} - 4.50 \text{ m/s}) = \boxed{-396 \text{ kg}\cdot\text{m/s};}$

$\Delta p_2 = m_2(v_2' - v_2) = (550 \text{ kg})(4.42 \text{ m/s} - 3.70 \text{ m/s}) = \boxed{+396 \text{ kg}\cdot\text{m/s}.}$

As expected, the changes are equal and opposite.

27. (*a*) For the elastic collision of the two balls, we use momentum conservation:

$m_1v_1 + m_2v_2 = m_1v_1' + m_2v_2'$;

$(0.280 \text{ kg})v_1 + m_2(0) = (0.280 \text{ kg})v_1' + m_2(\tfrac{1}{2}v_1)$.

Because the collision is elastic, the relative speed does not change:

$v_1 - v_2 = -(v_1' - v_2')$, or $v_1 - 0 = \tfrac{1}{2}v_1 - v_1'$, which gives $v_1' = -\tfrac{1}{2}v_1$.

Using this result in the momentum equation, we get $\boxed{m_2 = 0.840 \text{ kg}.}$

(*b*) The fraction transferred is

$\text{fraction} = \Delta\text{KE}_2/\text{KE}_1 = \tfrac{1}{2}m_2(v_2'^2 - v_2^2)/\tfrac{1}{2}m_1v_1^2$

$= \tfrac{1}{2}m_2[(\tfrac{1}{2}v_1)^2 - 0]/\tfrac{1}{2}m_1v_1^2 = \tfrac{1}{4}m_2/m_1 = \tfrac{1}{4}(0.840 \text{ kg})/(0.280 \text{ kg}) = \boxed{0.750.}$

28. We find the speed after falling a height h from energy conservation:

$\frac{1}{2}Mv^2 = Mgh$, or $v = (2gh)^{1/2}$.

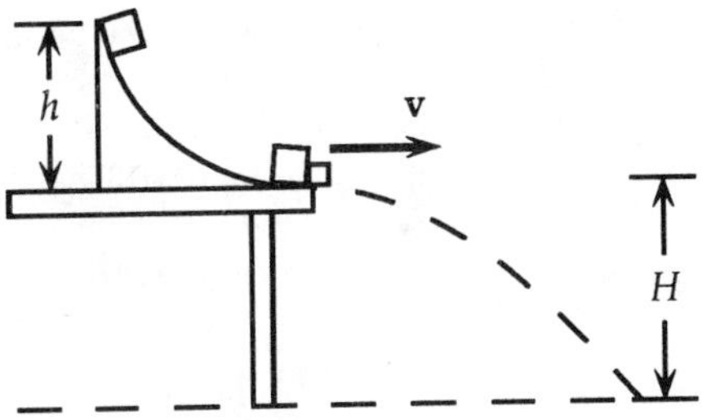

The speed of the first cube after sliding down the incline and just before the collision is

$v_1 = [2(9.80 \text{ m/s}^2)(0.20 \text{ m})]^{1/2} = 1.98 \text{ m/s}$.

For the elastic collision of the two cubes, we use momentum conservation:

$Mv_1 + mv_2 = Mv_1' + mv_2'$;

$M(1.98 \text{ m/s}) + \frac{1}{2}M(0) = Mv_1' + \frac{1}{2}Mv_2'$.

Because the collision is elastic, the relative speed does not change:

$v_1 - v_2 = -(v_1' - v_2')$, or $1.98 \text{ m/s} - 0 = v_2' - v_1'$.

Combining these two equations, we get

$v_1' = 0.660 \text{ m/s}$, and $v_2' = 2.64 \text{ m/s}$.

Because both cubes leave the table with a horizontal velocity, they will fall to the floor in the same time, which we find from

$H = \frac{1}{2}gt^2$;

$0.90 \text{ m} = \frac{1}{2}(9.80 \text{ m/s}^2)t^2$, which gives $t = 0.429$ s.

Because the horizontal motion has constant velocity, we have

$x_1 = v_1't = (0.660 \text{ m/s})(0.429 \text{ s}) = \boxed{0.28 \text{ m};}$

$x_2 = v_2't = (2.64 \text{ m/s})(0.429 \text{ s}) = \boxed{1.1 \text{ m.}}$

29. (*a*) For the elastic collision of the two masses, we use momentum conservation:

$m_1v_1 + m_2v_2 = m_1v_1' + m_2v_2'$;

$m_1v_1 + 0 = m_1v_1' + m_2v_2'$.

Because the collision is elastic, the relative speed does not change:

$v_1 - v_2 = -(v_1' - v_2')$, or $v_1 - 0 = v_2' - v_1'$.

If we multiply this equation by m_2 and subtract it from the momentum equation, we get

$(m_1 - m_2)v_1 = (m_1 + m_2)v_1'$, or $v_1' = [(m_1 - m_2)/(m_1 + m_2)]v_1$.

If we multiply the relative speed equation by m_1 and add it to the momentum equation, we get

$2m_1v_1 = (m_1 + m_2)v_2'$, or $v_2' = [2m_1/(m_1 + m_2)]v_1$.

(*b*) When $m_1 \ll m_2$ we have

$v_1' = [(m_1 - m_2)/(m_1 + m_2)]v_1 \approx [(-m_2)/(m_2)]v_1 = -v_1$;

$v_2' = [2m_1/(m_1 + m_2)]v_1 \approx [(2m_1/m_2)v_1 \approx 0$; so

$\boxed{v_1' \approx -v_1, v_2' \approx 0;}$ the small mass rebounds with the same speed; the large mass does not move. An example is throwing a ping pong ball against a concrete block.

(*c*) When $m_1 \gg m_2$ we have

$v_1' = [(m_1 - m_2)/(m_1 + m_2)]v_1 \approx [(m_1)/(m_1)]v_1 = v_1$;

$v_2' = [2m_1/(m_1 + m_2)]v_1 \approx (2m_1/m_1)v_1 = 2v_1$; so

$\boxed{v_1' \approx v_1, v_2' \approx 2v_1;}$ the large mass continues with the same speed; the small mass acquires a large velocity. An example is hitting a light stick with a bowling ball.

(*d*) When $m_1 = m_2$ we have

$v_1' = [(m_1 - m_2)/(m_1 + m_2)]v_1 = 0$;

$v_2' = [2m_1/(m_1 + m_2)]v_1 = (2m_1/2m_1)v_1 = v_1$; so

$\boxed{v_1' \approx 0, v_2' \approx v_1;}$ the striking mass stops; the hit mass acquires the striking mass's velocity. An example is one billiard ball hitting an identical one.

30. We let V be the speed of the block and bullet immediately after the collision and before the pendulum swings. For this perfectly inelastic collision, we use momentum conservation:

$mv + 0 = (M + m)V;$

$(0.018\ \text{kg})(230\ \text{m/s}) = (0.018\ \text{kg} + 3.6\ \text{kg})V,$

which gives $V = 1.14\ \text{m/s}$.

Because the tension does no work, we can use energy conservation for the swing:

$\frac{1}{2}(M + m)V^2 = (M + m)gh,$ or $V = (2gh)^{1/2};$

$1.14\ \text{m/s} = [2(9.80\ \text{m/s}^2)h]^{1/2}$, which gives $h = 0.0666\ \text{m}$.

We find the horizontal displacement from the triangle:

$L^2 = (L - h)^2 + x^2;$

$(2.8\ \text{m})^2 = (2.8\ \text{m} - 0.0666\ \text{m})^2 + x^2$, which gives $x =$ **0.61 m.**

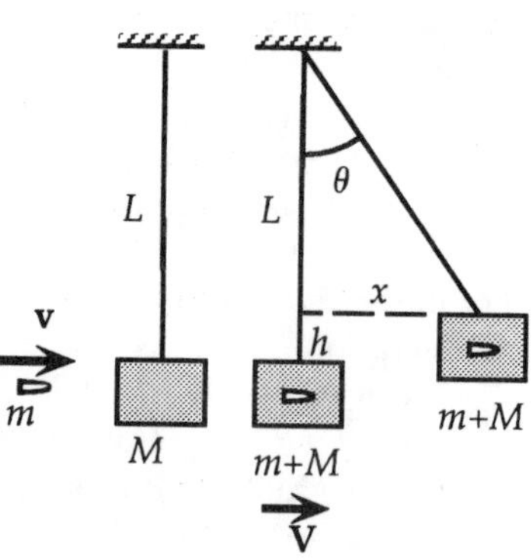

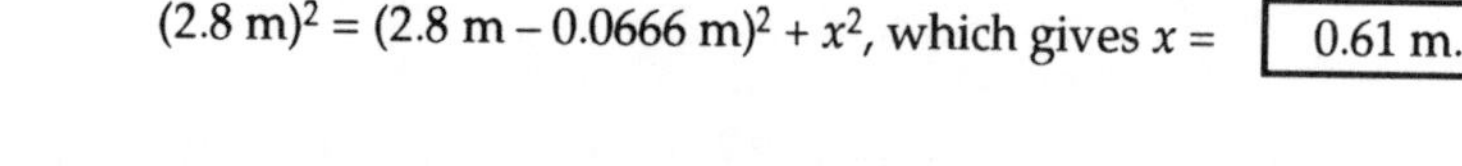

31. (*a*) The velocity of the block and projectile after the collision is

$v' = mv_1/(m + M).$

The fraction of kinetic energy lost is

$$\text{fraction lost} = -\Delta\text{KE}/\text{KE} = -[\tfrac{1}{2}(m + M)v'^2 - \tfrac{1}{2}mv_1^2]/\tfrac{1}{2}mv_1^2$$

$$= -\{(m + M)[mv_1/(m + M)]^2 - mv_1^2\}/mv_1^2$$

$$= -[m/(m + M)] + 1 = \boxed{+M/(m + M).}$$

(*b*) For the data given we have

$$\text{fraction lost} = M/(m + M) = (14.0\ \text{g})/(14.0\ \text{g} + 380\ \text{g}) = \boxed{0.964.}$$

32. Momentum conservation gives

$0 = m_1v_1' + m_2v_2';$

$0 = m_1v_1' + 1.5m_1v_2'$, or $v_1' = -1.5v_2'$.

The kinetic energy of each piece is

$\text{KE}_2 = \frac{1}{2}m_2v_2'^2;$

$\text{KE}_1 = \frac{1}{2}m_1v_1'^2 = \frac{1}{2}(m_2/1.5)(-1.5v_2')^2 = (1.5)\frac{1}{2}m_2v_2'^2 = 1.5\text{KE}_2.$

The energy supplied by the explosion produces the kinetic energy:

$E = \text{KE}_1 + \text{KE}_2 = 2.5\text{KE}_2;$

$7500\ \text{J} = 2.5\text{KE}_2$, which gives $\text{KE}_2 = 3000\ \text{J}$.

For the other piece we have

$\text{KE}_1 = E - \text{KE}_2 = 7500\ \text{J} - 3000\ \text{J} = 4500\ \text{J}.$

Thus

KE(heavier) = 3000 J; KE(lighter) = 4500 J.

33. On the horizontal surface after the collision, the normal force on the joined cars is $F_N = (m + M)g$. We find the common speed of the joined cars immediately after the collision by using the work-energy principle for the sliding motion:

$W_{fr} = \Delta\text{KE};$

$-\mu_k(m + M)gd = 0 - \frac{1}{2}(M + m)V^2;$

$0.40(9.80\ \text{m/s}^2)(2.8\ \text{m}) = \frac{1}{2}V^2$, which gives $V = 4.68\ \text{m/s}$.

For the collision, we use momentum conservation:

$mv + 0 = (m + M)V;$

$(1.0 \times 10^3\ \text{kg})v = (1.0 \times 10^3\ \text{kg} + 2.2 \times 10^3\ \text{kg})(4.68\ \text{m/s})$, which gives $\boxed{v = 15\ \text{m/s}}$ (54 km/h).

34. (*a*) For a perfectly elastic collision, we use momentum conservation:

$$m_1v_1 + m_2v_2 = m_1v_1' + m_2v_2', \text{ or } m_1(v_1 - v_1') = m_2(v_2' - v_2).$$

Kinetic energy is conserved, so we have

$$\tfrac{1}{2}m_1v_1^2 + \tfrac{1}{2}m_2v_2^2 = \tfrac{1}{2}m_1v_1'^2 + \tfrac{1}{2}m_2v_2'^2, \text{ or } m_1(v_1^2 - v_1'^2) = m_2(v_2'^2 - v_2^2),$$

which can be written as

$$m_1(v_1 - v_1')(v_1 + v_1') = m_2(v_2' - v_2)(v_2' + v_2).$$

When we divide this by the momentum result, we get

$$v_1 + v_1' = v_2' + v_2, \text{ or } v_1' - v_2' = v_2 - v_1.$$

If we use this in the definition of the coefficient of restitution, we get

$$e = (v_1' - v_2')/(v_2 - v_1) = (v_2 - v_1)/(v_2 - v_1) = 1.$$

For a completely inelastic collision, the two objects move together, so we have

$v_1' = v_2'$, which gives $e = 0$.

(*b*) We find the speed after falling a height h from energy conservation:

$$\tfrac{1}{2}mv_1^2 = mgh, \text{ or } v_1 = (2gh)^{1/2}.$$

The same expression holds for the height reached by an object moving upward:

$$v_1' = (2gh')^{1/2}.$$

Because the steel plate does not move, when we take into account the directions we have

$$e = (v_1' - v_2')/(v_2 - v_1) = [(2gh')^{1/2} - 0]/\{0 - [-(2gh)^{1/2}]\}, \text{ so } \boxed{e = (h'/h)^{1/2}.}$$

35. Momentum conservation for the explosion gives us

$$0 = m_1v_1' + m_2v_2';$$

$$0 = m_1v_1' + 3m_1v_2', \text{ or } v_1' = -3v_2'.$$

On the horizontal surface after the collision, the normal force on a block is $F_N = mg$.
We relate the speed of a block immediately after the collision to the distance it slides from the work-energy principle for the sliding motion:

$$W_{fr} = \Delta KE;$$

$$-\mu_k mgd = 0 - \tfrac{1}{2}mv^2, \text{ or } d = \tfrac{1}{2}v^2/\mu_k g.$$

If we use this for each block and form the ratio, we get

$$d_1/d_2 = (v_1/v_2)^2 = (-3)^2 = \boxed{9,}$$ with the lighter block traveling farther.

36. For the momentum conservation of this one-dimensional collision, we have

$$m_1v_1 + m_2v_2 = m_1v_1' + m_2v_2'.$$

(*a*) If the bodies stick together, $v_1' = v_2' = V$:

(5.0 kg)(5.5 m/s) + (3.0 kg)(– 4.0 m/s) = (5.0 kg + 3.0 kg)V, which gives V = $\boxed{v_1' = v_2' = 1.9 \text{ m/s.}}$

(*b*) If the collision is elastic, the relative speed does not change:

$v_1 - v_2 = -(v_1' - v_2')$, or 5.5 m/s – (– 4.0 m/s) = 9.5 m/s = $v_2' - v_1'$.

The momentum equation is

(5.0 kg)(5.5 m/s) + (3.0 kg)(– 4.0 m/s) = (5.0 kg)v_1' + (3.0 kg)v_2', or

(5.0 kg)v_1' + (3.0 kg)v_2' = 15.5 kg·m/s.

When we combine these two equations, we get $\boxed{v_1' = -1.6 \text{ m/s}, v_2' = 7.9 \text{ m/s.}}$

(*c*) If m_1 comes to rest, $v_1' = 0$.

(5.0 kg)(5.5 m/s) + (3.0 kg)(– 4.0 m/s) = 0 + (3.0 kg)v_2', which gives $\boxed{v_1' = 0, v_2' = 5.2 \text{ m/s.}}$

(*d*) If m_2 comes to rest, $v_2' = 0$.

(5.0 kg)(5.5 m/s) + (3.0 kg)(– 4.0 m/s) = (5.0 kg)v_1' + 0, which gives $\boxed{v_1' = 3.1 \text{ m/s}, v_2' = 0.}$

(*e*) The momentum equation is

(5.0 kg)(5.5 m/s) + (3.0 kg)(– 4.0 m/s) = (5.0 kg)(– 4.0 m/s) + (3.0 kg)v_2',

which gives $\boxed{v_1' = -4.0 \text{ m/s}, v_2' = 12 \text{ m/s.}}$

The result for (*c*) is reasonable. The 3.0-kg body rebounds.

The result for (*d*) is not reasonable. The 5.0-kg body would have to pass through the 3.0-kg body.

To check the result for (*e*) we find the change in kinetic energy:

$$\begin{aligned}\Delta KE &= (\tfrac{1}{2}m_1v_1'^2 + \tfrac{1}{2}m_2v_2'^2) - (\tfrac{1}{2}m_1v_1^2 + \tfrac{1}{2}m_2v_2^2)\\ &= \tfrac{1}{2}[(5.0 \text{ kg})(5.5 \text{ m/s})^2 + (3.0 \text{ kg})(-4.0 \text{ m/s})^2] - \tfrac{1}{2}[(5.0 \text{ kg})(-4.0 \text{ m/s})^2 + (3.0 \text{ kg})(12 \text{ m/s})^2]\\ &= +156 \text{ J}.\end{aligned}$$

Because the kinetic energy cannot increase in a simple collision, **the result for (*e*) is not reasonable.**

37. Because the initial momentum is zero, the momenta of the three products of the decay must add to zero. If we draw the vector diagram, we see that

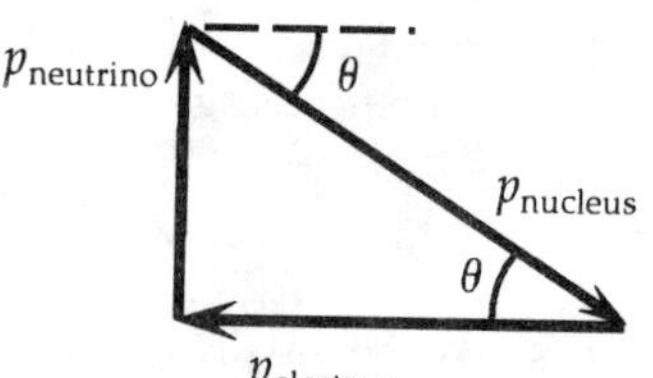

$$p_{nucleus} = (p_{electron}^2 + p_{neutrino}^2)^{1/2}$$
$$= [(9.30 \times 10^{-23}\ \text{kg}\cdot\text{m/s})^2 + (5.40 \times 10^{-23}\ \text{kg}\cdot\text{m/s})^2]^{1/2}$$
$$= \boxed{1.08 \times 10^{-22}\ \text{kg}\cdot\text{m/s.}}$$

We find the angle from

$$\tan\theta = p_{neutrino}/p_{electron}$$
$$= (5.40 \times 10^{-23}\ \text{kg}\cdot\text{m/s})/(9.30 \times 10^{-23}\ \text{kg}\cdot\text{m/s})$$
$$= 0.581, \text{ so the angle is } \boxed{30.1^\circ \text{ from the direction opposite to the electron's.}}$$

38. For the collision we use momentum conservation:

x-direction: $m_1v_1 + 0 = (m_1 + m_2)v' \cos\theta$;
(4.3 kg)(7.8 m/s) = (4.3 kg + 5.6 kg)$v' \cos\theta$, which gives
$v' \cos\theta = 3.39$ m/s.

y-direction: $0 + m_2v_2 = (m_1 + m_2)v' \sin\theta$;
(5.6 kg)(10.2 m/s) = (4.3 kg + 5.6 kg)$v' \sin\theta$, which gives
$v' \sin\theta = 5.77$ m/s.

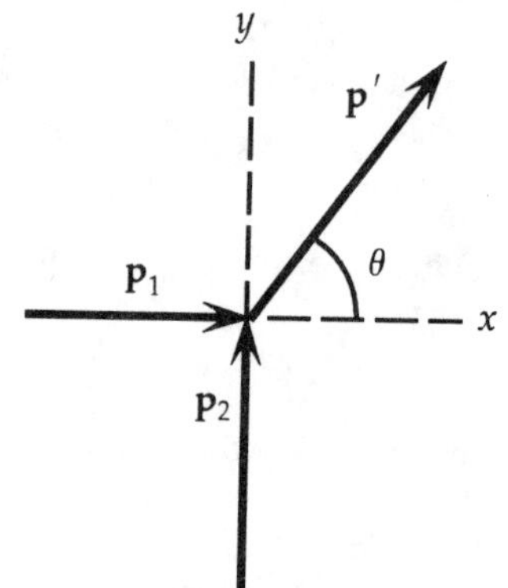

We find the direction by dividing the equations:

$\tan\theta = (5.77\ \text{m/s})/(3.39\ \text{m/s}) = 1.70$, so $\theta = \boxed{60^\circ.}$

We find the magnitude by squaring and adding the equations:

$v' = [(5.77\ \text{m/s})^2 + (3.39\ \text{m/s})^2]^{1/2} = \boxed{6.7\ \text{m/s.}}$

39. (*a*) Using the coordinate system shown, for momentum conservation we have

$$\boxed{\begin{array}{l} x\text{-momentum: } m_Av_A + 0 = m_Av_A' \cos\theta_A' + m_Bv_B' \cos\theta_B'; \\ y\text{-momentum: } 0 + 0 = m_Av_A' \sin\theta_A' - m_Bv_B' \sin\theta_B'. \end{array}}$$

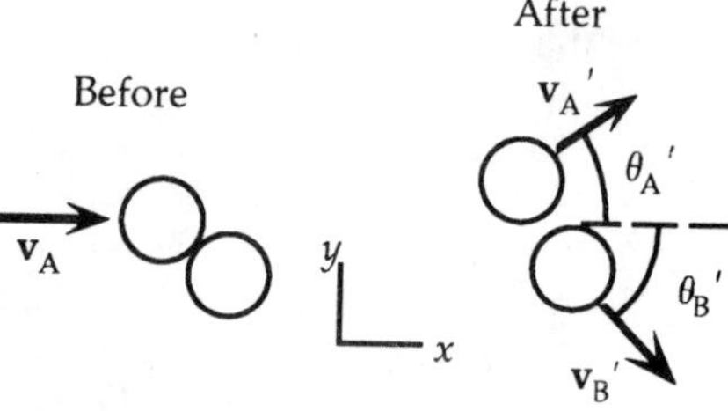

(*b*) With the given data, we have

x: (0.400 kg)(1.80 m/s) =
(0.400 kg)(1.10 m/s) cos 30° + (0.500 kg)$v_B' \cos\theta_B'$,
which gives $v_B' \cos\theta_B' = 0.678$ m/s;

y: 0 = (0.400 kg)(1.10 m/s) sin 30° – (0.500 kg)$v_B' \sin\theta_B'$,
which gives $v_B' \sin\theta_B' = 0.440$ m/s.

We find the magnitude by squaring and adding the equations:

$v_B' = [(0.440\ \text{m/s})^2 + (0.678\ \text{m/s})^2]^{1/2} = \boxed{0.808\ \text{m/s.}}$

We find the direction by dividing the equations:

$\tan\theta_B' = (0.440\ \text{m/s})/(0.678\ \text{m/s}) = 0.649$, so $\theta_B = \boxed{33.0^\circ.}$

40. (*a*) Using the coordinate system shown, for momentum conservation we have

x-momentum: $mv + 0 = 0 + 2mv_2' \cos\theta$, or
$2v_2' \cos\theta = v$;

y-momentum: $0 + 0 = -mv_1' + 2mv_2' \sin\theta$, or
$2v_2' \sin\theta = v_1'$.

If we square and add these two equations, we get
$v^2 + v_1'^2 = 4v_2'^2$.

For the conservation of kinetic energy, we have
$\frac{1}{2}mv^2 + 0 = \frac{1}{2}mv_1'^2 + \frac{1}{2}(2m)v_2'^2$, or
$v^2 - v_1'^2 = 2v_2'^2$.

When we add this to the previous result, we get
$v^2 = 3v_2'^2$.

Using this in the x-momentum equation, we get
$\cos\theta = \sqrt{3}/2$, or $\theta =$ $\boxed{30°.}$

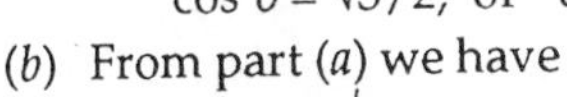

(*b*) From part (*a*) we have
$v_2' = v/\sqrt{3}$.

Using the energy result, we get
$v_1'^2 = v^2 - 2v_2'^2 = v^2 - 2v^2/3 = \frac{1}{3}v^2$, or $\boxed{v_1' = v/\sqrt{3}.}$

(*c*) The fraction of the kinetic energy transferred is
fraction = $\text{KE}_2/\text{KE}_1 = \frac{1}{2}(2m)v_2'^2/\frac{1}{2}mv^2 = m(v^2/3)/\frac{1}{2}mv^2 =$ $\boxed{\frac{2}{3}.}$

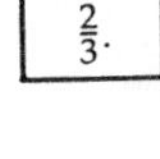

41. Using the coordinate system shown, for momentum conservation we have

y-momentum: $-mv\sin\theta_1 + mv\sin\theta_2 = 0$, or $\theta_1 = \theta_2$.

x-momentum: $mv\cos\theta_1 + mv\cos\theta_2 = 2mv_2'$;
$2mv\cos\theta_1 = 2mv/3$;
$\cos\theta_1 = \frac{1}{3}$, or $\theta_1 = 70.5° = \theta_2$.

The angle between their initial directions is
$\phi = \theta_1 + \theta_2 = 2(70.5°) =$ $\boxed{141°.}$

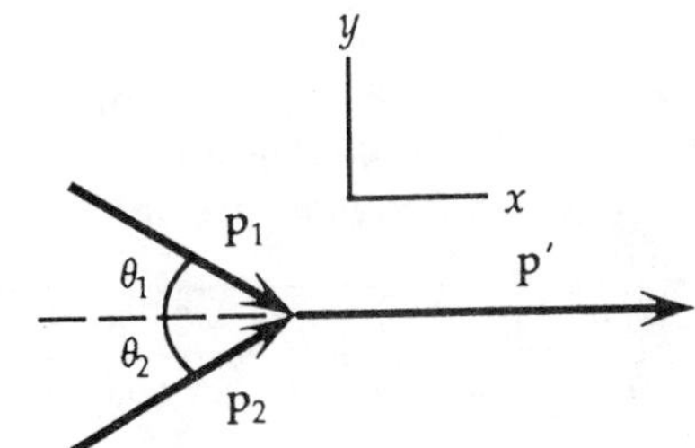

42. Using the coordinate system shown, for momentum conservation we have

y-momentum: $mv_1 + 0 = mv_1' \cos\alpha + Mv_2' \cos\theta_2$;
$5M(12.0 \text{ m/s}) = 5Mv_1' \cos\alpha + Mv_2' \cos 80°$, or
$5v_1' \cos\alpha = -v_2' \cos 80° + 60.0 \text{ m/s}$.

x-momentum: $0 = -mv_1' \sin\alpha + Mv_2' \sin\theta_2$;
$0 = -5Mv_1' \sin\alpha + Mv_2' \sin 80°$, or
$5v_1' \sin\alpha = v_2' \sin 80°$.

For the conservation of kinetic energy, we have
$\frac{1}{2}mv_1^2 + 0 = \frac{1}{2}mv_1'^2 + \frac{1}{2}Mv_2'^2$;
$5M(12.0 \text{ m/s})^2 = 5Mv_1'^2 + Mv_2'^2$, or
$5v_1'^2 + v_2'^2 = 720 \text{ m}^2/\text{s}^2$.

We have three equations in three unknowns: α, v_1', v_2'. We eliminate α by squaring and adding the two momentum results, and then combine this with the energy equation, with the results:

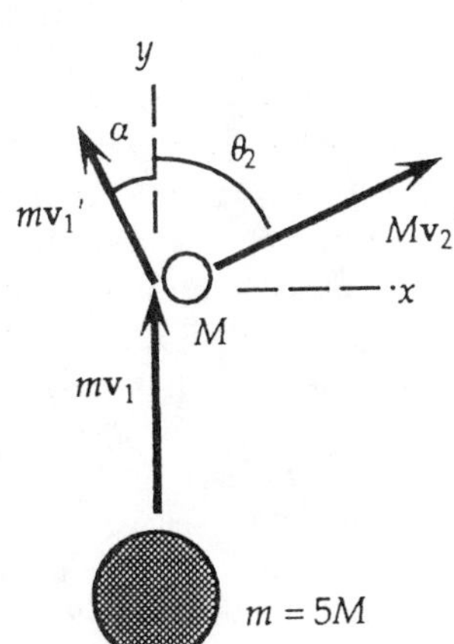

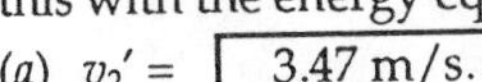

(*a*) $v_2' =$ $\boxed{3.47 \text{ m/s.}}$

(*b*) $v_1' =$ $\boxed{11.9 \text{ m/s.}}$

(*c*) $\alpha =$ $\boxed{3.29°.}$

43. Using the coordinate system shown, for momentum conservation we have

x: $m_n v_n + 0 = m_n v_n' \cos\theta_1' + m_{He} v_{He}' \cos\theta_2'$;
$m_n(6.2\times10^5 \text{ m/s}) = m_n v_n' \cos\theta_1' + 4m_n v_{He}' \cos 45°$, or
$v_n' \cos\theta_1' = (6.2\times10^5 \text{ m/s}) - 4v_{He}' \cos 45°$.

y: $0 + 0 = -m_n v_n' \sin\theta_1' + m_{He} v_{He}' \sin\theta_2'$;
$0 = -m_n v_n' \sin\theta_1' + 4m_n v_{He}' \sin 45°$, or
$v_n' \sin\theta_1' = 4v_{He}' \sin 45°$.

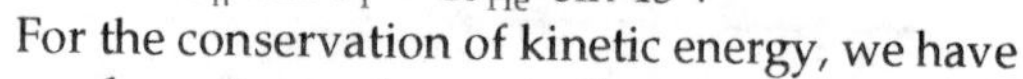

For the conservation of kinetic energy, we have

$\frac{1}{2}m_n v_n^2 + 0 = \frac{1}{2}m_n v_n'^2 + \frac{1}{2}m_{He} v_{He}'^2$;
$m_n(6.2\times10^5 \text{ m/s})^2 = m_n v_n'^2 + 4m_n v_{He}'^2$, or
$v_n'^2 + 4v_{He}'^2 = 3.84\times10^{11} \text{ m}^2/\text{s}^2$.

We have three equations in three unknowns: θ_1', v_n', v_{He}'. We eliminate θ_1' by squaring and adding the two momentum results, and then combine this with the energy equation, with the results:

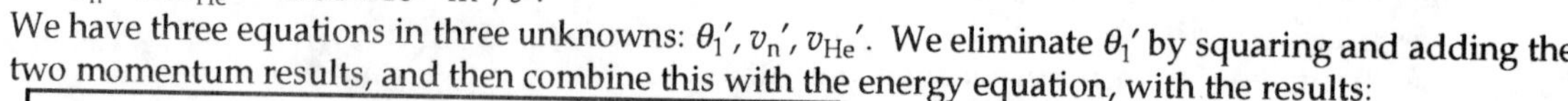

$\theta_1' = 76°$, $v_n' = 5.1\times10^5$ m/s, $v_{He}' = 1.8\times10^5$ m/s.

44. Using the coordinate system shown, for momentum conservation we have

x: $0 + mv_2 = mv_1' \cos\alpha + 0$;
$3.7 \text{ m/s} = v_1' \cos\alpha$;

y: $mv_1 + 0 = mv_1' \sin\alpha + mv_2'$;
$2.0 \text{ m/s} = v_1' \sin\alpha + v_2'$, or
$v_1' \sin\alpha = 2.0 \text{ m/s} - v_2'$.

For the conservation of kinetic energy, we have

$\frac{1}{2}mv_1^2 + \frac{1}{2}mv_2^2 = \frac{1}{2}mv_1'^2 + \frac{1}{2}mv_2'^2$;
$(2.0 \text{ m/s})^2 + (3.7 \text{ m/s})^2 = v_1'^2 + v_2'^2$.

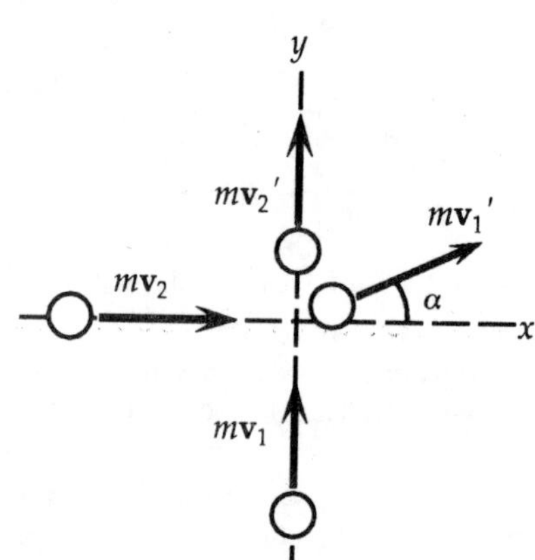

We have three equations in three unknowns: α, v_1', v_2'. We eliminate α by squaring and adding the two momentum results, and then combine this with the energy equation, with the results:

$\alpha = 0°$, $v_1' = 3.7$ m/s, $v_2' = 2.0$ m/s.

The two billiard balls exchange velocities.

45. Using the coordinate system shown, for momentum conservation we have

x: $mv_1 + 0 = mv_1' \cos\theta_1 + mv_2' \cos\theta_2$, or
$v_1 = v_1' \cos\theta_1 + v_2' \cos\theta_2$;

y: $0 + 0 = mv_1' \sin\theta_1 - mv_2' \sin\theta_2$, or
$0 = v_1' \sin\theta_1 - v_2' \sin\theta_2$.

For the conservation of kinetic energy, we have

$\frac{1}{2}mv_1^2 + 0 = \frac{1}{2}mv_1'^2 + \frac{1}{2}mv_2'^2$;
$v_1^2 = v_1'^2 + v_2'^2$.

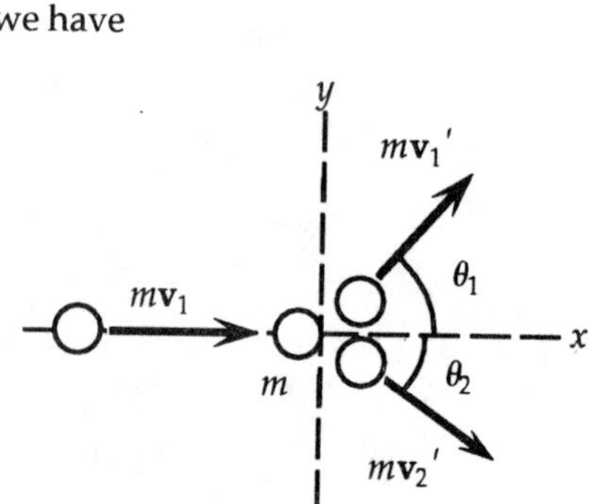

We square each of the momentum equations:

$v_1^2 = v_1'^2 \cos^2\theta_1 + 2v_1'v_2' \cos\theta_1 \cos\theta_2 + v_2'^2 \cos^2\theta_2$;
$0 = v_1'^2 \sin^2\theta_1 - 2v_1'v_2' \sin\theta_1 \sin\theta_2 + v_2'^2 \sin^2\theta_2$.

If we add these two equations and use $\sin^2\theta + \cos^2\theta = 1$, we get

$v_1^2 = v_1'^2 + 2v_1'v_2'(\cos\theta_1 \cos\theta_2 - \sin\theta_1 \sin\theta_2) + v_2'^2$.

If we subtract the energy equation, we get

$0 = 2v_1'v_2'(\cos\theta_1 \cos\theta_2 - \sin\theta_1 \sin\theta_2)$, or $\cos\theta_1 \cos\theta_2 - \sin\theta_1 \sin\theta_2 = 0$.

We reduce this with a trigonometric identity:

$\cos\theta_1 \cos\theta_2 - \sin\theta_1 \sin\theta_2 = \cos(\theta_1 + \theta_2) = 0$,

which means that $\theta_1 + \theta_2 = 90°$.

46. We choose the origin at the carbon atom. The center of mass will lie along the line joining the atoms:

$x_{CM} = (m_C x_C + m_O x_O)/(m_C + m_O)$
$= [0 + (16\text{ u})(1.13 \times 10^{-10}\text{ m})]/(12\text{ u} + 16\text{ u}) = \boxed{6.5 \times 10^{-11}\text{ m}}$ from the carbon atom.

5.332×10^{-6} 1.81×10^{-5}

47. We choose the origin at the front of the car:

$x_{CM} = (m_{car}x_{car} + m_{front}x_{front} + m_{back}x_{back})/(m_{car} + m_{front} + m_{back})$
$= [(1050\text{ kg})(2.50\text{ m}) + (140\text{ kg})(2.80\text{ m}) + (210\text{ kg})(3.90\text{ m})]/(1050\text{ kg} + 140\text{ kg} + 210\text{ kg})$
$= \boxed{2.74\text{ m}}$ from the front of the car.

48. Because the cubes are made of the same material, their masses will be proportional to the volumes:

$m_1, m_2 = 2^3 m_1 = 8m_1, m_3 = 3^3 m_1 = 27m_1$.

From symmetry we see that $y_{CM} = 0$.

We choose the x-origin at the outside edge of the small cube:

$x_{CM} = (m_1x_1 + m_2x_2 + m_3x_3)/(m_1 + m_2 + m_3)$
$= \{m_1(\frac{1}{2}\ell_0) + 8m_1[\ell_0 + \frac{1}{2}(2\ell_0)] +$
$27m_1[\ell_0 + 2\ell_0 + \frac{1}{2}(3\ell_0)]\}/(m_1 + 8m_1 + 27m_1)$
$= 138\ell_0/36$
$= \boxed{3.83\ell_0 \text{ from the outer edge of the small cube.}}$

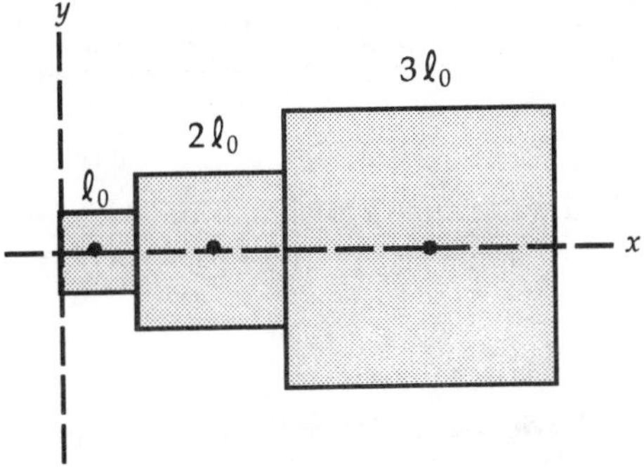

49. We choose the origin at the center of the raft, which is the CM of the raft:

$x_{CM} = (Mx_{raft} + m_1x_1 + m_2x_2 + m_3x_3)/(M + m_1 + m_2 + m_3)$
$= [0 + (1200\text{ kg})(9.0\text{ m}) + (1200\text{ kg})(9.0\text{ m}) +$
$(1200\text{ kg})(-9.0\text{ m})]/[6200\text{ kg} + 3(1200\text{ kg})]$
$= \boxed{1.10\text{ m (East).}}$

$y_{CM} = (My_{raft} + m_1y_1 + m_2y_2 + m_3y_3)/(M + m_1 + m_2 + m_3)$
$= [0 + (1200\text{ kg})(9.0\text{ m}) + (1200\text{ kg})(-9.0\text{ m}) +$
$(1200\text{ kg})(-9.0\text{ m})]/[6200\text{ kg} + 3(1200\text{ kg})]$
$= \boxed{-1.10\text{ m (South).}}$

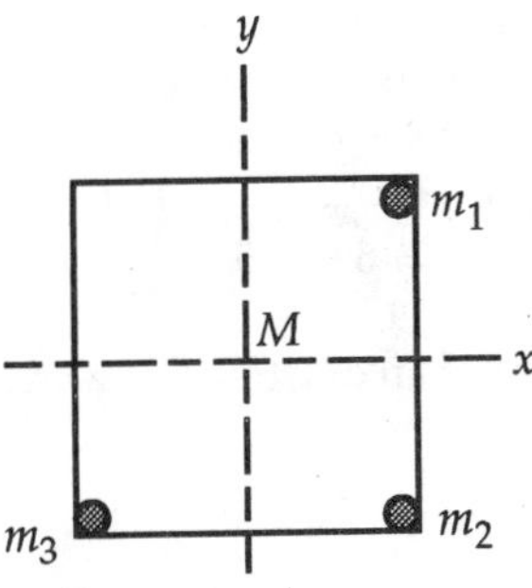

50. We choose the coordinate system shown. There are 10 cases.

$x_{CM} = (5mx_1 + 3mx_2 + 2mx_3)/(10m)$
$= [5(\frac{1}{2}\ell) + 3(\ell + \frac{1}{2}\ell) + 2(2\ell + \frac{1}{2}\ell)]/(10)$
$= 1.2\ell$.

$y_{CM} = (7my_1 + 2my_2 + my_3)/(10m)$
$= [7(\frac{1}{2}\ell) + 2(\ell + \frac{1}{2}\ell) + (2\ell + \frac{1}{2}\ell)]/(10)$
$= 0.9\ell$.

The CM is $\boxed{1.2\ell \text{ from the left, and } 0.9\ell \text{ from the back}}$ of the pallet.

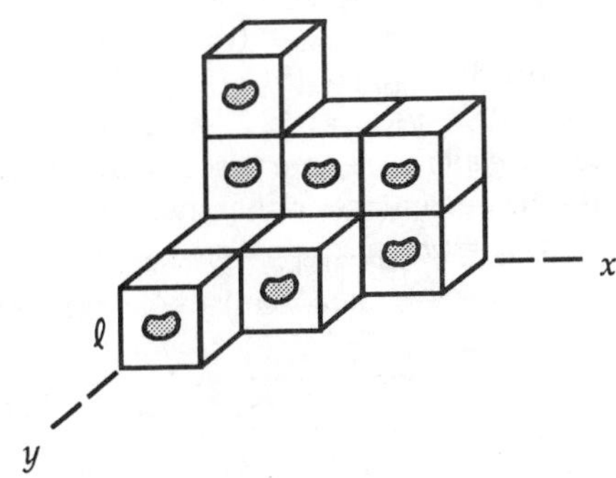

51. We know from the symmetry that the center of mass lies on a line containing the center of the plate and the center of the hole. We choose the center of the plate as origin and x along the line joining the centers. Then $y_{CM} = 0$. A uniform circle has its center of mass at its center. We can treat the system as two circles:

    a circle of radius $2R$, density ρ and mass $\rho\pi(2R)^2$ with $x_1 = 0$;

    a circle of radius R, density $-\rho$ and mass $-\rho\pi R^2$ with $x_2 = 0.80R$.

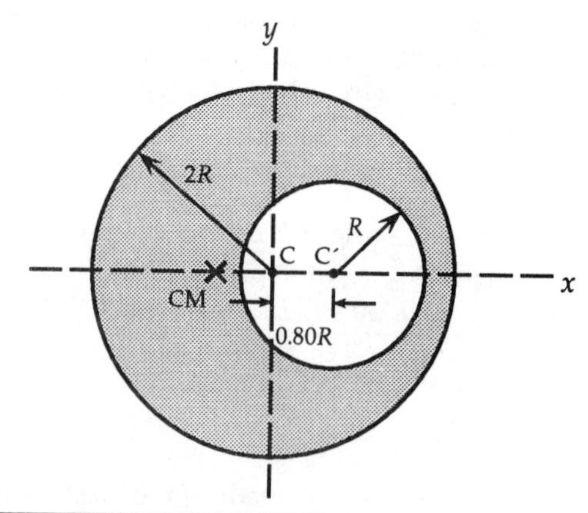

    We find the center of mass from

    $$x_{CM} = (m_1x_1 + m_2x_2)/(m_1 + m_2)$$
    $$= [4\rho\pi R^2 (0) - \rho\pi R^2 (0.80R)]/(4\rho\pi R^2 - \rho\pi R^2)$$
    $$= -0.27R.$$

    The center of mass is **along the line joining the centers 0.07R outside the hole.**

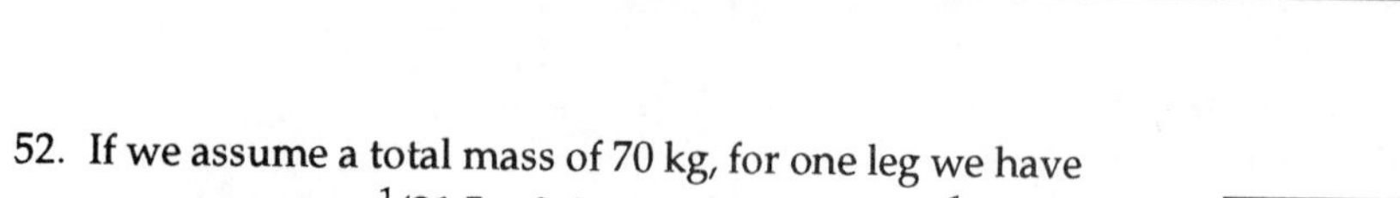

52. If we assume a total mass of 70 kg, for one leg we have

    $$m_{leg} = m_{body}\tfrac{1}{2}(21.5 + 9.6 + 3.4)/100 = (70\text{ kg})\tfrac{1}{2}(34.5)/100 = \boxed{12\text{ kg.}}$$

53. If we measure from the shoulder, the percentage of the height to the center of mass for each of the segments is

    upper arm: $81.2 - 71.7 = 9.5$;

    lower arm: $81.2 - 55.3 = 25.9$;

    hand: $81.2 - 43.1 = 38.1$.

    Because all masses are percentages of the body mass, we can use the percentages rather than the actual mass. Thus we have

    $$x_{CM} = (m_{upper}x_{upper} + m_{lower}x_{lower} + m_{hand}x_{hand})/(m_{upper} + m_{lower} + m_{hand})$$
    $$= [\tfrac{1}{2}(6.6)(9.5) + \tfrac{1}{2}(4.2)(25.9) + \tfrac{1}{2}(1.7)(38.1)]/[\tfrac{1}{2}(6.6) + \tfrac{1}{2}(4.2) + \tfrac{1}{2}(1.7)]$$
    $$= 19.$$

    The CM of an outstretched arm is **19% of the height.**

54. We choose the shoulder as the origin. The locations of the centers of mass for each of the segments are

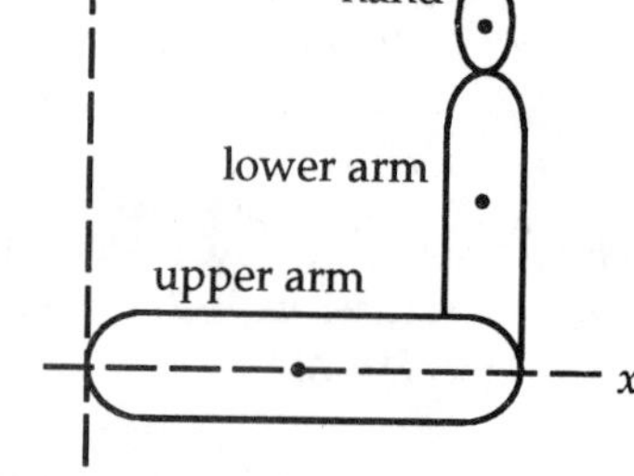

    upper arm: $x_{upper} = [(81.2 - 71.7)/100](155\text{ cm}) = 14.7\text{ cm}$;

    $y_{upper} = 0$;

    lower arm: $x_{lower} = [(81.2 - 62.2)/100](155\text{ cm}) = 29.5\text{ cm}$;

    $y_{lower} = [(62.2 - 55.3)/100](155\text{ cm}) = 10.7\text{ cm}$;

    hand: $x_{hand} = [(81.2 - 62.2)/100](155\text{ cm}) = 29.5\text{ cm}$;

    $y_{hand} = [(62.2 - 43.1)/100](155\text{ cm}) = 29.6\text{ cm}$.

    Because all masses are percentages of the body mass, we can use the percentages rather than the actual mass. Thus we have

    $$x_{CM} = (m_{upper}x_{upper} + m_{lower}x_{lower} + m_{hand}x_{hand})/(m_{upper} + m_{lower} + m_{hand})$$
    $$= [\tfrac{1}{2}(6.6)(14.7\text{ cm}) + \tfrac{1}{2}(4.2)(29.5\text{ cm}) + \tfrac{1}{2}(1.7)(29.5\text{ cm})]/[\tfrac{1}{2}(6.6) + \tfrac{1}{2}(4.2) + \tfrac{1}{2}(1.7)]$$
    $$= \boxed{22\text{ cm.}}$$

    $$y_{CM} = (m_{upper}y_{upper} + m_{lower}y_{lower} + m_{hand}y_{hand})/(m_{upper} + m_{lower} + m_{hand})$$
    $$= [\tfrac{1}{2}(6.6)(0) + \tfrac{1}{2}(4.2)(10.7\text{ cm}) + \tfrac{1}{2}(1.7)(29.6\text{ cm})]/[\tfrac{1}{2}(6.6) + \tfrac{1}{2}(4.2) + \tfrac{1}{2}(1.7)]$$
    $$= \boxed{7.6\text{ cm.}}$$

55. We use the line of the torso as the origin. The vertical locations of the centers of mass for each of the segments, as a percentage of the height, are

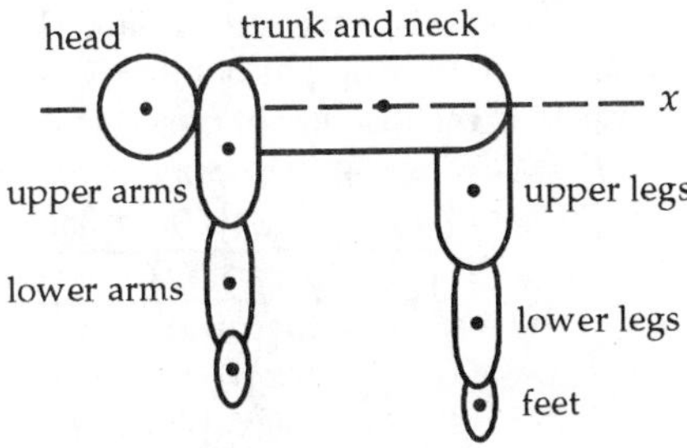

torso and head: 0;
upper arms: $y_{ua} = -(81.2 - 71.7) = -9.5$;
lower arms: $y_{la} = -(81.2 - 55.3) = -25.9$;
hands: $y_h = -(81.2 - 43.1) = -38.1$;
upper legs: $y_{ul} = -(52.1 - 42.5) = -9.6$;
lower legs: $y_{ll} = -(52.1 - 18.1) = -33.9$;
feet: $y_f = -(52.1 - 1.8) = -50.3$.

Because all masses are percentages of the body mass, we can use the percentages rather than the actual mass. Thus we have

$$y_{CM} = (m_{ua}y_{ua} + m_{la}y_{la} + m_h y_h + m_{ul}y_{ul} + m_{ll}y_{ll} + m_f y_f)/m_{body}$$
$$= [(6.6)(-9.5) + (4.2)(-25.9) + (1.7)(-38.1) + (21.5)(-9.6) + (9.6)(-33.9) + (3.4)(-50.3)]/100$$
$$= -9.4.$$

The CM will be **9.4% of the body height** below the line of the torso. For a height of 1.8 m, this is about 17 cm, so **yes,** this will most likely be outside the body.

56. (*a*) If we choose the origin at the center of the Earth, we have

$$x_{CM} = (m_{Earth}x_{Earth} + m_{Moon}x_{Moon})/(m_{Earth} + m_{Moon})$$
$$= [0 + (7.35 \times 10^{22}\text{ kg})(3.84 \times 10^{8}\text{ m})]/(5.98 \times 10^{24}\text{ kg} + 7.35 \times 10^{22}\text{ kg})$$
$$= \boxed{4.66 \times 10^{6}\text{ m.}}$$

Note that this is less than the radius of the Earth and thus is inside the Earth.

(*b*) The CM found in part (*a*) will move around the Sun on an elliptical path. The Earth and Moon will revolve about the CM. Because this is near the center of the Earth, the Earth will essentially be on the elliptical path around the Sun. The motion of the Moon about the Sun is more complicated.

57. We choose the origin of our coordinate system at the woman.

(*a*) For their CM we have

$$x_{CM} = (m_{woman}x_{woman} + m_{man}x_{man})/(m_{woman} + m_{man})$$
$$= [0 + (90\text{ kg})(10.0\text{ m})]/(55\text{ kg} + 90\text{ kg})$$
$$= \boxed{6.2\text{ m.}}$$

(*b*) Because the CM will not move, we find the location of the woman from

$$x_{CM} = (m_{woman}x_{woman}' + m_{man}x_{man}')/(m_{woman} + m_{man})$$
$$6.2\text{ m} = [(55\text{ kg})x_{woman}' + (90\text{ kg})(10.0\text{ m} - 2.5\text{ m})]/(55\text{ kg} + 90\text{ kg}),\text{ which gives}$$
$$x_{woman}' = 4.1\text{ m.}$$

The separation of the two will be 7.5 m – 4.1 m = **3.4 m.**

(*c*) The two will meet at the CM, so he will have moved 10.0 m – 6.2 m = **3.8 m.**

58. Because the two segments of the mallet are uniform, we know that the center of mass of each segment is at its midpoint. We choose the origin at the bottom of the handle. The mallet will spin about the CM, which is the point that will follow a parabolic trajectory:

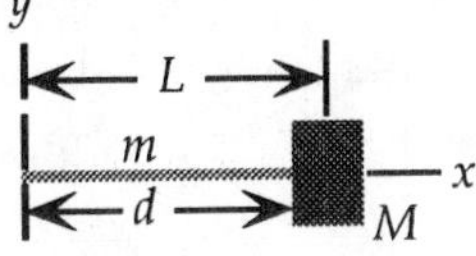

$$x_{CM} = (md + ML)/(m + M)$$
$$= [(0.500\text{ kg})(12.0\text{ cm}) + (2.00\text{ kg})(24.0\text{ cm} + 4.00\text{ cm})]/(0.500\text{ kg} + 2.00\text{ kg})$$
$$= \boxed{24.8\text{ cm.}}$$

59. The CM will land at the same point, $2D$ from the launch site. If part I is still stopped by the explosion, it will fall straight down, as before.

(*a*) We find the location of part II from the CM:

$$x_{CM} = (m_I x_I + m_{II} x_{II})/(m_I + m_{II})$$
$$2D = [m_I D + 3m_I x_{II}]/(m_I + 3m_I),\text{ which gives}$$
$$x_{II} = \boxed{7D/3,\text{ or } 2D/3 \text{ closer to the launch site.}}$$

(*b*) For the new mass distribution, we have

$$x_{CM} = (m_I x_I + m_{II} x_{II})/(m_I + m_{II})$$
$$2D = [3m_{II} D + m_{II} x_{II}]/(3m_{II} + m_{II}),\text{ which gives}$$
$$x_{II} = \boxed{5D,\text{ or } 2D \text{ farther from the launch site.}}$$

60. The forces on the balloon, gondola, and passenger are balanced, so the CM does not move relative to the Earth. As the passenger moves down at a speed v relative to the balloon, the balloon will move up. If the speed of the balloon is v' relative to the Earth, the passenger will move down at a speed $v - v'$ relative to the Earth. We choose the location of the CM as the origin and determine the positions after a time t:

$$x_{CM} = (m_{balloon} x_{balloon} + m_{passenger} x_{passenger})/(m_{balloon} + m_{passenger})$$
$$0 = [Mv't + m(v - v')t]/(M + m),\text{ which gives}$$
$$\boxed{v' = mv/(M + m)\text{ up.}}$$

If the passenger stops, the gondola and $\boxed{\text{the balloon will also stop.}}$ There will be equal and opposite impulses acting when the passenger grabs the rope to stop.

61. We find the force on the person from the magnitude of the force required to change the momentum of the air:

$$F = \Delta p/\Delta t = (\Delta m/\Delta t)v$$
$$= (40\text{ kg/s}\cdot\text{m}^2)(1.50\text{ m})(0.50\text{ m})(100\text{ km/h})/(3.6\text{ ks/h}) = \boxed{8.3 \times 10^2\text{ N.}}$$

The maximum friction force will be

$F_{fr} = \mu mg \approx (1.0)(70\text{ kg})(9.80\text{ m/s}^2) = 6.9 \times 10^2$ N, so the forces are $\boxed{\text{about the same.}}$

62. For the system of railroad car and snow, the horizontal momentum will be constant. For the horizontal motion, we take the direction of the car for the positive direction. The snow initially has no horizontal velocity. For this perfectly inelastic collision, we use momentum conservation:

$$M_1 v_1 + M_2 v_2 = (M_1 + M_2)V;$$

$(5800\text{ kg})(8.60\text{ m/s}) + 0 = [5800\text{ kg} + (3.50\text{ kg/min})(90.0\text{ min})]V$, which gives $\boxed{V = 8.16\text{ m/s.}}$

Note that there is a vertical impulse, so the vertical momentum is not constant.

63. We find the speed after being hit from the height h using energy conservation:

$$\tfrac{1}{2}mv'^2 = mgh,\text{ or } v' = (2gh)^{1/2} = [2(9.80\text{ m/s}^2)(55.6\text{ m})]1/2 = 33.0\text{ m/s.}$$

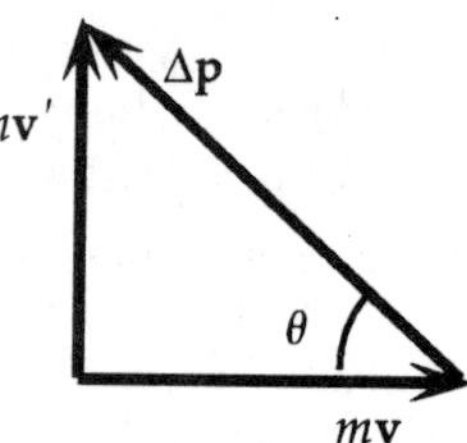

We see from the diagram that the magnitude of the change in momentum is

$$\Delta p = m(v^2 + v'^2)^{1/2}$$
$$= (0.145\text{ kg})[(35.0\text{ m/s})^2 + (33.0\text{ m/s})^2]^{1/2} = 6.98\text{ kg}\cdot\text{m/s.}$$

We find the force from

$$F\,\Delta t = \Delta p;$$
$$F(0.50 \times 10^{-3}\text{ s}) = 6.98\text{ kg}\cdot\text{m/s, which gives } F = \boxed{1.4 \times 10^4\text{ N.}}$$

We find the direction of the force from

$$\tan\theta = v'/v = (33.0\text{ m/s})/(35.0\text{ m/s}) = 0.943,\ \theta = \boxed{43.3°.}$$

64. For momentum conservation we have

x: $mv_0 = \tfrac{2}{3}mv_x'$, which gives $\boxed{v_x' = \tfrac{3}{2}v_0;}$

y: $0 = \tfrac{1}{3}m(2v_0) - \tfrac{2}{3}mv_y'$, which gives $\boxed{v_y' = -v_0.}$

The rocket's forward speed increases because the fuel is shot backward relative to the rocket.

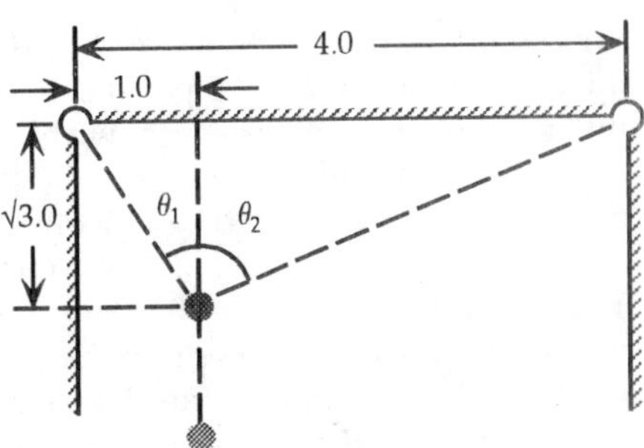

65. From the result of Problem 45, the angle between the two final directions will be 90° for an elastic collision. We take the initial direction of the cue ball to be parallel to the side of the table. The angles for the two balls after the collision are

$\tan\theta_1 = 1.0/\sqrt{3.0}$, which gives $\theta_1 = 30°$;

$\tan\theta_2 = (4.0 - 1.0)/\sqrt{3.0}$, which gives $\theta_2 = 60°$.

Because their sum is 90°, **this will be a "scratch shot".**

66. In the reference frame of the capsule before the push, we take the positive direction in the direction the capsule will move.

(a) Momentum conservation gives us

$mv_{\text{astronaut}} + Mv_{\text{satellite}} = mv_{\text{astronaut}}' + Mv_{\text{satellite}}'$,

$0 + 0 = (140\text{ kg})(-2.50\text{ m/s}) + (1800\text{ kg})v_{\text{satellite}}'$, which gives $v_{\text{satellite}}' =$ **0.194 m/s.**

(b) We find the force on the satellite from

$F_{\text{satellite}} = \Delta p_{\text{satellite}}/\Delta t = m_{\text{satellite}}\,\Delta v_{\text{satellite}}/\Delta t$

$=(1800\text{ kg})(0.194\text{ m/s} - 0)/(0.500\text{ s}) =$ **700 N.**

There will be an equal but opposite force on the astronaut.

67. For each of the elastic collisions with a step, conservation of kinetic energy means that the velocity reverses direction but has the same magnitude. Thus the golf ball always rebounds to the height from which it started. Thus, after five bounces, the bounce height will be **4.00 m.**

68. For the elastic collision of the two balls, we use momentum conservation:

$m_1v_1 + m_2v_2 = m_1v_1' + m_2v_2'$;

$mv_1 + 0 = m(-v_1/4) + m_2v_2'$, or $m_2v_2' = 5mv_1/4$.

Because the collision is elastic, the relative speed does not change:

$v_1 - 0 = -(v_1' - v_2')$; $v_1 = v_2' - (-v_1/4)$, or $v_2' = 3v_1/4$.

Combining these two equations, we get **$m_2 = 5m/3$.**

69. On the horizontal surface, the normal force on a car is $F_N = mg$. We find the speed of a car immediately after the collision by using the work-energy principle for the succeeding sliding motion:

$W_{\text{fr}} = \Delta\text{KE}$;

$-\mu_k mgd = 0 - \frac{1}{2}mv^2$.

We use this to find the speeds of the cars after the collision:

$0.60(9.80\text{ m/s}^2)(15\text{ m}) = \frac{1}{2}v_A'^2$, which gives $v_A' = 13.3\text{ m/s}$;

$0.60(9.80\text{ m/s}^2)(30\text{ m}) = \frac{1}{2}v_B'^2$, which gives $v_B' = 18.8\text{ m/s}$.

For the collision, we use momentum conservation:

$m_Av_A + m_Bv_B = m_Av_A' + m_Bv_B'$;

$(2000\text{ kg})v_A + 0 = (2000\text{ kg})(13.3\text{ m/s}) + (1000\text{ kg})(18.8\text{ m/s})$, which gives $v_A = 22.7\text{ m/s}$.

We find the speed of car A before the brakes were applied by using the work-energy principle for the preceding sliding motion:

$W_{\text{fr}} = \Delta\text{KE}$;

$-\mu_k m_A gd = \frac{1}{2}m_Av_A^2 - \frac{1}{2}m_Av_{A0}^2$;

$-0.60(9.80\text{ m/s}^2)(15\text{ m}) = \frac{1}{2}[(22.7\text{ m/s})^2 - v_{A0}^2]$,

which gives $v_{A0} = 26.3\text{ m/s} = 94.7\text{ km/h} =$ **59 mi/h.**

70. We choose the origin of our coordinate system at the initial position of the 75-kg person. For the location of the center of mass of the system we have

$$x_{CM} = (m_1x_1 + m_2x_2 + m_{boat}x_{boat})/(m_1 + m_2 + m_{boat})$$
$$= [(75 \text{ kg})(0) + (60 \text{ kg})(2.0 \text{ m}) + (80 \text{ kg})(1.0 \text{ m})]/(75 \text{ kg} + 60 \text{ kg} + 80 \text{ kg}) = 0.93 \text{ m}.$$

Thus the CM will be 2.0 m – 0.93 m = 1.07 m from the 60-kg person. When the two people exchange seats, the CM will not move. The end where the 75-kg person started, which was 0.93 m from the CM, is now 1.07 m from the CM, that is, the boat must have moved 1.07 m – 0.93 m =

0.14 m toward the initial position of the 75-kg person.

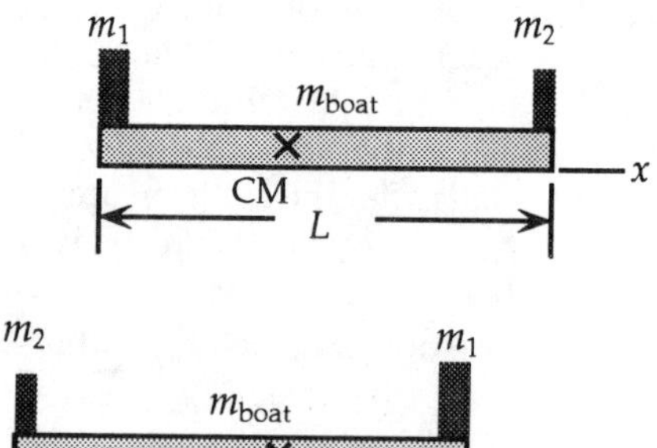

71. (*a*) We take the direction of the meteor for the positive direction.
For this perfectly inelastic collision, we use momentum conservation:

$$M_{meteor}v_{meteor} + M_{Earth}v_{Earth} = (M_{meteor} + M_{Earth})V;$$
$$(10^8 \text{ kg})(15 \times 10^3 \text{ m/s}) + 0 = (10^8 \text{ kg} + 6.0 \times 10^{24} \text{ kg})V, \text{ which gives } \boxed{V = 2.5 \times 10^{-13} \text{ m/s.}}$$

(*b*) The fraction transformed was

$$\text{fraction} = \Delta KE_{Earth}/KE_{meteor} = \tfrac{1}{2}m_{Earth}V^2/\tfrac{1}{2}m_{meteor}v_{meteor}^2$$
$$= (6.0 \times 10^{24} \text{ kg})(2.5 \times 10^{-13} \text{ m/s})^2/(10^8 \text{ kg})(15 \times 10^3 \text{ m/s})^2 = \boxed{1.7 \times 10^{-17}.}$$

(*c*) The change in the Earth's kinetic energy was

$$\Delta KE_{Earth} = \tfrac{1}{2}m_{Earth}V^2$$
$$= \tfrac{1}{2}(6.0 \times 10^{24} \text{ kg})(2.5 \times 10^{-13} \text{ m/s})^2 = \boxed{0.19 \text{ J.}}$$

72. Momentum conservation gives

$$0 = m_1v_1' + m_2v_2', \text{ or } v_2'/v_1' = -m_1/m_2.$$

The ratio of kinetic energies is

$$KE_2/KE_1 = \tfrac{1}{2}m_2v_2'^2/\tfrac{1}{2}m_1v_1'^2 = (m_2/m_1)(v_2'/v_1')^2 = 2.$$

When we use the result from momentum, we get

$$(m_2/m_1)(-m_1/m_2)^2 = 2, \text{ which gives } m_1/m_2 = \boxed{2.}$$

73. (*b*) The force would become zero at

$$t = 580/(1.8 \times 10^5) = 3.22 \times 10^{-3} \text{ s}.$$

At $t = 3.0 \times 10^{-3}$ s the force is

$$580 - (1.8 \times 10^5)(3.0 \times 10^{-3} \text{ s}) = +40 \text{ N}.$$

The impulse is the area under the F vs. t curve, and consists of a triangle and a rectangle:

$$\text{Impulse} = \tfrac{1}{2}(580 \text{ N} - 40 \text{ N})(3.0 \times 10^{-3} \text{ s}) +$$

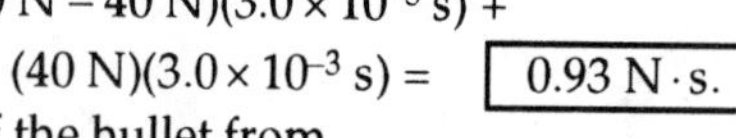

$$(40 \text{ N})(3.0 \times 10^{-3} \text{ s}) = \boxed{0.93 \text{ N}\cdot\text{s.}}$$

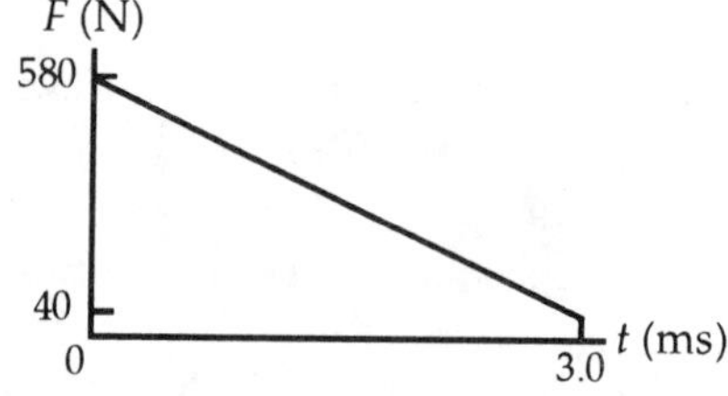

(*c*) We find the mass of the bullet from

$$\text{Impulse} = \Delta p = m\,\Delta v;$$
$$0.93 \text{ N}\cdot\text{s} = m(220 \text{ m/s} - 0), \text{ which gives } m = 4.23 \times 10^{-3} \text{ kg} = \boxed{4.2 \text{ g.}}$$

74. We find the speed for falling or rising through a height h from energy conservation:

$\frac{1}{2}mv^2 = mgh$, or $v^2 = 2gh$.

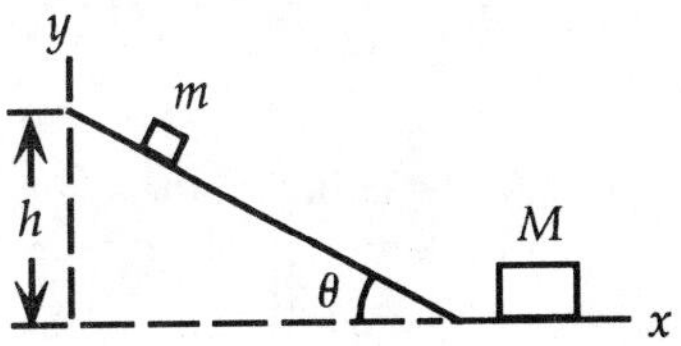

(a) The speed of the first block after sliding down the incline and just before the collision is

$v_1 = [2(9.80 \text{ m/s}^2)(3.60 \text{ m})]^{1/2} = 8.40 \text{ m/s}$.

For the elastic collision of the two blocks, we use momentum conservation:

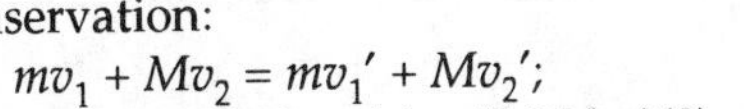

$mv_1 + Mv_2 = mv_1' + Mv_2'$;

$(2.20 \text{ kg})(8.40 \text{ m/s}) + (7.00 \text{ kg})(0) = (2.20 \text{ kg})v_1' + (7.00 \text{ kg})v_2'$.

Because the collision is elastic, the relative speed does not change:

$v_1 - v_2 = -(v_1' - v_2')$, or $8.40 \text{ m/s} - 0 = v_2' - v_1'$.

Combining these two equations, we get

$v_1' = -4.38 \text{ m/s}$, $v_2' = 4.02 \text{ m/s}$.

(b) We find the height of the rebound from

$v_1'^2 = 2gh'$;

$(-4.38 \text{ m/s})^2 = 2(9.80 \text{ m/s}^2)h'$, which gives $h' = 0.979$ m.

The distance along the incline is

$d = h'/\sin\theta = (0.979 \text{ m})/\sin 30° =$ 1.96 m.

75. Because energy is conserved for the motion up and down the incline, mass m will return to the level with the speed $-v_1'$. For a second collision to occur, mass m must be moving faster than mass M: $-v_1' \geq v_2'$.

In the first collision, the relative speed does not change:

$v_1 - 0 = -(v_1' - v_2')$, or $-v_1' = v_1 - v_2'$,

so the condition becomes $v_1 - v_2' \geq v_2'$, or $v_1 \geq 2v_2'$.

For the first collision, we use momentum conservation:

$mv_1 + 0 = mv_1' + Mv_2'$, or $v_1 - v_1' = (M/m)v_2'$.

When we use the two versions of the condition, we get $v_1 - v_1' \geq 3v_2'$, so we need

$(M/m) \geq 3$, or $m \leq M/3$.

76. (*a*) If the skeet were not hit by the pellet, the horizontal distance it would travel can be found from the range expression for projectile motion:

$R = (v_0^2/g) \sin 2\theta$

$= [(30 \text{ m/s})^2/(9.80 \text{ m/s}^2)] \sin 2(30°) = 79.5 \text{ m}.$

At the collision the skeet will have the x-component of the initial velocity:

$v_1 = v_0 \cos\theta = (30 \text{ m/s}) \cos 30° = 26.0 \text{ m/s}.$

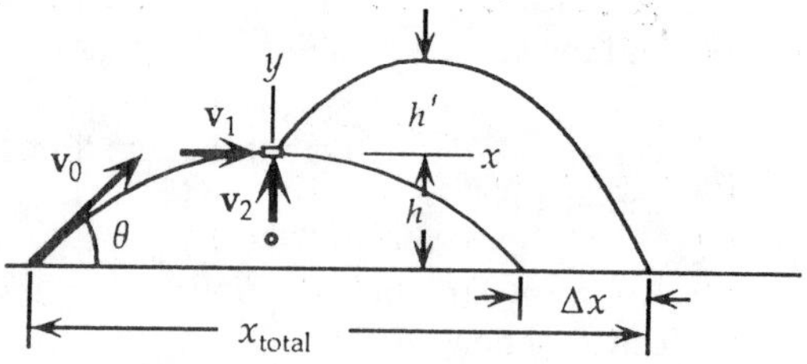

We use energy conservation to find the height attained by the skeet when the collision occurs:

$\frac{1}{2}Mv_0^2 = \frac{1}{2}Mv_1^2 + Mgh;$

$\frac{1}{2}(30 \text{ m/s})^2 = \frac{1}{2}(26.0 \text{ m/s})^2 + (9.80 \text{ m/s}^2)h$, which gives $h = 11.5$ m.

Using the coordinate system shown, for momentum conservation of the collision we have

x: $Mv_1 + 0 = (M + m)V_x;$

$(250 \text{ g})(26.0 \text{ m/s}) = (250 \text{ g} + 15 \text{ g})V_x$, which gives $V_x = 24.5$ m/s;

y: $0 + mv_2 = (M + m)V_y;$

$(15 \text{ g})(200 \text{ m/s}) = (250 \text{ g} + 15 \text{ g})V_y$, which gives $V_y = 11.3$ m/s.

We use energy conservation to find the additional height attained by the skeet after the collision:

$\frac{1}{2}(M + m)(V_x^2 + V_y^2) = \frac{1}{2}(M + m)V_x^2 + (M + m)gh';$

$\frac{1}{2}[(21.7 \text{ m/s})^2 + (11.3 \text{ m/s})^2] = \frac{1}{2}(21.7 \text{ m/s})^2 + (9.80 \text{ m/s}^2)h'$, which gives $\boxed{h' = 6.54 \text{ m.}}$

(*b*) We find the time for the skeet to reach the ground from the vertical motion:

$y = y_0 + V_y t + \frac{1}{2}(-g)t^2;$

$-11.5 \text{ m} = 0 + (11.3 \text{ m/s})t - \frac{1}{2}(9.80 \text{ m/s}^2)t^2.$

The positive solution to this quadratic equation is $t = 3.07$ s.

The horizontal distance from the collision is

$x = V_x t = (24.5 \text{ m/s})(3.07 \text{ s}) = 75 \text{ m}.$

The total horizontal distance covered is

$x_{\text{total}} = \frac{1}{2}R + x = \frac{1}{2}(79.5 \text{ m}) + 75 \text{ m} = 115 \text{ m}.$

Because of the collision, the skeet will have traveled an additional distance of

$\Delta x = x_{\text{total}} - R = 115 \text{ m} - 79.5 \text{ m} = \boxed{35 \text{ m.}}$

77. Obviously the spacecraft will have negligible effect on the motion of Saturn. In the reference frame of Saturn, we can treat this as the equivalent of a small mass "bouncing off" a massive object. The relative velocity of the spacecraft in this reference frame will be reversed.

The initial relative velocity of the spacecraft is

$v_{SpS} = v_{Sp} - v_S = 10.4 \text{ km/s} - (-9.6 \text{ km/s}) = 20.0 \text{ km/s}.$

so the final relative velocity is $v_{SpS}' = -20.0$ km/s. Therefore, we find the final velocity of the spacecraft from

$v_{SpS}' = v_{Sp}' - v_S;$

$-20.0 \text{ km/s} = v_{Sp}' - (-9.6 \text{ km/s})$, which gives $v_{Sp}' = -29.6$ km/s,

so the final speed of the spacecraft is $\boxed{29.6 \text{ km/s.}}$

CHAPTER 8

1. (*a*) $30° = (30°)(\pi \text{ rad}/180°) =$ $\boxed{\pi/6 \text{ rad} = 0.524 \text{ rad};}$
   (*b*) $57° = (57°)(\pi \text{ rad}/180°) =$ $\boxed{19\pi/60 = 0.995 \text{ rad};}$
   (*c*) $90° = (90°)(\pi \text{ rad}/180°) =$ $\boxed{\pi/2 = 1.571 \text{ rad};}$
   (*d*) $360° = (360°)(\pi \text{ rad}/180°) =$ $\boxed{2\pi = 6.283 \text{ rad};}$
   (*e*) $420° = (420°)(\pi \text{ rad}/180°) =$ $\boxed{7\pi/3 = 7.330 \text{ rad}.}$

2. The subtended angle in radians is the size of the object divided by the distance to the object:
   $\theta = 2R_{Sun}/r;$
   $(0.5°)(\pi \text{ rad}/180°) = 2R_{Sun}/(150 \times 10^6 \text{ km})$, which gives $\boxed{R_{Sun} = 6.5 \times 10^5 \text{ km}.}$

3. The subtended angle in radians is the size of the object divided by the distance to the object:
   $\theta_{Sun} = 2R_{Sun}/r_{Sun} = 2(6.96 \times 10^5 \text{ km})/(149.6 \times 10^6 \text{ km}) =$ $\boxed{9.30 \times 10^{-3} \text{ rad } (0.53°);}$
   $\theta_{Moon} = 2R_{Moon}/r_{Moon} = 2(1.74 \times 10^3 \text{ km})/(384 \times 10^3 \text{ km}) =$ $\boxed{9.06 \times 10^{-3} \text{ rad } (0.52°).}$

4. We find the distance from
   $\theta = h/r;$
   $(6°)(\pi \text{ rad}/180°) = (300 \text{ m})/r$; which gives $r =$ $\boxed{2.9 \times 10^3 \text{ m}.}$

5. We find the diameter of the spot from
   $\theta = D_{spot}/r;$
   $(1.8 \times 10^{-5} \text{ rad}) = D_{spot}/(380 \times 10^3 \text{ km})$, which gives $\boxed{D_{spot} = 6.8 \text{ km}.}$

6. $\omega = (1800 \text{ rev/min})(2\pi \text{ rad/rev})(1 \text{ min}/60 \text{ s}) =$ $\boxed{188 \text{ rad/s}.}$

7. The linear speed of the point on the edge is the tangential speed:
   $v = r\omega = (0.175 \text{ m})(188 \text{ rad/s}) =$ $\boxed{33 \text{ m/s}.}$
   Because the speed is constant, the tangential acceleration is zero. There will be a radial acceleration:
   $a_R = \omega^2 R = (188 \text{ rad/s})^2(0.175 \text{ m}) =$ $\boxed{6.2 \times 10^3 \text{ m/s}^2.}$

8. From the definition of angular acceleration, we have
   $\alpha = \Delta\omega/\Delta t = [(33 \text{ rev/min})(2\pi \text{ rad/rev})(1 \text{ min}/60 \text{ s}) - 0]/(1.8 \text{ s}) =$ $\boxed{1.9 \text{ rad/s}^2.}$

9. From the definition of angular velocity, we have
   $\omega = \Delta\theta/\Delta t$, and we use the time for each hand to turn through a complete circle, 2π rad.
   (*a*) $\omega_{second} = \Delta\theta/\Delta t$
   $= (2\pi \text{ rad})/(60 \text{ s}) =$ $\boxed{0.105 \text{ rad/s}.}$
   (*b*) $\omega_{minute} = \Delta\theta/\Delta t$
   $= (2\pi \text{ rad})/(60 \text{ min})(60 \text{ s/min}) =$ $\boxed{1.75 \times 10^{-3} \text{ rad/s}.}$
   (*c*) $\omega_{hour} = \Delta\theta/\Delta t$
   $= (2\pi \text{ rad})/(12 \text{ h})(60 \text{ min/h})(60 \text{ s/min}) =$ $\boxed{1.45 \times 10^{-4} \text{ rad/s}.}$
   (*d*) For each case, the angular velocity is constant, so the angular acceleration is $\boxed{\text{zero.}}$

10. From the definition of angular acceleration, we have
    $\alpha = \Delta\omega/\Delta t = [0 - (7500 \text{ rev/min})(2\pi \text{ rad/rev})(1 \text{ min}/60 \text{ s})]/(3.0 \text{ s}) =$ $\boxed{-2.6 \times 10^2 \text{ rad/s}^2.}$

11. In each revolution the ball rolls a distance equal to its circumference, so we have
$L = N(\pi D)$;
4.5 m = (15.0)πD, which gives D = 0.095 m = **9.5 cm.**

12. In each revolution the wheel rolls a distance equal to its circumference, so we have
$L = N(\pi D)$;
7.0×10^3 m = $N\pi$(0.68 m), which gives N = **3.3×10^3 revolutions.**

13. The subtended angle in radians is the size of the object divided by the distance to the object. A pencil with a diameter of 6 mm will block out the Moon if it is held about 60 cm from the eye. For the angle subtended we have $\theta_{Moon} = D_{pencil}/r_{pencil} \approx$ (0.6 cm)/(60 cm) **$\approx$ 0.01 rad.**
We estimate the diameter of the Moon from
$\theta_{Moon} = D_{Moon}/r_{Moon}$;
0.01 rad = $D_{Moon}/(3.8 \times 10^5$ km), which gives **$D_{Moon} \approx 3.8 \times 10^3$ km.**

14. (*a*) The Earth moves one revolution around the Sun in one year, so we have
$\omega_{orbit} = \Delta\theta/\Delta t$
= (2π rad)/(1 yr)(3.16×10^7 s/yr) = **1.99×10^{-7} rad/s.**
(*b*) The Earth rotates one revolution in one day, so we have
$\omega_{rotation} = \Delta\theta/\Delta t$
= (2π rad)/(1 day)(24 h/day)(3600 s/h) = **7.27×10^{-5} rad/s.**

15. All points will have the angular speed of the Earth:
$\omega = \Delta\theta/\Delta t$ = (2π rad)/(1 day)(24 h/day)(3600 s/h) = 7.27×10^{-5} rad/s.
Their linear speed will depend on the distance from the rotation axis.
(*a*) On the equator we have
$v = R_{Earth}\omega$ = (6.38×10^6 m)(7.27×10^{-5} rad/s) = **464 m/s.**
(*b*) At a latitude of 66.5° the distance is R_{Earth} cos 66.5°, so we have
$v = R_{Earth}$ cos 66.5° ω = (6.38×10^6 m)(cos 66.5°)(7.27×10^{-5} rad/s) = **185 m/s.**
(*c*) At a latitude of 45.0° the distance is R_{Earth} cos 45.0°, so we have
$v = R_{Earth}$ cos 45.0° ω = (6.38×10^6 m)(cos 45.0°)(7.27×10^{-5} rad/s) = **328 m/s.**

16. The particle will experience a radial acceleration:
$a_R = \omega^2 r$;
(100,000)(9.80 m/s^2) = ω^2(0.070 m), which gives
ω = (3740 rad/s)(60 s/min)/(2π rad/rev) = **3.6×10^4 rpm.**

17. The initial and final angular speeds are
ω_0 = (160 rpm)(2π rad/rev)/(60 s/min) = 16.8 rad/s;
ω = (280 rpm)(2π rad/rev)/(60 s/min) = 29.3 rad/s.
(*a*) We find the angular acceleration from
$\alpha = \Delta\omega/\Delta t$
= (29.3 rad/s – 16.8 rad/s)/(4.0 s) = **3.1 rad/s^2.**
(*b*) We find the angular speed after 2.0 s:
$\omega = \omega_0 + \alpha t$ = 16.8 rad/s + (3.13 rad/s^2)(2.0 s) = 23.1 rad/s.
At this time the radial acceleration of a point on the rim is
$a_R = \omega^2 r$ = (23.1 rad/s)^2(0.35 m) = **1.9×10^2 m/s^2.**
The tangential acceleration is
$a_{tan} = \alpha r$ = (3.13 rad/s^2)(0.35 m) = **1.1 m/s^2.**

18. If there is no slipping, the tangential speed of the outer edge of the turntable is the tangential speed of the outer edge of the roller:
$v = R_1\omega_1 = R_2\omega_2$, which gives $\boxed{\omega_1/\omega_2 = R_2/R_1.}$

19. The final angular speed is
$\omega = (1 \text{ rpm})(2\pi \text{ rad/rev})/(60 \text{ s/min}) = 0.105 \text{ rad/s}.$
(*a*) We find the angular acceleration from
$\alpha = \Delta\omega/\Delta t$
$= (0.105 \text{ rad/s} - 0)/(10.0 \text{ min})(60 \text{ s/min}) = \boxed{1.8 \times 10^{-4} \text{ rad/s}^2.}$
(*b*) We find the angular speed after 5.0 min:
$\omega = \omega_0 + \alpha t = 0 + (1.8 \times 10^{-4} \text{ rad/s}^2)(5.0 \text{ min})(60 \text{ s/min}) = 5.45 \times 10^{-2} \text{ rad/s}.$
At this time the radial acceleration of a point on the skin is
$a_R = \omega^2 r = (5.45 \times 10^{-2} \text{ rad/s})^2(4.25 \text{ m}) = \boxed{1.2 \times 10^{-2} \text{ m/s}^2.}$
The tangential acceleration is
$a_{tan} = \alpha r = (1.8 \times 10^{-4} \text{ rad/s}^2)(4.25 \text{ m}) = \boxed{7.7 \times 10^{-4} \text{ m/s}^2.}$

20. For motion with constant angular acceleration we use
$\omega^2 = \omega_0^2 + 2\alpha\theta;$
$[(33 \text{ rpm})(2\pi \text{ rad/rev})/(60 \text{ s/min})]^2 = 0 + 2\alpha(1.7 \text{ rev})(2\pi \text{ rad/rev})$, which gives $\alpha = \boxed{0.56 \text{ rad/s}^2.}$

21. For the angular displacement we use
$\theta = \frac{1}{2}(\omega_0 + \omega)t$
$= \frac{1}{2}[0 + (15{,}000 \text{ rpm})(2\pi \text{ rad/rev})/(60 \text{ s/min})](220 \text{ s}) = 1.73 \times 10^5 \text{ rad}.$
Thus $\theta = (1.73 \times 10^5 \text{ rad})/(2\pi \text{ rad/rev}) = \boxed{2.75 \times 10^4 \text{ rev.}}$

22. (*a*) For motion with constant angular acceleration we use
$\omega = \omega_0 + \alpha t;$
$(1200 \text{ rev/min})(2\pi \text{ rad/rev})(1 \text{ min}/60\text{s}) = (4000 \text{ rev/min})(2\pi \text{ rad/rev})(1 \text{ min}/60\text{s}) + \alpha(3.5 \text{ s}),$
which gives $\alpha = \boxed{-84 \text{ rad/s}^2.}$
(*b*) For the angular displacement we use
$\theta = \frac{1}{2}(\omega_0 + \omega)t$
$= \frac{1}{2}[(4000 \text{ rev/min} + 1200 \text{ rev/min})][(2\pi \text{ rad/rev})/(60 \text{ s/min})](3.5 \text{ s}) = 954 \text{ rad}.$
Thus $\theta = (954 \text{ rad})/(2\pi \text{ rad/rev}) = \boxed{1.5 \times 10^2 \text{ rev.}}$

23. (*a*) We find the angular acceleration from
$\theta = \omega_0 t + \frac{1}{2}\alpha t^2;$
$(20 \text{ rev})(2\pi \text{ rad/rev}) = 0 + \frac{1}{2}\alpha[(1 \text{ min})(60 \text{ s/min})]^2$, which gives $\boxed{\alpha = 0.070 \text{ rad/s}^2.}$
(*b*) We find the final angular speed from
$\omega = \omega_0 + \alpha t = 0 + (0.070 \text{ rad/s}^2)(60 \text{ s}) = 4.2 \text{ rad/s} = \boxed{40 \text{ rpm.}}$

24. We find the total angle the wheel turns from
$\theta = \frac{1}{2}(\omega_0 + \omega)t$
$= \frac{1}{2}[(210 \text{ rev/min} + 350 \text{ rev/min})][(2\pi \text{ rad/rev})/(60 \text{ s/min})](6.5 \text{ s}) = 191 \text{ rad} = 30.3 \text{ rev}.$
For each revolution the point on the edge will travel one circumference, so the total distance traveled is
$d = \theta\pi D = (30.3 \text{ rev})\pi(0.40 \text{ m}) = \boxed{38 \text{ m.}}$

25. We use the initial conditions of $t = 0$, $\theta_0 = 0$, and ω_0. If the angular acceleration is constant, the average angular acceleration is also the instantaneous angular acceleration. From the definition of angular acceleration, we have

$\alpha = \alpha_{av} = \Delta\omega/\Delta t = (\omega - \omega_0)/(t - 0)$, which gives $\omega = \omega_0 + \alpha t$; Eq. (8–9a).

Because the angular velocity is a linear function of the time, the average velocity will be

$\omega_{av} = \frac{1}{2}(\omega_0 + \omega)$; Eq. (8–9d).

From the definition of angular velocity, we have

$\omega = \omega_{av} = \Delta\theta/\Delta t$;

$\frac{1}{2}(\omega_0 + \omega) = (\theta - 0)/(t - 0)$, which gives $t = 2\theta/(\omega + \omega_0)$, or $\omega = -\omega_0 + 2\theta/t$.

When we substitute the expression for t in Eq. (8–9a), we get

$\omega - \omega_0 = \alpha[2\theta/(\omega + \omega_0)]$, which simplifies to $\omega^2 = \omega_0^2 + 2\alpha\theta$; Eq. (8–9c).

When we substitute the expression for ω in Eq. (8–9a), we get

$-\omega_0 + 2\theta/t = \omega_0 + \alpha t$, which simplifies to $\theta = \omega_0 t + \frac{1}{2}\alpha t^2$; Eq. (8–9b).

26. (*a*) If there is no slipping, the linear tangential acceleration of the pottery wheel and the rubber wheel must be the same:

$a_{tan} = R_1\alpha_1 = R_2\alpha_2$;

(2.0 cm)(7.2 rad/s²) = (25.0 cm)α_2, which gives $\alpha_2 =$ **0.58 rad/s².**

(*b*) We find the time from

$\omega = \omega_0 + \alpha t$;

(65 rev/min)(2π rad/rev)/(60 s/min) = 0 + (0.58 rad/s²)t, which gives $t =$ **12 s.**

27. We find the initial and final angular velocities of the wheel from the rolling condition:

$\omega_0 = v_0/r$ = [(100 km/h)/(3.6 ks/h)]/(0.40 m) = 69.5 rad/s;

$\omega = v/r$ = [(50 km/h)/(3.6 ks/h)]/(0.40 m) = 34.8 rad/s.

(*a*) We find the angular acceleration from

$\omega^2 = \omega_0^2 + 2\alpha\theta$;

(34.8 rad/s)² = (69.5 rad/s)² + 2α(65 rev)(2π rad/rev), which gives $\alpha =$ **– 4.4 rad/s².**

(*b*) We find the additional time from

$\omega_{final} = \omega + \alpha t$;

0 = 34.8 rad/s + (– 4.4 rad/s²)t, which gives $t =$ **7.9 s.**

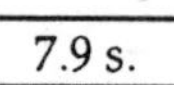

28. (*a*) The tangential acceleration is

$a_{tan} = \alpha r$.

The angular speed as a function of time is

$\omega = \omega_0 + \alpha t = 0 + \alpha t = \alpha t$.

The radial acceleration is

$a_R = \omega^2 r = (\alpha t)^2 r = \alpha^2 t^2 r$.

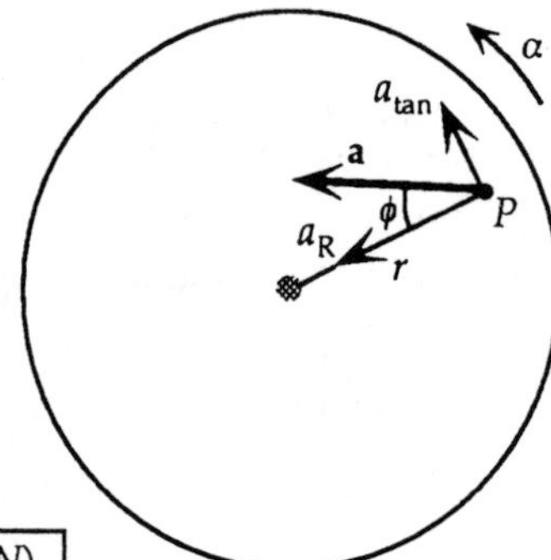

(*b*) Because each revolution corresponds to an angular displacement of 2π, the number of revolutions as a function of time is

$N = \theta/2\pi = (\omega_0 t + \frac{1}{2}\alpha t^2)/2\pi = \alpha t^2/4\pi$.

From the figure we see that

$\tan\phi = a_{tan}/a_R = \alpha r/\alpha^2 t^2 r = 1/\alpha t^2 = 1/4\pi N$, or **$\phi = \tan^{-1}(1/4\pi N)$.**

29. The force being applied by the rider is equal to his force of gravity. The maximum torque will be exerted when this force is perpendicular to the line from the axis to the pedal:

$\tau_{max} = rF$ = (0.17 m)(55 kg)(9.80 m/s²) = **92 m·N.**

30. If θ is the angle between the force and the surface of the door, we have

(a) $\tau = LF \sin \theta = (0.84 \text{ m})(45 \text{ N}) \sin 90° =$ $\boxed{38 \text{ m} \cdot \text{N}.}$

(b) $\tau = LF \sin \theta = (0.84 \text{ m})(45 \text{ N}) \sin 60° =$ $\boxed{33 \text{ m} \cdot \text{N}.}$

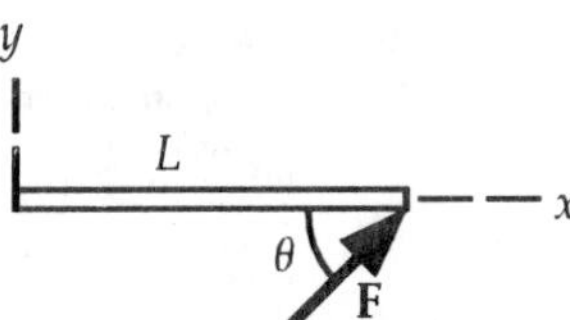

31. We assume clockwise motion, so the frictional torque is counterclockwise. If we take the clockwise direction as positive, we have

$$\tau_{\text{net}} = rF_1 - RF_2 + RF_3 - \tau_{\text{fr}}$$
$$= (0.10 \text{ m})(35 \text{ N}) - (0.20 \text{ m})(30 \text{ N}) + (0.20 \text{ m})(20 \text{ N}) - 0.40 \text{ m} \cdot \text{N}$$
$$= \boxed{1.1 \text{ m} \cdot \text{N (clockwise).}}$$

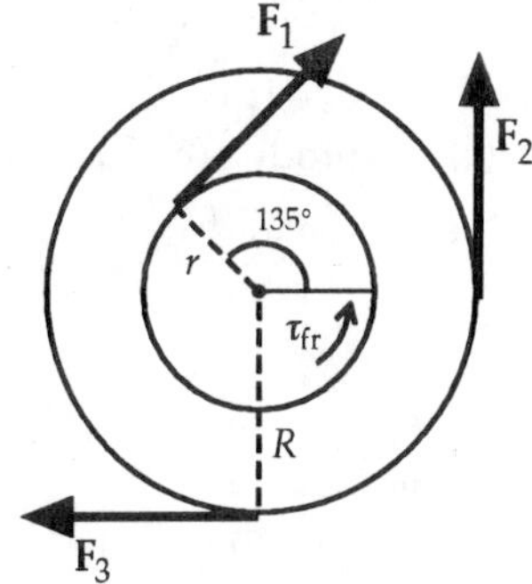

32. Each wheel supports one-quarter of the weight. For the wheels to spin, the applied torque must be greater than the maximum frictional torque produced by the static friction from the pavement:

$$\tau_{\text{applied}} \geq F_{\text{fr}}r = \mu_s F_N r = (0.75)\tfrac{1}{4}(1080 \text{ kg})(9.80 \text{ m/s}^2)(0.33 \text{ m})$$
$$\geq \boxed{6.5 \times 10^2 \text{ m} \cdot \text{N}.}$$

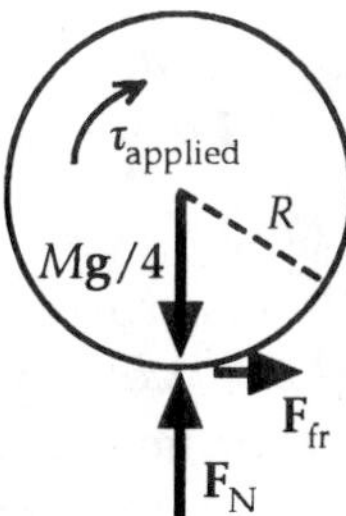

33. The force to produce the required torque is

$$F_{\text{wrench}} = \tau/L = (80 \text{ m} \cdot \text{N})/(0.30 \text{ m}) = \boxed{2.7 \times 10^2 \text{ N}.}$$

Because this torque is balanced by the torque produced by the bolt on the wrench, an equal torque is produced on the bolt. Because there are six points where a force is applied to the bolt, we have

$$F_{\text{bolt}} = (\tau/r)/6 = (80 \text{ m} \cdot \text{N})/6(0.0075 \text{ m}) = \boxed{1.8 \times 10^3 \text{ N}.}$$

34. The moment of inertia of a sphere about an axis through its center is

$$I = (2/5)MR^2 = (2/5)(12.2 \text{ kg})(0.623 \text{ m})^2 = \boxed{1.89 \text{ kg} \cdot \text{m}^2.}$$

35. Because all of the mass is the same distance from the axis, we have

$$I = MR^2 = (1.25 \text{ kg})[\tfrac{1}{2}(0.667 \text{ m})]^2 = \boxed{0.139 \text{ kg} \cdot \text{m}^2.}$$

The mass of the hub can be ignored because $\boxed{\text{the distance of its mass from the axis is so small.}}$

36. (a) For the moment of inertia about the y-axis, we have

$$I_a = \Sigma m_i R_i^2 = md_1^2 + Md_1^2 + m(d_2 - d_1)^2 + M(d_2 - d_1)^2$$
$$= (1.8 \text{ kg})(0.50 \text{ m})^2 + (3.1 \text{ kg})(0.50 \text{ m})^2 +$$
$$(1.8 \text{ kg})(1.00 \text{ m})^2 + (3.1 \text{ kg})(1.00 \text{ m})^2 = \boxed{6.1 \text{ kg} \cdot \text{m}^2.}$$

(b) For the moment of inertia about the x-axis, all the masses are the same distance from the axis, so we have

$$I_b = \Sigma m_i R_i^2 = (2m + 2M)(\tfrac{1}{2}h)^2$$
$$= [2(1.8 \text{ kg}) + 2(3.1 \text{ kg})](0.25 \text{ m})^2 = \boxed{0.61 \text{ kg} \cdot \text{m}^2.}$$

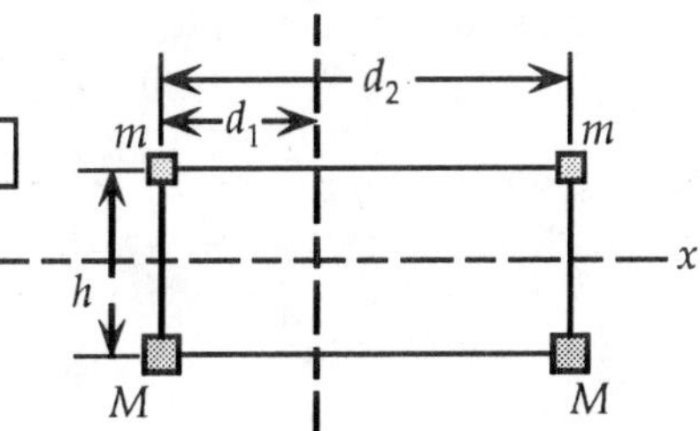

37. If M is the total mass and D is the effective separation, each atom has a mass $\frac{1}{2}M$ and is $\frac{1}{2}D$ from the axis. We find the distance D from

$I = 2(\frac{1}{2}M)(\frac{1}{2}D)^2 = \frac{1}{4}MD^2;$

$1.9 \times 10^{-46}\ \text{kg}\cdot\text{m}^2 = \frac{1}{4}(5.3 \times 10^{-26}\ \text{kg})D^2$, which gives $D =$ $\boxed{1.2 \times 10^{-10}\ \text{m.}}$

38. (*a*) Because we can ignore the mass of the rod, for the moment of inertia we have

$I = m_{\text{ball}}R^2$

$= (1.05\ \text{kg})(0.900\ \text{m})^2 =$ $\boxed{0.851\ \text{kg}\cdot\text{m}^2.}$

(*b*) To produce constant angular velocity, the net torque must be zero:

$\tau_{\text{net}} = \tau_{\text{applied}} - \tau_{\text{friction}} = 0,$ or

$\tau_{\text{applied}} = F_{\text{fr}}R = (0.0800\ \text{N})(0.900\ \text{m}) =$ $\boxed{0.0720\ \text{m}\cdot\text{N.}}$

39. We find the required constant angular acceleration from

$\omega = \omega_0 + \alpha t;$

$(30\ \text{rev/min})(2\pi\ \text{rad/rev})(1\ \text{min}/60\text{s}) = 0 + \alpha(5.0\ \text{min})(60\ \text{s/min})$, which gives $\alpha = 0.0105\ \text{rad/s}^2$.

The moment of inertia of the solid cylinder is $\frac{1}{2}MR^2$. Because we have four forces creating the torque that produces the required acceleration, we have

$\tau = I\alpha;$

$4FR = \frac{1}{2}MR^2\alpha,$ or

$F = MR\alpha/8 = (2600\ \text{kg})(3.0\ \text{m})(0.0105\ \text{rad/s}^2)/8 =$ $\boxed{10\ \text{N.}}$

40. (*a*) The moment of inertia of the solid cylinder is

$I = \frac{1}{2}MR^2 = \frac{1}{2}(0.550\ \text{kg})(0.0850\ \text{m})^2 =$ $\boxed{1.99 \times 10^{-3}\ \text{kg}\cdot\text{m}^2.}$

(*b*) We can find the frictional torque acting on the wheel from the slowing-down motion:

$\tau_{\text{fr}} = I\alpha_1 = I(\omega_1 - \omega_{01})/t_1$

$= (1.99 \times 10^{-3}\ \text{kg}\cdot\text{m}^2)[0 - (1500\ \text{rpm})(2\pi\ \text{rad/rev})/(60\ \text{s/min})]/(55.0\ \text{s}) = -5.67 \times 10^{-3}\ \text{m}\cdot\text{N}.$

For the accelerating motion, we have

$\tau_{\text{applied}} + \tau_{\text{fr}} = I\alpha_2 = I(\omega_2 - \omega_{02})/t_2;$

$\tau_{\text{applied}} - 5.67 \times 10^{-3}\ \text{m}\cdot\text{N} = (1.99 \times 10^{-3}\ \text{kg}\cdot\text{m}^2)[(1500\ \text{rpm})(2\pi\ \text{rad/rev})/(60\ \text{s/min}) - 0]/(5.00\ \text{s}),$

which gives

$\boxed{\tau_{\text{applied}} = 6.82 \times 10^{-2}\ \text{m}\cdot\text{N.}}$

41. For the accelerating motion, we have

$\tau_{\text{applied}} = I\alpha = I(\omega - \omega_0)/t = \frac{1}{3}mL^2(\omega - \omega_0)/t$

$= \frac{1}{3}(2.2\ \text{kg})(0.95\ \text{m})^2[(3.0\ \text{rev/s})(2\pi\ \text{rad/rev}) - 0]/(0.20\ \text{s}) =$ $\boxed{62\ \text{m}\cdot\text{N.}}$

42. The moment of inertia for the system of merry-go-round and children about the center is

$I = \frac{1}{2}MR^2 + 2m_{\text{child}}R^2 = (\frac{1}{2}M + 2m_{\text{child}})R^2 = [\frac{1}{2}(800\ \text{kg}) + 2(25\ \text{kg})](2.5\ \text{m})^2 = 2.81 \times 10^3\ \text{kg}\cdot\text{m}^2.$

We find the torque required from

$\tau_{\text{applied}} = I\alpha = I(\omega - \omega_0)/t$

$= (2.81 \times 10^3\ \text{kg}\cdot\text{m}^2)[(20\ \text{rpm})(2\pi\ \text{rad/rev})/(60\ \text{s/min}) - 0]/(10.0\ \text{s}) =$ $\boxed{5.9 \times 10^2\ \text{m}\cdot\text{N.}}$

Because the worker is pushing perpendicular to the radius, the required force is

$F = \tau_{\text{applied}}/R = (5.9 \times 10^2\ \text{m}\cdot\text{N})/(2.5\ \text{m}) =$ $\boxed{2.4 \times 10^2\ \text{N.}}$

43. We find the acceleration from

$\tau_{\text{friction}} = I\alpha = \frac{1}{2}MR^2\alpha;$

$-1.20 \text{ m}\cdot\text{N} = \frac{1}{2}(4.80 \text{ kg})(0.0710 \text{ m})^2\alpha$, which gives $\alpha = -99.2 \text{ rad/s}^2$.

We find the angle turned through from

$\omega^2 = \omega_0^2 + 2\alpha\theta;$

$0 = [(10{,}000 \text{ rpm})(2\pi \text{ rad/rev})/(60 \text{ s/min})]^2 + 2(-99.2 \text{ rad/s}^2)\theta$, which gives

$\theta = 5.53 \times 10^3 \text{ rad} =$ $\boxed{8.80 \times 10^2 \text{ rev.}}$

We can find the time from

$\omega = \omega_0 + \alpha t;$

$0 = (10{,}000 \text{ rev/min})(2\pi \text{ rad/rev})/(60 \text{ s/min}) + (-99.2 \text{ rad/s}^2)t$; which gives $t =$ $\boxed{10.6 \text{ s.}}$

44. (*a*) Because we ignore the mass of the arm, for the moment of inertia we have

$I = m_{\text{ball}}d_1^2 = (3.6 \text{ kg})(0.30 \text{ m})^2 = 0.324 \text{ kg}\cdot\text{m}^2.$

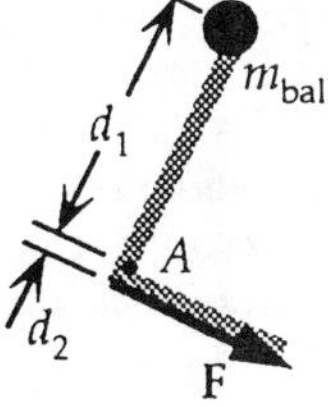

The angular acceleration of the ball-arm system is

$\alpha = a_{\tan}/d_1 = (7.0 \text{ m/s}^2)/(0.30 \text{ m}) = 23.3 \text{ rad/s}^2.$

Thus we find the required torque from

$\tau = I\alpha$

$= (0.324 \text{ kg}\cdot\text{m}^2)(23.3 \text{ rad/s}^2) =$ $\boxed{7.5 \text{ m}\cdot\text{N.}}$

(*b*) Because the force from the triceps muscle is perpendicular to the line from the axis, we find the force from

$F = \tau/d_2 = (7.5 \text{ m}\cdot\text{N})/(0.025 \text{ m}) =$ $\boxed{3.0 \times 10^2 \text{ N.}}$

45. (*a*) The final angular velocity of the arm and ball is

$\omega = v/d_1 = (10.0 \text{ m/s})/(0.30 \text{ m}) = 33.3 \text{ rad/s}.$

We find the angular acceleration from

$\omega = \omega_0 + \alpha t;$

$33.3 \text{ rad/s} = 0 + \alpha(0.350 \text{ s})$, which gives $\alpha =$ $\boxed{95.2 \text{ rad/s}^2.}$

(*b*) For the moment of inertia of the ball and arm we have

$I = m_{\text{ball}}d_1^2 + \frac{1}{3}m_{\text{arm}}d_1^2$

$= (1.50 \text{ kg})(0.30 \text{ m})^2 + \frac{1}{3}(3.70 \text{ kg})(0.30 \text{ m})^2 = 0.246 \text{ kg}\cdot\text{m}^2.$

Because the force from the triceps muscle is perpendicular to the line from the axis, we find the force from

$F = \tau/d_2 = I\alpha/d_2 = (0.246 \text{ kg}\cdot\text{m}^2)(95.2 \text{ rad/s}^2)/(0.025 \text{ m}) =$ $\boxed{9.4 \times 10^2 \text{ N.}}$

46. For the moment of inertia of the rotor blades we have

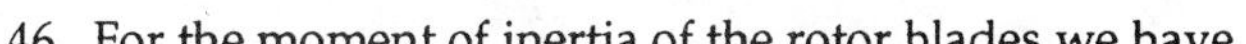

$I = 3(\frac{1}{3}m_{\text{blade}}L^2) = m_{\text{blade}}L^2 = (160 \text{ kg})(3.75 \text{ m})^2 =$ $\boxed{2.25 \times 10^3 \text{ kg}\cdot\text{m}^2.}$

We find the required torque from

$\tau = I\alpha = I(\omega - \omega_0)/t$

$= (2.25 \times 10^3 \text{ kg}\cdot\text{m}^2)[(5.0 \text{ rev/s})(2\pi \text{ rad/rev}) - 0]/(8.0 \text{ s}) =$ $\boxed{8.8 \times 10^3 \text{ m}\cdot\text{N.}}$

47. We choose the clockwise direction as positive.

(*a*) With the force acting, we write $\Sigma\tau = I\alpha$ about the axis from the force diagram for the roll:

$Fr - \tau_{fr} = I\alpha_1$;

$(3.2\ \text{N})(0.076\ \text{m}) - 0.11\ \text{m}\cdot\text{N} = (2.9\times10^{-3}\ \text{kg}\cdot\text{m}^2)\alpha_1$,

which gives $\alpha_1 = 45.9\ \text{rad/s}^2$.

We find the angle turned while the force is acting from

$\theta_1 = \omega_0 t + \frac{1}{2}\alpha_1 t_1^2$

$= 0 + \frac{1}{2}(45.9\ \text{rad/s}^2)(1.3\ \text{s})^2 = 38.8\ \text{rad}.$

The length of paper that unrolls during this time is

$s_1 = r\theta_1 = (0.076\ \text{m})(38.8\ \text{rad}) = \boxed{2.9\ \text{m.}}$

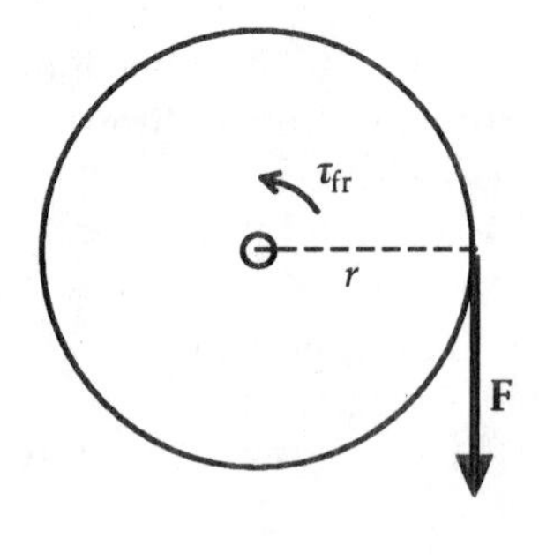

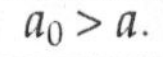

(*b*) With no force acting, we write $\Sigma\tau = I\alpha$ about the axis from the force diagram for the roll:

$-\tau_{fr} = I\alpha_2$;

$-0.11\ \text{m}\cdot\text{N} = (2.9\times10^{-3}\ \text{kg}\cdot\text{m}^2)\alpha_2$,

which gives $\alpha_2 = -37.9\ \text{rad/s}^2$.

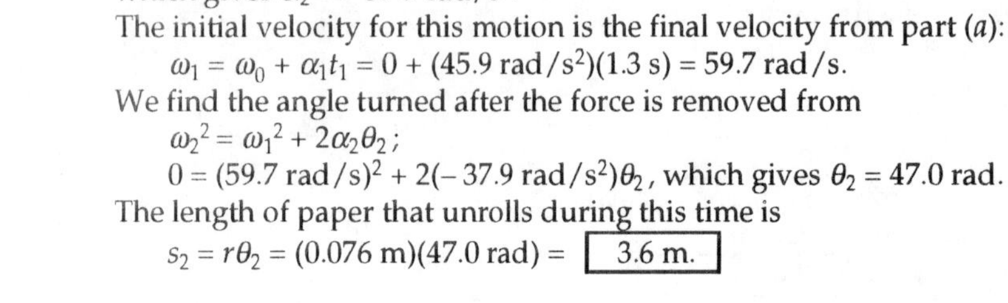

The initial velocity for this motion is the final velocity from part (*a*):

$\omega_1 = \omega_0 + \alpha_1 t_1 = 0 + (45.9\ \text{rad/s}^2)(1.3\ \text{s}) = 59.7\ \text{rad/s}.$

We find the angle turned after the force is removed from

$\omega_2^2 = \omega_1^2 + 2\alpha_2\theta_2$;

$0 = (59.7\ \text{rad/s})^2 + 2(-37.9\ \text{rad/s}^2)\theta_2$, which gives $\theta_2 = 47.0\ \text{rad}$.

The length of paper that unrolls during this time is

$s_2 = r\theta_2 = (0.076\ \text{m})(47.0\ \text{rad}) = \boxed{3.6\ \text{m.}}$

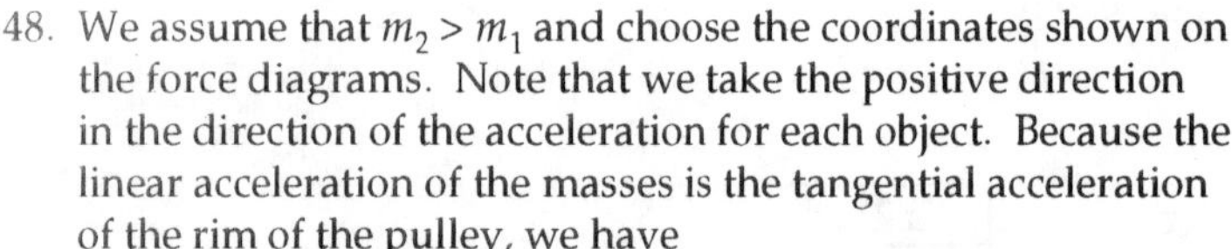

48. We assume that $m_2 > m_1$ and choose the coordinates shown on the force diagrams. Note that we take the positive direction in the direction of the acceleration for each object. Because the linear acceleration of the masses is the tangential acceleration of the rim of the pulley, we have

$a = a_{tan} = \alpha R.$

We write $\Sigma F_y = ma_y$ for m_2:

$m_2 g - F_{T2} = m_2 a.$

We write $\Sigma F_y = ma_y$ for m_1:

$F_{T1} - m_1 g = m_1 a.$

We write $\Sigma\tau = I\alpha$ for the pulley about its axle:

$F_{T2}R - F_{T1}R = I\alpha = Ia/R$, or $F_{T2} - F_{T1} = Ia/R^2$.

If we add the two force equations, we get

$F_{T1} - F_{T2} = (m_1 + m_2)a + (m_1 - m_2)g.$

When we add these two equations, we get

$\boxed{a = (m_2 - m_1)g/(m_1 + m_2 + I/R^2).}$

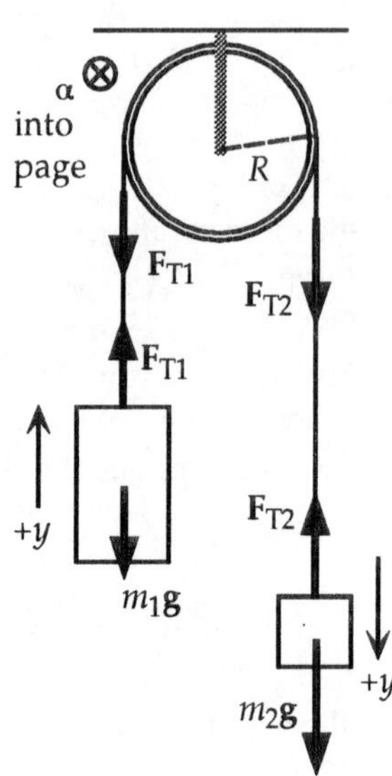

If the moment of inertia of the pulley is ignored, from the torque equation, we see that the two tensions will be equal.

For the acceleration, we set $I = 0$ and get

$\boxed{a_0 = (m_2 - m_1)g/(m_1 + m_2).}$

Thus we see that $\boxed{a_0 > a.}$

49. (*a*) The final angular velocity of the hammer is

$\omega = v/r = (28.0 \text{ m/s})/(1.20 \text{ m}) = 23.3 \text{ rad/s}.$

We find the angular acceleration from

$\omega^2 = \omega_0^2 + 2\alpha\theta;$

$(23.3 \text{ rad/s})^2 = 0 + 2\alpha (4 \text{ rev})(2\pi \text{ rad/rev}),$

which gives $\alpha =$ $\boxed{10.8 \text{ rad/s}^2.}$

(*b*) For the tangential acceleration we have

$a_{tan} = \alpha r = (10.8 \text{ rad/s}^2)(1.20 \text{ m}) =$ $\boxed{13.0 \text{ m/s}^2.}$

(*c*) For the radial acceleration at release we have

$a_R = \omega^2 r = (23.3 \text{ rad/s})^2(1.20 \text{ m}) =$ $\boxed{653 \text{ m/s}^2.}$

(*d*) The magnitude of the resultant acceleration of the hammer is

$a = (a_{tan}^2 + a_R^2)^{1/2} = [(13.0 \text{ m/s}^2)^2 + (653 \text{ m/s}^2)^2]^{1/2} = 653 \text{ m/s}^2.$

This acceleration is provided by the net force exerted on the hammer, so we have

$F_{net} = ma = (7.30 \text{ kg})(653 \text{ m/s}^2) =$ $\boxed{4.77 \times 10^3 \text{ N.}}$

(*e*) We find the angle from

$\tan\phi = a_{tan}/a_R = (13.0 \text{ m/s}^2)/(653 \text{ m/s}^2) = 0.0199$, which gives $\phi =$ $\boxed{1.14°.}$

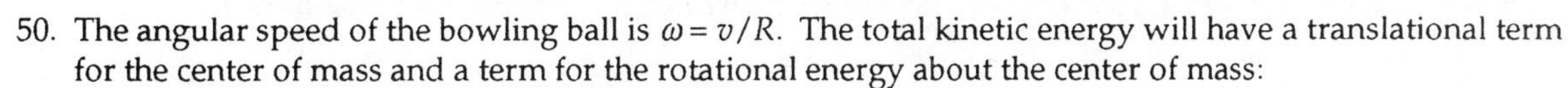

50. The angular speed of the bowling ball is $\omega = v/R$. The total kinetic energy will have a translational term for the center of mass and a term for the rotational energy about the center of mass:

$$\text{KE}_{total} = \text{KE}_{trans} + \text{KE}_{rot} = \tfrac{1}{2}Mv^2 + \tfrac{1}{2}I\omega^2 = \tfrac{1}{2}Mv^2 + \tfrac{1}{2}(\tfrac{2}{5}MR^2)(v/R)^2 = 7Mv^2/10$$

$$= 7(7.3 \text{ kg})(4.3 \text{ m})^2/10 = \boxed{94 \text{ J.}}$$

51. The work done increases the kinetic energy of the rotor:

$$W = \Delta\text{KE} = \tfrac{1}{2}I\omega^2 - 0$$

$$= \tfrac{1}{2}(3.15 \times 10^{-2} \text{ kg}\cdot\text{m}^2)[(8000 \text{ rev/min})(2\pi \text{ rad/rev})/(60 \text{ s/min})]^2 = \boxed{1.11 \times 10^4 \text{ J.}}$$

52. (*a*) The Earth rotates one revolution in one day, so we have

$$\omega_{rotation} = \Delta\theta/\Delta t$$

$$= (2\pi \text{ rad})/(1 \text{ day})(24 \text{ h/day})(3600 \text{ s/h}) = 7.27 \times 10^{-5} \text{ rad/s}.$$

The kinetic energy of rotation is

$$\text{KE}_{rotation} = \tfrac{1}{2}I_{rotation}\omega_{rotation}^2 = \tfrac{1}{2}(\tfrac{2}{5}Mr^2)\omega_{rotation}^2$$

$$= \tfrac{1}{2}(\tfrac{2}{5})(6.0 \times 10^{24} \text{ kg})(6.4 \times 10^6 \text{ m})^2(7.27 \times 10^{-5} \text{ rad/s})^2 = \boxed{2.6 \times 10^{29} \text{ J.}}$$

(*b*) The Earth moves one revolution around the Sun in one year, so we have

$$\omega_{orbit} = \Delta\theta/\Delta t$$

$$= (2\pi \text{ rad})/(1 \text{ yr})(3.16 \times 10^7 \text{ s/yr}) = 1.99 \times 10^{-7} \text{ rad/s}.$$

The kinetic energy of revolution is

$$\text{KE}_{revolution} = \tfrac{1}{2}I_{revolution}\omega_{revolution}^2 = \tfrac{1}{2}(MR^2)\omega_{revolution}^2$$

$$= \tfrac{1}{2}(6.0 \times 10^{24} \text{ kg})(1.5 \times 10^{11} \text{ m})^2(1.99 \times 10^{-7} \text{ rad/s})^2 = \boxed{2.7 \times 10^{33} \text{ J.}}$$

We see that the kinetic energy of revolution is much greater than that of rotation, so the total energy is $\boxed{\text{KE}_{total} = 2.7 \times 10^{33} \text{ J.}}$

53. The work done increases the kinetic energy of the merry-go-round:

$$W = \Delta\text{KE} = \tfrac{1}{2}I\omega^2 - 0 = \tfrac{1}{2}(\tfrac{1}{2}Mr^2)(\Delta\theta/\Delta t)^2$$

$$= \tfrac{1}{4}(1640 \text{ kg})(8.20 \text{ m})^2[(1 \text{ rev})(2\pi \text{ rad/rev})/(8.00 \text{ s})]^2 = \boxed{1.70 \times 10^4 \text{ J.}}$$

54. We choose the reference level for gravitational potential energy at the bottom. The kinetic energy will be the translational energy of the center of mass and the rotational energy about the center of mass.
    (*a*) Because there is no work done by friction while the cylinder is rolling, for the work-energy principle we have

    $W_{net} = \Delta KE + \Delta PE;$

    $0 = (\frac{1}{2}Mv^2 + \frac{1}{2}I\omega^2 - 0) + Mg(0 - d \sin \theta).$

    Because the cylinder is rolling, $v = R\omega$. The rotational inertia is $\frac{2}{5}MR^2$. Thus we get

    $\frac{1}{2}Mv^2 + \frac{1}{2}(\frac{2}{5}MR^2)(v/R^2)^2 = Mgd \sin \theta$, which gives

    $v = (\frac{4}{3}gd \sin \theta)^{1/2}$, and $\omega = (\frac{4}{3}gd \sin \theta)^{1/2}/R.$

    When we use the given data, we get

    $v = [\frac{4}{3}(9.80 \text{ m/s}^2)(10.0 \text{ m}) \sin 30°]^{1/2} =$ **8.37 m/s,** and

    $\omega = v/R = (8.37 \text{ m/s})/(0.200 \text{ m}) =$ **41.8 rad/s.**

    (*b*) For the ratio of kinetic energies we have

    $KE_{trans}/KE_{rot} = \frac{1}{2}Mv^2/\frac{1}{2}I\omega^2$

    $= Mv^2/(\frac{2}{5}MR^2)(v/R^2)^2 =$ **2.50.**

    (*c*) None of the answers depends on the mass; the rotational speed depends on the radius.

55. For the system of the two blocks and pulley, no work will be done by nonconservative forces. The rope ensures that each block has the same speed v and the angular speed of the pulley is $\omega = v/R_0$. We choose the reference level for gravitational potential energy at the floor.
    The rotational inertia of the pulley is $I = \frac{1}{2}MR_0^2$.
    For the work-energy principle we have

    $W_{net} = \Delta KE + \Delta PE;$

    $0 = [(\frac{1}{2}m_1v^2 + \frac{1}{2}m_2v^2 + \frac{1}{2}I\omega^2) - 0] + m_1g(h - 0) + m_2g(0 - h);$

    $\frac{1}{2}m_1v^2 + \frac{1}{2}m_2v^2 + \frac{1}{2}(\frac{1}{2}MR_0^2)(v/R_0)^2 = (m_2 - m_1)gh;$

    $\frac{1}{2}[m_1 + m_2 + \frac{1}{2}M]v^2 = (m_2 - m_1)gh;$

    $\frac{1}{2}[18.0 \text{ kg} + 26.5 \text{ kg} + \frac{1}{2}(7.50 \text{ kg})]v^2 =$

    $(26.5 \text{ kg} - 18.0 \text{ kg})(9.80 \text{ m/s}^2)(3.00 \text{ m})$, which gives

    $v =$ **3.22 m/s.**

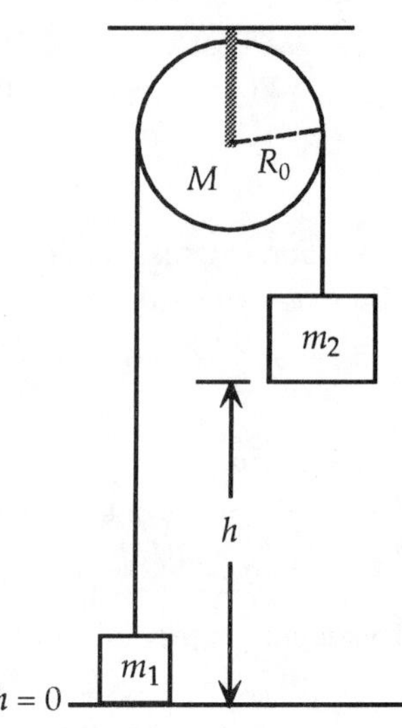

56. If the contact point does not move, no work is done by the friction force. With the reference level for potential energy at the ground, we use energy conservation to find the angular speed just before the pole hits the ground:

    $KE_i + PE_i = KE_f + PE_f;$

    $0 + Mg\frac{1}{2}L = \frac{1}{2}(\frac{1}{3}ML^2)\omega^2 + 0$, which gives $\omega = (3g/L)^{1/2}$.

    Because the pole is rotating about the contact point, the speed of the upper end is

    $v = \omega L = (3gL)^{1/2} = [3(9.80 \text{ m/s}^2)(3.30 \text{ m})]^{1/2} =$ **9.85 m/s.**

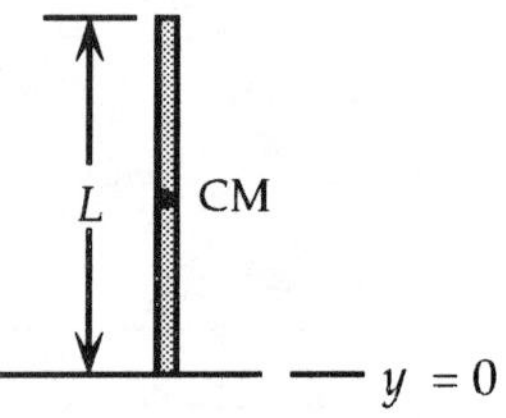

57. The angular momentum of rotation about the fixed end of the string is

    $L = I\omega = mR^2\omega$

    $= (0.210 \text{ kg})(1.10 \text{ m})^2(10.4 \text{ rad/s}) =$ **$2.64 \text{ kg} \cdot \text{m}^2/\text{s}$.**

58. (*a*) As the arms are raised some of the person's mass is farther from the axis of rotation, so the moment of inertia has increased. For the isolated system of platform and person, the angular momentum is conserved. As the moment of inertia increases, the angular velocity must decrease.

(*b*) If the mass and thus the moment of inertia of the platform can be neglected, for the conservation of angular momentum, we have

$L = I_1\omega_1 = I_2\omega_2$, or

$I_2/I_1 = \omega_1/\omega_2 = (1.30\text{ rad/s})/(0.80\text{ rad/s}) = \boxed{1.6.}$

59. Because the diver in the air is an isolated system, for the conservation of angular momentum we have

$L = I_1\omega_1 = I_2\omega_2$, or $I_2/I_1 = \omega_1/\omega_2$;

$1/3.5 = \omega_1/(2\text{ rev}/1.5\text{ s})$, which gives $\omega_1 = \boxed{0.38\text{ rev/s.}}$

60. Because the skater is an isolated system, for the conservation of angular momentum we have

$L = I_1\omega_1 = I_2\omega_2$;

$(4.6\text{ kg}\cdot\text{m}^2)(1.0\text{ rev}/2.0\text{ s}) = I_2(3.0\text{ rev/s})$, which gives $I_2 = \boxed{0.77\text{ kg}\cdot\text{m}^2.}$

She accomplishes this by pulling her arms closer to her body.

61. If we approximate the hurricane as a solid cylinder, we have

$M = \rho(\pi R^2 H) = (1.3\text{ kg/m}^3)\pi(100\times10^3\text{ m})^2(4.0\times10^3\text{ m}) = 1.63\times10^{14}\text{ kg}$;

$I = \frac{1}{2}MR^2 = \frac{1}{2}(1.63\times10^{14}\text{ kg})(100\times10^3\text{ m})^2 = 8.15\times10^{23}\text{ kg}\cdot\text{m}^2$.

The winds of a hurricane are obviously not uniform; they are generally higher near the eye. If we assume the highest winds are at half the radius, for the average angular speed we have

$\omega = v/\frac{1}{2}R = [(120\text{ km/h})/(3.6\text{ ks/h})]/\frac{1}{2}(100\times10^3\text{ m}) = 6.67\times10^{-4}\text{ rad/s}$.

(*a*) For the kinetic energy we have

$\text{KE} = \frac{1}{2}I\omega^2 = \frac{1}{2}(8.15\times10^{23}\text{ kg}\cdot\text{m}^2)(6.67\times10^{-4}\text{ rad/s})^2 \approx \boxed{2\times10^{17}\text{ J.}}$

(*b*) For the angular momentum we have

$L = I\omega = (8.15\times10^{23}\text{ kg}\cdot\text{m}^2)(6.67\times10^{-4}\text{ rad/s}) \approx \boxed{5\times10^{20}\text{ kg}\cdot\text{m}^2\text{/s.}}$

62. (*a*) We approximate the mass distribution as a solid cylinder. The angular momentum is

$L = I\omega = \frac{1}{2}mR^2\omega = \frac{1}{2}(55\text{ kg})(0.15\text{ m})^2[(3.5\text{ rev/s})(2\pi\text{ rad/rev})] = \boxed{14\text{ kg}\cdot\text{m}^2\text{/s.}}$

(*b*) If the arms do not move, the moment of inertia will not change. We find the torque from the change in angular momentum:

$\tau = \Delta L/\Delta t = (0 - 13.6\text{ kg}\cdot\text{m}^2\text{/s})/(5.0\text{ s}) = \boxed{-2.7\text{ m}\cdot\text{N.}}$

63. (*a*) The Earth rotates one revolution in one day, so we have

$$\begin{aligned}\omega_{\text{rotation}} &= \Delta\theta/\Delta t\\ &= (2\pi\text{ rad})/(1\text{ day})(24\text{ h/day})(3600\text{ s/h}) = 7.27\times10^{-5}\text{ rad/s}.\end{aligned}$$

If we assume the Earth is a uniform sphere, the angular momentum is

$$\begin{aligned}L_{\text{rotation}} &= I_{\text{rotation}}\omega_{\text{rotation}} = (\tfrac{2}{5}Mr^2)\omega_{\text{rotation}}\\ &= (\tfrac{2}{5})(6.0\times10^{24}\text{ kg})(6.4\times10^6\text{ m})^2(7.27\times10^{-5}\text{ rad/s}) = \boxed{7.1\times10^{33}\text{ kg}\cdot\text{m}^2\text{/s.}}\end{aligned}$$

(*b*) The Earth moves one revolution around the Sun in one year, so we have

$$\begin{aligned}\omega_{\text{orbit}} &= \Delta\theta/\Delta t\\ &= (2\pi\text{ rad})/(1\text{ yr})(3.16\times10^7\text{ s/yr}) = 1.99\times10^{-7}\text{ rad/s}.\end{aligned}$$

The angular momentum is

$$\begin{aligned}L_{\text{revolution}} &= I_{\text{revolution}}\omega_{\text{revolution}} = (MR^2)\omega_{\text{revolution}}\\ &= (6.0\times10^{24}\text{ kg})(1.5\times10^{11}\text{ m})^2(1.99\times10^{-7}\text{ rad/s}) = \boxed{2.7\times10^{40}\text{ kg}\cdot\text{m}^2\text{/s.}}\end{aligned}$$

64. If there are no external torques, angular momentum will be conserved:

$L = I_1\omega + I_2(0) = (I_1 + I_2)\omega_{final}$, or

$I\omega = (I + I)\omega_{final}$, which gives $\omega_{final} =$ $\boxed{\omega/2.}$

65. If there are no external torques, angular momentum will be conserved:

$L = I_{disk}\omega_1 + I_{rod}(0) = (I_{disk} + I_{rod})\omega_2$;

$(Mr^2/2)\omega_1 = [(Mr^2/2) + (ML^2/12)]\omega_2 = \{(Mr^2/2) + [M(2r)^2/12]\}\omega_2$, which gives

$\omega_2 = (3/5)\omega_1 = (3/5)(7.0 \text{ rev/s}) =$ $\boxed{4.2 \text{ rev/s.}}$

66. (*a*) By walking to the edge, the moment of inertia of the person changes. Because the system of person and platform is isolated, angular momentum will be conserved:

$L = (I_{platform} + I_{person1})\omega_1 = (I_{platform} + I_{person2})\omega_2$;

$[1000 \text{ kg}\cdot\text{m}^2 + (75 \text{ kg})(0)^2](2.0 \text{ rad/s}) = [1000 \text{ kg}\cdot\text{m}^2 + (75 \text{ kg})(3.0 \text{ m})^2]\omega_2$, which gives

$\omega_2 =$ $\boxed{1.2 \text{ rad/s.}}$

(*b*) For the kinetic energies, we have

$KE_1 = \frac{1}{2}(I_{platform} + I_{person1})\omega_1^2 = \frac{1}{2}(1000 \text{ kg}\cdot\text{m}^2 + 0)(2.0 \text{ rad/s})^2 =$ $\boxed{2.0 \times 10^3 \text{ J;}}$

$KE_2 = \frac{1}{2}(I_{platform} + I_{person2})\omega_2^2 = \frac{1}{2}[1000 \text{ kg}\cdot\text{m}^2 + (75 \text{ kg})(3.0 \text{ m})^2](1.2 \text{ rad/s})^2 =$ $\boxed{1.2 \times 10^3 \text{ J.}}$

Thus there is a $\boxed{\text{loss of } 8.0 \times 10^2 \text{ J, a decrease of 40\%.}}$

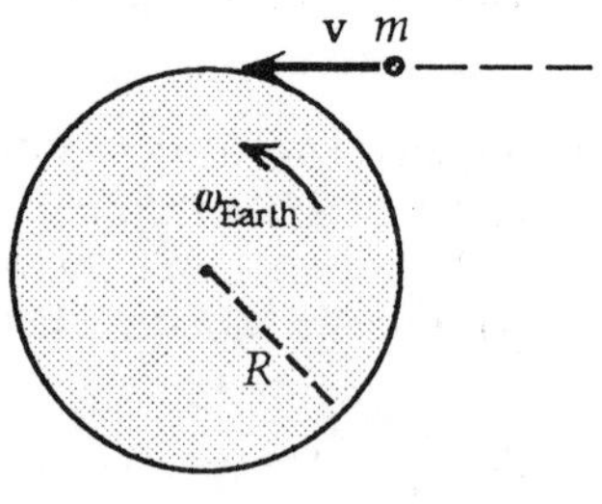

67. The initial angular speed of the asteroid about the center of the Earth is

$\omega_a = v_a/R = (30 \times 10^3 \text{ m/s})/(6.4 \times 10^6 \text{ m}) = 4.7 \times 10^{-3} \text{ rad/s}$.

The initial angular speed of the Earth is

$\omega_E = (2\pi \text{ rad})/(1 \text{ day})(24 \text{ h/day})(3600 \text{ s/h}) = 7.27 \times 10^{-5} \text{ rad/s}$.

For the system of asteroid and Earth, angular momentum is conserved:

$I_{asteroid}\omega_a + I_{Earth}\omega_E = (I_{asteroid} + I_{Earth})\omega \approx I_{Earth}\omega$,

because the mass of the Earth is much greater than the mass of the asteroid.

This gives

$\omega - \omega_E \approx (I_{asteroid}/I_{Earth})\omega_a$.

For the fractional change, we have

$(\omega - \omega_E)/\omega_E \approx (I_{asteroid}/I_{Earth})(\omega_a/\omega_E) = (m_aR^2)/(\frac{2}{5}M_ER^2)](\omega_a/\omega_E)$

$= [(1.0 \times 10^5 \text{ kg})/\frac{2}{5}(6.0 \times 10^{24} \text{ kg})][(4.7 \times 10^{-3} \text{ rad/s})/(7.27 \times 10^{-5} \text{ rad/s})] =$ $\boxed{3 \times 10^{-18}.}$

68. When the people step onto the merry-go-round, they have no initial angular momentum. For the system of merry-go-round and people, angular momentum is conserved:

$I_{merry\text{-}go\text{-}round}\omega_0 + I_{people}\omega_i = (I_{merry\text{-}go\text{-}round} + I_{people})\omega$;

$I_{merry\text{-}go\text{-}round}\omega_0 + 4mR^2(0) = (I_{merry\text{-}go\text{-}round} + 4mR^2)\omega$;

$(1760 \text{ kg}\cdot\text{m}^2)(0.80 \text{ rad/s}) = [(1760 \text{ kg}\cdot\text{m}^2) + 4(65 \text{ kg})(2.1 \text{ m})^2]\omega$, which gives $\omega =$ $\boxed{0.48 \text{ rad/s.}}$

If the people jump off in a radial direction with respect to the merry-go-round, they have the tangential velocity of the merry-go-round: $v = R\omega_0$. For the system of merry-go-round and people, angular momentum is conserved:

$(I_{merry\text{-}go\text{-}round} + I_{people})\omega_0 = I_{merry\text{-}go\text{-}round}\omega + I_{people}\omega_0$, which gives

$\omega = \omega_0$. $\boxed{\text{The angular speed of the merry-go-round does not change.}}$

Note that the angular momentum of the people will change when contact is made with the ground.

69. We assume that the lost mass does not carry away any angular momentum. For the Sun, angular momentum is conserved:

$I_0\omega_0 = I\omega;$

$\frac{2}{5}M_1R_1^2\omega_0 = \frac{2}{5}M_2R_2^2\omega;$ or

$\omega = (R_1/R_2)^2(M_1/M_2)\omega_0 = (1/0.01)^2(1/0.5)(2\pi \text{ rad})/(30 \text{ days})(24 \text{ h/day})(3600 \text{ s/h})$

$= \boxed{4.8 \times 10^{-2} \text{ rad/s.}}$

The new period is 130 s.

For the ratio of kinetic energies we have

$\text{KE}_2/\text{KE}_1 = \frac{1}{2}I\omega^2/\frac{1}{2}I_0\omega_0^2$

$= \frac{1}{2}(\frac{2}{5}M_2R_2^2)[(R_1/R_2)^2(M_1/M_2)]^2\omega_0^2/\frac{1}{2}(\frac{2}{5}M_1R_1^2)\omega_0^2$

$= (R_1/R_2)^2(M_1/M_2) = (1/0.01)^2(1/0.5) = 2.0 \times 10^4,$ or $\boxed{\text{KE}_2 = 2.0 \times 10^4 \text{ KE}_1.}$

70. We choose the direction the person walks for the positive rotation. The speed of the person with respect to the turntable is v. If the edge of the turntable acquires a speed v_t with respect to the ground, the speed of the person will be $v_t + v$. Because all speeds are the same distance from the axis, if we divide by R, we get

$\omega_p = \omega_t + (v/R).$

Using the speeds with respect to the ground, from the conservation of angular momentum of the system of turntable and person, we have

$L = 0 = I_t\omega_t + I_p\omega_p = I_t\omega_t + m_pR^2[\omega_t + (v/R)];$

$(1760 \text{ kg}\cdot\text{m}^2)\omega_t + (55 \text{ kg})(3.25 \text{ m})^2\{\omega_t + [(3.8 \text{ m/s})/(3.25 \text{ m})]\} = 0,$ which gives

$\boxed{\omega_t = -0.30 \text{ rad/s.}}$

The negative sign indicates a motion opposite to that of the person.

71. Initially there is no angular momentum about the vertical axis. Because there are no torques about this vertical axis for the system of platform and wheel, the angular momentum about the vertical axis is zero and conserved. We choose up for the positive direction.

(*a*) From the conservation of angular momentum about the vertical axis, we have

$L = 0 = I_P\omega_P + I_W\omega_W$, which gives $\boxed{\omega_P = -(I_W/I_P)\omega_W \text{ (down).}}$

(*b*) From the conservation of angular momentum about the vertical axis, we have

$L = 0 = I_P\omega_P + I_W\omega_W \cos 60°$, which gives $\boxed{\omega_P = -(I_W/2I_P)\omega_W \text{ (down).}}$

(*c*) From the conservation of angular momentum about the vertical axis, we have

$L = 0 = I_P\omega_P + I_W(-\omega_W)$, which gives $\boxed{\omega_P = (I_W/I_P)\omega_W \text{ (up).}}$

(*d*) Because the total angular momentum is zero, when the wheel stops, the platform and person must also stop.

Thus $\boxed{\omega_P = 0.}$

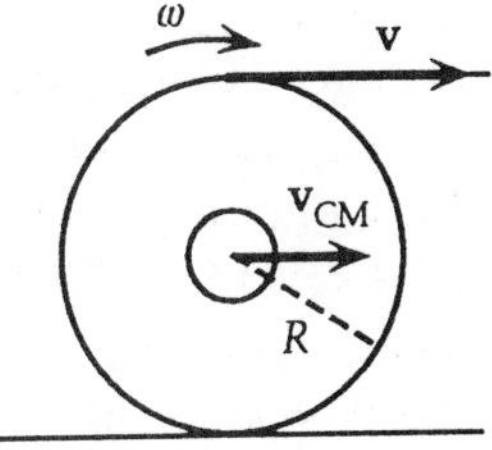

72. Because the spool is rolling, $v_{CM} = R\omega$. The velocity of the rope at the top of the spool, which is also the velocity of the person, is

$v = R\omega + v_{CM} = 2R\omega = 2v_{CM}.$

Thus in the time it takes for the person to walk a distance L, the CM will move a distance $L/2$. Therefore, the length of rope that unwinds is

$L_{rope} = L - (L/2) = \boxed{L/2.}$

The CM will move a distance $\boxed{L/2.}$

73. For the same side of the Moon to always face the Earth, the angular velocity of the orbital motion and the angular velocity of the spinning motion must be the same. We use r for the radius of the Moon and R for the distance from the Earth to the Moon. For the ratio of angular momenta, we have

$L_{spin}/L_{orbital} = \frac{2}{5}Mr^2\omega/MR^2\omega$

$= \frac{2}{5}(r/R)^2 = \frac{2}{5}[(1.74 \times 10^6 \text{ m})/(3.84 \times 10^8 \text{ m})]^2 = \boxed{8.21 \times 10^{-6}.}$

74. After 3.0 s the velocity of the CM will be

$v_{CM} = 0 + a_{CM}t = (1.00\text{ m/s}^2)(3.0\text{ s}) = 3.0\text{ m/s}.$

Because the wheel is rolling, $v_{CM} = R\omega$. The velocity at the top of the wheel is

$v = R\omega + v_{CM} = 2R\omega = 2v_{CM} = 2(3.0\text{ m/s}) =$ $\boxed{6.0\text{ m/s}.}$

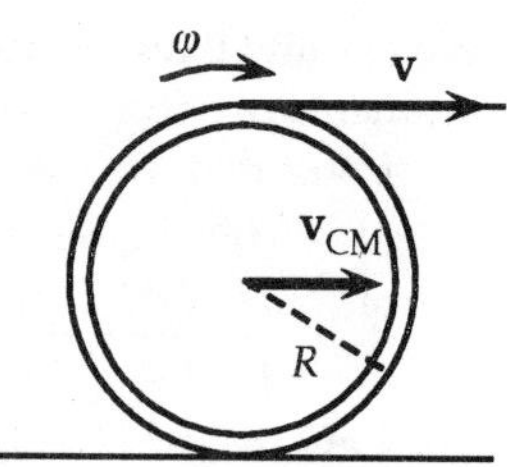

75. (*a*) The yo-yo is considered as three cylinders, with a total mass of

$M = 2M_{disk} + M_{hub} = 2(0.050\text{ kg}) + 0.0050\text{ kg} = 0.105\text{ kg},$

and a moment of inertia of the yo-yo about its axis of

$I = 2(\tfrac{1}{2}M_{disk}R_{disk}{}^2) + \tfrac{1}{2}M_{hub}R_{hub}{}^2$

$= (0.050\text{ kg})[\tfrac{1}{2}(0.075\text{ m})]^2 + \tfrac{1}{2}(0.0050\text{ kg})[\tfrac{1}{2}(0.010\text{ m})]^2$

$= 7.04 \times 10^{-5}\text{ kg}\cdot\text{m}^2.$

Because the yo-yo is rolling about a point on the rim of the hub,

$v_{CM} = R_{hub}\omega.$

The kinetic energy of the yo-yo is the translational kinetic energy of the CM and the rotational kinetic energy about the CM. Because the top of the string does not move, the tension in the string does no work. Thus energy is conserved:

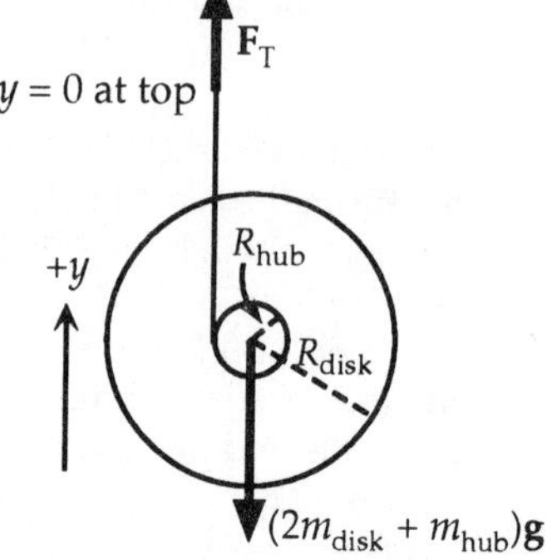

$KE_i + PE_i = KE_f + PE_f;$

$0 + 0 = \tfrac{1}{2}Mv_{CM}{}^2 + \tfrac{1}{2}I\omega^2 + Mg(-L);$

$\tfrac{1}{2}Mv_{CM}{}^2 + \tfrac{1}{2}I(v_{CM}/R_{hub})^2 = MgL;$

$\tfrac{1}{2}\{0.105\text{ kg} + [(7.04 \times 10^{-5}\text{ kg}\cdot\text{m}^2)/(0.0050\text{ m})^2]\}v_{CM}{}^2 = (0.105\text{ kg})(9.80\text{ m/s}^2)(1.0\text{ m}),$

which gives $v_{CM} =$ $\boxed{0.84\text{ m/s}.}$

(*b*) For the fraction of the kinetic energy that is rotational, we have

$KE_{rot}/(KE_{trans} + KE_{rot}) = \tfrac{1}{2}I\omega^2 / (\tfrac{1}{2}Mv_{CM}{}^2 + \tfrac{1}{2}I\omega^2)$

$= I\omega^2/[M(R_{hub}\omega)^2 + I\omega^2] = 1/[(MR_{hub}{}^2/I) + 1]$

$= 1/\{[(0.105\text{ kg})(0.0050\text{ m})^2/(7.04 \times 10^{-5}\text{ kg}\cdot\text{m}^2)] + 1\} =$ $\boxed{0.964.}$

76. (*a*) If we let d represent the spacing of the teeth, which is the same on both sprockets, we can relate the number of teeth to the radius for each wheel:

$N_Fd = 2\pi R_F,$ and $N_Rd = 2\pi R_R$, which gives $N_F/N_R = R_F/R_R$.

The linear speed of the chain is the tangential speed for each socket:

$v = R_F\omega_F = R_R\omega_R.$

Thus we have

$\boxed{\omega_R/\omega_F = R_F/R_R = N_F/N_R.}$

(*b*) For the given data we have

$\omega_R/\omega_F = 52/13 =$ $\boxed{4.0.}$

(*c*) For the given data we have

$\omega_R/\omega_F = 42/28 =$ $\boxed{1.5.}$

77. We assume that the lost mass does not carry away any angular momentum. For the star, angular momentum is conserved:

$I_0\omega_0 = I\omega;$

$\tfrac{2}{5}M_1R_1{}^2\omega_0 = \tfrac{2}{5}M_2R_2{}^2\omega;$ or

$\omega = (R_1/R_2)^2(M_1/M_2)\omega_0$

$= (6.96 \times 10^6\text{ km}/10\text{ km})^2(1/0.25)[(1.0\text{ rev})/(10\text{ days})] =$ $\boxed{2.0 \times 10^9\text{ rev/day}.}$

78. We convert the speed: (90 km/h)/(3.6 ks/h) = 25 m/s.

(*a*) We assume that the linear kinetic energy that the automobile acquires during each acceleration is not regained when the automobile slows down. For the work-energy principle we have

$W_{\text{net}} = \Delta\text{KE} + \Delta\text{PE};$

$-F_{\text{fr}}D = [20(\frac{1}{2}Mv^2) - \text{KE}_{\text{flywheel}}] + 0,$ or

$\text{KE}_{\text{flywheel}} = (20)\frac{1}{2}(1400\text{ kg})(25\text{ m/s})^2 + (500\text{ N})(300 \times 10^3\text{ m}) = 1.6 \times 10^8\text{ J}.$

(*b*) We find the angular velocity of the flywheel from

$\text{KE}_{\text{flywheel}} = \frac{1}{2}I\omega^2 = \frac{1}{2}(\frac{1}{2}mR^2)\omega^2;$

$1.6 \times 10^8\text{ J} = \frac{1}{4}(240\text{ kg})(0.75\text{ m})^2\omega^2,$ which gives $\omega =$ $\boxed{2.2 \times 10^3\text{ rad/s.}}$

(*c*) We find the time from

$t = \text{KE}_{\text{flywheel}}/P = (1.6 \times 10^8\text{ J})/(150\text{ hp})(746\text{ W/hp}) = 1.43 \times 10^3\text{ s} =$ $\boxed{24\text{ min.}}$

79. (*a*) We choose the reference level for gravitational potential energy at the initial position at the bottom of the incline. The kinetic energy will be the translational energy of the center of mass and the rotational energy about the center of mass. Because there is no work done by friction while the cylinder is rolling, for the work-energy principle we have

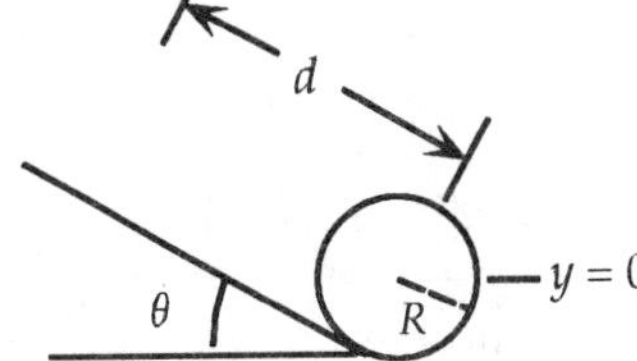

$W_{\text{net}} = \Delta\text{KE} + \Delta\text{PE};$

$0 = [0 - (\frac{1}{2}Mv^2 + \frac{1}{2}I\omega^2)] + Mg(d\sin\theta - 0).$

Because the cylinder is rolling, $v = R\omega$. For a hoop the moment of inertia is MR^2. Thus we get

$\frac{1}{2}Mv^2 + \frac{1}{2}(MR^2)(v^2/R^2) = Mv^2 = Mgd\sin\theta;$

$(4.3\text{ m/s})^2 = (9.80\text{ m/s}^2)d\sin 15°,$ which gives $d =$ $\boxed{7.3\text{ m.}}$

(*b*) We find the time to go up the incline from the linear motion (which has constant acceleration):

$d = \frac{1}{2}(v + 0)t;$

$7.3\text{ m} = \frac{1}{2}(4.3\text{ m/s})t,$ which gives $t = 3.4$ s.

Because there are no losses to friction, the time to go up the incline will be the same as the time to return. The total time will be

$T = 2t =$ $\boxed{6.8\text{ s.}}$

80. (*a*) For the angular acceleration of the rod about the pivot, we have

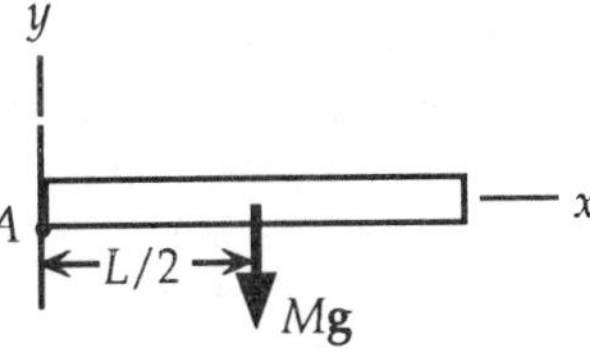

$\Sigma\tau = I\alpha;$

$MgL/2 = \frac{1}{3}ML^2\alpha,$ which gives $\boxed{\alpha = 3g/2L.}$

(*b*) The tangential acceleration of the end of the rod is

$a_{\tan} = \alpha L = (3g/2L)(L) =$ $\boxed{3g/2.}$

Note that there is no radial acceleration of the end of the rod at the moment of release because there is no tangential speed then.

81. The cylinder will roll about the contact point A.

We write $\Sigma\tau = I\alpha$ about the point A:

$F(R - h) + F_{N1}[R^2 - (R - h)^2]^{1/2} - Mg[R^2 - (R - h)^2]^{1/2} = I_A\alpha.$

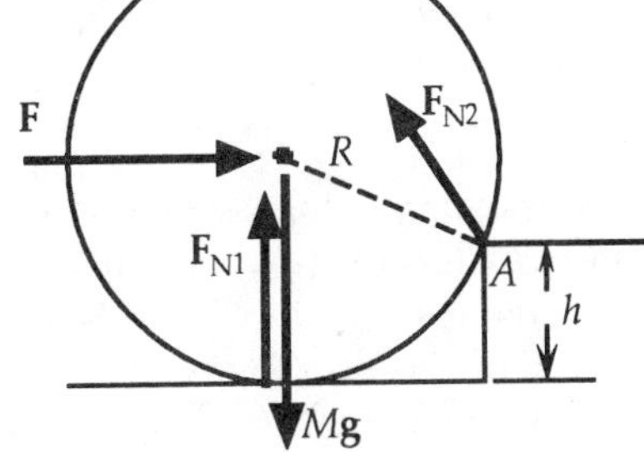

When the cylinder does roll over the curb, contact with the ground is lost and $F_{N1} = 0$. Thus we get

$F = \{I_A\alpha + Mg[R^2 - (R - h)^2]^{1/2}\}/(R - h)$

$= [I_A\alpha/(R - h)] + [Mg(2Rh - h^2)^{1/2}/(R - h)].$

The minimum force occurs when $\alpha = 0$:

$\boxed{F_{\min} = Mg[h(2R - h)]^{1/2}/(R - h).}$

82. (*a*) If we consider an axis through the CM parallel to the velocity vector (that is, parallel to the ground), there is no angular acceleration about this axis. If d is the distance from the CM to the contact point on the ground, we have

$$\Sigma\tau = I\alpha;$$

$$F_N d \sin\theta - F_{fr} d \cos\theta = 0, \text{ which gives } \tan\theta = F_{fr}/F_N.$$

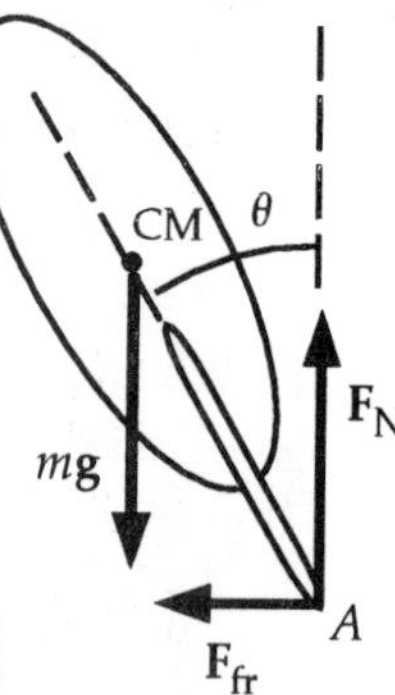

(*b*) The friction force is producing the necessary radial acceleration for the turn:

$$F_{fr} = F_N \tan\theta = mg\tan\theta = mv^2/r, \text{ or}$$

$$\tan\theta = v^2/gr = (4.2 \text{ m/s})^2/(9.80 \text{ m/s}^2)(6.4 \text{ m}) = 0.281, \text{ so } \boxed{\theta = 16°.}$$

(*c*) From $F_{fr} = mv^2/r$, we see that the minimum turning radius requires the maximum static friction force:

$$\mu_s mg = mv^2/r_{min}, \text{ or } r_{min} = v^2/\mu_s g = (4.2 \text{ m/s})^2/(0.70)(9.80 \text{ m/s}^2) = \boxed{2.6 \text{ m.}}$$

83. At the top of the loop, if the marble stays on the track, the normal force and the weight provide the radial acceleration:

$$F_N + mg = mv^2/R_0.$$

The minimum value of the normal force is zero, so we find the minimum speed at the top from

$$mg = mv_{min}^2/R_0, \text{ or } v_{min}^2 = gR_0.$$

Because the marble is rolling, the corresponding angular velocity at the top is $\omega_{min} = v_{min}/r$, so the minimum kinetic energy at the top is

$$\text{KE}_{min} = \tfrac{1}{2}mv_{min}^2 + \tfrac{1}{2}I\omega_{min}^2$$

$$= \tfrac{1}{2}mv_{min}^2 + \tfrac{1}{2}(\tfrac{2}{5}mr^2)(v_{min}/r)^2 = 7mv_{min}^2/10 = 7mgR_0/10.$$

If there are no frictional losses, we use energy conservation from the release point to the highest point of the loop:

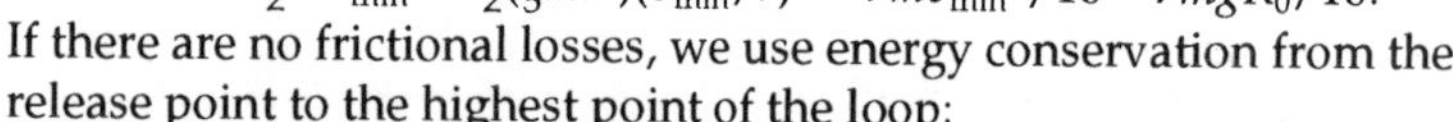

$$\text{KE}_i + \text{PE}_i = \text{KE}_f + \text{PE}_f;$$

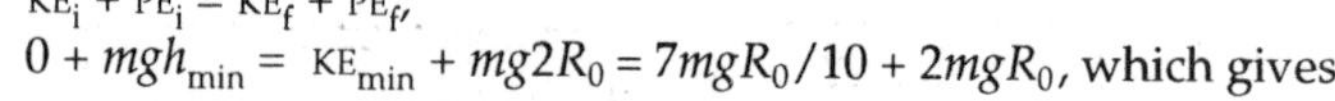

$$0 + mgh_{min} = \text{KE}_{min} + mg2R_0 = 7mgR_0/10 + 2mgR_0, \text{ which gives } \boxed{h_{min} = 2.7R_0.}$$

84. If r is not much smaller than R_0, the CM of the marble moves around the loop in a circle with radius $R_0 - r$. At the minimum speed at the top of the loop, the weight provides the radial acceleration:

$$mg = mv_{min}^2/(R_0 - r), \text{ or } v_{min}^2 = g(R_0 - r).$$

The corresponding angular velocity at the top is $\omega_{min} = v_{min}/r$, so the minimum kinetic energy is

$$\text{KE}_{min} = \tfrac{1}{2}mv_{min}^2 + \tfrac{1}{2}I\omega_{min}^2 = \tfrac{1}{2}mv_{min}^2 + \tfrac{1}{2}(\tfrac{2}{5}mr^2)(v_{min}/r)^2 = 7mv_{min}^2/10 = 7mg(R_0 - r)/10.$$

The distance h is to the bottom of the marble. We use energy conservation from the release point to the highest point of the loop:

$$\text{KE}_i + \text{PE}_i = \text{KE}_f + \text{PE}_f;$$

$$0 + mg(h_{min} + r) = \text{KE}_{min} + mg(2R_0 - r) = 7mg(R_0 - r)/10 + mg(2R_0 - r),$$

which gives $\boxed{h_{min} = 2.7(R_0 - r).}$

85. Because there is no friction, the CM must fall straight down. The vertical velocity of the right end of the stick must always be zero. If ω is the angular velocity of the stick just before it hits the table, the velocity of the right end with respect to the CM will be $\omega(L/2)$ up. Thus we have $\omega(L/2) - v_{CM} = 0$, or $\omega = 2v_{CM}/L$.

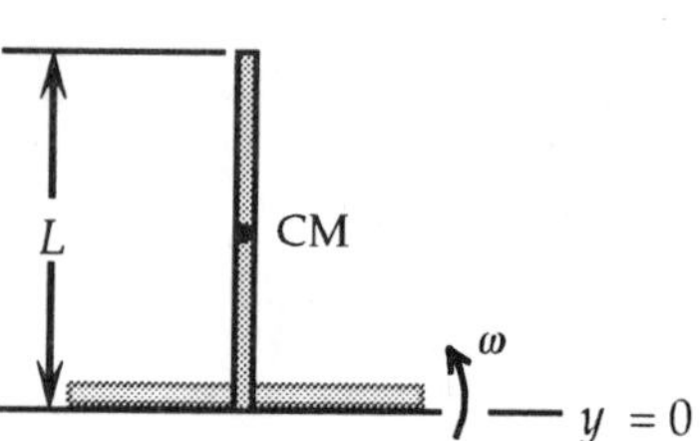

The kinetic energy will be the translational energy of the center of mass and the rotational energy about the center of mass. With the reference level for potential energy at the ground, we use energy conservation to find the speed of the CM just before the stick hits the ground:

$$\text{KE}_i + \text{PE}_i = \text{KE}_f + \text{PE}_f;$$

$$0 + Mg\tfrac{1}{2}L = \tfrac{1}{2}Mv_{CM}^2 + \tfrac{1}{2}(ML^2/12)\omega^2 + 0;$$

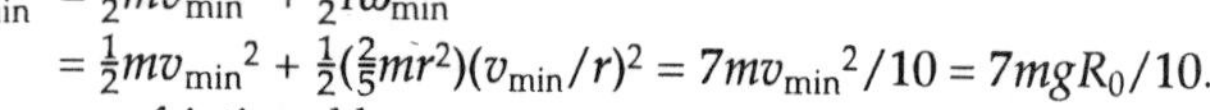

$$Mg\tfrac{1}{2}L = \tfrac{1}{2}Mv_{CM}^2 + \tfrac{1}{2}(ML^2/12)(2v_{CM}/L)^2 = \tfrac{2}{3}(Mv_{CM}^2/3), \text{ which gives } \boxed{v_{CM} = (3gL/4)^{1/2}.}$$

CHAPTER 9

1. From the force diagram for the sapling we can write
   $\Sigma F_x = F_1 - F_2 \sin 20° - F_3 \cos \alpha = 0;$
   $282 \text{ N} - (355 \text{ N}) \sin 20° - F_3 \cos \alpha = 0,$ or
   $F_3 \cos \alpha = 161 \text{ N}.$
   $\Sigma F_y = F_2 \cos 20° - F_3 \sin \alpha = 0;$
   $F_3 \sin \alpha = (355 \text{ N}) \cos 20° = 334 \text{ N}.$
   Thus we have
   $F_3 = [(161 \text{ N})^2 + (334 \text{ N})^2]^{1/2} =$ $\boxed{370 \text{ N.}}$
   $\tan \alpha = (334 \text{ N})/(161 \text{ N}) = 2.08,$ $\alpha = 64°.$ So $\theta = 180° - \alpha =$ $\boxed{116°.}$

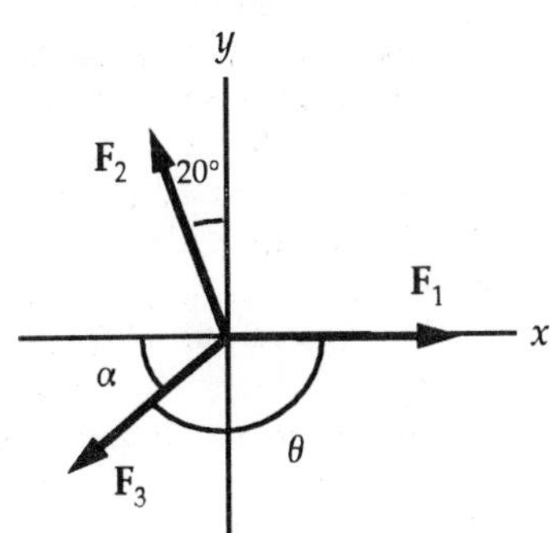

2. We choose the coordinate system shown, with the y-axis in the direction of the net force. From symmetry we know that the two tensions will be at the same angle from the y-axis.
   We write $\Sigma F_y = ma_y$ from the force diagram for the tooth:
   $F_{net} = 2F_T \cos (\frac{1}{2}\theta);$
   $0.75 \text{ N} = 2F_T \cos 77.5°,$ which gives $F_T =$ $\boxed{1.7 \text{ N.}}$

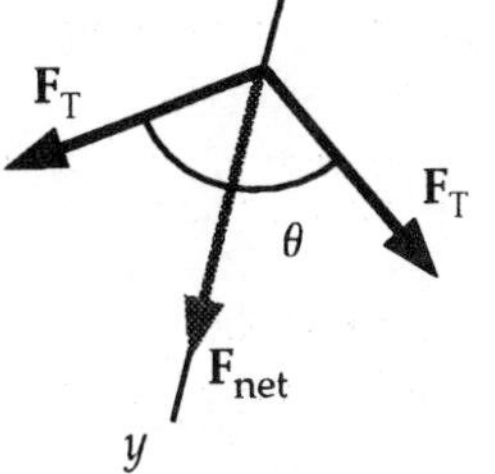

3. We choose the coordinate system shown, with positive torques clockwise. For the torque from the person's weight about the point B we have
   $\tau_B = MgL = (60 \text{ kg})(9.80 \text{ m/s}^2)(3.0 \text{ m}) =$ $\boxed{1.8 \times 10^3 \text{ m} \cdot \text{N.}}$

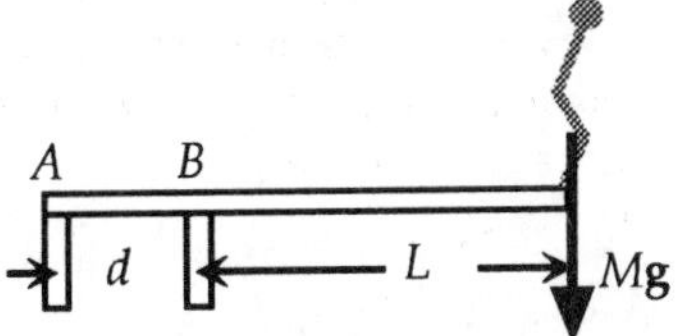

4. We choose the coordinate system shown, with positive torques clockwise. For the torque from the person's weight about the point A we have
   $\tau_A = Mgx;$
   $1000 \text{ m} \cdot \text{N} = (60 \text{ kg})(9.80 \text{ m/s}^2)x,$ which gives $x =$ $\boxed{1.7 \text{ m.}}$

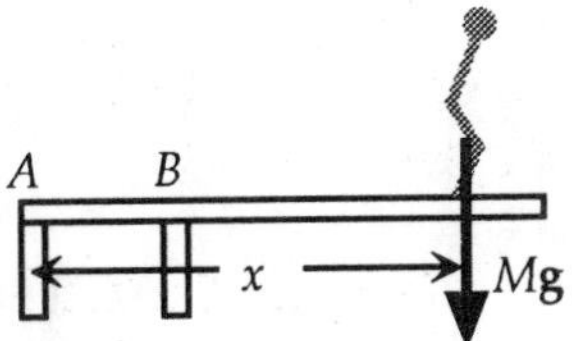

5. From the force diagram for the junction we can write
   $\Sigma F_x = F_{T1} - F_{T2} \cos 45° = 0.$
   This shows that $F_{T2} > F_{T1}$, so we take F_{T2} to be the maximum.
   $\Sigma F_y = F_{T2} \sin 45° - Mg = 0;$
   $Mg = (1300 \text{ N}) \sin 45° =$ $\boxed{9.2 \times 10^2 \text{ N.}}$

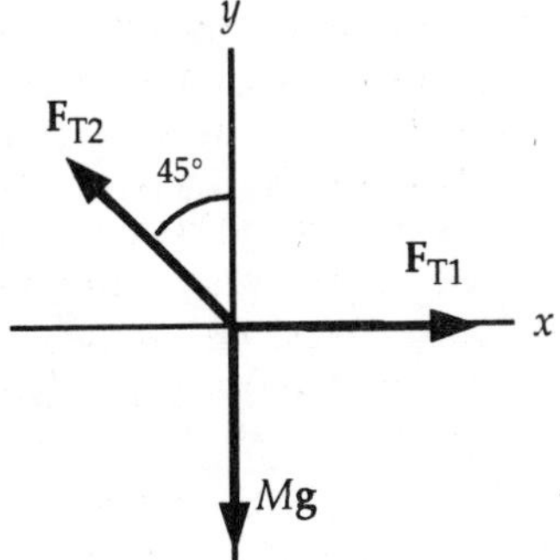

6. We choose the coordinate system shown, with positive torques clockwise. We write $\Sigma\tau = I\alpha$ about the point A from the force diagram for the leg:
$\Sigma\tau_A = MgD - F_T L = 0;$
$(15.0\text{ kg})(9.80\text{ m/s}^2)(0.350\text{ m}) - F_T(0.805\text{ m}),$
which gives $F_T = 63.9\text{ N}.$
Because there is no acceleration of the hanging mass, we have
$F_T = mg,$ or $m = F_T/g = (63.9\text{ N})/(9.80\text{ m/s}^2) =$ **6.52 kg.**

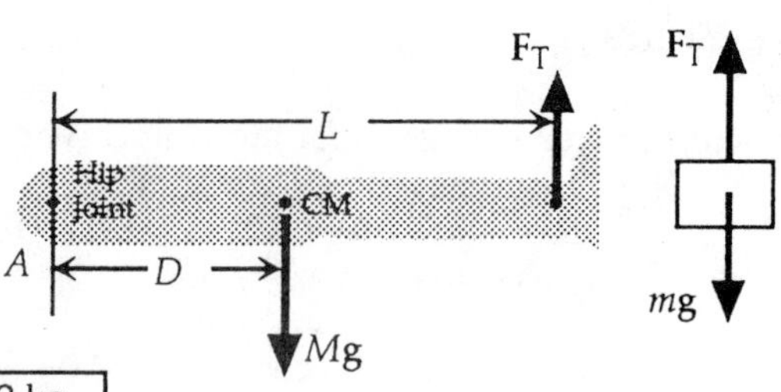

7. We choose the coordinate system shown, with positive torques clockwise. We write $\Sigma\tau = I\alpha$ about the point A from the force diagram for the beam and piano:
$\Sigma\tau_A = Mg\frac{1}{4}L + mg\frac{1}{2}L - F_{N2}L = 0,$ which gives
$F_{N2} = \frac{1}{4}Mg + \frac{1}{2}mg$
$= \frac{1}{4}(300\text{ kg})(9.80\text{ m/s}^2) + \frac{1}{2}(160\text{ kg})(9.80\text{ m/s}^2)$
$= 1.52 \times 10^3\text{ N}.$
We write $\Sigma F_y = ma_y$ from the force diagram for the beam and piano:
$F_{N1} + F_{N2} - Mg - mg = 0,$ which gives
$F_{N1} = Mg + mg - F_{N2}$
$= (300\text{ kg})(9.80\text{ m/s}^2) + (160\text{ kg})(9.80\text{ m/s}^2) - 1.52 \times 10^3\text{ N} = 2.94 \times 10^3\text{ N}.$
The forces on the supports are the reactions to these forces:
2.94×10^3 N down, and 1.52×10^3 N down.

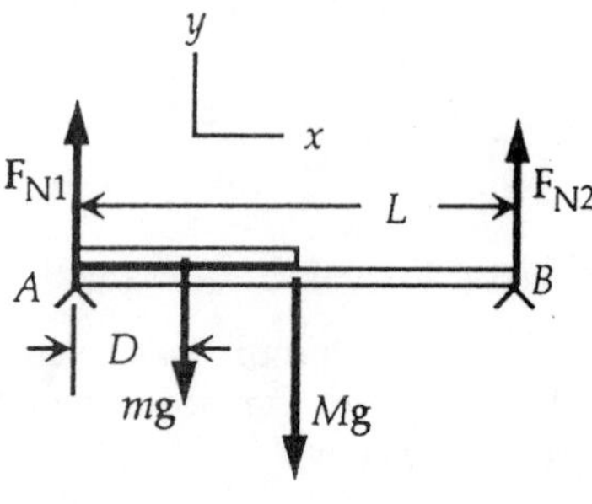

8. We choose the coordinate system shown, with positive torques clockwise. We write $\Sigma\tau = I\alpha$ about the point A from the force diagram for the two beams:
$\Sigma\tau_A = mg\frac{1}{4}L + Mg\frac{1}{2}L - F_{N2}L = 0,$ which gives
$F_{N2} = \frac{1}{4}Mg + \frac{1}{2}mg$
$= \frac{1}{4}[\frac{1}{2}(1000\text{ kg})](9.80\text{ m/s}^2) + \frac{1}{2}(1000\text{ kg})(9.80\text{ m/s}^2)$
$=$ **6.13×10^3 N.**
We write $\Sigma F_y = ma_y$ from the force diagram for the two beams:
$F_{N1} + F_{N2} - Mg - mg = 0,$ which gives
$F_{N1} = Mg + mg - F_{N2}$
$= (1000\text{ kg})(9.80\text{ m/s}^2) + \frac{1}{2}(1000\text{ kg})(9.80\text{ m/s}^2) - 6.13 \times 10^3\text{ N} =$ **8.57×10^3 N.**

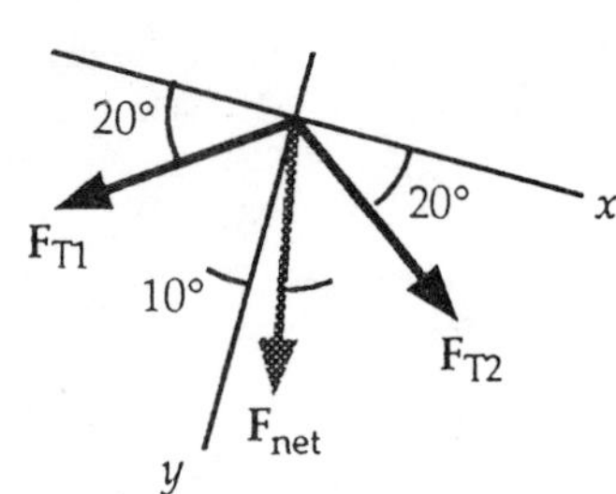

9. We must move the direction of the net force 10° to the right.
We choose the coordinate system shown, with the y-axis in the direction of the original net force.
We write $\Sigma\mathbf{F}$ from the force diagram for the tooth:
$\Sigma F_x = -F_{T1}\cos 20° + F_{T2}\cos 20° = F_{net}\sin 10°;$
$-(2.0\text{ N})\cos 20° + F_{T2}\cos 20° = F_{net}\sin 10°;$
$\Sigma F_y = +F_{T1}\sin 20° + F_{T2}\sin 20° = F_{net}\cos 10°;$
$+(2.0\text{ N})\sin 20° + F_{T2}\sin 20° = F_{net}\cos 10°.$
We have two equations for the two unknowns: F_{net}, and F_{T2}.
When we eliminate F_{net}, we get $F_{T2} =$ **2.3 N.**

20°
20°
x
F_{T1}
10°
F_{T2}
F_{net}
y

10. We choose the coordinate system shown, with positive torques clockwise. We write $\Sigma\tau = I\alpha$ about the support point A from the force diagram for the board and people:

$$\Sigma\tau_A = -m_1g(L-d) + m_2gd = 0;$$
$$-(30\text{ kg})(10\text{ m} - d) + (70\text{ kg})d = 0,$$

which gives $d =$ 3.0 m from the adult.

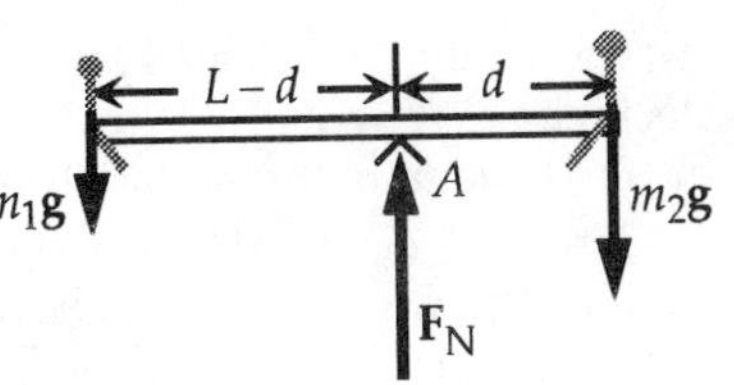

11. We choose the coordinate system shown, with positive torques clockwise. We write $\Sigma\tau = I\alpha$ about the support point A from the force diagram for the board and people:

$$\Sigma\tau_A = -m_1g(L-d) - Mg(\tfrac{1}{2}L - d) + m_2gd = 0;$$
$$-(30\text{ kg})(10\text{ m} - d) - (15\text{ kg})(5.0\text{ m} - d) + (70\text{ kg})d = 0,$$

which gives $d =$ 3.3 m from the adult.

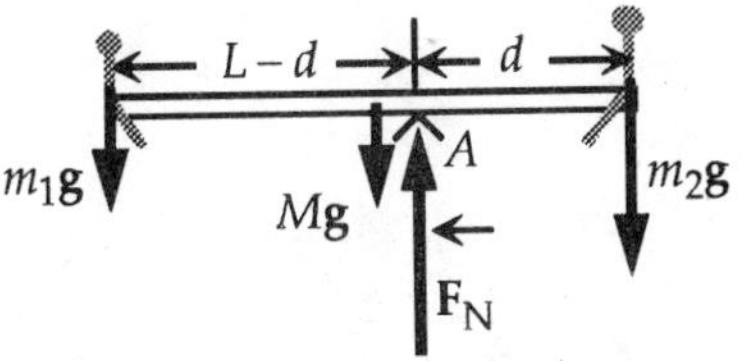

12. From the force diagram for the mass we can write

$$\Sigma F_x = F_{T1} - F_{T2}\cos\theta = 0, \text{ or}$$
$$F_{T1} = F_{T2}\cos 30°.$$
$$\Sigma F_y = F_{T2}\sin\theta - mg = 0, \text{ or}$$
$$F_{T2}\sin 30° = mg = (200\text{ kg})(9.80\text{ m/s}^2),$$

which gives $F_{T2} =$ 3.9×10^3 N.

Thus we have

$$F_{T1} = F_{T2}\cos 30° = (3.9 \times 10^3\text{ N})\cos 30° = 3.4 \times 10^3\text{ N}.$$

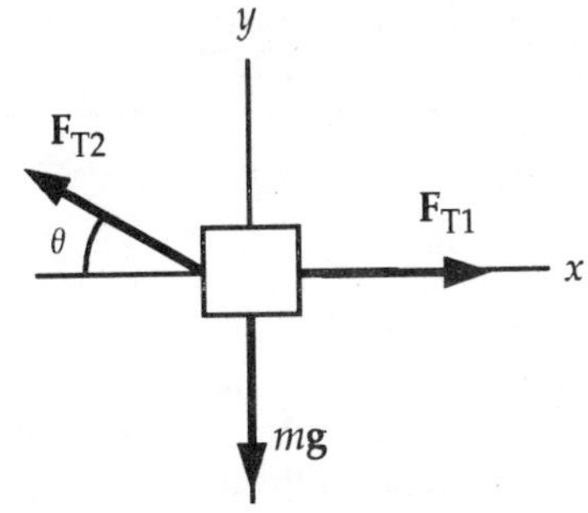

13. From the force diagram for the hanging light and junction we can write

$$\Sigma F_x = F_{T1}\cos\theta_1 - F_{T2}\cos\theta_2 = 0;$$
$$F_{T1}\cos 37° = F_{T2}\cos 53°;$$
$$\Sigma F_y = F_{T1}\sin\theta_1 + F_{T2}\sin\theta_2 - mg = 0;$$
$$F_{T1}\sin 37° + F_{T2}\sin 53° = (30\text{ kg})(9.80\text{ m/s}^2).$$

When we solve these two equations for the two unknowns, F_{T1}, and F_{T2}, we get $F_{T1} = 1.8 \times 10^2$ N, and $F_{T2} = 2.4 \times 10^2$ N.

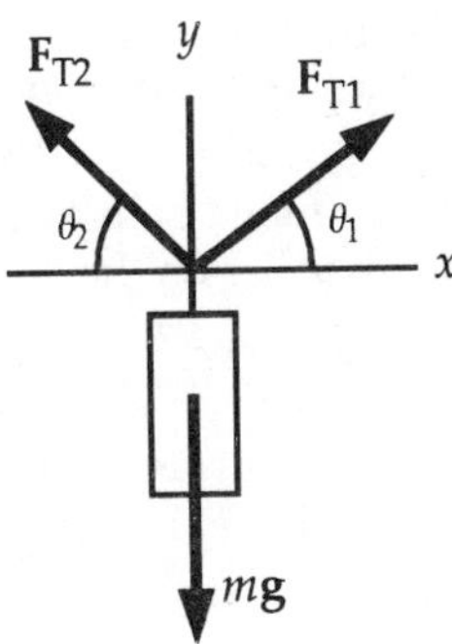

14. From the force diagram for the seesaw and children we can write

$$\Sigma F_y = F_N - m_1g - m_2g - Mg = 0, \text{ or}$$
$$F_N = (m_1 + m_2 + M)g$$
$$= (30\text{ kg} + 25\text{ kg} + 2.0\text{ kg})(9.80\text{ m/s}^2) = 5.6 \times 10^2\text{ N}.$$

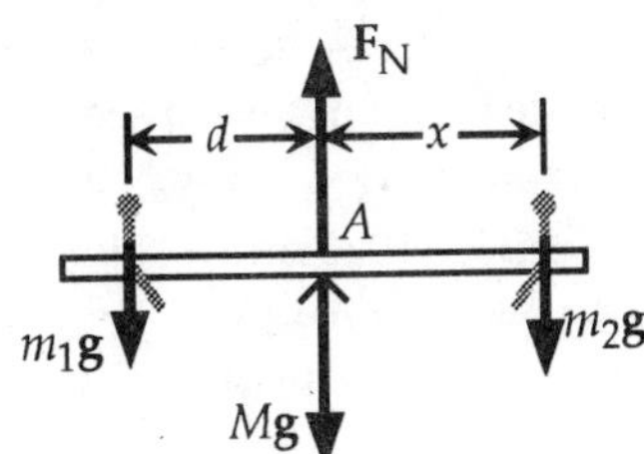

15. We choose the coordinate system shown, with positive torques clockwise. We write $\Sigma\tau = I\alpha$ about the support point A from the force diagram for the cantilever:

$\Sigma\tau_A = -F_2 d + Mg\frac{1}{2}L = 0;$

$-F_2(20.0\text{ m}) + (1200\text{ kg})(9.80\text{ m/s}^2)(25.0\text{ m}) = 0,$

which gives $F_2 =$ $\boxed{1.47 \times 10^4\text{ N.}}$

For the forces in the y-direction we have

$\Sigma F_y = F_1 + F_2 - Mg = 0,$ or

$F_1 = Mg - F_2 = (1200\text{ kg})(9.80\text{ m/s}^2) - (1.47 \times 10^4\text{ N}) =$ $\boxed{-2.94 \times 10^3\text{ N (down).}}$

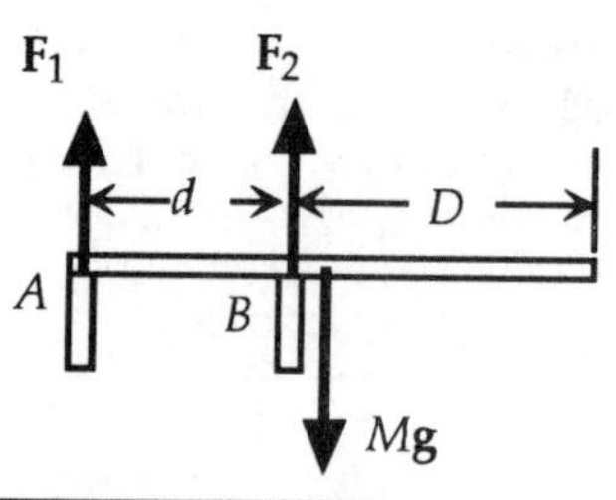

16. From the force diagram for the sheet we can write

$\Sigma F_x = F_{T2}\cos\theta - F_{T1}\cos\theta = 0,$ which gives

$F_{T2} = F_{T1}.$

$\Sigma F_y = F_{T1}\sin\theta + F_{T2}\sin\theta - mg = 0;$

$2F_{T1}\sin\theta = mg;$

$2F_{T1}\sin 3.5° = (0.60\text{ kg})(9.80\text{ m/s}^2),$

which gives $F_{T1} =$ $\boxed{48\text{ N.}}$

The tension is so much greater than the weight because $\boxed{\text{only the vertical components balance the weight.}}$

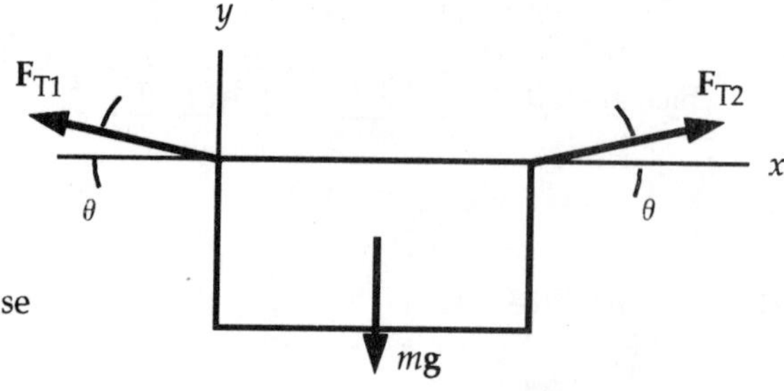

17. We choose the coordinate system shown, with positive torques clockwise. We write $\Sigma\tau = I\alpha$ about the lower hinge B from the force diagram for the door:

$\Sigma\tau_B = F_{Ax}(H - 2D) - Mg\frac{1}{2}w = 0;$

$F_{Ax}[2.30\text{ m} - 2(0.40\text{ m})] - (13.0\text{ kg})(9.80\text{ m/s}^2)\frac{1}{2}(1.30\text{ m}),$

which gives $F_{Ax} = 55.2$ N.

We write $\Sigma\mathbf{F} = m\mathbf{a}$ from the force diagram for the door:

$\Sigma F_x = F_{Ax} + F_{Bx} = 0;$

$55\text{ N} + F_{Bx} = 0,$ which gives $F_{Bx} = -55.2$ N.

The top hinge pulls away from the door, and the bottom hinge pushes on the door.

$\Sigma F_y = F_{Ay} + F_{By} - Mg = 0.$

Because each hinge supports half the weight, we have

$F_{Ay} = F_{By} = \frac{1}{2}(3.0\text{ kg})(9.80\text{ m/s}^2) = 63.7\text{ N}.$

Thus we have $\boxed{\text{top hinge: } F_{Ax} = 55.2\text{ N}, F_{Ay} = 63.7\text{ N; bottom hinge: } F_{Ax} = -55.2\text{ N}, F_{Ay} = 63.7\text{ N.}}$

18. We choose the coordinate system shown, with positive torques clockwise. We write $\Sigma\tau = I\alpha$ about the support point A from the force diagram for the seesaw and boys:

$\Sigma\tau_A = +m_2 g\frac{1}{2}L + m_3 g x - m_1 g\frac{1}{2}L = 0;$

$+(35\text{ kg})\frac{1}{2}(3.6\text{ m}) + (25\text{ kg})x - (50\text{ kg})\frac{1}{2}(3.6\text{ m}) = 0,$

which gives $x = 1.1$ m.

The third boy should be $\boxed{\text{1.1 m from pivot on side of lighter boy.}}$

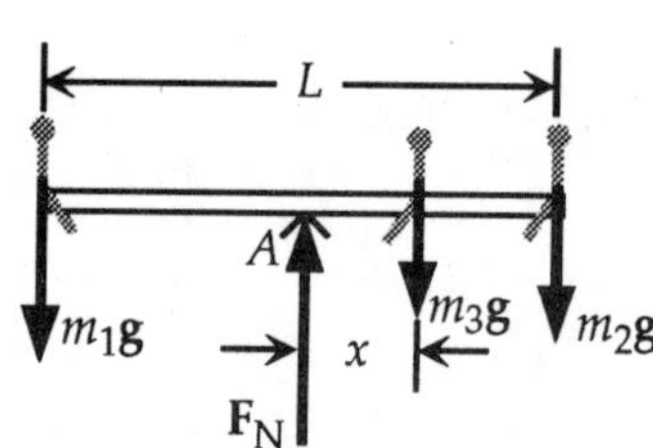

19. We choose the coordinate system shown, with positive torques clockwise. For the torques about the point B we have

$\Sigma\tau_B = F_1 d + MgD = 0;$

$F_1(1.0\text{ m}) + (60\text{ kg})(9.80\text{ m/s}^2)(3.0\text{ m}) = 0,$

which gives $F_1 = -1.8 \times 10^3\text{ N (down)}.$

For the torques about the point A we have

$\Sigma\tau_A = -F_2 d + Mg(D + d) = 0;$

$F_2(1.0\text{ m}) = (60\text{ kg})(9.80\text{ m/s}^2)(3.0\text{ m} + 1.0\text{ m}),$

which gives $F_2 = 2.4 \times 10^3\text{ N (up)}.$

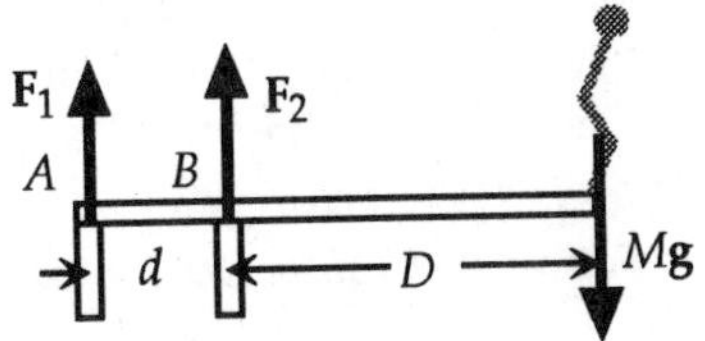

20. We choose the coordinate system shown, with positive torques clockwise. For the torques about the point B we have

$\Sigma\tau_B = F_1 d + MgD + mg[\tfrac{1}{2}(D + d) - d] = 0;$

$F_1(1.0\text{ m}) = -(60\text{ kg})(9.80\text{ m/s}^2)(3.0\text{ m}) - (35\text{ kg})(9.80\text{ m/s}^2)(1.0\text{ m}),$

which gives $F_1 = -2.1 \times 10^3\text{ N (down)}.$

For the torques about the point A we have

$\Sigma\tau_A = -F_2 d + Mg(D + d) + mg\tfrac{1}{2}(D + d) = 0;$

$F_2(1.0\text{ m}) = (60\text{ kg})(9.80\text{ m/s}^2)(4.0\text{ m}) + (35\text{ kg})(9.80\text{ m/s}^2)(2.0\text{ m}),$

which gives $F_2 = 3.0 \times 10^3\text{ N (up)}.$

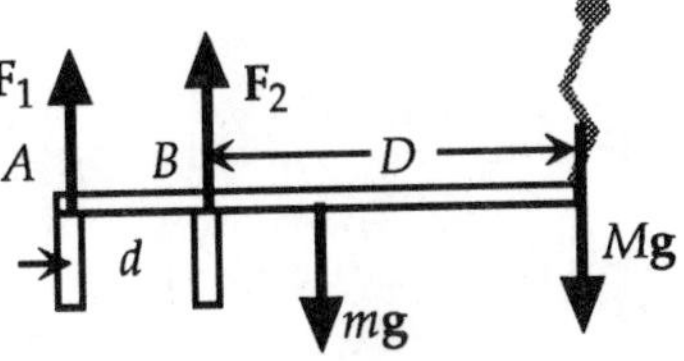

21. From Example 7–16 we have

$D = (20.4/100)(1.60\text{ m}) = 0.326\text{ m}.$

From Table 7–1, we have

$L = [(52.1 - 4.0)/100](1.60\text{ m}) = 0.770\text{ m};$

$M = \tfrac{1}{2}[(21.5 + 9.6 + 3.4)/100](60.0\text{ kg}) = 10.35\text{ kg}.$

We choose the coordinate system shown, with positive torques clockwise. We write $\Sigma\tau = I\alpha$ about the hip joint from the force diagram for the leg:

$\Sigma\tau = MgD - F_T L = 0;$

$(10.35\text{ kg})(9.80\text{ m/s}^2)(0.326\text{ m}) - F_T(0.770\text{ m}) = 0.$

which gives $F_T = 42.9\text{ N}.$

Because there is no acceleration of the hanging mass, we have

$F_T = mg$, or $m = F_T/g = (42.9\text{ N})/(9.80\text{ m/s}^2) = 4.38\text{ kg}.$

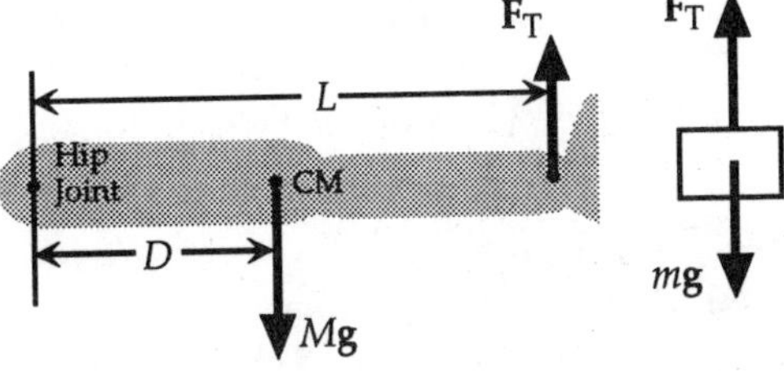

22. We choose the coordinate system shown, with positive torques clockwise. We write $\Sigma\tau = I\alpha$ about the support point B from the force diagram for the beam:

$\Sigma\tau_B = +F_1(d_1 + d_2 + d_3 + d_4) - F_3(d_2 + d_3 + d_4) -$
$F_4(d_3 + d_4) - F_5 d_4 - Mg\tfrac{1}{2}(d_1 + d_2 + d_3 + d_4) = 0;$

$F_1(10.0\text{ m}) - (4000\text{ N})(8.0\text{ m}) - (3000\text{ N})(4.0\text{ m}) -$
$(2000\text{ N})(1.0\text{ m}) - (250\text{ kg})(9.80\text{ m/s}^2)(5.0\text{ m}) = 0,$

which gives $F_1 = 5.8 \times 10^3\text{ N}.$

We write $\Sigma\tau = I\alpha$ about the support point A from the force diagram for the beam:

$\Sigma\tau_A = -F_2(d_1 + d_2 + d_3 + d_4) + F_3 d_1 - F_4(d_1 + d_2) -$
$F_5(d_1 + d_2 + d_3) - Mg\tfrac{1}{2}(d_1 + d_2 + d_3 + d_4) = 0;$

$-F_2(10.0\text{ m}) - (4000\text{ N})(2.0\text{ m}) - (3000\text{ N})(6.0\text{ m}) -$
$(2000\text{ N})(9.0\text{ m}) - (250\text{ kg})(9.80\text{ m/s}^2)(5.0\text{ m}) = 0,$

which gives $F_2 = 5.6 \times 10^3\text{ N}.$

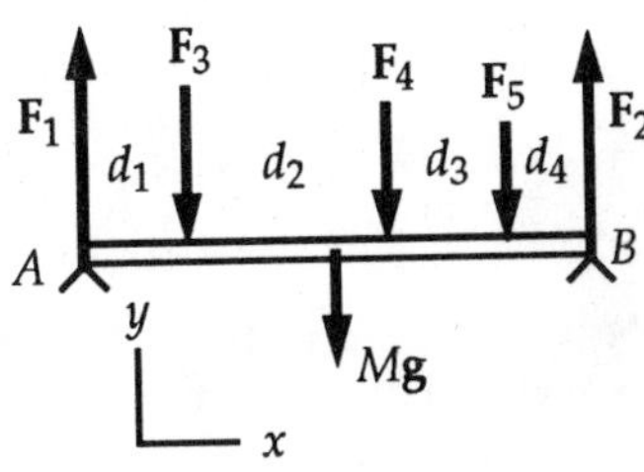

23. We choose the coordinate system shown, with positive torques clockwise. We write $\Sigma\tau = I\alpha$ about the point A from the force diagram for the beam:

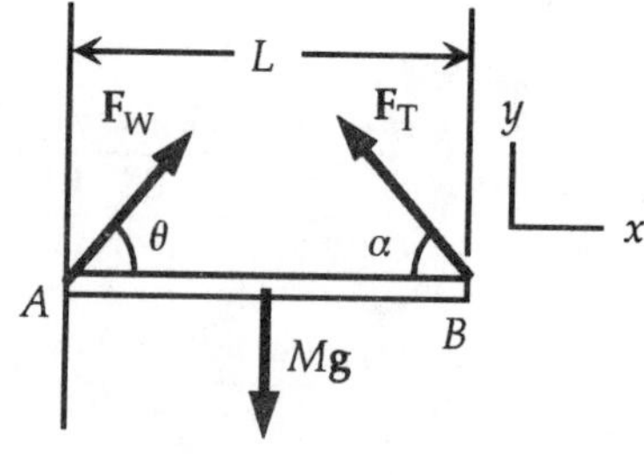

$\Sigma\tau_A = -(F_T \sin\alpha)L + Mg\frac{1}{2}L = 0;$

$-F_T \sin 50° + (30 \text{ kg})(9.80 \text{ m/s}^2)\frac{1}{2} = 0,$

which gives $F_T = 1.9 \times 10^2$ N.

Note that we find the torque produced by the tension by finding the torques produced by the components.

We write $\Sigma\mathbf{F} = m\mathbf{a}$ from the force diagram for the beam:

$\Sigma F_x = F_{Wx} - F_T \cos\alpha = 0;$

$F_{Wx} - (1.9 \times 10^2 \text{ N}) \cos 50° = 0$, which gives $F_{Wx} = 123$ N.

$\Sigma F_y = F_{Wy} + F_T \sin\alpha - Mg = 0;$

$F_{Wy} + (1.9 \times 10^2 \text{ N}) \sin 50° - (30 \text{ kg})(9.80 \text{ m/s}^2) = 0$, which gives $F_{Wy} = 147$ N.

For the magnitude of $\mathbf{F}_W$ we have

$F_W = (F_{Wx}{}^2 + F_{Wy}{}^2)^{1/2} = [(123 \text{ N})^2 + (147 \text{ N})^2]^{1/2} = 1.9 \times 10^2$ N.

We find the direction from

$\tan\theta = F_{Wy}/F_{Wx} = (147 \text{ N})/(123 \text{ N}) = 1.19$, which gives $\theta = 50°$.

Thus the force at the wall is $\mathbf{F}_W = 1.9 \times 10^2$ N, 50° above the horizontal.

24. Because the backpack is at the midpoint of the rope, the angles are equal. The force exerted by the backpacker is the tension in the rope. From the force diagram for the backpack and junction we can write

$\Sigma F_x = F_{T1} \cos\theta - F_{T2} \cos\theta = 0$, or $F_{T1} = F_{T2} = F;$

$\Sigma F_y = F_{T1} \sin\theta + F_{T2} \sin\theta - mg = 0$, or

$2F \sin\theta = mg.$

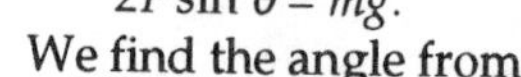

(*a*) We find the angle from

$\tan\theta = h/\frac{1}{2}L = (1.5 \text{ m})/\frac{1}{2}(7.6 \text{ m}) = 0.395$, or $\theta = 21.5°$.

When we put this in the force equation, we get

$2F \sin 21.5° = (16 \text{ kg})(9.80 \text{ m/s}^2)$, which gives $F = 2.1 \times 10^2$ N.

(*b*) We find the angle from

$\tan\theta = h/\frac{1}{2}L = (0.15 \text{ m})/\frac{1}{2}(7.6 \text{ m}) = 0.0395$, or $\theta = 2.26°$.

When we put this in the force equation, we get

$2F \sin 2.26° = (16 \text{ kg})(9.80 \text{ m/s}^2)$, which gives $F = 2.0 \times 10^3$ N.

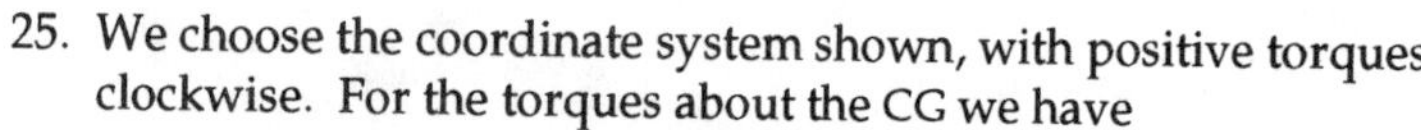

25. We choose the coordinate system shown, with positive torques clockwise. For the torques about the CG we have

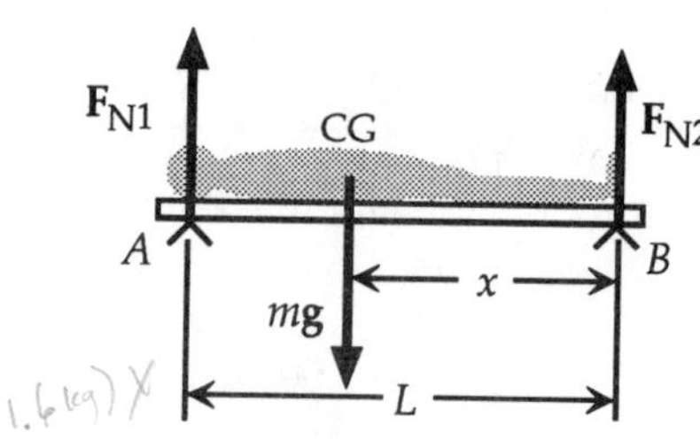

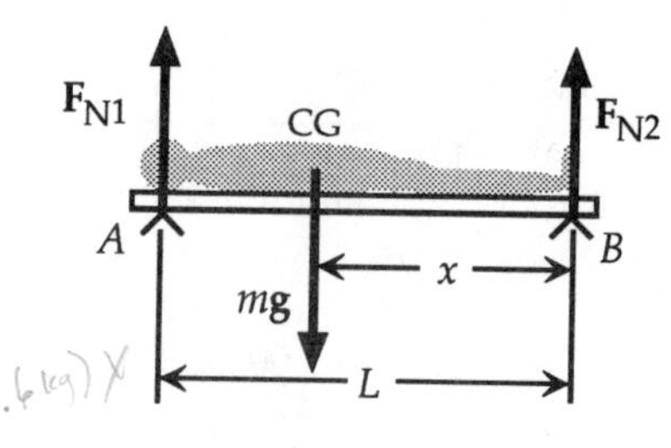

$\Sigma\tau_{CG} = F_{N1}(L - x) - F_{N2}x = 0;$

$(35.1 \text{ kg})g(170 \text{ cm} - x) - (31.6 \text{ kg})gx = 0,$

which gives $x = 89.5$ cm from the feet.

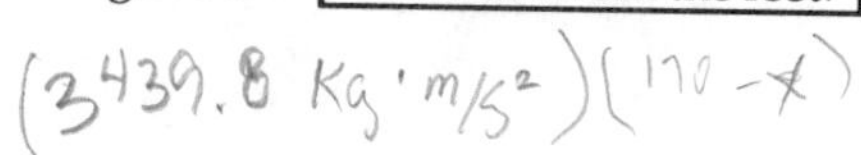

(3439.8 Kg·m/s²)(170 − x)

188.83

5967 kg·cm − 35.1Kgx = (31.6Kg)(31.6kg)x

26. We choose the coordinate system shown, with positive torques clockwise. We write $\Sigma\tau = I\alpha$ about the point A from the force diagram for the beam and sign:

$\Sigma\tau_A = -(F_T \sin\theta)D + MgL + mg\frac{1}{2}L = 0;$

$-F_T(\sin 41.0°)(1.35\text{ m}) + (215\text{ N})(1.70\text{ m}) + (135\text{ N})\frac{1}{2}(1.70\text{ m}) = 0,$

which gives $\boxed{F_T = 542\text{ N.}}$

Note that we find the torque produced by the tension by finding the torques produced by the components.

We write $\Sigma\mathbf{F} = m\mathbf{a}$ from the force diagram for the beam and sign:

$\Sigma F_x = F_{hingeH} - F_T\cos\theta = 0;$

$F_{hingeH} - (542\text{ N})\cos 41.0° = 0$, which gives $F_{hingeH} = \boxed{409\text{ N.}}$

$\Sigma F_y = F_{hingeV} + F_T\sin\theta - Mg - mg = 0;$

$F_{hingeV} + (542\text{ N})\cos 41.0° - 215\text{ N} - 135\text{ N} = 0,$

which gives $F_{hingeV} = \boxed{-59\text{ N (down).}}$

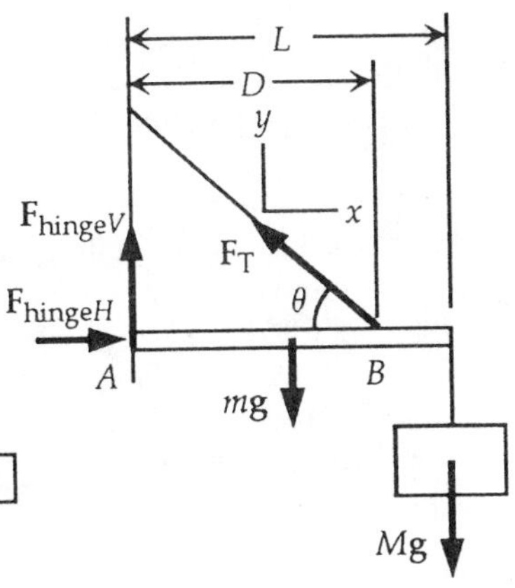

27. We choose the coordinate system shown, with positive torques clockwise. We write $\Sigma\tau = I\alpha$ about the point A from the force diagram for the pole and light:

$\Sigma\tau_A = -F_TH + MgL\cos\theta + mg\frac{1}{2}L\cos\theta = 0;$

$-F_T(3.80\text{ m}) + (12.0\text{ kg})(9.80\text{ m/s}^2)(7.5\text{ m})\cos 37° +$

$(8.0\text{ kg})(9.80\text{ m/s}^2)\frac{1}{2}(7.5\text{ m})\cos 37° = 0,$

which gives $\boxed{F_T = 2.5\times10^2\text{ N.}}$

We write $\Sigma\mathbf{F} = m\mathbf{a}$ from the force diagram for the pole and light:

$\Sigma F_x = F_{AH} - F_T = 0;$

$F_{AH} - 2.5\times10^2\text{ N} = 0$, which gives $F_{AH} = \boxed{2.5\times10^2\text{ N.}}$

$\Sigma F_y = F_{AV} - Mg - mg = 0;$

$F_{AV} - (12.0\text{ kg})(9.80\text{ m/s}^2) - (8.0\text{ kg})(9.80\text{ m/s}^2) = 0$, which gives $F_{AV} = \boxed{2.0\times10^2\text{ N.}}$

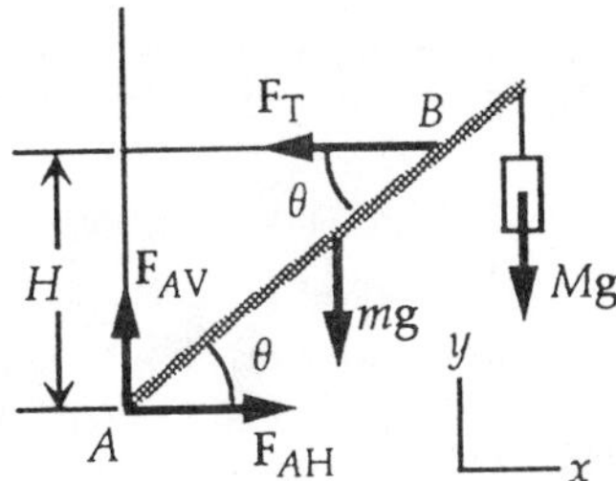

28. We choose the coordinate system shown, with positive torques counterclockwise. We write $\Sigma\tau = I\alpha$ about the point A from the force diagram for the ladder:

$\Sigma\tau_A = mg(\frac{1}{2}L)\cos\theta - F_{N2}L\sin\theta = 0$, which gives

$F_{N2} = mg/2\tan\theta.$

We write $\Sigma F_x = ma_x$ from the force diagram for the ladder:

$F_{fr} - F_{N2} = 0$, which gives $F_{fr} = F_{N2} = mg/2\tan\theta.$

We write $\Sigma F_y = ma_y$ from the force diagram for the ladder:

$F_{N1} - mg = 0$, which gives $F_{N1} = mg.$

For the bottom not to slip, we must have

$F_{fr} \le \mu F_{N1}$, or $mg/2\tan\theta \le \mu mg,$

from which we get $\tan\theta \ge 1/2\mu.$

The minimum angle is $\boxed{\theta_{min} = \tan^{-1}(1/2\mu).}$

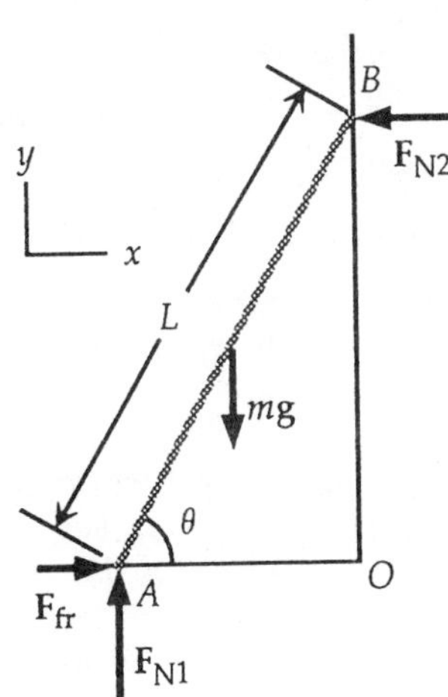

29. Because the backpack is at the midpoint of the rope, the angles are equal. From the force diagram for the backpack and junction we can write

$\Sigma F_x = F_{T1} \cos \theta - F_{T2} \cos \theta = 0$, or $F_{T1} = F_{T2} = F$;

$\Sigma F_y = F_{T1} \sin \theta + F_{T2} \sin \theta - mg - F_{bear} = 0$.

When the bear is not pulling, we have

$2F_1 \sin \theta = mg$;

$2F_1 \sin 15° = (23.0 \text{ kg})(9.80 \text{ m/s}^2)$, which gives $F_1 = 435$ N.

When the bear is pulling, we have

$2F_2 \sin \theta = mg + F_{bear}$;

$2(2)(435 \text{ N}) \sin 30° = (23.0 \text{ kg})(9.80 \text{ m/s}^2) + F_{bear}$,

which gives $F_{bear} =$ $\boxed{6.5 \times 10^2 \text{ N.}}$

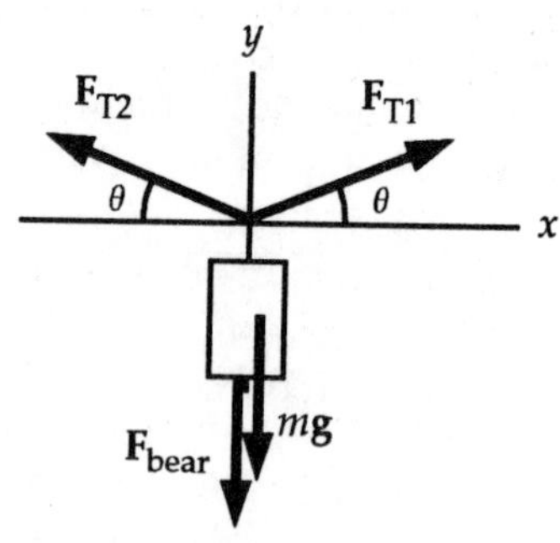

30. We choose the coordinate system shown, with positive torques clockwise.

(*a*) For the torques about the point *B* we have

$\Sigma \tau_B = F_{T1}(\frac{1}{2}L + D) - MgD = 0$;

$F_{T1}(0.500 \text{ m} + 0.400 \text{ m}) - (0.230 \text{ kg})(9.80 \text{ m/s}^2)(0.400 \text{ m}) = 0$,

which gives $F_{T1} =$ $\boxed{1.00 \text{ N.}}$

(*b*) For the torques about the point *A* we have

$\Sigma \tau_A = - F_{T2}(\frac{1}{2}L + D) + Mg\frac{1}{2}L = 0$;

$- F_{T2}(0.500 \text{ m} + 0.400 \text{ m}) + (0.230 \text{ kg})(9.80 \text{ m/s}^2)(0.500 \text{ m})$,

which gives $F_{T2} =$ $\boxed{1.25 \text{ N.}}$

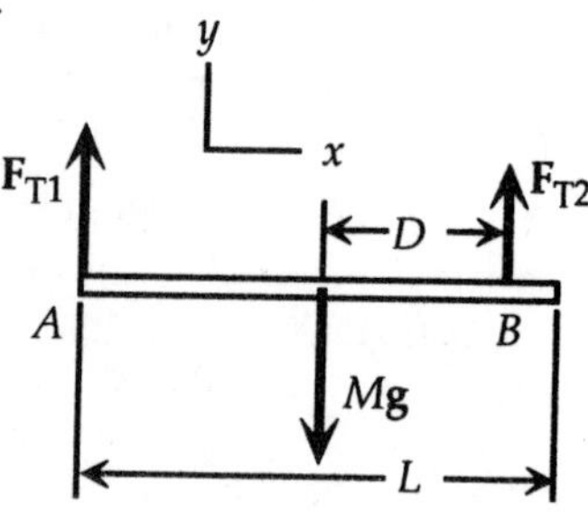

31. We choose the coordinate system shown, with positive torques clockwise. We write $\Sigma \tau = I\alpha$ about the point *A* from the force diagram for the ladder and painter:

$\Sigma \tau_A = mg(\frac{1}{2}d_L) + Mgd_P - F_W h_L = 0$;

$(12.0 \text{ kg})(9.80 \text{ m/s}^2)\frac{1}{2}(3.0 \text{ m}) + (60.0 \text{ kg})(9.80 \text{ m/s}^2)(2.1 \text{ m}) - F_W(4.0 \text{ m}) = 0$, which gives

$F_W = 353$ N.

We write $\Sigma F_x = ma_x$ from the force diagram:

$F_{Gx} - F_W = 0$, which gives

$F_{Gx} = F_W = 353$ N.

We write $\Sigma F_y = ma_y$ from the force diagram:

$F_{Gy} - mg - Mg = 0$, which gives

$F_{Gy} = (m + M)g = (12.0 \text{ kg} + 60.0 \text{ kg})(9.80 \text{ m/s}^2) = 706$ N.

Because the ladder is on the verge of slipping, we must have

$F_{Gx} = \mu F_{Gy}$, or $\mu = F_{Gx}/F_{Gy} = (353 \text{ N})/(706 \text{ N}) =$ $\boxed{0.50.}$

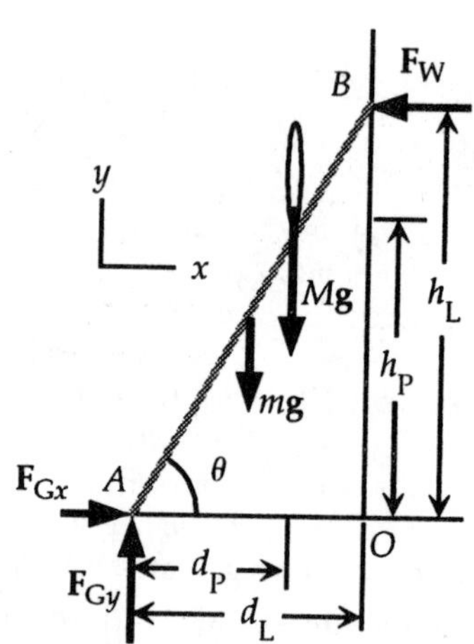

32. We choose the coordinate system shown, with positive torques clockwise.
We write $\Sigma F_x = ma_x$ from the force diagram for the lamp:
$F_P - F_{fr} = 0$, which gives $F_P = F_{fr}$.
We write $\Sigma F_y = ma_y$ from the force diagram for the lamp:
$F_N - mg = 0$, which gives $F_N = mg$.
(*a*) If we assume that the lamp slides, we have $F_P = F_{fr} = \mu F_N = \mu mg$.
The normal force would have to be inside the base, so we find the distance from the pole at which it acts. We write $\Sigma\tau = I\alpha$ about the center of the base from the force diagram for the lamp:
$\Sigma\tau = F_P H - F_N x = \mu mgH - mgx = 0$;
$(0.20)(0.60\ \text{m}) - x = 0$, which gives $x = 0.12$ m.
Because this is greater than 0.10 m, the lamp will **tip over.**
(*b*) If the lamp slides, we have $F_P = F_{fr} = \mu F_N = \mu mg$.
We write $\Sigma\tau = I\alpha$ about the center of the base from the force diagram for the lamp:
$\Sigma\tau = F_P H - F_N x = \mu mgH - mgx = 0$, or $H = x/\mu$.
The maximum height will be when x is maximum, which is d:
$H_{max} = d/\mu = (0.10\ \text{m})/(0.20) = 0.50\ \text{m} =$ **50 cm.**

33. From the symmetry of the wires, we see that the angle between a horizontal line on the ground parallel to the net and the line from the base of the pole to the anchoring point is $\theta = 30°$. We find the angle between the pole and a wire from
$\tan\alpha = d/H = (2.0\ \text{m})/(2.6\ \text{m}) = 0.769$, which gives $\alpha = 37.6°$.
Thus the horizontal component of each tension is $F_T \sin\alpha$.
We write $\Sigma\tau = I\alpha$ about the horizontal axis through the base A perpendicular to the net from the force diagram for the pole:
$\Sigma\tau_A = F_{net}H - 2(F_T \sin\alpha\cos\theta)H = 0$, or
$F_{net} = 2F_T \sin\alpha\cos\theta = 2(95\ \text{N})\sin 37.6°\cos 30° =$ **1.0×10^2 N.**

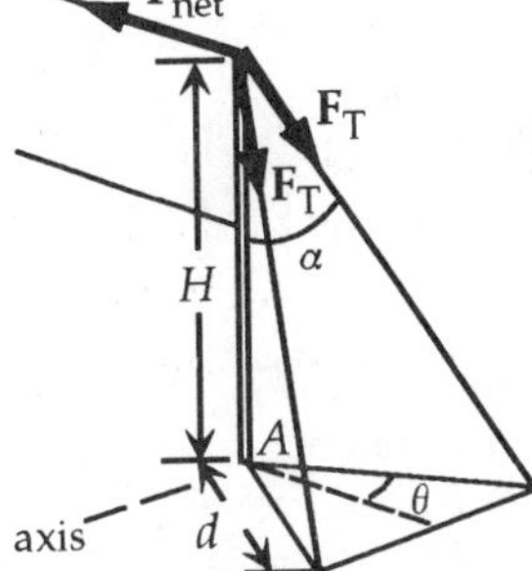

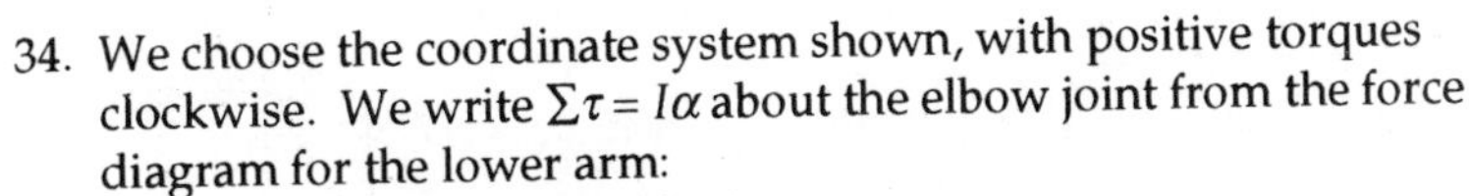

34. We choose the coordinate system shown, with positive torques clockwise. We write $\Sigma\tau = I\alpha$ about the elbow joint from the force diagram for the lower arm:
$\Sigma\tau = mgD + MgL - F_M d = 0$;
$(2.0\ \text{kg})(9.80\ \text{m/s}^2)(0.15\ \text{m}) + M(9.80\ \text{m/s}^2)(0.35\ \text{m}) - (400\ \text{N})(0.060\ \text{m}) = 0$,
which gives $M =$ **6.1 kg.**

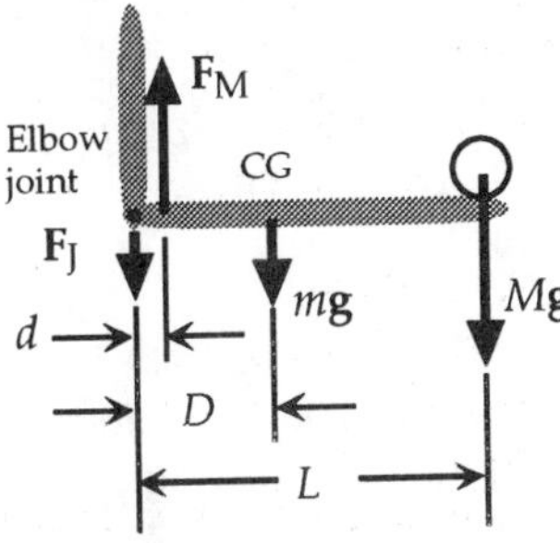

35. We choose the coordinate system shown, with positive torques clockwise. We write $\Sigma\tau = I\alpha$ about the elbow joint from the force diagram for the lower arm:
$\Sigma\tau = mgD + MgL - F_M d = 0$;
$(2.8\ \text{kg})(9.80\ \text{m/s}^2)(0.12\ \text{m}) + (7.3\ \text{kg})(9.80\ \text{m/s}^2)(0.300\ \text{m}) - F_M(0.025\ \text{m}) = 0$,
which gives $F_M =$ **9.9×10^2 N.**

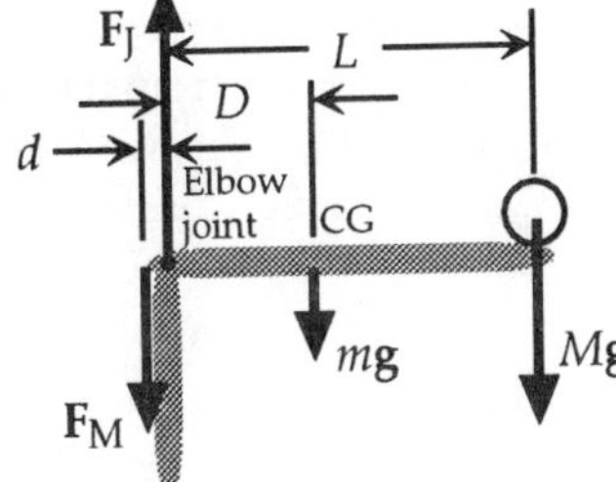

36. We choose the coordinate system shown, with positive torques clockwise. We write $\Sigma\tau = I\alpha$ about the shoulder joint A from the force diagram for the arm:

    $\Sigma\tau_A = mgD - (F_M \sin\theta)d = 0;$

    $(3.3 \text{ kg})(9.80 \text{ m/s}^2)(0.24 \text{ m}) - (F_M \sin 15°)(0.12 \text{ m}) = 0,$

    which gives $F_M =$ $\boxed{2.5 \times 10^2 \text{ N.}}$

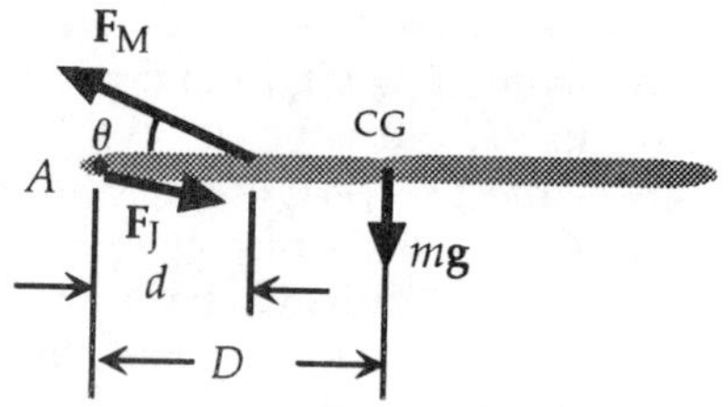

37. We choose the coordinate system shown, with positive torques clockwise. We write $\Sigma\tau = I\alpha$ about the shoulder joint A from the force diagram for the arm:

    $\Sigma\tau_A = mgD + MgL - (F_M \sin\theta)d = 0;$

    $(3.3 \text{ kg})(9.80 \text{ m/s}^2)(0.24 \text{ m}) +$

    $(15 \text{ kg})(9.80 \text{ m/s}^2)(0.52 \text{ m}) - (F_M \sin 15°)(0.12 \text{ m}) = 0,$

    which gives $F_M =$ $\boxed{2.7 \times 10^3 \text{ N.}}$

    Note that this is more than 10 times the result from Problem 36.

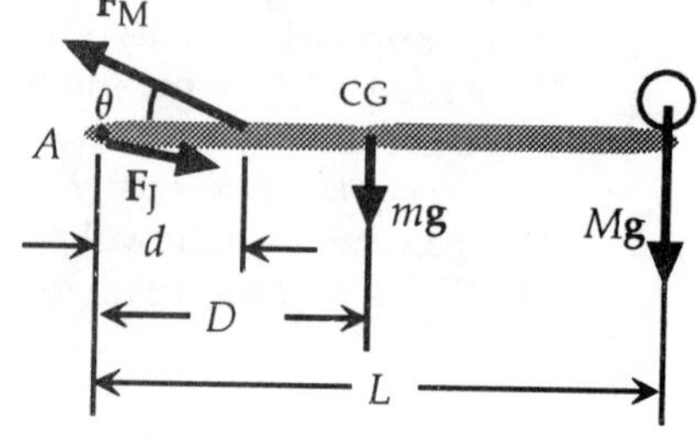

38. We write $\Sigma F_x = ma_x$ from the force diagram:

    $F_{Jx} - F_M \cos\theta = 0$, which gives $F_{Jx} = F_M \cos\theta$.

    We write $\Sigma F_y = ma_y$ from the force diagram:

    $F_{Jy} + F_M \sin\theta - mg - Mg = 0,$

    which gives $F_{Jy} = (m + M)g - F_M \sin\theta$.

    For Problem 36, $M = 0$, so we have

    $F_{Jx} = F_M \cos\theta = (2.5 \times 10^2 \text{ N}) \cos 15° = 241 \text{ N}.$

    $F_{Jy} = (m + M)g - F_M \sin\theta$

    $= (3.3 \text{ kg})(9.80 \text{ m/s}^2) - (2.5 \times 10^2 \text{ N}) \sin 15° = -32 \text{ N}.$

    $F_J = (F_{Jx}^2 + F_{Jy}^2)^{1/2} = [(241 \text{ N})^2 + (-32 \text{ N})^2]^{1/2} =$ $\boxed{2.4 \times 10^2 \text{ N.}}$

    For Problem 37, $M = 15$ kg, so we have

    $F_{Jx} = F_M \cos\theta = (2.7 \times 10^3 \text{ N}) \cos 15° = 2.61 \times 10^3 \text{ N}.$

    $F_{Jy} = (m + M)g - F_M \sin\theta = (3.3 \text{ kg} + 15 \text{ kg})(9.80 \text{ m/s}^2) - (2.7 \times 10^3 \text{ N}) \sin 15° = -517 \text{ N}.$

    $F_J = (F_{Jx}^2 + F_{Jy}^2)^{1/2} = [(2.61 \times 10^3 \text{ N})^2 + (-517 \text{ N})^2]^{1/2} =$ $\boxed{2.7 \times 10^3 \text{ N.}}$

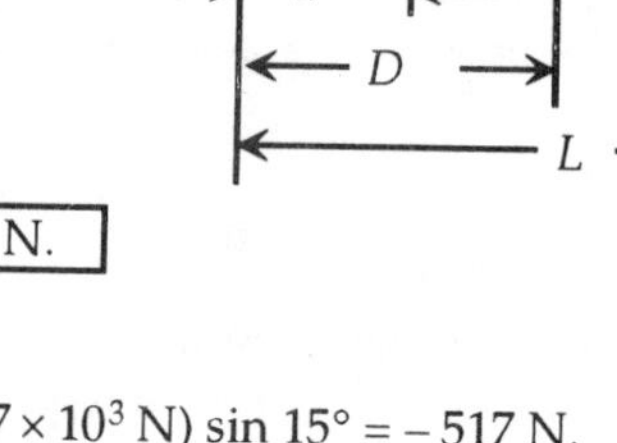

39. Because the person is standing on one foot, the normal force on the ball of the foot must support the weight: $F_N = Mg$. We choose the coordinate system shown, with positive torques clockwise. We write $\Sigma\tau = I\alpha$ about the point A from the force diagram for the foot:

    $\Sigma\tau_A = F_T d - F_N D = 0;$

    $F_T d - F_N(2d) = 0$, which gives

    $F_T = 2F_N = 2(70 \text{ kg})(9.80 \text{ m/s}^2) =$ $\boxed{1.4 \times 10^3 \text{ N (up).}}$

    We write $\Sigma F_y = ma_y$ from the force diagram:

    $F_T + F_N - F_{bone} = 0$, which gives

    $F_{bone} = F_T + F_N = 3F_N = 3(70 \text{ kg})(9.80 \text{ m/s}^2) =$ $\boxed{2.1 \times 10^3 \text{ N (down).}}$

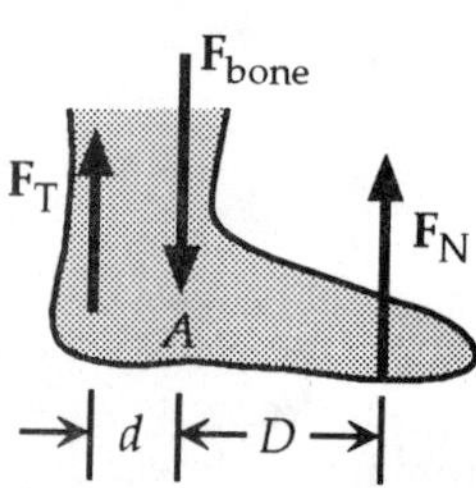

40. We choose the coordinate system shown, with positive torques clockwise. We write $\Sigma\tau = I\alpha$ about the point A from the force diagram for the torso:

$$\Sigma\tau_A = w_1(d_1 + d_2 + d_3)\cos\theta + w_2(d_2 + d_3)\cos\theta + w_3 d_3 \cos\theta - (F_M \sin\alpha)(d_2 + d_3) = 0;$$

$$0.07w(24\text{ cm} + 12\text{ cm} + 36\text{ cm})\cos 60° + 0.12w(12\text{ cm} + 36\text{ cm})\cos 60° + 0.46w(36\text{ cm})\cos 60° - F_M(\sin 12°)(12\text{ cm} + 36\text{ cm}) = 0,$$

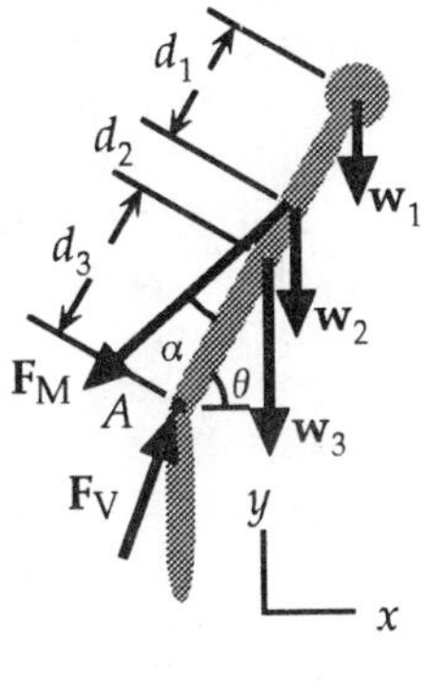

which gives $F_M = 1.37w$.

We write $\Sigma F_x = ma_x$ from the force diagram:

$$F_{Vx} - F_M\cos(\theta - \alpha) = 0, \text{ which gives}$$
$$F_{Vx} = F_M\cos(\theta - \alpha) = (1.37w)\cos 48° = 0.917w.$$

We write $\Sigma F_y = ma_y$ from the force diagram:

$$F_{Vy} - F_M\sin(\theta - \alpha) - w_1 - w_2 - w_3 = 0, \text{ which gives}$$
$$F_{Vy} = F_M\sin(\theta - \alpha) + w_1 + w_2 + w_3 = (1.37w)\sin 48° + 0.07w + 0.12w + 0.46w = 1.67w.$$

For the magnitude we have

$$F_V = (F_{Vx}^2 + F_{Vy}^2)^{1/2} = [(0.917w)^2 + (1.67w)^2]^{1/2} = \boxed{1.9w.}$$

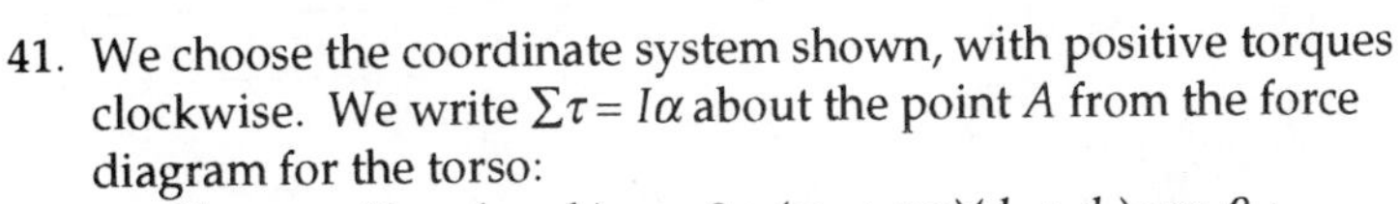

41. We choose the coordinate system shown, with positive torques clockwise. We write $\Sigma\tau = I\alpha$ about the point A from the force diagram for the torso:

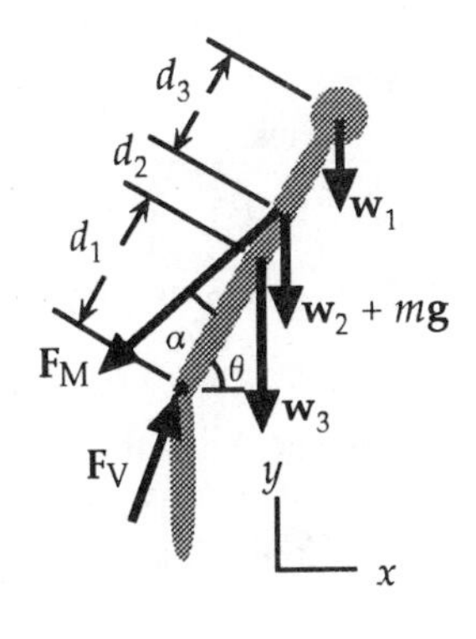

$$\Sigma\tau_A = w_1(d_1 + d_2 + d_3)\cos\theta + (w_2 + mg)(d_2 + d_3)\cos\theta + w_3 d_3\cos\theta - (F_M\sin\alpha)(d_2 + d_3) = 0;$$

$$0.07(70\text{ kg})(9.80\text{ m/s}^2)(24\text{ cm} + 12\text{ cm} + 36\text{ cm})\cos 30° + [0.12(70\text{ kg}) + 20\text{ kg}](9.80\text{ m/s}^2)(12\text{ cm} + 36\text{ cm})\cos 30° + 0.46(70\text{ kg})(9.80\text{ m/s}^2)(36\text{ cm})\cos 30° - F_M(\sin 12°)(12\text{ cm} + 36\text{ cm}) = 0,$$

which gives $F_M = 2.45 \times 10^3$ N.

We write $\Sigma F_x = ma_x$ from the force diagram:

$$F_{Vx} - F_M\cos(\theta - \alpha) = 0, \text{ which gives}$$
$$F_{Vx} = F_M\cos(\theta - \alpha) = (2.45 \times 10^3\text{ N})\cos 18° = 2.33 \times 10^3\text{ N}.$$

We write $\Sigma F_y = ma_y$ from the force diagram:

$$F_{Vy} - F_M\sin(\theta - \alpha) - w_1 - w_2 - w_3 - mg = 0, \text{ which gives}$$
$$F_{Vy} = F_M\sin(\theta - \alpha) + w_1 + w_2 + w_3 + mg$$
$$= (2.45 \times 10^3\text{ N})\sin 18° + (0.07 + 0.12 + 0.46)(70\text{ kg})(9.80\text{ m/s}^2) + (20\text{ kg})(9.80\text{ m/s}^2) = 1.40 \times 10^3\text{ N}.$$

For the magnitude we have

$$F_V = (F_{Vx}^2 + F_{Vy}^2)^{1/2} = [(2.33 \times 10^3\text{ N})^2 + (1.40 \times 10^3\text{ N})^2]^{1/2} = \boxed{2.7 \times 10^3\text{ N.}}$$

42. If the tower is of uniform composition, the center of gravity will be at the center of the tower. Because the top is 4.5 m off center, the CG will be $\frac{1}{2}$(4.5 m) = 2.25 m off center. This is less than the radius of the tower, 3.5 m, so the line of the weight is inside the base and the tower is in $\boxed{\text{stable equilibrium.}}$

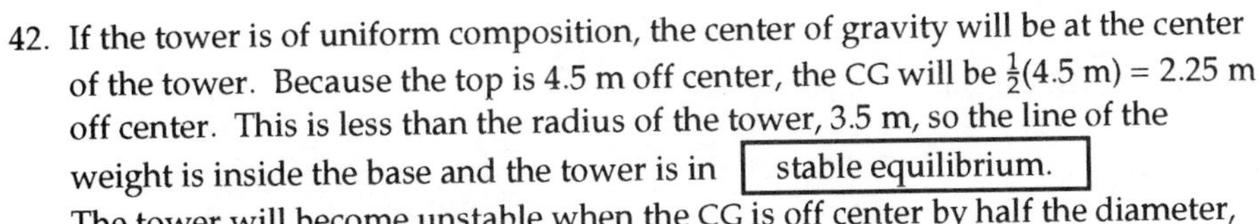

The tower will become unstable when the CG is off center by half the diameter, or the top is off center by the diameter. Thus the tower can lean an additional

$$7.0\text{ m} - 4.5\text{ m} = \boxed{2.5\text{ m.}}$$

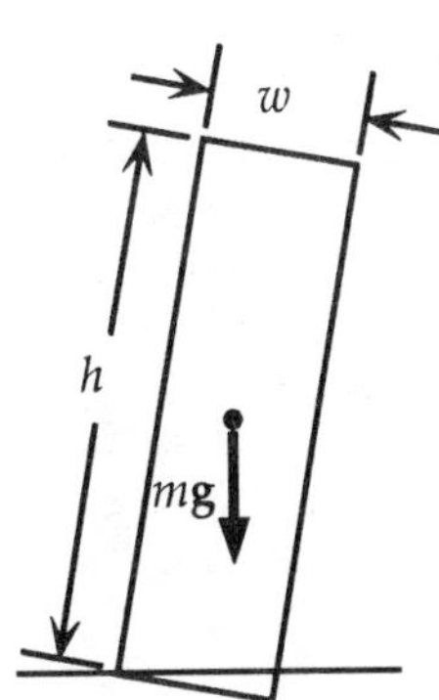

43. (*a*) The maximum distance for the top brick to remain on the next brick will be reached when its center of mass is directly over the edge of the next brick. Thus the top brick will overhang by $x_1 = L/2$.

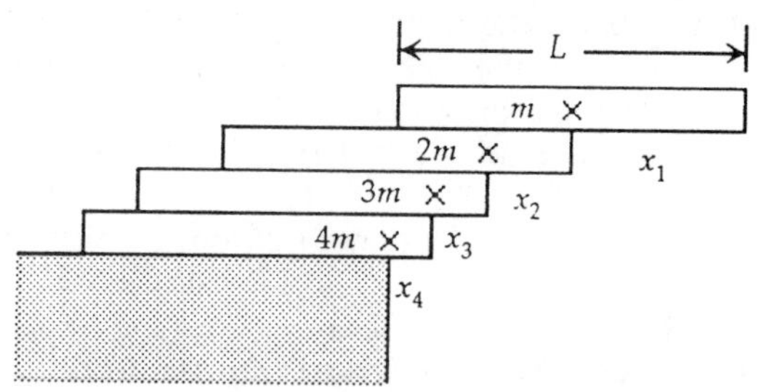

The maximum distance for the top two bricks to remain on the next brick will be reached when the center of mass of the top two bricks is directly over the edge of the third brick. If we take the edge of the third brick as the origin, we have

$x_{CM} = [m(x_2 - L/2) + mx_2]/2m = 0$, which gives $x_2 = L/4$.

The maximum distance for the top three bricks to remain on the next brick will be reached when the center of mass of the top three bricks is directly over the edge of the fourth brick. If we take the edge of the fourth brick as the origin, we have

$x_{CM} = [m(x_3 - L/2) + 2mx_3]/3m = 0$, which gives $x_3 = L/6$.

The maximum distance for the four bricks to remain on the table will be reached when the center of mass of the four bricks is directly over the edge of the table. If we take the edge of the table as the origin, we have

$x_{CM} = [m(x_4 - L/2) + 3mx_4]/3m = 0$, which gives $x_4 = L/8$.

(*b*) With the origin at the table edge, we find the position of the left edge of the top brick from

$D = x_1 + x_2 + x_3 + x_4 - L = (L/2) + (L/4) + (L/6) + (L/8) - L = L/24$, which is $\boxed{\text{beyond the table.}}$

(*c*) We can generalize our results by recognizing that the *i*th brick is a distance $L/2i$ beyond the edge of the brick below it. The total distance spanned by *n* bricks is

$$\boxed{D = L\sum_{i=1}^{n}\frac{1}{2i}.}$$

(*d*) Each side of the arch must span 0.50 m. From our general result, we have

$$0.50 \text{ m} = (0.30 \text{ m})\sum_{i=1}^{n}\frac{1}{2i}.$$

If we evaluate this numerically, such as by using a spreadsheet, we find that 16 bricks will create a span of 0.507 m, so a minimum of $\boxed{\text{32 bricks}}$ is necessary, not counting the one on top. If the bottom brick is flush with the opening, as shown in Fig. 9–32, then two more bricks will be needed.

44. We find the increase in length from the elastic modulus:

$E = \text{Stress}/\text{Strain} = (F/A)/(\Delta L/L_0)$;

$5 \times 10^9 \text{ N/m}^2 = [(250 \text{ N})/\pi(0.50 \times 10^{-3} \text{ m})^2]/[\Delta L/(30.0 \text{ cm})]$, which gives $\Delta L = \boxed{1.91 \text{ cm.}}$

45. (a) We find the stress from

Stress $= F/A = (25{,}000 \text{ kg})(9.80 \text{ m/s2})/(2.0 \text{ m}^2) = \boxed{1.2 \times 10^5 \text{ N/m}^2.}$

(*b*) We find the strain from

Strain $= \text{Stress}/E = (1.2 \times 10^5 \text{ N/m}^2)/(50 \times 10^9 \text{ N/m}^2) = \boxed{2.4 \times 10^{-6}.}$

46.. We use the strain to find how much the column is shortened:

Strain $= \Delta L/L_0$;

$2.4 \times 10^{-6} = \Delta L/(12 \text{ m})$, which gives $\Delta L = 2.9 \times 10^{-5} \text{ m} = \boxed{0.029 \text{ mm.}}$

47. (*a*) We find the stress from

Stress $= F/A = (2000 \text{ kg})(9.80 \text{ m/s}^2)/(0.15 \text{ m}^2) = \boxed{1.3 \times 10^5 \text{ N/m}^2.}$

(*b*) We find the strain from

Strain $= \text{Stress}/E = (1.3 \times 10^5 \text{ N/m}^2)/(200 \times 10^9 \text{ N/m}^2) = \boxed{6.5 \times 10^{-7}.}$

(*c*) We use the strain to find how much the girder is lengthened:

Strain $= \Delta L/L_0$;

$6.5 \times 10^{-7} = \Delta L/(9.50 \text{ m})$, which gives $\Delta L = 6.2 \times 10^{-6} \text{ m} = \boxed{0.0062 \text{ mm.}}$

48. The tension in each wire produces the stress. We find the strain from

$\text{Strain} = \text{Stress}/E = F_T/EA.$

For wire 1 we have

$\text{Strain}_1 = (1.8 \times 10^2\ \text{N})/(200 \times 10^9\ \text{N/m}^2)\pi(0.50 \times 10^{-3}\ \text{m})^2 = 1.15 \times 10^{-3} =$ $\boxed{0.12\%.}$

For wire 2 we have

$\text{Strain}_2 = (2.4 \times 10^2\ \text{N})/(200 \times 10^9\ \text{N/m}^2)\pi(0.50 \times 10^{-3}\ \text{m})^2 = 1.53 \times 10^{-3} =$ $\boxed{0.15\%.}$

49. We find the volume change from

$\Delta P = -B\,\Delta V/V_0;$

$(2.6 \times 10^6\ \text{N/m}^2 - 1.0 \times 10^5\ \text{N/m}^2) = -(1.0 \times 10^9\ \text{N/m}^2)\Delta V/(1000\ \text{cm}^3),$

which gives $\Delta V = -2.5\ \text{cm}^3$.

The new volume is $V_0 + \Delta V = 1000\ \text{cm}^3 + (-2.5\ \text{cm}^3) =$ $\boxed{997\ \text{cm}^3.}$

50. We find the elastic modulus from

$E = \text{Stress}/\text{Strain} = (F/A)/(\Delta L/L_0)$

$= [(13.4\ \text{N})/\tfrac{1}{4}\pi(8.5 \times 10^{-3}\ \text{m})^2]/[(0.37\ \text{cm})/(15\ \text{cm})] =$ $\boxed{9.6 \times 10^6\ \text{N/m}^2.}$

51. The pressure needed is determined by the bulk modulus:

$\Delta P = -B\,\Delta V/V_0 = -(90 \times 10^9\ \text{N/m}^2)(-0.10 \times 10^{-2}) =$ $\boxed{9.0 \times 10^7\ \text{N/m}^2.}$

This is $(9.0 \times 10^7\ \text{N/m}^2)/(1.0 \times 10^5\ \text{N/m}^2 \cdot \text{atm}) =$ $\boxed{9.0 \times 10^2\ \text{atm.}}$

52. We will take the change in pressure to be 200 atm. We find the volume change from

$\Delta P = -B\,\Delta V/V_0;$

$(200\ \text{atm})(1.0 \times 10^5\ \text{N/m}^2 \cdot \text{atm}) = -(90 \times 10^9\ \text{N/m}^2)\Delta V/V_0,$

which gives $\Delta V/V_0 = -2.2 \times 10^{-4} =$ $\boxed{-0.022\%.}$

53. If we treat the abductin as an elastic spring, we find the effective spring constant from

$k = F/\Delta L = EA/L_0 = (2.0 \times 10^6\ \text{N/m}^2)(0.50 \times 10^{-4}\ \text{m}^2)/(3.0 \times 10^{-3}\ \text{m}) = 3.33 \times 10^4\ \text{N/m}.$

We find the elastic potential energy stored in the abductin from

$\text{PE} = \tfrac{1}{2}kx^2 = \tfrac{1}{2}(3.33 \times 10^4\ \text{N/m})(1.0 \times 10^{-3}\ \text{m})^2 =$ $\boxed{0.017\ \text{J.}}$

54. (*a*) For the torque from the sign's weight about the point *A* we have

$\tau_A = Mgd = (5.1\ \text{kg})(9.80\ \text{m/s}^2)(2.2\ \text{m}) =$ $\boxed{1.1 \times 10^2\ \text{m}\cdot\text{N.}}$

(*b*) The balancing torque must be exerted by the $\boxed{\text{wall,}}$ which is the only other contact point. Because the torque from the sign is clockwise, this torque must be counterclockwise.

(*c*) If we think of the torque from the wall as a pull on the top of the pole and a push on the bottom of the pole, there is tension along the top of the pole and compression along the bottom. There must also be a vertical force at the wall which, in combination with the weight of the sign, will create a shear stress in the pole. Thus $\boxed{\text{all three}}$ play a part.

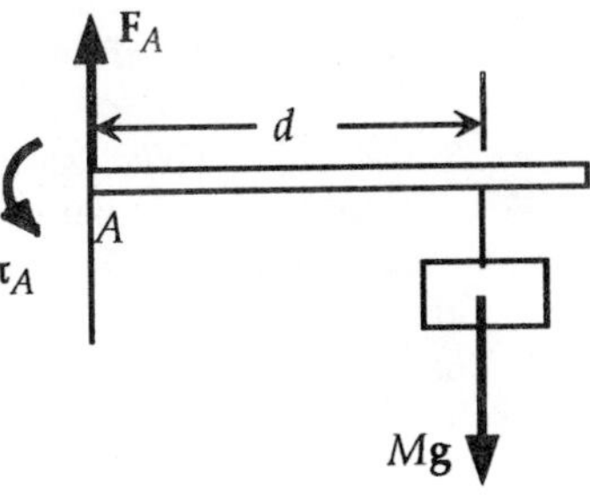

55. We find the maximum compressive force from the compressive strength of bone:

$F_{max} = (\text{Compressive strength})A = (170 \times 10^6\ \text{N/m}^2)(3.0 \times 10^{-4}\ \text{m}^2) =$ $\boxed{5.1 \times 10^4\ \text{N.}}$

56. We find the maximum tension from the tensile strength of nylon:

F_{Tmax} = (Tensile strength)A

= $(500 \times 10^6\ \text{N/m}^2)\pi(0.50 \times 10^{-3}\ \text{m})^2$ = **3.9×10^2 N.**

We can increase the maximum tension by increasing the area, so we use **thicker strings.**

The impulse on the ball that changes its momentum must be provided by an increased tension, **so that the maximum strength is exceeded.**

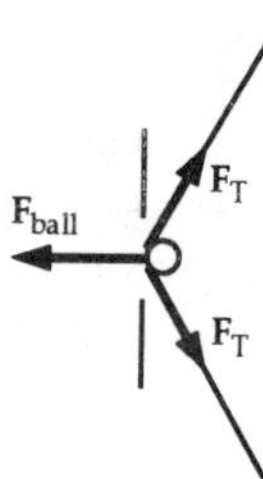

57. (*a*) We determine if the compressive strength, $1.7 \times 10^8\ \text{N/m}^2$, is exceeded:

Stress = $F/A = (3.6 \times 10^4\ \text{N})/(3.6 \times 10^{-4}\ \text{m}^2) = 1.0 \times 10^8\ \text{N/m}^2$.

Because this is less than the compressive strength, the bone will **not break.**

(*b*) We find the change in length from

Strain = $\Delta L/L_0$ = Stress/E, or

ΔL = (Stress)$L_0/E = (1.0 \times 10^8\ \text{N/m}^2)(0.20\ \text{m})/(15 \times 10^9\ \text{N/m}^2) = 1.3 \times 10^{-3}$ m = **1.3 mm.**

58. (*a*) We want the maximum stress to be (1/7.0) of the tensile strength:

$\text{Stress}_{max} = F/A_{min}$ = (Tensile strength)/7.0;

$(320\ \text{kg})(9.80\ \text{m/s}^2)/A_{min} = (500 \times 10^6\ \text{N/m}^2)/7.0$, which gives A_{min} = **$4.4 \times 10^{-5}\ \text{m}^2$.**

(*b*) We find the change in length from

Strain = $\Delta L/L_0$ = Stress/E, or

ΔL = (Stress)$L_0/E = [(500 \times 10^6\ \text{N/m}^2)/7.0](7.5\ \text{m})/(200 \times 10^9\ \text{N/m}^2) = 2.7 \times 10^{-3}$ m = **2.7 mm.**

59. We choose the coordinate system shown, with positive torques clockwise. We write $\Sigma\tau = I\alpha$ about the support point A from the force diagram for the cantilever:

$\Sigma\tau_A = -F_2 d + Mg\frac{1}{2}L = 0;$

$-F_2(20.0\ \text{m}) + (2600\ \text{kg})(9.80\ \text{m/s}^2)(25.0\ \text{m}) = 0,$

which gives $F_2 = 3.19 \times 10^4$ N.

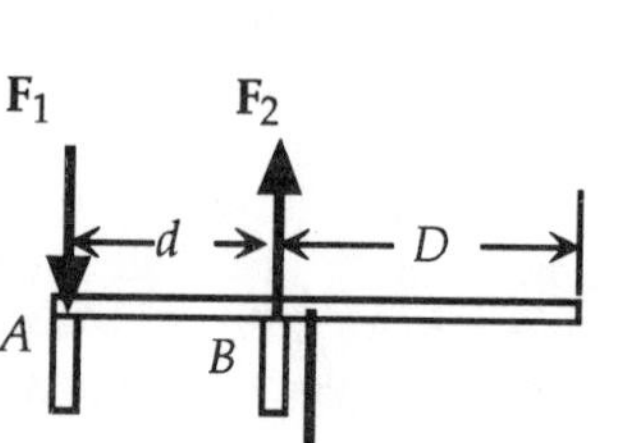

We assume the force is parallel to the grain. We want the maximum stress to be (1/8.5) of the compressive strength:

$\text{Stress}_{max2} = F/A_{min2}$ = (Compressive strength)/8.5;

$(3.19 \times 10^4\ \text{N})/A_{min2} = (35 \times 10^6\ \text{N/m}^2)/8.5$, which gives **$A_{min2} = 7.7 \times 10^{-3}\ \text{m}^2$.**

For the forces in the y-direction we have

$\Sigma F_y = F_1 + F_2 - Mg = 0$, or

$F_1 = Mg - F_2 = (2600\ \text{kg})(9.80\ \text{m/s}^2) - (3.19 \times 10^4\ \text{N}) = -6.42 \times 10^3$ N.

We assume the force is parallel to the grain. We want the maximum stress to be (1/8.5) of the tensile strength:

$\text{Stress}_{max1} = F/A_{min1}$ = (Tensile strength)/8.5;

$(6.42 \times 10^3\ \text{N})/A_{min1} = (40 \times 10^6\ \text{N/m}^2)/8.5$, which gives **$A_{min1} = 1.4 \times 10^{-3}\ \text{m}^2$.**

60. We want the maximum shear stress to be (1/6.0) of the shear strength:

$\text{Stress}_{max} = F/A_{min}$ = (Shear strength)/6.0;

$(3200\ \text{N})/\frac{1}{4}\pi d_{min}^2 = (170 \times 10^6\ \text{N/m}^2)/6.0$, which gives $d_{min} = 1.2 \times 10^{-2}$ m = **1.2 cm.**

61. We find the required tension from $\Sigma F_y = ma_y$:

$F_T - mg = ma$, or

$F_T = m(a + g) = (3100\text{ kg})(1.2\text{ m/s}^2 + 9.80\text{ m/s}^2) = 3.41 \times 10^4\text{ N}$.

We want the maximum stress to be (1/7.0) of the tensile strength:

$\text{Stress}_{max} = F/A_{min} = (\text{Tensile strength})/7.0$;

$(3.41 \times 10^4\text{ N})/\frac{1}{4}\pi d_{min}^2 = (500 \times 10^6\text{ N/m}^2)/7.0$,

which gives $d_{min} = 2.5 \times 10^{-2}\text{ m} =$ **2.5 cm.**

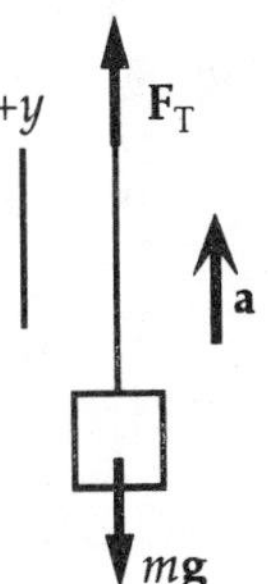

62. In each arch the horizontal force at the base must equal the horizontal force at the top. Because the two arches support the same load, we see from the force diagrams that the vertical forces will be the same and have the same moment arms. Thus the torque about the base of the horizontal force at the top must be the same for the two arches:

$\tau = F_{round}h_{round} = F_{round}R = F_{pointed}h_{pointed}$;

$F_{round}(4.0\text{ m}) = \frac{1}{3}F_{round}h_{pointed}$,

which gives $h_{pointed} =$ **12 m.**

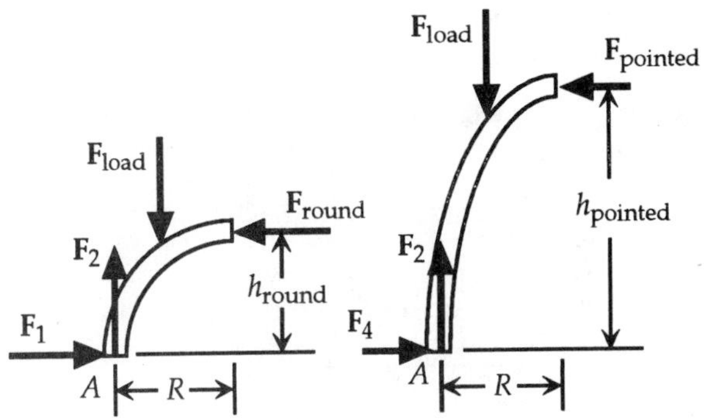

63. We find the required tension from $\Sigma F_y = ma_y$:

$2F\sin\theta - F_{load} = 0$;

$2F\sin 5° - 4.3 \times 10^5\text{ N} = 0$, which gives $F =$ **2.5×10^6 N.**

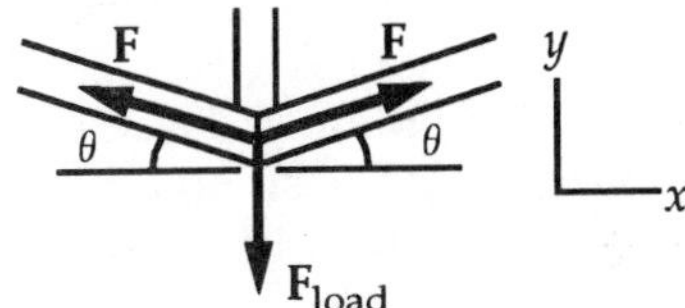

64. All elements are in equilibrium. For the C-D pair, we write $\Sigma\tau = I\alpha$ about the point c from the force diagram:

$\Sigma\tau_c = M_CgL_6 - M_DgL_5 = 0$;

$M_C(5.00\text{ cm}) - M_D(17.50\text{ cm}) = 0$, or $M_C = 3.500M_D$.

The center of mass of C and D must be under the point c.

We write $\Sigma\tau = I\alpha$ about the point b from the force diagram:

$\Sigma\tau_b = M_BgL_4 - (M_C + M_D)gL_3 = 0$;

$(0.735\text{ kg})(5.00\text{ cm}) - (M_C + M_D)(15.00\text{ cm}) = 0$, or

$M_C + M_D = 0.245\text{ kg}$.

When we combine these two results, we get

$M_C = 0.191$ kg, and **$M_D = 0.0544$ kg.**

The center of mass of B, C, and D must be under the point b. For the entire mobile, we write $\Sigma\tau = I\alpha$ about the point a from the force diagram:

$\Sigma\tau_a = (M_B + M_C + M_D)gL_2 - M_AgL_1 = 0$;

$(0.735\text{ kg} + 0.245\text{ kg})(7.50\text{ cm}) - M_A(30.00\text{ cm}) = 0$, which gives **$M_A = 0.245$ kg.**

65. We choose the coordinate system shown, with positive torques clockwise. We write $\Sigma\tau$ about the rear edge from the force diagram:

$$\Sigma\tau_{\text{edge}} = mg\tfrac{1}{2}L - F_A\tfrac{1}{2}H$$
$$= (1.8 \times 10^8\ \text{N})\tfrac{1}{2}(40\ \text{m}) - (950\ \text{N/m}^2)(200\ \text{m})(70\ \text{m})\tfrac{1}{2}(200\ \text{m})$$
$$= \boxed{+2.3 \times 10^9\ \text{m}\cdot\text{N}.}$$

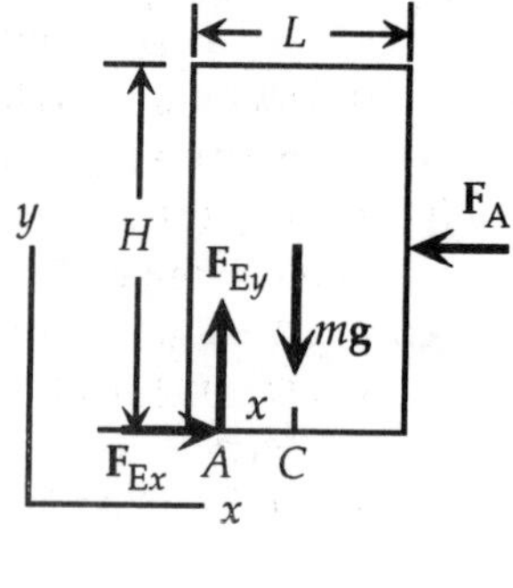

Because the result is positive, the torque is clockwise, so the building $\boxed{\text{will not topple.}}$

An alternative procedure is to find the location of the force $F_{Ey} = mg$. We write $\Sigma\tau = 0$ about the middle of the base from the force diagram:

$$\Sigma\tau_C = F_{Ey}x - F_A\tfrac{1}{2}H = 0;$$
$$(1.8 \times 10^8\ \text{N})x - (950\ \text{N/m}^2)(200\ \text{m})(70\ \text{m})\tfrac{1}{2}(200\ \text{m}) = 0,$$

which gives $x = 11$ m.

Because this is less than 20 m, the building will not topple.

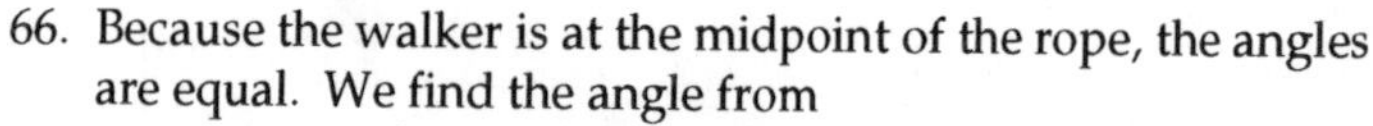

66. Because the walker is at the midpoint of the rope, the angles are equal. We find the angle from

$$\tan\theta = h/\tfrac{1}{2}L = (3.4\ \text{m})/\tfrac{1}{2}(46\ \text{m}) = 0.148, \text{ or } \theta = 8.41°.$$

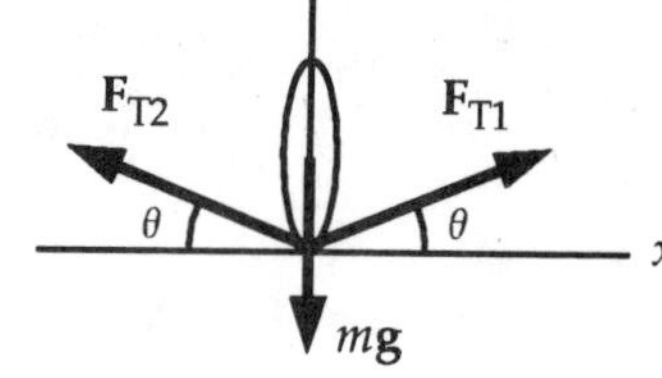

From the force diagram for the walker we can write

$$\Sigma F_x = F_{T1}\cos\theta - F_{T2}\cos\theta = 0, \text{ or } F_{T1} = F_{T2} = F_T;$$
$$\Sigma F_y = F_{T1}\sin\theta + F_{T2}\sin\theta - mg = 0, \text{ or}$$
$$2F_T\sin 8.41° = (60\ \text{kg})(9.80\ \text{m/s}^2), \text{ which gives}$$
$$F_T = \boxed{2.0 \times 10^3\ \text{N}.}$$

$\boxed{\text{No.}}$ There must always be an upward component of the tension to balance the weight.

67. (*a*) The cylinder will roll about the contact point *A*.

We write $\Sigma\tau = I\alpha$ about the point *A*:

$$F_a(2R - h) + F_{N1}[R^2 - (R - h)^2]^{1/2} - Mg[R^2 - (R - h)^2]^{1/2} = I_A\alpha.$$

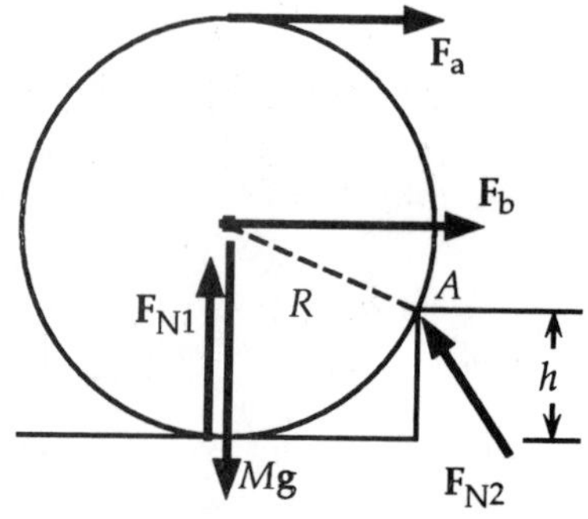

When the cylinder does roll over the curb, contact with the ground is lost and $F_{N1} = 0$. Thus we get

$$F_a = \{I_A\alpha + Mg[R^2 - (R - h)^2]^{1/2}\}/(2R - h)$$
$$= [I_A\alpha/(2R - h)] + [Mg(2Rh - h^2)^{1/2}/(2R - h)].$$

The minimum force occurs when $\alpha = 0$:

$$F_{a\min} = Mg[h(2R - h)]^{1/2}/(2R - h) = \boxed{Mg[h/(2R - h)]^{1/2}.}$$

(*b*) The cylinder will roll about the contact point *A*.

We write $\Sigma\tau = I\alpha$ about the point *A*:

$$F_b(R - h) + F_{N1}[R^2 - (R - h)^2]^{1/2} - Mg[R^2 - (R - h)^2]^{1/2} = I_A\alpha.$$

When the cylinder does roll over the curb, contact with the ground is lost and $F_{N1} = 0$. Thus we get

$$F_b = \{I_A\alpha + Mg[R^2 - (R - h)^2]^{1/2}\}/(R - h)$$
$$= [I_A\alpha/(R - h)] + [Mg(2Rh - h^2)^{1/2}/(R - h)].$$

The minimum force occurs when $\alpha = 0$:

$$F_{b\min} = \boxed{Mg[h(2R - h)]^{1/2}/(R - h).}$$

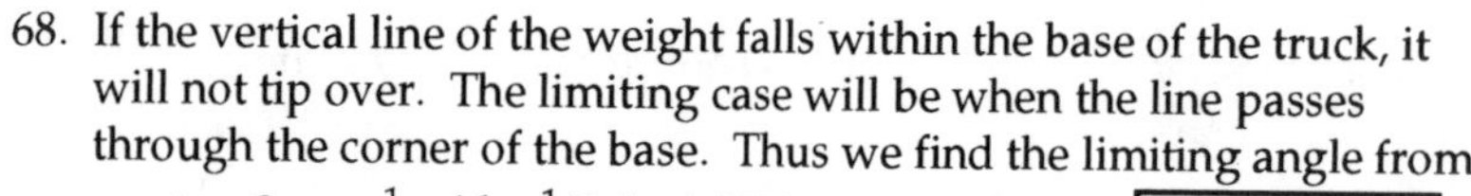

68. If the vertical line of the weight falls within the base of the truck, it will not tip over. The limiting case will be when the line passes through the corner of the base. Thus we find the limiting angle from

$$\tan\theta_{\max} = \tfrac{1}{2}w/d = \tfrac{1}{2}(2.4\ \text{m})/(2.2\ \text{m}) = 0.545, \text{ or } \boxed{\theta_{\max} = 29°.}$$

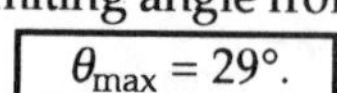

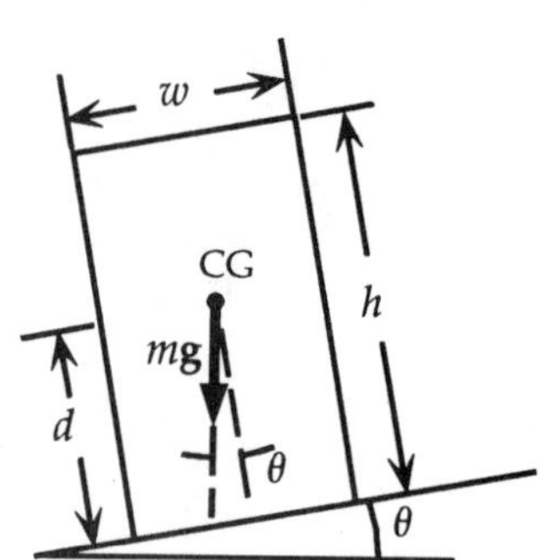

69. (*a*) From Example 7–7, the force of the ground on one leg is
$F_{leg} = \frac{1}{2}(2.1 \times 10^5 \text{ N}) = 1.05 \times 10^5 \text{ N}$.
We find the stress in the tibia bone from
Stress $= F_{leg}/A = (1.05 \times 10^5 \text{ N})/(3.0 \times 10^{-4} \text{ m}) =$ $\boxed{3.5 \times 10^8 \text{ N/m}^2.}$
(*b*) The compressive strength of bone is 1.7×10^8 N/m². Thus the bone $\boxed{\text{will break.}}$
(*c*) From Example 7–7, the force of the ground on one leg is
$F_{leg} = \frac{1}{2}(4.9 \times 10^3 \text{ N}) = 2.45 \times 10^3 \text{ N}$.
We find the stress in the tibia bone from
Stress $= F_{leg}/A = (2.45 \times 10^3 \text{ N})/(3.0 \times 10^{-4} \text{ m}) =$ $\boxed{8.2 \times 10^6 \text{ N/m}^2.}$
This is less than the compressive strength of bone, so the bone $\boxed{\text{will not break.}}$

70. The force is parallel to the grain. We want the maximum stress to be (1/12) of the compressive strength. For N studs we have
$\text{Stress}_{max} = (Mg/N)/A = (\text{Compressive strength})/12;$
$(12{,}600 \text{ kg})(9.80 \text{ m/s}^2)/N(0.040 \text{ m})(0.090 \text{ m}) = (35 \times 10^6 \text{ N/m}^2)/12$, which gives $N = 11.8$.
Thus we need $\boxed{6}$ studs on each side.
There are five spaces between the studs, so they will be
$(10.0 \text{ m})/5 =$ $\boxed{\text{2.0 m apart.}}$

71. From the force diagram for the section we can write

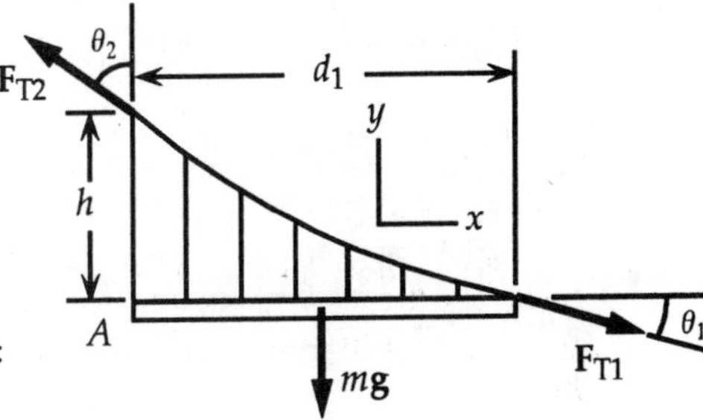

$\Sigma F_x = F_{T1} \cos \theta_1 - F_{T2} \sin \theta_2 = 0$, or
$F_{T1} \cos 19° - F_{T2} \sin 60° = 0;$
$\Sigma F_y = -F_{T1} \sin \theta_1 + F_{T2} \cos \theta_2 - mg = 0$, or
$-F_{T1} \sin 19° + F_{T2} \cos 60° = mg.$
When we combine these equations, we get
$\boxed{F_{T1} = 4.54mg, \text{ and } F_{T2} = 4.96mg.}$
We write $\Sigma\tau = I\alpha$ about the point A from the force diagram:
$\Sigma\tau_A = -(F_{T2} \sin \theta_2)h + (F_{T1} \sin \theta_1)d_1 + mg\frac{1}{2}d_1 = 0;$
$-(4.96mg \sin 60°)h + (4.54mg \sin 19°)(343 \text{ m}) + mg\frac{1}{2}(343 \text{ m}) = 0,$
which gives $\boxed{h = 158 \text{ m.}}$

72. We choose the coordinate system shown, with positive torques clockwise.
   (*a*) The maximum weight will cause the force F_A to be zero.
   We write $\Sigma\tau = I\alpha$ about the support point B from the force diagram for the beam and person:
   $\Sigma\tau_B = -W(\frac{1}{2}L - d_2) + wd_2 + F_AD = 0;$
   $-(600\text{ N})[\frac{1}{2}(20.0\text{ m}) - 5.0\text{ m}] + w_{max}(5.0\text{ m}) + 0 = 0,$
   which gives $w_{max} =$ $\boxed{600\text{ N.}}$
   (*b*) The maximum weight means the force $\boxed{F_A = 0.}$
   We write $\Sigma\tau = I\alpha$ about the support point A from the force diagram for the beam and person:
   $\Sigma\tau_A = +W(\frac{1}{2}L - d_1) + w(D - d_2) - F_BD = 0;$
   $+(600\text{ N})[\frac{1}{2}(20.0\text{ m}) - 3.0\text{ m}] + (600\text{ N})(5.0\text{ m}) - F_B(12.0\text{ m}) = 0,$ which gives $\boxed{F_B = 1200\text{ N.}}$
   (*c*) We write $\Sigma\tau = I\alpha$ about the support point B from the force diagram for the beam and person:
   $\Sigma\tau_B = -W(\frac{1}{2}L - d_2) + wx + F_AD = 0;$
   $-(600\text{ N})[\frac{1}{2}(20.0\text{ m}) - 5.0\text{ m}] + (600\text{ N})(2.0\text{ m}) + F_A(12.0\text{ m}) = 0,$
   which gives $\boxed{F_A = 150\text{ N.}}$
   We write $\Sigma\tau = I\alpha$ about the support point A from the force diagram for the beam and person:
   $\Sigma\tau_A = +W(\frac{1}{2}L - d_1) + w(D + x) - F_BD = 0;$
   $+(600\text{ N})[\frac{1}{2}(20.0\text{ m}) - 3.0\text{ m}] + (600\text{ N})(12.0\text{ m} + 2.0\text{ m}) - F_B(12.0\text{ m}) = 0,$
   which gives $\boxed{F_B = 1050\text{ N.}}$
   (*d*) We write $\Sigma\tau = I\alpha$ about the support point B from the force diagram for the beam and person:
   $\Sigma\tau_B = -W(\frac{1}{2}L - d_2) + wx + F_AD = 0;$
   $-(600\text{ N})[\frac{1}{2}(20.0\text{ m}) - 5.0\text{ m}] + (600\text{ N})(-10.0\text{ m}) + F_A(12.0\text{ m}) = 0,$
   which gives $\boxed{F_A = 750\text{ N.}}$
   We write $\Sigma\tau = I\alpha$ about the support point A from the force diagram for the beam and person:
   $\Sigma\tau_A = +W(\frac{1}{2}L - d_1) + w(D + x) - F_BD = 0;$
   $+(600\text{ N})[\frac{1}{2}(20.0\text{ m}) - 3.0\text{ m}] + (600\text{ N})(12.0\text{ m} - 10.0\text{ m}) - F_B(12.0\text{ m}) = 0,$
   which gives $\boxed{F_B = 450\text{ N.}}$

73. The minimum mass is placed symmetrically between two of the legs of the table, so the normal force on the opposite leg becomes zero, as shown in the top view of the table. We write $\Sigma\tau = I\alpha$ about a horizontal axis that passes through the two legs where there is a normal force:
   $\Sigma\tau = -MgR\cos\theta + mgR(1 - \cos\theta) = 0;$
   $-(36\text{ kg})\cos 60° + m(1 - \cos 60°) = 0,$
   which gives $m =$ $\boxed{36\text{ kg.}}$

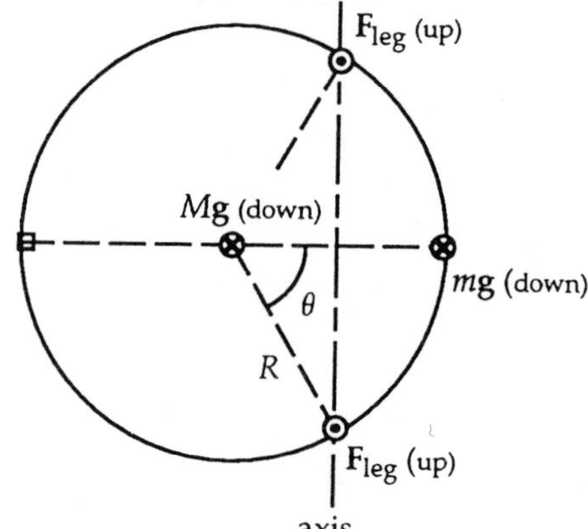

74. We select one-half of the cable for our system. From the force diagram for the section we can write

$\Sigma F_x = F_{T1} - F_{T2} \cos \theta = 0;$

$F_{T1} - F_{T2} \cos 60° = 0.$

$\Sigma F_y = + F_{T2} \sin \theta - \frac{1}{2}mg = 0;$

$F_{T2} \sin 60° - \frac{1}{2}mg = 0.$

When we combine these equations, we get

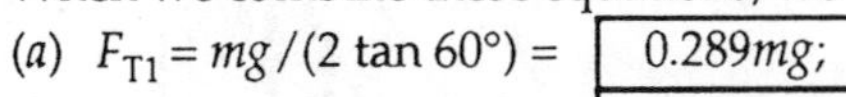

(*a*) $F_{T1} = mg/(2 \tan 60°) =$ $\boxed{0.289mg;}$

(*b*) $F_{T2} = mg/(2 \sin 60°) =$ $\boxed{0.577mg.}$

(c) The direction of the tension in each case is tangent to the cable:

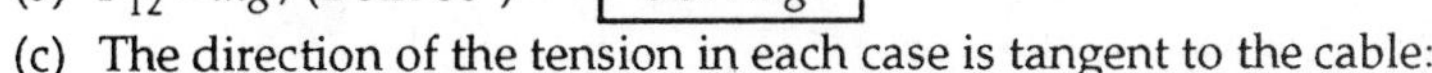

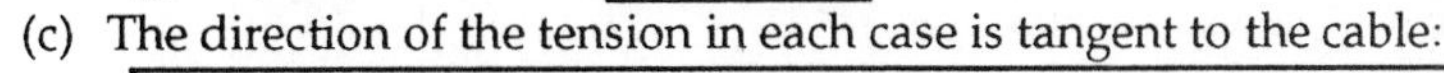

horizontal at the lowest point, and 60° above the horizontal at the attachment.

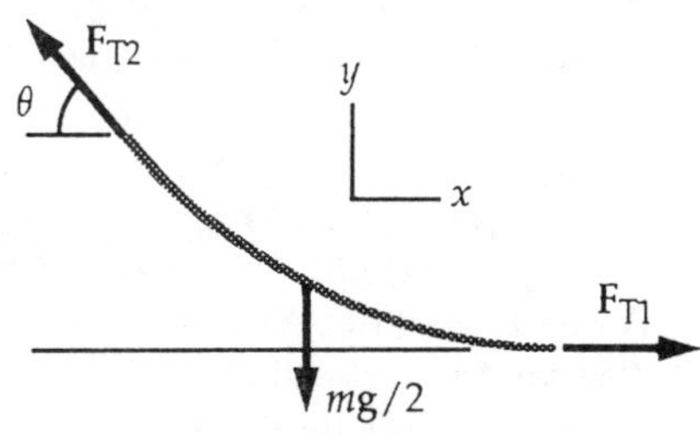

75. Because there is no net horizontal force on the tower, from the force diagram for the tower we can write

$\Sigma F_x = F_{T2} \sin \theta_2 - F_{T3} \sin \theta_3 = 0,$

or

$F_{T3} = F_{T2} (\sin \theta_2)/(\sin \theta_3).$

From the force diagram for the north span we can write

$\Sigma F_x = F_{T1} \cos \theta_1 - F_{T2} \sin \theta_2 = 0,$

or

$F_{T1} = F_{T2} (\sin \theta_2)/(\cos \theta_1).$

$\Sigma F_y = + F_{T2} \cos \theta_2 - F_{T1} \sin \theta_1 - mg = 0,$

or

$mg = F_{T2} \cos \theta_2 - F_{T1} \sin \theta_1.$

From the force diagram for one-half of the center span we can write

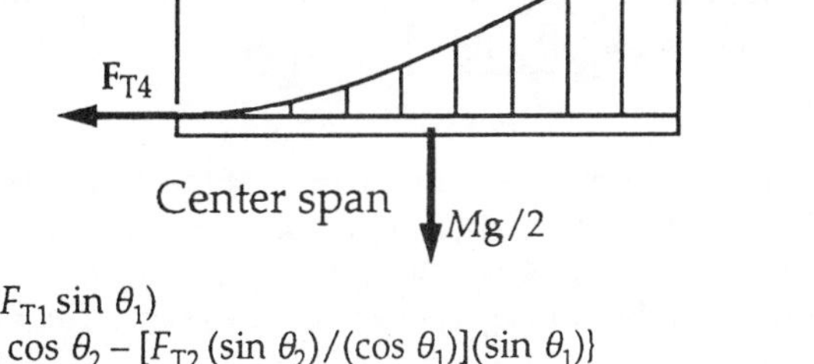

$\Sigma F_y = + F_{T3} \cos \theta_3 - \frac{1}{2}Mg = 0,$ or

$Mg = 2F_{T3} \cos \theta_3.$

Because the roadway is uniform, the length of each roadway is proportional to the mass:

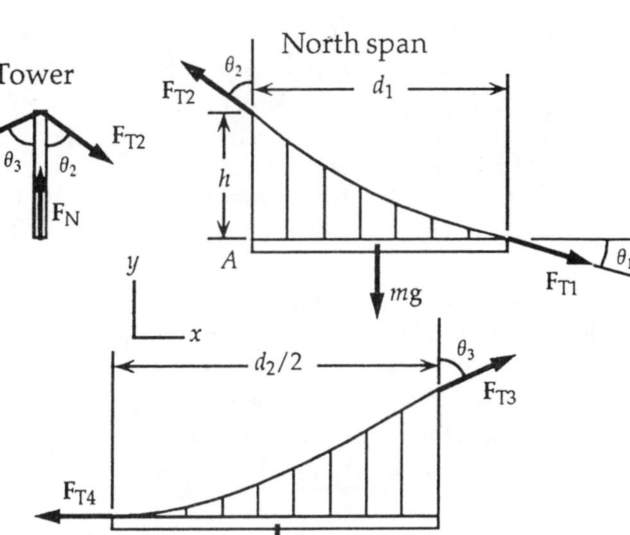

$d_2/d_1 = M/m = (2F_{T3} \cos \theta_3)/(F_{T2} \cos \theta_2 - F_{T1} \sin \theta_1)$

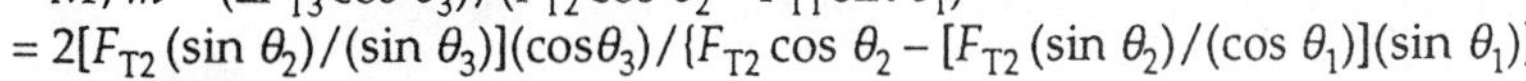

$= 2[F_{T2} (\sin \theta_2)/(\sin \theta_3)](\cos\theta_3)/\{F_{T2} \cos \theta_2 - [F_{T2} (\sin \theta_2)/(\cos \theta_1)](\sin \theta_1)\}$

$= 2(\cot \theta_3)/(\cot \theta_2 - \tan \theta_1)$

$= 2(\cot 66°)/(\cot 60° - \tan 19°) =$ $\boxed{3.8.}$

76. We choose the coordinate system shown, with positive torques clockwise. 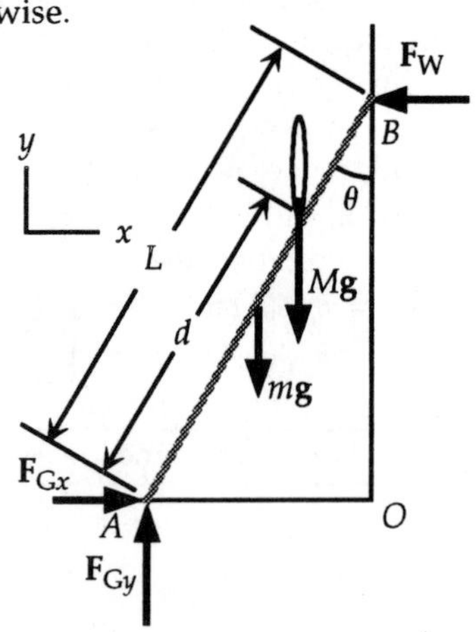
   (*a*) We write $\Sigma F_x = ma_x$ from the force diagram:
   $F_{Gx} - F_W = 0$, or $F_{Gx} = F_W$.
   We write $\Sigma F_y = ma_y$ from the force diagram:
   $F_{Gy} - mg = 0$, which gives
   $F_{Gy} = mg = (15.0 \text{ kg})(9.80 \text{ m/s}^2) = 147 \text{ N}$.
   We write $\Sigma\tau = I\alpha$ about the point A from the force diagram for the ladder:
   $\Sigma\tau_A = mg(\frac{1}{2}L \sin\theta) - F_W L \cos\theta = 0$;
   $(15.0 \text{ kg})(9.80 \text{ m/s}^2)\frac{1}{2}(\sin 20°) - F_W(\cos 20°) = 0$,
   which gives $F_W = 27$ N.
   Thus the components of the force at the ground are
   $F_{Gx} = 27 \text{ N}, F_{Gy} = 147 \text{ N}.$

   (*b*) We write $\Sigma F_x = ma_x$ from the force diagram:
   $F_{Gx} - F_W = 0$, or $F_{Gx} = F_W$.
   We write $\Sigma F_y = ma_y$ from the force diagram:
   $F_{Gy} - mg - Mg = 0$, which gives $F_{Gy} = (m + M)g = (15.0 \text{ kg} + 70 \text{ kg})(9.80 \text{ m/s}^2) = 833 \text{ N}$.
   We write $\Sigma\tau = I\alpha$ about the point A from the force diagram for the ladder and person:
   $\Sigma\tau_A = mg(\frac{1}{2}L \sin\theta) + Mg(d \sin\theta) - F_W L \cos\theta = 0$;
   $(15.0 \text{ kg})(9.80 \text{ m/s}^2)\frac{1}{2}(7.0 \text{ m})(\sin 20°) + (70 \text{ kg})(9.80 \text{ m/s}^2)\frac{3}{4}(7.0 \text{ m})(\sin 20°) - F_W(7.0 \text{ m})(\cos 20°) = 0$,
   which gives $F_W = 214 \text{ N} = F_{Gx}$.
   Because the ladder is on the verge of slipping, we must have
   $F_{Gx} = \mu F_{Gy}$, or $\mu = F_{Gx}/F_{Gy} = (214 \text{ N})/(833 \text{ N}) = 0.26.$

77. We have the same results from $\Sigma \mathbf{F} = m\mathbf{a}$: $F_{Gx} = F_W$; $F_{Gy} = (m + M)g = 833$ N.
   Because the ladder is on the verge of slipping, we must have $F_{Gx} = \mu F_{Gy} = (0.30)(833 \text{ N}) = 250 \text{ N} = F_W$.
   We write $\Sigma\tau = I\alpha$ about the point A from the force diagram for the ladder and person:
   $\Sigma\tau_A = mg(\frac{1}{2}L \sin\theta) + Mg(d \sin\theta) - F_W L \cos\theta = 0$;
   $(15.0 \text{ kg})(9.80 \text{ m/s}^2)\frac{1}{2}(7.0 \text{ m})(\sin 20°) + (70 \text{ kg})(9.80 \text{ m/s}^2)d(\sin 20°) - (250 \text{ N})(7.0 \text{ m})(\cos 20°) = 0$,
   which gives $d = 6.3 \text{ m.}$

78. The maximum stress in a column will be at the bottom, caused by the weight of the material. If the column has density ρ, height h, and area A, we have
   Stress $= F/A = mg/A = \rho Vg/A = \rho Ahg/A = \rho gh$, which is independent of area.
   The column will buckle when this stress exceeds the compressive strength:
   h_{max} = (Compressive strength)/ρg.
   (*a*) For steel we have
   $h_{max} = (500 \times 10^6 \text{ N/m}^2)/(7.8 \times 10^3 \text{ kg/m}^3)(9.80 \text{ m/s}^2) = 6.5 \times 10^3 \text{ m.}$
   (*b*) For granite we have
   $h_{max} = (170 \times 10^6 \text{ N/m}^2)/(2.7 \times 10^3 \text{ kg/m}^3)(9.80 \text{ m/s}^2) = 6.4 \times 10^3 \text{ m.}$

79. We assume when the brick strikes the floor there is an average force which produces an average stress in the brick, creating an average strain:
   $\Delta L/L_0 = (F/A)/E$.
   If we use this average strain for the distance the CM moves while the brick comes to rest, the work done by the average force is $-F\,\Delta L$. When we use the work-energy principle from the release point to the final resting point, we have
   $-F\,\Delta L = \Delta\text{KE} + \Delta\text{PE} = 0 + (0 - mgh)$, or
   $h = F\,\Delta L/mg = F(F/A)L_0/Emg = A(F/A)^2 L_0/Emg$.
   If we assume that the stress varies linearly from zero at the top of the brick to maximum at the bottom, the brick will break when the average stress exceeds one-half the compressive strength, so we have
   $h_{min} = (0.15 \text{ m})(0.060 \text{ m})[\frac{1}{2}(35 \times 10^6 \text{ N/m}^2)]^2(0.040 \text{ m})/(14 \times 10^9 \text{ N/m}^2)(1.2 \text{ kg})(9.80 \text{ m/s}^2) = 1.3 \text{ m.}$

80. We write $\Sigma F_x = ma_x$ from the force diagram:
$F - F_{fr} = 0$, or $F = F_{fr}$.
We write $\Sigma F_y = ma_y$ from the force diagram:
$F_N - Mg = 0$, or $F_N = Mg$.
We find the location of the force F_N when the static friction force reaches its maximum value:
$F = F_{fr} = \mu_s F_N = \mu_s Mg$.
We write $\Sigma\tau = 0$ about the edge A of the block from the force diagram:
$\Sigma\tau_A = -Mg\frac{1}{2}L + F_N x - Fh = 0$, which gives
$x = (\frac{1}{2}MgL - Fh)/F_N = (\frac{1}{2}MgL - \mu_s Mgh)/Mg = \frac{1}{2}L - \mu_s h$.

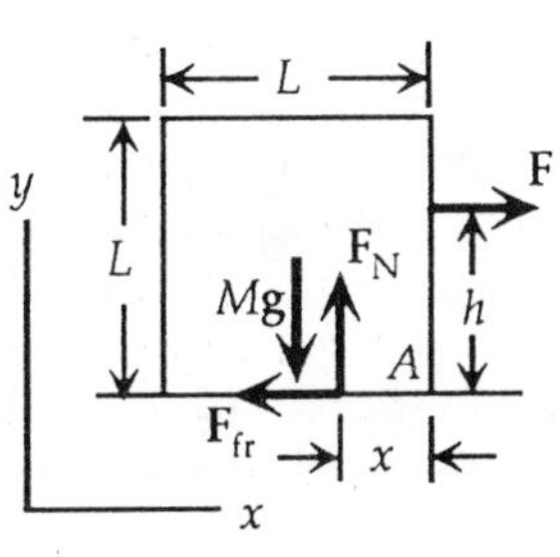

(*a*) For the block to slide, we must have $x > 0$, or
$\frac{1}{2}L > \mu_s h$, which gives $\boxed{\mu_s < L/2h.}$

(*b*) For the block to tip, we must have $x < 0$, or
$\frac{1}{2}L < \mu_s h$, which gives $\boxed{\mu_s > L/2h.}$

81. (*a*) We write $\Sigma\tau = 0$ about the feet from the force diagram:
$\Sigma\tau_{feet} = 2F_{hand}(d_1 + d_2) - mgd_2 = 0$, or
$F_{hand} = mgd_2/2(d_1 + d_2)$
$= (70\text{ kg})(9.80\text{ m/s}^2)(0.73\text{ m})/2(0.25\text{ m} + 0.73\text{ m})$
$= \boxed{2.6 \times 10^2\text{ N.}}$

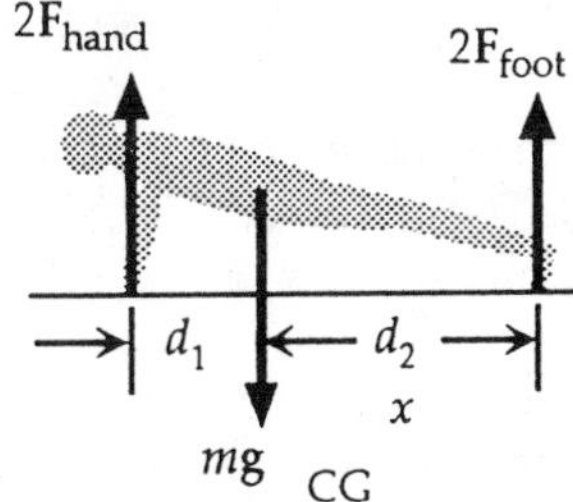

(*b*) We write $\Sigma\tau = 0$ about the hands from the force diagram:
$\Sigma\tau_{hands} = -2F_{foot}(d_1 + d_2) + mgd_1 = 0$, or
$F_{foot} = mgd_1/2(d_1 + d_2)$
$= (70\text{ kg})(9.80\text{ m/s}^2)(0.25\text{ m})/2(0.25\text{ m} + 0.73\text{ m})$
$= \boxed{88\text{ N.}}$

82. The ropes can only provide a tension, so the scaffold will be stable if F_{T1} and F_{T2} are greater than zero. The tension will be least in the rope farthest from the painter. To find how far the painter can walk from the right rope toward the right end, we set $F_{T1} = 0$.
We write $\Sigma\tau = 0$ about B from the force diagram:
$\Sigma\tau_B = Mgx_{right} + F_{T1}(D + 2d) - m_{pail}g(D + d) - mgD = 0$;
$(60\text{ kg})(9.80\text{ m/s}^2)x_{right} + 0 -$
$(4.0\text{ kg})(9.80\text{ m/s}^2)(2.0\text{ m} + 1.0\text{ m}) - (25\text{ kg})(9.80\text{ m/s}^2)(2.0\text{ m}) = 0$,
which gives $x_{right} = 1.03$ m.
Because this is greater than the distance to the end of the plank, 1.0 m, walking to the $\boxed{\text{right end is safe.}}$

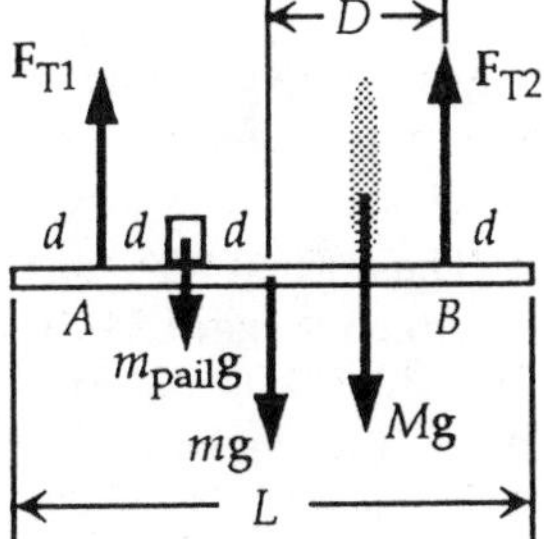

To find how far the painter can walk from the right rope toward the left end, we set $F_{T2} = 0$.
We write $\Sigma\tau = 0$ about A from the force diagram:
$\Sigma\tau_A = -Mgx_{left} - F_{T2}(D + 2d) + m_{pail}gd + mg2d = 0$;
$-(60\text{ kg})(9.80\text{ m/s}^2)x_{left} - 0 + (4.0\text{ kg})(9.80\text{ m/s}^2)(1.0\text{ m}) + (25\text{ kg})(9.80\text{ m/s}^2)2(1.0\text{ m}) = 0$,
which gives $x_{left} = 0.90$ m.
Because this is less than the distance to the end of the plank, 1.0 m, walking to the $\boxed{\text{left end is not safe.}}$
The painter can safely walk to within $\boxed{0.10\text{ m}}$ of the left end.

CHAPTER 10

1. When we use the density of granite, we have
   $m = \rho V = (2.7 \times 10^3\ \text{kg/m}^3)(1 \times 10^8\ \text{m}^3) = \boxed{2.7 \times 10^{11}\ \text{kg.}}$

2. When we use the density of air, we have
   $m = \rho V = \rho LWH = (1.29\ \text{kg/m}^3)(5.8\ \text{m})(3.8\ \text{m})(2.8\ \text{m}) = \boxed{80\ \text{kg.}}$

3. When we use the density of gold, we have
   $m = \rho V = \rho LWH = (19.3 \times 10^3\ \text{kg/m}^3)(0.60\ \text{m})(0.25\ \text{m})(0.15\ \text{m}) = \boxed{4.3 \times 10^2\ \text{kg}}$ (≈ 950 lb!).

4. If we assume a mass of 65 kg with the density of water, we have
   $m = \rho V;$
   $65\ \text{kg} = (1.0 \times 10^3\ \text{kg/m}^3)V$, which gives $V = \boxed{6.5 \times 10^{-2}\ \text{m}^3}$ (≈ 65 L).

5. From the masses we have
   $m_{\text{water}} = 98.44\ \text{g} - 35.00\ \text{g} = 63.44\ \text{g};$
   $m_{\text{fluid}} = 88.78\ \text{g} - 35.00\ \text{g} = 53.78\ \text{g}.$
   Because the water and the fluid occupy the same volume, we have
   $\text{SG}_{\text{fluid}} = \rho_{\text{fluid}}/\rho_{\text{water}} = m_{\text{fluid}}/m_{\text{water}} = (53.78\ \text{g})/(63.44\ \text{g}) = \boxed{0.8477.}$

6. The definition of the specific gravity of the mixture is
   $\text{SG}_{\text{mixture}} = \rho_{\text{mixture}}/\rho_{\text{water}}.$
   The density of the mixture is
   $\rho_{\text{mixture}} = m_{\text{mixture}}/V = (\text{SG}_{\text{antifreeze}}\rho_{\text{water}}V_{\text{antifreeze}} + \text{SG}_{\text{water}}\rho_{\text{water}}V_{\text{water}})/V$, so we get
   $\text{SG}_{\text{mixture}} = (\text{SG}_{\text{antifreeze}}V_{\text{antifreeze}} + \text{SG}_{\text{water}}V_{\text{water}})/V$
   $= [(0.80)(5.0\ \text{L}) + (1.0)(4.0\ \text{L})]/(9.0\ \text{L}) = \boxed{0.89.}$

7. The difference in pressure is caused by the difference in elevation:
   $\Delta P = \rho g\ \Delta h = (1.05 \times 10^3\ \text{kg/m}^3)(9.80\ \text{m/s}^2)(1.60\ \text{m}) = \boxed{1.65 \times 10^4\ \text{N/m}^2.}$

8. (*a*) The normal force on the heel must equal the weight. The pressure of the reaction to the normal force, which is exerted on the floor, is
   $P = F_N/A = mg/A = (50\ \text{kg})(9.80\ \text{m/s}^2)/(0.05 \times 10^{-4}\ \text{m}^2) = \boxed{9.8 \times 10^7\ \text{N/m}^2.}$
   (*b*) For the elephant standing on one foot, we have
   $P = F_N/A = mg/A = (1500\ \text{kg})(9.80\ \text{m/s}^2)/(800 \times 10^{-4}\ \text{m}^2) = \boxed{1.8 \times 10^5\ \text{N/m}^2.}$
   Note that this is a factor of ≈ 1000× less than that of the model!

9. (*a*) The force of the air on the table top is
   $F = PA = (1.013 \times 10^5\ \text{N/m}^2)(1.6\ \text{m})(1.9\ \text{m}) = \boxed{3.1 \times 10^5\ \text{N (down).}}$
   (*b*) Because the pressure is the same on the underside of the table, the upward force has the same magnitude: $\boxed{3.1 \times 10^5\ \text{N.}}$ This is why the table does not move!

10. The pressure difference on the lungs is the pressure change from the depth of water:
    $\Delta P = \rho g\ \Delta h;$
    $(80\ \text{mm-Hg})(133\ \text{N/m}^2 \cdot \text{mm-Hg}) = (1.00 \times 10^3\ \text{kg/m}^3)(9.80\ \text{m/s}^2)\Delta h$, which gives $\Delta h = \boxed{1.1\ \text{m.}}$

11. There is atmospheric pressure outside the tire, so we find the net force from the gauge pressure. Because the reaction to the force from the pressure on the four footprints of the tires supports the automobile, we have

$4PA = mg;$

$4(240 \times 10^3 \text{ N/m}^2)(200 \times 10^{-4} \text{ m}^2) = m(9.80 \text{ m/s}^2)$, which gives $m = \boxed{1.96 \times 10^3 \text{ kg.}}$

12. Because the force from the pressure on the cylinder supports the automobile, we have

$PA = mg;$

$(18 \text{ atm})(1.013 \times 10^5 \text{ N/m}^2 \cdot \text{atm})\pi(11 \times 10^{-2} \text{ m})^2 = m(9.8 \text{ m/s}^2)$, which gives $m = \boxed{7.1 \times 10^3 \text{ kg.}}$

Note that we use gauge pressure because there is atmospheric pressure on the outside of the cylinder.

13. The pressure from the height of alcohol must balance the atmospheric pressure:

$P = \rho gh;$

$1.013 \times 10^5 \text{ N/m}^2 = (0.79 \times 10^3 \text{ kg/m}^3)(9.80 \text{ m/s}^2)h$, which gives $h = \boxed{13 \text{ m.}}$

14. The pressure at a depth h is

$P = P_0 + \rho gh = 1.013 \times 10^5 \text{ N/m}^2 + (1.00 \times 10^3 \text{ kg/m}^3)(9.80 \text{ m/s}^2)(2.0 \text{ m}) = \boxed{1.21 \times 10^5 \text{ N/m}^2.}$

The force on the bottom is

$F = PA = (1.21 \times 10^5 \text{ N/m}^2)(22.0 \text{ m})(12.0 \text{ m}) = \boxed{3.2 \times 10^7 \text{ N (down).}}$

The pressure depends only on depth, so it will be the same: $\boxed{1.21 \times 10^5 \text{ N/m}^2.}$

15. The pressure is produced by a column of air:

$P = \rho gh;$

$1.013 \times 10^5 \text{ N/m}^2 = (1.29 \text{ kg/m}^3)(9.80 \text{ m/s}^2)h$, which gives $h = 8.0 \times 10^3 \text{ m} = \boxed{8.0 \text{ km.}}$

16. Because points a and b are at the same elevation of water, the pressures must be the same. Each pressure is due to the atmospheric pressure at the top of the column and the height of the liquid, so we have

$P_a = P_b$, or $P_0 + \rho_{\text{oil}} g h_{\text{oil}} = P_0 + \rho_{\text{water}} g h_{\text{water}};$

$\rho_{\text{oil}} g(27.2 \text{ cm}) = (1.00 \times 10^3 \text{ kg/m}^3)g(27.2 \text{ cm} - 9.41 \text{ cm}),$

which gives $\rho_{\text{oil}} = \boxed{6.54 \times 10^2 \text{ kg/m}^3.}$

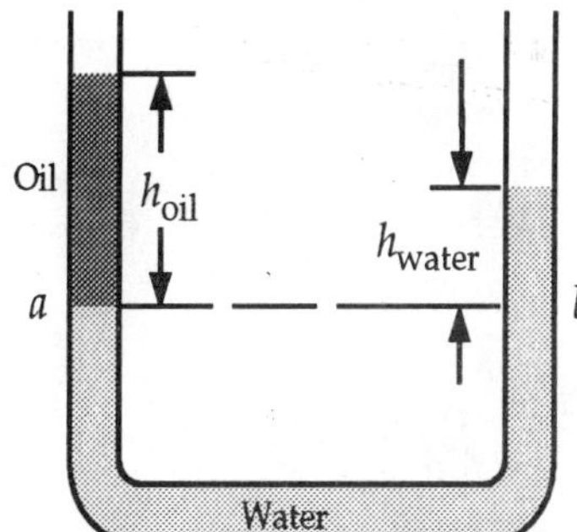

17. We find the gauge pressure from the water height:

$P = \rho_{\text{water}} g h_{\text{water}} = (1.00 \times 10^3 \text{ kg/m}^3)(9.80 \text{ m/s}^2)[5.0 \text{ m} + (100 \text{ m}) \sin 60°] = \boxed{9.0 \times 10^5 \text{ N/m}^2 \text{ (gauge).}}$

If we neglect turbulence and frictional effects, we know from energy considerations that the water would rise to the elevation at which it started:

$h = [5.0 \text{ m} + (100 \text{ m}) \sin 60°] = \boxed{92 \text{ m.}}$

18. The minimum gauge pressure would cause the water to come out of the faucet with very little speed. This means the gauge pressure must be enough to hold the water at this elevation:

$P_{\text{gauge}} = \rho gh;$

$= (1.00 \times 10^3 \text{ kg/m}^3)(9.80 \text{ m/s}^2)(41 \text{ m}) = \boxed{4.0 \times 10^5 \text{ N/m}^2.}$

19. (*a*) For the mass of water in the tube we have
$m = \rho V = \rho \pi R_T^2 H = (1.00 \times 10^3\ \mathrm{kg/m^3})\pi(0.0030\ \mathrm{m})^2(12\ \mathrm{m})$
$= \boxed{0.34\ \mathrm{kg}.}$
(*b*) The net force on the lid of the barrel is due to the gauge pressure of the water at the top of the barrel. Because this gauge pressure is from the mass of water in the tube, we have
$F = PA_B = (mg/A_T)A_B = mg(R_B/R_T)^2$
$= (0.34\ \mathrm{kg})(9.80\ \mathrm{m/s^2})[(20\ \mathrm{cm})/(0.30\ \mathrm{cm})]^2 = \boxed{1.5 \times 10^4\ \mathrm{N\ (up)}.}$

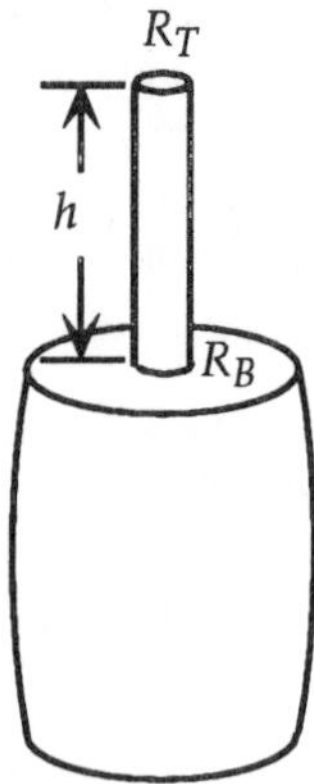

20. We consider a mass of water m that occupies a volume V_0 at the surface. The pressure increase at a depth h will decrease the volume to V, where $m = \rho_0 V_0 = \rho V$. We use the bulk modulus to find the volume change:
$\Delta V/V_0 = -\Delta P/B = -\rho_0 gh/B = -(1.025 \times 10^3\ \mathrm{kg/m^3})(9.80\ \mathrm{m/s^2})(6.0 \times 10^3\ \mathrm{m})/(2.0 \times 10^9\ \mathrm{N/m^2}) = -0.030.$
The fact that this is small justifies using the surface density for the pressure calculation.
The volume at the depth h is
$V = V_0 + \Delta V = (1 - 0.030)V_0 = 0.970V_0,$
so the density is
$\rho = \rho_0(V_0/V) = (1.025 \times 10^3\ \mathrm{kg/m^3})(1/0.970) = \boxed{1.057 \times 10^3\ \mathrm{kg/m^3}.}$
For the fractional change in density, we have
$\Delta\rho/\rho_0 = (V_0/V) - 1 = (1/0.970) - 1 = 1.030 - 1 = \boxed{+0.030\ (3\%).}$

21. Because the mass of the displaced liquid is the mass of the hydrometer, we have
$m = \rho_{\mathrm{water}} h_{\mathrm{water}} A = \rho_{\mathrm{liquid}} h_{\mathrm{liquid}} A;$
$(1000\ \mathrm{kg/m^3})(22.5\ \mathrm{cm}) = \rho_{\mathrm{liquid}}\ (22.9)$, which gives $\rho_{\mathrm{liquid}} = \boxed{983\ \mathrm{kg/m^3}.}$

22. Because the mass of the displaced water is the apparent change in mass of the rock, we have
$\Delta m = \rho_{\mathrm{water}} V.$
For the density of the rock we have
$\rho_{\mathrm{rock}} = m_{\mathrm{rock}}/V = (m_{\mathrm{rock}}/\Delta m)\rho_{\mathrm{water}}$
$= [(8.20\ \mathrm{kg})/(8.20\ \mathrm{kg} - 6.18\ \mathrm{kg})](1.00 \times 10^3\ \mathrm{kg/m^3}) = \boxed{4.06 \times 10^3\ \mathrm{kg/m^3}.}$

23. When the aluminum floats, the net force is zero. If the fraction of the aluminum that is submerged is f, we have
$F_{\mathrm{net}} = 0 = F_{\mathrm{buoy}} - m_{\mathrm{Al}}g = \rho_{\mathrm{Hg}} gfV - \rho_{\mathrm{Al}} gV;$
$(13.6 \times 10^3\ \mathrm{kg/m^3})gfV = (2.70 \times 10^3\ \mathrm{kg/m^3})gV,$
which gives $f = \boxed{0.199.}$

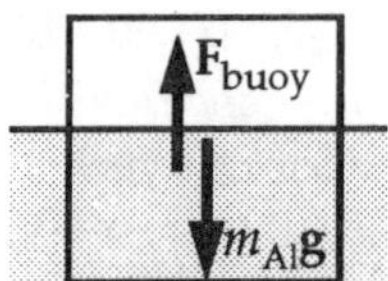

24. When the balloon and cargo float, the net force is zero, so we have

$F_{net} = 0 = F_{buoy} - m_{He}g - m_{balloon}g - m_{cargo}g;$

$0 = \rho_{air}gV_{balloon} - \rho_{He}gV_{balloon} - m_{balloon}g - m_{cargo}g;$

$0 = (1.29 \text{ kg/m}^3 - 0.179 \text{ kg/m}^3)\frac{4}{3}\pi(9.5 \text{ m})^3 - (1000 \text{ kg} + m_{cargo}),$

which gives $m_{cargo} =$ $\boxed{3.0 \times 10^3 \text{ kg.}}$

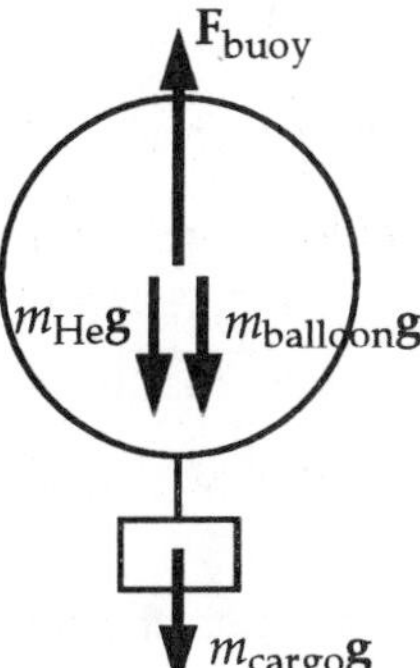

25. Because the mass of the displaced water is the apparent change in mass of the person, we have

$\Delta m = \rho_{water}V_{legs}.$

For the mass of one leg we have

$m_{leg} = SG_{body}\rho_{water}\frac{1}{2}V_{legs} = SG_{body}\rho_{water}\frac{1}{2}\Delta m/\rho_{water} = SG_{body}\frac{1}{2}\Delta m$

$= (1.00)\frac{1}{2}(78 \text{ kg} - 54 \text{ kg}) =$ $\boxed{12 \text{ kg.}}$

26. Because the mass of the displaced water is the apparent change in mass of the sample, we have

$\Delta m = \rho_{water}V.$

For the density of the sample we have

$\rho = m/V = (m/\Delta m)\rho_{water}$

$= [(63.5 \text{ g})/(63.5 \text{ g} - 56.4 \text{ g})](1.00 \times 10^3 \text{ kg/m}^3) = 8.94 \times 10^3 \text{ kg/m}^3.$

From the table of densities, the most likely metal is $\boxed{\text{copper.}}$

27. (*a*) Because the mass of the displaced liquid is the apparent change in mass of the ball, we have

$\Delta m = \rho_{liquid}V,$ or

$\rho_{liquid} = \Delta m/V = (\Delta m/m)\rho_{ball};$

$= [(3.40 \text{ kg} - 2.10 \text{ kg})/(3.40 \text{ kg})](2.70 \times 10^3 \text{ kg/m}^3) =$ $\boxed{1.03 \times 10^3 \text{ kg/m}^3.}$

(*b*) For a submerged object, from part (*a*) we have

$\boxed{\rho_{liquid} = [(m - m_{apparent})/m]\rho_{object}.}$

28. (*a*) The buoyant force is a measure of the net force on the partially submerged object due to the pressure in the fluid. In order to remove the object and have no effect on the fluid, we would have to fill the submerged volume with an equal volume of fluid. As expected, the buoyant force on this fluid is the weight of the fluid. To have no net torque on the fluid, the buoyant force and the weight of the fluid would have to act at the same point, the center of gravity.

(*b*) From the diagram we see that, if the center of buoyancy is above the center of gravity, when the ship tilts, the net torque will tend to restore the ship's position. From the diagram we see that, if the center of buoyancy is below the center of gravity, when the ship tilts, the net torque will tend to continue the tilt. If the center of buoyancy is at the center of gravity, there will be no net torque from these forces, so other torques (from the wind and waves) would determine the motion of the ship. Thus stability is achieved when the $\boxed{\text{center of buoyancy is above the center of gravity.}}$

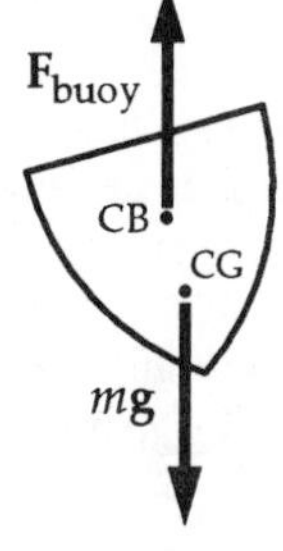

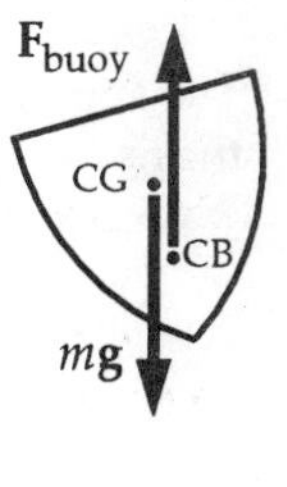

29. Because the mass of the displaced alcohol is the apparent change in mass of the wood, we have
    $\Delta m = \rho_{\text{alcohol}} V$.
    We find the specific gravity of the wood from
    $SG_{\text{wood}}\rho_{\text{water}} = m_{\text{wood}}/V = (m_{\text{wood}}/\Delta m)\rho_{\text{alcohol}}$;
    $SG_{\text{wood}}(1.00 \times 10^3 \text{ kg/m}^3) = [(0.48 \text{ kg})/(0.48 \text{ kg} - 0.035 \text{ kg})](0.79 \times 10^3 \text{ kg/m}^3)$, which gives
    $SG_{\text{wood}} = \boxed{0.85.}$
    As expected, the specific gravity is less than 1, so the wood floats in water.

30. When the ice floats, the net force is zero. If the fraction of the ice that is above water is f, we have
    $F_{\text{net}} = 0 = F_{\text{buoy}} - m_{\text{ice}}g = \rho_{\text{sw}}g(1 - f)V - \rho_{\text{ice}}gV$, or
    $(1.025)\rho_{\text{w}}g(1 - f)V = (0.917)\rho_{\text{w}}gV$, which gives $f = \boxed{0.105\ (10.5\%).}$

31. From Problem 30, we know that the initial volume out of the water, without the bear on the ice, is
    $V_1 = 0.105V_0 = 0.105(10 \text{ m}^3) = 1.05 \text{ m}^3$.
    Thus we find the submerged volume of ice with the bear on the ice from
    $V_2 = V_0 - \frac{1}{2}V_1 = 10 \text{ m}^3 - \frac{1}{2}(1.05 \text{ m}^3) = 9.48 \text{ m}^3$.
    Because the net force is zero, we have
    $F_{\text{net}} = 0 = F_{\text{bear}} + F_{\text{ice}} - m_{\text{bear}}g - m_{\text{ice}}g$, or
    $\rho_{\text{sea water}}g(0.30)V_{\text{bear}} + \rho_{\text{sea water}}gV_2 = m_{\text{bear}}g + \rho_{\text{ice}}gV_0$;
    $\rho_{\text{sea water}}g(0.30)(m_{\text{bear}}/\rho_{\text{bear}}) + \rho_{\text{sea water}}gV_2 = m_{\text{bear}}g + \rho_{\text{ice}}gV_0$;
    $(1.025)\rho_{\text{w}}(0.30)[m_{\text{bear}}/(1.00)\rho_{\text{w}}] + (1.025 \times 10^3 \text{ kg/m}^3)(9.48 \text{ m}^3) =$
    $m_{\text{bear}} + (0.917)(1.00 \times 10^3 \text{ kg/m}^3)(10 \text{ m}^3)$, which gives $m_{\text{bear}} = \boxed{7.9 \times 10^2 \text{ kg.}}$

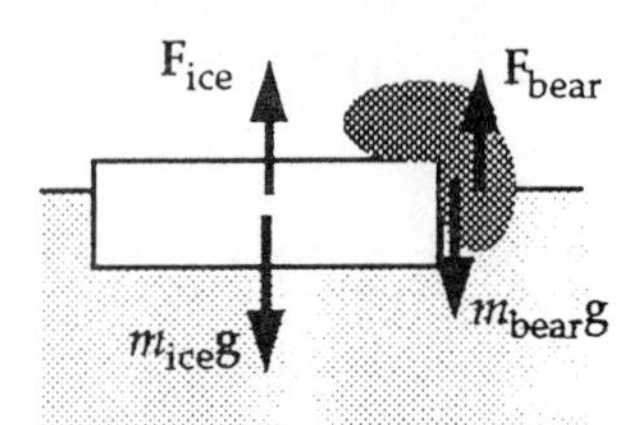

32. The minimum mass of lead will suspend it under water.
    Because the net force is zero, we have
    $F_{\text{net}} = 0 = F_{\text{lead}} + F_{\text{wood}} - m_{\text{lead}}g - m_{\text{wood}}g$, or
    $\rho_{\text{water}}gV_{\text{lead}} + \rho_{\text{water}}gV_{\text{wood}} = m_{\text{lead}}g + m_{\text{wood}}g$;
    $\rho_{\text{water}}g(m_{\text{lead}}/\rho_{\text{lead}}) + \rho_{\text{water}}g(m_{\text{wood}}/\rho_{\text{wood}}) = m_{\text{lead}}g + m_{\text{wood}}g$.
    We can rearrange this:
    $m_{\text{lead}}[1 - (1/SG_{\text{lead}})] = m_{\text{wood}}[(1/SG_{\text{wood}}) - 1]$;
    $m_{\text{lead}}[1 - (1/11.3)] = (2.52 \text{ kg})[(1/0.50) - 1]$, which gives
    $m_{\text{lead}} = \boxed{2.76 \text{ kg.}}$

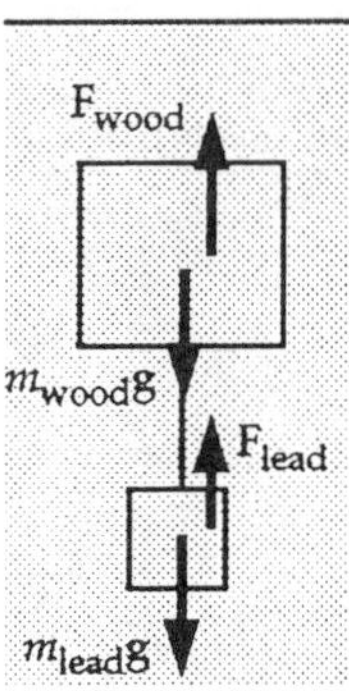

33. The apparent weight is the force required to hold the system, so the net force is zero. When only the sinker is submerged, we have
    $F_{\text{net}} = 0 = w_1 + F_{\text{buoy1}} - w - mg$, or
    $w_1 + \rho_{\text{water}}gV_{\text{sinker}} = w + mg$.
    When the sinker and object are submerged, we have
    $F_{\text{net}} = 0 = w_2 + F_{\text{buoy1}} + F_{\text{buoy2}} - w - mg$, or
    $w_2 + \rho_{\text{water}}gV_{\text{sinker}} + \rho_{\text{water}}gV_{\text{object}} = w + mg$.
    If we subtract the two results, we get
    $w_1 - w_2 - \rho_{\text{water}}gV_{\text{object}} = 0$, or
    $\rho_{\text{water}}g(m_{\text{object}}/\rho_{\text{object}}) = w/SG_{\text{object}} = w_1 - w_2$,
    so $SG_{\text{object}} = w/(w_1 - w_2)$.

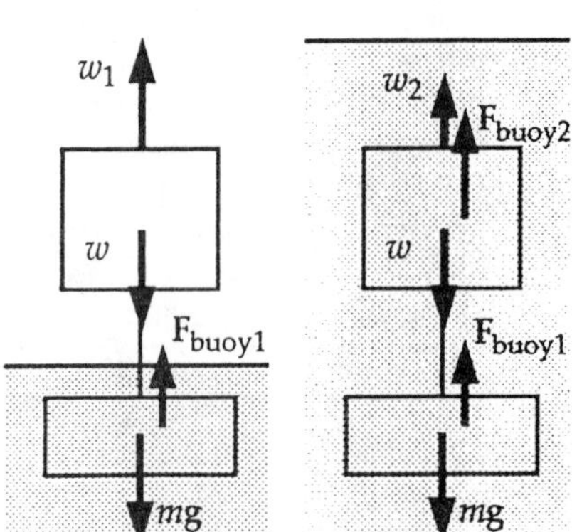

34. The flow rate in the major arteries must be the flow rate in the aorta:

$\rho v_1 A_1 = \rho v_2 A_2$;

$(30\text{ cm/s})\pi(1.0\text{ cm})^2 = v_2(2.0\text{ cm}^2)$, which gives $v_2 =$ **47 cm/s.**

35. From the equation of continuity we have

Flow rate $= Av$;

$(9.2\text{ m})(5.0\text{ m})(4.5\text{ m})/(10\text{ min})(60\text{ s/min}) = \pi(0.17\text{ cm})^2 v$, which gives $v =$ **3.8 m/s.**

36. From the equation of continuity we have

Flow rate $= A_{\text{pool}}h/t = A_{\text{hose}}v$;

$\frac{1}{4}\pi(7.2\text{ m})^2(1.5\text{ m})/t = \frac{1}{4}\pi[(0.625\text{ in})(0.0254\text{ m/in})]^2(0.28\text{ m/s})$, which gives $t = 1.10 \times 10^6\text{ s} =$ **13 days.**

37. If we choose the initial point at the water main where the water is not moving and the final point at the top of the spray, where the water also is not moving, from Bernoulli's equation we have

$P_1 + \frac{1}{2}\rho v_1^2 + \rho g h_1 = P_2 + \frac{1}{2}\rho v_2^2 + \rho g h_2$;

$P_1 + 0 + 0 = P_{\text{atm}} + 0 + (1.00 \times 10^3\text{ kg/m}^3)(9.80\text{ m/s}^2)(12\text{ m})$, which gives

$P_1 - P_{\text{atm}} = P_{\text{gauge}} =$ **$1.2 \times 10^5\text{ N/m}^2 = 1.2$ atm.**

38. If we choose the initial point at the pressure head where the water is not moving and the final point at the faucet, from Bernoulli's equation we have

$P_1 + \frac{1}{2}\rho v_1^2 + \rho g h_1 = P_2 + \frac{1}{2}\rho v_2^2 + \rho g h_2$;

$P_{\text{atm}} + 0 + (1.00 \times 10^3\text{ kg/m}^3)(9.80\text{ m/s}^2)(12.0\text{ m}) = P_{\text{atm}} + \frac{1}{2}(1.00 \times 10^3\text{ kg/m}^3)v_2^2$, which gives

$v_2 = 15.3\text{ m/s}$.

For the flow rate we have

Flow rate $= Av = \frac{1}{4}\pi(0.016\text{ m})^2\,(15.3\text{ m/s}) =$ **$3.1 \times 10^{-3}\text{ m}^3/\text{s}$.**

39. The pressure under the roof will be atmospheric. If we choose the initial point where the air is not moving and the final point above the roof, from Bernoulli's equation we have

$P_1 + \frac{1}{2}\rho v_1^2 + \rho g h_1 = P_2 + \frac{1}{2}\rho v_2^2 + \rho g h_2$;

$P_{\text{atm}} + 0 + 0 = P + \frac{1}{2}(1.29\text{ kg/m}^3)(30\text{ m/s})^2$, which gives $P - P_{\text{atm}} = 5.8 \times 10^2\text{ N/m}^2$.

The net upward force on the roof is

$F = (P - P_{\text{atm}})A = (5.8 \times 10^2\text{ N/m}^2)(240\text{ m}^2) =$ **1.4×10^5 N.**

40. If we consider the points at the top and bottom surfaces of the wing compared to the air flow in front of the wing, from Bernoulli's equation we have

$P_0 + \frac{1}{2}\rho v_0^2 + \rho g h_0 = P_1 + \frac{1}{2}\rho v_1^2 + \rho g h_1 = P_2 + \frac{1}{2}\rho v_2^2 + \rho g h_2$;

$P_1 + \frac{1}{2}(1.29\text{ kg/m}^3)(340\text{ m/s})^2 + 0 = P_2 + \frac{1}{2}(1.29\text{ kg/m}^3)(290\text{ m/s})^2$,

which gives $P_2 - P_1 = 2.03 \times 10^4\text{ N/m}^2$.

The net upward force on the wing is

$F = (P_2 - P_1)A = (2.03 \times 10^4\text{ N/m}^2)(80\text{ m}^2) =$ **1.6×10^6 N.**

41. If we consider the points far away from the center of the hurricane and at the center of the hurricane, from Bernoulli's equation we have

$P_1 + \frac{1}{2}\rho v_1^2 + \rho g h_1 = P_2 + \frac{1}{2}\rho v_2^2 + \rho g h_2$;

$1.013 \times 10^5\text{ N/m}^2 + 0 + 0 = P_2 + \frac{1}{2}(1.29\text{ kg/m}^3)[(300\text{ km/h})/(3.6\text{ ks/h})]^2 + 0$,

which gives $P_2 =$ **$9.7 \times 10^4\text{ N/m}^2$.**

42. If we consider a volume of fluid, at each end of the pipe there is a force toward the fluid of PA. If the area of the pipe is constant, the net force on the fluid is

$F_{net} = (P_1 - P_2)A.$

The required power is

$\text{Power} = F_{net}v = (P_1 - P_2)Av = (P_1 - P_2)Q.$

43. The flow rate in the pipe at street level must be the flow rate at the top floor:

$v_1A_1 = v_2A_2;$

$(0.60\text{ m/s})\frac{1}{4}\pi(5.0\text{ cm})^2 = v_2\frac{1}{4}\pi(2.6\text{ cm})^2$, which gives $v_2 =$ **2.22 m/s.**

If we use Bernoulli's equation between the street level and the top floor, we have

$P_1 + P_{atm} + \frac{1}{2}\rho v_1^2 + \rho g h_1 = P_2 + P_{atm} + \frac{1}{2}\rho v_2^2 + \rho g h_2;$

$3.8\text{ atm} + P_{atm} + \frac{1}{2}(1.00\times10^3\text{ kg/m}^3)(0.60\text{ m/s})^2 + 0 =$

$P_2 + P_{atm} + \frac{1}{2}(1.00\times10^3\text{ kg/m}^3)(2.22\text{ m/s})^2 + (1.00\times10^3\text{ kg/m}^3)(9.80\text{ m/s}^2)(20\text{ m}),$

which gives $P_2 =$ **1.8 atm (gauge).**

44. (*a*) The flow rate through the venturi meter is constant:

$v_1A_1 = v_2A_2.$

If we use Bernoulli's equation between the segments of the meter, we have

$P_1 + \frac{1}{2}\rho v_1^2 + \rho g h_1 = P_2 + \frac{1}{2}\rho v_2^2 + \rho g h_2;$

$P_1 + \frac{1}{2}\rho v_1^2 + 0 = P_2 + \frac{1}{2}\rho v_2^2 + 0$, or $P_1 - P_2 = \frac{1}{2}\rho(v_2^2 - v_1^2).$

When we substitute for v_2 from the flow rate, we get

$P_1 - P_2 = \frac{1}{2}\rho[(v_1A_1/A_2)^2 - v_1^2] = \frac{1}{2}\rho v_1^2[(A_1^2 - A_2^2)/A_2^2],$

which gives

$v_1 = A_2[2(P_1 - P_2)/\rho(A_1^2 - A_2^2)]^{1/2}.$

(*b*) With the given data, we have

$v_1 = A_2[2(P_1 - P_2)/\rho(A_1^2 - A_2^2)]^{1/2} = D_1^2[2(P_1 - P_2)/\rho(D_1^4 - D_2^4)]^{1/2}$

$= (1.0\text{ cm})^2\{2(18\text{ mm-Hg})(133\text{ N/m}^2\cdot\text{mm-Hg})/(1.00\times10^3\text{ kg/m}^3)[(3.0\text{ cm})^4 - (1.0\text{ cm})^4]\}^{1/2}$

$=$ **0.24 m/s.**

45. The torque corresponds to a tangential force on the inner cylinder:

$F = \tau/R_1.$

The layer of fluid has a thickness $R_2 - R_1$, the fluid next to the outer cylinder is at rest, and the fluid next to the inner cylinder has a speed $v = R_1\omega$, where

$\omega = (62\text{ rev/min})(2\pi\text{ rad/rev})/(60\text{ s/min}) = 6.49\text{ rad/s}.$

If we use the average radius for the area over which the torque acts in the definition of viscosity, we have

$\eta = FL/Av = \tau(R_2 - R_1)/R_1[2\pi\frac{1}{2}(R_1 + R_2)H]R_1\omega = \tau(R_2 - R_1)/\pi R_1^2(R_1 + R_2)H\omega$

$= (0.024\text{ m}\cdot\text{N})(0.0530\text{ m} - 0.0510\text{ m})/\pi(0.0510\text{ m})^2(0.0510\text{ m} + 0.0530\text{ m})(0.120\text{ m})(6.49\text{ rad/s})$

$=$ **0.072 Pa·s.**

46. The same volume of water is used, so the time is inversely proportional to the flow rate. Taking into account the viscosity of the water, with the only change the diameter of the hose, we have

$t \propto 1/Q \propto 1/r^4 \propto 1/d^4.$

For the two hoses we have.

$t_2/t_1 = (d_1/d_2)^4 = (3/5)^4 =$ **1/7.7 = 0.13.**

47. From Poiseuille's equation we have

$Q = \pi r^4(P_1 - P_2)/8\eta L;$

$450\times10^{-6}\text{ m}^3/\text{s} = \pi(0.145\text{ m})^4(P_1 - P_2)/8(0.20\text{ Pa}\cdot\text{s})(1.9\times10^3\text{ m}),$

which gives $P_1 - P_2 =$ **9.9×10^2 N/m².**

48. The pressure forcing the blood through the needle is produced from the elevation of the bottle:

$\Delta P = \rho gh.$

From Poiseuille's equation we have

$Q = \pi r^4(P_1 - P_2)/8\eta L = \pi r^4 \rho gh/8\eta L;$

$(4.1 \times 10^{-6}\ \text{m}^3/\text{min})/(60\ \text{s/min}) = \pi(0.20 \times 10^{-3}\ \text{m})^4(1.05 \times 10^3\ \text{kg/m}^3)(9.80\ \text{m/s}^2)(1.70\ \text{m})/8\eta(0.038\ \text{m}),$

which gives $\eta = \boxed{4.2 \times 10^{-3}\ \text{Pa}\cdot\text{s}.}$

49. From Poiseuille's equation we have

$Q = V/t = \pi r^4(P_1 - P_2)/8\eta L;$

$(9.0\ \text{m})(14\ \text{m})(4.0\ \text{m})/(10\ \text{min})(60\ \text{s/min}) =$

$\pi r^4(0.71 \times 10^{-3}\ \text{atm})(1.013 \times 10^5\ \text{N/m}^2\cdot\text{atm})/8(0.018 \times 10^{-3}\ \text{Pa}\cdot\text{s})(21\ \text{m}),$

which gives $r = 0.058$ m.

Thus the diameter needed is $\boxed{0.12\ \text{m}.}$

50. For the viscous flow of the blood, we have

$Q \propto r^4.$

For the two flows we have.

$Q_2/Q_1 = (r_2/r_1)^4;$

$0.25 = (r_2/r_1)^4$, which gives $r_2/r_1 = \boxed{0.71\ \text{(reduced by 29\%).}}$

51. From Poiseuille's equation we have

$Q = Av = \pi r^4\ \Delta P/8\eta L;$

$\pi(0.010\ \text{m})^2(0.30\ \text{m/s}) = \pi(0.010\ \text{m})^4\ \Delta P/8(4 \times 10^{-3}\ \text{Pa}\cdot\text{s})L,$

which gives $\Delta P/L = 96\ \text{N/m}^2/\text{m} = \boxed{0.96\ \text{N/m}^2/\text{cm}.}$

52. (*a*) We find the Reynolds number for the blood flow:

$Re = 2\bar{v}r\rho/\eta = 2(0.30\ \text{m/s})(0.010\ \text{m})(1.05 \times 10^3\ \text{kg/m}^3)/(4.0 \times 10^{-3}\ \text{Pa}\cdot\text{s}) = 1600.$

Thus the flow is $\boxed{\text{laminar}}$ but close to turbulent.

(*b*) If the only change is in the average speed, we have

$Re_2/Re_1 = \bar{v}_2/\bar{v}_1;$

$Re_2/1600 = 2$, or $Re_2 = \boxed{3200,}$ so the flow is $\boxed{\text{turbulent.}}$

53. Although the flow is not through a pipe, we approximate the thickness as the diameter. For the Reynolds number we have

$Re = 2\bar{v}r\rho/\eta = 2[(50 \times 10^{-3}\ \text{m/yr})/(3.16 \times 10^7\ \text{s/yr})]\tfrac{1}{2}(100 \times 10^3\ \text{m})(3200\ \text{kg/m}^3)/(4 \times 10^{19}\ \text{Pa}\cdot\text{s})$

$\approx \boxed{1 \times 10^{-20}.}$

This is definitely laminar flow.

54. We find the pressure difference required across the needle from Poiseuille's equation:

$Q = \pi r^4(P_1 - P_2)/8\eta L;$

$(4.0 \times 10^{-6}\ \text{m}^3/\text{min})/(60\ \text{s/min}) = \pi(0.20 \times 10^{-3}\ \text{m})^4(P_1 - P_2)/8(4.0 \times 10^{-3}\ \text{Pa}\cdot\text{s})(0.040\ \text{m}),$

which gives $P_1 - P_2 = 1.70 \times 10^4\ \text{N/m}^2.$

The pressure P_2 is the pressure in the vein. The pressure P_1 is produced from the elevation of the bottle:

$P_1 = P_{atm} + \rho gh.$

Thus we have

$P_{atm} + \rho gh - P_2 = 1.70 \times 10^4\ \text{N/m}^2$, or

$\rho gh = P_2 - P_{atm} + 1.70 \times 10^4\ \text{N/m}^2;$

$(1.05 \times 10^3\ \text{kg/m}^3)(9.80\ \text{m/s}^2)h = (18\ \text{torr})(133\ \text{N/m}^2\cdot\text{torr}) + 1.70 \times 10^4\ \text{N/m}^2$, which gives $h = \boxed{1.9\ \text{m}.}$

55. For the two surfaces, top and bottom, we have

$$\gamma = F/2\ell = (5.1 \times 10^{-3}\ \text{N})/2(0.070\ \text{m}) = \boxed{0.036\ \text{N/m.}}$$

56. For the two surfaces, top and bottom, we have

$$F = \gamma 2\ell = (0.025\ \text{N/m})2(0.182\ \text{m}) = \boxed{9.1 \times 10^{-3}\ \text{N.}}$$

57. From Example 10–14, the six upward surface tension forces must balance the weight. We find the required angle from

$$6(2\pi r\gamma \cos\theta) \approx w;$$

$$6(2)\pi(3.0 \times 10^{-5}\ \text{m})(0.072\ \text{N/m}) \cos\theta \approx (0.016 \times 10^{-3}\ \text{kg})(9.80\ \text{m/s}^2),\ \text{which gives } \cos\theta \approx 1.9.$$

Because the maximum possible value of $\cos\theta$ is 1, the insect $\boxed{\text{would not remain on the surface.}}$

58. (*a*) We assume that the net force from the weight and the buoyant force is much smaller than the surface tension. For the two surfaces, inner and outer circumferences, we have

$$F = \gamma 2L = \gamma 2(2\pi r),\ \text{or}\ \boxed{\gamma = F/4\pi r.}$$

(*b*) For the given data we have

$$\gamma = F/4\pi r = (840 \times 10^{-3}\ \text{N})/4\pi(0.028\ \text{m}) = \boxed{2.4\ \text{N/m.}}$$

59. The volume of water must remain the same. We find the surface area of the pool from

$$V = A_{\text{pool}}h = 100(\tfrac{1}{2})(\tfrac{4}{3}\pi h^3),\ \text{which gives } A_{\text{pool}} = 100(\tfrac{2}{3}\pi h^2).$$

The surface energy is equal to the work necessary to form the surface: $W = \gamma A$. For the ratio we have

$$W_{\text{droplets}}/W_{\text{pool}} = A_{\text{droplets}}/A_{\text{pool}} = 100(\tfrac{1}{2})(4\pi h^2)/100(\tfrac{2}{3}\pi h^2) = \boxed{3.}$$

60. We consider half of the soap bubble. The forces on the hemisphere will be the surface tensions on the two circles and the net force from the excess pressure between the inside and the outside of the bubble. This net force is the sum of all of the forces perpendicular to the surface of the hemisphere, but must be parallel to the surface tension. Therefore we can find it by finding the force on the circle that is the base of the hemisphere. The total force must be zero, so we have

$$2(2\pi r)\gamma = (\pi r^2)\,\Delta P,\ \text{which gives}\ \ \Delta P = 4\gamma/r.$$

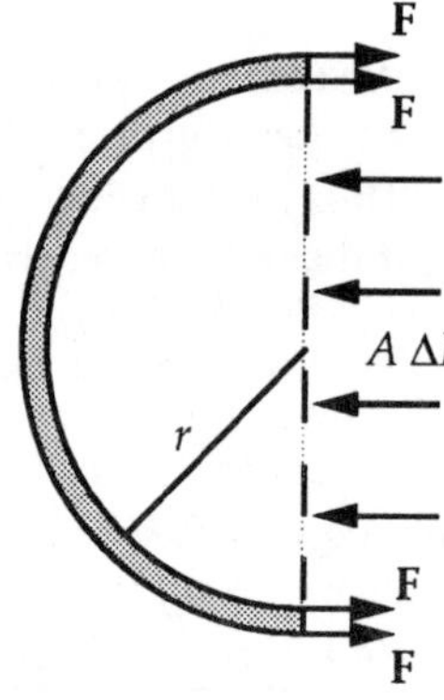

61. The liquid pressure is produced from the elevation of the bottle:

$$\Delta P = \rho g h.$$

(*a*) $(65\ \text{mm-Hg})(133\ \text{N/m}^2\cdot\text{mm-Hg}) = (1.00 \times 10^3\ \text{kg/m}^3)(9.80\ \text{m/s}^2)h$, which gives $h = \boxed{0.88\ \text{m.}}$

(*b*) $(550\ \text{mm-H}_2\text{O})(9.81\ \text{N/m}^2\cdot\text{mm-H}_2\text{O}) = (1.00 \times 10^3\ \text{kg/m}^3)(9.80\ \text{m/s}^2)h$, which gives $h = \boxed{0.55\ \text{m.}}$

(*c*) If we neglect viscous effects, we must produce a pressure to balance the blood pressure:

$(18\ \text{mm-Hg})(133\ \text{N/m}^2\cdot\text{mm-Hg}) = (1.00 \times 10^{-3}\ \text{kg/m}^3)(9.80\ \text{m/s}^2)h$, which gives $h = \boxed{0.24\ \text{m.}}$

62. (*a*) If the fluid is incompressible, the pressure must be constant, so we have

$$P = F_{\text{needle}}/A_{\text{needle}} = F_{\text{plunger}}/A_{\text{plunger}};$$

$F_{\text{needle}}/\tfrac{1}{4}\pi(0.020\ \text{cm})^2 = (2.4\ \text{N})/\tfrac{1}{4}\pi(1.3\ \text{cm})^2$, which gives $F_{\text{needle}} = \boxed{5.7 \times 10^{-4}\ \text{N.}}$

(*b*) Just before the fluid starts to move, the pressure must be the gauge pressure in the vein:

$$P = F_{\text{plunger}}/A_{\text{plunger}};$$

$(18\ \text{mm-Hg})(133\ \text{N/m}^2\cdot\text{mm-Hg}) = F_{\text{plunger}}/\tfrac{1}{4}\pi(1.3 \times 10^{-2}\ \text{cm})^2$, which gives $F_{\text{plunger}} = \boxed{0.32\ \text{N.}}$

63. If the motion is such that we can consider the air to be in equilibrium, we find the applied force from

$F = PA;$

$F_i = P_iA = (210 \times 10^3\ \text{N/m}^2)\frac{1}{4}\pi(0.030\ \text{m})^2 = 1.5 \times 10^2\ \text{N};$

$F_f = P_fA = (310 \times 10^3\ \text{N/m}^2)\frac{1}{4}\pi(0.030\ \text{m})^2 = 2.2 \times 10^2\ \text{N}.$

Thus the range of the applied force is $\boxed{1.5 \times 10^2\ \text{N} \leq F \leq 2.2 \times 10^2\ \text{N}.}$

64. The pressure difference is produced from the elevation:

$\Delta P = \rho gh$

$= (1.29\ \text{kg/m}^3)(9.80\ \text{m/s}^2)(410\ \text{m})/(1.013 \times 10^5\ \text{N/m}^2 \cdot \text{atm}) = \boxed{0.051\ \text{atm.}}$

65. The pressure is

$P = P_{atm} + \rho gh$

$= 1\ \text{atm} + (917\ \text{kg/m}^3)(9.80\ \text{m/s}^2)(4 \times 10^3\ \text{m})/(1.013 \times 10^5\ \text{N/m}^2 \cdot \text{atm}) \approx \boxed{4 \times 10^2\ \text{atm.}}$

66. The pressure difference in the blood is produced by the elevation change:

$\Delta P = \rho g\ \Delta h$

$= (1.05 \times 10^3\ \text{kg/m}^3)(9.80\ \text{m/s}^2)(6\ \text{m})/(1.013 \times 10^5\ \text{N/m}^2 \cdot \text{atm}) = \boxed{0.6\ \text{atm.}}$

67. The pressure difference on the ear drum is the change produced by the elevation change:

$\Delta P = \rho g\ \Delta h.$

The net force is

$F = A\ \Delta P = A\rho g\ \Delta h.$

$= (0.50 \times 10^{-4}\ \text{m}^2)(1.29\ \text{kg/m}^3)(9.80\ \text{m/s}^2)(1000\ \text{m}) = \boxed{0.63\ \text{N.}}$

68. Because the animal is suspended in the mixture, the mass of the animal is the mass of the displaced mixture, and the volume of the animal is the volume of the mixture:

$m_{animal} = m_{disp};$

$V = V_{alcohol} + V_{water}.$

We are given that

$m_{alcohol} = \rho_{alcohol}V_{alcohol} = 0.180m_{disp}$, and $m_{water} = \rho_{water}V_{water} = 0.820m_{disp}.$

For the density of the animal we have

$\rho_{animal} = m_{animal}/V = m_{disp}/(V_{alcohol} + V_{water}) = m_{disp}/[(0.180m_{disp}/\rho_{alcohol}) + (0.820m_{disp}/\rho_{water})]$

$= \rho_{water}\rho_{alcohol}/(0.180\rho_{water} + 0.820\rho_{alcohol})$

$= (1.00 \times 10^3\ \text{kg/m}^3)(0.79 \times 10^3\ \text{kg/m}^3)/[0.180(1.00 \times 10^3\ \text{kg/m}^3) + 0.820(0.79 \times 10^3\ \text{kg/m}^3)]$

$= \boxed{954\ \text{kg/m}^3.}$

69. The sum of the magnitudes of the forces on the ventricle wall is

$F = A\ \Delta P;$

$= (85 \times 10^{-4}\ \text{m}^2)(120\ \text{mm-Hg})(133\ \text{N/m}^2 \cdot \text{mm-Hg}) = \boxed{1.4 \times 10^2\ \text{N.}}$

Note that the forces on the wall are not all parallel, so this is not the vector sum.

70. We assume that ρ and g are constant. The pressure on a small area of the Earth's surface is produced by the weight of the air above it:

$P = mg/A = m_{total}g/A_{total} = m_{total}g/4\pi R_{Earth}{}^2;$

$1.013 \times 10^5\ \text{N/m}^2 = m_{total}(9.80\ \text{m/s}^2)/4\pi(6.37 \times 10^6\ \text{m})^2$, which gives $m_{total} = \boxed{5.3 \times 10^{18}\ \text{kg.}}$

71. The pressure difference on the water in the straw produces the elevation change:
    $\Delta P = \rho g \,\Delta h$
    (80 mm-Hg)(133 N/m$^2\cdot$mm-Hg) = (1.00 × 10^3 kg/m^3)(9.80 m/s^2)h, which gives h = **1.1 m.**

72. If we choose the initial point at the pressure head, where the water is not moving, and the final point at the faucet, from Bernoulli's equation we have
    $P_1 + \frac{1}{2}\rho v_1^2 + \rho g h_1 = P_2 + \frac{1}{2}\rho v_2^2 + \rho g h_2$;
    $P_{atm} + 0 + (1.00 \times 10^3\text{ kg/m}^3)(9.80\text{ m/s}^2)h_1 = P_{atm} + \frac{1}{2}(1.00 \times 10^3\text{ kg/m}^3)(7.2\text{ m/s})^2 + 0$,
    which gives h_1 = **2.6 m.**

73. The mass of the unloaded water must equal the mass of the displaced sea water:
    $m = \rho_{\text{sea water}}Ah = (1.025 \times 10^3\text{ kg/m}^3)(2800\text{ m}^2)(8.50\text{ m})$ = **2.44 × 10^7 kg.**
    Note that this is a volume of
    $V = m/\rho_{\text{water}} = (2.44 \times 10^7\text{ kg})/(1.00 \times 10^3\text{ kg/m}^3) = 2.44 \times 10^4\text{ m}^3$,
    which is greater than the volume change of the sea water.

74. We assume that for the maximum number of people, the logs are completely in the water and the people are not. Because the net force is zero, we have
    $F_{\text{buoy}} = m_{\text{people}}g + m_{\text{logs}}g$;
    $\rho_{\text{water}}g(\pi r^2)N_{\text{logs}}L = N_{\text{people}}mg + N_{\text{logs}}\rho_{\text{logs}}g(\pi r^2)L$.
    Because the specific gravity is the ratio of the density to the density of water, this can be written
    $(1 - SG_{\text{logs}})N_{\text{logs}}\rho_{\text{water}}\pi r^2 L = N_{\text{people}}m$;
    $(1 - 0.60)(10)(1.00 \times 10^3\text{ kg/m}^3)\pi(0.165\text{ m})^2(6.1\text{ m}) = N_{\text{people}}(70\text{ kg})$, which gives $N_{\text{people}} = 29.8$.
    Thus the raft can hold a maximum of **29 people.**

75. The work done in each heartbeat is
    $W = Fd = PAd = PV$.
    If n is the heart rate, the power output is
    Power = $nW = nPV$
    = [(70 beats/min)/(60 s/min)](105 mm-Hg)(133 N/m$^2\cdot$mm-Hg)(70 × 10^{-6} m^3) = **1.1 W.**

76. We find the effective g by considering a volume of the water. The net force must produce the acceleration:
    $F_{\text{buoy1}} - mg = ma$;
    $\rho_{\text{water}}g'V_{\text{water}} - \rho_{\text{water}}V_{\text{water}}g = \rho_{\text{water}}V_{\text{water}}a$;
    $g' = g + a = g + 3.5g = 4.5g$.
    The buoyant force on the rock is
    $F_{\text{buoy2}} = \rho_{\text{water}}g'V_{\text{rock}} = \rho_{\text{water}}g'(m_{\text{rock}}/\rho_{\text{rock}}) = g'm_{\text{rock}}/SG_{\text{rock}}$
    $= (4.5)(9.80\text{ m/s}^2)(3.0\text{ kg})/(2.7)$
    = **49 N.**

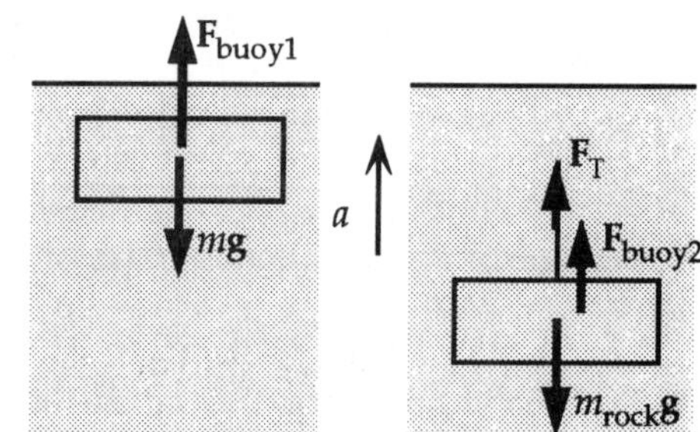

To see if the rock will float, we find the tension required to accelerate the rock:
    $F_T + F_{\text{buoy2}} = m_{\text{rock}}a$;
    $F_T + 49\text{ N} = (3.0\text{ kg})(3.5)(9.80\text{ m/s}^2)$, which gives $F_T = 54$ N.
Because the result is positive, tension is required, so the rock will **not float.**

77. We choose the reference level at the nozzle. If we apply Bernoulli's equation from the exit of the nozzle to the top of the spray, we have

$P_1 + \frac{1}{2}\rho v_1^2 + \rho g h_1 = P_2 + \frac{1}{2}\rho v_2^2 + \rho g h_2;$

$P_{atm} + \frac{1}{2}\rho v_1^2 + 0 = P_{atm} + 0 + \rho g h_2$, which gives $v_1^2 = 2gh_2$.

If we use the equation of continuity from the pump to the nozzle, we have

Flow rate $= A_3 v_3 = A_1 v_1$, or $v_3 = (A_1/A_3)v_1 = (D_1/D_3)^2 v_1$.

If we apply Bernoulli's equation from the pump to the exit of the nozzle, we have

$P_3 + \frac{1}{2}\rho v_3^2 + \rho g h_3 = P_1 + \frac{1}{2}\rho v_1^2 + \rho g h_1;$

$P_{pump} + \frac{1}{2}\rho[(D_1/D_3)^2 v_1]^2 + \rho g h_3 = P_{atm} + \frac{1}{2}\rho v_1^2 + 0$, or

$P_{pump} - P_{atm} = \frac{1}{2}\rho v_1^2[1 - (D_1/D_3)^4] - \rho g h_3 = \rho g\{h_2[1 - (D_1/D_3)^4] - h_3\}$

$= (1.00 \times 10^3\ \text{kg/m}^3)(9.80\ \text{m/s}^2)\{(0.12\ \text{m})[1 - (0.060\ \text{cm}/0.12\ \text{cm})^4] - (-1.1\ \text{m})\}$

$= \boxed{1.13 \times 10^4\ \text{N/m}^2.}$

78. We let d represent the diameter of the stream a distance y below the faucet. If we use the equation of continuity, we have

Flow rate $= A_0 v_0 = A_1 v_1$, or $v_1 = (A_0/A_1)v_0 = (D/d)^2 v_0$.

We choose the reference level at the faucet. If we apply Bernoulli's equation to the stream, we have

$P_0 + \frac{1}{2}\rho v_0^2 + \rho g h_0 = P_1 + \frac{1}{2}\rho v_1^2 + \rho g h_1;$

$P_{atm} + \frac{1}{2}\rho v_0^2 + 0 = P_{atm} + \frac{1}{2}\rho v_1^2 + \rho g(-y);$

$v_1^2 = v_0^2 + 2gy = (D/d)^4 v_0^2$, which gives $\boxed{d = D[v_0^2/(v_0^2 + 2gy)]^{1/4}.}$

79. (*a*) We use the flow rate to find the speed in an injector:

$Q = Av = N_{cylinders} N_{injectors} \pi r^2 v;$

$(7.5 \times 10^{-3}\ \text{m}^3/\text{rev})(3000\ \text{rev/min})/(60\ \text{s/min}) = (12)(4)\pi(3.0 \times 10^{-3}\ \text{m})^2 v,$

which gives $v = \boxed{2.8 \times 10^2\ \text{m/s.}}$

(*b*) If we apply Bernoulli's equation to the flow through the injectors, we have

$P_0 + \frac{1}{2}\rho v_0^2 + \rho g h_0 = P_1 + \frac{1}{2}\rho v_1^2 + \rho g h_1;$

$(1.5\ \text{atm})(1.013 \times 10^5\ \text{N/m}^2 \cdot \text{atm}) + 0 + 0 = P_{injector} + \frac{1}{2}(1.29\ \text{kg/m}^3)(2.8 \times 10^2\ \text{m/s})^2 + 0,$

which gives

$P_{injector} = \boxed{1.03 \times 10^5\ \text{N/m}^2\ (1.01\ \text{atm}).}$

80. (*a*) We choose the reference level at the bottom of the sink. If we apply Bernoulli's equation to the flow from the average depth of water in the sink to the pail, we have

$P_0 + \frac{1}{2}\rho v_0^2 + \rho g h_0 = P_1 + \frac{1}{2}\rho v_1^2 + \rho g h_1;$

$P_{atm} + 0 + \rho g h_0 = P_{atm} + \frac{1}{2}\rho v_1^2 + \rho g h_1$, or

$v_1 = [2g(h_0 - h_1)]^{1/2} = \{2(9.80\ \text{m/s}^2)[0.020\ \text{m} - (-0.50\ \text{m})]\}^{1/2} = \boxed{3.1\ \text{m/s.}}$

(*b*) We use the flow rate to find the time:

$Q = Av = V/t;$

$\pi(0.010\ \text{m})^2(3.07\ \text{m/s}) = (0.375\ \text{m}^2)(0.04\ 0\text{m})/t$, which gives $t = \boxed{16\ \text{s.}}$

CHAPTER 11

1. We find the spring constant from the compression caused by the increased weight:
   $k = mg/x = (65\text{ kg})(9.80\text{ m/s}^2)/(0.028\text{ m}) = 2.28 \times 10^4\text{ N/m}$.
   The frequency of vibration will be
   $f = (k/m)^{1/2}/2\pi = [(2.28 \times 10^4\text{ N/m})/(1065\text{ kg})]^{1/2}/2\pi =$ $\boxed{0.74\text{ Hz.}}$

2. We find the spring constant from the elongation caused by the increased weight:
   $k = \Delta mg/\Delta x = (80\text{ N} - 55\text{ N})/(0.85\text{ m} - 0.65\text{ m}) =$ $\boxed{1.3 \times 10^2\text{ N/m.}}$

3. In one period the particle will travel from one extreme position to the other (a distance of $2A$) and back again. The total distance traveled is
   $d = 4A = 4(0.25\text{ m}) =$ $\boxed{1.00\text{ m.}}$

4. (*a*) We find the spring constant from the elongation caused by the weight:
   $k = mg/\Delta x = (2.7\text{ kg})(9.80\text{ m/s}^2)/(0.039\text{ m}) =$ $\boxed{6.8 \times 10^2\text{ N/m.}}$
   (*b*) Because the fish will oscillate about the equilibrium position, the amplitude will be the distance the fish was pulled down from equilibrium:
   $\boxed{A = 2.5\text{ cm.}}$
   The frequency of vibration will be
   $f = (k/m)^{1/2}/2\pi = [(6.8 \times 10^2\text{ N/m})/(2.7\text{ kg})]^{1/2}/2\pi =$ $\boxed{2.5\text{ Hz.}}$

5. Because the mass starts at the maximum displacement, we have

t	0	$T/4$	$T/2$	$3T/4$	T	$5T/4$
x	A	0	$-A$	0	A	0

   We see that the curve resembles a $\boxed{\text{cosine wave.}}$

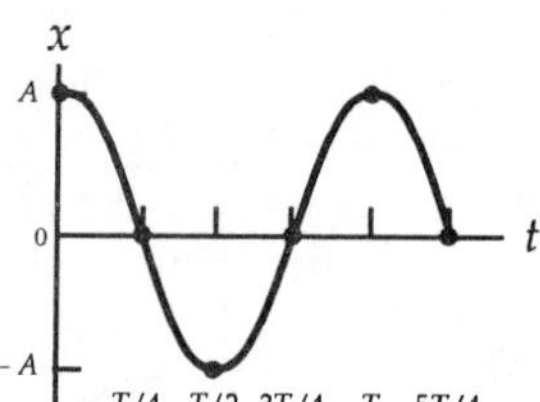

6. (*a*) We find the effective spring constant from the frequency:
   $f_1 = (k/m_1)^{1/2}/2\pi$;
   $4.0\text{ Hz} = [k/(0.15 \times 10^{-3}\text{ kg})]^{1/2}/2\pi$, which gives $k =$ $\boxed{9.5 \times 10^{-2}\text{ N/m.}}$
   (*b*) The new frequency of vibration will be
   $f_2 = (k/m_2)^{1/2}/2\pi = [(9.5 \times 10^{-2}\text{ N/m})/(0.50 \times 10^{-3}\text{ kg})]^{1/2}/2\pi =$ $\boxed{2.2\text{ Hz.}}$

7. (*a*) We find the effective spring constant from the frequency:
   $f_1 = (k/m_1)^{1/2}/2\pi$;
   $2.5\text{ Hz} = [k/(0.050\text{ kg})]^{1/2}/2\pi$, which gives $k =$ $\boxed{12\text{ N/m.}}$
   (*b*) Because the size and shape are the same, the spring constant will be the same.
   The new frequency of vibration will be
   $f_2 = (k/m_2)^{1/2}/2\pi = [(12\text{ N/m})/(0.25\text{ kg})]^{1/2}/2\pi =$ $\boxed{1.1\text{ Hz.}}$

8. The dependence of the frequency on the mass is
   $f = (k/m)^{1/2}/2\pi$.
   Because the spring constant does not change, we have
   $f_2/f_1 = (m_1/m_2)^{1/2}$;
   $f_2/(3.0\text{ Hz}) = [(0.60\text{ kg})/(0.38\text{ kg})]^{1/2}$, which gives $f_2 =$ $\boxed{3.8\text{ Hz.}}$

9. (*a*) The velocity will be maximum at the equilibrium position:
$v_0 = A\omega = 2\pi fA = 2\pi(3.0\ \text{Hz})(0.15\ \text{m}) =$ $\boxed{2.8\ \text{m/s.}}$
(*b*) We find the velocity at the position from
$v = v_0[1 - (x^2/A^2)]^{1/2}$
$= (2.8\ \text{m/s})\{1 - [(0.10\ \text{m})^2/(0.15\ \text{m})^2]\}^{1/2} =$ $\boxed{2.1\ \text{m/s.}}$
(*c*) We find the total energy from the maximum kinetic energy:
$E = KE_{max} = \frac{1}{2}mv_0^2 = \frac{1}{2}(0.50\ \text{kg})(2.8\ \text{m/s})^2 =$ $\boxed{2.0\ \text{J.}}$
(*d*) Because $x = A$ at $t = 0$, we have a cosine function:
$x = A\cos(\omega t) = A\cos(2\pi ft) =$ $\boxed{(0.15\ \text{m})\cos[2\pi(3.0\ \text{Hz})t].}$

10. The dependence of the frequency on the mass is
$f = (k/m)^{1/2}/2\pi.$
Because the spring constant does not change, we have
$f_2/f_1 = (m/m_2)^{1/2};$
$(0.60\ \text{Hz})/(0.88\ \text{Hz}) = [m/(m + 0.600\ \text{kg})]^{1/2}$, which gives $m =$ $\boxed{0.52\ \text{kg.}}$

11. In the equilibrium position, the net force is zero. When the mass is pulled down a distance x, the net restoring force is the sum of the additional forces from the springs, so we have
$F_{net} = \Delta F_2 + \Delta F_1 = -kx - kx = -2kx,$
which gives an effective force constant of $2k$.
We find the frequency of vibration from
$f = (k_{eff}/m)^{1/2}/2\pi =$ $\boxed{(2k/m)^{1/2}/2\pi.}$

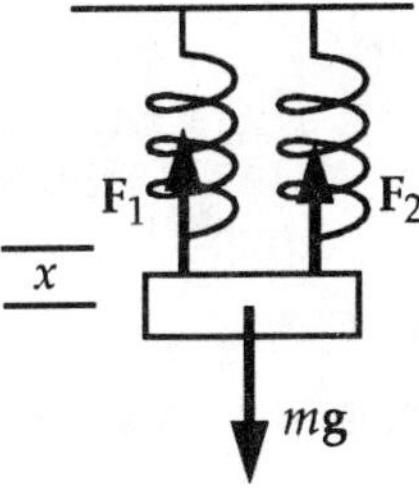

12. We find the spring constant from the elongation caused by the mass:
$k = \Delta mg/\Delta x = (1.62\ \text{kg})(9.80\ \text{m/s}^2)/(0.315\ \text{m}) = 50.4\ \text{N/m}.$
The period of the motion is independent of amplitude:
$T = 2\pi(m/k)^{1/2} = 2\pi[(1.62\ \text{kg})/(50.4\ \text{N/m})]^{1/2} = 1.13\ \text{s}.$
The time to return to the equilibrium position is one-quarter of a period:
$t = \frac{1}{4}T = \frac{1}{4}(1.13\ \text{s}) =$ $\boxed{0.282\ \text{s.}}$

13. We find the spring constant from the compression caused by the force:
$k = F/\Delta x = (80.0\ \text{N})/(0.200\ \text{m}) = 400\ \text{N/m}.$
The ball will leave at the equilibrium position, where the kinetic energy is maximum. Because this is also the maximum potential energy, we have
$KE_{max} = \frac{1}{2}mv_0^2 = PE_{max} = \frac{1}{2}kA^2;$
$\frac{1}{2}(0.150\ \text{kg})v_0^2 = \frac{1}{2}(400\ \text{N/m})(0.200\ \text{m})^2$, which gives $v_0 =$ $\boxed{10.3\ \text{m/s.}}$

14. (*a*) The period of the motion is independent of amplitude:
$T = 2\pi(m/k)^{1/2} = 2\pi[(0.750\text{ kg})/(124\text{ N/m})]^{1/2} =$ $\boxed{0.489\text{ s.}}$
The frequency is
$f = 1/T = 1/(0.489\text{ s}) =$ $\boxed{2.04\text{ Hz.}}$
(*b*) Because the mass is struck at the equilibrium position, the initial speed is the maximum speed. We find the amplitude from
$v_0 = A\omega = 2\pi fA;$
$2.76\text{ m/s} = 2\pi(2.04\text{ Hz})A$, which gives $A =$ $\boxed{0.215\text{ m.}}$
(*c*) The maximum acceleration is
$a_{max} = \omega^2 A = (2\pi f)^2 A = [2\pi(2.04\text{ Hz})]^2(0.215\text{ m}) =$ $\boxed{35.5\text{ m/s}^2.}$
(*d*) Because the mass starts at the equilibrium position, we have a sine function. If we take the positive x-direction in the direction of the initial velocity, we have
$x = A\sin(\omega t) = A\sin(2\pi ft) = (0.215\text{ m})\sin[2\pi(2.04\text{ Hz})t] =$ $\boxed{(0.215\text{ m})\sin[(12.8\text{ s}^{-1})t].}$
(*e*) We find the total energy from the maximum kinetic energy:
$E = KE_{max} = \frac{1}{2}mv_0^2 = \frac{1}{2}(0.750\text{ kg})(2.76\text{ m/s})^2 =$ $\boxed{2.86\text{ J.}}$

15. (*a*) The work done on the spring increases the potential energy:
$W = PE_{max} = \frac{1}{2}kA_0^2;$
$3.0\text{ J} = \frac{1}{2}k(0.12\text{ m})^2$, which gives $k =$ $\boxed{4.2\times10^2\text{ N/m.}}$
(*b*) The maximum acceleration is produced by the maximum restoring force:
$F_{max} = kA = ma;$
$(4.17\times10^2\text{ N/m})(0.12\text{ m}) = m(15\text{ m/s}^2)$, which gives $m =$ $\boxed{3.3\text{ kg.}}$

16. Because the mass is released at the maximum displacement, we have
$x = x_0\cos(\omega t);\quad v = -v_0\sin(\omega t);\quad a = -a_{max}\cos(\omega t).$
(*a*) We find ωt from
$v = -\frac{1}{2}v_0 = -v_0\sin(\omega t)$, which gives $\omega t = 30°$.
Thus the distance is
$x = x_0\cos(\omega t) = x_0\cos(30°) =$ $\boxed{0.866\,x_0.}$
(*b*) We find ωt from
$a = -\frac{1}{2}a_{max} = -a_{max}\cos(\omega t)$, which gives $\omega t = 60°$.
Thus the distance is
$x = x_0\cos(\omega t) = x_0\cos(60°) =$ $\boxed{0.500\,x_0.}$

17. (*a*) The amplitude is the maximum value of x: $\boxed{0.45\text{ m.}}$
(*b*) We find the frequency from the coefficient of t:
$2\pi f = 8.40\text{ s}^{-1}$, which gives $f =$ $\boxed{1.34\text{ Hz.}}$
(*c*) The maximum speed is
$v_0 = \omega A = (8.40\text{ s}^{-1})(0.45\text{ m}) = 3.78\text{ m/s}.$
We find the total energy from the maximum kinetic energy:
$E = KE_{max} = \frac{1}{2}mv_0^2 = \frac{1}{2}(0.50\text{ kg})(3.78\text{ m/s})^2 =$ $\boxed{3.6\text{ J.}}$
(*d*) We find the velocity at the position from
$v = v_0[1 - (x^2/A^2)]^{1/2}$
$= (3.78\text{ m/s})\{1 - [(0.30\text{ m})^2/(0.45\text{ m})^2]\}^{1/2} = 2.82\text{ m/s}.$
The kinetic energy is
$KE = \frac{1}{2}mv^2 = \frac{1}{2}(0.50\text{ kg})(2.82\text{ m/s})^2 =$ $\boxed{2.0\text{ J.}}$
The potential energy is
$PE = E - KE = 3.6\text{ J} - 2.0\text{ J} =$ $\boxed{1.6\text{ J.}}$

18. (*a*) The amplitude is the maximum value of x: **0.35 m.**
(*b*) We find the frequency from the coefficient of t:
$2\pi f = 5.50\ \text{s}^{-1}$, which gives $f =$ **0.875 Hz.**
(*c*) The period is
$T = 1/f = 1/(0.875\ \text{Hz}) =$ **1.14 s.**
(*d*) The maximum speed is
$v_0 = \omega A = (5.50\ \text{s}^{-1})(0.35\ \text{m}) = 1.93\ \text{m/s}$.
We find the total energy from the maximum kinetic energy:
$E = KE_{max} = \frac{1}{2}mv_0^2 = \frac{1}{2}(0.400\ \text{kg})(1.93\ \text{m/s})^2 =$ **0.74 J.**
(*e*) We find the velocity at the position from
$v = v_0[1 - (x^2/A^2)]^{1/2}$
$= (1.93\ \text{m/s})\{1 - [(0.10\ \text{m})^2/(0.35\ \text{m})^2]\}^{1/2} = 1.84\ \text{m/s}$.
The kinetic energy is
$KE = \frac{1}{2}mv^2 = \frac{1}{2}(0.400\ \text{kg})(1.84\ \text{m/s})^2 =$ **0.68 J.**
The potential energy is
$PE = E - KE = 0.74\ \text{J} - 0.68\ \text{J} =$ **0.06 J.**

x (m)
0.35
0
−0.35
0.25 0.50 0.75 1.00 1.25 1.50
t (s)

19. (*a*) We find the frequency from
$f = (k/m)^{1/2}/2\pi = [(210\ \text{N/m})/(0.250\ \text{kg})]^{1/2}/2\pi = 4.61\ \text{Hz}$, so
$\omega = 2\pi f = 2\pi(4.61\ \text{Hz}) = 29.0\ \text{s}^{-1}$.
Because the mass starts at the equilibrium position moving in the positive direction, we have a sine function:
$x = A \sin(\omega t) = (0.280\ \text{m}) \sin[(29.0\ \text{s}^{-1})t]$.
(*b*) The period of the motion is
$T = 1/f = 1/(4.61\ \text{Hz}) = 0.217\ \text{s}$.
It will take one-quarter period to reach the maximum extension, so the spring will have maximum extensions at **0.0542 s, 0.271 s, 0.488 s, … .**
It will take three-quarters period to reach the minimum extension, so the spring will have minimum extensions at **0.163 s, 0.379 s, 0.596 s, … .**

20. (*a*) We find the frequency from the period:
$f = 1/T = 1/(0.55\ \text{s}) = 1.82\ \text{Hz}$, so
$\omega = 2\pi f = 2\pi(1.82\ \text{Hz}) = 11.4\ \text{s}^{-1}$.
The amplitude is the compression: 0.10 m. Because the mass is released at the maximum displacement, we have a cosine function:
$y = A \cos(\omega t) = (0.10\ \text{m}) \cos[(11.4\ \text{s}^{-1})t]$.
(*b*) The time to return to the equilibrium position is one-quarter of a period:
$t = \frac{1}{4}T = \frac{1}{4}(0.55\ \text{s}) =$ **0.14 s.**
(*c*) The maximum speed is
$v_0 = \omega A = (11.4\ \text{s}^{-1})(0.10\ \text{m}) =$ **1.1 m/s.**
(*d*) The maximum acceleration is
$a_{max} = \omega^2 A = (11.4\ \text{s}^{-1})^2(0.10\ \text{m}) =$ **13 m/s².**
The maximum magnitude of the acceleration occurs at the endpoints of the motion, so it will be attained first at **the release point.**

21. Immediately after the collision, the block-bullet system will have its maximum velocity at the equilibrium position. We find this velocity from energy conservation:
$KE_i + PE_i = KE_f + PE_f$;
$\frac{1}{2}(M + m)v_0^2 + 0 = 0 + \frac{1}{2}kA^2$;
$\frac{1}{2}(0.600\ \text{kg} + 0.025\ \text{kg})v_0^2 = \frac{1}{2}(6.70 \times 10^3\ \text{N/m})(0.215\ \text{m})^2$, which gives $v_0 = 22.3\ \text{m/s}$.
We find the initial speed of the bullet from momentum conservation for the impact:
$mv + 0 = (M + m)v_0$;
$(0.025\ \text{kg})v = (0.600\ \text{kg} + 0.025\ \text{kg})(22.3\ \text{m/s})$, which gives $v =$ **557 m/s.**

22. Because the frequencies and masses are the same, the spring constant must be the same. We can compare the two maximum potential energies:

$$\text{PE}_1/\text{PE}_2 = \tfrac{1}{2}kA_1^2/\tfrac{1}{2}kA_2^2 = (A_1/A_2)^2;$$

$$10 = (A_1/A_2)^2, \text{ or } \boxed{A_1 = 3.16A_2.}$$

23. (*a*) The total energy is the maximum potential energy, so we have

$$\text{PE} = \tfrac{1}{2}\text{PE}_{\text{max}};$$

$$\tfrac{1}{2}kx^2 = \tfrac{1}{2}(\tfrac{1}{2}kA^2), \text{ which gives } x = \boxed{0.707A.}$$

(*b*) We find the position from

$$v = v_0[1 - (x^2/A^2)]^{1/2};$$

$$\tfrac{1}{2}v_0 = v_0[1 - (x^2/A^2)]^{1/2}, \text{ which gives } x = \boxed{0.866A.}$$

24. We use a coordinate system with down positive.
With x_0 a magnitude, at the equilibrium position we have

$$\Sigma F = -kx_0 + mg = 0.$$

If the spring is compressed a distance x from the equilibrium position, we have

$$\Sigma F = -k(x_0 + x) + mg = 0.$$

When we use the equilibrium condition, we get

$$\Sigma F = F = -kx.$$

Note that x is negative, so the force is positive.

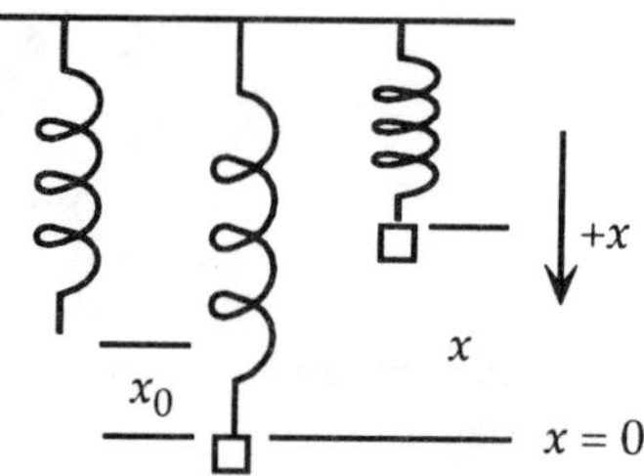

25. For the vertical spring there are both gravitational and elastic potential energy terms. We choose the unloaded position of the spring as the reference level with down positive.
With x_0 a magnitude, at the equilibrium position we have

$$\Sigma F = -kx_0 + mg = 0, \text{ or } kx_0 = mg.$$

When the spring is stretched a distance x from the equilibrium position, the stretch of the spring is $x + x_0$.
For the total energy we have

$$\text{E} = \text{KE} + \text{PE}_{\text{grav}} + \text{PE}_{\text{spring}}$$

$$= \tfrac{1}{2}mv^2 - mg(x + x_0) + \tfrac{1}{2}k(x + x_0)^2 = \tfrac{1}{2}mv^2 - mgx - mg\,x_0 + \tfrac{1}{2}kx^2 + kxx_0 + \tfrac{1}{2}kx_0^2.$$

When we use the equilibrium condition, we get

$$\text{E} = \tfrac{1}{2}mv^2 - mgx_0 + \tfrac{1}{2}kx^2 + \tfrac{1}{2}kx_0^2.$$

Because the terms containing x_0 are constant, we have

$$\text{E} + mgx_0 - \tfrac{1}{2}kx_0^2 = (\text{a constant}) = \text{E}' = \tfrac{1}{2}mv^2 + \tfrac{1}{2}kx^2.$$

If we had chosen a reference level for the gravitational potential energy halfway between the unloaded position and the equilibrium position, the terms added to E would cancel and E = E′.

26. We find the period from the time for N oscillations:

$$T = t/N = (34.7 \text{ s})/8 = 4.34 \text{ s}.$$

From this we can get the spring constant:

$$T = 2\pi(m/k)^{1/2};$$

$$4.34 \text{ s} = 2\pi[(65.0 \text{ kg})/k]^{1/2}, \text{ which gives } k = \boxed{136 \text{ N/m.}}$$

At the equilibrium position, we have

$$mg = kx_0;$$

$$(65.0 \text{ kg})(9.80 \text{ m/s}^2) = (136 \text{ N/m})x_0, \text{ which gives } x_0 = 4.7 \text{ m}.$$

Because this is how much the cord has stretched, we have

$$L = D - x_0 = 25.0 \text{ m} - 4.7 \text{ m} = \boxed{20.3 \text{ m.}}$$

27. (*a*) If we apply a force F to stretch the springs, the total displacement Δx is the sum of the displacements of the two springs: $\Delta x = \Delta x_1 + \Delta x_2$.
The effective spring constant is defined from $F = -k_{\text{eff}}\Delta x$.
Because they are in series, the force must be the same in each spring:
$$F_1 = F_2 = F = -k_1\,\Delta x_1 = -k_2\,\Delta x_2.$$
Then $\Delta x = \Delta x_1 + \Delta x_2$ becomes
$$-F/k_{\text{eff}} = -(F/k_1) - (F/k_2), \quad \text{or} \quad 1/k_{\text{eff}} = (1/k_1) + (1/k_2).$$
For the period we have
$$T = 2\pi(m/k_{\text{eff}})^{1/2} = 2\pi\{m[(1/k_1) + (1/k_2)]\}^{1/2}.$$

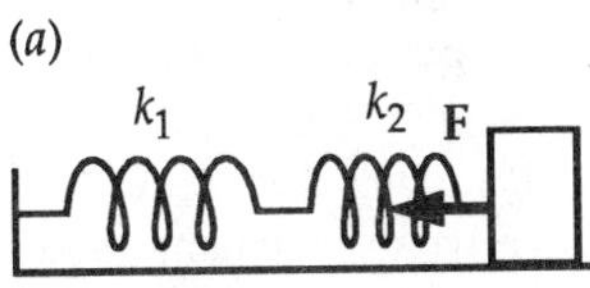

(*b*) In the equilibrium position, we have
$$F_{\text{net}} = F_{20} - F_{10} = 0, \quad \text{or} \quad F_{10} = F_{20}.$$
When the object is moved to the right a distance x, we have
$$F_{\text{net}} = F_{20} - k_2x - (F_{10} + k_1x) = -(k_1 + k_2)x.$$
The effective spring constant is $k_{\text{eff}} = k_1 + k_2$, so the period is
$$T = 2\pi(m/k_{\text{eff}})^{1/2} = 2\pi[m/(k_1 + k_2)]^{1/2}.$$

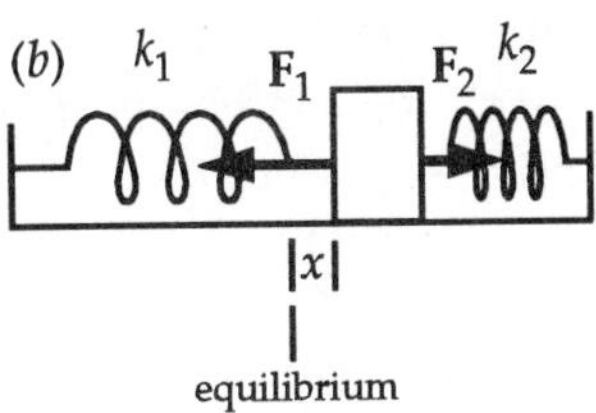

28. (*a*) We find the period from the time for N oscillations:
$$T = t/N = (50\text{ s})/36 = \boxed{1.4\text{ s.}}$$
(*b*) The frequency is
$$f = 1/T = 1/(1.39\text{ s}) = \boxed{0.72\text{ Hz.}}$$

29. (*a*) Because the period includes one "tick" and one "tock", the period is two seconds. We find the length from
$$T = 2\pi(L/g)^{1/2};$$
$$2.00\text{ s} = 2\pi[(L/(9.80\text{ m/s}^2)]^{1/2}, \text{ which gives } L = \boxed{0.993\text{ m.}}$$
(*b*) We see that an increase in L will cause an increase in T. This means that there will be fewer swings each hour, so the clock will run $\boxed{\text{slow.}}$

30. (*a*) For the period on Earth we have
$$T = 2\pi(L/g)^{1/2} = 2\pi[(0.50\text{ m})/(9.80\text{ m/s}^2)]^{1/2} = \boxed{1.4\text{ s.}}$$
(*b*) In a freely falling elevator, the effective g is zero, so the period would be $\boxed{\text{infinite (no swing).}}$

31. (*a*) For the frequency we have
$$f = (g/L)^{1/2}(1/2\pi) = [(9.80\text{ m/s}^2)/(0.66\text{ m})]^{1/2}(1/2\pi) = \boxed{0.61\text{ Hz.}}$$
(*b*) We use energy conservation between the release point and the lowest point:
$$\text{KE}_i + \text{PE}_i = \text{KE}_f + \text{PE}_f;$$
$$0 + mgh = \tfrac{1}{2}mv_0^2 + 0;$$
$$(9.80\text{ m/s}^2)(0.66\text{ m})(1 - \cos 12°) = \tfrac{1}{2}v_0^2,$$
which gives $v_0 = \boxed{0.53\text{ m/s.}}$
(*c*) The energy stored in the oscillation is the initial potential energy:
$$\text{PE}_i = mgh = (0.310\text{ kg})(9.80\text{ m/s}^2)(0.66\text{ m})(1 - \cos 12°) = \boxed{0.044\text{ J.}}$$

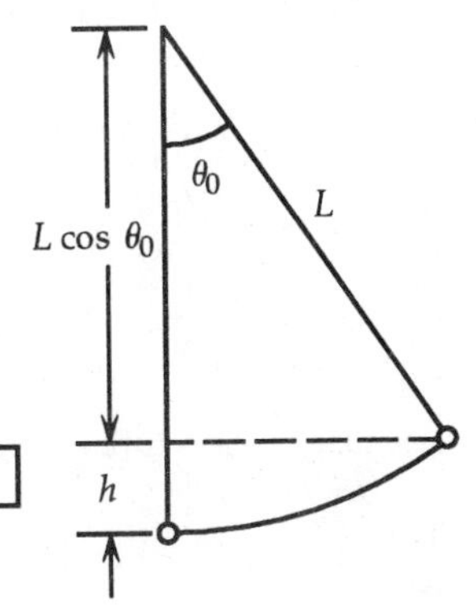

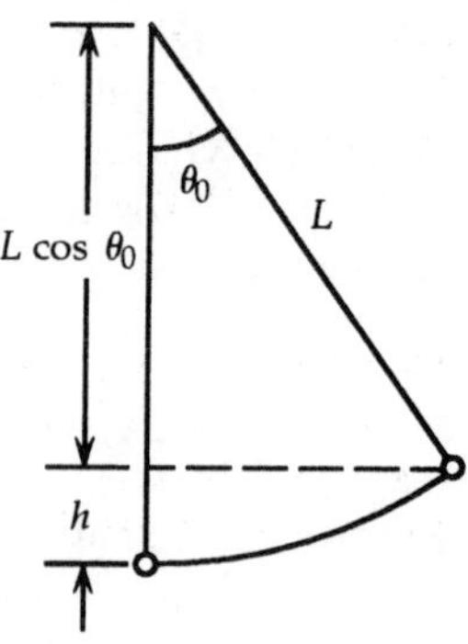

32. We use energy conservation between the release point and the lowest point:

$KE_i + PE_i = KE_f + PE_f$;

$0 + mgh = \frac{1}{2}mv_0^2 + 0$, or $v_0^2 = 2gh = 2gL(1 - \cos \theta_0)$.

When we use a trigonometric identity, we get

$v_0^2 = 2gL(2 \sin^2 \frac{1}{2}\theta_0)$.

For a simple pendulum θ_0 is small, so we have $\sin \frac{1}{2}\theta_0 \approx \frac{1}{2}\theta_0$.

Thus we get

$v_0^2 = 2gL2\,(\frac{1}{2}\theta_0)^2$, or $\boxed{v_0 = \theta_0(gL)^{1/2}.}$

33. We assume that 15° is small enough that we can consider this a simple pendulum, with a period

$T = 1/f = 1/(2.0 \text{ Hz}) = 0.50$ s.

Because the pendulum is released at the maximum angle, the angle will oscillate as a cosine function:

$\theta = \theta_0 \cos (2\pi ft) = (15°) \cos [2\pi(2.0 \text{ Hz})t] = (15°) \cos [(4.0\pi \text{ s}^{-1})t]$.

(*a*) $\theta = (15°) \cos [2\pi(2.0 \text{ Hz})(0.25 \text{ s})] = \boxed{-15°.}$

This is expected, since the time is half a period.

(*b*) $\theta = (15°) \cos [2\pi(2.0 \text{ Hz})(1.60 \text{ s})] = \boxed{+4.6°.}$

(*c*) $\theta = (15°) \cos [2\pi(2.0 \text{ Hz})(500 \text{ s})] = \boxed{+15°.}$

This is expected, since the time is one thousand periods.

34. The speed of the wave is

$v = f\lambda = \lambda/T = (8.5 \text{ m})/(3.0 \text{ s}) = \boxed{2.8 \text{ m/s.}}$

35. We find the wavelength from

$v = f\lambda$;

$330 \text{ m/s} = (262 \text{ Hz})\lambda$, which gives $\lambda = \boxed{1.26 \text{ m.}}$

36. For AM we find the wavelengths from

$\lambda_{\text{AMhigher}} = v/f_{\text{AMlower}} = (3.00 \times 10^8 \text{ m/s})/(550 \times 10^3 \text{ Hz}) = \boxed{545 \text{ m;}}$

$\lambda_{\text{AMlower}} = v/f_{\text{AMhigher}} = (3.00 \times 10^8 \text{ m/s})/(1600 \times 10^3 \text{ Hz}) = \boxed{188 \text{ m.}}$

For FM we have

$\lambda_{\text{FMhigher}} = v/f_{\text{FMlower}} = (3.00 \times 10^8 \text{ m/s})/(88.0 \times 10^6 \text{ Hz}) = \boxed{3.41 \text{ m;}}$

$\lambda_{\text{FMlower}} = v/f_{\text{FMhigher}} = (3.00 \times 10^8 \text{ m/s})/(108 \times 10^6 \text{ Hz}) = \boxed{2.78 \text{ m.}}$

37. We find the speed of the longitudinal (compression) wave from

$v = (B/\rho)^{1/2}$ for fluids and $v = (E/\rho)^{1/2}$ for solids.

(*a*) For water we have

$v = (B/\rho)^{1/2} = [(2.0 \times 10^9 \text{ N/m}^2)/(1.00 \times 10^3 \text{ kg/m}^3)]^{1/2} = \boxed{1.4 \times 10^3 \text{ m/s.}}$

(*b*) For granite we have

$v = (E/\rho)^{1/2} = [(45 \times 10^9 \text{ N/m}^2)/(2.7 \times 10^3 \text{ kg/m}^3)]^{1/2} = \boxed{4.1 \times 10^3 \text{ m/s.}}$

(*c*) For steel we have

$v = (E/\rho)^{1/2} = [(200 \times 10^9 \text{ N/m}^2)/(7.8 \times 10^3 \text{ kg/m}^3)]^{1/2} = \boxed{5.1 \times 10^3 \text{ m/s.}}$

38. Because the modulus does not change, the speed depends on the density:

$v \propto (1/\rho)^{1/2}$.

Thus we see that the speed will be greater in the $\boxed{\text{less dense rod.}}$

For the ratio of speeds we have

$v_1/v_2 = (\rho_2/\rho_1)^{1/2} = (2)^{1/2} = \boxed{1.41.}$

39. We find the speed of the wave from
$v = [F_T/(m/L)]^{1/2} = \{(150\text{ N})/[(0.55\text{ kg})/(30\text{ m})]\}^{1/2} = 90.5\text{ m/s}.$
We find the time from
$t = L/v = (30\text{ m})/(90.5\text{ m/s}) =$ **0.33 s.**

40. The speed of the longitudinal (compression) wave is
$v = (E/\rho)^{1/2},$
so the wavelength is
$\lambda = v/f = (E/\rho)^{1/2}/f = [(100 \times 10^9\text{ N/m}^2)/(7.8 \times 10^3\text{ kg/m}^3)]^{1/2}/(6{,}000\text{ Hz}) =$ **0.60 m.**

41. The speed of the longitudinal wave is
$v = (B/\rho)^{1/2},$
so the distance that the wave traveled is
$2D = vt = (B/\rho)^{1/2}t;$
$2D = [(2.0 \times 10^9\text{ N/m}^2)/(1.00 \times 10^3\text{ kg/m}^3)]^{1/2}(3.0\text{ s})$, which gives $D = 2.1 \times 10^3\text{ m} =$ **2.1 km.**

42. (*a*) Because both waves travel the same distance, we have
$\Delta t = (d/v_S) - (d/v_P) = d[(1/v_S) - (1/v_P)];$
$(2.0\text{ min})(60\text{ s/min}) = d\{[1/(5.5\text{ km/s})] - [1/(8.5\text{ km/s})]\}$, which gives $d =$ **1.9×10^3 km.**
(*b*) The direction of the waves is not known, thus the position of the epicenter **cannot be determined.** It would take at least one more station to find the intersection of the two circles.

43. For the surface wave we have
$\omega = 2\pi f = 2\pi(0.50\text{ Hz}) = \pi\text{ s}^{-1}.$
The object will leave the surface when the maximum acceleration of the SHM becomes greater than g, so the normal force becomes zero. Thus we have
$a_{max} = \omega^2 A > g;$
$(\pi\text{ s}^{-1})^2 A > 9.80\text{ m/s}^2$, which gives $A >$ **1.0 m.**

44. We assume that the wave spreads out uniformly in all directions.
(a) The intensity will decrease as $1/r^2$, so the ratio of intensities is
$I_2/I_1 = (r_1/r_2)^2 = [(10\text{ km})/(20\text{ km}]^2 =$ **0.25.**
(*b*) Because the intensity depends on A^2, the amplitude will decrease as $1/r$, so the ratio of amplitudes is
$A_2/A_1 = (r_1/r_2) = [(10\text{ km})/(20\text{ km}] =$ **0.50.**

45. We assume that the wave spreads out uniformly in all directions.
(*a*) The intensity will decrease as $1/r^2$, so the ratio of intensities is
$I_2/I_1 = (r_1/r_2)^2;$
$I_2/(2.0 \times 10^6\text{ J/m}^2\cdot\text{s}) = [(50\text{ km})/(1.0\text{ km})]^2$, which gives $I_2 =$ **5.0×10^9 J/m²·s.**
(*b*) We can take the intensity to be constant over the small area, so we have
$P_2 = I_2 S = (5.0 \times 10^9\text{ J/m}^2\cdot\text{s})(10.0\text{ m}^2) =$ **5.0×10^{10} W.**

46. If we consider two concentric circles around the spot where the waves are generated, the same energy must go past each circle in the same time. The intensity of a wave depends on A^2, so for the energy passing through a circle of radius r, we have
$E = I(2\pi r) = kA^2 2\pi r = \text{a constant}.$
Thus A must vary with r, in particular, we have $A \propto 1/\sqrt{r}$.

47. Because the speed and frequency are the same for the two waves, the intensity depends on the amplitude:

  $I \propto A^2$.

 For the ratio of intensities we have

  $I_2/I_1 = (A_2/A_1)^2$;

  $2 = (A_2/A_1)^2$, which gives $A_2/A_1 =$ **1.41.**

48. Because the speed and frequency are the same for the two waves, the intensity depends on the amplitude:

  $I \propto A^2$.

 For the ratio of intensities we have

  $I_2/I_1 = (A_2/A_1)^2$;

  $3 = (A_2/A_1)^2$, which gives $A_2/A_1 =$ **1.73.**

49. The bug will undergo SHM, so the maximum KE is also the maximum PE, which occurs at the maximum displacement. For the ratio of energies we have

  $KE_2/KE_1 = PE_2/PE_1 = (A_2/A_1)^2 = [\frac{1}{2}(4.5\text{ cm})/\frac{1}{2}(6.0\text{ cm})]^2 =$ **0.56.**

50. (*a*) (*b*)

 (*c*) Because all particles of the string are at equilibrium positions, there is no potential energy. Particles of the string will have transverse velocities, so they have **kinetic energy.**

51. All harmonics are present in a vibrating string. Because the harmonic specifies the multiple of the fundamental, we have $f_n = nf_1$, $n = 1, 2, 3, \ldots$:

  $f_1 = 1f_1 = (1)(440\text{ Hz}) =$ **440 Hz;**

  $f_2 = 2f_1 = (2)(440\text{ Hz}) =$ **880 Hz;**

  $f_3 = 3f_1 = (3)(440\text{ Hz}) =$ **1320 Hz;**

  $f_4 = 4f_1 = (4)(440\text{ Hz}) =$ **1760 Hz.**

52. From the diagram the initial wavelength is $2L$, and the final wavelength is $4L/3$. The tension has not changed, so the velocity has not changed:

  $v = f_1\lambda_1 = f_2\lambda_2$;

  $(294\text{ Hz})(2L) = f_2(4L/3)$, which gives $f_2 =$ **441 Hz.**

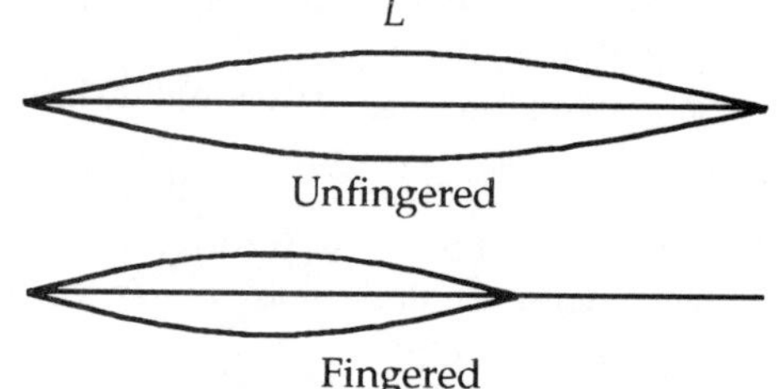

53. From the diagram the initial wavelength is $L/2$. We see that the other wavelengths are

  $\lambda_1 = 2L$, $\lambda_2 = L$ and $\lambda_3 = 2L/3$.

 The tension has not changed, so the velocity has not changed:

  $v = f\lambda = f_n\lambda_n$;

  $(280\text{ Hz})(L/2) = f_1(2L)$, which gives $f_1 =$ **70 Hz;**

  $(280\text{ Hz})(L/2) = f_2(L)$, which gives $f_2 =$ **140 Hz;**

  $(280\text{ Hz})(L/2) = f_3(2L/3)$, which gives $f_3 =$ **210 Hz.**

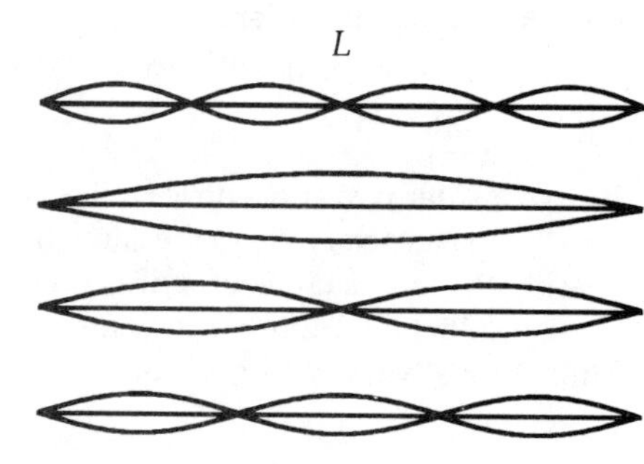

54. The oscillation corresponds to the fundamental with a frequency:

$f_1 = 1/T = 1/(2.5 \text{ s}) = 0.40 \text{ Hz}.$

This is similar to the vibrating string, so all harmonics are present:

$f_n = nf_1 = \boxed{n(0.40 \text{ Hz}), n = 1, 2, 3, \ldots .}$

We find the corresponding periods from

$T_n = 1/f_n = 1/nf_1 = T/n = \boxed{(2.5 \text{ s})/n, n = 1, 2, 3, \ldots .}$

55. We find the wavelength from

$v = f\lambda;$

$92 \text{ m/s} = (475 \text{ Hz})\lambda$, which gives $\lambda = 0.194$ m.

The distance between adjacent nodes is $\frac{1}{2}\lambda$, so we have

$d = \frac{1}{2}\lambda = \frac{1}{2}(0.194 \text{ m}) = \boxed{0.097 \text{ m.}}$

56. All harmonics are present in a vibrating string: $f_n = nf_1, n = 1, 2, 3, \ldots$. The difference in frequencies for two successive overtones is

$\Delta f = f_{n+1} - f_n = (n + 1)f_1 - nf_1 = f_1$, so we have $f_1 = 350 \text{ Hz} - 280 \text{ Hz} = \boxed{70 \text{ Hz.}}$

57. We find the speed of the wave from

$v = (F_T/\mu)^{1/2} = \{(520 \text{ N})/[(0.0036 \text{ kg})/(0.90 \text{ m})]\}^{1/2} = 361 \text{ m/s}.$

The wavelength of the fundamental for a string is $\lambda_1 = 2L$. We find the fundamental frequency from

$f_1 = v/\lambda_1 = (361 \text{ m/s})/2(0.60 \text{ m}) = \boxed{300 \text{ Hz.}}$

All harmonics are present so the first overtone is the second harmonic:

$f_2 = (2)f_1 = (2)300 \text{ Hz} = \boxed{600 \text{ Hz.}}$

The second overtone is the third harmonic:

$f_3 = (3)f_1 = (3)300 \text{ Hz} = \boxed{900 \text{ Hz.}}$

58. We assume that the change in tension does not change the mass density, so the velocity variation depends only on the tension. Because the wavelength does not change, we have

$\lambda = v_1/f_1 = v_2/f_2$, or $F_{T2}/F_{T1} = (f_2/f_1)^2$.

For the fractional change we have

$(F_{T2} - F_{T1})/F_{T1} = (F_{T2}/F_{T1}) - 1 = (f_2/f_1)^2 - 1 = [(200 \text{ Hz})/(205 \text{ Hz})]^2 - 1 = -0.048.$

Thus the tension should be $\boxed{\text{decreased by 4.8\%.}}$

59. The speed of the wave depends on the tension and the mass density:

$v = (F_T/\mu)^{1/2}.$

The wavelength of the fundamental for a string is $\lambda_1 = 2L$. We find the fundamental frequency from

$f_1 = v/\lambda_1 = (1/2L)(F_T/\mu)^{1/2}.$

All harmonics are present in a vibrating string, so we have

$f_n = nf_1 = (n/2L)(F_T/\mu)^{1/2}, n = 1, 2, 3, \ldots .$

60. The hanging weight creates the tension in the string: $F_T = mg$. The speed of the wave depends on the tension and the mass density:

$v = (F_T/\mu)^{1/2} = (mg/\mu)^{1/2}$.

The frequency is fixed by the vibrator, so the wavelength is

$\lambda = v/f = (1/f)(mg/\mu)^{1/2}$. With a node at each end, each loop corresponds to $\lambda/2$.

(*a*) For one loop, we have $\lambda_1/2 = L$, or

$2L = v_1/f = (1/f)(m_1 g/\mu)^{1/2}$;

$2(1.50\text{ m}) = (1/60\text{ Hz})[m_1(9.80\text{ m/s}^2)/(4.3 \times 10^{-4}\text{ kg/m})]^{1/2}$, which gives $m_1 =$ **1.4 kg.**

(*b*) For two loops, we have $\lambda_2/2 = L/2$, or

$L = v_2/f = (1/f)(m_2 g/\mu)^{1/2}$;

$1.50\text{ m} = (1/60\text{ Hz})[m_2(9.80\text{ m/s}^2)/(4.3 \times 10^{-4}\text{ kg/m})]^{1/2}$, which gives $m_2 =$ **0.36 kg.**

(*c*) For five loops, we have $\lambda_5/2 = L/5$, or

$2L/5 = v_5/f = (1/f)(m_5 g/\mu)^{1/2}$;

$2(1.50\text{ m})/5 = (1/60\text{ Hz})[m_5(9.80\text{ m/s}^2)/(4.3 \times 10^{-4}\text{ kg/m})]^{1/2}$, which gives $m_5 =$ **0.057 kg.**

The amplitude of the standing wave can be much greater than the vibrator amplitude because of the resonance built up from the reflected waves at the two ends of the string.

61. The hanging weight creates the tension in the string: $F_T = mg$. The speed of the wave depends on the tension and the mass density:

$v = (F_T/\mu)^{1/2} = (mg/\mu)^{1/2}$, and thus is constant.

The frequency is fixed by the vibrator, so the constant wavelength is

$\lambda = v/f = (1/f)(mg/\mu)^{1/2} = (1/60\text{ Hz})[(0.080\text{ kg})(9.80\text{ m/s}^2)/(5.6 \times 10^{-4}\text{ kg/m})]^{1/2} = 0.624\text{ m}$.

The different standing waves correspond to different integral numbers of loops, starting at one loop. With a node at each end, each loop corresponds to $\lambda/2$. The lengths of the string for the possible standing wavelengths are

$L_n = n\lambda/2 = n(0.624\text{ m})/2 = n(0.312\text{ m}), n = 1, 2, 3, \ldots$, or

$L_n = 0.312\text{ m}, 0.624\text{ m}, 0.924\text{ m}, 1.248\text{ m}, 1.560\text{ m}, \ldots$.

Thus we see that there are **4** standing waves for lengths between 0.10 m and 1.5 m.

62. (*a*) The wavelength of the fundamental for a string is $2L$, so the fundamental frequency is

$f = (1/2L)(F_T/\mu)^{1/2}$.

When the tension is changed, the change in frequency is

$\Delta f = f' - f = (1/2L)[(F_T'/\mu)^{1/2} - (F_T/\mu)^{1/2}]$

$= (1/2L)\{(F_T/\mu)^{1/2}[(F_T'/F_T)^{1/2} - 1]\} = f\{[(F_T + \Delta F_T)/F_T]^{1/2} - 1]\}$

$= f\{[1 + (\Delta F_T/F_T)]^{1/2} - 1]\}$.

If $\Delta F_T/F_T$ is small, we have

$[1 + (\Delta F_T/F_T)]^{1/2} \approx 1 + \frac{1}{2}(\Delta F_T/F_T)$, so we get

$\Delta f \approx f[1 + \frac{1}{2}(\Delta F_T/F_T) - 1] = \frac{1}{2}(\Delta F_T/F_T)f$.

(*b*) With the given data, we get

$\Delta f = \frac{1}{2}(\Delta F_T/F_T)f$;

$442\text{ Hz} - 438\text{ Hz} = \frac{1}{2}(\Delta F_T/F_T)(438\text{ Hz})$, which gives $\Delta F_T/F_T = 0.018 =$ **1.8% (increase).**

(*c*) For each overtone there will be a new wavelength, but the wavelength does not change when the tension changes, so the formula **will apply** to the overtones.

63. For the refraction of the waves we have

$v_2/v_1 = (\sin\theta_2)/(\sin\theta_1)$;

$v_2/(8.0\text{ km/s}) = (\sin 31°)/(\sin 50°)$, which gives $v_2 =$ **5.4 km/s.**

64. For the refraction of the waves we have

$v_2/v_1 = (\sin\theta_2)/(\sin\theta_1)$;

$(2.1\text{ km/s})/(2.8\text{ km/s}) = (\sin\theta_2)/(\sin 34°)$, which gives $\theta_2 =$ **25°.**

65. The speed of the longitudinal (compression) wave for the solid rock depends on the modulus and the density: $v = (E/\rho)^{1/2}$. The modulus does not change, so we have $v \propto (1/\rho)^{1/2}$.
For the refraction of the waves we have
$v_2/v_1 = (\rho_1/\rho_2)^{1/2} = (SG_1/SG_2)^{1/2} = (\sin\theta_2)/(\sin\theta_1)$;
$[(3.7)/(2.8)]^{1/2} = (\sin\theta_2)/(\sin 35°)$, which gives $\theta_2 =$ $\boxed{41°.}$

66. (*a*) For the refraction of the waves we have
$v_2/v_1 = (\sin\theta_2)/(\sin\theta_i)$;
Because $v_2 > v_1$, $\theta_2 > \theta_i$. When we use the maximum value of θ_2, we get
$v_2/v_1 = (\sin 90°)/(\sin\theta_{iM}) = 1/(\sin\theta_{iM})$, or $\boxed{\theta_{iM} = \sin^{-1}(v_1/v_2).}$
(*b*) We have
$\theta_{iM} = \sin^{-1}(v_1/v_2) = \sin^{-1}[(7.2\text{ km/s})/(8.4\text{ km/s})] = 59°$.
Thus for angles $\boxed{> 59°}$ there will be only reflection.

67. If we approximate the sloshing as a standing wave with the fundamental frequency, we have $\lambda = 2D$.
We find the speed of the waves from
$v = f\lambda = (1.0\text{ Hz})2(0.08\text{ m}) =$ $\boxed{0.16\text{ m/s.}}$

68. We choose $h = 0$ at the unstretched position of the net and let the stretch of the net be x. We use energy conservation between the release point and the lowest point to find the spring constant:
$KE_i + PE_i = KE_f + PE_f$;
$0 + mgh_i = 0 + mgh_f + \frac{1}{2}kx_1^2$, or $mg[h_i - (-x_1)] = \frac{1}{2}kx_1^2$;
$(70\text{ kg})(9.80\text{ m/s}^2)(20\text{ m} + 1.1\text{ m}) = \frac{1}{2}k(1.1\text{ m})^2$, which gives $k = 2.39 \times 10^4\text{ N/m}$.
When the person lies on the net, the weight causes the deflection:
$mg = kx_2$;
$(70\text{ kg})(9.80\text{ m/s}^2) = (2.39 \times 10^4\text{ N/m})x_2$, which gives $x_2 = 0.029\text{ m} =$ $\boxed{2.9\text{ cm.}}$
We use energy conservation between the release point and the lowest point to find the stretch:
$KE_i + PE_i = KE_f + PE_f$;
$0 + mgh_i = 0 + mgh_f + \frac{1}{2}kx_3^2$, or $mg[h_i - (-x_3)] = \frac{1}{2}kx_3^2$;
$(70\text{ kg})(9.80\text{ m/s}^2)(35\text{ m} + x_3) = \frac{1}{2}(2.39 \times 10^4\text{ N/m})x_3^2$.
This is a quadratic equation for x_3, for which the positive result is $\boxed{1.4\text{ m.}}$

69. The stress from the tension in the cable causes the strain. We find the effective spring constant from
$E = \text{stress/strain} = (F_T/A)/(\Delta L/L_0)$, or
$k = F_T/\Delta L = EA/L_0 = (200 \times 10^9\text{ N/m}^2)\pi(3.2 \times 10^{-3}\text{ m})^2/(20\text{ m}) = 3.22 \times 10^5\text{ N/m}$.
We find the period from
$T = 2\pi(m/k)^{1/2} = 2\pi[(1200\text{ kg})/(3.22 \times 10^5\text{ N/m})]^{1/2} =$ $\boxed{0.38\text{ s.}}$

70. We ignore any frictional losses and use energy conservation:
$KE_i + PE_i = KE_f + PE_f$;
$\frac{1}{2}mv_0^2 + 0 = 0 + \frac{1}{2}kx^2$;
$\frac{1}{2}(1500\text{ kg})(2\text{ m/s})^2 = \frac{1}{2}(500 \times 10^3\text{ N/m})x^2$, which gives $x = 0.11\text{ m} =$ $\boxed{11\text{ cm.}}$

71. We treat the oscillation of the Jell-O as a standing wave produced by shear waves traveling up and down. The speed of the shear waves is

$v = (G/\rho)^{1/2} = [(520\ \text{N/m}^2)/(1300\ \text{kg/m}^3)]^{1/2} = 0.632\ \text{m/s}.$

Because the maximum shear displacement is at the top, we estimate the wavelength as

$\lambda = 4h = 4(0.040\ \text{m}) = 0.16\ \text{m}.$

We find the frequency from

$v = f\lambda;$

$0.632\ \text{m/s} = f(0.16\ \text{m})$, which gives $f =$ $\boxed{4.0\ \text{Hz.}}$

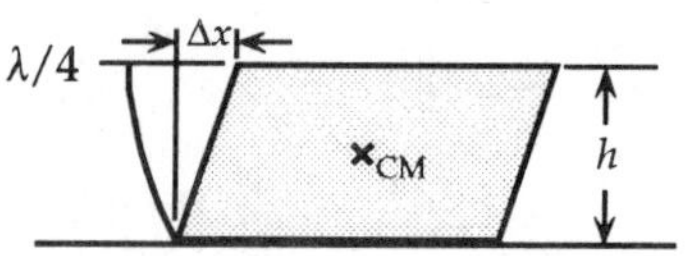

72. The effective value of g is increased when the acceleration is upward and decreased when the acceleration is downward. Because the length does not change, for the ratio of frequencies we have

$f'/f = (g'/g)^{1/2}.$

(*a*) For the upward acceleration we get

$f'/f = (g'/g)^{1/2} = [(g + \frac{1}{2}g)/g]^{1/2}$, which gives $f' =$ $\boxed{1.22f.}$

(*b*) For the downward acceleration we get

$f'/f = (g'/g)^{1/2} = [(g - \frac{1}{2}g)/g]^{1/2}$, which gives $f' =$ $\boxed{0.71f.}$

73. (*a*) We find the effective force constant from the displacement caused by the additional weight:

$k\,\Delta y = mg$, or $k = \Delta mg/\Delta y = (75\ \text{kg})(9.80\ \text{m/s}^2)/(0.040\ \text{m}) = 1.84 \times 10^4\ \text{N/m}.$

We find the frequency of vibration from

$f = (k/m)^{1/2}(1/2\pi) = [(1.84 \times 10^4\ \text{N/m})/(250\ \text{kg})]^{1/2}(1/2\pi) =$ $\boxed{1.4\ \text{Hz.}}$

(*b*) The total energy is the maximum potential energy, so we have

$E = PE_{max} = \frac{1}{2}kA^2 = \frac{1}{2}(1.84 \times 10^4\ \text{N/m})(0.040\ \text{m})^2 =$ $\boxed{15\ \text{J.}}$

Note that this is similar to the weight hanging on a spring. If we measure from the equilibrium position, we ignore changes in mgh.

74. The frequency of the sound will be the frequency of the needle passing over the ripples. The speed of the needle relative to the ripples is $v = r\omega$, so the frequency is

$f = v/\lambda = r\omega/\lambda$

$= (0.128\ \text{m})(33\ \text{rev/min})(2\pi\ \text{rad/rev})/(60\ \text{s/min})(1.70 \times 10^{-3}\ \text{m}) =$ $\boxed{260\ \text{Hz.}}$

75. (*a*) All harmonics are present in a vibrating string: $f_n = nf_1$, $n = 1, 2, 3, \ldots$.

The first overtone is f_2 and the second overtone is f_3.

For G we have

$f_2 = 2(392\ \text{Hz}) =$ $\boxed{784\ \text{Hz;}}$ $f_3 = 3(392\ \text{Hz}) =$ $\boxed{1176\ \text{Hz.}}$

For A we have

$f_2 = 2(440\ \text{Hz}) =$ $\boxed{880\ \text{Hz;}}$ $f_3 = 3(440\ \text{Hz}) =$ $\boxed{1320\ \text{Hz.}}$

(*b*) The speed of the wave in a string is $v = [F_T/(M/L)]^{1/2}$. Because the lengths are the same, the wavelengths of the fundamentals must be the same. For the ratio of frequencies we have

$f_A/f_G = v_A/v_G = (M_G/M_A)^{1/2};$

$(440\ \text{Hz})/(392\ \text{Hz}) = (M_G/M_A)^{1/2}$, which gives $M_G/M_A =$ $\boxed{1.26.}$

(*c*) Because the mass densities and the tensions are the same, the speeds must be the same. The wavelengths are proportional to the lengths, so for the ratio of frequencies we have

$f_A/f_G = \lambda_G/\lambda_A = L_G/L_A;$

$(440\ \text{Hz})/(392\ \text{Hz}) = L_G/L_A$, which gives $L_G/L_A =$ $\boxed{1.12.}$

(*d*) The speed of the wave in a string is $v = [F_T/(M/L)]^{1/2}$. Because the lengths are the same, the wavelengths of the fundamentals must be the same. For the ratio of frequencies we have

$f_A/f_G = v_A/v_G = (F_{TA}/F_{TG})^{1/2};$

$(440\ \text{Hz})/(392\ \text{Hz}) = (F_{TA}/F_{TG})^{1/2}$, which gives $F_{TG}/F_{TA} =$ $\boxed{0.794.}$

76. (*a*) We find the spring constant from energy conservation:

$KE_i + PE_i = KE_f + PE_f;$

$\frac{1}{2}Mv_0^2 + 0 = 0 + \frac{1}{2}kA^2;$

$\frac{1}{2}(900\text{ kg})(20\text{ m/s})^2 = \frac{1}{2}k(5.0\text{ m})^2$, which gives $k =$ $\boxed{1.4 \times 10^4\text{ N/m.}}$

(*b*) We find the period of the oscillation from

$T = 2\pi(m/k)^{1/2} = 2\pi[(900\text{ kg})/(1.44 \times 10^4\text{ N/m})]^{1/2} = 1.57\text{ s}.$

The car will be in contact with the spring for half a cycle, so the time is

$t = \frac{1}{2}T = \frac{1}{2}(1.57\text{ s}) = \boxed{0.79\text{ s.}}$

77. (*a*) The speed of the wave in a string is $v = [F_T/\mu]^{1/2}$. Because the tensions must be the same anywhere along the string, for the ratio of velocities we have

$\boxed{v_2/v_1 = (\mu_1/\mu_2)^{1/2}.}$

(*b*) Because the motion of one string is creating the motion of the other, the frequencies must be the same. For the ratio of wavelengths we have

$\boxed{\lambda_2/\lambda_1 = v_2/v_1 = (\mu_1/\mu_2)^{1/2}.}$

(*c*) From the result for part (*b*) we see that, if $\mu_2 > \mu_1$, we have $\lambda_2 < \lambda_1$, so the $\boxed{\text{lighter cord}}$ will have the greater wavelength.

78. The object will leave the surface when the maximum acceleration of the SHM becomes greater than g, so the normal force becomes zero. For the pebble to remain on the board, we have

$a_{max} = \omega^2 A = (2\pi f)^2 A < g;$

$[2\pi(3.5\text{ Hz})]^2 A < 9.80\text{ m/s}^2$, which gives $A < 2.0 \times 10^{-2}\text{ m} =$ $\boxed{2.0\text{ cm.}}$

79. In the equilibrium position, the net force is zero, so we have

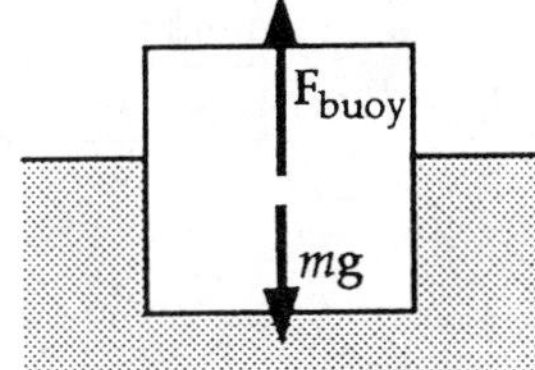

$F_{buoy} = mg.$

When the block is pushed into the water, there will be an additional buoyant force, equal to the weight of the additional water displaced, to bring the block back to the equilibrium position. When the block is pushed down a distance Δx, this net upward force is

$F_{net} = -\rho_{water}gA\,\Delta x.$

Because the net restoring force is proportional to the displacement, the block will oscillate with SHM. We find the effective force constant from the coefficient of Δx:

$\boxed{k = \rho_{water}gA.}$

80. The distance the mass falls is the distance the spring is stretched. We use energy conservation between the initial point, where the spring is unstretched, and the lowest point, our reference level for the gravitational potential energy, to find the spring constant:

$KE_i + PE_i = KE_f + PE_f;$

$0 + mgh = 0 + \frac{1}{2}kh^2$, which gives $k = 2mg/h$.

We find the frequency from

$f = (k/m)^{1/2}/2\pi = (2mg/hm)^{1/2}/2\pi = (2g/h)^{1/2}/2\pi = [2(9.80\text{ m/s}^2)/(0.30\text{ m})]^{1/2}/2\pi =$ $\boxed{1.3\text{ Hz.}}$

81. The increase in temperature will cause the length of the brass rod to increase. The period of the pendulum depends on the length,

$$T = 2\pi(L/g)^{1/2},$$

so the period will be greater. This means the pendulum will make fewer swings in a day, so the clock will be **slow.**

We use T_C for the temperature to distinguish it from the period.

For the length of the brass rod, we have

$$L = L_0(1 + \alpha \Delta T_C).$$

Thus the ratio of periods is

$$T/T_0 = (L/L_0)^{1/2} = (1 + \alpha \Delta T_C)^{1/2}.$$

Because $\alpha \Delta T_C$ is much less than 1, we have

$$T/T_0 \approx 1 + \tfrac{1}{2}\alpha \Delta T_C, \quad \text{or} \quad \Delta T/T_0 = \tfrac{1}{2}\alpha \Delta T_C.$$

The number of swings in a time t is $N = t/T$. For the same time t, the change in period will cause a change in the number of swings:

$$\Delta N = (t/T) - (t/T_0) = t(T_0 - T)/TT_0 \approx -t(\Delta T/T_0)/T_0,$$

because $T \approx T_0$. The time difference in one day is

$$\Delta t = T_0\,\Delta N = -t(\Delta T/T_0) = -t(\tfrac{1}{2}\alpha \Delta T_C)$$

$$= -(1\text{ day})(86{,}400\text{ s/day})\tfrac{1}{2}[19 \times 10^{-6}\,(\text{C}°)^{-1}](35°\text{C} - 20°\text{C}) = \boxed{-12.3\text{ s.}}$$

82. When the water is displaced a distance Δx from equilibrium, the net restoring force is the unbalanced weight of water in the height $2\,\Delta x$:

$$F_{net} = -2\rho g A\,\Delta x.$$

We see that the net restoring force is proportional to the displacement, so **the block will oscillate with SHM.**

We find the effective spring constant from the coefficient of Δx:

$$\boxed{k = 2\rho g A.}$$

From the formula for k, we see that the effective spring constant depends on **the density and the cross section.**

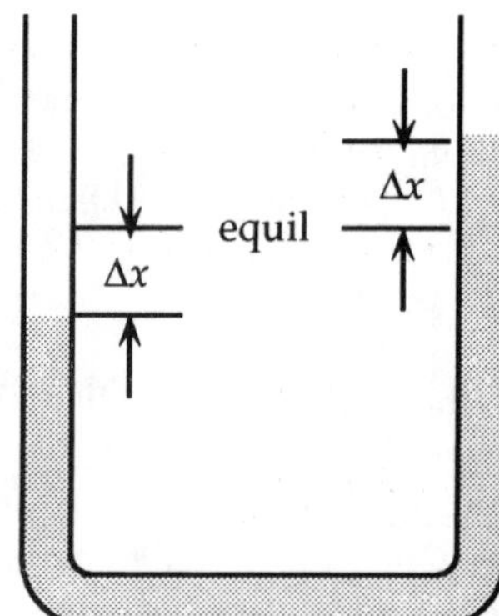

CHAPTER 12

1. Because the sound travels both ways across the lake, we have
$L = \frac{1}{2}vt = \frac{1}{2}(343 \text{ m/s})(1.5 \text{ s}) =$ $\boxed{2.6 \times 10^2 \text{ m.}}$

2. Because the sound travels down and back, we have
$L = \frac{1}{2}vt = \frac{1}{2}(1560 \text{ m/s})(2.0 \text{ s}) = 1.6 \times 10^3 \text{ m} =$ $\boxed{1.6 \text{ km.}}$

3. (*a*) We find the extreme wavelengths from
$\lambda_1 = v/f_1 = (343 \text{ m/s})/(20 \text{ Hz}) = 17 \text{ m};$
$\lambda_2 = v/f_2 = (343 \text{ m/s})/(20{,}000 \text{ Hz}) = 1.7 \times 10^{-2} \text{ m} = 1.7 \text{ cm}.$
The range of wavelengths is $\boxed{1.7 \text{ cm} \le \lambda \le 17 \text{ m.}}$
(*b*) We find the wavelength from
$\lambda = v/f = (343 \text{ m/s})/(10 \times 10^6 \text{ Hz}) =$ $\boxed{3.4 \times 10^{-5} \text{ m.}}$

4. For the travel times we have
$t_{\text{wire}} = d/v_{\text{wire}} = (8.4 \text{ m})/(5100 \text{ m/s}) =$ $\boxed{0.0016 \text{ s;}}$
$t_{\text{air}} = (d/v_{\text{air}}) = (8.4 \text{ m})/(343 \text{ m/s}) =$ $\boxed{0.024 \text{ s } (\approx 15\times t_{\text{wire}}).}$

5. The speed in the concrete is determined by the elastic modulus:
$v_{\text{concrete}} = (E/\rho)^{1/2} = [(20 \times 10^9 \text{ N/m}^2)/(2.3 \times 10^3 \text{ kg/m}^3)]^{1/2} = 2.95 \times 10^3 \text{ m/s}.$
For the time interval we have
$\Delta t = (d/v_{\text{air}}) - (d/v_{\text{concrete}});$
$1.4 \text{ s} = d\{[1/(343 \text{ m/s})] - [1/(2.95 \times 10^3 \text{ m/s})]\}$, which gives $d =$ $\boxed{5.4 \times 10^2 \text{ m.}}$

6. (*a*) For the time interval in sea water we have
$\Delta t = d/v_{\text{water}} = (1.0 \times 10^3 \text{ m})/(1560 \text{ m/s}) =$ $\boxed{0.64 \text{ s.}}$
(*b*) For the time interval in the air we have
$\Delta t = d/v_{\text{air}} = (1.0 \times 10^3 \text{ m})/(343 \text{ m/s}) =$ $\boxed{2.9 \text{ s.}}$

7. If we let L_1 represent the thickness of the top layer, the total transit time is
$t = (L_1/v_1) + [(L - L_1)/v_2)];$
$4.5 \text{ s} = [L_1/(331 \text{ m/s})] + [(1500 \text{ m} - L_1)/(343 \text{ m/s})]$, which gives $L_1 =$ $\boxed{1200 \text{ m.}}$
Thus the bottom layer is $1500 \text{ m} - 1200 \text{ m} =$ $\boxed{300 \text{ m.}}$

8. Because the distance is $d = vt$, the change in distance from the change in velocity is
$\Delta d = t\,\Delta v$; so the percentage change is
$(\Delta d/d)100 = (\Delta v/v)100 = (343 - 331 \text{ m/s})(100)/(343 \text{ m/s}) =$ $\boxed{3.5\%.}$

9. We find the intensity of the sound from
$\beta_1 = 10 \log_{10}(I_1/I_0);$
$120 \text{ dB} = 10 \log(I_1/10^{-12} \text{ W/m}^2)$, which gives $I_1 =$ $\boxed{1.0 \text{ W/m}^2.}$
For a whisper we have
$\beta_2 = 10 \log_{10}(I_2/I_0);$
$20 \text{ dB} = 10 \log(I_1/10^{-12} \text{ W/m}^2)$, which gives $I_2 =$ $\boxed{1.0 \times 10^{-10} \text{ W/m}^2.}$

10. We find the intensity level from
$\beta = 10 \log_{10}(I_1/I_0) = 10 \log_{10}[(2.0 \times 10^{-6} \text{ W/m}^2)/(10^{-12} \text{ W/m}^2)] =$ $\boxed{63 \text{ dB.}}$

11. We find the ratio of intensities of the sounds from
    $\beta = 10 \log_{10}(I_2/I_1)$;
    $2.0 \text{ dB} = 10 \log_{10}(I_2/I_1)$, which gives $I_2/I_1 = 1.58$.
    Because the intensity is proportional to the square of the amplitude, we have
    $I_2/I_1 = (A_2/A_1)^2$;
    $1.58 = (A_2/A_1)^2$, which gives $A_2/A_1 = \boxed{1.3.}$

12. When three of the four engines are shut down, the intensity is reduced by a factor of 4, so we have
    $\beta_2 - \beta_1 = 10 \log_{10}(I_2/I_0) - 10 \log_{10}(I_1/I_0) = 10 \log_{10}(I_2/I_1)$;
    $\beta_2 - 120 \text{ dB} = 10 \log(\frac{1}{4}I_1/I_1)$, which gives $\beta_2 = \boxed{114 \text{ dB.}}$

13. We find the ratio of intensities from
    $\beta = 10 \log_{10}(I_{\text{signal}}/I_{\text{noise}})$;
    $58 \text{ dB} = 10 \log_{10}(I_{\text{signal}}/I_{\text{noise}})$, which gives $I_{\text{signal}}/I_{\text{noise}} = \boxed{6.3 \times 10^5.}$

14. (*a*) Using the intensity from Table 12–2, we have
    $P = IA = I2\pi r^2 = (3 \times 10^{-6} \text{ W/m}^2)2\pi(0.50 \text{ m})^2 = \boxed{5 \times 10^{-6} \text{ W.}}$
    (*b*) We find the number of people required from
    $N = P_{\text{total}}/P = (100 \text{ W})/(5 \times 10^{-6} \text{ W}) = \boxed{2 \times 10^7.}$

15. (*a*) We find the intensity of the sound from
    $\beta = 10 \log_{10}(I/I_0)$;
    $50 \text{ dB} = 10 \log(I/10^{-12} \text{ W/m}^2)$, which gives $I = 1.0 \times 10^{-7} \text{ W/m}^2$.
    The rate at which energy is absorbed is the power of the sound wave:
    $P = IA = (1.0 \times 10^{-7} \text{ W/m}^2)(5.0 \times 10^{-5} \text{ m}^2) = \boxed{5.0 \times 10^{-12} \text{ W.}}$
    (*b*) We find the time from
    $t = E/P = (1.0 \text{ J})/(5.0 \times 10^{-12} \text{ W}) = 2.0 \times 10^{11} \text{ s} = \boxed{6.3 \times 10^3 \text{ yr.}}$

16. (*a*) If we assume that one channel is connected to the speaker, the rating is the power in the sound, and the sound spreads out uniformly, we have
    $I = P/4\pi r^2$, so we get
    $I_1 = (250 \text{ W})/4\pi(3.0 \text{ m})^2 = 2.21 \text{ W/m}^2$;
    $I_2 = (40 \text{ W})/4\pi(3.0 \text{ m})^2 = 0.354 \text{ W/m}^2$.
    For the intensity levels we have
    $\beta_1 = 10 \log_{10}(I_1/I_0) = 10 \log_{10}[(2.21 \text{ W/m}^2)/(10^{-12} \text{ W/m}^2)] = \boxed{123 \text{ dB;}}$
    $\beta_2 = 10 \log_{10}(I_2/I_0) = 10 \log_{10}[(0.354 \text{ W/m}^2)/(10^{-12} \text{ W/m}^2)] = \boxed{115 \text{ dB.}}$
    (*b*) The difference in intensity levels is 8 dB. A change of 10 dB corresponds to a doubling of the loudness, so the expensive amp is $\boxed{\text{almost twice as loud.}}$

17. (*a*) We find the intensity of the sound from
    $\beta = 10 \log_{10}(I/I_0)$;
    $130 \text{ dB} = 10 \log(I/10^{-12} \text{ W/m}^2)$, which gives $I = 10 \text{ W/m}^2$.
    The power output of the speaker is
    $P = IA = I4\pi r^2 = (10 \text{ W/m}^2)4\pi(2.5 \text{ m})^2 = \boxed{7.9 \times 10^2 \text{ W.}}$
    (*b*) We find the intensity of the sound from
    $\beta = 10 \log_{10}(I/I_0)$;
    $90 \text{ dB} = 10 \log(I/10^{-12} \text{ W/m}^2)$, which gives $I = 1.0 \times 10^{-3} \text{ W/m}^2$.
    We find the distance from
    $P = IA$;
    $7.9 \times 10^2 \text{ W} = (1.0 \times 10^{-3} \text{ W/m}^2)4\pi r^2$, which gives $r = \boxed{2.5 \times 10^2 \text{ m.}}$

18. (*a*) The intensity of the sound wave is given by

$I = 2\pi^2 \rho f^2 x_0^2 v.$

Because the frequency, density, and velocity are the same, for the ratio we have

$I_2/I_1 = (x_{02}/x_{01})^2 = 3^2 = \boxed{9.}$

(*b*) We find the change in intensity level from

$\Delta\beta = 10 \log_{10}(I_2/I_1) = 10 \log_{10}(9) = \boxed{9.5 \text{ dB.}}$

19. (*a*) We find the intensity of the sound from

$P = IA = I2\pi r^2;$

$5.0 \times 10^5 \text{ W} = I4\pi(30 \text{ m})^2$, which gives $I = 44.2 \text{ W/m}^2$.

We find the intensity level from

$\beta = 10 \log_{10}(I/I_0) = 10 \log_{10}[(44.2 \text{ W/m}^2)/(10^{-12} \text{ W/m}^2)] = \boxed{136 \text{ dB.}}$

(*b*) We find the intensity of the sound without air absorption from

$P = I_1A_1 = I_1 2\pi r_1^2;$

$5.0 \times 10^5 \text{ W} = I_1 4\pi(1000 \text{ m})^2$, which gives $I_1 = 3.98 \times 10^{-2} \text{ W/m}^2$.

We find the intensity level from

$\beta_{10} = 10 \log_{10}(I_1/I_0) = 10 \log_{10}[(3.98 \times 10^{-2} \text{ W/m}^2)/(10^{-12} \text{ W/m}^2)] = 106 \text{ dB}.$

When we consider air absorption, we have

$\beta_1 = \beta_{10} - (7.0 \text{ dB/km})r_1 = 106 \text{ dB} - (7.0 \text{ dB/km})(1.0 \text{ km}) = \boxed{99 \text{ dB.}}$

(*c*) We find the intensity of the sound without air absorption from

$P = I_2A_2 = I_2 2\pi r_2^2;$

$5.0 \times 10^5 \text{ W} = I_2 4\pi(5000 \text{ m})^2$, which gives $I_2 = 1.59 \times 10^{-3} \text{ W/m}^2$.

We find the intensity level from

$\beta_{20} = 10 \log_{10}(I_2/I_0) = 10 \log_{10}[(1.59 \times 10^{-3} \text{ W/m}^2)/(10^{-12} \text{ W/m}^2)] = 92 \text{ dB}.$

When we consider air absorption, we have

$\beta_2 = \beta_{20} - (7.0 \text{ dB/km})r_2 = 92 \text{ dB} - (7.0 \text{ dB/km})(5.0 \text{ km}) = \boxed{57 \text{ dB.}}$

20. The intensity of the sound wave is given by

$I = 2\pi^2 \rho f^2 x_0^2 v.$

Because the amplitude, density, and velocity are the same, for the ratio we have

$I_2/I_1 = (f_2/f_1)^2 = 2^2 = \boxed{4.}$

21. The intensity of the sound wave is given by

$I = 2\pi^2 \rho f^2 x_0^2 v$

$= 2\pi^2(1.29 \text{ kg/m}^3)(260 \text{ Hz})^2(1.3 \times 10^{-3} \text{ m})^2(343 \text{ m/s}) = 1.0 \times 10^3 \text{ W/m}^2.$

We find the intensity level from

$\beta = 10 \log_{10}(I/I_0) = 10 \log_{10}[(1.0 \times 10^3 \text{ W/m}^2)/(10^{-12} \text{ W/m}^2)] = \boxed{150 \text{ dB.}}$

22. We find the intensity of the sound from

$\beta = 10 \log_{10}(I/I_0);$

$120 \text{ dB} = 10 \log(I/10^{-12} \text{ W/m}^2)$, which gives $I = 1.00 \text{ W/m}^2$.

We find the maximum displacement from

$I = 2\pi^2 \rho f^2 x_0^2 v$

$1.00 \text{ W/m}^2 = 2\pi^2(1.29 \text{ kg/m}^3)(131 \text{ Hz})^2 x_0^2(343 \text{ m/s})$, which gives $x_0 = \boxed{8.17 \times 10^{-5} \text{ m.}}$

23. From Figure 12–7, we see that a 100-Hz tone with an intensity level of 50 dB has a loudness level of 20 phons. If we follow this loudness curve to 6000 Hz, we find that the intensity level must be $\boxed{25 \text{ dB.}}$

24. From Figure 12–7, we see that an intensity level of 30 dB intersects the loudness level of 0 phons at 150 Hz.

At the high frequency end, the highest frequency is reached, so the range is $\boxed{150 \text{ Hz to } 20{,}000 \text{ Hz.}}$

25. (*a*) From Figure 12–7, we see that at 100 Hz the lowest threshold is 5×10^{-9} W/m^2 and the threshold of pain is 1 W/m^2. For the ratio we have

$(1\ \text{W/m}^2)/(5 \times 10^{-9}\ \text{W/m}^2) = \boxed{2 \times 10^8.}$

(*b*) From Figure 12–7, we see that at 5000 Hz the lowest threshold is 5×10^{-13} W/m^2 and the threshold of pain is 1×10^{-1} W/m^2. For the ratio we have

$(1 \times 10^{-1}\ \text{W/m}^2)/(5 \times 10^{-13}\ \text{W/m}^2) = \boxed{2 \times 10^{11}.}$

26. The wavelength of the fundamental frequency for a string is $\lambda = 2L$, so the speed of a wave on the string is

$v = \lambda f = 2(0.32\ \text{m})(440\ \text{Hz}) = 282\ \text{m/s}.$

We find the tension from

$v = [F_T/(m/L)]^{1/2};$

$282\ \text{m/s} = \{F_T/[(0.35 \times 10^{-3}\ \text{kg})/(0.32\ \text{m})]\}^{1/2}$, which gives $F_T = \boxed{87\ \text{N.}}$

27. The wavelength of the fundamental frequency for a string is $\lambda = 2L$. Because the speed of a wave on the string does not change, we have

$v = \lambda_1 f_1 = \lambda_2 f_2;$

$2(0.70\ \text{m})(330\ \text{Hz}) = 2L_2(440\ \text{Hz})$, which gives $L_2 = 0.525$ m.

Thus the finger must be placed 0.70 m – 0.525 m = $\boxed{0.17\ \text{m}}$ from the end.

28. We find the speed of sound at the temperature from

$v = (331 + 0.60T)\ \text{m/s} = [331 + (0.60)(21°\text{C})]\ \text{m/s} = 344\ \text{m/s}.$

The fundamental wavelength for an open pipe has an antinode at each end, so the wavelength is $\lambda = 2L$. We find the length from

$v = \lambda f;$

$344\ \text{m/s} = 2L(262\ \text{Hz})$, which gives $L = \boxed{0.66\ \text{m.}}$

29. (*a*) The empty soda bottle is approximately a closed pipe with a node at the bottom and an antinode at the top. The wavelength of the fundamental frequency is $\lambda = 4L$. We find the frequency from

$v = \lambda f;$

$343\ \text{m/s} = 4(0.15\ \text{m})f$, which gives $f = \boxed{570\ \text{Hz.}}$

(*b*) The length of the pipe is now $\frac{2}{3}$ of the original length. We find the frequency from

$v = \lambda f;$

$343\ \text{m/s} = 4(\frac{2}{3})(0.15\ \text{m})f$, which gives $f = \boxed{860\ \text{Hz.}}$

30. For an open pipe the wavelength of the fundamental frequency is $\lambda = 2L$.

We find the required lengths from

$v = \lambda f = 2Lf;$

$343\ \text{m/s} = 2L_{\text{lowest}}(20\ \text{Hz})$, which gives $L_{\text{lowest}} = 8.6$ m;

$343\ \text{m/s} = 2L_{\text{highest}}(20{,}000\ \text{Hz})$, which gives $L_{\text{highest}} = 8.6 \times 10^{-3}\ \text{m} = 8.6$ mm.

Thus the range of lengths is $\boxed{8.6\ \text{mm} < L < 8.6\ \text{m.}}$

31\. (*a*) For a closed pipe the wavelength of the fundamental frequency is $\lambda_1 = 4L$. We find the fundamental frequency from

$v = \lambda_1 f_1$;

343 m/s = 4(1.12 m)f_1, which gives f_1 = 76.6 Hz.

Only the odd harmonics are present, so we have

$f_3 = 3f_1$ = (3)(76.6 Hz) = 230 Hz;

$f_5 = 5f_1$ = (5)(76.6 Hz) = 383 Hz;

$f_7 = 7f_1$ = (7)(76.6 Hz) = 536 Hz.

(*b*) For an open pipe the wavelength of the fundamental frequency is $\lambda_1 = 2L$. We find the fundamental frequency from

$v = \lambda_1 f_1$;

343 m/s = 2(1.12 m)f_1, which gives f_1 = 153 Hz.

All harmonics are present, so we have

$f_2 = 2f_1$ = (2)(153 Hz) = 306 Hz;

$f_3 = 3f_1$ = (3)(153 Hz) = 459 Hz;

$f_4 = 4f_1$ = (4)(153 Hz) = 612 Hz.

32\. The fundamental wavelength for a flute has an antinode at each end, so the wavelength is $\lambda = 2L$. Uncovering a hole shortens the length. We find the new length from

$v = \lambda' f$;

343 m/s = $2L'$(294 Hz), which gives L' = 0.583 m.

Thus the hole must be 0.655 m – 0.583 m = 0.072 m = 7.2 cm from the end.

33\. Without the support the bridge is like a string, so the wavelength is $\lambda_1 = 2L$. Adding the support creates a node at the middle, so the wavelength is $\lambda_2 = L$. The wave speed in the bridge has not changed. We find the new frequency from

$v = \lambda_1 f_1 = \lambda_2 f_2$;

$2L$(4.0 Hz) = Lf_2, which gives f_2 = 8.0 Hz.

Yes, because this is higher than the expected earthquake frequencies, the modification did some good.

34\. We assume that the length, and thus the wavelength, has not changed. The frequency change is due to the change in the speed of sound. We find the speed of sound at 5°C:

$v = (331 + 0.60T)$ m/s = [331 + (0.60)(5°C)] m/s = 334 m/s.

For the percent change in frequency we have

$(\Delta f/f)100 = (\Delta v/v)100$ = [(334 m/s – 343 m/s)/(343 m/s)]100 = –2.6%.

35\. (*a*) We find the speed of sound at 15°C:

$v = (331 + 0.60T)$ m/s = [331 + (0.60)(15°C)] m/s = 340 m/s.

The wavelength of the fundamental frequency is $\lambda_1 = 4L$. We find the length from

$v = \lambda_1 f_1 = 4Lf_1$;

340 m/s = $4L$(294 Hz), which gives L = 0.289 m.

(*b*) For helium we have

$v = \lambda_1 f_1 = 4Lf_1$;

1005 m/s = 4(0.289 m)f_1, which gives f_1 = 869 Hz.

Note that we have no correction for the 5 C° temperature change.

36. (*a*) For an open pipe all harmonics are present, the difference in frequencies is the fundamental frequency, and all frequencies will be integral multiples of the difference. For a closed pipe only odd harmonics are present, the difference in frequencies is twice the fundamental frequency, and frequencies will not be integral multiples of the difference but odd multiples of half the difference. For this pipe we have

$\Delta f = 616\text{ Hz} - 440\text{ Hz} = 440\text{ Hz} - 264\text{ Hz} = 176\text{ Hz}.$

Because we have frequencies that are not integral multiples of this, the pipe is **closed.**

(*b*) The fundamental frequency is

$f_1 = \frac{1}{2}\Delta f = \frac{1}{2}(176\text{ Hz}) =$ **88 Hz.**

Note that the given frequencies are the third, fifth, and seventh harmonics.

37. For an open pipe all harmonics are present, the difference in frequencies is the fundamental frequency, and all frequencies will be integral multiples of the difference. Thus we have

$f_1 = \Delta f = 330\text{ Hz} - 275\text{ Hz} = 55\text{ Hz}.$

The wavelength of the fundamental frequency is $\lambda_1 = 2L$. We find the speed of sound from

$v = \lambda_1 f_1 = 2Lf_1 = 2(1.80\text{ m})(55\text{ Hz}) =$ **198 m/s.**

38. For an open pipe all harmonics are present, the difference in frequencies is the fundamental frequency, and all frequencies will be integral multiples of the difference. For a closed pipe only odd harmonics are present, the difference in frequencies is twice the fundamental frequency, and frequencies will not be integral multiples of the difference but odd multiples of half the difference. For this pipe we have

$\Delta f = 280\text{ Hz} - 240\text{ Hz} = 40\text{ Hz}.$

Because we have frequencies that are integral multiples of this, the pipe is **open,** with a fundamental frequency of 40 Hz.

The wavelength of the fundamental frequency is $\lambda_1 = 2L$. We find the length from

$v = \lambda_1 f_1 = 2Lf_1$;

$343\text{ m/s} = 2L(40\text{ Hz})$, which gives $L =$ **4.3 m.**

39. (*a*) The wavelength of the fundamental frequency is $\lambda_1 = 2L$. We find the fundamental frequency from

$v = \lambda_1 f_1 = 2Lf_1$;

$343\text{ m/s} = 2(2.44\text{ m})f_1$, which gives $f_1 = 70.3\text{ Hz}$.

We find the highest harmonic from

$f_n = nf_1$;

$20{,}000\text{ Hz} = n(70.3\text{ Hz})$, which gives $n = 284.5$.

Because all harmonics are present in an open pipe and the fundamental frequency is within the audible range, 284 harmonics are present, which means **283 overtones.**

(*b*) The wavelength of the fundamental frequency is $\lambda_1 = 4L$. We find the fundamental frequency from

$v = \lambda_1 f_1 = 4Lf_1$;

$343\text{ m/s} = 4(2.44\text{ m})f_1$, which gives $f_1 = 35.1\text{ Hz}$.

We find the highest harmonic from

$f_n = nf_1$;

$20{,}000\text{ Hz} = n(35.1\text{ Hz})$, which gives $n = 569$.

Because only the odd harmonics are present in a closed pipe and the fundamental frequency is within the audible range, 569 harmonics are present, which means **284 overtones.**

40. If we consider the ear canal as a closed pipe, the wavelength of the fundamental frequency is $\lambda_1 = 4L$. We take the most sensitive frequency as the fundamental frequency to find the length of the ear canal:

$v = \lambda_1 f_1 = 4Lf_1$;

$343\text{ m/s} = 4L(3500\text{ Hz})$, which gives $L = 2.5 \times 10^{-2}\text{ m} =$ **2.5 cm.**

41. Because the intensity is proportional to the square of both the amplitude and the frequency, we have

$I_2/I_1 = (A_2/A_1)^2(f_2/f_1)^2 = (0.4)^2(2)^2 =$ **0.64;**

$I_3/I_1 = (A_3/A_1)^2(f_3/f_1)^2 = (0.15)^2(3)^2 =$ **0.20.**

We find the relative intensity levels from

$\beta_2 - \beta_1 = 10 \log_{10}(I_2/I_1) = 10 \log_{10}(0.64) =$ **– 2 dB;**

$\beta_3 - \beta_1 = 10 \log_{10}(I_3/I_1) = 10 \log_{10}(0.20) =$ **– 7 dB.**

42. The beat frequency is the difference in frequencies, so we have

$\Delta f = f_{\text{beat}} = 1/(2.0 \text{ s}) =$ **0.50 Hz.**

Note that the frequency of the second string could be higher or lower.

43. The beat frequency is the difference in frequencies, so we have

$f_{\text{beat}} = \Delta f = 277 \text{ Hz} - 262 \text{ Hz} =$ **15 Hz.**

Because the human ear can detect beats up to about 8 Hz, this **will not be audible.**

If each frequency is reduced by a factor of 4, the beat frequency will be reduced by the same factor:

$f_{\text{beat}}' = \frac{1}{4}f_{\text{beat}} = \frac{1}{4}(15 \text{ Hz}) =$ **3.8 Hz.**

This will be audible.

44. The beat frequency is the difference in frequencies, so we have

$f_{\text{beat}} = \Delta f = f_2 - f_1;$

$\pm 5.0 \text{ kHz} = f_2 - 23.5 \text{ kHz}$, which gives $f_2 = 18.5 \text{ kHz}, 28.5 \text{ kHz}$.

Because the second whistle cannot be heard by humans, its frequency is **28.5 kHz.**

45. Because the beat frequency increases for the fork with the higher frequency, the string frequency must be below 350 Hz. Thus we have

$f = f_1 - f_{\text{beat1}} = f_2 - f_{\text{beat2}} = 350 \text{ Hz} - 4 \text{ Hz} = 355 \text{ Hz} - 9 \text{ Hz} =$ **346 Hz.**

46. (*a*) The beat frequency will be the difference in frequencies. Because the second frequency could be higher or lower, we have

$f_{\text{beat}} = \Delta f = f_2 - f_1;$

$\pm (3 \text{ beats})/(2.0 \text{ s}) = f_2 - 132 \text{ Hz}$, which gives $f_2 =$ **130.5 Hz, 133.5 Hz.**

(*b*) We assume that the change in tension does not change the mass density, so the velocity variation depends only on the tension. Because the wavelength does not change, we have

$\lambda = v_1/f_1 = v_2/f_2$, or $F_{T2}/F_{T1} = (f_2/f_1)^2$.

For the fractional change we have

$(F_{T2} - F_{T1})/F_{T1} = (F_{T2}/F_{T1}) - 1 = (f_2/f_1)^2 - 1 = [(f_2 - f_1)/f_1]^2;$

$\Delta F_T/F_T = [(\pm 1.5 \text{ Hz})/(132 \text{ Hz})]^2$, which gives $\Delta F_T/F_T = \pm 0.023 =$ **± 2.3%.**

47. The beat frequency will be the difference in frequencies. We assume that the change in tension does not change the mass density, so the velocity variation depends only on the tension. Because the wavelength does not change, we have

$\lambda = v_1/f_1 = v_2/f_2$, or $(F_{T2}/F_{T1})^{1/2} = (f_2/f_1)^2$.

If we write the changes as

$F_{T2} = F_{T1} + \Delta F_T$, and $f_2 = f_1 + \Delta f$, we get

$[(F_{T1} + \Delta F_T)/F_{T1}]^{1/2} = [(f_1 + \Delta f)/f_1]$, or $[1 + (\Delta F_T/F_{T1})]^{1/2} = 1 + (\Delta f/f_1)$.

Because $(\Delta F_T/F_{T1}) \ll 1$, we can expand the left hand side to get

$1 + \frac{1}{2}(\Delta F_T/F_{T1}) = 1 + (\Delta f/f_1)$, or

$\frac{1}{2}(\Delta F_T/F_{T1}) = \Delta f/f_1;$

$\frac{1}{2}(-0.015) = \Delta f/(294 \text{ Hz})$, which gives $\Delta f =$ **2.2 Hz.**

48. For destructive interference, the path difference is an odd multiple of half the wavelength.
   (*a*) Because the path difference is fixed, the lowest frequency corresponds to the longest wavelength. We find this from $\Delta L = \lambda_1/2$, so we have
   $f_1 = v/\lambda_1 = (343 \text{ m/s})/2(3.5 \text{ m} - 3.0 \text{ m}) = \boxed{343 \text{ Hz.}}$
   (*b*) We find the wavelength for the next frequency from $\Delta L = 3\lambda_2/2$, so we have
   $f_2 = v/\lambda_2 = (343 \text{ m/s})/[2(3.5 \text{ m} - 3.0 \text{ m})/3] = \boxed{1030 \text{ Hz.}}$
   We find the wavelength for the next frequency from $\Delta L = 5\lambda_3/2$, so we have
   $f_3 = v/\lambda_3 = (343 \text{ m/s})/[2(3.5 \text{ m} - 3.0 \text{ m})/5] = \boxed{1715 \text{ Hz.}}$

49. (*a*) We find the two frequencies from
   $f_1 = v/\lambda_1 = (343 \text{ m/s})/(2.64 \text{ m}) = 129.9 \text{ Hz};$
   $f_2 = v/\lambda_2 = (343 \text{ m/s})/(2.76 \text{ m}) = 124.3 \text{ Hz}.$
   The beat frequency is
   $f_{\text{beat}} = \Delta f = f_2 - f_1 = 129.9 \text{ Hz} - 124.3 \text{ Hz} = \boxed{5.6 \text{ Hz.}}$
   (*b*) The intensity maxima will travel with the speed of sound, so the separation of regions of maximum intensity is the "wavelength" of the beats:
   $\lambda_{\text{beat}} = v/f_{\text{beat}} = (343 \text{ m/s})/(5.6 \text{ Hz}) = \boxed{61 \text{ m.}}$

50. We see that there is no path difference for a listener on the bisector of the two speakers. The maximum path difference occurs for a listener on the line of the speakers. If this path difference is less than $\lambda/2$, there will be no location where destructive interference can occur. Thus we have
   path difference $= d_B - d_A = d \geq \lambda/2.$

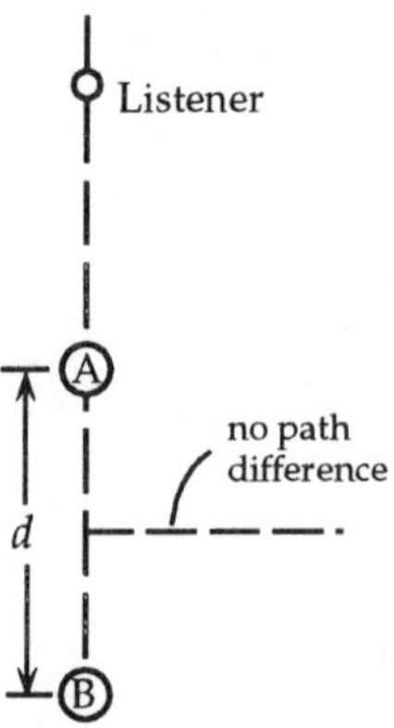

51. Because the police car is at rest, the wavelength traveling toward you is
   $\lambda_1 = v/f_0 = (343 \text{ m/s})/(1800 \text{ Hz}) = 0.191 \text{ m}.$
   (*a*) This wavelength approaches you at a relative speed of $v + v_L$. You hear a frequency
   $f_1 = (v + v_L)/\lambda_1 = (343 \text{ m/s} + 30.0 \text{ m/s})/(0.191 \text{ m}) = \boxed{1950 \text{ Hz.}}$
   (*b*) The wavelength approaches you at a relative speed of $v - v_L$. You hear a frequency
   $f_2 = (v - v_L)/\lambda_1 = (343 \text{ m/s} - 30.0 \text{ m/s})/(0.191 \text{ m}) = \boxed{1640 \text{ Hz.}}$

52. Because the bat is at rest, the wavelength traveling toward the object is
   $\lambda_1 = v/f_0 = (343 \text{ m/s})/(50{,}000 \text{ Hz}) = 6.86 \times 10^{-3} \text{ m}.$
   The wavelength approaches the object at a relative speed of $v - v_{\text{object}}$. The sound strikes and reflects from the object with a frequency
   $f_1 = (v - v_{\text{object}})/\lambda_1 = (343 \text{ m/s} - 25.0 \text{ m/s})/(6.86 \times 10^{-3} \text{ m}) = 46{,}360 \text{ Hz}.$
   This frequency can be considered emitted by the object, which is moving away from the bat. Because the wavelength behind a moving source increases, the wavelength approaching the bat is
   $\lambda_2 = (v + v_{\text{object}})/f_1 = (343 \text{ m/s} + 25.0 \text{ m/s})/(46{,}360 \text{ Hz}) = 7.94 \times 10^{-3} \text{ m}.$
   This wavelength approaches the bat at a relative speed of v, so the frequency received by the bat is
   $f_2 = v/\lambda_2 = (343 \text{ m/s})/(7.94 \times 10^{-3} \text{ m}) = \boxed{43{,}200 \text{ Hz.}}$

53. Because the wavelength in front of a moving source decreases, the wavelength from the approaching tuba is

$\lambda_1 = (v - v_1)/f_0 = (343 \text{ m/s} - 10.0 \text{ m/s})/(75 \text{ Hz}) = 4.44 \text{ m}.$

This wavelength approaches the stationary listener at a relative speed of v, so the frequency heard by the listener is

$f_1 = v/\lambda_1 = (343 \text{ m/s})/(4.44 \text{ m}) = 77 \text{ Hz}.$

Because the frequency from the stationary tube is unchanged, the beat frequency is

$f_{\text{beat}} = \Delta f = f_1 - f_0 = 77 \text{ Hz} - 75 \text{ Hz} =$ $\boxed{2 \text{ Hz.}}$

54. Because the wavelength in front of a moving source decreases, the wavelength from the approaching automobile is

$\lambda_1 = (v - v_1)/f_0 = (343 \text{ m/s} - 15 \text{ m/s})/f_0.$

This wavelength approaches the stationary listener at a relative speed of v, so the frequency heard by the listener is

$f_1 = v/\lambda_1 = (343 \text{ m/s})/[(343 \text{ m/s} - 15 \text{ m/s})/f_0] = (343 \text{ m/s})f_0/(328 \text{ m/s}).$

Because the frequency from the stationary automobile is unchanged, the beat frequency is

$f_{\text{beat}} = \Delta f = f_1 - f_0;$

$5.5 \text{ Hz} = [(343 \text{ m/s})f_0/(328 \text{ m/s})] - f_0$, which gives $f_0 =$ $\boxed{120 \text{ Hz.}}$

55. We convert the speed of the train: $(40 \text{ km/h})/(3.6 \text{ ks/h}) = 11.1 \text{ m/s}.$

Because the wavelength behind a moving source increases, the wavelength from the receding train is

$\lambda_1 = (v + v_1)/f_0 = (343 \text{ m/s} + 11.1 \text{ m/s})/(277 \text{ Hz}) = 1.28 \text{ m}.$

This wavelength approaches the stationary observer at a relative speed of v, so the frequency heard by the observer is

$f_1 = v/\lambda_1 = (343 \text{ m/s})/(1.28 \text{ m}) = 268 \text{ Hz}.$

Because the frequency from the stationary train is unchanged, the beat frequency is

$f_{\text{beat}} = |\Delta f| = |f_1 - f_0| = |268 \text{ Hz} - 277 \text{ Hz}| =$ $\boxed{9 \text{ Hz.}}$

56. Because the wavelength in front of a moving source decreases, the wavelength from the approaching source is

$\lambda_1 = (v - v_1)/f_0.$

This wavelength approaches the stationary listener at a relative speed of v, so the frequency heard by the listener is

$f_1 = v/\lambda_1 = v/[(v - v_1)/f_0] = vf_0/(v - v_1) = (343 \text{ m/s})(2000 \text{ Hz})/(343 \text{ m/s} - 15 \text{ m/s}) =$ $\boxed{2091 \text{ Hz.}}$

The wavelength from a stationary source is

$\lambda_2 = v/f_0.$

This wavelength approaches the moving receiver at a relative speed of $v + v_1$, so the frequency heard is

$f_2 = (v + v_1)/\lambda_2 = (v + v_1)/(v/f_0) = (v + v_1)f_0/v = (343 \text{ m/s} + 15 \text{ m/s})(2000 \text{ Hz})/(343 \text{ m/s}) =$ $\boxed{2087 \text{ Hz.}}$

The two frequencies are $\boxed{\text{not exactly the same, but close.}}$

For the other speeds we have,

at 150 m/s:

$f_3 = v/\lambda_1 = v/[(v - v_1)/f_0] = vf_0/(v - v_1) = (343 \text{ m/s})(2000 \text{ Hz})/(343 \text{ m/s} - 150 \text{ m/s}) =$ $\boxed{3554 \text{ Hz;}}$

$f_4 = (v + v_1)/\lambda_2 = (v + v_1)/(v/f_0) = (v + v_1)f_0/v = (343 \text{ m/s} + 150 \text{ m/s})(2000 \text{ Hz})/(343 \text{ m/s}) =$ $\boxed{2875 \text{ Hz.}}$

at 300 m/s:

$f_3 = v/\lambda_1 = v/[(v - v_1)/f_0] = vf_0/(v - v_1) = (343 \text{ m/s})(2000 \text{ Hz})/(343 \text{ m/s} - 1300 \text{ m/s}) =$ $\boxed{15950 \text{ Hz;}}$

$f_4 = (v + v_1)/\lambda_2 = (v + v_1)/(v/f_0) = (v + v_1)f_0/v = (343 \text{ m/s} + 300 \text{ m/s})(2000 \text{ Hz})/(343 \text{ m/s}) =$ $\boxed{3750 \text{ Hz.}}$

The Doppler formulas are not symmetric; the speed of the source creates a greater shift than the speed of the observer.

We can write the expression for the frequency from an approaching source as

$f_1 = v/\lambda_1 = vf_0/(v - v_1) = f_0/[1 - (v_1/v)].$

If $v_1 \ll v$, we use $1/[1 - (v_1/v)] \approx 1 + (v_1/v)$, so we have

$f_1 \approx f_0[1 + (v_1/v)],$

which is the expression $f_2 = (v + v_1)f_0/v$ for the frequency for a stationary source and a moving listener.

57. Because the source is at rest, the wavelength traveling toward the blood is

$$\lambda_1 = v/f_0.$$

This wavelength approaches the blood at a relative speed of $v - v_{\text{blood}}$. The ultrasound strikes and reflects from the blood with a frequency

$$f_1 = (v - v_{\text{blood}})/\lambda_1 = (v - v_{\text{blood}})/(v/f_0) = (v - v_{\text{blood}})f_0/v.$$

This frequency can be considered emitted by the blood, which is moving away from the source. Because the wavelength behind the moving blood increases, the wavelength approaching the source is

$$\lambda_2 = (v + v_{\text{blood}})/f_1 = (v + v_{\text{blood}})/[(v - v_{\text{blood}})f_0/v] = [1 + (v_{\text{blood}}/v)]v/f_0[1 - (v_{\text{blood}}/v)].$$

This wavelength approaches the source at a relative speed of v, so the frequency received by the source is

$$f_2 = v/\lambda_2 = v/\{[1 + (v_{\text{blood}}/v)]v/f_0[1 - (v_{\text{blood}}/v)]\} = [1 - (v_{\text{blood}}/v)]f_0/[1 + (v_{\text{blood}}/v)].$$

Because $v_{\text{blood}} \ll v$, we use $1/[1 + (v_{\text{blood}}/v)] \approx 1 - (v_{\text{blood}}/v)$, so we have

$$f_2 \approx f_0[1 - (v_{\text{blood}}/v)]^2 \approx f_0[1 - 2(v_{\text{blood}}/v)].$$

For the beat frequency we have

$$f_{\text{beat}} = f_0 - f_2 = 2(v_{\text{blood}}/v)f_0 = 2[(0.020 \text{ m/s})/(1540 \text{ m/s})](500 \times 10^3 \text{ Hz}) = \boxed{13 \text{ Hz.}}$$

58. Because the source is at rest, the wavelength traveling toward the heart is

$$\lambda_1 = v/f_0.$$

If we assume that the heart is moving away, this wavelength approaches the heart at a relative speed of $v - v_{\text{heart}}$. The ultrasound strikes and reflects from the heart with a frequency

$$f_1 = (v - v_{\text{heart}})/\lambda_1 = (v - v_{\text{heart}})/(v/f_0) = (v - v_{\text{heart}})f_0/v.$$

This frequency can be considered emitted by the heart, which is moving away from the source. Because the wavelength behind the moving heart increases, the wavelength approaching the source is

$$\lambda_2 = (v + v_{\text{heart}})/f_1 = (v + v_{\text{heart}})/[(v - v_{\text{heart}})f_0/v] = [1 + (v_{\text{heart}}/v)]v/f_0[1 - (v_{\text{heart}}/v)].$$

This wavelength approaches the source at a relative speed of v, so the frequency received by the source is

$$f_2 = v/\lambda_2 = v/\{[1 + (v_{\text{heart}}/v)]v/f_0[1 - (v_{\text{heart}}/v)]\} = [1 - (v_{\text{heart}}/v)]f_0/[1 + (v_{\text{heart}}/v)].$$

Because $v_{\text{heart}} \ll v$, we use $1/[1 + (v_{\text{heart}}/v)] \approx 1 - (v_{\text{heart}}/v)$, so we have

$$f_2 \approx f_0[1 - (v_{\text{heart}}/v)]^2 \approx f_0[1 - 2(v_{\text{heart}}/v)].$$

The maximum beat frequency occurs for the maximum heart velocity, so we have

$$f_{\text{beat}} = f_0 - f_2 = 2(v_{\text{heart}}/v)f_0;$$

$$600 \text{ Hz} = 2[v_{\text{heart}}/(1.54 \times 10^3 \text{ m/s})](2.25 \times 10^6 \text{ Hz}), \text{ which gives } v_{\text{heart}} = \boxed{0.205 \text{ m/s.}}$$

59. In Problem 58, we assumed that the heart is moving away from the source. Because the heart velocity is much less than the wave speed, the same beat frequency will occur when the heart is moving toward the source. Thus the maximum beat frequency will occur twice during each beat of the heart, so the heart-beat rate is

$$\tfrac{1}{2}(180 \text{ maxima/min}) = \boxed{90 \text{ beats/min.}}$$

60. From the diagram we see that

$$\sin\theta = (v_{\text{wave}}t)/(v_{\text{boat}}t) = v_{\text{wave}}/v_{\text{boat}};$$

$$\sin 20° = (2.0 \text{ m/s})/v_{\text{boat}}, \text{ which gives } v_{\text{boat}} = \boxed{5.8 \text{ m/s.}}$$

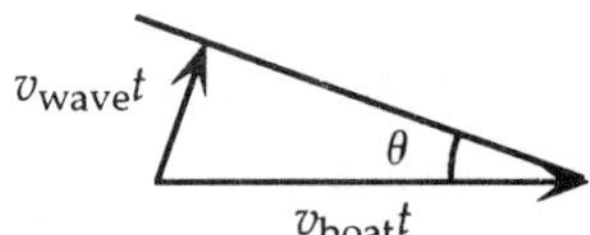

61. (*a*) From the definition of the Mach number, we have

$$v = (\text{Mach number})v_{\text{sound}} = (0.33)(343 \text{ m/s}) = \boxed{1.1 \times 10^2 \text{ m/s.}}$$

(*b*) We find the speed of sound from

$$v = (\text{Mach number})v_{\text{sound}};$$

$$(3000 \text{ km/h})/(3.6 \text{ ks/h}) = (3.2)v_{\text{sound}}, \text{ which gives } v_{\text{sound}} = \boxed{2.6 \times 10^2 \text{ m/s.}}$$

62. In a time t the shock wave moves a distance $v_{sound}t$ perpendicular to the wavefront. In the same time the object moves $v_{object}t$. We see from the diagram that

$$\sin\theta = (v_{sound}t)/(v_{object}t) = v_{sound}/v_{object}.$$

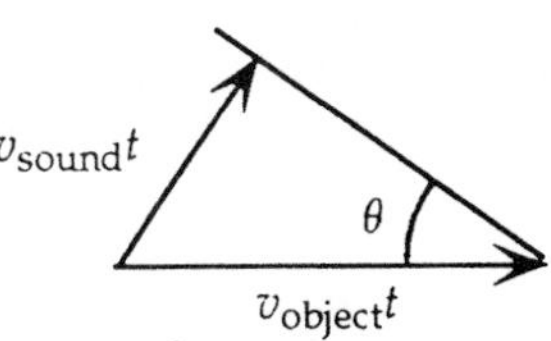

63. (*a*) We find the angle of the shock wave from

$$\sin\theta = v_{sound}/v_{object} = v_{sound}/(\text{Mach number})v_{sound}$$
$$= 1/(\text{Mach number}) = 1/2.3 = 0.435, \text{ so } \theta = \boxed{26°.}$$

(*b*) From the diagram we see that, when the shock wave hits the ground, we have

$$\tan\theta = h/v_{airplane}t = h/(\text{Mach number})v_{sound}t;$$
$$\tan 26° = (7100\text{ m})/(2.1)(310\text{ m/s})t, \text{ which gives } t = \boxed{23\text{ s.}}$$

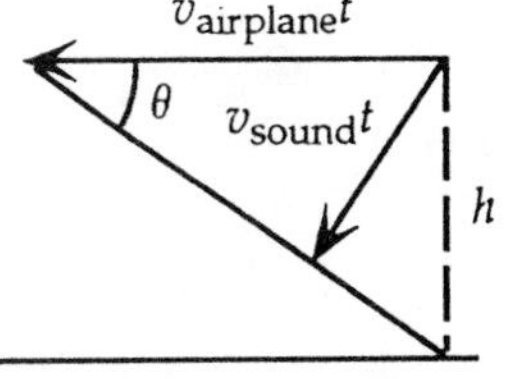

64. (*a*) From the definition of the Mach number, we have

$$v = (\text{Mach number})v_{sound};$$
$$(15{,}000\text{ km/h})/(3.6\text{ ks/h}) = (\text{Mach number})(35\text{ m/s}), \text{ which gives Mach number} = \boxed{120.}$$

(*b*) We find the angle of the shock wave from

$$\sin\theta = v_{sound}/v_{object} = (35\text{ m/s})/[(15{,}000\text{ km/h})/(3.6\text{ ks/h})] = 0.0084, \text{ so } \theta = 0.48°.$$

Thus the apex angle is $2\theta = \boxed{0.96°.}$

65. (*a*) We find the angle of the shock wave in the air from

$$\sin\theta = v_{sound}/v_{object} = (343\text{ m/s})/(8000\text{ m/s}) = 0.0429, \text{ so } \theta = \boxed{2.46°.}$$

(*b*) We find the angle of the shock wave in the ocean from

$$\sin\theta = v_{sound}/v_{object} = (1540\text{ m/s})/(8000\text{ m/s}) = 0.193, \text{ so } \theta = \boxed{11.1°.}$$

66. (*a*) From the diagram we see that, when the shock wave hits the ground, we have

$$\tan\theta = h/d_{plane} = (1.5\text{ km})/(2.0\text{ km}) = 0.75, \text{ so } \theta = \boxed{37°.}$$

(*b*) We find the speed of the plane from

$$\sin\theta = v_{sound}/v_{plane} = v_{sound}/(\text{Mach number})v_{sound} = 1/(\text{Mach number});$$
$$\sin 37° = 1/(\text{Mach number}), \text{ which gives Mach number} = \boxed{1.7.}$$

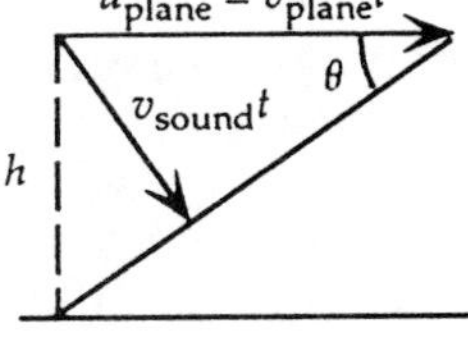

67. Because the frequency doubles for each octave, we have

$$f_{high}/f_{low} = 2^x;$$
$$20{,}000\text{ Hz})/(20\text{ Hz}) = 2^x, \text{ or}$$
$$\log_{10}(1000) = x\log_{10}(2). \text{ which gives } x \approx \boxed{10.}$$

68. We choose a coordinate system with origin at the top of the cliff, down positive, and $t = 0$ when the stone is dropped. If we call t_1 the time of fall for the stone, we have

$$y = y_{01} + v_{01}t_1 + \tfrac{1}{2}gt_1^2;$$
$$h = 0 + 0 + \tfrac{1}{2}gt_1^2, \text{ or } t_1 = (2h/g)^{1/2}.$$

For the time t_2 for the sound to reach the top of the cliff, we have

$$t_2 = h/v_{sound}.$$

Thus for the total time we have

$$t = t_1 + t_2 = (2h/g)^{1/2} + h/v_{sound};$$
$$3.5\text{ s} = [2h/(9.80\text{ m/s}^2)]^{1/2} + [h/(343\text{ m/s})].$$

This is a quadratic equation for $h^{1/2}$, which has the positive result $h^{1/2} = 7.39\text{ m}^{1/2}$, so the height of the cliff is $\boxed{55\text{ m.}}$

69. The intensity for a 0 dB sound is $I_0 = 10^{-12}$ W/m^2. For 1000 mosquitoes, we find the intensity level from
$\beta = 10 \log_{10}(I_1/I_0) = 10 \log_{10}(1000I_0/I_0) =$ **30 dB.**

70. Because the wavelength in front of a moving source decreases, the wavelength from the approaching car is
$\lambda_1 = (v - v_{car})/f_0$.
This wavelength approaches the stationary listener at a relative speed of v, so the frequency heard by the listener is
$f_1 = v/\lambda_1 = v/[(v - v_{car})/f_0] = vf_0/(v - v_{car})$.
Because the wavelength behind a moving source increases, the wavelength from the receding car is
$\lambda_2 = (v + v_{car})/f_0$.
This wavelength approaches the stationary listener at a relative speed of v, so the frequency heard by the listener is
$f_2 = v/\lambda_2 = v/[(v + v_{car})/f_0] = vf_0/(v + v_{car})$.
If the frequency drops by one octave, we have
$f_1/f_2 = [vf_0/(v - v_{car})]/[vf_0/(v + v_{car})] = (v + v_{car})/(v - v_{car}) = 2$, or
$2(v - v_{car}) = v + v_{car}$, which gives $v_{car} = \frac{1}{3}v = \frac{1}{3}(343 \text{ m/s}) = 114 \text{ m/s} =$ **410 km/h (257 mi/h).**

71. The wavelength of the fundamental frequency for a string is $\lambda = 2L$. Because the speed of a wave on the string does not change, we have
$v = \lambda_1 f_1 = \lambda_1' f_1'$;
$2L[(600 \text{ Hz})/3] = 2(0.60L)f_1'$, which gives $f_1' =$ **333 Hz.**

72. The tension and mass density of the string determines the velocity:
$v = (F_T/\mu)^{1/2}$.
Because the strings have the same length, the wavelengths are the same, so for the ratio of frequencies we have
$f_{n+1}/f_n = v_{n+1}/v_n = (\mu_n/\mu_{n+1})^{1/2} = 1.5$, or $\mu_{n+1}/\mu_n = (1/1.5)^2 = 0.444$.
With respect to the lowest string, we have
$\mu_{n+1}/\mu_1 = (0.444)^n$.
If we call the mass density of the lowest string 1, we have
1, 0.444, 0.198, 0.0878, 0.0389.

73. (*a*) The wavelength of the fundamental frequency for a string is $\lambda = 2L$, so the speed of a wave on the string is
$v = \lambda f = 2(0.32 \text{ m})(440 \text{ Hz}) =$ **2.8×10^2 m/s.**
We find the tension from
$v = (F_T/\mu)^{1/2}$;
$282 \text{ m/s} = [F_T/(5.5 \times 10^{-4} \text{ kg/m})]^{1/2}$, which gives $F_T =$ **44 N.**
(*b*) For a closed pipe the wavelength of the fundamental frequency is $\lambda = 4L$. We find the required length from
$v = \lambda f = 4Lf$;
$343 \text{ m/s} = 4L(440 \text{ Hz})$, which gives $L = 0.195 \text{ m} =$ **19.5 cm.**
(*c*) All harmonics are present in the string, so the first overtone is the second harmonic:
$f_2 = 2f_1 = 2(440 \text{ Hz}) =$ **880 Hz.**
Only the odd harmonics are present in a closed pipe, so the first overtone is the third harmonic:
$f_3 = 3f_1 = 3(440 \text{ Hz}) =$ **1320 Hz.**

74. We find the ratio of intensities from

$\Delta\beta = 10 \log_{10}(I_2/I_1)$;

$-10 \text{ dB} = 10 \log_{10}(I_2/I_1)$, which gives $I_2/I_1 = 0.10$.

If we assume uniform spreading of the sounds, the intensity is proportional to the power output, so we have

$P_2/P_1 = I_2/I_1$;

$P_2/(100 \text{ W}) = 0.10$, which gives $P_2 =$ $\boxed{10 \text{ W.}}$

75. At each resonant position the top of the water column is a node. Thus the distance between the two readings corresponds to the distance between adjacent nodes, or half a wavelength:

$\Delta L = \lambda/2$, or $\lambda = 2\,\Delta L$.

For the frequency we have

$f = v/\lambda = (343 \text{ m/s})/2(0.395 \text{ m} - 0.125 \text{ m}) =$ $\boxed{635 \text{ Hz.}}$

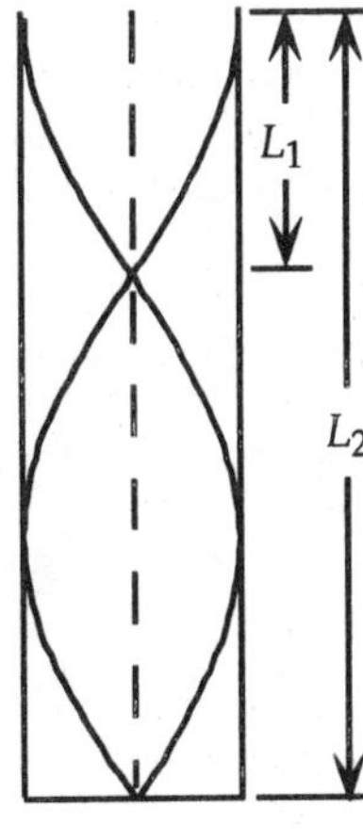

76. We find the gain from

$\beta = 10 \log_{10}(P_2/P_1) = 10 \log_{10}[(100 \text{ W})/(1 \times 10^{-3} \text{ W})] =$ $\boxed{50 \text{ dB.}}$

77. (*a*) Because both sources are moving toward the observer at the same speed, they will have the same Doppler shift, so the beat frequency will be $\boxed{0.}$

(*b*) Because the wavelength in front of a moving source decreases, the wavelength from the approaching loudspeaker is

$\lambda_1 = (v - v_{car})/f_0 = (343 \text{ m/s} - 10.0 \text{ m/s})/(200 \text{ Hz}) = 1.665 \text{ m}$.

This wavelength approaches the stationary listener at a relative speed of v, so the frequency heard by the listener is

$f_1 = v/\lambda_1 = (343 \text{ m/s})/(1.665 \text{ m}) = 206 \text{ Hz}$.

Because the wavelength behind a moving source decreases, the wavelength from the receding loudspeaker is

$\lambda_2 = (v + v_{car})/f_0 = (343 \text{ m/s} + 10.0 \text{ m/s})/(200 \text{ Hz}) = 1.765 \text{ m}$.

This wavelength approaches the stationary listener at a relative speed of v, so the frequency heard by the listener is

$f_2 = v/\lambda_2 = (343 \text{ m/s})/(1.765 \text{ m}) = 194 \text{ Hz}$.

The beat frequency is

$f_{beat} = \Delta f = 206 \text{ Hz} - 194 \text{ Hz} =$ $\boxed{12 \text{ Hz.}}$

Note that this frequency may be too high to be audible.

(*c*) Because both sources are moving away from the observer at the same speed, they will have the same Doppler shift, so the beat frequency will be $\boxed{0.}$

78. Because the wavelength in front of a moving source decreases, the wavelength from the approaching train whistle is

$\lambda_1 = (v - v_{train})/f_0$.

Because you are stationary, this wavelength approaches you at a relative speed of v, so the frequency heard by you is

$f_1 = v/\lambda_1 = v/[(v - v_{train})/f_0] = vf_0/(v - v_{train})$.

Because the wavelength behind a moving source increases, the wavelength from the receding train whistle is

$\lambda_2 = (v + v_{train})/f_0$.

This wavelength approaches you at a relative speed of v, so the frequency heard by you is

$f_2 = v/\lambda_2 = v/[(v + v_{train})/f_0] = vf_0/(v + v_{train})$.

For the ratio of frequencies, we get

$f_2/f_1 = (v - v_{train})/(v + v_{train})$;

$(486 \text{ Hz})/(522 \text{ Hz}) = (343 \text{ m/s} - v_{train})/(343 \text{ m/s} + v_{train})$, which gives $v_{train} = \boxed{12.3 \text{ m/s.}}$

79. For a closed pipe the wavelength of the fundamental frequency is $\lambda_{1pipe} = 4L$. Thus the frequency of the third harmonic is

$f_{3pipe} = 3f_{1pipe} = 3v/4L = 3(343 \text{ m/s})/4(0.75 \text{ m}) = 343 \text{ Hz}$.

This is the fundamental frequency for the string. The wavelength of the fundamental frequency for the string is $\lambda_{1string} = 2L$, so we have

$v_{string} = f_{1string}/2L = [F_T/(m/L)]^{1/2}$;

$(343 \text{ Hz})/2(0.75 \text{ m}) = \{F_T/[(0.00180 \text{ kg})/(0.75 \text{ m})]\}^{1/2}$, which gives $F_T = \boxed{6.4 \times 10^2 \text{ N.}}$

80. Because the source is at rest, the wavelength traveling toward the blood is

$\lambda_1 = v/f_0$.

This wavelength approaches the blood at a relative speed of $v - v_{blood}$. The ultrasound strikes and reflects from the blood with a frequency

$f_1 = (v - v_{blood})/\lambda_1 = (v - v_{blood})/(v/f_0) = (v - v_{blood})f_0/v$.

This frequency can be considered emitted by the blood, which is moving away from the source.

Because the wavelength behind the moving blood increases, the wavelength approaching the source is

$\lambda_2 = (v + v_{blood})/f_1 = (v + v_{blood})/[(v - v_{blood})f_0/v] = [1 + (v_{blood}/v)]v/f_0[1 - (v_{blood}/v)]$.

This wavelength approaches the source at a relative speed of v, so the frequency received by the source is

$f_2 = v/\lambda_2 = v/\{[1 + (v_{blood}/v)]v/f_0[1 - (v_{blood}/v)]\} = [1 - (v_{blood}/v)]f_0/[1 + (v_{blood}/v)]$.

Because $v_{blood} \ll v$, we use $1/[1 + (v_{blood}/v)] \approx 1 - (v_{blood}/v)$, so we have

$f_2 \approx f_0[1 - (v_{blood}/v)]^2 \approx f_0[1 - 2(v_{blood}/v)]$.

For the beat frequency we have

$f_{beat} = f_0 - f_2 = 2(v_{blood}/v)f_0 = 2[(0.32 \text{ m/s})/(1.54 \times 10^3 \text{ m/s})](5.50 \times 10^6 \text{ Hz}) = 2.29 \times 10^3 \text{ Hz} = \boxed{2.29 \text{ kHz.}}$

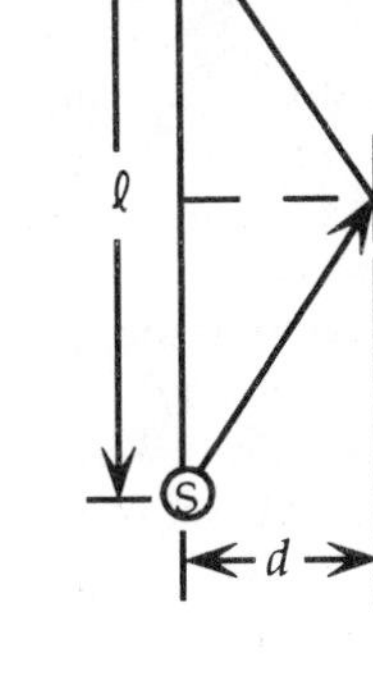

81. The initial path difference is

$$\Delta D = 2[(\ell/2)^2 + d^2]^{1/2} - \ell.$$

When the obstacle is moved Δd, the new path difference is

$$\Delta D' = 2[(\ell/2)^2 + (d + \Delta d)^2]^{1/2} - \ell.$$

Because $\Delta d \ll d$, we get

$$\Delta D' = 2[(\ell/2)^2 + d^2 + \Delta d(2d + \Delta d)]^{1/2} - \ell$$
$$\approx 2[(\ell/2)^2 + d^2 + 2d\,\Delta d]^{1/2} - \ell$$
$$= 2[(\ell/2)^2 + d^2]^{1/2}(1 + \{2d\,\Delta d/[(\ell/2)^2 + d^2]\}^{1/2}) - \ell.$$
$$\approx 2[(\ell/2)^2 + d^2]^{1/2}(1 + \tfrac{1}{2}\{2d\,\Delta d/[(\ell/2)^2 + d^2]\}) - \ell.$$
$$= 2[(\ell/2)^2 + d^2]^{1/2}(1 + \{d\,\Delta d/[(\ell/2)^2 + d^2]\}) - \ell.$$

To create destructive interference, the change in path difference must be $\lambda/2$, so we have

$$\Delta D' - \Delta D = \lambda/2;$$
$$2[(\ell/2)^2 + d^2]^{1/2}(1 + \{d\,\Delta d/[(\ell/2)^2 + d^2]\}) - \ell - \{2[(\ell/2)^2 + d^2]^{1/2} - \ell\} = \lambda/2;$$

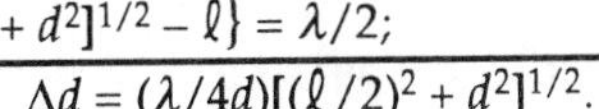

$2[(\ell/2)^2 + d^2]^{1/2} d\,\Delta d/[(\ell/2)^2 + d^2] = \lambda/2$, which gives $\boxed{\Delta d = (\lambda/4d)[(\ell/2)^2 + d^2]^{1/2}.}$

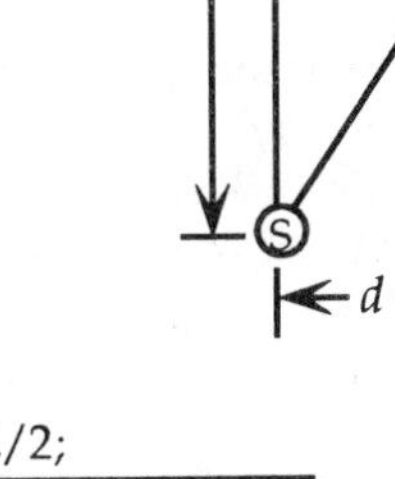

82. Because the wind velocity is a movement of the medium, it adds or subtracts from the speed of sound in the medium.

(*a*) Because the wind is blowing away from the observer, the effective speed of sound is $v - v_{\text{wind}}$. Therefore the wavelength traveling toward the observer is

$$\lambda_a = (v - v_{\text{wind}})/f_0.$$

This wavelength approaches the observer at a relative speed of $v - v_{\text{wind}}$. The observer will hear a frequency

$$f_a = (v - v_{\text{wind}})/\lambda_a = (v - v_{\text{wind}})/[(v - v_{\text{wind}})/f_0] = f_0 = \boxed{645\text{ Hz.}}$$

(*b*) Because the wind is blowing toward the observer, the effective speed of sound is $v + v_{\text{wind}}$. From the analysis in part (*a*), we see that there will be no change in the frequency: $\boxed{645\text{ Hz.}}$

(*c*) Because the wind is blowing perpendicular to the line toward the observer, the effective speed of sound is v. Because there is no relative motion of the whistle and the observer, there will be no change in the frequency: $\boxed{645\text{ Hz.}}$

(*d*) Because the wind is blowing perpendicular to the line toward the observer, the effective speed of sound is v. Because there is no relative motion of the whistle and the observer, there will be no change in the frequency: $\boxed{645\text{ Hz.}}$

(*e*) Because the wind is blowing toward the cyclist, the effective speed of sound is $v + v_{\text{wind}}$. Therefore the wavelength traveling toward the cyclist is

$$\lambda_e = (v + v_{\text{wind}})/f_0.$$

This wavelength approaches the cyclist at a relative speed of $v + v_{\text{wind}} + v_{\text{cycle}}$. The cyclist will hear a frequency

$$f_e = (v + v_{\text{wind}} + v_{\text{cycle}})/\lambda_e = (v + v_{\text{wind}} + v_{\text{cycle}})/[(v + v_{\text{wind}})/f_0]$$
$$= (v + v_{\text{wind}} + v_{\text{cycle}})f_0/(v + v_{\text{wind}})$$
$$= (343\text{ m/s} + 9.0\text{ m/s} + 13.0\text{ m/s})(645\text{ Hz})/(343\text{ m/s} + 9.0\text{ m/s}) = \boxed{669\text{ Hz.}}$$

(*f*) Because the wind is blowing perpendicular to the line toward the cyclist, the effective speed of sound is v. Therefore the wavelength traveling toward the cyclist is

$$\lambda_f = v/f_0.$$

This wavelength approaches the cyclist at a relative speed of $v + v_{\text{cycle}}$. The cyclist will hear a frequency

$$f_f = (v + v_{\text{cycle}})/\lambda_f = (v + v_{\text{cycle}})/(v/f_0)$$
$$= (v + v_{\text{cycle}})f_0/v = (343\text{ m/s} + 13.0\text{ m/s})(645\text{ Hz})/(343\text{ m/s}) = \boxed{669\text{ Hz.}}$$

83. When the person is equidistant from the sources, there is constructive interference. To move to a point where there is destructive interference, the path difference from the sources must be $\Delta L = n\lambda/2$. If we assume that $n = 1$, we find the frequency from

$f = v/\lambda = v/2\,\Delta L = (343\text{ m/s})/2(0.22\text{ m}) =$ **780 Hz.**

Because larger values of n would give a frequency out of the given range, this is the result.

84. (*a*) We find the wavelength from

$\lambda = v/f = (1560\text{ m/s})/(200{,}000\text{ Hz}) =$ **7.8×10^{-3} m.**

(*b*) We find the transit time from

$\Delta t = 2d/v = 2(100\text{ m})/(1560\text{ m/s}) =$ **0.13 s.**

85. If T is the period of the pulses and Δt is the duration of a pulse, and the moth must receive the entire reflected pulse, the maximum time for the signal to travel to the moth and return is $t_{max} = T - \Delta t$. We find the distance from

$2d_{max} = vt_{max} = (343\text{ m/s})(70.0\text{ ms} - 3.0\text{ ms})(10^{-3}\text{ s/ms})$, which gives $d_{max} =$ **11.5 m.**

86. Because the wavelength in front of a moving source decreases, the wavelength approaching the moth is

$\lambda_1 = (v - v_{bat})/f_0$.

The wavelength approaches the moth at a relative speed of $v + v_{moth}$. The sound strikes and reflects from the object with a frequency

$f_1 = (v + v_{moth})/\lambda_1 = (v + v_{moth})/[(v - v_{bat})/f_0] = (v + v_{moth})f_0/(v - v_{bat})$.

This frequency can be considered emitted by the moth, which is moving toward the bat. Because the wavelength in front of a moving source decreases, the wavelength approaching the bat is

$\lambda_2 = (v - v_{moth})/f_1 = (v - v_{moth})/[(v + v_{moth})f_0/(v - v_{bat})] = (v - v_{moth})(v - v_{bat})/(v + v_{moth})f_0$.

This wavelength approaches the bat at a relative speed of $v + v_{bat}$, so the frequency received by the bat is

$f_2 = (v + v_{bat})/\lambda_2 = (v + v_{bat})/[(v - v_{moth})/f_1] = (v + v_{bat})(v + v_{moth})f_0/(v - v_{moth})(v - v_{bat})$

$= (343.0\text{ m/s} + 8.0\text{ m/s})(343.0\text{ m/s} + 5.0\text{ m/s})(51.53\text{ kHz})/(343.0\text{ m/s} - 5.0\text{ m/s})(343.0\text{ m/s} - 8.0\text{ m/s})$

$=$ **55.39 kHz.**

CHAPTER 13

1. The number of atoms in a mass m is given by
   $N = m/Mm_{atom}$.
   Because the masses of the two rings are the same, for the ratio we have
   $N_{Au}/N_{Ag} = M_{Ag}/M_{Au} = 108/197 =$ $\boxed{0.548.}$

2. The number of atoms in a mass m is given by
   $N = m/Mm_{atom} = (3.4 \times 10^{-3}\text{ kg})/(63.5\text{ u})(1.66 \times 10^{-27}\text{ kg/u}) =$ $\boxed{3.2 \times 10^{22}\text{ atoms.}}$

3. (*a*) $T(°C) = (5/9)[T(°F) - 32] = (5/9)(68°F - 32) =$ $\boxed{20°C.}$
   (*b*) $T(°F) = (9/5)T(°C) + 32 = (9/5)(1800°C) + 32 =$ $\boxed{3272°F.}$

4. (*a*) $T(°F) = (9/5)T(°C) + 32 = (9/5)(-15°C) + 32 =$ $\boxed{5°F.}$
   (*b*) $T(°C) = (5/9)[T(°F) - 32] = (5/9)(-15°F - 32) =$ $\boxed{-26°C.}$

5. $T_1(°C) = (5/9)[T_1(°F) - 32] = (5/9)(136°F - 32) =$ $\boxed{58°C;}$
   $T_2(°C) = (5/9)[T_2(°F) - 32] = (5/9)(-129°F - 32) =$ $\boxed{-89°C.}$

6. Because the temperature and length are linearly related, we have
   $\Delta T/\Delta L = (100.0°C - 0.0°C)/(22.85\text{ cm} - 10.70\text{ cm}) = 8.23\text{ C°/cm}.$
   (*a*) $(T_1 - 0.0°C)/(16.70\text{ cm} - 10.70\text{ cm}) = 8.23\text{ C°/cm}$, which gives $T_1 =$ $\boxed{49.4°C.}$
   (*b*) $(T_2 - 0.0°C)/(20.50\text{ cm} - 10.70\text{ cm}) = 8.23\text{ C°/cm}$, which gives $T_2 =$ $\boxed{80.7°C.}$

7. Because 0° in the original scale corresponds to 100°C and 100° corresponds to 0°C, the conversion between the two scales is
   $[T(°C) - 0°C]/(100°C - 0°C) = (T - 100°)/(0° - 100°)$, or
   $T = 100° - T(°C) = 100° - 35° =$ $\boxed{65°.}$

8. We set $T(°F) = T(°C) = T$ in the conversion between the temperature scales:
   $T(°F) = (9/5)T(°C) + 32$
   $T = (9/5)T + 32$, which gives $T =$ $\boxed{-40°F = -40°C.}$

9. If the slabs are in contact at 20°C, at any temperature below this the expansion cracks will increase. Thus the expansion from 20°C to 50°C must eliminate the cracks. Any higher temperature will cause stress in the concrete. If the cracks have a width ΔL, we have
   $\Delta L = \alpha L_0 \Delta T = [12 \times 10^{-6}\,(\text{C°})^{-1}](14\text{ m})(50°C - 20°C) = 5.9 \times 10^{-3}\text{ m} =$ $\boxed{0.50\text{ cm.}}$

10. For the expansion ΔL, we have
    $\Delta L = \alpha L_0 \Delta T = [0.2 \times 10^{-6}\,(\text{C°})^{-1}](2.0\text{ m})(5.0°C) =$ $\boxed{2.0 \times 10^{-6}\text{ m.}}$

11. The unit for α is $(\text{C°})^{-1}$. In British units it will be $(\text{F°})^{-1}$, so we have
    $\alpha_B/\alpha = (\text{F°})^{-1}/(\text{C°})^{-1} = \text{C°/F°} =$ $\boxed{5/9.}$

12. We can treat the change in diameter as a simple change in length, so we have
$\Delta L = \alpha L_0 \Delta T;$
$1.869\text{ cm} - 1.871\text{ cm} = [12 \times 10^{-6}\,(\text{C}°)^{-1}](1.871\text{ cm})(T - 20°\text{C})$, which gives $T =$ $\boxed{-69°\text{C}.}$

13. For the expanded dimensions, we have
$\ell' = \ell(1 + \alpha\Delta T);\ w' = w(1 + \alpha\Delta T).$
Thus the change in area is
$\Delta A = A' - A = \ell'w' - \ell w = \ell w(1 + \alpha\Delta T)^2 - \ell w = \ell w[2\alpha\Delta T + (\alpha\Delta T)^2] = \ell w\alpha\Delta T(2 + \alpha\Delta T).$
Because $\alpha\Delta T \ll 2$, we have
$\Delta A = 2\alpha\ell w\,\Delta T.$

14. We consider a fixed mass of iron. The change in volume from the temperature change is
$\Delta V_T = \beta V_0\,\Delta T.$
The change in volume from the pressure change depends on the bulk modulus:
$\Delta V_P = -(V_0/B)\,\Delta P.$
Because the density is mass/volume, for the fractional change in the density we have
$$\begin{aligned}\Delta\rho/\rho_0 &= [(1/V) - (1/V_0)]/(1/V_0) = (V_0 - V)/V \approx (V_0 - V)/V_0 = -\Delta V/V_0\\ &= -\beta\,\Delta T + \Delta P/B\\ &= -[35 \times 10^{-6}\,(\text{C}°)^{-1}](2000°\text{C} - 20°\text{C}) + (5000\text{ atm})(1.013 \times 10^5\text{ N/m}^2\cdot\text{atm})/(90 \times 10^9\text{ N/m}^2)\\ &= -0.064 = \boxed{-6.4\%.}\end{aligned}$$

15. The contraction of the glass causes the enclosed volume to decrease as if it were glass. The volume of water that can be added is
$$\begin{aligned}\Delta V &= \Delta V_{\text{glass}} - \Delta V_{\text{water}} = V_0\beta_{\text{glass}}\,\Delta T - V_0\beta_{\text{water}}\,\Delta T = V_0(\beta_{\text{glass}} - \beta_{\text{water}})\Delta T\\ &= (350\text{ mL})[27 \times 10^{-6}\,(\text{C}°)^{-1} - 210 \times 10^{-6}\,(\text{C}°)^{-1}](20°\text{C} - 100°\text{C}) = \boxed{5.1\text{ mL.}}\end{aligned}$$

16. (*a*) The expansion of the container causes the enclosed volume to increase as if it were made of the same material as the container. The volume of water that was lost is
$\Delta V = \Delta V_{\text{water}} - \Delta V_{\text{container}} = V_0\beta_{\text{water}}\,\Delta T - V_0\beta_{\text{container}}\,\Delta T = V_0(\beta_{\text{water}} - \beta_{\text{container}})\Delta T;$
$(0.35\text{ g})/(0.98324\text{ g/mL}) = (65.50\text{ mL})[210 \times 10^{-6}\,(\text{C}°)^{-1} - \beta_{\text{container}}](60°\text{C} - 20°\text{C})$, which gives
$\beta_{\text{container}} = \boxed{74 \times 10^{-6}\,(\text{C}°)^{-1}.}$
(*b*) From Table 13–1, $\boxed{\text{aluminum}}$ is the most likely material.

17. We find the change in volume from
$$\begin{aligned}\Delta V &= V_0\beta\,\Delta T = \tfrac{4}{3}\pi r^3\beta\,\Delta T\\ &= \tfrac{4}{3}\pi(7.25\text{ cm})^3[1 \times 10^{-6}\,(\text{C}°)^{-1}](200°\text{C} - 30°\text{C}) = \boxed{0.27\text{ cm}^3.}\end{aligned}$$

18. We can treat the change in diameter as a simple change in length, so we have
$D = D_0(1 + \alpha\Delta T).$
The two objects must reach the same diameter:
$D = D_{0\text{brass}}(1 + \alpha_{\text{brass}}\,\Delta T) = D_{0\text{iron}}(1 + \alpha_{\text{iron}}\,\Delta T);$
$(8.753\text{ cm})\{1 + [19 \times 10^{-6}\,(\text{C}°)^{-1}](T - 20°\text{C})\} = (8.743\text{ cm})\{1 + [12 \times 10^{-6}\,(\text{C}°)^{-1}](T - 20°\text{C})\},$
which gives
$T = \boxed{-1.4 \times 10^2\,°\text{C}.}$

19. We assume that we can ignore the change in cross sectional area of the tube. The volume change of the fluid is the increased volume in the column:

$\Delta V = AL\beta\,\Delta T = A\,\Delta L$, or $\Delta L = L\beta\,\Delta T$.

When we compare this to the expression for linear expansion,

$\Delta L = L\alpha\,\Delta T$,

we see that $\alpha = \beta$.

20. (*a*) We consider a fixed mass of the substance. The change in volume from the temperature change is

$\Delta V = \beta V_0\,\Delta T$.

Because the density is mass/volume, for the fractional change in the density we have

$\Delta\rho/\rho = [(1/V) - (1/V_0)]/(1/V_0) = (V_0 - V)/V \approx (V_0 - V)/V_0 = -\Delta V/V_0 = -\beta\,\Delta T$,

which we can write

$\Delta\rho = -\beta\rho\,\Delta T$.

(*b*) For the lead sphere we have

$\Delta\rho/\rho = -[87 \times 10^{-6}\,(\mathrm{C°})^{-1}](40°\mathrm{C} - 25°\mathrm{C}) =$ $\boxed{-0.0057\ (0.57\%).}$

21. For the expanded dimensions of the rectangular solid, we have

$L = L_0(1 + \alpha\Delta T)$; $W = W_0(1 + \alpha\Delta T)$; $H = H_0(1 + \alpha\Delta T)$.

Thus the change in volume is

$$\begin{aligned}\Delta V &= V - V_0 = LWH - L_0W_0H_0\\ &= L_0(1 + \alpha\Delta T)W_0(1 + \alpha\Delta T)H_0(1 + \alpha\Delta T) - L_0W_0H_0 = L_0W_0H_0(1 + \alpha\Delta T)^3 - L_0W_0H_0\\ &= L_0W_0H_0[3\alpha\Delta T + 3(\alpha\Delta T)^2 + (\alpha\Delta T)^3] = L_0W_0H_0\alpha\Delta T[3 + 3(\alpha\Delta T) + (\alpha\Delta T)^2].\end{aligned}$$

Because $\alpha\Delta T \ll 1$, we have

$\Delta V = 3\alpha V_0\,\Delta T$.

When we compare this to the expression for volume expansion,

$\Delta V = \beta V_0\,\Delta T$,

we see that $\beta = 3\alpha$.

22. The increase in temperature will cause the length of the brass rod to increase. The period of the pendulum depends on the length,

$T = 2\pi(L/g)^{1/2}$,

so the period will be greater. This means the pendulum will make fewer swings in a day, so the clock will be slow and the clock will lose time.

We use T_C for the temperature to distinguish it from the period.

For the length of the brass rod, we have

$L = L_0(1 + \alpha\,\Delta T_C)$.

Thus the ratio of periods is

$T/T_0 = (L/L_0)^{1/2} = (1 + \alpha\,\Delta T_C)^{1/2}$.

Because $\alpha\,\Delta T_C$ is much less than 1, we have

$T/T_0 \approx 1 + \frac{1}{2}\alpha\,\Delta T_C$, or $\Delta T/T_0 = \frac{1}{2}\alpha\,\Delta T_C$.

The number of swings in a time t is $N = t/T$. For the same time t, the change in period will cause a change in the number of swings:

$\Delta N = (t/T) - (t/T_0) = t(T_0 - T)/TT_0 \approx -t(\Delta T/T_0)/T_0$,

because $T \approx T_0$. The time difference in one year is

$$\begin{aligned}\Delta t &= T_0\,\Delta N = -t(\Delta T/T_0) = -t(\tfrac{1}{2}\alpha\,\Delta T_C)\\ &= -(1\ \mathrm{yr})(3.16 \times 10^7\ \mathrm{s/yr})\tfrac{1}{2}[19 \times 10^{-6}\,(\mathrm{C°})^{-1}](25°\mathrm{C} - 17°\mathrm{C}) = -2.4 \times 10^3\ \mathrm{s} = \boxed{-40\ \mathrm{min.}}\end{aligned}$$

23. We find the change in radius from

$\Delta R = R\alpha\Delta T$.

Because the bearings are frictionless, angular momentum will be conserved:

$I_1\omega_1 = I_2\omega_2$, with $I = \frac{1}{2}mR^2$ for a solid cylinder.

For the fractional change in the angular velocity, we have

$\Delta\omega/\omega = (\omega_2 - \omega_1)/\omega_1 = [(I_1\omega_1/I_2) - \omega_1]/\omega_1 = (I_1 - I_2)/I_2$.

Because the mass is constant, we have

$\Delta\omega/\omega = [R^2 - (R + \Delta R)^2]/(R + \Delta R)^2 = -[2R\,\Delta R + (\Delta R)^2]/(R + \Delta R)^2 \approx -2R\,\Delta R\,/R^2 = -2\,\Delta R\,/R$.

From the temperature change, we have

$\Delta\omega/\omega = -2\alpha\Delta T = -2[25\times10^{-6}\,(\text{C}°)^{-1}](75.0°\text{C} - 20.0°\text{C}) = \boxed{-2.8\times10^{-3}\,(0.28\%).}$

24. The compressive strain must compensate for the thermal expansion. From the relation between stress and strain, we have

Stress = E(Strain) = $E\alpha\Delta T$.

When we use the ultimate strength of concrete, we have

$20\times10^6\,\text{N/m}^2 = (20\times10^9\,\text{N/m}^2)[12\times10^{-6}\,(\text{C}°)^{-1}](T - 10.0°\text{C})$, which gives $T = \boxed{93°\text{C.}}$

25. The compressive strain must compensate for the thermal expansion. From the relation between stress and strain, we have

Stress = E(Strain) = $E\alpha\Delta T$;

$F/A = (70\times10^9\,\text{N/m}^2)[25\times10^{-6}\,(\text{C}°)^{-1}](35°\text{C} - 15°\text{C}) = \boxed{3.5\times10^7\,\text{N/m}^2.}$

26. (*a*) The tensile strain must compensate for the thermal contraction. From the relation between stress and strain, we have

Stress = E(Strain) = $E\alpha\Delta T$;

$F/A = (200\times10^9\,\text{N/m}^2)[12\times10^{-6}\,(\text{C}°)^{-1}](-30°\text{C} - 30°\text{C}) = \boxed{-1.4\times10^8\,\text{N/m}^2.}$

(*b*) $\boxed{\text{No,}}$ because the ultimate strength of steel is $500\times10^6\,\text{N/m}^2 = 5.0\times10^8\,\text{N/m}^2$.

(*c*) For concrete we have

$F/A = (20\times10^9\,\text{N/m}^2)[12\times10^{-6}\,(\text{C}°)^{-1}](-30°\text{C} - 30°\text{C}) = \boxed{-1.4\times10^7\,\text{N/m}^2.}$

Because the ultimate tensile strength of concrete is $2\times10^6\,\text{N/m}^2$, it $\boxed{\text{will fracture.}}$

27. (*a*) As the iron band expands, the inside diameter will increase as if it were iron. We can treat the change in inside diameter as a simple change in length, so we have

$D = D_0(1 + \alpha\Delta T)$, or $D - D_0 = D_0\alpha\Delta T$;

134.122 cm – 134.110 cm = (134.110 cm)$[12\times10^{-6}\,(\text{C}°)^{-1}](T - 20°\text{C})$, which gives $T = \boxed{27°\text{C.}}$

(*b*) We assume that the barrel is rigid. The tensile strain in the circumference of the band is

$\pi\,\Delta D/\pi D_0 = \Delta D/D_0$,

which is the thermal strain. The tensile strain must compensate for the thermal contraction. From the relation between stress and strain, we have

Stress = E(Strain) = $E\alpha\Delta T$;

$F/(0.089\,\text{m})(0.0065\,\text{m}) = (100\times10^9\,\text{N/m}^2)(134.122\,\text{cm} - 134.110\,\text{cm})/(134.110\,\text{cm})$,

which gives

$F = \boxed{5.2\times10^3\,\text{N.}}$

28. (*a*) $T(\text{K}) = T(°\text{C}) + 273 = 86°\text{C} + 273 = \boxed{359\,\text{K.}}$

(*b*) $T(°\text{C}) = (5/9)[T(°\text{F}) - 32] = (5/9)(78°\text{F} - 32) = 26°\text{C}$.

$T(\text{K}) = T(°\text{C}) + 273 = 26°\text{C} + 273 = \boxed{299\,\text{K.}}$

(*c*) $T(\text{K}) = T(°\text{C}) + 273 = -100°\text{C} + 273 = \boxed{173\,\text{K.}}$

(*d*) $T(\text{K}) = T(°\text{C}) + 273 = 5500°\text{C} + 273 = \boxed{5773\,\text{K.}}$

29. On the Celsius scale, absolute zero is

$T(°C) = T(K) - 273.15 = 0\text{ K} - 273.15 = -273.15°C.$

On the Fahrenheit scale, we have

$T(°F) = (9/5)T(°C) + 32 = (9/5)(-273.15°C) + 32 =$ $\boxed{-459.7°F.}$

30. (*a*) $T_1(K) = T_1(°C) + 273 = 4000°C + 273 =$ $\boxed{4273\text{ K};}$

$T_2(K) = T_2(°C) + 273 = 15 \times 10^6\ °C + 273 =$ $\boxed{15 \times 10^6\text{ K.}}$

(*b*) The difference in each case is 273, so we have

Earth: $(273)(100)/(4273) =$ $\boxed{6.4\%;}$

Sun: $(273)(100)/(15 \times 10^6) =$ $\boxed{0.0018\%.}$

31. For the two states of the gas we can write

$P_1V_1 = nRT_1$ and $P_2V_2 = nRT_2$, which can be combined to give

$(P_2/P_1)(V_2/V_1) = T_2/T_1$;

$(4.00\text{ atm}/1.00\text{ atm})(V_2/3.00\text{ m}^3) = (311\text{ K}/273\text{ K})$, which gives $V_2 =$ $\boxed{0.854\text{ m}^3.}$

32. If we assume oxygen is an ideal gas, we have

$PV = nRT = (m/M)RT;$

$(1.013 \times 10^5\text{ Pa})V = [m/(32\text{ g/mol})(10^3\text{ kg/g})](8.315\text{ J/mol}\cdot\text{K})(273\text{ K})$, which gives

$m/V =$ $\boxed{1.43\text{ kg/m}^3.}$

33. The volume, temperature, and mass are constant. For the two states of the gas we can write

$P_1V = n_1RT = (m/M_1)RT$, and $P_2V = n_2RT = (m/M_2)RT$, which can be combined to give

$P_2/P_1 = M_1/M_2$;

$P_2/(3.65\text{ atm}) = (28\text{ g/mol})/(44\text{ g/mol})$, which gives $P_2 =$ $\boxed{2.32\text{ atm.}}$

34. (*a*) For the ideal gas we have

$PV = nRT = (m/M)RT;$

$(1.013 \times 10^5\text{ Pa})V = [(18.5\text{ kg})(10^3\text{ g/kg})/(28\text{ g/mol})](8.314\text{ J/mol}\cdot\text{K})(273\text{ K}),$

which gives $V =$ $\boxed{14.8\text{ m}^3.}$

(*b*) With the additional mass in the same volume, we have

$PV = nRT = (m/M)RT;$

$P(14.8\text{ m}^3) = [(18.5\text{ kg} + 15.0\text{ kg})(10^3\text{ g/kg})/(28\text{ g/mol})](8.314\text{ J/mol}\cdot\text{K})(273\text{ K}),$

which gives $P = 1.83 \times 10^5\text{ Pa} =$ $\boxed{1.81\text{ atm.}}$

35. (*a*) For the ideal gas we have

$PV = nRT;$

$(1.000\text{ atm} + 0.350\text{ atm})(1.013 \times 10^5\text{ Pa/atm})V = (25.50\text{ mol})(8.314\text{ J/mol}\cdot\text{K})(283\text{ K}),$

which gives $V =$ $\boxed{0.439\text{ m}^3.}$

(*b*) For the two states of the gas we can write

$P_1V_1 = nRT_1$ and $P_2V_2 = nRT_2$, which can be combined to give

$(P_2/P_1)(V_2/V_1) = T_2/T_1$;

$[(1.00\text{ atm} + 1.00\text{ atm})/(1.00\text{ atm} + 0.350\text{ atm})](1/2) = (T_2/283\text{ K}),$

which gives $T_2 = 210\text{ K} =$ $\boxed{-63°C.}$

36. If we assume argon is an ideal gas, we have

$PV = nRT = (m/M)RT;$

$P(50.0 \times 10^{-3}\text{ m}^3) = [(105.0\text{ kg})(10^3\text{ g/kg})/(40\text{ g/mol})](8.314\text{ J/mol}\cdot\text{K})(293\text{ K}),$

which gives $P = 1.28 \times 10^8\text{ Pa} =$ $\boxed{1.26 \times 10^3\text{ atm.}}$

37. The volume and pressure are constant. For the fractional change in the number of moles, we can write

$\Delta n/n = [(PV/RT_2) - (PV/RT_1)]/(PV/RT_1) = (T_1 - T_2)/T_2$

$= (288 \text{ K} - 311 \text{ K})/(311 \text{ K}) = \boxed{-0.074\ (7.4\%).}$

38. For the two states of the gas we can write

$P_1V_1 = nRT_1$ and $P_2V_2 = nRT_2$, which can be combined to give

$(P_2/P_1)(V_2/V_1) = T_2/T_1$;

$(P_2/2.45 \text{ atm})(48.8 \text{ L}/55.0 \text{ L}) = (323.2 \text{ K}/291.2 \text{ K})$, which gives $P_2 = \boxed{3.06 \text{ atm.}}$

39. If we assume water vapor is an ideal gas, we have

$PV = nRT = (m/M)RT$;

$(1.013 \times 10^5 \text{ Pa})V = [m/(18 \text{ g/mol})(10^3 \text{ kg/g})](8.315 \text{ J/mol}\cdot\text{K})(373 \text{ K})$, which gives

$m/V = \boxed{0.588 \text{ kg/m}^3.}$

This is less than the listed value of 0.598 kg/m^3. We expect a difference because the tendency of steam to form droplets indicates an attractive force, so $\boxed{\text{water vapor is not an ideal gas.}}$

40. For the two states of the gas we can write

$P_1V_1 = nRT_1$ and $P_2V_2 = nRT_2$, which can be combined to give

$(P_2/P_1)(V_2/V_1) = T_2/T_1$;

$(0.70 \text{ atm}/1.00 \text{ atm})(V_2/V_1) = (278.2 \text{ K}/293.2 \text{ K})$, which gives $V_2/V_1 = \boxed{1.4\times.}$

41. The pressure at the bottom of the lake is

$P_{\text{bottom}} = P_{\text{top}} + \rho gh = 1.013 \times 10^5 \text{ Pa} + (1000 \text{ kg/m}^3)(9.80 \text{ m/s}^2)(43.5 \text{ m}) = 5.28 \times 10^5 \text{ Pa}.$

For the two states of the gas we can write

$P_{\text{bottom}}V_{\text{bottom}} = nRT_{\text{bottom}}$, and $P_{\text{top}}V_{\text{top}} = nRT_{\text{top}}$, which can be combined to give

$(P_{\text{bottom}}/P_{\text{top}})(V_{\text{bottom}}/V_{\text{top}}) = T_{\text{bottom}}/T_{\text{top}}$;

$(5.28 \times 10^5 \text{ Pa}/1.013 \times 10^5 \text{ Pa})(1.00 \text{ cm}^3/V_{\text{top}}) = (278.7 \text{ K}/294.2 \text{ K})$, which gives $V_{\text{top}} = \boxed{5.50 \text{ cm}^3.}$

42. If we write the ideal gas law as $PV = NkT$, we have

$N/V = P/kT = (1.013 \times 10^5 \text{ Pa})/(1.38 \times 10^{-23} \text{ J/K})(273 \text{ K}) = \boxed{2.69 \times 10^{25} \text{ molecules/m}^3.}$

43. We find the number of moles from

$n = \rho V/M = (1.000 \text{ g/cm}^3)(1.000 \times 10^3 \text{ cm}^3)/(18 \text{ g/mol}) = \boxed{55.6 \text{ mol.}}$

For the number of molecules we have

$N = nN_A = (55.6 \text{ mol})(6.02 \times 10^{23} \text{ molecules/mol}) = \boxed{3.34 \times 10^{25} \text{ molecules.}}$

44. (*a*) We find the number of moles from

$n = \rho\tfrac{3}{4}V/M = \rho\tfrac{3}{4}4\pi R^2 d/M$

$= (1000 \text{ kg/m}^3)3\pi(6.4 \times 10^6 \text{ m})^2(3 \times 10^3 \text{ m})(10^3 \text{ g/kg})/(18 \text{ g/mol}) = \boxed{6 \times 10^{22} \text{ mol.}}$

(*b*) For the number of molecules we have

$N = nN_A = (6 \times 10^{22} \text{ mol})(6.02 \times 10^{23} \text{ molecules/mol}) = \boxed{4 \times 10^{46} \text{ molecules.}}$

45. The volume and mass are constant. For the two states of the gas we can write

$P_1V_1 = nRT_1$, and $P_2V_2 = nRT_2$, which can be combined to give

$P_2/P_1 = T_2/T_1$;

$(P_2/1.00 \text{ atm}) = (455 \text{ K}/293 \text{ K})$, which gives $P_2 = 1.55$ atm.

We find the length of a side of the box from

$V = L^3$;

$3.9 \times 10^{-2} \text{ m}^3 = L^3$, which gives $L = 0.339$ m.

The net force is the same on each side of the box. Because there is atmospheric pressure outside the box, the net force is

$F = A\,\Delta P = L^2(P_2 - P_1) = (0.339 \text{ m})^2(1.55 \text{ atm} - 1.00 \text{ atm})(1.013 \times 10^5 \text{ Pa/atm}) =$ **6.4 × 10³ N.**

Note that we have assumed no change in dimensions from the increased pressure.

46. We find the number of moles in one breath from

$PV = nRT$;

$(1.013 \times 10^5 \text{ Pa})(2.0 \times 10^{-3} \text{ m}^3) = n(8.315 \text{ J/mol}\cdot\text{K})(300 \text{ K})$, which gives $n = 8.02 \times 10^{-2}$ mol.

For the number of molecules in one breath we have

$N = nN_A = (8.02 \times 10^{-2} \text{ mol})(6.02 \times 10^{23} \text{ molecules/mol}) = 4.83 \times 10^{22}$ molecules.

We assume that all of the molecules from the last breath that Einstein took are uniformly spread throughout the atmosphere, so the fraction that are in one breath is given by $V/V_{\text{atmosphere}}$.

We find the number now in one breath from

$N'/N = V/V_{\text{atmosphere}} = V/4\pi R^2 h$;

$N'/(4.83 \times 10^{22} \text{ molecules}) = (2.0 \times 10^{-3} \text{ m}^3)/4\pi(6.4 \times 10^6 \text{ m})^2(10 \times 10^3 \text{ m})$,

which gives $N' =$ **20 molecules.**

47. The average kinetic energy depends on the temperature:

$\frac{1}{2}mv_{\text{rms}}^2 = \frac{3}{2}kT$, which gives

$v_{\text{rms}} = (3kT/m)^{1/2} = [3(1.38 \times 10^{-23} \text{ J/K})(6000 \text{ K})/(4 \text{ u})(1.66 \times 10^{-27} \text{ kg/u})]^{1/2} =$ **6.1 × 10³ m/s.**

48. (*a*) The average kinetic energy depends on the temperature:

$\frac{1}{2}mv_{\text{rms}}^2 = \frac{3}{2}kT = \frac{3}{2}(1.38 \times 10^{-23} \text{ J/K})(273 \text{ K}) =$ **5.65 × 10⁻²¹ J.**

(*b*) For the total translational kinetic energy we have

$\text{KE} = N(\frac{1}{2}mv_{\text{rms}}^2) = \frac{3}{2}nN_AkT$

$= \frac{3}{2}(1.00 \text{ mol})(6.02 \times 10^{23} \text{ molecules/mol})(1.38 \times 10^{-23} \text{ J/K})(293 \text{ K}) =$ **3.65 × 10³ J.**

49. We find the rms speed from

$v_{\text{rms}} = [(\Sigma v^2)/N]^{1/2} = [(6^2 + 2^2 + 4^2 + 6^2 + 0^2 + 4^2 + 1^2 + 8^2 + 5^2 + 3^2 + 7^2 + 8^2)/12]^{1/2} =$ **5.2.**

Note that this is greater than the average speed of 4.5.

50. The average kinetic energy depends on the temperature:

$\frac{1}{2}mv_{\text{rms}}^2 = \frac{3}{2}kT$.

We form the ratio at the two temperatures:

$(v_{\text{rms}2}/v_{\text{rms}1})^2 = T_2/T_1$;

$(2)^2 = T_2/273 \text{ K}$, which gives $T_2 = 1092 \text{ K} =$ **819°C.**

51. The average kinetic energy depends on the temperature:

$\frac{1}{2}mv_{rms}^2 = \frac{3}{2}kT.$

We form the ratio at the two temperatures:

$(v_{rms2}/v_{rms1})^2 = T_2/T_1;$

$(1.020)^2 = T_2/293.2$ K, which gives $T_2 = 304.9$ K = **31.7°C.**

52. The average kinetic energy depends on the temperature:

$\frac{1}{2}mv_{rms}^2 = \frac{3}{2}kT.$

We form the ratio at the two temperatures, and use the ideal gas law:

$(v_{rms2}/v_{rms1})^2 = T_2/T_1 = P_2V_2/P_1V_1 = P_2/P_1;$

$(v_{rms2}/v_{rms1})^2 = 2$, which gives $v_{rms2}/v_{rms1} =$ **$\sqrt{2}$.**

53. The average kinetic energy depends on the temperature:

$\frac{1}{2}mv_{rms}^2 = \frac{3}{2}kT.$

We form the ratio at the two temperatures, and use the ideal gas law:

$(m_2/m_1)(v_{rms2}/v_{rms1})^2 = T_2/T_1 = 1$, so we have

$(v_{rms2}/v_{rms1})^2 = (m_1/m_2)$, or $v_{rms2}/v_{rms1} = (m_1/m_2)^{1/2}$.

54. (*a*) We find the rms speed from

$v_{rms} = (3kT/m)^{1/2} = [3(1.38 \times 10^{-23}\text{ J/K})(310\text{ K})/(89\text{ u})(1.66 \times 10^{-27}\text{ kg/u})]^{1/2} =$ **2.9×10^2 m/s.**

(*b*) For the protein we have

$v_{rms} = (3kT/m)^{1/2} = [3(1.38 \times 10^{-23}\text{ J/K})(310\text{ K})/(50{,}000\text{ u})(1.66 \times 10^{-27}\text{ kg/u})]^{1/2}$

= **12 m/s.**

55. The average kinetic energy depends on the temperature:

$\frac{1}{2}mv_{rms}^2 = \frac{3}{2}kT$, or $kT/m = \frac{1}{3}v_{rms}^2$.

With M the mass of the gas and m the mass of a molecule, we write the ideal gas law as

$PV = NkT = (M/m)kT$, or

$P = (M/V)(kT/m) = \rho(\frac{1}{3}v_{rms}^2) = \frac{1}{3}\rho v_{rms}^2.$

56. The average kinetic energy depends on the temperature:

$\frac{1}{2}mv_{rms}^2 = \frac{3}{2}kT.$

We form the ratio for the two masses:

$(m_{235}/m_{238})(v_{rms235}/v_{rms238})^2 = T_2/T_1 = 1$, so we have

$(v_{rms235}/v_{rms238})^2 = (m_{238}/m_{235})$, or

$v_{rms235}/v_{rms238} = (m_{238}/m_{235})^{1/2} = \{[6(19\text{ u}) + 238\text{ u}]/[6(19\text{ u}) + 235\text{ u}]\}^{1/2} =$ **1.00429.**

57. (*a*) From Figure 13–23 we see that at atmospheric pressure CO_2 can exist as a **solid or vapor.**

(*b*) From Figure 13–23 we see that CO_2 may be a liquid when

5.11 atm < P < 73 atm, and – 56.6°C < T < 31°C.

58. From Figure 13–23 we see that CO_2 is a **vapor** at 30 atm and 30°C.

59. (a) From Figure 13–22 we see that the phase is **vapor.**

(*b*) From Figure 13–22 we see that the phase is **solid.**

60. The saturated vapor pressure at 25°C is 23.8 torr.
At 50% humidity the partial vapor pressure is
$P = 0.50P_s = 0.50(23.8 \text{ torr}) = 11.9 \text{ torr}.$
This corresponds to a saturated vapor pressure between 10°C and 15°C. We assume a linear change between values listed in Table 13–4 and use the values at 10°C and 15°C to find the dew point;
$T = 10°\text{C} + [(15°\text{C} - 10°\text{C})(11.9 \text{ torr} - 9.21 \text{ torr})/(12.8 \text{ torr} - 9.21 \text{ torr})] =$ **14°C.**

61. Water boils when the saturated vapor pressure equals the air pressure. From Table 13–4 we see that the saturated vapor pressure at 90°C is
$7.01 \times 10^4 \text{ Pa} =$ **0.69 atm.**

62. Water boils when the saturated vapor pressure equals the air pressure. From Table 13–4 we see that $0.85 \text{ atm} = 8.6 \times 10^4 \text{ Pa}$ lies between 90°C and 100°C. We use the values at 90°C and 100°C to find the temperature;
$T = 90°\text{C} + [(100°\text{C} - 90°\text{C})(8.6 \times 10^4 \text{ Pa} - 7.0 \times 10^4 \text{ Pa})/(10.1 \times 10^4 \text{ Pa} - 7.0 \times 10^4 \text{ Pa})] =$ **95°C.**

63. We find the saturated vapor pressure from
$P = (\text{RH})P_s;$
$530 \text{ Pa} = 0.40P_s$, which gives $P_s = 1325 \text{ Pa}$.
This saturated vapor pressure corresponds to a temperature between 10°C and 15°C. We use the values at 10°C and 15°C to find the dew point;
$T = 10°\text{C} + [(15°\text{C} - 10°\text{C})(1.325 \times 10^3 \text{ Pa} - 1.23 \times 10^3 \text{ Pa})/(1.71 \times 10^3 \text{ Pa} - 1.23 \times 10^3 \text{ Pa})] =$ **11°C.**

64. Water boils when the saturated vapor pressure equals the air pressure. From Table 13–4 we see that the saturated vapor pressure at 120°C is
$1.99 \times 10^4 \text{ Pa} =$ **1.96 atm.**

65. From Table 13–4 we see that the saturated vapor pressure at 25°C is 3.17×10^3 Pa. Water can evaporate until the saturated vapor pressure is reached. The initial pressure is (relative humidity)(saturated vapor pressure). Because the volume and temperature are constant, we use the ideal gas law to find the number of moles that can evaporate:
$\Delta n = \Delta P(V/RT) = (1 - \text{RH})P_s V/RT$
$= (1 - 0.80)(3.17 \times 10^3 \text{ Pa})(680 \text{ m}^3)/(8.315 \text{ J/mol}\cdot\text{K})(298 \text{ K}) = 174 \text{ mol}.$
We find the mass from
$m = M\,\Delta n = (18 \text{ g/mol})(174 \text{ mol}) = 3.1 \times 10^3 \text{ g} =$ **3.1 kg.**

66. From Table 13–4 we see that the saturated vapor pressure at 20°C is 2.33×10^3 Pa. The vapor pressure is (relative humidity)(saturated vapor pressure). Because the volume and temperature are constant, we use the ideal gas law to find the number of moles that must be removed:
$\Delta n = \Delta P(V/RT) = (\text{RH}_2 - \text{RH}_1)P_s V/RT$
$= (0.30 - 0.95)(2.33 \times 10^3 \text{ Pa})(95 \text{ m}^2)(2.8 \text{ m})/(8.315 \text{ J/mol}\cdot\text{K})(293 \text{ K}) = -165 \text{ mol}.$
We find the mass from
$m = M\,\Delta n = (18 \text{ g/mol})(165 \text{ mol}) = 3.0 \times 10^3 \text{ g} =$ **3.0 kg.**

67. Because there is only steam in the autoclave, the saturated vapor pressure is the gauge pressure plus atmospheric pressure:
$1.0 \text{ atm} + 1.0 \text{ atm} = 2.0 \text{ atm} = 2.03 \times 10^5 \text{ Pa}.$
From Table 13–4 we see this saturated vapor pressure occurs at **120°C.**

68. Because the outside air is at the dew point, its vapor pressure is the saturated vapor pressure at 5°C, which is 872 Pa. We consider a constant mass of gas, that is a fixed number of moles, that moves from outside to inside. Because the pressure is constant, we have

$V_2/T_2 = V_1/T_1$.

The vapor pressure inside is

$P_2 = nRT_2/V_2 = nRT_1/V_1 = P_1 = 872$ Pa.

The saturated vapor pressure at 25°C is 3170 Pa, so the relative humidity is

(872 Pa)/(3170 Pa) = 0.28 = $\boxed{28\%.}$

69. Because the air at the wet-bulb thermometer is at the dew point, the partial vapor pressure at 30°C is the saturated vapor pressure at 10°C, which is 1230 Pa. The saturated vapor pressure at 30°C is 4240 Pa, so the relative humidity is

(1230 Pa)/(4240 Pa) = 0.29 = $\boxed{29\%.}$

70. From the result of Example 13–19, we have

$$t = \left(\frac{\bar{C}}{\Delta C}\right)\frac{(\Delta x)^2}{D} = \frac{1}{2}\frac{(2.0\text{ m})^2}{4.0\times 10^{-5}\text{ m}^2/\text{s}} = 5.0\times 10^4\text{ /s} = \boxed{14\text{ h.}}$$

Our experience is that the odor is detected much sooner than this, which means that convection is much more important than diffusion.

71. From the result of Example 13–19, we have

$$t = \left(\frac{\bar{C}}{\Delta C}\right)\frac{(\Delta x)^2}{D} = \frac{(1.0\text{ mol/m}^3 + 0.40\text{ mol/m}^3)/2}{(1.0\text{ mol/m}^3 - 0.40\text{ mol/m}^3)}\frac{(15\times 10^{-6}\text{ m})^2}{95\times 10^{-11}\text{ m}^2/\text{s}} = \boxed{0.28\text{ s.}}$$

The diffusion speed is

$v = \Delta x/t = (15 \times 10^{-6}\text{ m})/(0.28\text{ s}) = \boxed{5.4 \times 10^{-5}\text{ m/s.}}$

We find the rms speed from

$v_{rms} = (3kT/m)^{1/2} = [3(1.38 \times 10^{-23}\text{ J/K})(293\text{ K})/(75\text{ u})(1.66 \times 10^{-27}\text{ kg/u})]^{1/2} = \boxed{3.1 \times 10^2\text{ m/s.}}$

72. (*a*) From the ideal gas law, we have

$C_0 = n/V = P/RT = (0.21)(1.013 \times 10^5\text{ Pa})/(8.315\text{ J/mol}\cdot\text{K})(293\text{ K}) = 8.7\text{ mol/m}^3$.

(*b*) The concentration change is

$\Delta C = C_0 - \frac{1}{2}C_0 = \frac{1}{2}C_0 = \frac{1}{2}(8.7\text{ mol/m}^3) = 4.35\text{ mol/m}^3$.

We find the diffusion rate from

$J = DA\,\Delta C/\Delta x = (1 \times 10^{-5}\text{ m}^2/\text{s})(2 \times 10^{-9}\text{ m}^2)(4.35\text{ mol/m}^3)/(2 \times 10^{-3}\text{ m}) = \boxed{4 \times 10^{-11}\text{ mol/s.}}$

(*c*) For the average concentration we have

$C_{av} = \frac{1}{2}(C_0 + \frac{1}{2}C_0) = \frac{3}{4}C_0 = 6.53\text{ mol/m}^3$.

We find the average time from

$t = N/J = C_{av}V/J = (6.53\text{ mol/m}^3)(2 \times 10^{-9}\text{ m}^2)(2 \times 10^{-3}\text{ m})/(4 \times 10^{-11}\text{ mol/s}) = \boxed{0.6\text{ s.}}$

73. (*a*) If we consider a cross-sectional area through which diffusion is taking place, the diffusion rate will be proportional to how frequently a molecule, during its random motion, will strike the area. This frequency is proportional to its average speed, for which we can use the rms speed. The temperature is related to the average kinetic energy of a molecule:

$\frac{1}{2}mv_{\rm rms}{}^2 = \frac{3}{2}kT,$ or $v_{\rm rms}{}^2 \propto 1/m.$

Thus we have

$J \propto v_{\rm rms} \propto 1/\sqrt{m}.$

(*b*) The molecular mass is proportional to the molecular weight, so the molecule with the lighter mass, which is **nitrogen,** will diffuse faster.

For the ratio of diffusion rates, we have

$J_{\rm N}/J_{\rm O} = (m_{\rm O}/m_{\rm N})^{1/2} = (M_{\rm O}/M_{\rm N})^{1/2} = (32/28)^{1/2} = 1.069,$ or **6.9%.**

74. (*a*) The tape will expand, so the numbers will be beyond the true length, so it will read **low.**

(*b*) The percentage error will be

$(\Delta L/L)(100) = (L\alpha\,\Delta T/L)(100) = (\alpha\,\Delta T)(100) = [12 \times 10^{-6}\,(\mathrm{C}°)^{-1}](34°\mathrm{C} - 20°\mathrm{C})(100) =$ **0.017%.**

75. Because we neglect the glass expansion, when the 300 mL cools to room temperature, the change in volume of the water will be

$\Delta V = \beta V_0\,\Delta T = [210 \times 10^{-6}\,(\mathrm{C}°)^{-1}](300\ \mathrm{mL})(20°\mathrm{C} - 80°\mathrm{C}) =$ **– 3.8 mL (1.3%).**

76. For the two conditions of the gas in the cylinder, we can write

$P_1V = n_1RT,$ and $P_2V = n_2RT,$ which can be combined to give

$P_2/P_1 = n_2/n_1\,;$

$(5\ \mathrm{atm}/28\ \mathrm{atm}) = n_2/n_1,$ which gives $n_2/n_1 =$ **0.18.**

77. The ideal gas law is

$P_1V = n_1RT = (m/M)RT,$

where m is the mass and M is the molecular weight. We write this as

$P = (m/V)RT/M = \rho RT/M.$

78. We use the ideal gas law:

$PV = NkT;$

$(1.013 \times 10^5\ \mathrm{Pa})(5\ \mathrm{m})(3\ \mathrm{m})(2.5\ \mathrm{m}) = N(1.38 \times 10^{-23}\ \mathrm{J/K})(293\ \mathrm{K}),$

which gives $N =$ **9.4×10^{26} molecules.**

We find the number of moles from

$n = N/N_A = (9.4 \times 10^{26}\ \text{molecules})/(6.02 \times 10^{23}\ \text{molecules/mol}) =$ **1.6×10^3 mol.**

79. We use the ideal gas law:

$PV = NkT,$ or

$N/V = P/kT = (1 \times 10^{-12}\ \mathrm{N/m^2})/(1.38 \times 10^{-23}\ \mathrm{J/K})(273\ \mathrm{K})(10^6\ \mathrm{cm^3/m^3}) =$ **3×10^2 molecules/cm³.**

80. The average kinetic energy depends on the temperature:

$\frac{1}{2}mv_{\rm rms}{}^2 = \frac{3}{2}kT,$ which gives

$v_{\rm rms} = (3kT/m)^{1/2} = [3(1.38 \times 10^{-23}\ \mathrm{J/K})(2.7\ \mathrm{K})/(1\ \mathrm{u})(1.66 \times 10^{-27}\ \mathrm{kg/u})]^{1/2} =$ **2.6×10^2 m/s.**

We find the pressure from

$PV = NkT,$ or

$P = (N/V)/kT = (1\ \mathrm{atom/cm^3})(10^6\ \mathrm{cm^3/m^3})/(1.38 \times 10^{-23}\ \mathrm{J/K})(2.7\ \mathrm{K}) =$ **$4 \times 10^{-17}\ \mathrm{N/m^2}$.**

81. The pressure at a depth h is

$P = P_0 + \rho gh = 1.013 \times 10^5$ Pa + (1000 kg/m^3)(9.80 m/s^2)(10 m) = 1.99×10^5 Pa.

For the two states of the gas we can write

$PV = nRT$, and $P_0V_0 = nRT$, which can be combined to give

$P/P_0 = V_0/V$;

(1.99×10^5 Pa/1.013×10^5 Pa) = (V_0/5.5 L), which gives V_0 = $\boxed{\text{11 L.}}$

This doubling of the volume is definitely $\boxed{\text{not advisable.}}$

82. (*a*) The volume of each gas is $\frac{1}{2}V_{\text{tank}}$. We use the ideal gas law:

$PV_O = N_OkT$;

(10 atm + 1 atm)(1.013×10^5 Pa/atm)$\frac{1}{2}$(3500 cm^3)/(10^6 cm^3/m^3) = N_O(1.38×10^{-23} J/K)(293 K),

which gives $\boxed{N_O = 4.8 \times 10^{23} \text{ molecules} = N_{He},}$ because the mass is not in the ideal gas law.

(*b*) The average kinetic energy depends on the temperature:

$\frac{1}{2}mv_{\text{rms}}^2 = \frac{3}{2}kT$.

Because the gases are at the same temperature, the ratio of average kinetic energies is $\boxed{1.}$

(c) We see that

$(v_{\text{rmsHe}}/v_{\text{rmsO}})^2 = m_O/m_{He}$ = (32 u)/(4 u), which gives $v_{\text{rmsHe}}/v_{\text{rmsO}}$ = $\boxed{2.8.}$

83. The rms speed is the speed of the nitrogen molecules, so we have

$\frac{1}{2}mv_{\text{rms}}^2 = \frac{3}{2}kT$;

(28 u)(1.66×10^{-27} kg/u)[(40,000 km/h)/(3.6 ks/h)]^2 = 3(1.38×10^{-23} J/K)T,

which gives T = $\boxed{1.4 \times 10^5 \text{ K.}}$

84. For the two conditions of the gas in the cylinder, we can write

$P_1V = nRT_1$, and $P_2V = nRT_2$, which can be combined to give

$P_2/P_1 = T_2/T_1$;

P_2/P_1 = (633 K/393 K) = $\boxed{1.61.}$

The average kinetic energy depends on the temperature:

$\frac{1}{2}mv_{\text{rms}}^2 = \frac{3}{2}kT$.

We form the ratio for the two temperatures:

$(v_{\text{rms2}}/v_{\text{rms1}})^2 = T_2/T_1$ = (633 K/393 K), which gives $v_{\text{rms2}}/v_{\text{rms1}}$ = $\boxed{1.27.}$

85. (*a*) The ideal gas law is

$PV = nRT = (m/M)RT$;

(1.013×10^5 Pa)(770 m^3) = [m/(29 g/mol)](8.315 J/mol·K)(293 K),

which gives $m = 9.2 \times 10^5$ g = $\boxed{9.2 \times 10^2 \text{ kg.}}$

(*b*) At the lower temperature we have

(1.013×10^5 Pa)(770 m^3) = [m/(29 g/mol)](8.315 J/mol·K)(263 K),

which gives $m = 1.03 \times 10^6$ g = 10.3×10^2 kg.

Thus the mass that has entered the house is 10.3×10^2 kg – 9.2×10^2 kg = $\boxed{1.1 \times 10^2 \text{ kg.}}$

86. (*a*) The ideal gas law is

$PV = nRT.$

At constant pressure we have

$P\,\Delta V = nR\,\Delta T$, or

$\Delta V/V = \Delta T/T.$

The thermal expansion is $\Delta V/V = \beta\,\Delta T$, so we see that $\beta = 1/T$.

At 293 K we have

$\beta = 1/293\text{ K} = 3.41 \times 10^{-3}\,(\text{C}°)^{-1}$, which agrees with the value of $3400 \times 10^{-6}\,(\text{C}°)^{-1}$ in Table 13–1.

(*b*) At constant temperature, when the pressure is changed we can write

$P_1V_1 = nRT = P_2V_2$, or $P_2 = P_1V_1/V_2$.

For the small fractional change in pressure, we have

$\Delta P/P = (P_2 - P_1)/P_1 = [(P_1V_1/V_2) - P_1]/P_1 = (V_1 - V_2)/V_2 = -\Delta V/V.$

From the definition of the bulk modulus, we have

$B = -\Delta P/(\Delta V/V) = -\Delta P/(-\Delta P/P) = P.$

87. The pressure on a small area of the surface can be considered to be due to the weight of the air column above the area:

$P = Mg/A.$

When we consider the total surface of the Earth, we have

$M_{\text{total}} = PA_{\text{total}}/g = P4\pi R^2/g$

$= (1.013 \times 10^5\text{ Pa})4\pi(6.37 \times 10^6\text{ m})^2/(9.80\text{ m/s}^2) = 5.27 \times 10^{18}\text{ kg}.$

If we use the average mass of an air molecule, we find the number of molecules from

$N = M_{\text{total}}/m = (5.27 \times 10^{18}\text{ kg})/(28.8\text{ u})(1.66 \times 10^{-27}\text{ kg}) =$ $\boxed{1.1 \times 10^{44}\text{ molecules.}}$

88. We use the ideal gas law to find the temperature:

$PV = nRT;$

$(4.2\text{ atm})(1.013 \times 10^5\text{ Pa/atm})(7.6\text{ m}^3) = (1800\text{ mol})(8.315\text{ J/mol}\cdot\text{K})T,$

which gives $T = 216$ K.

The average kinetic energy depends on the temperature:

$\frac{1}{2}mv_{\text{rms}}^2 = \frac{3}{2}kT$, which gives

$v_{\text{rms}} = (3kT/m)^{1/2} = [3(1.38 \times 10^{-23}\text{ J/K})(216\text{ K})/(28\text{ u})(1.66 \times 10^{-27}\text{ kg/u})]^{1/2} =$ $\boxed{439\text{ m/s.}}$

89. (*a*) The volume of the bulb is so much greater than the volume of mercury in the tube that we can ignore any changes in the tube dimensions. The additional length of the mercury column in the tube will be due to the increased expansion of the mercury in the bulb compared to the expansion of the glass bulb. The volume of mercury that adds to the length in the tube is

$\Delta V = \Delta V_{\text{mercury}} - \Delta V_{\text{glass}} = V_0\beta_{\text{mercury}}\,\Delta T - V_0\beta_{\text{glass}}\,\Delta T = V_0(\beta_{\text{mercury}} - \beta_{\text{glass}})\Delta T$

$= (0.255\text{ cm}^3)[180 \times 10^{-6}\,(\text{C}°)^{-1} - 9 \times 10^{-6}\,(\text{C}°)^{-1}](33.0°\text{C} - 11.5°\text{C}) = 9.43 \times 10^{-4}\text{ cm}^3.$

We find the additional length from

$L = \Delta V/A_{\text{tube}} = \Delta V/\frac{1}{4}\pi d^2 = 4(9.43 \times 10^{-4}\text{ cm}^3)/\pi(0.0140\text{ cm})^2 =$ $\boxed{6.09\text{ cm.}}$

(*b*) If we combine the two expressions from part (*a*), we get

$L\frac{1}{4}\pi d^2 = \Delta V = V_0(\beta_{\text{mercury}} - \beta_{\text{glass}})\Delta T$, which gives $\boxed{L = 4V_0(\beta_{\text{mercury}} - \beta_{\text{glass}})\Delta T/\pi d^2.}$

90. We find the molecular density from the ideal gas law:

$PV = NkT$, or

$N/V = P/kT = (1.013 \times 10^5\text{ Pa})/(1.38 \times 10^{-23}\text{ J/K})(273\text{ K})(10^6\text{ cm}^3/\text{m}^3) = 2.79 \times 10^{19}\text{ molecules/cm}^3.$

If we assume that each molecule occupies a cube of side a, we can find a, which is the average distance between molecules, from the volume occupied by a molecule:

$V/N = a^3;$

$1/(2.79 \times 10^{19}\text{ molecules/cm}^3) = a^3$, which gives $a =$ $\boxed{3.3 \times 10^{-7}\text{ cm.}}$

91. We find the total number of molecules from the total number of moles of water plus others:

$N = \{[(0.70)/(18 \text{ g/mol})] + [(0.30)/(10^5 \text{ g/mol})]\}(2.0 \times 10^{-12} \text{ kg})(6.02 \times 10^{23} \text{ molecules/mol})$
$= 4.7 \times 10^{10}$ molecules.

Because each molecule has an average kinetic energy of $\frac{3}{2}kT$, the total translational kinetic energy is

$\text{KE} = \frac{3}{2}NkT = \frac{3}{2}(4.7 \times 10^{10} \text{ molecules})(1.38 \times 10^{-23} \text{ J/K})(310 \text{ K}) = \boxed{3.0 \times 10^{-10} \text{ J.}}$

92. (*a*) If $V_{0\text{Fe}}$ is the volume of the iron and $V_{0\text{Hg}}$ is the volume of mercury that is displaced, the fraction of the volume of the iron that is submerged is

$f = V_{0\text{Hg}}/V_{0\text{Fe}}$.

Each volume will increase as the temperature is raised, so the new fraction will be

$f' = V_{0\text{Hg}}(1 + \beta_{\text{Hg}}\,\Delta T)/V_{0\text{Fe}}(1 + \beta_{\text{Fe}}\,\Delta T) = f(1 + \beta_{\text{Hg}}\,\Delta T)/(1 + \beta_{\text{Fe}}\,\Delta T)$.

Because $\beta\Delta T \ll 1$, we use the approximation

$1/(1 + \beta_{\text{Fe}}\,\Delta T) \approx 1 - \beta_{\text{Fe}}\,\Delta T$, so we have

$f' = f(1 + \beta_{\text{Hg}}\,\Delta T)(1 - \beta_{\text{Fe}}\,\Delta T) = f[1 + \beta_{\text{Hg}}\,\Delta T - \beta_{\text{Fe}}\,\Delta T - \beta_{\text{Hg}}\beta_{\text{Fe}}(\Delta T)^2]$, or

$(f' - f)/f = \beta_{\text{Hg}}\,\Delta T - \beta_{\text{Fe}}\,\Delta T - \beta_{\text{Hg}}\beta_{\text{Fe}}(\Delta T)^2$.

The last term is the product of two small numbers, so it can be neglected and we have

$\Delta f/f = \beta_{\text{Hg}}\,\Delta T - \beta_{\text{Fe}}\,\Delta T = (\beta_{\text{Hg}} - \beta_{\text{Fe}})\Delta T$.

Because $\beta_{\text{Hg}} > \beta_{\text{Fe}}$, the fraction that is submerged will increase, so the cube will float $\boxed{\text{lower.}}$

(*b*) For the percent change in the fraction submerged, we have

$(\Delta f/f)(100) = (\beta_{\text{Hg}} - \beta_{\text{Fe}})(\Delta T)(100)$
$= [180 \times 10^{-6}\,(\text{C}°)^{-1} - 35 \times 10^{-6}\,(\text{C}°)^{-1}](25°\text{C} - 0°\text{C})(100) = \boxed{0.36\%.}$

93. We treat the circumference of the band as a length, which will expand according to

$2\pi R = 2\pi R_0(1 + \alpha\Delta T)$, or

$R - R_0 = R_0\alpha\,\Delta T = (6.37 \times 10^6 \text{ m})[12 \times 10^{-6}\,(\text{C}°)^{-1}](30°\text{C} - 20°\text{C}) = \boxed{7.6 \times 10^2 \text{ m.}}$

CHAPTER 14

1. The required heat flow is
   $\Delta Q = mc\,\Delta T = (20.0\text{ kg})(4186\text{ J/kg}\cdot\text{C}°)(95°\text{C} - 15°\text{C}) = \boxed{6.7 \times 10^6\text{ J.}}$

2. For the work to equal the energy value of the food, we have
   $W = (750\text{ Cal})(4186\text{ J/Cal}) = \boxed{3.14 \times 10^6\text{ J.}}$

3. We find the temperature from
   $W = \Delta Q = mc\,\Delta T;$
   $7700\text{ J} = (3.0\text{ kg})(4186\text{ J/kg}\cdot\text{C}°)(T - 10.0°\text{C})$, which gives $T = \boxed{10.6°\text{C.}}$

4. (*a*) We convert the units:
   $(2500\text{ Cal/day})(4186\text{ J/Cal}) = \boxed{1.05 \times 10^7\text{ J/day.}}$
   (*b*) We convert the units:
   $(1.05 \times 10^7\text{ J/day})(1\text{ W}\cdot\text{s/J})/(3600\text{ s/h})(1000\text{ W/kW}) = \boxed{2.9\text{ kWh/day.}}$
   (*c*) $\text{Cost} = (\text{Rate})(\text{Energy}) = (\$0.10/\text{kWh})(2.9\text{ kWh/day}) = \boxed{\$0.29\text{ /day.}}$
   Difficult to feed yourself on 29 cents/day.

5. We convert the units:
   $\Delta Q = mc\,\Delta T;$
   $1\text{ Btu} = (1\text{ lb})(1.00\text{ kcal/kg}\cdot\text{C}°)(1\text{ F}°)(0.454\text{ kg/lb})(5\text{ C}°/9\text{ F}°) = 0.252\text{ kcal};$
   $1\text{ Btu} = (0.252\text{ kcal})(4186\text{ J/kcal}) = 1055\text{ J}.$

6. We find the time from
   $\Delta Q = mc\,\Delta T;$
   $(350\text{ W})t = (250\text{ mL})(1.00\text{ g/mL})(10^{-3}\text{ kg/g})(4186\text{ J/kg}\cdot\text{C}°)(50°\text{C} - 20°\text{C})$, which gives $t = \boxed{90\text{ s.}}$

7. We find the mass per hour from
   $\Delta Q/t = (m/t)c\,\Delta T;$
   $7200\text{ kcal/h} = (m/t)(1.00\text{ kcal/kg}\cdot\text{C}°)(50°\text{C} - 15°\text{C})$, which gives $m/t = \boxed{2.1 \times 10^2\text{ kg/h.}}$

8. The heat flow generated must equal the kinetic energy loss:
   $\Delta Q = \frac{1}{2}mv^2 = \frac{1}{2}(1000\text{ kg})[(100\text{ km/h})/(3.6\text{ ks/h})]^2(1\text{ kcal}/4186\text{ J}) = \boxed{92\text{ kcal.}}$

9. We find the specific heat from
   $\Delta Q = mc\,\Delta T;$
   $135 \times 10^3\text{ J} = (5.1\text{ kg})c\,(30°\text{C} - 20°\text{C})$, which gives $c = \boxed{2.6 \times 10^3\text{ J/kg}\cdot\text{C}°.}$

10. The required heat flow is
    $\Delta Q = mc\,\Delta T = (18\text{ L})(1.00\text{ kg/L})(4186\text{ J/kg}\cdot\text{C}°)(90°\text{C} - 20°\text{C}) = \boxed{5.3 \times 10^6\text{ J.}}$

11. Because ΔQ and ΔT are the same, from $\Delta Q = mc\,\Delta T$ we see that $m \propto 1/c$:
    $m_{\text{Cu}} : m_{\text{Al}} : m_{\text{w}} = 1/c_{\text{Cu}} : 1/c_{\text{Al}} : 1/c_{\text{w}}$
    $= 1/390 : 1/900 : 1/4186 = \boxed{10.7 : 4.65 : 1.}$

12. We convert the units:
$c = (4186\ \text{J/kg}\cdot\text{C}°)(1\ \text{Btu}/1055\ \text{J})(5\ \text{C}°/9\ \text{F}°)(0.454\ \text{kg/lb}) = \boxed{1.00\ \text{Btu/lb}\cdot\text{F}°.}$

13. To be equivalent the water and lead must have the same mc:
$m_{\text{lead}}c_{\text{lead}} = m_{\text{water}}c_{\text{water}};$
$(4.00\ \text{kg})(130\ \text{J/kg}\cdot\text{C}°) = m_{\text{water}}(4186\ \text{J/kg}\cdot\text{C}°)$, which gives $m_{\text{water}} = \boxed{0.124\ \text{kg.}}$

14. We find the temperature from
heat lost = heat gained;
$m_{\text{water}}c_{\text{water}}\,\Delta T_{\text{water}} = m_{\text{glass}}c_{\text{glass}}\,\Delta T_{\text{glass}};$
$(135\ \text{mL})(1.00\ \text{g/mL})(10^{-3}\ \text{kg/g})(4186\ \text{J/kg}\cdot\text{C}°)(T - 39.2°\text{C}) = (0.030\ \text{kg})(840\ \text{J/kg}\cdot\text{C}°)(39.2°\text{C} - 21.6°\text{C})$,
which gives $T = \boxed{40.0°\text{C.}}$

15. If all the kinetic energy in the hammer blows is absorbed by the nail, we have
$\text{KE} = 10(\tfrac{1}{2}mv^2) = mc\,\Delta T;$
$10[\tfrac{1}{2}(1.20\ \text{kg})(8.0\ \text{m/s})^2] = (0.014\ \text{kg})(450\ \text{J/kg}\cdot\text{C}°)\,\Delta T$, which gives $\Delta T = \boxed{61\ \text{C}°.}$

16. We find the temperature from
heat lost = heat gained;
$m_{\text{Cu}}c_{\text{Cu}}\,\Delta T_{\text{Cu}} = (m_{\text{Al}}c_{\text{Al}} + m_{\text{water}}c_{\text{water}})\,\Delta T_{\text{Al}};$
$(0.270\ \text{kg})(390\ \text{J/kg}\cdot\text{C}°)(300°\text{C} - T) =$
$[(0.150\ \text{kg})(900\ \text{J/kg}\cdot\text{C}°) + (0.820\ \text{kg})(4186\ \text{J/kg}\cdot\text{C}°)](T - 12.0°\text{C})$, which gives $T = \boxed{20.2°\text{C.}}$

17. We find the temperature from
heat lost = heat gained;
$m_{\text{poker}}c_{\text{poker}}\,\Delta T_{\text{poker}} = m_{\text{cider}}c_{\text{cider}}\,\Delta T_{\text{cider}};$
$(0.55\ \text{kg})(450\ \text{J/kg}\cdot\text{C}°)(700°\text{C} - T) = (0.50\ \text{L})(1.00\ \text{kg/L})(4186\ \text{J/kg}\cdot\text{C}°)(T - 15°\text{C})$,
which gives $T = \boxed{87°\text{C.}}$

18. We find the temperature from
heat lost = heat gained;
$m_{\text{shoe}}c_{\text{shoe}}\,\Delta T_{\text{shoe}} = (m_{\text{pot}}c_{\text{pot}} + m_{\text{water}}c_{\text{water}})\,\Delta T_{\text{pot}};$
$(0.40\ \text{kg})(450\ \text{J/kg}\cdot\text{C}°)(T - 25°\text{C}) =$
$[(0.30\ \text{kg})(450\ \text{J/kg}\cdot\text{C}°) + (1.60\ \text{L})(1.00\ \text{kg/L})(4186\ \text{J/kg}\cdot\text{C}°)](25°\text{C} - 20°\text{C})$,
which gives $T = \boxed{215°\text{C.}}$

19. We find the specific heat from
heat lost = heat gained;
$m_{\text{Fe}}c_{\text{Fe}}\,\Delta T_{\text{Fe}} = (m_{\text{Al}}c_{\text{Al}} + m_{\text{glycerin}}c_{\text{glycerin}})\,\Delta T_{\text{glycerin}};$
$(0.290\ \text{kg})(450\ \text{J/kg}\cdot\text{C}°)(180°\text{C} - 38°\text{C}) = [(0.100\ \text{kg})(900\ \text{J/kg}\cdot\text{C}°) + (0.250\ \text{kg})c_{\text{glycerin}}](38°\text{C} - 10°\text{C})$,
which gives $c_{\text{glycerin}} = \boxed{2.3 \times 10^3\ \text{J/kg}\cdot\text{C}°.}$

20. We find the specific heat from
heat lost = heat gained;
$m_x c_x\,\Delta T_x = (m_{\text{Al}}c_{\text{Al}} + m_{\text{water}}c_{\text{water}} + m_{\text{glass}}c_{\text{glass}})\,\Delta T_{\text{water}};$
$(0.195\ \text{kg})c_x\,(330°\text{C} - 35.0°\text{C}) =$
$[(0.100\ \text{kg})(900\ \text{J/kg}\cdot\text{C}°) + (0.150\ \text{kg})(4186\ \text{J/kg}\cdot\text{C}°) + (0.017\ \text{kg})(840\ \text{J/kg}\cdot\text{C}°)](35.0°\text{C} - 12.5°\text{C})$,
which gives $c_x = \boxed{286\ \text{J/kg}\cdot\text{C}°.}$

21. The water must be heated to the boiling temperature, 100°C. We find the time from

t = (heat gained)/P = [($m_{Al}c_{Al}$+ $m_{water}c_{water}$) ΔT_{water}]/P

= [(0.360 kg)(900 J/kg·C°) + (0.60 L)(1.00 kg/L)(4186 J/kg·C°)](100°C – 8.0C)/(750 W)

= 348 s = **5.8 min.**

22. The silver must be heated to the melting temperature, 961°C, and then melted. We find the heat required from

$Q = m_{silver}c_{silver}\,\Delta T_{silver} + m_{silver}L_{silver}$

= (16.50 kg)(230 J/kg·C°)(961°C – 20° C) + (16.50 kg)(0.88 × 10^5 J/kg) = **5.02 × 10^6 J.**

23. If we assume that the heat is required just to evaporate the water, we have

$Q = m_{water}L_{water}$;

180 kcal = m_{water}(539 kcal/kg), which gives m_{water} = **0.334 kg (0.334 L).**

24. The temperature of the liquid nitrogen will not change, so the ice will cool to 77 K. We find the amount of nitrogen that has evaporated from

heat lost = heat gained;

$m_{ice}c_{ice}\,\Delta T_{ice} = m_{nitrogen}L_{nitrogen}$;

(0.030 kg)(2100 J/kg·C°)(273 K – 77 K) = $m_{nitrogen}$(200 × 10^3 J/kg), which gives $m_{nitrogen}$ = **0.062 kg .**

Note that a change of 1 K is equal to a change of 1 C°.

25. The temperature of the ice will rise to 0°C, at which point melting will occur, and then the resulting water will rise to the final temperature. We find the mass of the ice cube from

heat lost = heat gained;

$(m_{Al}c_{Al} + m_{water}c_{water})\,\Delta T_{Al} = m_{ice}(c_{ice}\,\Delta T_{ice} + L_{ice} + c_{water}\,\Delta T_{water})$;

[(0.100 kg)(900 J/kg·C°) + (0.300 kg)(4186 J/kg·C°)](20°C – 17°C) =

m_{ice}{(2100 J/kg·C°)[0°C – (– 8.5°C)] + (3.33 × 10^5 J/kg) + (4186 J/kg·C°)(17°C – 0°C)},

which gives m_{ice} = 9.6 × 10^3 kg = **9.6 g.**

26. (*a*) We find the heat required to reach the boiling point from

$Q_1 = (m_{Fe}c_{Fe} + m_{water}c_{water})\Delta T$

= [(230 kg)(450 J/kg·C°) + (830 kg)(4186 J/kg·C°)](100°C – 20° C) = 2.86 × 10^8 J.

We find the time from

$t_1 = Q_1/P$ = (2.86 × 10^8 J)/(52,000 × 10^3 J/h) = **5.5 h.**

(*b*) There is no change in the temperature, so the additional heat required to change the water into steam is

$Q_2 = m_{water}L_{steam}$

= (830 kg)(22.6 × 10^5 J/kg) = 1.88 × 10^9 J.

We find the additional time from

$t_2 = Q_2/P$ = (1.88 × 10^9 J)/(52,000 × 10^3 J/h) = 36.1 h.

Thus the total time required is

$t = t_1 + t_2$ = 5.5 h + 36.1 h = **41.6 h.**

27. We use the heat of vaporization at body temperature: 585 kcal/kg.

If all of the energy supplied by the bicyclist evaporates the water, we have

$Q = m_{water}L_{water}$

= (8.0 L)(1.00 kg/L)(585 kcal/kg) = **4.7 × 10^3 kcal.**

28. The steam will condense at 100°C, and then the resulting water will cool to the final temperature. The ice will melt at 0°C, and then the resulting water will rise to the final temperature. We find the mass of the steam required from

heat lost = heat gained;

$m_{\text{steam}}(L_{\text{steam}} + c_{\text{water}}\,\Delta T_1) = m_{\text{ice}}(L_{\text{ice}} + c_{\text{water}}\,\Delta T_2)$;

$m_{\text{steam}}[(22.6 \times 10^5\ \text{J/kg}) + (4186\ \text{J/kg}\cdot\text{C°})](100°\text{C} - 20°\text{C}) =$
$(1.00\ \text{kg})[(3.33 \times 10^5\ \text{J/kg}) + (4186\ \text{J/kg}\cdot\text{C°})(20°\text{C} - 0°\text{C})]$,

which gives $m_{\text{steam}} =$ **0.16 kg.**

29. We find the latent heat of fusion from

heat lost = heat gained;

$(m_{\text{Al}}c_{\text{Al}} + m_{\text{water}}c_{\text{water}})\,\Delta T_{\text{water}} = m_{\text{Hg}}(L_{\text{Hg}} + c_{\text{Hg}}\,\Delta T_{\text{Hg}})$

$[(0.620\ \text{kg})(900\ \text{J/kg}\cdot\text{C°}) + (0.400\ \text{kg})(4186\ \text{J/kg}\cdot\text{C°})](12.80°\text{C} - 5.06°\text{C}) =$
$(1.00\ \text{kg})\{L_{\text{Hg}} + (138\ \text{J/kg}\cdot\text{C°})[5.06°\text{C} - (-39.0°\text{C})]\}$,

which gives $L_{\text{Hg}} =$ **1.12×10^4 J/kg.**

30. We assume that the water created by the melting of the ice stays at 0°C. Because the work done by friction, which decreases the kinetic energy, generates the heat flow, we have

$Q = \frac{1}{2}(\Delta\text{KE})$;

$m_{\text{ice}}L_{\text{ice}} = \frac{1}{2}(\frac{1}{2}mv^2)$;

$m_{\text{ice}}(3.33 \times 10^5\ \text{J/kg}) = \frac{1}{4}(54.0\ \text{kg})(6.4\ \text{m/s})^2$, which gives $m_{\text{ice}} = 1.7 \times 10^{-3}\ \text{kg} =$ **1.7 g.**

31. The work done by friction, which decreases the kinetic energy, generates the heat flow. In general a fraction of the heat flow is used to raise the temperature of the lead bullet and then melt the lead bullet. The larger this fraction is, the smaller the bullet velocity needed. We determine the minimum muzzle velocity by assuming that this fraction is 1:

$Q = \Delta\text{KE}$;

$m_{\text{lead}}(c_{\text{lead}}\,\Delta T + L_{\text{lead}}) = \frac{1}{2}m_{\text{lead}}v^2$;

$[(130\ \text{J/kg}\cdot\text{C°})(327°\text{C} - 20°\text{C}) + 0.25 \times 10^5\ \text{J/kg}] = \frac{1}{2}v^2$, which gives $v =$ **360 m/s.**

32. The strong gusty winds will provide heat convection, so the temperature at the outside of the window will be the external air temperature. We find the rate of heat flow from

$\Delta Q/\Delta t = kA(\Delta T/L)$

$= (0.84\ \text{J/s}\cdot\text{m}\cdot\text{C°})(3.0\ \text{m}^2)[15°\text{C} - (-5°\text{C})]/(3.2 \times 10^{-3}\ \text{m}) =$ **1.6×10^4 W.**

33. (*a*) We find the radiated power from

$\Delta Q/\Delta t = e\sigma AT^4$

$= (0.35)(5.67 \times 10^{-8}\ \text{W/m}^2\cdot\text{K}^4)4\pi(0.22\ \text{m})^2(298\ \text{K})^4 =$ **95 W.**

(*b*) We find the net flow rate from

$\Delta Q/\Delta t = e\sigma A(T_2^4 - T_1^4)$

$= (0.35)(5.67 \times 10^{-8}\ \text{W/m}^2\cdot\text{K}^4)4\pi(0.22\ \text{m})^2[(298\ \text{K})^4 - (268\ \text{K})^4] =$ **33 W.**

34. We find the distance for the conduction from

$\Delta Q/\Delta t = kA(\Delta T/L)$

$200\ \text{W} = (0.2\ \text{J/s}\cdot\text{m}\cdot\text{C°})(1.5\ \text{m}^2)(0.50\ \text{C°})/L$, which gives $L = 7.5 \times 10^{-4}\ \text{m} =$ **0.75 mm.**

35. We find the radiated power from

$\Delta Q/\Delta t = e_{\text{Betelgeuse}}\sigma A_{\text{Betelgeuse}}T_{\text{Betelgeuse}}^4 = e_{\text{Betelgeuse}}\sigma 4\pi r_{\text{Betelgeuse}}^2 T_{\text{Betelgeuse}}^4.$

$= (1.0)(5.67 \times 10^{-8}\ \text{W/m}^2\cdot\text{K}^4)4\pi(3.1 \times 10^{11}\ \text{m})^2(2800\ \text{K})^4 = \boxed{4.2 \times 10^{30}\ \text{W.}}$

If we form the ratio for Betelgeuse and the Sun, we have

$P_{\text{Betelgeuse}}/P_{\text{Sun}} = (r_{\text{Betelgeuse}}/r_{\text{Sun}})^2(T_{\text{Betelgeuse}}/T_{\text{Sun}})^4$, or

$P_{\text{Betelgeuse}} = (3.1 \times 10^{11}\ \text{m}/7.0 \times 10^8\ \text{m})^2(\tfrac{1}{2})^4 P_{\text{Sun}} = \boxed{1.2 \times 10^4\ P_{\text{Sun}}.}$

36. (*a*) The cross-sectional area of the beam that falls on an area A is $A \cos\theta$. Thus the rate at which energy is absorbed is

$P = IeA\cos\theta = (1000\ \text{W/m}^2)(0.75)(225 \times 10^{-4}\ \text{m}^2)\cos 40°$

$= \boxed{13\ \text{W.}}$

(*b*) With the new emissivity we have

$P = IeA\cos\theta = (1000\ \text{W/m}^2)(0.20)(225 \times 10^{-4}\ \text{m}^2)\cos 40°$

$= \boxed{3.4\ \text{W.}}$

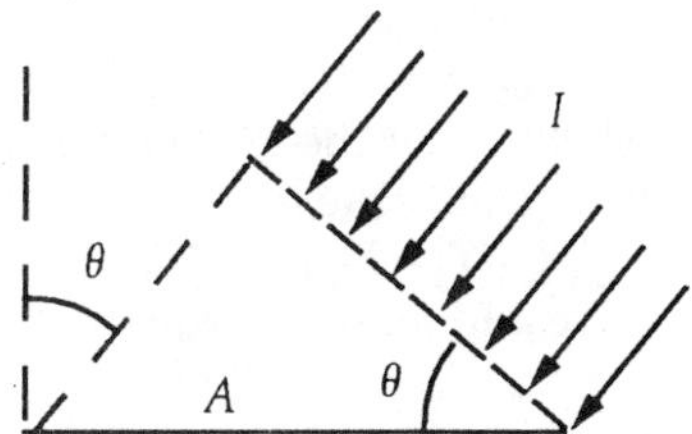

37. We find the rate of heat flow through the wall from

$\Delta Q/\Delta t = kA(\Delta T/L)$

$= (0.84\ \text{J/s}\cdot\text{m}\cdot\text{C°})(4.0\ \text{m})(4.0\ \text{m})(30°\text{C} - 10°\text{C})/(0.12\ \text{m}) = 2.24 \times 10^3\ \text{W}.$

We find the number of bulbs required to provide this heat flow from

$N = (\Delta Q/\Delta t)/P = (2.24 \times 10^3\ \text{W})/(100\ \text{W}) = 22.4 = \boxed{22\ \text{bulbs.}}$

38. The cross-sectional area of the beam that falls on an area A is $A\cos\theta$. Thus the rate at which energy is absorbed is

$P = IeA\cos\theta.$

There is no change in temperature of the ice. We find the time to provide the energy to melt the ice from

$Q = mL = \rho AhL = Pt = (IeA\cos\theta)t;$

$(917\ \text{kg/m}^3)(0.010\ \text{m})(3.33 \times 10^5\ \text{J/kg}) = (1000\ \text{W/m}^2)(0.050)(\cos 30°)t$, which gives

$t = 7.1 \times 10^4\ \text{s} = \boxed{20\ \text{h.}}$

Note that the result is independent of the area.

39. In the steady state, the intermediate temperature does not change, so the heat flow must be the same through the two rods:

$\Delta Q/\Delta t = k_{\text{Cu}}A_{\text{Cu}}(T_{\text{hot}} - T)/L_{\text{Cu}} = k_{\text{Al}}A_{\text{Al}}(T_{\text{hot}} - T)/L_{\text{Al}}.$

The rods have the same area and length, so we have

$k_{\text{Cu}}(T_{\text{hot}} - T) = k_{\text{Al}}(T_{\text{hot}} - T);$

$(380\ \text{J/s}\cdot\text{m}\cdot\text{C°})(250°\text{C} - T) = (200\ \text{J/s}\cdot\text{m}\cdot\text{C°})(T - 0.0°\text{C})$, which gives $T = \boxed{164°\text{C.}}$

40. We find the temperature from the power radiated into space:

$\Delta Q/\Delta t = e\sigma AT^4$, or $(\Delta Q/\Delta t)/A = e\sigma T^4;$

$430\ \text{W/m}^2 = (1.0)(5.67 \times 10^{-8}\ \text{W/m}^2\cdot\text{K}^4)T^4$, which gives $T = \boxed{295\ \text{K}\ (22°\text{C}).}$

41. We find the temperature difference from conduction through the glass:

$\Delta Q/\Delta t = kA(\Delta T/L);$

$95\ \text{W} = (0.84\ \text{J/s}\cdot\text{m}\cdot\text{C°})4\pi(0.030\ \text{m})^2\,\Delta T/(1.0 \times 10^{-3}\ \text{m})$, which gives $\Delta T = \boxed{10\ \text{C°.}}$

42. The total rate of thermal energy loss through the wall and windows must be the power output of the stove:

$$P = \Delta Q/\Delta t = (\Delta Q/\Delta t)_{\text{wall}} + (\Delta Q/\Delta t)_{\text{windows}}$$
$$= (k_1 A_1\,\Delta T/L_1) + (k_2 A_2\,\Delta T/L_2) = \boxed{[(k_1A_1/L_1) + (k_2A_2/L_2)]\Delta T.}$$

43. The rate of thermal energy flow is the same for the brick and the insulation:

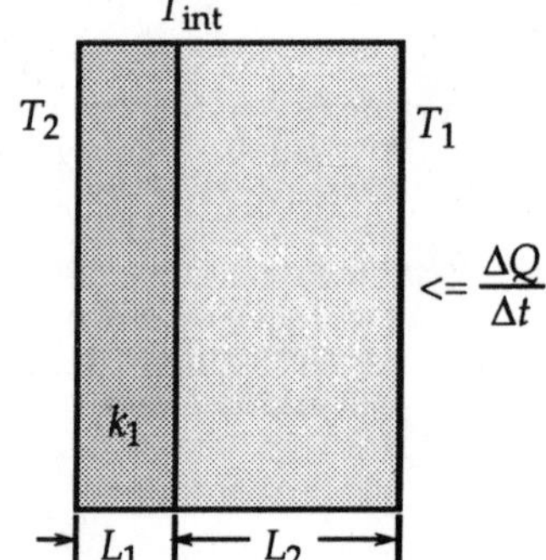

$$\Delta Q/\Delta t = A(T_1 - T_2)/R_{\text{eff}} = A(T_1 - T_{\text{int}})/R_2 = A(T_{\text{int}} - T_2)/R_1,$$

where T_{int} is the temperature at the brick-insulation interface.

By equating the first term to each of the others, we have

$$R_{\text{eff}}(T_1 - T_{\text{int}}) = R_2(T_1 - T_2), \quad \text{and} \quad R_{\text{eff}}(T_{\text{int}} - T_2) = R_1(T_1 - T_2).$$

If we add these two equations, we get

$$R_{\text{eff}}(T_1 - T_2) = R_1(T_1 - T_2) + R_2(T_1 - T_2), \text{ which gives}$$
$$R_{\text{eff}} = R_1 + R_2.$$

This shows the usefulness of the *R*-value.

For the *R*-value of the brick we have

$$R_1 = L_1/k_1$$
$$= [(4.0\text{ in})/(12\text{ in/ft})](3.28\text{ ft/m})/(0.84\text{ J/s}\cdot\text{m}\cdot\text{C°})(1\text{ Btu}/1055\text{ J})(3600\text{ s/h})(5\text{ C°}/9\text{ F°})$$
$$= 0.69\text{ ft}^2\cdot\text{h}\cdot\text{F°/Btu}.$$

Thus the total *R*-value of the wall is

$$R_{\text{eff}} = R_1 + R_2 = 0.69\text{ ft}^2\cdot\text{h}\cdot\text{F°/Btu} + 19\text{ ft}^2\cdot\text{h}\cdot\text{F°/Btu} = 19.7\text{ ft}^2\cdot\text{h}\cdot\text{F°/Btu}.$$

The rate of heat loss is

$$\Delta Q/\Delta t = A(T_1 - T_2)/R_{\text{eff}}$$
$$= [(240\text{ ft}^2)(10\text{ F°})/(19.7\text{ ft}^2\cdot\text{h}\cdot\text{F°/Btu})](1055\text{ J/Btu})/(3600\text{ s/h}) = \boxed{36\text{ W.}}$$

44. (*a*) We call the temperatures at the interfaces T_a and T_b, as shown. In the steady state, the rate of heat flow is the same for each layer:

$$\Delta Q/\Delta t = k_1A(T_2 - T_a)/\ell_1 = k_2A(T_a - T_b)/\ell_2 = k_3A(T_b - T_1)/\ell_3.$$

We treat this as three equations:

$$T_2 - T_a = [(\Delta Q/\Delta t)/A]\ell_1/k_1;$$
$$T_a - T_b = [(\Delta Q/\Delta t)/A]\ell_2/k_2;$$
$$T_b - T_1 = [(\Delta Q/\Delta t)/A]\ell_3/k_3.$$

If we add these equations, we get

$$T_2 - T_1 = [(\Delta Q/\Delta t)/A][(\ell_1/k_1) + (\ell_2/k_2) + (\ell_3/k_3)], \text{ which gives}$$
$$\Delta Q/\Delta t = A(T_2 - T_1)/[(\ell_1/k_1) + (\ell_2/k_2) + (\ell_3/k_3)].$$

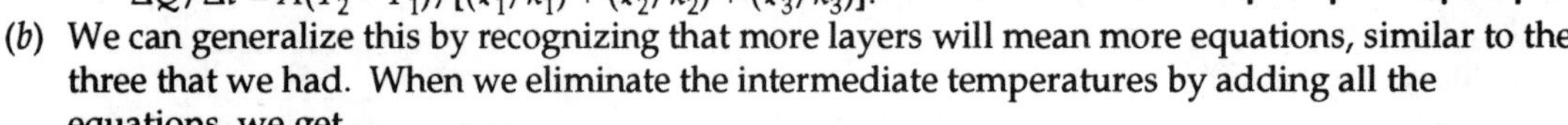

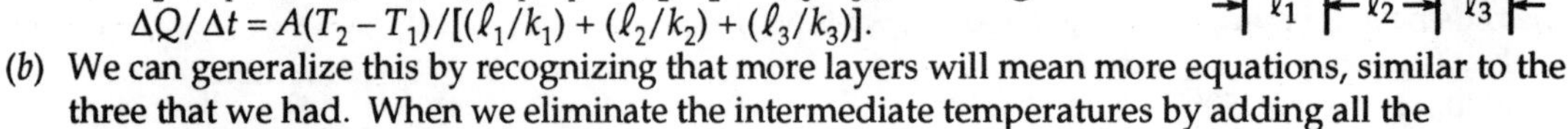

(*b*) We can generalize this by recognizing that more layers will mean more equations, similar to the three that we had. When we eliminate the intermediate temperatures by adding all the equations, we get

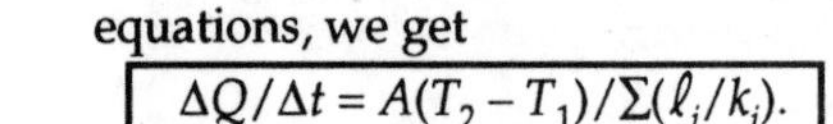

$$\boxed{\Delta Q/\Delta t = A(T_2 - T_1)/\Sigma(\ell_i/k_i).}$$

45. The total area of the six sides of the icebox is

$$A = 2[(0.25\text{ m})(0.35\text{ m}) + (0.25\text{ m})(0.50\text{ m}) + (0.35\text{ m})(0.50\text{ m})] = 0.775\text{ m}^2.$$

As the ice melts, the inside temperature of the icebox remains at 0°C. The rate at which heat flows through the sides of the icebox is

$$\Delta Q/\Delta t = m_{\text{ice}}L_{\text{ice}}/\Delta t = kA(\Delta T/L);$$
$$(11.0\text{ kg})(3.33\times10^5\text{ J/kg})/\Delta t = (0.023\text{ J/s}\cdot\text{m}\cdot\text{C°})(0.775\text{ m}^2)(30°\text{C} - 0°\text{C})/(0.015\text{ m}),$$

which gives $\Delta t = 5.14\times10^4\text{ s} = \boxed{14\text{ h.}}$

46. We assume that the rate of heat loss is proportional to the temperature difference, so the heat flow in a time t is

$Q = K(\Delta T)t.$

We assume that the constant K takes into account all forms of heat loss, but does not depend on the temperature difference, and thus is the same day and night. When the thermostat is turned down, we have

$Q_1 = K(\Delta T_{\text{day}})t_{\text{day}} + K(\Delta T_{1\text{night}})t_{\text{night}}$

$= K[(22°\text{C} - 8°\text{C})(17.0\ \text{h}) + (12°\text{C} - 0°\text{C})(7.0\ \text{h})] = (322\ \text{h}\cdot\text{C}°)K.$

When the thermostat is not turned down, we have

$Q_2 = K(\Delta T_{\text{day}})t_{\text{day}} + K(\Delta T_{2\text{night}})t_{\text{night}}$

$= K[(22°\text{C} - 8°\text{C})(17.0\ \text{h}) + (22°\text{C} - 0°\text{C})(7.0\ \text{h})] = (392\ \text{h}\cdot\text{C}°)K.$

For the percentage increase we have

$(\Delta Q/Q_1)(100) = \{[(392\ \text{h}\cdot\text{C}°)K - (322\ \text{h}\cdot\text{C}°)K]/[(322\ \text{h}\cdot\text{C}°)K]\}(100) = \boxed{22\%.}$

47. Because 70% of the heat generated by burning the coal heats the house, we have

$0.70m_{\text{coal}}(7000\ \text{kcal/kg}) = 4.8 \times 10^7\ \text{kcal}$, which gives $m_{\text{coal}} = \boxed{9.8 \times 10^3\ \text{kg.}}$

48. The work done by friction, which decreases the kinetic energy, generates the heat flow. If all the heat flow is absorbed by the lead bullet and wooden block, we have

$Q = \Delta\text{KE};$

$(m_{\text{lead}}c_{\text{lead}} + m_{\text{wood}}c_{\text{wood}})\Delta T = \frac{1}{2}m_{\text{lead}}v^2;$

$[(0.015\ \text{kg})(130\ \text{J/kg}\cdot\text{C}°) + (1.05\ \text{kg})(1700\ \text{J/kg}\cdot\text{C}°)](0.020\ \text{C}°) = \frac{1}{2}(0.015\ \text{kg})v^2,$

which gives $v = \boxed{69\ \text{m/s.}}$

49. (*a*) We find the power radiated from

$\Delta Q/\Delta t = e\sigma AT^4$

$= (1.0)(5.67 \times 10^{-8}\ \text{W/m}^2\cdot\text{K}^4)4\pi(7.0 \times 10^8\ \text{m})^2(5500\ \text{K})^4 = \boxed{3.2 \times 10^{26}\ \text{W.}}$

(*b*) Because this radiation passes through a sphere centered at the Sun, we have

$P/4\pi R^2 = (3.2 \times 10^{26}\ \text{W})/4\pi(1.5 \times 10^{11}\ \text{m})^2 = \boxed{1.1 \times 10^3\ \text{W/m}^2.}$

50. We find the temperature rise from

$Q = mc\,\Delta T;$

$(0.80)(200\ \text{kcal/h})(1.00\ \text{h}) = (70\ \text{kg})(0.83\ \text{kcal/kg}\cdot\text{C}°)\Delta T$, which gives $\Delta T = \boxed{2.8\ \text{C}°.}$

51. The heat generated in stopping the fall equals the decrease in kinetic energy. From energy conservation for the fall, this must equal the change in potential energy. We find the temperature rise from

$Q = (0.50)mgh = mc\,\Delta T;$

$(0.50)(340\ \text{kg})(9.80\ \text{m/s}^2)(140\ \text{m}) = (340\ \text{kg})(860\ \text{J/kg}\cdot\text{C}°)\Delta T$, which gives $\Delta T = \boxed{0.80\ \text{C}°.}$

52. (*a*) We use the two expressions for Q:

$Q = mc\,\Delta T = C\,\Delta T$, which gives $\boxed{C = mc.}$

(*b*) For 1 kg we have

$C = mc = (1.00\ \text{kg})(4186\ \text{J/kg}\cdot\text{C}°) = \boxed{4186\ \text{J/C}°.}$

(*c*) For 50 kg we have

$C = mc = (50.0\ \text{kg})(4186\ \text{J/kg}\cdot\text{C}°) = \boxed{2.09 \times 10^5\ \text{J/C}°.}$

53. We find the temperature rise in the rod of length L from

$\Delta Q = mc\,\Delta T = \rho ALc\,\Delta T;$

$320 \times 10^3\ \text{J} = (3.64\ \text{kg/m}^3)\pi(0.0100\ \text{m})^2L(130\ \text{J/kg}\cdot\text{C}^\circ)\,\Delta T$, which gives $\Delta T = (676\ \text{C}^\circ\cdot\text{m})/L$.

Because the rod is very long, the temperature rise will be small. We find the change in length from

$\Delta L = L\alpha\Delta T = L[29 \times 10^{-6}\ (\text{C}^\circ)^{-1}][(676\ \text{C}^\circ\cdot\text{m})/L] = 1.96 \times 10^{-2}\ \text{m} = \boxed{1.96\ \text{cm.}}$

If the rod is 2.0 cm long, the temperature rise will be

$\Delta T = (676\ \text{C}^\circ\cdot\text{m})/L = (676\ \text{C}^\circ\cdot\text{m})/(2.0 \times 10^{-2}\ \text{m}) = 3.4 \times 10^4\ \text{C}^\circ.$

This is so much greater than the boiling point for lead, 1750°C, that the $\boxed{\text{rod vaporizes.}}$

54. (*a*) We find the rate of heat flow through the clothing from

$\Delta Q/\Delta t = kA(\Delta T/L)$

$= (0.025\ \text{J/s}\cdot\text{m}\cdot\text{C}^\circ)(1.7\ \text{m}^2)[34^\circ\text{C} - (-20^\circ\text{C})]/(0.035\ \text{m}) = \boxed{66\ \text{W.}}$

(*b*) When the clothing is wet, we have

$\Delta Q/\Delta t = kA(\Delta T/L)$

$= (0.56\ \text{J/s}\cdot\text{m}\cdot\text{C}^\circ)(1.7\ \text{m}^2)[34^\circ\text{C} - (-20^\circ\text{C})]/(0.0050\ \text{m}) = \boxed{1.0 \times 10^4\ \text{W.}}$

55. If we assume that all the energy evaporates the water, with the latent heat at 20° given in the text, we have

$Q = m_{\text{water}}L_{\text{water}}$

$(1000\ \text{kcal/h})(2.5\ \text{h}) = m_{\text{water}}(585\ \text{kcal/kg})$, which gives $m_{\text{water}} = \boxed{4.3\ \text{kg.}}$

56. We find the rate of heat conduction from

$\Delta Q/\Delta t = kA(\Delta T/L)$

$= (0.2\ \text{J/s}\cdot\text{m}\cdot\text{C}^\circ)(1.5\ \text{m}^2)(37^\circ\text{C} - 34^\circ\text{C})/(0.040\ \text{m}) = \boxed{23\ \text{W.}}$

This is much less than the 230 W that must be dissipated, so the convection provided by the blood in carrying a heat flow to the skin is necessary.

57. The net heat flow rate from radiation, when the temperature difference is $\Delta T = T_1 - T_2$, is

$\Delta Q/\Delta t = e\sigma A(T_1^4 - T_2^4)$

$= e\sigma A[(T_2 + \Delta T)^4 - T_2^4] = e\sigma A\{T_2^4[1 + (\Delta T/T_2)]^4 - T_2^4\} = e\sigma AT_2^4\{[1 + (\Delta T/T_2)]^4 - 1\}.$

Because $\Delta T/T_2 \ll 1$, we use the binomial expansion for the first term:

$[1 + (\Delta T/T_2)]^4 \approx 1 + 4(\Delta T/T_2).$

When we substitute this, we get

$\Delta Q/\Delta t \approx e\sigma AT_2^4[1 + 4(\Delta T/T_2) - 1]$

$= 4e\sigma AT_2^4(\Delta T/T_2) = 4e\sigma AT_2^3\,\Delta T = 4e\sigma AT_2^3\,(T_1 - T_2) = \text{constant} \times (T_1 - T_2).$

58. (*a*) We find the rate of heat flow from

$$\Delta Q/\Delta t = k_1A_1(\Delta T/L_1) + k_2A_2(\Delta T/L_2) + k_3A_3(\Delta T/L_3)$$
$$= [(k_1A_1/L_1) + (k_2A_2/L_2) + (k_3A_3/L_3)]\Delta T$$
$$= \{[(0.023\ \text{J/s}\cdot\text{m}\cdot\text{C°})(410\ \text{m}^2)/(0.175\ \text{m})] + [(0.12\ \text{J/s}\cdot\text{m}\cdot\text{C°})(280\ \text{m}^2)/(0.065\ \text{m})] + [(0.84\ \text{J/s}\cdot\text{m}\cdot\text{C°})(33\ \text{m}^2)/(0.0065\ \text{m})]\}[23°\text{C} - (-10°\text{C})]$$
$$= 1.60\times10^5\ \text{W} = \boxed{160\ \text{kW.}}$$

(*b*) The heat flow needed to raise the temperature of the air is

$$Q_{air} = m_{air}c_{air}\,\Delta T = \rho_{air}V_{air}c_{air}\,\Delta T$$
$$= (1.29\ \text{kg/m}^3)(750\ \text{m}^3)(0.24\ \text{kcal/kg}\cdot\text{C°})(4186\ \text{J/kcal})(23°\text{C} - 10°\ \text{C}) = 1.26\times10^7\ \text{J}.$$

During the 30 minutes there will be heat loss from conduction. We use the average temperature difference:

$$\Delta T_{av} = \tfrac{1}{2}(23°\text{C} + 10°\text{C}) - (-10°\text{C}) = 26.5\ \text{C°}.$$

Because the conduction loss is proportional to the temperature difference, the loss during the 30 minutes is

$$Q_{loss} = [(1.60\times10^5\ \text{W})/(33\ \text{C°})](26.5\ \text{C°})(30\ \text{min})(60\ \text{s/min}) = 2.30\times10^8\ \text{J}.$$

The total heat required is

$$Q_{total} = Q_{air} + Q_{loss} = 1.26\times10^7\ \text{J} + 2.30\times10^8\ \text{J} = \boxed{2.4\times10^8\ \text{J.}}$$

(*c*) We find the amount of gas required from

$$0.90m_{gas}(5.4\times10^7\ \text{J/kg}) = (1.60\times10^5\ \text{W})(24\ \text{h/day})(30\ \text{day/month})(3600\ \text{s/h}),$$

which gives $m_{gas} = 8.48\times10^3$ kg/month.

We find the cost from

$$\text{Cost} = (\text{rate})m_{gas} = (\$0.080/\text{kg})(8.48\times10^3\ \text{kg/month}) = \boxed{\$6.8\times10^2/\text{month.}}$$

59. (*a*) The work done by friction, which decreases the kinetic energy, generates the heat flow. If 50% of the heat flow is absorbed by the lead bullet, we have

$$Q = \tfrac{1}{2}\Delta\text{KE};$$
$$m_{lead}c_{lead}\,\Delta T = \tfrac{1}{2}(\tfrac{1}{2}m_{lead})(v_f^2 - v_i^2)$$
$$[(0.015\ \text{kg})(130\ \text{J/kg}\cdot\text{C°})\Delta T = \tfrac{1}{4}(0.015\ \text{kg})[(220\ \text{m/s})^2 - (160\ \text{m/s})^2],$$

which gives $\Delta T = \boxed{44\ \text{C°.}}$

(*b*) With an ambient temperature of 20°C, we have T_{lead} = 20°C + 44 C° = 64°C. Because this is less than the melting point of lead, 327°C, there will be $\boxed{\text{no melting}}$ of the bullet.

60. (*a*) If we assume that all of the radiation is absorbed to raise the temperature of the leaf, we have

$$P = IeA = m_{leaf}c_{leaf}\,(\Delta T/\Delta t);$$
$$(1000\ \text{W/m}^2)(0.85)(40\times10^{-4}\ \text{m}^2) = (4.5\times10^{-4}\ \text{kg})(0.80\ \text{kcal/kg}\cdot\text{C°})(4186\ \text{J/kcal})(\Delta T/\Delta t),$$

which gives $\Delta T/\Delta t = \boxed{2.3\ \text{C°/s.}}$

(*b*) When the leaf reaches the temperature at which the absorbed energy is re-radiated to the surroundings from both sides of the leaf, we have

$$IeA = e\sigma 2A(T_2^4 - T_1^4),\quad \text{or}\quad I = 2\sigma(T_2^4 - T_1^4);$$
$$(1000\ \text{W/m}^2) = 2(5.67\times10^{-8}\ \text{W/m}^2\cdot\text{K}^4)[T_2^4 - (293\ \text{K})^4],\ \text{which gives}\ T_2 = 357\ \text{K} = \boxed{84°\text{C.}}$$

(*c*) The major ways that heat can be dissipated are by $\boxed{\text{convection}}$ from $\boxed{\text{conduction}}$ to the air in contact with the leaf, and $\boxed{\text{evaporation.}}$

61. When we consider radiation from both sides of the leaf, the net absorbtion rate is

$$P_{net} = IeA - e\sigma 2A(T_2^4 - T_1^4) = eA\,[I - 2\sigma(T_2^4 - T_1^4)].$$

To remove this energy by evaporation, we have

$$eA\,[I - 2\sigma(T_2^4 - T_1^4)] = (m/t)L;$$

$$(0.85)(40 \times 10^{-4}\ \text{m}^2)\{1000\ \text{W/m}^2 - 2(5.67 \times 10^{-8}\ \text{W/m}^2\cdot\text{K}^4)[(308\ \text{K})^4 - (293)^4]\} = (m/t)(585\ \text{kcal/kg})(4186\ \text{J/kcal}),$$

which gives $m/t = \boxed{1.1 \times 10^{-6}\ \text{kg/s}\ (4.1\ \text{g/h}).}$

Note that we have used the latent heat at 20°C given in the text.

62. The work done by friction, which decreases the kinetic energy, generates the heat flow. In general, some of the heat flow will heat the air and some will be radiated, so a fraction of the heat flow is used to raise the temperature of the iron meteorite and then melt the iron meteorite. The larger this fraction is, the smaller the necessary velocity. We determine the minimum velocity by assuming that this fraction is 1:

$$Q = \Delta\text{KE};$$

$$m_{iron}(c_{iron}\,\Delta T + L_{iron}) = \tfrac{1}{2}m_{iron}v_{min}^2;$$

$(450\ \text{J/kg}\cdot\text{C°})[1808°\text{C} - (-125°\text{C})] + 2.89 \times 10^5\ \text{J/kg} = \tfrac{1}{2}v_{min}^2$, which gives $v_{min} = \boxed{1.5 \times 10^3\ \text{m/s.}}$

CHAPTER 15

1. We use the first law of thermodynamics to find the change in internal energy:

$\Delta U = Q - W$

$= -3.42 \times 10^3\ \text{J} - (-1.6 \times 10^3\ \text{J}) = \boxed{-1.8 \times 10^3\ \text{J.}}$

2. (*a*) The internal energy of an ideal gas depends only on the temperature, $U = \frac{3}{2}nRT$, so we have

$\Delta U = \frac{3}{2}nR\,\Delta T = \quad 0.$

(*b*) We use the first law of thermodynamics to find the heat absorbed:

$\Delta U = Q - W;$

$0 = Q - 4.40 \times 10^3\ \text{J}$, which gives $Q = \boxed{4.40 \times 10^3\ \text{J.}}$

3.

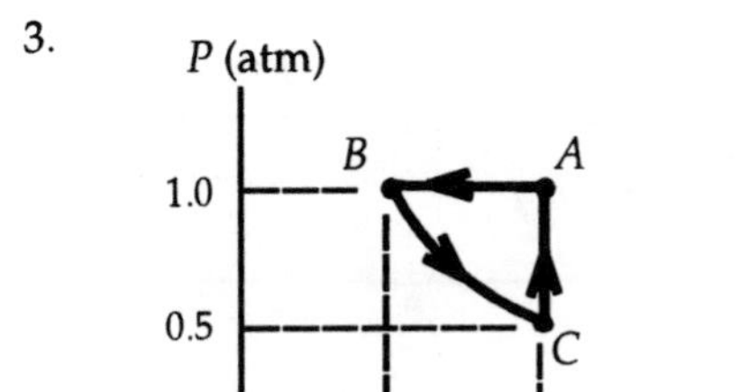

4. For the isothermal process, we have

$P_1V_1 = P_2V_2;$

$(6.5\ \text{atm})(1.0\ \text{L}) = (1.0\ \text{atm})V_2,$

which gives $V_2 = 6.5\ \text{L}.$

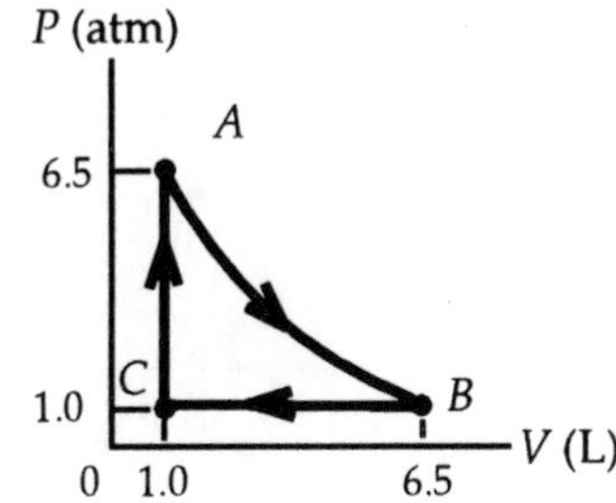

5.

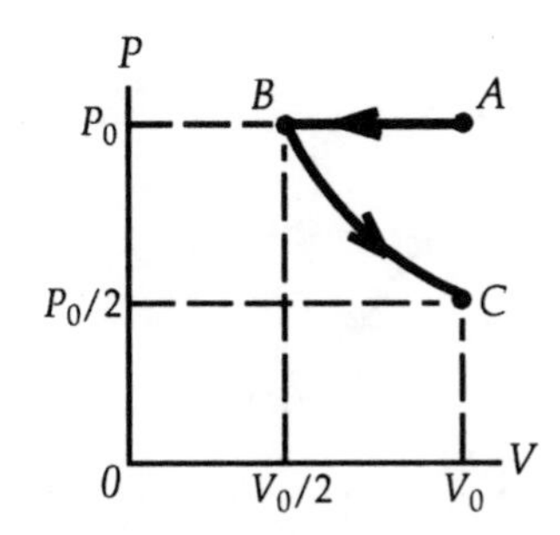

6. (*a*) Because there is no change in volume, we have

$W = \boxed{0.}$

(*b*) We use the first law of thermodynamics to find the change in internal energy:

$\Delta U = Q - W$

$= -265\ \text{kJ} - 0 = \boxed{-265\ \text{kJ.}}$

7. (*a*) In an adiabatic process there is no heat flow:

$Q = \boxed{0.}$

(*b*) We use the first law of thermodynamics to find the change in internal energy:

$\Delta U = Q - W$

$= 0 - (-1350 \text{ J}) - 0 = \boxed{+1350 \text{ J.}}$

(*c*) The internal energy of an ideal gas depends only on the temperature, $U = \frac{3}{2}nRT$, so we see that an increase in the internal energy means that the temperature must $\boxed{\text{rise.}}$

8. (*a*) Work is done only in the constant pressure process:

$W = p_a(V_b - V_a)$

$= (5.0 \text{ atm})(1.013 \times 10^5 \text{ N/m}^2 \cdot \text{atm})[(0.660 - 0.400) \times 10^{-3} \text{ m}^3]$

$= \boxed{1.3 \times 10^2 \text{ J.}}$

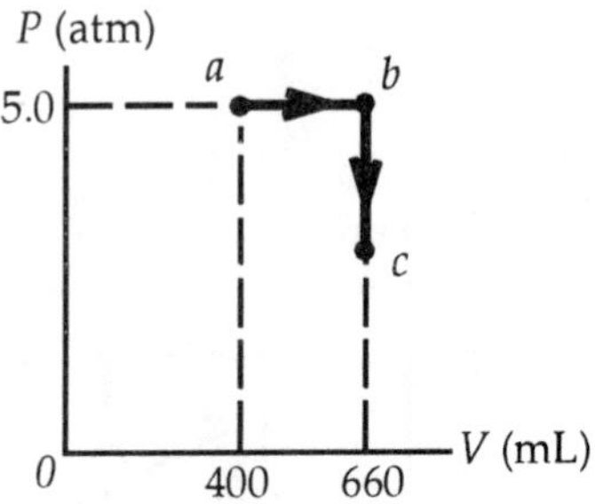

(*b*) Because the initial and final temperatures are the same, we know that $\Delta U = 0$. We use the first law of thermodynamics to find the heat flow for the entire process:

$\Delta U = Q - W;$

$0 = Q - (+1.3 \times 10^2 \text{ J})$, which gives $Q = \boxed{+1.3 \times 10^2 \text{ J.}}$

9. (*a*) Work is done only in the constant pressure process:

$W = p_b(V_c - V_b)$

$= (1.4 \text{ atm})(1.013 \times 10^5 \text{ N/m}^2 \cdot \text{atm})[(9.3 - 6.8) \times 10^{-3} \text{ m}^3]$

$= \boxed{3.5 \times 10^2 \text{ J.}}$

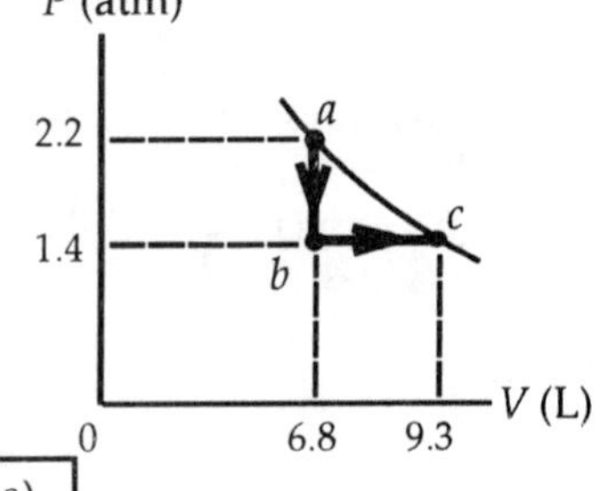

(*b*) Because the initial and final temperatures are the same, we know that $\boxed{\Delta U = 0.}$

(*c*) We use the first law of thermodynamics to find the heat flow for the entire process:

$\Delta U = Q - W;$

$0 = Q - (+3.5 \times 10^2 \text{ J})$, which gives $Q = \boxed{+3.5 \times 10^2 \text{ J (into the gas).}}$

10. (*b*) Work done in the constant pressure process is

$W = p_1(V_2 - V_1)$

$= (450 \text{ N/m}^2)(8.00 \text{ m}^3 - 2.00 \text{ m}^3) = \boxed{2.70 \times 10^3 \text{ J.}}$

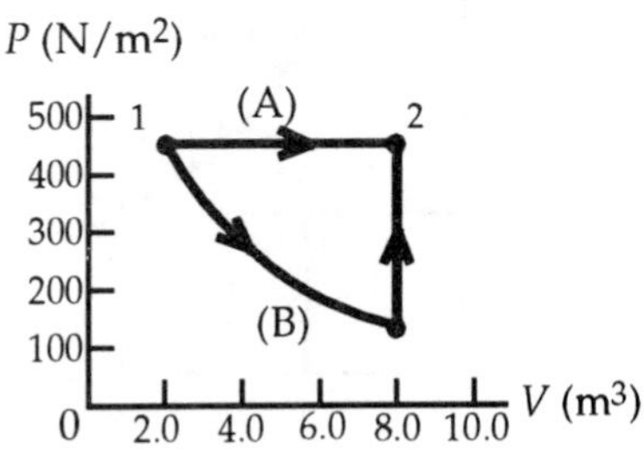

The internal energy of an ideal gas depends only on the temperature, so we have

$\Delta U = \frac{3}{2}nR\,\Delta T = \frac{3}{2}(nRT_2 - nRT_1) = \frac{3}{2}(p_2V_2 - p_1V_1)$

$= \frac{3}{2}p_1(V_2 - V_1) = \frac{3}{2}W = \frac{3}{2}(2.70 \times 10^3 \text{ J}) = \boxed{4.05 \times 10^3 \text{ J.}}$

(*d*) Because both paths have the same initial and final states, the change in internal energy is the same:

$\Delta U = \boxed{4.05 \times 10^3 \text{ J.}}$

11. (*a*) We can find the internal energy change $U_c - U_a$ from the information for the curved path $a \to c$:

$$U_c - U_a = Q_{a\to c} - W_{a\to c} = -63\text{ J} - (-35\text{ J}) = -28\text{ J}.$$

For the path $a \to b \to c$, we have

$$U_c - U_a = Q_{a\to b\to c} - W_{a\to b\to c} = Q_{a\to b\to c} - W_{a\to b};$$
$$-28\text{ J} = Q_{a\to b\to c} - (-48\text{ J}),\text{ which gives } Q_{a\to b\to c} = \boxed{-76\text{ J.}}$$

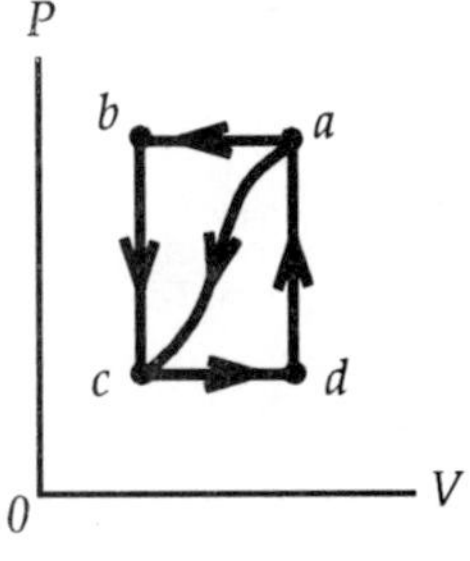

(*b*) For the path $c \to d \to a$, work is done only during the constant pressure process, $c \to d$, so we have

$$W_{c\to d\to a} = P_c(V_d - V_c) = \tfrac{1}{2}P_b(V_a - V_b)$$
$$= \tfrac{1}{2}W_{b\to a} = -\tfrac{1}{2}W_{a\to b} = -\tfrac{1}{2}(-48\text{ J}) = \boxed{+24\text{ J.}}$$

(*c*) We use the first law of thermodynamics for the path $c \to d \to a$ to find $Q_{c\to d\to a}$:

$$U_a - U_c = -(U_c - U_a) = Q_{c\to d\to a} - W_{c\to d\to a};$$
$$-(-28\text{ J}) = Q_{c\to d\to a} - (24\text{ J}),\text{ which gives } Q_{c\to d\to a} = \boxed{+52\text{ J.}}$$

(*d*) $U_a - U_c = -(U_c - U_a) = -(-28\text{ J}) = \boxed{+28\text{ J.}}$

(*e*) Because there is no work done for the path $d \to a$, we have

$$U_a - U_d = (U_a - U_c) + (U_c - U_d) = (U_a - U_c) - (U_d - U_c) = Q_{d\to a} - W_{d\to a};$$
$$+28\text{ J} - (+5\text{ J}) = Q_{d\to a} - 0,\text{ which gives } Q_{d\to a} = \boxed{+23\text{ J.}}$$

12. (*a*) We can find the internal energy change $U_a - U_c$ from the information for the curved path $a \to c$:

$$(U_c - U_a) = -(U_a - U_c) = Q_{a\to c} - W_{a\to c} = -80\text{ J} - (-55\text{ J}) = -25\text{ J},$$

so $U_a - U_c = \boxed{+25\text{ J.}}$

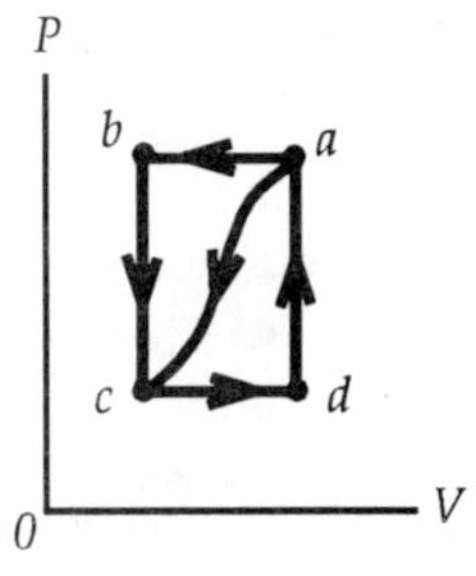

(*b*) We use the first law of thermodynamics for the path $c \to d \to a$ to find $Q_{c\to d\to a}$:

$$U_a - U_c = Q_{c\to d\to a} - W_{c\to d\to a};$$
$$+25\text{ J} = Q_{c\to d\to a} - (+38\text{ J}),\text{ which gives } Q_{c\to d\to a} = \boxed{+63\text{ J.}}$$

(*c*) For the path $a \to b \to c$, work is done only during the constant pressure process, $a \to b$, so we have

$$W_{a\to b\to c} = P_a(V_b - V_a) = (2.5)P_d(V_c - V_d)$$
$$= (2.5)W_{d\to c} = -(2.5)W_{c\to d} = -(2.5)(+38\text{ J}) = \boxed{-95\text{ J.}}$$

(*d*) We use the first law of thermodynamics for the path $a \to b \to c$ to find $Q_{a\to b\to c}$:

$$U_c - U_a = Q_{a\to b\to c} - W_{a\to b\to c};$$
$$-25\text{ J} = Q_{a\to b\to c} - (-95\text{ J}),\text{ which gives } Q_{a\to b\to c} = \boxed{-120\text{ J.}}$$

(*e*) Because there is no work done for the path $b \to c$, we have

$$U_c - U_b = (U_c - U_a) + (U_a - U_b) = Q_{b\to c} - W_{b\to c};$$
$$-25\text{ J} + (+10\text{ J}) = Q_{b\to c} - 0,\text{ which gives } Q_{b\to c} = \boxed{-15\text{ J.}}$$

13. Because the directions along the path are opposite to the directions in Problem 12, all terms for Q and W will have the opposite sign.

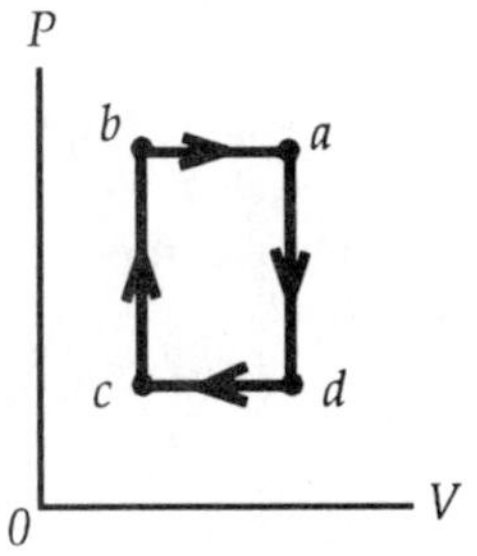

(*a*) For the work done around the cycle, we have

$$W_{\text{cycle}} = W_{c\to b\to a} + W_{a\to d\to c} = -W_{a\to b\to c} - W_{c\to d\to a}$$
$$= -(-95\text{ J}) - (+38\text{ J}) = \boxed{+57\text{ J.}}$$

(*b*) For the heat flow during the cycle, we have

$$Q_{\text{cycle}} = Q_{c\to b\to a} + Q_{a\to d\to c} = -Q_{a\to b\to c} - Q_{c\to d\to a}$$
$$= -(-120\text{ J}) - (+63\text{ J}) = \boxed{+57\text{ J.}}$$

(*c*) For the internal energy change of the cycle, we have

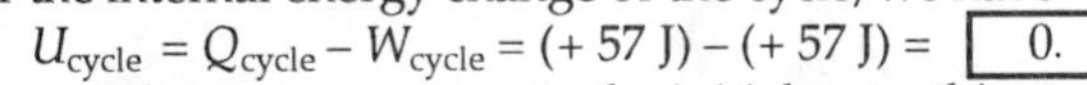

$$U_{\text{cycle}} = Q_{\text{cycle}} - W_{\text{cycle}} = (+57\text{ J}) - (+57\text{ J}) = \boxed{0.}$$

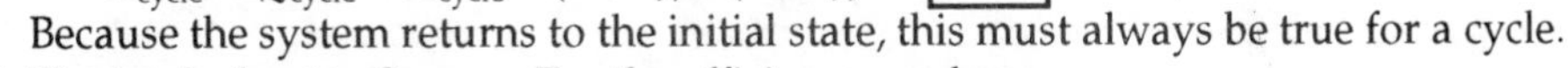

Because the system returns to the initial state, this must always be true for a cycle.

(*d*) The intake heat is $Q_{c\to b\to a}$. For the efficiency, we have

$$e = W_{\text{cycle}}/Q_{c\to b\to a} = (57\text{ J})/(120\text{ J}) = 0.475 = \boxed{48\%.}$$

14. Using the metabolic rates from Table 15–1, we have

$E = [(8.0\ \text{h})(70\ \text{W}) + (1.0\ \text{h})(460\ \text{W}) + (4.0\ \text{h})(230\ \text{W}) + (10.0\ \text{h})(115\ \text{W}) + (1.0\ \text{h})(1150\ \text{W})](3600\ \text{s/h}) = \boxed{1.5 \times 10^7\ \text{J}\ (3.6 \times 10^3\ \text{kcal}).}$

15. Using the metabolic rates from Table 15–1, we have

$\text{Rate} = E/t = [(8.0\ \text{h})(70\ \text{W}) + (8.0\ \text{h})(115\ \text{W}) + (4.0\ \text{h})(230\ \text{W}) + (2.0\ \text{h})(115\ \text{W}) + (1.5\ \text{h})(460\ \text{W}) + (0.5\ \text{h})(1150\ \text{W})]/(24\ \text{h}) = \boxed{1.6 \times 10^2\ \text{W}.}$

16. Using the metabolic rates from Table 15–1, we find the energy difference per year:

$\Delta E/t = [(1.0\ \text{h/day})(230\ \text{W} - 70\ \text{W})(365\ \text{days/yr})(3600\ \text{s/h}) = 2.1 \times 10^8\ \text{J/yr}.$

We find the equivalent amount of fat from

$\Delta m = (2.1 \times 10^8\ \text{J/yr})/(4186\ \text{J/kcal})(9000\ \text{kcal/kg}) = \boxed{5.6\ \text{kg}.}$

17. For the heat input, we have

$Q_H = Q_L + W = 8200\ \text{J} + 3200\ \text{J} = 11{,}400\ \text{J}.$

We find the efficiency from

$e = W/Q_H = (3200\ \text{J})/(11{,}400\ \text{J}) = 0.28 = \boxed{28\%.}$

18. We find the efficiency from

$e = W/Q_H = (7200\ \text{J})/(12.0\ \text{kcal})(4186\ \text{J/kcal}) = 0.143 = \boxed{14.3\%.}$

19. The maximum efficiency is the efficiency of the Carnot cycle:

$e = 1 - (T_L/T_H) = 1 - [(593\ \text{K})/(853\ \text{K})] = 0.305 = \boxed{30.5\%.}$

20. We find the temperature from the Carnot efficiency:

$e = 1 - (T_L/T_H);$

$0.28 = 1 - [(503\ \text{K})/T_H]$, which gives $T_H = 699\ \text{K} = \boxed{426°\text{C}.}$

21. The efficiency of the plant is

$e = 0.75e_{\text{Carnot}} = 0.75[1 - (T_L/T_H)] = 0.75\{1 - [(623\ \text{K})/(873\ \text{K})]\} = 0.215.$

We find the intake heat flow rate from

$e = W/Q_H;$

$0.215 = (1.3 \times 10^9\ \text{J/s})(3600\ \text{s/h})/(Q_H/t)$, which gives $Q_H/t = 2.18 \times 10^{13}\ \text{J/h}.$

We find the discharge heat flow from

$Q_L/t = (Q_H/t) - (W/t) = 2.18 \times 10^{13}\ \text{J/h} - (1.3 \times 10^9\ \text{J/s})(3600\ \text{s/h}) = \boxed{1.7 \times 10^{13}\ \text{J/h}.}$

22. We find the efficiency from

$e = W/Q_H = (W/t)/(Q_H/t) = (440 \times 10^3\ \text{J/s})/(680\ \text{kcal})(4186\ \text{J/kcal}) = 0.155.$

We find the temperature from the Carnot efficiency:

$e = 1 - (T_L/T_H);$

$0.155 = 1 - [T_L/(843\ \text{K})]$, which gives $T_L = 712\ \text{K} = \boxed{439°\text{C}.}$

23. (*a*) The efficiency of the engine is

$e = fe_{\text{Carnot}}$;

$0.15 = f[1 - (T_L/T_H)] = f\{1 - [(358\text{ K})/(773\text{ K})]\}$, which gives $f = 0.28 = \boxed{28\%.}$

(*b*) The power output of the engine is

$P = (100\text{ hp})(746\text{ W/hp}) = \boxed{7.46 \times 10^4\text{ W.}}$

We find the intake heat flow rate from

$e = P/(Q_H/t)$;

$0.15 = (7.46 \times 10^4\text{ W})(3600\text{ s/h})/(Q_H/t)$, which gives $Q_H/t = 1.79 \times 10^9\text{ J/h}$.

We find the discharge heat flow from

$Q_L/t = (Q_H/t) - (W/t) = 1.79 \times 10^9\text{ J/h} - (7.46 \times 10^4\text{ J/s})(3600\text{ s/h})$

$= \boxed{1.52 \times 10^9\text{ J/h } (3.64 \times 10^5\text{ kcal/h}).}$

24. We find the low temperature from the Carnot efficiency:

$e_1 = 1 - (T_L/T_{H1})$;

$0.30 = 1 - [T_L/(823\text{ K})]$, which gives $T_L = 576\text{ K}$.

Because the low temperature does not change, for the new efficiency we have

$e_2 = 1 - (T_L/T_{H2})$;

$0.40 = 1 - [(576\text{ K})/T_{H2}]$, which gives $T_{H2} = 960\text{ K} = \boxed{687°\text{C.}}$

25. We find the high temperature from the Carnot efficiency:

$e_1 = 1 - (T_{L1}/T_H)$;

$0.39 = 1 - [(623\text{ K})/T_H]$, which gives $T_H = 1021\text{ K}$.

Because the high temperature does not change, for the new efficiency we have

$e_2 = 1 - (T_{L2}/T_H)$;

$0.50 = 1 - [T_{L2}/(1021\text{ K})]$, which gives $T_{L2} = 511\text{ K} = \boxed{238°\text{C.}}$

26. For the efficiencies of the engines, we have

$e_1 = 0.60e_{\text{Carnot}} = 0.60[1 - (T_{L1}/T_{H1})]$

$= 0.60\{1 - [(713\text{ K})/(943\text{ K})]\} = 0.146$;

$e_2 = 0.60e_{\text{Carnot}} = 0.60[1 - (T_{L2}/T_{H2})]$

$= 0.60\{1 - [(563\text{ K})/(703\text{ K})]\} = 0.119$.

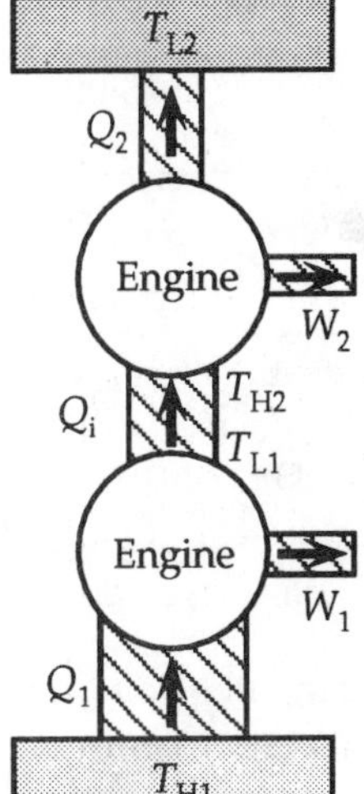

Because coal is burned to produce the input heat to the first engine, we need to find Q_{H1}. We relate W_1 to Q_{H1} from the efficiency:

$e_1 = W_1/Q_{H1}$, or $W_1 = 0.146Q_{H1}$.

Because the exhaust heat from the first engine is the input heat to the second engine, we have

$e_2 = W_2/Q_{H1} = W_2/Q_{L1}$, or $W_2 = 0.119Q_{L1}$.

For the first engine we know that

$Q_{L1} = Q_{H1} - W_1$, so we get

$W_2 = 0.119Q_{L1} = 0.119(Q_{H1} - 0.146Q_{H1}) = 0.102Q_{H1}$.

For the total work, we have

$W = W_1 + W_2 = 0.146Q_{H1} + 0.102Q_{H1} = 0.248Q_{H1}$.

When we use the rate at which this work is done, we get

$1000 \times 10^6\text{ W} = 0.248(Q_{H1}/t)$, which gives $Q_{H1}/t = 4.03 \times 10^9\text{ J/s}$.

We find the rate at which coal must be burned from

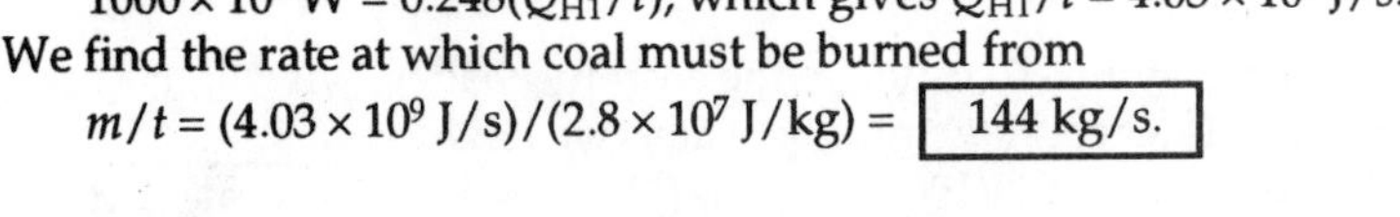

$m/t = (4.03 \times 10^9\text{ J/s})/(2.8 \times 10^7\text{ J/kg}) = \boxed{144\text{ kg/s.}}$

27. We find the discharge heat flow for the plant from

$Q_{L2}/t = (Q_{H1}/t) - (W/t) = 4.03 \times 10^9\text{ J/s} - 1000 \times 10^6\text{ J/s} = 3.03 \times 10^9\text{ J/s}$.

We find the rate at which water must pass through the plant from

$Q_{L2}/t = (m/t)c\,\Delta T$;

$(3.03 \times 10^9\text{ J/s})(3600\text{ s/h}) = (m/t)(4186\text{ J/kg}\cdot\text{C°})(6.0\text{ C°})$, which gives $m/t = \boxed{4.3 \times 10^8\text{ kg/h.}}$

28. The maximum coefficient of performance for a cooling coil is
$CP = Q_L/W = T_L/(T_H - T_L) = (258\text{ K})/(303\text{ K} - 258\text{ K}) = \boxed{5.7.}$

29. The coefficient of performance for the refrigerator is
$CP = Q_L/W = T_L/(T_H - T_L) = (258\text{ K})/(295\text{ K} - 258\text{ K}) = \boxed{7.0.}$

30. We find the low temperature from the coefficient of performance:
$CP = Q_L/W = T_L/(T_H - T_L)$;
$5.0 = T_L/(302\text{ K} - T_L)$, which gives $= 252\text{ K} = \boxed{-21°\text{C}.}$

31. (*a*) The coefficient of performance for the heat pump is
$CP_1 = T_H/(T_H - T_{L1}) = (295\text{ K})/(295\text{ K} - 273\text{ K}) = 13.4.$
We find the work from
$CP_1 = Q_H/W_1$;
$13.4 = (2800\text{ J})/W_1$, which gives $W_1 = \boxed{2.1 \times 10^2\text{ J.}}$
(*b*) The coefficient of performance for the heat pump is now
$CP_2 = T_H/(T_H - T_{L2}) = (295\text{ K})/(295\text{ K} - 258\text{ K}) = 7.97.$
We find the work from
$CP_2 = Q_H/W_2$;
$7.97 = (2800\text{ J})/W_2$, which gives $W_2 = \boxed{3.5 \times 10^2\text{ J.}}$

32. The efficiency of the engine is
$e = 1 - (T_L/T_H)$.
For the coefficient of performance, we have
$CP = T_H/(T_H - T_L) = 1/[1 - (T_L/T_H)] = 1/e = 1/0.35 = \boxed{2.9.}$

33. We assume a room temperature of 20°C. The efficiency of the engine is
$e = 1 - (T_L/T_H)$.
We find the coefficient of performance of the refrigerator from the efficiency of the engine:
$CP = Q_L/W = T_L/(T_H - T_L) = (T_L/T_H)/[1 - (T_L/T_H)] = (1 - e)/e = (1 - 0.30)/(0.30) = 2.33.$
The heat flow required to cool and freeze the water is
$Q_L = m(c\,\Delta T + L)$.
We find the time from
$t = W/P = (Q_L/CP)/P = m(c\,\Delta T + L)/(CP)P$
$= (2.33)(12)(0.040\text{ kg})\{[(4186\text{ J/kg}\cdot\text{C°})(20°\text{C} - 0°\text{C})] + 3.33 \times 10^5\text{ J/kg}\}/(450\text{ W})$
$= 190\text{ s} = \boxed{3.2\text{ min.}}$

34. The condensation occurs at constant temperature. Because there is a heat flow out of the steam, we have
$\Delta S = Q/T = -mL/T = -(0.100\text{ kg})(539\text{ kcal/kg})/(373\text{ K}) = \boxed{-0.145\text{ kcal/K.}}$

35. The heating does not occur at constant temperature. To estimate the entropy change, we will use the average temperature of 50°C. Because there is a heat flow into the water, we have
$\Delta S \approx Q/T = mc\,\Delta T/T_{av} = (1.00\text{ kg})(1.00\text{ kcal/kg}\cdot\text{C°})(100°\text{C} - 0°\text{C})/(323\text{ K}) = \boxed{+0.31\text{ kcal/K.}}$

36. The freezing occurs at constant temperature. Because there is a heat flow out of the water, we have
$\Delta S = Q/T = -mL/T = -(1.00\text{ m}^3)(1000\text{ kg/m}^3)(79.7\text{ kcal/kg})/(273\text{ K}) = \boxed{-292\text{ kcal/K.}}$

37. We assume that when the ice is formed at 0°C, it is removed, so its entropy change will be the same as in Problem 36. The heat flow from the water went into the great deal of ice at – 10°C. Because there is a great deal of ice, its temperature will not change. We find the entropy change of the ice from

$$\Delta S_{\text{ice}} = Q/T_{\text{ice}} = +\,mL/T = (1.00\text{ m}^3)(1000\text{ kg/m}^3)(79.7\text{ kcal/kg})/(263\text{ K}) = +\,303\text{ kcal/K}.$$

Thus the total entropy change is

$$\Delta S_{\text{ice}} = \Delta S_{\text{ice}} + \Delta S = +\,303\text{ kcal/K} + (-\,292\text{ kcal/K}) = \boxed{+\,11\text{ kcal/K.}}$$

If the new ice were not removed, we would include an additional heat term which would be negative for cooling the new ice and positive for the great deal of ice. The net additional entropy change would be small.

38. The total rate of the entropy change is

$$\Delta S_{\text{total}}/t = \Delta S_{\text{source}}/t + \Delta S_{\text{water}}/t = (-\,Q/t)/T_{\text{source}} + (+\,Q/t)/T_{\text{water}}$$
$$= (-\,7.50\text{ cal/s})/(513\text{ K}) + (+\,7.50\text{ cal/s})/(300\text{ K}) = \boxed{+\,0.0104\text{ cal/K}\cdot\text{s.}}$$

39. We find the final temperature from

heat lost = heat gained;

$$m_{\text{hot}}c_{\text{water}}\,\Delta T_{\text{hot}} = m_{\text{cold}}c_{\text{water}}\,\Delta T_{\text{cold}};$$

$(1.0\text{ kg})(1000\text{ cal/kg}\cdot\text{C°})(60\text{°C} - T) = (1.0\text{ kg})(1000\text{ cal/kg}\cdot\text{C°})(T - 30\text{°C})$, which gives $T = 45\text{°C}$.

The heat flow from the hot water to the cold water is

$$Q = m_{\text{cold}}c_{\text{water}}\,\Delta T_{\text{cold}} = (1.0\text{ kg})(1000\text{ cal/kg}\cdot\text{C°})(45\text{°C} - 30\text{°C}) = 1.5 \times 10^4\text{ cal}.$$

The heating and cooling do not occur at constant temperature. To estimate the entropy changes, we will use the average temperatures:

$$T_{\text{coldav}} = \tfrac{1}{2}(T_{\text{cold}} + T) = \tfrac{1}{2}(30\text{°C} + 45\text{°C}) = 37.5\text{°C};$$
$$T_{\text{hotav}} = \tfrac{1}{2}(T_{\text{hot}} + T) = \tfrac{1}{2}(60\text{°C} + 45\text{°C}) = 52.5\text{°C}.$$

The total entropy change is

$$\Delta S = \Delta S_{\text{hot}} + \Delta S_{\text{cold}} = (-\,Q/T_{\text{hotav}}) + (+\,Q/T_{\text{coldav}})$$
$$= [(-\,1.5 \times 10^4\text{ cal})/(325.7\text{ K})] + [(+\,1.5 \times 10^4\text{ cal})/(310.7\text{ K})] = \boxed{+\,2.2\text{ cal/K.}}$$

40. The aluminum and water are isolated, so we find the final temperature from

heat lost = heat gained;

$$m_{\text{Al}}c_{\text{Al}}\,\Delta T_{\text{Al}} = m_{\text{water}}c_{\text{water}}\,\Delta T_{\text{water}};$$

$(3.8\text{ kg})(0.22\text{ kcal/kg}\cdot\text{C°})(30\text{°C} - T) = (1.0\text{ kg})(1.00\text{ kcal/kg}\cdot\text{C°})(T - 20\text{°C})$, which gives $T = 24.55\text{°C}$.

The heat flow from the aluminum to the water is

$$Q = m_{\text{Al}}c_{\text{Al}}\,\Delta T_{\text{Al}} = (3.8\text{ kg})(0.22\text{ kcal/kg}\cdot\text{C°})(30\text{°C} - 24.55\text{°C}) = 4.55\text{ kcal}.$$

The heating and cooling do not occur at constant temperature. To estimate the entropy changes, we will use the average temperatures:

$$T_{\text{waterav}} = \tfrac{1}{2}(T_{\text{water}} + T) = \tfrac{1}{2}(20\text{°C} + 24.55\text{°C}) = 22.3\text{°C};$$
$$T_{\text{Alav}} = \tfrac{1}{2}(T_{\text{Al}} + T) = \tfrac{1}{2}(30\text{°C} + 24.55\text{°C}) = 27.3\text{°C}.$$

The total entropy change is

$$\Delta S = \Delta S_{\text{Al}} + \Delta S_{\text{water}} = (-\,Q/T_{\text{Alav}}) + (+\,Q/T_{\text{waterav}})$$
$$= [(-\,4.55 \times 10^3\text{ cal})/(300.5\text{ K})] + [(+\,4.55 \times 10^3\text{ cal})/(295.5\text{ K})] = \boxed{+\,0.26\text{ cal/K.}}$$

41. (*a*) The actual efficiency of the engine is

$e_{actual} = W/Q_H = (550\ J)/(2200\ J) = 0.250 = 25.0\%.$

The ideal efficiency is

$e_{ideal} = 1 - (T_L/T_H) = 1 - [(650\ K)/(970\ K)] = 0.330 = 33.0\%.$

Thus the engine is running at

$e_{actual}/e_{ideal} = (0.250)/(0.330) = 0.758 =$ $\boxed{75.8\% \text{ of ideal.}}$

(*b*) We find the heat exhausted in one cycle from

$Q_L = Q_H - W = 2200\ J - 550\ J = 1650\ J.$

In one cycle of the engine there is no entropy change for the engine. The input heat is taken from the universe at T_H and the exhaust heat is added to the universe at T_L. The total entropy change is

$\Delta S_{total} = (-Q_H/T_H) + (Q_L/T_L)$

$= [-(2200\ J)/(970\ K)] + [(1650\ J)/(650\ K)] =$ $\boxed{+0.270\ J/K.}$

(*c*) For a Carnot engine we have

$W = eQ_H = 0.330Q_H;$

$Q_L = Q_H - W = (1 - e)Q_H = (1 - 0.330)Q_H = 0.667\ Q_H.$

The total entropy change is

$\Delta S_{total} = (-Q_H/T_H) + (Q_L/T_L)$

$= [-Q_H/(970\ K)] + [+0.667\ Q_H/(650\ K)] =$ $\boxed{0,}$ as expected for an ideal engine.

42. Because each die has six possible results, there are (6)(6) = 36 possible microstates.

(*a*) We find the number of microstates that give a 5 by listing all the possibilities:

(1, 4), (2, 3), (3, 2), and (4, 1), so there are 4 microstates.

Thus the probability of obtaining a 5 is

$P_5 = (4\text{ microstates})/(36\text{ microstates}) =$ $\boxed{1/9.}$

(*b*) We find the number of microstates that give an 11 by listing all the possibilities:

(5, 6), and (6, 5), so there are 2 microstates.

Thus the probability of obtaining an 11 is

$P_{11} = (2\text{ microstates})/(36\text{ microstates}) =$ $\boxed{1/18.}$

43. We are not concerned with the order that the cards are placed in the hand.

(*a*) One of the possible microstates is the four aces and one of the four kings. Because the suit of the king is not specified, there are four different possibilities, and thus the macrostate of four aces and one king has 4 microstates.

(*b*) Because the deck contains only one of each card specified, there is only 1 microstate for the macrostate of six of hearts, eight of diamonds, queen of clubs, three of hearts, jack of spades.

(*c*) If we call the four jacks J1, J2, J3, J4, without regard to order we have the following possible pairs:

J1J2, J1J3, J1J4, J2J3, J2J4, J3J4, so there are 6 combinations for the jacks.

Similarly, there will be 6 combinations for the queens, but only 4 combinations for the ace.

Thus the total number of microstates for the macrostate of two jacks, two queens, and an ace is

(6)(6)(4) = 144.

(*d*) We construct the hand by considering the number of ways we can draw each card. Because we are not concerned with any specific values, there will be 52 possibilities for the first card. For the second draw, because we cannot have any of the cards equal in value to the first one, there will be only 48 possibilities. Similarly, there will be 44 possibilities for the third draw, 40 possibilities for the fourth draw, and 36 possibilities for the fifth draw. If we were concerned with order, the total number of possibilities would be the product of these. The number of microstates must be reduced by dividing by the number of ways of arranging five cards, which is

(5)(4)(3)(2)(1) = 120.

Thus the number of microstates for a hand with no two equal-value cards is

$(52)(48)(44)(40)(36)/120 = 1.32 \times 10^6.$

The probability will increase with the number of microstates, so the order is

$\boxed{(b), (a), (c), (d).}$

44. We use H for head, and T for tail. For the microstates we construct the following table:

Macrostate	Microstates	Number
6 heads	HHHHHH	1
5 heads, 1 tail	HHHHHT, HHHHTH, HHHTHH, HHTHHH, HTHHHH, THHHHH	6
4 heads, 2 tails	HHHHTT, HHHTHT, HHHTTH, HHTHHT, HHTHTH, HHTTHH, HTHHHT, HTHHTH, HTHTHH, HTTHHH, THHHHT, THHHTH, THHTHH, THTHHH,TTHHHH	15
3 heads, 3 tails	HHHTTT, HHTHTT, HHTTHT, HHTTTH, HTHHTT, HTHTHT, HTHTTH, HTTHHT, HTTHTH, HTTTHH, THHHTT, THHTHT, THHTTH, THTHHT, THTHTH, THTTHH, TTHHHT, TTHHTH, TTHTHH, TTTHHH	20
2 heads, 4 tails	HHTTTT, HTHTTT, HTTHTT, HTTTHT, HTTTTH, THHTTT, THTHTT, THTTHT, THTTTH, TTHHTT, TTHTHT, TTHTTH, TTTHHT, TTTHTH, TTTTHH	15
1 head, 5 tails	HTTTTT, THTTTT, TTHTTT, TTTHTT, TTTTHT, TTTTTH	6
6 tails	TTTTTT	1

There are a total of 64 microstates.

(*a*) The probability of obtaining three heads and three tails is

$P_{33} = (20 \text{ microstates})/(64 \text{ microstates}) = \boxed{5/16.}$

(*b*) The probability of obtaining six heads is

$P_{60} = (1 \text{ microstate})/(64 \text{ microstates}) = \boxed{1/64.}$

45. (*a*) We assume that our hand is the first dealt and we receive the aces in the first four cards. The probability of the first card being an ace is 4/52. For the next three cards dealt to the other players, there must be no aces, so the probabilities are 48/51, 47/50, 46/49. The next card is ours; the probability of it being an ace is 3/48. For the next three cards, there must be no aces, so the probabilities are 45/47, 44/46, 43/45. For the next cards, we have

2/44, 42/43, 41/42, 40/41, 1/40, 39/39, 38/38, … .

For the product of all these, we have

$P_1 = (4/52)(48/51)(47/50)(46/49)(3/48)(45/47)(44/46)(43/45)(2/44)(42/43)(41/42)(40/41)(1/40)$
$= (4)(3)(2)(1)/(52)(51)(50)(49).$

Because we do not have to receive the aces as the first four cards, we multiply this probability by the number of ways 4 cards can be drawn from a total of 13, which is

$13!/4!9! = (13)(12)(11)(10)/(4)(3)(2)(1).$

Thus the probability of being dealt four aces is

$P = [(4)(3)(2)(1)/(52)(51)(50)(49)][(13)(12)(11)(10)/(4)(3)(2)(1)]$
$= (13)(12)(11)(10)/(52)(51)(50)(49) = 0.00264 = \boxed{1/379.}$

[Note that this is (48! 4!/52!)(13!/4! 9!).]

(*b*) We assume that our hand is the first dealt. Because the suit is not specified, any first card is acceptable, so the probability is (52/52) = 1. For the next three cards dealt to the other players, there must be no cards in the suit we were dealt, so the probabilities are

39/51, 38/50, 37/49.

The next card is ours; the probability of it being in the same suit is 12/48.

For the next three cards, there must be no cards in the suit we were dealt, so the probabilities are

36/47, 35/46, 34/45.

For the next cards, we have

11/44, 33/43, 32/42, 31/41, 10/40, … .

For the product of all these, we have

$P_1 = (1)(39/51)(38/50)(37/49)(12/48)(36/47)(35/46)(34/45)(11/44))(33/43)(32/42)(31/41)(10/40)\ldots$
$= 12!\,39!/51! = 6.3 \times 10^{-12} = \boxed{1/1.59 \times 10^{11}.}$

46. We find the area from

$E = (IA)t;$

22×10^3 Wh/day = (40 W/m^2)A(12 h/day), which gives A = 46 m^2 ≈ 490 ft^2.

Although this area (about 20 ft ×25 ft) would probably fit, it would have to be larger because it is not directly facing the sun for the 12 hours.

47. If we assume 100% efficiency, the electrical power is the rate at which the gravitational energy of the water is changed:

$P = (\Delta m/\Delta t)gh = (35\ \text{m}^3/\text{s})(1000\ \text{kg/m}^3)(9.80\ \text{m/s}^2)(45\ \text{m}) = 1.5 \times 10^7\ \text{W} = 15\ \text{MW}.$

48. (*a*) The energy required to increase the gravitational potential energy of the water is

$E = (\Delta m/\Delta t)ght$

$= (1.00 \times 10^5\ \text{kg/s})(9.80\ \text{m/s}^2)(135\ \text{m})(10.0\ \text{h}) = 1.32 \times 10^9\ \text{Wh} = 1.32 \times 10^6\ \text{kWh}.$

(*b*) If the stored energy is released at 75% efficiency, we have

$P = (0.75)E/t = (0.75)(1.32 \times 10^6\ \text{kWh})/(14\ \text{h}) = 7.09 \times 10^4\ \text{kW}.$

49. Because the change in elevation of the water is not constant during the tidal flow, we take half the difference in height as the average change. In one pass through the turbines, for the work done by the falling water, we have

$W_1 = mgh_{av} = \rho Ahg(\tfrac{1}{2}h).$

If there are two high and two low tides each day, there will be four passes through the turbines, so the total work in one day is

$W = 4W_1 = 2\rho Agh^2 = 2(1000\ \text{kg/m}^3)(23 \times 10^6\ \text{m}^2)(9.80\ \text{m/s}^2)(8.5\ \text{m})^2 = 3.3 \times 10^{13}\ \text{J}.$

50. We find the heat released from

$Q = mc\,\Delta T = \rho Vc\,\Delta T = (1000\ \text{kg/m}^3)(1000\ \text{m})^3(4186\ \text{J/kg}\cdot\text{C}°)(1\ \text{K}) = 4.2 \times 10^{15}\ \text{J}.$

51. The internal energy of an ideal gas depends only on the temperature, $U = \tfrac{3}{2}nRT$. For an adiabatic process there is no heat flow. For the first law of thermodynamics we have

$\Delta U = \tfrac{3}{2}nR\,\Delta T = Q - W;$

$\tfrac{3}{2}(1.5\ \text{mol})(8.31\ \text{J/mol}\cdot\text{K})\,\Delta T = 0 - (7500\ \text{J})$, which gives $\Delta T = -401\ \text{K} = -401\ \text{C}°.$

52. First we check to see if energy is conserved:

$Q_L + W = 1.50\ \text{MW} + 1.50\ \text{MW} = 3.00\ \text{MW} = Q_H,$

so energy is conserved.

The efficiency of the engine is

$e = W/Q_H = (1.50\ \text{MW})/(3.00\ \text{MW}) = 0.500 = 50.0\%.$

The maximum possible efficiency is the efficiency of the Carnot cycle:

$e = 1 - (T_L/T_H) = 1 - [(215\ \text{K})/(425\ \text{K})] = 0.494 = 49.4\%.$

Yes, there is something fishy, because his claimed efficiency is not possible.

53. (*a*) Because the pressure is constant, we find the work from

$W = p(V_2 - V_1)$

$= (1.013 \times 10^5\ \text{N/m}^2)(4.1\ \text{m}^3 - 1.9\ \text{m}^3) = 2.2 \times 10^5\ \text{J}.$

(*b*) We use the first law of thermodynamics to find the change in internal energy:

$\Delta U = Q - W$

$= +5.30 \times 10^4\ \text{J} - 2.2 \times 10^5\ \text{J} = -1.7 \times 10^5\ \text{J}.$

(*c*)

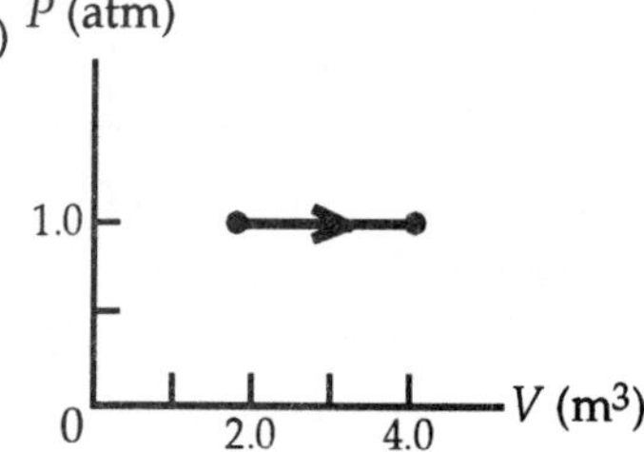

54. (*a*) We find the rate at which work is done from

$P = W/t = (200\text{ J/cycle/cyl})(4\text{ cyl})(25\text{ cycles/s}) =$ $\boxed{2.0 \times 10^4\text{ J/s.}}$

(*b*) We find the heat input rate from

$e = W/Q_H;$

$0.25 = (2.0 \times 10^4\text{ J/s})/(Q_H/t)$, which gives $Q_H/t =$ $\boxed{8.0 \times 10^4\text{ J/s.}}$

(*c*) We find the time to use one gallon from

$t = E/(Q_H/t) = (130 \times 10^6\text{ J})/(8.0 \times 10^4\text{ J/s}) = 1.6 \times 10^3\text{ s} =$ $\boxed{27\text{ min.}}$

55. The maximum possible efficiency is the efficiency of the Carnot cycle:

$e = 1 - (T_L/T_H) = 1 - [(90\text{ K})/(293\text{ K})] = 0.69 =$ $\boxed{69\%.}$

56. The maximum possible efficiency is the efficiency of the Carnot cycle:

$e = 1 - (T_L/T_H) = 1 - [(277\text{ K})/(300\text{ K})] = 0.077 =$ $\boxed{7.7\%.}$

The engine might be feasible because the great amount of water in the ocean could allow a large flow rate through the engine. Possible adverse environmental effects would be that mixing the waters on a large scale would change the environment for those creatures that live in the cooler water, and change in the surface temperature of the ocean over a large area could cause atmospheric changes.

57. We assume that the loss in kinetic energy is transferred to the environment as a heat flow. For the entropy change we have

$\Delta S = Q/T = 2(\tfrac{1}{2}mv^2)/T = mv^2/T$

$= (1100\text{ kg})[(100\text{ km/h})/(3.6\text{ ks/h})]^2/(293\text{ K}) =$ $\boxed{2.9 \times 10^3\text{ J/K.}}$

58. (*a*) We find the final temperature from

heat lost = heat gained;

$m_{\text{water}}c_{\text{water}}\,\Delta T_{\text{water}} = m_{\text{Al}}c_{\text{Al}}\,\Delta T_{\text{Al}};$

$(0.140\text{ kg})(4186\text{ J/kg}\cdot\text{C}^\circ)(50^\circ\text{C} - T) = (0.120\text{ kg})(900\text{ J/kg}\cdot\text{C}^\circ)(T - 15^\circ\text{C}),$

which gives $T =$ $\boxed{44.6^\circ\text{C.}}$

(*b*) The heat flow from the water to the aluminum is

$Q = m_{\text{Al}}c_{\text{Al}}\,\Delta T_{\text{Al}} = (0.120\text{ kg})(900\text{ J/kg}\cdot\text{C}^\circ)(44.6^\circ\text{C} - 15^\circ\text{C}) = 3.20 \times 10^3\text{ J}.$

The heating and cooling do not occur at constant temperature. To estimate the entropy changes, we will use the average temperatures:

$T_{\text{waterav}} = \tfrac{1}{2}(T_{\text{water}} + T) = \tfrac{1}{2}(50^\circ\text{C} + 44.6^\circ\text{C}) = 47.3^\circ\text{C} = 320.5\text{ K};$

$T_{\text{Alav}} = \tfrac{1}{2}(T_{\text{Al}} + T) = \tfrac{1}{2}(15^\circ\text{C} + 44.6^\circ\text{C}) = 29.8^\circ\text{C} = 303.0\text{ K}.$

The total entropy change is

$\Delta S = \Delta S_{\text{Al}} + \Delta S_{\text{water}} = (+\,Q/T_{\text{Alav}}) + (-\,Q/T_{\text{waterav}})$

$= [(+\,3.20 \times 10^3\text{ J})/(303.0\text{ K})] + [(-\,3.20 \times 10^3\text{ J})/(320.5\text{ K})] =$ $\boxed{+0.58\text{ J/K.}}$

59. (*a*) The coefficient of performance for the heat pump is

$\text{CP} = T_H/(T_H - T_L) = (297\text{ K})/(297\text{ K} - 279\text{ K}) =$ $\boxed{17.}$

(*b*) We find the heat delivered at the high temperature from

$\text{CP} = Q_H/W;$

$17 = Q_H/(1000\text{ J/s})(3600\text{ s/h})$, which gives $Q_H =$ $\boxed{6.1 \times 10^7\text{ J/h.}}$

60. We use the units to help us find the rate of heat input to the engine from the burning of the gasoline:

$Q_H/t = [(3.0 \times 10^4\text{ kcal/gal})/(41\text{ km/gal})](90\text{ km/h})(4186\text{ J/kcal}) = 2.76 \times 10^8\text{ J/h}.$

The horsepower is the work done by the engine. We find the efficiency from

$e = W/Q_H = (25\text{ hp})(746\text{ W/hp})(3600\text{ s/h})/(2.76 \times 10^8\text{ J/h}) = 0.24 =$ $\boxed{24\%.}$

61. We assume that the loss in kinetic energy is transferred to the environment as a heat flow. For the entropy change we have

$\Delta S = Q/T = \frac{1}{2}mv^2/T$

$= \frac{1}{2}(15 \text{ kg})(2.4 \text{ m/s})^2/(293 \text{ K}) = \boxed{0.15 \text{ J/K.}}$

62. We find the change in length of the steel I-beam from

$\Delta L = L\alpha\,\Delta T = (7.5 \text{ m})[11 \times 10^{-6} (\text{C°})^{-1}](-6.0 \text{ C°}) = -4.95 \times 10^{-4} \text{ m}.$

The heat flow from the steel beam is

$Q = mc\,\Delta T = (300 \text{ kg})(0.11 \text{ cal/kg}\cdot\text{C°})(-6.0 \text{ C°})(4186 \text{ J/kcal}) = -8.29 \times 10^5 \text{ J}.$

Work is done on the beam as the load at the top moves down a distance ΔL and the center of gravity moves down a distance $\frac{1}{2}\Delta L$. If we take the beam as our system, the work done by the system is negative:

$W = F\,\Delta L + mg\frac{1}{2}\Delta L = (F + \frac{1}{2}mg)\,\Delta L = [4.3 \times 10^5 \text{ N} + \frac{1}{2}(300 \text{ kg})(9.80 \text{ m/s}^2)](-4.95 \times 10^{-4} \text{ m}) = -214 \text{ J}.$

We use the first law of thermodynamics to find the change in internal energy:

$\Delta U = Q - W$

$= -8.29 \times 10^5 \text{ J} - (-214 \text{ J}) = \boxed{-8.3 \times 10^5 \text{ J.}}$

63. We find the heat input rate from

$e = W/Q_H;$

$0.33 = (800 \text{ MW})/(Q_H/t)$, which gives $Q_H/t = 2424 \text{ MW}$.

We find the discharge heat flow from

$Q_L/t = (Q_H/t) - (W/t) = 2424 \text{ MW} - 800 \text{ MW} = 1624 \text{ MW}.$

If this heat flow warms the air, we have

$Q_L/t = (n/t)c\,\Delta T;$

$(1624 \times 10^6 \text{ W})(3600 \text{ s/h})(24 \text{ h/day}) = (n/t)(7.0 \text{ cal/mol}\cdot\text{C°})(7.0 \text{ C°})(4.186 \text{ J/cal}),$

which gives $n/t = 6.84 \times 10^{11} \text{ mol/day}$.

To find the volume rate, we use the ideal gas law:

$P(V/t) = (n/t)RT;$

$(1.013 \times 10^5 \text{ Pa})(V/t) = (6.84 \times 10^{11} \text{ mol/day})(8.315 \text{ J/mol}\cdot\text{K})(293 \text{ K}),$

which gives $V/t = 1.65 \times 10^{10} \text{ m}^3\text{/day} = \boxed{17 \text{ km}^3\text{/day.}}$

Depending on the dispersal by the winds, the local climate could be heated significantly.

We find the area from

$A = (V/t)t/h = (16.5 \text{ km}^3\text{/day})(1 \text{ day})/(0.200 \text{ km}) = \boxed{83 \text{ km}^2.}$

64. (*a*) The efficiency of the plant is

$e = 1 - (T_L/T_H) = 1 - [(285 \text{ K})/(600 \text{ K})] = 0.525 = 52.5\%.$

We find the heat input rate from

$e = W/Q_H;$

$0.525 = (900 \text{ MW})/(Q_H/t)$, which gives $Q_H/t = 1714 \text{ MW}$.

We find the discharge heat flow from

$Q_L/t = (Q_H/t) - (W/t) = 1714 \text{ MW} - 900 \text{ MW} = 814 \text{ MW}.$

If this heat flow warms some river water, which mixes with the rest of the river water, we have

$Q_L/t = (n/t)c\,\Delta T;$

$(814 \times 10^6 \text{ W}) = (37 \text{ m}^3\text{/s})(1000 \text{ kg/m}^3)(4186 \text{ J/kg}\cdot\text{C°})\,\Delta T$, which gives $\Delta T = \boxed{5.3 \text{ C°.}}$

(*b*) The heat flow does not occur at constant temperature. To estimate the entropy changes, we will use the average temperatures:

$T_{waterav} = T_{water} + \frac{1}{2}\,\Delta T = 285 \text{ K} + \frac{1}{2}(5.3 \text{ C°}) = 287.6 \text{ K};$

The entropy change is

$\Delta S/m = (Q_L/t)/(m/t)T_{waterav}$

$= (814 \times 10^6 \text{ W})/(37 \text{ m}^3\text{/s})(1000 \text{ kg/m}^3)(287.6 \text{ K}) = \boxed{+77 \text{ J/kg}\cdot\text{K.}}$

CHAPTER 16

1. The number of electrons is
   $N = Q/e = (-30.0\times10^{-6}\text{ C})/(-1.60\times10^{-19}\text{ C/electrons}) =$ $\boxed{1.88\times10^{14}\text{ electrons.}}$

2. The magnitude of the Coulomb force is
   $F = kQ_1Q_2/r^2.$
   If we divide the expressions for the two forces, we have
   $F_2/F_1 = (r_1/r_2)^2;$
   $F_2/(4.2\times10^{-2}\text{ N}) = (8)^2$, which gives $F_2 =$ $\boxed{2.7\text{ N.}}$

3. The magnitude of the Coulomb force is
   $F = kQ_1Q_2/r^2.$
   If we divide the expressions for the two forces, we have
   $F_2/F_1 = (r_1/r_2)^2;$
   $3 = [(20.0\text{ cm})/r_2]^2$, which gives $r_2 =$ $\boxed{11.5\text{ cm.}}$

4. The magnitude of the Coulomb force is
   $F = kQ_1Q_2/r^2.$
   If we divide the expressions for the two forces, we have
   $F_2/F_1 = (r_1/r_2)^2;$
   $F_2/(0.0200\text{ N}) = (150\text{ cm}/30.0\text{ cm})^2$, which gives $F_2 =$ $\boxed{0.500\text{ N.}}$

5. The magnitude of the Coulomb force is
   $F = kQ_1Q_2/r^2$
   $= (9.0\times10^9\text{ N}\cdot\text{m}^2/\text{C}^2)(26)(1.60\times10^{-19}\text{ C})(1.60\times10^{-19}\text{ C})/(1.5\times10^{-12}\text{ m})^2 =$ $\boxed{2.7\times10^{-3}\text{ N.}}$

6. The magnitude of the Coulomb force is
   $F = kQ_1Q_2/r^2$
   $= (9.0\times10^9\text{ N}\cdot\text{m}^2/\text{C}^2)(1.60\times10^{-19}\text{ C})(1.60\times10^{-19}\text{ C})/(5.0\times10^{-15}\text{ m})^2 =$ $\boxed{9.2\text{ N.}}$

7. The magnitude of the Coulomb force is
   $F = kQ_1Q_2/r^2$
   $= (9.0\times10^9\text{ N}\cdot\text{m}^2/\text{C}^2)(15\times10^{-6}\text{ C})(3.00\times10^{-3}\text{ C})/(0.40\text{ m})^2 =$ $\boxed{2.5\times10^3\text{ N.}}$

8. The number of excess electrons is
   $N = Q/e = (-60\times10^{-6}\text{ C})/(-1.60\times10^{-19}\text{ C/electrons}) =$ $\boxed{3.8\times10^{14}\text{ electrons.}}$
   The mass increase is
   $\Delta m = Nm_e = (3.8\times10^{14}\text{ electrons})(9.11\times10^{-31}\text{ kg/electron}) =$ $\boxed{3.4\times10^{-16}\text{ kg.}}$

9. Because the charge on the Earth can be considered to be at the center, we can use the expression for the force between two point charges. For the Coulomb force to be equal to the weight, we have
   $kQ^2/R^2 = mg;$
   $(9.0\times10^9\text{ N}\cdot\text{m}^2/\text{C}^2)Q^2/(6.38\times10^6\text{ m})^2 = (1050\text{ kg})(9.80\text{ m/s}^2)$, which gives $Q =$ $\boxed{6.8\times10^3\text{ C.}}$

10. The number of molecules in 1.0 kg is

$N = [(1.0 \text{ kg})(10^3 \text{ g/kg})/(18 \text{ g/mol})](6.02 \times 10^{23} \text{ molecules/mol}) = 3.34 \times 10^{25} \text{ molecules}.$

Each molecule of H_2O contains 2(1) + 8 = 10 electrons. The charge of the electrons in 1.0 kg is

$q = (3.34 \times 10^{25} \text{ molecules})(10 \text{ electrons/molecule})(-1.60 \times 10^{-19} \text{ C/electron})$

$= \boxed{-5.4 \times 10^7 \text{ C.}}$

11. Using the symbols in the figure, we find the magnitudes of the three individual forces:

$F_{12} = F_{21} = kQ_1Q_2/r_{12}^2 = kQ_1Q_2/L^2$
$= (9.0 \times 10^9 \text{ N}\cdot\text{m}^2/\text{C}^2)(70 \times 10^{-6} \text{ C})(48 \times 10^{-6} \text{ C})/(0.35 \text{ m})^2$
$= 2.47 \times 10^2 \text{ N}.$

$F_{13} = F_{31} = kQ_1Q_3/r_{12}^2 = kQ_1Q_3/(2L)^2$
$= (9.0 \times 10^9 \text{ N}\cdot\text{m}^2/\text{C}^2)(70 \times 10^{-6} \text{ C})(80 \times 10^{-6} \text{ C})/[2(0.35 \text{ m})]^2$
$= 1.03 \times 10^2 \text{ N}.$

$F_{23} = F_{32} = kQ_2Q_3/r_{12}^2 = kQ_2Q_3/L^2$
$= (9.0 \times 10^9 \text{ N}\cdot\text{m}^2/\text{C}^2)(48 \times 10^{-6} \text{ C})(80 \times 10^{-6} \text{ C})/(0.35 \text{ m})^2$
$= 2.82 \times 10^2 \text{ N}.$

The directions of the forces are determined from the signs of the charges and are indicated on the diagram. For the net forces, we get

$F_1 = F_{13} - F_{12} = 1.03 \times 10^2 \text{ N} - 2.47 \times 10^2 \text{ N} = \boxed{-1.4 \times 10^2 \text{ N (left).}}$

$F_2 = F_{21} - F_{23} = 2.47 \times 10^2 \text{ N} + 2.82 \times 10^2 \text{ N} = \boxed{+5.3 \times 10^2 \text{ N (right).}}$

$F_3 = -F_{31} - F_{32} = -1.03 \times 10^2 \text{ N} - 2.82 \times 10^2 \text{ N} = \boxed{-3.9 \times 10^2 \text{ N (left).}}$

Note that the sum for the three charges is zero.

12. Because all the charges and their separations are equal, we find the magnitude of the individual forces:

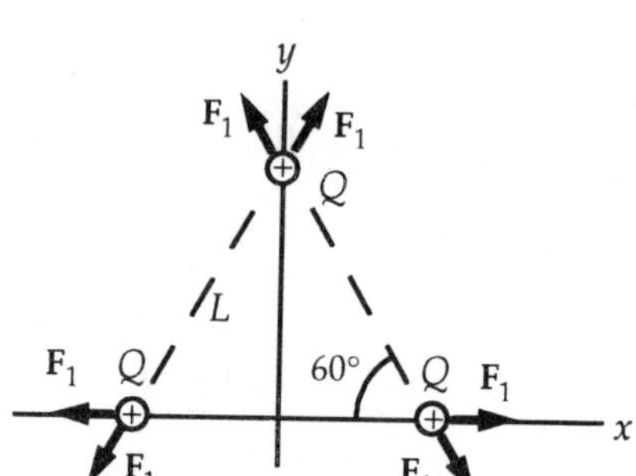

$F_1 = kQQ/L^2 = kQ^2/L^2$
$= (9.0 \times 10^9 \text{ N}\cdot\text{m}^2/\text{C}^2)(11.0 \times 10^{-6} \text{ C})^2/(0.150 \text{ m})^2$
$= 48.4 \text{ N}.$

The directions of the forces are determined from the signs of the charges and are indicated on the diagram.

For the forces on the top charge, we see that the horizontal components will cancel. For the net force, we have

$F = F_1 \cos 30° + F_1 \cos 30° = 2F_1 \cos 30°$
$= 2(48.4 \text{ N}) \cos 30°$
$= 83.8 \text{ N}$ up, or away from the center of the triangle.

From the symmetry each of the other forces will have the same magnitude and a direction away from the center: The net force on each charge is $\boxed{\text{83.8 N away from the center of the triangle.}}$

Note that the sum for the three charges is zero.

13. We find the magnitudes of the individual forces on the charge at the upper right corner:

$F_1 = F_2 = kQQ/L^2 = kQ^2/L^2$
$= (9.0 \times 10^9\ \text{N}\cdot\text{m}^2/\text{C}^2)(6.00 \times 10^{-3}\ \text{C})^2/(1.00\ \text{m})^2$
$= 3.24 \times 10^5$ N.

$F_3 = kQQ/(L\sqrt{2})^2 = kQ^2/2L^2$
$= (9.0 \times 10^9\ \text{N}\cdot\text{m}^2/\text{C}^2)(6.00 \times 10^{-3}\ \text{C})^2/2(1.00\ \text{m})^2$
$= 1.62 \times 10^5$ N.

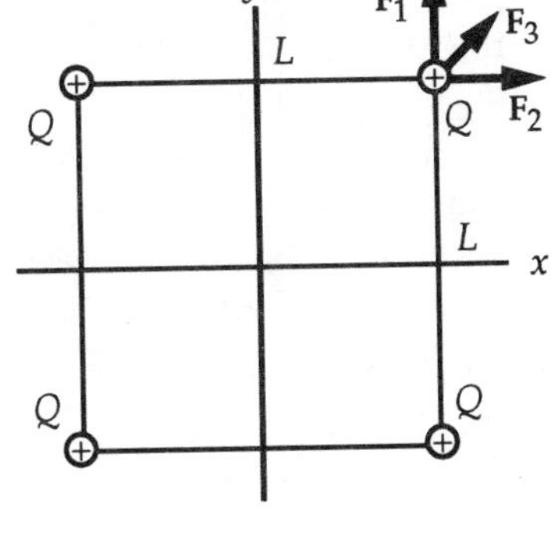

The directions of the forces are determined from the signs of the charges and are indicated on the diagram.
For the forces on the top charge, we see that the net force will be along the diagonal. For the net force, we have

$F = F_1 \cos 45° + F_2 \cos 45° + F_3$
$= 2(3.24 \times 10^5\ \text{N}) \cos 45° + 1.62 \times 10^5$ N
$= 6.20 \times 10^5$ N along the diagonal, or away from the center of the square.

From the symmetry, each of the other forces will have the same magnitude and a direction away from the center: The net force on each charge is **6.20×10^5 N away from the center of the square.**
Note that the sum for the three charges is zero.

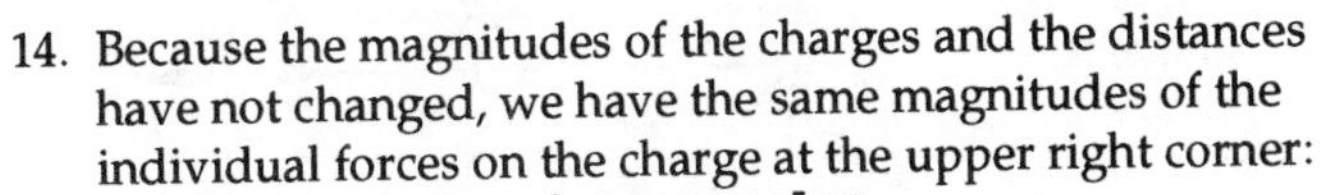

14. Because the magnitudes of the charges and the distances have not changed, we have the same magnitudes of the individual forces on the charge at the upper right corner:

$F_1 = F_2 = kQQ/L^2 = 3.24 \times 10^5$ N.
$F_3 = kQ^2/2L^2 = 1.62 \times 10^5$ N.

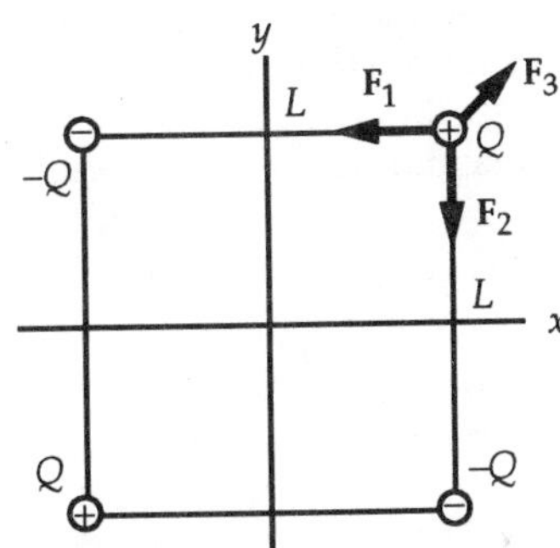

The directions of the forces are determined from the signs of the charges and are indicated on the diagram.
For the forces on the top charge, we see that the net force will be along the diagonal. For the net force, we have

$F = -F_1 \cos 45° - F_2 \cos 45° + F_3$
$= -2(3.24 \times 10^5\ \text{N}) \cos 45° + 1.62 \times 10^5$ N
$= -2.96 \times 10^5$ N along the diagonal, or toward the center of the square.

From the symmetry, each of the other forces will have the same magnitude and a direction toward the center: The net force on each charge is **2.96×10^5 N toward the center of the square.**
Note that the sum for the three charges is zero.

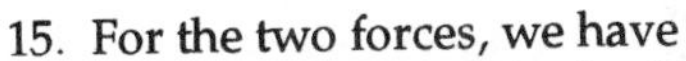

15. For the two forces, we have

$F_{\text{electric}} = kq_1q_2/r_{12}^2 = ke^2/r^2$
$= (9.0 \times 10^9\ \text{N}\cdot\text{m}^2/\text{C}^2)(1.6 \times 10^{-19}\ \text{C})^2/(0.53 \times 10^{-10}\ \text{m})^2 =$ **8.2×10^{-8} N.**

$F_{\text{gravitational}} = Gm_1m_2/r^2$
$= (6.67 \times 10^{-11}\ \text{N}\cdot\text{m}^2/\text{kg}^2)(9.11 \times 10^{-31}\ \text{kg})(1.67 \times 10^{-27}\ \text{kg})/(0.53 \times 10^{-10}\ \text{m})^2$
$=$ **3.6×10^{-47} N.**

The ratio of the forces is

$F_{\text{electric}}/F_{\text{gravitational}} = (8.2 \times 10^{-8}\ \text{N})/(3.6 \times 10^{-47}\ \text{N}) =$ **2.3×10^{39}.**

16. Because the electrical attraction must provide the same force as the gravitational attraction, we equate the two forces:

$kQQ/r^2 = Gm_1m_2/r^2$;
$(9.0 \times 10^9\ \text{N}\cdot\text{m}^2/\text{C}^2)Q^2 = (6.67 \times 10^{-11}\ \text{N}\cdot\text{m}^2/\text{kg}^2)(5.97 \times 10^{24}\ \text{kg})(7.35 \times 10^{22}\ \text{kg})$, which gives
$Q =$ **5.71×10^{13} C.**

17. If the separation is r and one charge is Q_1, the other charge will be $Q_2 = Q_T - Q_1$. For the repulsive force, we have

$$F = kQ_1Q_2/r^2 = kQ_1(Q_T - Q_1)/r^2.$$

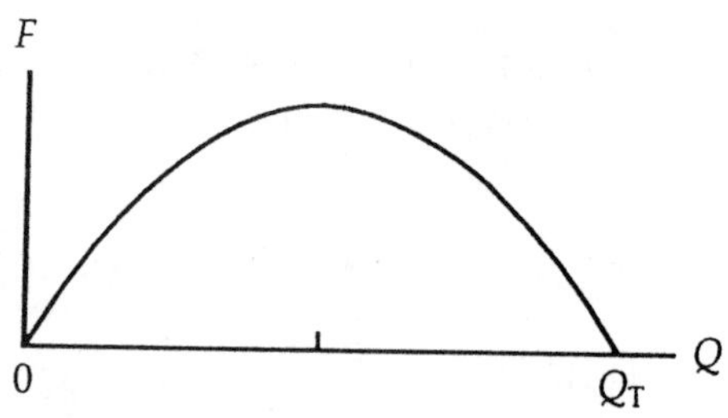

(*a*) If we plot the force as a function of Q_1, we see that the maximum occurs when

$$\boxed{Q_1 = \tfrac{1}{2}Q_T,}$$

which we would expect from symmetry, since we could interchange the two charges without changing the force.

(*b*) We see from the plot that the minimum occurs when either charge is zero:

$$\boxed{Q_1 \text{ (or } Q_2) = 0.}$$

18. The attractive Coulomb force provides the centripetal acceleration of the electron:

$$ke^2/r^2 = mv^2/r, \quad \text{or} \quad r = ke^2/mv^2;$$

$$r = (9.0 \times 10^9\ \text{N}\cdot\text{m}^2/\text{C}^2)(1.60 \times 10^{-19}\ \text{C})^2/(9.11 \times 10^{-31}\ \text{kg})(1.1 \times 10^6\ \text{m/s})^2 = \boxed{2.1 \times 10^{-10}\ \text{m.}}$$

19. If we place a positive charge, it will be repelled by the positive charge and attracted by the negative charge. Thus the third charge must be placed along the line of the charges, but not between them. For the net force to be zero, the magnitudes of the individual forces must be equal:

$$F = kQ_1Q/r_1^2 = kQ_2Q/r_2^2, \quad \text{or} \quad Q_1/(L + x)^2 = Q_2/x^2;$$

$$(5.7\ \mu\text{C})/(0.25\ \text{m} + x)^2 = (3.5\ \mu\text{C})/x^2, \text{ which gives } x = 0.91\ \text{m}, -0.11\ \text{m}.$$

The negative result corresponds to the position between the charges where the magnitudes and the directions are the same. Thus the third charge should be placed $\boxed{\text{0.91 m beyond the negative charge.}}$ Note that we would have the same analysis if we used a negative charge.

20. If one charge is Q_1, the other charge will be $Q_2 = Q - Q_1$. For the force to be repulsive, the two charges must have the same sign. Because the total charge is positive, each charge will be positive. We account for this by considering the force to be positive:

$$F = kQ_1Q_2/r^2 = kQ_1(Q - Q_1)/r^2;$$

$$12.0\ \text{N} = (9.0 \times 10^9\ \text{N}\cdot\text{m}^2/\text{C}^2)Q_1(80.0 \times 10^{-6}\ \text{C} - Q_1)/(1.06\ \text{m})^2, \text{ which is a quadratic equation:}$$

$$Q_1^2 - (80.0 \times 10^{-6}\ \text{C})Q_1 + 1.50 \times 10^{-9}\ \text{C}^2 = 0, \text{ which gives } Q_1 = \boxed{50.0 \times 10^{-6}\ \text{C}, 30.0 \times 10^{-6}\ \text{C.}}$$

Note that, because the labels are arbitrary, we get the value of both charges.

For an attractive force, the charges must have opposite signs, so their product will be negative. We account for this by considering the force to be negative:

$$F = kQ_1Q_2/r^2 = kQ_1(Q - Q_1)/r^2;$$

$$-12.0\ \text{N} = (9.0 \times 10^9\ \text{N}\cdot\text{m}^2/\text{C}^2)Q_1(80.0 \times 10^{-6}\ \text{C} - Q_1)/(1.06\ \text{m})^2, \text{ which is a quadratic equation:}$$

$$Q_1^2 - (80.0 \times 10^{-6}\ \text{C})Q_1 - 1.50 \times 10^{-9}\ \text{C}^2 = 0, \text{ which gives } Q_1 = \boxed{-15.7 \times 10^{-6}\ \text{C}, 95.7 \times 10^{-6}\ \text{C.}}$$

21. The acceleration is produced by the force from the electric field:

$$F = qE = ma;$$

$$(1.60 \times 10^{-19}\ \text{C})(600\ \text{N/C}) = (9.11 \times 10^{-31}\ \text{kg})a, \text{ which gives } a = \boxed{1.05 \times 10^{14}\ \text{m/s}^2.}$$

Because the charge on the electron is negative, the direction of force, and thus the acceleration, is $\boxed{\text{opposite to the direction of the electric field.}}$

The direction of the acceleration is $\boxed{\text{independent of the velocity.}}$

22. If we take the positive direction to the east, we have

$$F = qE = (-1.60 \times 10^{-19}\ \text{C})(+3500\ \text{N/C}) = -5.6 \times 10^{-16}\ \text{N}, \text{ or } \boxed{5.6 \times 10^{-16}\ \text{N (west).}}$$

23. If we take the positive direction to the south, we have

$F = qE;$

$3.2 \times 10^{-14}\ \text{N} = (+\ 1.60 \times 10^{-19}\ \text{C})E$, which gives $E =$ $\boxed{+\ 2.0 \times 10^{5}\ \text{N/C (south).}}$

24. If we take the positive direction up, we have

$F = qE;$

$+\ 8.4\ \text{N} = (-\ 8.8 \times 10^{-6}\ \text{C})E$, which gives $E =$ $\boxed{+\ 9.5 \times 10^{5}\ \text{N/C (up).}}$

25. The electric field above a positive charge will be away from the charge, or up.
We find the magnitude from

$E = kQ/r^2$

$= (9.0 \times 10^{9}\ \text{N}\cdot\text{m}^2/\text{C}^2)(33.0 \times 10^{-6}\ \text{C})/(0.300\ \text{m})^2 =$ $\boxed{3.30 \times 10^{6}\ \text{N/C (up).}}$

26. The directions of the fields are determined from the signs of the charges and are indicated on the diagram. The net electric field will be to the left. We find its magnitude from

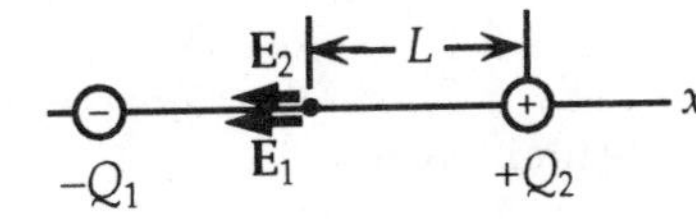

$E = kQ_1/L^2 + kQ_2/L^2 = k(Q_1 + Q_2)/L^2$

$= (9.0 \times 10^{9}\ \text{N}\cdot\text{m}^2/\text{C}^2)(8.0 \times 10^{-6}\ \text{C} + 6.0 \times 10^{-6}\ \text{C})/(0.020\ \text{m})^2$

$= 3.2 \times 10^{8}\ \text{N/C}.$

Thus the electric field is $\boxed{3.2 \times 10^{8}\ \text{N/C toward the negative charge.}}$

27. The acceleration is produced by the force from the electric field:

$F = qE = ma;$

$(-\ 1.60 \times 10^{-19}\ \text{C})E = (9.11 \times 10^{-31}\ \text{kg})(125\ \text{m/s}^2)$, which gives $E = -\ 7.12 \times 10^{-10}\ \text{N/C}$.

Because the charge on the electron is negative, the direction of force, and thus the acceleration, is opposite to the direction of the electric field, so the electric field is $\boxed{7.12 \times 10^{-10}\ \text{N/C (south).}}$

28. The directions of the fields are determined from the signs of the charges and are in the same direction, as indicated on the diagram.
The net electric field will be to the left. We find its magnitude from

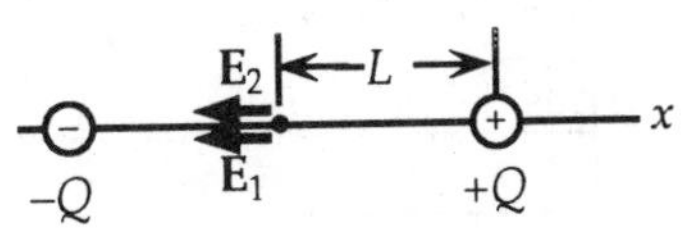

$E = kQ_1/L^2 + kQ_2/L^2 = k(Q + Q)/L^2 = 2kQ/L^2$

$1750\ \text{N/C} = 2(9.0 \times 10^{9}\ \text{N}\cdot\text{m}^2/\text{C}^2)Q/(0.080\ \text{m})^2$, which gives

$\boxed{Q = 6.2 \times 10^{-10}\ \text{C.}}$

29. From the diagram, we see that the electric fields produced by the charges will have the same magnitude, and the resultant field will be down.
The distance from the origin is x, so we have

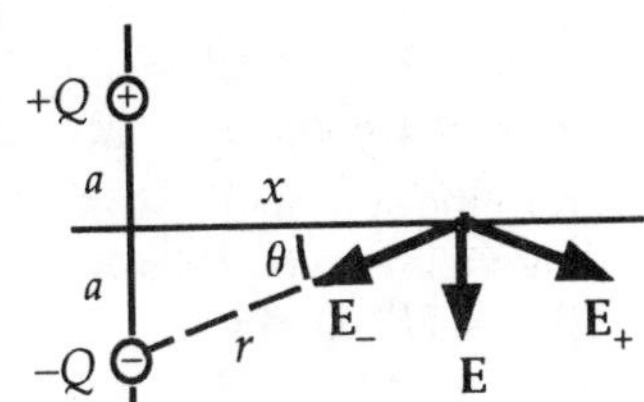

$E_+ = E_- = kQ/r^2 = kQ/(a^2 + x^2).$

From the symmetry, for the magnitude of the electric field we have

$E = 2E_+ \sin\theta = 2[kQ/(a^2 + x^2)][a/(a^2 + x^2)^{1/2}]$

$=$ $\boxed{2kQa/(a^2 + x^2)^{3/2}\ \text{parallel to the line of the charges.}}$

30. At point A, from the diagram, we see that the electric fields produced by the charges will have the same magnitude, and the resultant field will be up. We find the angle θ from

$$\tan \theta = (0.050 \text{ m})/(0.100 \text{ m}) = 0.500, \text{ or } \theta = 26.6°.$$

For the magnitudes of the individual fields we have

$$E_{1A} = E_{2A} = kQ/r_A^2$$
$$= (9.0 \times 10^9 \text{ N}\cdot\text{m}^2/\text{C}^2)(9.0 \times 10^{-6} \text{ C})/[(0.100 \text{ m})^2 + (0.050 \text{ m})^2]$$
$$= 6.48 \times 10^6 \text{ N/C}.$$

From the symmetry, the resultant electric field is

$$E_A = 2E_{1A} \sin \theta = 2(6.48 \times 10^6 \text{ N/C}) \sin 26.6° = \boxed{5.8 \times 10^6 \text{ N/C up.}}$$

For point B we find the angles for the directions of the fields from

$$\tan \theta_1 = (0.050 \text{ m})/(0.050 \text{ m}) = 1.00, \text{ or } \theta_1 = 45.0°.$$
$$\tan \theta_2 = (0.050 \text{ m})/(0.150 \text{ m}) = 0.333, \text{ or } \theta_2 = 18.4°.$$

For the magnitudes of the individual fields we have

$$E_{1B} = kQ/r_{1B}^2$$
$$= (9.0 \times 10^9 \text{ N}\cdot\text{m}^2/\text{C}^2)(9.0 \times 10^{-6} \text{ C})/[(0.050 \text{ m})^2 + (0.050 \text{ m})^2]$$
$$= 1.62 \times 10^7 \text{ N/C}.$$
$$E_{2B} = kQ/r_{2B}^2$$
$$= (9.0 \times 10^9 \text{ N}\cdot\text{m}^2/\text{C}^2)(9.0 \times 10^{-6} \text{ C})/[(0.150 \text{ m})^2 + (0.050 \text{ m})^2] = 3.24 \times 10^6 \text{ N/C}.$$

For the components of the resultant field we have

$$E_{Bx} = E_{1B} \cos \theta_1 - E_{2B} \cos \theta_2 = (1.62 \times 10^7 \text{ N/C}) \cos 45.0° - (3.24 \times 10^6 \text{ N/C}) \cos 18.4° = 8.38 \times 10^6 \text{ N/C};$$
$$E_{By} = E_{1B} \sin \theta_1 - E_{2B} \sin \theta_2 = (1.62 \times 10^7 \text{ N/C}) \sin 45.0° - (3.24 \times 10^6 \text{ N/C}) \sin 18.4° = 1.25 \times 10^7 \text{ N/C}.$$

We find the direction from

$$\tan \theta_B = E_{By}/E_{Bx} = (1.25 \times 10^7 \text{ N/C})/(8.38 \times 10^6 \text{ N/C}) = 1.49, \text{ or } \theta_1 = 56.2°.$$

We find the magnitude from

$$E_B = E_{Bx}/\cos \theta_B = (8.38 \times 10^6 \text{ N/C})/\cos 56.2° = 1.51 \times 10^7 \text{ N/C}.$$

Thus the field at point B is $\boxed{1.5 \times 10^7 \text{ N/C } 56° \text{ above the horizontal.}}$

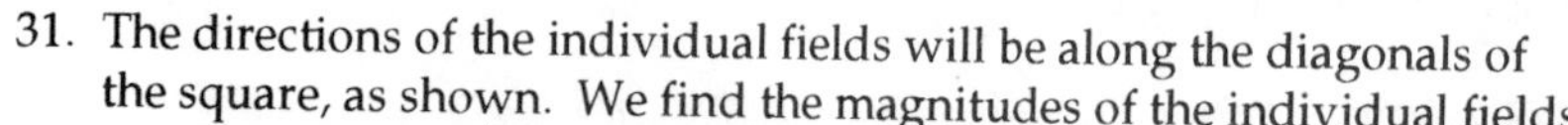

31. The directions of the individual fields will be along the diagonals of the square, as shown. We find the magnitudes of the individual fields:

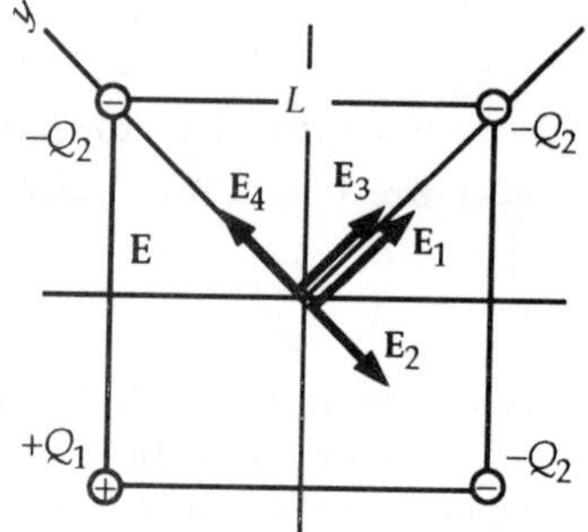

$$E_1 = kQ_1/(L/\sqrt{2})^2 = 2kQ_1/L^2$$
$$= 2(9.0 \times 10^9 \text{ N}\cdot\text{m}^2/\text{C}^2)(45.0 \times 10^{-6} \text{ C})/(0.60 \text{ m})^2 = 2.25 \times 10^6 \text{ N/C}.$$
$$E_2 = E_3 = E_4 = kQ_2/(L/\sqrt{2})^2 = 2kQ_2/L^2$$
$$= 2(9.0 \times 10^9 \text{ N}\cdot\text{m}^2/\text{C}^2)(31.0 \times 10^{-6} \text{ C})/(0.60 \text{ m})^2 = 1.55 \times 10^6 \text{ N/C}.$$

From the symmetry, we see that the resultant field will be along the diagonal shown as the x-axis. For the net field, we have

$$E = E_1 + E_3 = 2.25 \times 10^6 \text{ N/C} + 1.55 \times 10^6 \text{ N/C} = 3.80 \times 10^6 \text{ N/C}.$$

Thus the field at the center is

$$\boxed{3.80 \times 10^6 \text{ N/C away from the positive charge.}}$$

32. The directions of the individual fields are shown in the figure. We find the magnitudes of the individual fields:

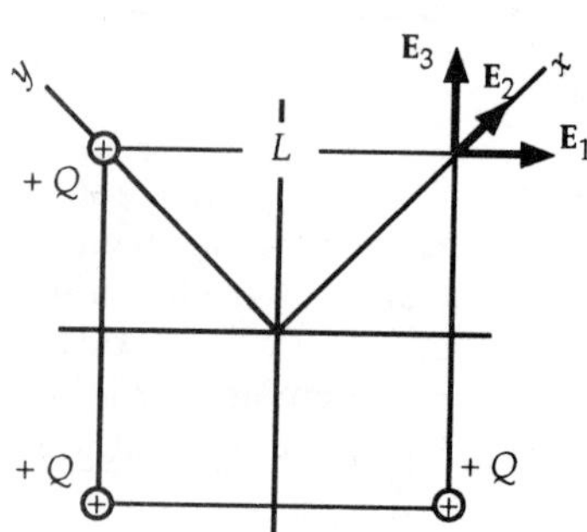

$$E_1 = E_3 = kQ/L^2$$
$$= (9.0 \times 10^9 \text{ N}\cdot\text{m}^2/\text{C}^2)(2.80 \times 10^{-6} \text{ C})/(1.00 \text{ m})^2 = 2.52 \times 10^4 \text{ N/C}.$$
$$E_2 = kQ/(L\sqrt{2})^2 = \tfrac{1}{2}kQ_2/L^2$$
$$= \tfrac{1}{2}(9.0 \times 10^9 \text{ N}\cdot\text{m}^2/\text{C}^2)(2.80 \times 10^{-6} \text{ C})/(1.00 \text{ m})^2 = 1.26 \times 10^4 \text{ N/C}.$$

From the symmetry, we see that the resultant field will be along the diagonal shown as the x-axis. For the net field, we have

$$E = 2E_1 \cos 45° + E_2 = 2(2.52 \times 10^4 \text{ N/C}) \cos 45° + 1.26 \times 10^4 \text{ N/C}$$
$$= 4.82 \times 10^4 \text{ N/C}.$$

Thus the field at the unoccupied corner is

$$\boxed{4.82 \times 10^4 \text{ N/C away from the opposite corner.}}$$

33. (*a*) The directions of the individual fields are shown in the figure.
We find the magnitudes of the individual fields:
$E_1 = E_2 = kQ/L^2$.
For the components of the resultant field we have
$E_x = -E_2 \sin 60° = -0.866kQ/L^2$;
$E_y = -E_1 - E_2 \cos 60° = -kQ/L^2 - 0.500kQ/L^2 = -1.50kQ/L^2$.
We find the direction from
$\tan\theta = E_y/E_x = (-1.50kQ/L^2)/(-0.866kQ/L^2) = 1.73$, or $\theta = 60°$.
We find the magnitude from
$E = E_x/\cos\theta = (0.866kQ/L^2)/\cos 60° = 1.73kQ/L^2$.
Thus the field is $1.73kQ/L^2$ 60° below the – x-axis.

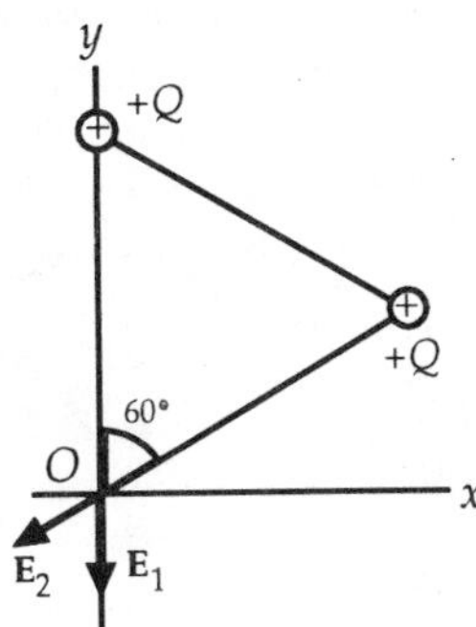

(*b*) The directions of the individual fields are shown in the figure.
The magnitudes of the individual fields will be the same:
$E_1 = E_2 = kQ/L^2$.
For the components of the resultant field we have
$E_x = +E_2 \sin 60° = +0.866kQ/L^2$;
$E_y = -E_1 + E_2 \cos 60° = -kQ/L^2 + 0.500kQ/L^2 = -0.500kQ/L^2$.
We find the direction from
$\tan\theta = E_y/E_x = (-0.500kQ/L^2)/(+0.866kQ/L^2) = -0.577$, or $\theta = -30°$.
We find the magnitude from
$E = E_x/\cos\theta = (0.866kQ/L^2)/\cos 30° = kQ/L^2$.
Thus the field is kQ/L^2 30° below the + x-axis.

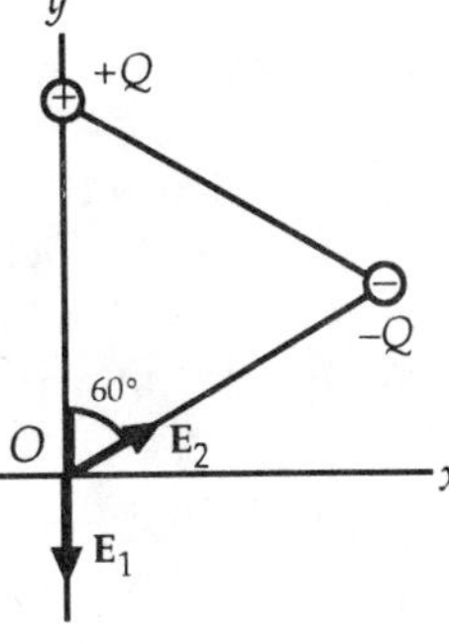

34.

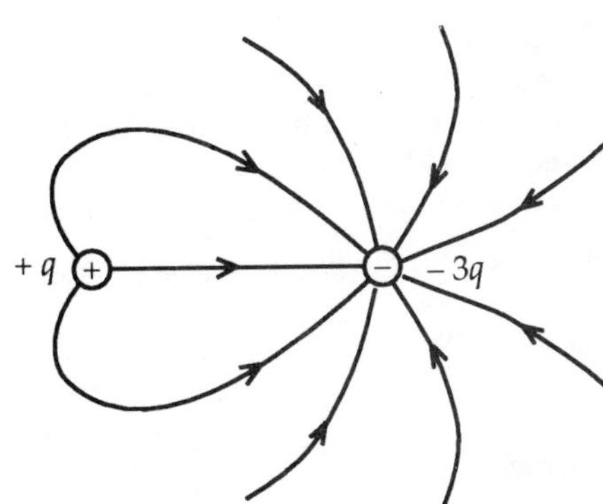

35. The acceleration is produced by the force from the electric field:
$F = qE = ma$;
$(1.60 \times 10^{-19}\text{ C})E = (1.67 \times 10^{-27}\text{ kg})(1 \times 10^6)(9.80\text{ m/s}^2)$, which gives $E =$ 0.10 N/C.

36. If we let x be the distance from the center of the Earth, we have
$GM_{\text{Moon}}/(D - x)^2 = GM_{\text{Earth}}/x^2 = 81GM_{\text{Moon}}/x^2$, or $81(D - x)^2 = x^2$.
When we take the square root of both sides, we get
$x = 9D/10 = 9(3.80 \times 10^5\text{ km})/10 =$ 3.42×10^5 km from the center of the Earth.
Note that taking a negative square root gives $x = 9D/8$, the point on the other side of the Moon where the magnitudes are equal, but the fields have the same direction.

37. For the electric field to be zero, the individual fields must have opposite directions, so the two charges must have the same sign. For the net field to be zero, the magnitudes of the individual fields must be equal:

$E = kQ_1/r_1^2 = kQ_2/r_2^2$, or $Q_1/(\frac{1}{3}L)^2 = Q_2/(\frac{2}{3}L)^2$,

which gives $\boxed{Q_2 = 4Q_1.}$

L/3
x
$\mathbf{E}_2$ $\mathbf{E}_1$
Q_1 Q_2

38. (*a*) We find the acceleration produced by the electric field:

$F = qE = ma;$

$(1.60 \times 10^{-19}\ \text{C})(1.85 \times 10^4\ \text{N/C}) = (9.11 \times 10^{-31}\ \text{kg})a$, which gives $a = 3.24 \times 10^{15}\ \text{m/s}^2$.

Because the field is constant, the acceleration is constant, so we find the speed from

$v^2 = v_0^2 + 2ax = 0 + 2(3.24 \times 10^{15}\ \text{m/s}^2)(0.0120\ \text{m})$, which gives $v = \boxed{8.83 \times 10^6\ \text{m/s.}}$

(*b*) For the ratio of the two forces, we have

$mg/qE = (9.11 \times 10^{-31}\ \text{kg})(9.80\ \text{m/s}^2)/(1.60 \times 10^{-19}\ \text{C})(1.85 \times 10^4\ \text{N/C}) = 3.0 \times 10^{-15}$.

Thus $mg \ll qE$.

39. (*a*) The acceleration of the electron, and thus the force produced by the electric field, must be opposite its velocity. Because the electron has a negative charge, the direction of the electric field will be opposite that of the force, so the direction of the electric field is

$\boxed{\text{in the direction of the velocity, to the right.}}$

(*b*) Because the field is constant, the acceleration is constant, so we find the required acceleration from

$v^2 = v_0^2 + 2ax;$

$0 = [0.01(3.0 \times 10^8\ \text{m/s})]^2 + 2a(0.050\ \text{m})$, which gives $a = -9.00 \times 10^{13}\ \text{m/s}^2$.

We find the electric field from

$F = qE = ma;$

$(1.60 \times 10^{-19}\ \text{C})E = (9.11 \times 10^{-31}\ \text{kg})(9.00 \times 10^{13}\ \text{m/s}^2)$, which gives $E = \boxed{5.1 \times 10^2\ \text{N/C.}}$

40. (*a*) To estimate the force between a thymine and an adenine, we assume that only the atoms with an indicated charge make a contribution. Because all charges are fractions of the electronic charge, we let

$Q_H = Q_N = f_1 e$, and $Q_O = Q_C = f_2 e$.

A convenient numerical factor will be

$ke^2/(10^{-10}\ \text{m/Å})^2$
$= (9.0 \times 10^9\ \text{N}\cdot\text{m}^2/\text{C}^2)(1.60 \times 10^{-19}\ \text{C})^2/(10^{-10}\ \text{m/Å})^2$
$= 2.30 \times 10^{-8}\ \text{N}\cdot\text{Å}^2$.

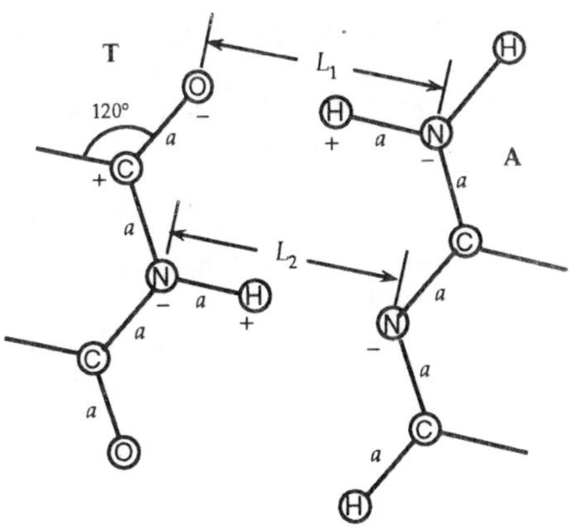

T, A, 120°, L_1, L_2, a, C, O, N, H, +, –

For the first contribution we find the force for the bond of the oxygen on the thymine with the H-N pair on the adenine. From Newton's third law, we know that the force on one must equal the force on the other. We find the attractive force on the oxygen:

$F_O = kQ_O\{[Q_H/(L_1 - a)^2] - (Q_N/L_1^2)\}$
$= ke^2 f_2 f_1\{[1/(L_1 - a)^2] - (1/L_1^2)\}$
$= (2.30 \times 10^{-8}\ \text{N}\cdot\text{Å}^2)(0.4)(0.2)\{[1/(2.80\ \text{Å} - 1.00\ \text{Å})^2] - [1/(2.80\ \text{Å})^2]\} = 3.33 \times 10^{-10}\ \text{N}$.

For the force for the lower bond of the H-N pair on the thymine with the nitrogen on the adenine, we find the attractive force on the nitrogen:

$F_N = kQ_N\{[Q_H/(L_2 - a)^2] - (Q_N/L_2^2)\} = ke^2 f_1 f_1\{[1/(L_2 - a)^2] - (1/L_2^2)\}$
$= (2.30 \times 10^{-8}\ \text{N}\cdot\text{Å}^2)(0.2)(0.2)\{[1/(3.00\ \text{Å} - 1.00\ \text{Å})^2] - [1/(3.00\ \text{Å})^2]\} = 1.28 \times 10^{-10}\ \text{N}$.

There will be a repulsive force between the oxygen of the first bond and the nitrogen of the second bond. To find the separation of the two, we note that the distance between the two nitrogens of the adenine, which is approximately perpendicular to L_1, is $2a \cos 30° = 1.73a$. We find the magnitude of this force from

$F_{O\text{-}N} = kQ_O\{Q_N/[L_1^2 + (1.73a)^2]\} = ke^2 f_2 f_1\{1/[L_1^2 + (1.73a)^2]\}$
$= (2.30 \times 10^{-8}\ \text{N}\cdot\text{Å}^2)(0.4)(0.2)\{1/[(2.80\ \text{Å})^2 + (1.73\ \text{Å})^2]\} = 1.7 \times 10^{-10}\ \text{N}$.

We find the angle that this force makes with the line of the other bonds from

$\tan\theta = 1.73a/L_1 = 1.73\ \text{Å}/2.89\ \text{Å} = 0.62$, or $\theta = 32°$.

Thus the component that contributes to the bond is $(1.7 \times 10^{-10}\ \text{N}) \cos 32° = 1.4 \times 10^{-10}\ \text{N}$.

The other contribution will be from the carbon atom on the thymine. Because the distance is slightly greater and there will be attraction to the nitrogens and repulsion from the hydrogen, we neglect this contribution.

Thus the estimated net bond is $3.33 \times 10^{-10}\ \text{N} + 1.28 \times 10^{-10}\ \text{N} - 1.4 \times 10^{-10}\ \text{N}$ $\boxed{\approx 3 \times 10^{-10}\ \text{N.}}$

(*b*) To estimate the net force between a cytosine and a guanine, we note that there are two oxygen bonds, one nitrogen bond, and one repulsive O-N force. We neglect the other forces because they involve cancellation from the involvement of both hydrogen and nitrogen. If we ignore the small change in distances, we have

$2(3.33 \times 10^{-10}\ \text{N}) + 1.28 \times 10^{-10}\ \text{N} - 1.4 \times 10^{-10}\ \text{N}$ $\boxed{\approx 7 \times 10^{-10}\ \text{N.}}$

(*c*) The total force for the DNA molecule is

$(3 \times 10^{-10}\ \text{N} + 7 \times 10^{-10}\ \text{N})(10^5\ \text{pairs})$ $\boxed{\approx 10^{-4}\ \text{N.}}$

41. When we equate the two forces, we have

$mg = ke^2/r^2$;
$(9.11 \times 10^{-31}\ \text{kg})(9.80\ \text{m/s}^2) = (9.0 \times 10^9\ \text{N}\cdot\text{m}^2/\text{C}^2)(1.60 \times 10^{-19}\ \text{C})^2/r^2$, which gives $r =$ $\boxed{5.08\ \text{m.}}$

42. Because a copper atom has 29 electrons, we find the number of electrons in the penny from

$N = [(3.0\ \text{g})/(63.5\ \text{g/mol})](6.02 \times 10^{23}\ \text{atoms/mol})(29\ \text{electrons/atom}) = 8.24 \times 10^{23}$ electrons.

We find the fractional loss from

$\Delta q/q = (42 \times 10^{-6}\ \text{C})/(8.24 \times 10^{23}\ \text{electrons})(1.6 \times 10^{-19}\ \text{C/electron}) =$ $\boxed{3.2 \times 10^{-10}.}$

43. The weight must be balanced by the force from the electric field:

$mg = qE;$

$(1.67 \times 10^{-27}\ \text{kg})(9.80\ \text{m/s}^2) = (1.60 \times 10^{-19}\ \text{C})E$, which gives $E = \boxed{1.02 \times 10^{-7}\ \text{N/C (up).}}$

44. Because we can treat the charge on the Earth as a point charge at the center, we have

$E = kQ/r^2;$

$150\ \text{N/C} = (9.0 \times 10^9\ \text{N}\cdot\text{m}^2/\text{C}^2)Q/(6.38 \times 10^6\ \text{m})^2$, which gives $Q = \boxed{6.8 \times 10^5\ \text{C.}}$

Because the field points toward the center, the charge must be $\boxed{\text{negative.}}$

45. The weight must be balanced by the force from the electric field:

$mg = \rho\frac{4}{3}\pi r^3 g = NeE\,;$

$(1000\ \text{kg/m}^3)\frac{4}{3}\pi(1.8 \times 10^{-5}\ \text{m})^3(9.80\ \text{m/s}^2) = N(1.60 \times 10^{-19}\ \text{C})(150\ \text{N/C})$,

which gives $N = \boxed{1.0 \times 10^7\ \text{electrons.}}$

46. The directions of the individual fields will be along the diagonals of the square, as shown. All distances are the same. We find the magnitudes of the individual fields:

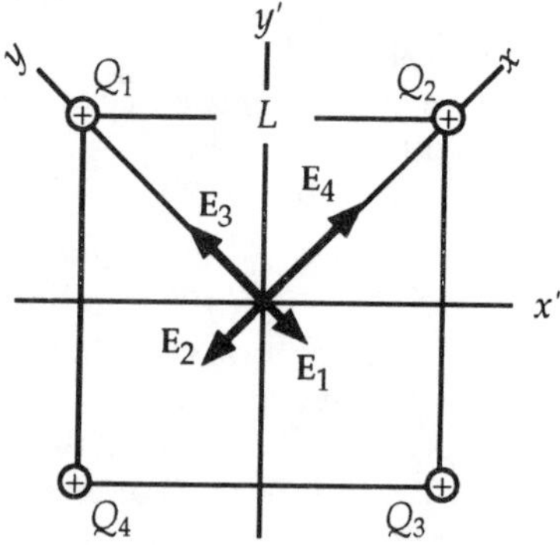

$E_1 = kQ_1/(L/\sqrt{2})^2 = 2kQ_1/L^2$

$= 2(9.0 \times 10^9\ \text{N}\cdot\text{m}^2/\text{C}^2)(1.0 \times 10^{-6}\ \text{C})/(0.25\ \text{m})^2 = 2.88 \times 10^5\ \text{N/C}.$

$E_2 = kQ_2/(L/\sqrt{2})^2 = 2E_1 = 2(2.88 \times 10^5\ \text{N/C}) = 5.76 \times 10^5\ \text{N/C}.$

$E_3 = kQ_3/(L/\sqrt{2})^2 = 3E_1 = 3(2.88 \times 10^5\ \text{N/C}) = 8.64 \times 10^5\ \text{N/C}.$

$E_4 = kQ_4/(L/\sqrt{2})^2 = 4E_1 = 4(2.88 \times 10^5\ \text{N/C}) = 11.52 \times 10^5\ \text{N/C}.$

We simplify the vector addition by using the xy-coordinate system shown. For the components of the resultant field we have

$E_x = E_4 - E_2 = 11.52 \times 10^5\ \text{N/C} - 5.76 \times 10^5\ \text{N/C} = 5.76 \times 10^5\ \text{N/C};$

$E_y = E_3 - E_1 = 8.64 \times 10^5\ \text{N/C} - 2.88 \times 10^5\ \text{N/C} = 5.76 \times 10^5\ \text{N/C}.$

Thus we see that the resultant will be in the y'-direction:

$E = 2E_x \cos 45° = 2(5.76 \times 10^5\ \text{N/C}) \cos 45° = \boxed{8.1 \times 10^5\ \text{N/C up.}}$

47. We find the force between the groups by finding the force on the CO group from the HN group. A convenient numerical factor will be

$ke^2/(10^{-9}\ \text{m/nm})^2$

$= (9.0 \times 10^9\ \text{N}\cdot\text{m}^2/\text{C}^2)(1.60 \times 10^{-19}\ \text{C})^2/(10^{-9}\ \text{m/nm})^2$

$= 2.30 \times 10^{-10}\ \text{N}\cdot\text{nm}^2.$

For the forces on the atoms, we have

$F_O = kQ_O\{[Q_H/(L - d_2)^2] - (Q_N/L^2)\} = ke^2 f_O f_H\{[1/(L - d_2)^2] - (1/L^2)\}$

$= (2.30 \times 10^{-10}\ \text{N}\cdot\text{nm}^2)(0.4)(0.2)\{[1/(0.28\ \text{nm} - 0.10\ \text{nm})^2] - [1/(0.28\ \text{nm})^2]\} = 3.33 \times 10^{-10}\ \text{N}.$

$F_C = kQ_C\{[Q_N/(L + d_1)^2] - Q_H/(L + d_1 - d_2)^2]\} = ke^2 f_C f_N\{[1/(L + d_1)^2] - [1/(L + d_1 - d_2)^2]\}$

$= (2.30 \times 10^{-10}\ \text{N}\cdot\text{nm}^2)(0.4)(0.2)\{[1/(0.28\ \text{nm} + 0.12\ \text{nm})^2] - [1/(0.28\ \text{nm} + 0.12\ \text{nm} - 0.10\ \text{nm})^2]\}$

$= -8.94 \times 10^{-11}\ \text{N}.$

Thus the net force is

$F = F_O + F_C = 3.33 \times 10^{-10}\ \text{N} - 8.94 \times 10^{-11}\ \text{N} = \boxed{2.4 \times 10^{-10}\ \text{N (attraction).}}$

48. Because the charges have the same sign, they repel each other. The force from the third charge must balance the repulsive force for each charge, so the third charge must be positive and between the two negative charges. For each of the negative charges, we have

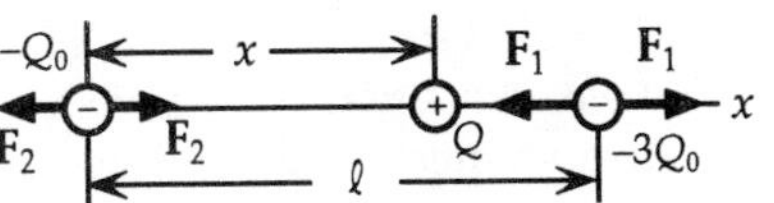

Q_0: $kQ_0Q/x^2 = kQ_0(3Q_0)/\ell^2$, or $\ell^2Q = 3x^2Q_0$;

$3Q_0$: $k3Q_0Q/(\ell - x)^2 = kQ_0(3Q_0)/\ell^2$, or $\ell^2Q = (\ell - x)^2Q_0$.

Thus we have

$3x^2 = (\ell - x)^2$, which gives $x = -1.37\ell, + 0.366\ell$.

Because the positive charge must be between the charges, it must be 0.366ℓ from Q_0. When we use this value in one of the force equations, we get

$Q = 3(0.366\ell)^2Q_0/\ell^2 = 0.402Q_0$.

Thus we place a charge of **$0.402Q_0$ 0.366ℓ from Q_0.**

Note that the force on the middle charge is also zero.

49. Because the charge moves in the direction of the electric field, it must be **positive.**

We find the angle of the string from the dimensions:

$\cos\theta = (0.49\text{ m})/(0.50\text{ m}) = 0.98$, or $\theta = 11.5°$.

Because the charge is in equilibrium, the resultant force is zero.

We see from the force diagram that

$\tan\theta = QE/mg$;

$\tan 11.5° = Q(9200\text{ N/C})/(1.0 \times 10^{-3}\text{ kg})(9.80\text{ m/s}^2)$,

which gives $Q =$ **2.2×10^{-7} C.**

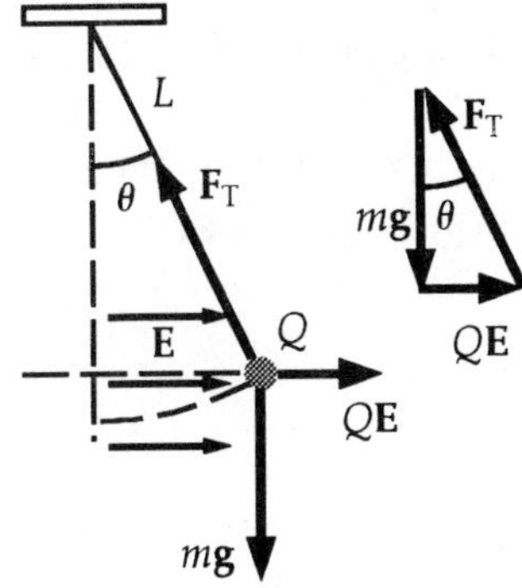

50. Because the charges have opposite signs, the location where the electric field is zero must be outside the two charges, as shown.

The fields from the two charges must balance:

$kQ_1/x^2 = kQ_2/(L - x)^2$;

$(2.5 \times 10^{-5}\text{ C})/x^2 = (5.0 \times 10^{-6}\text{ C})/(2.0\text{ m} - x)^2$,

which gives $x = 1.4$ m, 3.6 m.

Because 1.4 m is between the charges, the location is

3.6 m from the positive charge, and 1.6 m from the negative charge.

51. (*a*) The force is opposite to the direction of the electron. We find the acceleration produced by the electric field:

$-qE = ma$;

$-(1.60 \times 10^{-19}\text{ C})(7.7 \times 10^3\text{ N/C}) = (9.11 \times 10^{-31}\text{ kg})a$, which gives $a = -1.35 \times 10^{15}\text{ m/s}^2$.

Because the field is constant, the acceleration is constant, so we find the distance from

$v^2 = v_0^2 + 2ax$;

$0 = (1.5 \times 10^6\text{ m/s})^2 + 2(-1.35 \times 10^{15}\text{ m/s}^2)x$, which gives $x = 8.3 \times 10^{-4}\text{ m} =$ **0.83 mm.**

(*b*) We find the time from

$x = v_0t + \frac{1}{2}at^2$;

$0 = (1.5 \times 10^6\text{ m/s})t + \frac{1}{2}(-1.35 \times 10^{15}\text{ m/s}^2)t^2$,

which gives $t = 0$ (the starting time), and $2.2 \times 10^{-9}\text{ s} =$ **2.2 ns.**

52. The angular frequency of the SHM is
$\omega = (k/m)^{1/2} = [(126\text{ N/m})/(0.800\text{ kg})]^{1/2} = 12.5\text{ s}^{-1}$.
If we take down as positive, with respect to the equilibrium position, the ball will start at maximum displacement, so the position as a function of time is
$x = A\cos(\omega t) = (0.0500\text{ m})\cos[(12.5\text{ s}^{-1})t]$.
Because the charge is negative, the electric field at the table will be up and the distance from the table is
$r = H - x = 0.150\text{ m} - (0.0500\text{ m})\cos[(12.5\text{ s}^{-1})t]$.
The electric field is
$E = kQ/r^2 = (9.0\times10^9\text{ N}\cdot\text{m}^2/\text{C}^2)(3.00\times10^{-6}\text{ C})/\{0.150\text{ m} - (0.0500\text{ m})\cos[(12.5\text{ s}^{-1})t]\}^2$
$= \boxed{(1.08\times10^7\text{ N/C})/\{3 - \cos[(12.5\text{ s}^{-1})t]\}^2\text{ up.}}$

53. We consider the forces on one ball. (The other will be the same except for the reversal.) The separation of the charges is
$r = 2L\sin 30° = 2(0.70\text{ m})\sin 30° = 0.70\text{ m}$.
From the equilibrium force diagram, we have
$\tan\theta = F/mg = [k(\tfrac{1}{2}Q)(\tfrac{1}{2}Q)/r^2]/mg;$
$\tan 30° = \tfrac{1}{4}(9.0\times10^9\text{ N}\cdot\text{m}^2/\text{C}^2)Q^2/(0.70\text{ m})^2(24\times10^{-3}\text{ kg})(9.80\text{ m/s}^2),$
which gives $Q = 5.4\times10^{-6}\text{ C} = \boxed{5.4\ \mu\text{C.}}$

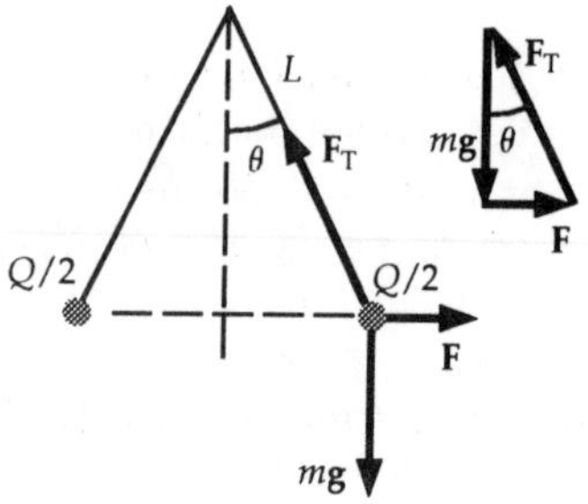

54. The pea will discharge when the electric field at the surface exceeds the breakdown field. Because we can treat the charge on the pea as a point charge at the center, we have
$E = kQ/r^2;$
$3\times10^6\text{ N/C} = (9.0\times10^9\text{ N}\cdot\text{m}^2/\text{C}^2)Q/(0.375\times10^{-2}\text{ m})^2$, which gives $Q = \boxed{5\times10^{-9}\text{ C.}}$

55. We find the electric field at the location of Q_1 due to the plates and Q_2.
For the field of Q_2 we have
$E_2 = kQ_2/x^2$
$= (9.0\times10^9\text{ N}\cdot\text{m}^2/\text{C}^2)(1.3\times10^{-6}\text{ C})/(0.34\text{ m})^2$
$= 1.01\times10^5\text{ N/C (left)}.$
The field from the plates is to the right, so we have
$E_{net} = E_{plates} - E_2$
$= 73{,}000\text{ N/C} - 1.01\times10^5\text{ N/C} = -2.8\times10^5\text{ N/C (left)}.$
For the force on Q_1, we have
$F_1 = Q_1E_{net} = (-6.7\times10^{-6}\text{ C})(-2.8\times10^5\text{ N/C}) = \boxed{+0.19\text{ N (right).}}$

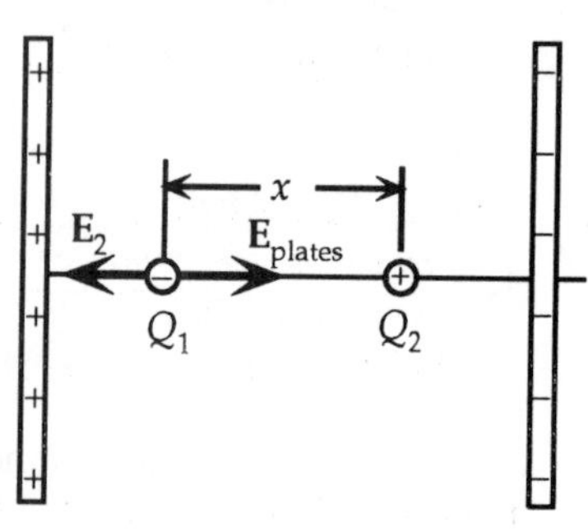

56. We take up as the positive direction and assume that E is up.
From the equilibrium force diagram, we have
$F_T + QE = mg;$
$5.67\text{ N} + (0.340\times10^{-6}\text{ C})E = (0.210\text{ kg})(9.80\text{ m/s}^2),$
which gives $E = \boxed{-1.06\times10^7\text{ N/C (down).}}$

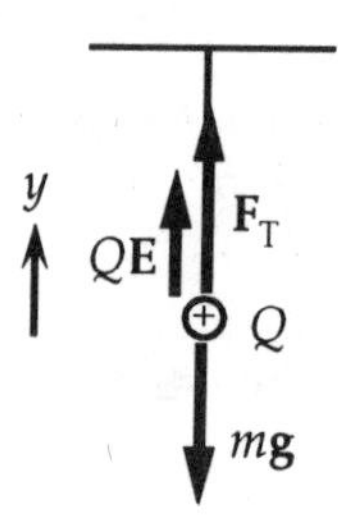

CHAPTER 17

1. We find the work done by an external agent from the work-energy principle:
$W = \Delta\text{KE} + \Delta\text{PE} = 0 + q(V_b - V_a)$
$= (-8.6 \times 10^{-6}\ \text{C})(+75\ \text{V} - 0) =$ **-6.5×10^{-4} J (done by the field).**

2. We find the work done by an external agent from the work-energy principle:
$W = \Delta\text{KE} + \Delta\text{PE} = 0 + q(V_b - V_a)$
$= (1.60 \times 10^{-19}\ \text{C})[(-50\ \text{V}) - (+100\ \text{V})] =$ **-2.40×10^{-17} J (done by the field);**
$W = q(V_b - V_a)$
$= (+1\ \text{e})[(-50\ \text{V}) - (+100\ \text{V})] =$ **-150 eV.**

3. Because the total energy of the electron is conserved, we have
$\Delta\text{KE} + \Delta\text{PE} = 0$, or
$\Delta\text{KE} = -q(V_B - V_A) = -(-1.60 \times 10^{-19}\ \text{C})(21{,}000\ \text{V}) =$ **3.4×10^{-15} J;**
$\Delta\text{KE} = -(-1\ \text{e})(21{,}000\ \text{V}) =$ **21 keV.**

4. Because the total energy of the electron is conserved, we have
$\Delta\text{KE} + \Delta\text{PE} = 0;$
$\Delta\text{KE} + q(V_B - V_A) = 0;$
$3.45 \times 10^{-15}\ \text{J} + (-1.60 \times 10^{-19}\ \text{C})(V_B - V_A)$; which gives $V_B - V_A =$ **2.16×10^3 V.**
Plate B is at the higher potential.

5. For the uniform electric field between two large, parallel plates, we have
$E = \Delta V/d = (220\ \text{V})/(5.2 \times 10^{-3}\ \text{m}) =$ **4.2×10^4 V/m.**

6. For the uniform electric field between two large, parallel plates, we have
$E = \Delta V/d;$
$640\ \text{V/m} = \Delta V/(11.0 \times 10^{-3}\ \text{m})$, which gives $\Delta V =$ **7.04 V.**

7. Because the total energy of the helium nucleus is conserved, we have
$\Delta\text{KE} + \Delta\text{PE} = 0;$
$\Delta\text{KE} + q(V_B - V_A) = 0;$
$65.0\ \text{keV} + (+2\text{e})(V_B - V_A)$; which gives $V_B - V_A =$ **-32.5 kV.**

8. For the uniform electric field between two large, parallel plates, we have
$E = \Delta V/d;$
$3 \times 10^6\ \text{V/m} = (100\ \text{V})/d$, which gives $d =$ **3×10^{-5} m.**

9. We use the work-energy principle:
$W = \Delta\text{KE} + \Delta\text{PE} = \Delta\text{KE} + q(V_b - V_a);$
$25.0 \times 10^{-4}\ \text{J} = 4.82 \times 10^{-4}\ \text{J} + (-7.50 \times 10^{-6}\ \text{C})(V_b - V_a)$, which gives $V_b - V_a = -269\text{V}$, or
$V_a - V_b = 269$ V.

10. The data given are the kinetic energies, so we find the speed from

(*a*) $\text{KE}_a = \frac{1}{2}mv_a^2;$

$(750\text{ eV})(1.60\times10^{-19}\text{ J/eV}) = \frac{1}{2}(9.11\times10^{-31}\text{ kg})v_a^2$, which gives $v_a =$ $\boxed{1.62\times10^7\text{ m/s.}}$

(*b*) $\text{KE}_b = \frac{1}{2}mv_b^2;$

$(3.5\times10^3\text{ eV})(1.60\times10^{-19}\text{ J/eV}) = \frac{1}{2}(9.11\times10^{-31}\text{ kg})v_b^2$, which gives $v_b =$ $\boxed{3.5\times10^7\text{ m/s.}}$

11. We find the speed from

$\text{KE} = \frac{1}{2}mv^2;$

$(28.0\times10^6\text{ eV})(1.60\times10^{-19}\text{ J/eV}) = \frac{1}{2}(1.67\times10^{-27}\text{ kg})v^2$, which gives $v =$ $\boxed{7.32\times10^7\text{ m/s.}}$

12. We find the speed from

$\text{KE} = \frac{1}{2}mv^2;$

$(5.53\times10^6\text{ eV})(1.60\times10^{-19}\text{ J/eV}) = \frac{1}{2}(6.64\times10^{-27}\text{ kg})v^2$, which gives $v =$ $\boxed{1.63\times10^7\text{ m/s.}}$

13. We find the electric potential of the point charge from

$V = kq/r = (9.0\times10^9\text{ N}\cdot\text{m}^2/\text{C}^2)(4.00\times10^{-6}\text{ C})/(15.0\times10^{-2}\text{ m}) =$ $\boxed{2.40\times10^5\text{ V.}}$

14. We find the charge from

$V = kQ/r;$

$125\text{ V} = (9.0\times10^9\text{ N}\cdot\text{m}^2/\text{C}^2)Q/(15\times10^{-2}\text{ m})$, which gives $Q = 2.1\times10^{-9}\text{ C} =$ $\boxed{2.1\text{ nC.}}$

15. We find the electric potentials of the stationary charges at the initial and final points:

$V_a = k[(Q_1/r_{1a}) + (Q_2/r_{2a})]$

$= (9.0\times10^9\text{ N}\cdot\text{m}^2/\text{C}^2)\{[(30\times10^{-6}\text{ C})/(0.16\text{ m})] + [(30\times10^{-6}\text{ C})/(0.16\text{ m})]\} = 3.38\times10^6\text{ V}.$

$V_b = k[(Q_1/r_{1b}) + (Q_2/r_{2b})]$

$= (9.0\times10^9\text{ N}\cdot\text{m}^2/\text{C}^2)\{[(30\times10^{-6}\text{ C})/(0.26\text{ m})] + [(30\times10^{-6}\text{ C})/(0.06\text{ m})]\} = 5.54\times10^6\text{ V}.$

Because there is no change in kinetic energy, we have

$W_{a\to b} = \Delta K + \Delta U = 0 + q(V_b - V_a)$

$= (0.50\times10^{-6}\text{ C})(5.54\times10^6\text{ V} - 3.38\times10^6\text{ V}) =$ $\boxed{+1.08\text{ J.}}$

16. (*a*) We find the electric potential of the proton from

$V = kq/r = (9.0\times10^9\text{ N}\cdot\text{m}^2/\text{C}^2)(1.60\times10^{-19}\text{ C})/(2.5\times10^{-15}\text{ m}) =$ $\boxed{5.8\times10^5\text{ V.}}$

(*b*) We find the electric potential energy of the system by considering one of the charges to be at the potential created by the other charge:

$\text{PE} = qV = (1.60\times10^{-19}\text{ C})(5.76\times10^5\text{ V}) =$ $\boxed{9.2\times10^{-14}\text{ J}}$ $= 0.58\text{ MeV}.$

17. When the proton is accelerated by a potential, it acquires a kinetic energy:

$\text{KE} = Q_pV_{\text{accel}}.$

If it is far from the silicon nucleus, its potential is zero. It will slow as it approaches the positive charge of the nucleus, because the potential produced by the silicon nucleus is increasing. At the proton's closest point the kinetic energy will be zero. We find the required accelerating potential from

$\Delta\text{KE} + \Delta\text{PE} = 0;$

$0 - \text{KE} + Q_p(V_{\text{Si}} - 0) = 0$, or

$Q_pV_{\text{accel}} = Q_pkQ_{\text{Si}}/(R_p + R_{\text{Si}});$

$V_{\text{accel}} = (9.0\times10^9\text{ N}\cdot\text{m}^2/\text{C}^2)(14)(1.60\times10^{-19}\text{ C})/(1.2\times10^{-15}\text{ m} + 3.6\times10^{-15}\text{ m})$

$= 4.2\times10^6\text{ V} =$ $\boxed{4.2\text{ MV.}}$

18. We find the potential energy of the system of charges by adding the work required to bring the three electrons in from infinity successively. Because there is no potential before the electrons are brought in, for the first electron we have

$$W_1 = (-e)V_0 = 0.$$

When we bring in the second electron, there will be a potential from the first:

$$W_2 = (-e)V_1 = (-e)k(-e)/r_{12} = ke^2/d.$$

When we bring in the third electron, there will be a potential from the first two:

$$W_3 = (-e)V_2 = (-e)\{[k(-e)/r_{13}] + [k(-e)/r_{23}]\} = 2ke^2/d.$$

The total work required is

$$W = W_1 + W_2 + W_3 = (ke^2/d) + (2ke^2/d) = 3ke^2/d$$

$$= 3(9.0 \times 10^9\ \text{N}\cdot\text{m}^2/\text{C}^2)(1.60 \times 10^{-19}\ \text{C})^2/(1.0 \times 10^{-10}\ \text{m}) = \boxed{6.9 \times 10^{-18}\ \text{J}} = 43\ \text{eV}.$$

19. (*a*) We find the electric potentials at the two points:

$$V_a = kQ/r_a$$
$$= (9.0 \times 10^9\ \text{N}\cdot\text{m}^2/\text{C}^2)(-3.8 \times 10^{-6}\ \text{C})/(0.70\ \text{m})$$
$$= -4.89 \times 10^4\ \text{V}.$$
$$V_b = kQ/r_b$$
$$= (9.0 \times 10^9\ \text{N}\cdot\text{m}^2/\text{C}^2)(-3.8 \times 10^{-6}\ \text{C})/(0.80\ \text{m})$$
$$= -4.28 \times 10^4\ \text{V}.$$

Thus the difference is

$$V_{ba} = V_b - V_a = -4.28 \times 10^4\ \text{V} - (-4.89 \times 10^4\ \text{V}) = \boxed{+6.1 \times 10^3\ \text{V}.}$$

(*b*) We find the electric fields at the two points:

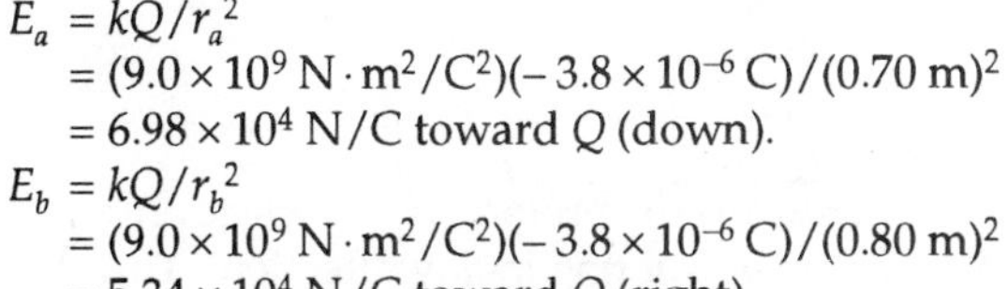

$$E_a = kQ/r_a^2$$
$$= (9.0 \times 10^9\ \text{N}\cdot\text{m}^2/\text{C}^2)(-3.8 \times 10^{-6}\ \text{C})/(0.70\ \text{m})^2$$
$$= 6.98 \times 10^4\ \text{N/C toward } Q \text{ (down).}$$

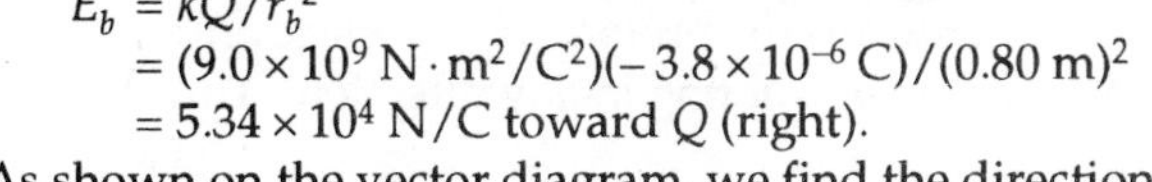

$$E_b = kQ/r_b^2$$
$$= (9.0 \times 10^9\ \text{N}\cdot\text{m}^2/\text{C}^2)(-3.8 \times 10^{-6}\ \text{C})/(0.80\ \text{m})^2$$
$$= 5.34 \times 10^4\ \text{N/C toward } Q \text{ (right).}$$

As shown on the vector diagram, we find the direction of $\mathbf{E}_b - \mathbf{E}_a$ from

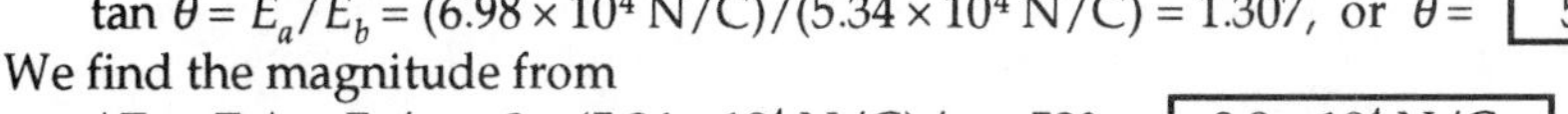

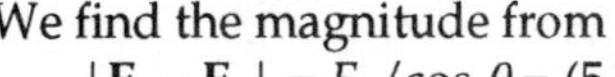

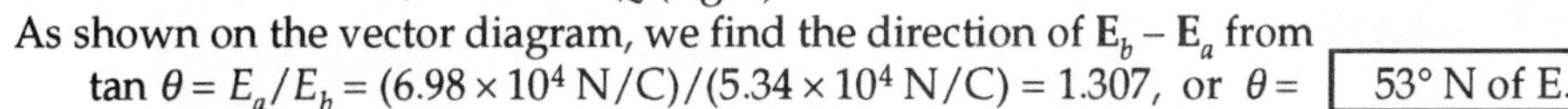

$$\tan\theta = E_a/E_b = (6.98 \times 10^4\ \text{N/C})/(5.34 \times 10^4\ \text{N/C}) = 1.307,\ \text{or}\ \theta = \boxed{53^\circ\ \text{N of E.}}$$

We find the magnitude from

$$|\mathbf{E}_b - \mathbf{E}_a| = E_b/\cos\theta = (5.34 \times 10^4\ \text{N/C})/\cos 53^\circ = \boxed{8.8 \times 10^4\ \text{N/C.}}$$

20. When the electron is far away, the potential from the fixed charge is zero. Because energy is conserved, we have

$$\Delta\text{KE} + \Delta\text{PE} = 0;$$
$$\tfrac{1}{2}mv^2 - 0 + (-e)(0 - V) = 0,\ \text{or}$$
$$\tfrac{1}{2}mv^2 = e(kQ/r)$$
$$\tfrac{1}{2}(9.11 \times 10^{-31}\ \text{kg})v^2 = (1.60 \times 10^{-19}\ \text{C})(9.0 \times 10^9\ \text{N}\cdot\text{m}^2/\text{C}^2)(-0.125 \times 10^{-6}\ \text{C})/(0.725\ \text{m}),$$

which gives $v = \boxed{2.33 \times 10^7\ \text{m/s.}}$

21. We find the electric potential energy of the system by considering one of the charges to be at the potential created by the other charge. This will be zero when they are far away. Because the masses are equal, the speeds will be equal. From energy conservation we have

$$\Delta\text{KE} + \Delta\text{PE} = 0;$$
$$\tfrac{1}{2}mv^2 + \tfrac{1}{2}mv^2 - 0 + Q(0 - V) = 0,\ \text{or}$$
$$2(\tfrac{1}{2}mv^2) = mv^2 = Q(kQ/r) = kQ^2/r;$$
$$(1.0 \times 10^{-6}\ \text{kg})v^2 = (9.0 \times 10^9\ \text{N}\cdot\text{m}^2/\text{C}^2)(7.5 \times 10^{-6}\ \text{C})^2/(0.055\ \text{m}),$$

which gives $v = \boxed{3.0 \times 10^3\ \text{m/s.}}$

22. (*a*) We find the electric potential of the proton from
$V = ke/r = (9.0 \times 10^9\ \text{N}\cdot\text{m}^2/\text{C}^2)(1.60 \times 10^{-19}\ \text{C})/(0.53 \times 10^{-10}\ \text{m}) =$ $\boxed{27.2\ \text{V.}}$

(*b*) For the electron orbiting the nucleus, the attractive Coulomb force provides the centripetal acceleration:
$ke^2/r^2 = mv^2/r$, which gives $\text{KE} = \frac{1}{2}mv^2 = \frac{1}{2}ke^2/r = \frac{1}{2}eV = \frac{1}{2}(1\ \text{e})(27.2\ \text{V}) =$ $\boxed{13.6\ \text{eV.}}$

(*c*) For the total energy we have
$\text{E} = \text{KE} + \text{PE} = (\frac{1}{2}ke^2/r) + (-e)(ke/r) = -\frac{1}{2}ke^2/r =$ $\boxed{-13.6\ \text{eV.}}$

(*d*) Because the final energy of the electron is zero, for the ionization energy we have
$\text{E}_{\text{ionization}} = -\text{E} =$ $\boxed{13.6\ \text{eV}}$ $(2.2 \times 10^{-18}\ \text{J})$.

23. We find the electric potentials from the charges at the two points:

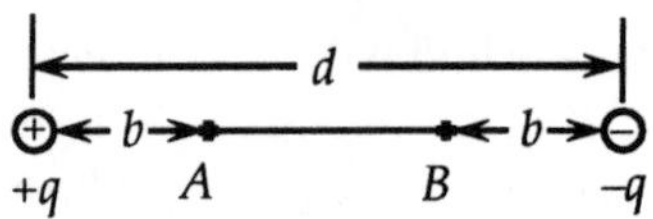

$$V_A = k\{(+q/b) + [-q/(d-b)]\}$$
$$= kq\{(1/b) - [1/(d-b)]\} = kq(d-2b)/b(d-b).$$
$$V_B = k\{[+q/(d-b)] + (-q/b)\}$$
$$= kq\{[1/(d-b)] - (1/b)\} = kq(2b-d)/b(d-b).$$

Thus we have
$V_{BA} = V_B - V_A = [kq(2b-d)/b(d-b)] - [kq(d-2b)/b(d-b)] =$ $\boxed{2kq(2b-d)/b(d-b).}$

Note that, as expected, $V_{BA} = 0$ when $b = \frac{1}{2}d$.

24. (*a*) We find the dipole moment from
$p = eL = (1.60 \times 10^{-19}\ \text{C})(0.53 \times 10^{-10}\ \text{m}) =$ $\boxed{8.5 \times 10^{-30}\ \text{C}\cdot\text{m.}}$

(*b*) The dipole moment will point from the electron toward the proton. As the electron revolves about the proton, the dipole moment will spend equal times pointing in any direction. Thus the average over time will be $\boxed{\text{zero.}}$

25. With the dipole pointing along the axis, the potential at a point a distance r from the dipole which makes an angle θ with the axis is

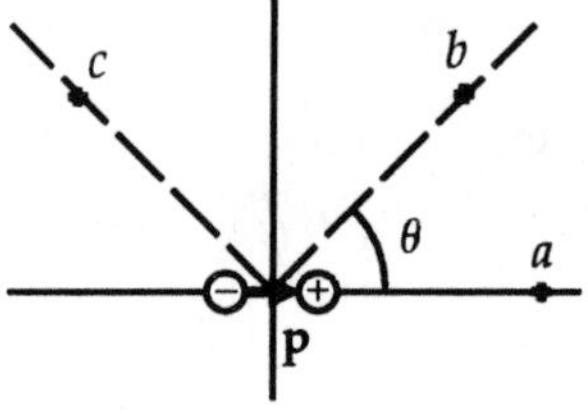

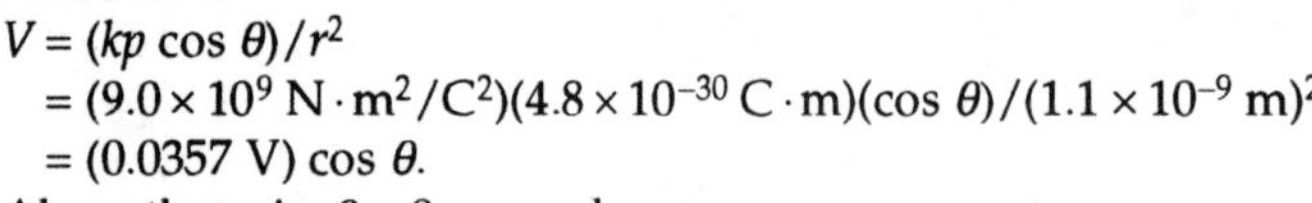

$$V = (kp \cos \theta)/r^2$$
$$= (9.0 \times 10^9\ \text{N}\cdot\text{m}^2/\text{C}^2)(4.8 \times 10^{-30}\ \text{C}\cdot\text{m})(\cos \theta)/(1.1 \times 10^{-9}\ \text{m})^2$$
$$= (0.0357\ \text{V}) \cos \theta.$$

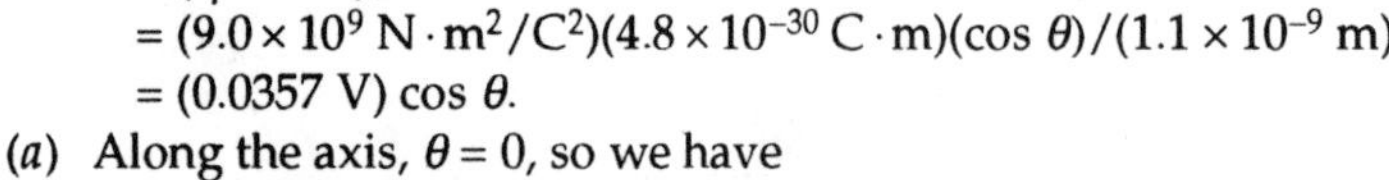

(*a*) Along the axis, $\theta = 0$, so we have
$V = (kp \cos \theta)/r^2$
$= (0.0357\ \text{V}) \cos 0° =$ $\boxed{0.036\ \text{V.}}$

(*b*) Above the axis near the positive charge, $\theta = 45°$, so we have
$V = (kp \cos \theta)/r^2$
$= (0.0357\ \text{V}) \cos 45° =$ $\boxed{0.025\ \text{V.}}$

(*c*) Above the axis near the negative charge, $\theta = 135°$, so we have
$V = (kp \cos \theta)/r^2$
$= (0.0357\ \text{V}) \cos 135° =$ $\boxed{-0.025\ \text{V.}}$

26. (*a*) With the distance measured from the center of the dipole, we find the potential from each charge:

$V_O = kQ_O/r_O$
$= (9.0 \times 10^9\ \text{N}\cdot\text{m}^2/\text{C}^2)(-6.6 \times 10^{-20}\ \text{C})/(9.0 \times 10^{-10}\ \text{m} - 0.6 \times 10^{-10}\ \text{m})$
$= -0.707\ \text{V}.$

$V_C = kQ_C/r_C$
$= (9.0 \times 10^9\ \text{N}\cdot\text{m}^2/\text{C}^2)(+6.6 \times 10^{-20}\ \text{C})/(9.0 \times 10^{-10}\ \text{m} + 0.6 \times 10^{-10}\ \text{m})$
$= +0.619\ \text{V}.$

Thus the total potential is

$V = V_O + V_C = -0.707\ \text{V} + 0.619\ \text{V} = \boxed{-0.088\ \text{V}.}$

(*b*) The percent error introduced by the dipole approximation is

% error $= (100)(0.089\ \text{V} - 0.088\ \text{V})/(0.088\ \text{V}) = \boxed{1\%.}$

27. Because $p_1 = p_2$, from the vector addition we have

$p = 2p_1 \cos(\frac{1}{2}\theta) = 2qL \cos(\frac{1}{2}\theta);$

$6.1 \times 10^{-30}\ \text{C}\cdot\text{m} = 2q(0.96 \times 10^{-10}\ \text{m}) \cos[\frac{1}{2}(104°)]$, which gives $q = \boxed{5.2 \times 10^{-20}\ \text{C}.}$

28. We find the potential energy of the system by considering each of the charges of the dipole on the right to be in the potential created by the other dipole. The potential of the dipole on the left along its axis is

$V_1 = (kp_1 \cos\theta)/r^2 = kp_1/r^2.$

If r is the distance between centers of the dipoles, the potential energy is

$\text{PE} = (q_2)[kp_1/(r + \frac{1}{2}d)^2] + (-q_2)[kp_1/(r - \frac{1}{2}d)^2]$
$= q_2kp_1\{[1/(r + \frac{1}{2}d)^2] - [1/(r - \frac{1}{2}d)^2]\} = (q_2kp_1/r^2)(\{1/[1 + (d/2r)]^2\} - \{1/[1 - (d/2r)]^2\}).$

Because $d \ll r$, we can use the approximation $1/(1 \pm x)^2 \approx 1 \mp 2x$, when $x \ll 1$:

$\text{PE} \approx (q_2kp_1/r^2)\{[1 - (d/r)] - [1 + (d/r)]\} = -2q_2dkp_1/r^3 = -2kp_1p_2/r^3.$

29. Because the field is uniform, the magnitudes of the forces on the charges of the dipole will be equal:

$F_+ = F_- = QE.$

If the separation of the charges is ℓ, the dipole moment will be $p = Q\ell$.
If we choose the center of the dipole for the axis of rotation, both forces create a CCW torque with a net torque of

$\tau = F_+(\frac{1}{2}\ell) \sin\phi + F_-(\frac{1}{2}\ell) \sin\phi = 2QE(\frac{1}{2}\ell) \sin\phi = pE \sin\phi.$

Because the forces are in opposite directions, the net force is $\boxed{\text{zero.}}$
If the field is nonuniform, there would be a torque produced by the average field. The magnitudes of the forces would not be the same, so there would be a $\boxed{\text{resultant force}}$ that would cause a translation of the dipole.

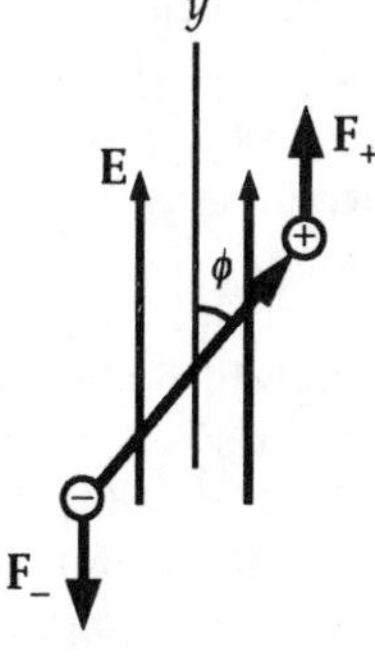

30. From $Q = CV$, we have

$2500\ \mu\text{C} = C(950\ \text{V})$, which gives $C = \boxed{2.6\ \mu\text{F}.}$

31. From $Q = CV$, we have

$95\ \text{pC} = C(120\ \text{V})$, which gives $C = \boxed{0.79\ \text{pF}.}$

32. From $Q = CV$, we have

$16.5 \times 10^{-8}\ \text{C} = (7500 \times 10^{-12}\ \text{F})V$, which gives $V = \boxed{22.0\ \text{V}.}$

33. The final potential on the capacitor will be the voltage of the battery. Positive charge will move from one plate to the other, so the charge that flows through the battery is
    $Q = CV = (9.00\ \mu\text{F})(12.0\ \text{V}) = \boxed{108\ \mu\text{C}.}$

34. For a parallel-plate capacitor, we find the area from
    $C = \epsilon_0 A/d;$
    $0.20\ \text{F} = (8.85 \times 10^{-12}\ \text{C}^2/\text{N}\cdot\text{m}^2)A/(2.2 \times 10^{-3}\ \text{m})$, which gives $A = \boxed{5.0 \times 10^7\ \text{m}^2.}$
    If the area were a square, it would be ≈ 7 km on a side.

35. We find the capacitance from
    $C = K\epsilon_0 A/d = K\epsilon_0 \pi r^2/d$
    $= (7)(8.85 \times 10^{-12}\ \text{C}^2/\text{N}\cdot\text{m}^2)\pi(0.050\ \text{m})^2/(3.2 \times 10^{-3}\ \text{m}) = \boxed{1.5 \times 10^{-10}\ \text{F}.}$

36. From $Q = CV$, we see that
    $\Delta Q = C\,\Delta V;$
    $15\ \mu\text{C} = C(121\ \text{V} - 97\ \text{V})$, which gives $C = \boxed{0.63\ \mu\text{F}.}$

37. The uniform electric field between the plates is related to the potential difference across the plates:
    $E = V/d.$
    For a parallel-plate capacitor, we have
    $Q = CV = (\epsilon_0 A/d)(Ed) = \epsilon_0 AE$
    $= (8.85 \times 10^{-12}\ \text{C}^2/\text{N}\cdot\text{m}^2)(35.0 \times 10^{-4}\ \text{m}^2)(8.50 \times 10^5\ \text{V/m})$
    $= 2.63 \times 10^{-8}\ \text{C} = \boxed{26.3\ \text{nC}.}$

38. The uniform electric field between the plates is related to the potential difference across the plates:
    $E = V/d.$
    For a parallel-plate capacitor, we have
    $Q = CV = (\epsilon_0 A/d)(Ed) = \epsilon_0 AE;$
    $4.2 \times 10^{-6}\ \text{C} = (8.85 \times 10^{-12}\ \text{C}^2/\text{N}\cdot\text{m}^2)A(2.0 \times 10^3\ \text{V/mm})(10^3\ \text{mm/m}),$
    which gives $A = \boxed{0.24\ \text{m}^2.}$

39. We find the potential difference across the plates from
    $Q = CV;$
    $72\ \mu\text{C} = (0.80\ \mu\text{F})V$, which gives $V = 90\ \text{V}$.
    We find the uniform electric field between the plates from
    $E = V/d = (90\ \text{V})/(2.0 \times 10^{-3}\ \text{m}) = \boxed{4.5 \times 10^4\ \text{V/m}.}$

40. The uniform electric field between the plates is related to the potential difference across the plates:
    $E = V/d.$
    For a parallel-plate capacitor, we have
    $Q = CV = CEd;$
    $0.775 \times 10^{-6}\ \text{C} = C(9.21 \times 10^4\ \text{V/m})(1.95 \times 10^{-3}\ \text{m})$, which gives $C = \boxed{4.32 \times 10^{-9}\ \text{F}.}$
    We find the area of the plates from
    $C = K\epsilon_0 A/d;$
    $4.32 \times 10^{-9}\ \text{F} = (3.75)(8.85 \times 10^{-12}\ \text{C}^2/\text{N}\cdot\text{m}^2)A/(1.95 \times 10^{-3}\ \text{m})$, which gives $A = \boxed{0.254\ \text{m}^2.}$

41. We find the initial charge on the 7.7-μF capacitor when it is connected to the battery:

$Q = C_1V = (7.7\ \mu\text{F})(125\ \text{V}) = 962.5\ \mu\text{C}.$

When C_1 is disconnected from the battery and then connected to C_2, some charge will flow from C_1 to C_2. The flow will stop when the voltage across the two capacitors is the same:

$V_1 = V_2 = 15\ \text{V}.$

Because charge is conserved, we have

$Q = Q_1 + Q_2.$

We find the charge remaining on C_1 from

$Q_1 = C_1V_1 = (7.7\ \mu\text{F})(15\ \text{V}) = 115.5\ \mu\text{C}.$

The charge on C_2 is

$Q_2 = Q - Q_1 = 962.5\ \mu\text{C} - 115.5\ \mu\text{C} = 847\ \mu\text{C}.$

We find the value of C_2 from

$Q_2 = C_2V_2;$

$847\ \mu\text{C} = C_2(15\ \text{V})$, which gives $C_2 = \boxed{56\ \mu\text{F}.}$

42. We find the initial charges on the capacitors:

$Q_1 = C_1V_1 = (2.50\ \mu\text{F})(1000\ \text{V}) = 2500\ \mu\text{C};$

$Q_2 = C_2V_2 = (6.80\ \mu\text{F})(650\ \text{V}) = 4420\ \mu\text{C}.$

When the capacitors are connected, some charge will flow from C_2 to C_1 until the potential difference across the two capacitors is the same:

$V_1' = V_2' = V.$

Because charge is conserved, we have

$Q = Q_1' + Q_2' = Q_1 + Q_2 = 2500\ \mu\text{C} + 4420\ \mu\text{C} = 6920\ \mu\text{C}.$

For the two capacitors we have

$Q_1' = C_1V$, and $Q_2' = C_2V.$

When we add these, we get

$Q_1' + Q_2' = Q = (C_1 + C_2)V;$

$6920\ \mu\text{C} = (2.50\ \mu\text{F} + 6.80\ \mu\text{F})V$, which gives $V = \boxed{744\ \text{V}.}$

The charge on C_1 is

$Q_1' = C_1V = (2.50\ \mu\text{F})(744\ \text{V}) = 1.86 \times 10^3\ \mu\text{C} = \boxed{1.86 \times 10^{-3}\ \text{C}.}$

The charge on C_2 is

$Q_2' = C_2V = (6.80\ \mu\text{F})(744\ \text{V}) = 5.06 \times 10^3\ \mu\text{C} = \boxed{5.06 \times 10^{-3}\ \text{C}.}$

43. The energy stored in the capacitor is

$U = \frac{1}{2}CV^2 = \frac{1}{2}(7200 \times 10^{-12}\ \text{F})(550\ \text{V})^2 = \boxed{1.09 \times 10^{-3}\ \text{J}.}$

44. We find the capacitance from

$U = \frac{1}{2}CV^2;$

$200\ \text{J} = \frac{1}{2}C(6000\ \text{V})^2$, which gives $C = 1.1 \times 10^{-5}\ \text{F} = \boxed{11\ \mu\text{F}.}$

45. (*a*) The radius of the pie plate is

$r = \frac{1}{2}(9.0 \text{ in})(2.54 \times 10^{-2} \text{ m/in}) = 0.114 \text{ m}.$

If we assume that it approximates a parallel-plate capacitor, we have

$C = \epsilon_0 A/d = \epsilon_0 \pi r^2/d$

$= (8.85 \times 10^{-12} \text{ C}^2/\text{N}\cdot\text{m}^2)\pi(0.114 \text{ m})^2/(0.10 \text{ m}) = 3.6 \times 10^{-12} \text{ F} = \boxed{3.6 \text{ pF.}}$

(*b*) We find the charge on each plate from

$Q = CV = (3.6 \text{ pF})(9.0 \text{ V}) = \boxed{32 \text{ pC.}}$

(*c*) We assume that the electric field is uniform, so we have

$E = V/d = (9.0 \text{ V})/(0.10 \text{ m}) = \boxed{90 \text{ V/m.}}$

(*d*) The work done by the battery is the energy stored in the capacitor:

$W = U = \frac{1}{2}CV^2 = \frac{1}{2}(3.6 \times 10^{-12} \text{ F})(9.0 \text{ V})^2 = \boxed{1.5 \times 10^{-10} \text{ J.}}$

(*e*) Because the battery is still connected, the electric field will not change. Insertion of the dielectric will change $\boxed{\text{capacitance, charge, and work done by the battery.}}$

46. From $C = \epsilon_0 A/d$, we see that separating the plates will change C. For the stored energy we have

$U = \frac{1}{2}CV^2 = \frac{1}{2}Q^2/C.$

Because the charge is constant, for the two conditions we have

$U_2/U_1 = C_1/C_2 = d_2/d_1 = \boxed{2.}$

47. (*a*) For the stored energy we have $U = \frac{1}{2}CV^2$. Because the capacitance does not change, we have

$U_2/U_1 = (V_2/V_1)^2 = (2)^2 = \boxed{4\times.}$

(*b*) For the stored energy we have $U = \frac{1}{2}CV^2 = \frac{1}{2}Q^2/C$. The capacitance does not change, so we have

$U_2/U_1 = (Q_2/Q_1)^2 = (2)^2 = \boxed{4\times.}$

(*c*) Because the battery is still connected, the potential difference will not change.

From $C = \epsilon_0 A/d$, we see that separating the plates will change C.

For the stored energy we have $U = \frac{1}{2}CV^2$, so we get

$U_2/U_1 = C_2/C_1 = (2)^2 = d_1/d_2 = \boxed{\tfrac{1}{2}\times.}$

48. Because the capacitor is isolated, the charge will not change. The initial stored energy is

$U_1 = \frac{1}{2}C_1V_1^2 = \frac{1}{2}Q^2/C_1$, with $C_1 = \epsilon_0 A/d_1$.

The changes will change the capacitance:

$C_2 = K\epsilon_0 A/d_2.$

For the ratio of stored energies, we have

$U_2/U_1 = C_1/C_2 = (\epsilon_0 A/d_1)/(K\epsilon_0 A/d_2) = d_2/Kd_1 = \frac{1}{2}/K = \boxed{1/2K.}$

The stored energy decreases from two factors. Because the plates attract each other, when the separation is halved, work is done by the field, so the energy decreases. When the dielectric is inserted, the induced charges on the dielectric are attracted to the plates; again work is done by the field and the energy decreases.

The uniform electric field between the plates is related to the potential difference across the plates:

$E = V/d.$

For a parallel-plate capacitor, we have

$Q = C_1V_1 = C_1E_1d_1 = C_2E_2d_2$, or

$E_2/E_1 = C_1d_1/C_2d_2 = \epsilon_0 A/K\epsilon_0 A = \boxed{1/K.}$

49. (*a*) Because there is no stored energy on the uncharged 4.00-μF capacitor, the total stored energy is

$$U_a = \tfrac{1}{2}C_1V_0^2 = \tfrac{1}{2}(2.70 \times 10^{-6}\ \text{F})(45.0\ \text{V})^2 = \boxed{2.73 \times 10^{-3}\ \text{J.}}$$

(*b*) We find the initial charge on the 2.70-μF capacitor when it is connected to the battery;

$$Q = C_1V_0 = (2.70\ \mu\text{F})(45.0\ \text{V}) = 121.5\ \mu\text{C}.$$

When the capacitors are connected, some charge will flow from C_1 to C_2 until the potential difference across the two capacitors is the same:

$$V_1 = V_2 = V.$$

Because charge is conserved, we have

$$Q = Q_1 + Q_2 = 121.5\ \mu\text{C}.$$

For the two capacitors we have

$$Q_1 = C_1V, \quad \text{and} \quad Q_2 = C_2V.$$

When we form the ratio, we get

$Q_2/Q_1 = (121.5\ \mu\text{C} - Q_1)/Q_1 = C_2/C_1 = (4.00\ \mu\text{F})/(2.70\ \mu\text{F})$, which gives $Q_1 = 49.0\ \mu\text{C}$.

We find V from

$$Q_1 = C_1V;$$

$49.0\ \mu\text{C} = (2.70\ \mu\text{F})V$, which gives $V = 18.1\ \text{V}$.

For the stored energy we have

$$U_b = \tfrac{1}{2}C_1V^2 + \tfrac{1}{2}C_2V^2 = \tfrac{1}{2}(C_1 + C_2)V^2 = \tfrac{1}{2}[(2.70 + 4.00) \times 10^{-6}\ \text{F}](18.1\ \text{V})^2 = \boxed{1.10 \times 10^{-3}\ \text{J.}}$$

(*c*) The change in stored energy is

$$\Delta U = U_b - U_a = 1.10 \times 10^{-3}\ \text{J} - 2.73 \times 10^{-3}\ \text{J} = \boxed{-1.63 \times 10^{-3}\ \text{J.}}$$

(*d*) **The stored potential energy is not conserved.** During the flow of charge before the final steady state, some of the stored energy is dissipated as thermal and radiant energy.

50. We find the rms speed from

$$\text{KE} = \tfrac{1}{2}mv_{\text{rms}}^2 = \tfrac{3}{2}kT;$$

$(9.11 \times 10^{-31}\ \text{kg})v_{300}^2 = 3(1.38 \times 10^{-23}\ \text{J/K})(300\ \text{K})$, which gives $v_{300} = \boxed{1.17 \times 10^5\ \text{m/s.}}$

$(9.11 \times 10^{-31}\ \text{kg})v_{2500}^2 = 3(1.38 \times 10^{-23}\ \text{J/K})(2500\ \text{K})$, which gives $v_{2500} = \boxed{3.37 \times 10^5\ \text{m/s.}}$

51. We find the horizontal velocity of the electron as it enters the electric field from the accelerating voltage:

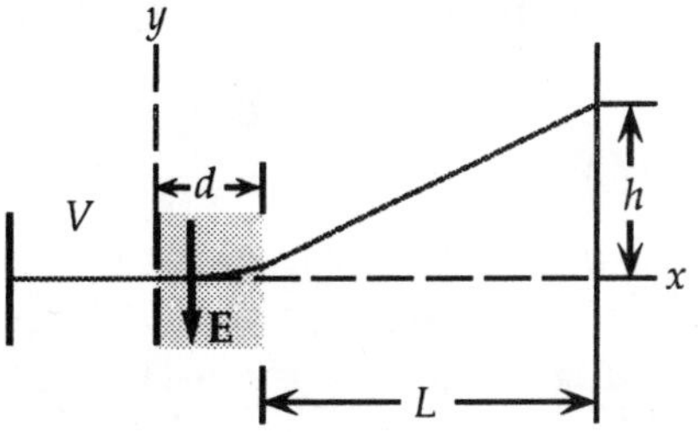

$$\tfrac{1}{2}mv_0^2 = eV;$$

$$\tfrac{1}{2}(9.11 \times 10^{-31}\ \text{kg})v_0^2 = (1.60 \times 10^{-19}\ \text{C})(15 \times 10^3\ \text{V}),$$

which gives $v_0 = 7.26 \times 10^7\ \text{m/s}$.

Because the force from the electric field is vertical, the horizontal velocity is constant. The time to pass through the field is

$$t_1 = d/v_0 = (0.028\ \text{m})/(7.26 \times 10^7\ \text{m/s}) = 3.86 \times 10^{-10}\ \text{s}.$$

The time for the electron to go from the field to the screen is

$$t_2 = L/v_0 = (0.22\ \text{m})/(7.26 \times 10^7\ \text{m/s}) = 3.03 \times 10^{-9}\ \text{s}.$$

If we neglect the small deflection during the passage through the field, we find the vertical velocity when the electron leaves the field from the vertical displacement:

$$v_y = h/t_2 = (0.11\ \text{m})/(3.03 \times 10^{-9}\ \text{s}) = 3.63 \times 10^7\ \text{m/s}.$$

This velocity was produced by the acceleration in the electric field:

$$F = eE = ma_y, \quad \text{or} \quad a_y = eE/m.$$

From the vertical motion in the field, we have

$$v_y = v_{0y} + a_yt_1;$$

$$3.63 \times 10^7\ \text{m/s} = 0 + [(1.60 \times 10^{-19}\ \text{C})E/(9.11 \times 10^{-31}\ \text{kg})](3.86 \times 10^{-10}\ \text{s}),$$

which gives $E = \boxed{5.4 \times 10^5\ \text{V/m.}}$

52. We find the horizontal velocity of the electron as it enters the electric field from the accelerating voltage:

$\frac{1}{2}mv_0^2 = eV;$

$\frac{1}{2}(9.11 \times 10^{-31}\ \text{kg})v_0^2 = (1.60 \times 10^{-19}\ \text{C})(14 \times 10^3\ \text{V}),$

which gives $v_0 = 7.01 \times 10^7$ m/s.

Because the force from the electric field is vertical, the horizontal velocity is constant. The time to pass through the field is

$t_1 = d/v_0 = (0.026\ \text{m})/(7.01 \times 10^7\ \text{m/s}) = 3.71 \times 10^{-10}$ s.

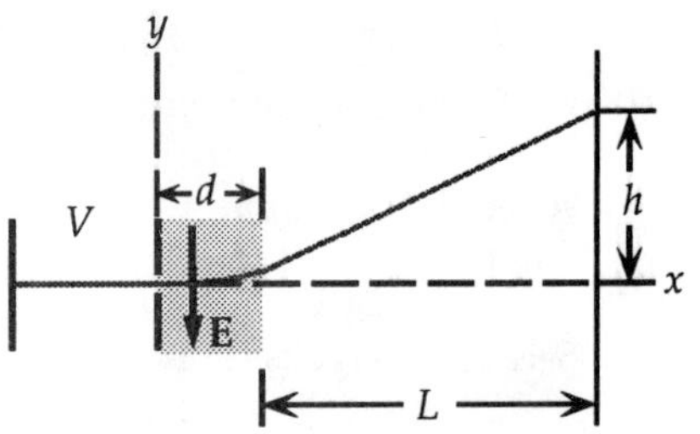

The time for the electron to go from the field to the screen is

$t_2 = L/v_0 = (0.34\ \text{m})/(7.01 \times 10^7\ \text{m/s}) = 4.85 \times 10^{-9}$ s.

The electron will sweep up and down across the screen. If we neglect the small deflection during the passage through the deflecting plates, when the electron leaves the plates the vertical velocity required to reach the edge of the screen is

$v_{y\text{max}} = h/t_2 = (0.15\ \text{m})/(4.85 \times 10^{-9}\ \text{s}) = 3.10 \times 10^7$ m/s.

This velocity was produced by the acceleration in the electric field:

$F = eE_{\text{max}} = ma_{y\text{max}},$ or $a_{y\text{max}} = eE_{\text{max}}/m.$

From the vertical motion in the field, we have

$v_{y\text{max}} = v_{0y} + a_{y\text{max}}t_1;$

$3.10 \times 10^7\ \text{m/s} = 0 + [(1.60 \times 10^{-19}\ \text{C})E_{\text{max}}/(9.11 \times 10^{-31}\ \text{kg})](3.71 \times 10^{-10}\ \text{s}),$

which gives $E_{\text{max}} = 4.8 \times 10^5$ V/m.

Thus the range for the electric field is $\boxed{-4.8 \times 10^5\ \text{V/m} < E < 4.8 \times 10^5\ \text{V/m}.}$

53. The energy density in the field is

$u = \frac{1}{2}\epsilon_0 E^2 = \frac{1}{2}(8.85 \times 10^{-12}\ \text{C}^2/\text{N}\cdot\text{m}^2)(150\ \text{V/m})^2 = \boxed{9.96 \times 10^{-8}\ \text{J/m}^3.}$

54. (*a*) We find the potential difference from

$U = Q\,\Delta V;$

$4.2\ \text{MJ} = (4.0\ \text{C})\,\Delta V,$ which gives $\Delta V = \boxed{1.1\ \text{MV}.}$

(*b*) We find the amount of water that can have its temperature raised to the boiling point from

$U = mc\,\Delta T;$

$4.2 \times 10^6\ \text{J} = m(4186\ \text{J/kg}\cdot\text{C}°)(100°\text{C} - 20°\text{C}),$ which gives $m = \boxed{13\ \text{kg}.}$

55. (a) We find the average translational kinetic energy from

$\text{KE}_\text{O} = \frac{3}{2}kT = \frac{3}{2}(1.38 \times 10^{-23}\ \text{J/K})(293\ \text{K})/(1.60 \times 10^{-19}\ \text{J/eV}) = \boxed{0.039\ \text{eV}.}$

(*b*) The average translational kinetic energy depends only on the temperature, so we have

$\text{KE}_\text{N} = \boxed{0.039\ \text{eV}.}$

(*c*) For the iron atom we have

$\text{KE}_\text{Fe} = \frac{3}{2}kT = \frac{3}{2}(1.38 \times 10^{-23}\ \text{J/K})(2 \times 10^6\ \text{K})/(1.60 \times 10^{-19}\ \text{J/eV}) = 3 \times 10^2\ \text{eV} = \boxed{0.3\ \text{keV}.}$

(*d*) For the carbon dioxide molecule we have

$\text{KE}_{\text{CO}_2} = \frac{3}{2}kT = \frac{3}{2}(1.38 \times 10^{-23}\ \text{J/K})(223\ \text{K})/(1.60 \times 10^{-19}\ \text{J/eV}) = \boxed{0.029\ \text{eV}.}$

56. The acceleration produced by a potential difference of 1000 V over a distance of 1 cm is

$a = eE/m = eV/md = (1.60 \times 10^{-19}\ \text{C})(1000\ \text{V})/(9.11 \times 10^{-31}\ \text{kg})(0.01\ \text{m}) = 2 \times 10^{16}\ \text{m/s}^2.$

Because this is so much greater than g, $\boxed{\text{yes,}}$ the electron can easily move upward.

To find the potential difference to hold the electron stationary, we have

$a = g = eE/m;$

$9.80\ \text{m/s}^2 = (1.60 \times 10^{-19}\ \text{C})V/(9.11 \times 10^{-31}\ \text{kg})(0.030\ \text{m}),$ which gives $V = \boxed{1.7 \times 10^{-12}\ \text{V}.}$

57. If the plates initially have a charge Q on each plate, the energy to move a charge ΔQ will increase the stored energy:

$$\Delta U = U_2 - U_1 = (\tfrac{1}{2}Q_2{}^2/C) - (\tfrac{1}{2}Q_1{}^2/C)$$
$$= [(Q + \Delta Q)^2 - Q^2]/2C = [(2Q\,\Delta Q + (\Delta Q)^2]/2C = (2Q + \Delta Q)\,\Delta Q/2C;$$
$$8.5\text{ J} = (2Q + 3.0\times10^{-3}\text{ C})(3.0\times10^{-3}\text{ C})/2(9.0\times10^{-6}\text{ F}),$$ which gives $Q = 0.024\text{ C} =$ **24 mC.**

58. (*a*) The kinetic energy of the electron ($q = -e$) is

$$\text{KE}_\text{e} = -qV_{BA} = -(-e)V_{BA} = eV_{BA}.$$

The kinetic energy of the proton ($q = +e$) is

$$\text{KE}_\text{p} = -qV_{AB} = -(+e)(-V_{BA}) = eV_{BA} =$$ **5.2 keV.**

(*b*) We find the ratio of their speeds, starting from rest, from

$$\tfrac{1}{2}m_\text{e}v_\text{e}{}^2 = \tfrac{1}{2}m_\text{p}v_\text{p}{}^2,\ \text{or}\ v_\text{e}/v_\text{p} = (m_\text{p}/m_\text{e})^{1/2} = [(1.67\times10^{-27}\text{ kg})/(9.11\times10^{-31}\text{ kg})]^{1/2} =$$ **42.8.**

59. The mica will change the capacitance. The potential difference is constant, so we have

$$\Delta Q = Q_2 - Q_1 = (C_2 - C_1)V = (K - 1)C_1V$$
$$= (7 - 1)(2600\times10^{-12}\text{ F})(9.0\text{ V}) = 1.4\times10^{-7}\text{ C} =$$ **0.14 μC.**

60. If we equate the heat flow to the stored energy, we have

$$U = \tfrac{1}{2}CV^2 = mc\,\Delta T;$$
$$\tfrac{1}{2}(4.0\text{ F})V^2 = (2.5\text{ kg})(4186\text{ J/kg}\cdot\text{C}°)(95°\text{C} - 20°\text{C}),$$ which gives $V =$ **6.3×10^2 V.**

61. Because the charged capacitor is disconnected from the plates, the charge must be constant. The paraffin will change the capacitance, so we have

$$Q = C_1V_1 = C_2V_2 = KC_1V_2;$$
$$24.0\text{ V} = (2.2)V_2,$$ which gives $V_2 =$ **10.9 V.**

62. The uniform electric field between the plates is related to the potential difference across the plates:

$$E = V/d.$$

For a parallel-plate capacitor, we have

$$Q = CV = (\epsilon_0A/d)(Ed) = \epsilon_0AE$$
$$= (8.85\times10^{-12}\text{ C}^2/\text{N}\cdot\text{m}^2)(56\times10^{-4}\text{ m}^2)(3.0\times10^6\text{ V/mm}) = 1.5\times10^{-7}\text{ C} =$$ **0.15 μC.**

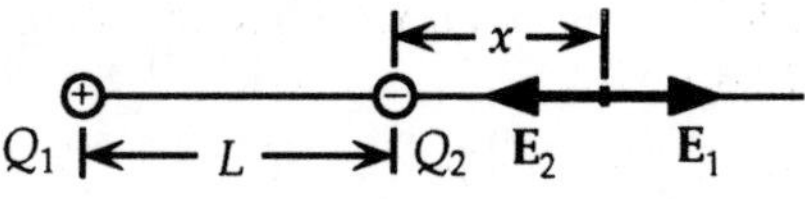

63. (*a*) Because the charges have opposite signs, the location where the electric field is zero must be outside the negative charge, as shown.
The fields from the two charges must balance:
$kQ_1/(x + L)^2 = kQ_2/x^2$;
$(3.4\ \mu C)/(x + 1.5\ cm)^2 = (2.0\ \mu C)/x^2$,
which gives $x = -0.65$ cm, 4.9 cm.

Because -0.65 cm is between the charges, the location is
4.9 cm from the negative charge, and 6.4 cm from the positive charge.

(*b*) The potential is a scalar that depends only on the distance. If the potential is 0 at the point x from the negative charge, the potential for the two charges is
$V = k[(Q_1/|(x + L)|) + (Q_2/|x|)]$;
$0 = [(3.4\ \mu C)/|(x + 1.5\ cm)|] + [(-2.0\ \mu C)/|x|]$,
which gives $3.4|x| = 2.0|(x + 1.5\ cm)|$.
For a point between the two charges, x is negative, so we have
$3.4(-x_1) = 2.0(x_1 + 1.5\ cm)$, which gives $x_1 = -0.56$ cm.
For a point outside the two charges, x is positive, so we have
$3.4(x_2) = 2.0(x_2 + 1.5\ cm)$, which gives $x_2 = 2.1$ cm.
Thus there are two positions: **0.56 cm from the negative charge toward the positive charge, and 2.1 cm from the negative charge away from the positive charge.**

64. The distances from the midpoint of a side to the three charges are $\ell/2$, $\ell/2$, and $\ell \cos 30°$.
At point a, we have
$V_a = k\{[(-Q)/(\ell/2)] + [(+Q)/(\ell/2)] + [(-3Q)/(\ell \cos 30°)]\}$
$= (kQ/\ell)[(-2) + (+2) + (-3/\cos 30°)] =$ **$-3.5\ kQ/\ell$.**
At point b, we have
$V_b = k\{[(+Q)/(\ell/2)] + [(-3Q)/(\ell/2)] + [(-Q)/(\ell \cos 30°)]\}$
$= (kQ/\ell)[(+2) + (-6) + (-1/\cos 30°)] =$ **$-5.2\ kQ/\ell$.**
At point c, we have
$V_c = k\{[(-3Q)/(\ell/2)] + [(-Q)/(\ell/2)] + [(+Q)/(\ell \cos 30°)]\}$
$= (kQ/\ell)[(-6) + (-2) + (+1/\cos 30°)] =$ **$-6.8\ kQ/\ell$.**

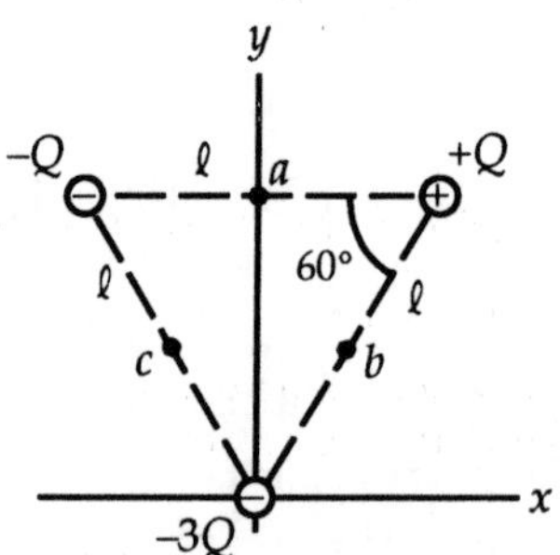

65. When the capacitors are connected, some charge will flow from C_1 to C_2 until the potential difference across the two capacitors is the same:
$V_1 = V_2 = V$.
Because charge is conserved, we have
$Q_0 = Q_1 + Q_2$.
For the two capacitors we have
$Q_1 = C_1V$, and $Q_2 = C_2V$.
When we form the ratio, we get
$Q_2/Q_1 = (Q_0 - Q_1)/Q_1 = C_2/C_1$, which gives **$Q_1 = Q_0C_1/(C_1 + C_2)$.**
For Q_2 we have
$Q_2 = Q_0 - Q_1 = Q_0\{1 - [C_1/(C_1 + C_2)]\}$, thus **$Q_2 = Q_0C_2/(C_1 + C_2)$.**
We find the potential difference from
$Q_1 = C_1V$;
$Q_0C_1/(C_1 + C_2) = C_1V$, which gives **$V = Q_0/(C_1 + C_2)$.**

66. We find the horizontal velocity of the electron as it enters the electric field from the accelerating voltage:

$\frac{1}{2}mv_0^2 = eV_{\text{accel}}$;

$\frac{1}{2}(9.11 \times 10^{-31}\text{ kg})v_0^2 = (1.60 \times 10^{-19}\text{ C})(25 \times 10^3\text{ V})$,

which gives $v_0 = 9.37 \times 10^7\text{ m/s}$.

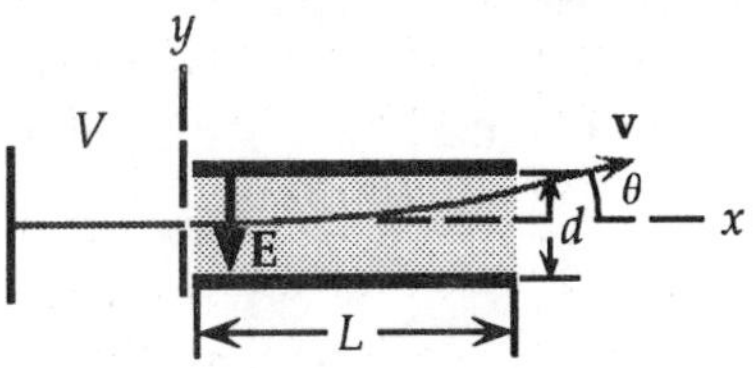

We find the vertical acceleration due to the electric field from

$F = eE = ma_y$, or $a_y = eE/m = eV/md$;

$a_y = (1.60 \times 10^{-19}\text{ C})(250\text{ V})/(9.11 \times 10^{-31}\text{ kg})(0.013\text{ m}) = 3.38 \times 10^{15}\text{ m/s}^2$.

Because the force from the electric field is vertical, the horizontal velocity is constant. The time to pass through the field is

$t_1 = L/v_0 = (0.065\text{ m})/(9.37 \times 10^7\text{ m/s}) = 6.94 \times 10^{-10}\text{ s}$.

From the vertical motion in the field, we have

$v_y = v_{0y} + a_y t_1 = 0 + (3.38 \times 10^{15}\text{ m/s}^2)(6.94 \times 10^{-10}\text{ s}) = 2.34 \times 10^6\text{ m/s}$.

The angle θ is the direction of the velocity, which we find from

$\tan\theta = v_y/v_0 = (2.34 \times 10^6\text{ m/s})/(9.37 \times 10^7\text{ m/s}) = 0.025$, or $\boxed{\theta = 1.4°.}$

67. For the motion of the electron from emission to the plate, the energy of the electron is conserved, so we have

$\Delta\text{KE} + \Delta\text{PE} = 0$, or $0 - \frac{1}{2}mv^2 + (-e)\,\Delta V = 0$;

$-\frac{1}{2}(9.11 \times 10^{-31}\text{ kg})v^2 + (-1.60 \times 10^{-19}\text{ C})(-3.02\text{ V} - 0) = 0$, which gives $v = \boxed{1.03 \times 10^6\text{ m/s.}}$

68. (*a*) For a parallel-plate capacitor, we find the gap from

$C = \epsilon_0 A/d$;

$1\text{ F} = (8.85 \times 10^{-12}\text{ C}^2/\text{N}\cdot\text{m}^2)(1.0 \times 10^{-4}\text{ m}^2)/d$, which gives $d = \boxed{9 \times 10^{-16}\text{ m.}}$

Because this is many orders of magnitude less than the size of an atom, it is not practical.

(*b*) We find the area from

$C = \epsilon_0 A/d$;

$1\text{ F} = (8.85 \times 10^{-12}\text{ C}^2/\text{N}\cdot\text{m}^2 A/(1.0 \times 10^{-3}\text{ m})$, which gives $A = \boxed{1.1 \times 10^8\text{ m}^2.}$

Because this corresponds to a square $\approx$ 10 km on a side, it is not practical.

69. Because the electric field points downward, the potential is greater at the higher elevation. For the potential difference, we have

$\Delta V = -(150\text{ V/m})(2.00\text{ m}) = -300\text{ V}$.

For the motion of the falling charged balls, the energy is conserved:

$\Delta\text{KE} + \Delta\text{PE} = 0$, or $\frac{1}{2}mv^2 - 0 + q\,\Delta V + mg(0 - h) = 0$, which gives

$v^2 = 2gh - 2(q/m)\,\Delta V$.

For the positive charge, we have

$v_1^2 = 2gh - 2(q_1/m)\,\Delta V$

$= 2(9.80\text{ m/s}^2)(2.00\text{ m}) - 2[(550 \times 10^{-6}\text{ C})/(0.540\text{ kg})](-300\text{ V})$, which gives $v_1 = 6.31\text{ m/s}$.

For the negative charge, we have

$v_2^2 = 2gh - 2(q_2/m)\,\Delta V$

$= 2(9.80\text{ m/s}^2)(2.00\text{ m}) - 2[(-550 \times 10^{-6}\text{ C})/(0.540\text{ kg})](-300\text{ V})$, which gives $v_2 = 6.21\text{ m/s}$.

Thus the difference in speeds is $6.31\text{ m/s} - 6.21\text{ m/s} = \boxed{0.10\text{ m/s.}}$

70. (*a*) The energy stored in the capacitor is

$$U = \tfrac{1}{2}CV^2 = \tfrac{1}{2}(0.050 \times 10^{-6}\ \text{F})(30 \times 10^3\ \text{V})^2 = \boxed{23\ \text{J.}}$$

(*b*) We find the power of the pulse from

$$P = 0.10\ U/t = (0.10)(23\ \text{J})/(10 \times 10^{-6}\ \text{s}) = 2.3 \times 10^5\ \text{W} = \boxed{0.23\ \text{MW.}}$$

71. (*a*) We find the capacitance from

$$C = \epsilon_0 A/d$$

$$= (8.85 \times 10^{-12}\ \text{C}^2/\text{N}\cdot\text{m}^2)(110 \times 10^6\ \text{m}^2)/(1500\ \text{m}) = 6.49 \times 10^{-7}\ \text{F} = \boxed{0.649\ \mu\text{F.}}$$

(*b*) We find the stored charge from

$$Q = CV = (6.49 \times 10^{-7}\ \text{F})(35 \times 10^6\ \text{V}) = \boxed{23\ \text{C.}}$$

(*c*) For the stored energy we have

$$U = \tfrac{1}{2}CV^2 = \tfrac{1}{2}(6.49 \times 10^{-7}\ \text{F})(35 \times 10^6\ \text{V})^2 = \boxed{4.0 \times 10^8\ \text{J.}}$$

CHAPTER 18

1. The charge that passes through the battery is
   $\Delta Q = I\,\Delta t = (5.7\text{ A})(7.0\text{ h})(3600\text{ s/h}) = \boxed{1.4 \times 10^5\text{ C.}}$

2. The rate at which electrons pass any point in the wire is the current:
   $I = 1.00\text{ A} = (1.00\text{ C/s})/(1.60 \times 10^{-19}\text{ C/electron}) = \boxed{6.25 \times 10^{18}\text{ electron/s.}}$

3. We find the current from
   $I = \Delta Q/\Delta t = (1000\text{ ions})(1.60 \times 10^{-19}\text{ C/ion})/(6.5 \times 10^{-6}\text{ s}) = \boxed{2.5 \times 10^{-11}\text{ A.}}$

4. We find the voltage from
   $V = IR = (0.25\text{ A})(3000\ \Omega) = \boxed{7.5 \times 10^2\text{ V.}}$

5. We find the resistance from
   $V = IR;$
   $110\text{ V} = (3.1\text{ A})R$, which gives $R = \boxed{35\ \Omega.}$

6. For the device we have $V = IR$.
   (*a*) If we assume constant resistance and divide expressions for the two conditions, we get
   $V_2/V_1 = I_2/I_1;$
   $0.90 = I_2/(5.50\text{ A})$, which gives $I_2 = \boxed{4.95\text{ A.}}$
   (*b*) With the voltage constant, if we divide expressions for the two conditions, we get
   $I_2/I_1 = R_1/R_2;$
   $I_2/(5.50\text{ A}) = 1/0.90$, which gives $I_2 = \boxed{6.11\text{ A.}}$

7. The rate at which electrons leave the battery is the current:
   $I = V/R = [(9.0\text{ V})/(1.6\ \Omega)](60\text{ s/min})/(1.60 \times 10^{-19}\text{ C/electron}) = \boxed{2.1 \times 10^{21}\text{ electron/min.}}$

8. (*a*) We find the resistance from
   $V = IR;$
   $120\text{ V} = (9.0\text{ A})R$, which gives $R = \boxed{13\ \Omega.}$
   (*b*) The charge that passes through the hair dryer is
   $\Delta Q = I\,\Delta t = (9.0\text{ A})(15\text{ min})(60\text{ s/min}) = \boxed{8.1 \times 10^3\text{ C.}}$

9. We find the resistance from
   $V = IR;$
   $12\text{ V} = (0.50\text{ A})R$, which gives $R = \boxed{24\ \Omega.}$
   The energy taken out of the battery is
   $E = Pt = IVt = (0.50\text{ A})(12\text{ V})(1\text{ min})(60\text{ s/min}) = \boxed{3.6 \times 10^2\text{ J.}}$

10. We find the voltage across the bird's feet from
    $V = IR = (2500\text{ A})(2.5 \times 10^{-5}\ \Omega/\text{m})(4.0 \times 10^{-2}\text{ m}) = \boxed{2.5 \times 10^{-3}\text{ V.}}$

11. We find the resistance from

$$R = \rho L/A = \rho L/\tfrac{1}{4}\pi d^2$$
$$= (1.68 \times 10^{-8}\ \Omega\cdot\text{m})(3.0\ \text{m})/\tfrac{1}{4}\pi(1.5 \times 10^{-3}\ \text{m})^2 = \boxed{0.029\ \Omega.}$$

12. We find the radius from

$$R = \rho L/A = \rho L/\pi r^2;$$
$$0.22\ \Omega = (5.6 \times 10^{-8}\ \Omega\cdot\text{m})(1.00\ \text{m})/\pi r^2, \text{ which gives } r = 2.85 \times 10^{-4}\ \text{m},$$

so the diameter is $5.7 \times 10^{-4}\ \text{m} = \boxed{0.57\ \text{mm.}}$

13. From the expression for the resistance, $R = \rho L/A$, we form the ratio

$$R_{Al}/R_{Cu} = (\rho_{Al}/\rho_{Cu})(L_{Al}/L_{Cu})(A_{Cu}/A_{Al}) = (\rho_{Al}/\rho_{Cu})(L_{Al}/L_{Cu})(d_{Cu}/d_{Al})^2$$
$$= [(2.65 \times 10^{-8}\ \Omega\cdot\text{m})/(1.68 \times 10^{-8}\ \Omega\cdot\text{m})][(10.0\ \text{m})/(15.0\ \text{m})][(2.5\ \text{mm})/(2.0\ \text{mm})]^2$$
$$= 1.6, \text{ or } \boxed{R_{Al} = 1.6R_{Cu}.}$$

14. $\boxed{\text{Yes,}}$ if we select the appropriate diameter. From the expression for the resistance, $R = \rho L/A$, we form the ratio

$$R_T/R_{Cu} = (\rho_T/\rho_{Cu})(L_T/L_{Cu})(A_{Cu}/A_T) = (\rho_T/\rho_{Cu})(d_{Cu}/d_T)^2;$$
$$1 = [(5.6 \times 10^{-8}\ \Omega\cdot\text{m})/(1.68 \times 10^{-8}\ \Omega\cdot\text{m})][(2.5\ \text{mm})/d_T]^2, \text{ which gives } d_T = \boxed{4.6\ \text{mm.}}$$

15. Because the material and area of the two pieces are the same, from the expression for the resistance, $R = \rho L/A$, we see that the resistance is proportional to the length:

$$R_1/R_2 = L_1/L_2 = 7.$$

Because $L_1 + L_2 = L$, we have

$$7L_2 + L_2 = L, \text{ or } L_2 = L/8, \text{ and } L_1 = 7L/8, \text{ so the wire should be cut at } \boxed{1/8 \text{ the length.}}$$

We find the resistance of each piece from

$$R_1 = (L_1/L)R = (7/8)(10.0\ \Omega) = \boxed{8.75\ \Omega;}$$
$$R_2 = (L_2/L)R = (1/8)(10.0\ \Omega) = \boxed{1.25\ \Omega.}$$

16. We find the temperature change from

$$R = R_0(1 + \alpha\Delta T), \text{ or } \Delta R = R_0\alpha\Delta T;$$
$$0.20R_0 = R_0[0.0068\ (\text{C}°)^{-1}]\,\Delta T, \text{ which gives } \Delta T = \boxed{29\ \text{C}°.}$$

17. For the wire we have $R = V/I$. We find the temperature from

$$R = R_0(1 + \alpha\Delta T);$$
$$(V/I) = (V/I_0)(1 + \alpha\Delta T);$$
$$(10.00\ \text{V}/0.3618\ \text{A}) = (10.00\ \text{V}/0.4212\ \text{A})\{1 + [0.00429\ (\text{C}°)^{-1}](T - 20.0°\text{C})\}, \text{ which gives } T = \boxed{58.3°\text{C.}}$$

18. We find the temperature from

$$\rho_{0,T} = \rho_{Cu} = \rho_{0,Cu}(1 + \alpha_{Cu}\Delta T);$$
$$5.6 \times 10^{-8}\ \Omega\cdot\text{m} = (1.68 \times 10^{-8}\ \Omega\cdot\text{m})\{1 + [0.0068\ (\text{C}°)^{-1}](T - 20.0°\text{C})\}, \text{ which gives } T = \boxed{363°\text{C.}}$$

19. We find the temperature from

$$R = R_0(1 + \alpha_{Cu}\Delta T);$$
$$140\ \Omega = (12\ \Omega)\{1 + [0.0060\ (\text{C}°)^{-1}](T - 20.0°\text{C})\}, \text{ which gives } T = \boxed{1.8 \times 10^3\ °\text{C.}}$$

20. For each direction through the solid, the length and area will be constant, so we have $R = \rho L/A$.

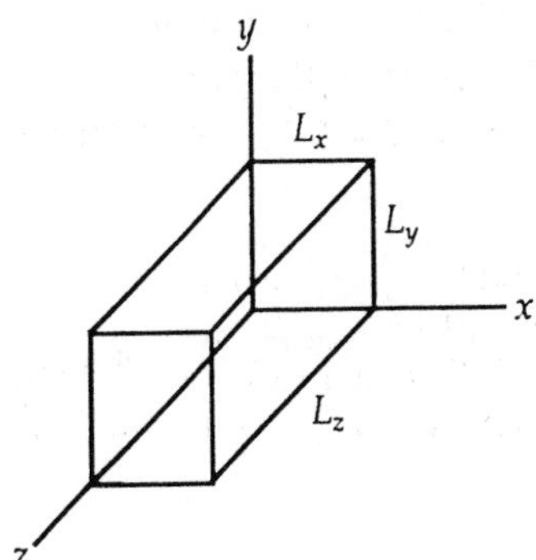

(a) In the x-direction, we have

$$R_x = \rho L_x/L_yL_z = (3.0 \times 10^{-5}\ \Omega\cdot\text{m})(0.010\ \text{m})/(0.020\ \text{m})(0.040\ \text{m}) = \boxed{3.8 \times 10^{-4}\ \Omega.}$$

(b) In the y-direction, we have

$$R_y = \rho L_y/L_xL_z = (3.0 \times 10^{-5}\ \Omega\cdot\text{m})(0.020\ \text{m})/(0.010\ \text{m})(0.040\ \text{m}) = \boxed{1.5 \times 10^{-3}\ \Omega.}$$

(c) In the z-direction, we have

$$R_z = \rho L_z/L_xL_y = (3.0 \times 10^{-5}\ \Omega\cdot\text{m})(0.040\ \text{m})/(0.010\ \text{m})(0.020\ \text{m}) = \boxed{6.0 \times 10^{-3}\ \Omega.}$$

21. (a) For the resistance of each wire we have

$$R_{\text{Cu}} = \rho_{\text{Cu}}L_{\text{Cu}}/A = (1.68 \times 10^{-8}\ \Omega\cdot\text{m})(5.0\ \text{m})/\pi(0.50 \times 10^{-3}\ \text{m})^2 = 0.107\ \Omega.$$

$$R_{\text{Al}} = \rho_{\text{Al}}L_{\text{Al}}/A = (2.65 \times 10^{-8}\ \Omega\cdot\text{m})(5.0\ \text{m})/\pi(0.50 \times 10^{-3}\ \text{m})^2 = 0.169\ \Omega.$$

Thus the total resistance is

$$R = R_{\text{Cu}} + R_{\text{Al}} = 0.107\ \Omega + 0.169\ \Omega = \boxed{0.28\ \Omega.}$$

(b) We find the current from

$$V = IR;$$

$$80\ \text{V} = I(0.28\ \Omega),\ \text{which gives}\ I = \boxed{2.9 \times 10^2\ \text{A.}}$$

(c) The current must be the same for the two wires, so we have

$$V_{\text{Cu}} = IR_{\text{Cu}} = (2.9 \times 10^2\ \text{A})(0.107\ \Omega) = \boxed{31\ \text{V.}}$$

$$V_{\text{Al}} = IR_{\text{Al}} = (2.9 \times 10^2\ \text{A})(0.169\ \Omega) = \boxed{49\ \text{V.}}$$

22. We use the temperature coefficients at 20°C. For the total resistance we have

$$R = R_{\text{C}} + R_{\text{N}} = R_{\text{C0}}(1 + \alpha_{\text{C}}\,\Delta T) + R_{\text{N0}}(1 + \alpha_{\text{N}}\,\Delta T) = R_{\text{C0}} + R_{\text{N0}} + R_{\text{C0}}\alpha_{\text{C}}\,\Delta T + R_{\text{N0}}\alpha_{\text{N}}\,\Delta T.$$

If the total resistance does not change, we have

$$R_{\text{C}} + R_{\text{N}} = R_{\text{C0}} + R_{\text{N0}},\ \text{or}$$

$$R_{\text{C0}}\alpha_{\text{C}}\,\Delta T = -\,R_{\text{N0}}\alpha_{\text{N}}\,\Delta T;$$

$$R_{\text{C0}}[-0.0005\ (\text{C}°)^{-1}] = -\,R_{\text{N0}}[0.0004\ (\text{C}°)^{-1}],\ \text{which gives}\ R_{\text{C0}} = 0.8R_{\text{N0}}.$$

For the total resistance we have

$$4.70\ \text{k}\Omega = R_{\text{C0}} + R_{\text{N0}} = 0.8R_{\text{N0}} + R_{\text{N0}},\ \text{which gives}\ \boxed{R_{\text{N0}} = 2.61\ \text{k}\Omega,}\ \text{and}\ \boxed{R_{\text{C0}} = 2.09\ \text{k}\Omega.}$$

23. For an ohmic resistor, we have

$$P = IV = V^2/R,\ \text{or}\ R = V^2/P = (240\ \text{V})^2/(3.3 \times 10^3\ \text{W}) = \boxed{17\ \Omega.}$$

24. We find the power from

$$P = IV = (0.350\ \text{A})(9.0\ \text{V}) = \boxed{3.2\ \text{W.}}$$

25. From $P = V^2/R$, we see that the maximum voltage will produce the maximum power, so we have

$$\tfrac{1}{4}\ \text{W} = V_{\text{max}}^2/(2.7 \times 10^3\ \Omega),\ \text{which gives}\ V_{\text{max}} = \boxed{26\ \text{V.}}$$

26. (*a*) From $P = V^2/R$, we see that the lower power setting, $\boxed{600\text{ W,}}$ must have the higher resistance.
    (*b*) At the lower setting, we have
    $P_1 = V^2/R_1$;
    $600\text{ W} = (120\text{ V})^2/R_1$, which gives $R_1 = \boxed{24\ \Omega.}$
    (*c*) At the higher setting, we have
    $P_2 = V^2/R_2$;
    $1200\text{ W} = (120\text{ V})^2/R_2$, which gives $R_2 = \boxed{12\ \Omega.}$

27. (*a*) We find the resistance from
    $P_1 = V^2/R_1$;
    $60\text{ W} = (120\text{ V})^2/R_1$, which gives $R_1 = \boxed{240\ \Omega.}$
    The current is
    $I_1 = V/R_1 = (120\text{ V})/(240\ \Omega) = \boxed{0.50\text{ A.}}$
    (*b*) We find the resistance from
    $P_2 = V^2/R_2$;
    $440\text{ W} = (120\text{ V})^2/R_2$, which gives $R_2 = \boxed{32.7\ \Omega.}$
    The current is
    $I_2 = V/R_2 = (120\text{ V})/(32.7\ \Omega) = \boxed{3.67\text{ A.}}$

28. We find the operating resistance from
    $P = V^2/R$;
    $60\text{ W} = (240\text{ V})^2/R$, which gives $R = 9.6 \times 10^2\ \Omega$.
    If we assume that the resistance stays the same, for the lower voltage we have
    $P = V^2/R = (120\text{ V})^2/(9.6 \times 10^2\ \Omega) = \boxed{15\text{ W.}}$
    At one-quarter the power, the bulb will be much dimmer.

29. We find the energy used by the toaster from
    $E = Pt = (0.550\text{ kW})(10\text{ min})/(60\text{ min/h}) = \boxed{0.092\text{ kWh.}}$
    The cost for a month would be
    $\text{Cost} = E(\text{rate}) = (0.092\text{ kWh/day})(4\text{ days/wk})(4\text{ wk/month})(12¢/\text{kWh}) = \boxed{18¢/\text{month.}}$

30. The cost for a year would be
    $\text{Cost} = E(\text{rate}) = Pt(\text{rate})$
    $= (40 \times 10^{-3}\text{ kW})(1\text{ yr})(365\text{ days/yr})(24\text{ h/day})(\$0.110/\text{kWh}) = \boxed{\$38.50.}$

31. (*a*) We find the resistance from
    $V = IR$;
    $2(1.5\text{ V}) = (0.350\text{ A})/R$, which gives $R = \boxed{8.6\ \Omega.}$
    The power dissipated is
    $P = IV = (0.350\text{ A})(3.0\text{ V}) = \boxed{1.1\text{ W.}}$
    (*b*) We assume that the resistance does not change, so we have
    $P_2/P_1 = (V_2/V_1)^2 = (6.0\text{ V}/3.0\text{ V})^2 = \boxed{4\times.}$
    The increased power would last for a short time, until the increased temperature of the filament would burn out the bulb.

32. 90 A · h is the total charge that passed through the battery when it was charged.
    We find the energy from
    $E = Pt = VIt = VQ = (12\text{ V})(90\text{ A}\cdot\text{h})(10^{-3}\text{ kW/W}) = \boxed{1.1\text{ kWh}} = 3.9 \times 10^6\text{ J}.$

33. (*a*) We find the maximum power output from

$P_{max} = I_{max}V = (0.025\text{ A})(9.0\text{ V}) = \boxed{0.23\text{ W.}}$

(*b*) The power output eventually becomes thermal energy. The circuit is designed to allow the dissipation of the maximum power, which we assume is the same. Thus we have

$P_{max} = I_{max}V = I_{max}'V'$;

$0.225\text{ W} = I_{max}'(7.0\text{ V})$, which gives $I_{max}' = 0.032\text{ A} = \boxed{32\text{ mA.}}$

34. The total power will be the sum, so we have

$P_{total} = I_{total}V$;

$N(100\text{ W}) = (2.5\text{ A})(120\text{ V})$, which gives $N = \boxed{3.}$

35. The power rating is the mechanical power output, so we have

efficiency $= \text{output}/\text{input} = P_{mechanical}/P_{electrical}$

$= (0.50\text{ hp})(746\text{ W/hp})/(4.4\text{ A})(120\text{ V}) = 0.706 = \boxed{71\%.}$

36. The required current to deliver the power is $I = P/V$, and the wasted power (thermal losses in the wires) is $P_{loss} = I^2R$. For the two conditions we have

$I_1 = (520\text{ kW})/(12\text{ kV}) = 43.3\text{ A};\ P_{loss1} = (43.3\text{ A})^2(3.0\ \Omega)(10^{-3}\text{ kW/W}) = 5.63\text{ kW}$;

$I_2 = (520\text{ kW})/(50\text{ kV}) = 10.4\text{ A};\ P_{loss2} = (10.4\text{ A})^2(3.0\ \Omega)(10^{-3}\text{ kW/W}) = 0.324\text{ kW}$.

Thus the decrease in power loss is

$\Delta P_{loss} = P_{loss1} - P_{loss2} = 5.63\text{ kW} - 0.324\text{ kW} = \boxed{5.3\text{ kW.}}$

37. (*a*) We find the resistance from

$P = V^2/R$;

$2200\text{ W} = (240\text{ V})^2/R$, which gives $R = \boxed{26.2\ \Omega.}$

(*b*) If 80% of the electrical energy is used to heat the water to the boiling point, we have

$0.80E_{elec} = 0.80\,P_{elec}t = mc\,\Delta T$;

$(0.80)(2200\text{ W})t = (100\text{ mL})(1\text{ g/mL})(10^{-3}\text{ kg/g})(4186\text{ J/kg}\cdot\text{C°})(100\text{°C} - 20\text{°C})$,

which gives $t = \boxed{19\text{ s.}}$

(*c*) We find the cost from

Cost $= E(\text{rate}) = P_{elec}t\ (\text{rate})$

$= [(2.20\text{ kW})(19\text{ s})/(3600\text{ s/h})](10¢/\text{kWh}) = \boxed{0.12¢.}$

38. For the water to remove the thermal energy produced, we have

$P = IV = (m/t)c\,\Delta T$;

$(14.5\text{ A})(240\text{ V}) = (m/t)(4186\text{ J/kg}\cdot\text{C°})(7.50\text{ C°})$, which gives $m/t = \boxed{0.111\text{ kg/s.}}$

39. If 60% of the electrical energy is used to heat the water, we have

$0.60E_{elec} = 0.60\,IVt = mc\,\Delta T$;

$(0.60)I(12\text{ V})(5.0\text{ min})(60\text{ s/min}) = (150\text{ mL})(1\text{ g/mL})(10^{-3}\text{ kg/g})(4186\text{ J/kg}\cdot\text{C°})(95\text{°C} - 5\text{°C})$,

which gives $I = \boxed{26\text{ A.}}$

The resistance of the heating coil is

$R = V/I = (12\text{ V})/(26\text{ A}) = \boxed{0.46\ \Omega.}$

40. We find the peak current from the peak voltage:

$V_0 = \sqrt{2}\,V_{rms} = I_0R$;

$\sqrt{2}(120\text{ V}) = I_0(2.2\times10^3\ \Omega)$, which gives $I_0 = \boxed{0.077\text{ A.}}$

41. We find the peak current from the peak voltage:
    $V_0 = I_0R$;
    180 V = I_0(330 Ω), which gives I_0 = 0.545 A.
    The rms current is
    $I_{rms} = I_0/\sqrt{2}$ = (0.545 A)/$\sqrt{2}$ = 0.386 A.

42. (*a*) Because the total resistance is
    $R_{total} = V/I$, when $I = 0$, the resistance is infinite.
    (*b*) With one lightbulb on, we have
    $P = I_{rms}V_{rms} = V_{rms}^2/R$;
    75 W = (120 V)$^2/R$, which gives R = 1.9×10^2 Ω.

43. We find the rms voltage from
    $P = I_{rms}V_{rms}$;
    1500 W = [(4.0 A)/$\sqrt{2}$]V_{rms}, which gives V_{rms} = 5.3×10^2 V.

44. The peak voltage is
    $V_0 = \sqrt{2}\,V_{rms} = \sqrt{2}$(450 V) = 636 V.
    We find the peak current from
    $P = I_{rms}V_{rms} = (I_0/\sqrt{2})V_{rms}$;
    1800 W = $(I_0/\sqrt{2})$(450 V), which gives I_0 = 5.66 A.

45. The maximum instantaneous power is
    $P_0 = I_0V_0 = (\sqrt{2}\,I_{rms})(\sqrt{2}\,V_{rms}) = 2P$ = 2(3.0 hp) = 6.0 hp (4.5 kW).
    For the maximum current, we have
    $P = I_{rms}V_{rms} = (I_0/\sqrt{2})V_{rms}$;
    (3.0 hp)(746 W/hp) = $(I_0/\sqrt{2})$(240 V), which gives I_0 = 13 A.

46. For the average power, we have
    $P = I_{rms}V_{rms} = V_{rms}^2/R$ = (240 V)^2/(34 W) = 1.7×10^3 W = 1.7 kW.
    The maximum power is
    $P_0 = I_0V_0 = (\sqrt{2}\,I_{rms})(\sqrt{2}\,V_{rms}) = 2P$ = 2(1.7 kW) = 3.4 kW.
    Because the power is always positive, the minimum power is zero.

47. From Example 18–14 we know that the density of free electrons in copper is
    $n = 8.4 \times 10^{28}\ \text{m}^{-3}$.
    We find the drift speed from
    $I = neAv_d = ne(\frac{1}{4}\pi D^2)v_d$;
    2.5×10^{-6} A = $(8.4 \times 10^{28}\ \text{m}^{-3})(1.60 \times 10^{-19}\ \text{C})[\frac{1}{4}\pi(0.55 \times 10^{-3}\ \text{m})^2]v_d$,
    which gives v_d = 7.8×10^{-10} m/s.

48. (*a*) We find the resistance from

$R = V/I = (22.0 \text{ mV})/(750 \text{ mA}) = \boxed{0.0293 \ \Omega.}$

(*b*) We find the resistivity from

$R = \rho L/A;$

$0.0293 \ \Omega = \rho(5.00 \text{ m})/\pi(1.0 \times 10^{-3} \text{ m})^2$, which gives $\rho = \boxed{1.8 \times 10^{-8} \ \Omega \cdot \text{m.}}$

(*c*) We find the density of free electrons from the drift speed:

$I = neAv_d = ne(\pi r^2)v_d;$

$750 \text{ mA} = n(1.60 \times 10^{-19} \text{ C})\pi(1.0 \times 10^{-3} \text{ m})^2(1.7 \times 10^{-5} \text{ m/s})$, which gives $n = \boxed{8.8 \times 10^{28} \text{ m}^{-3}.}$

49. For the total current we have

$I = n_+(+e)Av_{d+} + n_-(-2e)Av_{d-} = 0;$

$(5.00 \text{ mol/m}^3)(5.00 \times 10^{-4} \text{ m/s}) - 2n_-(-2.00 \times 10^{-4} \text{ m/s}) = 0$, which gives $n_- = \boxed{6.25 \text{ mol/m}^3.}$

50. For the net current density we have

$I/A = n_+(+2e)v_{d+} + n_-(-e)v_{d-}$

$= (2.8 \times 10^{12} \text{ m}^{-3})(2)(1.60 \times 10^{-19} \text{ C})(2.0 \times 10^6 \text{ m/s}) +$

$(8.0 \times 10^{11} \text{ m}^{-3})(-1.60 \times 10^{-19} \text{ C})(-7.2 \times 10^6 \text{ m/s}) = \boxed{2.7 \text{ A/m}^2 \text{ north.}}$

51. We find the magnitude of the electric field from

$E = V/d = (70 \times 10^{-3} \text{ V})/(1.0 \times 10^{-8} \text{ m}) = \boxed{7.0 \times 10^6 \text{ V/m.}}$

Note that the direction of the field will be into the cell.

52. We find the speed of the pulse from

$v = \Delta x/\Delta t = (7.20 \text{ cm} - 3.40 \text{ cm})/(100 \text{ cm/m})(0.0063 \text{ s} - 0.0052 \text{ s}) = \boxed{35 \text{ m/s.}}$

Two measurements are necessary to eliminate uncertainty over the exact location of the stimulation and the effects of the initial creation of the stimulation, including any initial delay in producing the change in concentrations.

53. From the data of Example 18–15, we can find the required energy from the energy stored in the capacitor during a pulse:

$U = \frac{1}{2}CV^2 = \frac{1}{2}(10^{-8} \text{ F})(0.10 \text{ V})^2 = \boxed{5 \times 10^{-11} \text{ J.}}$

If the time between pulses is

$t = 1/(100 \text{ pulses/s}) = 0.010 \text{ s},$

for N neurons the average power is

$P_{av} = NU/t = (10^4)(5 \times 10^{-11} \text{ J})/(0.010 \text{ s}) = \boxed{5 \times 10^{-5} \text{ W.}}$

54. Because each ion carries the electron charge, the effective current of the Na^+ ions through the surface of the axon is

$I = jN_A e(\pi dL)$

$= (3 \times 10^{-7} \text{ mol/m}^2 \cdot \text{s})(6.02 \times 10^{23} \text{ ion/mol})(1.60 \times 10^{-19} \text{ C/ion})\pi(20 \times 10^{-6} \text{ m})(0.10 \text{ m})$

$= 1.8 \times 10^{-7} \text{ A}.$

The power required to move these charges through the potential difference is

$P = IV = (1.8 \times 10^{-7} \text{ A})(30 \times 10^{-3} \text{ V}) = \boxed{5.4 \times 10^{-9} \text{ W.}}$

55. The charge is

$\Delta Q = I\,\Delta t = (1.00 \text{ A} \cdot \text{h})(3600 \text{ s/h}) = \boxed{3.60 \times 10^3 \text{ C.}}$

56. We find the current from

$P = IV;$

$(1.0\text{ hp})(746\text{ W/hp}) = I(120\text{ V})$, which gives $I =$ $\boxed{6.2\text{ A.}}$

57. We find the current when the lights are on from

$P = IV;$

$(92\text{ W}) = I(12\text{ V})$, which gives $I = 7.67\text{ A}$.

90 A · h is the total charge that passes through the battery when it is completely discharged. Thus the time for complete discharge is

$t = Q/I = (90\text{ A}\cdot\text{h})/(7.67\text{ A}) =$ $\boxed{12\text{ h.}}$

58. We find the resistance of the heating element from

$P = IV = V^2/R;$

$1500\text{ W} = (110\text{ V})^2/R$, which gives $R = 8.07\ \Omega$.

We find the diameter from

$R = \rho L/A = \rho L/\frac{1}{4}\pi D^2 = 4\rho L/\pi D^2;$

$8.07\ \Omega = 4(9.71 \times 10^{-8}\ \Omega\cdot\text{m})(5.4\text{ m})/\pi D^2$, which gives $D = 2.9 \times 10^{-4}\text{ m} =$ $\boxed{0.29\text{ mm.}}$

59. We find the conductance from

$G = 1/R = I/V = (0.700\text{ A})/(3.0\text{ V}) =$ $\boxed{0.23\text{ S.}}$

60. (*a*) The daily energy use is

$E = (1.8\text{ kW})(3.0\text{ h/day}) + 4(0.10\text{ kW})(6.0\text{ h/day}) + (3.0\text{ kW})(1.4\text{ h/day}) + 2.0\text{ kWh/day}$
$= 14\text{ kWh/day}.$

The cost for a month is

$\text{Cost} = E(\text{rate}) = (14\text{ kWh/day})(30\text{ days/month})(\$0.105/\text{kWh}) =$ $\boxed{\$44.}$

(*b*) For a 35-percent efficient power plant, we find the amount of coal from

$0.35m(7000\text{ kcal/kg})(4186\text{ J/kcal}) = (14\text{ kWh/day})(365\text{ days/yr})(10^3\text{ W/kW})(3600\text{ s/yr})$,

which gives $m =$ $\boxed{1.8 \times 10^3\text{ kg.}}$

61. The dependence of the resistance on the dimensions is $R = \rho L/A$. When we form the ratio for the two wires, we get

$R_2/R_1 = (L_2/L_1)(A_1/A_2) = (\frac{1}{2})(\frac{1}{2}) = \frac{1}{4}$, so $\boxed{R_2 = \frac{1}{4}R_1.}$

62. (*a*) The dependence of the power output on the voltage is $P = V^2/R$. When we form the ratio for the two conditions, we get

$P_2/P_1 = (V_2/V_1)^2.$

For the percentage change we have

$[(P_2 - P_1)/P_1](100) = [(V_2/V_1)^2 - 1](100) = [(105\text{ V}/117\text{ V})^2 - 1](100) =$ $\boxed{-19.5\%.}$

(*b*) The decreased power output would cause a decrease in the temperature, so the resistance would decrease. This means for the reduced voltage, the

$\boxed{\text{percentage decrease in the power output would be less}}$ than calculated.

63. The maximum current will produce the maximum rate of heating. We can find the resistance per meter from

$P/L = I^2R/L;$

$1.6 \text{ W/m} = (35 \text{ A})^2(R/L)$, which gives $R/L = 1.31 \times 10^{-3}\ \Omega/\text{m}$.

From the dependence of the resistance on the dimensions, $R = \rho L/A$, we get

$R/L = \rho/\frac{1}{4}\pi D^2 = 4\rho/\pi D^2;$

$1.31 \times 10^{-3}\ \Omega/\text{m} = 4(1.68 \times 10^{-8}\ \Omega\cdot\text{m})/\pi D^2$, which gives $D = 4.0 \times 10^{-3}$ m = **4.0 mm.**

64. (*a*) We find the frequency from the coefficient of t:

$2\pi f = 210\ \text{s}^{-1}$, which gives $f =$ **33.4 Hz.**

(*b*) The maximum current is 1.80 A, so the rms current is

$I_{rms} = I_0/\sqrt{2} = (1.80 \text{ A})/\sqrt{2} =$ **1.27 A.**

(*c*) For the voltage we have

$V = IR = (1.80 \text{ A})(42.0\ \Omega) \sin (210\ \text{s}^{-1})t =$ **(75.6 V) sin (210 s⁻¹)t.**

65. The dependence of the power output on the voltage is $P = V^2/R$. When we form the ratio for the two conditions, we get

$P_2/P_1 = (V_2/V_1)^2.$

For the percentage change we have

$(\Delta P/P)(100) = [(P_2 - P_1)/P_1](100) = [(V_2/V_1)^2 - 1](100)$

$= \{[(V + \Delta V)^2/V]^2 - 1\}(100) = \{[1 + (\Delta V/V)]^2 - 1\}(100).$

If $\Delta V/V$ is small, we can use the approximation: $[1 + (\Delta V/V)]^2 \approx 1 + 2\,\Delta V/V$:

$(\Delta P/P)(100) \approx (2\,\Delta V/V)(100)$, so the power drop is twice the voltage drop.

If we assume that the drop from 60 W to 50 W is small, we have

$(\Delta V/V)(100) = \frac{1}{2}[(50 \text{ W} - 60 \text{ W})/(60 \text{ W})](100) = -8.4\%.$

Thus the required voltage drop is **8.4%.**

66. (*a*) We find the input power from

$P_{output} = (\text{efficiency})P_{input};$

$900 \text{ W} = (0.60)P_{input}$, which gives $P_{input} = 1500$ W = **1.5 kW.**

(*b*) We find the current from

$P = IV;$

$1500 \text{ W} = I(120 \text{ V})$, which gives $I =$ **12.5 A.**

67. Because the volume is constant, we have

$AL = A'L'$, or $A'/A = L/L' = 1/3.00.$

The dependence of the resistance on the dimensions is $R = \rho L/A$. When we form the ratio for the two wires, we get

$R'/R = (L'/L)(A/A')$

$R'/(1.00\ \Omega) = (3.00)(3.00)$, which gives $R' =$ **9.00 Ω.**

68. The dependence of the resistance on the dimensions is $R = \rho L/A$. When we form the ratio for the two wires, we get

$R_1/R_2 = (L_1/L_2)(A_2/A_1) = (L_1/L_2)(D_2/D_1)^2 = (2)(\frac{1}{2})^2 = \frac{1}{2}.$

For a fixed voltage, the power dissipation is

$P = V^2/R.$

When we form the ratio for the two wires, we get

$P_1/P_2 = R_2/R_1 = 1/\frac{1}{2} =$ **2.**

69. Heat must be provided to replace the heat loss through the walls and to raise the temperature of the air brought in:

$$P = P_{\text{loss}} + mc\,\Delta T = 850\text{ kcal/h} + 2(1.29\text{ kg/m}^3)(62\text{ m}^3)(0.17\text{ kcal/kg}\cdot\text{C°})(20°\text{C} - 5°\text{C})$$
$$= 1.26 \times 10^3\text{ kcal/h} = (1.26 \times 10^3\text{ kcal/h})(4186\text{ J/kcal})/(3600\text{ s/h}) = 1.46 \times 10^3\text{ W} = \boxed{1.5\text{ kW.}}$$

70. (*a*) We find the average power required to provide the force to balance the average friction force from

$$P = Fv = (240\text{ N})(40\text{ km/h})/(3.6\text{ ks/h})(746\text{ W/hp}) = \boxed{3.6\text{ hp.}}$$

(*b*) We find the average current from

$$P = IV;$$
$$(3.6\text{ hp})(746\text{ W/hp}) = I(12\text{ V}),\text{ which gives } I = 222\text{ A.}$$

52 A·h is the total charge that passes through the battery when it is completely discharged. Thus the time for complete discharge is

$$t = Q/I = (26)(52\text{ A}\cdot\text{h})/(222\text{ A}) = 6.09\text{ h.}$$

In this time the car can travel

$$x = vt = (40\text{ km/h})(6.09\text{ h}) = \boxed{2.4 \times 10^2\text{ km.}}$$

71. For the resistance, we have

$$R = \rho L/A = \rho L/\tfrac{1}{4}\pi d^2;$$
$$12.5\ \Omega = 4(1.68 \times 10^{-8}\ \Omega\cdot\text{m})L/\pi d^2.$$

The mass of the wire is

$$m = \rho_m AL;$$
$$0.0180\text{ kg} = (8.9 \times 10^3\text{ kg/m}^3)\tfrac{1}{4}\pi d^2 L.$$

This gives us two equations with two unknowns, L and d. When we solve them, we get

$$d = 2.58 \times 10^{-4}\text{ m} = \boxed{0.258\text{ mm,}}\quad\text{and}\quad L = \boxed{38.8\text{ m.}}$$

72. (*a*) We find the initial power consumption from

$$P = V^2/R_0 = (120\text{ V})^2/(12\ \Omega) = 1.2 \times 10^3\text{ W} = \boxed{1.2\text{ kW.}}$$

(*b*) The power consumption when the bulb is hot is

$$P = V^2/R = (120\text{ V})^2/(140\ \Omega) = \boxed{100\text{ W.}}$$

The designated power is the operating power.

73. The stored energy in the capacitor must provide the energy used during the lapse:

$$U = \tfrac{1}{2}CV^2 = Pt;$$
$$\tfrac{1}{2}C(120\text{ V})^2 = (150\text{ W})(0.10\text{ s}),\text{ which gives } C = \boxed{2.1 \times 10^{-3}\text{ F.}}$$

74. The time for a proton to travel completely around the accelerator is

$$t = L/v.$$

In this time all the protons stored in the beam will pass a point, so the current is

$$I = Ne/t = Nev/L;$$
$$11 \times 10^{-3}\text{ A} = N(1.60 \times 10^{-19}\text{ C})(3.0 \times 10^8\text{ m/s})/(6300\text{ m}),\text{ which gives } N = \boxed{1.4 \times 10^{12}\text{ protons.}}$$

75. From Example 18–14 we know that the density of free electrons in copper is

$n = 8.4 \times 10^{28}\ \text{m}^{-3}$.

With the alternating current an electron will oscillate with SHM. We find the rms current from

$P = I_{rms}V_{rms}$;

$300\ \text{W} = I_{rms}(120\ \text{V})$, which gives $I_{rms} = 2.50\ \text{A}$, so the peak current is

$I_0 = \sqrt{2}\, I_{rms} = \sqrt{2}(2.50\ \text{A}) = 3.54\ \text{A}$.

The peak current corresponds to the maximum drift speed, which we find from

$I_0 = neAv_{dmax} = ne(\frac{1}{4}\pi D^2)v_{dmax}$;

$3.54\ \text{A} = (8.4 \times 10^{28}\ \text{m}^{-3})(1.60 \times 10^{-19}\ \text{C})[\frac{1}{4}\pi(1.8 \times 10^{-3}\ \text{m})^2]v_{dmax}$,

which gives $v_{dmax} = 1.03 \times 10^{-4}\ \text{m/s}$.

For SHM the maximum speed is related to the maximum displacement:

$v_{dmax} = A\omega = A(2\pi f)$;

$1.03 \times 10^{-4}\ \text{m/s} = A[2\pi(60\ \text{Hz})]$, which gives $A = 2.73 \times 10^{-7}\ \text{m}$.

For SHM the electron will move from one extreme to the other, so the total distance covered is $2A$:

$\boxed{5.5 \times 10^{-7}\ \text{m.}}$

CHAPTER 19

1. When the bulbs are connected in series, the equivalent resistance is
   $R_{\text{series}} = \Sigma R_i = 4R_{\text{bulb}} = 4(140\ \Omega) = \boxed{560\ \Omega.}$
   When the bulbs are connected in parallel, we find the equivalent resistance from
   $1/R_{\text{parallel}} = \Sigma(1/R_i) = 4/R_{\text{bulb}} = 4/(140\ \Omega)$, which gives $R_{\text{parallel}} = \boxed{35\ \Omega.}$

2. (*a*) When the bulbs are connected in series, the equivalent resistance is
   $R_{\text{series}} = \Sigma R_i = 3R_1 + 3R_2 = 3(40\ \Omega) + 3(80\ \Omega) = \boxed{360\ \Omega.}$
   (*b*) When the bulbs are connected in parallel, we find the equivalent resistance from
   $1/R_{\text{parallel}} = \Sigma(1/R_i) = (3/R_1) + (3/R_2) = [3/(40\ \Omega)] + [3/(80\ \Omega)]$, which gives $R_{\text{parallel}} = \boxed{8.9\ \Omega.}$

3. If we use them as single resistors, we have
   $R_1 = \boxed{30\ \Omega;}\quad R_2 = \boxed{50\ \Omega.}$
   When the bulbs are connected in series, the equivalent resistance is
   $R_{\text{series}} = \Sigma R_i = R_1 + R_2 = 30\ \Omega + 50\ \Omega = \boxed{80\ \Omega.}$
   When the bulbs are connected in parallel, we find the equivalent resistance from
   $1/R_{\text{parallel}} = \Sigma(1/R_i) = (1/R_1) + (1/R_2) = [1/(30\ \Omega)] + [1/(50\ \Omega)]$, which gives $R_{\text{parallel}} = \boxed{19\ \Omega.}$

4. Because resistance increases when resistors are connected in series, the maximum resistance is
   $R_{\text{series}} = R_1 + R_2 + R_3 = 500\ \Omega + 900\ \Omega + 1400\ \Omega = 2800\ \Omega = \boxed{2.80\ \text{k}\Omega.}$
   Because resistance decreases when resistors are connected in parallel, we find the minimum resistance from
   $1/R_{\text{parallel}} = (1/R_1) + (1/R_2) + (1/R_3) = [1/(500\ \Omega)] + [1/(900\ \Omega)] + [1/(1400\ \Omega)]$,
   which gives $R_{\text{parallel}} = \boxed{261\ \Omega.}$

5. The voltage is the same across resistors in parallel, but is less across a resistor in a series connection. We connect three 1.0-Ω resistors in series as shown in the diagram. Each resistor has the same current and the same voltage:
   $V_i = \frac{1}{3}V = \frac{1}{3}(6.0\ \text{V}) = 2.0\ \text{V}.$
   Thus we can get a 4.0-V output between *a* and *c*.

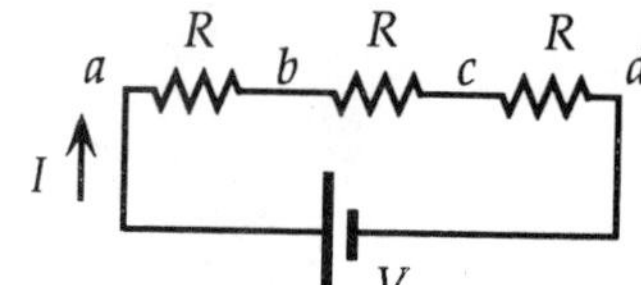

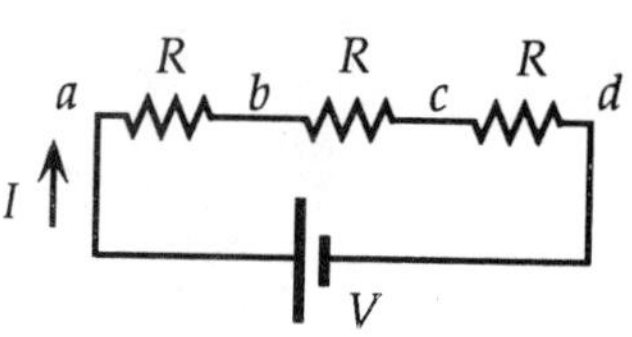

6. When the resistors are connected in series, as shown in A, we have
   $R_A = \Sigma R_i = 3R = 3(240\ \Omega) = \boxed{720\ \Omega.}$
   When the resistors are connected in parallel, as shown in B, we have
   $1/R_B = \Sigma(1/R_i) = 3/R = 3/(240\ \Omega)$, so $R_B = \boxed{80\ \Omega.}$
   In circuit C, we find the equivalent resistance of the two resistors in parallel:
   $1/R_1 = \Sigma(1/R_i) = 2/R = 2/(240\ \Omega)$, so $R_1 = 120\ \Omega$.
   This resistance is in series with the third resistor, so we have
   $R_C = R_1 + R = 120\ \Omega + 240\ \Omega = \boxed{360\ \Omega.}$
   In circuit D, we find the equivalent resistance of the two resistors in series:
   $R_2 = R + R = 240\ \Omega + 240\ \Omega = 480\ \Omega$.
   This resistance is in parallel with the third resistor, so we have
   $1/R_D = (1/R_2) + (1/R) = (1/480\ \Omega) + (1/240\ \Omega)$, so $R_D = \boxed{160\ \Omega.}$

7. We can reduce the circuit to a single loop by successively combining parallel and series combinations.
We combine R_1 and R_2, which are in series:
$R_7 = R_1 + R_2 = 2.8\ \text{k}\Omega + 2.8\ \text{k}\Omega = 5.6\ \text{k}\Omega.$
We combine R_3 and R_7, which are in parallel:
$1/R_8 = (1/R_3) + (1/R_7) = [1/(2.8\ \text{k}\Omega)] + [1/(5.6\ \text{k}\Omega)],$
which gives $R_8 = 1.87\ \text{k}\Omega$.
We combine R_4 and R_8, which are in series:
$R_9 = R_4 + R_8 = 2.8\ \text{k}\Omega + 1.87\ \text{k}\Omega = 4.67\ \text{k}\Omega.$
We combine R_5 and R_9, which are in parallel:
$1/R_{10} = (1/R_5) + (1/R_9) = [1/(2.8\ \text{k}\Omega)] + [1/(4.67\ \text{k}\Omega)],$
which gives $R_{10} = 1.75\ \text{k}\Omega$.
We combine R_{10} and R_6, which are in series:
$R_{eq} = R_{10} + R_6 = 1.75\ \text{k}\Omega + 2.8\ \text{k}\Omega = \boxed{4.6\ \text{k}\Omega.}$

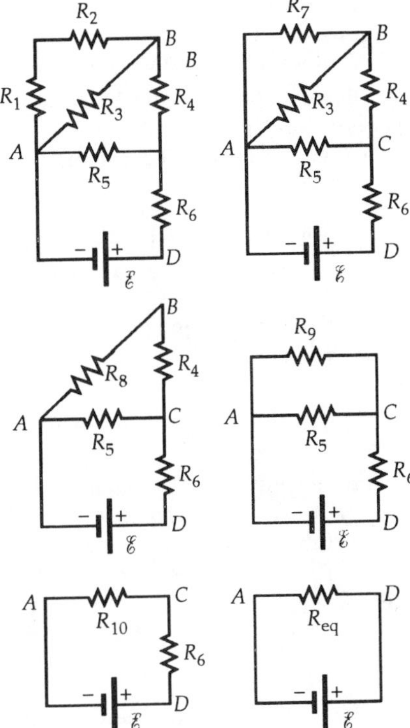

8. (*a*) In series the current must be the same for all bulbs. If all bulbs have the same resistance, they will have the same voltage:
$V_{bulb} = V/N = (110\ \text{V})/8 = \boxed{13.8\ \text{V.}}$
(*b*) We find the resistance of each bulb from
$R_{bulb} = V_{bulb}/I = (13.8\ \text{V})/(0.40\ \text{A}) = \boxed{34\ \Omega.}$
The power dissipated in each bulb is
$P_{bulb} = IV_{bulb} = (0.40\ \text{A})(13.8\ \text{V}) = \boxed{5.5\ \text{W.}}$

9. For the parallel combination, the total current from the source is
$I = NI_{bulb} = 8(0.240\ \text{A}) = 1.92\ \text{A}.$
The voltage across the leads is
$V_{leads} = IR_{leads} = (1.92\ \text{A})(1.5\ \Omega) = 2.9\ \text{V}.$
The voltage across each of the bulbs is
$V_{bulb} = V - V_{leads} = 110\ \text{V} - 2.88\ \text{V} = 107\ \text{V}.$
We find the resistance of a bulb from
$R_{bulb} = V_{bulb}/I_{bulb} = (107\ \text{V})/(0.240\ \text{A}) = \boxed{450\ \Omega.}$
The power dissipated in the leads is IV_{leads} and the total power used is IV, so the fraction wasted is
$IV_{leads}/IV = V_{leads}/V = (2.9\ \text{V})/(110\ \text{V}) = 0.026 = \boxed{2.6\%.}$

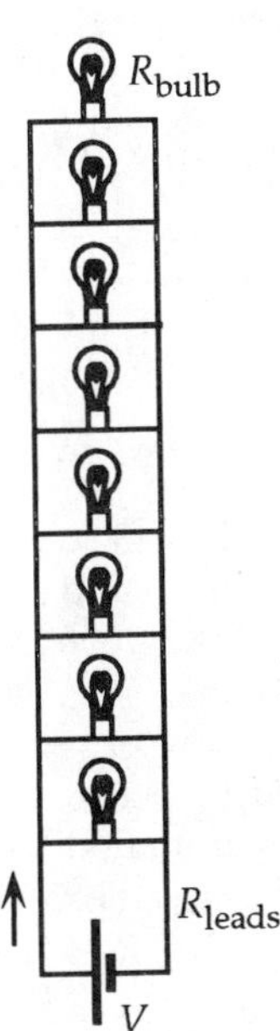

10. In series the current must be the same for all bulbs. If all bulbs have the same resistance, they will have the same voltage:
$V_{bulb} = V/N = (110\text{ V})/8 = 13.8\text{ V}.$
We find the resistance of each bulb from
$P_{bulb} = V_{bulb}^2/R_{bulb};$
$7.0\text{ W} = (13.8\text{ V})^2/R_{bulb}$, which gives $R_{bulb} =$ $\boxed{27\ \Omega.}$

11. Fortunately the required resistance is less. We can reduce the resistance by adding a parallel resistor, which does not require breaking the circuit. We find the necessary resistance from
$1/R = (1/R_1) + (1/R_2);$
$1/(320\ \Omega) = [1/(480\ \Omega)] + (1/R_2)$, which gives $R_2 =$ $\boxed{960\ \Omega \text{ in parallel.}}$

12. The equivalent resistance of the two resistors connected in series is
$R_s = R_1 + R_2.$
We find the equivalent resistance of the two resistors connected in parallel from
$1/R_p = (1/R_1) + (1/R_2)$, or $R_p = R_1R_2/(R_1 + R_2).$
The power dissipated in a resistor is $P = V^2/R$, so the ratio of the two powers is
$P_p/P_s = R_s/R_p = (R_1 + R_2)^2/R_1R_2 = 4.$
When we expand the square, we get
$R_1^2 + 2R_1R_2 + R_2^2 = 4R_1R_2$, or $R_1^2 - 2R_1R_2 + R_2^2 = (R_1 - R_2)^2 = 0$, which gives $R_2 = R_1 =$ $\boxed{2.20\text{ k}\Omega.}$

13. With the two bulbs connected in parallel, there will be 110 V across each bulb, so the total power will be 75 W + 40 W = 115 W. We find the net resistance of the bulbs from
$P = V^2/R;$
$115\text{ W} = (110\text{ V})^2/R$, which gives $R =$ $\boxed{105\ \Omega.}$

14. We can reduce the circuit to a single loop by successively combining parallel and series combinations.
We combine R_1 and R_2, which are in series:
$R_7 = R_1 + R_2 = 2.20\text{ k}\Omega + 2.20\text{ k}\Omega = 4.40\text{ k}\Omega.$
We combine R_3 and R_7, which are in parallel:
$1/R_8 = (1/R_3) + (1/R_7) = [1/(2.20\text{ k}\Omega)] + [1/(4.40\text{ k}\Omega)],$
which gives $R_8 = 1.47\text{ k}\Omega.$
We combine R_4 and R_8, which are in series:
$R_9 = R_4 + R_8 = 2.20\text{ k}\Omega + 1.47\text{ k}\Omega = 3.67\text{ k}\Omega.$
We combine R_5 and R_9, which are in parallel:
$1/R_{10} = (1/R_5) + (1/R_9) = [1/(2.20\text{ k}\Omega)] + [1/(3.67\text{ k}\Omega)],$
which gives $R_{10} = 1.38\text{ k}\Omega.$
We combine R_{10} and R_6, which are in series:
$R_{eq} = R_{10} + R_6 = 1.38\text{ k}\Omega + 2.20\text{ k}\Omega = 3.58\text{ k}\Omega.$
The current in the single loop is the current through R_6:
$I_6 = I = \mathscr{E}/R_{eq} = (12\text{ V})/(3.58\text{ k}\Omega) =$ $\boxed{3.36\text{ mA.}}$
For V_{AC} we have
$V_{AC} = IR_{10} = (3.36\text{ mA})(1.38\text{ k}\Omega) = 4.63\text{ V}.$
This allows us to find I_5 and I_4;
$I_5 = V_{AC}/R_5 = (4.63\text{ V})/(2.20\text{ k}\Omega) =$ $\boxed{2.11\text{ mA;}}$
$I_4 = V_{AC}/R_9 = (4.63\text{ V})/(3.67\text{ k}\Omega) =$ $\boxed{1.26\text{ mA.}}$
For V_{AB} we have
$V_{AB} = I_4R_8 = (1.26\text{ mA})(1.47\text{ k}\Omega) = 1.85\text{ V}.$
This allows us to find I_3, I_2, and I_1;
$I_3 = V_{AB}/R_3 = (1.85\text{ V})/(2.20\text{ k}\Omega) =$ $\boxed{0.84\text{ mA;}}$
$I_1 = I_2 = V_{AB}/R_7 = (1.85\text{ V})/(4.40\text{ k}\Omega) =$ $\boxed{0.42\text{ mA.}}$
From above, we have $\boxed{V_{AB} = 1.85\text{ V.}}$

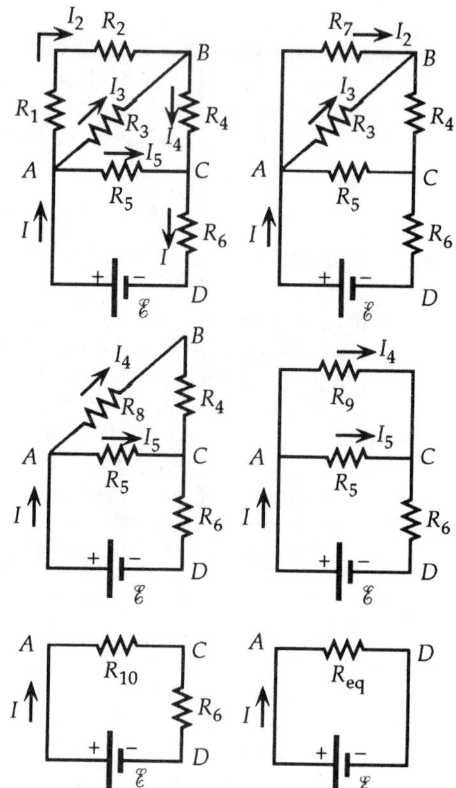

15. (a) When the switch is closed the addition of R_2 to the parallel set will decrease the equivalent resistance, so the current from the battery will increase. This causes an increase in the voltage across R_1, and a corresponding decrease across R_3 and R_4. The voltage across R_2 increases from zero. Thus we have

V_1 and V_2 increase; V_3 and V_4 decrease.

(b) The current through R_1 has increased. This current is now split into three, so currents through R_3 and R_4 decrease. Thus we have

I_1 (= I) and I_2 increase; I_3 and I_4 decrease.

(c) The current through the battery has increased, so the power output of the battery increases.

(d) Before the switch is closed, $I_2 = 0.$ We find the resistance for R_3 and R_4 in parallel from

$1/R_A = \Sigma(1/R_i) = 2/R_3 = 2/(100\ \Omega)$,

which gives $R_A = 50\ \Omega$.

For the single loop, we have

$I = I_1 = V/(R_1 + R_A)$
$= (45.0\ \text{V})/(100\ \Omega + 50\ \Omega) =$ 0.300 A.

This current will split evenly through R_3 and R_4:

$I_3 = I_4 = \frac{1}{2}I = \frac{1}{2}(0.300\ \text{A}) =$ 0.150 A.

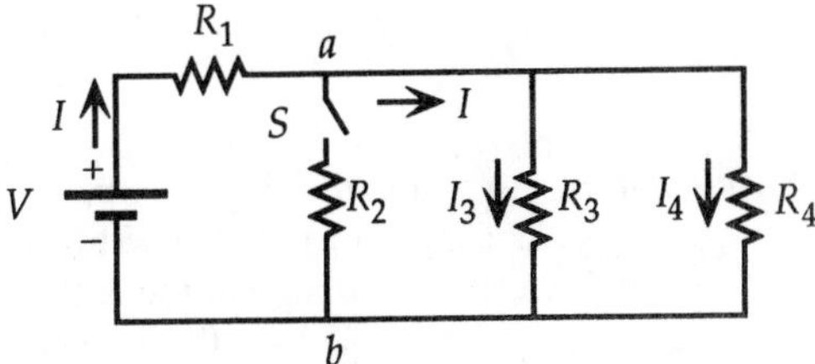

After the switch is closed, we find the resistance for R_2, R_3, and R_4 in parallel from

$1/R_B = \Sigma(1/R_i) = 3/R_3 = 3/(100\ \Omega)$,

which gives $R_B = 33.3\ \Omega$.

For the single loop, we have

$I = I_1 = V/(R_1 + R_B)$
$= (45.0\ \text{V})/(100\ \Omega + 33.3\ \Omega) =$ 0.338 A.

This current will split evenly through R_2, R_3, and R_4:

$I_2 = I_3 = I_4 = \frac{1}{3}I = \frac{1}{3}(0.338\ \text{A}) =$ 0.113 A.

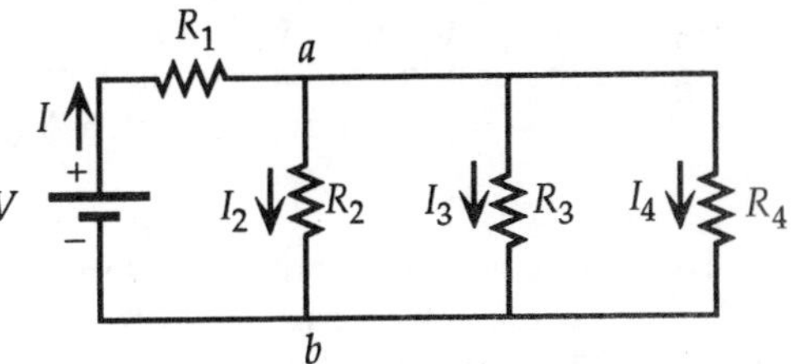

16. (a) When the switch is opened, the removal of a resistor from the parallel set will increase the equivalent resistance, so the current from the battery will decrease. This causes a decrease in the voltage across R_1, and a corresponding increase across R_2. The voltage across R_3 decreases to zero. Thus we have

V_1 and V_3 decrease; V_2 increases.

(b) The current through R_1 has decreased. The current through R_2 has increased. The current through R_3 has decreased to zero. Thus we have

I_1 (= I) and I_3 decrease; I_2 increases.

(c) Because the current through the battery decreases, the Ir term decreases, so the terminal voltage of the battery will increase.

(d) When the switch is closed, we find the resistance for R_2 and R_3 in parallel from

$1/R_A = \Sigma(1/R_i) = 2/R = 2/(5.50\ \Omega)$,

which gives $R_A = 2.75\ \Omega$.

For the single loop, we have

$I = V/(R_1 + R_A + r)$
$= (18.0\ \text{V})/(5.50\ \Omega + 2.75\ \Omega + 0.50\ \Omega) = 2.06\ \text{A}$.

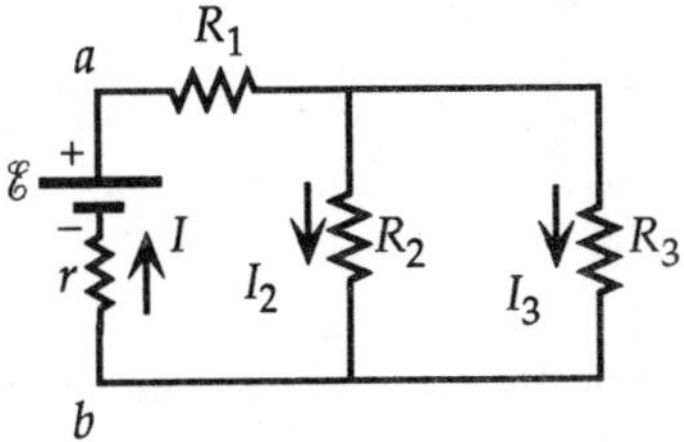

For the terminal voltage of the battery, we have

$V_{ab} = \mathscr{E} - Ir = 18.0\ \text{V} - (2.06\ \text{A})(0.50\ \Omega) =$ 17.0 V.

When the switch is open, for the single loop, we have

$I' = V/(R_1 + R_2 + r)$
$= (18.0\ \text{V})/(5.50\ \Omega + 5.50\ \Omega + 0.50\ \Omega) = 1.57\ \text{A}$.

For the terminal voltage of the battery, we have

$V_{ab}' = \mathscr{E} - I'r = 18.0\ \text{V} - (1.57\ \text{A})(0.50\ \Omega) =$ 17.2 V.

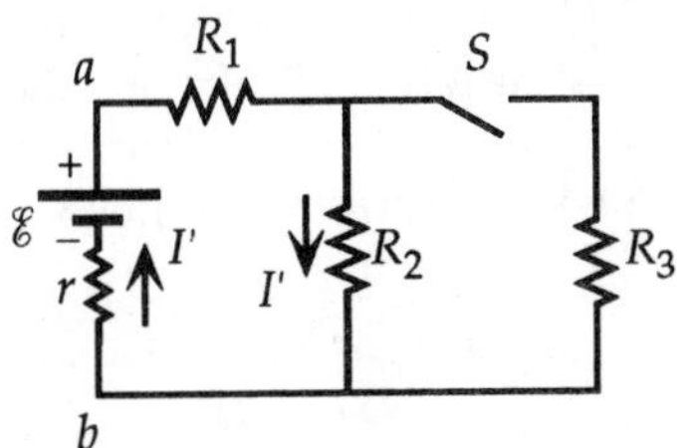

17. We find the resistance for R_1 and R_2 in parallel from

$$1/R_p = (1/R_1) + (1/R_2) = [1/(2.8\ k\Omega)] + [1/(2.1\ k\Omega)],$$

which gives $R_p = 1.2\ k\Omega$.

Because the same current passes through R_p and R_3, the higher resistor will have the higher power dissipation, so the limiting resistor is R_3, which will have a power dissipation of $\frac{1}{2}$ W. We find the current from

$$P_{3max} = I_{3max}{}^2 R_3;$$

$$0.50\ W = I_{3max}{}^2(1.8 \times 10^3\ \Omega), \text{ which gives } I_{3max} = 0.0167\ A.$$

The maximum voltage for the network is

$$V_{max} = I_{3max}(R_p + R_3)$$

$$= (0.0167\ A)(1.2 \times 10^3\ \Omega + 1.8 \times 10^3\ \Omega) = \boxed{50\ V.}$$

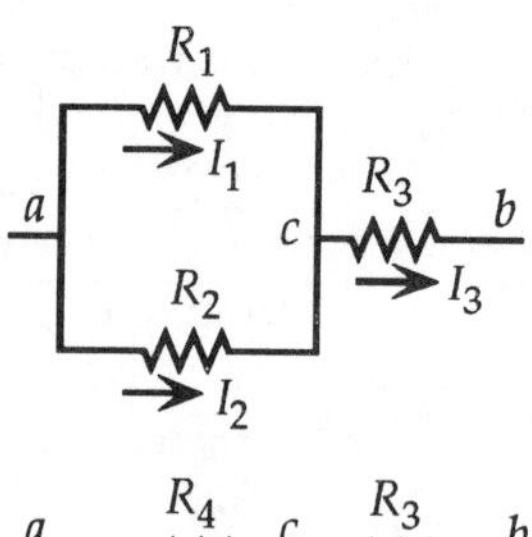

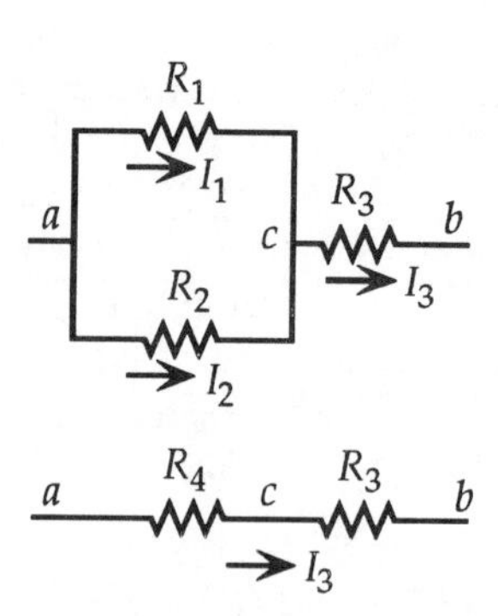

18. (*a*) For the current in the single loop, we have

$$I_a = V/(R_a + r) = (8.50\ V)/(81.0\ \Omega + 0.900\ \Omega) = 0.104\ A.$$

For the terminal voltage of the battery, we have

$$V_a = \mathscr{E} - I_a r = 8.50\ V - (0.104\ A)(0.900\ \Omega) = \boxed{8.41\ V.}$$

(*b*) For the current in the single loop, we have

$$I_b = V/(R_b + r) = (8.50\ V)/(810\ \Omega + 0.900\ \Omega) = 0.0105\ A.$$

For the terminal voltage of the battery, we have

$$V_b = \mathscr{E} - I_b r = 8.50\ V - (0.0105\ A)(0.900\ \Omega) = \boxed{8.49\ V.}$$

19. The voltage across the bulb is the terminal voltage of the four cells:

$$V = IR_{bulb} = 4(\mathscr{E} - Ir);$$

$$(0.62\ A)(12\ \Omega) = 4[2.0\ V - (0.62\ A)r], \text{ which gives } r = \boxed{0.23\ \Omega.}$$

20. We find the resistance of a bulb from the nominal rating:

$$R_{bulb} = V_{nominal}{}^2/P_{nominal} = (12.0\ V)^2/(3.0\ W) = 48\ \Omega.$$

We find the current through each bulb when connected to the battery from:

$$I_{bulb} = V/R_{bulb} = (11.8\ V)/(48\ \Omega) = 0.246\ A.$$

Because the bulbs are in parallel, the current through the battery is

$$I = 2I_{bulb} = 2(0.246\ A) = 0.492\ A.$$

We find the internal resistance from

$$V = \mathscr{E} - Ir;$$

$$11.8\ V = [12.0\ V - (0.492\ A)r], \text{ which gives } r = \boxed{0.4\ \Omega.}$$

21. If we can ignore the resistance of the ammeter, for the single loop we have

$$I = \mathscr{E}/r;$$

$$25\ A = (1.5\ V)/r, \text{ which gives } r = \boxed{0.060\ \Omega.}$$

22. We find the internal resistance from

$$V = \mathscr{E} - Ir;$$

$$8.8\ V = [12.0\ V - (60\ A)r], \text{ which gives } r = \boxed{0.053\ \Omega.}$$

Because the terminal voltage is the voltage across the starter, we have

$$V = IR;$$

$$8.8\ V = (60\ A)R, \text{ which gives } R = \boxed{0.15\ \Omega.}$$

23.

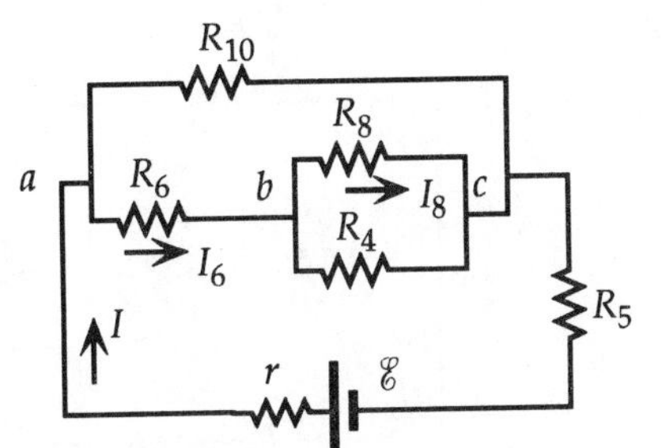

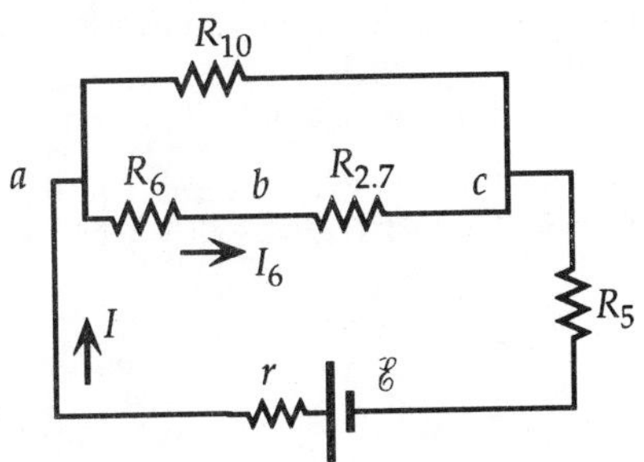

From the results of Example 19–7, we know that the current through the 6.0-Ω resistor, I_6, is 0.48 A.
We find V_{bc} from

$$V_{bc} = I_6 R_{2.7} = (0.48 \text{ mA})(2.7\ \Omega) = 1.30 \text{ V}.$$

This allows us to find I_8;

$$I_8 = V_{bc}/R_8 = (1.30 \text{ V})/(8.0 \text{ k}\Omega) = \boxed{0.16 \text{ A.}}$$

24. For the current in the single loop, we have

$$I = V/(R_1 + R_2 + r)$$
$$= (9.0 \text{ V})/(8.0\ \Omega + 12.0\ \Omega + 2.0\ \Omega) = \boxed{0.41 \text{ A.}}$$

For the terminal voltage of the battery, we have

$$V_{ab} = \mathcal{E} - Ir = 9.0 \text{ V} - (0.41 \text{ A})(2.0\ \Omega) = 8.18 \text{ V}.$$

The current in a resistor goes from high to low potential.
For the voltage changes across the resistors, we have

$$V_{bc} = -IR_2 = -(0.41 \text{ A})(12.0\ \Omega) = -4.91 \text{ V};$$
$$V_{ca} = -IR_1 = -(0.41 \text{ A})(8.0\ \Omega) = -3.27 \text{ V}.$$

For the sum of the voltage changes, we have

$$V_{ab} + V_{bc} + V_{ca} = 8.18 \text{ V} - 4.91 \text{ V} - 3.27 \text{ V} = 0.$$

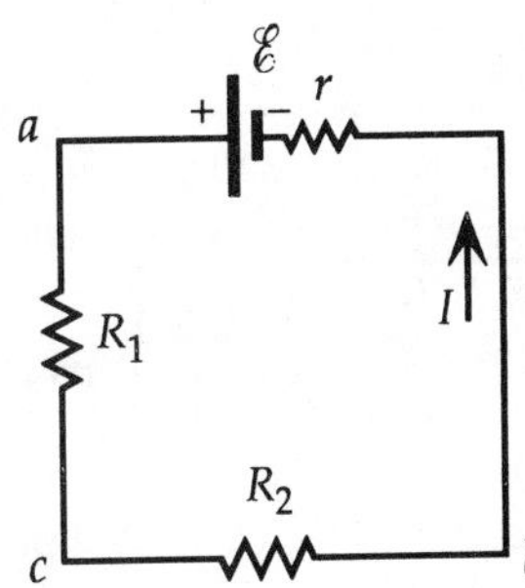

25. For the loop, we start at point a:

$$-IR - \mathcal{E}_2 - Ir_2 + \mathcal{E}_1 - Ir_1 = 0;$$
$$-I(6.6\ \Omega) - 12 \text{ V} - I(2\ \Omega) + 18 \text{ V} - I(1\ \Omega) = 0,$$

which gives $I = 0.625$ A.
The top battery is discharging, so we have

$$V_1 = \mathcal{E}_1 - Ir_1 = 18 \text{ V} - (0.625 \text{ A})(1\ \Omega) = \boxed{17.4 \text{ V.}}$$

The bottom battery is charging, so we have

$$V_2 = \mathcal{E}_2 + Ir_2 = 12 \text{ V} + (0.625 \text{ A})(2\ \Omega) = \boxed{13.3 \text{ V.}}$$

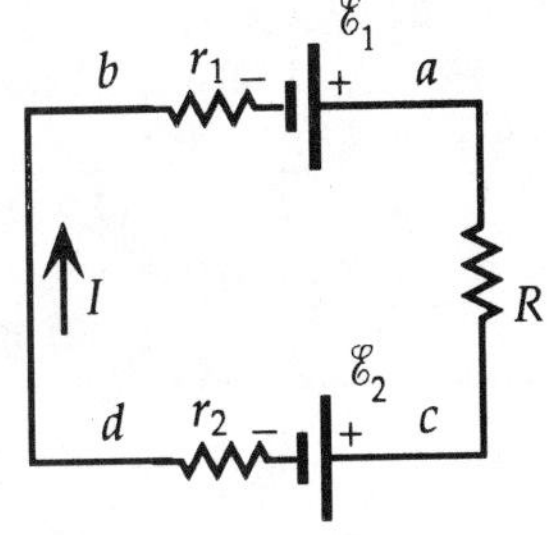

26. To find the potential difference between points a and d, we can use any path. The simplest one is through the top resistor R_1. From the results of Example 19–8, we know that the current I_1 is – 0.87 A.
We find V_{ad} from

$$V_{ad} = V_a - V_d = I_1 R_1 = (-0.87 \text{ A})(30\ \Omega) = \boxed{-26 \text{ V.}}$$

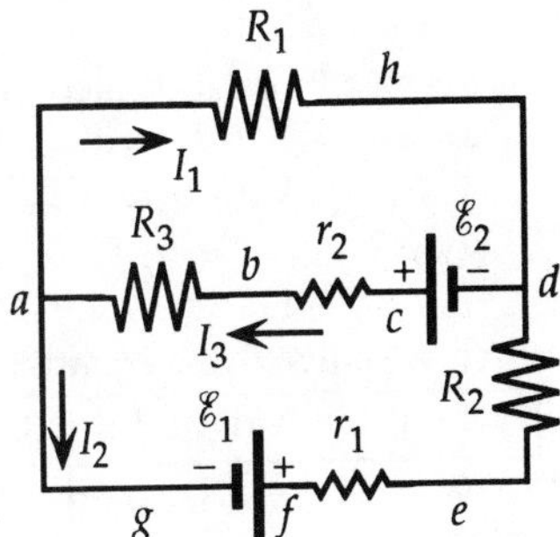

27. From the results of Example 19–8, we know that the currents are
$I_1 = -0.87$ A, $I_2 = 2.6$ A, $I_3 = 1.7$ A.
On the circuit diagram, both batteries are discharging. so we have
$V_{bd} = \mathscr{E}_2 - I_3 r_2 = 45\text{ V} - (1.7\text{ A})(1\ \Omega) =$ $\boxed{43\text{ V.}}$
$V_{eg} = \mathscr{E}_1 - I_2 r_1 = 80\text{ V} - (2.6\text{ A})(1\ \Omega) =$ $\boxed{77\text{ V.}}$

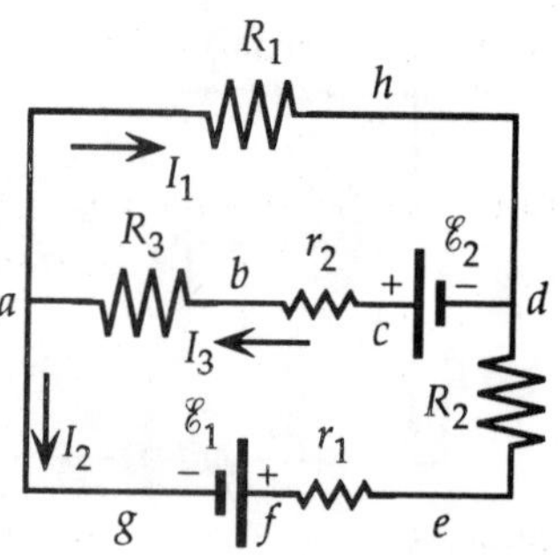

28. For the conservation of current at point c, we have
$I_{in} = I_{out}$;
$I_1 = I_2 + I_3$.
For the two loops indicated on the diagram, we have
loop 1: $V_1 - I_2R_2 - I_1R_1 = 0$;
$+9.0\text{ V} - I_2(15\ \Omega) - I_1(22\ \Omega) = 0$;
loop 2: $V_3 + I_2R_2 = 0$;
$+6.0\text{ V} + I_2(15\ \Omega) = 0$.
When we solve these equations, we get
$\boxed{I_1 = 0.68\text{ A}, I_2 = -0.40\text{ A},}$ $I_3 = 1.08$ A.
Note that I_2 is opposite to the direction shown.

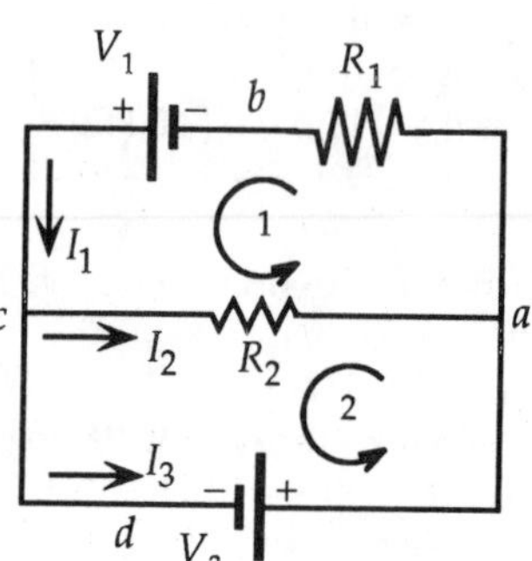

29. For the conservation of current at point c, we have
$I_{in} = I_{out}$;
$I_1 = I_2 + I_3$.
When we add the internal resistance terms for the two loops indicated on the diagram, we have
loop 1: $V_1 - I_2R_2 - I_1R_1 - I_1r_1 = 0$;
$+9.0\text{ V} - I_2(15\ \Omega) - I_1(22\ \Omega) - I_1(1.2\ \Omega) = 0$;
loop 2: $V_3 + I_2R_2 + I_3r_3 = 0$;
$+6.0\text{ V} + I_2(15\ \Omega) + I_2(1.2\ \Omega) = 0$.
When we solve these equations, we get
$\boxed{I_1 = 0.60\text{ A}, I_2 = -0.33\text{ A},}$ $I_3 = 0.93$ A.

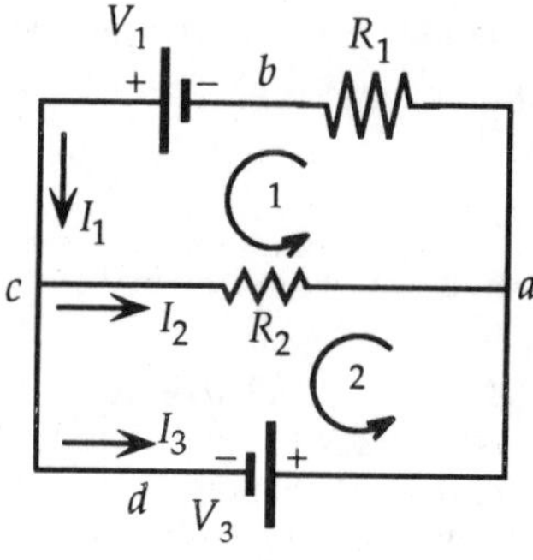

30. For the conservation of current at point b, we have
$I_{in} = I_{out}$;
$I_1 + I_3 = I_2$.
For the two loops indicated on the diagram, we have
loop 1: $\mathscr{E}_1 - I_1R_1 - I_2R_2 = 0$;
$+9.0\text{ V} - I_1(15\ \Omega) - I_2(20\ \Omega) = 0$;
loop 2: $-\mathscr{E}_2 + I_2R_2 + I_3R_3 = 0$;
$-12.0\text{ V} + I_2(20\ \Omega) + I_2(30\ \Omega) = 0$.
When we solve these equations, we get
$\boxed{I_1 = 0.156\text{ A right}, I_2 = 0.333\text{ A left}, I_3 = 0.177\text{ A up.}}$
We have carried an extra decimal place to show the agreement with the junction equation.

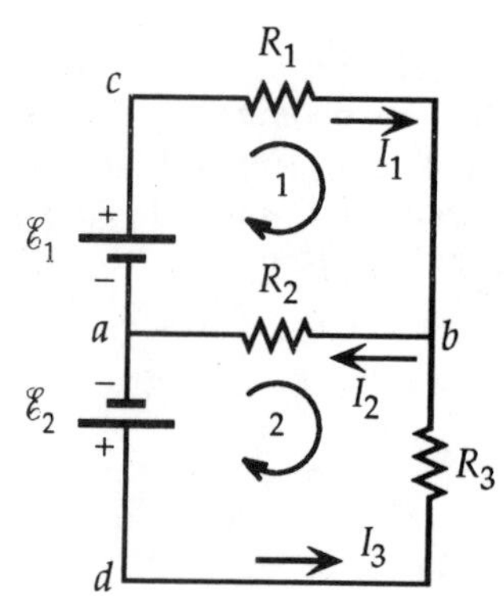

31. For the conservation of current at point b, we have

$I_{in} = I_{out}$;

$I_1 + I_3 = I_2$.

When we add the internal resistance terms for the two loops indicated on the diagram, we have

loop 1: $\mathcal{E}_1 - I_1R_1 - I_1r_1 - I_2R_2 = 0$;

$+ 9.0 \text{ V} - I_1(1.0 \ \Omega) - I_1(15 \ \Omega) - I_2(20 \ \Omega) = 0$;

loop 2: $- \mathcal{E}_2 + I_3r_2 + I_2R_2 + I_3R_3 = 0$;

$- 12.0 \text{ V} + I_3(1.0 \ \Omega) + I_2(20 \ \Omega) + I_2(30 \ \Omega) = 0$.

When we solve these equations, we get

$I_1 = 0.15$ A right, $I_2 = 0.33$ A left, $I_3 = 0.18$ A up.

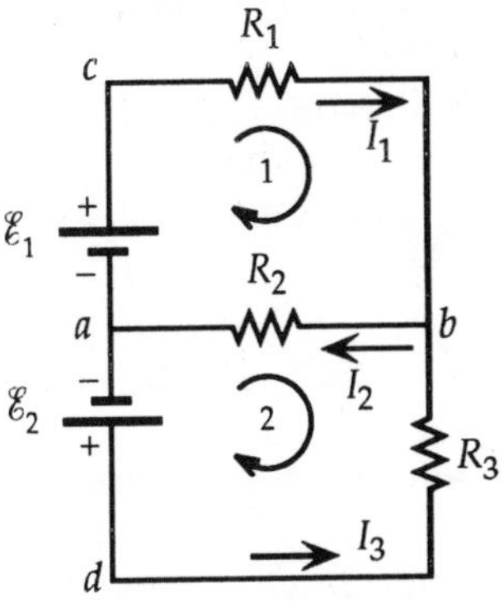

32. When we include the current through the battery, we have six unknowns. For the conservation of current, we have

junction a: $I = I_1 + I_2$;

junction b: $I_1 = I_3 + I_5$;

junction d: $I_2 + I_5 = I_4$.

For the three loops indicated on the diagram, we have

loop 1: $- I_1R_1 - I_5R_5 + I_2R_2 = 0$;

$- I_1(20 \ \Omega) - I_5(10 \ \Omega) + I_2(25 \ \Omega) = 0$;

loop 2: $- I_3R_3 + I_4R_4 + I_5R_5 = 0$;

$- I_3(2 \ \Omega) + I_4(2 \ \Omega) + I_5(10 \ \Omega) = 0$;

loop 3: $+ \mathcal{E} - I_2R_2 - I_4R_4 = 0$.

$+ 6.0 \text{ V} - I_2(25 \ \Omega) - I_4(2 \ \Omega) = 0$;

When we solve these six equations, we get

$I_1 = 0.274$ A, $I_2 = 0.222$ A, $I_3 = 0.266$ A,
$I_4 = 0.229$ A, $I_5 = 0.007$ A, $I = 0.496$ A.

We have carried an extra decimal place to show the agreement with the junction equations.

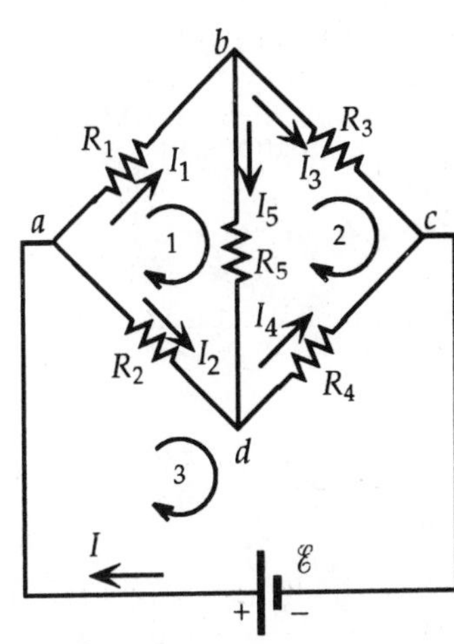

33. When the 25-Ω resistor is shorted, points a and d become the same point and we lose I_2.

For the conservation of current, we have

junction a: $I + I_5 = I_1 + I_4$;

junction b: $I_1 = I_3 + I_5$;

For the three loops indicated on the diagram, we have

loop 1: $- I_1R_1 - I_5R_5 = 0$;

$- I_1(20 \ \Omega) - I_5(10 \ \Omega) = 0$;

loop 2: $- I_3R_3 - I_4R_4 + I_5R_5 = 0$;

$- I_3(2 \ \Omega) + I_4(2 \ \Omega) + I_5(10 \ \Omega) = 0$;

loop 3: $+ \mathcal{E} - I_4R_4 = 0$.

$+ 6.0 \text{ V} - I_4(2 \ \Omega) = 0$;

When we solve these six equations, we get

$I_1 = 0.23$ A, $I_3 = 0.69$ A, $I_4 = 3.00$ A, $I_5 = - 0.46$ A, $I = 3.69$ A.

Thus the current through the 10-Ω resistor is 0.46 A up.

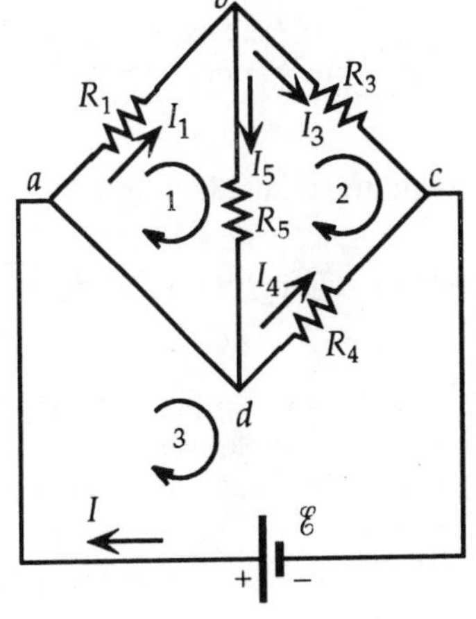

34. For the conservation of current at point *a*, we have
$I_2 + I_3 = I_1$.
For the two loops indicated on the diagram, we have
loop 1: $\mathscr{E}_1 - I_1r_1 - I_1R_3 + \mathscr{E}_2 - I_2r_2 - I_2R_2 - I_1R_1 = 0;$
$+ 12.0 \text{ V} - I_1(1.0\ \Omega) - I_1(8.0\ \Omega) + 12.0 \text{ V} - I_2(1.0\ \Omega) - I_2(10\ \Omega) - I_1(12\ \Omega) = 0;$
loop 2: $\mathscr{E}_3 - I_3r_3 - I_3R_5 + I_2R_2 - \mathscr{E}_2 + I_2r_2 - I_3R_4 = 0;$
$+ 6.0 \text{ V} - I_3(1.0\ \Omega) - I_3(18\ \Omega) - 12.0 \text{ V} + I_2(1.0\ \Omega) - I_3(15\ \Omega) = 0.$
When we solve these equations, we get
$\boxed{I_1 = 0.77 \text{ A}, I_2 = 0.71 \text{ A}, I_3 = 0.055 \text{ A}.}$
For the terminal voltage of the 6.0-V battery, we have
$V_{fe} = \mathscr{E}_3 - I_3r_3 = 6.0 \text{ V} - (0.055 \text{ A})(1.0\ \Omega) = \boxed{5.95 \text{ V}.}$

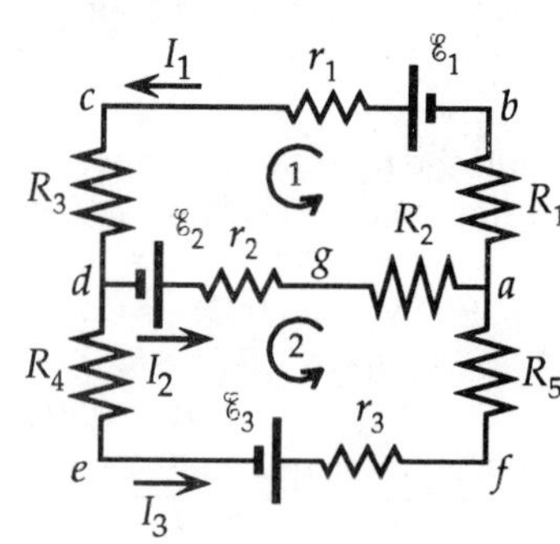

35. The upper loop equation becomes
loop 1: $\mathscr{E}_1 - I_1r_1 - I_1R_3 + \mathscr{E}_2 - I_2r_2 - I_2R_2 - I_1R_1 = 0;$
$+ 12.0 \text{ V} - I_1(1.0\ \Omega) - I_1(8.0\ \Omega) + 12.0 \text{ V} - I_2(1.0\ \Omega) - I_2(10\ \Omega) - I_1(12\ \Omega) = 0.$
The other equations are the same:
$I_2 + I_3 = I_1$.
loop 2: $\mathscr{E}_3 - I_3r_3 - I_3R_5 + I_2R_2 - \mathscr{E}_2 + I_2r_2 - I_3R_4 = 0;$
$+ 6.0 \text{ V} - I_3(1.0\ \Omega) - I_3(18\ \Omega) - 12.0 \text{ V} + I_2(1.0\ \Omega) - I_3(15\ \Omega) = 0.$
When we solve these equations, we get $\boxed{I_1 = 1.30 \text{ A},}$ $I_2 = 1.12 \text{ A}, I_3 = 0.18 \text{ A}.$

36. For the conservation of current at point *b*, we have
$I = I_1 + I_2$.
For the two loops indicated on the diagram, we have
loop 1: $\mathscr{E}_1 - I_1r_1 - IR = 0;$
$+ 2.0 \text{ V} - I_1(0.10\ \Omega) - I(4.0\ \Omega) = 0;$
loop 2: $\mathscr{E}_2 - I_2r_2 - IR = 0;$
$+ 3.0 \text{ V} - I_2(0.10\ \Omega) - I(4.0\ \Omega) = 0.$
When we solve these equations, we get
$I_1 = 5.31 \text{ A}, I_2 = -4.69 \text{ A}, I = 0.62 \text{ A}.$
For the voltage across *R* we have
$V_{ab} = IR = (0.62 \text{ A})(4.0\ \Omega) = \boxed{2.5 \text{ V}.}$
Note that one battery is charging the other with a significant current.

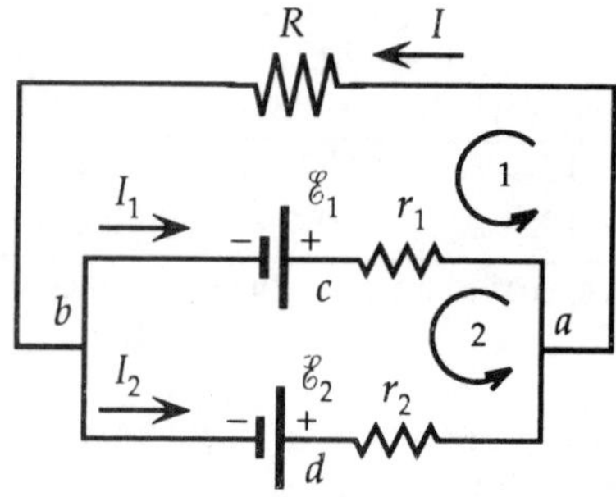

37. We find the equivalent capacitance for a parallel connection from
$C_{\text{parallel}} = \Sigma C_i = 6(3.7\ \mu\text{F}) = \boxed{22\ \mu\text{F}.}$
When the capacitors are connected in series, we find the equivalent capacitance from
$1/C_{\text{series}} = \Sigma(1/C_i) = 6/(3.7\ \mu\text{F})$, which gives $C_{\text{series}} = \boxed{0.62\ \mu\text{F}.}$

38. Fortunately the required capacitance is greater. We can increase the capacitance by adding a parallel capacitor, which does not require breaking the circuit. We find the necessary capacitor from
$C = C_1 + C_2;$
$16\ \mu\text{F} = 5.0\ \mu\text{F} + C_2$, which gives $C_2 = \boxed{11\ \mu\text{F in parallel.}}$

39. We can decrease the capacitance by adding a series capacitor. We find the necessary capacitor from
$1/C = (1/C_1) + (1/C_2);$
$1/(3300 \text{ pF}) = [1/(4800 \text{ pF})] + (1/C_2)$, which gives $C_2 = \boxed{10{,}560 \text{ pF}.}$
$\boxed{\text{Yes,}}$ it is necessary to break a connection to add a series component.

40. The capacitance increases with a parallel connection, so the maximum capacitance is

$$C_{max} = C_1 + C_2 + C_3 = 2000 \text{ pF} + 7500 \text{ pF} + 0.0100\ \mu\text{F} = 0.0020\ \mu\text{F} + 0.0075\ \mu\text{F} + 0.0100\ \mu\text{F}$$
$$= \boxed{0.0195\ \mu\text{F.}}$$

The capacitance decreases with a series connection, so we find the minimum capacitance from

$$1/C_{min} = (1/C_1) + (1/C_2) + (1/C_3) = [1/(2000 \text{ pF})] + [1/(7500 \text{ pF})] + [1/(0.0100\ \mu\text{F})]$$
$$= [1/(2.0 \text{ nF})] + [1/(7.5 \text{ nF})] + [1/(10.0 \text{ nF})], \text{ which gives } C_{min} = \boxed{1.4 \text{ nF.}}$$

41. The energy stored in a capacitor is

$$U = \tfrac{1}{2}CV^2.$$

To increase the energy we must increase the capacitance, which means adding a parallel capacitor. Because the potential is constant, to have three times the energy requires three times the capacitance:

$$C = 3C_1 = C_1 + C_2, \text{ or } C_2 = 2C_1 = 2(150 \text{ pF}) = \boxed{300 \text{ pF in parallel.}}$$

42. (*a*) From the circuit, we see that C_2 and C_3 are in series and find their equivalent capacitance from

$$1/C_4 = (1/C_2) + (1/C_3), \text{ which gives } C_4 = C_2C_3/(C_2 + C_3).$$

From the new circuit, we see that C_1 and C_4 are in parallel, with an equivalent capacitance

$$C_{eq} = C_1 + C_4 = C_1 + [C_2C_3/(C_2 + C_3)]$$
$$= \boxed{(C_1C_2 + C_1C_3 + C_2C_3)/(C_2 + C_3).}$$

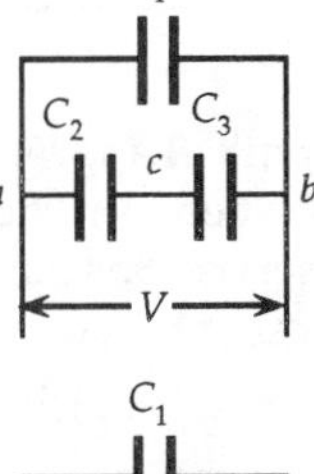

(*b*) Because V is across C_1, we have

$$Q_1 = C_1V = (12.5\ \mu\text{F})(45.0 \text{ V}) = \boxed{563\ \mu\text{C.}}$$

Because C_2 and C_3 are in series, the charge on each is the charge on their equivalent capacitance:

$$Q_2 = Q_3 = C_4V = C_2C_3/(C_2 + C_3)V$$
$$= [(12.5\ \mu\text{F})(6.25\ \mu\text{F})/(12.5\ \mu\text{F} + 6.25\ \mu\text{F})](45.0 \text{ V}) = \boxed{188\ \mu\text{C.}}$$

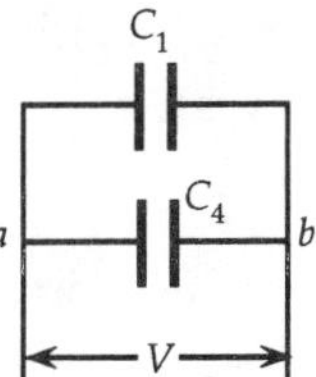

43. From Problem 42 we know that the equivalent capacitance is

$$C_{eq} = (C_1C_2 + C_1C_3 + C_2C_3)/(C_2 + C_3) = 3C_1/2 = 3(8.8\ \mu\text{F})/2 = 13.2\ \mu\text{F}.$$

Because this capacitance is equivalent to the three, the energy stored in it is the energy stored in the network:

$$U = \tfrac{1}{2}C_{eq}V^2 = \tfrac{1}{2}(13.2 \times 10^{-6} \text{ F})(90 \text{ V})^2 = \boxed{0.053 \text{ J.}}$$

44. (*a*) We find the equivalent capacitance from

$$1/C_{series} = (1/C_1) + (1/C_2) = [1/(0.40\ \mu\text{F})] + [1/(0.50\ \mu\text{F})],$$

which gives $C_{series} = 0.222\ \mu\text{F}$.

The charge on the equivalent capacitor is the charge on each capacitor:

$$Q_1 = Q_2 = Q_{series} = C_{series}V = (0.222\ \mu\text{F})(9.0 \text{ V}) = 2.00\ \mu\text{C}.$$

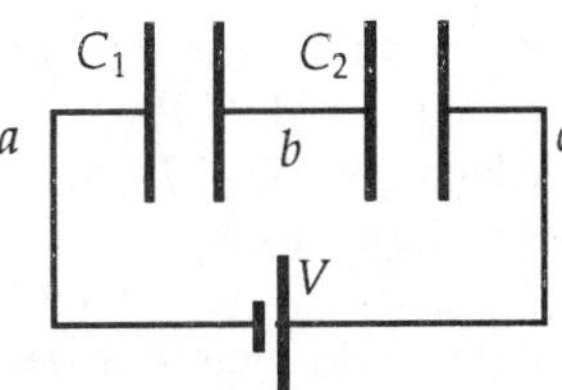

We find the potential differences from

$$Q_1 = C_1V_1;$$
$$2.00\ \mu\text{C} = (0.40\ \mu\text{F})V_1, \text{ which gives } \boxed{V_1 = 5.0 \text{ V.}}$$
$$Q_2 = C_2V_2;$$
$$2.00\ \mu\text{C} = (0.50\ \mu\text{F})V_2, \text{ which gives } \boxed{V_2 = 4.0 \text{ V.}}$$

(*b*) As we found above

$$Q_1 = Q_2 = \boxed{2.0\ \mu\text{C.}}$$

(*c*) For the parallel network, we have

$$V_1 = V_2 = \boxed{9.0 \text{ V.}}$$

We find the two charges from

$$Q_1 = C_1V_1 = (0.40\ \mu\text{F})(9.0 \text{ V}) = \boxed{3.6\ \mu\text{C;}}$$
$$Q_2 = C_2V_2 = (0.50\ \mu\text{F})(9.0 \text{ V}) = \boxed{4.5\ \mu\text{C.}}$$

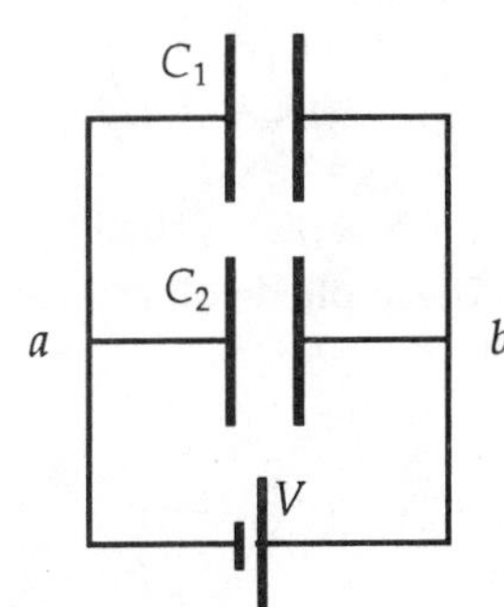

45. For the parallel network the potential difference is the same for all capacitors, and the total charge is the sum of the individual charges. We find the charge on each from
$Q_1 = C_1V = (\epsilon_0A_1/d_1)V;\ Q_2 = C_2V = (\epsilon_0A_2/d_2)V;\ Q_3 = C_3V = (\epsilon_0A_3/d_3)V.$
Thus the sum of the charges is
$Q = Q_1 + Q_2 + Q_3 = [(\epsilon_0A_1/d_1)V] + [(\epsilon_0A_2/d_2)V] + [(\epsilon_0A_3/d_3)V].$
The definition of the equivalent capacitance is
$C_{eq} = Q/V = [(\epsilon_0A_1/d_1)] + [(\epsilon_0A_2/d_2)] + [(\epsilon_0A_3/d_3)] = C_1 + C_2 + C_3.$

46. The potential difference must be the same on each half of the capacitor, so we can treat the system as two capacitors in parallel:
$C = C_1 + C_2 = [K_1\epsilon_0(\frac{1}{2}A)/d] + [K_2\epsilon_0(\frac{1}{2}A)/d]$
$= (\epsilon_0\frac{1}{2}A/d)(K_1 + K_2) = \frac{1}{2}(K_1 + K_2)(\epsilon_0A/d)$
$= \boxed{\frac{1}{2}(K_1 + K_2)C_0.}$

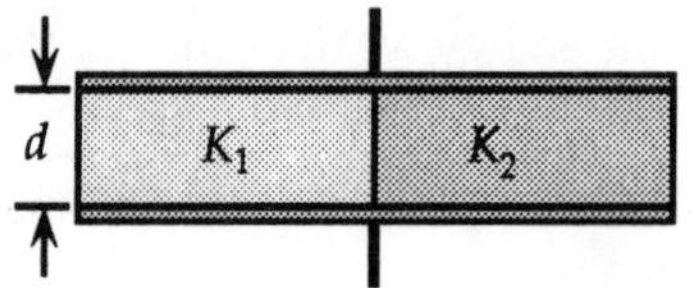

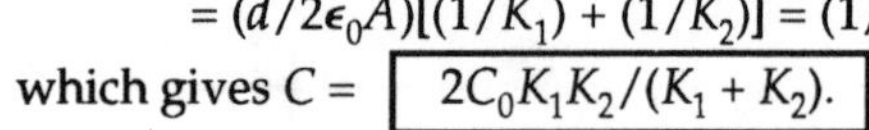

47. If we think of a layer of equal and opposite charges on the interface between the two dielectrics, we see that they are in series. For the equivalent capacitance, we have
$1/C= (1/C_1) + (1/C_2) = (\frac{1}{2}d/K_1\epsilon_0A) + (\frac{1}{2}d/K_2\epsilon_0A)$
$= (d/2\epsilon_0A)[(1/K_1) + (1/K_2)] = (1/2C_0)[(K_1 + K_2)/K_1K_2],$
which gives $C = \boxed{2C_0K_1K_2/(K_1 + K_2).}$

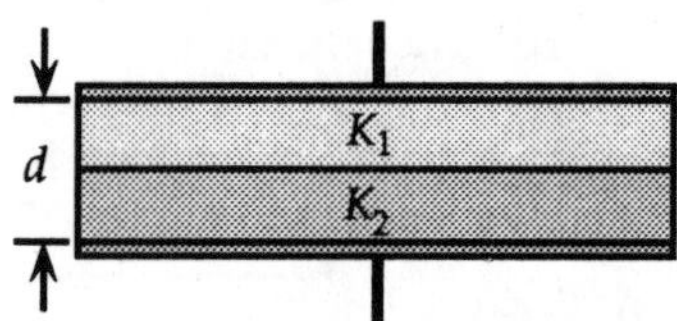

48. For a series network, we have
$Q_1 = Q_2 = Q_{series} = 125$ pC.
We find the equivalent capacitance from
$Q_{series} = C_{series}V;$
125 pC $= C_{series}(25.0$ V), which gives $C_{series} = 5.00$ pF.
We find the unknown capacitance from
$1/C_{series} = (1/C_1) + (1/C_2);$
$1/(5.00$ pF$) = [1/(200$ pF$)] + (1/C_2)$, which gives $C_2 = \boxed{5.13 \text{ pF.}}$

49. (*a*) We find the equivalent capacitance of the two in series from
$1/C_4 = (1/C_1) + (1/C_2) = [1/(7.0\ \mu F)] + [1/(3.0\ \mu F)],$
which gives $C_4 = 2.1\ \mu$F.
This is in parallel with C_3, so we have
$C_{eq} = C_3 + C_4 = 4.0\ \mu F + 2.1\ \mu F = \boxed{6.1\ \mu F.}$
(*b*) We find the charge on each of the two in series:
$Q_1 = Q_2 = Q_4 = C_4V = (2.1\ \mu F)(24\ V) = 50.4\ \mu C.$
We find the voltages from
$Q_1 = C_1V_1;$
$50.4\ \mu C = (7.0\ \mu F)V_1$, which gives $V_1 = \boxed{7.2\ V;}$
$Q_2 = C_2V_2;$
$50.4\ \mu C = (3.0\ \mu F)V_2$, which gives $V_2 = \boxed{16.8\ V.}$
The applied voltage is across C_3: $V_3 = \boxed{24\ V.}$

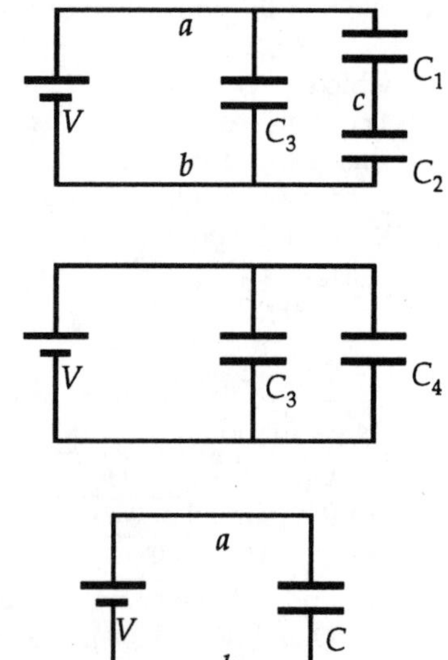

50. Because the two sides of the circuit are identical, we find the resistance from the time constant:

$\tau = RC;$

$3.0\ \text{s} = R(3.0\ \mu\text{F})$, which gives $R =$ **1.0 MΩ.**

51. (*a*) We know from Example 19–7 that the equivalent resistance of the two resistors in parallel is 2.7 Ω. There can be no steady current through the capacitor, so we can find the current in the series resistor circuit:

$I = \mathscr{E}/(R_6 + R_{2.7} + R_5 + r)$

$= (9.0\ \text{V})/(6.0\ \Omega + 2.7\ \Omega + 5.0\ \Omega + 0.50\ \Omega) = 0.634\ \text{A}.$

We use this current to find the potential difference across the capacitor:

$V_{ac} = I(R_6 + R_{2.7}) = (0.634\ \text{A})(6.0\ \Omega + 2.7\ \Omega) = 5.52\ \text{V}.$

The charge on the capacitor is

$Q = CV_{ac} = (7.5\ \mu\text{F})(5.52\ \text{V}) =$ **41 μC.**

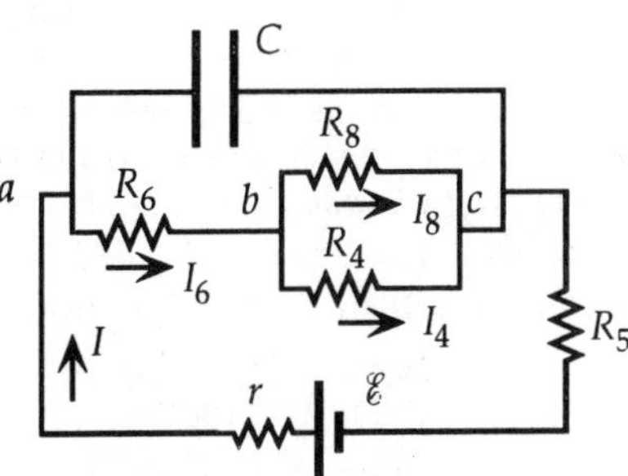

(*b*) As we found above, the steady state current through the 6.0-Ω and 5.0-Ω resistors is **0.63 A.**

The potential difference across the 2.7-Ω resistor is

$V_{bc} = IR_{2.7} = (0.634\ \text{A})(2.7\ \Omega) = 1.7\ \text{V}.$

We find the currents through the 8.0-Ω and 4.0-Ω resistors from

$V_{bc} = I_8 R_8;$

$1.7\ \text{V} = I_8(8.0\ \Omega)$, which gives $I_8 =$ **0.21 A;**

$V_{bc} = I_4 R_4;$

$1.7\ \text{V} = I_4(4.0\ \Omega)$, which gives $I_4 =$ **0.42 A.**

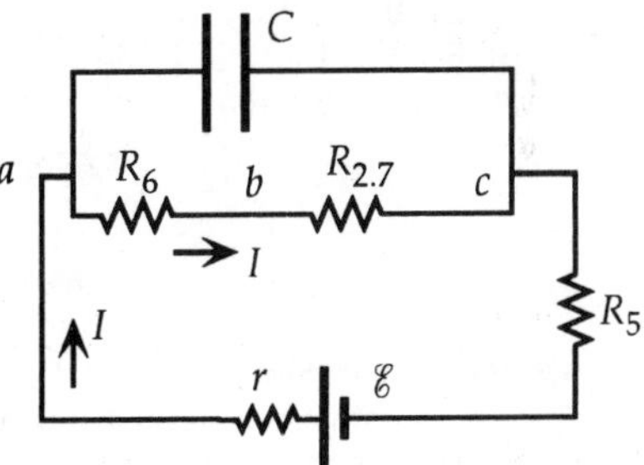

52. (*a*) We find the capacitance from

$\tau = RC;$

$3.5 \times 10^{-6}\ \text{s} = (15 \times 10^3\ \Omega)C,$

which gives $C = 2.3 \times 10^{-9}\ \text{F} =$ **2.3 nF.**

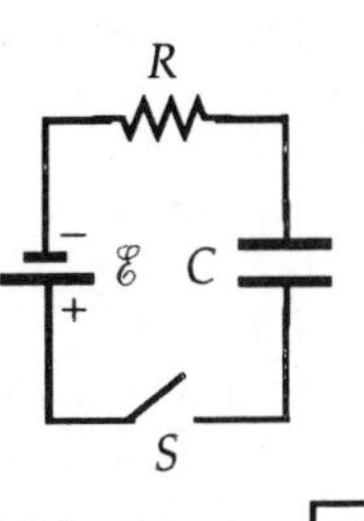

(*b*) The voltage across the capacitance will increase to the final steady state value. The voltage across the resistor will start at the battery voltage and decrease exponentially:

$V_R = \mathscr{E}\, e^{-t/\tau};$

$16\ \text{V} = (24\ \text{V})e^{-t/(35\ \mu\text{s})}$, or $t/(35\ \mu\text{s}) = \ln(24\ \text{V}/16\ \text{V}) = 0.405$, which gives $t =$ **14 μs.**

53. The time constant of the circuit is

$\tau = RC = (6.7 \times 10^3\ \Omega)(3.0 \times 10^{-6}\ \text{F}) = 0.0201\ \text{s} = 20.1\ \text{ms}.$

The capacitor voltage will decrease exponentially:

$V_C = V_0 e^{-t/\tau};$

$0.01V_0 = V_0 e^{-t/(20.1\ \text{ms})}$, or $t/(20.1\ \text{ms}) = \ln(100) = 4.61,$

which gives $t =$ **93 ms.**

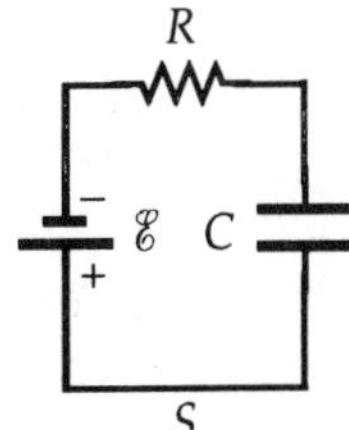

54. (*a*) In the steady state there is no current through the capacitors. Thus the current through the resistors is

$I = V_{cd}/(R_1 + R_2) = (24\ \text{V})/(8.8\ \Omega + 4.4\ \Omega) = 1.82\ \text{A}.$

The potential at point *a* is

$V_a = V_{ad} = IR_2 = (1.82\ \text{A})(4.4\ \Omega) =$ **8.0 V.**

(*b*) We find the equivalent capacitance of the two series capacitors:

$1/C = (1/C_1) + (1/C_2) = [1/(0.48\ \mu\text{F})] + [1/(0.24\ \mu\text{F})],$

which gives $C = 0.16\ \mu\text{F}$.

We find the charge on each of the two in series:

$Q_1 = Q_2 = Q = CV_{cd} = (0.16\ \mu\text{F})(24\ \text{V}) = 3.84\ \mu\text{C}.$

The potential at point *b* is

$V_b = V_{bd} = Q_2/C_2 = (3.84\ \mu\text{C})/(0.24\ \mu\text{F}) =$ **16 V.**

(*c*) With the switch closed, the current is the same. Point *b* must have the same potential as point *a*:

$V_b = V_a =$ **8.0 V.**

(*d*) We find the charge on each of the two capacitors, which are no longer in series:

$Q_1 = C_1V_{cb} = (0.48\ \mu\text{F})(24\ \text{V} - 8.0\ \text{V}) = 7.68\ \mu\text{C};$

$Q_2 = C_2V_{bd} = (0.24\ \mu\text{F})(8.0\ \text{V}) = 1.92\ \mu\text{C}.$

When the switch was open, the net charge at point *b* was zero, because the charge on the negative plate of C_1 had the same magnitude as the charge on the positive plate of C_2. With the switch closed, these charges are not equal. The net charge at point *b* is

$Q_b = -Q_1 + Q_2 = -7.68\ \mu\text{C} + 1.92\ \mu\text{C} =$ **$-5.8\ \mu\text{C}$,** which flowed through the switch.

55. We find the resistance on the voltmeter from

$R = (\text{sensitivity})(\text{scale}) = (30{,}000\ \Omega/\text{V})(250\ \text{V}) = 7.50 \times 10^6\ \Omega =$ **7.50 MΩ.**

56. We find the current for full-scale deflection of the ammeter from

$I = V_{\text{max}}/R = V_{\text{max}}/(\text{sensitivity})V_{\text{max}} = 1/(10{,}000\ \Omega/\text{V}) = 1.00 \times 10^{-4}\ \text{A} =$ **100 μA.**

57. (*a*) We make an ammeter by putting a resistor in parallel with the galvanometer. For full-scale deflection, we have

$V_{\text{meter}} = I_{\text{G}}r = I_sR_s;$

$(50 \times 10^{-6}\ \text{A})(30\ \Omega) = (30\ \text{A} - 50 \times 10^{-6}\ \text{A})R_s,$

which gives $R_s =$ **$50 \times 10^{-6}\ \Omega$ in parallel.**

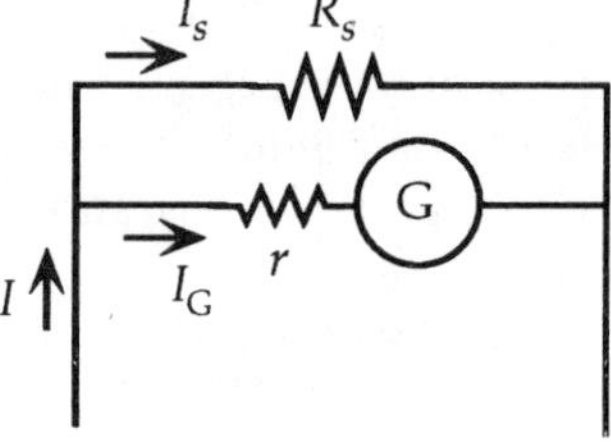

(*b*) We make a voltmeter by putting a resistor in series with the galvanometer. For full-scale deflection, we have

$V_{\text{meter}} = I(R_x + r) = I_{\text{G}}(R_x + r);$

$1000\ \text{V} = (50 \times 10^{-6}\ \text{A})(R_x + 30\ \Omega),$

which gives $R_x = 20 \times 10^6\ \Omega =$ **20 MΩ in series.**

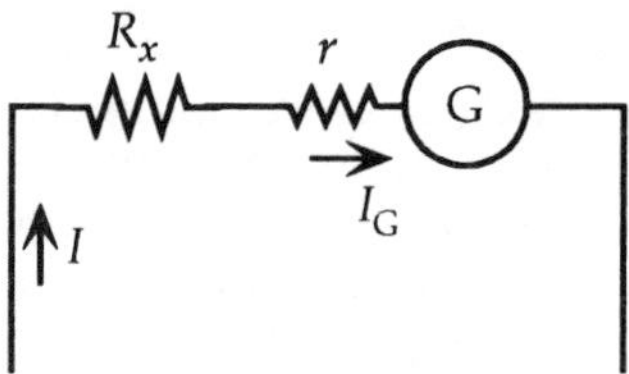

58. (*a*) The current for full-scale deflection of the galvanometer is

$I = 1/(\text{sensitivity}) = 1/(35\ \text{k}\Omega/\text{V}) = 2.85 \times 10^{-2}\ \text{mA} = 28.5\ \mu\text{A}.$

We make an ammeter by putting a resistor in parallel with the galvanometer. For full-scale deflection, we have

$V_{\text{meter}} = I_G r = I_s R_s;$

$(28.5 \times 10^{-6}\ \text{A})(20.0\ \Omega) = (2.0\ \text{A} - 28.5 \times 10^{-6}\ \text{A})R_s,$

which gives $R_s =$ **$2.9 \times 10^{-5}\ \Omega$ in parallel.**

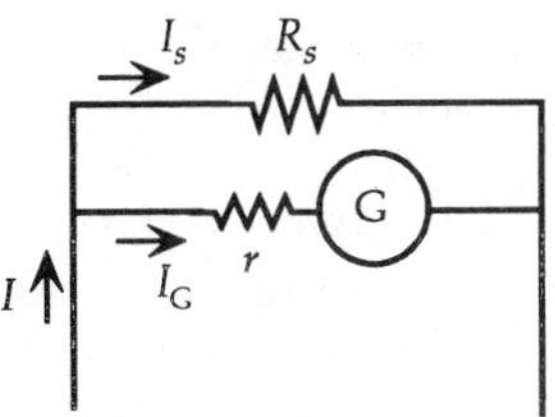

(*b*) We make a voltmeter by putting a resistor in series with the galvanometer. For full-scale deflection, we have

$V_{\text{meter}} = I(R_x + r) = I_G(R_x + r);$

$1.00\ \text{V} = (28.5 \times 10^{-6}\ \text{A})(R_x + 20\ \Omega),$

which gives $R_x = 3.5 \times 10^4\ \Omega =$ **$35\ \text{k}\Omega$ in series.**

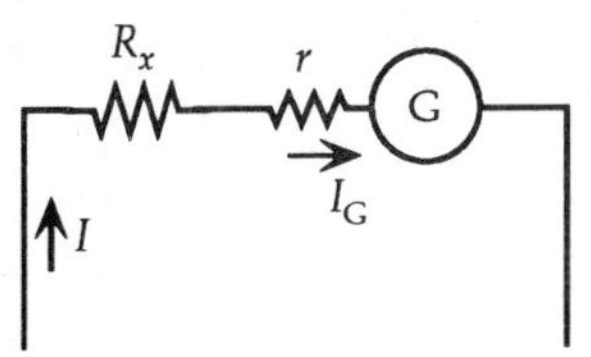

59. We can treat the milliammeter as a galvanometer coil, and find its resistance from

$1/R_A = (1/R_s) + (1/r) = (1/0.20\ \Omega) + (1/30\ \Omega),$

which gives $R_A = 0.199\ \Omega.$

We make a voltmeter by putting a resistor in series with the galvanometer. For full-scale deflection, we have

$V_{\text{meter}} = I(R_x + R_A) = I_G(R_x + R_A);$

$10\ \text{V} = (10 \times 10^{-3}\ \text{A})(R_x + 0.199\ \Omega),$

which gives $R_x = 1.0 \times 10^3\ \Omega =$ **$1.0\ \text{k}\Omega$ in series.**

The sensitivity of the voltmeter is

Sensitivity $= (1000\ \Omega)/(10\ \text{V}) =$ **$100\ \Omega/\text{V}.$**

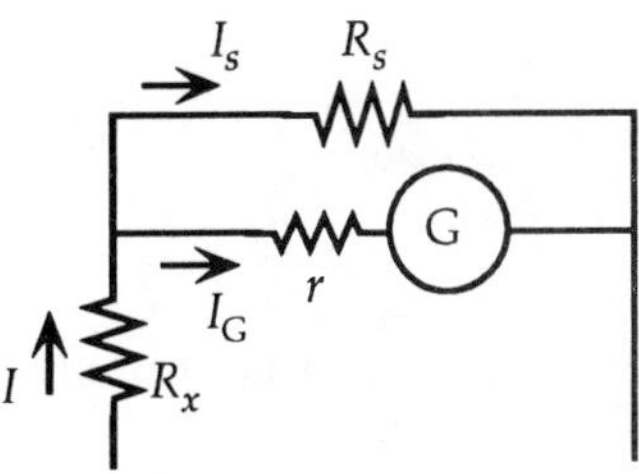

60. Before connecting the voltmeter, the current in the series circuit is

$I_0 = \mathscr{E}/(R_1 + R_2) = (45\ \text{V})/(37\ \text{k}\Omega + 28\ \text{k}\Omega) = 0.692\ \text{mA}.$

The voltages across the resistors are

$V_{01} = I_0 R_1 = (0.692\ \text{mA})(37\ \text{k}\Omega) = 25.6\ \text{V};$

$V_{02} = I_0 R_2 = (0.692\ \text{mA})(28\ \text{k}\Omega) = 19.4\ \text{V}.$

When the voltmeter is across R_1, we find the equivalent resistance of the pair in parallel:

$1/R_A = (1/R_1) + (1/R_V) = (1/37\ \text{k}\Omega) + (1/100\ \text{k}\Omega),$

which gives $R_A = 27.0\ \text{k}\Omega.$

The current in the circuit is

$I_1 = \mathscr{E}/(R_A + R_2) = (45\ \text{V})/(27\ \text{k}\Omega + 28\ \text{k}\Omega) = 0.818\ \text{mA}.$

The reading on the voltmeter is

$V_{V1} = I_1 R_A = (0.818\ \text{mA})(27.0\ \text{k}\Omega) = 22.1\ \text{V} =$ **$22\ \text{V}.$**

When the voltmeter is across R_2, we find the equivalent resistance of the pair in parallel:

$1/R_B = (1/R_2) + (1/R_V) = (1/28\ \text{k}\Omega) + (1/100\ \text{k}\Omega),$

which gives $R_B = 21.9\ \text{k}\Omega.$

The current in the circuit is

$I_2 = \mathscr{E}/(R_1 + R_B) = (45\ \text{V})/(37\ \text{k}\Omega + 22\ \text{k}\Omega) = 0.764\ \text{mA}.$

The reading on the voltmeter is

$V_{V1} = I_1 R_B = (0.764\ \text{mA})(21.9\ \text{k}\Omega) = 16.7\ \text{V} =$ **$17\ \text{V}.$**

We find the percent inaccuracies introduced by the meter:

$(25.6\ \text{V} - 22.1\ \text{V})(100)/(25.6\ \text{V}) =$ **14% low;**

$(19.4\ \text{V} - 16.7\ \text{V})(100)/(19.4\ \text{V}) =$ **14% low.**

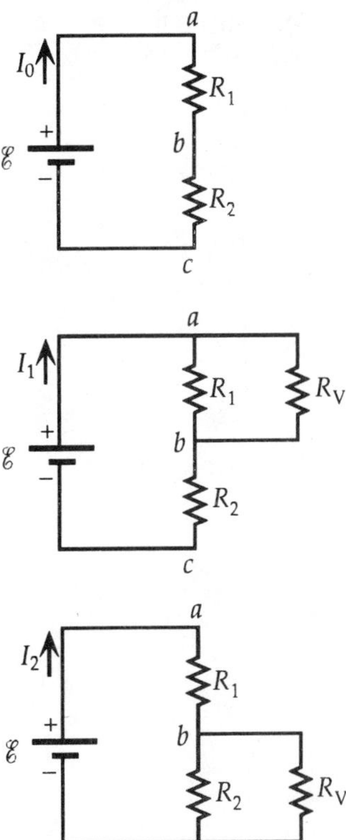

61. We find the voltage of the battery from the series circuit with the ammeter in it:

$\mathcal{E} = I_A(R_A + R_1 + R_2)$
$= (5.25 \times 10^{-3}\ \text{A})(60\ \Omega + 700\ \Omega + 400\ \Omega) = 6.09\ \text{V}.$

Without the meter in the circuit, we have

$\mathcal{E} = I_0(R_1 + R_2);$
$6.09\ \text{V} = I_0(700\ \Omega + 400\ \Omega),$

which gives $I_0 = 5.54 \times 10^{-3}\ \text{A} =$ 5.54 mA.

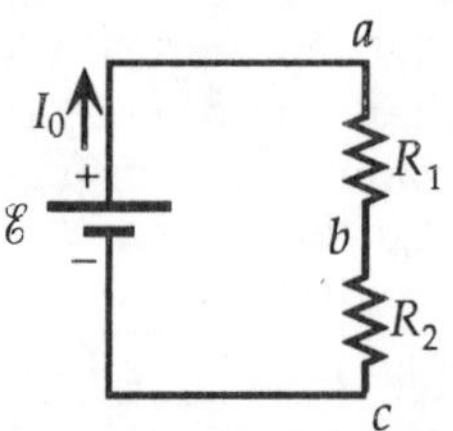

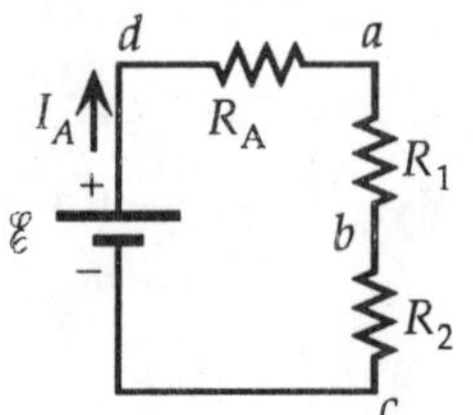

62. We find the equivalent resistance of the voltmeter in parallel with one of the resistors:

$1/R = (1/R_1) + (1/R_V) = (1/9.0\ \text{k}\Omega) + (1/15\ \text{k}\Omega),$

which gives $R = 5.63\ \text{k}\Omega$.

The current in the circuit, which is read by the ammeter, is

$I = \mathcal{E}/(r + R_A + R + R_2)$
$= (12.0\ \text{V})/(1.0\ \Omega + 0.50\ \Omega + 5.63\ \text{k}\Omega + 9.0\ \text{k}\Omega)$
$=$ 0.82 mA.

The reading on the voltmeter is

$V_{ab} = IR = (0.82\ \text{mA})(5.63\ \text{k}\Omega) =$ 4.6 V.

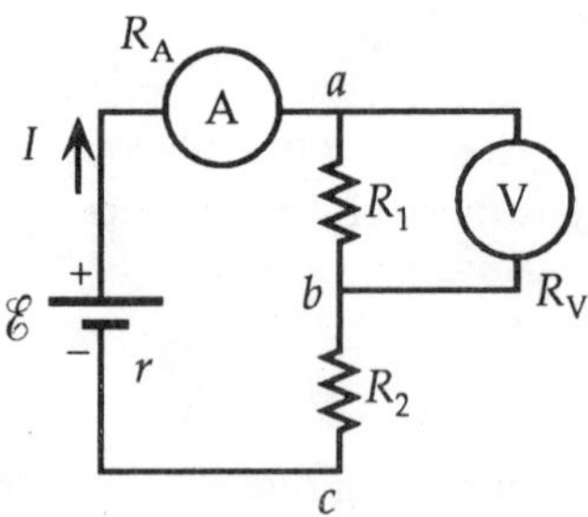

63. In circuit 1 the voltmeter is placed in parallel with the resistor, so we find their equivalent resistance from

$1/R_{eq1} = (1/R) + (1/R_V)$, or $R_{eq1} = RR_V/(R + R_V)$.

The ammeter measures the current through this equivalent resistance and the voltmeter measures the voltage across this equivalent resistance, so we have

$R_1 = V/I = R_{eq1} = RR_V/(R + R_V)$.

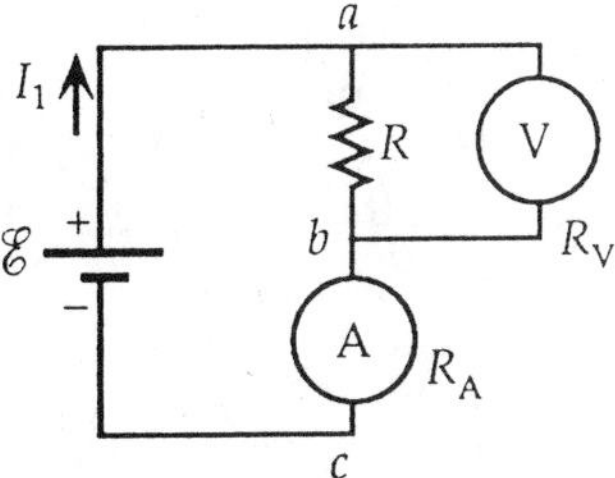

In circuit 2 the ammeter is placed in series with the resistor, so we find their equivalent resistance from

$R_{eq2} = R + R_A$.

The ammeter measures the current through this equivalent resistance and the voltmeter measures the voltage across this equivalent resistance, so we have

$R_2 = V/I = R_{eq2} = R + R_A$.

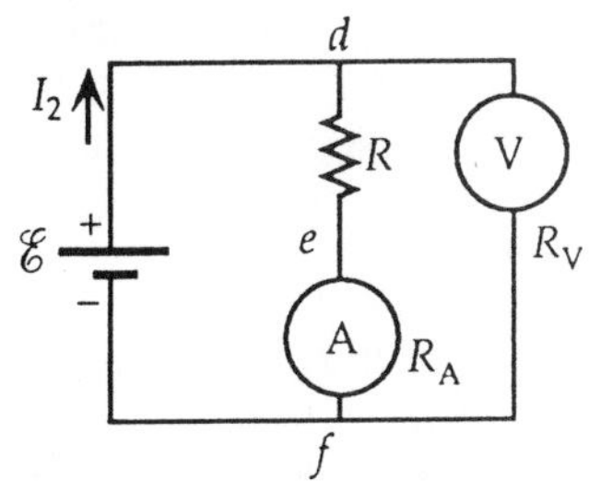

(*a*) For circuit 1 we get

$R_{1a} = (2.00\ \Omega)(10.0 \times 10^3\ \Omega)/(2.00\ \Omega + 10.0 \times 10^3\ \Omega)$

$= \boxed{2.00\ \Omega.}$

For circuit 2 we get

$R_{2a} = (2.00\ \Omega + 1.00\ \Omega) = \boxed{3.00\ \Omega.}$

Thus circuit 1 is better.

(*b*) For circuit 1 we get

$R_{1b} = (100\ \Omega)(10.0 \times 10^3\ \Omega)/(100\ \Omega + 10.0 \times 10^3\ \Omega)$

$= \boxed{99\ \Omega.}$

For circuit 2 we get

$R_{2b} = (100\ \Omega + 1.00\ \Omega) = \boxed{101\ \Omega.}$

Thus both circuits give about the same inaccuracy.

(*c*) For circuit 1 we get

$R_{1c} = (5.0\ \text{k}\Omega)(10.0\ \text{k}\Omega)/(5.0\ \text{k}\Omega + 10.0\ \text{k}\Omega)$

$= \boxed{3.3\ \text{k}\Omega.}$

For circuit 2 we get

$R_{2c} = (5.0\ \text{k}\Omega + 1.00\ \Omega) = \boxed{5.0\ \text{k}\Omega.}$

Thus circuit 2 is better.

Circuit 1 is better when the resistance is small compared to the voltmeter resistance.
Circuit 2 is better when the resistance is large compared to the ammeter resistance.

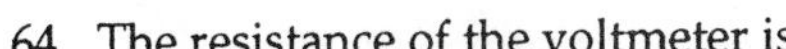

64. The resistance of the voltmeter is

$R_V = (\text{sensitivity})(\text{scale}) = (1000\ \Omega/\text{V})(3.0\ \text{V}) = 3.0 \times 10^3\ \Omega = 3.0\ \text{k}\Omega$.

We find the equivalent resistance of the resistor and the voltmeter from

$1/R_{eq} = (1/R) + (1/R_V)$, or

$R_{eq} = RR_V/(R + R_V) = (7.4\ \text{k}\Omega)(3.0\ \text{k}\Omega)/(7.4\ \text{k}\Omega + 3.0\ \text{k}\Omega) = 2.13\ \text{k}\Omega$.

The voltmeter measures the voltage across this equivalent resistance, so the current in the circuit is

$I = V_{ab}/R_{eq} = (2.0\ \text{V})/(2.13\ \text{k}\Omega) = 0.937\ \text{mA}$.

For the series circuit, we have

$\mathcal{E} = I(R + R_{eq}) = (0.937\ \text{mA})(7.4\ \text{k}\Omega + 2.13\ \text{k}\Omega) = \boxed{8.9\ \text{V.}}$

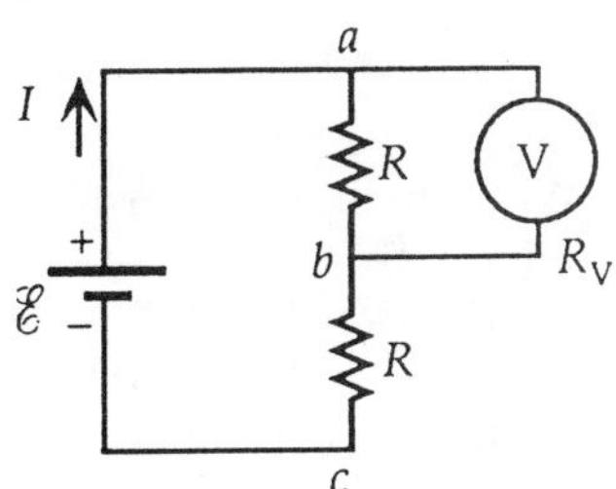

65. We know from Example 19–14 that the voltage across the resistor without the voltmeter connected is 4.0 V.
Thus the minimum voltmeter reading is
$V_{ab} = (0.97)(4.0\ \text{V}) = 3.88\ \text{V}.$
We find the maximum current in the circuit from
$I = V_{bc}/R_2 = (8.0\ \text{V} - 3.88\ \text{V})/(15\ \text{k}\Omega) = 0.275\ \text{mA}.$
Now we can find the minimum equivalent resistance for the voltmeter and R_1:
$R_{eq} = V_{ab}/I = (3.88\ \text{V})/(0.275\ \text{mA}) = 14.1\ \text{k}\Omega.$
For the equivalent resistance, we have
$1/R_{eq} = (1/R_1) + (1/R_V);$
$1/(14.1\ \text{k}\Omega) = [1/(15\ \text{k}\Omega)] + (1/R_V)$, which gives $R_V = 240\ \text{k}\Omega.$
We see that the minimum R_{eq} gives the minimum R_V, so we have $R_V \geq 240\ \text{k}\Omega.$

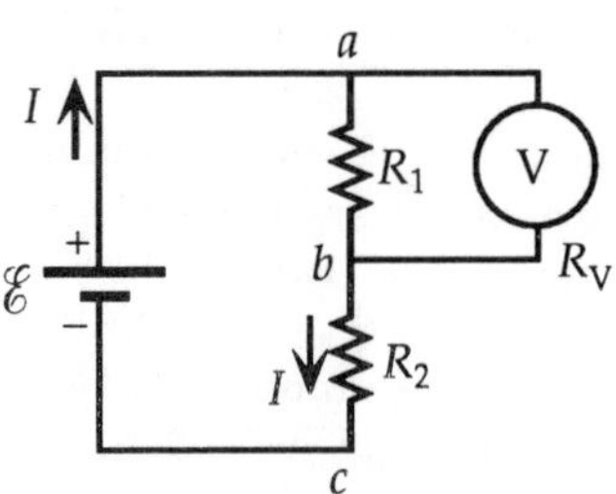

66. We find the resistances of the voltmeter scales:
R_{V100} = (sensitivity)(scale)
$= (20{,}000\ \Omega/\text{V})(100\ \text{V}) = 2.0 \times 10^6\ \Omega = 2000\ \text{k}\Omega;$
R_{V30} = (sensitivity)(scale)
$= (20{,}000\ \Omega/\text{V})(30\ \text{V}) = 6.0 \times 10^5\ \Omega = 600\ \text{k}\Omega.$
The current in the circuit is
$I = (V_{ab}/R_V) + (V_{ab}/R_1).$
For the series circuit, we have
$\mathscr{E} = V_{ab} + IR_2.$
When the 100-volt scale is used, we have
$I = [(25\ \text{V})/(2000\ \text{k}\Omega)] + [(25\ \text{V})/(120\ \text{k}\Omega)] = 0.221\ \text{mA}.$
$\mathscr{E} = 25\ \text{V} + (0.221\ \text{mA})R_2.$
When the 30-volt scale is used, we have
$I = [(23\ \text{V})/(600\ \text{k}\Omega)] + [(23\ \text{V})/(120\ \text{k}\Omega)] = 0.230\ \text{mA}.$
$\mathscr{E} = 23\ \text{V} + (0.230\ \text{mA})R_2.$
We have two equations for two unknowns, with the results: $\mathscr{E} = 74.1$ V, and

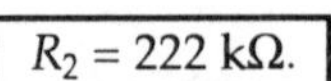
$R_2 = 222\ \text{k}\Omega.$

Without the voltmeter in the circuit, we find the current:
$\mathscr{E} = I'(R_1 + R_2);$
$74.1\ \text{V} = I'(120\ \text{k}\Omega + 222\ \text{k}\Omega)$, which gives $I' = 0.217\ \text{mA}.$
Thus the voltage across R_1 is
$V_{ab}' = I'R_1 = (0.217\ \text{mA})(120\ \text{k}\Omega) =$ 26 V.

67. When the voltmeter is across R_1, for the junction b, we have
$I_{1A} + I_{1V} = I_1;$
$[(5.5\ \text{V})/R_1] + [(5.5\ \text{V})/(15.0\ \text{k}\Omega)] = (12.0\ \text{V} - 5.5\ \text{V})/R_2;$
$[(5.5\ \text{V})/R_1] + 0.367\ \text{mA} = (6.5\ \text{V})/R_2.$
When the voltmeter is across R_2, for the junction e, we have
$I_2 = I_{2A} + I_2;$
$(12.0\ \text{V} - 4.0\ \text{V})/R_1 = [(4.0\ \text{V})/R_2] + [(4.0\ \text{V})/(15.0\ \text{k}\Omega)];$
$[(8.0\ \text{V})/R_1] = [(4.0\ \text{V})/R_2] + 0.267\ \text{mA}.$
We have two equations for two unknowns, with the results:
$R_1 = 9.4\ \text{k}\Omega,$ and $R_2 = 6.8\ \text{k}\Omega.$

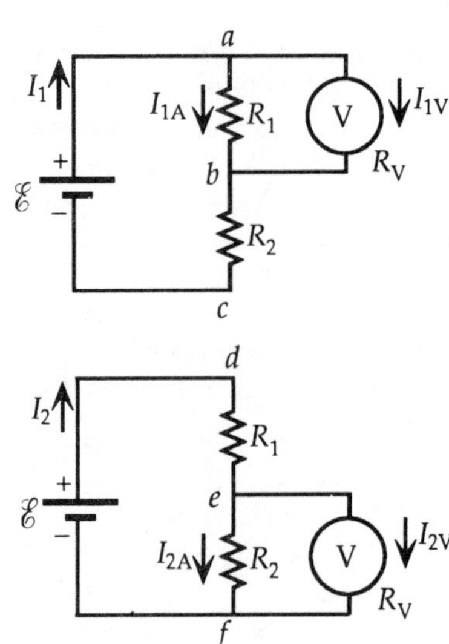

68. The voltage is the same across resistors in parallel, but is less across a resistor in a series connection. We connect two resistors in series as shown in the diagram. Each resistor has the same current:

$I = V/(R_1 + R_2) = (9.0\text{ V})/(R_1 + R_2).$

If the desired voltage is across R_1, we have

$V_{ab} = IR_1 = (9.0\text{ V})R_1/(R_1 + R_2);$

$0.25\text{ V} = (9.0\text{ V})R_1/(R_1 + R_2) = (9.0\text{ V})/[1 + (R_2/R_1)],$

which gives $R_2/R_1 = 35.$

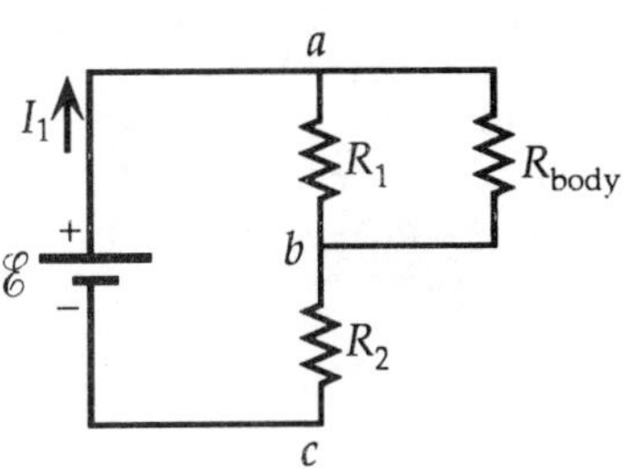

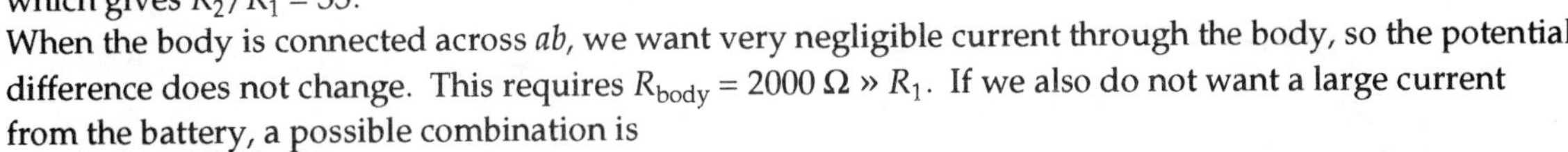

When the body is connected across ab, we want very negligible current through the body, so the potential difference does not change. This requires $R_{body} = 2000\ \Omega \gg R_1$. If we also do not want a large current from the battery, a possible combination is

$\boxed{R_1 = 2\ \Omega, R_2 = 70\ \Omega.}$

69. Because the voltage is constant and the power is additive, we can use two resistors in parallel. For the lower ratings, we use the resistors separately; for the highest rating, we use them in parallel. The rotary switch shown allows the B contact to successively connect to C and D. The A contact connects to C and D for the parallel connection. We find the resistances for the three settings from

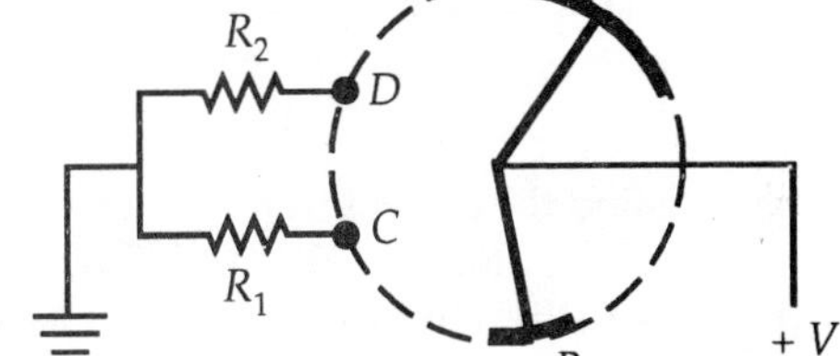

$P = V^2/R;$

$50\text{ W} = (120\text{ V})^2/R_1$, which gives $R_1 = 288\ \Omega;$

$100\text{ W} = (120\text{ V})^2/R_2$, which gives $R_2 = 144\ \Omega;$

$150\text{ W} = (120\text{ V})^2/R_3$, which gives $R_3 = 96\ \Omega.$

As expected, for the parallel arrangement we have

$1/R_{eq} = (1/R_1) + (1/R_2);$

$1/R_{eq} = [1/(288\ \Omega)] + [1/(144\ \Omega)]$, which gives $R_{eq} = 96\ \Omega = R_3.$

Thus the two required resistors are $\boxed{288\ \Omega, 144\ \Omega.}$

70. The voltage drop across the two wires is

$V_{drop} = IR = (3.0\text{ A})(0.0065\ \Omega/\text{m})(2)(95\text{ m}) = \boxed{3.7\text{ V.}}$

The applied voltage at the apparatus is

$V = V_0 - V_{drop} = 120\text{ V} - 3.7\text{ V} = \boxed{116\text{ V.}}$

71. We find the current through the patient (and nurse) from the series circuit:

$I = V/(R_{motor} + R_{bed} + R_{nurse} + R_{patient})$

$= (220\text{ V})/(10^4\ \Omega + 0 + 10^4\ \Omega + 10^4\ \Omega) = 7.3 \times 10^{-3}\text{ A} = \boxed{7.3\text{ mA.}}$

72. (*a*) When the capacitors are connected in parallel, we find the equivalent capacitance from

$C_{parallel} = C_1 + C_2 = 0.40\ \mu\text{F} + 0.60\ \mu\text{F} = 1.00\ \mu\text{F}.$

The stored energy is

$U_{parallel} = \frac{1}{2}C_{parallel}V^2 = \frac{1}{2}(1.00 \times 10^{-6}\text{ F})(45\text{ V})^2 = \boxed{1.0 \times 10^{-3}\text{ J.}}$

(*b*) When the capacitors are connected in series, we find the equivalent capacitance from

$1/C_{series} = (1/C_1) + (1/C_2) = [1/(0.40\ \mu\text{F})] + [1/(0.60\ \mu\text{F})]$, which gives $C_{series} = 0.24\ \mu\text{F}.$

The stored energy is

$U_{series} = \frac{1}{2}C_{series}V^2 = \frac{1}{2}(0.24 \times 10^{-6}\text{ F})(45\text{ V})^2 = \boxed{2.4 \times 10^{-4}\text{ J.}}$

(*c*) We find the charges from

$Q = C_{eq}V;$

$Q_{parallel} = C_{parallel}V = (1.00\ \mu\text{F})(45\text{ V}) = \boxed{45\ \mu\text{C.}}$

$Q_{series} = C_{series}V = (0.24\ \mu\text{F})(45\text{ V}) = \boxed{11\ \mu\text{C.}}$

73. The time between firings is

$t = (60\text{ s})/(72\text{ beats}) = 0.833\text{ s}.$

We find the time for the capacitor to reach 63% of maximum from

$V = V_0(1 - e^{-t/\tau}) = 0.63V_0$, which gives $e^{-t/\tau} = 0.37$, or

$t = \tau = RC;$

$0.833\text{ s} = R(7.5\ \mu\text{C})$, which gives $R =$ $\boxed{0.11\text{ M}\Omega.}$

74. We find the required current for the hearing aid from

$P = IV;$

$2\text{ W} = I(4.0\text{ V})$, which gives $I = 0.50\text{ A}$.

With this current the terminal voltage of the three mercury cells would be

$V_{\text{mercury}} = 3(\mathscr{E}_{\text{mercury}} - Ir_{\text{mercury}}) = 3[1.35\text{ V} - (0.50\text{ A})(0.030\ \Omega)] = 4.01\text{ V}.$

With this current the terminal voltage of the three dry cells would be

$V_{\text{dry}} = 3(\mathscr{E}_{\text{dry}} - Ir_{\text{dry}}) = 3[1.5\text{ V} - (0.50\text{ A})(0.35\ \Omega)] = 3.98\text{ V}.$

Thus the $\boxed{\text{mercury cells would have a higher terminal voltage.}}$

75. (*a*) We find the current from

$I_1 = V/R_{\text{body}} = (110\text{V})/(900\ \Omega) =$ $\boxed{0.12\text{ A.}}$

(*b*) Because the alternative path is in parallel, the current is the same: $\boxed{0.12\text{ A.}}$

(*c*) The current restriction means that the voltage will change. Because the voltage will be the same across both resistances, we have

$I_2R_2 = I_{\text{body}}R_{\text{body}};$

$I_2(40\ ) = I_{\text{body}}(900\ )$, or $I_2 = 22.5I_{\text{body}}$.

For the sum of the currents, we have

$I_2 + I_{\text{body}} = 23.5I_{\text{body}} = 1.5\text{ A}$, which gives $I_{\text{body}} = 6.4 \times 10^{-2}\text{ A} =$ $\boxed{64\text{ mA.}}$

76. (*a*) When there is no current through the galvanometer, we have $V_{BD} = 0$, a current I_1 through R_1 and R_2, and a current I_3 through R_3 and R_x. Thus we have

$V_{AD} = V_{AB};\ I_1R_1 = I_3R_3$, and

$V_{BC} = V_{DC};\ I_1R_2 = I_3R_x$.

When we divide these two equations, we get

$R_2/R_1 = R_x/R_3$, or $R_x = (R_2/R_1)R_3$.

(*b*) The unknown resistance is

$R_x = (R_2/R_1)R_3 = (972\ \Omega/630\ \Omega)(42.6\ \Omega) =$ $\boxed{65.7\ \Omega.}$

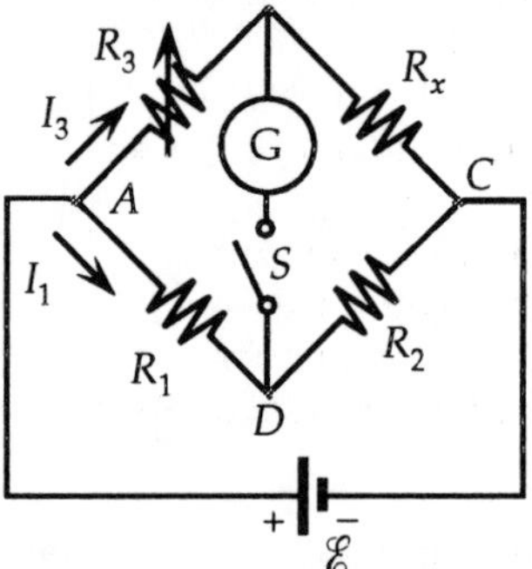

77. The resistance of the platinum wire is

$R_x = (R_2/R_1)R_3 = (46.0\ \Omega/38.0\ \Omega)(3.48\ \Omega) = 4.21\ \Omega.$

We find the length from

$R = \rho L/A;$

$= (10.6 \times 10^{-8}\ \Omega\cdot\text{m})L/\pi(0.460 \times 10^{-3}\text{ m})2$, which gives $L =$ $\boxed{26.4\text{ m.}}$

78. (*a*) When there is no current through the galvanometer, the current I must pass through the long resistor R', so the potential difference between A and C is

$$V_{AC} = IR.$$

Because there is no current through the measured emf, for the bottom loop we have

$$\mathscr{E} = IR.$$

When different emfs are balanced, the current I is the same, so we have

$$\mathscr{E}_s = IR_s, \text{ and } \mathscr{E}_x = IR_x.$$

When we divide the two equations, we get

$$\mathscr{E}_s/\mathscr{E}_x = R_s/R_x, \text{ or } \mathscr{E}_x = (R_x/R_s)\mathscr{E}_s.$$

(*b*) Because the resistance is proportional to the length, we have

$$\mathscr{E}_x = (R_x/R_s)\mathscr{E}_s = (45.8 \text{ cm}/25.4 \text{ cm})(1.0182 \text{ V}) = \boxed{1.836 \text{ V.}}$$

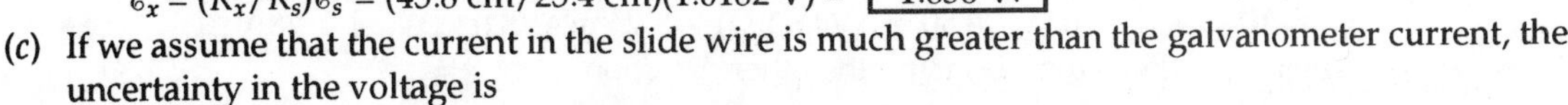

(*c*) If we assume that the current in the slide wire is much greater than the galvanometer current, the uncertainty in the voltage is

$$\Delta V = \pm I_G R_G = \pm (30\ \Omega)(0.015 \text{ mA}) = \pm 0.45 \text{ mV}.$$

Because this can occur for each setting and there will be uncertainties in measuring the distances, the minimum uncertainty is $\boxed{\pm 0.90 \text{ mV.}}$

(*d*) The advantage of this method is that there is $\boxed{\text{no effect of the internal resistance,}}$ because there is no current through the cell.

79. (*a*) We see from the diagram that all positive plates are connected to the positive side of the battery, and that all negative plates are connected to the negative side of the battery, so the capacitors are connected in $\boxed{\text{parallel.}}$

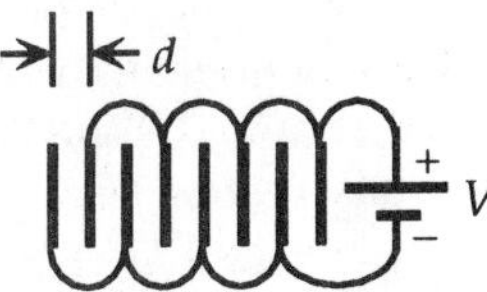

(*b*) For parallel capacitors, the total capacitance is the sum, so we have

$$C_{min} = 7(\epsilon_0 A_{min}/d)$$
$$= 7(8.85 \times 10^{-12} \text{ C}^2/\text{N}\cdot\text{m}^2)(2.0 \times 10^{-4} \text{ m}^2)/(2.0 \times 10^{-3} \text{ m}) = 6.2 \times 10^{-12} \text{ F} = 6.2 \text{ pF}.$$
$$C_{max} = 7(\epsilon_0 A_{min}/d)$$
$$= 7(8.85 \times 10^{-12} \text{ C}^2/\text{N}\cdot\text{m}^2)(12.0 \times 10^{-4} \text{ m}^2)/(2.0 \times 10^{-3} \text{ m}) = 3.7 \times 10^{-11} \text{ F} = 37 \text{ pF}.$$

Thus the range is $\boxed{6.2 \text{ pF} \le C \le 37 \text{ pF.}}$

80. The terminal voltage of a discharging battery is

$$V = \mathscr{E} - Ir.$$

For the two conditions, we have

$$40.8 \text{ V} = \mathscr{E} - (7.40 \text{ A})r;$$
$$47.3 \text{ V} = \mathscr{E} - (2.20 \text{ A})r.$$

We have two equations for two unknowns, with the solutions: $\mathscr{E} = \boxed{50.1 \text{ V,}}$ and $r = \boxed{1.25\ \Omega.}$

81. One arrangement is to connect N resistors in series. Each resistor will have the same power, so we need

$$N = P_{total}/P = 5 \text{ W}/\tfrac{1}{2} \text{ W} = 10 \text{ resistors}.$$

We find the required value of resistance from

$$R_{total} = NR_{series};$$
$$2.2 \text{ k}\Omega = 10R_{series}, \text{ which gives } R_{series} = 0.22 \text{ k}\Omega.$$

Thus we have $\boxed{10\ \ 0.22\text{-k}\Omega \text{ resistors in series.}}$

Another arrangement is to connect N resistors in series. Each resistor will again have the same power, so we need the same number of resistors: 10.

We find the required value of resistance from

$$1/R_{total} = \Sigma(1/R_i) = N/R_{parallel};$$
$$1/2.2 \text{ k}\Omega = 10/R_{parallel}, \text{ which gives } R_{parallel} = 22 \text{ k}\Omega.$$

Thus we have $\boxed{10\ \ 22\text{-k}\Omega \text{ resistors in parallel.}}$

82. If we assume the current in R_4 is to the right, we have
$V_{cd} = I_4R_4 = (3.50 \text{ mA})(4.0 \text{ k}\Omega) = 14.0 \text{ V}$.
We can now find the current in R_8:
$I_8 = V_{cd}/R_8 = (14.0 \text{ V})/(8.0 \text{ k}\Omega) = 1.75 \text{ mA}$.
From conservation of current at the junction c, we have
$I = I_4 + I_8 = 3.50 \text{ mA} + 1.75 \text{ mA} = 5.25 \text{ mA}$.
If we go clockwise around the outer loop, starting at a, we have

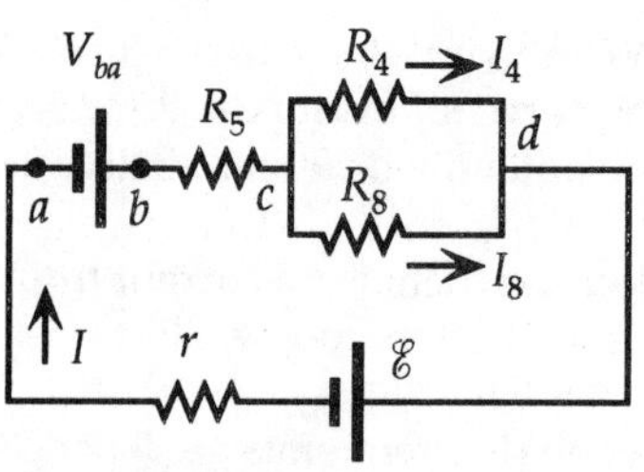

$V_{ba} - IR_5 - I_4R_4 - \mathscr{E} - Ir = 0$, or
$V_{ba} = (5.25 \text{ mA})(5.0 \text{ k}\Omega) - (3.50 \text{ mA})(4.0 \text{ k}\Omega) - 12.0 \text{ V} - (5.25 \text{ mA})(1.0 \ \Omega) = \boxed{52 \text{ V.}}$
If we assume the current in R_4 is to the left, all currents are reversed, so we have
$V_{dc} = 14.0 \text{ V}; \quad I_8 = 1.75 \text{ mA}, \quad \text{and} \quad I = 5.25 \text{ mA}$.
If we go counterclockwise around the outer loop, starting at a, we have
$-Ir + \mathscr{E} - I_4R_4 - IR_5 - Ir + V_{ab} = 0$, or $V_{ba} = -V_{ab} = -Ir + \mathscr{E} - I_4R_4 - IR_5 - Ir$;
$V_{ba} = -(5.25 \text{ mA})(1.0 \ \Omega) + 12.0 \text{ V} - (3.50 \text{ mA})(4.0 \text{ k}\Omega) - (5.25 \text{ mA})(5.0 \text{ k}\Omega) = \boxed{-28 \text{ V.}}$
The negative value means the battery is facing the other direction.

83. When the leads are shorted ($R_x = 0$), there will be maximum current in the circuit, including the galvanometer. We have
$V_{ab,\text{max}} = I_{G,\text{max}}r = (35 \times 10^{-6} \text{ A})(25 \ \Omega) = 8.75 \times 10^{-4} \text{ V}$.
We can now find the current in the shunt, R_{sh}:
$I_{sh,\text{max}} = V_{ab,\text{max}}/R_{sh} = (8.75 \times 10^{-4} \text{ V})/R_{sh}$.
From conservation of current at the junction a, we have
$I_{\text{max}} = I_{G,\text{max}} + I_{sh,\text{max}} = 35 \times 10^{-6} \text{ A} + (8.75 \times 10^{-4} \text{ V})/R_{sh}$.
If we go clockwise around the outer loop, starting at a, we have

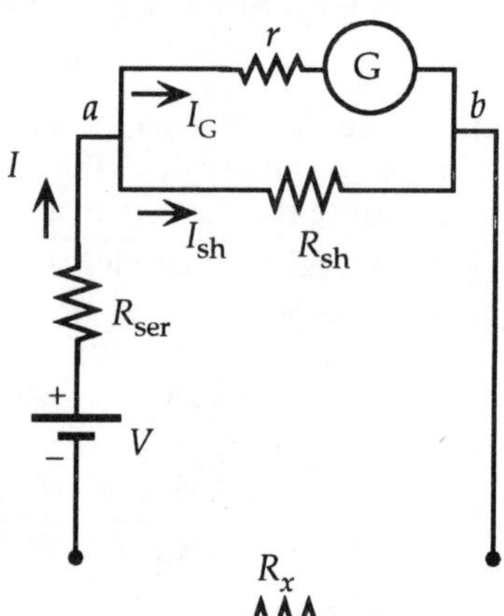

$V_{ba,\text{max}} + V - I_{\text{max}}R_{ser} = 0$;
$-8.75 \times 10^{-4} \text{ V} + 3.0 \text{ V} - [35 \times 10^{-6} \text{ A} + (8.75 \times 10^{-4} \text{ V})/R_{sh}]R_{ser} = 0$;
$35 \times 10^{-6} \text{ A} + (8.75 \times 10^{-4} \text{ V})/R_{sh} = (3.0 \text{ V} - 8.75 \times 10^{-4} \text{ V})/R_{ser}$;
$35 \times 10^{-6} \text{ A} + (8.75 \times 10^{-4} \text{ V})/R_{sh} \approx (3.0 \text{ V})/R_{ser}$.
When the leads are across $R_x = 30 \text{ k}\Omega$, all currents will be one-half their maximum values. We have
$V_{ab} = I_Gr = \frac{1}{2}(35 \times 10^{-6} \text{ A})(25 \ \Omega) = 4.375 \times 10^{-4} \text{ V}$.
The current in the shunt is
$I_{sh} = V_{ab}/R_{sh} = (4.375 \times 10^{-4} \text{ V})/R_{sh}$.
From conservation of current at the junction a, we have
$I = I_G + I_{sh} = 17.5 \times 10^{-6} \text{ A} + (4.375 \times 10^{-4} \text{ V})/R_{sh}$.
If we go clockwise around the outer loop, starting at a, we have
$V_{ba} + V - I(R_x + R_{ser}) = 0$;
$-4.375 \times 10^{-4} \text{ V} + 3.0 \text{ V} - [17.5 \times 10^{-6} \text{ A} + (4.375 \times 10^{-4} \text{ V})/R_{sh}](30 \times 10^3 \ \Omega + R_{ser}) = 0$;
$17.5 \times 10^{-6} \text{ A} + (4.375 \times 10^{-4} \text{ V})/R_{sh} = (3.0 \text{ V} - 4.375 \times 10^{-4} \text{ V})/(30 \times 10^3 \ \Omega + R_{ser})$;
$17.5 \times 10^{-6} \text{ A} + (4.375 \times 10^{-4} \text{ V})/R_{sh} \approx (3.0 \text{ V})/(30 \times 10^3 \ \Omega + R_{ser})$.
We have two equations for two unknowns, with the solutions:
$R_{sh} = \boxed{13 \ \Omega,}$ and $R_{ser} = 30 \times 10^3 \ \Omega = \boxed{30 \text{ k}\Omega.}$

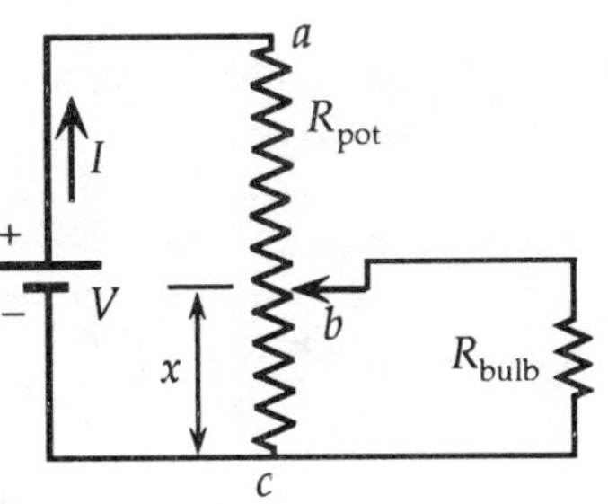

84. The resistance along the potentiometer is proportional to the length, so we find the equivalent resistance between points b and c:

$1/R_{eq} = (1/xR_{pot}) + (1/R_{bulb})$, or

$R_{eq} = xR_{pot}R_{bulb}/(xR_{pot} + R_{bulb})$.

We find the current in the loop from

$I = V/[(1 - x)R_{pot} + R_{eq}]$.

The potential difference across the bulb is

$V_{bc} = IR_{eq}$, so the power expended in the bulb is

$P = V_{bc}^2/R_{bulb}$.

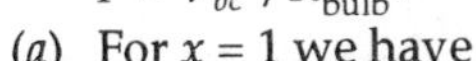

(a) For $x = 1$ we have

$R_{eq} = (1)(100\ \Omega)(200\ \Omega)/[(1)(100\ \Omega) + 200\ \Omega] = 66.7\ \Omega$.

$I = (120\text{ V})/[(1 - 1)(100\ \Omega) + 66.7\ \Omega] = 1.80\text{ A}$.

$V_{bc} = (1.80\text{ A})(66.7\ \Omega) = 120\text{ V}$.

$P = (120\text{ V})^2/(200\ \Omega) = \boxed{72.0\text{ W.}}$

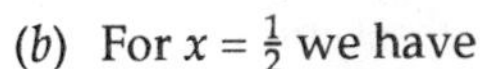

(b) For $x = \frac{1}{2}$ we have

$R_{eq} = (\frac{1}{2})(100\ \Omega)(200\ \Omega)/[(\frac{1}{2})(100\ \Omega) + 200\ \Omega] = 40.0\ \Omega$.

$I = (120\text{ V})/[(1 - \frac{1}{2})(100\ \Omega) + 40.0\ \Omega] = 1.33\text{ A}$.

$V_{bc} = (1.33\text{ A})(40.0\ \Omega) = 53.3\text{ V}$.

$P = (53.3\text{ V})^2/(200\ \Omega) = \boxed{14.2\text{ W.}}$

(c) For $x = \frac{1}{4}$ we have

$R_{eq} = (\frac{1}{4})(100\ \Omega)(200\ \Omega)/[(\frac{1}{4})(100\ \Omega) + 200\ \Omega] = 22.2\ \Omega$.

$I = (120\text{ V})/[(1 - \frac{1}{4})(100\ \Omega) + 66.7\ \Omega] = 1.23\text{ A}$.

$V_{bc} = (1.23\text{ A})(22.2\ \Omega) = 27.4\text{ V}$.

$P = (27.4\text{ V})^2/(200\ \Omega) = \boxed{3.75\text{ W.}}$

85. (a) Normally there is no DC current in the circuit, so the voltage of the battery is across the capacitor. When there is an interruption, the capacitor voltage will decrease exponentially:

$V_C = V_0 e^{-t/\tau}$.

We find the time constant from the need to maintain 70% of the voltage for 0.20 s:

$0.70V_0 = V_0 e^{-(0.20\text{ s})/\tau}$, or $(0.20\text{ s})/\tau = \ln(1.43) = 3.57$,

which gives $\tau = 0.56$ s.

We find the required resistance from

$\tau = RC$;

$0.56\text{ s} = R(22 \times 10^{-6}\text{ F})$, which gives $R = 2.5 \times 10^4\ \Omega = \boxed{25\text{ k}\Omega.}$

(b) In normal operation, there will be no voltage across the resistor, so the device should be connected

$\boxed{\text{between } b \text{ and } c.}$

86. (a) Because the capacitor is disconnected from the power supply, the charge is constant. We find the new voltage from

$Q = C_1V_1 = C_2V_2$;

$(10\text{ pF})(10{,}000\text{ V}) = (1\text{ pF})V_2$, which gives $V_2 = 1.0 \times 10^5\text{ V} = \boxed{0.10\text{ MV.}}$

(b) A major disadvantage is that, when the stored energy is used, the

$\boxed{\text{voltage will decrease exponentially,}}$ so it can be used for only short bursts.

CHAPTER 20

1. (*a*) The maximum force will be produced when the wire and the magnetic field are perpendicular, so we have
$F_{max} = ILB$, or
$F_{max}/L = IB = (9.80\text{ A})(0.80\text{ T}) =$ 7.8 N/m.
(*b*) We find the force per unit length from
$F/L = IB \sin 45.0° = (F_{max}/L) \sin 45.0° = (7.8\text{ N/m}) \sin 45.0° =$ 5.5 N/m.

2. The force on the wire is produced by the component of the magnetic field perpendicular to the wire:
$F = ILB \sin 40°$
$= (6.5\text{ A})(1.5\text{ m})(5.5 \times 10^{-5}\text{ T}) \sin 40° =$ 3.4×10^{-4} N perpendicular to the wire and to **B**.

3. For the maximum force the wire is perpendicular to the field, so we find the current from
$F = ILB$;
$0.900\text{ N} = I(4.20\text{ m})(0.0800\text{ T})$, which gives $I =$ 2.68 A.

4. The maximum force will be produced when the wire and the magnetic field are perpendicular, so we have
$F_{max} = ILB$;
$4.14\text{ N} = (25.0\text{ A})(0.220\text{ m})B$, which gives $B =$ 0.753 T.

5. To find the direction of the force on the electron, we point our fingers west and curl them upward into the magnetic field. Our thumb points north, which would be the direction of the force on a positive charge. Thus the force on the electron is south.
$F = qvB = (1.60 \times 10^{-19}\text{ C})(3.58 \times 10^{6}\text{ m/s})(1.30\text{ T}) =$ 7.45×10^{-13} N south.

6. To find the direction of the force on the electron, we point our fingers upward and curl them forward into the magnetic field. Our thumb points left, which would be the direction of the force on a positive charge. Thus the force on the electron is right. As the electron deflects to the right, the force will always be perpendicular, so the electron will travel in a clockwise vertical circle.
The magnetic force provides the radial acceleration, so we have
$F = qvB = mv^2/r$, so the radius of the path is
$r = mv/qB$;
$= (9.11 \times 10^{-31}\text{ kg})(1.80 \times 10^{6}\text{ m/s})/(1.60 \times 10^{-19}\text{ C})(0.250\text{ T}) =$ 4.10×10^{-5} m.

7. To find the direction of the force on the electron, we point our fingers in the direction of **v** and curl them into the magnetic field **B**. Our thumb points in the direction of the force on a positive charge. Thus the force on the electron is opposite to our thumb.
(*a*) Fingers out, curl down, thumb right, force left.
(*b*) Fingers down, curl back, thumb right, force left.
(*c*) Fingers in, curl right, thumb down, force up.
(*d*) Fingers right, curl up, thumb out, force in.
(*e*) Fingers left, but cannot curl into B, so force is zero.
(*f*) Fingers left, curl out, thumb up, force down.

8. We assume that we want the direction of **B** that produces the maximum force, i. e., perpendicular to **v**. Because the charge is positive, we point our thumb in the direction of **F** and our fingers in the direction of **v**. To find the direction of **B**, we note which way we should curl our fingers, which will be the direction of the magnetic field **B**.
   (*a*) Thumb out, fingers left, curl **down.**
   (*b*) Thumb up, fingers right, curl **in.**
   (*c*) Thumb down, fingers in, curl **right.**

9. To produce a circular path, the magnetic field is perpendicular to the velocity. The magnetic force provides the centripetal acceleration:
   $qvB = mv^2/r$, or
   $r = mv/qB$;
   $0.25 \text{ m} = (6.6 \times 10^{-27} \text{ kg})(1.6 \times 10^{7} \text{ m/s})/2(1.60 \times 10^{-19} \text{ C})B$, which gives $B =$ **1.3 T.**

10. The greatest force will be produced when the velocity and the magnetic field are perpendicular. We point our thumb down (a negative charge!), and our fingers south. We must curl our fingers to the west, which will be the direction of the magnetic field. We find the magnitude from
    $F = qvB$;
    $2.2 \times 10^{-12} \text{ N} = (1.60 \times 10^{-19} \text{ C})(1.8 \times 10^{6} \text{ m/s})B$, which gives $B =$ **7.6 T west.**

11. The force is maximum when the current and field are perpendicular:
    $F_{max} = ILB$.
    When the current makes an angle θ with the field, the force is
    $F = ILB \sin \theta$.
    Thus we have
    $F/F_{max} = \sin \theta = 0.45$, or $\theta =$ **27°.**

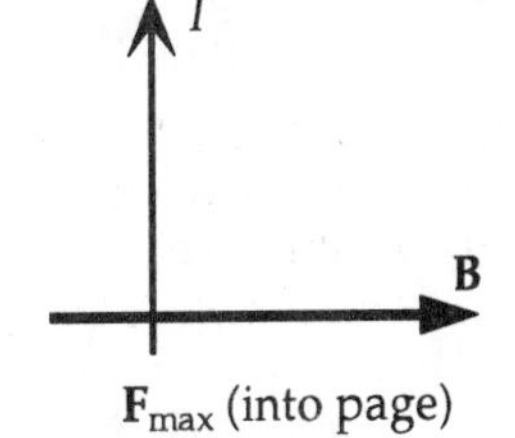

$\mathbf{F}_{max}$ (into page)

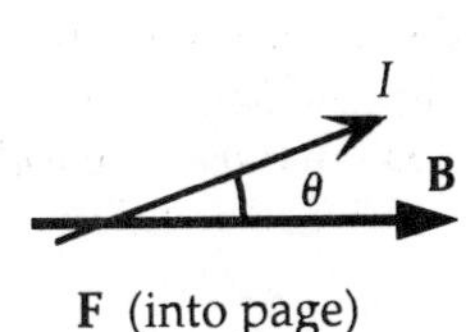

F (into page)

12. (*a*) We see from the diagram that the magnetic field is up, so the top pole face is a **south pole.**
    (*b*) We find the current from
    $F = ILB$;
    $5.30 \text{ N} = I(0.10 \text{ m})(0.15 \text{ T})$, which gives $I =$ **3.5×10^{2} A.**
    (*c*) The new force is
    $F' = ILB \sin \theta = F \sin \theta = (5.30 \text{ N}) \sin 80° =$ **5.22 N.**
    Note that the wire could be tipped either way.

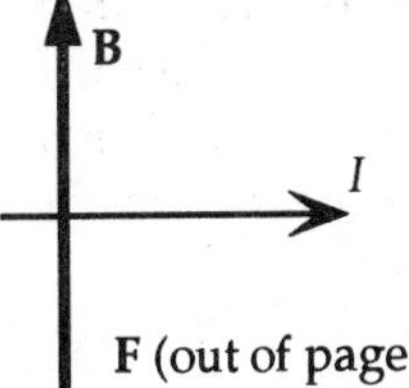

F (out of page)

13. The magnetic force provides the centripetal acceleration:
    $qvB = mv^2/r$, or $mv = qBr$.
    The kinetic energy of the electron is
    $\text{KE} = \frac{1}{2}mv^2 = \frac{1}{2}(qBr)^2/m$
    $= \frac{1}{2}[(1.60 \times 10^{-19} \text{ C})(1.15 \text{ T})(8.40 \times 10^{-3} \text{ m})]^2/(1.67 \times 10^{-27} \text{ kg})$
    $= 7.15 \times 10^{-16} \text{ J} = (7.15 \times 10^{-16} \text{ J})/(1.60 \times 10^{-19} \text{ J/eV}) = 4.47 \times 10^{3} \text{ eV} =$ **4.47 keV.**

14. The magnetic force provides the centripetal acceleration:
    $qvB = mv^2/r$, or $mv = qBr$.
    The kinetic energy of the electron is
    $\text{KE} = \frac{1}{2}mv^2 = \frac{1}{2}(qBr)^2/m = (q^2B^2/2m)r^2$.

15. The magnetic force provides the centripetal acceleration:
$qvB = mv^2/r$, or $mv = p = qBr$.

16. The magnetic force provides the centripetal acceleration:
$qvB = mv^2/r$, or $mv = p = qBr$.
The angular momentum is
$L = mvr = qBr^2$.

17. We find the required acceleration from
$v^2 = v_0^2 + 2ax$;
$(30\text{ m/s})^2 = 0 + 2a(1.0\text{ m})$, which gives $a = 450\text{ m/s}^2$.
This acceleration is provided by the force from the magnetic field:
$F = ILB = ma$;
$I(0.20\text{ m})(1.7\text{ T}) = (1.5 \times 10^{-3}\text{ kg})(450\text{ m/s}^2)$, which gives $I =$ $\boxed{2.0\text{ A.}}$
The force is away from the battery, fingers in the direction of I would have to curl down; thus the field points $\boxed{\text{down.}}$

18. The magnetic force produces an acceleration perpendicular to the original motion:
$a_\perp = qvB/m = (8.10 \times 10^{-9}\text{ C})(180\text{ m/s})(5.00 \times 10^{-5}\text{ T})/(3.80 \times 10^{-3}\text{ kg}) = 1.92 \times 10^{-8}\text{ m/s}^2$.
The time the bullet takes to travel 1.00 km is
$t = L/v = (1.00 \times 10^3\text{ m})/(180\text{ m/s}) = 5.56\text{ s}$.
The small acceleration will produce a small deflection, so we assume the perpendicular acceleration is constant. We find the deflection of the electron from
$y = v_{0y}t + \frac{1}{2}at^2 = 0 + \frac{1}{2}(1.92 \times 10^{-8}\text{ m/s}^2)(5.56\text{ s})^2 =$ $\boxed{3.0 \times 10^{-7}\text{ m.}}$
This justifies our assumption of constant acceleration.

19. The magnetic field of a long wire depends on the distance from the wire:
$B = (\mu_0/4\pi)2I/r$
$= (10^{-7}\text{ T}\cdot\text{m/A})2(15\text{ A})/(0.15\text{ m}) =$ $\boxed{2.0 \times 10^{-5}\text{ T.}}$
When we compare this to the Earth's field, we get
$B/B_{\text{Earth}} = (2.0 \times 10^{-5}\text{ T})/(5.5 \times 10^{-5}\text{ T}) = 0.36 =$ $\boxed{36\%.}$

20. We find the current from
$B = (\mu_0/4\pi)2I/r$;
$5.5 \times 10^{-5}\text{ T} = (10^{-7}\text{ T}\cdot\text{m/A})2I/(0.30\text{ m})$, which gives $I =$ $\boxed{83\text{ A.}}$

21. The two currents in the same direction will be attracted with a force of
$F = I_1(\mu_0 I_2/2\pi d)L = \mu_0 I_1 I_2 L/2\pi d$
$= (4\pi \times 10^{-7}\text{ T}\cdot\text{m/A})(35\text{ A})(35\text{ A})(45\text{ m})/2\pi(0.060\text{ m}) =$ $\boxed{0.18\text{ N attraction.}}$

22. Because the force is attractive, the second current must be in the same direction as the first. We find the magnitude from
$F/L = \mu_0 I_1 I_2/2\pi d$
$8.8 \times 10^{-4}\text{ N/m} = (4\pi \times 10^{-7}\text{ T}\cdot\text{m/A})(12\text{ A})I_2/2\pi(0.070\text{ m})$, which gives $I_2 =$ $\boxed{26\text{ A upward.}}$

23. The magnetic field produced by the wire must be less than 1% of the magnetic field of the Earth. We find the current from

$B = (\mu_0/4\pi)2I/r;$

$0.01(5.5 \times 10^{-5}\ \text{T}) > (10^{-7}\ \text{T}\cdot\text{m/A})2I/(1.0\ \text{m})$, which gives $I <$ **3 A.**

24. The magnetic field produced by the wire is

$B = (\mu_0/4\pi)2I/r = (10^{-7}\ \text{T}\cdot\text{m/A})2(30\ \text{A})/(0.086\ \text{m}) = 6.98 \times 10^{-5}\ \text{T}$,

and will be perpendicular to the motion of the airplane.

We find the acceleration produced by the magnetic force from

$F = qvB = ma;$

$(18.0\ \text{C})(1.8\ \text{m/s})(6.98 \times 10^{-5}\ \text{T}) = (175 \times 10^{-3}\ \text{kg})a$, which gives $a = 1.29 \times 10^{-2}\ \text{m/s}^2 =$ **$1.3 \times 10^{-3}\,g$.**

25. The magnetic field from the wire at a point south of a downward current will be to the west, with a magnitude:

$B_{\text{wire}} = (\mu_0/4\pi)2I/r$

$= (10^{-7}\ \text{T}\cdot\text{m/A})2(30\ \text{A})/(0.20\ \text{m}) = 3.0 \times 10^{-5}\ \text{T}.$

Because this is perpendicular to the Earth's field, we find the direction of the resultant field, and thus of the compass needle, from

$\tan\theta = B_{\text{wire}}/B_{\text{Earth}} = (3.0 \times 10^{-5}\ \text{T})/(0.45 \times 10^{-4}\ \text{T}) = 0.667$, or

$\theta = 34°$ W of N.

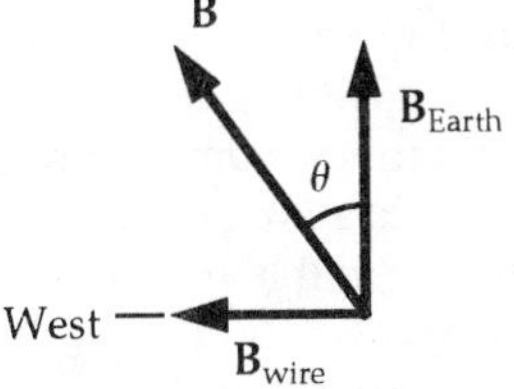

26. The magnetic field to the west of a wire with a current to the north will be up, with a magnitude:

$B_{\text{wire}} = (\mu_0/4\pi)2I/r$

$= (10^{-7}\ \text{T}\cdot\text{m/A})2(12.0\ \text{A})/(0.200\ \text{m}) = 1.20 \times 10^{-5}\ \text{T}.$

The net downward field is

$B_{\text{down}} = B_{\text{Earth}} \sin 40° - B_{\text{wire}} = (5.0 \times 10^{-5}\ \text{T}) \sin 40° - 1.20 \times 10^{-5}\ \text{T} = 2.01 \times 10^{-5}\ \text{T}.$

The northern component is $B_{\text{north}} = B_{\text{Earth}} \cos 40° = 3.83 \times 10^{-5}\ \text{T}.$

We find the magnitude from

$B = [(B_{\text{down}})^2 + (B_{\text{north}})^2]^{1/2} = [(2.01 \times 10^{-5}\ \text{T})^2 + (3.83 \times 10^{-5}\ \text{T})^2]^{1/2} =$ **4.3×10^{-5} T.**

We find the direction from

$\tan\theta = B_{\text{down}}/B_{\text{north}} = (2.01 \times 10^{-5}\ \text{T})/(3.83 \times 10^{-5}\ \text{T}) = 0.525$, or $\theta =$ **28° below the horizontal.**

27. Because a current represents the amount of charge that passes a given point, the effective current of the proton beam is

$I = \Delta q/\Delta t = (10^9\ \text{protons/s})(1.60 \times 10^{-19}\ \text{C/proton}) = 1.60 \times 10^{-10}\ \text{A}.$

The magnetic field from this current will be

$B = (\mu_0/4\pi)2I/r$

$= (10^{-7}\ \text{T}\cdot\text{m/A})2(1.60 \times 10^{-10}\ \text{A})/(2.0\ \text{m}) =$ **1.6×10^{-17} T.**

28. (*a*) When the currents are in the same direction, the fields between the currents will be in opposite directions, so at the midpoint we have

$B_a = B_2 - B_1 = [(\mu_0/4\pi)2I_2/r] - [(\mu_0/4\pi)2I/r] = [(\mu_0/4\pi)2/r](I_2 - I)$

$= (10^{-7}\ \text{T}\cdot\text{m/A})2/(0.010\ \text{m})(15\ \text{A} - I)$

$=$ **$(2.0 \times 10^{-5}\ \text{T/A})(15\ \text{A} - I)$ up,** with the currents as shown.

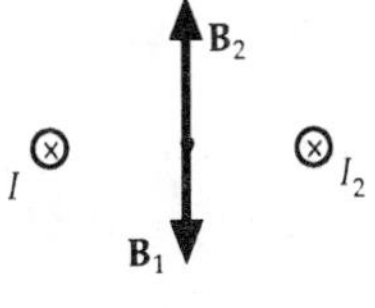

(*b*) When the currents are in opposite directions, the fields between the currents will be in the same direction, so at the midpoint we have

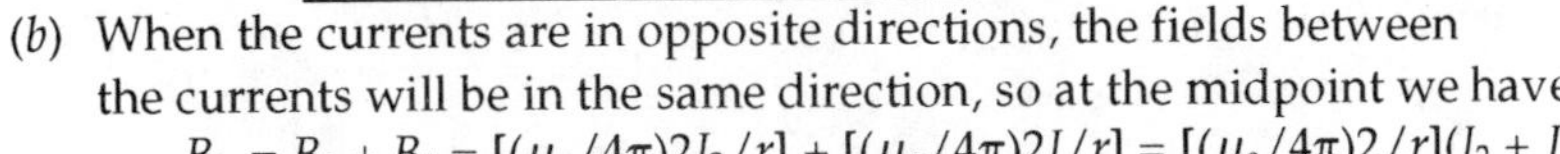

$B_b = B_2 + B_1 = [(\mu_0/4\pi)2I_2/r] + [(\mu_0/4\pi)2I/r] = [(\mu_0/4\pi)2/r](I_2 + I)$

$= (10^{-7}\ \text{T}\cdot\text{m/A})2/(0.010\ \text{m})(15\ \text{A} + I)$

$=$ **$(2.0 \times 10^{-5}\ \text{T/A})(15\ \text{A} + I)$ down,** with the currents as shown.

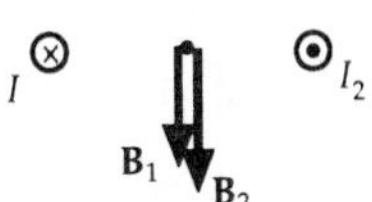

29. Because the currents are in opposite directions, the fields will be in opposite directions. For the net field we have

$$B = B_1 - B_2 = [(\mu_0/4\pi)2I_1/r_1] - [(\mu_0/4\pi)2I_2/r_2]$$
$$= [(\mu_0/4\pi)2I]\{[1/(L - \tfrac{1}{2}d)] - [1/(L + \tfrac{1}{2}d)]\}$$
$$= [(\mu_0/4\pi)2I/L]\{[1/(1 - \tfrac{1}{2}d/L)] - [1/(1 + \tfrac{1}{2}d/L)]\}.$$

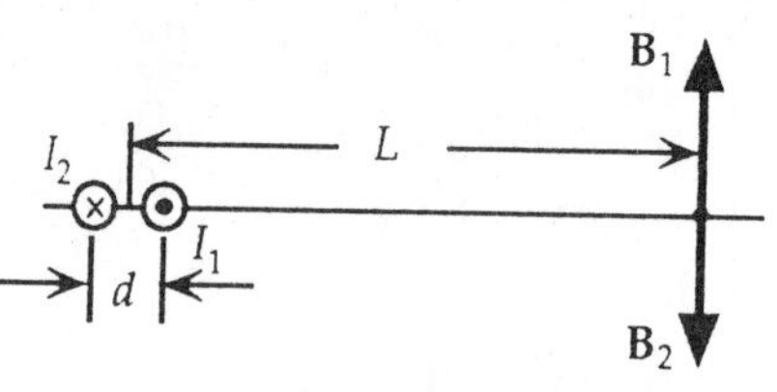

Because $d \ll L$, we can use the approximation $1/(1 \pm x) \approx 1 \mp x$:

$$B = [(\mu_0/4\pi)2I/L][(1 + \tfrac{1}{2}d/L) - (1 - \tfrac{1}{2}d/L)]$$
$$= [(\mu_0/4\pi)2I/L](d/L) = (\mu_0/4\pi)2Id/L^2$$
$$= [(10^{-7}\ \mathrm{T \cdot m/A})2(25\ \mathrm{A})(2.0 \times 10^{-3}\ \mathrm{m})/(0.100\ \mathrm{m})^2$$
$$= \boxed{1.0 \times 10^{-6}\ \mathrm{T\ up,}}$$ with the currents as shown.

This is

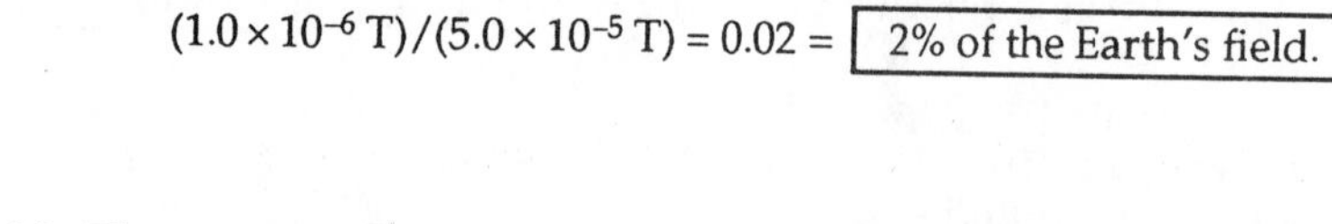

$$(1.0 \times 10^{-6}\ \mathrm{T})/(5.0 \times 10^{-5}\ \mathrm{T}) = 0.02 = \boxed{2\%\ \text{of the Earth's field.}}$$

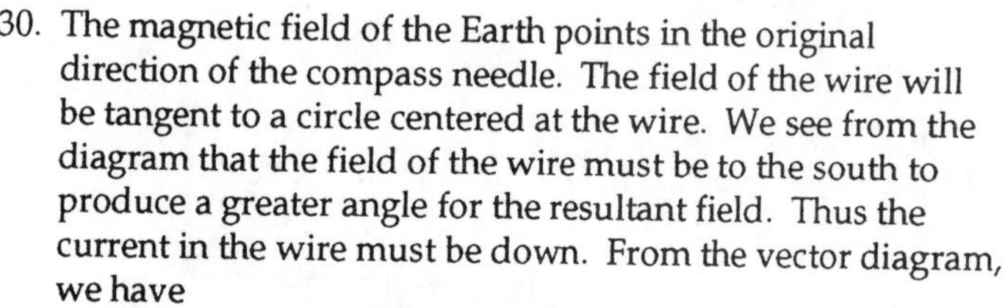

30. The magnetic field of the Earth points in the original direction of the compass needle. The field of the wire will be tangent to a circle centered at the wire. We see from the diagram that the field of the wire must be to the south to produce a greater angle for the resultant field. Thus the current in the wire must be down. From the vector diagram, we have

$$B \sin\theta_2 = B_{\mathrm{Earth}} \sin\theta_1;$$
$$B \cos\theta_2 = B_{\mathrm{Earth}} \cos\theta_1 - B_{\mathrm{wire}}.$$

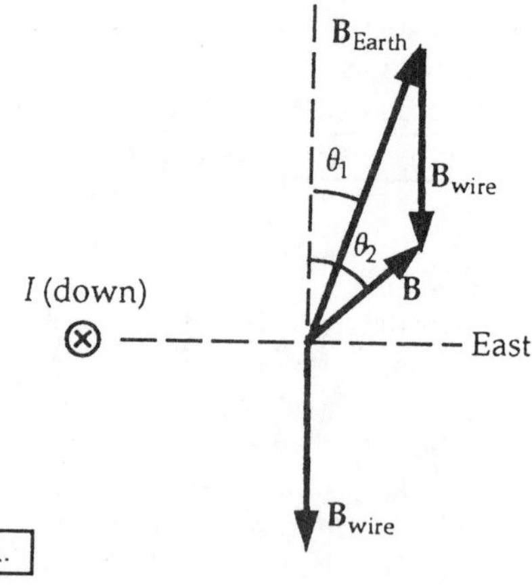

When we divide the two equations, we get

$$\tan\theta_2 = (B_{\mathrm{Earth}} \sin\theta_1)/(B_{\mathrm{Earth}} \cos\theta_1 - B_{\mathrm{wire}});$$
$$\tan 55° = [(0.50 \times 10^{-4}\ \mathrm{T}) \sin 20°]/[(0.50 \times 10^{-4}\ \mathrm{T}) \cos 20° - B_{\mathrm{wire}}],$$

which gives $B_{\mathrm{wire}} = 3.50 \times 10^{-4}$ T.

We find the current from

$$B_{\mathrm{wire}} = (\mu_0/4\pi)2I/r;$$
$$3.50 \times 10^{-4}\ \mathrm{T} = (10^{-7}\ \mathrm{T \cdot m/A})2I/(0.080\ \mathrm{m}),\ \text{which gives}\ I = \boxed{14\ \mathrm{A.}}$$

31. Because the currents and the separations are the same, we find the force per unit length between any two wires from

$$F/L = I_1(\mu_0 I_2/2\pi d) = \mu_0 I^2/2\pi d$$
$$= (4\pi \times 10^{-7}\ \mathrm{T \cdot m/A})(8.0\ \mathrm{A})^2/2\pi(0.380\ \mathrm{m})$$
$$= 3.37 \times 10^{-5}\ \mathrm{N/m}.$$

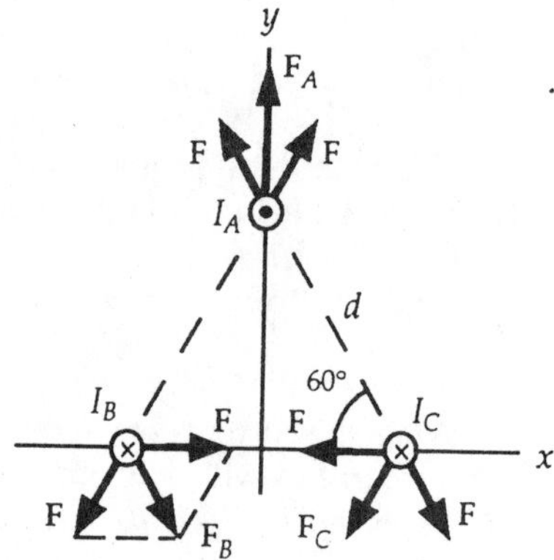

The directions of the forces are shown on the diagram. The symmetry of the force diagrams simplifies the vector addition, so we have

$$F_A/L = 2(F/L) \cos 30°$$
$$= 2(3.37 \times 10^{-5}\ \mathrm{N/m}) \cos 30° = \boxed{5.8 \times 10^{-5}\ \text{up.}}$$
$$F_B/L = F/L$$
$$= \boxed{3.4 \times 10^{-5}\ \mathrm{N/m}\ 60°\ \text{below the line toward}\ C.}$$
$$F_C/L = F/L$$
$$= \boxed{3.4 \times 10^{-5}\ \mathrm{N/m}\ 60°\ \text{below the line toward}\ B.}$$

32. The Coulomb force between the charges provides the centripetal acceleration:
$ke^2/r^2 = mv^2/r$, which gives $v = (ke^2/mr)^{1/2}$.
The period of the electron's orbit is
$T = 2\pi r/v = 2\pi r/(ke^2/mr)^{1/2} = 2\pi(mr^3/ke^2)^{1/2}$
$= 2\pi[(9.11 \times 10^{-31}\text{ kg})(5.3 \times 10^{-11}\text{ m})^3/(9.0 \times 10^9\text{ N}\cdot\text{m}^2/\text{C}^2)(1.60 \times 10^{-19}\text{ C})^2]^{1/2}$
$= 1.52 \times 10^{-16}$ s.
Thus the effective current of the electron is
$I = e/T = (1.60 \times 10^{-19}\text{ C})/(1.52 \times 10^{-16}\text{ s}) = 1.05 \times 10^{-3}$ A.
The magnitude of the magnetic field is
$B = (\mu_0/2\pi)I/r = (\mu_0/4\pi)2I/r$
$= (10^{-7}\text{ T}\cdot\text{m/A})2(1.05 \times 10^{-3}\text{ A})/(5.3 \times 10^{-11}\text{ m}) =$ **12 T.**

33. We find the direction of the field for each wire from the tangent to the circle around the wire, as shown.
For their magnitudes, we have

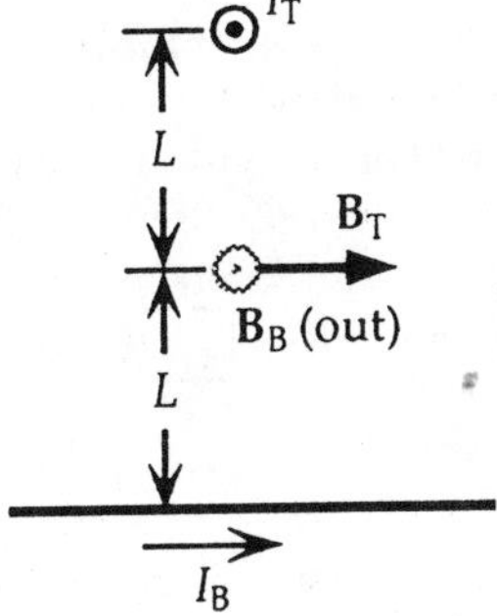

$B_T = (\mu_0/4\pi)2I_T/L$
$= (10^{-7}\text{ T}\cdot\text{m/A})2(20.0\text{ A})/(0.100\text{ m}) = 4.00 \times 10^{-5}$ T.
$B_B = (\mu_0/4\pi)2I_B/L$
$= (10^{-7}\text{ T}\cdot\text{m/A})2(5.0\text{ A})/(0.100\text{ m}) = 1.00 \times 10^{-5}$ T.
Because the fields are perpendicular, we find the magnitude from
$B = (B_T^2 + B_B^2)^{1/2}$
$= [(4.00 \times 10^{-5}\text{ T})^2 + (1.00 \times 10^{-5}\text{ T})^2]^{1/2} =$ **4.1×10^{-5} T.**

34. (*a*) For the force produced by the magnetic field of the upper wire to balance the weight, it must be up, i. e., an attractive force. Thus the currents must be in the same direction. When we equate the magnitudes of the two forces for a length *L*, we get

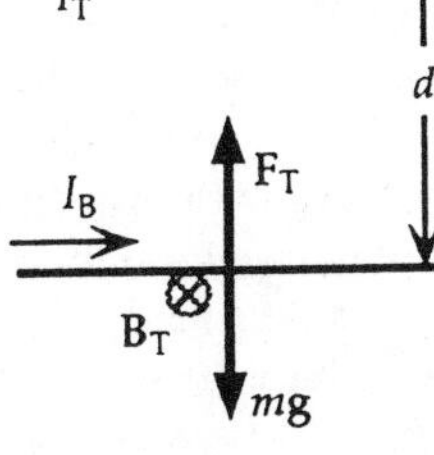

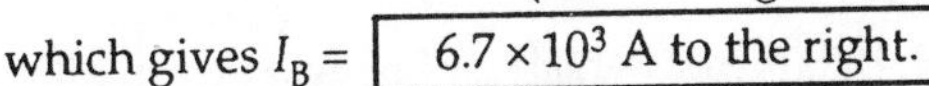

$I_B B_T L = mg$;
$(\mu_0/4\pi)2I_B I_T L/d = \rho(\pi r^2 L)g$;
$(10^{-7}\text{ T}\cdot\text{m/A})2I_B(48\text{ A})/(0.15\text{ m}) =$
$(8.9 \times 10^3\text{ kg/m}^3)\pi(1.25 \times 10^{-3}\text{ m})^2(9.80\text{ m/s}^2)$,
which gives $I_B =$ **6.7×10^3 A to the right.**

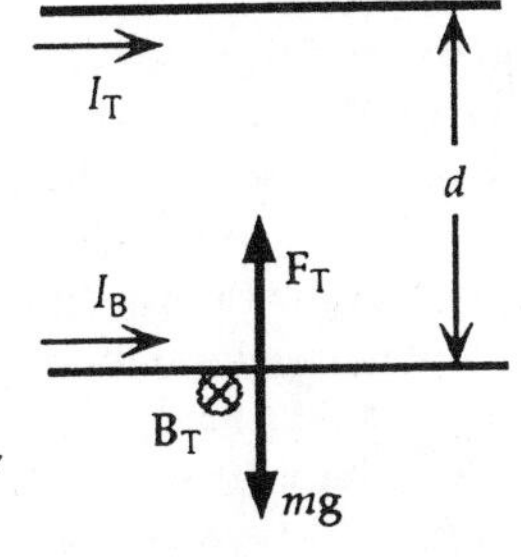

(*b*) The magnetic force will decrease with increasing separation. If the wire is moved a small distance above or below the equilibrium position, there will be a net force, away from equilibrium, and the wire will be **unstable.**

(*c*) If the second wire is above the first, there must be a repulsive magnetic force between the two wires to balance the weight, which means the currents must be opposite. Because the separation is the same, the magnitude of the current is the same:
$I_2 =$ **6.7×10^3 A to the left.**
The magnetic force will decrease with increasing separation. If the wire is moved a small distance above or below the equilibrium position, there will be a net force back toward equilibrium, and the wire will be **stable for vertical displacements.**

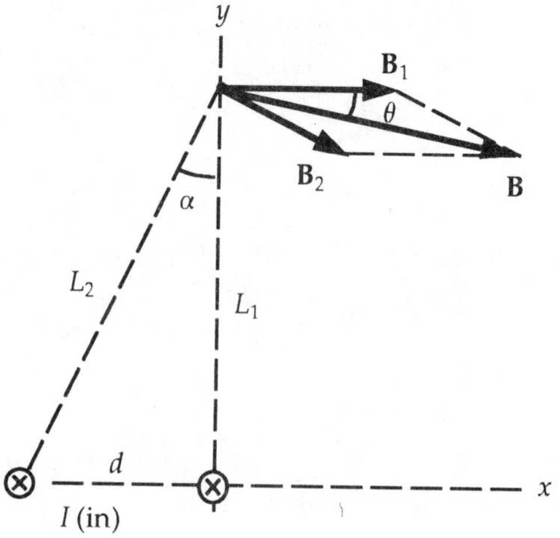

35. Because $(12)^2 + (5)^2 = (13)^2$, we have a right triangle, so
$\tan \alpha = d/L_1 = (5.00 \text{ cm})/(12.0 \text{ cm}) = 0.417, \ \alpha = 22.6°.$
We find the direction of the field for each wire from the tangent to the circle around the wire, as shown.
For their magnitudes, we have
$B_1 = (\mu_0/4\pi)2I/L_1$
$= (10^{-7} \text{ T}\cdot\text{m/A})2(16.5 \text{ A})/(0.120 \text{ m}) = 2.75 \times 10^{-5} \text{ T}.$
$B_2 = (\mu_0/4\pi)2I_B/L_2$
$= (10^{-7} \text{ T}\cdot\text{m/A})2(16.5 \text{ A})/(0.130 \text{ m}) = 2.54 \times 10^{-5} \text{ T}.$
From the vector diagram, we have
$B_x = B_1 + B_2 \cos \alpha$
$= 2.75 \times 10^{-5} \text{ T} + (2.54 \times 10^{-5} \text{ T}) \cos 22.6°$
$= 5.09 \times 10^{-5} \text{ T}.$
$B_y = -B_2 \cos \alpha;$
$= -(2.54 \times 10^{-5} \text{ T}) \cos 22.6° = -0.976 \times 10^{-5} \text{ T}.$
For the direction of the field, we have
$\tan \theta = B_y/B_x = (0.976 \times 10^{-5} \text{ T})/(5.09 \times 10^{-5} \text{ T}) = 0.192, \ \theta = 10.9°.$
We find the magnitude from
$B_x = B \cos \theta;$
$5.09 \times 10^{-5} \text{ T} = B \cos 10.9°,$
which gives $B =$ 5.18×10^{-5} T 10.9° below the plane parallel to the two wires.

36. We find the current in the solenoid from
$B = \mu_0 nI = \mu_0 NI/L;$
$0.385 \text{ T} = (4\pi \times 10^{-7} \text{ T}\cdot\text{m/A})[(1000 \text{ turns})/(0.300 \text{ m})]I$, which gives $I =$ 91.9 A.

37. The mass, and thus the volume, of the wire is fixed, so we have $\pi r^2 L = k$; a smaller radius will give a greater length. If we assume a given current (a variable voltage supply), the magnetic field of the solenoid will be determined by the density of turns: $B = \mu_0 nI$. The greatest density will be when the wires are closely wound. In this case, the separation of turns is $2r$, so the density of turns is $1/2r$, which would indicate that the radius should be very small.
If D is the diameter of the solenoid, the number of turns is
$N = L/\pi D,$
so the length of the solenoid is $N2r = 2Lr/\pi D = 2k/\pi^2 rD$.
The length of the solenoid must be much greater than the diameter, which will be true for small r, as long as the diameter is not large, which is the only restriction on the diameter. These considerations indicate that a long and thin wire should be used. However, we must be concerned with the resistance of the wire, because the thermal power generation, I^2R, must be dissipated in the solenoid. The resistance is
$R = \rho L/\pi r^2 = \rho k/\pi r^4.$
Thus a very thin wire will create thermal dissipation problems, which means that the insulation and/or wire could melt. Thus the wire should be something between long, thin and short, fat.

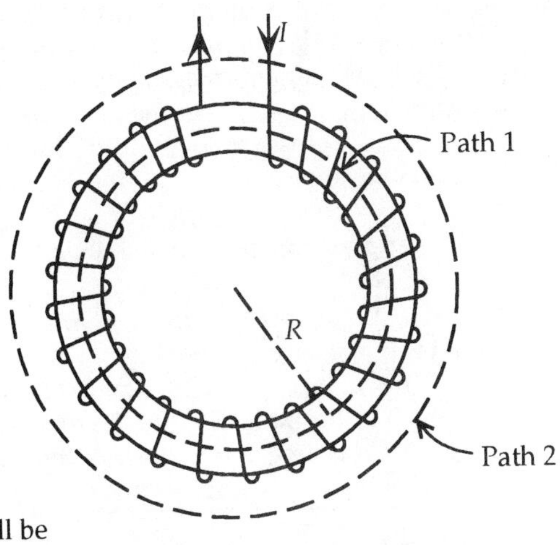

38. (*a*) Each loop will produce a field along its axis. For path 1, the symmetry means that the magnetic field will have the same magnitude anywhere on the path and be circular, so that B is parallel to the path. We apply Ampere's law to path 1, which is a circle with radius R. One side of every turn of the coil passes through the area enclosed by this path, so we have

$$\Sigma B_{\parallel} \Delta \ell = \mu_0 \Sigma I.$$

Because $B_{\parallel} = B$ is the same for all segments of the path, and each turn has the same current, we get

$$B \Sigma \Delta \ell = B(2\pi R) = \mu_0 NI, \quad \text{or} \quad B = \mu_0 NI/2\pi R.$$

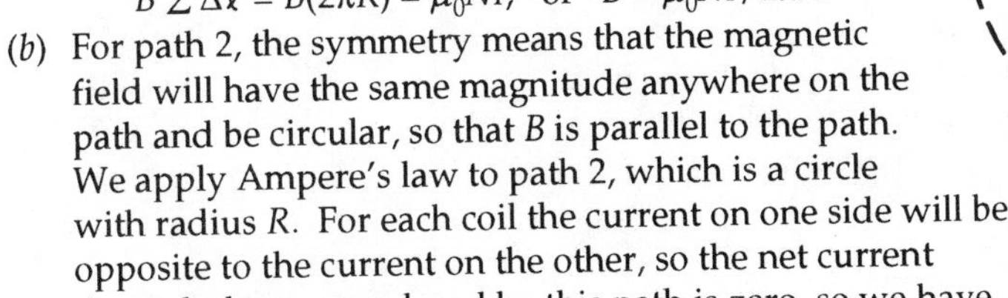

(*b*) For path 2, the symmetry means that the magnetic field will have the same magnitude anywhere on the path and be circular, so that B is parallel to the path. We apply Ampere's law to path 2, which is a circle with radius R. For each coil the current on one side will be opposite to the current on the other, so the net current through the area enclosed by this path is zero, so we have

$$\Sigma B_{\parallel} \Delta \ell = \mu_0 \Sigma I;$$

$$B \Sigma \Delta \ell = B(2\pi R) = 0, \quad \text{or} \quad B = 0.$$

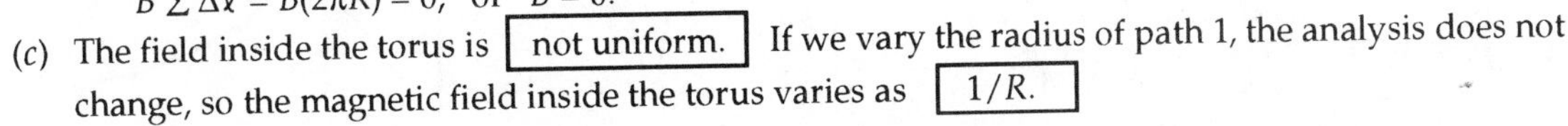

(*c*) The field inside the torus is **not uniform.** If we vary the radius of path 1, the analysis does not change, so the magnetic field inside the torus varies as **$1/R$.**

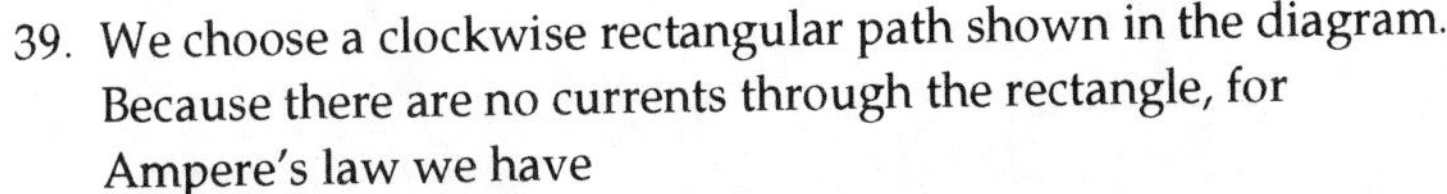

39. We choose a clockwise rectangular path shown in the diagram. Because there are no currents through the rectangle, for Ampere's law we have

$$\Sigma B_{\parallel} \Delta \ell = \mu_0 \Sigma I = 0.$$

The sum on the left-hand side consists of four parts:

$$(\Sigma B_{\parallel} \Delta \ell)_{\text{left}} + (\Sigma B_{\parallel} \Delta \ell)_{\text{top}} + (\Sigma B_{\parallel} \Delta \ell)_{\text{right}} + (\Sigma B_{\parallel} \Delta \ell)_{\text{bottom}} = 0.$$

The field on the left side is constant and parallel to the path; the field on the right is zero. Thus we have

$$Bh + (\Sigma B_{\parallel} \Delta \ell)_{\text{top}} + 0 + (\Sigma B_{\parallel} \Delta \ell)_{\text{bottom}} = 0, \quad \text{or}$$

$$Bh = -[(\Sigma B_{\parallel} \Delta \ell)_{\text{top}} + (\Sigma B_{\parallel} \Delta \ell)_{\text{bottom}}].$$

Thus there must be a component of B parallel to the paths on top and bottom, as shown, so there must be a fringing field.

Note that the contributions to the sum from the top and bottom have the same sign.

40. (*a*) We choose a circular path with radius r, centered on the axis of the cylinder, so the symmetry means that the magnetic field will have the same magnitude anywhere on the path and be circular; B is parallel to the path. Because the current is uniform across the cross section, we find the current through the path from the area:

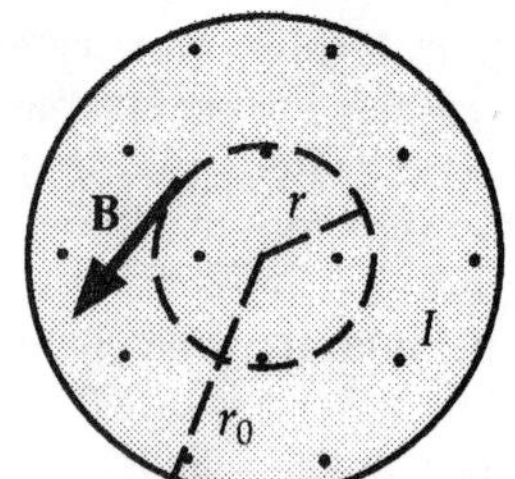

$I' = (\pi r^2/\pi r_0^2)I = (r^2/r_0^2)I.$

We apply Ampere's law to the path:

$\Sigma B_{\parallel} \Delta \ell = \mu_0 \Sigma I.$

$B \Sigma \Delta \ell = B(2\pi r) = \mu_0(r^2/r_0^2)I$, or $B = \mu_0 Ir/2\pi r_0^2.$

(*b*) At the surface of the wire, $r = r_0$, so we have

$B = \mu_0 Ir/2\pi r_0^2 = \mu_0 Ir_0/2\pi r_0^2 = \mu_0 I/2\pi r_0,$

which is the expression for the magnetic field outside the wire.

(*c*) The field is zero at the center of the wire, and outside the wire it decreases with distance, so the maximum is

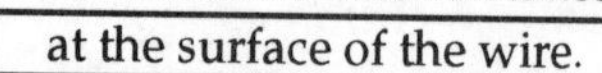

For the given data, we have

$B_{max} = \mu_0 I/2\pi r_0 = (\mu_0/4\pi)2I/r_0$

$= (10^{-7}\ \text{T}\cdot\text{m/A})2(15.0\ \text{A})/(0.50 \times 10^{-3}\ \text{m})$

$= \boxed{6.0 \times 10^{-3}\ \text{T.}}$

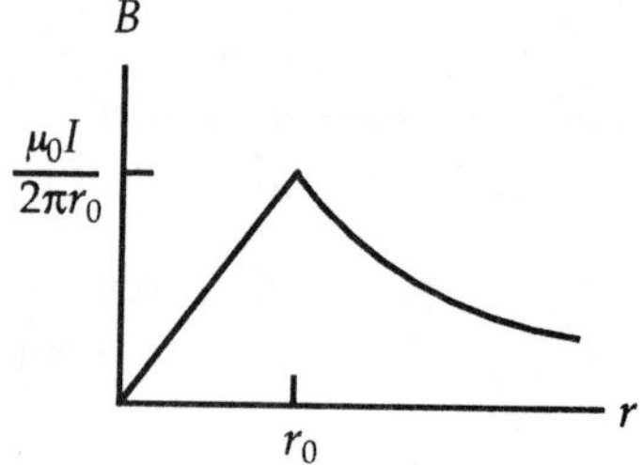

(*d*) Inside the wire we have

$B/B_{max} = r/r_0 = 0.10$, or $\boxed{r = 0.10r_0.}$

Outside the wire we have

$B/B_{max} = r_0/r = 0.10$, or $\boxed{r = 10r_0.}$

41. When the coil comes to rest, the magnetic torque is balanced by the restoring torque:

$NIAB = k\phi.$

Because the deflection is the same, we have

$I_1B_1 = I_2B_2;$

$(63.0\ \mu\text{A})B_1 = I_2(0.860B_1)$, which gives $I_2 = \boxed{73.3\ \mu\text{A.}}$

42. When the coil comes to rest, the magnetic torque is balanced by the restoring torque:

$NIAB = k\phi.$

Because the deflection is the same, we have

$I_1/k_1 = I_2/k_2;$

$(36\ \mu\text{A})/k_1 = I_2/(0.80k_1)$, which gives $I_2 = \boxed{29\ \mu\text{A.}}$

43. If we assume that the magnetic field is constant, we have

$\tau_2/\tau_1 = I_2/I_1 = 0.85,$

so the torque will decrease by 15%.

If we assume that the magnetic field is produced by the current, it will be proportional to the current and will also decrease by 15%, so we have

$\tau_2/\tau_1 = (I_2/I_1)(B_2/B_1) = (0.85)(0.85) = 0.72,$

so the torque will $\boxed{\text{decrease by 28\%.}}$

44. When the coil is parallel to the magnetic field, the torque is maximum, so we have

$\tau = NIAB;$

$0.325\ \text{m}\cdot\text{N} = (1)(6.30\ \text{A})(0.220)^2B$, which gives $B = \boxed{1.07\ \text{T.}}$

45. The angular momentum of the electron for the circular orbit is
$L = mvr.$
The time for the electron to go once around the orbit is
$T = 2\pi r/v,$
so the effective current is
$I = e/T = ev/2\pi r.$
The magnetic dipole moment is
$M = IA = (ev/2\pi r)\pi r^2 = evr/2 = (e/2m)L.$

46. (*a*) The angle between the normal to the coil and the field is 34.0°, so the torque is
$\tau = NIAB \sin\theta$
$= (11)(7.70 \text{ A})\pi(0.090 \text{ m})^2(5.50 \text{ T}) \sin 34.0°$
$= \boxed{6.63 \times 10^{-5} \text{ m}\cdot\text{N}.}$
(*b*) From the directions of the forces shown on the diagram, the $\boxed{\text{south edge}}$ of the coil will rise.

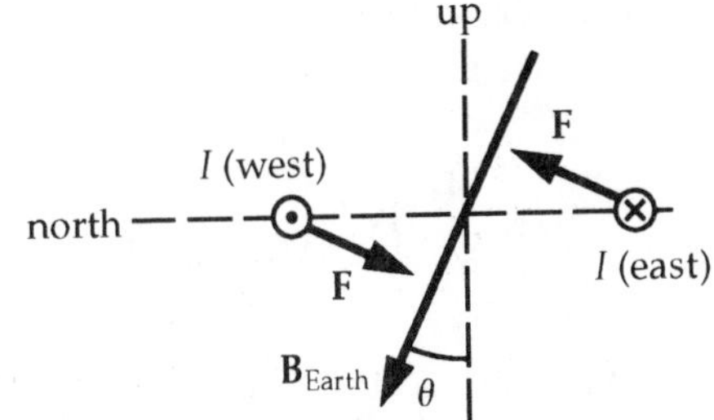

47. (*a*) The Hall emf is across the width of the sample, so the Hall field is
$E_H = \mathcal{E}_H/w = (6.5 \times 10^{-6} \text{ V})/(0.030 \text{ m}) = \boxed{2.2 \times 10^{-4} \text{ V/m}.}$
(*b*) The forces from the electric field and the magnetic field balance. We find the drift speed from
$E_H = v_d B;$
$2.2 \times 10^{-4} \text{ V/m} = v_d(0.80 \text{ T})$, which gives $v_d = \boxed{2.7 \times 10^{-4} \text{ m/s}.}$
(*c*) We find the density from
$I = neAv_d;$
$30 \text{ A} = n(1.60 \times 10^{-19} \text{ C})(0.030 \text{ m})(500 \times 10^{-6} \text{ m})(2.7 \times 10^{-4} \text{ m/s}),$
which gives $n = \boxed{4.6 \times 10^{28} \text{ electrons/m}^3.}$

48. The Hall field is
$E_H = \mathcal{E}_H/w = (2.42 \times 10^{-6} \text{ V})/(0.015 \text{ m}) = 1.613 \times 10^{-4} \text{ V/m}.$
To determine the drift speed, we first find the density of free electrons:
$n = [(0.971)(1000 \text{ kg/m}^3)(10^3 \text{ g/kg})/(23 \text{ g/mol})](6.02 \times 10^{23} \text{ free electrons/mol})$
$= 2.543 \times 10^{28} \text{ electrons/m}^3.$
We find the drift speed from
$I = neAv_d;$
$12.0 \text{ A} = (2.543 \times 10^{28} \text{ electrons/m}^3)(1.60 \times 10^{-19} \text{ C})(0.015 \text{ m})(1.00 \times 10^{-3} \text{ m})(2.7 \times 10^{-4} \text{ m/s})v_d,$
which gives $v_d = 1.967 \times 10^{-4} \text{ m/s}.$
The forces from the electric field and the magnetic field balance, so we have
$E_H = v_d B;$
$1.613 \times 10^{-4} \text{ V/m} = (1.967 \times 10^{-4} \text{ m/s})B$, which gives $B = \boxed{0.820 \text{ T}.}$

49. (*a*) The sign of the ions will not change the magnitude of the Hall emf, but will $\boxed{\text{determine the polarity of the emf.}}$
(*b*) The forces from the electric field and the magnetic field balance. We find the flow velocity from
$\mathcal{E}_H = v_d Bw;$
$0.10 \times 10^{-3} \text{ V} = v_d(0.070 \text{ T})(3.3 \times 10^{-3} \text{ m})$, which gives $v_d = \boxed{0.43 \text{ m/s}.}$

50. For the circular motion, the magnetic force provides the centripetal acceleration:
$qvB = mv^2/r$, or $v = qBr/m$.
To make the path straight, the forces from the electric field and the magnetic field balance:
$qE = qvB = q(qBr/m)B$, or
$E = qB^2r/m = (1.60 \times 10^{-19}\text{ C})(0.566\text{ T})^2(0.0510\text{ m})/(1.67 \times 10^{-27}\text{ kg}) = \boxed{1.57 \times 10^6\text{ V/m.}}$

51. For the circular motion, the magnetic force provides the centripetal acceleration:
$qvB = mv^2/r$, or $r = mv/qB$.
The velocities are the same because of the velocity selector, so we have
$m/m_0 = r/r_0$;
$m_1/(76\text{ u}) = (21.0\text{ cm})/(22.8\text{ cm}) = \boxed{70\text{ u;}}$
$m_2/(76\text{ u}) = (21.6\text{ cm})/(22.8\text{ cm}) = \boxed{72\text{ u;}}$
$m_3/(76\text{ u}) = (21.9\text{ cm})/(22.8\text{ cm}) = \boxed{73\text{ u;}}$
$m_4/(76\text{ u}) = (22.2\text{ cm})/(22.8\text{ cm}) = \boxed{74\text{ u.}}$

52. We find the velocity of the velocity selector from
$v = E/B = (2.48 \times 10^4\text{ V/m})/(0.68\text{ T}) = 3.65 \times 10^4\text{ m/s}$.
For the radius of the path, we have
$r = mv/qB' = [(3.65 \times 10^4\text{ m/s})/(1.60 \times 10^{-19}\text{ C})(0.68\text{ T})]m = (3.35 \times 10^{23}\text{ m/kg})m$.
If we let A represent the mass number, we can write this as
$r = (3.35 \times 10^{23}\text{ m/kg})(1.67 \times 10^{-27}\text{ kg})A = (5.60 \times 10^{-4}\text{ m})A = (0.560\text{ mm})A$.
The separation of the lines is the difference in the diameter, or
$\Delta D = 2\,\Delta r = 2(0.560\text{ mm})\,\Delta A = (1.12\text{ mm})(1) = \boxed{1.12\text{ mm.}}$
If the ions were doubly charged, all radii would be reduced by one-half, so the separation would be $\boxed{0.56\text{ mm.}}$

53. (*a*) To make the path straight, the forces from the electric field and the magnetic field balance:
$qE = qvB$;
$10{,}000\text{ V/m} = (4.8 \times 10^6\text{ m/s})B$, which gives $B = \boxed{2.1 \times 10^{-3}\text{ T.}}$

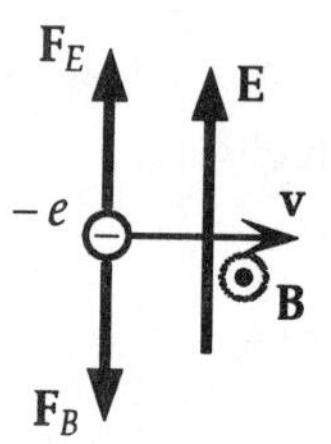

(*b*) Because the electric force is up, the magnetic force must be down, so the magnetic field is $\boxed{\text{out of the page.}}$

(*c*) If there is only the magnetic field, the radius of the circular orbit is
$r = mv/qB$.
The time to complete a circle is
$T = 2\pi r/v = 2\pi m/qB$, so the frequency is
$f = 1/T = qB/2\pi m = (1.60 \times 10^{-19}\text{ C})(2.08 \times 10^{-3}\text{ T})/2\pi(9.11 \times 10^{-31}\text{ kg}) = \boxed{5.8 \times 10^7\text{ Hz.}}$

F_E E $-e$ v B F_B

54. The velocity of the velocity selector is
$v = E/B$.
For the radius of the path, we have
$r = mv/qB' = mE/qB'B$, so
$r = km$, and $\Delta r = k\,\Delta m$.
If we form the ratio, we get
$\Delta m/m = \Delta r/r$;
$(28.0134\text{ u} - 28.0106\text{ u})/(28.0134\text{ u}) = \frac{1}{2}(0.50 \times 10^{-3}\text{ m})/r$, which gives $r = \boxed{2.5\text{ m.}}$

55. We find the speed acquired from the accelerating voltage from energy conservation:

$0 = \Delta\text{KE} + \Delta\text{PE};$

$0 = \frac{1}{2}mv^2 - 0 + q(-V)$, which gives $v = (2qV/m)^{1/2}$.

We combine this with the expression for the radius of the path:

$R = mv/qB = m(2qV/m)^{1/2}/qB$, or $m = qB^2R^2/2V$.

56. We find the permeability from

$B = \mu nI;$

$1.8\ \text{T} = \mu[(600\ \text{turns})/(0.36\ \text{m})](40\ \text{A})$, which gives $\mu = \boxed{2.7 \times 10^{-5}\ \text{T}\cdot\text{m/A}.}$

57. The magnetic force must be toward the center of the circular path, so the magnetic field must be up.

The magnetic force provides the centripetal acceleration:

$qvB = mv^2/r$, or $mv = qBr;$

$4.8 \times 10^{-16}\ \text{kg}\cdot\text{m/s} = (1.60 \times 10^{-19}\ \text{C})B(1.0 \times 10^3\ \text{m})$,

which gives $B = \boxed{3.0\ \text{T up.}}$

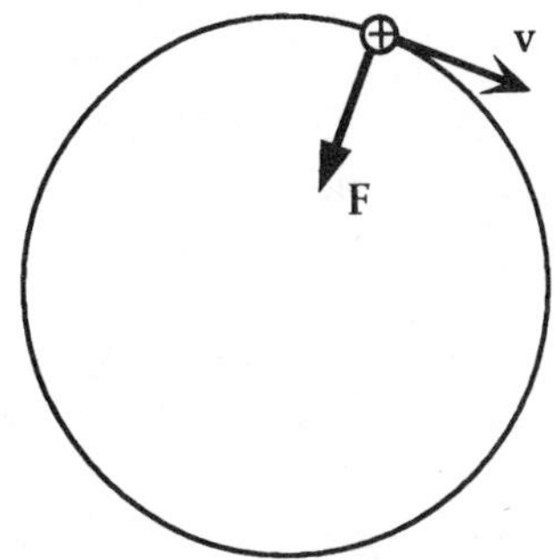

58. The magnetic field at the loop from the wire will be into the page, and will depend only on the distance from the wire r:

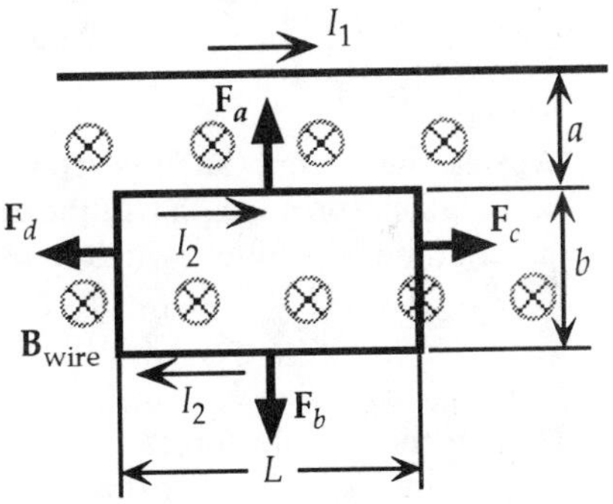

$B = (\mu_0/4\pi)2I_1/r,$

For the two vertical sides of the loop, the currents are in opposite directions so their forces will be in opposite directions. Because the current will be in the same average field, the magnitudes of the forces will be equal, so $\mathbf{F}_c + \mathbf{F}_d = 0$.

For the sum of the two forces on the top and bottom of the loop, we have

$$F_{\text{net}} = F_a - F_b = I_2B_aL - I_2B_bL$$
$$= I_2[(\mu_0/4\pi)2I_1/a]L - I_2[(\mu_0/4\pi)2I_1/(a + b)]L$$
$$= (\mu_0/4\pi)(2I_1I_2L)\{(1/a) - [1/(a + b)]\}$$
$$= (10^{-7}\ \text{T}\cdot\text{m/A})2(2.5\ \text{A})(2.5\ \text{A})(0.100\ \text{m})[(1/0.030\ \text{m}) - (1/0.080\ \text{m})]$$
$$= \boxed{2.6 \times 10^{-6}\ \text{N toward the wire.}}$$

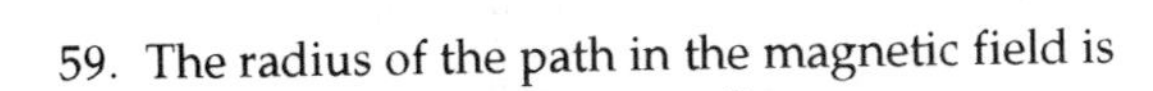

59. The radius of the path in the magnetic field is

$r = mv/eB$, or $mv = eBr$.

The kinetic energy is

$\text{KE} = \frac{1}{2}mv^2 = \frac{1}{2}(eBr)^2/m.$

If we form the ratio for the two particles, we have

$\text{KE}_p/\text{KE}_e = (r_p/r_e)^2(m_e/m_p);$

$1 = (r_p/r_e)^2[(9.11 \times 10^{-31}\ \text{kg})/(1.67 \times 10^{-27}\ \text{kg})]$, which gives $r_p/r_e = \boxed{42.8.}$

60. The magnetic force on the electron must be up, so the velocity must be toward the west.

For the balanced forces, we have

$mg = qvB;$

$(9.11 \times 10^{-31}\ \text{kg})(9.80\ \text{m/s}^2) = (1.60 \times 10^{-19}\ \text{C})v(0.50 \times 10^{-4}\ \text{T})$,

which gives $v = \boxed{1.1 \times 10^{-6}\ \text{m/s west.}}$

61. The force on the airplane is
$F = qvB = (155\ \text{C})(120\ \text{m/s})(5.0 \times 10^{-5}\ \text{T}) =$ $\boxed{0.93\ \text{N.}}$

62. Even though the Earth's field dips, the current and the field are perpendicular. The direction of the force will be perpendicular to both the cable and the Earth's field, so it will be 68° above the horizontal toward the north. For the magnitude, we have
$F = ILB$
$= (330\ \text{A})(10\ \text{m})(5.0 \times 10^{-5}\ \text{T}) =$ $\boxed{0.17\ \text{N}\ 68° \text{ above the horizontal toward the north.}}$

63. (*a*) We find the speed acquired from the accelerating voltage from energy conservation:
$0 = \Delta\text{KE} + \Delta\text{PE};$
$0 = \frac{1}{2}mv^2 - 0 + q(-V)$, which gives
$v = (2qV/m)^{1/2} = [2(2)(1.60 \times 10^{-19}\ \text{C})(2400\ \text{V})/(6.6 \times 10^{-27}\ \text{kg})]^{1/2} = 4.82 \times 10^5\ \text{m/s}.$
For the radius of the path, we have
$r = mv/qB = (6.6 \times 10^{-27}\ \text{kg})(4.82 \times 10^5\ \text{m/s})/(2)(1.60 \times 10^{-19}\ \text{C})(0.240\ \text{T})$
$= 4.1 \times 10^{-2}\ \text{m} =$ $\boxed{4.1\ \text{cm.}}$
(*b*) The period of revolution is
$T = 2\pi r/v = 2\pi m/qB = 2\pi(6.6 \times 10^{-27}\ \text{kg})/(2)(1.60 \times 10^{-19}\ \text{C})(0.240\ \text{T}) =$ $\boxed{5.4 \times 10^{-7}\ \text{s.}}$

64. If we consider a length L of the wire, for the balanced forces, we have
$mg = \rho\pi r^2Lg = ILB;$
$(8.9 \times 10^3\ \text{kg/m}^3)\pi(0.500 \times 10^{-3}\ \text{m})^2(9.80\ \text{m/s}^2) = I(5.00 \times 10^{-5}\ \text{T}),$
which gives $I =$ $\boxed{1.37 \times 10^3\ \text{A.}}$

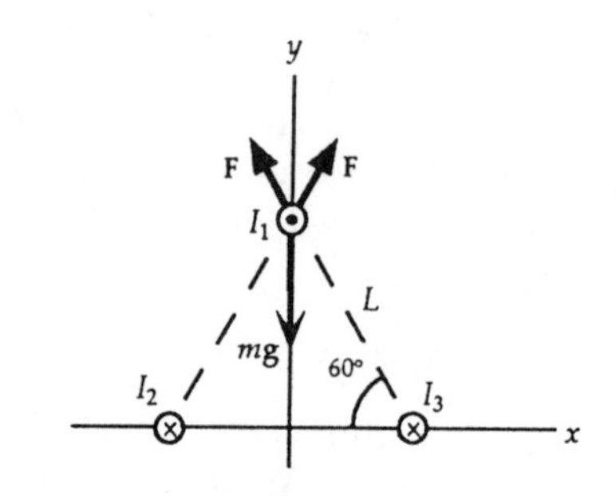

65. Because the currents and the separations are the same, we find the force on a length L of the top wire from either of the two bottom wires from
$F = I_1(\mu_0 I_2/2\pi d)L = \mu_0 I_1 I_2 L/2\pi d$
$= (4\pi \times 10^{-7}\ \text{T}\cdot\text{m/A})(20.0\ \text{A})I_1L/2\pi(0.380\ \text{m})$
$= (1.05 \times 10^{-5}\ \text{N/A}\cdot\text{m})I_1L.$
The directions of the forces are shown on the diagram. The symmetry of the force diagrams simplifies the vector addition, so for the net force to be zero, we have
$F_A = 2F\cos 30° = mg = \rho\pi r^2Lg;$
$2(1.05 \times 10^{-5}\ \text{N/A}\cdot\text{m})I_1L\cos 30° = (8.9 \times 10^3\ \text{kg/m}^3)\pi(1.00 \times 10^{-3}\ \text{m})^2(9.80\ \text{m/s}^2)L,$
which gives $I_1 =$ $\boxed{1.50 \times 10^4\ \text{A.}}$

66. (*a*) The force from the magnetic field will accelerate the rod:
$F = ILB = ma$, which gives $a = ILB/m$.
Because the rod starts from rest and the acceleration is constant, we have
$v = v_0 + at = 0 + (ILB/m)t =$ $\boxed{ILBt/m.}$

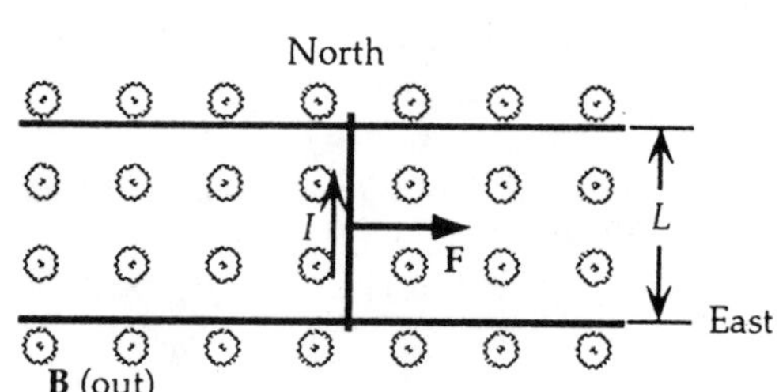

(*b*) The total normal force on the rod is mg, so there is a friction force of $\mu_k mg$. For the acceleration, we have
$\Sigma F = ILB - \mu_k mg = ma$, which gives $a = (ILB/m) - \mu_k g$.
For the speed we have
$v = at =$ 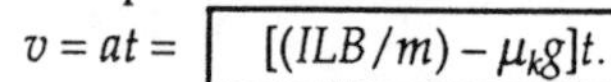 $\boxed{[(ILB/m) - \mu_k g]t.}$
(*c*) For a current toward the north in an upward field, the force will be to the $\boxed{\text{east.}}$

67. We find the speed acquired from the accelerating voltage from energy conservation:

$0 = \Delta\text{KE} + \Delta\text{PE};$

$0 = \frac{1}{2}mv^2 - 0 + (-e)(V)$, or $v = (2eV/m)^{1/2}$.

If we assume that the deflection is small, the time the electron takes to reach the screen is

$t = L/v = L(m/2eV)^{1/2}$.

The magnetic force produces an acceleration perpendicular to the original motion:

$a_\perp = evB/m$.

For a small deflection, we can take the force to be constant, so the deflection of the electron is

$d = \frac{1}{2}a_\perp t^2 = \frac{1}{2}(evB/m)(L/v)^2 = \frac{1}{2}eBL^2/mv = \frac{1}{2}BL^2(e/2mV)^{1/2}$

$= \frac{1}{2}(5.0 \times 10^{-5}\text{ T})(0.20\text{ m})^2[(1.60 \times 10^{-19}\text{ C})/2(9.11 \times 10^{-31}\text{ kg})V]^{1/2} = (0.296\text{ m} \cdot \text{V}^{1/2})/V^{1/2}$.

(*a*) For a voltage of 2.0 kV, we have

$d = (0.296\text{ m} \cdot \text{V}^{1/2})/V^{1/2} = (0.296\text{ m} \cdot \text{V}^{1/2})/(2.0 \times 10^3\text{ V})^{1/2} = 6.6 \times 10^{-3}\text{ m} =$ **6.6 mm.**

(*b*) For a voltage of 30 kV, we have

$d = (0.296\text{ m} \cdot \text{V}^{1/2})/V^{1/2} = (0.296\text{ m} \cdot \text{V}^{1/2})/(30 \times 10^3\text{ V})^{1/2} = 1.7 \times 10^{-3}\text{ m} =$ **1.7 mm.**

These results justify our assumption of small deflection.

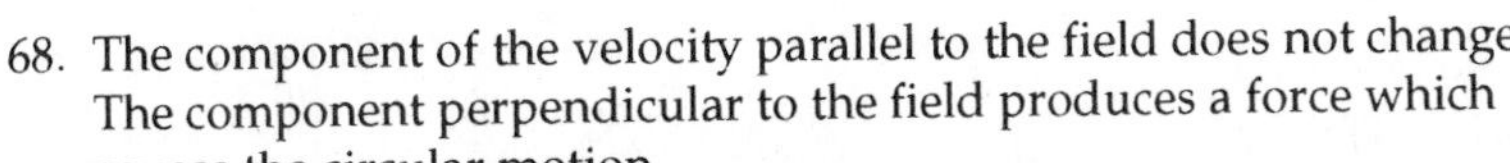

68. The component of the velocity parallel to the field does not change. The component perpendicular to the field produces a force which causes the circular motion.

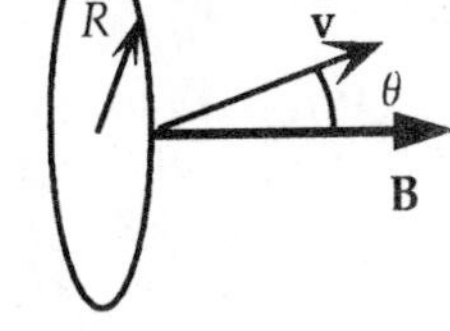

We find the radius of the circular motion from

$R = mv_\perp/qB$

$= (9.11 \times 10^{-31}\text{ kg})(1.8 \times 10^7\text{ m/s})(\sin 7.0°)/(1.60 \times 10^{-19}\text{ C})(3.3 \times 10^{-2}\text{ T})$

$= 3.78 \times 10^{-4}\text{ m} =$ **0.38 mm.**

We find the time for one revolution:

$T = 2\pi R/v_\perp = 2\pi(3.78 \times 10^{-4}\text{ m})/(1.8 \times 10^7\text{ m/s}) \sin 7.0° = 1.08 \times 10^{-9}\text{ s}$.

In this time, the distance the electron travels along the field is

$d = v_\parallel T = (1.8 \times 10^7\text{ m/s})(\cos 7.0°)(1.08 \times 10^{-9}\text{ s}) = 1.9 \times 10^{-2}\text{ m} =$ **1.9 cm.**

69. (*a*) The radius of the circular orbit is

$r = mv/qB$.

The time to complete a circle is

$T = 2\pi r/v = 2\pi m/qB$, so the frequency is

$f = 1/T = qB/2\pi m$.

Note that this is independent of r.

Because we want the ac voltage to be maximum when the proton reaches the gap and minimum (reversed) when the proton has made half a circle, the frequency of the ac voltage must be the same: $f = 1/T = qB/2\pi m$.

(*b*) In a full circle, the proton crosses the gap twice. If the gap is small, the ac voltage will not change significantly from its maximum value while the proton is in the gap.

The energy gain from the two crossings is

$\Delta\text{KE} = 2qV_0$.

(*c*) From $r = mv/qB$, we see that the maximum speed, and thus the maximum kinetic energy occurs at the maximum radius of the path. The maximum kinetic energy is

$\text{KE}_{max} = \frac{1}{2}mv_{max}^2 = \frac{1}{2}m(qBr_{max}/m)^2 = (qBr_{max})^2/2m$

$= [(1.60 \times 10^{-19}\text{ C})(0.50\text{ T})(2.0\text{ m})]^2/2(1.67 \times 10^{-27}\text{ kg})$

$= 7.66 \times 10^{-12}\text{ J} = (7.66 \times 10^{-12}\text{ J})/(1.60 \times 10^{-13}\text{ J/MeV}) =$ **48 MeV.**

(*d*) The cyclotron is like a swing because a small push is given in resonance with the natural frequency of the motion.

70. If the beam is perpendicular to the magnetic field, the force from the magnetic field is always perpendicular to the velocity, so it will change the direction of the velocity, but not its magnitude. The radius of the path in the magnetic field is

$R = mv/qB.$

Protons with different speeds will have paths of different radii. Thus slower protons will deflect more, and faster protons will deflect less, than those with the design speed.

We find the radius of the path from

$R = mv/qB$

$= (1.67 \times 10^{-27}\ \text{kg})(1.0 \times 10^{7}\ \text{m/s})/(1.60 \times 10^{-19}\ \text{C})(0.33\ \text{T})$

$= 0.316\ \text{m}.$

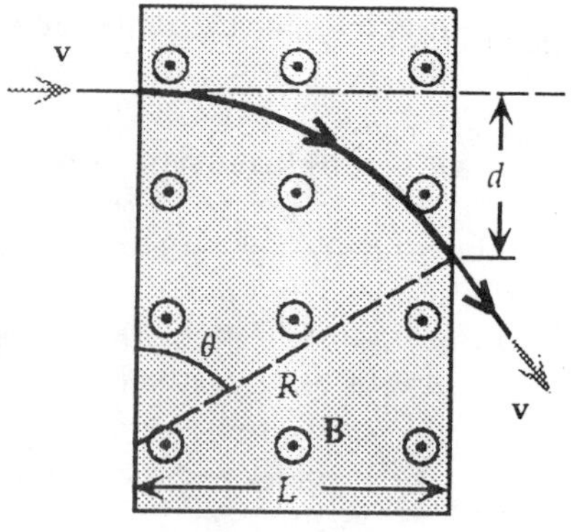

Because the exit velocity is perpendicular to the radial line from the center of curvature, the exit angle is also the angle the radial line makes with the boundary of the field:

$\sin\theta = L/R = (0.050\ \text{m})/(0.316\ \text{m}) = 0.158,$ so $\boxed{\theta = 9.1°.}$

71. (*a*) The force on one side of the loop is

$F = ILB = (25.0\ \text{A})(0.200\ \text{m})(1.65\ \text{T}) = 8.25\ \text{N}.$

When the loop is perpendicular to the magnetic field, the forces at top and bottom will create a tension in the other two sides of $\frac{1}{2}F$. This produces a tensile stress of $\frac{1}{2}F/A = F/2\pi r^2$.

When the loop is parallel to the magnetic field, the forces on right and left will create a shear in the other two sides. This produces a shear stress of $\frac{1}{2}F/A = F/2\pi r^2$.

Because the magnitudes are the same and the tensile strength of aluminum is equal to the shear strength, we can use either condition to determine the minimum diameter. With a safety factor of 10, we have

$10(F/2\pi r^2) < \text{Strength};$

$10(8.25\ \text{N})/2\pi r^2 < 200 \times 10^{6}\ \text{N/m}^2$, which gives $r > 2.56 \times 10^{-4}\ \text{m} = 0.256\ \text{mm}.$

Thus the minimum diameter is $\boxed{0.512\ \text{mm.}}$

(*b*) The resistance of a single loop is

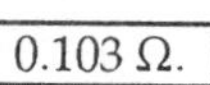

$R = \rho L/A = \rho L/\pi r^2 = (2.65 \times 10^{-8}\ \Omega\cdot\text{m})(4)(0.200\ \text{m})/\pi(2.56 \times 10^{-4}\ \text{m})^2 = \boxed{0.103\ \Omega.}$

72. (*a*) We find the resistance of the coil from

$P = V^2/R;$

$4.0 \times 10^{3}\ \text{W} = (120\ \text{V})^2/R$, which gives $R = 3.60\ \Omega.$

If the coil is tightly wound, each turn will have a length of πD. We find the number of turns from the length of wire required to give this resistance:

$R = \rho L/A = \rho N\pi D/w^2;$

$3.60\ \Omega = (1.65 \times 10^{-8}\ \Omega\cdot\text{m})N\pi(1.2\ \text{m})/(1.6 \times 10^{-3}\ \text{m})^2$, which gives $N = \boxed{1.5 \times 10^{2}\ \text{turns.}}$

(*b*) We find the current in the coil from

$I = V/R = (120\ \text{V})/(3.60\ \Omega) = 33.3\ \text{A}.$

We find the magnetic field strength from

$B = \mu_0 NI/2r$

$= (4\pi \times 10^{-7}\ \text{T}\cdot\text{m/A})(1.5 \times 10^{2}\ \text{turns})(33.3\ \text{A})/2(0.60\ \text{m}) = \boxed{5.2 \times 10^{-3}\ \text{T.}}$

(*c*) If we increase the number of turns by a factor *k*, the resistance will increase by this factor. Because the voltage is constant, the current will decrease by this factor, so the product *NI* will not change. Thus the magnetic field strength $\boxed{\text{will not change.}}$

CHAPTER 21

1. The magnitude of the average induced emf is
$\mathcal{E} = -\Delta\Phi_B/\Delta t = -A\,\Delta B/\Delta t = -\pi(0.046\text{ m})^2(0 - 1.5\text{ T})/(0.20\text{ s}) =$ **0.050 V.**

2. We assume the plane of the coil is perpendicular to the magnetic field. The magnitude of the average induced emf is
$\mathcal{E} = -\Delta\Phi_B/\Delta t = -A\,\Delta B/\Delta t = -\pi(0.080\text{ m})^2(0 - 1.10\text{ T})/(0.15\text{ s}) =$ **0.15 V.**

3. As the coil is pushed into the field, the magnetic flux increases into the page. To oppose this increase, the flux produced by the induced current must be out of the page, so the induced current is **counterclockwise.**

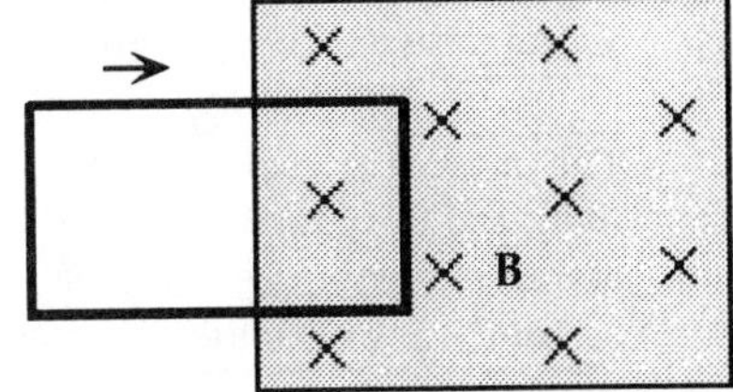

4. As the magnet is pushed into the coil, the magnetic flux increases to the right. To oppose this increase, the flux produced by the induced current must be to the left, so the induced current in the resistor will be **from right to left.**

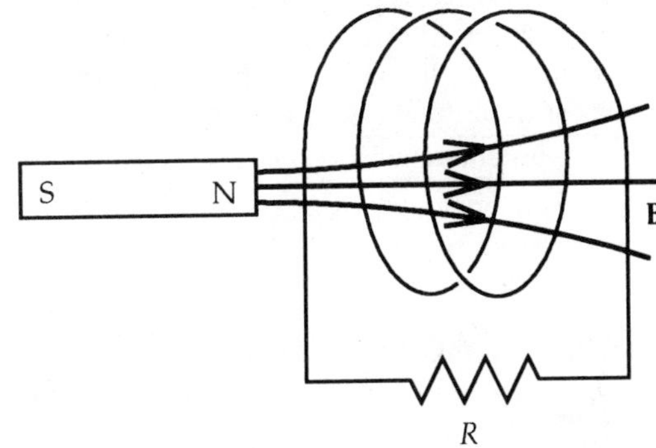

5. The average induced emf is
$\mathcal{E} = -N\,\Delta\Phi_B/\Delta t = -(2)[(+38\text{ Wb}) - (-30\text{ Wb})]/(0.42\text{ s}) =$ **-3.2×10^2 V.**

6. We choose up as the positive direction. The average induced emf is
$\mathcal{E} = -\Delta\Phi_B/\Delta t = -A\,\Delta B/\Delta t = -\pi(0.036\text{ m})^2[(-0.25\text{ T}) - (+0.63\text{ T})]/(0.15\text{ s}) = 2.4 \times 10^{-2}\text{ V} =$ **24 mV.**

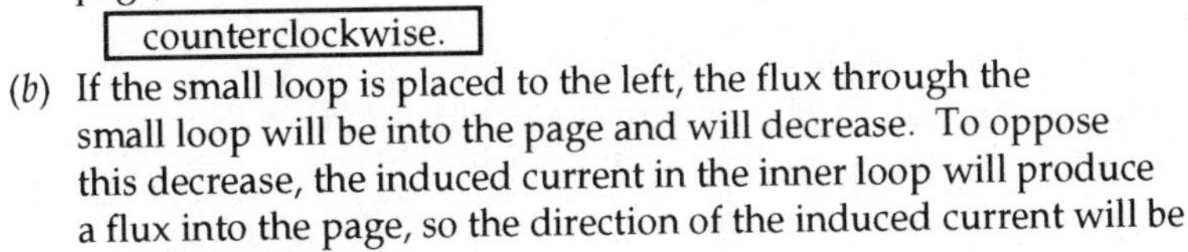

7. (*a*) As the resistance is increased, the current in the outer loop will decrease. Thus the flux through the inner loop, which is out of the page, will decrease. To oppose this decrease, the induced current in the inner loop will produce a flux out of the page, so the direction of the induced current will be **counterclockwise.**
(*b*) If the small loop is placed to the left, the flux through the small loop will be into the page and will decrease. To oppose this decrease, the induced current in the inner loop will produce a flux into the page, so the direction of the induced current will be **clockwise.**

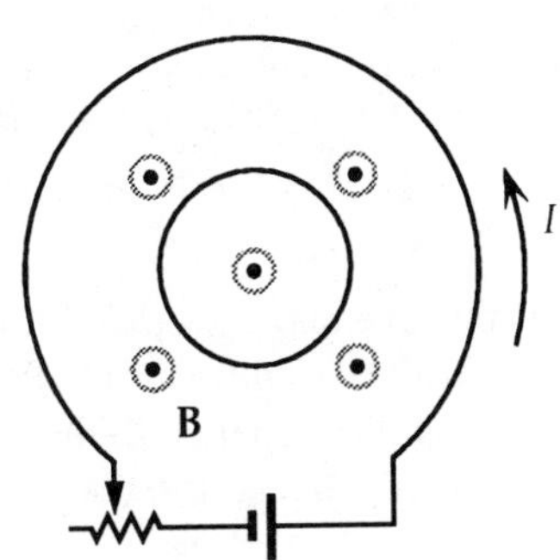

8. (*a*) The increasing current in the wire will cause an increasing field into the page through the loop. To oppose this increase, the induced current in the loop will produce a flux out of the page, so the direction of the induced current will be counterclockwise.

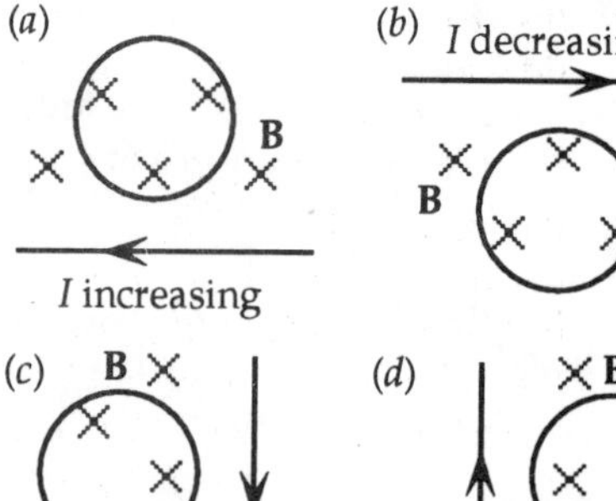

(*b*) The decreasing current in the wire will cause a decreasing field into the page through the loop. To oppose this decrease, the induced current in the loop will produce a flux into the page, so the direction of the induced current will be clockwise.

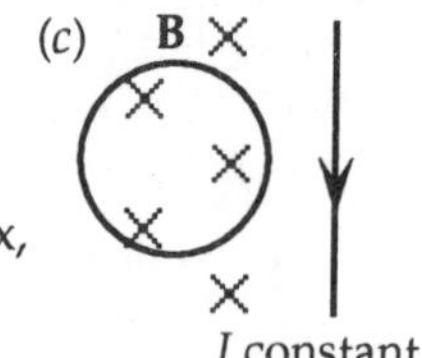

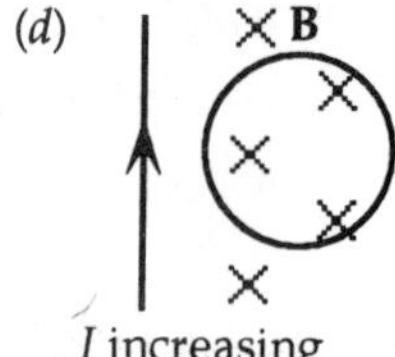

(*c*) Because the current is constant, there will be no change in flux, so the induced current will be zero.

(*d*) The increasing current in the wire will cause an increasing field into the page through the loop. To oppose this increase, the induced current in the loop will produce a flux out of the page, so the direction of the induced current will be counterclockwise.

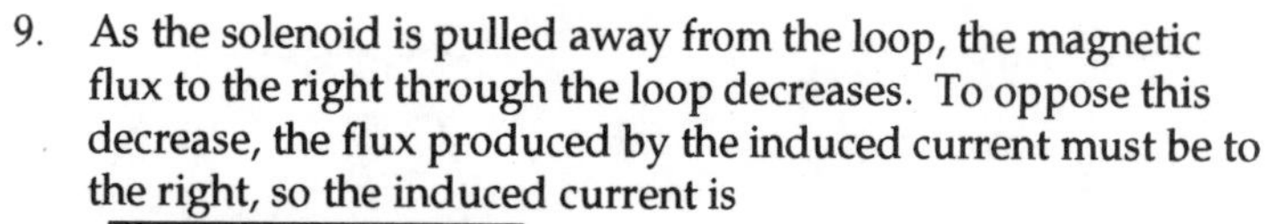

9. As the solenoid is pulled away from the loop, the magnetic flux to the right through the loop decreases. To oppose this decrease, the flux produced by the induced current must be to the right, so the induced current is counterclockwise as viewed from the solenoid.

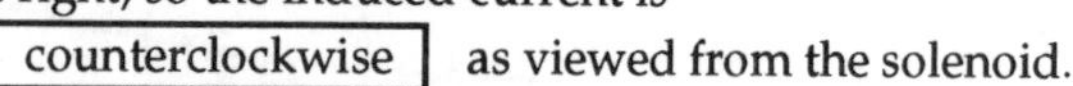

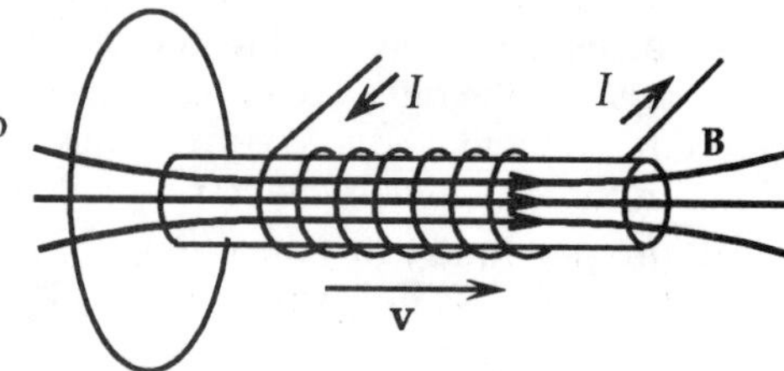

10. (*a*) The average induced emf is

$$\mathscr{E} = -\Delta\Phi_B/\Delta t = -A\,\Delta B/\Delta t = -\pi(0.10\text{ m})^2[(-0.45\text{ T}) - (+0.52\text{ T})]/(0.180\text{ s}) = \boxed{0.17\text{ V.}}$$

(*b*) The positive result for the induced emf means the induced field is away from the observer, so the induced current is clockwise.

11. (*a*) Because the velocity is perpendicular to the magnetic field and the rod, we find the induced emf from

$$\mathscr{E} = BLv$$
$$= (0.800\text{ T})(0.120\text{ m})(0.150\text{ m/s})$$
$$= 1.44\times10^{-2}\text{ V} = \boxed{14.4\text{ mV.}}$$

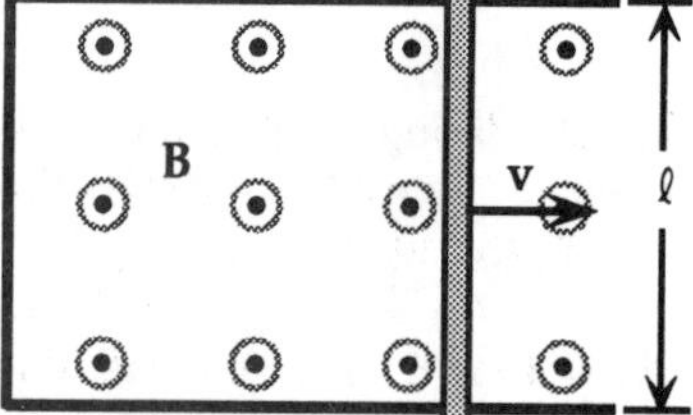

(*b*) Because the upward flux is increasing, the induced flux will be into the page, so the induced current is clockwise. Thus the induced emf in the rod is down, which means that the electric field will be down. We find its magnitude from

$$E = \mathscr{E}/\ell = (1.44\times10^{-2}\text{ V})/(0.120\text{ m}) = \boxed{0.120\text{ V/m down.}}$$

12. (*a*) The magnetic flux through the loop is into the paper and decreasing, because the area is decreasing. To oppose this decrease, the induced current in the loop will produce a flux into the paper, so the direction of the induced current will be clockwise.

(*b*) We choose into the paper as the positive direction. The average induced emf is

$$\mathscr{E} = -\Delta\Phi_B/\Delta t = -B\,\Delta A/\Delta t = -\pi B\,\Delta(r^2)/\Delta t$$
$$= -\pi(0.75\text{ T})[(0.030\text{ m})^2 - (0.100\text{ m})^2]/(0.50\text{ s}) = 4.3\times10^{-2}\text{ V} = \boxed{43\text{ mV.}}$$

(*c*) We find the average induced current from

$$I = \mathscr{E}/R = (43\text{ mV})/(2.5\ \Omega) = \boxed{17\text{ mA.}}$$

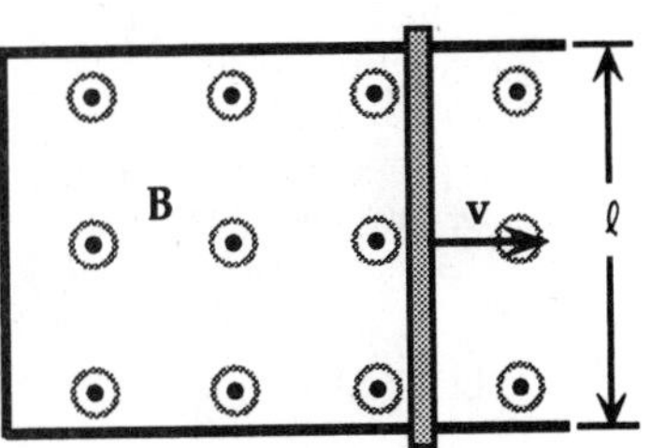

13. (*a*) Because the velocity is perpendicular to the magnetic field and the rod, we find the speed from

$\mathcal{E} = BLv;$

$100 \times 10^{-3}\ \text{V} = (0.90\ \text{T})(0.132\ \text{m})v,$

which gives $v =$ **0.84 m/s.**

(*b*) Because the outward flux is increasing, the induced flux will be into the page, so the induced current is clockwise. Thus the induced emf in the rod is down, which means that the electric field will be down. We find its magnitude from

$E = \mathcal{E}/\ell = (100 \times 10^{-3}\ \text{V})/(0.132\ \text{m}) =$ **0.758 V/m down.**

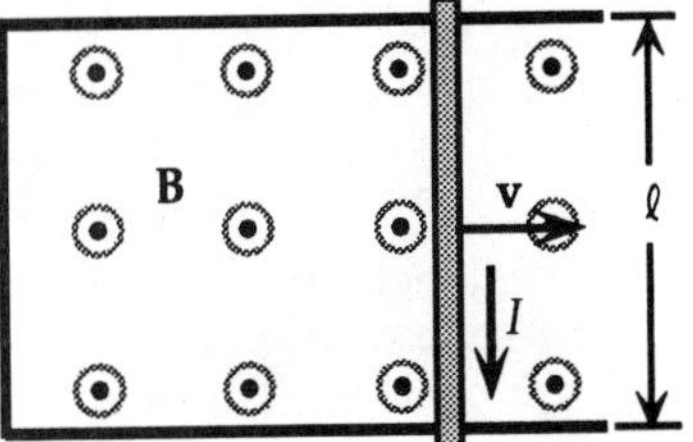

14. (*a*) Because the velocity is perpendicular to the magnetic field and the rod, we find the induced emf from

$\mathcal{E} = BLv$

$= (0.75\ \text{T})(0.300\ \text{m})(1.9\ \text{m/s}) =$ **0.43 V.**

(*b*) We find the induced current from

$I = \mathcal{E}/R = (0.43\ \text{V})/(2.5\ \Omega + 25.0\ \Omega) =$ **0.016 A.**

(*c*) The induced current in the rod will be down. Because this current is in an upward magnetic field, there will be a magnetic force to the left. To keep the rod moving, there must be an equal external force to the right, which we find from

$F = ILB = (0.016\ \text{A})(0.300\ \text{m})(0.75\ \text{T}) =$ **3.5×10^{-3} N.**

15. As the loop is pulled from the field, the flux through the loop will decrease. We find the induced emf from

$\mathcal{E} = -\Delta\Phi_B/\Delta t = -B\,\Delta A/\Delta t = -B\ell\,\Delta x/\Delta t = -B\ell\,(-v) = B\ell v.$

Because the inward flux is decreasing, the induced flux will be into the page, so the induced current is clockwise, given by

$I = \mathcal{E}/R.$

Because this current in the left hand side of the loop is in a downward magnetic field, there will be a magnetic force to the left. To keep the rod moving, there must be an equal external force to the right, which we find from

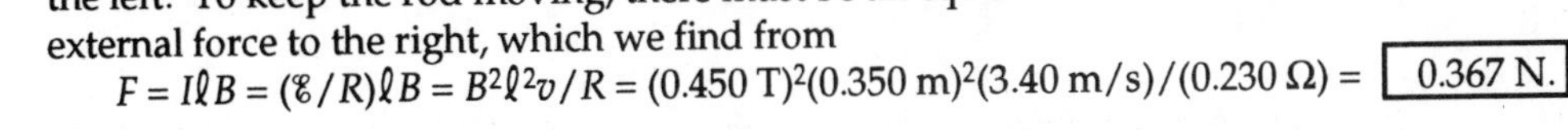

$F = I\ell B = (\mathcal{E}/R)\ell B = B^2\ell^2 v/R = (0.450\ \text{T})^2(0.350\ \text{m})^2(3.40\ \text{m/s})/(0.230\ \Omega) =$ **0.367 N.**

16. (*a*) For the resistance of the loop, we have

$R = \rho L/A = (1.68 \times 10^{-8}\ \Omega\cdot\text{m})(20)\pi(0.310\ \text{m})/\pi(1.3 \times 10^{-3}\ \text{m})^2 = 0.0616\ \Omega.$

The induced emf is

$\mathcal{E} = -\Delta\Phi_B/\Delta t = -A\,\Delta B/\Delta t = -(20)\pi(0.155\ \text{m})^2(8.65 \times 10^{-3}\ \text{T/s}) = -0.0131\ \text{V}.$

Thus the induced current is

$I = \mathcal{E}/R = (0.0131\ \text{V})/(0.0616\ \Omega) =$ **0.21 A.**

(*b*) Thermal energy is produced in the wire at the rate of

$P = I^2R = (0.21\ \text{A})^2(0.0616\ \Omega) = 2.8 \times 10^{-3}\ \text{W} =$ **2.8 mW.**

17. If we assume that the movable rod starts at the bottom of the U, in a time t it will have moved a distance $x = vt$. For the resistance of the U, we have

$R = \rho L/A = \rho(2vt + \ell)/A.$

The induced emf is

$\mathcal{E} = BLv;$

so the induced current is

$I = \mathcal{E}/R = B\ell v/[\rho(2vt + \ell)/A] =$ **$B\ell vA/\rho(2vt + \ell)$.**

18. For the resistance of the loop, we have

$$R = \rho L/A = \rho\pi D/\tfrac{1}{4}\pi d^2 = 4\rho D/d^2.$$

The induced emf is

$$\mathscr{E} = -\Delta\Phi_B/\Delta t = -\tfrac{1}{4}\pi D^2\,\Delta B/\Delta t;$$

so the induced current is

$$I = \mathscr{E}/R = -(\pi D d^2/16\rho)\,\Delta B/\Delta t.$$

In the time Δt the amount of charge that will pass a point is

$$Q = I\,\Delta t = -(\pi D d^2/16\rho)\,\Delta B = -[\pi(0.132\text{ m})(2.25\times10^{-3}\text{ m})^2/16(1.68\times10^{-8}\,\Omega\cdot\text{m})](0 - 0.750\text{ T}) = \boxed{5.86\text{ C.}}$$

19. The maximum induced emf is

$$\mathscr{E} = NBA\omega.$$

If the only change is in the rotation speed, for the two conditions we have

$$\mathscr{E}_2/\mathscr{E}_1 = \omega_2/\omega_1;$$

$\mathscr{E}_2/(12.4\text{ V}) = (2500\text{ rpm})/(1000\text{ rpm})$ which gives $\mathscr{E}_2 = \boxed{31.0\text{ V.}}$

20. We find the number of turns from

$$\mathscr{E}_{peak} = NBA\omega;$$

$24.0\text{ V} = N(0.420\text{ T})(0.050\text{ m})^2(60\text{ rev/s})(2\pi\text{ rad/rev})$, which gives $N = \boxed{61\text{ turns.}}$

21. The induced emf is

$$\mathscr{E} = NBA\omega\sin\omega t.$$

For the rms value of the output, we have

$$V_{rms} = [(\mathscr{E}^2)_{av}]^{1/2} = [(NBA\omega)^2(\sin^2\omega t)_{av}]^{1/2} = NBA\omega\,[(\sin^2\omega t)_{av}]^{1/2}.$$

The $\sin^2\omega t$ function varies from 0 to 1, with an average value of 1/2, so we get

$$V_{rms} = NBA\omega(1/\sqrt{2}) = NBA\omega/\sqrt{2}.$$

22. We find the rotation speed from

$$\mathscr{E}_{peak} = NBA\omega;$$

$120\text{ V} = (720\text{ turns})(0.650\text{ T})(0.210\text{ m})^2\omega$, which gives $\omega = 5.81\text{ rad/s} = \boxed{0.925\text{ rev/s.}}$

23. We find the peak emf from

$$\mathscr{E}_{peak} = NBA\omega = (450\text{ turns})(0.75\text{ T})\pi(0.050\text{ m})^2(60\text{ rev/s})(2\pi\text{ rad/rev}) = 999\text{ V}.$$

The rms voltage output is

$$V_{rms} = \mathscr{E}_{peak}/\sqrt{2} = (999\text{ V})/\sqrt{2} = 707\text{ V} = \boxed{0.71\text{ kV.}}$$

If only the rotation frequency changes, to double the rms voltage output, we must double the rotation speed, so we have

$$f_2 = 2f_1 = 2(60\text{ rev/s}) = \boxed{120\text{ rev/s.}}$$

24. We find the counter emf from

$$\mathscr{E} - \mathscr{E}_{back} = IR;$$

$120\text{ V} - \mathscr{E}_{back} = (9.20\text{ A})(3.75\,\Omega)$, which gives $\mathscr{E}_{back} = \boxed{86\text{ V.}}$

25. Because the counter emf is proportional to the rotation speed, for the two conditions we have

$$\mathscr{E}_{back2}/\mathscr{E}_{back1} = \omega_2/\omega_1;$$

$\mathscr{E}_{back2}/(72\text{ V}) = (2500\text{ rpm})/(1800\text{ rpm})$, which gives $\mathscr{E}_{back2} = \boxed{100\text{ V.}}$

26. Because the counter emf is proportional to both the rotation speed and the magnetic field, for the two conditions we have

$\mathcal{E}_{back2}/\mathcal{E}_{back1} = (\omega_2/\omega_1)(B_2/B_1)$;

$(65\text{ V})/(100\text{ V}) = [(2500\text{ rpm})/(1000\text{ rpm})](B_2/B_1)$, which gives $\boxed{B_2/B_1 = 0.26.}$

27. Because the counter emf is proportional to the rotation speed, we find the new value from

$\mathcal{E}_{back2}/\mathcal{E}_{back1} = \omega_2/\omega_1$;

$\mathcal{E}_{back2}/(108\text{ V}) = (1/2)$, which gives $\mathcal{E}_{back2} = 54$ V.

We find the new current from

$\mathcal{E} - \mathcal{E}_{back} = IR$;

$120\text{ V} - 54\text{ V} = I(5.0\ \Omega)$, which gives $I = \boxed{13\text{ A.}}$

28. (*b*) At start up there will be no induced emf in the armature. Because the line voltage is across each resistor, we find the currents from

$I_{field0} = \mathcal{E}/R_{field} = (115\text{ V})/(66.0\ \Omega) = 1.74$ A;

$I_{armature0} = \mathcal{E}/R_{armature} = (115\text{ V})/(5.00\ \Omega) = 23.0$ A.

We use the junction condition to find the total current:

$I_0 = I_{field0} + I_{armature0} = 1.74\text{ A} + 23.0\text{ A} = \boxed{24.7\text{ A.}}$

(*c*) At full speed, the back emf is maximum. Because the line voltage is across the field resistor, we find the field current from

$I_{field0} = \mathcal{E}/R_{field} = (115\text{ V})/(66.0\ \Omega) = 1.74$ A.

We find the armature current from

$\mathcal{E} - \mathcal{E}_{back} = I_{armature}R_{armature}$;

$115\text{ V} - 105\text{ V} = I_{armature}(5.00\ \Omega)$,

which gives $I_{armature} = 2.0$ A.

Thus the total current is

$I_0 = I_{field} + I_{armature} = 1.74\text{ A} + 2.0\text{ A} = \boxed{3.7\text{ A.}}$

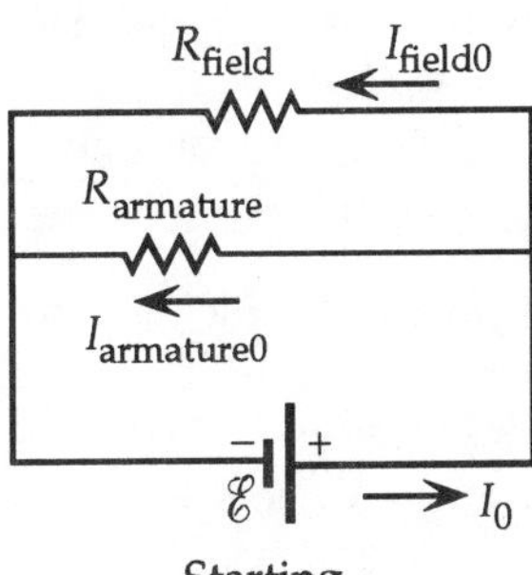

Starting

Full Speed

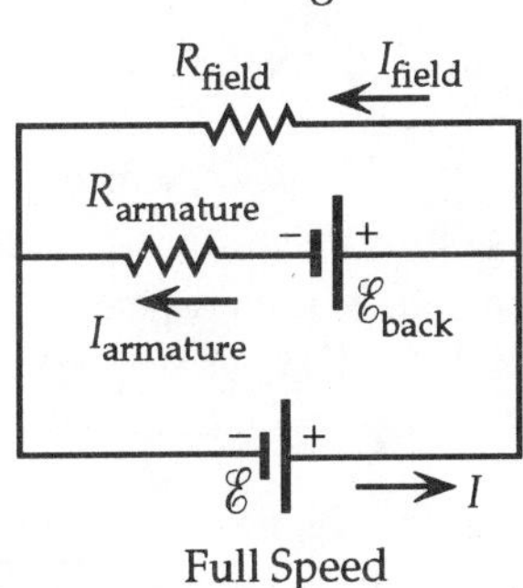

29. (*a*) We find the emf of the generator from the load conditions:

$V = \mathcal{E} - I_{armature}R_{armature}$;

$200\text{ V} = \mathcal{E} - (50\text{ A})(0.40\ \Omega)$, which gives $\mathcal{E} = 220$ V.

When there is no load on the generator, the current is zero, so the voltage will be the emf: $\boxed{220\text{ V.}}$

Note that no-load means little torque required to turn the generator.

(*b*) Because the generator emf is proportional to the rotation speed, we find the new value from

$\mathcal{E}_2/\mathcal{E}_1 = \omega_2/\omega_1$;

$\mathcal{E}_2/(220\text{ V}) = (800\text{ rpm}/1000\text{ rpm})$, which gives $\mathcal{E}_2 = 176$ V.

We find the new load voltage from

$V_2 = \mathcal{E}_2 - I_{armature}R_{armatrer} = 176\text{ V} - (50\text{ A})(0.40\text{ W}) = \boxed{156\text{ V.}}$

Note that the power output is reduced to 7.8 kW.

30. We find the number of turns in the secondary from

$V_S/V_P = N_S/N_P$;

$(10{,}000\text{V})/(120\text{ V}) = N_S/(5000\text{ turns})$, which gives $N_S = \boxed{4.17 \times 10^5\text{ turns.}}$

31. Because $N_S < N_P$, this is a step-down transformer.
   We find the ratio of the voltages from
   $V_S/V_P = N_S/N_P = (120 \text{ turns})/(420 \text{ turns}) =$ 0.285.
   For the ratio of currents, we have
   $I_S/I_P = N_P/N_S = (420 \text{ turns})/(120 \text{ turns}) =$ 3.50.

32. With 100% efficiency, the power on each side of the transformer is the same:
   $I_P V_P = I_S V_S$, so we have
   $I_S/I_P = V_P/V_S = (16 \text{ V})/(120) =$ 0.13.

33. We find the ratio of the number of turns from
   $N_S/N_P = V_S/V_P = (12 \times 10^3 \text{ V})/(220 \text{ V}) =$ 55.
   If the transformer is connected backward, the role of the turns will be reversed:
   $V_S/V_P = N_S/N_P$;
   $V_S/(220 \text{ V}) = 1/55$, which gives $V_S =$ 4.0 V.

34. (*a*) We assume 100% efficiency, so we have
   $I_P V_P = I_S V_S$;
   $(15 \text{ A})/(0.75 \text{ A}) = (120 \text{ V})/V_S$, which gives $V_S =$ 6.0 V.
   (*b*) Because $V_S < V_P$, this is a step-down transformer.

35. (*a*) We assume 100% efficiency, so we find the input voltage from
   $P = I_P V_P$;
   $100 \text{ W} = (20 \text{ A})V_P$, which gives $V_P = 5.0 \text{ V}$.
   Because $V_S > V_P$, this is a step-up transformer.
   (*b*) For the voltage ratio we have
   $V_S/V_P = (12 \text{ V})/(5.0 \text{ V}) =$ 2.4.

36. (*a*) Because $V_S < V_P$, this is a step-down transformer.
   (*b*) We assume 100% efficiency, so we find the current in the secondary from
   $P = I_S V_S$;
   $40 \text{ W} = I_S(12 \text{ V})$, which gives $I_S =$ 3.3 A.
   (*c*) We find the current in the primary from
   $P = I_P V_P$;
   $40 \text{ W} = I_P(120 \text{ V})$, which gives $I_S =$ 0.33 A.
   (*d*) We find the resistance of the bulb from
   $V_S = I_S R_S$;
   $120 \text{ V} = (3.33 \text{ A})R_S$, which gives $R_S =$ 3.6 Ω.

37. We find the output voltage from
   $V_S/V_P = N_S/N_P$;
   $V_S/(120 \text{ V}) = (1240 \text{ turns})/(330 \text{ turns})$, which gives $V_S =$ 451 V.
   We find the input current from
   $I_S/I_P = N_P/N_S$;
   $(15.0 \text{ A})/I_P = (330 \text{ turns})/(1240 \text{ turns})$, which gives $I_P =$ 56.4 A.

38. (*a*) We can find the current in the transmission lines from the output emf:

$P_{out} = IV_{out}$;

30×10^6 W = $I(45 \times 10^3$ V), which gives $I = 667$ A.

We find the input emf from

$\mathscr{E}_{in} = V_{out} + IR_{lines} = 45 \times 10^3$ V + (667 A)(4.0 Ω) = 48×10^3 = **48 kV.**

(*b*) The power loss in the lines is

$P_{loss} = I^2R_{lines} = (667\ A)^2(4.0\ \Omega) = 1.78 \times 10^6$ W = 1.78 MW.

The total power is

$P_{total} = P_{out} + P_{loss}$ = 30 MW + 1.78 MW = 31.8 MW,

so the fraction lost is

(1.78 MW)/(31.8 MW) = **0.056 (5.6%).**

39. We can find the current in the transmission lines from the power transmitted to the user:

$P_T = IV$, or $I = P_T/V$.

The power loss in the lines is

$P_L = I^2R_L = (P_T/V)^2R_L = (P_T)^2R_L/V^2$.

40. Without the transformers, we can find the delivered current, which is the current in the transmission lines, from the delivered power:

$P_{out} = IV_{out}$;

50×10^3 W = I(120V), which gives $I = 417$ A.

The power loss in the lines is

$P_{L0} = I^2R_{lines} = (417\ A)^2(2)(0.100\ \Omega) = 3.48 \times 10^4$ W = 34.8 kW.

With the transformers, to deliver the same power at 120 V, the delivered current from the step-down transformer is still 417 A.

If the step-down transformer is 99% efficient, we have

$(0.99)I_{P2}V_{P2} = I_{S2}V_{S2}$;

$(0.99)I_{P2}(1200\ V) = (417\ A)(120\ V)$, which gives $I_{P2} = 42.1$ A.

Because this is the current in the lines, the power loss in the lines is

$P_{L2} = I_{P2}^2R_{lines} = (42.1\ A)^2(2)(0.100\ \Omega) = 3.55 \times 10^2$ W = 0.355 kW.

For the 1% losses in the transformers, we approximate the power in each transformer:

P_{Lt} = (0.01)(50 kW + 50.36 kW) = 1.00 kW.

The total power loss using the transformers is

P_L = 0.355 kW + 1.00 kW = 1.4 kW.

The power saved by using the transformers is

$P_{saved} = P_{L0} - P_L$ = 34.8 kW – 1.4 kW = **33.4 kW.**

41. We can find the delivered current, which is the current in the transmission lines, from the delivered power:

$P_{out} = IV_{out}$;

300×10^6 W = $I(600 \times 10^3$ V), which gives $I = 500$ A.

We can approximate the power at the input to the lines as 102% of the delivered power to account for the loss. Thus the power loss in the lines is

$P_L = (0.02)(1.02)P_{out} = I^2R_{lines}$;

$(0.02)(1.02)(300 \times 10^6\ W) = (500\ A)^2R_{lines}$, which gives $R_{lines} = 24.5\ \Omega$.

We find the radius of the 2 lines, each 200 km long, from

$R = \rho L/A$;

$24.5\ \Omega = (2.65 \times 10^{-8}\ \Omega \cdot m)(2)(200 \times 10^3\ m)/\pi r^2$, which gives $r = 1.174 \times 10^{-2}$ m.

Thus the diameter of the lines should be **2.35 cm.**

42. We find the induced emf from
$\mathcal{E} = -L\,\Delta I/\Delta t = -(0.120\text{ H})(10.0\text{ A} - 25.0\text{ A})/(0.350\text{ s}) = \boxed{5.14\text{ V.}}$
The emf is in the $\boxed{\text{direction of the current,}}$ to oppose the decrease in the current.

43. We use the inductance of a solenoid:
$L = \mu_0 AN^2/\ell = (4\pi\times10^{-7}\text{ T}\cdot\text{m/A})\pi(1.45\times10^{-2}\text{ m})^2(10{,}000\text{ turns})^2/(0.60\text{ m}) = \boxed{0.14\text{ H.}}$

44. Because the current in increasing, the emf is negative. We find the self-inductance from
$\mathcal{E} = -L\,\Delta I/\Delta t;$
$-8.50\text{ V} = -L[31.0\text{ mA} - (-28.0\text{ mA})]/(42.0\text{ ms})$, which gives $L = \boxed{6.05\text{ H.}}$

45. We find the number of turns from the inductance of a solenoid:
$L = \mu_0 AN^2/\ell;$
$0.100\text{ H} = (4\pi\times10^{-7}\text{ T}\cdot\text{m/A})\pi(2.6\times10^{-2}\text{ m})^2N^2/(0.300\text{ m})$, which gives $N = \boxed{3.4\times10^3\text{ turns.}}$

46. We use the inductance of a solenoid:
$L = \mu_0 AN^2/\ell = (4\pi\times10^{-7}\text{ T}\cdot\text{m/A})\pi(1.25\times10^{-2}\text{ m})^2(1000\text{ turns})^2/(0.282\text{ m}) = 2.2\times10^{-3}\text{ H} = \boxed{2.2\text{ mH.}}$
If we form the ratio of inductances for the two conditions, we have
$L_2/L_1 = (\mu/\mu_0)(N_2/N_1)^2;$
$1 = (1000)[N_2/(1000\text{ turns})]^2$, which gives $N_2 = \boxed{32\text{ turns.}}$

47. We find the induced emf from
$\mathcal{E}_{\text{induced}} = -L\,\Delta I/\Delta t = -(0.440\text{ H})(3.50\text{ A/s}) = -1.54\text{ V}.$
The negative sign indicates a direction opposite to the current.
If we start at point b and add the potential changes, we get
$V_b + IR + |\mathcal{E}_{\text{induced}}| = V_a$, or
$V_{ab} = (3.00\text{ A})(2.25\ \Omega) + 1.54\text{ V} = \boxed{8.29\text{ V.}}$

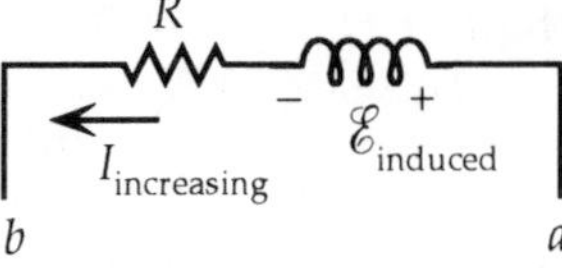

48. If D represents the diameter of the solenoid, the length of the wire is $N(\pi D)$. Because this is constant, we have
$N_1\pi D_1 = N_2\pi D_2$, or $N_2/N_1 = D_1/D_2 = \frac{1}{2}$.
The solenoid is tightly wound, so the length of the solenoid is $\ell = Nd$, where d is the diameter of the wire. Thus we have
$\ell_2/\ell_1 = N_2/N_1 = \frac{1}{2}$.
We use the inductance of a solenoid:
$L = \mu_0 AN^2/\ell$, and form the ratio of inductances for the two conditions, so we have
$L_2/L_1 = (D_2/D_1)^2(N_2/N_1)^2/(\ell_2/\ell_1) = (2)^2(\frac{1}{2})^2/(\frac{1}{2}) = \boxed{2.}$

49. (*a*) We know from Problem 20–38 that the magnetic field at the radius R of a torus is

$B = \mu_0 NI/2\pi R.$

If the current changes by ΔI, the magnetic field change is

$\Delta B = (\mu_0 N/2\pi R)\,\Delta I.$

If we assume that the field is uniform inside the torus (because the field varies with the distance from the center of the torus, this is not actually true, but will be a good approximation if $R \gg r$), the emf induced by a current change is

$\mathcal{E} = -N\,\Delta\Phi_B/\Delta t = -NA\,\Delta B/\Delta t = -N(\mu_0 N\pi r^2/2\pi R)\,\Delta I/\Delta t = -(\mu_0 N^2 r^2/2R)\,\Delta I/\Delta t.$

When we compare this to

$\mathcal{E} = -L\,\Delta I/\Delta t$, we see that $L = \mu_0 N^2 r^2/2R.$

Because the circumference of the torus is $\ell = 2\pi R$, we can write this as

$L = \mu_0 N^2 \pi r^2/\ell$, which is **consistent** with that of a solenoid.

(*b*) For the given data, we have

$L = (4\pi\times10^{-7}\,\mathrm{T\cdot m/A})(1000\ \mathrm{turns})^2(0.75\times10^{-2}\,\mathrm{m})^2/2(0.20\,\mathrm{m}) = 1.76\times10^{-4}\,\mathrm{H} =$ **0.18 mH.**

50. (*a*) For two inductors placed in series, the current through each inductor is the same. This current is also the current through the equivalent inductor, so the total emf is

$\mathcal{E} = \mathcal{E}_1 + \mathcal{E}_2$

$= (-L_1\,\Delta I/\Delta t) + (-L_2\,\Delta I/\Delta t) = -(L_1 + L_2)\,\Delta I/\Delta t = -L_{\text{series}}\,\Delta I/\Delta t$, which gives $\boxed{L_{\text{series}} = L_1 + L_2.}$

(*b*) For two inductors placed in parallel, the potential difference across each inductor, which is the emf, is the same:

$\mathcal{E} = \mathcal{E}_1 = \mathcal{E}_2 = -L_1\,\Delta I_1/\Delta t = -L_2\,\Delta I_2/\Delta t = -L_{\text{parallel}}\,\Delta I/\Delta t.$

The total current through the equivalent inductor is

$I = I_1 + I_2$, so we have

$\Delta I/\Delta t = \Delta I_1/\Delta t + \Delta I_2/\Delta t;$

$-\mathcal{E}/L_{\text{parallel}} = -\mathcal{E}/L_1 - \mathcal{E}/L_2$, which gives $1/L_{\text{parallel}} = (1/L_1) + (1/L_2)$, or $\boxed{L_{\text{parallel}} = L_1 L_2/(L_1 + L_2).}$

(*c*) Each inductor will have an additional emf term:

$\mathcal{E}_1 = -L_1\,\Delta I_1/\Delta t \pm M\,\Delta I_2/\Delta t$, and $\mathcal{E}_2 = -L_2\,\Delta I_2/\Delta t \pm M\,\Delta I_1/\Delta t,$

where the sign of the mutual inductance term is determined by the orientation of one winding with respect to the other.

For the series arrangement, if we assume that the windings are such that the fields of each inductor are in the same direction, the induced emf from the mutual inductance will be in the same direction (have the same sign) as the self-induced emfs. Because $I_1 = I_2 = I$, we have

$\mathcal{E} = \mathcal{E}_1 + \mathcal{E}_2;$

$-L_{\text{series}}\,\Delta I/\Delta t = (-L_1\,\Delta I/\Delta t - M\,\Delta I/\Delta t) + (-L_2\,\Delta I/\Delta t - M\,\Delta I/\Delta t) = -(L_1 + L_2 + 2M)\,\Delta I/\Delta t,$

which gives $\boxed{L_{\text{series}} = L_1 + L_2 + 2M.}$

For the parallel arrangement, if we assume that the windings are such that the internal fields of each inductor are in the same direction, the induced emf from the mutual inductance will be in the opposite direction (have the opposite sign) from the self-induced emfs:

$\mathcal{E} = \mathcal{E}_1 = -L_1\,\Delta I_1/\Delta t + M\,\Delta I_2/\Delta t = \mathcal{E}_2 = -L_2\,\Delta I_2/\Delta t + M\,\Delta I_1/\Delta t.$

When these equations are combined, we get

$\Delta I_1/\Delta t = -(L_2 + M)\mathcal{E}/(L_1L_2 - M^2)$, and $\Delta I_2/\Delta t = -(L_1 + M)\mathcal{E}/(L_1L_2 - M^2).$

Because $I = I_1 + I_2$, we have

$\Delta I/\Delta t = \Delta I_1/\Delta t + \Delta I_2/\Delta t;$

$-\mathcal{E}/L_{\text{parallel}} = -(L_2 + M)\mathcal{E}/(L_1L_2 - M^2) - (L_1 + M)\mathcal{E}/(L_1L_2 - M^2),$

which gives

$\boxed{L_{\text{parallel}} = (L_1L_2 - M^2)/(L_1 + L_2 + 2M).}$

51. The magnetic field of the solenoid, which passes through the coil is

$B = \mu_0 N_1 I_1/\ell.$

When the current in the solenoid changes, the induced emf in the coil is

$\mathcal{E} = -N_2 A\,\Delta B/\Delta t = -N_2(\mu_0 N_1 A/\ell)\,\Delta I_1/\Delta t = -M\,\Delta I_1/\Delta t.$

Thus we get

$\boxed{M = \mu_0 N_1 N_2 A/\ell.}$

52. (*a*) We find the magnetic field from the first coil using the expression for a solenoid:

$$B = \mu N_1 I_1/\ell = (3000)(4\pi \times 10^{-7}\ \mathrm{T \cdot m/A})(1500\ \mathrm{turns})I_1/(0.30\ \mathrm{m}) = (18.85\ \mathrm{T/A})I_1.$$

The induced emf in the second coil is

$$\mathscr{E} = -N_2 A\ \Delta B/\Delta t$$
$$= -(800\ \mathrm{turns})\pi(0.020\ \mathrm{m})^2(18.85\ \mathrm{T/A})\ \Delta I_1/\Delta t$$
$$= -(18.95\ \mathrm{T \cdot m^2/A})(0 - 3.0\ \mathrm{A})/(0.0080\ \mathrm{s}) = \boxed{7.1 \times 10^3\ \mathrm{V.}}$$

(*b*) We find the mutual inductance from

$$\mathscr{E} = -M\ \Delta I_1/\Delta t;$$

$7.1 \times 10^3\ \mathrm{V} = -M(0 - 3.0\ \mathrm{A})/0.0080\ \mathrm{s})$, which gives $M = \boxed{19.0\ \mathrm{H.}}$

53. The direction of the induced emf will oppose the change in the current, as shown in the diagrams.

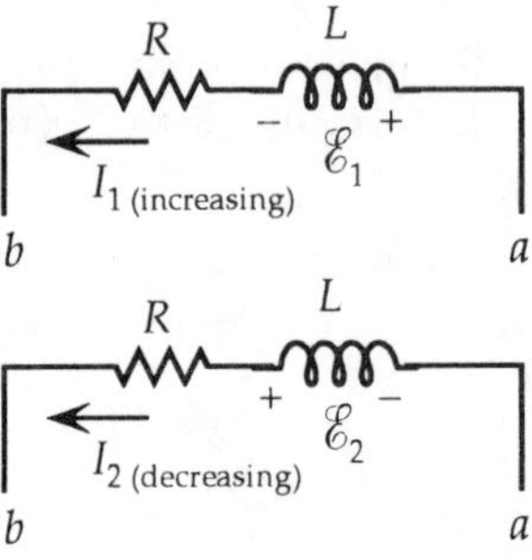

If we start at point *b* and add the potential changes, we get

$$V_b + IR \pm |\mathscr{E}_{\mathrm{induced}}| = V_a, \quad \text{or}$$
$$V_{ab} = IR \pm |\mathscr{E}_{\mathrm{induced}}|;$$
$$22.5\ \mathrm{V} = (0.860\ \mathrm{A})R + L\,(3.40\ \mathrm{A/s});$$
$$16.2\ \mathrm{V} = (0.700\ \mathrm{A})R + L\,(-1.80\ \mathrm{A/s}).$$

We have two equations for two unknowns, with the results:

$R = \boxed{24.3\ \Omega,}$ and $L = \boxed{0.463\ \mathrm{H.}}$

54. The magnetic energy in the field is

$$U = u_B V = \tfrac{1}{2}(B^2/\mu_0)L\pi r^2$$
$$= \tfrac{1}{2}[(0.80\ \mathrm{T})^2/(4\pi \times 10^{-7}\ \mathrm{T \cdot m/A})](0.36\ \mathrm{m})\pi(1.0 \times 10^{-2}\ \mathrm{m})^2 = \boxed{29\ \mathrm{J.}}$$

55. The initial energy stored in the inductor is

$$U_0 = \tfrac{1}{2}LI_0^2 = \tfrac{1}{2}(60.0 \times 10^{-3}\ \mathrm{H})(50.0 \times 10^{-3}\ \mathrm{A})^2 = \boxed{7.5 \times 10^{-5}\ \mathrm{J.}}$$

For the increase in energy, we have

$$U/U_0 = (I/I_0)^2;$$
$$10 = (I/50.0\ \mathrm{mA})^2, \text{ which gives } I = 158\ \mathrm{mA}.$$

We find the time from

$$\Delta I/\Delta t = 100\ \mathrm{mA/s} = (158\ \mathrm{mA} - 50.0\ \mathrm{mA})/\Delta t, \text{ which gives } \Delta t = \boxed{1.08\ \mathrm{s.}}$$

56. The magnetic energy in the field is

$$U = u_B V = \tfrac{1}{2}(B^2/\mu_0)h\pi r^2$$
$$= \tfrac{1}{2}[(0.50 \times 10^{-4}\ \mathrm{T})^2/(4\pi \times 10^{-7}\ \mathrm{T \cdot m/A})](10 \times 10^3\ \mathrm{m})\pi(6.38 \times 10^6\ \mathrm{m})^2 = \boxed{5.1 \times 10^{15}\ \mathrm{J.}}$$

57. (*a*) For an *LR* circuit, we have

$$I = I_{\max}(1 - e^{-t/\tau});$$
$$0.80I_{\max} = (1 - e^{-t/\tau}), \text{ or } -t/\tau = -(7.20\ \mathrm{ms})/\tau = \ln 0.20, \text{ which gives } \tau = \boxed{4.47\ \mathrm{ms.}}$$

(*b*) We find the inductance from

$$\tau = L/R;$$
$$4.47 \times 10^{-3}\ \mathrm{s} = L/(250\ \Omega), \text{ which gives } L = \boxed{1.12\ \mathrm{H.}}$$

58. At $t = 0$ there is no voltage drop across the resistor, so we have

$V = L(\Delta I/\Delta t)_0$, or $\boxed{(\Delta I/\Delta t)_0 = V/L.}$

The maximum value of the current is reached after a long time, when there is no voltage across the inductor:

$V = I_{max}R$, or $I_{max} = V/R$.

We find the time to reach maximum if the initial rate were maintained from

$V/R = (\Delta I/\Delta t)_0 t = (V/L)t$, which gives $t = L/R = \tau$.

59. For an LR circuit, we have

$I = I_{max}(1 - e^{-t/\tau})$, which we can write as

$e^{-t/\tau} = 1 - (I/I_{max})$, or $t/\tau = -\ln[1 - (I/I_{max})] = -\ln[(Imax - I)/I_{max}]$.

(*a*) $t_a/\tau = -\ln(0.10)$, which gives $t_a/\tau = \boxed{2.3.}$

(*b*) $t_b/\tau = -\ln(0.010)$, which gives $t_b/\tau = \boxed{4.6.}$

(*c*) $t_c/\tau = -\ln(0.001)$, which gives $t_c/\tau = \boxed{6.9.}$

60. (*a*) We use the inductance of a solenoid:

$L = \mu_0 AN^2/\ell$.

Because they are tightly wound, the number of turns is determined by the diameter of the wire:

$N = \ell/d$.

If we form the ratio of inductances for the two conditions, we have

$L_1/L_2 = (N_1/N_2)^2 = (d_2/d_1)^2 = 2^2 = \boxed{4.}$

(*b*) The length of wire used for the turns is $\ell_{wire} = N(\pi D)$, where D is the diameter of the solenoid. Thus for the ratio of resistances, we have

$R_1/R_2 = (\ell_{wire1}/\ell_{wire2})(d_2/d_1)^2 = (N_1/N_2)(d_2/d_1)^2 = (d_2/d_1)^3$.

For the ratio of the time constants, we get

$\tau_1/\tau_2 = (L_1/L_2)(R_2/R_1) = (L_1/L_2)(d_1/d_2)^3 = (4)(\tfrac{1}{2})^3 = \boxed{\tfrac{1}{2}.}$

61. We find the frequency from

$X_L = \omega L = 2\pi fL$;

$1.5\ k\Omega = 2\pi f(0.160\ H)$, which gives $f = \boxed{1.5\ kHz.}$

62. We find the frequency from

$X_C = 1/2\pi fC$;

$250\ \Omega = 1/2\pi f(9.20 \times 10^{-6}\ F)$, which gives $f = \boxed{69.2\ Hz.}$

63. The impedance is $Z = X_C = 1/2\pi fC$.

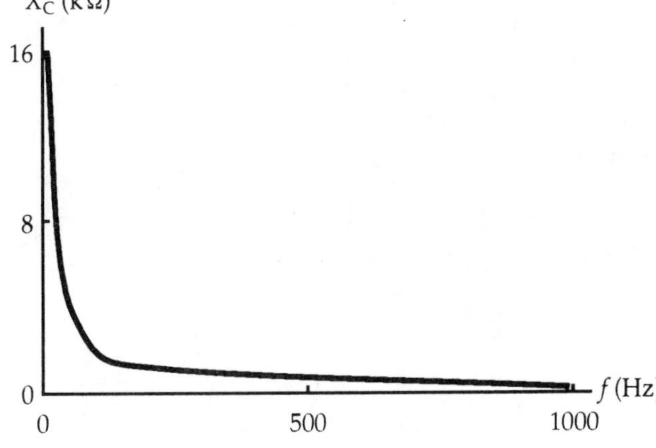

64. The impedance is $Z = X_L = \omega L = 2\pi f L$.

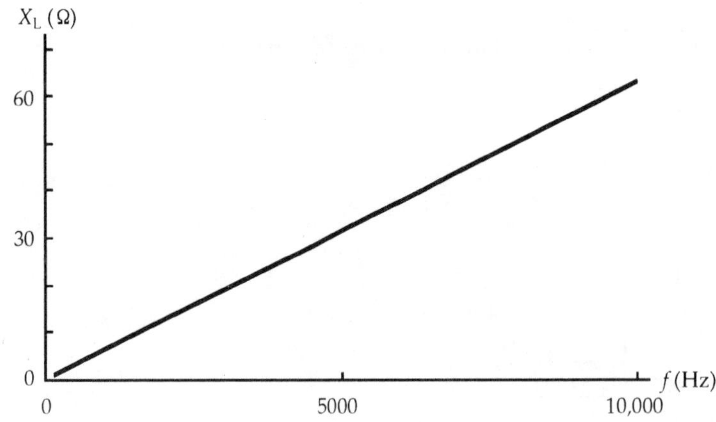

65. We find the impedance from
$Z = X_L = \omega L = 2\pi f L = 2\pi(10.0 \text{ kHz})(0.160 \text{ H}) = \boxed{10.1 \text{ k}\Omega.}$
For the rms current we have
$I_{rms} = V_{rms}/X_L = (240 \text{ V})/(10.1 \text{ k}\Omega) = \boxed{23.9 \text{ mA.}}$

66. We find the reactance from
$X_L = V_{rms}/I_{rms} = (240 \text{ V})/(12.8 \text{ A}) = 18.8\ \Omega.$
We find the inductance from
$X_L = 2\pi f L;$
$18.8\ \Omega = 2\pi(60 \text{ Hz})L$, which gives $L = 0.050 \text{ H} = \boxed{50 \text{ mH.}}$

67. (*a*) We find the impedance from
$Z = X_C = 1/2\pi f C = 1/2\pi(700 \text{ Hz})(0.030 \times 10^{-6} \text{ F}) = 7.6 \times 10^3\ \Omega = \boxed{7.6 \text{ k}\Omega.}$
(*b*) We find the peak value of the current from
$I_{peak} = \sqrt{2}\, I_{rms} = \sqrt{2}(V_{rms}/Z) = \sqrt{2}(2.0 \text{ kV})/(7.6 \text{ k}\Omega) = \boxed{0.37 \text{ A.}}$

68. Because the capacitor and resistor are in parallel, their currents are
$I_C = V/X_C$, and $I_R = V/R$.
The total current is $I = I_C + I_R$.
(*a*) The reactance of the capacitor is
$X_{C1} = 1/2\pi f_1 C = 1/2\pi(60 \text{ Hz})(0.60 \times 10^{-6} \text{ F}) = 4.42 \times 10^3\ \Omega = 4.42 \text{ k}\Omega.$
For the fraction of current that passes through C, we have
fraction1 $= I_{C1}/(I_{C1} + I_R) = (1/X_{C1})/[(1/X_{C1}) + (1/R)]$
$= (1/4.42 \text{ k}\Omega)/[(1/4.42 \text{ k}\Omega) + (1/300 \text{ k}\Omega)] = 0.064 = \boxed{6.4\%.}$
(*b*) The reactance of the capacitor is
$X_{C2} = 1/2\pi f_2 C = 1/2\pi(60{,}000 \text{ Hz})(0.60 \times 10^{-6} \text{ F}) = 4.42\ \Omega.$
For the fraction of current that passes through C, we have
fraction2 $= I_{C2}/(I_{C2} + I_R) = (1/X_{C2})/[(1/X_{C2}) + (1/R)]$
$= (1/4.42\ \Omega)/[(1/4.42\ \Omega) + (1/300 \text{ k}\Omega)] = 0.99 = \boxed{99\%.}$
Thus most of the high-frequency current passes through the capacitor.

69. Because the capacitor and resistor are in series, the impedance of the circuit is

$Z = (R^2 + X_C^2)^{1/2}$,

so the current is

$I = V/Z$,

and the voltage across the resistor is

$V_R = IR$.

(*a*) The reactance of the capacitor is

$X_{C1} = 1/2\pi f_1 C = 1/2\pi(60\text{ Hz})(2.0 \times 10^{-6}\text{ F}) = 1.33 \times 10^3\ \Omega = 1.33\text{ k}\Omega$.

The impedance of the circuit is

$Z_1 = (R^2 + X_{C1}^2)^{1/2} = [(0.500\text{ k}\Omega)^2 + (1.33\text{ k}\Omega)^2]^{1/2} = 1.42\text{ k}\Omega$.

The current is

$I_1 = V/Z_1 = (50\text{ mV})/(1.42\text{ k}\Omega) = 35.2\ \mu\text{A}$.

The voltage across the resistor is

$V_{R1} = I_1 R = (35.2\ \mu\text{A})(0.500\text{ k}\Omega) = \boxed{18\text{ mV.}}$

(*b*) The reactance of the capacitor is

$X_{C1} = 1/2\pi f_1 C = 1/2\pi(60{,}000\text{ Hz})(2.0 \times 10^{-6}\text{ F}) = 1.33\ \Omega$.

The impedance of the circuit is

$Z_1 = (R^2 + X_{C1}^2)^{1/2} = [(0.500\text{ k}\Omega)^2 + (0.00133\text{ k}\Omega)^2]^{1/2} = 0.500\text{ k}\Omega$.

The current is

$I_1 = V/Z_1 = (50\text{ mV})/(0.500\text{ k}\Omega) = 100\ \mu\text{A}$.

The voltage across the resistor is

$V_{R1} = I_1 R = (100\ \mu\text{A})(0.500\text{ k}\Omega) = \boxed{50\text{ mV.}}$

Thus the high-frequency signal passes to the resistor.

70. (*a*) The reactance of the inductor is

$X_{L1} = 2\pi f_1 L = 2\pi(60\text{ Hz})(0.50\text{ H}) = 188\ \Omega$.

The impedance of the circuit is

$Z_1 = (R^2 + X_{L1}^2)^{1/2} = [(30\text{ k}\Omega)^2 + (0.188\text{ k}\Omega)^2]^{1/2} = \boxed{30\text{ k}\Omega.}$

(*b*) The reactance of the inductor is

$X_{L2} = 2\pi f_2 L = 2\pi(30\text{ kHz})(0.50\text{ H}) = 94.2\text{ k}\Omega$.

The impedance of the circuit is

$Z_2 = (R^2 + X_{L2}^2)^{1/2} = [(30\text{ k}\Omega)^2 + (94.2\text{ k}\Omega)^2]^{1/2} = \boxed{99\text{ k}\Omega.}$

71. (*a*) The reactance of the capacitor is

$X_{C1} = 1/2\pi f_1 C = 1/2\pi(100\text{ Hz})(4.0 \times 10^{-6}\text{ F}) = 398\ \Omega = 0.398\text{ k}\Omega$.

The impedance of the circuit is

$Z_1 = (R^2 + X_{C1}^2)^{1/2} = [(1.5\text{ k}\Omega)^2 + (0.398\text{ k}\Omega)^2]^{1/2} = \boxed{1.6\text{ k}\Omega.}$

(*b*) The reactance of the capacitor is

$X_{C2} = 1/2\pi f_2 C = 1/2\pi(10{,}000\text{ Hz})(4.0 \times 10^{-6}\text{ F}) = 3.98\ \Omega = 0.00398\text{ k}\Omega$.

The impedance of the circuit is

$Z_2 = (R^2 + X_{C2}^2)^{1/2} = [(1.5\text{ k}\Omega)^2 + (0.00398\text{ k}\Omega)^2]^{1/2} = \boxed{1.5\text{ k}\Omega.}$

72. We find the impedance from

$Z = V_{rms}/I_{rms} = (120\text{ V})/(70\text{ mA}) = \boxed{1.7\text{ k}\Omega.}$

73. At 60 Hz, the reactance of the inductor is
$X_{L1} = 2\pi f_1 L = 2\pi(60\text{ Hz})(0.420\text{ H}) = 158\ \Omega.$
The impedance of the circuit is
$Z_1 = (R^2 + X_{L1}^2)^{1/2} = [(2.5\text{ k}\Omega)^2 + (0.158\text{ k}\Omega)^2]^{1/2} = 2.51\text{ k}\Omega.$
Thus the impedance at the new frequency is
$Z_2 = 2Z_1 = 2(2.51\text{ k}\Omega) = 5.02\text{ k}\Omega.$
We find the new reactance from
$Z_2 = (R^2 + X_{L2}^2)^{1/2};$
$5.02\text{ k}\Omega = [(2.5\text{ k}\Omega)^2 + X_{L2}^2]^{1/2}$, which gives $X_{L2} = 4.34\text{ k}\Omega.$
We find the new frequency from
$X_{L2} = 2\pi f_2 L;$
$4.34\text{ k}\Omega = 2\pi f_2(0.420\text{ H})$, which gives $f_2 = \boxed{1.6\text{ kHz.}}$

74. (*a*) The reactance of the capacitor is
$X_C = 1/2\pi fC = 1/2\pi(60\text{ Hz})(0.80 \times 10^{-6}\text{ F}) = 3.32 \times 10^3\ \Omega = 3.32\text{ k}\Omega.$
The impedance of the circuit is
$Z = (R^2 + X_C^2)^{1/2} = [(28.8\text{ k}\Omega)^2 + (3.32\text{ k}\Omega)^2]^{1/2} = 29.0\text{ k}\Omega.$
The rms current is
$I_{rms} = V_{rms}/Z = (120\text{ V})/(29.0\text{ k}\Omega) = \boxed{4.14\text{ mA.}}$
(*b*) We find the phase angle from
$\cos\phi = R/Z = (28.8\text{ k}\Omega)/(29.0\text{ k}\Omega) = 0.993.$
In an *RC* circuit, the current leads the voltage, so $\phi = \boxed{-6.6°.}$
(*c*) The power dissipated is
$P = I_{rms}^2 R = (4.14 \times 10^{-3}\text{ A})^2(28.8 \times 10^3\ \Omega) = \boxed{0.49\text{ W.}}$
(*d*) The rms readings across the elements are
$V_R = I_{rms}R = (4.14\text{ mA})(28.8\text{ k}\Omega) = \boxed{119\text{ V;}}$
$V_C = I_{rms}X_C = (4.14\text{ mA})(3.32\text{ k}\Omega) = \boxed{14\text{ V.}}$
Note that, because the maximum voltages occur at different times, the two readings do not add to the applied voltage of 120 V.

75. (*a*) The reactance of the inductor is
$X_L = 2\pi fL = 2\pi(60\text{ Hz})(0.900\text{ H}) = 339\ \Omega = 0.339\text{ k}\Omega.$
The impedance of the circuit is
$Z = (R^2 + X_L^2)^{1/2} = [(1.80\text{ k}\Omega)^2 + (0.339\text{ k}\Omega)^2]^{1/2} = 1.83\text{ k}\Omega.$
The rms current is
$I_{rms} = V_{rms}/Z = (120\text{ V})/(1.83\text{ k}\Omega) = \boxed{65.5\text{ mA.}}$
(*b*) We find the phase angle from
$\cos\phi = R/Z = (1.80\text{ k}\Omega)/(1.83\text{ k}\Omega) = 0.984.$
In an *RL* circuit, the current lags the voltage, so $\phi = \boxed{+10.4°.}$
(*c*) The power dissipated is
$P = I_{rms}^2 R = (65.5 \times 10^{-3}\text{ A})^2(1.80 \times 10^3\ \Omega) = \boxed{7.73\text{ W.}}$
(*d*) The rms readings across the elements are
$V_R = I_{rms}R = (65.5\text{ mA})(1.80\text{ k}\Omega) = \boxed{118\text{ V;}}$
$V_L = I_{rms}X_L = (65.5\text{ mA})(0.339\text{ k}\Omega) = \boxed{22\text{ V.}}$
Note that, because the maximum voltages occur at different times, the two readings do not add to the applied voltage of 120 V.

76. The reactance of the capacitor is

$X_C = 1/2\pi fC = 1/2\pi(10.0 \times 10^3\ \text{Hz})(5000 \times 10^{-12}\ \text{F}) = 3.18 \times 10^3\ \Omega = 3.18\ \text{k}\Omega.$

The reactance of the inductor is

$X_L = 2\pi fL = 2\pi(10.0\ \text{kHz})(0.0220\ \text{H}) = 1.38\ \text{k}\Omega.$

The impedance of the circuit is

$Z = [R^2 + (X_L - X_C)^2]^{1/2} = [(8.70\ \text{k}\Omega)^2 + (1.38\ \text{k}\Omega - 3.18\ \text{k}\Omega)^2]^{1/2} =$ **8.88 kΩ.**

We find the phase angle from

$\tan \phi = (X_L - X_L)/R = (1.38\ \text{k}\Omega - 3.18\ \text{k}\Omega)/(8.70\ \text{k}\Omega) = -0.207$, so $\phi =$ **– 11.7°.**

The rms current is

$I_{rms} = V_{rms}/Z = (300\ \text{V})/(8.70\ \text{k}\Omega) =$ **34.5 mA.**

77. We find the reactance from

$Z = [R^2 + X_L^2]^{1/2}$;

$35\ \Omega = [R^2 + (30\ \Omega)^2]^{1/2}$, which gives $R =$ **18 Ω.**

78. From the expression for V, we see that $V_0 = 4.8$ V, and $2\pi f = 754\ \text{s}^{-1}$.

For the reactances, we have

$X_C = 1/2\pi fC = 1/(754\ \text{s}^{-1})(3.0 \times 10^{-6}\ \text{F}) = 442\ \Omega.$

$X_L = 2\pi fL = (754\ \text{s}^{-1})(0.0030\ \text{H}) = 2.26\ \Omega.$

The impedance of the circuit is

$Z = [R^2 + (X_L - X_C)^2]^{1/2} = [(1.40\ \text{k}\Omega)^2 + (0.00226\ \text{k}\Omega - 0.442\ \text{k}\Omega)^2]^{1/2} = 1.47\ \text{k}\Omega.$

The power dissipated is

$P = I_{rms}^2R = (V_{rms}/Z)^2R = (V_0/Z\sqrt{2})^2R = \frac{1}{2}(V_0/Z)^2R$

$= \frac{1}{2}(4.8\ \text{V}/1.47 \times 10^3\ \Omega)^2(1.40 \times 10^3\ \Omega) = 7.5 \times 10^{-3}\ \text{W} =$ **7.5 mW.**

79. We find the rms current from

$P = I_{rms}^2R;$

$9.50\ \text{W} = I_{rms}^2(250\ \Omega)$, which gives $I_{rms} = 0.195$ A.

We find the impedance from

$Z = V_{rms}/I_{rms} = (50.0\ \text{V})/(0.195\ \text{A}) = 256\ \Omega.$

From this we can find the reactance of the coil:

$Z^2 = R^2 + X_L^2;$

$(256\ \Omega)^2 = (250\ \Omega)^2 + X_L^2$, which gives $X_L = 57.4\ \Omega$.

We find the frequency that will produce this reactance from

$X_L = 2\pi fL$;

$57.4\ \Omega = 2\pi f(0.040\ \text{H})$, which gives $f =$ **228 Hz.**

80. The voltage across the resistor is in phase with the current:

$V_R = IR = (I_0 \cos \omega t) = I_0R \cos \omega t.$

The voltage across the inductor leads the current by 90° or $\pi/2$:

$V_L = I_0 \cos (\omega t + \pi/2)X_L = I_0\omega L \cos (\omega t + \pi/2).$

The voltage across the capacitor lags the current by 90° or $\pi/2$:

$V_C = I_0 \cos (\omega t - \pi/2)X_C = (I_0/\omega C) \cos (\omega t - \pi/2).$

81. The resonant frequency is

$f_0 = (1/2\pi)(1/LC)^{1/2} = [1/(50 \times 10^{-6}\ \text{H})(3500 \times 10^{-12}\ \text{F})]^{1/2} =$ **3.8 ×10⁵ Hz.**

82. (*a*) The resonant frequency is given by
$f_0^2 = (1/2\pi)^2(1/LC)$.
When we form the ratio for the two stations, we get
$(f_{02}/f_{01})^2 = C_1/C_2$;
$(1600 \text{ kHz}/580 \text{ kHz})^2 = (2800 \text{ pF})/C_2$, which gives $C_2 =$ **368 pF.**

(*b*) We find the inductance from the first frequency:
$f_{01} = (1/2\pi)(1/LC_1)^{1/2}$;
$580 \times 10^3 \text{ Hz} = (1/2\pi)[1/L(2800 \times 10^{-12} \text{ F})]^{1/2}$, which gives $L = 2.68 \times 10^{-5} \text{ H} =$ **26.8 μH.**

83. (*a*) We find the capacitance from the frequency:
$f_0 = (1/2\pi)(1/LC)^{1/2}$;
$3600 \text{ Hz} = (1/2\pi)[1/(4.8 \times 10^{-3} \text{ H})C]^{1/2}$, which gives $C = 4.07 \times 10^{-7} \text{ F} =$ **0.41 μF.**

(*b*) At resonance the impedance is the resistance, so the current is
$I_0 = V_0/R = (50 \text{ V})/(4.4\ \Omega) =$ **11A.**

84. (*a*) The frequency of oscillation is the resonance frequency. We find the inductance from
$f_0 = (1/2\pi)(1/LC)^{1/2}$;
$20 \times 10^3 \text{ Hz} = (1/2\pi)[1/L(3000 \times 10^{-12} \text{ F})]^{1/2}$, which gives $L = 2.11 \times 10^{-2} \text{ H} =$ **21 mH.**

(*b*) The energy initially stored in the capacitor will oscillate between the capacitor and the inductor. We find the maximum current, when all of the energy is stored in the inductor, by equating the maximum energies:
$\frac{1}{2}CV_0^2 = \frac{1}{2}LI_{max}^2$;
$(3000 \times 10^{-12} \text{ F})(120 \text{ V})^2 = (2.11 \times 10^{-2} \text{ H})I_{max}^2$, which gives $I_{max} = 4.5 \times 10^{-2} \text{ A} =$ **45 mA.**

(*b*) The maximum energy stored in the inductor is
$U_{Lmax} = \frac{1}{2}LI_{max}^2 = \frac{1}{2}(2.11 \times 10^{-2} \text{ H})(4.5 \times 10^{-2} \text{ A})^2 =$ **2.2×10^{-5} J.**

85. We find the equivalent resistance of the two speakers in parallel from
$1/R_{eq} = (1/R_1) + (1/R_2) = (1/8\ \Omega) + (1/8\ \Omega)$, which gives $R_{eq} = 4\ \Omega$.
Thus the speakers should be connected to the **4 Ω** terminals.

86. Because the loudspeaker is connected to the secondary side of the transformer, we have
$Z_P/Z_S = (N_P/N_S)^2$;
$(30 \times 10^3\ \Omega)/(8.0\ \Omega) = (N_P/N_S)^2$, which gives $N_P/N_S =$ **61.**

87. (*a*) Because the increasing current in the first loop produces an increasing magnetic field in the second loop, **yes,** there will be an induced current in the second loop.

(*b*) Because the current in the first loop increases immediately, the induced current will increase **immediately.**

(*c*) The induced current will stop **when the current in the first loop reaches the steady state.**

(*d*) To oppose the increase in the magnetic field away from you from the first loop, the induced current will produce a magnetic field toward you. Thus the induced current will be **counterclockwise.**

(*e*) While there is an induced current, the loops will have parallel opposite currents, so, **yes,** there will be a force.

(*f*) Because we have opposite currents, the force is repulsive, so the force on each loop is **away from the other loop.**

88. (*a*) The clockwise current in the left-hand loop produces a magnetic field which is into the page within the loop and out of the page outside the loop. Thus the right-hand loop is in a magnetic field out of the page. Before the current in the left-hand loop reaches its steady state, there will be an induced current in the right-hand loop that will produce a magnetic field into the page to oppose the increase of the field from the left-hand loop. Thus the induced current will be **clockwise.**

(*b*) After a long time, the current in the left-hand loop will be constant, so there will be **no induced current.**

(*c*) If the second loop is pulled to the right, the magnetic field out of the page from the left-hand loop through the second loop will decrease. During the motion, there will be an induced current in the right-hand loop that will produce a magnetic field out of the page to oppose the decrease of the field from the left-hand loop. Thus the induced current will be **counterclockwise.**

89. The average induced emf is

$$\mathcal{E} = -\Delta\Phi_B/\Delta t = -A\,\Delta B/\Delta t = -(0.240\text{ m})^2(0 - 0.755\text{ T})/(0.0400\text{ s}) = 1.09\text{ V}.$$

The average current is

$$I = \mathcal{E}/R = (1.09\text{ V})/(6.50\ \Omega) = 0.168\text{ A}.$$

The energy dissipated is

$$E = I^2R\,\Delta t = (0.168\text{ A})^2(6.50\ \Omega)(0.0400\text{ s}) = \boxed{7.33\times10^{-3}\text{ J}.}$$

90. A side view of the rail and bar is shown in the figure. The component of the velocity of the bar that is perpendicular to the magnetic field is $v\cos\theta$, so the induced emf is

$$\mathcal{E} = BLv\cos\theta.$$

This produces a current in the wire

$$I = \mathcal{E}/R = (BLv\cos\theta)/R \text{ into the page.}$$

Because the current is perpendicular to the magnetic field, the force on the wire from the magnetic field will be horizontal, as shown, with magnitude

$$F_B = I\ell B = (B^2\ell^2 v\cos\theta)/R.$$

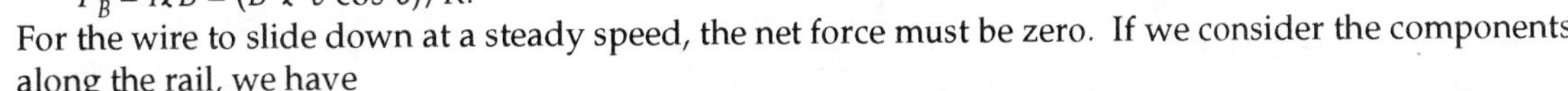

For the wire to slide down at a steady speed, the net force must be zero. If we consider the components along the rail, we have

$$F_B\cos\theta - mg\sin\theta = 0, \quad \text{or}$$

$$[(B^2\ell^2 v\cos\theta)/R]\cos\theta = (B^2\ell^2 v\cos^2\theta)/R = mg\sin\theta;$$

$(0.55\text{ T})^2(0.30\text{ m})^2v(\cos^2 5.0°)/(0.60\ \Omega) = (0.040\text{ kg})(9.80\text{ m/s}^2)\sin 5.0°$, which gives $\boxed{v = 0.76\text{ m/s}.}$

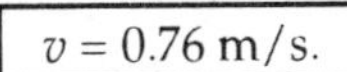

91. The average induced emf is

$$\mathcal{E} = -\Delta\Phi_B/\Delta t = -NA\,\Delta B/\Delta t = -NA[(-B) - (+B)]/\Delta t = 2NAB/\Delta t.$$

The average current is

$$I = \mathcal{E}/R = 2NAB/R\,\Delta t,$$

so the total charge that passes through the galvanometer is

$$Q = I\,\Delta t = (2NAB/R\,\Delta t)\,\Delta t = 2NAB/R, \quad \text{or} \quad B = QR/2NA.$$

92. (*a*) The velocity of the bar is perpendicular to the magnetic field, so the induced emf is
$\mathcal{E} = BLv$.
This produces a current in the wire
$I = \mathcal{E}/R = BLv/R$ downward.
Because the current is perpendicular to the magnetic field, the force on the wire from the magnetic field will be horizontal to the left with magnitude
$F_B = I\ell B = B^2\ell^2v/R$.
An external force of this magnitude must be exerted to the right to maintain the motion. The power expended by this force is
$P = Fv = (B^2\ell^2v/R)v = B^2\ell^2v^2/R$.
(*b*) The power dissipated in the resistance is
$P = I^2R = (BLv/R)^2R = B^2\ell^2v^2/R$,
which is the power expended by the external force. We expect this, since there is no increase in kinetic energy.

93. (*a*) From the efficiency of the transformer, we have $P_S = 0.80P_P$.
For the power input to the transformer, we have
$P_P = I_PV_P$;
$(58\text{ W})/0.80 = I_P(110\text{ V})$, which gives $I_P =$ **0.66 A.**
(*b*) We find the secondary voltage from
$P_P = V_S^2/R_S$;
$58\text{ W} = V_S^2/(2.4\ \Omega)$, which gives $V_S = 11.8$ V.
We find the ratio of the number of turns from
$N_P/N_S = V_P/V_S = (110\text{ V})/(11.8\text{ V}) =$ **9.3.**

94. (*a*) The voltage drop across the lines is
$\Delta V = 2IR = 2(700\text{ A})(0.80\ \Omega) = 1.12 \times 10^3\text{ V} = 1.12\text{ kV}$.
Thus the voltage at the other end is
$V_{out} = V_{in} - \Delta V = 42\text{ kV} - 1.12\text{ kV} =$ **41 kV.**
(*b*) The power input is
$P_{in} = IV_{in} = (0.700\text{ kA})(42\text{ kV}) =$ **29.4 MW.**
(*c*) The power loss in the lines is
$P_{loss} = 2I^2R = 2(0.700\text{ kA})^2(0.80\ \Omega) =$ **0.78 MW.**
(*d*) The power output is
$P_{out} = IV_{out} = (0.700\text{ kA})(40.9\text{ kV}) =$ **28.6 MW.**

95. We find the peak emf from
$\mathcal{E}_{peak} = NBA\omega = (125\text{ turns})(0.200\text{ T})(0.066\text{ m})^2(120\text{ rev/s})(2\pi\text{ rad/rev}) =$ **82 V.**

96. (*a*) Because we have direct coupling, the torque provided by the motor balances the torque of the friction force:
$NIAB = Fr$;
$(300\text{ turns})I(0.10\text{ m})(0.15\text{ m})(0.60\text{ T}) = (250\text{ N})(0.25\text{ m})$, which gives $I =$ **23 A.**
(*b*) To maintain the speed, we have a force equal to the friction force, so the power required is
$Fv = (250\text{ N})(30\text{ km/h})/(3.6\text{ ks/h}) = 2.08 \times 10^3$ W.
This must be provided by the net power from the motor, which is
$P_{net} = IV_{in} - I^2R = I(V_{in} - IR) = I\mathcal{E}_{back} = Fv$;
$(23\text{ A})\mathcal{E}_{back} = 2.08 \times 10^3$ W, which gives $\mathcal{E}_{back} =$ **90 V.**
(*c*) The power dissipation in the coils is
$P_{loss} = P_{in} - P_{net} = (23\text{ A})(10)(12\text{ V}) - 2.08 \times 10^3\text{ W} =$ **6.9 × 10² W.**
(*d*) The useful power percentage is
$(P_{net}/P_{in})(100) = (I\mathcal{E}_{back}/IV_{in})(100) = (90\text{ V}/120\text{ V})(100) =$ **75%.**

97. We find the impedance from

$Z = X_L = V_{rms}/I_{rms} = (220\text{ V})/(5.8\text{ A}) = 37.9\ \Omega.$

We find the inductance from

$X_L = 2\pi fL;$

$37.9\ \Omega = 2\pi(60\text{ Hz})L$, which gives $L =$ $\boxed{0.10\text{ H.}}$

98. For the current and voltage to be in phase, the net reactance of the capacitor and inductor must be zero, which means that we have resonance. Thus we have

$f_0 = (1/2\pi)(1/LC)^{1/2};$

$3360\text{ Hz} = (1/2\pi)[1/(0.230\text{ H})C]^{1/2}$, which gives $C = 9.76 \times 10^{-9}\text{ F} =$ $\boxed{9.76\text{ nF.}}$

99. Because the current lags the voltage, this must be an $\boxed{RL\text{ circuit.}}$

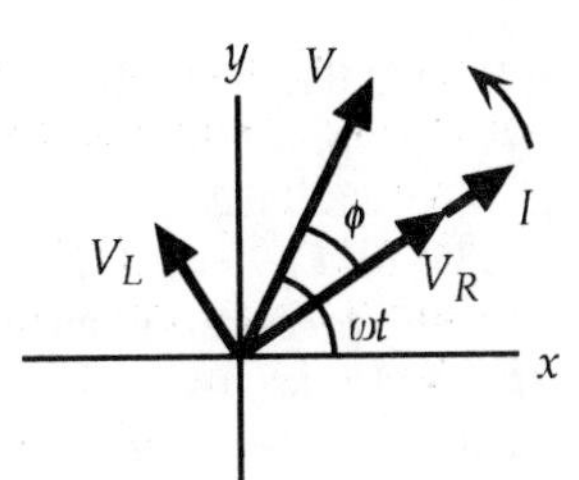

We find the impedance of the circuit from

$Z = V_{rms}/I_{rms} = (120\text{ V})/(5.6\text{ A}) = 21.4\ \Omega.$

We find the resistance from the phase:

$\cos\phi = R/Z;$

$\cos 50° = R/(21.4\ \Omega)$, which gives $R =$ $\boxed{13.8\ \Omega.}$

We find the reactance from the phase:

$\sin\phi = X_L/Z;$

$\sin 50° = X_L/(21.4\ \Omega)$, which gives $X_L = 16.4\ \Omega.$

We find the inductance from

$X_L = 2\pi fL;$

$16.4\ \Omega = 2\pi(60\text{ Hz})L$, which gives $L = 4.35 \times 10^{-2}\text{ H} =$ $\boxed{43.5\text{ mH.}}$

100. Because the circuit is in resonance, we find the inductance from

$f_0 = (1/2\pi)(1/LC)^{1/2};$

$48.0 \times 10^6\text{ Hz} = (1/2\pi)[1/L(220 \times 10^{-12}\text{ F})]^{1/2}$, which gives $L = 5.0 \times 10^{-8}\text{ H}.$

If r is the radius of the coil, the number of turns is

$N = \ell_{wire}/2\pi r.$

If d is the diameter of the wire, for closely-wound turns, the length of the coil is

$\ell = Nd.$

Thus the inductance of the coil is

$L = \mu_0 AN^2/\ell = \mu_0\pi r^2(\ell_{wire}/2\pi r)^2/Nd = \mu_0\ell_{wire}^2/4\pi Nd;$

$5.0 \times 10^{-8}\text{ H} = (4\pi \times 10^{-7}\text{ T}\cdot\text{m/A})(14.0\text{ m})^2/4\pi N(1.1 \times 10^{-3}\text{ m})$, which gives $N =$ $\boxed{3.56 \times 10^5\text{ turns.}}$

101. We find the resistance of the coil from the dc current:

$R = V_{dc}/I_{dc} = (36\text{ V})/(2.5\text{ A}) =$ $\boxed{14\ \Omega.}$

We find the impedance from the ac current:

$Z = V_{rms}/I_{rms} = (120\text{ V})/(3.8\text{ A}) = 31.6\ \Omega.$

We find the reactance from

$Z = [R^2 + X_L^2]^{1/2};$

$31.6\ \Omega = [(14.4\ \Omega)^2 + X_L^2]^{1/2}$, which gives $X_L = 28.1\ \Omega.$

We find the inductance from

$X_L = 2\pi fL;$

$28.1\ \Omega = 2\pi(60\text{ Hz})L$, which gives $L = 0.075\text{ H} =$ $\boxed{75\text{ mH.}}$

102. (*a*) At resonance we have

$2\pi f_0 = (1/LC)^{1/2}$.

The Q factor is

$Q = V_C/V_R = I_{rms}X_C/I_{rms}R = 1/2\pi f_0 CR = (LC)^{1/2}/CR = (1/R)(L/C)^{1/2}$.

(*b*) We find the inductance from the resonance frequency:

$2\pi f_0 = (1/LC)^{1/2}$;

$2\pi(1.0 \times 10^6 \text{ Hz}) = [1/L(0.010 \times 10^{-6} \text{ F})]$, which gives $L = 2.5 \times 10^{-6}$ H = $\boxed{2.5\ \mu\text{H.}}$

We find the resistance required from

$Q = (1/R)(L/C)^{1/2}$;

$1000 = (1/R)[(2.5 \times 10^{-6} \text{ H})/(0.010 \times 10^{-6} \text{ F})]$, which gives $R = \boxed{1.6 \times 10^{-2}\ \Omega.}$

103. We find the impedance from the power factor:

$\cos\phi = R/Z$;

$0.17 = (200\ \Omega)/Z$, which gives $Z = 1.18 \times 10^3\ \Omega$.

We get an expression for the reactances from

$Z^2 = R^2 + (X_L - X_C)^2$;

$(1.18 \times 10^3\ \Omega)^2 = (200\ \Omega)^2 + (X_L - X_C)^2$, which gives $X_L - X_C = \pm 1.16 \times 10^3\ \Omega$.

When we express this in terms of the inductance and capacitance, we get

$2\pi fL - (1/2\pi fC) = \pm 1.16 \times 10^3\ \Omega$;

$2\pi f(0.020 \text{ H}) - [1/2\pi f(50 \times 10^{-9} \text{ F})] = \pm 1.16 \times 10^3\ \Omega$, which reduces to two quadratic equations:

$0.126f^2 \pm (1.16 \times 10^3 \text{ Hz})f - 3.183 \times 10^6 \text{ Hz}^2$,

which have positive solutions of $f = \boxed{2.2 \times 10^3 \text{ Hz}, 1.1 \times 10^4 \text{ Hz.}}$

CHAPTER 22

1. The displacement current is
   $I_D = \epsilon_0 A\,(\Delta E/\Delta t) = (8.85\times 10^{-12}\text{ F/m})(0.028\text{ m})^2(2.0\times 10^6\text{ V/m}\cdot\text{s}) = \boxed{1.4\times 10^{-8}\text{ A.}}$

2. The electric field between the plates depends on the voltage:
   $E = V/d$, so $\Delta E/\Delta t = (1/d)\,\Delta V/\Delta t$.
   The displacement current is
   $I_D = \epsilon_0 A\,(\Delta E/\Delta t) = (8.85\times 10^{-12}\text{ F/m})\pi(0.030\text{ m})^2(120\text{ V/s})/(1.3\times 10^{-3}\text{ m}) = \boxed{2.3\times 10^{-9}\text{ A.}}$

3. The current in the wires must also be the displacement current in the capacitor. We find the rate at which the electric field is changing from
   $I_D = \epsilon_0 A\,(\Delta E/\Delta t)$;
   $3.8\text{ A} = (8.85\times 10^{-12}\text{ F/m})(0.0190\text{ m})^2(\Delta E/\Delta t)$, which gives $\Delta E/\Delta t = \boxed{1.2\times 10^{15}\text{ V/m}\cdot\text{s.}}$

4. The current in the wires is the rate at which charge is accumulating on the plates and must also be the displacement current in the capacitor. Because the location is outside the capacitor, we can use the expression for the magnetic field of a long wire:
   $B = (\mu_0/4\pi)2I/R = (10^{-7}\text{ T}\cdot\text{m/A})2(25.0\times 10^{-3}\text{ A})/(0.100\text{ m}) = \boxed{5.00\times 10^{-8}\text{ T.}}$
   After the capacitor is fully charged, all currents will be zero, so the magnetic field will be $\boxed{\text{zero.}}$

5. The electric field between the plates depends on the voltage:
   $E = V/d$, so $\Delta E/\Delta t = (1/d)\,\Delta V/\Delta t$.
   Thus the displacement current is
   $I_D = \epsilon_0 A\,(\Delta E/\Delta t) = (\epsilon_0 A/d)(\Delta V/\Delta t) = C\,\Delta V/\Delta t$.

6. (*a*) From the cylindrical symmetry, we know that the magnetic field will be circular, centered on the axis of the current. We choose a circular path with radius $r < R$ to apply Ampere's law. The total displacement current between the plates is
   $I_D = \epsilon_0 A\,(\Delta E/\Delta t)$.
   If the charge is uniformly distributed on the plates, we find the displacement current through the circular path from the area of the circle:
   $I_D' = \epsilon_0 \pi r^2(\Delta E/\Delta t)$.
   From the symmetry, the magnitude of **B** is constant on the path, so for Ampere's law, we have
   $\Sigma\, B\,\Delta L = B\,\Sigma\,\Delta L = \mu_0 I_D'$;
   $B(2\pi r) = \mu_0\epsilon_0\pi r^2\,(\Delta E/\Delta t)$, which gives $B = \frac{1}{2}\mu_0\epsilon_0 r\,(\Delta E/\Delta t)$.
   For a point outside the capacitor, the same symmetry tells us to choose a circular path with radius $r > R$. The total displacement current passes through the circle, so we have
   $\Sigma\, B\,\Delta L = B\,\Sigma\,\Delta L = \mu_0 I_D$;
   $B(2\pi r) = \mu_0\epsilon_0\pi R^2\,(\Delta E/\Delta t)$, which gives $B = (\mu_0\epsilon_0 R^2/2r)(\Delta E/\Delta t)$.
   (*b*) From the preceding result, we have
   $B(2\pi r) = \mu_0 I_D$, or $B = \mu_0 I_D/2\pi r$.
   (*c*) The conduction current between the plates is equal to the current in the wires, so it is as if there were a single current.

7. The electric field is
   $E = cB = (3.00\times 10^8\text{ m/s})(17.5\times 10^{-9}\text{ T}) = \boxed{5.25\text{ V/m.}}$

8. We find the magnetic field from
   $E = cB$;
   0.43×10^{-4} V/m = $(3.00 \times 10^{8}$ m/s$)B$, which gives B = $\boxed{1.4 \times 10^{-13} \text{ T.}}$

9. The frequency of the two fields must be the same: $\boxed{80.0 \text{ kHz.}}$
   The rms strength of the electric field is
   $E_{rms} = cB_{rms} = (3.00 \times 10^{8}\text{ m/s})(7.75 \times 10^{-9}\text{ T}) = \boxed{2.33 \text{ V/m.}}$
   The electric field is perpendicular to both the direction of travel and the magnetic field, so the electric field oscillates along the $\boxed{\text{horizontal north-south line.}}$

10. The wavelength of the radar signal is
    $\lambda = c/f = (3.00 \times 10^{8}\text{ m/s})/(27.75 \times 10^{9}\text{ Hz}) = 1.08 \times 10^{-2}\text{ m} = \boxed{1.08 \text{ cm.}}$

11. The frequency of the X-ray is
    $f = c/\lambda = (3.00 \times 10^{8}\text{ m/s})/(0.10 \times 10^{-9}\text{ m}) = \boxed{3.0 \times 10^{18}\text{ Hz.}}$

12. The frequency of the microwave is
    $f = c/\lambda = (3.000 \times 10^{8}\text{ m/s})/(1.50 \times 10^{-2}\text{ m}) = \boxed{2.00 \times 10^{10}\text{ Hz.}}$

13. The frequency of the wave is
    $f = c/\lambda = (3.00 \times 10^{8}\text{ m/s})/(900 \times 10^{-9}\text{ m}) = \boxed{3.33 \times 10^{14}\text{ Hz.}}$
    This frequency is just outside the red end of the visible region, so it is $\boxed{\text{infrared.}}$

14. The wavelength of the wave is
    $\lambda = c/f = (3.00 \times 10^{8}\text{ m/s})/(9.56 \times 10^{14}\text{ Hz}) = 3.14 \times 10^{-7}\text{ m} = \boxed{314 \text{ nm.}}$
    This wavelength is just outside the violet end of the visible region, so it is $\boxed{\text{ultraviolet.}}$

15. The time for light to travel from the Sun to the Earth is
    $\Delta t = L/c = (1.50 \times 10^{11}\text{ m})/(3.00 \times 10^{8}\text{ m/s}) = 5.00 \times 10^{2}\text{ s} = \boxed{8.33 \text{ min.}}$

16. We convert the units:
    $d = (4.2\text{ ly})(3.00 \times 10^{8}\text{ m/s})(3.16 \times 10^{7}\text{ s/yr}) = \boxed{4.0 \times 10^{16}\text{ m.}}$

17. The eight-sided mirror would have to rotate 1/8 of a revolution for the succeeding mirror to be in position to reflect the light in the proper direction. During this time the light must travel to the opposite mirror and back. Thus the angular speed required is
    $\omega = \Delta\theta/\Delta t = (2\pi\text{ rad}/8)/(2L/c) = (\pi\text{ rad})c/8L$
    $= (\pi\text{ rad})(3.00 \times 10^{8}\text{ m/s})/8(3.5 \times 10^{3}\text{ m}) = \boxed{3.4 \times 10^{3}\text{ rad/s}}$ $(3.2 \times 10^{4}\text{ rev/min})$.

18. (*a*) If we assume the closest approach of Mars to Earth, we have
    $\Delta t = L/c = [(227.9 - 149.6) \times 10^{9}\text{ m}]/(3.00 \times 10^{8}\text{ m/s}) = 2.6 \times 10^{2}\text{ s} = \boxed{4.3 \text{ min.}}$
    (*b*) If we assume the closest approach of Saturn to Earth, we have
    $\Delta t = L/c = [(1427 - 149.6) \times 10^{9}\text{ m}]/(3.00 \times 10^{8}\text{ m/s}) = 4.3 \times 10^{3}\text{ s} = \boxed{71 \text{ min.}}$

19. The distance that light travels is one year is
    $d = (3.00 \times 10^{8}\text{ m/s})(3.156 \times 10^{7}\text{ s/yr}) = \boxed{9.47 \times 10^{15}\text{ m.}}$

20. If we ignore the time for the sound to travel to the microphone, the time difference is
$\Delta t = t_{radio} - t_{sound} = (d_{radio}/c) - (d_{sound}/v_{sound}) = (3000\ \text{m}/3.00 \times 10^8\ \text{m/s}) - (50\ \text{m}/343\ \text{m/s}) = -0.14\ \text{s},$
so the **person at the radio hears the voice 0.14 s sooner.**

21. The length of the pulse is $\Delta d = c\,\Delta t$, so the number of wavelengths in this length is
$N = (c\,\Delta t)/\lambda = (3.00 \times 10^8\ \text{m/s})(30 \times 10^{-12}\ \text{s})/(1062 \times 10^{-9}\ \text{m}) =$ **8.5×10^3 wavelengths.**
The time for the length of the pulse to be one wavelength is
$\Delta t' = \lambda/c = (1062 \times 10^{-9}\ \text{m})/(3.00 \times 10^8\ \text{m/s}) = 3.5 \times 10^{-15}\ \text{s} =$ **3.5 fs.**

22. The energy per unit area per unit time is
$S = \frac{1}{2}c\epsilon_0 E_0^2$
$= \frac{1}{2}(3.00 \times 10^8\ \text{m/s})(8.85 \times 10^{-12}\ \text{C}^2/\text{N}\cdot\text{m}^2)(26.5 \times 10^{-3}\ \text{V/m})^2 =$ **$9.32 \times 10^{-7}\ \text{W/m}^2$.**

23. The energy per unit area per unit time is
$S = cB_{rms}^2/\mu_0$
$= (3.00 \times 10^8\ \text{m/s})(22.5 \times 10^{-9}\ \text{T})^2/(4\pi \times 10^{-7}\ \text{T}\cdot\text{m/A}) = 0.121\ \text{W/m}^2.$
We find the time from
$t = U/AS = (135\ \text{J})/(1.00 \times 10^{-4}\ \text{m}^2)(0.121\ \text{J/s}\cdot\text{m}^2) = 1.12 \times 10^7\ \text{s} = 0.353\ \text{yr} =$ **4.24 months.**

24. The energy per unit area per unit time is
$S = c\epsilon_0 E_{rms}^2$
$= (3.00 \times 10^8\ \text{m/s})(8.85 \times 10^{-12}\ \text{C}^2/\text{N}\cdot\text{m}^2)(18.6 \times 10^{-3}\ \text{V/m})^2 = 9.19 \times 10^{-7}\ \text{W/m}^2.$
We find the energy transported from
$U = AS = (1.00 \times 10^{-4}\ \text{m}^2)(9.19 \times 10^{-7}\ \text{W/m}^2)(3600\ \text{s/h}) =$ **3.31×10^{-7} J/h.**

25. Because the wave spreads out uniformly over the surface of the sphere, the power flux is
$S = P/A = (1000\ \text{W})/4\pi(10.0\ \text{m})^2 =$ **$0.796\ \text{W/m}^2$.**
We find the rms value of the electric field from
$S = c\epsilon_0 E_{rms}^2;$
$0.796\ \text{W/m}^2 = (3.00 \times 10^8\ \text{m/s})(8.85 \times 10^{-12}\ \text{C}^2/\text{N}\cdot\text{m}^2)E_{rms}^2$, which gives $E_{rms} =$ **17.3 V/m.**

26. For the energy density, we have
$u = \frac{1}{2}\epsilon_0 E_0^2 = S/c = (1350\ \text{W/m}^2)/(3.00 \times 10^8\ \text{m/s}) = 4.50 \times 10^{-6}\ \text{J/m}^3.$
The radiant energy is
$U = uV = (4.50 \times 10^{-6}\ \text{J/m}^3)(1.00\ \text{m}^3) =$ **4.50×10^{-6} J.**

27. The energy per unit area per unit time is
$S = P/A = c\epsilon_0 E_{rms}^2;$
$(12.8 \times 10^{-3}\ \text{W})/\pi(1.00 \times 10^{-3}\ \text{m})^2 = (3.00 \times 10^8\ \text{m/s})(8.85 \times 10^{-12}\ \text{C}^2/\text{N}\cdot\text{m}^2)E_{rms}^2,$
which gives $E_{rms} =$ **1.24×10^3 V/m.**
The rms value of the magnetic field is
$B_{rms} = E_{rms}/c = (1.24 \times 10^3\ \text{V/m})/(3.00 \times 10^8\ \text{m/s}) =$ **4.13×10^{-6} T.**

28. The radiation from the Sun has the same intensity in all directions, so the rate at which it reaches the Earth is the rate at which it passes through a sphere centered at the Sun:
$P = S4\pi R^2 = (1350\ \text{W/m}^2)4\pi(1.5 \times 10^{11}\ \text{m})^2 =$ **3.8×10^{26} W.**

29. (*a*) The energy emitted in each pulse is

$$U = Pt = (2.5 \times 10^{11}\ \text{W})(1.0 \times 10^{-9}\ \text{s}) = \boxed{2.5 \times 10^2\ \text{J}.}$$

(*b*) We find the rms electric field from

$$S = P/A = c\epsilon_0 E_{\text{rms}}^2;$$

$$(2.5 \times 10^{11}\ \text{W})/\pi(2.2 \times 10^{-3}\ \text{m})^2 = (3.00 \times 10^8\ \text{m/s})(8.85 \times 10^{-12}\ \text{C}^2/\text{N}\cdot\text{m}^2)E_{\text{rms}}^2,$$

which gives $E_{\text{rms}} = \boxed{2.5 \times 10^9\ \text{V/m}.}$

30. The resonant frequency is given by

$$f_0^2 = (1/2\pi)^2(1/LC).$$

When we form the ratio for the two stations, we get

$$(f_{02}/f_{01})^2 = C_1/C_2;$$

$$(1550\ \text{kHz}/550\ \text{kHz})^2 = (2400\ \text{pF})/C_2, \text{ which gives } C_2 = \boxed{302\ \text{pF}.}$$

31. We find the capacitance from the resonant frequency:

$$f_0 = (1/2\pi)(1/LC)^{1/2};$$

$$96.1 \times 10^6\ \text{Hz} = (1/2\pi)[1/(1.8 \times 10^{-6}\ \text{H})C]^{1/2}, \text{ which gives } C = 1.5 \times 10^{-12}\ \text{F} = \boxed{1.5\ \text{pF}.}$$

32. We find the inductance for the first frequency:

$$f_{01} = (1/2\pi)(1/L_1C)^{1/2};$$

$$8 \times 10^6\ \text{Hz} = (1/2\pi)[1/L_1(840 \times 10^{-12}\ \text{F})]^{1/2}, \text{ which gives } L_1 = 3.89 \times 10^{-9}\ \text{H} = 3.89\ \text{nH}.$$

For the second frequency we have

$$f_{02} = (1/2\pi)(1/L_2C)^{1/2};$$

$$108 \times 10^6\ \text{Hz} = (1/2\pi)[1/L_2(840 \times 10^{-12}\ \text{F})]^{1/2}, \text{ which gives } L_2 = 2.59 \times 10^{-9}\ \text{H} = 2.59\ \text{nH}.$$

Thus the range of inductances is

$$\boxed{2.59\ \text{nH} \le L \le 3.89\ \text{nH}.}$$

33. (*a*) The minimum value of C corresponds to the higher frequency, so we have

$$f_{01} = (1/2\pi)(1/LC_1)^{1/2};$$

$$15.0 \times 10^6\ \text{Hz} = (1/2\pi)[1/L(92 \times 10^{-12}\ \text{F})]^{1/2}, \text{ which gives } L = 1.22 \times 10^{-6}\ \text{H} = \boxed{1.22\ \mu\text{H}.}$$

(*b*) The maximum value of C corresponds to the lower frequency, so we have

$$f_{02} = (1/2\pi)(1/LC_2)^{1/2};$$

$$14.0 \times 10^6\ \text{Hz} = (1/2\pi)[1/(1.22 \times 10^{-6}\ \text{H})C_2]^{1/2}, \text{ which gives } C_2 = 1.06 \times 10^{-10}\ \text{F} = \boxed{106\ \text{pF}.}$$

34. To produce the voltage over the length of the antenna, we have

$$E_{\text{rms}} = V_{\text{rms}}/d = (1.00 \times 10^{-3}\ \text{V})/(1.40\ \text{m}) = \boxed{7.14 \times 10^{-4}\ \text{V/m}.}$$

The rate of energy transport is

$$S = c\epsilon_0 E_{\text{rms}}^2$$

$$= (3.00 \times 10^8\ \text{m/s})(8.85 \times 10^{-12}\ \text{C}^2/\text{N}\cdot\text{m}^2)(7.14 \times 10^{-4}\ \text{V/m})^2 = \boxed{1.35 \times 10^{-9}\ \text{W/m}^2.}$$

35. The wavelength of the AM station is

$$\lambda = c/f = (3.00 \times 10^8\ \text{m/s})/(680 \times 10^3\ \text{Hz}) = \boxed{441\ \text{m}.}$$

36. The wavelength of the FM station is

$$\lambda = c/f = (3.00 \times 10^8\ \text{m/s})/(100.7 \times 10^6\ \text{Hz}) = \boxed{2.979\ \text{m}.}$$

37. The frequencies are 940 kHz on the AM dial and 94 MHz on the FM dial. From $c = f\lambda$, we see that the lower frequency will have the longer wavelength: **the AM station.**
When we form the ratio of wavelengths, we get
$\lambda_2/\lambda_1 = f_1/f_2 = (94 \times 10^6 \text{ Hz})/(940 \times 10^3 \text{ Hz}) =$ **100×.**

38. The wavelength of Channel 2 is
$\lambda_2 = c/f_2 = (3.00 \times 10^8 \text{ m/s})/(54.0 \times 10^6 \text{ Hz}) =$ **5.56 m.**
The wavelength of Channel 69 is
$\lambda_{69} = c/f_{69} = (3.00 \times 10^8 \text{ m/s})/(806 \times 10^6 \text{ Hz}) =$ **0.372 m.**

39. After the change occurred, we would find out when the change in radiation reached the Earth:
$\Delta t = L/c = (1.50 \times 10^{14} \text{ m})/(3.00 \times 10^8 \text{ m/s}) = 5.00 \times 10^2 \text{ s} =$ **8.33 min.**

40. (*a*) The time for a signal to travel to the Moon is
$\Delta t = L/c = (3.84 \times 10^8 \text{ m})/(3.00 \times 10^8 \text{ m/s}) =$ **1.28 s.**
(*b*) The time for a signal to travel to Mars at the closest approach is
$\Delta t = L/c = (78 \times 10^9 \text{ m})/(3.00 \times 10^8 \text{ m/s}) = 260 \text{ s} =$ **4.3 min.**

41. The time consists of the time for the radio signal to travel to Earth and the time for the sound to travel from the loudspeaker:
$t = t_{\text{radio}} + t_{\text{sound}} = (d_{\text{radio}}/c) + (d_{\text{sound}}/v_{\text{sound}})$
$= (3.84 \times 10^8 \text{ m}/3.00 \times 10^8 \text{ m/s}) + (50 \text{ m}/343 \text{ m/s}) =$ **1.43 s.**
Note that about 10% of the time is for the sound wave.

42. The light has the same intensity in all directions, so the energy per unit area per unit time over a sphere centered at the source is
$S = P_0/A = P_0/4\pi r^2 = \frac{1}{2}c\epsilon_0 E_0^2 = \frac{1}{2}c(1/c^2\mu_0)E_0$, which gives $E_0 = (\mu_0 c P_0/2\pi r^2)^{1/2}$.

43. The light has the same intensity in all directions, so the energy per unit area per unit time over a sphere centered at the source is
$S = P_0/A = P_0/4\pi r^2 = (100 \text{ W})/4\pi(2.00 \text{ m})^2 = 1.99 \text{ W/m}^2$.
We find the electric field from
$S = \frac{1}{2}c\epsilon_0 E_0^2$;
$1.99 \text{ W/m}^2 = \frac{1}{2}(3.00 \times 10^8 \text{ m/s})(8.85 \times 10^{-12} \text{ C}^2/\text{N}\cdot\text{m}^2)E_0^2$, which gives $E_0 =$ **38.7 V/m.**
The magnetic field is
$B_0 = E_0/c = (38.7 \text{ V/m})/(3.00 \times 10^8 \text{ m/s}) =$ **1.29×10^{-7} T.**

44. The radiation from the Sun has the same intensity in all directions, so the rate at which it passes through a sphere centered at the Sun is
$P = S4\pi R^2$.
The rate must be the same for the two spheres, one containing the Earth and one containing Mars. When we form the ratio, we get
$P_{\text{Mars}}/P_{\text{Earth}} = (S_{\text{Mars}}/S_{\text{Earth}})(R_{\text{Mars}}/R_{\text{Earth}})^2$;
$1 = (S_{\text{Mars}}/1350 \text{ W/m}^2)(1.52)^2$, which gives $S_{\text{Mars}} = 584.3 \text{ W/m}^2$.
We find the rms value of the electric field from
$S_{\text{Mars}} = c\epsilon_0 E_{\text{rms}}^2$;
$584.3 \text{ W/m}^2 = (3.00 \times 10^8 \text{ m/s})(8.85 \times 10^{-12} \text{ C}^2/\text{N}\cdot\text{m}^2)E_{\text{rms}}^2$, which gives $E_{\text{rms}} =$ **469 V/m.**

45. If we curl the fingers of our right hand from the direction of the electric field (south) into the direction of the magnetic field (west), our thumb points down, so the direction of the wave is downward.

We find the electric field from

$S = \frac{1}{2}c\epsilon_0 E_0^2$;

$500\ \text{W/m}^2 = \frac{1}{2}(3.00 \times 10^8\ \text{m/s})(8.85 \times 10^{-12}\ \text{C}^2/\text{N}\cdot\text{m}^2)E_0^2$, which gives $E_0 =$ 614 V/m.

The magnetic field is

$B_0 = E_0/c = (614\ \text{V/m})/(3.00 \times 10^8\ \text{m/s}) =$ 2.05×10^{-6} T.

46. The length in space of a burst is

$d = ct = (3.00 \times 10^8\ \text{m/s})(10^{-8}\ \text{s}) =$ 3 m.

47. (*a*) The radio waves have the same intensity in all directions, so the energy per unit area per unit time over a sphere centered at the source with a radius of 100 m is

$S = P_0/A = P_0/4\pi r^2 = (500 \times 10^3\ \text{W})/4\pi(100\ \text{m})^2 = 0.398\ \text{W/m}^2$.

Thus the power through the area is

$P = SA = (0.398\ \text{W/m}^2)(1.0\ \text{m}^2) =$ 0.40 W.

(*b*) We find the rms value of the electric field from

$S = c\epsilon_0 E_{rms}^2$;

$0.398\ \text{W/m}^2 = (3.00 \times 10^8\ \text{m/s})(8.85 \times 10^{-12}\ \text{C}^2/\text{N}\cdot\text{m}^2)E_{rms}^2$, which gives $E_{rms} =$ 12 V/m.

(*c*) The voltage over the length of the antenna is

$E_{rms} = V_{rms} = E_{rms}d = (12\ \text{V/m})(1.0\ \text{m}) =$ 12 V.

48. (*a*) The radio waves have the same intensity in all directions, so the energy per unit area per unit time over a sphere centered at the source with a radius of 100 km is

$S = P_0/A = P_0/4\pi r^2 = (50 \times 10^3\ \text{W})/4\pi(100 \times 10^3\ \text{m})^2 = 3.98 \times 10^{-7}\ \text{W/m}^2$.

Thus the power through the area is

$P = SA = (3.98 \times 10^{-7}\ \text{W/m}^2)(1.0\ \text{m}^2) = 3.98 \times 10^{-7}\ \text{W} =$ 0.40 μW.

(*b*) We find the rms value of the electric field from

$S = c\epsilon_0 E_{rms}^2$;

$3.98 \times 10^{-7}\ \text{W/m}^2 = (3.00 \times 10^8\ \text{m/s})(8.85 \times 10^{-12}\ \text{C}^2/\text{N}\cdot\text{m}^2)E_{rms}^2$,

which gives $E_{rms} =$ 0.012 V/m.

(*c*) The voltage over the length of the antenna is

$E_{rms} = V_{rms} = E_{rms}d = (0.012\ \text{V/m})(1.0\ \text{m}) =$ 0.012 V.

49. The energy per unit area per unit time is

$S = \frac{1}{2}c\epsilon_0 E_0^2$;

$= \frac{1}{2}(3.00 \times 10^8\ \text{m/s})(8.85 \times 10^{-12}\ \text{C}^2/\text{N}\cdot\text{m}^2)(3 \times 10^6\ \text{V/m})^2 = 1.20 \times 10^{10}\ \text{W/m}^2$.

The power output is

$P = S4\pi r^2 = (1.20 \times 10^{10}\ \text{W/m2})4\pi(1.0\ \text{m})^2 =$ 1.5×10^{11} W.

50. We find the magnetic field from

$S = \frac{1}{2}(c/\mu_0)B_0^2$;

$1.0 \times 10^{-4}\ \text{W/m}^2 = \frac{1}{2}[(3.00 \times 10^8\ \text{m/s})/(4\pi \times 10^{-7}\ \text{T}\cdot\text{m/A})]B_0^2$, which gives $B_0 = 9.15 \times 10^{-10}$ T.

Because this field oscillates through the coil at $\omega = 2\pi f$, the maximum emf is

$\mathcal{E}_0 = NAB_0\omega = (480\ \text{turns})\pi(0.0085\ \text{m})^2(9.15 \times 10^{-10}\ \text{T})2\pi(810 \times 10^3\ \text{Hz}) = 5.07 \times 10^{-4}\ \text{V} = 0.507\ \text{mV}$.

The rms emf is

$\mathcal{E}_{rms} = \mathcal{E}_0/\sqrt{2} = (0.507\ \text{mV})/\sqrt{2} =$ 0.36 mV.

51. (a) We see from the diagram that all positive plates are connected to the positive side of the battery, and that all negative plates are connected to the negative side of the battery, so the 11 capacitors are connected in **parallel.**

(b) For parallel capacitors, the total capacitance is the sum, so we have

$C_{min} = 11(\epsilon_0 A_{min}/d)$
$= 11(8.85 \times 10^{-12}\ C^2/N\cdot m^2)(1.0 \times 10^{-4}\ m^2)/(1.1 \times 10^{-3}\ m) = 8.9 \times 10^{-12}\ F = 8.9\ pF;$

$C_{max} = 11(\epsilon_0 A_{max}/d)$
$= 11(8.85 \times 10^{-12}\ C^2/N\cdot m^2)(9.0 \times 10^{-4}\ m^2)/(1.1 \times 10^{-3}\ m) = 80 \times 10^{-11}\ F = 80\ pF.$

Thus the range is $8.9\ pF \le C \le 80\ pF.$

(c) The lowest resonant frequency requires the maximum capacitance.
We find the inductance for the lowest frequency:

$f_{01} = (1/2\pi)(1/L_1C_{max})^{1/2};$
$550 \times 10^3\ Hz = (1/2\pi)[1/L_1(80 \times 10^{-12}\ F)]^{1/2}$, which gives $L_1 = 1.05 \times 10^{-3}\ H = 1.05\ mH.$

We must check to make sure that the highest frequency can be reached.
We find the resonant frequency using this inductance and the minimum capacitance:

$f_{0max} = (1/2\pi)(1/L_1C_{min})^{1/2}$
$= (1/2\pi)[1/(1.05 \times 10^{-3}\ H)(8.9 \times 10^{-12}\ F)]^{1/2} = 1.64 \times 10^6\ Hz = 1640\ kHz.$

Because this is greater than the highest frequency desired, the inductor will work.
We could also start with the highest frequency.
We find the inductance for the highest frequency:

$f_{02} = (1/2\pi)(1/L_2C_{min})^{1/2};$
$1600 \times 10^3\ Hz = (1/2\pi)[1/L_2(8.9 \times 10^{-12}\ F)]^{1/2}$, which gives $L_2 = 1.22 \times 10^{-3}\ H = 1.22\ mH.$

We must check to make sure that the lowest frequency can be reached.
We find the resonant frequency using this inductance and the maximum capacitance:

$f_{0min} = (1/2\pi)(1/L_2C_{max})^{1/2}$
$= (1/2\pi)[1/(1.22 \times 10^{-3}\ H)(80 \times 10^{-12}\ F)]^{1/2} = 509 \times 10^5\ Hz = 509\ kHz.$

Because this is less than the lowest frequency desired, this inductor will also work.
Thus the range of inductances is

$1.05\ mH \le L \le 1.22\ mH.$

CHAPTER 23

1. For a flat mirror the image is as far behind the mirror as the object is in front, so the distance from object to image is

   $d_o + d_i = 1.5 \text{ m} + 1.5 \text{ m} =$ 3.0 m.

2. Because the angle of incidence must equal the angle of reflection, we see from the ray diagrams that the ray that reflects to the top of the object must be as far below the horizontal line to the reflection point on the mirror as the top is above the line, regardless of object position.

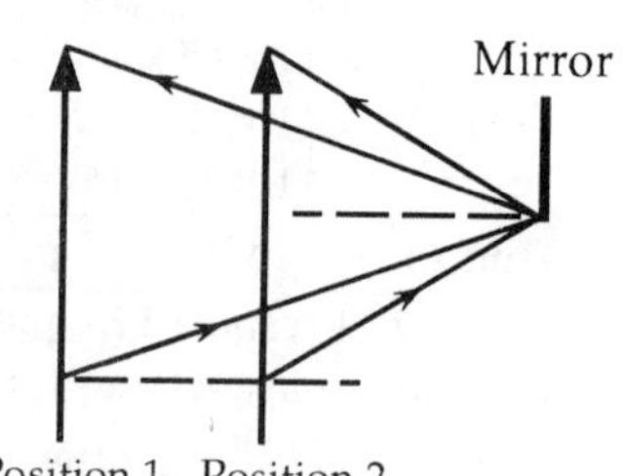

3. We show some of the rays from the tip of the arrow that form the three images. Single reflections form the two side images. Double reflections form the third image. The two reflections have reversed the orientation of the image.

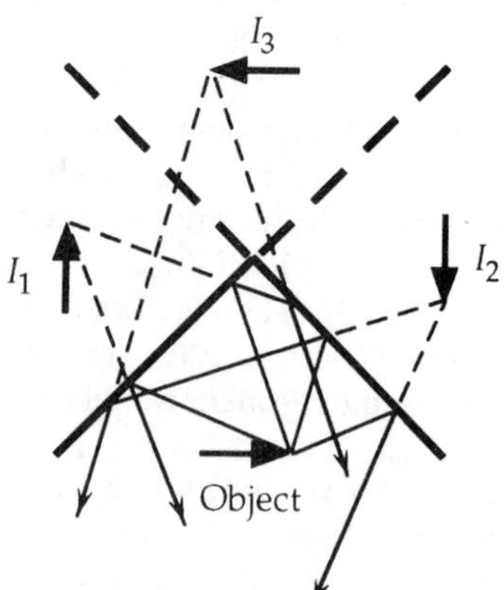

4. The angle of incidence is the angle of reflection. Thus we have

   $\tan \theta = (H - h)/L = h/x;$

   $(1.62 \text{ m} - 0.43 \text{ m})/(2.10 \text{ m}) = (0.43 \text{ m})/x,$

   which gives $x = 0.76 \text{ m} =$ 76 cm.

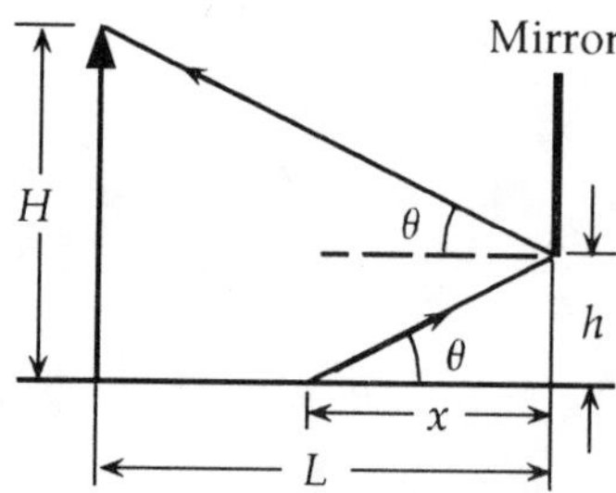

5. From the triangle formed by the mirrors and the first reflected ray, we have

   $\theta + \alpha + \phi = 180°;$

   $40° + 135° + \phi = 180°$, which gives $\phi =$ 5°.

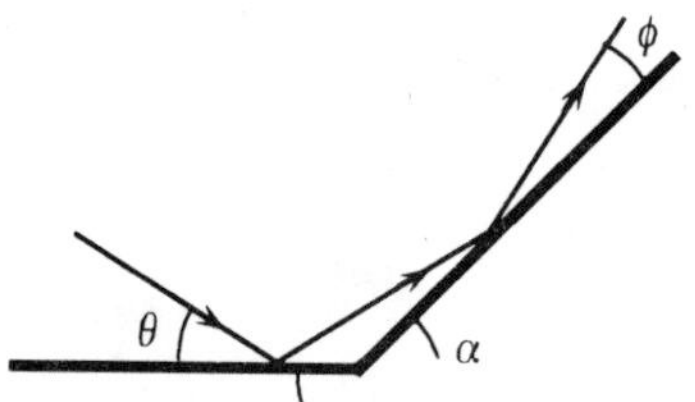

6. Because the rays entering your eye are diverging from the image position behind the mirror, the diameter of the area on the mirror and the diameter of your pupil must subtend the same angle from the image:

$D_{\text{mirror}}/d_i = D_{\text{pupil}}/(d_o + d_i)$;

$D_{\text{mirror}}/(70\text{ cm}) = (5.5\text{ mm})/(70\text{ cm} + 70\text{ cm})$, which gives $D_{\text{mirror}} = 2.75\text{ mm}$.

Thus the area on the mirror is

$A_{\text{mirror}} = \frac{1}{4}\pi D_{\text{mirror}}^2 = \frac{1}{4}\pi(2.75 \times 10^{-3}\text{ m})^2 = \boxed{5.9 \times 10^{-6}\text{ m}^2.}$

7. For the first reflection at A the angle of incidence θ_1 is the angle of reflection. For the second reflection at B the angle of incidence θ_2 is the angle of reflection. We can relate these angles to the angle at which the mirrors meet, ϕ, by using the sum of the angles of the triangle ABC:

$\phi + (90° - \theta_1) + (90° - \theta_2) = 180°$, or $\phi = \theta_1 + \theta_2$.

In the same way, for the triangle ABD, we have

$\alpha + 2\theta_1 + 2\theta_2 = 180°$, or $\alpha = 180° - 2(\theta_1 + \theta_2) = 180° - 2\phi$.

At point D we see that the deflection is

$\beta = 180° - \alpha = 180° - (180° - 2\phi) = 2\phi$.

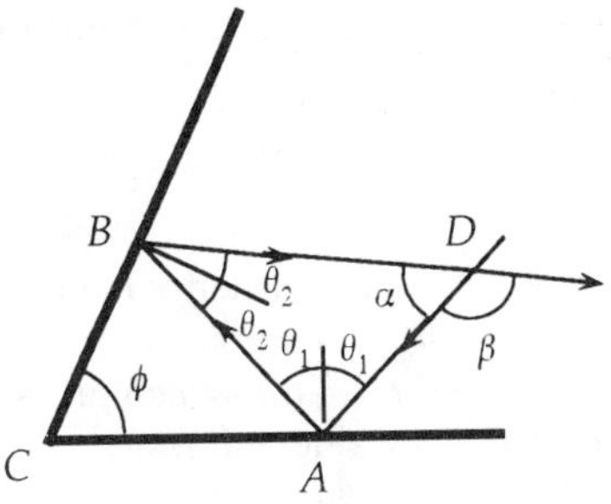

8. (a) The velocity of the wave, which specifies the direction of the light wave, is in the direction of the ray. If we consider a single reflection, because the angle of incidence is equal to the angle of reflection, the component of the velocity perpendicular to the mirror is reversed, while the component parallel to the mirror is unchanged. When we have three mirrors at right angles, we can choose the orientation of the mirrors for our axes. In each of the three reflections, the component of velocity perpendicular to the mirror will reverse. Thus after three reflections, all three components of the velocity will have been reversed, so the wave will return along the line of the original direction.

(b) Assuming the mirrors are large enough, the incident ray will make two reflections only if its direction is parallel to one of the mirrors. In that case, it has only two components, both of which will be reversed, so the ray will be $\boxed{\text{reversed along the line of the original direction.}}$

9. The rays from the Sun will be parallel, so the image will be at the focal point. The radius is

$r = 2f = 2(17.0\text{ cm}) = \boxed{34.0\text{ cm.}}$

10. To produce an image at infinity, the object will be at the focal point:

$d_o = f = r/2 = (27.0\text{ cm})/2 = \boxed{13.5\text{ cm.}}$

11.

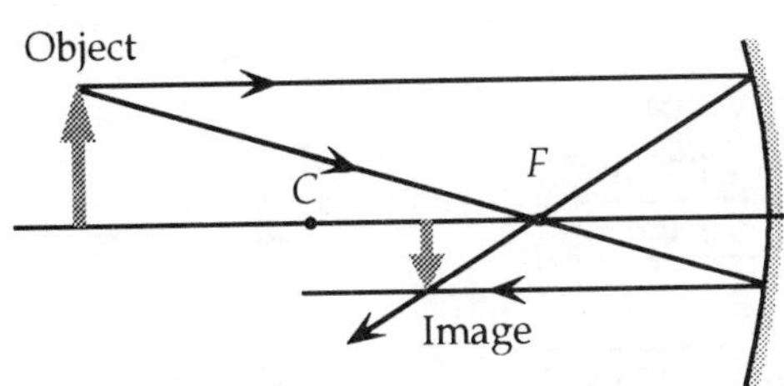

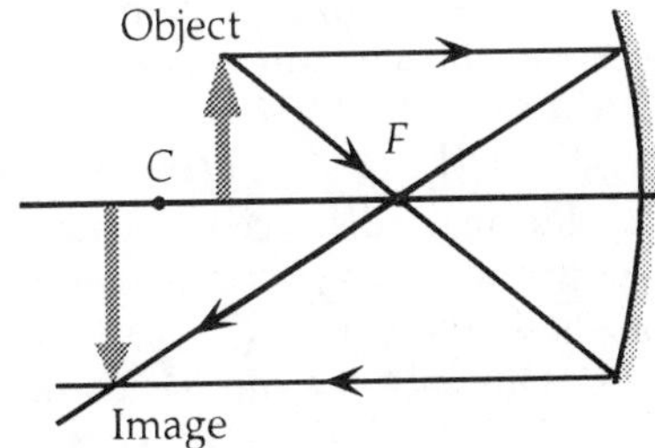

12. The ball is a convex mirror with a focal length
    $f = r/2 = (-4.5\text{ cm})/2 = -2.25\text{ cm}.$
    We locate the image from
    $(1/d_o) + (1/d_i) = 1/f;$
    $[1/(30.0\text{ cm})] + (1/d_i) = 1/(-2.25\text{ cm})$, which gives $d_i = -2.09\text{ cm}$.
    The image is **2.09 cm behind the surface, virtual.**
    The magnification is
    $m = -d_i/d_o = -(-2.09\text{ cm})/(30.0\text{ cm}) = +0.070.$
    Because the magnification is positive, the image is **upright.**

13. We find the image distance from the magnification:
    $m = h_i/h_o = -d_i/d_o;$
    $+3 = -d_i/(1.3\text{ m})$, which gives $d_i = -3.9\text{ m}$.
    We find the focal length from
    $(1/d_o) + (1/d_i) = 1/f;$
    $[1/(1.3\text{ m})] + [1/(-3.9\text{ m})] = 1/f$, which gives $f = 1.95\text{ m}$.
    The radius of the concave mirror is
    $r = 2f = 2(1.95\text{ m}) =$ **3.9 m.**

14. We find the image distance from the magnification:
    $m = h_i/h_o = -d_i/d_o;$
    $+4.5 = -d_i/(2.20\text{ cm})$, which gives $d_i = -9.90\text{ cm}$.
    We find the focal length from
    $(1/d_o) + (1/d_i) = 1/f;$
    $[1/(2.20\text{ cm})] + [1/(-9.90\text{ cm})] = 1/f$, which gives $f = 2.83\text{ cm}$.
    Because the focal length is positive, the mirror is **concave** with a radius of
    $r = 2f = 2(2.83\text{ cm}) =$ **5.7 cm.**

15. To produce an upright image, we have $d_i < 0$. A smaller image means $|d_i| < d_o$, so $f < 0$, which means the mirror is **convex.**
    The focal length of the mirror is
    $f = r/2 = (-3.2\text{ m})/2 = -1.6\text{ m}.$
    We locate the image from
    $(1/d_o) + (1/d_i) = 1/f;$
    $[1/(15.0\text{ m})] + (1/d_i) = 1/(-1.6\text{ m})$, which gives $d_i = -1.45\text{ m}$.
    Because the image distance is negative, the image is **virtual.**
    We find the image height from the magnification:
    $m = h_i/h_o = -d_i/d_o;$
    $h_i/(1.3\text{ m}) = -(-1.45\text{ m})/(15.0\text{ m})$, which gives $h_i =$ **0.13 m.**

16. (*a*) We see from the ray diagram that the image is behind the mirror, so it is virtual. We estimate the image distance as $\boxed{-7 \text{ cm.}}$

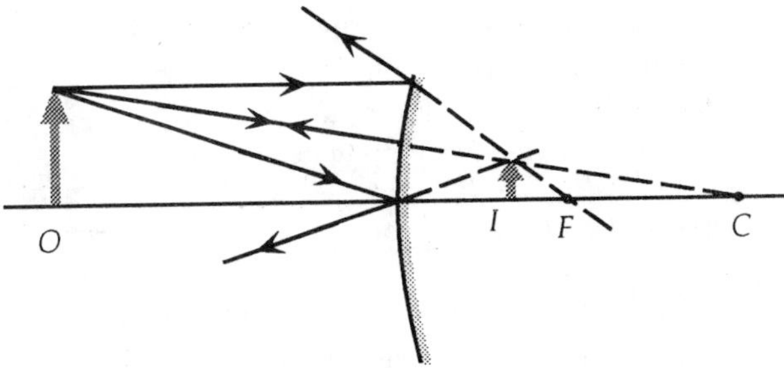

(*b*) If we use a focal length of – 10 cm, we locate the image from

$(1/d_o) + (1/d_i) = 1/f;$

$[1/(20 \text{ cm})] + (1/d_i) = 1/(-10 \text{ cm})$, which gives $d_i = -6.7$ cm.

(*c*) We find the image size from the magnification:

$m = h_i/h_o = -d_i/d_o;$

$h_i/(3.00 \text{ mm}) = -(-6.7 \text{ cm})/(20 \text{ cm})$, which gives $h_i = \boxed{1.0 \text{ mm.}}$

17. (*a*) With $d_i = d_o$, we locate the object from

$(1/d_o) + (1/d_i) = 1/f;$

$(1/d_o) + (1/d_o) = 1/f$, which gives $d_o = 2f = r$.

The object should be placed at the $\boxed{\text{center of curvature.}}$

(*b*) Because the image is in front of the mirror, $d_i > 0$, it is $\boxed{\text{real.}}$

(*c*) The magnification is

$m = -d_i/d_o = -d_o/d_o = -1.$

Because the magnification is negative, the image is $\boxed{\text{inverted.}}$

(*d*) As found in part (*c*), $m = \boxed{-1.}$

18. We take the object distance to be ∞, and find the focal length from

$(1/d_o) + (1/d_i) = 1/f;$

$(1/\infty) + [1/(-14.0 \text{ cm})] = 1/f$, which gives $f = -14.0$ cm.

Because the focal length is negative, the mirror is $\boxed{\text{convex.}}$

The radius is

$r = 2f = 2(-14.0 \text{ cm}) = \boxed{-28.0 \text{ cm.}}$

19. We find the image distance from

$(1/d_o) + (1/d_i) = 1/f = 2/r$, which we can write as $d_i = rd_o/(2d_o - r)$.

The magnification is

$m = -d_i/d_o = -r/(2d_o - r).$

If $d_o > r$, then $(2d_o - r) > r$, so

$|m| = r/(2d_o - r) = r/(>r) < 1.$

If $d_o < r$, then $(2d_o - r) < r$, so

$|m| = r/(2d_o - r) = r/(<r) > 1.$

20. From the ray that reflects from the center of the mirror, we have

$\tan\theta = h_o/d_o = h_i/d_i;$

$|m| = h_o/h_i = d_i/d_o.$

Because the image distance on the ray diagram is negative, we get

$m = -d_i/d_o.$

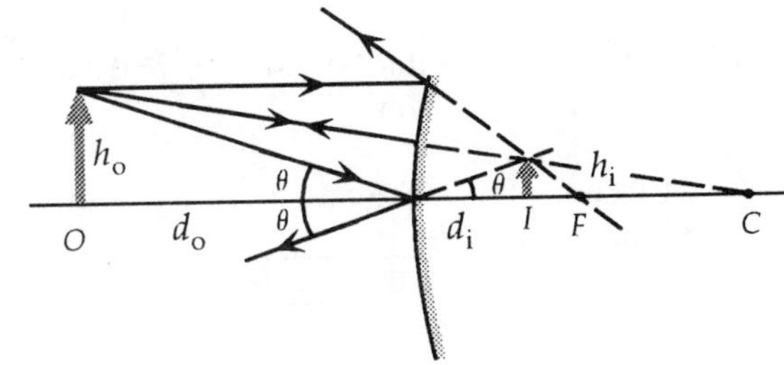

21. From the ray diagram, we see that
   $\tan \theta = h_o/d_o = h_i/d_i$;
   $\tan \alpha = h_o/(d_o + r) = h_i/(r - d_i)$.
   When we divide the two equations. we get
   $(d_o + r)/d_o = (r - d_i)/d_i$;
   $1 + (r/d_o) = (r/d_i) - 1$, or
   $(r/d_o) - (r/d_i) = -2$;
   $(1/d_o) - (1/d_i) = -2/r = -1/f$, with $f = r/2$.
   From the ray diagram, we see that $d_i < 0$.
   If we consider f to be negative, we have
   $(1/d_o) + (1/d_i) = 1/f$.

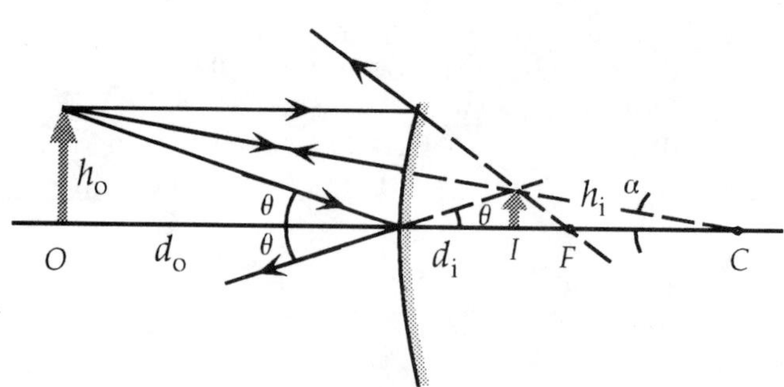

22. We find the image distance from the magnification:
   $m = h_i/h_o = -d_i/d_o$;
   $+0.80 = -d_i/(2.2\text{ m})$, which gives $d_i = -1.76$ m.
   We find the focal length from
   $(1/d_o) + (1/d_i) = 1/f$;
   $[1/(2.2\text{ m})] + [1/(-1.76\text{ m})] = 1/f$, which gives $f =$ $\boxed{-8.8\text{ m.}}$

23. (*a*) As the radius of a sphere gets larger, the surface is flatter. The plane mirror can be considered a spherical mirror with an infinite radius, and thus $\boxed{f = \infty.}$
   (*b*) When we use the mirror equation, we get
   $(1/d_o) + (1/d_i) = 1/f$;
   $(1/d_o) + (1/d_i) = 1/\infty = 0$, or $\boxed{d_i = -d_o.}$
   (*c*) For the magnification, we have
   $m = -d_i/d_o = -(-d_o)/d_o =$ $\boxed{+1.}$
   (*d*) $\boxed{\text{Yes,}}$ these are consistent with the discussion on plane mirrors.

24. (*a*) To produce a smaller image located behind the surface of the mirror requires a $\boxed{\text{convex mirror.}}$
   (*b*) We find the image distance from the magnification:
   $m = h_i/h_o = -d_i/d_o$;
   $(3.5\text{ cm})/(4.5\text{ cm}) = -d_i/(28\text{ cm})$, which gives $d_i = -21.8$ cm.
   As expected, $d_i < 0$. The image is located $\boxed{\text{22 cm behind the surface.}}$
   (*c*) We find the focal length from
   $(1/d_o) + (1/d_i) = 1/f$;
   $[1/(28\text{ cm})] + [1/(-21.8\text{ cm})] = 1/f$, which gives $f =$ $\boxed{-98\text{ cm.}}$
   (*d*) The radius of curvature is
   $r = 2f = 2(-98\text{ cm}) =$ $\boxed{-196\text{ cm.}}$

25. (*a*) To produce a larger image requires a $\boxed{\text{concave mirror.}}$
   (*b*) The image will be $\boxed{\text{erect and virtual.}}$
   (*c*) We find the image distance from the magnification:
   $m = h_i/h_o = -d_i/d_o$;
   $1.3 = -d_i/(20.0\text{ cm})$, which gives $d_i = -26.0$ cm.
   We find the focal length from
   $(1/d_o) + (1/d_i) = 1/f$;
   $[1/(20.0\text{ cm})] + [1/(-26.0\text{ cm})] = 1/f$, which gives $f = 86.7$ cm.
   The radius of curvature is
   $r = 2f = 2(86.7\text{ cm}) =$ $\boxed{173\text{ cm.}}$

26. (*a*) The speed in crown glass is
$v = c/n = (3.00 \times 10^8\ \text{m/s})/(1.52) =$ $\boxed{1.97 \times 10^8\ \text{m/s.}}$
(*b*) The speed in Lucite is
$v = c/n = (3.00 \times 10^8\ \text{m/s})/(1.51) =$ $\boxed{1.99 \times 10^8\ \text{m/s.}}$

27. We find the index of refraction from
$v = c/n;$
$2.29 \times 10^8\ \text{m/s} = (3.00 \times 10^8\ \text{m/s})/n$, which gives $n =$ $\boxed{1.31.}$

28. We find the index of refraction from
$v = c/n;$
$0.85v_{\text{water}} = 0.85c/1.33 = c/n$, which gives $n =$ $\boxed{1.56.}$

29. The % uncertainty in the index is
$\pm [(0.000010)/(1.00030)](100) = \pm 9.997 \times 10^{-4}\ \%$,
which will be the % uncertainty in the speed. Thus we have
$v = c/n = [(2.997925 \times 10^8\ \text{m/s})/(1.00030)] \pm 9.997 \times 10^{-4}\ \%$
$= 2.997026 \times 10^8\ \text{m/s} \pm 9.997 \times 10^{-4}\ \% =$ $\boxed{(2.997026 \pm 0.000030) \times 10^8\ \text{m/s.}}$

30. We find the angle of refraction in the glass from
$n_1 \sin \theta_1 = n_2 \sin \theta_2;$
$(1.00) \sin 63° = (1.50) \sin \theta_2$, which gives $\theta_2 =$ $\boxed{36°.}$

31. We find the angle of refraction in the water from
$n_1 \sin \theta_1 = n_2 \sin \theta_2;$
$(1.33) \sin 42.5° = (1.00) \sin \theta_2$, which gives $\theta_2 =$ $\boxed{64.0°.}$

32. We find the incident angle in the water from
$n_1 \sin \theta_1 = n_2 \sin \theta_2;$
$(1.33) \sin \theta_1 = (1.00) \sin 60°$, which gives $\theta_1 =$ $\boxed{41°.}$

33. We find the incident angle in the air from
$n_1 \sin \theta_1 = n_2 \sin \theta_2;$
$(1.00) \sin \theta_1 = (1.33) \sin 21.0°$, which gives $\theta_1 = 28.5°$.
Thus the angle above the horizon is
$90.0° - \theta_1 = 90.0° - 28.5° =$ $\boxed{61.5°.}$

34. For the refraction at the first surface, we have
$n_{\text{air}} \sin \theta_1 = n \sin \theta_2;$
$(1.00) \sin 45.0° = (1.52) \sin \theta_2$, which gives $\theta_2 = 27.7°$.
We find the angle of incidence at the second surface from
$(90° - \theta_2) + (90° - \theta_3) + A = 180°$, which gives
$\theta_3 = A - \theta_2 = 60° - 27.7° = 32.3°$.
For the refraction at the second surface, we have
$n \sin \theta_3 = n_{\text{air}} \sin \theta_4;$
$(1.52) \sin 32.3° = (1.00) \sin \theta_4$, which gives
$\theta_4 =$ $\boxed{54.3° \text{ from the normal.}}$

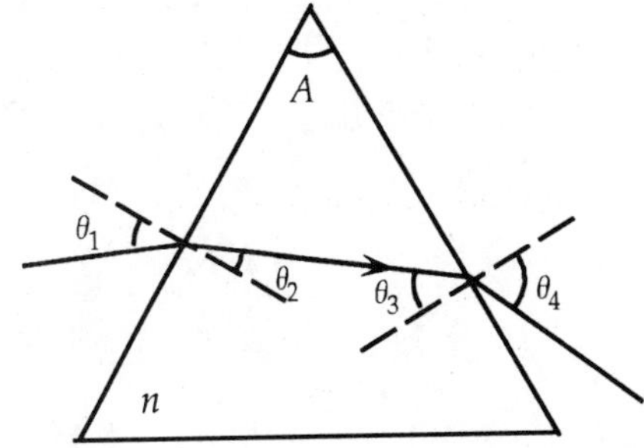

35. We find the angle of incidence from the distances:

$\tan \theta_1 = L_1/h_1 = (2.7 \text{ m})/(1.3 \text{ m}) = 2.076$, so $\theta_1 = 64.3°$.

For the refraction from air into water, we have

$n_{air} \sin \theta_1 = n_{water} \sin \theta_2$;

$(1.00) \sin 64.3° = (1.33) \sin \theta_2$, which gives $\theta_2 = 42.6°$.

We find the horizontal distance from the edge of the pool from

$L = L_1 + L_2 = L_1 + h_2 \tan \theta_2$

$= 2.7 \text{ m} + (2.1 \text{ m}) \tan 42.6° =$ **4.6 m.**

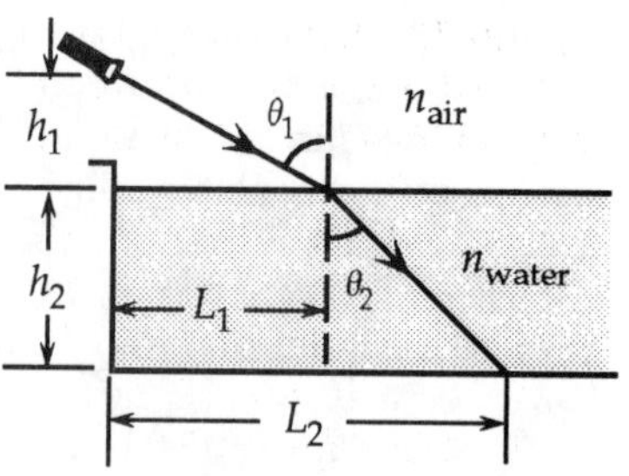

36. The angle of reflection is equal to the angle of incidence:

$\theta_{refl} = \theta_1 = 2\theta_2$.

For the refraction we have

$n_{air} \sin \theta_1 = n_{glass} \sin \theta_2$;

$(1.00) \sin 2\theta_2 = (1.52) \sin \theta_2$.

We use a trigonometric identity for the left-hand side:

$\sin 2\theta_2 = 2 \sin \theta_2 \cos \theta_2 = (1.52) \sin \theta_2$, or $\cos \theta_2 = 0.760$, so $\theta_2 = 40.5°$.

Thus the angle of incidence is $\theta_1 = 2\theta_2 =$ **81.0°.**

37. (*a*) We find the angle in the glass from the refraction at the air–glass surface:

$n_1 \sin \theta_1 = n_2 \sin \theta_2$;

$(1.00) \sin 43.5° = (1.52) \sin \theta_2$, which gives $\theta_2 =$ **26.9°.**

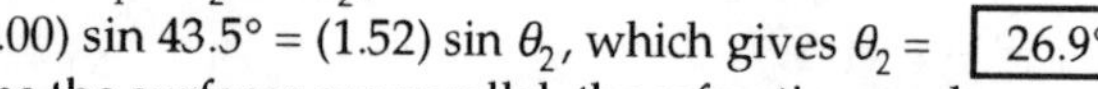

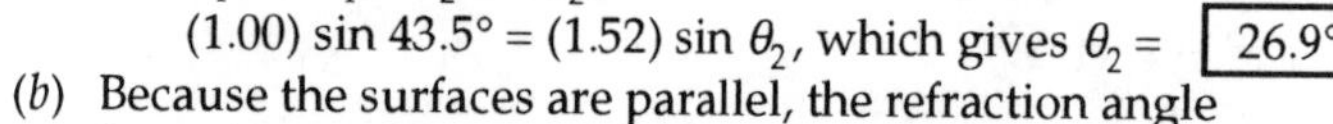

(*b*) Because the surfaces are parallel, the refraction angle from the first surface is the incident angle at the second surface. We find the angle in the water from the refraction at the glass–water surface:

$n_2 \sin \theta_2 = n_3 \sin \theta_3$;

$(1.52) \sin 26.9° = (1.33) \sin \theta_3$, which gives $\theta_3 =$ **31.2°.**

(*c*) If there were no glass, we would have

$n_1 \sin \theta_1 = n_3 \sin \theta_3'$;

$(1.00) \sin 43.5° = (1.33) \sin \theta_3'$, which gives $\theta_3' =$ **31.2°.**

Note that, because the sides are parallel, θ_3 is independent of the presence of the glass.

38. Because the surfaces are parallel, the angle of refraction from the first surface is the angle of incidence at the second, Thus for the refractions, we have

$n_1 \sin \theta_1 = n_2 \sin \theta_2$;

$n_2 \sin \theta_2 = n_3 \sin \theta_3$.

When we add the two equations, we get

$n_1 \sin \theta_1 = n_1 \sin \theta_3$, which gives $\theta_3 = \theta_1$.

Because the ray emerges in the same index of refraction, it is undeviated.

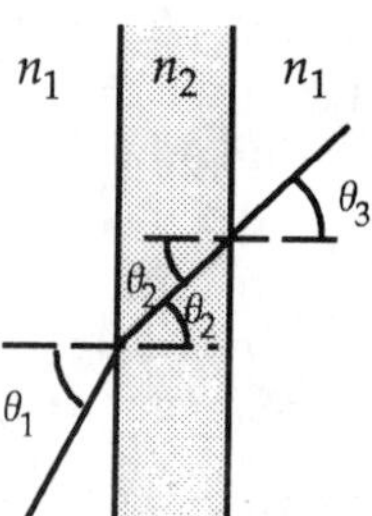

39. Because the glass surfaces are parallel, the exit beam will be traveling in the same direction as the original beam.
We find the angle inside the glass from
$n_{air} \sin \theta = n \sin \phi$.
If the angles are small, we use
$\cos \phi \approx 1$, and $\sin \phi \approx \phi$, where ϕ is in radians.
$(1.00)\ \theta = n\phi$, or $\phi = \theta/n$.
We find the distance along the ray in the glass from
$L = t/\cos \phi \approx t$.
We find the perpendicular displacement from the original direction from
$d = L \sin (\theta - \phi) \approx t(\theta - \phi) = t[\theta - (\theta/n)] = t\theta(n-1)/n$.

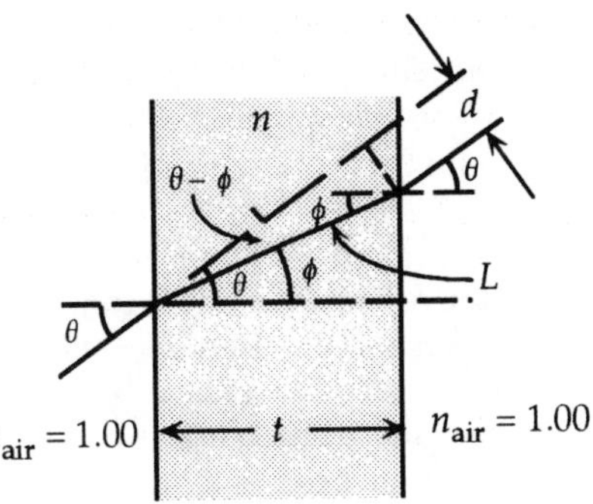

40. When the light in the material with a higher index is incident at the critical angle, the refracted angle is 90°:
$n_{Lucite} \sin \theta_1 = n_{water} \sin \theta_2$;
$(1.51) \sin \theta_1 = (1.33) \sin 90°$, which gives $\theta_1 =$ 61.7°.
Because Lucite has the higher index, the light must start in Lucite.

41. When the light in the liquid is incident at the critical angle, the refracted angle is 90°:
$n_{liquid} \sin \theta_1 = n_{air} \sin \theta_2$;
$n_{liquid} \sin 44.7° = (1.00) \sin 90°$, which gives $n_{liquid} =$ 1.42.

42. We find the critical angle for light leaving the water:
$n \sin \theta_1 = \sin \theta_2$;
$(1.33) \sin \theta_C = \sin 90°$, which gives $\theta_C = 48.8°$.
If the light is incident at a greater angle than this, it will totally reflect. We see from the diagram that
$R > H \tan \theta_C = (62.0\ \text{cm}) \tan 48.8° =$ 70.7 cm.

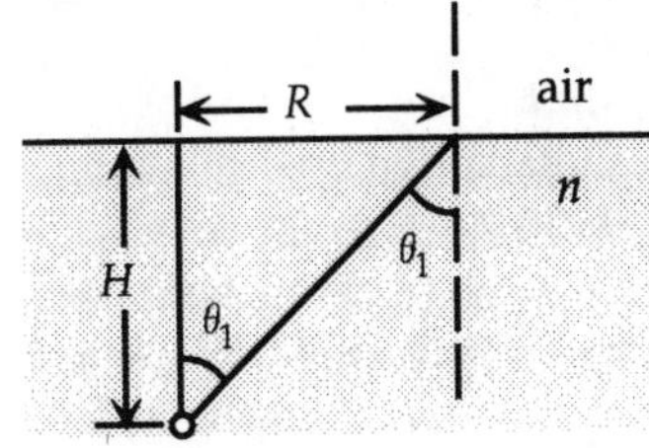

43. We find the distance L between reflections from
$d = L \sin \theta$;
$10^{-4}\ \text{m} = L \sin 15°$, which gives $L =$ 3.9×10^{-4} m.

44. We find the angle of incidence from the distances:
$\tan \theta_1 = L/h = (7.0\ \text{cm})/(8.0\ \text{cm}) = 0.875$, so $\theta_1 = 41.2°$.
For the maximum incident angle for the refraction from liquid into air, we have
$n_{liquid} \sin \theta_1 = n_{air} \sin \theta_2$;
$n_{liquid} \sin \theta_{1max} = (1.00) \sin 90°$, which gives $\sin \theta_{1max} = 1/n_{liquid}$.
Thus we have
$\sin \theta_1 \geq \sin \theta_{1max} = 1/n_{liquid}$;
$\sin 41.2° = 0.659 \geq 1/n_{liquid}$, or $n_{liquid} \geq 1.5$.

45. For the refraction at the first surface, we have

$n_{air} \sin \theta_1 = n \sin \theta_2$;

$(1.00) \sin \theta_1 = n \sin \theta_2$, which gives $\sin \theta_2 = \sin \theta_1 / n$.

We find the angle of incidence at the second surface from

$(90° - \theta_2) + (90° - \theta_3) + A = 180°$, which gives

$\theta_3 = A - \theta_2 = 75° - \theta_2$.

For the refraction at the second surface, we have

$n \sin \theta_3 = n_{air} \sin \theta_4 = (1.00) \sin \theta_4$.

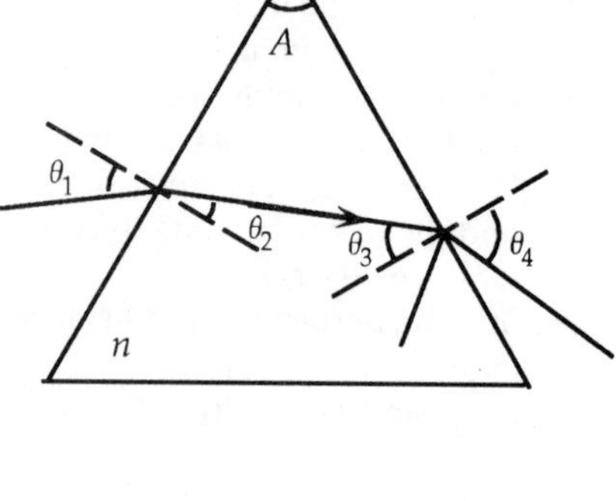

The maximum value of θ_4 before internal reflection takes place at the second surface is 90°. Thus for internal reflection to occur, we have

$n \sin \theta_3 = n \sin (A - \theta_2) \geq 1$.

When we expand the left-hand side, we get

$n(\sin A \cos \theta_2 - \cos A \sin \theta_2) \geq 1$.

If we use the result from the first surface to eliminate n, we get

$\sin \theta_1 (\sin A \cos \theta_2 - \cos A \sin \theta_2)/(\sin \theta_2) = \sin \theta_1 (\sin A / \tan \theta_2 - \cos A) \geq 1$, or

$1/\tan \theta_2 \geq [(1/\sin \theta_1) + \cos A]/\sin A = [(1/\sin 45°) + \cos 75°]/\sin 75° = 1.732$, which gives

$\tan \theta_2 \leq 0.577$, so $\theta_2 \leq 30°$.

From the result for the first surface, we have

$n_{min} = \sin \theta_1 / \sin \theta_{2max} = \sin 45°/\sin 30° = 1.414$, so $\boxed{n \geq 1.414.}$

46. For the refraction at the side of the rod, we have

$n_2 \sin \gamma = n_1 \sin \delta$.

The minimum angle for total reflection γ_{min} occurs when $\delta = 90°$:

$n_2 \sin \gamma_{min} = (1.00)(1) = 1$, or $\sin \gamma_{min} = 1/n_2$.

We find the maximum angle of refraction at the end of the rod from

$\beta_{max} = 90° - \gamma_{min}$.

Because the sine function increases with angle, for the refraction at the end of the rod, we have

$n_1 \sin \alpha_{max} = n_2 \sin \beta_{max}$;

$(1.00) \sin \alpha_{max} = n_2 \sin (90° - \gamma_{min}) = n_2 \cos \gamma_{min}$.

If we want total internal reflection to occur for any incident angle at the end of the fiber, the maximum value of α is 90°, so

$n_2 \cos \gamma_{min} = 1$.

When we divide this by the result for the refraction at the side, we get $\tan \gamma_{min} = 1$, or $\gamma_{min} = 45°$.

Thus we have

$n_2 \geq 1/\sin \gamma_{min} = 1/\sin 45° = 1.414$.

47. (*a*) The ray enters normal to the first surface, so there is no deviation there. The angle of incidence is 45° at the second surface. When there is air outside the surface, we have

$n_1 \sin \theta_1 = n_2 \sin \theta_2$;

$n_1 \sin 45° = (1.00) \sin \theta_2$.

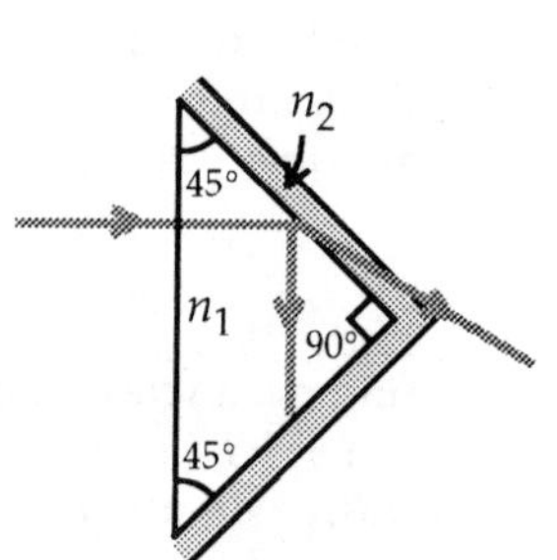

For total internal reflection to occur, $\sin \theta_2 \geq 1$, so we have

$n_1 \geq 1/\sin 45° = \boxed{1.414.}$

(*b*) When there is water outside the surface, we have

$n_1 \sin \theta_1 = n_2 \sin \theta_2$;

$(1.50) \sin 45° = (1.33) \sin \theta_2$, which gives $\sin \theta_2 = 0.80$.

Because $\sin \theta_2 < 1$, $\boxed{\text{the prism will not be totally reflecting.}}$

(*c*) For total reflection when there is water outside the surface, we have

$n_1 \sin \theta_1 = n_2 \sin \theta_2$;

$n_1 \sin 45° = (1.33) \sin \theta_2$.

For total internal reflection to occur, $\sin \theta_2 \geq 1$, so we have

$n_1 \geq 1.33/\sin 45° = \boxed{1.88.}$

48. (*a*) From the ray diagram, the object distance is about six focal lengths, or **390 mm.**

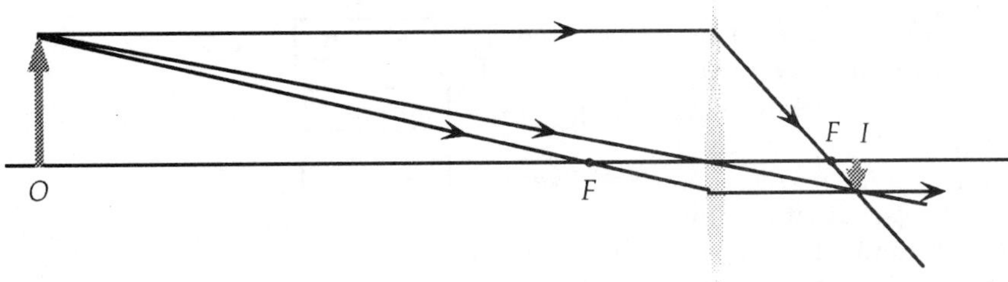

(*b*) We find the object distance from

$(1/d_o) + (1/d_i) = 1/f;$

$(1/d_o) + (1/78.0\text{ mm}) = 1/65.0\text{ mm}$, which gives $d_o = 390\text{ mm} =$ **39.0 cm.**

49. (*a*) To form a real image from parallel rays requires a **converging lens.**

(*b*) We find the power of the lens from

$(1/d_o) + (1/d_i) = 1/f = P$, when f is in meters;

$(1/\infty) + (1/0.185\text{ m}) = P =$ **5.41 D.**

50. To form a real image from a real object requires a **converging lens.**

We find the focal length of the lens from

$(1/d_o) + (1/d_i) = 1/f;$

$(1/225\text{ cm}) + (1/48.3\text{ cm}) = 1/f$, which gives $f =$ **+ 39.8 cm.**

Because $d_i > 0$, the image is **real.**

51. (*a*) The power of the lens is

$P = 1/f = 1/0.295\text{ m} =$ **3.39 D, converging.**

(*b*) We find the focal length of the lens from

$P = 1/f;$

$-6.25\text{ D} = 1/f$, which gives $f = -0.160\text{ m} =$ **– 16.0 cm, diverging.**

52. (*a*) We locate the image from

$(1/d_o) + (1/d_i) = 1/f;$

$(1/18\text{ cm}) + (1/d_i) = 1/24\text{ cm}$, which gives $d_i = -72\text{ cm}$.

The negative sign means the image is **72 cm behind the lens (virtual).**

(*b*) We find the magnification from

$m = -d_i/d_o = -(-72\text{ cm})/(18\text{ cm}) =$ **+ 4.0.**

53. (*a*) Because the Sun is very far away, the image will be at the focal point. We find the size of the image from

$m = h_i/h_o = -d_i/d_o;$

$h_i/(1.4 \times 10^6\text{ km}) = -(28\text{ mm})/(1.5 \times 10^8\text{ km})$, which gives $h_i =$ **– 0.26 mm.**

(*b*) For a 50 mm lens, we have

$h_i/(1.4 \times 10^6\text{ km}) = -(50\text{ mm})/(1.5 \times 10^8\text{ km})$, which gives $h_i =$ **– 0.47 mm.**

(*c*) For a 200 mm lens, we have

$h_i/(1.4 \times 10^6\text{ km}) = -(200\text{ mm})/(1.5 \times 10^8\text{ km})$, which gives $h_i =$ **– 1.9 mm.**

(*d*) The 28-mm lens simulates being farther away, so it would be a **wide-angle lens.**

The 200 mm lens simulates being closer, so it would be a **telephoto lens.**

54. (*a*) We find the image distance from
$(1/d_o) + (1/d_i) = 1/f;$
$(1/10.0 \times 10^3 \text{ mm}) + (1/d_i) = 1/80 \text{ mm}$, which gives $d_i =$ **81 mm.**
(*b*) For an object distance of 3.0 m, we have
$(1/3.0 \times 10^3 \text{ mm}) + (1/d_i) = 1/80 \text{ mm}$, which gives $d_i =$ **82 mm.**
(*c*) For an object distance of 1.0 m, we have
$(1/1.0 \times 10^3 \text{ mm}) + (1/d_i) = 1/80 \text{ mm}$, which gives $d_i =$ **87 mm.**
(*d*) We find the smallest object distance from
$(1/d_{omin}) + (1/120 \text{ mm}) = 1/80 \text{ mm}$, which gives $d_i = 240 \text{ mm} =$ **24 cm.**

55. We find the image distance from
$(1/d_o) + (1/d_i) = 1/f;$
$(1/0.140 \text{ m}) + (1/d_i) = -6.0 \text{ D},$
which gives $d_i = -0.076 \text{ m} =$ **– 7.6 cm (virtual image behind the lens).**
We find the height of the image from
$m = h_i/h_o = -d_i/d_o;$
$h_i/(1.0 \text{ mm}) = -(-7.6 \text{ cm})/(14.0 \text{ cm})$, which gives $h_i =$ **0.54 mm (upright).**

56. (*a*) We see that the image is behind the lens,
so it is **virtual.**
(*b*) From the ray diagram we see that we need a
converging lens.
(*c*) We find the image distance from the magnification:
$m = -d_i/d_o;$
$+2.5 = -d_i/(8.0 \text{ cm})$, which gives $d_i = -20 \text{ cm}$.
We find power of the lens from
$(1/d_o) + (1/d_i) = 1/f = P$, when f is in meters;
$(1/0.080 \text{ m}) + [1/(-0.20 \text{ m})] = P =$ **7.5 D.**

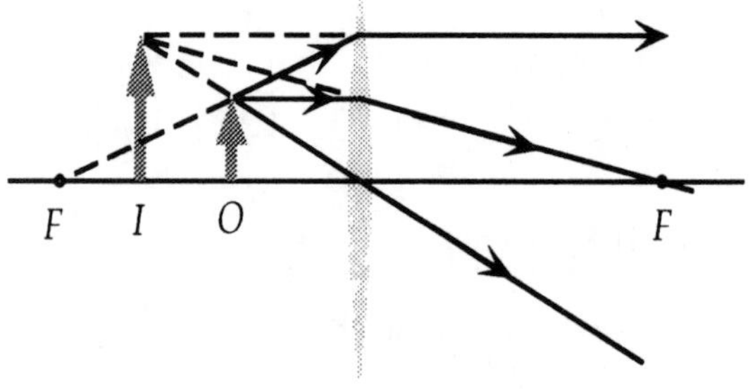

57. We can relate the image and object distance from the magnification:
$m = -d_i/d_o$, or $d_o = -d_i/m$.
We use this in the lens equation:
$(1/d_o) + (1/d_i) = 1/f;$
$-(m/d_i) + (1/d_i) = 1/f$, which gives $d_i = (1 - m)f$.
(*a*) If the image is real, $d_i > 0$. With $f > 0$, we see that $m < 1$; thus $m = -2.00$. The image distance is
$d_i = [1 - (-2.00)](50.0 \text{ mm}) = 150 \text{ mm}.$
The object distance is
$d_o = -d_i/m = -(150 \text{ mm})/(-2.00) =$ **75.0 mm.**
(b) If the image is virtual, $d_i < 0$. With $f > 0$, we see that $m > 1$; thus $m = +2.00$. The image distance is
$d_i = [1 - (+2.00)](50.0 \text{ mm}) = -50 \text{ mm}.$
The object distance is
$d_o = -d_i/m = -(-50 \text{ mm})/(+2.00) =$ **25.0 mm.**

58. We can relate the image and object distance from the magnification:

$m = -d_i/d_o$, or $d_o = -d_i/m$.

We use this in the lens equation:

$(1/d_o) + (1/d_i) = 1/f$;

$-(m/d_i) + (1/d_i) = 1/f$, which gives $d_i = (1 - m)f$.

(*a*) If the image is real, $d_i > 0$. With $f < 0$, we see that $m > 1$; thus $m = +2.00$. The image distance is

$d_i = [1 - (+2.00)](-50.0 \text{ mm}) = 50.0 \text{ mm}$.

The object distance is

$d_o = -d_i/m = -(50.0 \text{ mm})/(+2.00) =$ **-25.0 mm.**

The negative sign means the object is beyond the lens, so it would have to be an object formed by a preceding optical device.

(*b*) If the image is virtual, $d_i < 0$. With $f < 0$, we see that $m < 1$; thus $m = -2.00$. The image distance is

$d_i = [1 - (-2.00)](-50.0 \text{ mm}) = -150 \text{ mm}$.

The object distance is

$d_o = -d_i/m = -(-150 \text{ mm})/(-2.00) =$ **-75.0 mm.**

The negative sign means the object is beyond the lens, so it would have to be an object formed by a preceding optical device.

59. (*a*) We find the focal length of the lens from

$(1/d_o) + (1/d_i) = 1/f$;

$(1/31.5 \text{ cm}) + [1/(-8.20 \text{ cm})] = 1/f$, which gives $f =$ **-11.1 cm (diverging).**

The image is in front of the lens, so it is **virtual.**

(*b*) We find the focal length of the lens from

$(1/d_o) + (1/d_i) = 1/f$;

$(1/31.5 \text{ cm}) + [1/(-38.0 \text{ cm})] = 1/f$, which gives $f =$ **$+184$ cm (converging).**

The image is in front of the lens, so it is virtual.

60. (*a*) We find the image distance from

$(1/d_o) + (1/d_i) = 1/f$;

$(1/1.20 \times 10^3 \text{ mm}) + (1/d_i) = 1/135 \text{ mm}$, which gives $d_i =$ **152 mm (real, behind the lens).**

We find the height of the image from

$m = h_i/h_o = -d_i/d_o$;

$h_i/(2.20 \text{ cm}) = -(152 \text{ mm})/(1.20 \times 10^3 \text{ mm})$, which gives $h_i =$ **-0.279 cm (inverted).**

(*b*) We find the image distance from

$(1/d_o) + (1/d_i) = 1/f$;

$(1/1.20 \times 10^3 \text{ mm}) + (1/d_i) = 1/(-135 \text{ mm})$,

which gives $d_i =$ **-121 mm (virtual, in front of the lens).**

We find the height of the image from

$m = h_i/h_o = -d_i/d_o$;

$h_i/(2.20 \text{ cm}) = -(-121 \text{ mm})/(1.20 \times 10^3 \text{ mm})$, which gives $h_i =$ **$+0.222$ cm (upright).**

61. The sum of the object and image distances must be the distance between object and screen:

$d_o + d_i = L = 60 \text{ cm}$.

For the lens we have

$(1/d_o) + (1/d_i) = 1/f$;

$(1/d_o) + [1/(60 \text{ cm} - d_o)] = 1/(15 \text{ cm})$,

which gives a quadratic equation:

$d_o^2 - (60 \text{ cm})d_o + 900 \text{ cm}^2 = 0$, or $(d_o - 30 \text{ cm})^2 = 0$.

Thus the two answer are the same: 30 cm, so the lens should be placed **midway between the object and screen.**

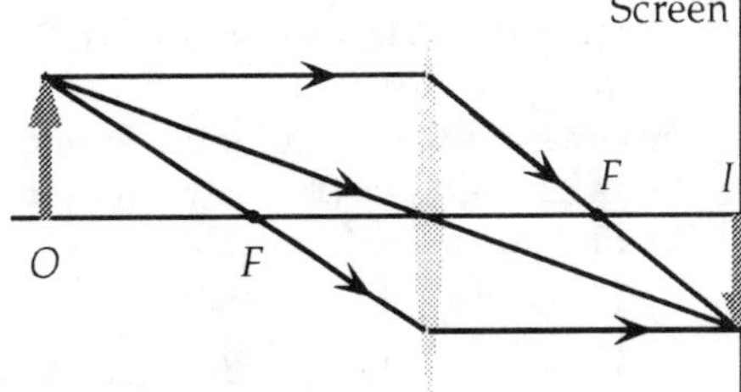

Note that in general the screen must be at least $4f$ from the object for an image to be formed on the screen.

62. For a real object and image, both d_o and d_i must be positive, so the magnification will be negative:

$m = -d_i/d_o;$

$-2.75 = -d_i/d_o,$ or $d_i = 2.75d_o.$

We find the object distance from

$(1/d_o) + (1/d_i) = 1/f;$

$(1/d_o) + (1/2.75d_o) = 1/(+ 75\text{ mm})$, which gives $d_o = 102$ cm.

The image distance is

$d_i = 2.75d_o = 2.75(102\text{ cm}) = 281$ cm.

The distance between object and image is

$L = d_o + d_i = 102\text{ cm} + 281\text{ cm} =$ **383 cm.**

63. (*a*) For the thin lens we have

$(1/d_o) + (1/d_i) = 1/f;$

$[1/(f + x)] + [1/(f + x')] = 1/f,$

which can be written as

$2f + x + x' = (f + x)(f + x')/f$

$= f + (x + x') + (xx'/f),$ or $xx' = f^2.$

(*b*) For the standard form we have

$(1/d_o) + (1/d_i) = 1/f;$

$(1/45.0\text{ cm}) + (1/d_i) = 1/32.0\text{ cm}$, which gives $d_i =$ 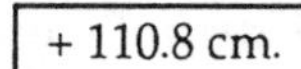

(*c*) For the Newtonian form we have

$xx' = f^2;$

$(45.0\text{ cm} - 32.0\text{ cm})x' = (32.0\text{ cm})^2$, which gives $x' = 78.7$ cm.

Thus the distance from the lens is

$d_i = x' + f = 78.7\text{ cm} + 32.0\text{ cm} =$ **110.8 cm.**

64. We find the image formed by the refraction of the first lens:

$(1/d_{o1}) + (1/d_{i1}) = 1/f_1;$

$(1/35.0\text{ cm}) + (1/d_{i1}) = 1/27.0\text{ cm}$, which gives $d_{i1} = + 118.1$ cm.

This image is the object for the second lens. Because it is beyond the second lens, it has a negative object distance: $d_{o2} = 16.5\text{ cm} - 118.1\text{ cm} = - 101.5$ cm.

We find the image formed by the refraction of the second lens:

$(1/d_{o2}) + (1/d_{i2}) = 1/f_2;$

$[1/(- 101.5\text{ cm})] + (1/d_{i2}) = 1/27.0\text{ cm}$, which gives $d_{i2} = + 21.3$ cm.

Thus the final image is **real, 21.3 cm beyond second lens.**

The total magnification is the product of the magnifications for the two lenses:

$m = m_1m_2 = (- d_{i1}/d_{o1})(- d_{i2}/d_{o2}) = d_{i1}d_{i2}/d_{o1}d_{o2}$

$= (+ 118.1\text{ cm})(+ 21.3\text{ cm})/(+ 35.0\text{ cm})(- 101.5\text{ cm}) =$ **– 0.708 (inverted).**

65. The image of an infinite object formed by the refraction of the first lens will be at the focal point:

$d_{i1} = f_1 = + 20.0$ cm.

This image is the object for the second lens. Because it is beyond the second lens, it has a negative object distance: $d_{o2} = 14.0\text{ cm} - 20.0\text{ cm} = - 6.0$ cm.

We find the image formed by the refraction of the second lens:

$(1/d_{o2}) + (1/d_{i2}) = 1/f_2;$

$[1/(- 6.0\text{ cm})] + (1/d_{i2}) = 1/(- 31.5\text{ cm})$, which gives $d_{i2} = + 7.4$ cm.

Thus the final image is **real, 7.4 cm beyond second lens.**

66. We see from the ray diagram that the image from the first lens will be a virtual image at its focal point. This is a real object for the second lens, and must be at the focal point of the second lens. If L is the separation of the lenses, the focal length of the first lens is

$f_1 = L - f_2 = 21.0 \text{ cm} - 31.0 \text{ cm} =$ **– 10.0 cm.**

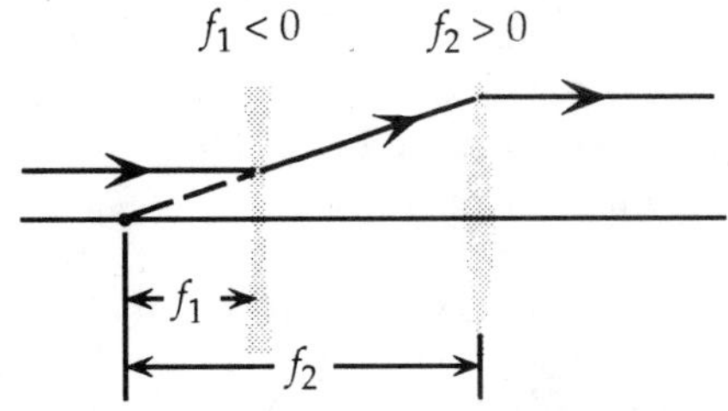

67. We find the focal length by finding the image distance for an object very far away.
For the first converging lens, we have

$(1/d_{o1}) + (1/d_{i1}) = 1/f_C$;

$(1/\infty) + (1/d_{i1}) = 1/f_C$, or, as expected, $d_{i1} = f_C$.

The first image is the object for the second lens. If the first image is real, the second object distance is negative:

$d_{o2} = -d_{i1} = -f_C$.

For the second diverging lens, we have

$(1/d_{o2}) + (1/d_{i2}) = 1/f_D$;

$[1/(-f_C)] + (1/d_{i2}) = 1/f_D$.

Because the second image must be at the focal point of the combination, we have

$(-1/f_C) + (1/f_T) = 1/f_D$, which gives $1/f_D = (1/f_T) - (1/f_C)$.

68. We find the focal length of the lens from

$1/f = (n-1)[(1/R_1) + (1/R_2)]$

$= (1.52-1)\{[1/(-31.2 \text{ cm})] + [1/(-23.8 \text{ cm})]\}$, which gives **$f = -26.0$ cm.**

69. We find the index from the lensmaker's equation:

$1/f = (n-1)[(1/R_1) + (1/R_2)]$;

$1/28.9 \text{ cm} = (n-1)[(1/31.0 \text{ cm}) + (1/31.0 \text{ cm})]$, which gives $n =$ **1.54.**

70. When the surfaces are reversed, we get

$1/f = (n-1)[(1/R_1) + (1/R_2)]$

$= (1.51-1)\{[1/(-18.4 \text{ cm})] + (1/\infty)\}$,

which gives $f = -36.0$ cm, which is the result from Example 23–17.

71. We find the radius from the lensmaker's equation:

$1/f = (n-1)[(1/R_1) + (1/R_2)]$;

$1/28.5 \text{ cm} = (1.46-1)[(1/\infty) + (1/R_2)]$, which gives $R_2 =$ **13.1 cm.**

72. We find the radius from the lensmaker's equation:

$1/f = (n-1)[(1/R_1) + (1/R_2)]$;

$1/(-25.4 \text{ cm}) = (1.50-1)[(1/\infty) + (1/R_2)]$, which gives $R_2 =$ **– 12.7 cm.**

The negative sign indicates concave.

73. We find the radius from the lensmaker's equation:

$1/f = (n-1)[(1/R_1) + (1/R_2)]$;

$+1.50 \text{ D} = (1.56-1)[(1/20.0 \text{ cm}) + (1/R_2)]$, which gives $R_2 =$ **– 43.1 cm (concave).**

74. We find the focal length of the lens from

$1/f = (n - 1)[(1/R_1) + (1/R_2)]$

$= (1.56 - 1)\{[1/(-21.0 \text{ cm})] + [1/(+18.5 \text{ cm})]\}$, which gives $f = 277.5$ cm.

We find the image distance from

$(1/d_o) + (1/d_i) = 1/f$;

$(1/100 \text{ cm}) + (1/d_i) = 1/277.5$ cm, which gives $d_i =$ **– 156 cm (in front of the lens).**

The magnification is

$m = -d_i/d_o = -(-156 \text{ cm})/(100 \text{ cm}) =$ **+ 1.56 (upright).**

75. The refraction equations are all based on

$n_1 \sin \theta_1 = n_2 \sin \theta_2$.

The lensmaker's equation is derived assuming air ($n = 1.00$) on the left-hand side. If we have some other material with n different from 1.00, we can make the equation equivalent to this by using an effective index:

$\sin \theta_1 = (n_2/n_1) \sin \theta_2 = n_{eff} \sin \theta_2$, where $n_{eff} = n_2/n_1$.

Thus we have

$P_{air} = 1/f_{air} = (n_{glass} - 1)[(1/R_1) + (1/R_2)]$;

$P_{water} = 1/f_{water} = [(n_{glass}/n_{water}) - 1][(1/R_1) + (1/R_2)]$.

If we divide the two equations, we get

$P_{water}/P_{air} = [(n_{glass}/n_{water}) - 1]/(n_{glass} - 1)$;

$P_{water}/(+5.2 \text{ D}) = [(1.50/1.33) - 1]/(1.50 - 1)$, which gives $P_{water} =$ **+ 1.3 D.**

76. For a plane mirror each image is as far behind the mirror as the object is in front. Each reflection produces a front-to-back reversal. We show the three images and the two intermediate images that are not seen.

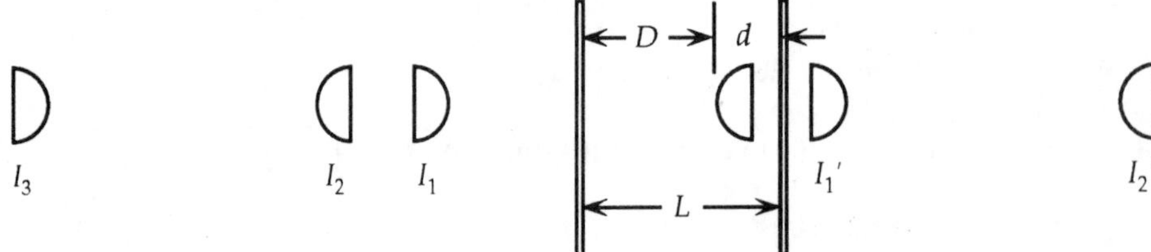

(*a*) The first image is from a single reflection, so it is

$d_1 = 2D = 2(1.5 \text{ m}) =$ **3.0 m** away.

The second image is from two reflections, so it is

$d_2 = L + d + D = 2.0 \text{ m} + 0.5 \text{ m} + 1.5 \text{ m} =$ **4.0 m** away.

The third image is from three reflections, so it is

$d_3 = 2L + D + D = 2(2.0 \text{ m}) + 1.5 \text{ m} + 1.5 \text{ m} =$ **7.0 m** away.

(*b*) We see from the diagram that

the first image is facing **toward you;**

the second image is facing **away from you;**

the third image is facing **toward you.**

77. We find the angle of incidence for the refraction from water into air:

$n_{water} \sin \theta_1 = n_{air} \sin \theta_2$;

$(1.33) \sin \theta_1 = (1.00) \sin (90° - 14°)$,

which gives $\theta_1 = 47°$.

We find the depth of the pool from

$\tan \theta_1 = x/h$;

$\tan 47° = (5.50 \text{ m})/h =$, which gives $h =$ **5.2 m.**

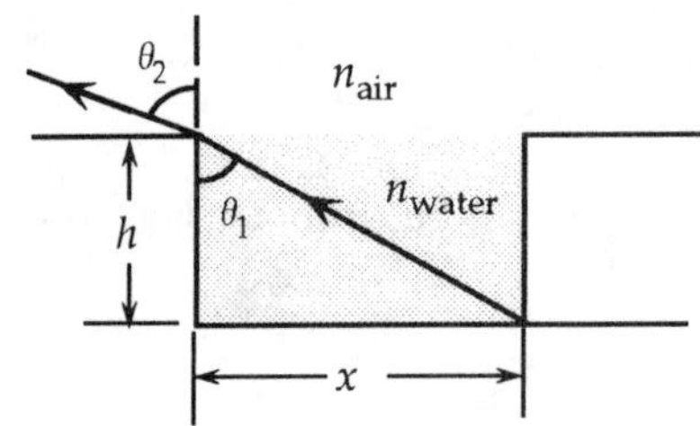

78. At the critical angle, the refracted angle is 90°. For the refraction from plastic to air, we have

$n_{plastic} \sin \theta_{plastic} = n_{air} \sin \theta_{air}$;

$n_{plastic} \sin 37.3° = (1.00) \sin 90°$, which gives $n_{plastic} = 1.65$.

For the refraction from plastic to water, we have

$n_{plastic} \sin \theta_{plastic}' = n_{water} \sin \theta_{water}$;

$(1.65) \sin \theta_{plastic}' = (1.33) \sin 90°$, which gives $\theta_{plastic}' = \boxed{53.7°.}$

79. We find the object distance from

$(1/d_o) + (1/d_i) = 1/f$;

$(1/d_o) + (1/7.50 \times 10^3 \text{ mm}) = 1/100 \text{ mm}$, which gives $d_o = 101 \text{ mm} = \boxed{0.101 \text{ m.}}$

We find the size of the image from

$m = h_i/h_o = -d_i/d_o$;

$h_i/(0.036 \text{ m}) = -(7.50 \text{ m})/(0.101 \text{ m})$, which gives $h_i = \boxed{-2.7 \text{ m.}}$

80.

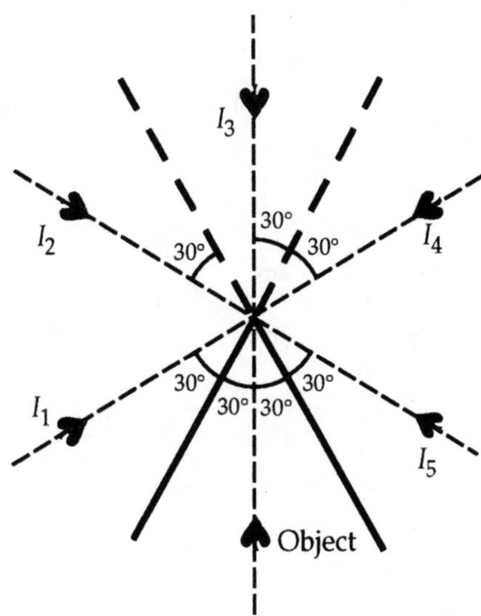

81. We find the object distance from the required magnification (which is negative for a real object and a real image):

$m = h_i/h_o = -d_i/d_o$;

$-(2.7 \times 10^3 \text{ mm})/(36 \text{ mm}) = -(9.00 \text{ m})/d_o$, which gives $d_o = 0.120$ m.

We find the focal length of the lens from

$(1/d_o) + (1/d_i) = 1/f$;

$(1/0.120 \text{ m}) + (1/9.00 \text{ m}) = 1/f$, which gives $f = 0.118 \text{ m} = \boxed{+12 \text{ cm.}}$

82. We get an expression for the image distance from the lens equation:
    $(1/d_o) + (1/d_i) = 1/f$;
    $1/d_i = (1/f) - (1/d_o)$, or $d_i = fd_o/(d_o - f)$.
    The magnification is
    $m = -d_i/d_o = -f/(d_o - f)$.
    If the lens is converging, $f > 0$.
    For a real object, $d_o > 0$.
    When $d_o > f$, we have $(d_o - f) > 0$, so all factors in the expressions for d_i and m are positive; thus $d_i > 0$ (real), and $m < 0$ (inverted).
    When $d_o < f$, we have $(d_o - f) < 0$, so the denominator in the expressions for d_i and m are negative; thus $d_i < 0$ (virtual), and $m > 0$ (upright).
    For an object beyond the lens, $d_o < 0$.
    When $-d_o > f$, we have $(d_o - f) < 0$, so both numerator and denominator in the expression for d_i are negative; thus $d_i > 0$, so the image is **real.** The numerator in the expression for m is negative; thus $m > 0$, so the image is **upright.**
    When $0 < -d_o < f$, we have $(d_o - f) < 0$, so we get the same result: **real and upright.**

83. We get an expression for the image distance from the lens equation:
    $(1/d_o) + (1/d_i) = 1/f$;
    $1/d_i = (1/f) - (1/d_o)$, or $d_i = fd_o/(d_o - f)$.
    If the lens is diverging, $f < 0$. If we write $f = -|f|$, we get $d_i = -|f|d_o/(d_o + |f|)$.
    For a real object, $d_o > 0$.
    All factors in the expression for d_i are positive, thus $d_i < 0$, so the image is always virtual.
    We can have a real image, $d_i > 0$, if $d_o < 0$, and $|d_o| < |f|$, so the denominator is still positive. Thus to have a real image from a diverging lens, the condition is
    $\boxed{0 < -d_o < -f.}$

84. For the refraction at the second surface, we have
    $n \sin\theta_3 = n_{air} \sin\theta_4$;
    $(1.52) \sin\theta_3 = (1.00) \sin\theta_4$.
    The maximum value of θ_4 before internal reflection takes place at the second surface is 90°. Thus for internal reflection not to occur, we have
    $(1.52) \sin\theta_3 \leq 1.00$;
    $\sin\theta_3 \leq 0.658$, so $\theta_3 \leq 41.1°$.
    We find the refraction angle at the second surface from
    $(90° - \theta_2) + (90° - \theta_3) + A = 180°$, which gives
    $\theta_2 = A - \theta_3 = 72° - \theta_3$.
    Thus $\theta_2 \geq 72° - 41.1° = 30.9°$.
    For the refraction at the first surface, we have
    $n_{air} \sin\theta_1 = n \sin\theta_2$;
    $(1.00) \sin\theta_1 = (1.52) \sin\theta_2$; which gives $\sin\theta_1 = (1.52) \sin\theta_2$.
    For the limiting condition, we have
    $\sin\theta_1 \geq (1.52) \sin 30.9° = 0.781$, so θ_1 $\boxed{\geq 51.3°.}$

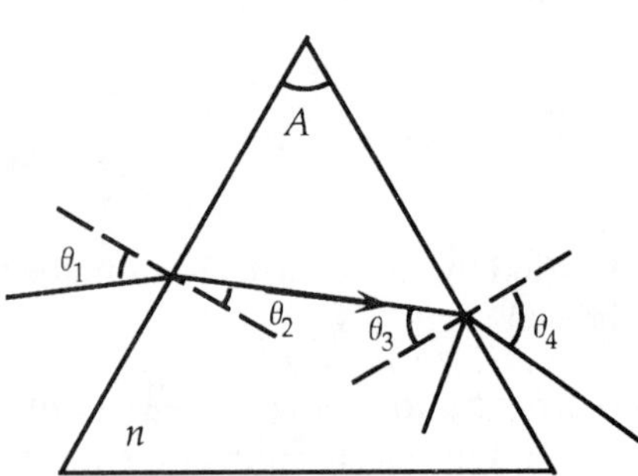

85. (*a*)

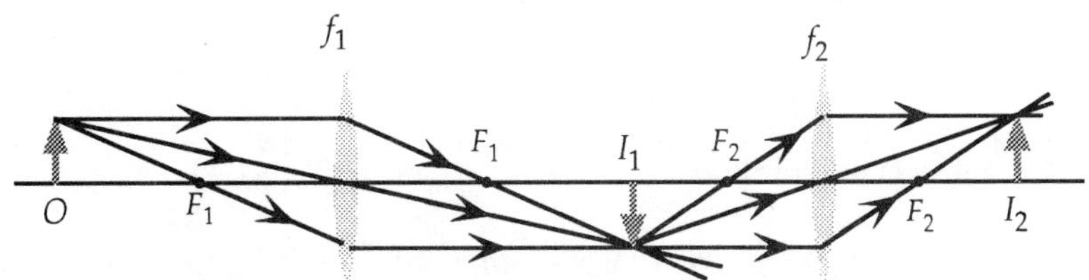

(b) We see that the mage is real and upright, and estimate that it is 20 cm beyond the second lens.
We find the image formed by the refraction of the first lens:

$(1/d_{o1}) + (1/d_{i1}) = 1/f_1;$

$(1/30\text{ cm}) + (1/d_{i1}) = 1/15\text{ cm}$, which gives $d_{i1} = +30\text{ cm}$.

This image is the object for the second lens. Because it is in front of the second lens, it is a real object, with an object distance of $d_{o2} = 50\text{ cm} - 30\text{ cm} = 20\text{ cm}$.

We find the image formed by the refraction of the second lens:

$(1/d_{o2}) + (1/d_{i2}) = 1/f_2;$

$(1/20\text{ cm}) + (1/d_{i2}) = 1/10\text{ cm}$, which gives $d_{i2} = +20\text{ cm}$.

Thus the final image is **real, 20 cm beyond second lens.**

The total magnification is the product of the magnifications for the two lenses:

$m = m_1m_2 = (-d_{i1}/d_{o1})(-d_{i2}/d_{o2}) = d_{i1}d_{i2}/d_{o1}d_{o2}$

$= (+30\text{ cm})(+30\text{ cm})/(+20\text{ cm})(+20\text{ cm}) = +1.0.$

Thus the final image is **upright, same size as object.**

86. (*a*) We find the focal length by finding the image distance for an object very far away.
For the first lens, we have

$(1/d_{o1}) + (1/d_{i1}) = 1/f_1;$

$(1/\infty) + (1/d_{i1}) = 1/f_1$, or, as expected, $d_{i1} = f_1$.

The first image is the object for the second lens. If the first image is real, the second object is virtual:

$d_{o2} = -d_{i1} = -f_1$.

For the second lens, we have

$(1/d_{o2}) + (1/d_{i2}) = 1/f_2;$

$[1/(-f_1)] + (1/d_{i2}) = 1/f_2$.

Because the second image must be at the focal point of the combination, we have

$(-1/f_1) + (1/f_T) = 1/f_2$, which gives $1/f_T = (1/f_1) + (1/f_2)$.

When we solve for f_T, we get

$f_T = f_1f_2/(f_1 + f_2)$.

(*b*) If we use the intermediate result $1/f_T = (1/f_1) + (1/f_2)$, we see that

$P = P_1 + P_1$.

87. (*a*) We use the lens equation with $d_o + d_i = d_T$:

$(1/d_o) + (1/d_i) = 1/f;$

$(1/d_o) + [1/(d_T - d_o) = 1/f.$

When we rearrange this, we get a quadratic equation for d_o:

$d_o^2 - d_T d_o + d_T f = 0$, which has the solution

$d_o = \frac{1}{2}[d_T \pm (d_T^2 - 4d_T f)^{1/2}].$

If $d_T > 4f$, we see that the term inside the square root $d_T^2 - 4d_T f > 0$, and $(d_T^2 - 4d_T f)^{1/2} < d_T$, so we get two real, positive solutions for d_o.

(*b*) If $d_T < 4f$, we see that the term inside the square root $d_T^2 - 4d_T f < 0$, so there are no real solutions for d_o.

(*c*) When there are two solutions, the distance between them is

$\Delta d = d_{o1} - d_{o2} = \frac{1}{2}[d_T + (d_T^2 - 4d_T f)^{1/2}] - \frac{1}{2}[d_T - (d_T^2 - 4d_T f)^{1/2}] = \boxed{(d_T^2 - 4d_T f)^{1/2}.}$

The image positions are given by

$d_i = d_T - d_o = \frac{1}{2}[d_T \mp (d_T^2 - 4d_T f)^{1/2}].$

The ratio of image sizes is the ratio of magnifications:

$m = m_2/m_1 = (d_{i2}/d_{o2})/(d_{i1}/d_{o1}) = (d_{i2}/d_{o2})(d_{o1}/d_{i1})$

$= \{\frac{1}{2}[d_T + (d_T^2 - 4d_T f)^{1/2}]/\frac{1}{2}[d_T - (d_T^2 - 4d_T f)^{1/2}]\}^2$

$= \boxed{\{[d_T + (d_T^2 - 4d_T f)^{1/2}]/[d_T - (d_T^2 - 4d_T f)^{1/2}]\}^2.}$

88. We find the focal length by finding the image distance for an object very far away.
For the first lens, we have

$(1/d_{o1}) + (1/d_{i1}) = 1/f_1;$

$(1/\infty) + (1/d_{i1}) = 1/(10.0 \text{ cm})$, or, as expected, $d_{i1} = 10.0$ cm.

The first image is the object for the second lens. The first image is real, so the second object has a negative object distance:

$d_{o2} = - d_{i1} = - 10.0$ cm.

For the second lens, we have

$(1/d_{o2}) + (1/d_{i2}) = 1/f_2;$

$[1/(- 10.0 \text{ cm})] + (1/d_{i2}) = 1/(- 20.0 \text{ cm})$, which gives $d_{i2} = + 20.0$ cm.

Because the second image must be at the focal point of the combination, we have

$\boxed{f = + 20.0 \text{ cm (converging).}}$

89. For both mirrors the image is virtual (behind the mirror), so the image distances are negative.
The image distance for the plane mirror is $d_{i1} = - d_o$, and the image is upright and the same size.
Because the angle subtended by the image is small, it is

$\theta_1 = h_1/(d_o - d_{i1}) = h/2d_o.$

The image distance for the convex mirror is d_{i2}, and the image is upright and smaller.
Because the angle subtended by the image is small, it is

$\theta_2 = h_2/(d_o - d_{i2}) = \theta_1/2 = h/4d_o$, or $h_2/h = (d_o - d_{i2})/4d_o.$

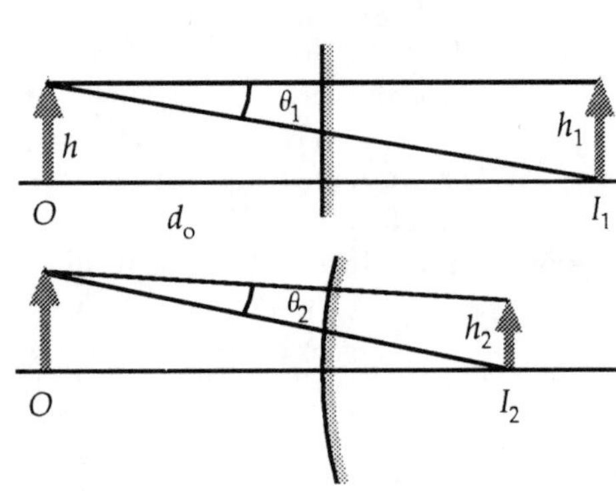

We find the image distance from the magnification:

$m_2 = h_2/h = - d_{i2}/d_o = (d_o - d_{i2})/4d_o,$

which gives $d_{i2} = - d_o/3$.

We find the required focal length of the convex mirror from

$(1/d_o) + (1/d_{i2}) = 1/f_2;$

$(1/d_o) + [1/(- d_o/3)] = 1/f_2$, which gives $f_2 = - d_o/2$.

Thus the radius of curvature is

$R_2 = 2f_2 = 2(- d_o/2) = - d_o = \boxed{- 3.25 \text{ m.}}$

90. The two students chose different signs for the magnification, i. e., one upright and one inverted.
The focal length of the concave mirror is $f = R/2 = (40\text{ cm})/2 = 20\text{ cm}$.
We relate the object and image distances from the magnification:
$m = -d_i/d_o$;
$\pm 3 = -d_i/d_o$, which gives $d_i = \mp 3d_o$.
When we use this in the mirror equation, we get
$(1/d_o) + (1/d_i) = 1/f$;
$(1/d_o) + [1/(\mp 3d_o)] = 1/f$ which gives $d_o = 2f/3, 4f/3 = 13.3\text{ cm}, 26.7\text{ cm}$.
The image distances are $= -40$ cm (virtual, upright), and $+80$ cm (real, inverted).

91. The refraction equations for the two surfaces are all based on
$n' \sin \theta_1 = n \sin \theta_2$, and $n \sin \theta_3 = n' \sin \theta_4$.
Eq. 23–10 was derived based on having air ($n' = 1$) outside the lens. We can make these equivalent to having air outside the lens, if we write them as
$\sin \theta_1 = (n/n') \sin \theta_2$, and $(n/n') \sin \theta_3 = \sin \theta_4$.
Thus we see that we have an effective index, $n_{eff} = n/n'$, to use in Eq. 23–10:
$1/f' = [(n/n') - 1][(1/R_1) + (1/R_2)]$.
The focal length in air is
$1/f = (n - 1)[(1/R_1) + (1/R_2)]$;
so we have
$1/f' = [(n/n') - 1]/f(n - 1)$.
Eq. 23–8 is derived using the focal points. Thus, if we use the above focal length, the derivation is the same, so we have the same equation:
$(1/d_o) + (1/d_i) = 1/f'$, where $1/f' = [(n/n') - 1]/f(n - 1)$.
Eq. 23–9 is derived by comparing heights and distances, so it is unchanged:
$m = h_i/h_o = -d_i/d_o$.

CHAPTER 24

1. We draw the wavelets and see that the incident wave fronts are parallel, with the angle of incidence θ_1 being the angle between the wave fronts and the surface. The reflecting wave fronts are parallel, with the angle of reflection θ_2 being the angle between the wave fronts and the surface. Both sets of wave fronts are in the same medium, so they travel at the same speed. The perpendicular distance between wave fronts is $BC = AD = c\,\Delta t$.
From the triangles, we see that
$AB = BC/\sin\theta_1 = AD/\sin\theta_2$.
Thus we have
$\sin\theta_1 = \sin\theta_2$, or $\theta_1 = \theta_2$.

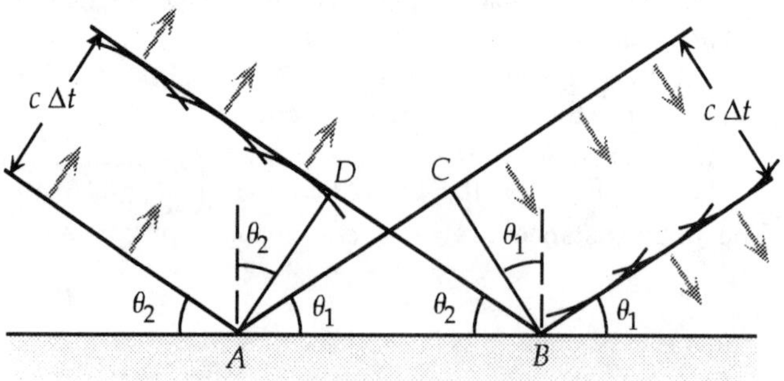

2. For constructive interference, the path difference is a multiple of the wavelength:
$d \sin\theta = m\lambda$, $m = 0, 1, 2, 3, \ldots$.
For the fifth order, we have
$(4.2 \times 10^{-5}\text{ m}) \sin 7.8° = (5)\lambda$, which gives $\lambda = 1.1 \times 10^{-6}$ m = **1.1 μm.**

3. For constructive interference, the path difference is a multiple of the wavelength:
$d \sin\theta = m\lambda$, $m = 0, 1, 2, 3, \ldots$.
For the third order, we have
$d \sin 15° = (3)(650 \times 10^{-9}\text{ m})$, which gives $d = 7.5 \times 10^{-6}$ m = **7.5 μm.**

4. For constructive interference, the path difference is a multiple of the wavelength:
$d \sin\theta = m\lambda$, $m = 0, 1, 2, 3, \ldots$.
We find the location on the screen from
$y = L \tan\theta$.
For small angles, we have
$\sin\theta \approx \tan\theta$, which gives
$y = L(m\lambda/d) = mL\lambda/d$.
For adjacent fringes, $\Delta m = 1$, so we have
$\Delta y = L\lambda\,\Delta m/d$;
$0.055\text{ m} = (5.00\text{ m})\lambda(1)/(0.040 \times 10^{-3}\text{ m})$, which gives $\lambda = 4.40 \times 10^{-7}$ m = **440 nm.**
The frequency is
$f = c/\lambda = (3.00 \times 10^8\text{ m/s})/(4.40 \times 10^{-7}\text{ m})$ = **6.82×10^{14} Hz.**

5. For constructive interference, the path difference is a multiple of the wavelength:
$d \sin\theta = m\lambda$, $m = 0, 1, 2, 3, \ldots$.
We find the location on the screen from
$y = L \tan\theta$.
For small angles, we have
$\sin\theta \approx \tan\theta$, which gives
$y = L(m\lambda/d) = mL\lambda/d$.
For adjacent fringes, $\Delta m = 1$, so we have
$\Delta y = L\lambda\,\Delta m/d$;
$= (2.6\text{ m})(656 \times 10^{-9}\text{ m})(1)/(0.050 \times 10^{-3}\text{ m}) = 3.4 \times 10^{-2}$ m = **3.4 cm.**

6. For destructive interference, the path difference is

$d \sin \theta = (m + \frac{1}{2})\lambda, \quad m = 0, 1, 2, 3, \ldots$; or

$\sin \theta = (m + \frac{1}{2})(2.5 \text{ cm})/(5.0 \text{ cm}) = (m + \frac{1}{2})(0.50), \quad m = 0, 1, 2, 3, \ldots$.

The angles for the first three regions of complete destructive interference are

$\sin \theta_0 = (m + \frac{1}{2})\lambda/d = (0 + \frac{1}{2})(0.50) = 0.25, \theta_0 = 15°$;

$\sin \theta_1 = (m + \frac{1}{2})\lambda/d = (1 + \frac{1}{2})(0.50) = 0.75, \theta_1 = 49°$;

$\sin \theta_2 = (m + \frac{1}{2})\lambda/d = (2 + \frac{1}{2})(0.50) = 1.25$, therefore, no third region.

We find the locations at the end of the tank from

$y = L \tan \theta$;

$y_0 = (2.0 \text{ m}) \tan 15° = 0.52 \text{ m}$;

$y_1 = (2.0 \text{ m}) \tan 49° = 2.3 \text{ m}$.

Thus you could stand

0.52 m, or 2.3 m away from the line perpendicular to the board midway between the openings.

7. For constructive interference, the path difference is a multiple of the wavelength:

$d \sin \theta = m\lambda, \quad m = 0, 1, 2, 3, \ldots$.

We find the location on the screen from

$y = L \tan \theta$.

For small angles, we have

$\sin \theta \approx \tan \theta$, which gives

$y = L(m\lambda/d) = mL\lambda/d$.

For the fourth order we have

$48 \times 10^{-3} \text{ m} = (1.5 \text{ m})(680 \times 10^{-9} \text{ m})(4)/d$, which gives $d = 8.5 \times 10^{-5} \text{ m} =$ 0.085 mm.

8. For constructive interference, the path difference is a multiple of the wavelength:

$d \sin \theta = m\lambda, \quad m = 0, 1, 2, 3, \ldots$.

We find the location on the screen from

$y = L \tan \theta$.

For small angles, we have

$\sin \theta \approx \tan \theta$, which gives

$y = L(m\lambda/d) = mL\lambda/d$.

For the second order of the two wavelengths, we have

$y_a = L\lambda_a m/d = (1.6 \text{ m})(480 \times 10^{-9} \text{ m})(2)/(0.54 \times 10^{-3} \text{ m}) = 2.84 \times 10^{-3} \text{ m} = 2.84 \text{ mm}$;

$y_b = L\lambda_b m/d = (1.6 \text{ m})(620 \times 10^{-9} \text{ m})(2)/(0.54 \times 10^{-3} \text{ m}) = 3.67 \times 10^{-3} \text{ m} = 3.67 \text{ mm}$.

Thus the two fringes are separated by 3.67 mm – 2.84 mm = 0.8 mm.

9. The 180° phase shift produced by the glass is equivalent to a path length of $\frac{1}{2}\lambda$. For constructive interference on the screen, the total path difference is a multiple of the wavelength:

$\frac{1}{2}\lambda + d \sin \theta_{max} = m\lambda, \quad m = 0, \pm 1, \pm 2, \pm 3, \ldots$; or $d \sin \theta = (m - \frac{1}{2})\lambda, \quad m = 0, \pm 1, \pm 2, \pm 3, \ldots$.

For destructive interference on the screen, the total path difference is

$\frac{1}{2}\lambda + d \sin \theta_{max} = (m + \frac{1}{2})\lambda, \quad m = 0, \pm 1, \pm 2, \pm 3, \ldots$; or $d \sin \theta = m\lambda, \quad m = 0, \pm 1, \pm 2, \pm 3, \ldots$.

Thus the pattern is just the reverse of the usual double-slit pattern.

10. For constructive interference of the second order for the blue light, we have
$d \sin \theta = m\lambda_b = (2)(460 \text{ nm}) = 920 \text{ nm}.$
For destructive interference of the other light, we have
$d \sin \theta = (m' + \frac{1}{2})\lambda, \quad m' = 0, 1, 2, 3, \ldots .$
When the two angles are equal, we get
$920 \text{ nm} = (m' + \frac{1}{2})\lambda, \quad m' = 0, 1, 2, 3, \ldots .$
For the first three values of m', we get
$920 \text{ nm} = (0 + \frac{1}{2})\lambda$, which gives $\lambda_{,} = 1.84 \times 10^3$ nm;
$920 \text{ nm} = (1 + \frac{1}{2})\lambda$, which gives $\lambda_{,} = 613$ nm;
$920 \text{ nm} = (2 + \frac{1}{2})\lambda$, which gives $\lambda_{,} = 368$ nm.
The only one of these that is visible light is **613 nm.**

11. The presence of the water changes the wavelength: $\lambda_{\text{water}} = \lambda/n_{\text{water}} = 400 \text{ nm}/1.33 = 300 \text{ nm}.$
For constructive interference, the path difference is a multiple of the wavelength in the water:
$d \sin \theta = m\lambda_{\text{water}}, \quad m = 0, 1, 2, 3, \ldots .$
We find the location on the screen from
$y = L \tan \theta.$
For small angles, we have
$\sin \theta \approx \tan \theta$, which gives
$y = L(m\lambda_{\text{water}}/d) = mL\lambda_{\text{water}}/d.$
For adjacent fringes, $\Delta m = 1$, so we have
$\Delta y = L\lambda_{\text{water}}\,\Delta m/d;$
$= (0.400 \text{ m})(300 \times 10^{-9} \text{ m})(1)/(5.00 \times 10^{-5} \text{ m}) = 2.40 \times 10^{-3} \text{ m} =$ **2.40 mm.**

12. To change the center point from constructive interference to destructive interference, the phase shift produced by the introduction of the plastic must be an odd multiple of half a wavelength, corresponding to the change in the number of wavelengths in the distance equal to the thickness of the plastic. The minimum thickness will be for a shift of a half wavelength:
$N = (t/\lambda_{\text{plastic}}) - (t/\lambda) = (tn_{\text{plastic}}/\lambda) - (t/\lambda) = (t/\lambda)(n_{\text{plastic}} - 1) = 1/2;$
$[t/(540 \text{ nm})](1.60 - 1) = 1/2$, which gives $t =$ **450 nm.**

13. We find the speed of light from the index of refraction, $v = c/n$. For the change, we have
$(v_{\text{red}} - v_{\text{violet}})/v_{\text{violet}} = [(c/n_{\text{red}}) - (c/n_{\text{violet}})]/(c/n_{\text{violet}})$
$= (n_{\text{violet}} - n_{\text{red}})/n_{\text{red}} = (1.665 - 1.617)/(1.617) = 0.030 =$ **3.0%.**

14. We find the speed of light from the index of refraction, $v = c/n$. For the change, we have
$(v_{\text{blue}} - v_{\text{red}})/v_{\text{red}} = [(c/n_{\text{blue}}) - (c/n_{\text{red}})]/(c/n_{\text{red}})$
$= (n_{\text{red}} - n_{\text{blue}})/n_{\text{blue}} = (1.617 - 1.645)/(1.645) = -0.017 =$ **– 1.7%.**

15. We find the angles of refraction in the glass from
$n_1 \sin \theta_1 = n_2 \sin \theta_2;$
$(1.00) \sin 60.00° = (1.4820) \sin \theta_{2,450}$, which gives $\theta_{2,450} = 35.76°;$
$(1.00) \sin 60.00° = (1.4742) \sin \theta_{2,700}$, which gives $\theta_{2,700} = 35.98°.$
Thus the angle between the refracted beams is
$\theta_{2,700} - \theta_{2,450} = 35.98° - 35.76° =$ **0.22°.**

16. For the refraction at the first surface, we have

$n_{air} \sin \theta_a = n \sin \theta_b$;

$(1.00) \sin 45° = (1.665) \sin \theta_{b1}$, which gives $\theta_{b1} = 25.13°$;

$(1.00) \sin 45° = (1.619) \sin \theta_{b2}$, which gives $\theta_{b2} = 25.90°$.

We find the angle of incidence at the second surface from

$(90° - \theta_b) + (90° - \theta_c) + A = 180°$, which gives

$\theta_{c1} = A - \theta_{b1} = 60.00° - 25.13° = 34.87°$;

$\theta_{c2} = A - \theta_{b2} = 60.00° - 25.90° = 34.10°$.

For the refraction at the second surface, we have

$n \sin \theta_c = n_{air} \sin \theta_d$;

$(1.665) \sin 34.87° = (1.00) \sin \theta_{d1}$, which gives $\theta_{d1} = 72.2°$ from the normal;

$(1.619) \sin 34.10° = (1.00) \sin \theta_{d2}$, which gives $\theta_{d2} = 65.2°$ from the normal.

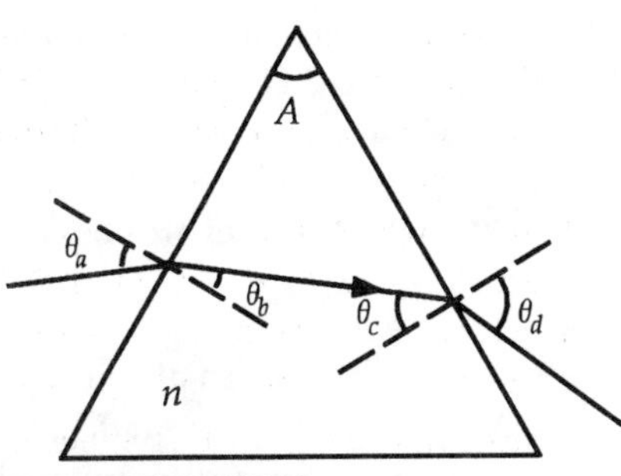

17. We find the focal length of the lens from

$1/f = (n - 1)[(1/R_1) + (1/R_2)]$;

$1/f_{400} = (1.518 - 1)\{[1/(17.5 \text{ cm})] + [1/(17.5 \text{ cm})]\}$, which gives $f_{400} = 16.89$ cm;

$1/f_{700} = (1.500 - 1)\{[1/(17.5 \text{ cm})] + [1/(17.5 \text{ cm})]\}$, which gives $f_{700} = 17.50$ cm.

Thus the distance between focal points is

$f_{700} - f_{400} = 17.50 \text{ cm} - 16.89 \text{ cm} =$ 0.61 cm.

18. We find the angle to the first minimum from

$\sin \theta_{1min} = m\lambda/D = (1)(520 \times 10^{-9} \text{ m})/(0.0440 \times 10^{-3} \text{ m}) = 0.0118$, so $\theta_{1min} = 0.677°$.

Thus the angular width of the central diffraction peak is

$\Delta\theta_1 = 2\theta_{1min} = 2(0.677°) =$ 1.35°.

19. The angle from the central maximum to the first minimum is 18.5°.

We find the wavelength from

$D \sin \theta_{1min} = m\lambda$;

$(3.00 \times 10^{-6} \text{ m}) \sin (18.5°) = (1)\lambda$, which gives $\lambda = 9.52 \times 10^{-7}$ m = 952 nm.

20. For constructive interference from the single slit, the path difference is

$D \sin \theta = (m + \frac{1}{2})\lambda, \quad m = 1, 2, 3, \ldots$.

For the first fringe away from the central maximum, we have

$(3.50 \times 10^{-6} \text{ m}) \sin \theta_1 = (\frac{3}{2})(550 \times 10^{-9} \text{ m})$, which gives $\theta_1 = 13.7°$.

We find the distance on the screen from

$y_1 = L \tan \theta_1 = (10.0 \text{ m}) \tan 13.7° =$ 2.4 m.

21. The angle from the central maximum to the first bright fringe is 9.75°.

For constructive interference from the single slit, the path difference is

$D \sin \theta = (m + \frac{1}{2})\lambda, \quad m = 1, 2, 3, \ldots$.

For the first fringe away from the central maximum, we have

$D \sin (9.75°) = (\frac{3}{2})(633 \times 10^{-9} \text{ m})$, which gives $D = 5.61 \times 10^{-6}$ m = 5.61 μm.

22. We find the angle to the first minimum from

$\sin \theta_{1min} = m\lambda/D = (1)(589 \times 10^{-9} \text{ m})/(0.0348 \times 10^{-3} \text{ m}) = 0.0169$, so $\theta_{1min} = 0.970°$.

We find the distance on the screen from

$y_1 = L \tan \theta_1 = (2.50 \text{ m}) \tan 0.970° = 4.23 \times 10^{-2} \text{ m} = 4.23$ cm.

Thus the width of the peak is

$\Delta y_1 = 2y_1 = 2(4.23 \text{ cm}) =$ 8.46 cm.

23. We find the angle to the first minimum from the distances:
    $\tan\theta_{1min} = \frac{1}{2}(9.20 \text{ cm})/(255 \text{ cm}) = 0.0180 = \sin\theta_{1min}$, because the angle is small.
    We find the slit width from
    $D \sin\theta_{1min} = m\lambda$;
    $D\,(0.0180) = (1)(415 \times 10^{-9} \text{ m})$, which gives $D = 2.30 \times 10^{-5} \text{ m} =$ $\boxed{0.0230 \text{ mm.}}$

24. Because the angles are small, we have
    $\tan\theta_{1min} = \frac{1}{2}(\Delta y_1)/L = \sin\theta_{1min}$.
    The condition for the first minimum is
    $D \sin\theta_{1min} = \frac{1}{2}D\,\Delta y_1/L = \lambda$.
    If we form the ratio of the expressions for the two wavelengths, we get
    $\Delta y_{1b}/\Delta y_{1a} = \lambda_b/\lambda_a$;
    $\Delta y_{1b}/(3.0 \text{ cm}) = (400 \text{ nm})/(550 \text{ nm})$, which gives $\Delta y_{1b} =$ $\boxed{2.2 \text{ cm.}}$

25. (*a*) There will be no diffraction minima if the angle for the first minimum is greater than 90°.
    Thus the limiting condition is
    $D \sin\theta_{1min} = m\lambda$;
    $D_{max} \sin 90° = (1)\lambda$, or $\boxed{D_{max} = \lambda.}$
    (*b*) Visible light has wavelengths from 400 nm to 700 nm, so the maximum slit width for no diffraction minimum for all of these wavelengths is the one for the smallest wavelength: $\boxed{400 \text{ nm.}}$

26. We find the angle for the second order from
    $d \sin\theta = m\lambda$;
    $(1.15 \times 10^{-5} \text{ m}) \sin\theta = (2)(650 \times 10^{-9} \text{ m})$, which gives $\sin\theta = 0.113$, so $\theta =$ $\boxed{6.49°.}$

27. We find the wavelength from
    $d \sin\theta = m\lambda$;
    $[1/(3500 \text{ lines/cm})](10^{-2} \text{ m/cm}) \sin 22.0° = 3\lambda$, which gives $\lambda = 3.57 \times 10^{-7} \text{ m} =$ $\boxed{357 \text{ nm.}}$

28. We find the slit separation from
    $d \sin\theta = m\lambda$;
    $d \sin 15.5° = (1)(589 \times 10^{-9} \text{ m})$, which gives $d = 2.20 \times 10^{-6} \text{ m} =$ $\boxed{2.20\ \mu\text{m.}}$
    We find the angle for the fourth order from
    $d \sin\theta = m\lambda$;
    $(2.20 \times 10^{-6} \text{ m}) \sin\theta_4 = (4)(589 \times 10^{-9} \text{ m})$, which gives $\sin\theta_4 = 1.069$, so there is $\boxed{\text{no fourth order.}}$

29. We find the wavelengths from
    $d \sin\theta = m\lambda$;
    $[1/(10{,}000 \text{ lines/cm})](10^{-2} \text{ m/cm}) \sin 31.2° = (1)\lambda_1$, which gives $\lambda_1 = 5.18 \times 10^{-7} \text{ m} =$ $\boxed{518 \text{ nm;}}$
    $[1/(10{,}000 \text{ lines/cm})](10^{-2} \text{ m/cm}) \sin 36.4° = (1)\lambda_2$, which gives $\lambda_2 = 5.93 \times 10^{-7} \text{ m} =$ $\boxed{593 \text{ nm;}}$
    $[1/(10{,}000 \text{ lines/cm})](10^{-2} \text{ m/cm}) \sin 47.5° = (1)\lambda_3$, which gives $\lambda_3 = 7.37 \times 10^{-7} \text{ m} =$ $\boxed{737 \text{ nm.}}$

30. We find the slit separation from
    $d \sin\theta = m\lambda$;
    $d \sin 23.0° = (3)(630 \times 10^{-9} \text{ m})$, which gives $d = 4.84 \times 10^{-6} \text{ m} = 4.84 \times 10^{-4} \text{ cm}$.
    The number of lines/cm is
    $1/d = 1/(4.84 \times 10^{-4} \text{ cm}) =$ $\boxed{2.07 \times 10^3 \text{ lines/cm.}}$

31. Because the angle increases with wavelength, to have a complete order we use the largest wavelength. The maximum angle is 90°, so we have

$d \sin \theta = m\lambda;$

$[1/(7000 \text{ lines/cm})](10^{-2} \text{ m/cm}) \sin 90° = m(750 \times 10^{-9} \text{ m})$, which gives $m = 1.90$.

Thus only $\boxed{\text{one full order}}$ can be seen on each side of the central white line.

32. The maximum angle is 90°, so we have

$d \sin \theta = m\lambda;$

$[1/(6000 \text{ lines/cm})](10^{-2} \text{ m/cm}) \sin 90° = m(633 \times 10^{-9} \text{ m})$, which gives $m = 2.63$.

Thus $\boxed{\text{two orders}}$ can be seen on each side of the central white line.

33. Because the angle increases with wavelength, to have a full order we use the largest wavelength. The maximum angle is 90°, so we find the minimum separation from

$d \sin \theta = m\lambda;$

$d_{\text{min}} \sin 90° = (2)(750 \times 10^{-9} \text{ m})$, which gives $d_{\text{min}} = 1.50 \times 10^{-6} \text{ m} = 1.50 \times 10^{-4} \text{ cm}$.

The maximum number of lines/cm is

$1/d_{\text{min}} = 1/(1.50 \times 10^{-4} \text{ cm}) = \boxed{6.67 \times 10^{3} \text{ lines/cm.}}$

34. We find the angles for the first order from

$d \sin \theta = m\lambda = \lambda;$

$[1/(7500 \text{ lines/cm})](10^{-2} \text{ m/cm}) \sin \theta_{400} = (400 \times 10^{-9} \text{ m})$, which gives $\sin \theta_{400} = 0.300$, so $\theta_{400} = 17.5°$;

$[1/(7500 \text{ lines/cm})](10^{-2} \text{ m/cm}) \sin \theta_{700} = (700 \times 10^{-9} \text{ m})$, which gives $\sin \theta_{700} = 0.563$, so $\theta_{700} = 34.2°$.

The distances from the central white line on the screen are

$y_{400} = L \tan \theta_{400} = (2.30 \text{ m}) \tan 17.5° = 0.723 \text{ m};$

$y_{700} = L \tan \theta_{700} = (2.30 \text{ m}) \tan 34.2° = 1.56 \text{ m}.$

Thus the width of the spectrum is

$y_{700} - y_{400} = 1.56 \text{ m} - 0.723 \text{ m} = \boxed{0.84 \text{ m.}}$

35. We find the angles for the first order from

$d \sin \theta = m\lambda = \lambda;$

$[1/(6600 \text{ lines/cm})](10^{-2} \text{ m/cm}) \sin \theta_{\alpha} = 656 \times 10^{-9} \text{ m}$, which gives $\sin \theta_{\alpha} = 0.433$, so $\theta_{\alpha} = 25.7°$;

$[1/(6600 \text{ lines/cm})](10^{-2} \text{ m/cm}) \sin \theta_{\delta} = 410 \times 10^{-9} \text{ m}$, which gives $\sin \theta_{\delta} = 0.271$, so $\theta_{\delta} = 15.7°$.

Thus the angular separation is

$\theta_{\alpha} - \theta_{\delta} = 25.7° - 15.7° = \boxed{10.0°.}$

36. Because the angles on each side of the central line are not the same, the incident light is not normal to the grating. We use the average angles:

$\theta_1 = (26°38' + 26°48')/2 = 26°43' = 26.72°;$

$\theta_2 = (41°08' + 41°19')/2 = 41°14' = 41.23°.$

We find the wavelengths from

$d \sin \theta = m\lambda;$

$[1/(8500 \text{ lines/cm})](10^{-2} \text{ m/cm}) \sin 26.72° = (1)\lambda_1$, which gives $\lambda_1 = 5.29 \times 10^{-7} \text{ m} = \boxed{529 \text{ nm;}}$

$[1/(8500 \text{ lines/cm})](10^{-2} \text{ m/cm}) \sin 41.23° = (1)\lambda_2$, which gives $\lambda_2 = 7.75 \times 10^{-7} \text{ m} = \boxed{775 \text{ nm.}}$

Note that the second wavelength is not visible.

37. We have the same average angles, but the path differences causing the interference must be measured in terms of the wavelengths in water. Thus the wavelengths calculated in Problem 36 are those in water. The wavelengths in air are

$\lambda_{1\text{air}} = \lambda_1 n_{\text{water}} = (529 \text{ nm})(1.33) = \boxed{704 \text{ nm;}}$

$\lambda_{2\text{air}} = \lambda_2 n_{\text{water}} = (775 \text{ nm})(1.33) = \boxed{1030 \text{ nm.}}$

Note that the second wavelength is not visible.

38. We equate a path difference of one wavelength with a phase difference of 2π. With respect to the incident wave, the wave that reflects at the top surface from the higher index of the soap bubble has a phase change of

$\phi_1 = \pi$.

$\phi_1 = \pi$

$\phi_2 = (2t/\lambda_{film})2\pi + 0$

t

n

With respect to the incident wave, the wave that reflects from the air at the bottom surface of the bubble has a phase change due to the additional path-length but no phase change on reflection:

$\phi_2 = (2t/\lambda_{film})2\pi + 0$.

For constructive interference, the net phase change is

$\phi = (2t/\lambda_{film})2\pi - \pi = m2\pi$, $m = 0, 1, 2, \ldots,$ or $t = \frac{1}{2}\lambda_{film}(m + \frac{1}{2})$, $m = 0, 1, 2, \ldots$.

The wavelengths in air that produce strong reflection are given be

$\lambda = n\lambda_{film} = 2nt/(m + \frac{1}{2}) = 4(1.34)(120 \text{ nm})/(2m + 1) = (643 \text{ nm})/(2m + 1)$.

Thus we see that, for the light to be in the visible spectrum, the only value of m is 0:

$\lambda = (643 \text{ nm})/(0 + 1) = 643$ nm, which is an **orange-red.**

39. Between the 25 dark lines there are 24 intervals. When we add the half-interval at the wire end, we have 24.5 intervals, so the separation is

26.5 cm/24.5 intervals = **1.08 cm.**

40. We equate a path difference of one wavelength with a phase difference of 2π. With respect to the incident wave, the wave that reflects at the top surface from the higher index of the soap bubble has a phase change of

$\phi_1 = \pi$.

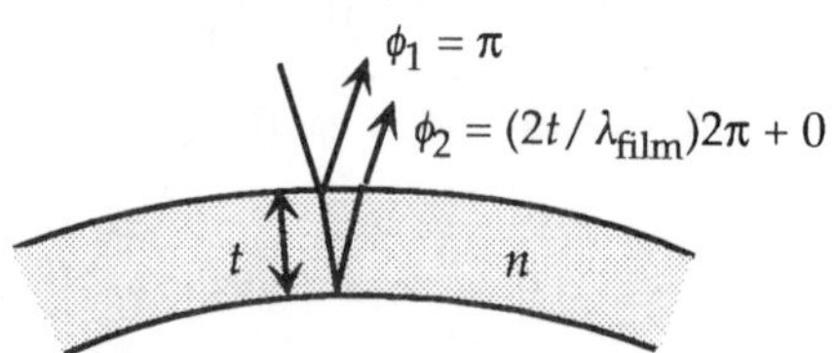

With respect to the incident wave, the wave that reflects from the air at the bottom surface of the bubble has a phase change due to the additional path-length but no phase change on reflection:

$\phi_2 = (2t/\lambda_{film})2\pi + 0$.

For destructive interference, the net phase change is

$\phi = (2t/\lambda_{film})2\pi - \pi = (m - \frac{1}{2})2\pi$, $m = 0, 1, 2, \ldots,$ or $t = \frac{1}{2}\lambda_{film}m = \frac{1}{2}(\lambda/n)m$, $m = 0, 1, 2, \ldots$.

The minimum non-zero thickness is

$t_{min} = \frac{1}{2}[(480 \text{ nm})/(1.42)](1) =$ **169 nm.**

41. With respect to the incident wave, the wave that reflects from the top surface of the coating has a phase change of

$\phi_1 = \pi$.

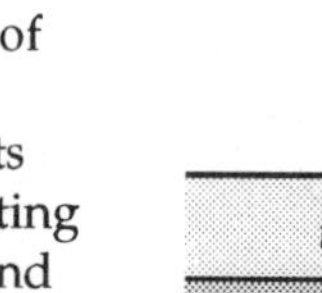

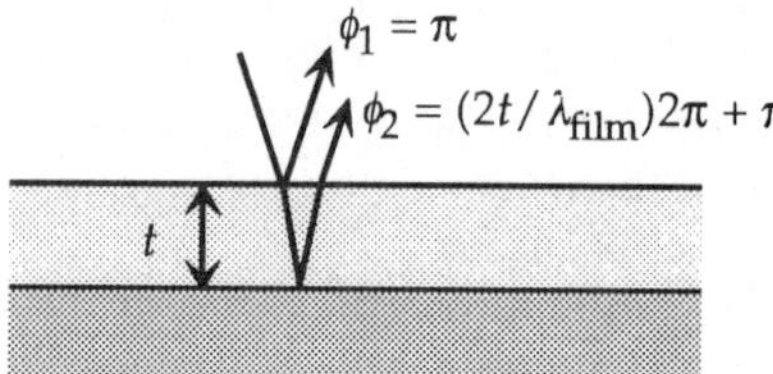

With respect to the incident wave, the wave that reflects from the glass ($n \approx 1.5$) at the bottom surface of the coating has a phase change due to the additional path-length and a phase change of π on reflection:

$\phi_2 = (2t/\lambda_{film})2\pi + \pi$.

For constructive interference, the net phase change is

$\phi = (2t/\lambda_{film})2\pi + \pi - \pi = m2\pi$, $m = 1, 2, 3, \ldots,$ or $t = \frac{1}{2}\lambda_{film}m = \frac{1}{2}(\lambda/n_{film})m$, $m = 1, 2, 3, \ldots$.

The minimum non-zero thickness occurs for $m = 1$:

$t_{min} = \lambda/2n_{film} = (570 \text{ nm})/2(1.25) =$ **228 nm.**

570 nm is in the middle of the visible spectrum. The transmitted light will be stronger in the wavelengths at the ends of the spectrum, so the lens would emphasize the red and violet wavelengths.

42. The phase difference for the reflected waves from the path-length difference and the reflection at the bottom surface is

$\phi = (2t/\lambda)2\pi + \pi.$

For the dark rings, this phase difference must be an odd multiple of π, so we have

$\phi = (2t/\lambda)2\pi + \pi = (2m + 1)\pi,\ m = 0, 1, 2, \ldots,$ or

$t = \frac{1}{2}m\lambda,\ m = 0, 1, 2, \ldots .$

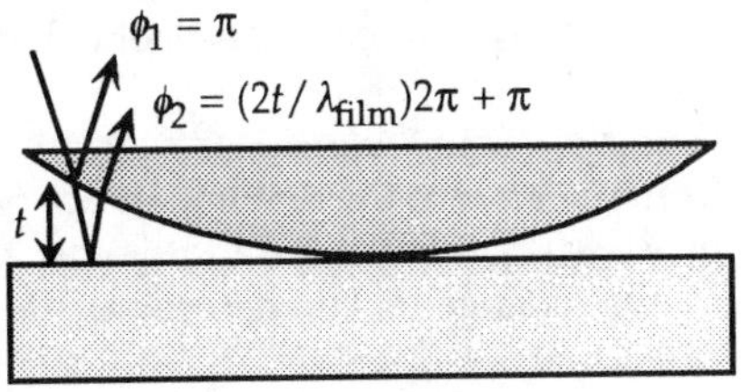

Because $m = 0$ corresponds to the dark center, m represents the number of the ring. Thus the thickness of the lens is the thickness of the air at the edge of the lens:

$t = \frac{1}{2}(31)(550\text{ nm}) = 8.5 \times 10^3\text{ nm} = \boxed{8.5\ \mu\text{m}.}$

$\phi_1 = \pi$

$\phi_2 = (2t/\lambda_{\text{film}})2\pi + \pi$

t

43. There is a phase difference for the reflected waves from the path-length difference, $(2t/\lambda)2\pi$, and the reflection at the bottom surface, π.

For destructive interference, this phase difference must be an odd multiple of π, so we have

$\phi = (2t/\lambda)2\pi + \pi = (2m + 1)\pi,\ m = 0, 1, 2, \ldots,$ or

$t = \frac{1}{2}m\lambda,\ m = 0, 1, 2, \ldots .$

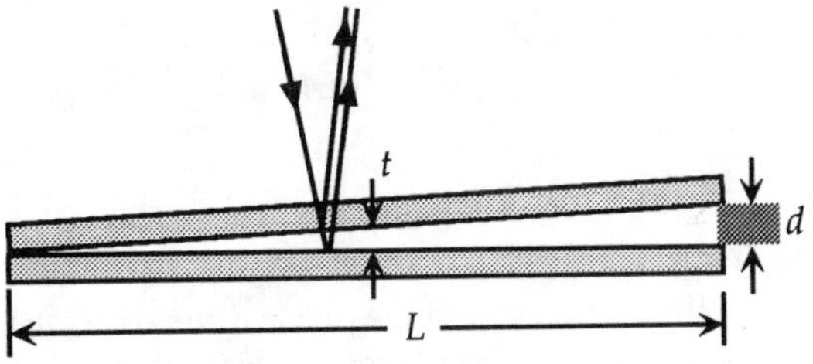

Because $m = 0$ corresponds to the edge where the glasses touch, $m + 1$ represents the number of the fringe. Thus the thickness of the foil is

$d = \frac{1}{2}(27)(670\text{ nm}) = 9.05 \times 10^3\text{ nm} = \boxed{9.05\ \mu\text{m}.}$

t d L

44. With respect to the incident wave, the wave that reflects from the air at the top surface of the air layer has a phase change of

$\phi_1 = 0.$

With respect to the incident wave, the wave that reflects from the glass at the bottom surface of the air layer has a phase change due to the additional path-length and a change on reflection:

$\phi_2 = (2t/\lambda)2\pi + \pi.$

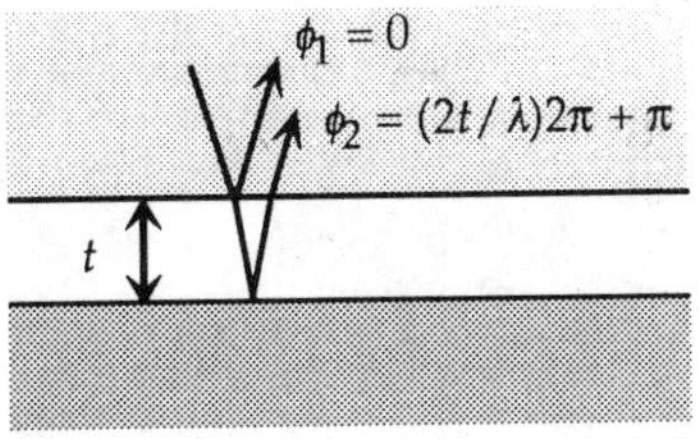

For constructive interference, the net phase change is

$\phi = (2t/\lambda)2\pi + \pi - 0 = m2\pi,\ m = 1, 2, 3, \ldots,$ or $t = \frac{1}{2}\lambda(m - \frac{1}{2}),\ m = 1, 2, 3, \ldots .$

The minimum thickness is

$t_{\text{min}} = \frac{1}{2}(450\text{ nm})(1 - \frac{1}{2}) = \boxed{113\text{ nm}.}$

For destructive interference, the net phase change is

$\phi = (2t/\lambda)2\pi + \pi - 0 = (2m + 1)\pi,\ m = 0, 1, 2, \ldots,$ or $t = \frac{1}{2}m\lambda,\ m = 0, 1, 2, \ldots .$

The minimum non-zero thickness is

$t_{\text{min}} = \frac{1}{2}(450\text{ nm})(1) = \boxed{225\text{ nm}.}$

$\phi_1 = 0$

$\phi_2 = (2t/\lambda)2\pi + \pi$

t

45. With respect to the incident wave, the wave that reflects from the top surface of the alcohol has a phase change of
$\phi_1 = \pi$.
With respect to the incident wave, the wave that reflects from the glass at the bottom surface of the alcohol has a phase change due to the additional path-length and a phase change of π on reflection:
$\phi_2 = (2t/\lambda_{film})2\pi + \pi$.

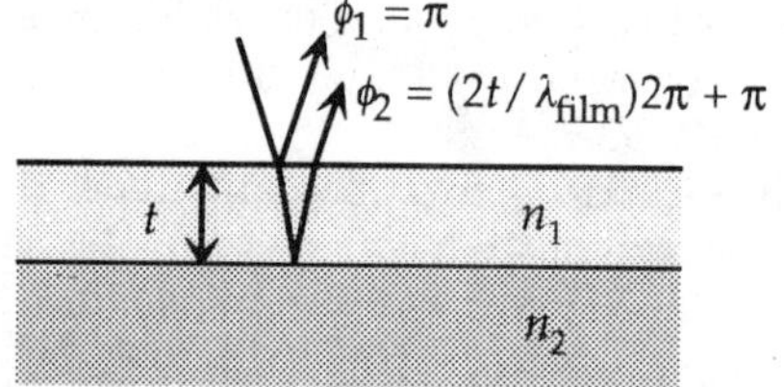

For constructive interference, the net phase change is
$\phi = (2t/\lambda_{1film})2\pi + \pi - \pi = m_1 2\pi, m_1 = 1, 2, 3, \ldots,$ or $t = \frac{1}{2}\lambda_{1film}(m_1) = \frac{1}{2}(\lambda_1/n_{film})(m_1), m_1 = 1, 2, 3, \ldots$.
For destructive interference, the net phase change is
$\phi = (2t/\lambda_{2film})2\pi + \pi - \pi = (2m_2 + 1)\pi,\ m_2 = 0, 1, 2, \ldots,$ or $t = \frac{1}{4}(\lambda_2/n_{film})(2m_2 + 1),\ m_2 = 0, 1, 2, \ldots$.
When we combine the two equations, we get
$\frac{1}{2}(\lambda_1/n_{film})(m_1) = \frac{1}{4}(\lambda_2/n_{film})(2m_2 + 1),$ or $(2m_2 + 1)/2m_1 = \lambda_1/\lambda_2 = (640 \text{ nm})/(512 \text{ nm}) = 1.25 = 5/4$.
Thus we see that $m_1 = m_2 = 2$, and the thickness of the film is
$t = \frac{1}{2}(\lambda_1/n_{film})(m_1) = \frac{1}{2}[(640 \text{ nm})/(1.36)](2) = \boxed{471 \text{ nm.}}$

46. At a distance r from the center of the lens, the thickness of the air space is y, and the phase difference for the reflected waves from the path-length difference and the reflection at the bottom surface is
$\phi = (2y/\lambda)2\pi + \pi$.
For the dark rings, we have
$\phi = (2y/\lambda)2\pi + \pi = (2m + 1)\pi,\ m = 0, 1, 2, \ldots,$ or
$y = \frac{1}{2}m\lambda,\ m = 0, 1, 2, \ldots$.
Because $m = 0$ corresponds to the dark center, m represents the number of the ring. From the triangle in the diagram, we have
$r^2 + (R - y)^2 = R^2,$ or $r^2 = 2yR - y^2 \approx 2yR$, when $y \ll R$,
which becomes
$r^2 = 2(\frac{1}{2}m\lambda)R = m\lambda R,\ m = 0, 1, 2, \ldots$.
When the apparatus is immersed in the liquid, the same analysis holds, if we use the wavelength in the liquid.
If we form the ratio for the two conditions, we get
$(r_1/r_2)^2 = \lambda_1/\lambda_2 = n,$ so
$n = (2.92 \text{ cm}/2.48 \text{ cm})^2 = \boxed{1.39.}$

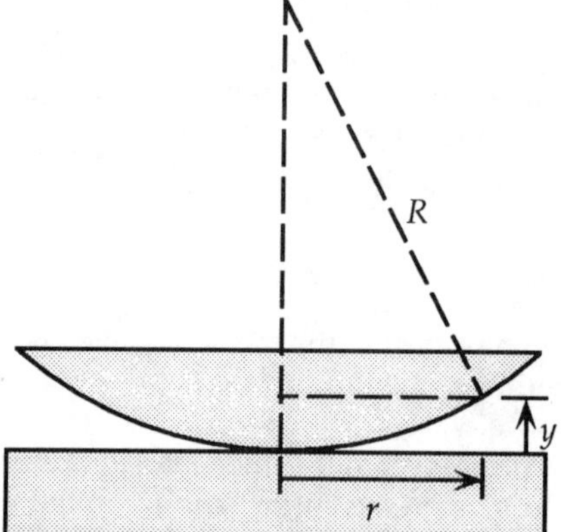

47. At a distance r from the center of the lens, the thickness of the air space is y, and the phase difference for the reflected waves from the path-length difference and the reflection at the bottom surface is
$\phi = (2y/\lambda)2\pi + \pi$.
For the dark rings, we have
$\phi = (2y/\lambda)2\pi + \pi = (2m + 1)\pi,\ m = 0, 1, 2, \ldots,$ or
$y = \frac{1}{2}m\lambda,\ m = 0, 1, 2, \ldots$.
Because $m = 0$ corresponds to the dark center, m represents the number of the ring. From the triangle in the diagram, we have
$r^2 + (R - y)^2 = R^2,$ or $r^2 = 2yR - y^2 \approx 2yR$, when $y \ll R$,
which becomes
$r^2 = 2(\frac{1}{2}m\lambda)R = m\lambda R,\ m = 0, 1, 2, \ldots,$ or $r = (m\lambda R)^{1/2}$.

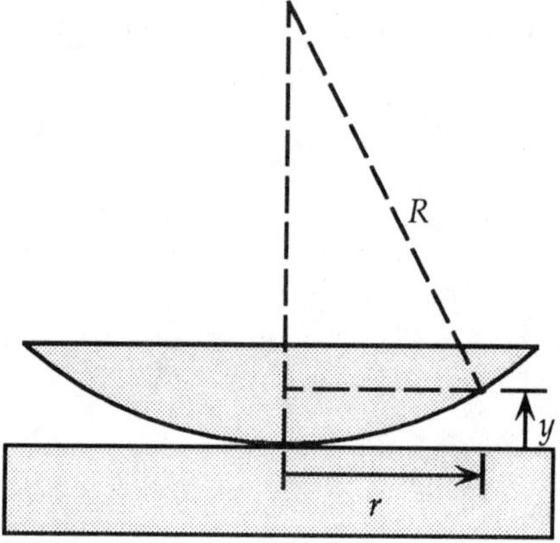

48. At a distance r from the center of the lens, the thickness of the air space is y, and the phase difference for the reflected waves from the path-length difference and the reflection at the bottom surface is

$\phi = (2y/\lambda)2\pi + \pi$.

For the bright rings, we have

$\phi = (2y/\lambda)2\pi + \pi = m2\pi$, $m = 1, 2, 3, \ldots$, or

$y = \frac{1}{2}\lambda(m - \frac{1}{2})$, $m = 1, 2, 3, \ldots$,

where m represents the number of the ring. From the triangle in the diagram, we have

$r^2 + (R - y)^2 = R^2$, or $r^2 = 2yR - y^2 \approx 2yR$, when $y \ll R$,

which becomes

$r^2 = 2[\frac{1}{2}\lambda(m - \frac{1}{2})]R = \lambda(m - \frac{1}{2})R$.

For the 48th ring, we have

$(3.05 \times 10^{-2}\text{ m})^2 = (610 \times 10^{-9}\text{ m})(48 - \frac{1}{2})R$, which gives $R = \boxed{32\text{ m.}}$

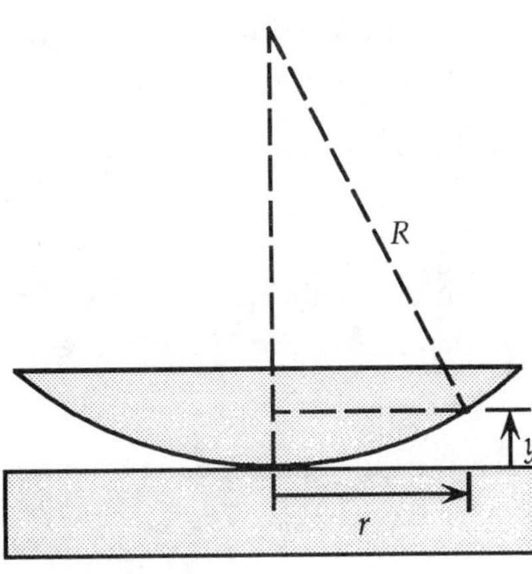

49. One fringe shift corresponds to a change in path length of λ. The number of fringe shifts produced by a mirror movement of ΔL is

$N = 2\,\Delta L/\lambda$;

$644 = 2(0.225 \times 10^{-3}\text{ m})/\lambda$, which gives $\lambda = 6.99 \times 10^{-7}\text{ m} = \boxed{699\text{ nm.}}$

50. One fringe shift corresponds to a change in path length of λ. The number of fringe shifts produced by a mirror movement of ΔL is

$N = 2\,\Delta L/\lambda$;

$272 = 2\,\Delta L/(589\text{ nm})$, which gives $\Delta L = 8.01 \times 10^4\text{ nm} = \boxed{80.1\ \mu\text{m.}}$

51. One fringe shift corresponds to a change in path length of λ. The number of fringe shifts produced by a mirror movement of ΔL is

$N = 2\,\Delta L/\lambda$;

$850 = 2\,\Delta L/(589 \times 10^{-9}\text{ m})$, which gives $\Delta L = 2.50 \times 10^{-4}\text{ m} = \boxed{0.250\text{ mm.}}$

52. One fringe shift corresponds to an effective change in path length of λ. The actual distance has not changed, but the number of wavelengths in the depth of the cavity has. If the cavity has a depth d, the number of wavelengths in vacuum is d/λ, and the number with the gas present is $d/\lambda_{gas} = n_{gas}d/\lambda$. Because the light passes through the cavity twice, the number of fringe shifts is

$N = 2[(n_{gas}d/\lambda) - (d/\lambda)] = 2(d/\lambda)(n_{gas} - 1)$;

$236 = 2[(1.30 \times 10^{-2}\text{ m})/(610 \times 10^{-9}\text{ m})](n_{gas} - 1)$, which gives $n_{gas} = \boxed{1.00554.}$

53. The two fringe patterns overlap. When the bright fringe of one occurs where there is a dark fringe of the other, there will be a region without fringes. When the next region occurs, the mirror movement must produce an integer number of fringe shifts for each wavelength:

$N_1 = 2\,\Delta L/\lambda_1$; $N_2 = 2\,\Delta L/\lambda_2$;

and the difference in the number of fringe shifts must be 1. Thus we have

$N_1\lambda_2 = N_2\lambda_1$;

$N_1(589.6\text{ nm}) = (N_1 + 1)(589.0\text{ nm})$, which gives $N_1 = 982$.

We find the mirror movement from

$N_1 = 2\,\Delta L/\lambda_1$;

$982 = 2\,\Delta L/(589.0\text{ nm})$, which gives $\Delta L = 2.89 \times 10^5\text{ nm} = \boxed{0.289\text{ mm.}}$

54. If the initial intensity is I_0, through the two sheets we have

$I_1 = \frac{1}{2}I_0,$

$I_2 = I_1 \cos^2\theta = \frac{1}{2}I_0 \cos^2\theta$, which gives

$I_2/I_0 = \frac{1}{2}\cos^2\theta = \frac{1}{2}\cos^2 70° =$ $\boxed{0.058.}$

55. Because the light is coming from air to glass, we find the angle from the vertical from

$\tan\theta_p = n_{glass} = 1.52$, which gives $\boxed{\theta_p = 56.7°.}$

56. Because the light is coming from water to diamond, we find the angle from the vertical from

$\tan\theta_p = n_{diamond}/n_{water} = 2.42/1.33 = 1.82$, which gives $\boxed{\theta_p = 61.2°.}$

57. For the refraction at the critical angle, with n_2 the higher index, we have

$n_1 \sin\theta_1 = n_2 \sin\theta_2;$

$n_1 \sin 90° = n_2 \sin 52°$, which gives $n_2/n_1 = 1.27$.

If the light is coming from lower index to higher index, we find the angle from

$\tan\theta_p = n_2/n_1 = 1.27$, which gives $\boxed{\theta_p = 52°.}$

If the light is coming from higher index to lower index, we find the angle from

$\tan\theta_p = n_1/n_2 = 1/1.27$, which gives $\boxed{\theta_p = 38°.}$

58. If I_0 is the intensity passed by the first Polaroid, the intensity passed by the second will be I_0 when the two axes are parallel. To reduce the intensity by half, we have

$I = I_0 \cos^2\theta = \frac{1}{2}I_0$, which gives $\theta =$ $\boxed{45°.}$

59. If the original intensity is I_0, the first Polaroid sheet will reduce the intensity of the original beam to

$I_1 = \frac{1}{2}I_0.$

If the axis of the second Polaroid sheet is oriented at an angle θ, the intensity is

$I_2 = I_1 \cos^2\theta.$

(*a*) $I_2 = I_1 \cos^2\theta = 0.75I_1$, which gives $\theta =$ $\boxed{30°.}$

(*b*) $I_2 = I_1 \cos^2\theta = 0.90I_1$, which gives $\theta =$ $\boxed{18°.}$

(*c*) $I_2 = I_1 \cos^2\theta = 0.99I_1$, which gives $\theta =$ $\boxed{5.7°.}$

60. If the initial intensity is I_0, through the two sheets we have

$I_1 = I_0 \cos^2\theta_1;$

$I_2 = I_1 \cos^2\theta_2 = I_0 \cos^2\theta_1 \cos^2\theta_2;$

$0.15I_0 = I_0 \cos^2\theta_1 \cos^2 40°$, which gives $\theta_1 =$ $\boxed{60°.}$

61. Through the successive sheets we have

$I_1 = I_0 \cos^2\theta_1,$

$I_2 = I_1 \cos^2\theta_2$, which gives

$I_2 = I_0 \cos^2\theta_1 \cos^2\theta_2 = I_0 (\cos^2 29.0°)(\cos^2 58.0°) = 0.215I_0.$

Thus the reduction is $\boxed{78.5\%.}$

62. If the light is coming from water to air, we find Brewster's angle from

$\tan \theta_p = n_{air}/n_{water} = 1.00/1.33 = 0.752$, which gives $\boxed{\theta_p = 36.9°.}$

For the refraction at the critical angle from water to air, we have

$n_{air} \sin \theta_1 = n_{water} \sin \theta_2$;

$(1.00) \sin 90° = (1.33) \sin \theta_c$, which gives $\boxed{\theta_c = 48.8°.}$

If the light is coming from air to water, we find Brewster's angle from

$\tan \theta_p' = n_{water}/n_{air} = 1.33/1.00 = 1.33$, which gives $\boxed{\theta_p' = 53.1°.}$

Thus $\theta_p + \theta_p' = 90.0°$.

63. If we have N polarizers, we set the angle between adjacent polarizers as θ, so that $N\theta = 90°$.
Through the successive polarizers, we have

$I_1 = I_0 \cos^2 \theta$,

$I_2 = I_1 \cos^2 \theta = I_0 \cos^2 \theta \cos^2 \theta = I_0 \cos^4 \theta$;

$I_3 = I_2 \cos^2 \theta = I_0 \cos^4 \theta \cos^2 \theta = I_0 \cos^6 \theta; \ldots .$

Thus for N polarizers, we have

$I_N = I_0 \cos^{2N} \theta = I_0 \cos^{2N} (90°/N) = 0.90 I_0$.

By using a numerical method, such as a spreadsheet, or trial and error, we find $N = 24$, $\theta = 3.75°$.

Thus we use $\boxed{\text{24 polarizers, with each at an angle of 3.75° with the next.}}$

64. The phase difference caused by the path back and forth through the coating must correspond to half a wavelength to produce destructive interference:

$2t = \lambda/2$, so $t = \lambda/4 = (2 \text{ cm})/4 = \boxed{0.5 \text{ cm.}}$

65. The wavelength of the sound is

$\lambda = v/f = (343 \text{ m/s})/(750 \text{ Hz}) = 0.457 \text{ m}$.

We find the angles of the minima from

$D \sin \theta = m\lambda, \quad m = 1, 2, 3, \ldots;$

$(0.88 \text{ m}) \sin \theta_1 = (1)(0.457 \text{ m})$, which gives $\sin \theta_1 = 0.520$, so $\theta_1 = 31°$;

$(0.88 \text{ m}) \sin \theta_2 = (2)(0.457 \text{ m})$, which gives $\sin \theta_2 = 1.04$, so there is no θ_2.

Thus the whistle would not be heard clearly at angles of $\boxed{\text{31° on either side of the normal.}}$

66. The lines act like a grating. Assuming the first order, we find the separation of the lines from

$d \sin \theta = m\lambda$;

$d \sin 50° = (1)(460 \times 10^{-9} \text{ m})$, which gives $d = 6.0 \times 10^{-7} \text{ m} = \boxed{600 \text{ nm.}}$

67. Because the angle increases with wavelength, to miss a complete order we use the smallest wavelength.
The maximum angle is 90°. We find the slit separation from

$d \sin \theta = m\lambda$;

$d \sin 90° = (2)(400 \times 10^{-9} \text{ m})$, which gives $d = 8.00 \times 10^{-7} \text{ m} = 8.00 \times 10^{-5} \text{ cm}$.

The number of lines/cm is

$1/d = 1/(8.00 \times 10^{-5} \text{ cm}) = \boxed{12{,}500 \text{ lines/cm.}}$

68. Because the angle increases with wavelength, we compare the maximum angle for the second order with the minimum angle for the third order:

$d \sin \theta = m\lambda$, or $\sin \theta = \lambda/d$;

$\sin \theta_{2max} = (2)(750 \text{ nm})/d$;

$\sin \theta_{3min} = (3)(400 \text{ nm})/d$.

When we divide the two equations, we get

$\sin \theta_{3min}/\sin \theta_{2max} = (1200 \text{ nm})/(1500 \text{ nm}) = 0.8$.

Because the value of the sine increases with angle, this means $\theta_{3min} < \theta_{2max}$, so the orders overlap. To determine the overlap, we find the second-order wavelength that coincides with θ_{3min}:

$(2)\lambda_2 = (3)(400 \text{ nm})$, which gives $\lambda_2 = 600$ nm.

We find the third-order wavelength that coincides with θ_{2max} from

$(2)(750 \text{ nm}) = (3)\lambda_3$, which gives $\lambda_3 = 500$ nm.

Thus 600 nm to 750 nm of the second order overlaps with 400 nm to 500 nm of the third order.

69. The wavelength of the signal is

$\lambda = v/f = (3.00 \times 10^8 \text{ m/s})/(75 \times 10^6 \text{ Hz}) = 4.00$ m.

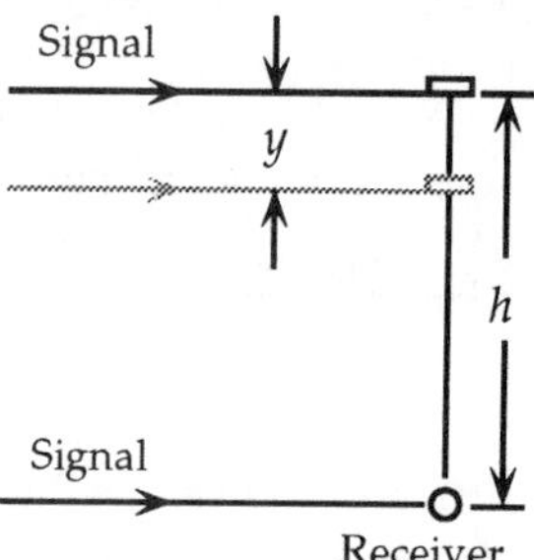

(*a*) There is a phase difference between the direct and reflected signals from the path difference, $(h/\lambda)2\pi$, and the reflection, π.

The total phase difference is

$\phi = (h/\lambda)2\pi + \pi = [(118 \text{ m})/(4.00 \text{ m})]2\pi + \pi = 30(2\pi)$.

Thus the interference is constructive.

(*b*) When the plane is 22 m closer to the receiver, the phase difference is

$\phi = [(h - y)/\lambda]2\pi + \pi$

$= [(118 \text{ m} - 22 \text{ m})/(4.00 \text{ m})]2\pi + \pi = 24(2\pi) + \pi$.

Thus the interference is destructive.

70. The wavelength of the signal is

$\lambda = v/f = (3.00 \times 10^8 \text{ m/s})/(102.1 \times 10^6 \text{ Hz}) = 2.94$ m.

Because measurements are made far from the antennae, we can use the analysis for the double slit. For constructive interference, the path difference is a multiple of the wavelength:

$d \sin \theta = m\lambda$, $m = 0, 1, 2, 3, \ldots$;

$(9.0 \text{ m}) \sin \theta_{1max} = (1)(2.94 \text{ m})$, which gives $\theta_{1max} = 19°$;

$(9.0 \text{ m}) \sin \theta_{2max} = (2)(2.94 \text{ m})$, which gives $\theta_{2max} = 41°$;

$(9.0 \text{ m}) \sin \theta_{3max} = (3)(2.94 \text{ m})$, which gives $\theta_{3max} = 78°$;

$(9.0 \text{ m}) \sin \theta_{4max} = (4)(2.94 \text{ m})$, which gives $\sin \theta_{4max} > 1$, so there is no fourth maximum.

Because the interference pattern will be symmetrical above and below the midline and on either side of the antennae, the angles for maxima are 19°, 41°, 78°, 102°, 139°, 161° above and below the midline.

For destructive interference, the path difference is an odd multiple of half a wavelength:

$d \sin \theta = (m - \frac{1}{2})\lambda$, $m = 1, 2, 3, \ldots$; or

$(9.0 \text{ m}) \sin \theta_{1min} = (1 - \frac{1}{2})(2.94 \text{ m})$, which gives $\theta_{1min} = 9.4°$;

$(9.0 \text{ m}) \sin \theta_{2min} = (2 - \frac{1}{2})(2.94 \text{ m})$, which gives $\theta_{2min} = 29°$;

$(9.0 \text{ m}) \sin \theta_{3min} = (3 - \frac{1}{2})(2.94 \text{ m})$, which gives $\theta_{3min} = 55°$;

$(9.0 \text{ m}) \sin \theta_{4min} = (4 - \frac{1}{2})(2.94 \text{ m})$, which gives $\sin \theta_{4min} > 1$, so there is no fourth minimum.

Because the interference pattern will be symmetrical above and below the midline and on either side of the antennae, the angles for minima are 9.4°, 29°, 55°, 125°, 151°, 171° above and below the midline.

71. For constructive interference, the path difference is a multiple of the wavelength:
$d \sin \theta = m\lambda, \quad m = 0, 1, 2, 3, \ldots .$
We find the location on the screen from
$y = L \tan \theta.$
For small angles, we have
$\sin \theta \approx \tan \theta$, which gives
$y = L(m\lambda/d) = mL\lambda/d.$
For the second-order fringes we have
$y_1 = 2L\lambda_1/d;$
$y_2 = 2L\lambda_2/d.$
When we subtract the two equations, we get
$\Delta y = y_1 - y_2 = (2L/d)(\lambda_1 - \lambda_2);$
1.33×10^{-3} m = [2(1.70 m)/(0.60 × 10^{-3} m)](590 nm – λ_2), which gives λ_2 = **355 nm.**

72. Because the light is coming from air to water, we find the angle from the vertical from
$\tan \theta_p = n_{water} = 1.33$, which gives $\theta_p = 53.1°$.
Thus the angle above the horizon is 90.0° – 53.1° = **36.9°.**

73. (*a*) If the initial intensity is I_0, through the two sheets we have
$I_1 = I_0/2;$
$I_2 = I_1 \cos^2 \theta = (I_0/2) \cos^2 90° =$ **0.**
(*b*) With the third polarizer inserted, we have
$I_1 = I_0/2;$
$I_2 = I_1 \cos^2 \theta_1 = (I_0/2) \cos^2 60°;$
$I_3 = I_2 \cos^2 \theta_2 = (I_0/2) \cos^2 60° \cos^2 30° =$ **$0.094I_0$.**
(*c*) If the third polarizer is placed in front of the other two, we have the same situation as in (*a*), with I_0 being less. Thus **no light gets transmitted.**

74. We equate a path difference of one wavelength with a phase difference of 2π. With respect to the incident wave, the wave that reflects at the top surface of the film has a phase change of

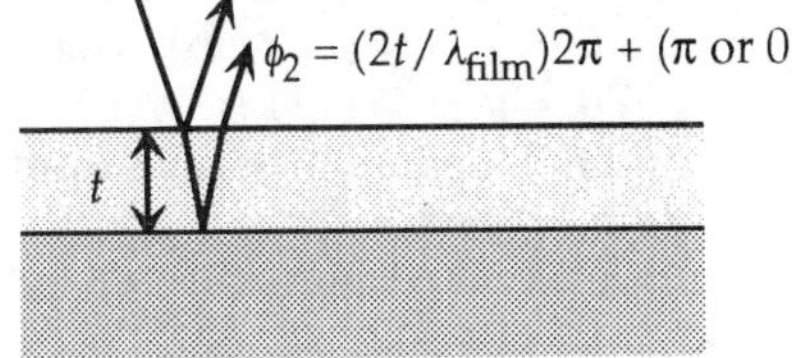

$\phi_1 = \pi.$
If we assume that the film has an index less than glass, the wave that reflects from the glass has a phase change due to the additional path-length and a phase change on reflection:
$\phi_2 = (2t/\lambda_{film})2\pi + \pi.$
For destructive interference, the net phase change is
$\phi = (2t/\lambda_{film})2\pi + \pi - \pi = (m - \frac{1}{2})2\pi, m = 1, 2, \ldots,$ or $t = \frac{1}{2}\lambda_{film}(m - \frac{1}{2}) = \frac{1}{2}(\lambda/n)(m - \frac{1}{2}), m = 1, 2, \ldots .$
For the minimum thickness, $m = 1$, we have
150 nm = $\frac{1}{2}$[(600 nm)/(n)](1 – $\frac{1}{2}$), which gives $n = 1$, so **no film with $n < n_{glass}$ is possible.**
If we assume that the film has an index greater than glass, the wave that reflects from the glass has a phase change due to the additional path-length and no phase change on reflection:
$\phi_2 = (2t/\lambda_{film})2\pi + 0.$
For destructive interference, the net phase change is
$\phi = (2t/\lambda_{film})2\pi - \pi = (m - \frac{1}{2})2\pi, m = 1, 2, \ldots,$ or $t = \frac{1}{2}\lambda_{film}m = \frac{1}{2}(m\lambda/n), m = 1, 2, \ldots .$
For the minimum thickness, $m = 1$, we have
150 nm = $\frac{1}{2}$(1)(600 nm)/n, which gives n = **2.00.**

75. With respect to the incident wave, the wave that reflects at the top surface of the film has a phase change of

$\phi_1 = \pi.$

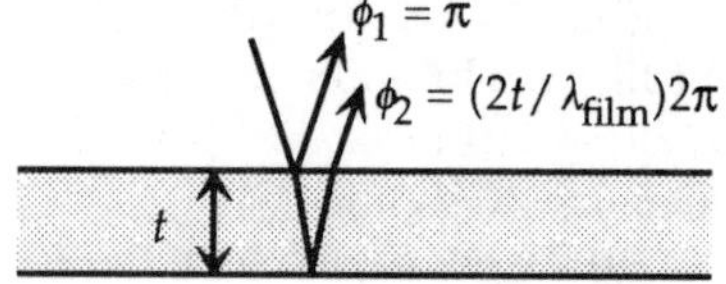

The wave that reflects from the bottom surface has a phase change due to the additional path-length and no phase change on reflection:

$\phi_2 = (2t/\lambda_{film})2\pi + 0.$

For destructive interference, the net phase change is

$\phi = (2t/\lambda_{film})2\pi - \pi = (m - \frac{1}{2})2\pi,\ m = 1, 2, \ldots,$ or $t = \frac{1}{2}\lambda_{film}m = \frac{1}{2}(m\lambda/n),\ m = 1, 2, \ldots.$

For the two wavelengths we have

$t = \frac{1}{2}(m_1\lambda_1/n) = \frac{1}{2}(m_2\lambda_2/n),$ or $m_1/m_2 = \lambda_2/\lambda_1 = 640\ \text{nm}/512\ \text{nm} = 1.25 = 5/4.$

Thus $m_1 = 5$, and $m_2 = 4$. For the thickness we have

$t = \frac{1}{2}(m_1\lambda_1/n) = \frac{1}{2}(5)(512\ \text{nm})/1.58 =$ **810 nm.**

76. The path difference between the top and bottom of the slit for the incident wave is

$D \sin \theta_i.$

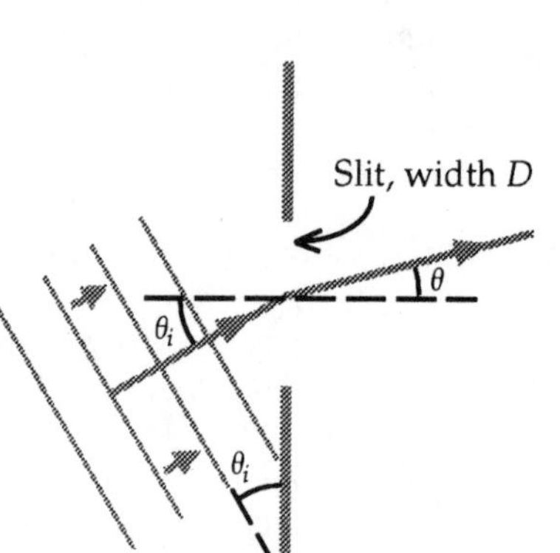

The path difference between the top and bottom of the slit for the diffracted wave is

$D \sin \theta.$

When $\theta = \theta_i$, the net path difference is zero, and there will be constructive interference. **There is a central maximum at $\theta = 30°$.**

When the net path difference is a multiple of a wavelength, there will be minima given by

$(D \sin \theta_i) - (D \sin \theta) = m\lambda,\ m = \pm 1, \pm 2, \ldots,$ or

$\sin \theta = \sin 30° - (m\lambda/D),$ where $m = \pm 1, \pm 2, \ldots.$

77. We find the angles for the first order from the distances:

$\tan \theta_1 = y_1/L = (3.32\ \text{cm})/(60.0\ \text{cm}) = 0.0553$, so $\theta_1 = 3.17°$;

$\tan \theta_2 = y_2/L = (3.71\ \text{cm})/(60.0\ \text{cm}) = 0.0618$, so $\theta_2 = 3.54°$.

We find the separation of lines from

$d \sin \theta_1 = m\lambda_1;$

$d \sin 3.17° = (1)(589 \times 10^{-9}\ \text{m})$, which gives $d = 1.066 \times 10^{-5}\ \text{m} = 1.066 \times 10^{-3}\ \text{cm}.$

For the second wavelength we have

$d \sin \theta_2 = m\lambda_2;$

$(1.06 \times 10^{-5}\ \text{m}) \sin 3.54° = (1)\lambda_2$, which gives $\lambda_2 = 6.58 \times 10^{-7}\ \text{m} =$ **658 nm.**

The number of lines/cm is

$1/d = 1/(1.066 \times 10^{-3}\ \text{cm}) =$ **938 lines/cm.**

78. With respect to the incident wave, the wave that reflects from the top surface of the coating has a phase change of

$\phi_1 = \pi$.

With respect to the incident wave, the wave that reflects from the glass ($n \approx 1.5$) at the bottom surface of the coating has a phase change due to the additional path-length and a phase change of π on reflection:

$\phi_2 = (2t/\lambda_{film})2\pi + \pi$.

$\phi_1 = \pi$

$\phi_2 = (2t/\lambda_{film})2\pi + \pi$

t

For destructive interference, this phase difference must be an odd multiple of π, so we have

$\phi = (2t/\lambda_{film})2\pi + \pi - \pi = (2m + 1)\pi,\ m = 0, 1, 2, \ldots,$ or $t = \frac{1}{4}(2m + 1)\lambda_{film},\ m = 0, 1, 2, \ldots$.

Thus the minimum thickness is

$t_{min} = \frac{1}{4}\lambda/n$.

(*a*) For the blue light we get

$t_{min} = \frac{1}{4}(450\text{ nm})/(1.38) =$ **81.5 nm.**

(*b*) For the red light we get

$t_{min} = \frac{1}{4}(700\text{ nm})/(1.38) =$ **127 nm.**

79. If we consider the two rays shown in the diagram, we see that the second ray has reflected twice. If $n_{film} < n_{glass}$, the first reflection from the glass produces a shift equivalent to $\frac{1}{2}\lambda_{film}$, while the second reflection from the air produces no shift. When we compare the two rays at the film-glass surface, we see that the second ray has a total shift of

$d_2 - d_1 = 2t + \frac{1}{2}\lambda_{film}$.

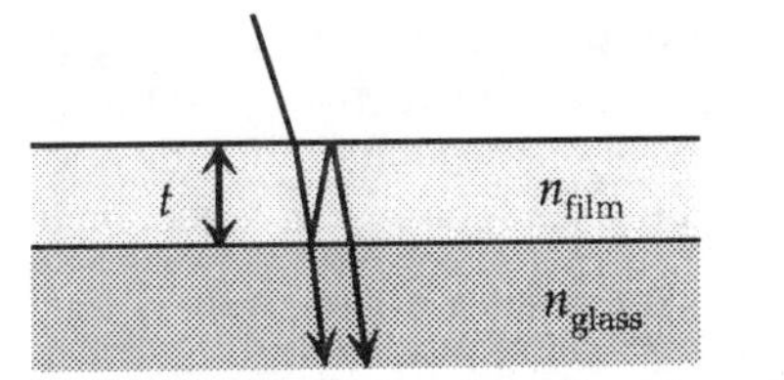

$d_1 = 0 \quad d_2 = 2t + (0 \text{ or } \lambda_{film}/2)$

For maxima, we have

$2t + \frac{1}{2}\lambda_{film} = m\lambda_{film},\ m = 1, 2, 3, \ldots,$ or $\boxed{t = \frac{1}{2}(m - \frac{1}{2})\lambda/n_{film},\ m = 1, 2, 3, \ldots}$.

For minima, we have

$2t + \frac{1}{2}\lambda_{film} = (m + \frac{1}{2})\lambda_{film},\ m = 0, 1, 2, 3, \ldots,$ or $\boxed{t = \frac{1}{2}m\lambda/n_{film},\ m = 0, 1, 2, 3, \ldots}$.

We see that for a film of zero thickness, that is, $t \ll \lambda_{film}$, there will be a minimum.

If $n_{film} > n_{glass}$, the first reflection from the glass produces no shift, while the second reflection from the air also produces no shift. When we compare the two rays at the film-glass surface, we see that the second ray has a total shift of

$d_2 - d_1 = 2t$.

For maxima, we have

$2t = m\lambda_{film},\ m = 0, 1, 2, 3, \ldots,$ or $\boxed{t = \frac{1}{2}m\lambda/n_{film},\ m = 0, 1, 2, 3, \ldots}$.

For minima, we have

$2t = (m - \frac{1}{2})\lambda_{film},\ m = 1, 2, 3, \ldots,$ or $\boxed{t = \frac{1}{2}(m - \frac{1}{2})\lambda/n_{film},\ m = 1, 2, 3, \ldots}$.

We see that for a film of zero thickness, that is, $t \ll \lambda_{film}$, there will be a maximum.

80. There will be a phase difference between the waves at the two slits because the wave at the upper slit will have traveled farther. The path difference at the two slits for the incident wave is

$d \sin \theta_i$.

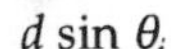

The path difference between the two slits for the diffracted wave is

$d \sin \theta$.

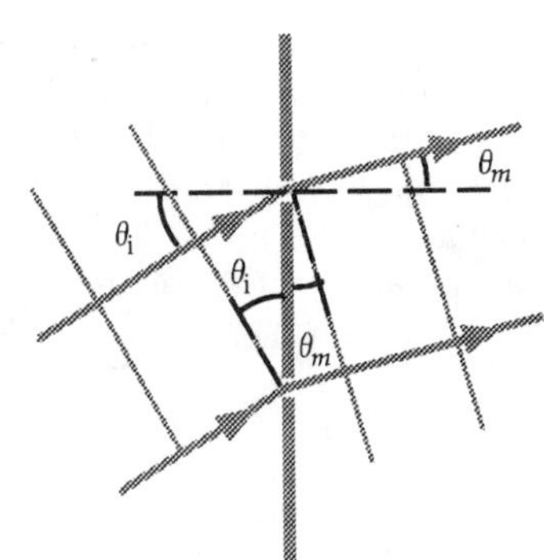

When the net path difference is a multiple of a wavelength, there will be maxima given by

$(d \sin \theta_i) - (d \sin \theta_m) = m\lambda,\ m = 0, \pm 1, \pm 2, \ldots,$ or

$$\boxed{\sin \theta_m = \sin \theta_i \pm (m\lambda/d), \text{ where } m = 0, 1, 2, \ldots}$$

81. (*a*) Through the successive polarizers we have

$I_1 = \frac{1}{2}I_0$;

$I_2 = I_1 \cos^2 \theta_2 = \frac{1}{2}I_0 \cos^2 \theta_2$;

$I_3 = I_2 \cos^2 \theta_3 = \frac{1}{2}I_0 \cos^2 \theta_2 \cos^2 \theta_3$;

$I_4 = I_3 \cos^2 \theta_4 = \frac{1}{2}I_0 \cos^2 \theta_2 \cos^2 \theta_3 \cos^2 \theta_4 = \frac{1}{2}I_0 \cos^2 30° \cos^2 30° \cos^2 30° = \boxed{0.21I_0.}$

(*b*) If we remove the second polarizer, we get

$I_1 = \frac{1}{2}I_0$;

$I_3 = I_1 \cos^2 \theta_3' = \frac{1}{2}I_0 \cos^2 \theta_3'$;

$I_4 = I_3 \cos^2 \theta_4 = \frac{1}{2}I_0 \cos^2 \theta_3' \cos^2 \theta_4 = \frac{1}{2}I_0 \cos^2 60° \cos^2 30° = 0.094I_0$.

Thus we can decrease the intensity by removing either the $\boxed{\text{second or third polarizer.}}$

(*c*) If we remove the $\boxed{\text{second and third polarizers,}}$ we will have two polarizers with their axes perpendicular, so no light will be transmitted.

82. To maximize reflection, we want the three rays shown on the diagram to be in phase. We first compare rays 2 and 3. We want them to be in phase when leaving the boundary between n_1 and n_2. Ray 2 reflects from $n_2 > n_1$, so there will be a phase shift of π. Ray 3 will have a phase change due to the additional path-length and no phase change on reflection from the next n_1 layer:

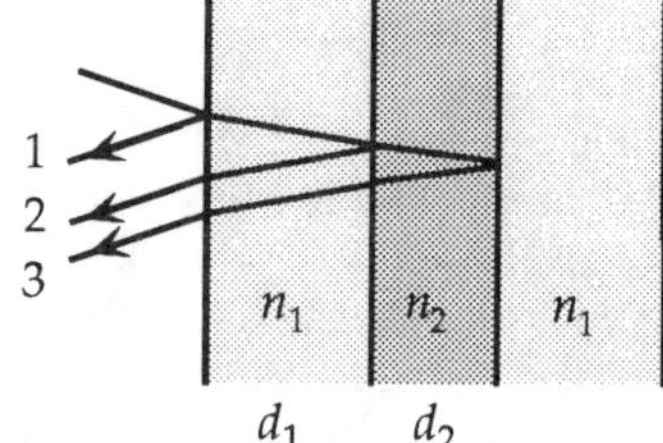

$\phi_3 = (2d_2/\lambda_2)2\pi + 0.$

For constructive interference, the net phase change is

$\phi = (2d_2/\lambda_2)2\pi - \pi = m2\pi, m = 0, 1, 2, \dots,$ or $d_2 = \frac{1}{2}\lambda_2(m + \frac{1}{2}) = \frac{1}{2}(\lambda/n_2)(m + \frac{1}{2}), m = 0, 1, 2, \dots .$

Thus for the minimum thickness ($m = 0$), we get

$\boxed{d_2 = \frac{1}{4}(\lambda/n_2).}$

We want rays 1 and 2 to be in phase when leaving the first surface. Ray 1 reflects from $n_1 > 1$, so there will be a phase shift of π. Ray 2 will have a phase change due to the additional path-length and a phase change on reflection from the n_2 layer:

$\phi_2 = (2d_1/\lambda_1)2\pi + \pi.$

For constructive interference, the net phase change is

$\phi = (2d_1/\lambda_1)2\pi + \pi - \pi = m2\pi, m = 1, 2, 3, \dots,$ or $d_1 = \frac{1}{2}\lambda_1 m, m = 1, 2, 3, \dots .$

Thus for the minimum thickness ($m = 1$), we get

$\boxed{d_1 = \frac{1}{2}(\lambda/n_1).}$

83. We find the angles for the first order from

$d \sin \theta = m\lambda = \lambda;$

$[1/(7500 \text{ lines/cm})](10^{-2} \text{ m/cm}) \sin \theta_1 = 4.4 \times 10^{-7}$ m, which gives $\sin \theta_1 = 0.330$, so $\theta_1 = 19.3°$;

$[1/(7500 \text{ lines/cm})](10^{-2} \text{ m/cm}) \sin \theta_2 = 6.3 \times 10^{-7}$ m, which gives $\sin \theta_2 = 0.473$, so $\theta_2 = 28.2°$.

The distances from the central white line on the screen are

$y_1 = L \tan \theta_1 = (2.5 \text{ m}) \tan 19.3° = 0.87$ m;

$y_2 = L \tan \theta_2 = (2.5 \text{ m}) \tan 28.2° = 1.34$ m.

Thus the separation of the lines is

$y_2 - y_1 = 1.34 \text{ m} - 0.87 \text{ m} = \boxed{0.47 \text{ m.}}$

CHAPTER 25

1. From the definition of the f-stop, we have
   $f\text{-stop} = f/D$;
   $1.4 = (55 \text{ mm})/D_{1.4}$, which gives $D_{1.4} = 39$ mm;
   $22 = (55 \text{ mm})/D_{22}$, which gives $D_{22} = 2.5$ mm.
   Thus the range of diameters is $\boxed{2.5 \text{ mm} \le D \le 39 \text{ mm.}}$

2. We find the f-number from
   $f\text{-stop} = f/D = (14 \text{ cm})/(6.0 \text{ cm}) = \boxed{f/2.3.}$

3. The exposure is proportional to the area and the time:
   Exposure $\propto At \propto D^2t \propto t/(f\text{-stop})^2$.
   Because we want the exposure to be the same, we have
   $t_1/(f\text{-stop}_1)^2 = t_2/(f\text{-stop}_2)^2$;
   $[(1/250) \text{ s}]/(5.6)^2 = t_2/(11)^2$, which gives $t_2 = \boxed{(1/60) \text{ s.}}$

4. The exposure is proportional to the area and the time:
   Exposure $\propto At \propto D^2t \propto t/(f\text{-stop})^2$.
   Because we want the exposure to be the same, we have
   $t_1/(f\text{-stop}_1)^2 = t_2/(f\text{-stop}_2)^2$;
   $[(1/125) \text{ s}]/(11)^2 = [(1/1000) \text{ s}]/(f\text{-stop}_2)^2$, which gives $f\text{-stop}_2 = \boxed{f/4.}$

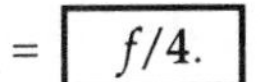

5. From the similar triangles on the ray diagram, we see that
   $H/L_1 = h/L_2$;
   $(2.0 \text{ cm})/(100 \text{ cm}) = h/(7.0 \text{ cm})$,
   which gives $h = 0.014$ cm $= 1.4$ mm.
   To find the radius of each spot of the image, we consider the light going through a slit and find the distance from the central bright spot to the first dark spot. For destructive interference, the path-length difference to the first dark spot is
   $d \sin \theta = \frac{1}{2}\lambda$;
   We find the locations on the screen from
   $y = L_2 \tan \theta$.
   For small angles, we have
   $\sin \theta \approx \tan \theta$, which gives
   $y = L_2(\frac{1}{2}\lambda/d) = \frac{1}{2}L_2\lambda/d = (7.0 \times 10^{-2} \text{ m})(550 \times 10^{-9} \text{ m})/(1.0 \times 10^{-3} \text{ m}) = 1.9 \times 10^{-5} \text{ m} = 0.019$ mm.
   Thus the diameter of the image spot is about 0.04 mm, which is much smaller than the separation of the spots, so they are easily resolvable.

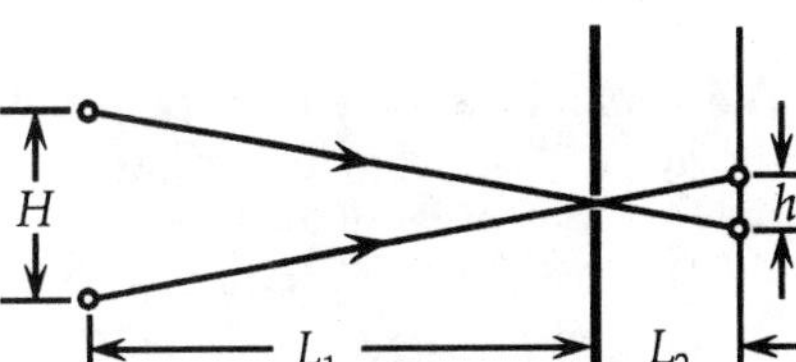

6. We find the effective f-number for the pinhole:
   $f\text{-stop}_2 = f/D = (70 \text{ mm})/(1.0 \text{ mm}) = f/70$.
   The exposure is proportional to the area and the time:
   Exposure $\propto At \propto D^2t \propto t/(f\text{-stop})^2$.
   Because we want the exposure to be the same, we have
   $t_1/(f\text{-stop}_1)^2 = t_2/(f\text{-stop}_2)^2$;
   $[(1/250) \text{ s}]/(11)^2 = t_2/(70)^2$, which gives $t_2 = \boxed{(1/6) \text{ s.}}$

7. For an object at infinity, the image will be in the focal plane, so we have $d_1 = f = 135$ mm.
When the object is at 1.2 m, we locate the image from
$(1/d_o) + (1/d_i) = 1/f$;
$(1/1200 \text{ mm}) + (1/d_i) = 1/135$ cm, which gives $d_i = 152$ mm.
Thus the distance from the lens to the film must change by
$d_i - f = 152 \text{ mm} - 135 \text{ mm} = \boxed{17 \text{ mm.}}$

8. We find the object distances from
$(1/d_o) + (1/d_i) = 1/f$;
$(1/d_{o1}) + (1/200.0 \text{ mm}) = 1/200$ mm, which gives $d_{o1} = \infty$;
$(1/d_{o2}) + (1/200.6 \text{ mm}) = 1/200$ mm, which gives $d_{o1} = 6.87 \times 10^3 \text{ mm} = 6.87$ m.
Thus the range of object distances is $\boxed{6.87 \text{ m} \le d_o \le \infty.}$

9. The converging camera lens will form a real, inverted image. For the magnification, we have
$m = h_i/h_o = -d_i/d_o$;
$-(24 \times 10^{-3} \text{ m})/(22 \text{ m}) = -d_i/(50 \text{ m})$, or $d_i = 5.45 \times 10^{-2}$ m.
We find the focal length of the lens from
$(1/d_o) + (1/d_i) = 1/f$;
$(1/50 \text{ m}) + (1/5.45 \times 10^{-2} \text{ m}) = 1/f$, which gives $f = 5.4 \times 10^{-2} \text{ m} = \boxed{54 \text{ mm.}}$

10. The length of the eyeball is the image distance for a far object, i. e., the focal length of the lens.
We find the *f*-number from
f-stop $= f/D = (20 \text{ mm})/(5.0 \text{ mm}) = \boxed{f/4.0.}$

11. The actual near point of the person is 50 cm. With the lens, an object placed at the normal near point, 25 cm, or 23 cm from the lens, is to produce a virtual image 50 cm from the eye, or 48 cm from the lens.
We find the power of the lens from
$(1/d_o) + (1/d_i) = 1/f = P$, when distances are in m;
$(1/0.23 \text{ m}) + (1/-0.48 \text{ m}) = P = \boxed{+2.3 \text{ D.}}$

12. With the lens, the screen placed 50 cm from the eye, or 48.2 cm from the lens, is to produce a virtual image 120 cm from the eye, or 118.2 cm from the lens.
We find the power of the lens from
$(1/d_o) + (1/d_i) = 1/f = P$, when distances are in m;
$(1/0.482 \text{ m}) + (1/-1.182 \text{ m}) = P = \boxed{+1.2 \text{ D.}}$

13. With the contact lens, an object at infinity would have a virtual image at the far point of the eye.
We find the power of the lens from
$(1/d_o) + (1/d_i) = 1/f = P$, when distances are in m;
$(1/\infty) + (1/-0.17 \text{ m}) = P = -5.9$ D.
To find the new near point, we have
$(1/d_o) + (1/d_i) = 1/f = P$;
$(1/d_o) + (1/-0.12 \text{ m}) = -5.9$ D, which gives $d_o = 0.41$ m.
$\boxed{\text{Glasses would be better,}}$ because they give a near point of 32 cm from the eye.

14. (*a*) The lens is diverging, so it produces images closer than the object, thus the person is **nearsighted.**
(*b*) We find the far point by finding the image distance for an object at infinity:
$(1/d_o) + (1/d_i) = 1/f = P;$
$(1/\infty) + (1/d_i) = -5.0$ D, which gives $d_i = -0.20$ m $= -20$ cm.
Because this is the distance from the lens, the far point without glasses is 20 cm + 2.0 cm = **22 cm.**

15. (*a*) We find the power of the lens for an object at infinity:
$(1/d_o) + (1/d_i) = 1/f = P;$
$(1/\infty) + (1/-0.70 \text{ m}) = P =$ **– 1.43 D.**
(*b*) To find the near point with the lens, we have
$(1/d_o) + (1/d_i) = 1/f = P;$
$(1/d_o) + (1/-0.20 \text{ m}) = -1.43$ D, which gives $d_o = 0.28$ m = **28 cm.**

16. The 2.0 cm of a normal eye is the image distance for an object at infinity, thus it is the focal length of the lens of the eye. To find the length of the nearsighted eye, we find the image distance (distance from lens to retina) for an object at the far point of the eye:
$(1/d_o) + (1/d_i) = 1/f;$
$(1/17 \text{ cm}) + (1/d_i) = 1/20$ cm, which gives $d_i = 2.27$ cm.
Thus the difference is 2.27 cm – 2.0 cm = **0.3 cm.**

17. We find the far point of the eye by finding the image distance with the lens for an object at infinity:
$(1/d_{o1}) + (1/d_{i1}) = 1/f_1;$
$(1/\infty) + (1/d_{i1}) = 1/-25.0$ cm, which gives $d_{i1} = -25.0$ cm from the lens, or 26.8 cm from the eye.
We find the focal length of the contact lens from
$(1/d_{o2}) + (1/d_{i2}) = 1/f_2;$
$(1/\infty) + (1/-26.8 \text{ cm}) = 1/f_2$, which gives $f_2 =$ **– 26.8 cm.**

18. (*a*) We find the focal length of the lens for an object at infinity and the image on the retina:
$(1/d_{o1}) + (1/d_{i1}) = 1/f_1;$
$(1/\infty) + (1/2.0 \text{ cm}) = 1/f_1$, which gives $f_1 =$ **2.0 cm.**
(*b*) With the object 30 cm from the eye, we have
$(1/d_{o2}) + (1/d_{i2}) = 1/f_2;$
$(1/30 \text{ cm}) + (1/2.0 \text{ cm}) = 1/f_2$, which gives $f_2 =$ **1.9 cm.**

19. The magnification with the image at infinity is
$M = N/f = (25 \text{ cm})/(12 \text{ cm}) =$ **2.1×.**

20. We find the focal length from
$M = N/f;$
$3.0 = (25 \text{ cm})/f$, which gives $f =$ **8.3 cm.**

21. (*a*) We find the focal length with the image at the near point from
$M = 1 + N/f_1;$
$2.5 = 1 + (25 \text{ cm})/f_1$, which gives $f_1 =$ **17 cm.**
(*b*) If the eye is relaxed, the image is at infinity, so we have
$M = N/f_2;$
$2.5 = (25 \text{ cm})/f_2$, which gives $f_2 =$ **10 cm.**

22. Maximum magnification is obtained with the image at the near point. We find the object distance from
$(1/d_o) + (1/d_i) = 1/f;$
$(1/d_o) + (1/-25.0 \text{ cm}) = 1/10.0 \text{ cm}$, which gives $d_o =$ $\boxed{7.1 \text{ cm from the lens.}}$
The magnification is
$M = 1 + N/f = 1 + (25.0 \text{ cm})/(10.0 \text{ cm}) =$ $\boxed{3.5\times.}$

23. (*a*) The angular magnification with the image at the near point is
$M = 1 + N/f = 1 + (25.0 \text{ cm})/(9.00 \text{ cm}) =$ $\boxed{3.78\times.}$
(*b*) Because the object without the lens and the image with the lens are at the near point, the angular magnification is also the ratio of widths:
$M = h_i/h_o;$
$3.78 = h_i/(3.10 \text{ mm})$, which gives $h_i =$ $\boxed{11.7 \text{ mm.}}$
(*c*) We find the object distance from
$(1/d_o) + (1/d_i) = 1/f;$
$(1/d_o) + (1/-25.0 \text{ cm}) = 1/9.00 \text{ cm}$, which gives $d_o =$ $\boxed{6.62 \text{ cm from the lens.}}$

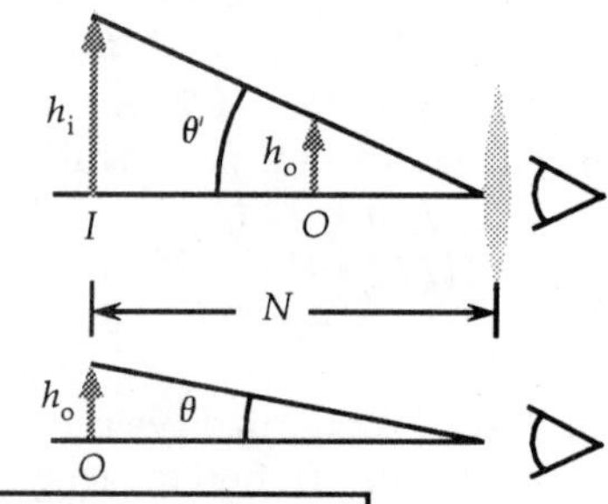

24. (*a*) We find the image distance from
$(1/d_o) + (1/d_i) = 1/f;$
$(1/5.35 \text{ cm}) + (1/d_i) = 1/6.00 \text{ cm},$
which gives $d_i =$ $\boxed{-49.4 \text{ cm.}}$
(*b*) From the diagram we see that the angular magnification is
$M = \theta'/\theta = (h_o/d_o)/(h_o/N) = N/d_o$
$= (25 \text{ cm})/(5.35 \text{ cm}) =$ $\boxed{4.67\times.}$

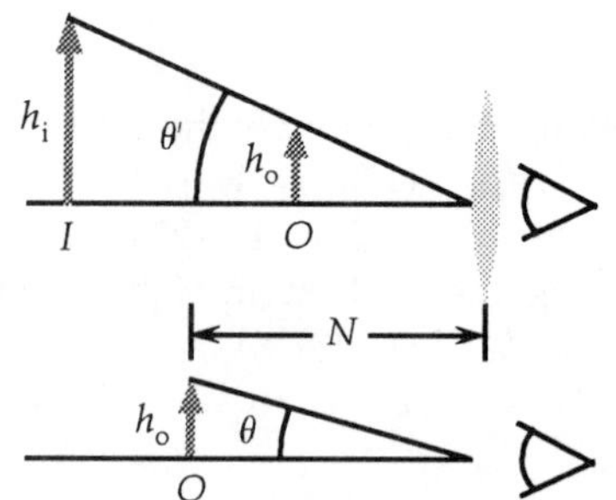

25. (*a*) We find the image distance from
$(1/d_o) + (1/d_i) = 1/f;$
$(1/8.5 \text{ cm}) + (1/d_i) = 1/9.5 \text{ cm}$, which gives $d_i =$ $\boxed{-81 \text{ cm.}}$
(*b*) The linear magnification is
$m = -d_i/d_o = -(-81 \text{ cm})/(8.5 \text{ cm}) =$ $\boxed{9.5\times.}$
(*c*) The angular magnification is
$M = \theta'/\theta = (h_o/d_o)/(h_o/N) = N/d_o$
$= (25 \text{ cm})/(8.5 \text{ cm}) =$ $\boxed{2.9\times.}$
Note that this is less than the linear magnification because the image is farther away.

26. We find the focal length of the lens from
$M = N/f;$
$3.0 = (25 \text{ cm})/f$, which gives $f = 8.3$ cm.
(*a*) The magnification with the image at infinity is
$M_1 = N_1/f = (50 \text{ cm})/(8.3 \text{ cm}) =$ $\boxed{6.0\times.}$
(*b*) The magnification with the image at infinity is
$M_2 = N_2/f = (16 \text{ cm})/(8.3 \text{ cm}) =$ $\boxed{1.9\times.}$
Without the lens, the closest an object can be placed is the near point. A farther near point means a smaller angle subtended by the object without the lens, and thus greater magnification.

27. We find the object distance for an image at her eye's near point:

$(1/d_o) + (1/d_i) = 1/f = P;$

$(1/d_o) + (1/-0.100\text{ m}) = -4.0\text{ D}$, which gives $d_o = 1.7 \times 10^{-2}\text{ m} =$ **17 cm.**

We find the object distance for an image at her eye's far point:

$(1/d_o) + (1/d_i) = 1/f = P;$

$(1/d_o) + (1/-0.200\text{ m}) = -4.0\text{ D}$, which gives $d_o = 1.0\text{ m} =$ **100 cm.**

28. The magnification of the telescope is given by

$M = -f_o/f_e = -(80\text{ cm})/(2.8\text{ cm}) =$ **$-29\times$.**

For both object and image far away, the separation of the lenses is

$L = f_o + f_e = 80\text{ cm} + 2.8\text{ cm} =$ **83 cm.**

29. We find the focal length of the eyepiece from the magnification:

$M = -f_o/f_e;$

$-25 = -(80\text{ cm})/f_e$, which gives $f_e =$ **3.2 cm.**

For both object and image far away, the separation of the lenses is

$L = f_o + f_e = 80\text{ cm} + 3.2\text{ cm} =$ **83 cm.**

30. We find the focal length of the objective from the magnification:

$M = f_o/f_e;$

$8.0 = f_o/(3.0\text{ cm})$, which gives $f_o =$ **24 cm.**

31. We find the focal length of the eyepiece from the power:

$f_e = 1/P = 1/40\text{ D} = 0.025\text{ m} = 2.5\text{ cm}.$

The magnification of the telescope is given by

$M = -f_o/f_e = -(85\text{ cm})/(2.5\text{ cm}) =$ **$-34\times$.**

32. For both object and image far away, we find the focal length of the eyepiece from the separation of the lenses:

$L = f_o + f_e;$

$76\text{ cm} = 74.5\text{ cm} + f_e$, which gives $f_e = 1.5\text{ cm}.$

The magnification of the telescope is given by

$M = -f_o/f_e = -(74.5\text{ cm})/(1.5\text{ cm}) =$ **$-50\times$.**

33. For both object and image far away, we find the (negative) focal length of the eyepiece from the separation of the lenses:

$L = f_o + f_e;$

$33\text{ cm} = 36\text{ cm} + f_e$, which gives $f_e = -3.0\text{ cm}.$

The magnification of the telescope is given by

$M = -f_o/f_e = -(36\text{ cm})/(-3.0\text{ cm}) =$ **$12\times$.**

34. The reflecting mirror acts as the objective, with a focal length

$f_o = r/2 = (5.0\text{ m})/2 = 2.5\text{ m}.$

The magnification of the telescope is given by

$M = -f_o/f_e = -(250\text{ cm})/(3.2\text{ cm}) =$ **$-78\times$.**

35. We find the focal length of the mirror from

$M = -f_o/f_e$;

$-120 = -f_o/(3.5 \text{ cm})$, which gives $f_o = 4.2 \times 10^2 \text{ cm} =$ $\boxed{4.2 \text{ m.}}$

The radius is

$r = 2f_o = 2(4.2 \text{ m}) =$ $\boxed{8.4 \text{ m.}}$

36. For the magnification we have

$M = -f_o/f_e = -180$, or $f_o = 180f_e$.

For both object and image far away, we have

$L = f_o + f_e$;

$1.25 \text{ m} = 180f_e + f_e$, which gives $f_e = 6.91 \times 10^{-3} \text{ m} =$ $\boxed{6.91 \text{ mm.}}$

The focal length of the objective is

$f_o = 180f_e = 180(6.91 \times 10^{-3} \text{ m}) =$ $\boxed{1.24 \text{ m.}}$

37. We assume a prism binocular so the magnification is positive, but simplify the diagram by ignoring the prisms. We find the focal length of the eyepiece from the design magnification:

$M = f_o/f_e$;

$6.0 = (26 \text{ cm})/f_e$, which gives $f_e = 4.33$ cm.

We find the intermediate image formed by the objective:

$(1/d_{oo}) + (1/d_{oi}) = 1/f_o$;

$(1/400 \text{ cm}) + (1/d_{oi}) = 1/26$ cm, which gives $d_{oi} = 27.8$ cm.

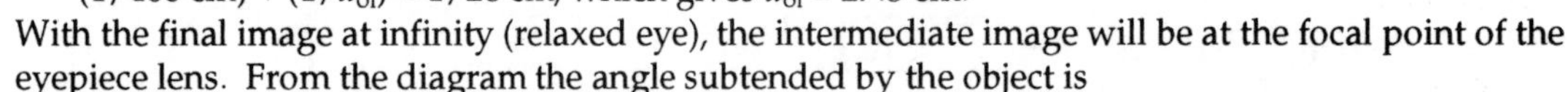

With the final image at infinity (relaxed eye), the intermediate image will be at the focal point of the eyepiece lens. From the diagram the angle subtended by the object is

$\theta = h/d_{oi}$,

while the angle subtended by the image is

$\theta' = h/f_e$.

Thus the angular magnification is

$M = \theta'/\theta = (h/f_e)/(h/d_{oi}) = d_{oi}/f_e = (27.8 \text{ cm})/(4.33 \text{ cm}) =$ $\boxed{6.4\times.}$

38. The magnification of the microscope is

$M = N\ell/f_o f_e = (25 \text{ cm})(17.5 \text{ cm})/(0.65 \text{ cm})(1.5 \text{ cm}) =$ $\boxed{450\times.}$

39. We find the focal length of the eyepiece from the magnification of the microscope:

$M = N\ell/f_o f_e$;

$720 = (25 \text{ cm})(17.5 \text{ cm})/(0.40 \text{ cm})f_e$, which gives $f_e =$ $\boxed{1.5 \text{ cm.}}$

40. (*a*) The total magnification is

$M = M_o M_e = (62.0)(12.0) =$ $\boxed{744\times.}$

(*b*) With the final image at infinity, we find the focal length of the eyepiece from

$M_e = N/f_e$;

$12.0 = (25.0 \text{ cm})/f_e$, which gives $f_e =$ $\boxed{2.08 \text{ cm.}}$

Because the image from the objective is at the focal point of the eyepiece, the image distance for the objective is

$d_i = \ell - f_e = 20.0 \text{ cm} - 2.08 \text{ cm} = 17.9$ cm.

We find the object distance from the magnification:

$M_o = d_i/d_o$;

$62.0 = (17.9 \text{ cm})/d_o$, which gives $d_o = 0.289$ cm.

We find the focal length of the objective from the lens equation:

$(1/d_o) + (1/d_i) = 1/f_o$;

$(1/0.289 \text{ cm}) + (1/17.9 \text{ cm}) = 1/f_o$, which gives $f_o =$ $\boxed{0.284 \text{ cm.}}$

(*c*) We found the object distance: $d_o =$ $\boxed{0.289 \text{ cm.}}$

41. (*a*) Because the image from the objective is at the focal point of the eyepiece, the image distance for the objective is

$d_{io} = \ell - f_e = 16.0 \text{ cm} - 1.8 \text{ cm} = 14.2 \text{ cm}.$

We find the object distance from the lens equation for the objective:

$(1/d_{oo}) + (1/d_{io}) = 1/f_o;$

$(1/d_{oo}) + (1/14.2 \text{ cm}) = 1/0.80 \text{ cm}$, which gives $d_{oo} =$ $\boxed{0.85 \text{ cm.}}$

(*b*) With the final image at infinity, the magnification of the eyepiece is

$M_e = N/f = (25.0 \text{ cm})/(1.8 \text{ cm}) = 13.9\times.$

The magnification of the objective is

$M_o = d_{io}/d_{oo} = (14.2 \text{ cm})/(0.85 \text{ cm}) = 16.7\times.$

The total magnification is

$M = M_oM_e = (16.7)(13.9) =$ $\boxed{230\times.}$

42. (*a*) We find the object distance from the lens equation for the eyepiece:

$(1/d_{oe}) + (1/d_{ie}) = 1/f_e;$

$(1/d_{oe}) + (1/-25 \text{ cm}) = 1/1.8 \text{ cm}$, which gives $d_{oe} = 1.7 \text{ cm}.$

The image distance for the objective is

$d_{io} = \ell - d_{oe} = 16.0 \text{ cm} - 1.7 \text{ cm} = 14.3 \text{ cm}.$

We find the object distance from the lens equation for the objective:

$(1/d_{oo}) + (1/d_{io}) = 1/f_o;$

$(1/d_{oo}) + (1/14.3 \text{ cm}) = 1/0.80 \text{ cm}$, which gives $d_{oo} =$ $\boxed{0.85 \text{ cm.}}$

(*b*) With the final image at the near point, the magnification of the eyepiece is

$M_e = 1 + N/f = 1 + (25.0 \text{ cm})/(1.8 \text{ cm}) = 14.9\times.$

The magnification of the objective is

$M_o = d_{io}/d_{oo} = (14.3 \text{ cm})/(0.85 \text{ cm}) = 16.9\times.$

The total magnification is

$M = M_oM_e = (16.9)(14.9) =$ $\boxed{250\times.}$

43. (*a*) We find the image distance from the lens equation for the objective:

$(1/d_{oo}) + (1/d_{io}) = 1/f_o;$

$(1/0.790 \text{ cm}) + (1/d_{io}) = 1/0.740 \text{ cm}$, which gives $d_{io} = 11.7 \text{ cm}.$

For the relaxed eye, the image from the objective is at the focal point of the eyepiece: $d_{oe} = 2.70 \text{ cm}.$

The distance between lenses is

$\ell = d_{io} + d_{oe} = 11.7 \text{ cm} + 2.70 \text{ cm} =$ $\boxed{14.4 \text{ cm.}}$

(*b*) With the final image at infinity, the magnification of the eyepiece is

$M_e = N/f = (25.0 \text{ cm})/(2.70 \text{ cm}) = 9.26\times.$

The magnification of the objective is

$M_o = d_{io}/d_{oo} = (11.7 \text{ cm})/(0.790 \text{ cm}) = 14.8\times.$

The total magnification is

$M = M_oM_e = (14.8)(9.26) =$ $\boxed{137\times.}$

44. (*a*) When lenses are in contact, the powers add:

$P = P_1 + P_2 = (1/-0.28 \text{ m}) + (1/0.23 \text{ m}) = +0.776 \text{ D}.$

It is a positive lens, and thus $\boxed{\text{converging.}}$

(*b*) The focal length is

$f = 1/P = 1/0.776 \text{ D} =$ $\boxed{1.3 \text{ m.}}$

45. (*a*) We find the incident angle from

$\sin \theta_1 = h_1/R = (1.0\text{ cm})/(12.0\text{ cm}) = 0.0833$, so $\theta_1 = 4.78°$.

For the refraction at the curved surface, we have

$\sin \theta_1 = n \sin \theta_2$;

$\sin 4.78° = (1.50) \sin \theta_2$, which gives $\sin \theta_2 = 0.0556$,

so $\theta_2 = 3.18°$.

We see from the diagram that

$\theta_3 = \theta_1 - \theta_2 = 4.78° - 3.18° = 1.60°$.

For the refraction at the flat face, we have

$n \sin \theta_3 = \sin \theta_4$;

$(1.50) \sin 1.60° = \sin \theta_4$, which gives $\sin \theta_4 = 0.0419$,

so $\theta_4 = 2.40°$.

We see from the diagram that

$h_1' = h_1 - R \sin \theta_3 = 1.0\text{ cm} - (12.0\text{ cm}) \sin 1.60° = 0.665\text{ cm}$,

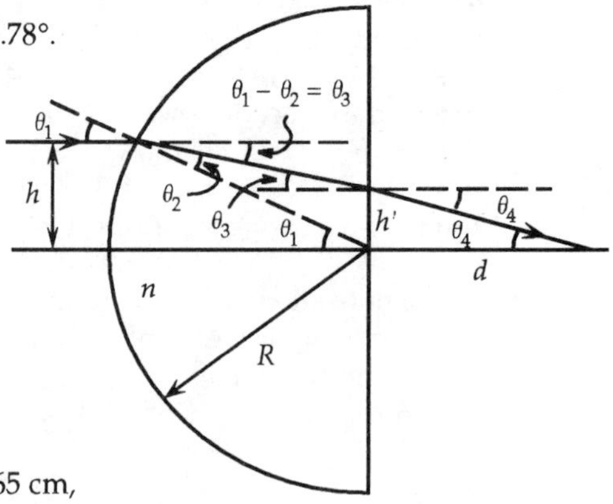

so the distance from the flat face to the point where the ray crosses the axis is

$d_1 = h_1'/\tan \theta_4 = (0.665\text{ cm})/\tan 2.40° =$ **15.9 cm.**

(*b*) We find the incident angle from

$\sin \theta_1 = h_2/R = (4.0\text{ cm})/(12.0\text{ cm}) = 0.333$, so $\theta_1 = 19.47°$.

For the refraction at the curved surface, we have

$\sin \theta_1 = n \sin \theta_2$;

$\sin 19.47° = (1.50) \sin \theta_2$, which gives $\sin \theta_2 = 0.222$, so $\theta_2 = 12.84°$.

We see from the diagram that

$\theta_3 = \theta_1 - \theta_2 = 19.47° - 12.84° = 6.63°$.

For the refraction at the flat face, we have

$n \sin \theta_3 = \sin \theta_4$;

$(1.50) \sin 6.63° = \sin \theta_4$, which gives $\sin \theta_4 = 0.173$, so $\theta_4 = 9.97°$.

We see from the diagram that

$h_2' = h_2 - R \sin \theta_3 = 4.0\text{ cm} - (12.0\text{ cm}) \sin 6.63° = 2.61\text{ cm}$,

so the distance from the flat face to the point where the ray crosses the axis is

$d_2 = h_2'/\tan \theta_4 = (2.61\text{ cm})/\tan 9.97° =$ **14.8 cm.**

(*c*) The separation of the "focal points" is

$\Delta d = d_1 - d_2 = 15.9\text{ cm} - 14.8\text{ cm} =$ **1.1 cm.**

(*d*) When $h_2 = 4.0$ cm, the rays focus closer to the lens, so they will form a circle at the "focal point" for $h_1 = 1.0$ cm. We find the radius of this circle from similar triangles:

$h_2'/d_2 = r/(d_1 - d_2)$;

$(2.61\text{ cm})/(14.8\text{ cm}) = r/(1.1\text{ cm})$, which gives $r =$ **0.19 cm.**

46. The minimum angular resolution is

$\theta = 1.22\lambda/D = (1.22)(550 \times 10^{-9}\text{ m})/(100\text{ in})(0.0254\text{ m/in}) =$ **2.64×10^{-7} rad = $(1.51 \times 10^{-5})°$.**

47. We find the angle of acceptance from

$\text{NA} = n \sin \alpha$;

$1.41 = (1.80) \sin \alpha$, which gives $\sin \alpha = 0.783$, so $\alpha =$ **51.6°.**

The resolving power is

$\text{RP} = 0.61\lambda/\text{NA} = 0.61(500\text{ nm})/(1.41) =$ **216 nm.**

48. (*a*) The numerical aperture is

$\text{NA} = n \sin \alpha = (1.60) \sin 60° =$ **1.39.**

(*b*) The resolving power is

$\text{RP} = 0.61\lambda/\text{NA} = 0.61(550\text{ nm})/(1.39) =$ **240 nm.**

49. The resolution of the telescope is
$\theta = 1.22\lambda/D = (1.22)(500 \times 10^{-9}\text{ m})/(12\text{ in})(0.0254\text{ m/in}) = 2.00 \times 10^{-6}\text{ rad}.$
The separation of the stars is
$d = L\theta = (20\text{ ly})(9.46 \times 10^{15}\text{ m/ly})(2.00 \times 10^{-6}\text{ rad}) = \boxed{3.8 \times 10^{11}\text{ m.}}$

50. If the lines are barely resolved, their separation is the resolving power:
$\text{RP} = 0.61\lambda/\text{NA} = 0.61(540\text{ nm})/(0.95) = \boxed{347\text{ nm.}}$
Because RP $\propto \lambda$, 400 nm requires a lower RP, so they $\boxed{\text{will be resolved by violet light.}}$
700 nm requires a higher RP, so they $\boxed{\text{will not be resolved by red light.}}$

51. The maximum resolving power is the spacing, so we find the minimum NA from
$\text{RP}_{max} = 0.61\lambda/\text{NA}_{min};$
$0.63 \times 10^{-6}\text{ m} = (0.61)(480 \times 10^{-9}\text{ m})/\text{NA}_{min}$, which gives $\text{NA}_{min} = \boxed{0.46.}$

52. The normal human eye can resolve about 10^{-4} m, so the required minimum magnification is
$M = (10^{-4}\text{ m})/(0.63 \times 10^{-6}\text{ m}) = \boxed{160\times.}$

53. (*a*) The resolution of the eye is
$\theta = 1.22\lambda/D = (1.22)(500 \times 10^{-9}\text{ m})/(5.0 \times 10^{-3}\text{ m}) = 1.22 \times 10^{-4}\text{ rad}.$
We find the maximum distance from
$d = L\theta;$
$2.0\text{ m} = L(1.22 \times 10^{-4}\text{ rad})$, which gives $L = 1.6 \times 10^4\text{ m} = \boxed{16\text{ km.}}$
(*b*) The angular separation is the resolution:
$\theta = 1.22 \times 10^{-4}\text{ rad} = (6.99 \times 10^{-3})^\circ = \boxed{0.42'.}$
Our answer is less than the real resolution because of the presence of aberrations.

54. A path difference of one wavelength corresponds to a phase difference of 2π. The actual distance has not changed, but the number of wavelengths is different for each path. If the object has a thickness t, the number of wavelengths in the medium is t/λ_1, and the number in the object is t/λ_2. Thus the phase difference is
$\delta = 2\pi[(t/\lambda_2) - (t/\lambda_1)] = 2\pi[(n_2t/\lambda) - (n_1t/\lambda)] = (2\pi/\lambda)(n_2 - n_1)t.$

55. For the diffraction from the crystal, we have
$m\lambda = 2d\sin\phi;\ m = 1, 2, 3, \ldots .$
For the first maximum, we get
$(1)(0.135\text{ nm}) = 2(0.280\text{ nm})\sin\phi$, which gives $\phi = \boxed{14.0^\circ.}$

56. We find the spacing from
$m\lambda = 2d\sin\phi;\ m = 1, 2, 3, \ldots .$
$(2)(0.0973\text{ nm}) = 2d\sin 23.4^\circ$, which gives $d = \boxed{0.245\text{ nm.}}$

57. (*a*) For the diffraction from the crystal, we have
$m\lambda = 2d\sin\phi;\ m = 1, 2, 3, \ldots .$
When we form the ratio for the two orders, we get
$m_2/m_1 = (\sin\phi_2)/(\sin\phi_1);$
$2/1 = (\sin\phi_2)/(\sin 23.4^\circ)$, which gives $\phi_2 = \boxed{52.6^\circ.}$
(*b*) We find the wavelength from
$m_1\lambda = 2d\sin\phi_1;$
$(1)\lambda = 2(0.24\text{ nm})\sin 23.4^\circ$, which gives $\lambda = \boxed{0.19\text{ nm.}}$

58. For the diffraction from the crystal, we have
$m\lambda = 2d \sin \phi;\ m = 1, 2, 3, \ldots;$
$(1)\lambda = 2d \sin \phi_1;$
$(2)\lambda = 2d \sin \phi_2;$
$(3)\lambda = 2d \sin \phi_3.$
We see that each equation contains the ratio λ/d, so the wavelength and lattice spacing
cannot be separately determined.

59. (*a*) Because X-ray images are shadows, the image will be the same size as the object, so the magnification is 1.
(*b*) The rays from the point source will not refract, so we can use similar triangles to compare the image size to the object size for the front of the body:
$m_1 = h_i/h_{o1} = (d_1 + d_2)/d_1 = (15 \text{ cm} + 25 \text{ cm})/(15 \text{ cm}) =$ 2.7.
For the back of the body, the image and object have the same size, so the magnification is 1.

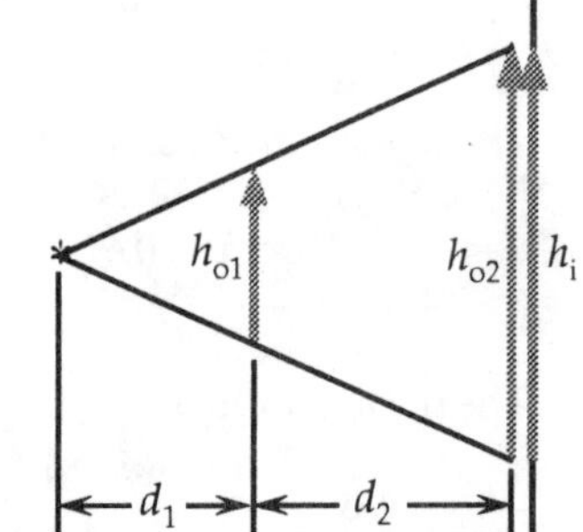

60. (*a*) With a positive (converging) lens, Sam is farsighted.
(*b*) The focal length of the lens is
$f = 1/P = 1/3.2 \text{ D} = 0.313 \text{ m} =$ 31 cm.
(*c*) The lens produces a virtual image at his near point:
$(1/d_{o1}) + (1/d_{i1}) = 1/f = P;$
$(1/0.23 \text{ m}) + (1/d_{i1}) = +3.2 \text{ D}$, which gives $d_{i1} = -0.87$ m, so his near point is 89 cm.
(*d*) For Pam, we find the object distance that will have an image at her near point:
$(1/d_{o2}) + (1/d_{i2}) = 1/f = P;$
$(1/d_{o2}) + (1/-0.23 \text{ m}) = +3.2 \text{ D}$, which gives $d_{o2} = 0.13$ m, which is 15 cm from her eyes.

61. The exposure is proportional to the intensity, the area and the time:
Exposure $\propto IAt \propto ID^2t \propto It/(f\text{-stop})^2$.
With the same shutter speed, the time is constant.
Because we want the exposure to be the same, we have
$I_1/(f\text{-stop}_1)^2 = I_2/(f\text{-stop}_2)^2;$
$I_1/(5.6)^2 = I_2/(22)^2$, which gives $I_2 = 16I_1$.
Note that we have followed convention to use multiples of 2.

62. When an object is very far away, the image will be at the focal point $d_i = f$. Thus the magnification is $m = -d_i/d_o = -f/d_o$, that is, proportional to f.

63. For the magnification, we have
$m = -h_i/h_o = -d_i/d_o = -1$, so $d_i = d_o$.
We find the object distance from
$(1/d_o) + (1/d_i) = 1/f;$
$(1/d_o) + (1/d_o) = 1/50$ mm, which gives $d_o =$ 100 mm.
The distance between the object and the film is
$d = d_o + d_i = 100 \text{ mm} + 100 \text{ mm} =$ 200 mm.

64. The actual far point of the person is 180 cm. With the lens, an object far away is to produce a virtual image 180 cm from the eye, or 178 cm from the lens.
We find the power of the lens from
$(1/d_{o1}) + (1/d_{i1}) = 1/f_1 = P_1$, when distances are in m;
$(1/\infty) + (1/-1.78\text{ m}) = P_1 =$ $\boxed{-0.56\text{ D (upper part).}}$
The actual near point of the person is 40 cm. With the lens, an object placed at the normal near point, 25 cm, or 23 cm from the lens, is to produce a virtual image 40 cm from the eye, or 38 cm from the lens.
We find the power of the lens from
$(1/d_{o2}) + (1/d_{i2}) = 1/f_2 = P_2$;
$(1/0.23\text{ m}) + (1/-0.38\text{ m}) = P_2 =$ $\boxed{+1.7\text{ D (lower part).}}$

65. The maximum magnification is achieved with the image at the near point:
$M_1 = 1 + N_1/f = 1 + (15.0\text{ cm})/(8.0\text{ cm}) =$ $\boxed{2.9\times.}$
For an adult we have
$M_2 = 1 + N_2/f = 1 + (25.0\text{ cm})/(8.0\text{ cm}) =$ $\boxed{4.1\times.}$

66. The magnification for a relaxed eye is
$M = N/f = NP = (0.25\text{ m})(+4.0\text{ D}) =$ $\boxed{1.0\times.}$

67. (*a*) The magnification of the telescope is given by
$M = -f_o/f_e = -P_e/P_o = -(4.5\text{ D})/(2.0\text{ D}) =$ $\boxed{-2.25\times.}$
(*b*) To get a magnification greater than 1, for the eyepiece we use the lens with the smaller focal length, or greater power: $\boxed{4.5\text{ D.}}$

68. The minimum angular resolution is
$\theta = 1.22\lambda/D.$
The distance between lines is the resolving power:
$\text{RP} = f\theta = 1.22\lambda f/D = 1.22\lambda(f\text{-stop}).$
For $f/2$ we have
$\text{RP}_2 = (1.22)(500 \times 10^{-9}\text{ m})(2) = 1.22 \times 10^{-6}\text{ m} = 1.22 \times 10^{-3}\text{ mm}$, so the resolution is
$1/\text{RP}_2 = 1/(1.22 \times 10^{-3}\text{ mm}) =$ $\boxed{820\text{ lines/mm.}}$
For $f/16$ we have
$\text{RP}_2 = (1.22)(500 \times 10^{-9}\text{ m})(16.7) = 1.02 \times 10^{-5}\text{ m} = 1.02 \times 10^{-2}\text{ mm}$, so the resolution is
$1/\text{RP}_2 = 1/(1.02 \times 10^{-2}\text{ mm}) =$ $\boxed{98\text{ lines/mm.}}$

69. To find the new near point, we have
$(1/d_{o1}) + (1/d_{i1}) = 1/f_1 = P_1$, when distances are in m;
$(1/0.35\text{ m}) + (1/d_{i1}) = +2.5\text{ D}$, which gives $d_{i1} = -2.8\text{ m}$.
To give him a normal near point, we have
$(1/d_{o2}) + (1/d_{i2}) = 1/f_2 = P_2$;
$(1/0.25\text{ m}) + (1/-2.8\text{ m}) = P_2 =$ $\boxed{+3.6\text{ D.}}$

70. For the minimum aperture the angle subtended at the lens by the smallest feature is the angular resolution:
$\theta = d/L = 1.22\lambda/D$;
$(5 \times 10^{-2}\text{ m})/(25 \times 10^{3}\text{ m}) = (1.22)(550 \times 10^{-9}\text{ m})/D$, which gives $D = 0.34\text{ m} =$ $\boxed{34\text{ cm.}}$

71. The angular resolution of the eye, which is the required resolution using the telescope, is

$\theta_{eye} = d_{eye}/L_{eye} = (0.10 \times 10^{-3}\text{ m})/(25 \times 10^{-2}\text{ m}) = 4.0 \times 10^{-4}\text{ rad}.$

The resolution without the telescope is

$\theta = d/L = (10\text{ km})/(3.84 \times 10^{5}\text{ km}) = 2.6 \times 10^{-5}\text{ rad}.$

If we ignore the inverted image, the magnification is

$M = \theta_{eye}/\theta = f_o/f_e;$

$(4.0 \times 10^{-4}\text{ rad})/(2.6 \times 10^{-5}\text{ rad}) = (2.2\text{ m})/f_e$, which gives $f_e = 0.14\text{ m} =$ $\boxed{14\text{ cm.}}$

The resolution limit is

$\theta = 1.22\lambda/D = (1.22)(500 \times 10^{-9}\text{ m})/(0.12\text{ m}) =$ $\boxed{5.1 \times 10^{-6}\text{ rad.}}$

This is a distance of 2 km on the surface of the Moon.

72. (*a*) When an object is very far away, the image will be at the focal point $d_{i1} = f$.

From the magnification, we have

$m = -h_i/h_o = -d_i/d_o$, so we see that h_i is proportional to d_i.

When the object and image are the same size, we get

$-h_i/h_o = -1 = -d_{i2}/d_{o2}$, so $d_{o2} = d_{i2}$.

From the lens equation, we get

$(1/d_{o2}) + (1/d_{i2}) = 1/f;$

$(1/d_{i2}) + (1/d_{i2}) = 1/f$, which gives $d_{i2} = 2f$.

The required exposure time is proportional to the area of the image on the film:

$t \propto A \propto (h_i)^2 \propto (d_i)^2.$

When we form the ratio, w get

$t_2/t_1 = (d_{i2}/d_{i1})^2 = (2f/f)^2 = 4.$

(*b*) For the increased object distances, we have

$(1/d_{o3}) + (1/d_{i3}) = 1/f;$

$(1/4f) + (1/d_{i3}) = 1/f$, which gives $d_{i3} = 4f/3$; and $t_3/t_1 = (4/3)^2 = 1.78.$

$(1/d_{o4}) + (1/d_{i4}) = 1/f;$

$(1/5f) + (1/d_{i4}) = 1/f$, which gives $d_{i4} = 5f/4$; and $t_4/t_1 = (5/4)^2 = 1.56.$

These increased exposures are less than the minimal adjustment on a typical camera, so they are negligible.

73. The focal length of the eyepiece is

$f_e = 1/P_e = 1/20\text{ D} = 5.0 \times 10^{-2}\text{ m} = 5.0\text{ cm}.$

For both object and image far away, we find the focal length of the objective from the separation of the lenses:

$L = f_o + f_e;$

$85\text{ cm} = f_o + 5.0\text{ cm}$, which gives $f_o = 80\text{ cm}$.

The magnification of the telescope is given by

$M = -f_o/f_e = -(80\text{ cm})/(5.0\text{ cm}) =$ $\boxed{-16\times.}$

74. We find the focal lengths of the lens for the two colors:

$1/f_{red} = (n_{red} - 1)(1/R_1 + 1/R_2)$

$= (1.5106 - 1)[(1/18.4 \text{ cm}) + (1/\infty)]$ which gives $f_{red} = 36.04$ cm.

$1/f_{yellow} = (n_{yellow} - 1)(1/R_1 + 1/R_2)$

$= (1.5226 - 1)[(1/18.4 \text{ cm}) + (1/\infty)]$ which gives $f_{yellow} = 35.21$ cm.

We find the image distances from

$(1/d_o) + (1/d_{ired}) = 1/f_{red}$;

$(1/66.0 \text{ cm}) + (1/d_{ired}) = 1/36.04$ cm, which gives $d_{ired} =$ 79.4 cm.

$(1/d_o) + (1/d_{iyellow}) = 1/f_{yellow}$;

$(1/66.0 \text{ cm}) + (1/d_{iorange}) = 1/35.21$ cm, which gives $d_{iyellow} =$ 75.5 cm.

The images are 3.9 cm apart, an example of chromatic aberration.

75. We can relate the object and image distances from the magnification:

$m = -h_i/h_o = -d_i/d_o$;

$(8.25 \times 10^{-3} \text{ m})/(1.75 \text{ m}) = d_i/d_o$, or $d_i = (4.71 \times 10^{-3})d_o$.

We find the object distance from

$(1/d_o) + (1/d_i) = 1/f$;

$(1/d_o) + [1/(4.71 \times 10^{-3})d_o] = 1/200 \times 10^{-3}$ m, which gives $d_o = 42.7$ m.

Thus the reporter was standing 42.7 m from his subject.

CHAPTER 26

1. (*a*) $[1 - (v/c)^2]^{1/2} = \{1 - [(20{,}000 \text{ m/s})/(3.00 \times 10^8 \text{ m/s})]^2\}^{1/2} =$ **1.00.**
   (*b*) $[1 - (v/c)^2]^{1/2} = [1 - (0.0100)^2]^{1/2} =$ **0.99995.**
   (*c*) $[1 - (v/c)^2]^{1/2} = [1 - (0.100)^2]^{1/2} =$ **0.995.**
   (*d*) $[1 - (v/c)^2]^{1/2} = [1 - (0.900)^2]^{1/2} =$ **0.436.**
   (*e*) $[1 - (v/c)^2]^{1/2} = [1 - (0.990)^2]^{1/2} =$ **0.141.**
   (*f*) $[1 - (v/c)^2]^{1/2} = [1 - (0.999)^2]^{1/2} =$ **0.0447.**

2. You measure the contracted length. We find the rest length from
   $L = L_0[1 - (v/c)^2]^{1/2}$;
   $48.2 \text{ m} = L_0[1 - (0.850)^2]^{1/2}$, which gives $L_0 =$ **91.5 m.**

3. We find the lifetime at rest from
   $\Delta t = \Delta t_0/[1 - (v^2/c^2)]^{1/2}$;
   $4.76 \times 10^{-6} \text{ s} = \Delta t_0/\{1 - [(2.70 \times 10^8 \text{ m/s})/(3.00 \times 10^8 \text{ m/s})]^2\}^{1/2}$, which gives $\Delta t_0 =$ **2.07×10^{-6} s.**

4. You measure the contracted length:
   $L = L_0[1 - (v/c)^2]^{1/2}$
   $= (100 \text{ ly})\{1 - [(2.60 \times 10^8 \text{ m/s})/(3.00 \times 10^8 \text{ m/s})]^2\}^{1/2} =$ **49.9 ly.**

5. The rest length of his car is 6.00 m. For his car you measure the contracted length:
   $L_1 = L_{01}[1 - (v/c)^2]^{1/2}$
   $= (6.00 \text{ m})[1 - (0.37)2]^{1/2} =$ **5.57 m.**
   He measured the contracted length of your car. We find the rest length from
   $L_2 = L_{02}[1 - (v/c)^2]^{1/2}$;
   $6.21 \text{ m} = L_{02}[1 - (0.37)^2]^{1/2}$, which gives $L_{02} =$ **6.68 m.**

6. We determine the speed from the time dilation:
   $\Delta t = \Delta t_0/[1 - (v^2/c^2)]^{1/2}$;
   $4.10 \times 10^{-8} \text{ s} = (2.60 \times 10^{-8} \text{ s})/[1 - (v/c)^2]^{1/2}$, which gives $v =$ **$0.773c$.**

7. We determine the speed from the length contraction:
   $L = L_0[1 - (v/c)^2]^{1/2}$;
   $25 \text{ ly} = (90 \text{ ly})[1 - (v/c)^2]^{1/2}$, which gives $v =$ **$0.96c$.**

8. For a 1.00 per cent change, the factor in the expressions for time dilation and length contraction must equal $1 - 0.0100 = 0.9900$:
   $[1 - (v/c)^2]^{1/2} = 0.9900$, which gives $v =$ **$0.141c$.**

9. In the Earth frame, the clock on the Enterprise will run slower.
   (*a*) We find the elapsed time on the ship from
   $\Delta t = \Delta t_0/[1 - (v^2/c^2)]^{1/2}$;
   $5.0 \text{ yr} = \Delta t_0/[1 - (0.89)^2]^{1/2}$, which gives $\Delta t_0 =$ **2.3 yr.**
   (*b*) We find the elapsed time on the Earth from
   $\Delta t = \Delta t_0/[1 - (v^2/c^2)]^{1/2}$
   $= (5.0 \text{ yr})/[1 - (0.89)^2]^{1/2} =$ **11 yr.**

10. (*a*) To an observer on Earth, 75.0 ly is the rest length, so the time will be
$t_{Earth} = L_0/v = (75.0 \text{ ly})/0.950c = \boxed{78.9 \text{ yr.}}$
(*b*) We find the dilated time on the spacecraft from
$\Delta t = \Delta t_0/[1 - (v^2/c^2)]^{1/2}$;
$78.9 \text{ yr} = \Delta t_0/[1 - (0.950)^2]^{1/2}$, which gives $\Delta t_0 = \boxed{24.6 \text{ yr.}}$
(*c*) To the spacecraft observer, the distance to the star is contracted:
$L = L_0[1 - (v/c)^2]^{1/2} = (75.0 \text{ ly})[1 - (0.950)^2]^{1/2} = \boxed{23.4 \text{ ly.}}$
(*d*) To the spacecraft observer, the speed of the spacecraft is
$v = L/\Delta t = (23.4 \text{ ly})/24.6 \text{ yr} = \boxed{0.950c,}$ as expected.

11. (*a*) You measure the contracted length. We find the rest length from
$L = L_0[1 - (v/c)^2]^{1/2}$;
$5.80 \text{ m} = L_0[1 - (0.580)^2]^{1/2}$, which gives $L_0 = \boxed{7.12 \text{ m.}}$
Distances perpendicular to the motion do not change, so the rest height is $\boxed{1.20 \text{ m.}}$
(*b*) We find the dilated time in the sports vehicle from
$\Delta t = \Delta t_0/[1 - (v^2/c^2)]^{1/2}$;
$20.0 \text{ s} = \Delta t_0/[1 - (0.580)^2]^{1/2}$, which gives $\Delta t_0 = \boxed{16.3 \text{ s.}}$
(*c*) To your friend, you moved at the same relative speed: $\boxed{0.580c.}$
(*d*) She would measure the same time dilation: $\boxed{16.3 \text{ s.}}$

12. In the Earth frame, the average lifetime of the pion will be dilated:
$\Delta t = \Delta t_0/[1 - (v^2/c^2)]^{1/2}$.
The speed as a fraction of the speed of light is
$v/c = d/c\,\Delta t = d[1 - (v^2/c^2)]^{1/2}/c\,\Delta t_0$;
$v/c = (10.0 \text{ m})[1 - (v^2/c^2)]^{1/2}/(3.00 \times 10^8 \text{ m/s})(2.60 \times 10^{-8} \text{ s})$,
which gives $v = \boxed{0.789c = 2.37 \times 10^8 \text{ m/s.}}$

13. The mass of the proton is
$m = m_0/[1 - (v^2/c^2)]^{1/2} = (1.67 \times 10^{-27} \text{ kg})/[1 - (0.90)^2]^{1/2} = \boxed{3.8 \times 10^{-27} \text{ kg.}}$

14. We find the speed from
$m = m_0/[1 - (v^2/c^2)]^{1/2}$;
$2m_0 = m_0/[1 - (v^2/c^2)]^{1/2}$, which gives $v = \boxed{0.866c.}$

15. We find the speed from
$m = m_0/[1 - (v^2/c^2)]^{1/2}$;
$1.10m_0 = m_0/[1 - (v^2/c^2)]^{1/2}$, which gives $v = \boxed{0.417c.}$

16. We convert the speed: $(40{,}000 \text{ km/h})/(3.6 \text{ ks/h}) = 1.11 \times 10^4 \text{ m/s}$.
Because this is much smaller than c, we can simplify the factor in the mass equation:
$1/[1 - (v^2/c^2)]^{1/2} \approx 1/[1 - \frac{1}{2}(v/c)^2] \approx 1 + \frac{1}{2}(v/c)^2$.
For the fractional change in mass, we have
$(m - m_0)/m_0 = \{1/[1 - (v^2/c^2)]^{1/2}\} - 1 \approx 1 + \frac{1}{2}(v/c)^2 - 1 = \frac{1}{2}(v/c)^2$;
$(m - m_0)/m_0 = \frac{1}{2}[(1.11 \times 10^4 \text{ m/s})/(3.00 \times 10^8 \text{ m/s})]^2 = 6.8 \times 10^{-10} = \boxed{6.8 \times 10^{-8}\%.}$

17. (*a*) We find the speed from

$m = m_0/[1 - (v^2/c^2)]^{1/2}$;

$10{,}000m_0 = m_0/[1 - (v^2/c^2)]^{1/2}$, which gives

$[1 - (v^2/c^2)]^{1/2} = 1.00 \times 10^{-4}$, or $(v/c)^2 = 1 - 1.00 \times 10^{-8}$.

When we take the square root, we get

$v/c = (1 - 1.00 \times 10^{-8})^{1/2} \approx 1 - \frac{1}{2}(1.00 \times 10^{-8}) = 1 - 0.50 \times 10^{-8}$.

Thus the speed is $\boxed{1.5\ \text{m/s less than } c.}$

(*b*) The contracted length of the tube is

$L = L_0[1 - (v/c)^2]^{1/2} = (3.0\ \text{km})(1.00 \times 10^{-4}) = 3.0 \times 10^{-4}\ \text{km} = \boxed{30\ \text{cm.}}$

18. The kinetic energy is

$\text{KE} = (m - m_0)c^2 = (3 - 1)m_0c^2 = 2m_0c^2 = 2(9.11 \times 10^{-31}\ \text{kg})(3.00 \times 10^8\ \text{m/s})^2 = \boxed{1.6 \times 10^{-13}\ \text{J (1.0 MeV).}}$

19. We find the increase in mass from

$\Delta m = \Delta E/c^2 = (4.82 \times 10^4\ \text{J})/(3.00 \times 10^8\ \text{m/s})^2 = \boxed{5.36 \times 10^{-13}\ \text{kg.}}$

Note that this is so small, most chemical reactions are considered to have mass conserved.

20. We find the loss in mass from

$\Delta m = \Delta E/c^2 = (200\ \text{MeV})(1.60 \times 10^{-13}\ \text{J/MeV})/(3.00 \times 10^8\ \text{m/s})^2 = \boxed{3.56 \times 10^{-28}\ \text{kg.}}$

21. The rest energy of the electron is

$E = m_0c^2 = (9.11 \times 10^{-31}\ \text{kg})(3.00 \times 10^8\ \text{m/s})^2 = \boxed{8.20 \times 10^{-14}\ \text{J}}$

$= (8.20 \times 10^{-14}\ \text{J})/(1.60 \times 10^{-13}\ \text{J/MeV}) = \boxed{0.511\ \text{MeV.}}$

22. The rest mass of the proton is

$m_0 = E/c^2 = (1.67 \times 10^{-27}\ \text{kg})(3.00 \times 10^8\ \text{m/s})^2/(1.60 \times 10^{-13}\ \text{J/MeV})c^2 = \boxed{939\ \text{MeV}/c^2.}$

23. We find the necessary mass conversion from

$\Delta m = \Delta E/c^2 = (8 \times 10^{19}\ \text{J})/(3.00 \times 10^8\ \text{m/s})^2 = \boxed{9 \times 10^2\ \text{kg.}}$

24. We find the energy equivalent of the mass from

$E = mc^2 = (1.0 \times 10^{-3}\ \text{kg})(3.00 \times 10^8\ \text{m/s})^2 = \boxed{9.0 \times 10^{13}\ \text{J.}}$

If this energy increases the gravitational energy, we have

$E = mgh$;

$9.0 \times 10^{13}\ \text{J} = m(9.80\ \text{m/s}^2)(100\ \text{m})$, which gives $m = \boxed{9.2 \times 10^{10}\ \text{kg.}}$

25. If the kinetic energy is equal to the rest energy, we have

$\text{KE} = (m - m_0)c^2 = m_0c^2$, or $m = 2m_0$.

We find the speed from

$m = m_0/[1 - (v^2/c^2)]^{1/2}$;

$2m_0 = m_0/[1 - (v^2/c^2)]^{1/2}$, which gives $v = 0.866c$.

26. (*a*) We find the work required from

$W = \Delta\text{KE} = (m - m_0)c^2 = m_0c^2(\{1/[1 - (v/c)^2]^{1/2}\} - 1)$

$= (939\text{ MeV})(\{1/[1 - (0.998)^2]^{1/2}\} - 1) = 13.9 \times 10^3\text{ MeV} =$ **13.9 GeV (2.23×10^{-9} J).**

(*b*) The momentum of the proton is

$p = mv = m_0v/[1 - (v/c)^2]^{1/2}$

$= (1.67 \times 10^{-27}\text{ kg})(0.998)(3.00 \times 10^8\text{ m/s})/[1 - (0.998)^2]^{1/2} =$ **7.91×10^{-18} kg·m/s.**

27. (*a*) The radiation falls on a circle with the Earth's radius. We find the increase in mass from

$\Delta m = \Delta E/c^2 = (1400\text{ W/m}^2)\pi(6.38 \times 10^6\text{ m})^2(3.16 \times 10^7\text{ s})/(3.00 \times 10^8\text{ m/s})^2 =$ **6.3×10^7 kg.**

Note that there will be mass loss from re-radiation into space.

(*b*) If the Sun radiates uniformly, the rate reaching the Earth is the rate through a sphere with a radius equal to the distance from the Sun to the Earth. We find the loss in mass from

$\Delta m = \Delta E/c^2 = (1400\text{ W/m}^2)4\pi(1.50 \times 10^{11}\text{ m})^2(3.16 \times 10^7\text{ s})/(3.00 \times 10^8\text{ m/s})^2 =$ **1.4×10^{17} kg.**

28. The speed of the proton is

$v = (2.50 \times 10^8\text{ m/s})/(3.00 \times 10^8\text{ m/s}) = 0.833c.$

The kinetic energy is

$\text{KE} = (m - m_0)c^2 = m_0c^2(\{1/[1 - (v/c)^2]^{1/2}\} - 1)$

$= (939\text{ MeV})(\{1/[1 - (0.833)^2]^{1/2}\} - 1) =$ **760 MeV (1.22×10^{-10} J).**

The momentum of the proton is

$p = mv = m_0v\{1/[1 - (v/c)^2]^{1/2}\}$

$= (1.67 \times 10^{-27}\text{ kg})(2.50 \times 10^8\text{ m/s})\{1/[1 - (0.833)^2]^{1/2}\} =$ **7.55×10^{-19} kg·m/s.**

29. The total energy of the proton is

$E = \text{KE} + m_0c^2 = 750\text{ MeV} + 939\text{ MeV} = 1689\text{ MeV}.$

The relation between the momentum and energy is

$(pc)^2 = E^2 - (m_0c^2)^2;$

$p^2(3.00 \times 10^8\text{ m/s})^2 = [(1689\text{ MeV})^2 - (939\text{ MeV})^2](1.60 \times 10^{-13}\text{ J/MeV})^2,$

which gives $p =$ **7.49×10^{-19} kg·m/s.**

30. The kinetic energy acquired by the proton is

$\text{KE} = qV = (1\text{ e})(75\text{ MV}) = 75\text{ MeV}.$

The mass of the proton is

$m = m_0 + \text{KE}/c^2 = 939\text{ MeV}/c^2 + (75\text{ MeV})/c^2 = 1014\text{ MeV}/c^2.$

We find the speed from

$m = m_0/[1 - (v^2/c^2)]^{1/2};$

$1014\text{ MeV}/c^2 = 939\text{ MeV}/c^2/[1 - (v^2/c^2)]^{1/2}$, which gives $v =$ **$0.377c$.**

31. The mass of the electron is

$m = m_0 + \text{KE}/c^2 = 0.511\text{ MeV}/c^2 + (1.00\text{ MeV})/c^2 = 1.51\text{ MeV}/c^2.$

We find the speed from

$m = m_0/[1 - (v^2/c^2)]^{1/2};$

$1.51\text{ MeV}/c^2 = 0.511\text{ MeV}/c^2/[1 - (v^2/c^2)]^{1/2}$, which gives $v =$ **$0.941c$.**

32. The kinetic energy acquired by the electron is
$\text{KE} = qV = (1\text{ e})(0.025\text{ MV}) = 0.025\text{ MeV}$.
The mass of the electron is
$m = m_0 + \text{KE}/c^2 = 0.511\text{ MeV}/c^2 + (0.025\text{ MeV})/c^2 = 0.536\text{ MeV}/c^2 = \boxed{1.05m_0.}$
We find the speed from
$m = m_0/[1-(v^2/c^2)]^{1/2}$;
$0.536\text{ MeV}/c^2 = (0.511\text{ MeV}/c^2)/[1-(v^2/c^2)]^{1/2}$, which gives $v = \boxed{0.302c.}$

33. If M_0 is the rest mass of the new particle, for conservation of energy we have
$2mc^2 = 2m_0c^2/[1-(v^2/c^2)]^{1/2} = M_0c^2$, which gives $M_0 = \boxed{2m_0/[1-(v^2/c^2)]^{1/2}.}$
Because energy is conserved, there was $\boxed{\text{no loss.}}$
The final particle is at rest, so the kinetic energy loss is the initial kinetic energy of the two colliding particles:
$\text{KE}_{\text{loss}} = 2(m-m_0)c^2 = \boxed{2m_0c^2(\{1/[1-(v^2/c^2)]^{1/2}\}-1).}$

34. The total energy of the proton is
$E = mc^2 = \text{KE} + m_0c^2 = \frac{1}{2}mc^2 + m_0c^2$, which gives $m = 2m_0 = 2(1.67\times10^{-27}\text{ kg}) = \boxed{3.34\times10^{-27}\text{ kg.}}$
We find the speed from
$m = m_0/[1-(v^2/c^2)]^{1/2}$;
$2m_0 = m_0/[1-(v^2/c^2)]^{1/2}$, which gives $v = \boxed{0.866c.}$

35. The total energy of the electron is
$E = mc^2 = \text{KE} + m_0c^2 = m_0c^2 + m_0c^2 = 2m_0c^2$, which gives $m = 2m_0$.
We find the speed from
$m = m_0/[1-(v^2/c^2)]^{1/2}$;
$2m_0 = m_0/[1-(v^2/c^2)]^{1/2}$, which gives $[1-(v^2/c^2)]^{1/2} = \frac{1}{2}$, so $v = \boxed{0.866c.}$
The momentum of the electron is
$p = mv = m_0v/[1-(v/c)^2]^{1/2}$
$= (9.11\times10^{-31}\text{ kg})(0.866)(3.00\times10^8\text{ m/s})/(\frac{1}{2}) = \boxed{4.73\times10^{-22}\text{ kg}\cdot\text{m/s.}}$

36. (*a*) The kinetic energy is
$\text{KE} = (m-m_0)c^2 = m_0c^2(\{1/[1-(v/c)^2]^{1/2}\}-1)$
$= (37{,}000\text{ kg})(3.00\times10^8\text{ m/s})^2(\{1/[1-(0.21)^2]^{1/2}\}-1) = \boxed{7.6\times10^{19}\text{ J.}}$
(*b*) When we use the classical expression, we get
$\text{KE}_c = \frac{1}{2}mv^2 = \frac{1}{2}(37{,}000\text{ kg})[(0.21)(3.00\times10^8\text{ m/s})]^2 = 7.34\times10^{19}\text{ J}$.
The error is
$(7.34-7.6)/(7.6) = -0.04 = \boxed{-4\%.}$

37. The speed of the proton is

$v = (9.8 \times 10^7 \text{ m/s})/(3.00 \times 10^8 \text{ m/s}) = 0.327c.$

The kinetic energy is

$\text{KE} = (m - m_0)c^2 = m_0c^2(\{1/[1 - (v/c)^2]^{1/2}\} - 1)$

$= (939 \text{ MeV})(\{1/[1 - (0.327)^2]^{1/2}\} - 1) =$ $\boxed{55 \text{ MeV } (8.7 \times 10^{-12} \text{ J}).}$

The momentum of the proton is

$p = mv = m_0v/[1 - (v/c)^2]^{1/2}$

$= (1.67 \times 10^{-27} \text{ kg})(9.8 \times 10^7 \text{ m/s})\{1/[1 - (0.327)^2]^{1/2}\} =$ $\boxed{1.7 \times 10^{-19} \text{ kg}\cdot\text{m/s.}}$

From the classical expressions, we get

$\text{KE}_c = \frac{1}{2}mv^2 = \frac{1}{2}(1.67 \times 10^{-27} \text{ kg})(9.8 \times 10^7 \text{ m/s})^2 = 8.02 \times 10^{-12}$ J, with an error of $(8.0 - 8.7)/(8.7) = -0.08 =$ $\boxed{-8\%.}$

$p = mv = (1.67 \times 10^{-27} \text{ kg})(9.8 \times 10^7 \text{ m/s}) = 1.6 \times 10^{-19} \text{ kg}\cdot\text{m/s}$, with an error of $(1.6 - 1.7)/(1.7) = -0.06 =$ $\boxed{-6\%.}$

38. If we ignore the recoil of the neptunium nucleus, the increase in kinetic energy is the kinetic energy of the alpha particle;

$\text{KE}_\alpha = [m_{\text{Am}} - (m_{\text{Np}} + m_\alpha)]c^2;$

$5.5 \text{ MeV} = [241.05682 \text{ u} - (m_{\text{Np}} + 4.00260 \text{ u})]c^2(931.5 \text{ MeV}/\text{u}c^2)$, which gives $m_{\text{Np}} =$ $\boxed{237.04832 \text{ u.}}$

39. The increase in kinetic energy comes from the decrease in potential energy:

$\text{KE} = (m - m_0)c^2 = m_0c^2(\{1/[1 - (v/c)^2]^{1/2}\} - 1);$

$7.60 \times 10^{-14} \text{ J} = (9.11 \times 10^{-31} \text{ kg})(3.00 \times 10^8 \text{ m/s})^2(\{1/[1 - (v/c)^2]^{1/2}\} - 1)$, which gives $v =$ $\boxed{0.855c.}$

40. (a)

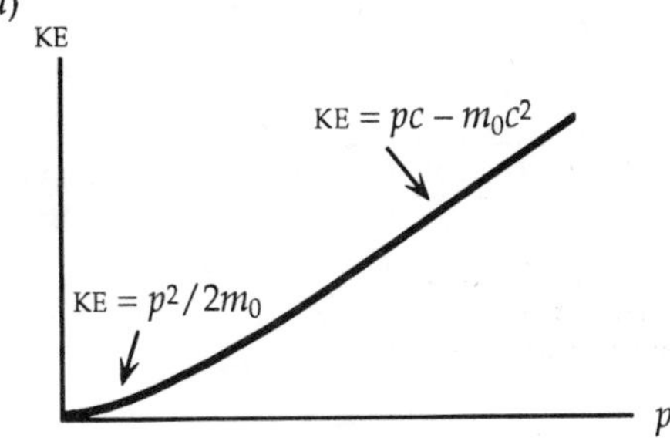

(b)

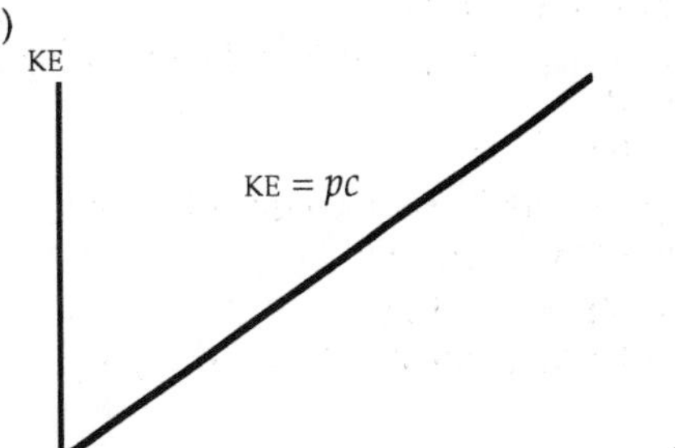

41. The total energy of the proton is

$E = mc^2 = \text{KE} + m_0c^2 = 900 \text{ GeV} + 0.938 \text{ GeV} = 901 \text{ GeV}$, so the mass is $901 \text{ GeV}/c^2$.

We find the speed from

$m = m_0/[1 - (v^2/c^2)]^{1/2};$

$901 \text{ GeV}/c^2 = (0.938 \text{ GeV}/c^2)/[1 - (v^2/c^2)]^{1/2}$, which gives $[1 - (v^2/c^2)]^{1/2} = 1.04 \times 10^{-3}$, so $v = 1.00c$.

The magnetic force provides the radial acceleration:

$qvB = mv^2/r$, or

$B = mv/qr = m_0v/qr[1 - (v^2/c^2)]^{1/2}$

$= (1.67 \times 10^{-27} \text{ kg})(3.00 \times 10^8 \text{ m/s})/(1.6 \times 10^{-19} \text{ C})(1.0 \times 10^3 \text{ m})(1.04 \times 10^{-3}) =$ $\boxed{3.0 \text{ T.}}$

Note that the mass is constant during the revolution.

42. Because the total energy of the muons becomes electromagnetic energy, we have

$E = m_1c^2 + m_2c^2 = m_0/[1 - (v_1^2/c^2)]^{1/2} + m_0/[1 - (v_2^2/c^2)]^{1/2}$

$= (105.7 \text{ MeV}/c^2)(c^2)/[1 - (0.33)^2]^{1/2} + (105.7 \text{ MeV}/c^2)(c^2)/[1 - (0.50)^2]^{1/2} =$ $\boxed{234 \text{ MeV.}}$

43. The magnetic force provides the radial acceleration:
$qvB = mv^2/r$, or
$m = qBr/v = E/c^2$.
With $v \approx c$, and $q = 1$ e, we get
$E\text{ (eV)} = (1)Brc^2/c = Brc$.
Note that the mass is constant during the revolution.

44. If we use the mass-speed relation,
$m = m_0/[1 - (v^2/c^2)]^{1/2}$,
and solve for the speed, we get
$v = c[1 - (m_0^2/m^2)]^{1/2}$.
Thus for the momentum we get
$p = mv = mc[1 - (m_0^2/m^2)]^{1/2} = [(mc)^2 - (m_0c)^2]^{1/2} = [(mc^2)^2 - (m_0c^2)^2]^{1/2}/c$.
When we use $\text{KE} = mc^2 - m_0c^2$, we get
$p = [(\text{KE} + m_0c^2)^2 - (m_0c^2)^2]^{1/2}/c = [(\text{KE})^2 + 2(\text{KE})m_0c^2]^{1/2}/c$.

45. We find the speed in the frame of the Earth from
$u = (v + u')/(1 + vu'/c^2) = (0.50c + 0.50c)/[1 + (0.50)(0.50)] =$ $\boxed{0.80c.}$

46. (*a*) In the reference frame of the second spaceship, the Earth is moving at $0.50c$, and the first spaceship is moving at $0.50c$ relative to the Earth. Thus the speed of the first spaceship relative to the second is
$u = (v + u')/(1 + vu'/c^2) = (0.50c + 0.50c)/[1 + (0.50)(0.50)] =$ $\boxed{0.80c.}$
(*b*) In the reference frame of the first spaceship, the Earth is moving at $-0.50c$, and the second spaceship is moving at $-0.50c$ relative to the Earth. Thus the speed of the second spaceship relative to the first is
$u = (v + u')/(1 + vu'/c^2) = [-0.50c + (-0.50c)]/[1 + (-0.50)(-0.50)] =$ $\boxed{-0.80c,}$ as expected.

47. We take the positive direction in the direction of the *Enterprise*. In the reference frame of the alien vessel, the Earth is moving at $-0.60c$, and the *Enterprise* is moving at $+0.90c$ relative to the Earth. Thus the speed of the *Enterprise* relative to the alien vessel is
$u = (v + u')/(1 + vu'/c^2) = (-0.60c + 0.90c)/[1 + (-0.60)(+0.90)] =$ $\boxed{0.65c.}$
Note that the relative speed of the two vessels as seen on Earth is $0.90c - 0.60c = 0.30c$.

48. We take the positive direction in the direction of the first spaceship.
(*a*) In the reference frame of the Earth, the first spaceship is moving at $+0.65c$, and the second spaceship is moving at $+0.91c$ relative to the first. Thus the speed of the second spaceship relative to the Earth is
$u = (v + u')/(1 + vu'/c^2) = (+0.65c + 0.91c)/[1 + (0.65)(0.91)] =$ $\boxed{0.98c.}$
(*b*) In the reference frame of the Earth, the first spaceship is moving at $+0.65c$, and the second spaceship is moving at $-0.91c$ relative to the first. Thus the speed of the second spaceship relative to the Earth is
$u = (v + u')/(1 + vu'/c^2) = [+0.65c + (-0.91c)]/[1 + (0.65)(-0.91)] =$ $\boxed{-0.64c.}$

49. The electrostatic force provides the radial acceleration:
$ke^2/r^2 = mv^2/r$.
Thus we find the speed from
$v^2 = (9.0 \times 10^9\ \text{N}\cdot\text{m}^2/\text{C}^2)(1.6 \times 10^{-19}\ \text{C})^2/(9.11 \times 10^{-31}\ \text{kg})(0.5 \times 10^{-10}\ \text{m})$,
which gives $v = 2 \times 10^6$ m/s.
Because this is less than $0.1c$, the electron is $\boxed{\text{not relativistic.}}$

50. Because the North Pole is has no tangential velocity, the clock there will measure a year (3.16×10^7 s). The clock at the equator has the tangential velocity of the equator:

$v = r\omega = (6.38 \times 10^6 \text{ m})(2\pi \text{ rad})/(24 \text{ h})(3600 \text{ s/h}) = 464 \text{ m/s}.$

The clock at the equator will run slow:

$t_{\text{equator}} = t_{\text{North}}[1 - (v^2/c^2)]^{1/2} \approx t_{\text{North}}[1 - \frac{1}{2}(v/c)^2].$

Thus the difference in times is

$t_{\text{North}} - t_{\text{equator}} = t_{\text{North}}\frac{1}{2}(v/c)^2 = (3.16 \times 10^7 \text{ s})\frac{1}{2}[(464 \text{ m/s})/(3.00 \times 10^8 \text{ m/s})]^2 =$ $\boxed{3.8 \times 10^{-5} \text{ s.}}$

51. (*a*) To travelers on the spacecraft, the distance to the star is contracted:

$L = L_0[1 - (v/c)^2]^{1/2} = (4.3 \text{ ly})[1 - (v/c)^2]^{1/2}.$

Because the star is moving toward the spacecraft, to cover this distance in 4.0 yr, the speed of the star must be

$v = L/t = (4.3 \text{ ly}/4.0 \text{ yr})[1 - (v/c)^2]^{1/2} = (1.075c)[1 - (v/c)^2]^{1/2}$, which gives $v = 0.73c$.

Thus relative to the Earth-star system, the speed of the spacecraft is $\boxed{0.73c.}$

(*b*) According to observers on Earth, clocks on the spacecraft run slow:

$t_{\text{Earth}} = t/[1 - (v^2/c^2)]^{1/2} = (4.0 \text{ yr})/[1 - (0.73)^2]^{1/2} =$ $\boxed{5.9 \text{ yr.}}$

Note that this agrees with the time found from distance and speed:

$t_{\text{Earth}} = L_0/v = (4.3 \text{ ly})/(0.73c) = 5.9 \text{ yr}.$

52. The dependence of the mass on the speed is

$m = m_0/[1 - (v^2/c^2)]^{1/2}.$

If we consider a box with sides x_0, y_0, and z_0, dimensions perpendicular to the motion, which we take to be the x-axis, do not change, but the length in the direction of motion will contract:

$x = x_0[1 - (v/c)^2]^{1/2}.$

Thus the density is

$\rho = m/xyz = m_0/[1 - (v^2/c^2)]^{1/2}x_0[1 - (v^2/c^2)]^{1/2}y_0z_0 =$ $\boxed{\rho_0/[1 - (v^2/c^2)].}$

53. We convert the speed: $(1500 \text{ km/h})/(3.6 \text{ ks/h}) = 417 \text{ m/s}$.

The flight time as observed on Earth is

$t_{\text{Earth}} = 2\pi r/v = 2\pi(6.38 \times 10^6 \text{ m})/(417 \text{ m/s}) = 9.62 \times 10^4 \text{ s}.$

The clock on the plane will run slow:

$t_{\text{plane}} = t_{\text{Earth}}[1 - (v^2/c^2)]^{1/2} \approx t_{\text{Earth}}[1 - \frac{1}{2}(v/c)^2].$

Thus the difference in times is

$t_{\text{Earth}} - t_{\text{plane}} = t_{\text{Earth}}\frac{1}{2}(v/c)^2 = (9.62 \times 10^4 \text{ s})\frac{1}{2}[(417 \text{ m/s})/(3.00 \times 10^8 \text{ m/s})]^2 =$ $\boxed{9.3 \times 10^{-8} \text{ s.}}$

54. We find the mass change from the required energy:

$E = Pt = m_0c^2;$

$(100 \text{ W})(3.16 \times 10^7 \text{ s}) = m_0(3.00 \times 10^8 \text{ m/s})^2$, which gives $m_0 =$ $\boxed{3.5 \times 10^{-8} \text{ kg.}}$

55. The minimum energy is required to produce the pair at rest:

$E = 2m_0c^2 = 2(0.511 \text{ MeV}) =$ $\boxed{1.02 \text{ MeV } (1.64 \times 10^{13} \text{ J}).}$

56. (*a*) Because the spring is at rest on the spaceship, its period is

$T = 2\pi(m/k)^{1/2} = 2\pi[(1.68 \text{ kg})/(48.7 \text{ N/m})]^{1/2} =$ $\boxed{1.17 \text{ s.}}$

(*b*) The oscillating mass is a clock. According to observers on Earth, clocks on the spacecraft run slow:

$T_{\text{Earth}} = T/[1 - (v^2/c^2)]^{1/2} = (1.17 \text{ s})/[1 - (0.900)^2]^{1/2} =$ $\boxed{2.68 \text{ s.}}$

57. The magnetic force provides the radial acceleration:

$qvB = mv^2/r$, or

$r = mv/qB = m_0v/qB[1 - (v^2/c^2)]^{1/2}$

$= (9.11 \times 10^{-31} \text{ kg})(0.92)(3.00 \times 10^8 \text{ m/s})/(1.6 \times 10^{-19} \text{ C})(1.8 \text{ T})[1 - (0.92)^2]^{1/2}$

$= 2.2 \times 10^{-3} \text{ m} =$ **2.2 mm.**

58. The kinetic energy comes from the decrease in rest mass:

$\text{KE} = [m_n - (m_p + m_e + m_\nu)]c^2$

$= [1.008665 \text{ u} - (1.00728 \text{ u} + 0.000549 \text{ u} + 0)]c^2(931.5 \text{ MeV}/uc^2) =$ **0.78 MeV.**

59. (*a*) We find the rate of mass loss from

$\Delta m/\Delta t = (\Delta E/\Delta t)/c^2$

$= (4 \times 10^{26} \text{ W})/(3 \times 10^8 \text{ m/s})^2 =$ **4.4×10^9 kg/s.**

(*b*) We find the time from

$\Delta t = \Delta m/\text{rate} = (5.98 \times 10^{24} \text{ kg})/(4.4 \times 10^9 \text{ kg/s})(3.16 \times 10^7 \text{ s/yr}) =$ **4.3×10^7 yr.**

(*c*) We find the time for the Sun to lose all of its mass at this rate from

$\Delta t = \Delta m/\text{rate} = (2.0 \times 10^{30} \text{ kg})/(4.4 \times 10^9 \text{ kg/s})(3.16 \times 10^7 \text{ s/yr}) =$ **1.4×10^{13} yr.**

60. The speed of the particle is

$v = (2.24 \times 10^7 \text{ m/s})/(3.00 \times 10^8 \text{ m/s}) = 0.747c.$

We use the momentum to find the rest mass:

$p = mv = m_0v/[1 - (v/c)^2]^{1/2};$

$3.0 \times 10^{-22} \text{ kg}\cdot\text{m/s} = m_0(0.747)(3.00 \times 10^8 \text{ m/s})/[1 - (0.747)^2]^{1/2},$

which gives $m_0 = 9.11 \times 10^{-31}$ kg.

Because the particle has a negative charge, it is **an electron.**

61. The binding energy is the energy required to provide the increase in rest mass:

$\text{KE} = [(2m_p + 2m_e) - m_{\text{He}}]c^2$

$= [2(1.00783 \text{ u}) + 2(1.00867 \text{ u}) - 4.00260 \text{ u}]c^2(931.5 \text{ MeV}/uc^2) =$ **28.3 MeV.**

62. We convert the speed: (110 km/h)/(3.6 ks/h) = 30.6 m/s.

Because this is much smaller than c, the mass of the car is

$m = m_0/[1 - (v^2/c^2)]^{1/2} \approx m_0[1 + \tfrac{1}{2}(v/c)^2].$

The fractional change in mass is

$(m - m_0)/m_0 = [1 + \tfrac{1}{2}(v/c)^2] - 1 = \tfrac{1}{2}(v/c)^2$

$= \tfrac{1}{2}[(30.6 \text{ m/s})/(3.00 \times 10^8 \text{ m/s})]^2 = 5.19 \times 10^{-15} =$ **5.19×10^{-13} %.**

63. (*a*) The magnitudes of the momenta are equal:

$p = mv = m_0v/[1 - (v/c)^2]^{1/2}$

$= (1.67 \times 10^{-27} \text{ kg})(0.935)(3.00 \times 10^8 \text{ m/s})/[1 - (0.935)^2]^{1/2} =$ **1.32×10^{-18} kg·m/s.**

(*b*) Because the protons are moving in opposite directions, the sum of the momenta is **0.**

(*c*) In the reference frame of one proton, the laboratory is moving at 0.935c. The other proton is moving at + 0.935c relative to the laboratory. Thus the speed of the other proton relative to the first is

$u = (v + u')/(1 + vu'/c^2) = [+ 0.935c + (+ 0.935c)]/[1 + (+ 0.935)(+ 0.935)] = 0.998c.$

The magnitude of the momentum of the other proton is

$p = mv = m_0v/[1 - (v/c)^2]^{1/2}$

$= (1.67 \times 10^{-27} \text{ kg})(0.998)(3.00 \times 10^8 \text{ m/s})/[1 - (0.998)^2]^{1/2} =$ **7.45×10^{-18} kg·m/s.**

64. The neutrino has no rest mass, so we have

$E_\nu = (p_\nu^2c^2 + m_\nu^2c^4)^{1/2} = p_\nu c.$

Because the pi meson decays at rest, momentum conservation tells us that the muon and neutrino have equal and opposite momenta:

$p_\mu = p_\nu = p.$

For energy conservation, we have

$E_\pi = E_\mu + E_\nu;$

$m_\pi c^2 = (p_\mu^2c^2 + m_\mu^2c^4)^{1/2} + p_\nu c = (p^2c^2 + m_\mu^2c^4)^{1/2} + pc.$

If we rearrange and square, we get

$(m_\pi c^2 - pc)^2 = m_\pi^2c^4 - 2m_\pi c^2pc + p^2c^2 = p^2c^2 + m_\mu^2c^4,$ or $pc = (m_\pi^2c^2 - m_\mu^2c^2)/2m_\pi.$

The kinetic energy of the muon is

$\text{KE}_\mu = E_\mu - m_\mu c^2 = (m_\pi c^2 - pc) - m_\mu c^2 = m_\pi c^2 - m_\mu c^2 - (m_\pi^2c^2 - m_\mu^2c^2)/2m_\pi$

$= (2m_\pi^2 - 2m_\mu m_\pi - m_\pi^2 + m_\mu^2)c^2/2m_\pi = (m_\pi^2 - 2m_\mu m_\pi + m_\mu^2)c^2/2m_\pi = (m_\pi - m_\mu)^2c^2/2m_\pi.$

65. To an observer in the barn reference frame, if the boy runs fast enough, the measured contracted length will be less than 12.0 m, so the observer can say that the two ends of the pole were simultaneously inside the barn. We find the necessary speed from

$L = L_0[1 - (v/c)^2]^{1/2};$

$12.0\text{ m} = (15.0\text{ m})[1 - (v/c)^2]^{1/2}$, which gives $v = 0.60c$.

To the boy, the barn is moving and thus the length of the barn as he would measure it is less than the length of the pole:

$L = L_0[1 - (v/c)^2]^{1/2} = (12.0\text{ m})[1 - (0.60)^2]^{1/2} = 7.68\text{ m}.$

However, simultaneity is relative. Thus when the two ends are simultaneously inside the barn to the barn observer, those two events are not simultaneous to the boy. Thus he would claim that the observer in the barn determined that the ends of the pole were inside the barn at different times, which is also what the boy would say. It is **not possible** in the boy's frame to have both ends of the pole inside the barn simultaneously.

66. The relation between energy and momentum is

$E = (m_0^2c^4 + p^2c^2)^{1/2} = c(m_0^2c^2 + p^2)^{1/2}.$

For the momentum, we have

$p = mv = Ev/c^2,$ or

$v = pc^2/E = pc/(m_0^2c^2 + p^2)^{1/2}.$

CHAPTER 27

Note: At the atomic scale, it is most convenient to have energies in electron-volts and wavelengths in nanometers. A useful expression for the energy of a photon in terms of its wavelength is

$E = hf = hc/\lambda = (6.63 \times 10^{-34}\ \text{J}\cdot\text{s})(3.00 \times 10^{8}\ \text{m/s})(10^{-9}\ \text{nm/m})/(1.60 \times 10^{-19}\ \text{J/eV})\lambda;$

$E = (1.24 \times 10^{3}\ \text{eV}\cdot\text{nm})/\lambda.$

1. The velocity of the electron in the crossed fields is given by

$v = E/B.$

The radius of the path in the magnetic field is

$r = mv/eB = mE/eB^2,$ or

$e/m = E/rB^2 = (300\ \text{V/m})/(7.0 \times 10^{-3}\ \text{m})(0.86\ \text{T})^2 = \boxed{5.8 \times 10^{4}\ \text{C/kg.}}$

2. The velocity of the electron in the crossed fields is given by

$v = E/B = (1.38 \times 10^{4}\ \text{V/m})/(2.90 \times 10^{-3}\ \text{T}) = \boxed{4.76 \times 10^{6}\ \text{m/s.}}$

For the radius of the path in the magnetic field, we have

$r = mv/qB$

$= (9.11 \times 10^{-31}\ \text{kg})(4.76 \times 10^{6}\ \text{m/s})/(1.60 \times 10^{-19}\ \text{C})(2.90 \times 10^{-3}\ \text{T}) = 9.34 \times 10^{-3}\ \text{m} = \boxed{9.34\ \text{mm.}}$

3. The force from the electric field must balance the weight:

$qE = neV/d = mg;$

$n(1.60 \times 10^{-19}\ \text{C})(340\ \text{V})/(1.0 \times 10^{-2}\ \text{m}) = (2.8 \times 10^{-15}\ \text{kg})(9.80\ \text{m/s}^2)$, which gives $n = \boxed{5.}$

4. When there is no charge on the drop, the forces acting on the drop are the effective downward force of gravity and the upward drag force;

$m_{\text{eff}}g - F_{\text{drag}} = 0,$ or

$(\rho - \rho_A)(\tfrac{4}{3}\pi r^3)g = 6\pi\eta r v_T$, which gives $r = [9\eta v_T/2(\rho - \rho_A)g]^{1/2}$.

With a positive charge, the electric force is up. If the drop is stationary, there is no drag force, so we have

$m_{\text{eff}}g - qE = 0,$ or

$qV/d = (\rho - \rho_A)(\tfrac{4}{3}\pi r^3)g.$

If we use the result for r, we get

$q = \tfrac{4}{3}\pi(\rho - \rho_A)g(d/V)[9\eta v_T/2(\rho - \rho_A)g]^{3/2}$

$= \tfrac{4}{3}\pi(\rho - \rho_A)g(d/V)(27/2)[(\rho - \rho_A)^2 g^2 \eta^3 v_T^{\,3}/2(\rho - \rho_A)^3 g]^{1/2} = (18\pi d/V)[\eta^3 v_T^{\,3}/2(\rho - \rho_A)g]^{1/2}.$

5. We find the temperature for a peak wavelength of 410 nm:

$T = (2.90 \times 10^{-3}\ \text{m}\cdot\text{K})/\lambda_P = (2.90 \times 10^{-3}\ \text{m}\cdot\text{K})/(410 \times 10^{-9}\ \text{m}) = \boxed{7.07 \times 10^{3}\ \text{K.}}$

6. (*a*) The temperature for a peak wavelength of 15.0 nm is

$T = (2.90 \times 10^{-3}\ \text{m}\cdot\text{K})/\lambda_P = (2.90 \times 10^{-3}\ \text{m}\cdot\text{K})/(15.0 \times 10^{-9}\ \text{m}) = \boxed{1.93 \times 10^{5}\ \text{K.}}$

(*b*) We find the peak wavelength from

$\lambda_P = (2.90 \times 10^{-3}\ \text{m}\cdot\text{K})/T = (2.90 \times 10^{-3}\ \text{m}\cdot\text{K})/(2000\ \text{K}) = 1.45 \times 10^{-6}\ \text{m} = \boxed{1.45\ \mu\text{m.}}$

Note that this is not in the visible range.

7. (*a*) We find the peak wavelength from
$\lambda_P = (2.90 \times 10^{-3}\ \text{m}\cdot\text{K})/T = (2.90 \times 10^{-3}\ \text{m}\cdot\text{K})/(273\ \text{K}) = 1.06 \times 10^{-5}\ \text{m} =$ $\boxed{10.6\ \mu\text{m}.}$
This wavelength is in the $\boxed{\text{near infrared.}}$
(*b*) We find the peak wavelength from
$\lambda_P = (2.90 \times 10^{-3}\ \text{m}\cdot\text{K})/T = (2.90 \times 10^{-3}\ \text{m}\cdot\text{K})/(3000\ \text{K}) = 9.67 \times 10^{-7}\ \text{m} =$ $\boxed{967\ \text{nm}.}$
This wavelength is in the $\boxed{\text{infrared.}}$
(*c*) We find the peak wavelength from
$\lambda_P = (2.90 \times 10^{-3}\ \text{m}\cdot\text{K})/T = (2.90 \times 10^{-3}\ \text{m}\cdot\text{K})/(4\ \text{K}) = 7.25 \times 10^{-4}\ \text{m} =$ $\boxed{0.73\ \text{mm}.}$
This wavelength is in the $\boxed{\text{far infrared.}}$

8. Because the energy is quantized, $E = nhf$, the difference in energy between adjacent levels is
$\Delta E = hf = (6.63 \times 10^{-34}\ \text{J}\cdot\text{s})(8.1 \times 10^{13}\ \text{Hz}) =$ $\boxed{5.4 \times 10^{-20}\ \text{J} = 0.34\ \text{eV}.}$

9. (*a*) Because the energy is quantized, $E = nhf$, the difference in energy between adjacent levels is
$\Delta E = hf = (6.63 \times 10^{-34}\ \text{J}\cdot\text{s})(0.90\text{Hz}) =$ $\boxed{6.0 \times 10^{-34}\ \text{J}.}$
(*b*) The total energy will be the maximum potential energy, so we have
$E = mgh = nhf;$
$(20\ \text{kg})(9.80\ \text{m/s}^2)(0.45\ \text{m}) = n(6.0 \times 10^{-34}\ \text{J})$, which gives $n =$ $\boxed{1.5 \times 10^{35}.}$
(*c*) The fractional change in energy is
$\Delta E/E = hf/nhf = 1/n = 1/(1.5 \times 10^{35}) =$ $\boxed{6.8 \times 10^{-36}, \text{ not measurable.}}$

10. (*a*) The potential energy on the first step is
$\text{PE}_1 = mgh = (58.0\ \text{kg})(9.80\ \text{m/s}^2)(0.200\ \text{m}) =$ $\boxed{114\ \text{J}.}$
(*b*) The potential energy on the second step is
$\text{PE}_2 = mg2h = 2\text{PE}_1 = 2(114\ \text{J}) =$ $\boxed{228\ \text{J}.}$
(*c*) The potential energy on the third step is
$\text{PE}_3 = mg3h = 3\text{PE}_1 = 3(114\ \text{J}) =$ $\boxed{342\ \text{J}.}$
(*d*) The potential energy on the nth step is
$\text{PE}_n = mgnh = n\text{PE}_1 = n(114\ \text{J}) =$ $\boxed{114n\ \text{J}.}$
(*e*) The change in energy is
$\Delta E = \text{PE}_2 - \text{PE}_6 = (2 - 6)(114\ \text{J}) =$ $\boxed{-456\ \text{J}.}$

11. We use a body temperature of 98°F = 37°C = 310 K. We find the peak wavelength from
$\lambda_P = (2.90 \times 10^{-3}\ \text{m}\cdot\text{K})/T = (2.90 \times 10^{-3}\ \text{m}\cdot\text{K})/(310\ \text{K}) = 9.4 \times 10^{-6}\ \text{m} =$ $\boxed{9.4\ \mu\text{m}.}$

12. The energy of the photons with wavelengths at the ends of the visible spectrum are
$E_1 = hf_1 = hc/\lambda_1 = (1.24 \times 10^3\ \text{eV}\cdot\text{nm})/(400\ \text{nm}) = 3.10\ \text{eV};$
$E_2 = hf_2 = hc/\lambda_2 = (1.24 \times 10^3\ \text{eV}\cdot\text{nm})/(700\ \text{nm}) = 1.77\ \text{eV}.$
Thus the range of energies is $\boxed{1.77\ \text{eV} < E < 3.10\ \text{eV}.}$

13. The energy of the photon is
$E = hf = (6.63 \times 10^{-34}\ \text{J}\cdot\text{s})(102.1 \times 10^6\ \text{Hz}) = 6.77 \times 10^{-26}\ \text{J} =$ $\boxed{4.23 \times 10^{-7}\ \text{eV}.}$

14. We find the wavelength from
$\lambda = c/f = hc/E = (1.24 \times 10^3\ \text{eV}\cdot\text{nm})/(200 \times 10^3\ \text{eV}) =$ $\boxed{6.22 \times 10^{-3}\ \text{nm}.}$
Significant diffraction occurs when the opening is on the order of the wavelength. Thus there would be $\boxed{\text{insignificant diffraction}}$ through the doorway.

15. We find the minimum frequency from

$E_{min} = hf_{min}$;

$(0.1 \text{ eV})(1.60 \times 10^{-19} \text{ J/eV}) = (6.63 \times 10^{-34} \text{ J}\cdot\text{s})f_{min}$, which gives $f_{min} = \boxed{2.4 \times 10^{13} \text{ Hz.}}$

The maximum wavelength is

$\lambda_{max} = c/f_{min} = (3.00 \times 10^{8} \text{ m/s})/(2.4 \times 10^{13} \text{ Hz}) = \boxed{1.2 \times 10^{-5} \text{ m.}}$

16. At the threshold frequency, the kinetic energy of the photoelectrons is zero, so we have

$\text{KE} = hf - W = 0$;

$hf_{min} = W_0$;

$(6.63 \times 10^{-34} \text{ J}\cdot\text{s})f_{min} = 4.3 \times 10^{-19} \text{ J}$, which gives $f_{min} = \boxed{6.5 \times 10^{14} \text{ Hz.}}$

17. At the threshold wavelength, the kinetic energy of the photoelectrons is zero, so we have

$\text{KE} = hf - W_0 = 0$;

$hc/\lambda_{max} = W_0$, or

$\lambda_{max} = (1.24 \times 10^{3} \text{ eV}\cdot\text{nm})/(3.10 \text{ eV}) = \boxed{400 \text{ nm.}}$

18. The energy of the photon is

$E = hf = (1.24 \times 10^{3} \text{ eV}\cdot\text{nm})/\lambda = (1.24 \times 10^{3} \text{ eV}\cdot\text{nm})/(390 \text{ nm}) = 3.18 \text{ eV}.$

The maximum kinetic energy of the photoelectrons is

$\text{KE}_{max} = hf - W_0 = 3.18 \text{ eV} - 2.48 \text{ eV} = \boxed{0.70 \text{ eV.}}$

We find the speed from

$\text{KE}_{max} = \frac{1}{2}mv^2$;

$(0.70 \text{ eV})(1.60 \times 10^{-19} \text{ J/eV}) = \frac{1}{2}(9.11 \times 10^{-31} \text{ kg})v^2$, which gives $v = \boxed{5.0 \times 10^{5} \text{ m/s.}}$

19. The photon of visible light with the maximum energy has the least wavelength:

$hf_{max} = (1.24 \times 10^{3} \text{ eV}\cdot\text{nm})/\lambda_{min} = (1.24 \times 10^{3} \text{ eV}\cdot\text{nm})/(400 \text{ nm}) = 3.10 \text{ eV}.$

Electrons will not be emitted if this energy is less than the work function.

The metals with work functions greater than 3.10 eV are $\boxed{\text{copper and iron.}}$

20. (*a*) At the threshold wavelength, the kinetic energy of the photoelectrons is zero, so we have

$\text{KE} = hf - W_0 = 0$;

$W_0 = hc/\lambda_{max} = (1.24 \times 10^{3} \text{ eV}\cdot\text{nm})/(570 \text{ nm}) = \boxed{2.18 \text{ eV.}}$

(*b*) The stopping voltage is the voltage that gives a potential energy change equal to the maximum kinetic energy:

$\text{KE}_{max} = eV_0 = hf - W_0$;

$(1 \text{ e})V_0 = [(1.24 \times 10^{3} \text{ eV}\cdot\text{nm})/(400 \text{ nm})] - 2.18 \text{ eV} = 3.10 \text{ eV} - 2.18 \text{ eV} = 0.92 \text{ eV},$

so the stopping voltage is $\boxed{0.92 \text{ V.}}$

21. The photon of visible light with the maximum energy has the minimum wavelength:

$hf_{max} = (1.24 \times 10^{3} \text{ eV}\cdot\text{nm})/\lambda_{min} = (1.24 \times 10^{3} \text{ eV}\cdot\text{nm})/(400 \text{ nm}) = 3.10 \text{ eV}.$

The maximum kinetic energy of the photoelectrons is

$\text{KE}_{max} = hf - W_0 = 3.10 \text{ eV} - 2.48 \text{ eV} = \boxed{0.62 \text{ eV.}}$

22. The energy of the photon is

$hf = (1.24 \times 10^{3} \text{ eV}\cdot\text{nm})/\lambda = (1.24 \times 10^{3} \text{ eV}\cdot\text{nm})/(225 \text{ nm}) = 5.51 \text{ eV}.$

We find the work function from

$\text{KE}_{max} = hf - W_0$;

$1.40 \text{ eV} = 5.51 \text{ eV} - W_0$, which gives $W_0 = \boxed{4.11 \text{ eV.}}$

23. The threshold wavelength determines the work function:
$W_0 = hc/\lambda_{max} = (1.24 \times 10^3 \text{ eV}\cdot\text{nm})/(320 \text{ nm}) = 3.88 \text{ eV}$.
(*a*) The energy of the photon is
$hf = (1.24 \times 10^3 \text{ eV}\cdot\text{nm})/\lambda = (1.24 \times 10^3 \text{ eV}\cdot\text{nm})/(250 \text{ nm}) = 4.96 \text{ eV}$.
The maximum kinetic energy of the photoelectrons is
$\text{KE}_{max} = hf - W_0 = 4.96 \text{ eV} - 3.88 \text{ eV} = \boxed{1.08 \text{ eV.}}$
(*b*) Because the wavelength is greater than the threshold wavelength, the photon energy is less than the work function, so there will be $\boxed{\text{no ejected electrons.}}$

24. The energy required for the chemical reaction is provided by the photon:
$E = hf = hc/\lambda = (1.24 \times 10^3 \text{ eV}\cdot\text{nm})/\lambda = (1.24 \times 10^3 \text{ eV}\cdot\text{nm})/(660 \text{ nm}) = \boxed{1.88 \text{ eV.}}$
Each reaction takes place in a molecule, so we have
$E = (1.88 \text{ eV/molecule})(6.02 \times 10^{23} \text{ molecules/mol})(1.60 \times 10^{-19} \text{ J/eV})/(4186 \text{ J/kcal})$
$= \boxed{43.3 \text{ kcal/mol.}}$

25. The reverse voltage is the voltage that gives a potential energy change equal to the maximum kinetic energy:
$\text{KE}_{max} = eV_0 = hf - W_0;$
$(1 \text{ e})(1.64 \text{ V}) = [(1.24 \times 10^3 \text{ eV}\cdot\text{nm})/(230 \text{ nm})] - W_0$, which gives $W_0 = \boxed{3.75 \text{ eV.}}$

26. The kinetic energy of the pair is
$\text{KE} = hf - 2m_0c^2 = 2.54 \text{ MeV} - 2(0.511 \text{ MeV}) = \boxed{1.52 \text{ MeV.}}$

27. The momentum of the photon is
$p = h/\lambda = (6.63 \times 10^{-34} \text{ J}\cdot\text{s})/(0.35 \times 10^{-9} \text{ m}) = \boxed{1.9 \times 10^{-24} \text{ kg}\cdot\text{m/s.}}$
Because the photon's speed is c, the effective mass is
$m = p/c = (1.9 \times 10^{-24} \text{ kg}\cdot\text{m/s})/(3.00 \times 10^8 \text{ m/s}) = \boxed{6.3 \times 10^{-33} \text{ kg.}}$

28. The photon with the longest wavelength has the minimum energy to create the masses:
$hf_{min} = hc/\lambda_{max} = 2m_0c^2;$
$(6.63 \times 10^{-34} \text{ J}\cdot\text{s})/\lambda_{max} = 2(1.67 \times 10^{-27} \text{ kg})(3.00 \times 10^8 \text{ m/s})$, which gives $\lambda_{max} = \boxed{6.62 \times 10^{-16} \text{ m.}}$

29. The photon with minimum energy to create the masses is
$hf_{min} = 2m_0c^2 = 2(207)(0.511 \text{ MeV}) = \boxed{212 \text{ MeV.}}$
The wavelength is
$\lambda = (1.24 \times 10^3 \text{ eV}\cdot\text{nm})/(212 \times 10^6 \text{ eV}) = 5.85 \times 10^{-6} \text{ nm} = \boxed{5.85 \times 10^{-15} \text{ m.}}$

30. The energy of the photon is
$hf = 2(\text{KE} + m_0c^2) = 2(0.345 \text{ MeV} + 0.511 \text{ MeV}) = \boxed{1.71 \text{ MeV.}}$
The wavelength is
$\lambda = (1.24 \times 10^3 \text{ eV}\cdot\text{nm})/(1.71 \times 10^6 \text{ eV}) = 7.24 \times 10^{-4} \text{ nm} = \boxed{7.24 \times 10^{-13} \text{ m.}}$

31. (*a*) $h/m_0c = (6.63 \times 10^{-34} \text{ J}\cdot\text{s})/(9.11 \times 10^{-31} \text{ kg})(3.00 \times 10^8 \text{ m/s}) = \boxed{2.43 \times 10^{-12} \text{ m.}}$
(*b*) $h/m_0c = (6.63 \times 10^{-34} \text{ J}\cdot\text{s})/(1.67 \times 10^{-27} \text{ kg})(3.00 \times 10^8 \text{ m/s}) = \boxed{1.32 \times 10^{-15} \text{ m.}}$
(*c*) For the energy of the photon we have
$E = hf = hc/\lambda = hc/(h/m_0c) = m_0c^2$.

32. We find the Compton wavelength shift for a photon scattered from an electron:

$$\lambda' - \lambda = (h/m_0c)(1 - \cos\theta).$$

(*a*) $\lambda'_a - \lambda = (2.43 \times 10^{-12}\text{ m})(1 - \cos 45°) = \boxed{7.12 \times 10^{-13}\text{ m.}}$

(*b*) $\lambda'_b - \lambda = (2.43 \times 10^{-12}\text{ m})(1 - \cos 90°) = \boxed{2.43 \times 10^{-12}\text{ m.}}$

(*c*) $\lambda'_c - \lambda = (2.43 \times 10^{-12}\text{ m})(1 - \cos 180°) = \boxed{4.86 \times 10^{-12}\text{ m.}}$

33. (*a*) The energy of a photon is

$$E = hf = hc/\lambda.$$

For the fractional loss, we have

$$(E - E')/E = [(1/\lambda) - (1/\lambda')]/(1/\lambda) = (\lambda' - \lambda)/\lambda'.$$

For 45° we get

$$(E - E'_a)/E = (\lambda'_a - \lambda)/\lambda'_a = (7.12 \times 10^{-13}\text{ m})/(0.120 \times 10^{-9}\text{ m} + 7.12 \times 10^{-13}\text{ m}) = \boxed{5.90 \times 10^{-3}.}$$

For 90° we get

$$(E - E'_b)/E = (\lambda'_b - \lambda)/\lambda'_b = (2.43 \times 10^{-12}\text{ m})/(0.120 \times 10^{-9}\text{ m} + 2.43 \times 10^{-12}\text{ m}) = \boxed{1.98 \times 10^{-2}.}$$

For 180° we get

$$(E - E'_c)/E = (\lambda'_c - \lambda)/\lambda'_c = (4.86 \times 10^{-12}\text{ m})/(0.120 \times 10^{-9}\text{ m} + 4.86 \times 10^{-12}\text{ m}) = \boxed{3.89 \times 10^{-2}.}$$

(*b*) The energy of the incident photon is

$$E = hf = hc/\lambda = (1.24 \times 10^3\text{ eV}\cdot\text{nm})/\lambda = (1.24 \times 10^3\text{ eV}\cdot\text{nm})/(0.120\text{ nm}) = 10.3 \times 10^3\text{ eV}.$$

From conservation of energy, the energy given to the scattered electron is the energy lost by the photon:

$$\text{KE} = (E - E') = [(E - E')/E]E.$$

For 45° we get

$$\text{KE}_a = [(E - E'_a)/E]E = (5.90 \times 10^{-3})(10.3 \times 10^3\text{ eV}) = \boxed{60.8\text{ eV.}}$$

For 90° we get

$$\text{KE}_b = [(E - E'_b)/E]E = (1.98 \times 10^{-2})(10.3 \times 10^3\text{ eV}) = \boxed{204\text{ eV.}}$$

For 180° we get

$$\text{KE}_c = [(E - E'_c)/E]E = (3.89 \times 10^{-2})(10.3 \times 10^3\text{ eV}) = \boxed{401\text{ eV.}}$$

34. For the conservation of momentum, we have

x: $h/\lambda + 0 = (h/\lambda')\cos\phi + p_e\cos\theta$, or

$$p_e\cos\theta = (h/\lambda) - (h/\lambda')\cos\phi;$$

y: $0 + 0 = (h/\lambda')\sin\phi - p_e\sin\theta$, or

$$p_e\sin\theta = (h/\lambda')\sin\phi.$$

If we square and add, to eliminate θ, we get

$$p_e^2 = (h/\lambda)^2 + (h/\lambda')^2 - (2h^2/\lambda\lambda')\cos\phi.$$

For energy conservation, we have

$$(hc/\lambda) + m_0c^2 = (hc/\lambda') + [p_e^2c^2 + (m_0c^2)^2]^{1/2},\text{ or}$$

$$(hc/\lambda) + m_0c^2 - (hc/\lambda') = [p_e^2c^2 + (m_0c^2)^2]^{1/2}.$$

When we substitute the result from momentum conservation and square, we get

$$(hc/\lambda)^2 + (m_0c^2)^2 + (hc/\lambda')^2 + 2(h^2c^2/\lambda\lambda') - 2(m_0c^2hc/\lambda) - 2(m_0c^2hc/\lambda') =$$
$$(hc/\lambda)^2 + (hc/\lambda')^2 - (2h^2c^2/\lambda\lambda')\cos\phi + (m_0c^2)^2,$$

which reduces to

$$2m_0c^2hc[(1/\lambda) - (1/\lambda')] = (2h^2c^2/\lambda\lambda')(1 - \cos\phi),\text{ or }\lambda' - \lambda = (h/m_0c)(1 - \cos\phi).$$

35. (*a*) For the conservation of momentum for the head-on collision, we have

$h/\lambda + 0 = -(h/\lambda') + p_e$, or $h/\lambda' = p_e - (h/\lambda)$.

	Before			After	
energy	hf	0		hf'	$p_e^2/2m$
momentum	h/λ	0		$-h/\lambda'$	p_e

For energy conservation, we have

$(hc/\lambda) + m_0c^2 = (hc/\lambda') + [(m_0c^2)^2 + p_e^2c^2]^{1/2}$
$= (hc/\lambda') + m_0c^2 + (p_e^2/2m_0)$,

where we have used the approximation $(1 + x)^{1/2} \approx 1 + x/2$, for $x \ll 1$.

When we use the result from momentum conservation, we get

$(hc/\lambda) = p_ec - (hc/\lambda) + (p_e^2/2m_0)$, or

$[p_e^2/2(9.11 \times 10^{-31}\ \text{kg})] + p_e(3.00 \times 10^8\ \text{m/s}) -$
$2[(6.63 \times 10^{-34}\ \text{J}\cdot\text{s})(3.00 \times 10^8\ \text{m/s})/(0.100 \times 10^{-9}\ \text{m})] = 0.$

When we solve this quadratic for p_e, the positive solution is

$p_e = 1.295 \times 10^{-23}\ \text{kg}\cdot\text{m/s}$.

The kinetic energy of the electron is

$\text{KE} = p_e^2/2m_0 = (1.295 \times 10^{-23}\ \text{kg}\cdot\text{m/s})^2/2(9.11 \times 10^{-31}\ \text{kg}) =$ $\boxed{9.21 \times 10^{-17}\ \text{J}}$ (576 eV).

Because this is much less than $m_0c^2 = 0.511$ MeV, we are justified in using the expansion.

(*b*) For the wavelength of the recoiling photon, we have

$h/\lambda' = p_e - (h/\lambda)$;

$(6.63 \times 10^{-34}\ \text{J}\cdot\text{s})/\lambda' = 1.295 \times 10^{-23}\ \text{kg}\cdot\text{m/s} - [(6.63 \times 10^{-34}\ \text{J}\cdot\text{s})/(0.100 \times 10^{-9}\ \text{m})]$,

which gives $\lambda' = 1.05 \times 10^{-10}\ \text{m} =$ $\boxed{0.105\ \text{nm.}}$

36. We find the wavelength from

$\lambda = h/p = h/mv = (6.63 \times 10^{-34}\ \text{J}\cdot\text{s})/(0.21\ \text{kg})(0.10\ \text{m/s}) =$ $\boxed{3.2 \times 10^{-32}\ \text{m.}}$

37. We find the wavelength from

$\lambda = h/p = h/mv = (6.63 \times 10^{-34}\ \text{J}\cdot\text{s})/(1.67 \times 10^{-27}\ \text{kg})(5.5 \times 10^4\ \text{m/s}) =$ $\boxed{7.2 \times 10^{-12}\ \text{m.}}$

38. We find the speed from

$\lambda = h/p = h/mv$;

$0.23 \times 10^{-9}\ \text{m} = (6.63 \times 10^{-34}\ \text{J}\cdot\text{s})/(9.11 \times 10^{-31}\ \text{kg})v$, which gives $v = 3.16 \times 10^6\ \text{m/s}$.

Because this is much less than c, we can use the classical expression for the kinetic energy. The kinetic energy is equal to the potential energy change:

$eV = \text{KE} = \frac{1}{2}mv^2 = \frac{1}{2}(9.11 \times 10^{-31}\ \text{kg})(3.16 \times 10^6\ \text{m/s})^2 = 4.56 \times 10^{-18}\ \text{J} = 28.5\ \text{eV}.$

Thus the required potential difference is $\boxed{29\ \text{V.}}$

39. The kinetic energy is equal to the potential energy change:

$\text{KE} = eV = (1\ \text{e})(20.0 \times 10^3\ \text{V}) = 20.0 \times 10^3\ \text{eV} = 0.0200\ \text{MeV}.$

Because this is 4% of m_0c^2, the electron is relativistic. We find the momentum from

$E^2 = [(\text{KE}) + m_0c^2]^2 = p^2c^2 + m_e^2c^4$, or

$p^2c^2 = (\text{KE})^2 + 2(\text{KE})m_0c^2$;

$p^2c^2 = (0.0200\ \text{MeV})^2 + 2(0.0200\ \text{MeV})(0.511\ \text{MeV})$, which gives $pc = 0.144$ MeV, or

$p = (0.144\ \text{MeV})(1.60 \times 10^{-13}\ \text{J/MeV})/(3.00 \times 10^8\ \text{m/s}) = 7.68 \times 10^{-23}\ \text{kg}\cdot\text{m/s}.$

The wavelength is

$\lambda = h/p = (6.63 \times 10^{-34}\ \text{J}\cdot\text{s})/(7.68 \times 10^{-23}\ \text{kg}\cdot\text{m/s}) =$ $\boxed{8.6 \times 10^{-12}\ \text{m.}}$

Because $\lambda \ll 5$ cm, $\boxed{\text{diffraction effects are negligible.}}$

40. Because all the energies are much less than m_0c^2, we can use $\text{KE} = p^2/2m_0$, so
    $\lambda = h/p = h/[2m_0(\text{KE})]^{1/2} = hc/[2m_0c^2(\text{KE})]^{1/2}$.
    (*a*) $\lambda = hc/[2m_0c^2(\text{KE})]^{1/2} = (1.24 \times 10^3\ \text{eV} \cdot \text{nm})/[2(0.511 \times 10^6\ \text{eV})(10\ \text{eV})]^{1/2} =$ $\boxed{0.39\ \text{nm.}}$
    (*b*) $\lambda = hc/[2m_0c^2(\text{KE})]^{1/2} = (1.24 \times 10^3\ \text{eV} \cdot \text{nm})/[2(0.511 \times 10^6\ \text{eV})(100\ \text{eV})]^{1/2} =$ $\boxed{0.12\ \text{nm.}}$
    (*c*) $\lambda = hc/[2m_0c^2(\text{KE})]^{1/2} = (1.24 \times 10^3\ \text{eV} \cdot \text{nm})/[2(0.511 \times 10^6\ \text{eV})(1.0 \times 10^3\ \text{eV})]^{1/2} =$ $\boxed{0.039\ \text{nm.}}$

41. With $\text{KE} = p^2/2m_0$, we have
    $\lambda = h/p = h/[2m_0(\text{KE})]^{1/2}$.
    If we form the ratio for the two particles with equal kinetic energies, we get
    $\lambda_p/\lambda_e = (m_{0e}/m_{0p})^{1/2}$.
    Because $m_{0p} > m_{0e}$, $\lambda_p < \lambda_e$.

42. With $\text{KE} = p^2/2m_0$, we have
    $\lambda = h/p = h/[2m_0(\text{KE})]^{1/2}$.
    If we form the ratio for the two particles with equal wavelengths, we get
    $1 = [m_{0e}(\text{KE}_e)/m_{0p}(\text{KE}_p)]^{1/2}$, or $\text{KE}_e/\text{KE}_p = m_{0p}/m_{0e} = (1.67 \times 10^{-27}\ \text{kg})/(9.11 \times 10^{-31}\ \text{kg}) =$ $\boxed{1.84 \times 10^3.}$

43. We find the speed from
    $\frac{1}{2}mv^2 = \frac{3}{2}kT$;
    $\frac{1}{2}(32\ \text{u})(1.66 \times 10^{-27}\ \text{kg/u})v^2 = \frac{3}{2}(1.38 \times 10^{-23}\ \text{J/K})(300\ \text{K})$, which gives $v = 484\ \text{m/s}$.
    The wavelength is
    $\lambda = h/p = h/mv = (6.63 \times 10^{-34}\ \text{J} \cdot \text{s})/(32\ \text{u})(1.66 \times 10^{-27}\ \text{kg/u})(484\ \text{m/s}) =$ $\boxed{2.6 \times 10^{-11}\ \text{m.}}$

44. For diffraction, the wavelength must be on the order of the opening. We find the speed from
    $\lambda = h/p = h/mv$;
    $10\ \text{m} = (6.63 \times 10^{-34}\ \text{J} \cdot \text{s})/(2000\ \text{kg})v$, which gives $v =$ $\boxed{3.3 \times 10^{-38}\ \text{m/s.}}$
    Not a good speed if you want to get somewhere.
    At a speed of 30 m/s, $\lambda \ll 10$ m, so there will be $\boxed{\text{no diffraction.}}$

45. If we assume that $E = hf$, then we get
    $E = hf = hv/\lambda = hv/(h/mv) = mv^2$.
    (*a*) If E is the kinetic energy, we consider the low speed and relativistic expressions:
    Low speed: $E = \frac{1}{2}mv^2 \neq mv^2$, so it is not possible to have $E = hf$.
    Relativistic: $E = (m - m_0)c^2$.
    If this is to equal mv^2, we get $m(c^2 - v^2) = m_0c^2$, or $m = m_0/(1 - v^2/c^2)$.
    Because this is not the mass-speed relation, it is not possible to have $E = hf$.
    (*b*) If E is the kinetic energy plus rest mass energy, we have
    $E = \text{KE} + m_0c^2 = mc^2 = hf = mv^2$, or $v = c$.
    Because this is not possible for a particle with a rest mass, it is not possible to have $E = hf$.

46. The kinetic energy of the electron is
    $\text{KE} = p^2/2m_0 = h^2/2m_0\lambda^2 = (6.63 \times 10^{-34}\ \text{J} \cdot \text{s})^2/2(9.11 \times 10^{-31}\ \text{kg})(0.10\ \text{m})^2 = 2.41 \times 10^{-17}\ \text{J} = 150\ \text{eV}$.
    Because this must equal the potential energy change, the required voltage is $\boxed{150\ \text{V.}}$

47. The wavelength of the electron is

$\lambda = h/p = h/[2m_0(\text{KE})]^{1/2} = hc/[2m_0c^2(\text{KE})]^{1/2}$

$= (1.24 \times 10^3 \text{ eV}\cdot\text{nm})/[2(0.511 \times 10^6 \text{ eV})(2250 \text{ eV})]^{1/2} = 2.59 \times 10^{-2} \text{ nm}.$

For a resolution of 5.0 nm, we find the numerical aperture from

$\text{RP} = 0.61\lambda/\text{NA};$

$5.0 \text{ nm} = 0.61(2.59 \times 10^{-2} \text{ nm})/\text{NA}$, which gives NA = $\boxed{3.2 \times 10^{-3}.}$

48. The energy of a level is $E_n = -(13.6 \text{ eV})/n^2$.

(*a*) The transition from $n = 1$ to $n' = 3$ is an $\boxed{\text{absorption,}}$ because the $\boxed{\text{final state,}}$ $n' = 3$, has a higher energy. The energy of the photon is

$hf = E_{n'} - E_n = -(13.6 \text{ eV})[(1/3^2) - (1/1^2)] = 12.1 \text{ eV}.$

(*b*) The transition from $n = 6$ to $n' = 2$ is an $\boxed{\text{emission,}}$ because the $\boxed{\text{initial state,}}$ $n' = 2$, has a higher energy. The energy of the photon is

$hf = -(E_{n'} - E_n) = (13.6 \text{ eV})[(1/2^2) - (1/6^2)] = 3.0 \text{ eV}.$

(*c*) The transition from $n = 4$ to $n' = 5$ is an $\boxed{\text{absorption,}}$ because the $\boxed{\text{final state,}}$ $n' = 5$, has a higher energy. The energy of the photon is

$hf = E_{n'} - E_n = -(13.6 \text{ eV})[(1/5^2) - (1/4^2)] = 0.31 \text{ eV}.$

The photon for the transition from $\boxed{n = 1 \text{ to } n' = 3}$ has the largest energy.

49. To ionize the atom means removing the electron, or raising it to zero energy:

$E_{\text{ion}} = 0 - E_n = (13.6 \text{ eV})/n^2 = (13.6 \text{ eV})/2^2 = \boxed{3.4 \text{ eV.}}$

50. From $\Delta E = hc/\lambda$, we see that the third longest wavelength comes from the transition with the third smallest energy: $\boxed{n = 6 \text{ to } n' = 3.}$

51. Doubly ionized lithium is like hydrogen, except that there are three positive charges (Z = 3) in the nucleus. The square of the product of the positive and negative charges appears in the energy term for the energy levels. We can use the results for hydrogen, if we replace e^2 by Ze^2:

$E_n = -Z^2(13.6 \text{ eV})/n^2 = -3^2(13.6 \text{ eV})/n^2 = -(122 \text{ eV})/n^2.$

We find the energy needed to remove the remaining electron from

$E = 0 - E_1 = 0 - [-(122 \text{ eV})/(1)^2] = \boxed{122 \text{ eV.}}$

52. For the Rydberg constant we have

$R = 2\pi^2e^4mk^2/h^3c$

$= 2\pi(1.602177 \times 10^{-19} \text{ C})^4(9.109390 \times 10^{-31} \text{ kg})(8.988 \times 10^9 \text{ N}\cdot\text{m}^2/\text{C}^2)^2/$
$(6.626076 \times 10^{-34} \text{ J}\cdot\text{s})^3(2.997925 \times 10^8 \text{ m/s})$

$= 1.0974 \times 10^7 \text{ m}^{-1}.$

53. The energy of the photon is

$hf = E_{\text{ion}} + \text{KE} = 13.6 \text{ eV} + 10.0 \text{ eV} = 23.6 \text{ eV}.$

We find the wavelength from

$\lambda = (1.24 \times 10^3 \text{ eV}\cdot\text{nm})/hf = (1.24 \times 10^3 \text{ eV}\cdot\text{nm})/(23.6 \text{ eV}) = \boxed{52.5 \text{ nm.}}$

54. (*a*) For the jump from $n = 4$ to $n = 2$, we have
$\lambda = (1.24 \times 10^3\ \text{eV}\cdot\text{nm})/(E_4 - E_2) = (1.24 \times 10^3\ \text{eV}\cdot\text{nm})/[-0.85\ \text{eV} - (-3.4\ \text{eV})] =$ **486 nm.**
(*b*) For the jump from $n = 3$ to $n = 1$, we have
$\lambda = (1.24 \times 10^3\ \text{eV}\cdot\text{nm})/(E_3 - E_1) = (1.24 \times 10^3\ \text{eV}\cdot\text{nm})/[-1.5\ \text{eV} - (-13.6\ \text{eV})] =$ **102 nm.**
(*c*) The energy of the $n = 5$ level is
$E_5 = -(13.6\ \text{eV})/5^2 = -0.54\ \text{eV}.$
For the jump from $n = 5$ to $n = 2$, we have
$\lambda = (1.24 \times 10^3\ \text{eV}\cdot\text{nm})/(E_5 - E_2) = (1.24 \times 10^3\ \text{eV}\cdot\text{nm})/[-0.54\ \text{eV} - (-3.4\ \text{eV})] =$ **434 nm.**

55. Singly ionized helium is like hydrogen, except that there are two positive charges ($Z = 2$) in the nucleus. The square of the product of the positive and negative charges appears in the energy term for the energy levels. We can use the results for hydrogen, if we replace e^2 by Ze^2:
$E_n = -Z^2(13.6\ \text{eV})/n^2 = -2^2(13.6\ \text{eV})/n^2 = -(54.4\ \text{eV})/n^2.$
We find the energy of the photon from the $n = 6$ to $n = 2$ transition:
$E = E_6 - E_2 = -(54.4\ \text{eV})[(1/6^2) - (1/2^2)] = 12.1\ \text{eV}.$
Because this is the energy difference for the $n = 1$ to $n = 3$ transition in hydrogen, the photon can be absorbed by a hydrogen atom which will jump from **$n = 1$ to $n = 3$.**

56. The longest wavelength corresponds to the minimum energy, which is the ionization energy:
$\lambda = (1.24 \times 10^3\ \text{eV}\cdot\text{nm})/E_{\text{ion}} = (1.24 \times 10^3\ \text{eV}\cdot\text{nm})/(13.6\ \text{eV}) =$ **91.2 nm.**

57. Singly ionized helium is like hydrogen, except that there are two positive charges ($Z = 2$) in the nucleus. The square of the product of the positive and negative charges appears in the energy term for the energy levels. We can use the results for hydrogen, if we replace e^2 by Ze^2:
$E_n = -Z^2(13.6\ \text{eV})/n^2 = -2^2(13.6\ \text{eV})/n^2 = -(54.4\ \text{eV})/n^2.$

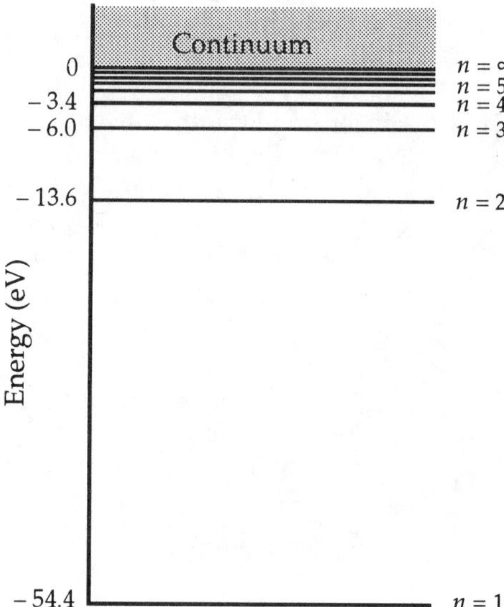

58. Doubly ionized lithium is like hydrogen, except that there are three positive charges ($Z = 3$) in the nucleus. The square of the product of the positive and negative charges appears in the energy term for the energy levels. We can use the results for hydrogen, if we replace e^2 by Ze^2:

$$E_n = -Z^2(13.6 \text{ eV})/n^2 = -3^2(13.6 \text{ eV})/n^2 = -(122 \text{ eV})/n^2.$$

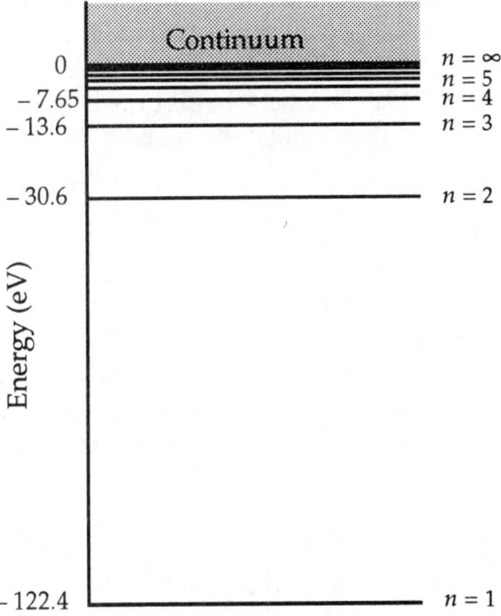

59. The potential energy for the ground state is

$$\text{PE} = -ke^2/r_1 = (9.00 \times 10^9 \text{ N}\cdot\text{m}^2/\text{C}^2)(1.60 \times 10^{-19} \text{ C})^2/(0.529 \times 10^{-10} \text{ m}) = -4.35 \times 10^{-18} \text{ J} = \boxed{-27.2 \text{ eV.}}$$

The kinetic energy is

$$\text{KE} = E_1 - \text{PE} = -13.6 \text{ eV} - (-27.2 \text{ eV}) = \boxed{+13.6 \text{ eV.}}$$

60. We find the value of n from

$$r_n = n^2 r_1;$$

$$1.00 \times 10^{-3} \text{ m} = n^2(0.529 \times 10^{-10} \text{ m}), \text{ which gives } n = \boxed{4.35 \times 10^3.}$$

The energy of this orbit is

$$E = -(13.6 \text{ eV})/n^2 = -(13.6 \text{ eV})/(4.35 \times 10^3)^2 = \boxed{-7.2 \times 10^{-7} \text{ eV.}}$$

61. We find the velocity from the quantum condition:

$$mvr_1 = nh/2\pi;$$

$$(9.11 \times 10^{-31} \text{ kg})v(0.529 \times 10^{-10} \text{ m}) = (1)(6.63 \times 10^{-34} \text{ J}\cdot\text{s})/2\pi,$$

which gives $v = 2.18 \times 10^6 \text{ m/s} = 7.3 \times 10^{-3}c$.

The relativistic factor is

$$[1 - (v/c)^2]^{1/2} \approx 1 - \tfrac{1}{2}(v/c)^2 = 1 - 2.7 \times 10^{-5}.$$

Because this is essentially 1, the use of nonrelativistic formulas is $\boxed{\text{justified.}}$

62. If we compare the two forces:

$$F_e = ke^2/r^2, \quad \text{and} \quad F_g = Gm_e m_p/r^2,$$

we see that we can use the hydrogen expressions if we replace ke^2 with $Gm_e m_p$. For the radius we get

$$r_1 = h^2/4\pi^2 G m_e^2 m_p$$

$$= (6.63 \times 10^{-34} \text{ J}\cdot\text{s})^2/4\pi^2(6.67 \times 10^{-11} \text{ N}\cdot\text{m}^2/\text{kg}^2)(9.11 \times 10^{-31} \text{ kg})^2(1.67 \times 10^{-27} \text{ kg})$$

$$= \boxed{1.20 \times 10^{29} \text{ m.}}$$

The ground state energy is

$$E_1 = 2\pi^2 G^2 m_e^3 m_p^2/h^2$$

$$= 2\pi^2(6.67 \times 10^{-11} \text{ N}\cdot\text{m}^2/\text{kg}^2)^2(9.11 \times 10^{-31} \text{ kg})^3(1.67 \times 10^{-27} \text{ kg})^2/(6.63 \times 10^{-34} \text{ J}\cdot\text{s})^2$$

$$= \boxed{-4.23 \times 10^{-97} \text{ J.}}$$

63. The potential energy for the nth state is
$$\text{PE} = -ke^2/r_n.$$
The Coulomb force provides the radial acceleration, so we have
$$ke^2/r_n{}^2 = mv_n{}^2/r_n, \quad \text{or} \quad mv_n{}^2 = ke^2/r_n.$$
Thus the kinetic energy is
$$\text{KE} = \tfrac{1}{2}mv_n{}^2 = \tfrac{1}{2}ke^2/r_n = \tfrac{1}{2}|\text{PE}|.$$

64. For the difference in radius for adjacent orbits, we have
$$\Delta r = r_n - r_{n-1} = [n^2 - (n-1)^2]r_1 = (2n-1)r_1.$$
When $n \gg 1$, we get
$$\Delta r \approx 2nr_1 = 2r_n/n.$$
In the expressions for r_n and E_n, we see that n and h are always paired, so letting $h \to 0$ is equivalent to letting $n \to \infty$.

65.

(*a*)

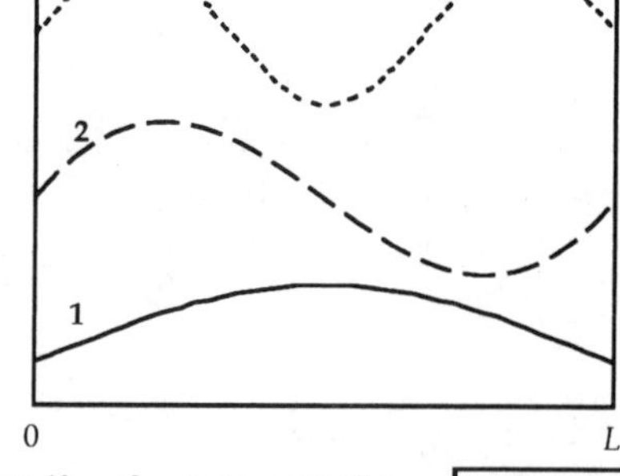

(*b*) From the diagram we see that the wavelengths are given by
$$\lambda_n = 2L/n, \quad n = 1, 2, 3, \ldots .$$
so the momentum is
$$p_n = h/\lambda = nh/2L, \quad n = 1, 2, 3, \ldots .$$
Thus the kinetic energy is
$$\text{KE}_n = p_n{}^2/2m = n^2h^2/8mL^2, \quad n = 1, 2, 3, \ldots .$$

(*c*) Because the potential energy is zero inside the box, the total energy is the kinetic energy. For the ground state energy, we get
$$E_1 = \text{KE}_1 = n^2h^2/8mL^2$$
$$= (1)^2(6.63 \times 10^{-34}\ \text{J}\cdot\text{s})^2/8(9.11 \times 10^{-31}\ \text{kg})(0.50 \times 10^{-10}\ \text{m})^2 = 2.41 \times 10^{-17}\ \text{J} = \boxed{150\ \text{eV.}}$$

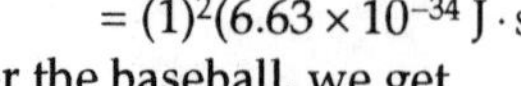

(*d*) For the baseball, we get
$$E_1 = \text{KE}_1 = n^2h^2/8mL^2$$
$$= (1)^2(6.63 \times 10^{-34}\ \text{J}\cdot\text{s})^2/8(0.140\ \text{kg})(0.50\ \text{m})^2 = \boxed{1.6 \times 10^{-66}\ \text{J.}}$$
We find the speed from
$$\text{KE}_1 = \tfrac{1}{2}mv^2;$$
$$1.6 \times 10^{-66}\ \text{J} = \tfrac{1}{2}(0.140\ \text{kg})v^2, \text{ which gives } v = \boxed{4.8 \times 10^{-33}\ \text{m/s.}}$$

(*e*) We find the width of the box from
$$E_1 = n^2h^2/8mL^2$$
$$(20\ \text{eV})(1.60 \times 10^{-19}\ \text{J/eV}) = (1)^2(6.63 \times 10^{-34}\ \text{J}\cdot\text{s})^2/8(9.11 \times 10^{-31}\ \text{kg})L^2,$$
which gives $L = 1.4 \times 10^{-10}\ \text{m} = \boxed{0.14\ \text{nm.}}$

66. We find the peak wavelength from
$$\lambda_P = (2.90 \times 10^{-3}\ \text{m}\cdot\text{K})/T = (2.90 \times 10^{-3}\ \text{m}\cdot\text{K})/(2.7\ \text{K}) = 1.1 \times 10^{-3}\ \text{m} = \boxed{1.1\ \text{mm.}}$$

67. To produce a photoelectron, the hydrogen atom must be ionized, so the minimum energy of the photon is 13.6 eV. We find the minimum frequency of the photon from
$$E_{\text{min}} = hf_{\text{min}};$$
$$(13.6\ \text{eV})(1.60 \times 10^{-19}\ \text{J/eV}) = (6.63 \times 10^{-34}\ \text{J}\cdot\text{s})f_{\text{min}}, \text{ which gives } f_{\text{min}} = \boxed{3.28 \times 10^{15}\ \text{Hz.}}$$

68. Because the energy is much less than m_0c^2, we can use KE = $p^2/2m_0$, so the wavelength of the electron is

$\lambda = h/p = h/[2m_0(\text{KE})]^{1/2} = hc/[2m_0c^2(\text{KE})]^{1/2}$

$= (1.24 \times 10^3 \text{ eV}\cdot\text{nm})/[2(0.511 \times 10^6 \text{ eV})(85 \text{ eV})]^{1/2} = 0.133 \text{ nm}.$

We find the spacing of the planes from

$2d \sin\theta = m\lambda;$

$2d \sin 38° = (1)(0.133 \text{ nm})$, which gives $d =$ **0.108 nm.**

69. For the energy of the photon, we have

$E = hf = hc/\lambda$

$= (6.63 \times 10^{-34} \text{ J}\cdot\text{s})(3.00 \times 10^8 \text{ m/s})/(1.60 \times 10^{-19} \text{ J/eV})\lambda = (1.24 \times 10^{-6} \text{ eV}\cdot\text{m})/\lambda.$

70. The energy of the photon is

$E = hf = hc/\lambda$

$= (6.63 \times 10^{-34} \text{ J}\cdot\text{s})(3.00 \times 10^8 \text{ m/s})/(12.2 \times 10^{-2} \text{ m}) = 1.63 \times 10^{-24} \text{ J}.$

Thus the rate at which photons are produced in the oven is

$N = P/E = (760 \text{ W})/(1.63 \times 10^{-24} \text{ J}) =$ **4.66×10^{26} photons/s.**

71. The energy of the photon is

$hf = (1.24 \times 10^3 \text{ eV}\cdot\text{nm})/\lambda = (1.24 \times 10^3 \text{ eV}\cdot\text{nm})/(550 \text{ nm}) = 2.25 \text{ eV}.$

We find the intensity of photons from

$I_{\text{photons}} = I/hf = (1400 \text{ W/m}^2)/(2.25 \text{ eV})(1.60 \times 10^{-19} \text{ J/eV}) =$ **2.78×10^{21} photons/s·m².**

72. The impulse on the wall is due to the change in momentum of the photons:

$F\,\Delta t = \Delta p = np = nh/\lambda$, or

$n/\Delta t = F\lambda/h = (5.5 \times 10^{-9} \text{ N})(633 \times 10^{-9} \text{ m})/(6.63 \times 10^{-34} \text{ J}\cdot\text{s}) =$ **5.3×10^{18} photons/s.**

73. The energy of the photon is

$hf = (1.24 \times 10^3 \text{ eV}\cdot\text{nm})/\lambda = (1.24 \times 10^3 \text{ eV}\cdot\text{nm})/(550 \text{ nm}) = 2.25 \text{ eV}.$

Because the light radiates uniformly, the intensity at a distance L is

$I = P/4\pi L^2$, so the rate at which energy enters the pupil is

$E/t = I\pi r^2 = Pr^2/4L^2.$

Thus the rate at which photons enter the pupil is

$n/t = (E/t)/hf = Pr^2/4L^2hf$

$= (0.030)(100 \text{ W})(2.0 \times 10^{-3} \text{ m})^2/4(1.0 \times 10^3 \text{ m})^2(2.25 \text{ eV})(1.60 \times 10^{-19} \text{ J/eV})$

$=$ **8.3×10^6 photons/s.**

74. The required momentum is

$p = h/\lambda$, or

$pc = hc/\lambda = (1.24 \times 10^3 \text{ eV}\cdot\text{nm})/(5.0 \times 10^{-3} \text{ nm}) = 2.48 \times 10^5 \text{ eV} = 0.248 \text{ MeV}.$

(*a*) For the proton, $pc \ll m_0c^2$, so we can find the required kinetic energy from

KE $= p^2/2m_0 = (pc)^2/2m_0c^2 = (0.248 \text{ MeV})^2/2(938 \text{ MeV}) = 3.3 \times 10^{-5} \text{ MeV} = 33 \text{ eV}.$

The potential difference to produce this kinetic energy is

$V = \text{KE}/e = (33 \text{ eV})/(1 \text{ e}) =$ **33 V.**

(*b*) For the electron, pc is on the order of m_0c^2, so we can find the required kinetic energy from

KE $= [(pc)^2 + (m_0c^2)^2]^{1/2} - m_0c^2$

$= [(0.248 \text{ MeV})^2 + (0.511 \text{ MeV})^2]^{1/2} - 0.511 \text{ MeV} = 0.057 \text{ MeV} = 57 \text{ keV}.$

The potential difference to produce this kinetic energy is

$V = \text{KE}/e = (57 \text{ keV})/(1 \text{ e}) =$ **57 kV.**

75. If we ignore the recoil motion, at the closest approach the kinetic energy of both particles is zero. The potential energy of the two charges must equal the initial kinetic energy of the α particle:

$\text{KE} = kZ_{\alpha}Z_{\text{Au}}e^2/r_{\text{min}}$;

$(4.8\text{ MeV})(1.60 \times 10^{-13}\text{ J/MeV}) = (9.00 \times 10^9\text{ N}\cdot\text{m}^2/\text{C}^2)(2)(79)(1.60 \times 10^{-19}\text{ C})^2/r_{\text{min}}$,

which gives $r_{\text{min}} =$ $\boxed{4.7 \times 10^{-14}\text{ m.}}$

76. The decrease in mass occurs because a photon has been emitted:

$\Delta m/m_0 = (\Delta E/c^2)/m_0 = \Delta E/m_0c^2 = (-13.6\text{ eV})[(1/1^2) - (1/4^2)]/(939 \times 10^6\text{ eV}) =$ $\boxed{-1.36 \times 10^{-8}.}$

77. The collision must be elastic as long as the electron does not have enough energy to raise the hydrogen atom to the $n = 2$ level. Thus we have

$\text{KE} < E_2 - E_1 = (-13.6\text{ eV})[(1/2^2) - (1/1^2)] =$ $\boxed{10.2\text{ eV.}}$

78. The Coulomb force provides the radial acceleration, so we have

$ke^2/r_n^2 = mv_n^2/r_n$, or $v_n = (ke^2/mr_n)^{1/2}$.

For the angular velocity, we get

$\omega_n = v_n/r_n = (ke^2/mr_n^3)^{1/2} = (64\pi^6m^2k^4e^8/n^6h^6)^{1/2} =$ $\boxed{8\pi^3mk^2e^4/n^3h^3.}$

where we have used the expression $r_n = n^2h^2/4\pi^2mke^2$ for the radius of the orbit.

For the frequency, we get

$f_n = \omega_n/2\pi =$ $\boxed{4\pi^2mk^2e^4/n^3h^3.}$

(*a*) For the ground state, we get

$\omega_1 = 8\pi^3(9.11 \times 10^{-31}\text{ kg})(9.00 \times 10^9\text{ N}\cdot\text{m}^2/\text{C}^2)^2(1.60 \times 10^{-19}\text{ C})^4/(1)^3(6.63 \times 10^{-34}\text{ J}\cdot\text{s})^3$

$= \boxed{4.12 \times 10^{16}\text{ rad/s.}}$

The frequency is

$f_1 = \omega_1/2\pi = (4.12 \times 10^{16}\text{ rad/s})/2\pi =$ $\boxed{6.55 \times 10^{15}\text{ Hz.}}$

(*b*) For the first excited state, we get

$\omega_2 = 8\pi^3(9.11 \times 10^{-31}\text{ kg})(9.00 \times 10^9\text{ N}\cdot\text{m}^2/\text{C}^2)^2(1.60 \times 10^{-19}\text{ C})^4/(2)^3(6.63 \times 10^{-34}\text{ J}\cdot\text{s})^3$

$= \boxed{5.15 \times 10^{15}\text{ rad/s.}}$

The frequency is

$f_2 = \omega_2/2\pi = (5.15 \times 10^{15}\text{ rad/s})/2\pi =$ $\boxed{8.19 \times 10^{14}\text{ Hz.}}$

79. The ratio of the forces is

$F_g/F_e = (Gm_em_p/r^2)/(ke^2/r^2) = Gm_em_p/ke^2$

$= (6.67 \times 10^{-11}\text{ N}\cdot\text{m}^2/\text{kg}^2)(9.11 \times 10^{-31}\text{ kg})^2(1.67 \times 10^{-27}\text{ kg})/$

$(9.00 \times 10^9\text{ N}\cdot\text{m}^2/\text{C}^2)(1.60 \times 10^{-19}\text{ C})^2 =$ $\boxed{4.4 \times 10^{-40}.}$

$\boxed{\text{Yes,}}$ the gravitational force may be safely ignored.

80. The potential difference produces a kinetic energy of 12.3 eV, so it is possible to provide this much energy to the hydrogen atom through collisions. From the ground state, the maximum energy of the atom is – 13.6 eV + 12.3 eV = – 1.3 eV. From the energy level diagram, we see that this means the atom could be excited to the $n = 3$ state, so the possible transitions when the atom returns to the ground state are $n = 3$ to $n = 2$, $n = 3$ to $n = 1$, and $n = 2$ to $n = 1$. For the wavelengths we have

$\lambda_{3\to2} = (1.24 \times 10^3\text{ eV}\cdot\text{nm})/(E_3 - E_2) = (1.24 \times 10^3\text{ eV}\cdot\text{nm})/[-1.5\text{ eV} - (-3.4\text{ eV})] =$ $\boxed{653\text{ nm;}}$

$\lambda_{3\to1} = (1.24 \times 10^3\text{ eV}\cdot\text{nm})/(E_3 - E_1) = (1.24 \times 10^3\text{ eV}\cdot\text{nm})/[-1.5\text{ eV} - (-13.6\text{ eV})] =$ $\boxed{102\text{ nm;}}$

$\lambda_{2\to1} = (1.24 \times 10^3\text{ eV}\cdot\text{nm})/(E_2 - E_1) = (1.24 \times 10^3\text{ eV}\cdot\text{nm})/[-3.4\text{ eV} - (-13.6\text{ eV})] =$ $\boxed{122\text{ nm.}}$

81. The energy levels are

$E_n = 2\pi^2 Z^2 e^4 mk^2/n^2h^2 = (82)^2(207)2\pi^2 e^4 m_e k^2/n^2h^2$

$= (82)^2(207)(-13.6\text{ eV})/n^2 = -(1.89 \times 10^7\text{ eV})/n^2 = -(18.9\text{ MeV})/n^2.$

For the $n = 2$ to $n = 1$ transition, the photon energy is

$hf = E_2 - E_1 = (-18.9\text{ MeV})[(1/2^2) - (1/1^2)] =$ **14.2 MeV.**

82. (*a*) The electron has a charge e, so the potential difference produces a kinetic energy of eV. The shortest wavelength photon is produced when all the kinetic energy is lost and a photon emitted:

$hf_{max} = hc/\lambda_0 = eV$, which gives $\lambda_0 = hc/eV$.

(*b*) $\lambda_0 = hc/eV = (1.24 \times 10^3\text{ eV}\cdot\text{nm})/(27\text{ eV}) =$ **0.046 nm.**

83. We find the momentum from

$E^2 = (\text{KE} + m_0c^2)^2 = p^2c^2 + m_e^2c^4$, or

$p^2c^2 = (\text{KE})^2 + 2(\text{KE})m_0c^2.$

The wavelength is

$\lambda = h/p = hc/pc = hc/[(\text{KE})^2 + 2(\text{KE})m_0c^2]^{1/2}.$

84. The kinetic energy of a thermal neutron is

$\text{KE} = \frac{3}{2}kT = \frac{3}{2}(1.38 \times 10^{-23}\text{ J/K})(300\text{ K}) = 6.21 \times 10^{-21}\text{ J} =$ **0.039 eV.**

We find the speed from

$\text{KE} = \frac{1}{2}mv^2;$

$6.21 \times 10^{-21}\text{ J} = \frac{1}{2}(1.67 \times 10^{-27}\text{ kg})v^2$, which gives $v = 2.73 \times 10^3\text{ m/s}$.

The wavelength is

$\lambda = h/p = h/mv = (6.63 \times 10^{-34}\text{ J}\cdot\text{s})/(1.67 \times 10^{-27}\text{ kg})(2.73 \times 10^3\text{ m/s}) = 1.5 \times 10^{-10}\text{ m} =$ **0.15 nm.**

85. We find the momentum from

$E^2 = (\text{KE} + m_0c^2)^2 = p^2c^2 + m_e^2c^4$, or

$p^2c^2 = (\text{KE})^2 + 2(\text{KE})m_0c^2.$

The wavelength is

$\lambda = h/p = hc/pc = hc/[(\text{KE})^2 + 2(\text{KE})m_0c^2]^{1/2}$

$= (1.24 \times 10^3\text{ eV}\cdot\text{nm})/[(60 \times 10^{-3}\text{ eV})^2 + 2(60 \times 10^{-3}\text{ eV})(0.511 \times 10^6\text{ eV})]^{1/2} = 4.9 \times 10^{-3}\text{ nm}.$

The theoretical resolution limit is on the order of the wavelength, or **5×10^{-12} m.**

86. The energy of a photon in terms of the momentum is

$E = hf = hc/\lambda = pc.$

The rate at which photons are striking the sail is

$N/\Delta t = IA/E = IA/pc.$

Because the photons reflect from the sail, the change in momentum of a photon is

$\Delta p = 2p.$

The impulse on the sail is due to the change in momentum of the photons:

$F\,\Delta t = N\,\Delta p$, or

$F = (N/\Delta t)\,\Delta p = (IA/pc)(2p) = 2IA/c = 2(1000\text{ W/m}^2)(1 \times 10^3\text{ m})^2/(3.00 \times 10^8\text{ m/s}) =$ **7 N.**

CHAPTER 28

Note: At the atomic scale, it is most convenient to have energies in electron-volts and wavelengths in nanometers. A useful expression for the energy of a photon in terms of its wavelength is
$E = hf = hc/\lambda = (6.63 \times 10^{-34}\ \text{J}\cdot\text{s})(3 \times 10^{8}\ \text{m/s})(10^{-9}\ \text{nm/m})/(1.60 \times 10^{-19}\ \text{J/eV})\lambda;$
$E = (1.24 \times 10^{3}\ \text{eV}\cdot\text{nm})/\lambda.$

1. We find the wavelength of the neutron from
$\lambda = h/p = h/[2m_0(\text{KE})]^{1/2}$
$= (6.63 \times 10^{-34}\ \text{J}\cdot\text{s})/[2(1.67 \times 10^{-27}\ \text{kg})(0.025\ \text{eV})(1.6 \times 10^{-19}\ \text{J/eV})]^{1/2} = 1.81 \times 10^{-10}\ \text{m}.$
The peaks of the interference pattern are given by
$d \sin \theta = n\lambda,\ n = 1, 2, \ldots .$
and the positions on the screen are
$y = L \tan \theta.$
For small angles, $\sin \theta = \tan \theta$, so we have
$y = nL\lambda/d.$
Thus the separation is
$\Delta y = L\lambda/d = (1.0\ \text{m})(1.81 \times 10^{-10}\ \text{m})/(0.50\ \text{m}) = \boxed{3.6 \times 10^{-7}\ \text{m.}}$

2. We find the wavelength of the bullet from
$\lambda = h/p = h/mv$
$= (6.63 \times 10^{-34}\ \text{J}\cdot\text{s})/(3.0 \times 10^{-3}\ \text{kg})(200\ \text{m/s}) = 1.1 \times 10^{-33}\ \text{m}.$
The half-angle for the central circle of the diffraction pattern is given by
$\sin \theta = 1.22\lambda/D.$
For small angles, $\sin \theta \approx \tan \theta$, so we have
$r = L \tan \theta \approx L \sin \theta = 1.22L\lambda/D;$
$0.50 \times 10^{-2}\ \text{m} = 1.22L(1.1 \times 10^{-33}\ \text{m})/(3.0 \times 10^{-3}\ \text{m})$, which gives $L = \boxed{1.1 \times 10^{28}\ \text{m.}}$
Diffraction effects are negligible for macroscopic objects.

3. We find the uncertainty in the momentum:
$\Delta p = m\,\Delta v = (1.67 \times 10^{-27}\ \text{kg})(0.024 \times 10^{5}\ \text{m/s}) = 4.00 \times 10^{-24}\ \text{kg}\cdot\text{m/s}.$
We find the uncertainty in the proton's position from
$\Delta x \geq \hbar/\Delta p = (1.055 \times 10^{-34}\ \text{J}\cdot\text{s})/(4.00 \times 10^{-24}\ \text{kg}\cdot\text{m/s}) = 2.6 \times 10^{-11}\ \text{m}.$
Thus the accuracy of the position is $\boxed{\pm 1.3 \times 10^{-11}\ \text{m.}}$

4. We find the uncertainty in the momentum:
$\Delta p \geq \hbar/\Delta x = (1.055 \times 10^{-34}\ \text{J}\cdot\text{s})/(2.0 \times 10^{-8}\ \text{m}) = 5.28 \times 10^{-27}\ \text{kg}\cdot\text{m/s}.$
We find the uncertainty in the velocity from
$\Delta p = m\,\Delta v;$
$5.28 \times 10^{-27}\ \text{kg}\cdot\text{m/s} = (9.11 \times 10^{-31}\ \text{kg})\,\Delta v$, which gives $\Delta v = \boxed{5.8 \times 10^{3}\ \text{m/s.}}$

5. We find the minimum uncertainty in the energy of the state from
$\Delta E \geq \hbar/\Delta t = (1.055 \times 10^{-34}\ \text{J}\cdot\text{s})/(10^{-8}\ \text{s}) = 1.1 \times 10^{-26}\ \text{J} = \boxed{6.6 \times 10^{-8}\ \text{eV.}}$

6. We find the lifetime of the particle from
$\Delta t \geq \hbar/\Delta E = (1.055 \times 10^{-34}\ \text{J}\cdot\text{s})/(2.49\ \text{GeV})(1.60 \times 10^{-10}\ \text{J/GeV}) = \boxed{2.65 \times 10^{-25}\ \text{s.}}$

7. We find the uncertainty in the energy of the muon from

$\Delta E \geq \hbar/\Delta t = (1.055 \times 10^{-34}\ \text{J}\cdot\text{s})/(2.2 \times 10^{-6}\ \text{s}) = 4.8 \times 10^{-29}\ \text{J} = 3.0 \times 10^{-10}\ \text{eV}.$

Thus the uncertainty in the mass is

$\Delta m = \Delta E/c^2 = \boxed{3.0 \times 10^{-10}\ \text{eV}/c^2.}$

8. We find the uncertainty in the energy of the free neutron from

$\Delta E \geq \hbar/\Delta t = (1.055 \times 10^{-34}\ \text{J}\cdot\text{s})/(900\ \text{s}) = 1.17 \times 10^{-37}\ \text{J}.$

Thus the uncertainty in the mass is

$\Delta m = \Delta E/c^2 = (1.17 \times 10^{-37}\ \text{J})/(3.00 \times 10^8\ \text{m/s})^2 = \boxed{1.30 \times 10^{-54}\ \text{kg.}}$

9. The uncertainty in the velocity is

$\Delta v = (0.055/100)(150\ \text{m/s}) = 0.0825\ \text{m/s}.$

For the electron, we have

$\Delta x \geq \hbar/m\,\Delta v = (1.055 \times 10^{-34}\ \text{J}\cdot\text{s})/(9.11 \times 10^{-31}\ \text{kg})(0.0825\ \text{m/s}) = \boxed{1.4 \times 10^{-3}\ \text{m.}}$

For the baseball, we have

$\Delta x \geq \hbar/m\,\Delta v = (1.055 \times 10^{-34}\ \text{J}\cdot\text{s})/(0.140\ \text{kg})(0.0825\ \text{m/s}) = \boxed{9.1 \times 10^{-33}\ \text{m.}}$

The uncertainty for the electron is greater by a factor of 1.5×10^{29}.

10. We use the radius as the uncertainty in position for the neutron. We find the uncertainty in the momentum from

$\Delta p \geq \hbar/\Delta x = (1.055 \times 10^{-34}\ \text{J}\cdot\text{s})/(1.0 \times 10^{-15}\ \text{m}) = 1.055 \times 10^{-19}\ \text{kg}\cdot\text{m/s}.$

If we assume that the lowest value for the momentum is the least uncertainty, we estimate the lowest possible kinetic energy as

$E = (\Delta p)^2/2m = (1.055 \times 10^{-19}\ \text{kg}\cdot\text{m/s})^2/2(1.67 \times 10^{-27}\ \text{kg}) = 3.33 \times 10^{-12}\ \text{J} = \boxed{21\ \text{MeV.}}$

11. We use the radius as the uncertainty in position for the electron. We find the uncertainty in the momentum from

$\Delta p \geq \hbar/\Delta x = (1.055 \times 10^{-34}\ \text{J}\cdot\text{s})/(1.0 \times 10^{-15}\ \text{m}) = 1.055 \times 10^{-19}\ \text{kg}\cdot\text{m/s}.$

If we assume that the lowest value for the momentum is the least uncertainty, we estimate the lowest possible energy as

$$E = (\text{KE}) + m_0c^2 = (p^2c^2 + m_0^2c^4)^{1/2} = [(\Delta p)^2c^2 + m_0^2c^4]^{1/2}$$
$$= [(1.055 \times 10^{-19}\ \text{kg}\cdot\text{m/s})^2(3.00 \times 10^8\ \text{m/s})^2 + (9.11 \times 10^{-31}\ \text{kg})^2(3.00 \times 10^8\ \text{m/s})^4]^{1/2}$$
$$= 3.175 \times 10^{-11}\ \text{J} \approx 200\ \text{MeV}.$$

12. The momentum of the electron is

$p = [2m(\text{KE})]^{1/2} = [2(9.11 \times 10^{-31}\ \text{kg})(3.00\ \text{keV})(1.60 \times 10^{-16}\ \text{J/keV})]^{1/2} = 2.96 \times 10^{-23}\ \text{kg}\cdot\text{m/s}.$

When the energy is increased by 1.00%, the new momentum is

$p' = [2m(1.0100\text{KE})]^{1/2} = [2m(\text{KE})]^{1/2}(1 + 0.0100)^{1/2} \approx p[1 + \tfrac{1}{2}(0.0100)].$

Thus the uncertainty in momentum is

$\Delta p = p' - p = p\tfrac{1}{2}(0.0100) = (2.96 \times 10^{-23}\ \text{kg}\cdot\text{m/s})\tfrac{1}{2}(0.0100) = 1.48 \times 10^{-25}\ \text{kg}\cdot\text{m/s}.$

We find the uncertainty in the electron's position from

$\Delta x \geq \hbar/\Delta p = (1.055 \times 10^{-34}\ \text{J}\cdot\text{s})/(1.48 \times 10^{-25}\ \text{kg}\cdot\text{m/s}) = \boxed{7.13 \times 10^{-10}\ \text{m.}}$

13. (*a*) We consider a particle of mass m moving in a circle with speed v, so the angular momentum is along the axis of the circle. The angular momentum is

$$L = rmv = rp,$$

so the uncertainty in the angular momentum is related to the uncertainty in momentum:

$$\Delta L = r\,\Delta p.$$

If the position on the circle is uncertain by Δx, this corresponds to an uncertainty in the angle:

$$\Delta x = r\,\Delta\phi.$$

If we use the uncertainty principle, we get

$$(\Delta x)(\Delta p) \geq \hbar;$$

$$(r\,\Delta\phi)(\Delta L/r) = (\Delta L)(\Delta\phi) \geq \hbar.$$

(*b*) If there is no uncertainty in the angular momentum, there is **infinite uncertainty** in the angular position of the electron, which is to say, the angular position is not known. Thus the concept of electron orbits is not valid.

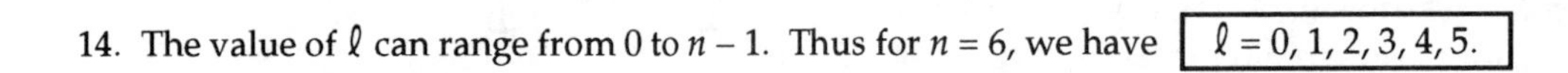

14. The value of ℓ can range from 0 to $n - 1$. Thus for $n = 6$, we have $\boxed{\ell = 0, 1, 2, 3, 4, 5.}$

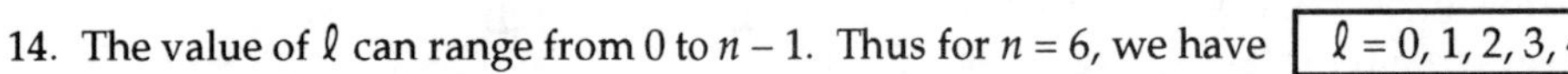

15. The value of m_ℓ can range from $-\ell$ to $+\ell$. Thus for $\ell = 4$, we have $\boxed{m_\ell = -4, -3, -2, -1, 0, 1, 2, 3, 4.}$ The possible values of m_s are $\boxed{-\frac{1}{2}, +\frac{1}{2}.}$

16. The number of electrons in the subshell is determined by the value of ℓ. For each ℓ the value of m_ℓ can range from $-\ell$ to $+\ell$, or $2\ell + 1$ values. For each of these there are two values of m_s. Thus the total number for $\ell = 3$ is

$$N = 2(2\ell + 1) = 2[2(3) + 1] = \boxed{14 \text{ electrons.}}$$

17. We start with the $n = 1$ shell, and list the quantum numbers in the order (n, ℓ, m_ℓ, m_s);

$\boxed{(1, 0, 0, -\frac{1}{2}), (1, 0, 0, +\frac{1}{2}), (2, 0, 0, -\frac{1}{2}), (2, 0, 0, +\frac{1}{2}), (2, 1, -1, -\frac{1}{2}), (2, 1, -1, +\frac{1}{2}), (2, 1, 0, -\frac{1}{2}).}$

Note that, without additional information, there are other possibilities for the last three electrons.

18. We start with the $n = 1$ shell, and list the quantum numbers in the order (n, ℓ, m_ℓ, m_s);

$(1, 0, 0, -\frac{1}{2}), (1, 0, 0, +\frac{1}{2}), (2, 0, 0, -\frac{1}{2}), (2, 0, 0, +\frac{1}{2}), (2, 1, -1, -\frac{1}{2}), (2, 1, -1, +\frac{1}{2}), (2, 1, 0, -\frac{1}{2}),$
$(2, 1, 0, +\frac{1}{2}), (2, 1, -1, -\frac{1}{2}), (2, 1, -1, +\frac{1}{2}), (3, 0, 0, -\frac{1}{2}), (3, 0, 0, +\frac{1}{2}).$

19. The value of ℓ can range from 0 to $n - 1$. Thus for $n = 4$, we have

$$\ell = 0, 1, 2, 3.$$

For each ℓ the value of m_ℓ can range from $-\ell$ to $+\ell$, or $2\ell + 1$ values. For each of these there are two values of m_s. Thus the total number for each ℓ is $2(2\ell + 1)$.

The number of states is

$$N = 2(0 + 1) + 2(2 + 1) + 2(4 + 1) + 2(6 + 1) = \boxed{32 \text{ states.}}$$

We start with $\ell = 0$, and list the quantum numbers in the order (n, ℓ, m_ℓ, m_s);

$(4, 0, 0, -\frac{1}{2}), (4, 0, 0, +\frac{1}{2}), (4, 1, -1, -\frac{1}{2}), (4, 1, -1, +\frac{1}{2}), (4, 1, 0, -\frac{1}{2}), (4, 1, 0, +\frac{1}{2}), (4, 1, 1, -\frac{1}{2}),$
$(4, 1, 1, +\frac{1}{2}), (4, 2, -2, -\frac{1}{2}), (4, 2, -2, +\frac{1}{2}), (4, 2, -1, -\frac{1}{2}), (4, 2, -1, +\frac{1}{2}), (4, 2, 0, -\frac{1}{2}), (4, 2, 0, +\frac{1}{2}),$
$(4, 2, 1, -\frac{1}{2}), (4, 2, 1, +\frac{1}{2}), (4, 2, 2, -\frac{1}{2}), (4, 2, 2, +\frac{1}{2}), (4, 3, -3, -\frac{1}{2}), (4, 3, -3, +\frac{1}{2}), (4, 3, -2, -\frac{1}{2}),$
$(4, 3, -2, +\frac{1}{2}), (4, 3, -1, -\frac{1}{2}), (4, 3, -1, +\frac{1}{2}), (4, 3, 0, -\frac{1}{2}), (4, 3, 0, +\frac{1}{2}), (4, 3, 1, -\frac{1}{2}), (4, 3, 1, +\frac{1}{2}),$
$(4, 3, 2, -\frac{1}{2}), (4, 3, 2, +\frac{1}{2}), (4, 3, 3, -\frac{1}{2}), (4, 3, 3, +\frac{1}{2}).$

20. The value of ℓ can range from 0 to $n - 1$. Thus for $\ell = 4$, we have $\boxed{n \geq 5.}$
For each ℓ the value of m_ℓ can range from $-\ell$ to $+\ell$: $\boxed{m_\ell = -4, -3, -2, -1, 0, 1, 2, 3, 4.}$
There are two values of m_s: $\boxed{m_s = -\frac{1}{2}, +\frac{1}{2}.}$

21. The value of m_ℓ can range from $-\ell$ to $+\ell$, so we have $\boxed{\ell \geq 3.}$
The value of ℓ can range from 0 to $n - 1$. Thus we have $\boxed{n \geq \ell + 1 \text{ (minimum 4).}}$
There are two values of m_s: $\boxed{m_s = -\frac{1}{2}, +\frac{1}{2}.}$

22. The magnitude of the angular momentum depends on ℓ only:
$L = \hbar[\ell(\ell + 1)]^{1/2} = (1.055 \times 10^{-34}\text{ J}\cdot\text{s})[(3)(3 + 1)]^{1/2} = \boxed{3.66 \times 10^{-34}\text{ kg}\cdot\text{m}^2/\text{s}.}$

23. From *spdfg*, we see that the "*g*" subshell has $\ell = 4$, so the number of electrons is
$N = 2(2\ell + 1) = 2[2(4) + 1] = 18$ electrons.

24. (*a*) We start with hydrogen and fill the levels as indicated in the periodic table:
$\boxed{1s^2 2s^2 2p^6 3s^2 3p^6 3d^7 4s^2.}$
Note that the $4s^2$ level is filled before the $3d$ level is started.
(*b*) For Z = 36 we have
$\boxed{1s^2 2s^2 2p^6 3s^2 3p^6 3d^{10} 4s^2 4p^6.}$
(*c*) For Z = 38 we have
$\boxed{1s^2 2s^2 2p^6 3s^2 3p^6 3d^{10} 4s^2 4p^6 5s^2.}$
Note that the $5s^2$ level is filled before the $4d$ level is started.

25. (*a*) Selenium has Z = 34:
$\boxed{1s^2 2s^2 2p^6 3s^2 3p^6 3d^{10} 4s^2 4p^4.}$
(*b*) Gold has Z = 79:
$\boxed{1s^2 2s^2 2p^6 3s^2 3p^6 3d^{10} 4s^2 4p^6 4d^{10} 4f^{14} 5s^2 5p^6 5d^{10} 6s^1.}$
(*c*) Uranium has Z = 92:
$\boxed{1s^2 2s^2 2p^6 3s^2 3p^6 3d^{10} 4s^2 4p^6 4d^{10} 4f^{14} 5s^2 5p^6 5d^{10} 6s^2 6p^6 5f^3 6d^1 7s^2.}$

26. (*a*) The $4p \to 3p$ transition is $\boxed{\text{forbidden,}}$ because $\Delta\ell \neq \pm 1$.
(*b*) The $2p \to 1s$ transition is $\boxed{\text{allowed,}}$ $\Delta\ell = -1$.
(*c*) The $3d \to 2d$ transition is $\boxed{\text{forbidden,}}$ because $\Delta\ell \neq \pm 1$.
(*d*) The $4d \to 3s$ transition is $\boxed{\text{forbidden,}}$ because $\Delta\ell \neq \pm 1$.
(*e*) The $4s \to 3p$ transition is $\boxed{\text{allowed,}}$ $\Delta\ell = +1$.

27. (*a*) The principal quantum number is $n = \boxed{6.}$
(*b*) The energy of the state is
$E_6 = -(13.6\text{ eV})/n^2 = -(13.6\text{ eV})/6^2 = \boxed{-0.378\text{ eV}.}$
(*c*) From *spdfgh*, we see that the "*h*" subshell has $\ell = 5$. The magnitude of the angular momentum depends on ℓ only:
$L = \hbar[\ell(\ell + 1)]^{1/2} = (1.055 \times 10^{-34}\text{ J}\cdot\text{s})[(5)(5 + 1)]^{1/2} = \boxed{5.78 \times 10^{-34}\text{ kg}\cdot\text{m}^2/\text{s}.}$
(*d*) For each ℓ the value of m_ℓ can range from $-\ell$ to $+\ell$: $\boxed{m_\ell = -5, -4, -3, -2, -1, 0, 1, 2, 3, 4, 5.}$

28. The third electron in lithium is in the 2*s* subshell, which is outside the more tightly bound filled 1*s* shell. This makes it appear as if there is a "nucleus" with a net charge of + 1e. Thus we use the energy of the hydrogen atom:

$E_2 = -(13.6 \text{ eV})/n^2 = -(13.6 \text{ eV})/2^2 = -3.4 \text{ eV},$

so the binding energy is $\boxed{3.4 \text{ eV.}}$

Our assumption of complete shielding of the nucleus by the 2*s* electrons is probably not correct.

29. (*a*) Boron has Z = 4, so the outermost electron has $n = 2$. We use the Bohr result with an effective Z:

$E_2 = -(13.6 \text{ eV})(Z_{\text{eff}})^2/n^2;$

$-8.26 \text{ eV} = -(13.6 \text{ eV})(Z_{\text{eff}})^2/2^2$, which gives $Z_{\text{eff}} = \boxed{1.56.}$

Note that this indicates some shielding by the second electron in the $n = 2$ shell.

(*b*) We find the average radius from

$r = n^2 r_1/Z_{\text{eff}} = 2^2(0.529 \times 10^{-10} \text{ m})/(1.56) = \boxed{1.4 \times 10^{-10} \text{ m.}}$

30. In a filled subshell, we have $2(2\ell + 1)$ electrons. All of the m_ℓ values

$-\ell, -\ell + 1, \ldots, 0, \ldots, \ell - 1, \ell$

are filled, so their sum is zero. For each m_ℓ value, both values of m_s are filled, so their sum is also zero. Thus the total angular momentum is zero.

31. For each ℓ the value of m_ℓ can range from $-\ell$ to $+\ell$, or $2\ell + 1$ values. For each of these there are two values of m_s. Thus the total number of electrons allowed in a subshell is

$N = 2(2\ell + 1).$

32. The shortest wavelength X-ray has the most energy, which is the maximum kinetic energy of the electron in the tube:

$\lambda = (1.24 \times 10^3 \text{ eV} \cdot \text{nm})/E = (1.24 \times 10^3 \text{ eV} \cdot \text{nm})/(30 \times 10^3 \text{ eV}) = \boxed{0.041 \text{ nm.}}$

The longest wavelength of the continuous spectrum would be at the limit of the X-ray region of the electromagnetic spectrum, generally on the order of $\boxed{1 \text{ nm.}}$

33. The shortest wavelength X-ray has the most energy, which is the maximum kinetic energy of the electron in the tube:

$E = (1.24 \times 10^3 \text{ eV} \cdot \text{nm})/\lambda = (1.24 \times 10^3 \text{ eV} \cdot \text{nm})/(0.030 \text{ nm}) = 4.1 \times 10^4 \text{ eV} = 41 \text{ keV}.$

Thus the operating voltage of the tube is $\boxed{41 \text{ kV.}}$

34. The energy of the photon with the shortest wavelength must equal the maximum kinetic energy of an electron:

$hf_0 = hc/\lambda_0 = eV$, or

$\lambda_0 = hc/eV = (6.63 \times 10^{-34} \text{ J} \cdot \text{s})(3 \times 10^8 \text{ m/s})(10^{-9} \text{ nm/m})/(1.60 \times 10^{-19} \text{ J/eV})(1 \text{ e})V$

$= (1.24 \times 10^3 \text{ V} \cdot \text{nm})/V.$

35. With the shielding provided by the remaining $n = 1$ electron, we use the energies of the hydrogen atom with Z replaced by Z – 1. The energy of the photon is

$hf = \Delta E = -(13.6 \text{ eV})(27 - 1)^2[(1/2^2) - (1/1^2)] = 6.90 \times 10^3 \text{ eV}.$

The wavelength of the photon is

$\lambda = (1.24 \times 10^3 \text{ eV} \cdot \text{nm})/\Delta E = (1.24 \times 10^3 \text{ eV} \cdot \text{nm})/(6.90 \times 10^3 \text{ eV}) = \boxed{0.18 \text{ nm.}}$

36. With the shielding provided by the remaining $n = 1$ electron, we use the energies of the hydrogen atom with Z replaced by Z – 1. The energy of the photon is

$hf = \Delta E = -(13.6\text{ eV})(24 - 1)^2[(1/2^2) - (1/1^2)] = 5.40 \times 10^3\text{ eV}.$

The wavelength of the photon is

$\lambda = (1.24 \times 10^3\text{ eV}\cdot\text{nm})/\Delta E = (1.24 \times 10^3\text{ eV}\cdot\text{nm})/(5.40 \times 10^3\text{ eV}) =$ **0.23 nm.**

37. If we assume that the shielding is provided by the remaining $n = 1$ electron, we use the energies of the hydrogen atom with Z replaced by Z – 1. The energy of the photon is

$hf = \Delta E = -(13.6\text{ eV})(42 - 1)^2[(1/3^2) - (1/1^2)] = 2.03 \times 10^4\text{ eV}.$

The wavelength of the photon is

$\lambda = (1.24 \times 10^3\text{ eV}\cdot\text{nm})/\Delta E = (1.24 \times 10^3\text{ eV}\cdot\text{nm})/(2.03 \times 10^4\text{ eV}) =$ **0.061 nm.**

We do not expect perfect agreement because there is some **partial shielding provided by the $n = 2$ shell,** which was ignored when we replaced Z by Z – 1.

38. The K_α line is from the $n = 2$ to $n = 1$ transition. We use the energies of the hydrogen atom with Z replaced by Z – 1. Thus we have

$hf = \Delta E \propto (Z - 1)^2$, so $\lambda \propto 1/(Z - 1)^2$.

When we form the ratio for the two materials, we get

$\lambda_X/\lambda_{\text{iron}} = (Z_{\text{iron}} - 1)^2/(Z_X - 1)^2;$

$(229\text{ nm})/(194\text{ nm}) = (26 - 1)^2/(Z_X - 1)^2$, which gives $Z_X = 24$,

so the material is **chromium.**

39. The energy of a pulse is

$E = P\,\Delta t = (0.60\text{ W})(30 \times 10^{-3}\text{ s}) =$ **0.018 J.**

The energy of a photon is

$hf = hc/\lambda = (1.24 \times 10^3\text{ eV}\cdot\text{nm})/\lambda = (1.24 \times 10^3\text{ eV}\cdot\text{nm})/(640\text{ nm}) = 1.94\text{ eV}.$

Thus the number of photons in a pulse is

$N = E/hf = (0.018\text{ J})/(1.94\text{ eV})(1.6 \times 10^{-19}\text{ J/eV}) =$ **5.8×10^{16} photons.**

40. We find the angular half width of the beam from

$\Delta\theta = 1.22\lambda/d = 1.22(694 \times 10^{-9}\text{ m})/(3.0 \times 10^{-3}\text{ m}) = 2.8 \times 10^{-4}\text{ rad},$

so the angular width is $\theta =$ **5.6×10^{-4} rad.**

The diameter of the beam when it reaches the satellite is

$D = r\theta = (300 \times 10^3\text{ m})(5.6 \times 10^{-4}\text{ rad}) =$ **1.7×10^2 m.**

41. We can relate the momentum to the radius of the orbit from the quantum condition:

$L = mvr = pr = n\hbar$, so $p = n\hbar/r = \hbar/r_1$ for the ground state.

If we assume that this is the uncertainty of the momentum, the uncertainty of the position is

$\Delta x \geq \hbar/\Delta p = \hbar/(\hbar/r_1) =$ **r_1, which is the Bohr radius.**

42. (*a*) We find the minimum uncertainty in the energy of the state from

$\Delta E \geq \hbar/\Delta t = (1.055 \times 10^{-34}\ \text{J}\cdot\text{s})/(10^{-8}\ \text{s}) = 1.1 \times 10^{-26}\ \text{J} = \boxed{6.6 \times 10^{-8}\ \text{eV.}}$

Note that, because the ground state is stable, we associate the uncertainty with the excited state.

(*b*) The transition energy is

$E = -(13.6\ \text{eV})[(1/2^2) - (1/1^2)] = 10.2\ \text{eV},$

so we have

$\Delta E/E = (6.6 \times 10^{-8}\ \text{eV})/(10.2\ \text{eV}) = \boxed{6.5 \times 10^{-9}.}$

(*c*) The wavelength of the line is

$\lambda = (1.24 \times 10^3\ \text{eV}\cdot\text{nm})E = (1.24 \times 10^3\ \text{eV}\cdot\text{nm})/(10.2\ \text{eV}) = \boxed{122\ \text{nm.}}$

We find the width of the line from

$\lambda + \Delta\lambda = (1.24 \times 10^3\ \text{eV}\cdot\text{nm})/(E + \Delta E)$

$= [(1.24 \times 10^3\ \text{eV}\cdot\text{nm})/E]/[1 + (\Delta E/E)] \approx \lambda[1 - (\Delta E/E)].$

If we ignore the sign, we get

$\Delta\lambda = \lambda(\Delta E/E) = (122\ \text{nm})(6.5 \times 10^{-9}) = \boxed{7.9 \times 10^{-7}\ \text{nm.}}$

43. The value of ℓ can range from 0 to $n - 1$. Thus for $n = 5$, we have $\ell \leq 4$.

The smallest value of L is $\boxed{0.}$

The largest magnitude of L is

$L = \hbar[\ell(\ell + 1)]^{1/2} = (1.055 \times 10^{-34}\ \text{J}\cdot\text{s})[(4)(4 + 1)]^{1/2} = \boxed{4.72 \times 10^{-34}\ \text{kg}\cdot\text{m}^2/\text{s.}}$

44. (*a*) We find the quantum number for the orbital angular momentum from

$L = M_{\text{earth}}vR = M_{\text{earth}}2\pi R^2/T = \hbar[\ell(\ell + 1)]^{1/2};$

$(5.98 \times 10^{24}\ \text{kg})2\pi(1.50 \times 10^{11}\ \text{m})^2/(3.16 \times 10^7\ \text{s}) = (1.055 \times 10^{-34}\ \text{J}\cdot\text{s})[\ell(\ell + 1)]^{1/2},$

which gives $\ell = \boxed{2.5 \times 10^{74}.}$

(*b*) The value of m_ℓ can range from $-\ell$ to $+\ell$, or $2\ell + 1$ values, so the number of orientations is

$N = 2\ell + 1 = 2(2.5 \times 10^{74}) + 1 = \boxed{5.0 \times 10^{74}.}$

45. (*a*) We find the wavelength of the bullet from

$\lambda = h/p = h/mv$

$= (6.63 \times 10^{-34}\ \text{J}\cdot\text{s})/(12 \times 10^{-3}\ \text{kg})(180\ \text{m/s}) = \boxed{3.1 \times 10^{-34}\ \text{m.}}$

(*b*) We find the uncertainty in the momentum component perpendicular to the motion:

$\Delta p_y \geq \hbar/\Delta y = (1.055 \times 10^{-34}\ \text{J}\cdot\text{s})/(0.60 \times 10^{-8}\ \text{m}) = \boxed{1.8 \times 10^{-32}\ \text{kg}\cdot\text{m/s.}}$

(*c*) We find the possible uncertainty in the y-position at the target from

$y/L = \Delta v_y/v_x = \Delta p_y/p_x;$

$y/(200\ \text{m}) = (1.8 \times 10^{-32}\ \text{kg}\cdot\text{m/s})/(12 \times 10^{-3}\ \text{kg})(180\ \text{m/s})$, which gives $y = \boxed{1.6 \times 10^{-30}\ \text{m.}}$

46. From the Bohr formula for the radius, we see that

$r \propto 1/Z$, so $r_{\text{lead}} = r_1/Z = (0.529 \times 10^{-10}\ \text{m})/(82) = \boxed{6.5 \times 10^{-13}\ \text{m.}}$

The innermost electron would "see" a nucleus with charge Ze.

Thus we use the energy of the hydrogen atom:

$E_1 = -(13.6\ \text{eV})Z^2/n^2 = -(13.6\ \text{eV})(82)^2/1^2 = -9.1 \times 10^4\ \text{eV},$

so the binding energy is $\boxed{91\ \text{keV.}}$

47. If we assume that the rate at which heat is produced is 100% of the electrical power input, we have

$P = IV = Q/t = mc(\Delta T/t);$

$(25\ \text{mA})(100\ \text{kV})(60\ \text{s/min})/(4186\ \text{J/kcal}) = (0.085\ \text{kg})(0.11\ \text{kcal/kg C}^\circ)(\Delta T/t),$

which gives $\Delta T/t = \boxed{3.8 \times 10^3\ \text{C}^\circ/\text{min.}}$

48. The wavelength of the photon emitted for the transition from n' to n is

$$1/\lambda = (2\pi^2 Z^2 e^4 mk^2/h^3c)[(1/n'^2) - (1/n^2)] = RZ^2[(1/n'^2) - (1/n^2)].$$

The K_α line is from the $n' = 2$ to $n = 1$ transition, and the other $n = 1$ electron shields the nucleus, so the effective Z is Z – 1:

$$1/\lambda = RZ_{\text{eff}}^2[(1/n'^2) - (1/n^2)] = R(Z-1)^2[(1/2^2) - (1/1^2)] = \tfrac{3}{4}R(Z-1)^2.$$

Thus we have

$1/\sqrt{\lambda} = (\tfrac{3}{4}R)^{1/2}(Z-1)$, which is the equation of a straight line, as in the Moseley plot, with $b = 1$.

The value of a is

$$a = (\tfrac{3}{4}R)^{1/2} = [\tfrac{3}{4}(1.0974 \times 10^7\ \text{m}^{-1})]^{1/2} = \boxed{2869\ \text{m}^{-1/2}.}$$

49. The magnitude of the angular momentum is given by

$$L = [\ell(\ell + 1)]^{1/2}(h/2\pi),$$

and the z-component is given by

$$L_z = m_\ell(h/2\pi).$$

Thus the angle between **L** and the z-axis is given by

$$\cos\theta = L_z/L = m_\ell(h/2\pi)/[\ell(\ell+1)]^{1/2}(h/2\pi) = m_\ell/[\ell(\ell+1)]^{1/2}.$$

For each ℓ the value of m_ℓ can range from $-\ell$ to $+\ell$, so there are $2\ell + 1$ values for θ.

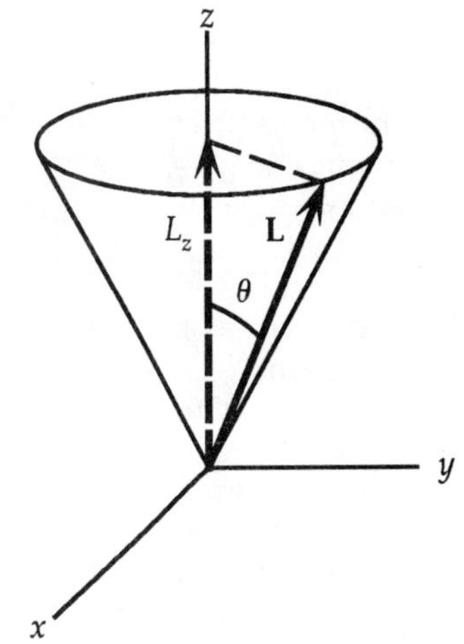

(*a*) For $\ell = 1$ the magnitude of **L** is

$$L = [(1)(1+1)]^{1/2}(h/2\pi) = (2)^{1/2}(h/2\pi).$$

The angles for the 3 values of m_ℓ are

$$\cos\theta_{1,3} = (\pm 1)/(2)^{1/2} = \pm 1/(2)^{1/2}, \quad\text{or}\quad \theta_1 = 45°,\ \theta_3 = 135°.$$

$$\cos\theta_2 = (0)/(2)^{1/2} = 0, \quad\text{or}\quad \theta_2 = 90°.$$

Thus the possible values for $\ell = 1$ are $\boxed{45°, 90°, 135°.}$

(*b*) For $\ell = 2$ the magnitude of **L** is

$$L = [(2)(2+1)]^{1/2}(h/2\pi) = (6)^{1/2}(h/2\pi).$$

The angles for the 5 values of m_ℓ are

$$\cos\theta_{1,5} = (\pm 2)/(6)^{1/2} = \pm(2/3)^{1/2}, \quad\text{or}\quad \theta_1 = 35.3°,\ \theta_5 = 144.7°;$$

$$\cos\theta_{2,4} = (\pm 1)/(6)^{1/2} = \pm(1/6)^{1/2}, \quad\text{or}\quad \theta_2 = 65.9°,\ \theta_4 = 114.1°;$$

$$\cos\theta_3 = (0)/(6)^{1/2} = 0, \quad\text{or}\quad \theta_3 = 90°.$$

Thus the possible values for $\ell = 2$ are $\boxed{35.3°, 65.9°, 90°, 114.1°, 144.7°.}$

(*c*) For $\ell = 3$ the magnitude of **L** is

$$L = [(3)(3+1)]^{1/2}(h/2\pi) = (12)^{1/2}(h/2\pi).$$

The angles for the 7 values of m_ℓ are

$$\cos\theta_{1,7} = (\pm 3)/(12)^{1/2} = \pm(3/4)^{1/2}, \quad\text{or}\quad \theta_1 = 30.0°,\ \theta_7 = 150.0°;$$

$$\cos\theta_{2,6} = (\pm 2)/(12)^{1/2} = \pm(1/3)^{1/2}, \quad\text{or}\quad \theta_2 = 54.7°,\ \theta_6 = 125.3°;$$

$$\cos\theta_{3,5} = (\pm 1)/(12)^{1/2} = \pm(1/12)^{1/2}, \quad\text{or}\quad \theta_3 = 73.2°,\ \theta_5 = 106.8°;$$

$$\cos\theta_4 = (0)/(6)^{1/2} = 0, \quad\text{or}\quad \theta_4 = 90°.$$

Thus the possible values for $\ell = 3$ are $\boxed{30.0°, 54.7°, 73.2°, 90°, 106.8°, 125.3°, 150.0°.}$

(*d*) The minimum value of θ has the largest value of m_ℓ. For $\ell = 100$, we have

$$\cos\theta_{\min} = (100)/[(100)(100+1)]^{1/2} = 0.995, \quad\text{or}\quad \theta_{\min} = \boxed{5.71°.}$$

For $\ell = 10^6$, we have

$$\cos\theta_{\min} = (10^6)/[(10^6)(10^6+1)]^{1/2} = 1/(1+10^{-6})^{1/2} \approx 1 - \tfrac{1}{2}(10^{-6}) = 0.9999995, \quad\text{or}\quad \theta_{\min} = \boxed{0.0573°.}$$

50. We find the uncertainty in the momentum component perpendicular to the motion, when the width of the beam is constrained to a dimension D:

$$\Delta p_y \geq (h/2\pi)/\Delta y = (h/2\pi)/D.$$

The half angle of the beam is given by the direction of the velocity:

$$\sin\theta = v_y/c.$$

We assume that the angle is small: $\sin\theta \approx \theta$, and we take the minimum uncertainty to be the perpendicular momentum, so we have

$$\theta \approx v_y/v_x = p_y/p = [(h/2\pi)/D]/(h/\lambda) = \lambda/2\pi D.$$

The angular spread is

$$2\theta \approx \lambda/\pi D \approx \lambda/D.$$

CHAPTER 29

Note: At the atomic scale, it is most convenient to have energies in electron-volts and wavelengths in nanometers. A useful expression for the energy of a photon in terms of its wavelength is

$E = hf = hc/\lambda = (6.63 \times 10^{-34}\ \text{J}\cdot\text{s})(3.00 \times 10^{8}\ \text{m/s})(10^{-9}\ \text{nm/m})/(1.60 \times 10^{-19}\ \text{J/eV})\lambda;$

$E = (1.24 \times 10^{3}\ \text{eV}\cdot\text{nm})/\lambda.$

A factor that appears in the analysis of electron energies is

$ke^2 = (9.00 \times 10^{9}\ \text{N}\cdot\text{m}^2/\text{C}^2)(1.60 \times 10^{-19}\ \text{C})^2 = 2.30 \times 10^{-28}\ \text{J}\cdot\text{m}.$

1. With the reference level at infinity, the binding energy of the two ions is

Binding energy $= -\text{PE} = ke^2/r$

$= (2.30 \times 10^{-28}\ \text{J}\cdot\text{m})/(0.28 \times 10^{-9}\ \text{m}) = 8.21 \times 10^{-19}\ \text{J} = \boxed{5.13\ \text{eV.}}$

2. With the repulsion of the electron clouds, the binding energy is

Binding energy $= -\text{PE} - \text{PE}_{\text{clouds}};$

$4.43\ \text{eV} = 5.13\ \text{eV} - \text{PE}_{\text{clouds}}$, which gives $\text{PE}_{\text{clouds}} = \boxed{0.70\ \text{eV.}}$

3. When the electrons are midway between the protons, each electron will have a potential energy due to the two protons:

$\text{PE}_{\text{ep}} = -(2)(0.33)ke^2/(r/2) = -(4)(0.33)(2.30 \times 10^{-28}\ \text{J}\cdot\text{m})/(0.074 \times 10^{-9}\ \text{m})(1.60 \times 10^{-19}\ \text{J/eV})$

$= -25.6\ \text{eV}.$

The protons have a potential energy:

$\text{PE}_{\text{pp}} = +ke^2/r = +(2.30 \times 10^{-28}\ \text{J}\cdot\text{m})/(0.074 \times 10^{-9}\ \text{m})(1.60 \times 10^{-19}\ \text{J/eV}) = +19.4\ \text{eV}.$

When the bond breaks, each hydrogen atom will be in the ground state with an energy $E_1 = -13.6$ eV.

Thus the binding energy is

Binding energy $= 2E_1 - (2\text{PE}_{\text{ep}} + \text{PE}_{\text{pp}}) = 2(-13.6\ \text{eV}) - [2(-25.6\ \text{eV}) + 19.4\ \text{eV}] = \boxed{4.6\ \text{eV.}}$

4. We convert the units:

$1\ \text{kcal/mol} = (1\ \text{kcal/mol})(4186\ \text{J/kcal})/(6.02 \times 10^{23}\ \text{molecules/mol})(1.60 \times 10^{-19}\ \text{J/eV})$

$= \boxed{0.0435\ \text{eV/molecule.}}$

For KCl we have

$(4.43\ \text{eV/molecule})[(1\ \text{kcal/mol})/(0.0434\ \text{eV/molecule})] = \boxed{102\ \text{kcal/mol.}}$

5. There are two bonds between O and the H–N dipole and one bond between N and the H–N dipole. The total binding energy is

Binding energy $= -(2\text{PE}_{\text{O}} + \text{PE}_{\text{N}})$

$= -2\{(-1.0e)[k(0.19e)/r_{\text{H1}}] + [k(-0.19e)/r_{\text{N1}}]\} - \{(-1.0e)[k(0.19e)/r_{\text{H2}}] + [k(-0.19e)/r_{\text{N2}}]\}$

$= (0.19ke^2)[(2/r_{\text{H1}}) - (2/r_{\text{N1}}) + (1/r_{\text{H2}}) - (1/r_{\text{N2}})]$

$= (0.19)(2.30 \times 10^{-28}\ \text{J}\cdot\text{m})[(2/1.90\ \text{Å}) - (2/2.90\ \text{Å}) + (1/2.00\ \text{Å}) - (1/3.00\ \text{Å})]/(10^{-10}\ \text{m/Å})$

$= 2.31 \times 10^{-19}\ \text{J} = \boxed{1.4\ \text{eV.}}$

6. The neutral He atom has two electrons in the ground state, $n = 1$, $\ell = 0$, $m_\ell = 0$. Thus the two electrons have opposite spins, $m_s = \pm\frac{1}{2}$. If we try to form a covalent bond, we see that an electron from one of the atoms will have the same quantum numbers as one of the electrons on the other atom. From the exclusion principle, this is not allowed, so the electrons cannot be shared.

We consider the He_2^+ molecular ion to be formed from a neutral He atom and an He^+ ion. If the electron on the ion has a certain spin value, it is possible for one of the electrons on the atom to have the opposite spin. Thus the electron can be in the same spatial region as the electron on the ion, so a bond can be formed.

7. The units of $\hbar^2/I$ are

$(\text{J}\cdot\text{s})^2/(\text{kg}\cdot\text{m}^2) = \text{J}^2/(\text{kg}\cdot\text{m/s}^2)\text{m} = \text{J}^2/(\text{N}\cdot\text{m}) = \text{J}^2/\text{J} = \text{J}.$

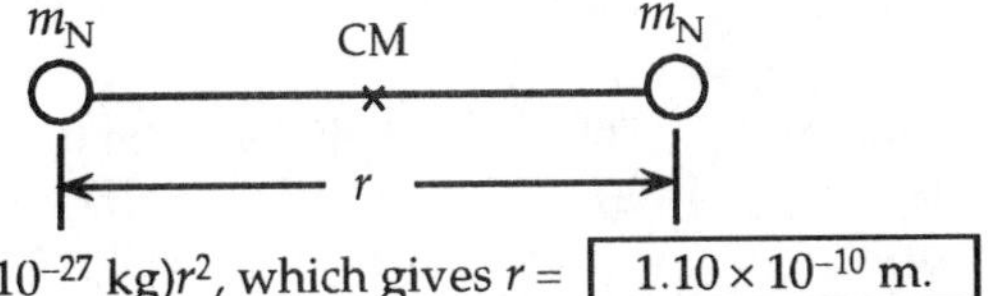

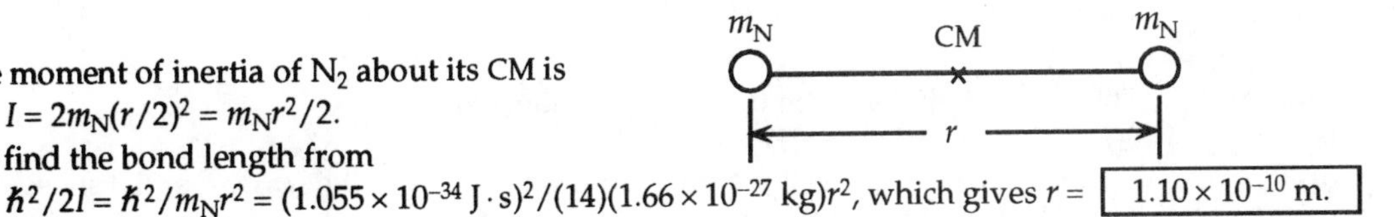

8. The moment of inertia of N_2 about its CM is

$I = 2m_N(r/2)^2 = m_N r^2/2.$

We find the bond length from

$\hbar^2/2I = \hbar^2/m_N r^2 = (1.055 \times 10^{-34}\ \text{J}\cdot\text{s})^2/(14)(1.66 \times 10^{-27}\ \text{kg})r^2$, which gives $r = \boxed{1.10 \times 10^{-10}\ \text{m.}}$

9. (*a*) The moment of inertia of O_2 about its CM is

$I = 2m_O(r/2)^2 = m_O r^2/2.$

We find the characteristic rotational energy from

$\hbar^2/2I = \hbar^2/m_O r^2$

$= (1.055 \times 10^{-34}\ \text{J}\cdot\text{s})^2/(16)(1.66 \times 10^{-27}\ \text{kg})(0.121 \times 10^{-9}\ \text{m})^2$

$= 2.86 \times 10^{-23}\ \text{J} = \boxed{1.79 \times 10^{-4}\ \text{eV.}}$

(*b*) The rotational energy is

$E_{rot} = L(L + 1)(\hbar^2/2I).$

Thus the energy of the emitted photon from the $L = 2$ to $L = 1$ transition is

$hf = \Delta E_{rot} = [(2)(2 + 1) - (1)(1 + 1)](\hbar^2/2I) = 4(\hbar^2/2I) = 4(1.79 \times 10^{-4}\ \text{eV}) = \boxed{7.16 \times 10^{-4}\ \text{eV.}}$

The wavelength is

$\lambda = c/f = hc/hf = (1.24 \times 10^3\ \text{eV}\cdot\text{nm})/(7.16 \times 10^{-4}\ \text{eV}) = 1.73 \times 10^6\ \text{nm} = \boxed{1.73\ \text{mm.}}$

10. The moment of inertia of H_2 about its CM is

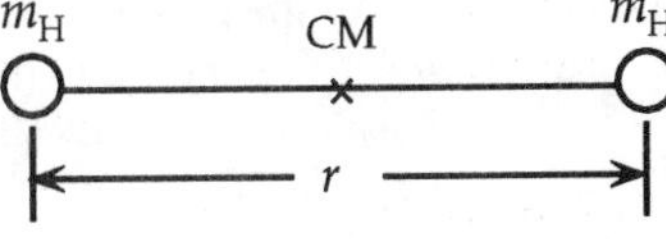

$I = 2m_H(r/2)^2 = m_H r^2/2.$

We find the characteristic rotational energy from

$\hbar^2/2I = \hbar^2/m_H r^2$

$= (1.055 \times 10^{-34}\ \text{J}\cdot\text{s})^2/(1.67 \times 10^{-27}\ \text{kg})(0.074 \times 10^{-9}\ \text{m})^2$

$= 1.22 \times 10^{-23}\ \text{J} = 7.61 \times 10^{-3}\ \text{eV}.$

The rotational energy is

$E_{rot} = L(L + 1)(\hbar^2/2I).$

Thus the energy of the emitted photon from the L to $L - 1$ transition is

$hf = \Delta E_{rot} = [(L)(L + 1) - (L - 1)(L)](\hbar^2/2I) = 2L(\hbar^2/2I).$

(*a*) For the $L = 1$ to $L = 0$ transition, we get

$hf = 2L(\hbar^2/2I) = 2(1)(7.61 \times 10^{-3}\ \text{eV}) = \boxed{1.52 \times 10^{-2}\ \text{eV.}}$

The wavelength is

$\lambda = c/f = hc/hf = (1.24 \times 10^3\ \text{eV}\cdot\text{nm})/(1.52 \times 10^{-2}\ \text{eV}) = 8.16 \times 10^4\ \text{nm} = \boxed{0.082\ \text{mm.}}$

(*b*) For the $L = 2$ to $L = 1$ transition, we get

$hf = 2L(\hbar^2/2I) = 2(2)(7.61 \times 10^{-3}\ \text{eV}) = \boxed{3.04 \times 10^{-2}\ \text{eV.}}$

The wavelength is

$\lambda = c/f = hc/hf = (1.24 \times 10^3\ \text{eV}\cdot\text{nm})/(3.04 \times 10^{-2}\ \text{eV}) = 4.08 \times 10^4\ \text{nm} = \boxed{0.041\ \text{mm.}}$

(*c*) For the $L = 3$ to $L = 2$ transition, we get

$hf = 2L(\hbar^2/2I) = 2(3)(7.61 \times 10^{-3}\ \text{eV}) = \boxed{4.56 \times 10^{-2}\ \text{eV.}}$

The wavelength is

$\lambda = c/f = hc/hf = (1.24 \times 10^3\ \text{eV}\cdot\text{nm})/(4.56 \times 10^{-2}\ \text{eV}) = 2.72 \times 10^4\ \text{nm} = \boxed{0.027\ \text{mm.}}$

11. We find the energies for the transitions from

$\Delta E = hf = hc/\lambda = (1.24 \times 10^3 \text{ eV}\cdot\text{nm})/\lambda$:

$\Delta E_1 = (1.24 \times 10^3 \text{ eV}\cdot\text{nm})/(23.1 \times 10^6 \text{ nm}) = 5.37 \times 10^{-5} \text{ eV}$;

$\Delta E_2 = (1.24 \times 10^3 \text{ eV}\cdot\text{nm})/(11.6 \times 10^6 \text{ nm}) = 10.7 \times 10^{-5} \text{ eV}$;

$\Delta E_3 = (1.24 \times 10^3 \text{ eV}\cdot\text{nm})/(7.71 \times 10^6 \text{ nm}) = 16.1 \times 10^{-5} \text{ eV}$.

The rotational energy is

$E_{\text{rot}} = L(L + 1)(\hbar^2/2I)$.

Thus the energy of the emitted photon from the L to $L - 1$ transition is

$hf = \Delta E_{\text{rot}} = [(L)(L + 1) - (L - 1)(L)](\hbar^2/2I) = 2L(\hbar^2/2I)$.

Because $\Delta E_3 = 3\,\Delta E_1$, and $\Delta E_2 = 2\,\Delta E_1$, the three transitions must be from the $L = 1, 2$, and 3 states.

We find the moment of inertia about the CM from

$\Delta E_3 = 2(3)(1.055 \times 10^{-34} \text{ J}\cdot\text{s})^2/2I$, which gives $I = 1.29 \times 10^{-45} \text{ kg}\cdot\text{m}^2$.

The positions of the atoms from the CM are

$r_1 = [m_{\text{Na}}(0) + m_{\text{Cl}}r]/(m_{\text{Na}} + m_{\text{Cl}})$

$= (35.5 \text{ u})r/(23.0 \text{ u} + 35.5 \text{ u}) = 0.607r$;

$r_2 = r - r_1 = 0.393r$.

We find the bond length from

$I = m_{\text{Na}}r_1^2 + m_{\text{Cl}}r_2^2$;

$1.29 \times 10^{-45} \text{ kg}\cdot\text{m}^2 = [(23.0 \text{ u})(0.607r)^2 + (35.5 \text{ u})(0.393r)^2](1.66 \times 10^{-27} \text{ kg/u})$,

which gives $r =$ $\boxed{2.35 \times 10^{-10} \text{ m.}}$

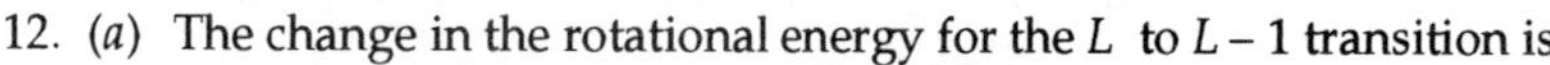

12. (*a*) The change in the rotational energy for the L to $L - 1$ transition is

$\Delta E_{\text{rot}} = [(L)(L + 1) - (L - 1)(L)](\hbar^2/2I) = 2L(\hbar^2/2I)$.

Thus the separation of adjacent lines is

$\Delta(\Delta E_{\text{rot}}) = 2\,\Delta L(\hbar^2/2I) = \hbar^2/I$;

$2.6 \times 10^{-3} \text{ eV} = (1.055 \times 10^{-34} \text{ J}\cdot\text{s})^2/I$,

which gives $I =$ $\boxed{2.68 \times 10^{-47} \text{ kg}\cdot\text{m}^2.}$

(*b*) The positions of the atoms from the CM are

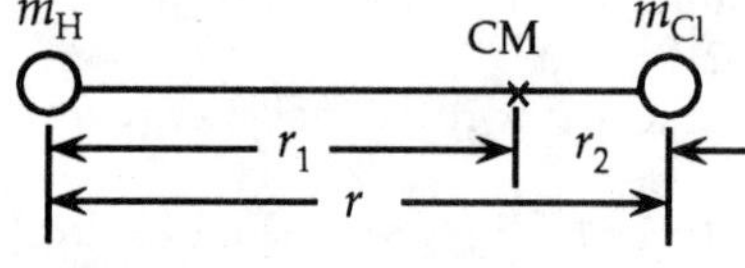

$r_1 = [m_{\text{H}}(0) + m_{\text{Cl}}r]/(m_{\text{H}} + m_{\text{Cl}})$

$= (35 \text{ u})r/(1 \text{ u} + 35 \text{ u}) = 0.972r$;

$r_2 = r - r_1 = 0.028r$.

We find the bond length from

$I = m_{\text{H}}r_1^2 + m_{\text{Cl}}r_2^2$;

$2.68 \times 10^{-47} \text{ kg}\cdot\text{m}^2 = [(1 \text{ u})(0.972r)^2 + (35 \text{ u})(0.028r)^2](1.66 \times 10^{-27} \text{ kg/u})$,

which gives $r =$ $\boxed{1.3 \times 10^{-10} \text{ m.}}$

13. The vibrational energy levels are

$E_{\text{vib}} = (v + \tfrac{1}{2})hf$.

The rotational energy levels are

$E_{\text{rot}} = L(L + 1)(\hbar^2/2I)$.

For an absorption from level v, L to $v + 1, L + 1$, the absorbed energy is

$$\begin{aligned}\Delta E &= \Delta E_{\text{vib}} + \Delta E_{\text{rot}}\\ &= [(v + 1 + \tfrac{1}{2}) - (v + \tfrac{1}{2})]hf + [(L + 1)(L + 2) - (L)(L + 1)](\hbar^2/2I)\\ &= hf + 2(L + 1)(\hbar^2/2I) = hf + (L + 1)(\hbar^2/I), \quad L = 0, 1, 2, \ldots .\end{aligned}$$

For an absorption from level v, L to $v + 1, L - 1$, the absorbed energy is

$$\begin{aligned}\Delta E &= \Delta E_{\text{vib}} + \Delta E_{\text{rot}}\\ &= [(v + 1 + \tfrac{1}{2}) - (v + \tfrac{1}{2})]hf + [(L - 1)(L) - (L)(L + 1)](\hbar^2/2I)\\ &= hf - 2L(\hbar^2/2I) = hf - L(\hbar^2/I), \quad L = 1, 2, 3, \ldots .\end{aligned}$$

14. (*a*) The positions of the atoms from the CM are

$r_1 = [m_1(0) + m_2r_0]/(m_1 + m_2) = m_2r_0/(m_1 + m_2);$

$r_2 = r_0 - r_1 = r - [m_2r_0/(m_1 + m_2)] = m_1r_0/(m_1 + m_2)$

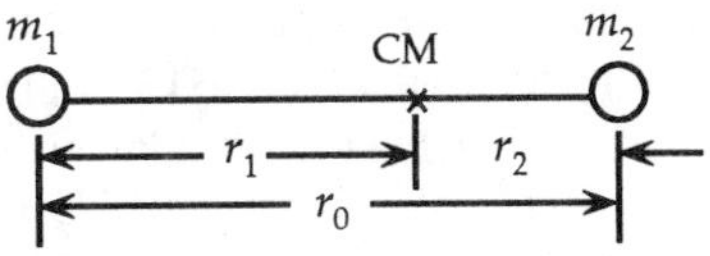

The moment of inertia is

$I = m_1r_1^2 + m_2r_2^2 = m_1[m_2r_0/(m_1 + m_2)]^2 + m_2[m_1r_0/(m_1 + m_2)]^2$

$= m_1m_2(m_2 + m_1)r_0^2/(m_1 + m_2)^2 = \mu r_0^2$, where $\mu = m_1m_2/(m_1 + m_2)$.

(*b*) For H_2 we have

$\mu = m_1m_2/(m_1 + m_2) = (1\text{ u})(1\text{ u})/(1\text{ u} + 1\text{ u}) = 0.50\text{ u} = 8.3 \times 10^{-28}\text{ kg}.$

$I = \mu r_0^2 = (8.3 \times 10^{-28}\text{ kg})(0.074 \times 10^{-9}\text{ m})^2 = \boxed{4.5 \times 10^{-48}\text{ kg}\cdot\text{m}^2.}$

(*c*) For O_2 we have

$\mu = m_1m_2/(m_1 + m_2) = (16\text{ u})(16\text{ u})/(16\text{ u} + 16\text{ u}) = 8\text{ u} = 1.33 \times 10^{-26}\text{ kg}.$

$I = \mu r_0^2 = (1.33 \times 10^{-26}\text{ kg})(0.121 \times 10^{-9}\text{ m})^2 = \boxed{1.94 \times 10^{-46}\text{ kg}\cdot\text{m}^2.}$

(*d*) For NaCl we have

$\mu = m_1m_2/(m_1 + m_2) = (23.0\text{ u})(35.5\text{ u})/(23.0\text{ u} + 35.5\text{ u}) = 14.0\text{ u} = 2.32 \times 10^{-26}\text{ kg}.$

$I = \mu r_0^2 = (2.32 \times 10^{-26}\text{ kg})(0.24 \times 10^{-9}\text{ m})^2 = \boxed{1.3 \times 10^{-45}\text{ kg}\cdot\text{m}^2.}$

(*e*) For CO we have

$\mu = m_1m_2/(m_1 + m_2) = (12\text{ u})(16\text{ u})/(12\text{ u} + 16\text{ u}) = 6.86\text{ u} = 1.14 \times 10^{-26}\text{ kg}.$

$I = \mu r_0^2 = (1.14 \times 10^{-26}\text{ kg})(0.113 \times 10^{-9}\text{ m})^2 = \boxed{1.45 \times 10^{-46}\text{ kg}\cdot\text{m}^2.}$

15. (*a*) The curve for $\text{PE} = \frac{1}{2}kx^2$ is shown in Figure 29–17 as a dotted line. This line crosses the PE = 0 axis at 0.120 nm. If we take the energy to be zero at the lowest point, the energy at the axis is 4.5 eV. Thus we have

$\text{PE} = \frac{1}{2}kx^2;$

$(4.5\text{ eV})(1.60 \times 10^{-19}\text{ J/eV}) = \frac{1}{2}k[(0.120\text{ nm} - 0.074\text{ nm})(10^{-9}\text{ m/nm})]^2$, which gives $k = \boxed{680\text{ N/m.}}$

(*b*) For H_2 the reduced mass is

$\mu = m_1m_2/(m_1 + m_2) = (1\text{ u})(1\text{ u})/(1\text{ u} + 1\text{ u}) = 0.50\text{ u} = 8.3 \times 10^{-28}\text{ kg}.$

The fundamental wavelength is

$\lambda = c/f = (2\pi c)(\mu/k)^{1/2}$

$= 2\pi(3.00 \times 10^8\text{ m/s})[(8.3 \times 10^{-28}\text{ kg})/(680\text{ N/m})]^{1/2} = 2.1 \times 10^{-6}\text{ m} = \boxed{2.1\ \mu\text{m.}}$

16. From the figure we see that the distance between nearest neighbor Na ions is the diagonal of the cube:

$D = d\sqrt{2} = (0.24\text{ nm})\sqrt{2} = \boxed{0.34\text{ nm.}}$

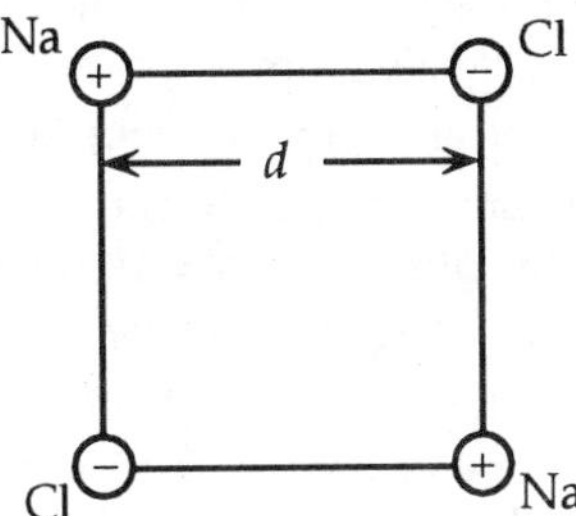

17. Because each ion occupies a cell of side s, a molecule occupies two cells. Thus the density is

$\rho = m_{\text{NaCl}}/2s^3;$

$2.165 \times 10^3\text{ kg/m}^3 = (58.44\text{ u})(1.66 \times 10^{-27}\text{ kg/u})/2s^3$, which gives $s = 2.83 \times 10^{-10}\text{ m} = \boxed{0.283\text{ nm.}}$

18. Because each ion occupies a cell of side s, a molecule occupies two cells. Thus the density is

$\rho = m_{\text{KCl}}/2s^3;$

$1.99 \times 10^3\text{ kg/m}^3 = (39.1\text{ u} + 35.5\text{ u})(1.66 \times 10^{-27}\text{ kg/u})/2s^3$, which gives $s = 3.15 \times 10^{-10}\text{ m} = \boxed{0.315\text{ nm.}}$

19. The partially filled shell in Na is the 3*s* shell, which has 1 electron in it. The partially filled shell in Cl is the 2*p* shell which has 5 electrons in it. In NaCl the electron from the 3*s* shell in Na is transferred to the 2*p* shell in Cl, which results in filled shells for both ions. Thus when many ions are considered, the resulting bands are either completely filled (the valence band) or completely empty (the conduction band). Thus a large energy is required to create a conduction electron by raising an electron from the valence band to the conduction band.

20. The photon with the minimum frequency for conduction must have an energy equal to the energy gap:
$E_g = hf = hc/\lambda = (1.24 \times 10^3 \text{ eV}\cdot\text{nm})/(640 \text{ nm}) = \boxed{1.94 \text{ eV.}}$

21. The photon with the longest wavelength or minimum frequency for conduction must have an energy equal to the energy gap:
$\lambda = c/f = hc/hf = hc/E_g = (1.24 \times 10^3 \text{ eV}\cdot\text{nm})/(1.1 \text{ eV}) = 1.1 \times 10^3 \text{ nm} = \boxed{1.1\ \mu\text{m.}}$

22. The energy of the photon must be greater than or equal to the energy gap. Thus the longest wavelength that will excite an electron is
$\lambda = c/f = hc/hf = hc/E_g = (1.24 \times 10^3 \text{ eV}\cdot\text{nm})/(0.72 \text{ eV}) = 1.7 \times 10^3 \text{ nm} = 1.7\ \mu\text{m}.$
Thus the wavelength range is $\boxed{\lambda \leq 1.7\ \mu\text{m.}}$

23. To use silicon to filter the wavelengths, we want wavelengths below the IR to be able to cause the electron to be raised to the conduction band, so the photon is absorbed in the silicon. We find the shortest wavelength from
$\lambda = c/f = hc/hf > hc/E_g = (1.24 \times 10^3 \text{ eV}\cdot\text{nm})/(1.14 \text{ eV}) = 1.09 \times 10^3 \text{ nm} = \boxed{1.09\ \mu\text{m.}}$
Because this is in the IR, the shorter wavelengths of visible light will excite the electron, so silicon $\boxed{\text{could be used}}$ as a window.

24. (*a*) In the 2*s* shell of an atom, $\ell = 0$, so there are two states: $m_s = \pm\frac{1}{2}$. When *N* atoms form bands, each atom provides 2 states, so the total number of states in the band is $\boxed{2N.}$
(*b*) In the 2*p* shell of an atom, $\ell = 1$, so there are three states from the m_ℓ values: $m_\ell = 0, \pm 1$; each of which has two states from the m_s values: $m_s = \pm\frac{1}{2}$, for a total of 6 states. When *N* atoms form bands, each atom provides 6 states, so the total number of states in the band is $\boxed{6N.}$
(*c*) In the 3*p* shell of an atom, $\ell = 1$, so there are three states from the m_ℓ values: $m_\ell = 0, \pm 1$; each of which has two states from the m_s values: $m_s = \pm\frac{1}{2}$, for a total of 6 states. When *N* atoms form bands, each atom provides 6 states, so the total number of states in the band is $\boxed{6N.}$
(*d*) In general, for a value of ℓ, there are $2\ell + 1$ states from the m_ℓ values: $m_\ell = 0, \pm 1, \ldots, \pm \ell$. For each of these there are two states from the m_s values: $m_s = \pm\frac{1}{2}$, for a total of $2(2\ell + 1)$ states. When *N* atoms form bands, each atom provides $2(2\ell + 1)$ states, so the total number of states in the band is $\boxed{2N(2\ell + 1).}$

25. The minimum energy provided to an electron must be equal to the energy gap:
$E_g = 0.72 \text{ eV}.$
Thus the maximum number of electrons is
$N = hf/E_g = (660 \times 10^3 \text{ eV})/(0.72 \text{ eV}) = \boxed{9.2 \times 10^5.}$

26. If we consider a mole of pure silicon (28 g or 6.02×10^{23} atoms), the number of conduction electrons is
$N_{\text{Si}} = [(28 \times 10^{-3}\text{ kg})/(2330\text{ kg/m}^3)](10^{16}\text{ electrons/m}^3) = 1.20 \times 10^{11}$ conduction electrons.
The additional conduction electrons provided by the doping is
$N_{\text{doping}} = (6.02 \times 10^{23}\text{ atoms})/10^6 = 6.02 \times 10^{17}$ added conduction electrons.
Thus the density of conduction electrons has increased by
$N_{\text{doping}}/N_{\text{Si}} = (6.02 \times 10^{17})/(1.20 \times 10^{11}) = \boxed{5 \times 10^6.}$

27. The photon will have an energy equal to the energy gap:
$\lambda = c/f = hc/hf = hc/E_g = (1.24 \times 10^3\text{ eV}\cdot\text{nm})/(1.4\text{ eV}) = 8.9 \times 10^2\text{ nm} = \boxed{0.89\ \mu\text{m.}}$

28. The photon will have an energy equal to the energy gap:
$E_g = hf = hc/\lambda = (1.24 \times 10^3\text{ eV}\cdot\text{nm})/(650\text{ nm}) = \boxed{1.91\text{ eV.}}$

29. From the current-voltage characteristic, we see that a current of 12 mA means a voltage of 0.68 V across the diode. Thus the battery voltage is
$V_{\text{battery}} = V_{\text{diode}} + V_{\text{R}} = 0.68\text{ V} + (12 \times 10^{-3}\text{ A})(660\ \Omega) = \boxed{8.6\text{ V.}}$

30. The battery voltage is
$V_{\text{battery}} = V_{\text{diode}} + V_{\text{R}};$
$2.0\text{ V} = V(I) + I(0.100\text{ k}\Omega)$, or $V(I) = 2.0\text{ V} - I(0.100\text{ k}\Omega)$.
This is a straight line which passes through the points
(20 mA, 0 V) and (12 mA, 0.8 V),
as drawn in the figure. Because $V(I)$ is represented by both curves, the intersection will give the current, which we see is $\boxed{13\text{ mA.}}$

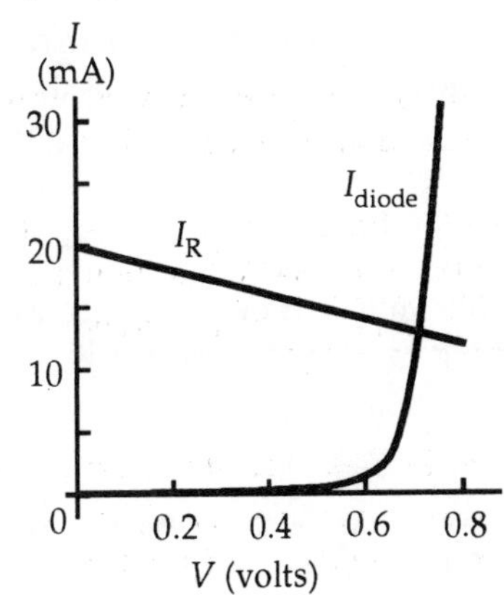

31.

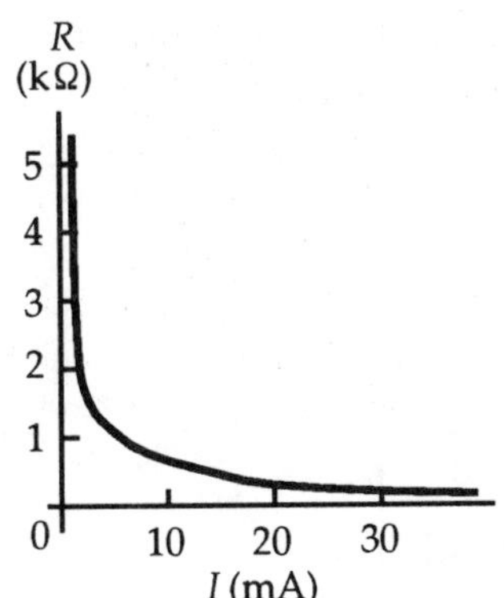

32. (*a*) For a half-wave rectifier without a capacitor, the current is zero for half the time. Thus the average current is
$I_{\text{av}} = \frac{1}{2}V_{\text{rms}}/R = \frac{1}{2}(120\text{ V})/(18\text{ k}\Omega) = \boxed{3.3\text{ mA.}}$
(*b*) For a full-wave rectifier without a capacitor, the current is positive all the time. Thus the average current is
$I_{\text{av}} = V_{\text{rms}}/R = (120\text{ V})/(18\text{ k}\Omega) = \boxed{6.7\text{ mA.}}$

33. There will be a current in the resistor while the ac voltage varies from 0.6 V to 12.0 V rms. Because the 0.6 V is small, the voltage across the resistor will be almost sinusoidal, so the rms voltage across the resistor will be close to 12.0 V – 0.6 V = 11.4 V.
    (*a*) For a half-wave rectifier without a capacitor, the current is zero for half the time. If we ignore the short times it takes to reach 0.6 V, this will also be true for the resistor. Thus the average current is
    $I_{av} = \frac{1}{2}V_{rms}/R = \frac{1}{2}(11.4\text{ V})/(0.150\text{ k}\Omega) =$ **38 mA.**
    (*b*) For a full-wave rectifier without a capacitor, the current is positive all the time. If we ignore the short times it takes to reach 0.6 V, this will also be true for the resistor. Thus the average current is
    $I_{av} = V_{rms}/R = (11.4\text{ V})/(0.150\text{ k}\Omega) =$ **76 mA.**

34. (*a*) The time constant for the circuit is
    $\tau_1 = RC_1 = (18 \times 10^3\ \Omega)(25 \times 10^{-6}\text{ F}) = 0.45\text{ s}.$
    Because there are two peaks per cycle, the period of the rectified voltage is
    $T = 1/2f = 1/2(60\text{ Hz}) = 0.0083\text{ s}.$
    Because $\tau_1 \gg T$, the voltage across the capacitor will be essentially constant during a cycle, so the voltage will be the peak voltage. Thus the average current is
    $I_{av} = V_0/R = \sqrt{2}(120\text{ V})/(18\text{ k}\Omega) =$ **9.4 mA (smooth).**
    (*b*) The time constant for the circuit is
    $\tau_2 = RC_2 = (18 \times 10^3\ \Omega)(0.10 \times 10^{-6}\text{ F}) = 0.0018\text{ s}.$
    Because $\tau_2 < T$, the voltage across the capacitor will be rippled, so the average voltage will be close to the rms voltage. Thus the average current is
    $I_{av} = V_{rms}/R = (120\text{ V})/(18\text{ k}\Omega) =$ **6.7 mA (rippled).**

35. The output voltage is the voltage across the resistor:
    $V = I_C R_C = \beta I_B R_C = (80)(2.0 \times 10^{-6}\text{ A})(3.3 \times 10^3\ \Omega) =$ **0.53 V.**

36. The output voltage is the voltage across the resistor:
    $V = I_C R_C = \beta I_B R_C;$
    $0.40\text{ V} = (100)(1.0 \times 10^{-6}\text{ A})R_C$, which gives $R_C = 4.0 \times 10^3\ \Omega =$ **4.0 kΩ.**

37. (*a*) The output voltage is the voltage across R_C:
    $V_C = I_C R_C,$
    while the input voltage is the voltage across R_B:
    $V_B = I_B R_B.$
    Thus the voltage gain is
    $V_C/V_B = I_C R_C/I_B R_B = \beta R_C/R_B = (90)(6.2\text{ k}\Omega)/(2.2\text{ k}\Omega) =$ **2.5×10^2.**
    (*b*) The power amplification is
    $P_{output}/P_{input} = I_C V_C/I_B V_B = \beta V_C/V_C = (90)(2.5 \times 10^2) =$ **2.3×10^4.**

38. The output current is
    $I_C = V_C/R_C = (70)V_B/R_C = (70)(0.080\text{ V})/(14\text{ k}\Omega) =$ **0.40 mA.**

39. For an electron confined within Δx, we find the uncertainty in the momentum from

$\Delta p = \hbar/\Delta x$,

which we take to be the momentum of the particle. The kinetic energy of the electron is

$\text{KE} = p^2/2m = \hbar^2/2m(\Delta x)^2$.

When the two electrons are in separated atoms, we get

$\text{KE}_1 = 2\hbar^2/2m(\Delta x)^2$

$= 2(1.055 \times 10^{-34}\ \text{J}\cdot\text{s})^2/2(9.11 \times 10^{-31}\ \text{kg})(0.053 \times 10^{-9}\ \text{m})^2 = 4.35 \times 10^{-18}\ \text{J} = 27.2\ \text{eV}$.

When the electrons are in the molecule, we get

$\text{KE}_2 = 2\hbar^2/2m(\Delta x)^2$

$= 2(1.055 \times 10^{-34}\ \text{J}\cdot\text{s})^2/2(9.11 \times 10^{-31}\ \text{kg})(0.074 \times 10^{-9}\ \text{m})^2 = 2.23 \times 10^{-18}\ \text{J} = 14.0\ \text{eV}$.

Thus the binding energy is

$\text{KE}_1 - \text{KE}_2 = 27.2\ \text{eV} - 14.0\ \text{eV} = \boxed{13\ \text{eV}.}$

40. (*a*) We find the temperature from

$\text{KE} = \frac{3}{2}kT$;

$(4.5\ \text{eV})(1.60 \times 10^{-19}\ \text{J/eV}) = \frac{3}{2}(1.38 \times 10^{-23}\ \text{J/K})T$, which gives $T = \boxed{3.5 \times 10^4\ \text{K}.}$

(*b*) We find the temperature from

$\text{KE} = \frac{3}{2}kT$;

$(0.15\ \text{eV})(1.60 \times 10^{-19}\ \text{J/eV}) = \frac{3}{2}(1.38 \times 10^{-23}\ \text{J/K})T$, which gives $T = \boxed{1.2 \times 10^3\ \text{K}.}$

41. (*a*) The potential energy for the point charges is

$\text{PE} = -ke^2/r = -(2.30 \times 10^{-28}\ \text{J}\cdot\text{m})/(0.27 \times 10^{-9}\ \text{m}) = -8.52 \times 10^{-19}\ \text{J} = \boxed{-5.3\ \text{eV}.}$

(*b*) Because the potential energy of the ions is negative, 5.3 eV is released when the ions are brought together. A release of energy means that energy must be provided to return the ions to the state of free atoms. Thus the total binding energy of the KF ions is

Binding energy $= 5.3\ \text{eV} + 4.07\ \text{eV} - 4.34\ \text{eV} = \boxed{5.0\ \text{eV}.}$

42. The photon with the longest wavelength has the minimum energy, so the energy gap must be

$E_g = hc/\lambda = (1.24 \times 10^3\ \text{eV}\cdot\text{nm})/(1000\ \text{nm}) = \boxed{1.24\ \text{eV}.}$

43. In a dielectric, Coulomb's law becomes

$F = ke^2/r^2 = e^2/4\pi K\epsilon_0 r^2$.

Thus where e^2 appears in an equation, we divide by K. If the "extra" electron is outside the arsenic ion, the effective Z will be 1, and we can use the hydrogen results.

(*a*) The energy of the electron is

$E = -2\pi^2 Z^2 e^4 mk^2/K^2h^2n^2 = -(13.6\ \text{eV})Z^2/K^2n^2 = -(13.6\ \text{eV})(1)^2/(12)^2(1)^2 = -0.094\ \text{eV}$.

Thus the binding energy is $\boxed{0.094\ \text{eV}.}$

(*b*) The radius of the electron orbit is

$r = K^2h^2n^2/4\pi^2 Ze^2mk = K^2n^2(0.0529\ \text{nm})/Z = (12)(1)^2(0.0529\ \text{nm})/(1)^2 = \boxed{0.63\ \text{nm}.}$

Note that this result justifies the assumption that the electron is outside the arsenic ion.

44. With a full-wave rectifier, there are two peaks for each cycle of the input voltage, so the time between peaks is
$T = 1/2f = 1/2(60 \text{ Hz}) = 0.00833 \text{ s}.$
The time constant of the rectifier is
$\tau = RC = (10 \times 10^3 \ \Omega)(30 \times 10^{-6} \text{ F}) = 0.30 \text{ s}.$
Because $T \ll \tau$, we assume that the exponential discharge of the capacitor voltage is linear:
$V_C = V_0(1 - t/\tau).$
We approximate the lowest voltage of the capacitor by finding the value reached in the time from one peak to the next:
$V = V_0[1 - (0.00833 \text{ s})/(0.30 \text{ s})] = 0.972\ V_0.$
Thus the ripple about the mean value is
$\pm \frac{1}{2}(V_0 - V)/V_0 = \pm \frac{1}{2}(1 - 0.972) = \pm 0.014 = \boxed{\pm 1.4\%.}$

45. (*a*) The current through the load resistor is
$I_{load} = V_{output}/R_{load} = (130 \text{ V})/(15.0 \text{ k}\Omega) = 8.67 \text{ mA}.$
At the minimum supply voltage the current through the diode will be zero, so the current through R is 8.67 mA, and the voltage across R is
$V_{R,min} = I_{R,min}R = (8.67 \text{ mA})(1.80 \text{ k}\Omega) = 15.6 \text{ V}.$
The minimum supply voltage is
$V_{min} = V_{R,min} + V_{output} = 15.6 \text{ V} + 130 \text{ V} = 146 \text{ V}.$
At the maximum supply voltage the current through the diode will be 100 mA, so the current through R is 100 mA + 8.67 mA = 108.7 mA, and the voltage across R is
$V_{R,max} = I_{R,max}R = (108.7 \text{ mA})(1.80 \text{ k}\Omega) = 196 \text{ V}.$
The maximum supply voltage is
$V_{max} = V_{R,max} + V_{output} = 196 \text{ V} + 130 \text{ V} = 326 \text{ V}.$
Thus the range of supply voltages is $\boxed{146 \text{ V} \le V \le 326 \text{ V}.}$

(*b*) At a constant supply voltage the voltage across R is 200 V – 130 V = 70 V, so the current in R is
$I_R = (70 \text{ V})/(1.80 \text{ k}\Omega) = 38.9 \text{ mA}.$
If there is no current through the diode, this current must be in the load resistor, so we have
$R_{load} = (130 \text{ V})/(38.9 \text{ mA}) = 3.34 \text{ k}\Omega.$
If R_{load} is less than this, there will be a greater current through R, and thus the voltage across the load will drop and regulation will be lost. If R_{load} is greater than 3.34 kΩ, the current through R_{load} will decrease and there will be current through the diode. The current through the diode is 38.9 mA when R_{load} is infinite, which is less than the maximum of 100 mA.
Thus the range for load resistance is $\boxed{3.34 \text{ k}\Omega \le R_{load} < \infty.}$

46. The moment of inertia of the baton is
$I = (ML^2/12) + [2m(L/2)^2] = [(0.200 \text{ kg})(0.30 \text{ m})^2/12] + [2(0.300 \text{ kg})(0.15)^2] = 0.015 \text{ kg}\cdot\text{m}^2.$
The rotational energy of the baton is
$E = \frac{1}{2}I\omega^2 = \frac{1}{2}(0.015 \text{ kg}\cdot\text{m}^2)[(2 \text{ rev/s})(2\pi \text{ rad/rev})]^2 = 1.18 \text{ J}.$
We find the rotational quantum number from
$L^2 = (I\omega)^2 = \ell(\ell + 1)\hbar^2;$
$[(0.015 \text{ kg}\cdot\text{m}^2)(2 \text{ rev/s})(2\pi \text{ rad/rev})]^2 = \ell(\ell + 1)(1.055 \times 10^{-34} \text{ J}\cdot\text{s})^2$, which gives $\ell = 1.8 \times 10^{33}$.
Thus the difference in rotational energy levels is
$\Delta E_{rot} = \ell\hbar^2/I = (1.8 \times 10^{33})(1.055 \times 10^{-34} \text{ J}\cdot\text{s})^2/(0.015 \text{ kg}\cdot\text{m}^2) = \boxed{1.3 \times 10^{-33} \text{ J}.}$
$\boxed{\text{No,}}$ because $\Delta E_{rot} \ll E$, we do not consider quantum effects for the baton.

47. The Hall voltage is produced by the drifting electrons:

$\mathcal{E}_H = v_d BL$;

18×10^{-3} V = v_d(1.5 T)(1.5×10^{-2} m), which gives v_d = 0.80 m/s.

We find the density of drifting electrons from the current:

$I = neAv_d$;

0.20×10^{-3} A = n(1.60×10^{-19} C)(1.5×10^{-2} m)(1.0×10^{-3} m)(0.80 m/s),

which gives $n = 1.04 \times 10^{20}$ electrons/m^3.

The density of silicon atoms is

N = [(2.33×10^6 g/m^3)/(28 g/mol)](6.02×10^{23} atoms/mol) = 5.0×10^{28} atoms/m^3.

Thus the ratio of electrons to atoms is

n/N = (1.04×10^{20} electrons/m^3)/(5.0×10^{28} atoms/m^3) = $\boxed{2.0 \times 10^{-9}.}$

48. From the diagram of the cubic lattice, we see that an atom inside the cube is bonded to the six nearest neighbors. Because each bond is shared by two atoms, the number of bonds per atom is 3. We find the heat of fusion for argon from the energy required to break the bonds:

L_{fusion} = (number of bonds/atom)(number of atoms/kg)E_{bond}

= (3)[(6.02×10^{23} atoms/mol)/(39.95×10^{-3} kg/mol)] ×

(3.9×10^{-3} eV)(1.60×10^{-19} J/eV)

= $\boxed{2.8 \times 10^4 \text{ J/kg.}}$

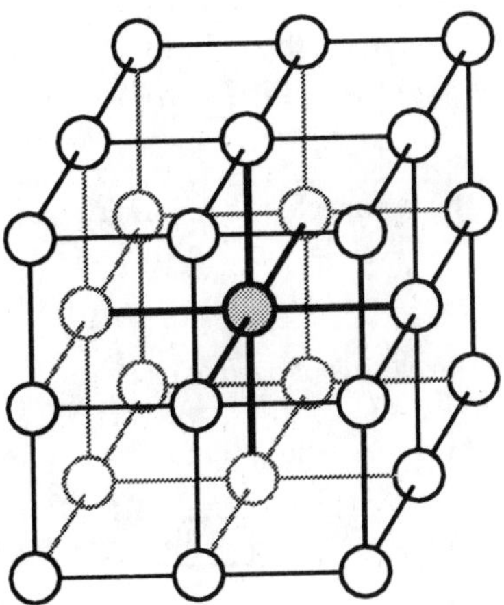

CHAPTER 30

Note: A useful expression for the energy of a photon in terms of its wavelength is
$E = hf = hc/\lambda = (6.63 \times 10^{-34}\ \text{J}\cdot\text{s})(3.00 \times 10^{8}\ \text{m/s})(10^{-9}\ \text{nm/m})/(1.60 \times 10^{-19}\ \text{J/eV})\lambda;$
$E = (1.24 \times 10^{3}\ \text{eV}\cdot\text{nm})/\lambda.$

A factor that appears in the analysis of energies is
$ke^2 = (9.00 \times 10^{9}\ \text{N}\cdot\text{m}^2/\text{C}^2)(1.60 \times 10^{-19}\ \text{C})^2 = 2.30 \times 10^{-28}\ \text{J}\cdot\text{m} = 1.44\ \text{MeV}\cdot\text{fm}.$

1. To find the rest mass of an α particle, we subtract the rest mass of the two electrons from the rest mass of a helium atom:
$m_\alpha = m_{\text{He}} - 2m_e$
$= (4.002602\ \text{u})(931.5\ \text{MeV}/\text{u}c^2) - 2(0.511\ \text{MeV}/c^2) = \boxed{3727\ \text{MeV}/c^2.}$

2. We convert the units:
$m = (139\ \text{MeV}/c^2)/(931.5\ \text{MeV}/\text{u}c^2) = \boxed{0.149\ \text{u}.}$

3. The α particle is a helium nucleus:
$r = (1.2 \times 10^{-15}\ \text{m})A^{1/3} = (1.2 \times 10^{-15}\ \text{m})(4)^{1/3} = \boxed{1.9 \times 10^{-15}\ \text{m}} = 1.9\ \text{fm}.$

4. The radius of a nucleus is
$r = (1.2 \times 10^{-15}\ \text{m})A^{1/3}.$
If we form the ratio for the two isotopes, we get
$r_{14}/r_{12} = (14/12)^{1/3} = 1.053.$
Thus the ^{14}C is $\boxed{5.3\%}$ greater than ^{12}C.

5. (*a*) The fraction of mass is
$m_p/(m_p + m_e) = (1.67 \times 10^{-27}\ \text{kg})/(1.67 \times 10^{-27}\ \text{kg} + 9.11 \times 10^{-31}\ \text{kg}) = \boxed{0.99945.}$
(*b*) The fraction of volume is
$(r_{\text{nucleus}}/r_{\text{atom}})^3 = [(1.2 \times 10^{-15}\ \text{m})/(0.53 \times 10^{-10}\ \text{m})]^3 = \boxed{1.2 \times 10^{-14}.}$
(*c*) If we take the density of the hydrogen nucleus as the density of nuclear matter, we get
$\rho = m_p/\tfrac{4}{3}\pi r^3 = (1.67 \times 10^{-27}\ \text{kg})/\tfrac{4}{3}\pi(1.2 \times 10^{-15}\ \text{m})^3 = \boxed{2.3 \times 10^{17}\ \text{kg/m}^3.}$
The density of water is 1000 kg/m^3, so nuclear matter is $\boxed{10^{14}\times}$ greater.

6. (*a*) The radius of ^{64}Cu is
$r = (1.2 \times 10^{-15}\ \text{m})A^{1/3} = (1.2 \times 10^{-15}\ \text{m})(64)^{1/3} = \boxed{4.8 \times 10^{-15}\ \text{m}} = 4.8\ \text{fm}.$
(*b*) We find the value of A from
$r = (1.2 \times 10^{-15}\ \text{m})A^{1/3};$
$3.7 \times 10^{-15}\ \text{m} = (1.2 \times 10^{-15}\ \text{m})A^{1/3}$, which gives $A = \boxed{29.}$

7. We find the radii of the two nuclei from
$R = r_0 A^{1/3};$
$R_\alpha = (1.2\ \text{fm})(4)^{1/3} = 1.9\ \text{fm};$
$R_U = (1.2\ \text{fm})(238)^{1/3} = 7.4\ \text{fm}.$
If the two nuclei are just touching, the Coulomb potential energy must be the initial kinetic energy of the α particle:
$\text{KE} = \text{PE} = Z_\alpha Z_U ke^2/(R_\alpha + R_U)$
$= (2)(92)(1.44\ \text{MeV}\cdot\text{fm})/(1.9\ \text{fm} + 7.4\ \text{fm}) = \boxed{28\ \text{MeV}.}$

8. We find the radii of the two nuclei from

$R = r_0 A^{1/3}$;

$R_\alpha = (1.2 \text{ fm})(4)^{1/3} = 1.9 \text{ fm}$;

$R_{Am} = (1.2 \text{ fm})(243)^{1/3} = 7.5 \text{ fm}$.

We assume that the nucleus is so much heavier than the α particle that the nucleus does not recoil. We find the kinetic energy of the α particle from the conservation of energy:

$KE_i + PE_i = KE_f + PE_f$;

$0 + Z_\alpha Z_{Am} ke^2/(R_\alpha + R_{Am}) = KE_f + 0$;

$(2)(95)(1.44 \text{ MeV} \cdot \text{fm})/(1.9 \text{ fm} + 7.5 \text{ fm}) = KE_f$, which gives $KE_f =$ **29 MeV.**

9. The radius of a nucleus is

$r = (1.2 \times 10^{-15} \text{ m})A^{1/3}$.

If we form the ratio for the two nuclei, we get

$r_X/r_U = (A_X/A_U)^{1/3}$;

$\frac{1}{2} = (A_X/238)^{1/3}$, which gives $A_X = 30$.

From the Appendix, we see that the stable nucleus could be **$^{31}_{15}P$.**

10. From Figure 30–1, we see that the average binding energy per nucleon at $A = 40$ is 8.6 MeV. Thus the total binding energy for ^{40}Ca is

$(40)(8.6 \text{ MeV}) =$ **340 MeV.**

11. (*a*) From Figure 30–1, we see that the average binding energy per nucleon at $A = 238$ is 7.5 MeV. Thus the total binding energy for ^{238}U is

$(238)(7.5 \text{ MeV}) =$ **1.8×10^3 MeV.**

(*b*) From Figure 30–1, we see that the average binding energy per nucleon at $A = 84$ is 8.7 MeV. Thus the total binding energy for ^{84}Kr is

$(84)(8.7 \text{ MeV}) =$ **730 MeV.**

12. Deuterium consists of one proton and one neutron. We find the binding energy from the masses:

Binding energy $= [m(^1H) + m(^1n) - m(^2H)]c^2$

$= [(1.007825 \text{ u}) + (1.008665 \text{ u}) - (2.014102 \text{ u})]c^2(931.5 \text{ MeV}/uc^2) =$ **2.22 MeV.**

13. ^{14}N consists of seven protons and seven neutrons. We find the binding energy from the masses:

Binding energy $= [7m(^1H) + 7m(^1n) - m(^{14}N)]c^2$

$= [7(1.007825 \text{ u}) + 7(1.008665 \text{ u}) - (14.003074 \text{ u})]c^2(931.5 \text{ MeV}/uc^2) = 104.7 \text{ MeV}$.

Thus the binding energy per nucleon is

$(104.7 \text{ MeV})/14 =$ **7.48 MeV.**

14. ^6Li consists of three protons and three neutrons. We find the binding energy from the masses:

Binding energy $= [3m(^1H) + 3m(^1n) - m(^6Li)]c^2$

$= [3(1.007825 \text{ u}) + 3(1.008665 \text{ u}) - (6.015121 \text{ u})]c^2(931.5 \text{ MeV}/uc^2) =$ **32.0 MeV.**

Thus the binding energy per nucleon is

$(32.0 \text{ MeV})/6 =$ **5.33 MeV.**

15. We find the binding energy of the last neutron from the masses:

Binding energy $= [m(^{11}C) + m(^1n) - m(^{12}C)]c^2$

$= [(11.011433 \text{ u}) + (1.008665 \text{ u}) - (12.000000 \text{ u})]c^2(931.5 \text{ MeV}/uc^2) =$ **18.7 MeV.**

16. We find the binding energy of the last neutron from the masses:

$$\text{Binding energy}(^{23}\text{Na}) = [m(^{22}\text{Na}) + m(^{1}\text{n}) - m(^{23}\text{Na})]c^2$$
$$= [(21.994434\text{ u}) + (1.008665\text{ u}) - (22.989767\text{ u})]c^2(931.5\text{ MeV}/\text{u}c^2)$$
$$= \boxed{12.4\text{ MeV.}}$$
$$\text{Binding energy}(^{24}\text{Na}) = [m(^{23}\text{Na}) + m(^{1}\text{n}) - m(^{24}\text{Na})]c^2$$
$$= [(22.989767\text{ u}) + (1.008665\text{ u}) - (23.990961\text{ u})]c^2(931.5\text{ MeV}/\text{u}c^2)$$
$$= \boxed{7.0\text{ MeV.}}$$

17. We find the required energy for separation from the masses.

(*a*) Removal of a proton creates an isotope of nitrogen:

$$\text{Energy(p)} = [m(^{15}\text{N}) + m(^{1}\text{H}) - m(^{16}\text{O})]c^2$$
$$= [(15.000108\text{ u}) + (1.007825\text{ u}) - (15.994915\text{ u})]c^2(931.5\text{ MeV}/\text{u}c^2) = \boxed{12.1\text{ MeV.}}$$

(*b*) Removal of a neutron creates another isotope of oxygen:

$$\text{Energy(n)} = [m(^{15}\text{O}) + m(^{1}\text{n}) - m(^{16}\text{O})]c^2$$
$$= [(15.003065\text{ u}) + (1.008665\text{ u}) - (15.994915\text{ u})]c^2(931.5\text{ MeV}/\text{u}c^2) = \boxed{15.7\text{ MeV.}}$$

The nucleons are held by the attractive strong nuclear force. It takes less energy to remove the proton because there is also the $\boxed{\text{repulsive electric force}}$ from the other protons.

18. (*a*) We find the binding energy from the masses:

$$\text{Binding energy} = [2m(^{4}\text{He}) - m(^{8}\text{Be})]c^2$$
$$= [2(4.002602\text{ u}) - (8.005305\text{ u})]c^2(931.5\text{ MeV}/\text{u}c^2) = -0.094\text{ MeV.}$$

Because the binding energy is negative, the nucleus is unstable.

(*b*) We find the binding energy from the masses:

$$\text{Binding energy} = [3m(^{4}\text{He}) - m(^{12}\text{C})]c^2$$
$$= [3(4.002602\text{ u}) - (12.000000\text{ u})]c^2(931.5\text{ MeV}/\text{u}c^2) = +7.3\text{ MeV.}$$

Because the binding energy is positive, the nucleus is $\boxed{\text{stable.}}$

19. The total mass of the excited atom includes the mass of the γ ray.

$$m = (59.933820) + [(1.33\text{ MeV})/(931.5\text{ MeV}/\text{u}c^2)] = \boxed{59.93525\text{ u.}}$$

20. The decay is ${}^{3}_{1}\text{H} \rightarrow {}^{3}_{2}\text{He} + {}^{0}_{-1}\text{e} + \bar{\nu}$. When we add an electron to both sides to use atomic masses, we see that the mass of the emitted β particle is included in the atomic mass of ^{3}He.
Thus the energy released is

$$Q = [m(^{3}\text{H}) - m(^{3}\text{He})]c^2 = [(3.016049\text{ u}) - (3.016029\text{ u})]c^2(931.5\text{ MeV}/\text{u}c^2) = 0.0186\text{ MeV} = \boxed{18.6\text{ keV.}}$$

21. The decay is ${}^{1}_{0}\text{n} \rightarrow {}^{1}_{1}\text{p} + {}^{0}_{-1}\text{e} + \bar{\nu}$. We take the electron mass to use the atomic mass of ^{1}H. The kinetic energy of the electron will be maximum if no neutrino is emitted. If we ignore the recoil of the proton, the maximum kinetic energy is

$$\text{KE} = [m(^{1}\text{n}) - m(^{1}\text{H})]c^2 = [(1.008665\text{ u}) - (1.007825\text{ u})]c^2(931.5\text{ MeV}/\text{u}c^2) = \boxed{0.783\text{ MeV.}}$$

22. For the decay ${}^{11}_{6}\text{C} \rightarrow {}^{10}_{5}\text{B} + {}^{1}_{1}\text{p}$, we find the difference of the initial and the final masses:

$$\Delta m = m(^{11}\text{C}) - m(^{10}\text{B}) - m(^{1}\text{H})$$
$$= (11.011433\text{ u}) - (10.012936\text{ u}) - (1.007825\text{ u}) = -0.0099318\text{ u.}$$

Thus some additional energy would have to be added.

23. If $^{22}_{11}\text{Na}$ were a β^- emitter, the resulting nucleus would be $^{22}_{12}\text{Mg}$, which has too few neutrons relative to the number of protons to be stable. Thus we have a **β^+ emitter.**

For the reaction $^{22}_{11}\text{Na} \rightarrow ^{22}_{10}\text{Ne} + \beta^+ + \nu$, if we add 11 electrons to both sides in order to use atomic masses, we see that we have two extra electron masses on the right. The kinetic energy of the β^+ will be maximum if no neutrino is emitted. If we ignore the recoil of the neon, the maximum kinetic energy is

$$\text{KE} = [m(^{22}\text{Na}) - m(^{22}\text{Ne}) - 2m(\text{e})]c^2$$
$$= [(21.994434\text{ u}) - (21.991383\text{ u}) - 2(0.00054858\text{ u})]c^2(931.5\text{ MeV}/uc^2) = \boxed{1.82\text{ MeV.}}$$

24. For each decay, we find the difference of the initial and the final masses:

(*a*) $\Delta m = m(^{236}\text{U}) - m(^{235}\text{U}) - m(^1\text{n})$
$= (236.045562\text{ u}) - (235.043924\text{ u}) - (1.008665\text{ u}) = -0.00703\text{ u}.$

Because an increase in mass is required, the decay is **not possible.**

(*b*) $\Delta m = m(^{16}\text{O}) - m(^{15}\text{O}) - m(^1\text{n})$
$= (15.994915\text{ u}) - (15.003065\text{ u}) - (1.008665\text{ u}) = -0.0168\text{ u}.$

Because an increase in mass is required, the decay is **not possible.**

(*c*) $\Delta m = m(^{23}\text{Na}) - m(^{22}\text{Na}) - m(^1\text{n})$
$= (22.989767\text{ u}) - (21.994434\text{ u}) - (1.008665\text{ u}) = -0.0133\text{ u}.$

Because an increase in mass is required, the decay is **not possible.**

25. We find the final nucleus by balancing the mass and charge numbers:

$Z(\text{X}) = Z(\text{U}) - Z(\text{He}) = 92 - 2 = 90;$

$A(\text{X}) = A(\text{U}) - A(\text{He}) = 232 - 4 = 228$, so the final nucleus is $\boxed{^{228}_{90}\text{Th.}}$

If we ignore the recoil of the thorium, the kinetic energy of the α particle is

$$\text{KE} = [m(^{232}\text{U}) - m(^{228}\text{Th}) - m(^4\text{He})]c^2;$$
$$5.32\text{ MeV} = [(232.037131\text{ u}) - m(^{228}\text{Th}) - (4.002602\text{ u})]c^2(931.5\text{ MeV}/uc^2),$$

which gives $m(^{228}\text{Th}) = \boxed{228.02883\text{ u.}}$

26. The kinetic energy of the electron will be maximum if no neutrino is emitted. If we ignore the recoil of the sodium, the maximum kinetic energy of the electron is

$$\text{KE} = [m(^{23}\text{Ne}) - m(^{23}\text{Na})]c^2$$
$$= [(22.9945\text{ u}) - (22.989767\text{ u})]c^2(931.5\text{ MeV}/uc^2) = \boxed{4.4\text{ MeV.}}$$

When the neutrino has all of the kinetic energy, the minimum kinetic energy of the electron is $\boxed{0.}$

The sum of the kinetic energy of the electron and the energy of the neutrino must be from the mass difference, so the energy range of the neutrino will be $\boxed{0 \leq E_\nu \leq 4.4\text{ MeV.}}$

27. (*a*) We find the final nucleus by balancing the mass and charge numbers:

$Z(\text{X}) = Z(\text{P}) - Z(\text{e}) = 15 - (-1) = 16;$

$A(\text{X}) = A(\text{P}) - A(\text{e}) = 32 - 0 = 32$, so the final nucleus is $\boxed{^{32}_{16}\text{S.}}$

(*b*) If we ignore the recoil of the sulfur, the maximum kinetic energy of the electron is

$$\text{KE} = [m(^{32}\text{P}) - m(^{32}\text{S})]c^2;$$
$$1.71\text{ MeV} = [(31.973908\text{ u}) - m(^{32}\text{S})]c^2(931.5\text{ MeV}/uc^2),$$

which gives $m(^{32}\text{S}) = \boxed{31.97207\text{ u.}}$

28. For alpha decay we have ${}^{218}_{84}\text{Po} \rightarrow {}^{214}_{82}\text{Pb} + {}^{4}_{2}\text{He}$. The Q value is

$Q = [m(^{218}\text{Po}) - m(^{214}\text{Pb}) - m(^{4}\text{He})]c^2$

$= [(218.008965 \text{ u}) - (213.999798 \text{ u}) - (4.002602 \text{ u})]c^2(931.5 \text{ MeV}/\text{u}c^2) = \boxed{6.12 \text{ MeV.}}$

For beta decay we have ${}^{218}_{84}\text{Po} \rightarrow {}^{218}_{85}\text{At} + {}^{0}_{-1}\text{e}$. The Q value is

$Q = [m(^{218}\text{Po}) - m(^{218}\text{At})]c^2$

$= [(218.008965 \text{ u}) - (218.00868 \text{ u})]c^2(931.5 \text{ MeV}/\text{u}c^2) = \boxed{0.27 \text{ MeV.}}$

29. For the electron capture ${}^{7}_{4}\text{Be} + {}^{0}_{-1}\text{e} \rightarrow {}^{7}_{3}\text{Li} + \nu$, we see that if we add three electron masses to both sides to use the atomic mass for Li, we use the captured electron for the atomic mass of Be.

We find the Q value from

$Q = [m(^{7}\text{Be}) - m(^{7}\text{Li})]c^2$

$= [(7.016928 \text{ u}) - (7.016003 \text{ u})]c^2(931.5 \text{ MeV}/\text{u}c^2) = \boxed{0.861 \text{ MeV.}}$

30. If we ignore the recoil of the lead, the kinetic energy of the α particle is

$\text{KE} = [m(^{210}\text{Po}) - m(^{206}\text{Pb}) - m(^{4}\text{He})]c^2$

$= [(209.982848 \text{ u}) - (205.974440 \text{ u}) - (4.002602 \text{ u})]c^2(931.5 \text{ MeV}/\text{u}c^2) = \boxed{5.41 \text{ MeV.}}$

31. The decay is ${}^{238}_{92}\text{U} \rightarrow {}^{234}_{90}\text{Th} + {}^{4}_{2}\text{He}$. If the uranium nucleus is at rest when it decays, for momentum conservation we have

$p_\alpha = p_{\text{Th}}.$

Thus the kinetic energy of the thorium nucleus is

$\text{KE}_{\text{Th}} = p_{\text{Th}}^2/2m_{\text{Th}} = p_\alpha^2/2m_{\text{Th}} = (m_\alpha/m_{\text{Th}})(\text{KE}_\alpha) = (4 \text{ u}/234 \text{ u})(4.20 \text{ MeV}) = \boxed{0.0718 \text{ MeV.}}$

The Q value is the total kinetic energy produced:

$Q = \text{KE}_\alpha + \text{KE}_{\text{Th}} = 4.20 \text{ MeV} + 0.0718 = \boxed{4.27 \text{ MeV.}}$

32. For the positron-emission process

${}^{A}_{Z+1}\text{X} \rightarrow {}^{A}_{Z}\text{X}' + \text{e}^+ + \nu,$

we need to add $Z + 1$ electrons to the nuclear mass of X to be able to use the atomic mass. On the right-hand side we use Z electrons to be able to use the atomic mass of X′. Thus we have 1 electron mass and the β-particle mass, which means that we must include 2 electron masses on the right-hand side.

The Q value will be

$Q = [M_{\text{P}} - (M_{\text{D}} + 2m_{\text{e}})]c^2 = (M_{\text{P}} - M_{\text{D}} - 2m_{\text{e}})c^2.$

33. The kinetic energy of the β^+ particle will be maximum if no neutrino is emitted. If we ignore the recoil of the boron, the maximum kinetic energy is

$\text{KE} = [m(^{11}\text{C}) - m(^{11}\text{B}) - 2m_{\text{e}}]c^2$

$= [(11.011433 \text{ u}) - (11.009305 \text{ u}) - 2(0.00054858)]c^2(931.5 \text{ MeV}/\text{u}c^2) = \boxed{0.960 \text{ MeV.}}$

The sum of the kinetic energy of the β^+ particle and the energy of the neutrino must be from the mass difference, so the kinetic energy of the neutrino will range from $\boxed{0.960 \text{ MeV to 0.}}$

34. If the initial nucleus is at rest when it decays, for momentum conservation we have

$p_\alpha = p_D$.

Thus the kinetic energy of the daughter is

$KE_D = p_D^2/2m_D = p_\alpha^2/2m_D = (m_\alpha/m_D)(KE_\alpha) = (A_\alpha/A_D)(KE_\alpha) = (4/A_D)(KE_\alpha)$.

Thus the fraction carried away by the daughter is

$(KE_D)/(KE_\alpha + KE_D) = (4/A_D)(KE_\alpha)/[(KE_\alpha) + (4/A_D)(KE_\alpha)] = 1/[1 + (A_D/4)]$.

For the decay of ^{226}Ra, the daughter has $A_D = 222$, so we get

$\text{fraction}_D = 1/[1 + (222/4)] = 0.018$.

Thus the α particle carries away $1 - 0.018 = 0.982 =$ 98.2%.

35. We find the decay constant from

$\lambda N = \lambda N_0 e^{-\lambda t}$;

320 decays/min = (1280 decays/min) $e^{-\lambda(6\text{ h})}$, which gives $\lambda = 0.231/\text{h}$.

Thus the half-life is

$T_{1/2} = 0.693/\lambda = 0.693/(0.231/\text{h}) =$ 3.0 h.

36. The decay constant is

$\lambda = 0.693/T_{1/2} = 0.693/(4.468 \times 10^9 \text{ yr})(3.16 \times 10^7 \text{ s/yr}) =$ $4.91 \times 10^{-18} \text{ s}^{-1}$.

37. The half-life is

$T_{1/2} = 0.693/\lambda = 0.693/(5.4 \times 10^{-3} \text{ s}^{-1}) = 128 \text{ s} =$ 2.1 min.

38. The activity of the sample is

$\Delta N/\Delta t = \lambda N = (0.693/T_{1/2})N = [0.693/(5730 \text{ yr})(3.16 \times 10^7 \text{ s/yr})](4.1 \times 10^{20}) =$ 1.6×10^9 decays/s.

39. We find the fraction remaining from

$N = N_0 e^{-\lambda t}$;

$N/N_0 = e^{-\lambda t} = e^{-[(0.693)(3.0 \text{ yr})(12 \text{ mo/yr})/9 \text{ mo}]} =$ 0.0625.

40. We find the number of nuclei from the activity of the sample:

$\Delta N/\Delta t = \lambda N$;

275 decays/s = $[(0.693)/(4.468 \times 10^9 \text{ yr})(3.16 \times 10^7 \text{ s/yr})]N$, which gives $N =$ 5.60×10^{19} nuclei.

41. (*a*) The fraction left is

$N/N_0 = (\frac{1}{2})^n = (\frac{1}{2})^4 =$ 0.0625.

(*b*) The fraction left is

$N/N_0 = (\frac{1}{2})^n = (\frac{1}{2})^{4.5} =$ 0.0442.

42. Because only α particle decay changes the mass number (by 4), we have

$N_\alpha = (235 - 207)/4 =$ 7 α particles.

An α particle decreases the atomic number by 2, while a β^- particle increases the atomic number by 1, so we have

$N_\beta = [92 - 82 - 7(2)]/(-1) =$ 4 β^- particles.

43. The decay constant is

$\lambda = 0.693/T_{1/2} = 0.693/(8.04\text{ days})(24\text{ h/day})(3600\text{ s/h}) = 9.976 \times 10^{-7}\text{ s}^{-1}$.

The initial number of nuclei is

$N_0 = [(532 \times 10^{-6}\text{ g})/(131\text{ g/mol})](6.02 \times 10^{23}\text{ atoms/mol}) = 2.445 \times 10^{18}$ nuclei.

(*a*) When $t = 0$, we get

$\lambda N = \lambda N_0 e^{-\lambda t} = (9.976 \times 10^{-7}\text{ s}^{-1})(2.445 \times 10^{18})\, e^0 = \boxed{2.44 \times 10^{12}\text{ decays/s.}}$

(*b*) When $t = 1.0$ h, the exponent is

$\lambda t = (9.976 \times 10^{-7}\text{ s}^{-1})(1.0\text{ h})(3600\text{ s/h}) = 3.591 \times 10^{-3}$,

so we get

$\lambda N = \lambda N_0 e^{-\lambda t} = (9.976 \times 10^{-7}\text{ s}^{-1})(2.445 \times 10^{18})\, e^{-0.003591} = \boxed{2.43 \times 10^{12}\text{ decays/s.}}$

(*c*) When $t = 6$ months, the exponent is

$\lambda t = (9.976 \times 10^{-7}\text{ s}^{-1})(6\text{ mo})(30\text{ days/mo})(24\text{ h/day})(3600\text{ s/h}) = 15.51$,

so we get

$\lambda N = \lambda N_0 e^{-\lambda t} = (9.976 \times 10^{-7}\text{ s}^{-1})(2.445 \times 10^{18})\, e^{-15.51} = \boxed{4.48 \times 10^{5}\text{ decays/s.}}$

44. The decay constant is

$\lambda = 0.693/T_{1/2} = 0.693/(30.8\text{ s}) = 0.0225\text{ s}^{-1}$.

(*a*) The initial number of nuclei is

$N_0 = (7.8 \times 10^{-6}\text{ g})/(124\text{ g/mol})](6.02 \times 10^{23}\text{ atoms/mol}) = \boxed{3.8 \times 10^{16}\text{ nuclei.}}$

(*b*) When $t = 2.0$ min, the exponent is

$\lambda t = (0.0225\text{ s}^{-1})(2.0\text{ min})(60\text{ s/min}) = 2.7$,

so we get

$N = N_0 e^{-\lambda t} = (3.8 \times 10^{16})\, e^{-2.7} = \boxed{2.5 \times 10^{15}\text{ nuclei.}}$

(*c*) The activity is

$\lambda N = (0.0225\text{ s}^{-1})(2.5 \times 10^{15}) = \boxed{5.7 \times 10^{13}\text{ decays/s.}}$

(*d*) We find the time from

$\lambda N = \lambda N_0 e^{-\lambda t}$;

1 decay/s $= (0.0225\text{ s}^{-1})(3.8 \times 10^{16})\, e^{-(0.0225\text{ /s})t}$, which gives $t = 1.53 \times 10^3\text{ s} = \boxed{2.5\text{ min.}}$

45. The number of nuclei is

$N = [(4.7 \times 10^{-6}\text{ g})/(32\text{ g/mol})](6.02 \times 10^{23}\text{ atoms/mol}) = 8.85 \times 10^{16}$ nuclei.

The activity is

$\lambda N = [(0.693)/(1.23 \times 10^{6}\text{ s})](8.85 \times 10^{16}) = \boxed{5.0 \times 10^{10}\text{ decays/s.}}$

46. We find the number of nuclei from

Activity $= \lambda N$;

3.55×10^5 decays/s $= [(0.693)/(7.56 \times 10^{6}\text{ s})]N$, which gives $N = 3.87 \times 10^{12}$ nuclei.

The mass is

$m = [(3.87 \times 10^{12}\text{ nuclei})/(6.02 \times 10^{23}\text{ atoms/mol})](35\text{ g/mol}) = \boxed{2.25 \times 10^{-10}\text{ g.}}$

47. (a) The decay constant is

$\lambda = 0.693/T_{1/2} = 0.693/(1.59 \times 10^5\text{ yr})(3.16 \times 10^7\text{ s/yr}) = \boxed{1.38 \times 10^{-13}\text{ s}^{-1}.}$

(b) The activity is

$\lambda N = (1.38 \times 10^{-13}\text{ s}^{-1})(6.50 \times 10^{19}) = 8.97 \times 10^{6}\text{ decays/s} = \boxed{5.387 \times 10^{8}\text{ decays/min.}}$

48. We find the number of half-lives from

$(\Delta N/\Delta t)/(\Delta N/\Delta t)_0 = (\tfrac{1}{2})^n$;

$1/10 = (\tfrac{1}{2})^n$, or $n \log 2 = \log 10$, which gives $n = 3.32$.

Thus the half-life is

$T_{1/2} = t/n = (9.6\text{ min})/3.32 = \boxed{2.9\text{ min.}}$

49. Because the fraction of atoms that are ^{14}C is so small, we use the atomic weight of ^{12}C to find the number of carbon atoms in 135 g:

$N = [(135 \text{ g})/(12 \text{ g/mol})](6.02 \times 10^{23} \text{ atoms/mol}) = 6.78 \times 10^{24}$ atoms.

The number of ^{14}C nuclei is

$N_{14} = (1.3/10^{12})(6.78 \times 10^{24}) = 8.81 \times 10^{12}$ nuclei.

The activity is

$\lambda N = [0.693/(5730 \text{ yr})(3.16 \times 10^7 \text{ s/yr})](8.81 \times 10^{12}) = \boxed{34 \text{ decays/s.}}$

50. We find the number of half-lives from

$(\Delta N/\Delta t)/(\Delta N/\Delta t)_0 = (\frac{1}{2})^n;$

$(820 \text{ decays/s})/(2880 \text{ decays /s}) = (\frac{1}{2})^n$, or $n \log 2 = \log 3.51$, which gives $n = 1.81$.

Thus the half-life is

$T_{1/2} = t/n = (1.6 \text{ h})/1.81 = 0.882 \text{ h} = \boxed{53 \text{ min.}}$

51. We find the number of nuclei from

Activity $= \lambda N;$

$8.70 \times 10^2 \text{ decays/s} = [(0.693)/(1.28 \times 10^9 \text{ s})(3.16 \times 10^7 \text{ s/yr})]N$, which gives $N = 5.08 \times 10^{19}$ nuclei.

The mass is

$m = [(5.08 \times 10^{19} \text{ nuclei})/(6.02 \times 10^{23} \text{ atoms/mol})](40 \text{ g/mol}) = 3.4 \times 10^{-3} \text{ g} = \boxed{3.4 \text{ mg.}}$

52. We assume that the elapsed time is much smaller than the half-life, so we can use a constant decay rate. Because ^{87}Sr is stable, and there was none present when the rocks were formed, every atom of ^{87}Rb that decayed is now an atom of ^{87}Sr. Thus we have

$N_{Sr} = -\Delta N_{Rb} = \lambda N_{Rb} \Delta t$, or

$N_{Sr}/N_{Rb} = (0.693/T_{1/2}) \Delta t;$

$0.0160 = [0.693/(4.75 \times 10^{10} \text{ yr})]\Delta t$, which gives $\Delta t = \boxed{1.1 \times 10^9 \text{ yr.}}$

This is $\approx 2\%$ of the half-life, so our original assumption is valid.

53. The decay rate is

$\Delta N/\Delta t = \lambda N.$

If we assume equal numbers of nuclei decaying by α emission, we have

$(\Delta N/\Delta t)_{218}/(\Delta N/\Delta t)_{214} = \lambda_{218}/\lambda_{214} = T_{1/2,214}/T_{1/2,218}$

$= (1.6 \times 10^{-4} \text{ s})/(3.1 \text{ min})(60 \text{ s/min}) = \boxed{8.6 \times 10^{-7}.}$

54. The decay constant is

$\lambda = 0.693/T_{1/2} = 0.693/(53 \text{ days}) = 0.0131 \text{ /day} = 1.52 \times 10^{-7} \text{ s}^{-1}.$

We find the number of half-lives from

$(\Delta N/\Delta t)/(\Delta N/\Delta t)_0 = (\frac{1}{2})^n;$

$(10 \text{ decays/s})/(350 \text{ decays/s}) = (\frac{1}{2})^n$, or $n \log 2 = \log 35$, which gives $n = 5.13$.

Thus the elapsed time is

$\Delta t = nT_{1/2} = (5.13)(53 \text{ days}) = \boxed{272 \text{ days}} \approx 9$ months.

We find the number of nuclei from

Activity $= \lambda N;$

$350 \text{ decays/s} = (1.52 \times 10^{-7} \text{ s}^{-1})N$, which gives $N = 2.31 \times 10^9$ nuclei.

The mass is

$m = [(2.31 \times 10^9 \text{ nuclei})/(6.02 \times 10^{23} \text{ atoms/mol})](7 \text{ g/mol}) = \boxed{2.7 \times 10^{-17} \text{ kg.}}$

55. We find the number of half-lives from

$(\Delta N/\Delta t)/(\Delta N/\Delta t)_0 = (\frac{1}{2})^n$;

$1.050 \times 10^{-2} = (\frac{1}{2})^n$, or $n \log 2 = \log(1/1.050 \times 10^{-2})$, which gives $n = 6.57$.

Thus the half-life is

$T_{1/2} = t/n = (4.00 \text{ h})/6.57 = 0.609 \text{ h} = 36.5 \text{ min}$.

From the Appendix we see that the isotope is $\boxed{{}^{211}_{82}\text{Pb}}$.

56. The decay constant is

$\lambda = 0.693/T_{1/2} = 0.693/(5730 \text{ yr}) = 1.209 \times 10^{-4} \text{ /yr}$.

Because the fraction of atoms that are ^{14}C is so small, we use the atomic weight of ^{12}C to find the number of carbon atoms in 170 g:

$N = [(170 \text{ g})/(12 \text{ g/mol})](6.02 \times 10^{23} \text{ atoms/mol}) = 8.53 \times 10^{24} \text{ atoms}$,

so the number of ^{14}C nuclei in a sample from a living tree is

$N_{14} = (1.3 \times 10^{-12})(8.53 \times 10^{24}) = 1.11 \times 10^{13}$ nuclei.

Because the carbon is being replenished in living trees, we assume that this number produced the activity when the club was made. We determine its age from

$\lambda N = \lambda N_{14} e^{-\lambda t}$;

$5.0 \text{ decays/s} = [(1.209 \times 10^{-4} \text{ /yr})/(3.16 \times 10^7 \text{ s/yr})](1.11 \times 10^{13} \text{ nuclei}) e^{-(1.209 \times 10^{-4} \text{ /yr})t}$,

which gives $t = \boxed{1.8 \times 10^4 \text{ yr.}}$

57. The number of radioactive nuclei decreases exponentially:

$N = N_0 e^{-\lambda t}$.

Every radioactive nucleus that decays becomes a daughter nucleus, so we have

$N_D = N_0 - N = \boxed{N_0(1 - e^{-\lambda t}).}$

58. The radius of a nucleus is

$r = (1.2 \times 10^{-15} \text{ m})A^{1/3}$.

If we form the ratio for the two isotopes, we get

$r_U/r_H = (238/1)^{1/3} = 6.2$.

59. (*a*) The mass of a nucleus with mass number A is A u and its radius is

$r = (1.2 \times 10^{-15} \text{ m})A^{1/3}$.

Thus the density is

$$\rho = m/V$$
$$= A(1.66 \times 10^{-27} \text{ kg/u})/\tfrac{4}{3}\pi r^3 = A(1.66 \times 10^{-27} \text{ kg/u})/\tfrac{4}{3}\pi(1.2 \times 10^{-15} \text{ m})^3 A$$
$$= \boxed{2.29 \times 10^{17} \text{ kg/m}^3,} \text{ independent of } A.$$

(*b*) We find the radius from

$M = \rho V$;

$5.98 \times 10^{24} \text{ kg} = (2.29 \times 10^{17} \text{ kg/m}^3)\tfrac{4}{3}\pi R^3$, which gives $R = \boxed{180 \text{ m.}}$

(*c*) For equal densities, we have

$\rho = M_{\text{Earth}}/\tfrac{4}{3}\pi R_{\text{Earth}}^3 = m_U/\tfrac{4}{3}\pi r_U^3$;

$(5.98 \times 10^{24} \text{ kg})/(6.38 \times 10^6 \text{ m})^3 = (238 \text{ u})(1.66 \times 10^{-27} \text{ kg/u})/r_U^3$, which gives $r_U = \boxed{2.58 \times 10^{-10} \text{ m.}}$

60. The radius of the iron nucleus is

$r = (1.2 \times 10^{-15}\text{ m})A^{1/3} = (1.2 \times 10^{-15}\text{ m})(56)^{1/3} = 4.6 \times 10^{-15}\text{ m}.$

We use twice the radius as the uncertainty in position for the nucleon. We find the uncertainty in the momentum from

$\Delta p \geq \hbar/\Delta x = (1.055 \times 10^{-34}\text{ J}\cdot\text{s})/2(4.6 \times 10^{-15}\text{ m}) = 1.15 \times 10^{-20}\text{ kg}\cdot\text{m/s}.$

If we assume that the lowest value for the momentum is the least uncertainty, we estimate the lowest possible kinetic energy as

$E = (\Delta p)^2/2m = (1.15 \times 10^{-20}\text{ kg}\cdot\text{m/s})^2/2(1.66 \times 10^{-27}\text{ kg}) = 3.96 \times 10^{-14}\text{ J} = \boxed{0.25\text{ MeV.}}$

61. If the ^{40}K nucleus in the excited state is at rest, the gamma ray and the nucleus must have equal and opposite momenta:

$p_K = p_\gamma = E_\gamma/c$, or $p_K c = E_\gamma = 1.46$ MeV.

The kinetic energy of the nucleus is

$\text{KE} = p_K^2/2m = (p_K c)^2/2mc^2 = (1.46\text{ MeV})^2/2(40\text{ u})(931.5\text{ MeV}/\text{u}c^2)c^2 = \boxed{28.6\text{ MeV.}}$

62. Because the carbon is being replenished in living trees, we assume that the amount of ^{14}C is constant until the wood is cut, and then it decays. We find the number of half-lives from

$N/N_0 = (\frac{1}{2})^n;$

$0.10 = (\frac{1}{2})^n$, or $n \log 2 = \log (10)$, which gives $n = 3.32$.

Thus the time is

$t = nT_{1/2} = (3.32)(5730\text{ yr}) = \boxed{1.9 \times 10^4\text{ yr.}}$

63. Because the tritium in water is being replenished, we assume that the amount is constant until the wine is made, and then it decays. We find the number of half-lives from

$N/N_0 = (\frac{1}{2})^n;$

$0.10 = (\frac{1}{2})^n$, or $n \log 2 = \log (10)$, which gives $n = 3.32$.

Thus the time is

$t = nT_{1/2} = (3.32)(12.33\text{ yr}) = \boxed{41\text{ yr.}}$

64. (a) We find the mass number from its radius:

$r = (1.2 \times 10^{-15}\text{ m})A^{1/3};$

$5.0 \times 10^3\text{ m} = (1.2 \times 10^{-15}\text{ m})A^{1/3}$, which gives $A = \boxed{7.2 \times 10^{55}.}$

(b) The mass of the neutron star is

$m = A(1.66 \times 10^{-27}\text{ kg/u}) = (7.2 \times 10^{55}\text{ u})(1.66 \times 10^{-27}\text{ kg/u}) = \boxed{1.2 \times 10^{29}\text{ kg.}}$

Note that this is about 6% of the mass of the Sun.

(c) The acceleration of gravity on the surface of the neutron star is

$g = Gm/r^2 = (6.67 \times 10^{-11}\text{ N}\cdot\text{m}^2/\text{kg}^2)(1.2 \times 10^{29}\text{ kg})/(5.0 \times 10^3\text{ m})^2 = \boxed{3.2 \times 10^{11}\text{ m/s}^2.}$

65. If we assume a body has 70 kg of water, the number of water molecules is

$N_{\text{water}} = [(70 \times 10^3\text{ g})/(18\text{ g/mol})](6.02 \times 10^{23}\text{ atoms/mol}) = 2.34 \times 10^{27}$ molecules.

The number of protons in a water molecule (H_2O) is 2 + 8 = 10, so the number of protons is

$N_0 = 2.34 \times 10^{28}$ protons.

If we assume that the time is much less than the half-life, the rate of decay is constant, so we have

$\Delta N/\Delta t = \lambda N = (0.693/T_{1/2})N;$

$(1\text{ proton})/\Delta t = [(0.693)/(10^{28}\text{ yr})](2.34 \times 10^{28}\text{ protons})$, which gives $\Delta t = \boxed{6 \times 10^3\text{ yr.}}$

66. The capture is ${}_1^1\text{H} + {}_0^1\text{n} \rightarrow {}_1^2\text{H} + \gamma$.
Because the kinetic energies of the particles are small, the gamma energy is the energy released:
$Q = [m(^1\text{H}) + m(^1\text{n}) - m(^2\text{H})]c^2$
$= [(1.007825 \text{ u}) + (1.008665 \text{ u}) - (2.014102 \text{ u})]c^2(931.5 \text{ MeV}/uc^2) =$ **2.22 MeV.**

67. We find the number of half-lives from
$(\Delta N/\Delta t)/(\Delta N/\Delta t)_0 = (\frac{1}{2})^n$;
$1.00 \times 10^{-2} = (\frac{1}{2})^n$, or $n \log 2 = \log (100)$, which gives $n = 6.65$, so the time is **$6.65T_{1/2}$.**

68. We find the number of ^{40}K nuclei from
Activity $= \lambda N_{40}$;
50 decays/s $= [(0.693)/(1.28 \times 10^9 \text{ s})(3.16 \times 10^7 \text{ s/yr})]N_{40}$, which gives $N_{40} = 2.9 \times 10^{18}$ nuclei.
The mass of ^{40}K is
$m_{40} = (2.9 \times 10^{18})(40 \text{ u})(1.66 \times 10^{-27} \text{ kg/u}) = 1.94 \times 10^{-7} \text{ kg} =$ **0.19 mg.**
From the Appendix we have
$N_{40} = (0.0117\%)N$, and $N_{39} = (93.2581\%)N$.
Thus the number of ^{39}K nuclei is
$N_{39} = [(93.2581\%)/(0.0117\%)](2.9 \times 10^{18} \text{ nuclei}) = 2.3 \times 10^{22}$ nuclei.
The mass of ^{39}K is
$m_{39} = (2.3 \times 10^{22})(39 \text{ u})(1.66 \times 10^{-27} \text{ kg/u}) = 1.51 \times 10^{-3} \text{ kg} =$ **1.5 g.**

69. The mass number changes only with an α decay for which the change is -4.
If the mass number is $4n$, then the new number is $4n - 4 = 4(n - 1) = 4n'$. Thus for each family, we have
$4n \rightarrow 4n - 4 \rightarrow 4n'$;
$4n + 1 \rightarrow 4n - 4 + 1 \rightarrow 4n' + 1$;
$4n + 2 \rightarrow 4n - 4 + 2 \rightarrow 4n' + 2$;
$4n + 3 \rightarrow 4n - 4 + 3 \rightarrow 4n' + 3$.
Thus the daughter nuclides are always in the same family.

70. We see from the periodic chart that Sr is in the same column as **calcium.**
If strontium is ingested, the body will treat it chemically as if it were calcium, which means it will be **stored by the body in bones.**
We find the number of half-lives to reach a 1% level from
$N/N_0 = (\frac{1}{2})^n$;
$0.01 = (\frac{1}{2})^n$, or $n \log 2 = \log (100)$, which gives $n = 6.64$.
Thus the time is
$t = nT_{1/2} = (6.64)(29 \text{ yr}) =$ **193 yr.**
The decay reactions are

${}_{38}^{90}\text{Sr} \rightarrow {}_{39}^{90}\text{Y} + {}_{-1}^{0}\text{e} + \bar{\nu}$, ${}_{39}^{90}\text{Y}$ is radioactive;
${}_{39}^{90}\text{Y} \rightarrow {}_{40}^{90}\text{Zr} + {}_{-1}^{0}\text{e} + \bar{\nu}$, ${}_{40}^{90}\text{Zr}$ is stable.

71. (*a*) We find the daughter nucleus by balancing the mass and charge numbers:
$Z(\text{X}) = Z(\text{Os}) - Z(\text{e}^-) = 76 - (-1) = 77$;
$A(\text{X}) = A(\text{Os}) - A(\text{e}^-) = 191 - 0 = 191$,
so the daughter nucleus is **${}_{77}^{191}\text{Ir}$.**
(*b*) Because there is only one β energy, the β decay must be to the higher excited state.

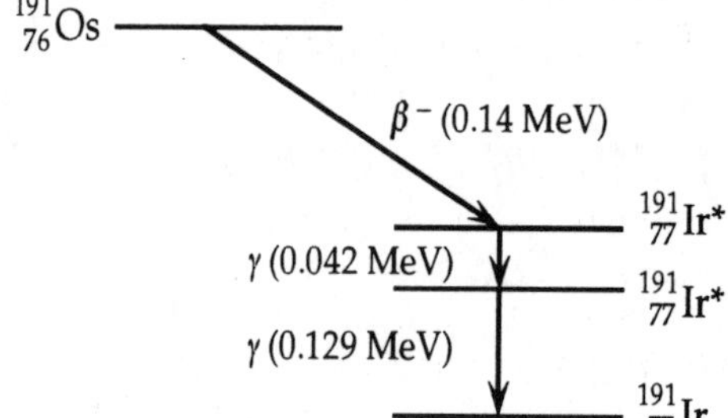

72. We use an average nuclear radius of 5×10^{-15} m.
We use twice the radius as the uncertainty in position for the electron to find the uncertainty in the momentum from
$\Delta p \geq \hbar/\Delta x = (1.055 \times 10^{-34}\ \text{J}\cdot\text{s})/2(5 \times 10^{-15}\ \text{m}) = 1.0 \times 10^{-20}\ \text{kg}\cdot\text{m/s}.$
If we assume that the lowest value for the momentum is the least uncertainty, we estimate the lowest possible energy as
$E = (\text{KE}) + m_0c^2 = (p^2c^2 + m_0{}^2c^4)^{1/2} = [(\Delta p)^2c^2 + m_0{}^2c^4]^{1/2}$
$= [(1.0 \times 10^{-19}\ \text{kg}\cdot\text{m/s})^2(3.00 \times 10^8\ \text{m/s})^2 + (9.11 \times 10^{-31}\ \text{kg})^2(3.00 \times 10^8\ \text{m/s})^4]^{1/2}$
$= 3.0 \times 10^{-11}\ \text{J} \approx 200\ \text{MeV}.$
This is much too high for the electron to be contained in the nucleus.

73. From Figure 30–1, we see that the average binding energy per nucleon at $A = 29$ is 8.6 MeV.
If we use the average atomic weight as the average number of nucleons for the two stable isotopes of copper, the total binding energy is
$(63.5)(8.6\ \text{MeV}) = \boxed{550\ \text{MeV.}}$
The number of atoms in a penny is
$N = [(3 \times 10^3\ \text{g})/(63.5\ \text{g/mol})](6.02 \times 10^{23}\ \text{atoms/mol}) = 2.84 \times 10^{22}\ \text{atoms}.$
Thus the total energy needed is
$(2.84 \times 10^{22})(550\ \text{MeV}) = 1.57 \times 10^{25}\ \text{MeV} = \boxed{2.5 \times 10^{12}\ \text{J.}}$

74. (*a*) $\Delta(^4\text{He}) = m(^4\text{He}) - A(^4\text{He}) = 4.002602\ \text{u} - 4 = \boxed{0.002602\ \text{u}}$
$= (0.002602\ \text{u})(931.5\ \text{MeV/uc}^2) = \boxed{2.424\ \text{MeV}/c^2.}$
(*b*) $\Delta(^{12}\text{C}) = m(^{12}\text{C}) - A(^{12}\text{C}) = 12.000000\ \text{u} - 12 = \boxed{0.}$
(*c*) $\Delta(^{107}\text{Ag}) = m(^{107}\text{Ag}) - A(^{107}\text{Ag}) = 106.905091\ \text{u} - 107 = \boxed{-0.094909\ \text{u}}$
$= (-0.094909\ \text{u})(931.5\ \text{MeV/uc}^2) = \boxed{-88.41\ \text{MeV}/c^2.}$
(*d*) $\Delta(^{235}\text{U}) = m(^{235}\text{U}) - A(^{235}\text{U}) = 235.043924\ \text{u} - 235 = \boxed{0.043924\ \text{u}}$
$= (0.043924\ \text{u})(931.5\ \text{MeV/uc}^2) = \boxed{40.92\ \text{MeV}/c^2.}$
(*e*) From the Appendix we see that
$\boxed{\Delta \geq 0 \text{ for } 0 \leq Z \leq 8 \text{ and } Z \geq 85; \quad \Delta < 0 \text{ for } 9 \leq Z \leq 84.}$

75. (*a*) The usual fraction of ^{14}C is 1.3×10^{-12}. Because the fraction of atoms that are ^{14}C is so small, we use the atomic weight of ^{12}C to find the number of carbon atoms in 100 g:
$N = [(100\ \text{g})/(12\ \text{g/mol})](6.02 \times 10^{23}\ \text{atoms/mol}) = 5.02 \times 10^{24}\ \text{atoms}.$
The number of ^{14}C nuclei in the sample is
$N_{14} = (1.3 \times 10^{-12})(5.02 \times 10^{24}) = 6.53 \times 10^{12}\ \text{nuclei}.$
We find the number of half-lives from
$N/N_0 = (\tfrac{1}{2})^n;$
$1/6.537 \times 10^{12} = (\tfrac{1}{2})^n$, or $n \log 2 = \log(6.53 \times 10^{12})$, which gives $n = 42.6$.
Thus the time is
$t = nT_{1/2} = (42.6)(5730\ \text{yr}) = \boxed{2.4 \times 10^5\ \text{yr.}}$
(*b*) It would take one half-life for the activity of 200 g to decay to the activity of 100 g.
This is so much smaller than the time in (*a*) that there is $\boxed{\text{no change.}}$
Thus carbon dating cannot be used for times much greater than the half-life.

76. Because there are so many low-energy electrons available, this reaction would turn most of the protons into neutrons, which would eliminate chemistry, and thus life.
The Q-value of the reaction is
$Q = [m(^1\text{H}) - m(^1\text{n})]c^2 = [(1.007825\ \text{u}) - (1.008665\ \text{u})]c^2(931.5\ \text{MeV/uc}^2) = -0.782\ \text{MeV}.$
The percentage increase in the proton's mass to make the Q-value = 0 is
$(\Delta m/m)(100) = [(0.782\ \text{MeV}/c^2)/(938.3\ \text{MeV}/c^2)](100) = \boxed{0.083\%.}$

CHAPTER 31

Note: A factor that appears in the analysis of energies is

$$ke^2 = (9.00 \times 10^9\ \text{N}\cdot\text{m}^2/\text{C}^2)(1.60 \times 10^{-19}\ \text{C})^2 = 2.30 \times 10^{-28}\ \text{J}\cdot\text{m} = 1.44\ \text{MeV}\cdot\text{fm}.$$

1. We find the product nucleus by balancing the mass and charge numbers:

$$Z(\text{X}) = Z(^{27}\text{Al}) + Z(\text{n}) = 13 + 0 = 13;$$

$$A(\text{X}) = A(^{27}\text{Al}) + A(\text{n}) = 27 + 1 = 28, \text{ so the product nucleus is } \boxed{^{28}_{13}\text{Al}.}$$

If $^{28}_{13}\text{Al}$ were a β^+ emitter, the resulting nucleus would be $^{28}_{12}\text{Mg}$, which has too many neutrons relative to the number of protons to be stable. Thus we have a $\boxed{\beta^- \text{ emitter.}}$

The reaction is $^{28}_{13}\text{Al} \rightarrow {}^{28}_{14}\text{Si} + {}^{\ 0}_{-1}\text{e} + \bar{\nu}$, so the product is $\boxed{^{28}_{14}\text{Si}.}$

2. For the reaction $^2_1\text{H}(\text{d, n})^3_2\text{He}$, we find the difference of the initial and the final masses:

$$\Delta m = m(^2\text{H}) + m(^2\text{H}) - m(\text{n}) - m(^3\text{He})$$

$$= 2(2.014102\ \text{u}) - (1.0086655\ \text{u}) - (3.016029\ \text{u}) = +\,0.003510\ \text{u}.$$

Thus $\boxed{\text{no threshold energy}}$ is required.

3. For the reaction $^{238}_{92}\text{U}(\text{n}, \gamma)^{239}_{92}\text{U}$ with slow neutrons, whose kinetic energy is negligible, we find the difference of the initial and the final masses:

$$\Delta m = m(^{238}\text{U}) + m(\text{n}) - m(^{239}\text{U})$$

$$= (238.050784\ \text{u}) + (1.0086655\ \text{u}) - (239.054289\ \text{u}) = +\,0.005160\ \text{u}.$$

Thus no threshold energy is required, so the reaction is $\boxed{\text{possible.}}$

4. For the reaction $^7_3\text{Li}(\text{p}, \alpha)^4_2\text{He}$, we determine the Q-value:

$$Q = [m(^7\text{Li}) + m(^1\text{H}) - m(^4\text{He}) - m(^4\text{He})]c^2$$

$$= [(7.016003\ \text{u}) + (1.007825\ \text{u}) - 2(4.002602\ \text{u})]c^2(931.5\ \text{MeV}/uc^2) = +\,17.35\ \text{MeV}.$$

Thus $\boxed{17.35\ \text{MeV is released.}}$

5. For the reaction $^{18}_8\text{O}(\text{p, n})^{18}_9\text{F}$, we find the mass from the Q-value:

$$Q = [m(^{18}\text{O}) + m(^1\text{H}) - m(\text{n}) - m(^{18}\text{F})]c^2;$$

$$-\,2.453\ \text{MeV} = [(17.999160\ \text{u}) + (1.007825\ \text{u}) - (1.008665\ \text{u}) - m(^{18}\text{F})]c^2(931.5\ \text{MeV}/uc^2),$$

which gives $m(^{18}\text{F}) = \boxed{+\,18.000953\ \text{u}.}$

6. For the reaction $^9_4\text{Be}(\alpha, \text{n})^{12}_6\text{C}$, we determine the Q-value:

$$Q = [m(^9\text{Be}) + m(^4\text{He}) - m(\text{n}) - m(^{12}\text{C})]c^2$$

$$= [(9.012182\ \text{u}) + (4.002602\ \text{u}) - (1.008665\ \text{u}) - (12.000000\ \text{u})]c^2(931.5\ \text{MeV}/uc^2) = +\,5.700\ \text{MeV}.$$

Thus $\boxed{5.700\ \text{MeV is released.}}$

7. (*a*) For the reaction $^{24}_{12}\text{Mg}(\text{n, d})^{23}_{11}\text{Na}$, we determine the Q-value:

$$Q = [m(^{24}\text{Mg}) + m(\text{n}) - m(^2\text{H}) - m(^{23}\text{Na})]c^2$$

$$= [(23.985042\ \text{u}) + (1.008665\ \text{u}) - (2.014102\ \text{u}) - (22.989767\ \text{u})]c^2(931.5\ \text{MeV}/uc^2) = -\,9.466\ \text{MeV}.$$

Because $(\text{KE} + Q) > 0$, the reaction $\boxed{\text{can occur.}}$

(*b*) The energy released is

$$\text{KE} + Q = 10.00\ \text{MeV} - 9.466\ \text{MeV} = \boxed{0.53\ \text{MeV.}}$$

8. (*a*) For the reaction ${}^{7}_{3}\text{Li}(\text{p}, \alpha){}^{4}_{2}\text{He}$, we determine the Q-value:

$$Q = [m(^7\text{Li}) + m(^1\text{H}) - m(^4\text{He}) - m(^4\text{He})]c^2$$
$$= [(7.016003\ \text{u}) + (1.007825\ \text{u}) - 2(4.002602\ \text{u})]c^2(931.5\ \text{MeV}/\text{u}c^2) = +17.35\ \text{MeV}.$$

Because $Q > 0$, the reaction **can occur.**

(*b*) The kinetic energy of the products is

$$\text{KE} = \text{KE}_i + Q = 2.500\ \text{MeV} + 17.35\ \text{MeV} = \boxed{19.85\ \text{MeV}.}$$

9. (*a*) For the reaction ${}^{14}_{7}\text{N}(\alpha, \text{p}){}^{17}_{8}\text{O}$, we determine the Q-value:

$$Q = [m(^{14}\text{N}) + m(^4\text{He}) - m(^1\text{H}) - m(^{17}\text{O})]c^2$$
$$= [(14.003074\ \text{u}) + (4.002602\ \text{u}) - (1.007825\ \text{u}) - (16.999131\ \text{u})]c^2(931.5\ \text{MeV}/\text{u}c^2) = -1.192\ \text{MeV}.$$

Because $(\text{KE} + Q) > 0$, the reaction **can occur.**

(*b*) The kinetic energy of the products is

$$\text{KE} = \text{KE}_i + Q = 7.68\ \text{MeV} - 1.192\ \text{MeV} = \boxed{6.49\ \text{MeV}.}$$

10. For the reaction ${}^{16}_{8}\text{O}(\alpha, \gamma){}^{20}_{10}\text{Ne}$, we determine the Q-value:

$$Q = [m(^{16}\text{O}) + m(^4\text{He}) - m(^{20}\text{Ne})]c^2$$
$$= [(15.994915\ \text{u}) + (4.002602\ \text{u}) - (19.992435\ \text{u})]c^2(931.5\ \text{MeV}/\text{u}c^2) = \boxed{+4.734\ \text{MeV}.}$$

11. For the reaction ${}^{13}_{6}\text{C}(\text{d}, \text{n}){}^{14}_{7}\text{N}$, we determine the Q-value:

$$Q = [m(^{13}\text{C}) + m(^2\text{H}) - m(\text{n}) - m(^{14}\text{N})]c^2$$
$$= [(13.003355\ \text{u}) + (2.014102\ \text{u}) - (1.008665\ \text{u}) - (14.003074\ \text{u})]c^2(931.5\ \text{MeV}/\text{u}c^2) = +5.326\ \text{MeV}.$$

The kinetic energy of the products is

$$\text{KE} = \text{KE}_i + Q = 36.3\ \text{MeV} + 5.326\ \text{MeV} = \boxed{41.6\ \text{MeV}.}$$

12. (*a*) We find the product nucleus by balancing the mass and charge numbers:

$$Z(\text{X}) = Z(^6\text{Li}) + Z(^2\text{H}) - Z(^1\text{H}) = 3 + 1 - 1 = 3;$$
$$A(\text{X}) = A(^6\text{Li}) + A(^2\text{H}) - A(^1\text{H}) = 6 + 2 - 1 = 7, \text{ so the product nucleus is } \boxed{{}^{7}_{3}\text{Li}.}$$

(*b*) It is a "stripping" reaction because a **neutron is stripped from the deuteron.**

(*c*) For the reaction ${}^{6}_{3}\text{Li}(\text{d}, \text{p}){}^{7}_{3}\text{Li}$, we determine the Q-value:

$$Q = [m(^6\text{Li}) + m(^2\text{H}) - m(^1\text{H}) - m(^7\text{Li})]c^2$$
$$= [(6.015121\ \text{u}) + (2.014102\ \text{u}) - (1.007825\ \text{u}) - (7.016003\ \text{u})]c^2(931.5\ \text{MeV}/\text{u}c^2) = \boxed{+5.025\ \text{MeV}.}$$

Because $Q > 0$, the reaction is **exothermic.**

13. (*a*) It is a "pickup" reaction because the **^{3}He picks up a neutron.**

(*b*) We find the product nucleus by balancing the mass and charge numbers:

$$Z(\text{X}) = Z(^{12}\text{C}) + Z(^3\text{He}) - Z(^4\text{He}) = 6 + 2 - 2 = 6;$$
$$A(\text{X}) = A(^{12}\text{C}) + A(^3\text{He}) - A(^4\text{He}) = 12 + 3 - 4 = 11, \text{ so the product nucleus is } \boxed{{}^{11}_{6}\text{C}.}$$

(*c*) For the reaction ${}^{12}_{6}\text{C}({}^{3}_{2}\text{He}, \alpha){}^{11}_{6}\text{C}$, we determine the Q-value:

$$Q = [m(^{12}\text{C}) + m(^3\text{He}) - m(^4\text{He}) - m(^{11}\text{C})]c^2$$
$$= [(12.000000\ \text{u}) + (3.016029\ \text{u}) - (4.002602\ \text{u}) - (11.011433\ \text{u})]c^2(931.5\ \text{MeV}/\text{u}c^2)$$
$$= \boxed{+1.857\ \text{MeV}.}$$

Because $Q > 0$, the reaction is **exothermic.**

14. (*a*) We find the initial nucleus by balancing the mass and charge numbers:

$Z(X) = Z(^{32}S) - Z(^{1}H) = 16 - 1 = 15;$

$A(X) = A(^{32}S) - A(^{1}H) = 32 - 1 = 31$, so the initial nucleus is $^{31}_{15}P$.

The reaction is $\boxed{^{31}_{15}P(p, \gamma)^{32}_{16}S.}$

(*b*) For the reaction, we determine the Q-value:

$Q = [m(^{31}P) + m(^{1}H) - m(^{32}S)]c^2$

$= [(30.973762\text{ u}) + (1.007825\text{ u}) - (31.972071\text{ u})]c^2(931.5\text{ MeV}/uc^2) = \boxed{+8.864\text{ MeV.}}$

15. For the reaction $^{13}_{6}C(p, n)^{13}_{7}N$, we determine the Q-value:

$Q = [m(^{13}C) + m(^{1}H) - m(n) - m(^{13}N)]c^2$

$= [(13.003355\text{ u}) + (1.007825\text{ u}) - (1.008665\text{ u}) - (13.005738\text{ u})]c^2(931.5\text{ MeV}/uc^2) = -3.002\text{ MeV}.$

The kinetic energy of the products is

$KE_n + KE_N = KE_p + Q.$

Because the kinetic energies « mc^2, we can use a non relativistic treatment: $KE = mv^2/2 = p^2/2m$.
The least kinetic energy is required when the product particles move together with the same speed.
With the target at rest, for momentum conservation we have

$p_p = p_n + p_N = (m_n + m_N)v$, or

$KE_p = p_p^2/2m_p = [(m_n + m_N)^2/2m_p]v^2 = [(m_n + m_N)/m_p](KE_n + KE_N)$, or

$KE_n + KE_N = [m_p/(m_n + m_N)]KE_p$.

When we use this in the kinetic energy equation, we get

$[m_p/(m_n + m_N)]KE_p = KE_p + Q;$

$\{[(1\text{ u})/(1\text{ u} + 13\text{ u})] - 1\}KE_p = -3.002\text{ MeV}$, which gives $KE_p = 3.23\text{ MeV}$.

16. The kinetic energy of the products is

$KE_{pr} = KE_b + Q.$

Because the kinetic energies « mc^2, we can use a non relativistic treatment: $KE = mv^2/2 = p^2/2m$.
The least kinetic energy is required when the product particles move together with the same speed.
With the target at rest, for momentum conservation we have

$p_b = p_{pr} = m_{pr}v$, or

$KE_b = p_b^2/2m_b = (m_{pr}^2/2m_b)v^2 = (m_{pr}/m_b)(KE_{pr})$, or $KE_{pr} = (m_b/m_{pr})(KE_b)$.

When we use this in the kinetic energy equation, we get

$(m_b/m_{pr})(KE_b) = (KE_b) + Q;$

$[(m_b/m_{pr}) - 1](KE_b) = Q$, which gives $KE_p = -Qm_{pr}/(m_{pr} - m_b)$.

17. If we neglect the kinetic energy of the neutron, the released energy is the Q-value:

$Q = [m(^{235}U) + m(n) - 12m(n) - m(^{88}Sr) - m(^{136}Xe)]c^2$

$= [(235.043924\text{ u}) - 11(1.008665\text{ u}) - (87.905618\text{ u}) - (135.90721\text{ u})]c^2(931.5\text{ MeV}/uc^2) = \boxed{126.5\text{ MeV.}}$

18. For the reaction $n + ^{235}_{92}U \to ^{141}_{56}Ba + ^{92}_{36}Kr + 3n$, we determine the Q-value:

$Q = [m(^{235}U) + m(n) - 3m(n) - m(^{141}Ba) - m(^{92}Kr)]c^2$

$= [(235.043924\text{ u}) - 2(1.008665\text{ u}) - (140.91440\text{ u}) - (91.92630\text{ u})]c^2(931.5\text{ MeV}/uc^2) = \boxed{173.2\text{ MeV.}}$

19. If we assume 100% efficiency, we have

$P = E/t;$

$200\text{ MW} = (200\text{ MeV})(1.60 \times 10^{-19}\text{ J/eV})(n/t)$, which gives $n/t = \boxed{6.3 \times 10^{18}\text{ fissions/s.}}$

20. We find the number of fissions from

$P = E/t$;

300 W = (200 MeV)(1.60×10^{-13} J/MeV)n/(3.16×10^{7} s/yr), which gives $n = 2.96 \times 10^{20}$ fissions.

Each fission uses one uranium atom, so the required mass is

$m = [(2.96 \times 10^{20}\text{ atoms})/(6.02 \times 10^{23}\text{ atoms/mol})](235\text{ g/mol}) = 1.16 \times 10^{-4}\text{ kg} = \boxed{0.116\text{ g.}}$

21. We find the number of fissions from

$P = E/t$;

(500 MW)/(0.40) = (200 MeV)(1.60×10^{-19} J/eV)n/(3.16×10^{7} s),

which gives $n = 1.23 \times 10^{27}$ fissions.

Each fission uses one uranium atom, so the required mass is

$m = [(1.23 \times 10^{27}\text{ atoms})/(6.02 \times 10^{23}\text{ atoms/mol})](235\text{ g/mol}) = 4.82 \times 10^{5}\text{ g} = \boxed{482\text{ kg.}}$

22. We find the number of collisions from

$E_n = E_0(\frac{1}{2})^n$;

$0.040\text{ eV} = (1.0 \times 10^{6}\text{ eV})(\frac{1}{2})^n$, which gives $n = \boxed{25.}$

23. The number of fissions in one second is

$n = t/\Delta t = (1.0\text{ s})/(1.0 \times 10^{-3}\text{ s}) = 1.0 \times 10^{3}$.

For each fission the number of neutrons is 1.0004 times the number from the previous fission. Thus the reaction rate will increase by

$(1.0004)^n = (1.0004)^{1000} = \boxed{1.49.}$

24. We find the average kinetic energy from

$\text{KE} = \frac{3}{2}kT = \frac{3}{2}(1.38 \times 10^{-23}\text{ J/K})(10^{7}\text{ K})/(1.60 \times 10^{-19}\text{ J/eV}) = 1.29 \times 10^{3}\text{ eV} = \boxed{1.3\text{ keV.}}$

25. For the reaction ${}^{2}_{1}\text{H} + {}^{3}_{1}\text{H} \rightarrow {}^{4}_{2}\text{He} + \text{n}$, we determine the Q-value:

$Q = [m({}^{2}\text{H}) + m({}^{3}\text{H}) - m(\text{n}) - m({}^{4}\text{He})]c^2$

$= [(2.014102\text{ u}) + (3.016049\text{ u}) - (1.008665\text{ u}) - (4.002602\text{ u})]c^2(931.5\text{ MeV}/uc^2) = +17.59\text{ MeV}.$

Thus 17.59 MeV is released.

26. For the reaction ${}^{2}_{1}\text{H} + {}^{2}_{1}\text{H} \rightarrow {}^{3}_{2}\text{He} + \text{n}$, we determine the Q-value:

$Q = [m({}^{2}\text{H}) + m({}^{2}\text{H}) - m(\text{n}) - m({}^{3}\text{He})]c^2$

$= [2(2.014102\text{ u}) - (1.008665\text{ u}) - (3.016029\text{ u})]c^2(931.5\text{ MeV}/uc^2) = +3.27\text{ MeV}.$

Thus 3.27 MeV is released.

27. For the reaction ${}^1_1\text{H} + {}^1_1\text{H} \rightarrow {}^2_1\text{H} + {}^{\;0}_{+1}\text{e} + \nu$, we must add two electron masses to the left hand side to use atomic masses. Thus we have two extra electron masses on the right hand side.
We determine the Q-value:
$Q = [m({}^1\text{H}) + m({}^1\text{H}) - m({}^2\text{H}) - 2m(\text{e})]c^2$
$= [2(1.007825\text{ u}) - (2.014102\text{ u}) - 2(0.0005486\text{ u})]c^2(931.5\text{ MeV/u}c^2) = 0.42\text{ MeV}.$
For the reaction ${}^1_1\text{H} + {}^2_1\text{H} \rightarrow {}^3_2\text{He} + \gamma$, we determine the Q-value:
$Q = [m({}^1\text{H}) + m({}^2\text{H}) - m({}^3\text{He})]c^2$
$= [(1.007825\text{ u}) + (2.014102\text{ u}) - (3.016029\text{ u})]c^2(931.5\text{ MeV/u}c^2) = 5.49\text{ MeV}.$
For the reaction ${}^3_2\text{He} + {}^3_2\text{He} \rightarrow {}^4_2\text{He} + {}^1_1\text{H} + {}^1_1\text{H}$, we determine the Q-value:
$Q = [m({}^3\text{He}) + m({}^3\text{He}) - m({}^4\text{He}) - m({}^1\text{H}) - m({}^1\text{H})]c^2$
$= [2(3.016029\text{ u}) - (4.002602\text{ u}) - 2(1.007825\text{ u})]c^2(931.5\text{ MeV/u}c^2) = 12.86\text{ MeV}.$

28. For the reaction ${}^2_1\text{H} + {}^2_1\text{H} \rightarrow {}^3_1\text{H} + {}^1_1\text{H}$, two atoms of ${}^2\text{H}$, or 4 u, are used as fuel. The energy release is
$(4.03\text{ MeV/reaction})/(4\text{ u/reaction})(1.66 \times 10^{-27}\text{ kg/u}) = 6.1 \times 10^{26}\text{ MeV/kg} = \boxed{6.1 \times 10^{23}\text{ MeV/g.}}$
For the reaction ${}^2_1\text{H} + {}^2_1\text{H} \rightarrow {}^3_2\text{He} + \text{n}$, two atoms of ${}^2\text{H}$, or 4 u, are used as fuel. The energy release is
$(3.27\text{ MeV/reaction})/(4\text{ u/reaction})(1.66 \times 10^{-27}\text{ kg/u}) = 4.9 \times 10^{26}\text{ MeV/kg} = \boxed{4.9 \times 10^{23}\text{ MeV/g.}}$
For the reaction ${}^2_1\text{H} + {}^3_1\text{H} \rightarrow {}^4_2\text{He} + \text{n}$, one atom of ${}^2\text{H}$ and one atom of ${}^3\text{H}$, or 5 u, are used as fuel. The energy release is
$(17.59\text{ MeV/reaction})/(5\text{ u/reaction})(1.66 \times 10^{-27}\text{ kg/u}) = 2.1 \times 10^{27}\text{ MeV/kg} = \boxed{2.1 \times 10^{24}\text{ MeV/g.}}$
In the fission reaction, one atom of ${}^{235}\text{U}$, or 235 u, is used as fuel. The energy release is
$(200\text{ MeV/reaction})/(235\text{ u/reaction})(1.66 \times 10^{-27}\text{ kg/u}) = 5.1 \times 10^{26}\text{ MeV/kg} = \boxed{5.1 \times 10^{23}\text{ MeV/g.}}$
Thus most fusion reactions yield more energy per unit mass.

29. We find the number of fusions from
$P = E/t;$
$300\text{ W} = (3.27\text{ MeV})(1.60 \times 10^{-13}\text{ J/MeV})n/(3.16 \times 10^7\text{ s/yr}),$
which gives $n = 1.81 \times 10^{28}$ fusions/yr.
Each fusion uses two deuterium atoms, so the required mass is
$m = [(1.81 \times 10^{28}\text{ atoms/yr})/(6.02 \times 10^{23}\text{ atoms/mol})](2)(2\text{ g/mol}) = \boxed{0.120\text{ g/yr.}}$

30. Because the kinetic energies « mc^2, we can use a non relativistic treatment: $\text{KE} = mv^2/2 = p^2/2m$.
We assume the kinetic energy of the deuterium and the tritium can be neglected, so for momentum conservation we have
$p_{\text{He}} = p_{\text{n}}.$
The kinetic energy of the particles is
$Q = \text{KE}_{\text{He}} + \text{KE}_{\text{n}} = (p_{\text{He}}{}^2/2m_{\text{He}}) + (p_{\text{n}}{}^2/2m_{\text{n}}) = p_{\text{He}}{}^2(m_{\text{He}} + m_{\text{n}})/2m_{\text{He}}m_{\text{n}};$
$17.59\text{ MeV} = p_{\text{He}}{}^2(4\text{ u} + 1\text{ u})/2(4\text{ u})(1\text{ u})$, which gives $p_{\text{He}}{}^2 = 28.1\text{ MeV}\cdot\text{u}.$
The kinetic energy of ${}^4\text{He}$ is
$\text{KE}_{\text{He}} = p_{\text{He}}{}^2/2m_{\text{He}} = (28.1\text{ MeV}\cdot\text{u})/2(4\text{ u}) = 3.5\text{ MeV}.$
The kinetic energy of n is
$\text{KE}_{\text{n}} = p_{\text{n}}{}^2/2m_{\text{n}} = (28.1\text{ MeV}\cdot\text{u})/2(1\text{ u}) = 14\text{ MeV}.$
This result is $\boxed{\text{not independent}}$ of the plasma temperature, which is a measure of the initial kinetic energies.

31. (*a*) For the reaction ${}^{12}_{6}\text{C} + {}^{12}_{6}\text{C} \rightarrow {}^{24}_{12}\text{Mg}$, we determine the Q-value:

$$Q = [m({}^{12}\text{C}) + m({}^{12}\text{C}) - m({}^{24}\text{Mg})]c^2$$
$$= [2(12.000000\text{ u}) - (23.985042\text{ u})]c^2(931.5\text{ MeV}/uc^2) = +13.93\text{ MeV}.$$

Thus **13.93 MeV is released.**

(*b*) If the two nuclei are just touching, the Coulomb potential energy must be the initial kinetic energies of the two nuclei:

$$2\text{KE} = \text{PE} = Z_C Z_C ke^2/R = (6)(6)(1.44\text{ MeV}\cdot\text{fm})/(6.0\text{ fm}) = 8.6\text{ MeV}.$$

Thus each carbon nucleus has a kinetic energy of **4.3 MeV.**

(*c*) We find the temperature from

$$\text{KE} = \tfrac{3}{2}kT;$$
$$(4.3\text{ MeV})(1.60\times10^{-13}\text{ J/MeV}) = \tfrac{3}{2}(1.38\times10^{-23}\text{ J/K})T,$$ which gives $T =$ **3.3×10^{10} K.**

32. The rate at which input energy is needed by the reactor is

$$P = [(1000\times10^{6}\text{ W})/(0.30)](3600\text{ s/h})/(1.60\times10^{-13}\text{ J/MeV}) = 7.50\times10^{25}\text{ MeV/h}.$$

If we assume equal contributions from the two equations, four deuterium nuclei release

$$4.03\text{ MeV} + 3.27\text{ MeV} = 7.30\text{ MeV}.$$

Each water molecule has two H atoms. If 0.015% of them are deuterium nuclei, the number of water molecules needed for fuel is

$$N = (7.50\times10^{25}\text{ MeV/h})/[(7.30\text{ MeV})/(4\text{ nuclei})](2\text{ nuclei/molecule})(0.015\times10^{-2})$$
$$= 1.37\times10^{29}\text{ molecules/h}.$$

The required mass of water is

$$m = [(1.37\times10^{29}\text{ molecules/h})/(6.02\times10^{23}\text{ molecules/mol})](18\text{ g/mol})$$
$$= 4.1\times10^{6}\text{ g/h} =$$ **4.1×10^{3} kg/h.**

33. The number of deuterium nuclei in 1.00 kg of water is

$$N = [(1.00\times10^{3}\text{ g})/(18\text{ g/mol})](6.02\times10^{23}\text{ molecules/mol})(2\text{ nuclei/molecule})(0.015\times10^{-2})$$
$$= 1.00\times10^{22}\text{ nuclei}.$$

Two deuterium nuclei release 4.03 MeV, so the total energy is

$$E = (1.00\times10^{22}\text{ nuclei})(4.03\text{ MeV})(1.60\times10^{-13}\text{ J/MeV})/(2\text{ nuclei}) =$$ **3.23×10^{9} J.**

This is **65×** the energy from burning gasoline.

34. (*a*) If we look at the reactions for the carbon cycle, we see that one carbon atom is used in the first reaction and one carbon atom is produced in the last reaction. If we add all of the reactions, we get

$4{}_1^1\text{H} \rightarrow {}_2^4\text{He} + 2e^+ + 2\nu$.

(*b*) If we ignore the gamma rays and the pair annihilation, the Q-value is

$Q = [4m(^1\text{H}) - m(^4\text{He}) - 2m(\text{e})]c^2$

$= [4(41.007825 \text{ u}) - (4.002602 \text{ u}) - 2(0.0005486 \text{ u})]c^2(931.5 \text{ MeV}/uc^2) = +24.69 \text{ MeV}$.

Thus 24.69 MeV is released.

(*c*) For the reaction ${}_6^{12}\text{C} + {}_1^1\text{H} \rightarrow {}_7^{13}\text{N}$, we determine the Q-value:

$Q = [m(^{12}\text{C}) + m(^1\text{H}) - m(^{13}\text{N})]c^2$

$= [(12.000000 \text{ u}) + (1.007825 \text{ u}) - (13.005738 \text{ u})]c^2(931.5 \text{ MeV}/uc^2) =$ $+1.94$ MeV.

For the decay ${}_7^{13}\text{N} \rightarrow {}_6^{13}\text{C} + e^+$, we determine the Q-value:

$Q = [m(^{13}\text{N}) - m(^{13}\text{C}) - 2m(\text{e})]c^2$

$= [(13.005738 \text{ u}) + (13.003355 \text{ u}) - 2(0.0005486 \text{ u})]c^2(931.5 \text{ MeV}/uc^2) =$ $+1.20$ MeV.

For the reaction ${}_6^{13}\text{C} + {}_1^1\text{H} \rightarrow {}_7^{14}\text{N}$, we determine the Q-value:

$Q = [m(^{13}\text{C}) + m(^1\text{H}) - m(^{14}\text{N})]c^2$

$= [(13.003355 \text{ u}) + (1.007825 \text{ u}) - (14.003074 \text{ u})]c^2(931.5 \text{ MeV}/uc^2) =$ $+7.55$ MeV.

For the reaction ${}_7^{14}\text{N} + {}_1^1\text{H} \rightarrow {}_8^{15}\text{O}$, we determine the Q-value:

$Q = [m(^{14}\text{N}) + m(^1\text{H}) - m(^{15}\text{O})]c^2$

$= [(14.003074 \text{ u}) + (1.007825 \text{ u}) - (15.003065 \text{ u})]c^2(931.5 \text{ MeV}/uc^2) =$ $+7.30$ MeV.

For the decay ${}_8^{15}\text{O} \rightarrow {}_7^{15}\text{N} + e^+$, we determine the Q-value:

$Q = [m(^{15}\text{O}) - m(^{15}\text{N}) - 2m(\text{e})]c^2$

$= [(15.003065 \text{ u}) + (15.000108 \text{ u}) - 2(0.0005486 \text{ u})]c^2(931.5 \text{ MeV}/uc^2) =$ $+1.73$ MeV.

For the reaction ${}_7^{15}\text{N} + {}_1^1\text{H} \rightarrow {}_6^{12}\text{C} + {}_2^4\text{He}$, we determine the Q-value:

$Q = [m(^{15}\text{N}) + m(^1\text{H}) - m(^{12}\text{C}) - m(^4\text{He})]c^2$

$= [(15.000108 \text{ u}) + (1.007825 \text{ u}) - (12.000000 \text{ u}) - (4.002602 \text{ u})]c^2(931.5 \text{ MeV}/uc^2) =$ $+4.97$ MeV.

(*d*) For the nuclei with higher atomic number, there is a greater repulsion from the positive charges, so higher kinetic energies are required. Thus higher temperatures are required.

35. (*a*) If n is the density of particles, for equal numbers of ^{2}H and ^{3}H atoms we have

$\rho = 200 \times 10^3 \text{ kg/m}^3 = \frac{1}{2}n(2 \text{ u})(1.66 \times 10^{-27} \text{ kg/u}) + \frac{1}{2}n(3 \text{ u})(1.66 \times 10^{-27} \text{ kg/u})$,

which gives $n =$ 4.8×10^{31} particles/m^3.

(*b*) To meet the Lawson criterion, we have

$n\tau \geq 3 \times 10^{20} \text{ s/m}^3$;

$(4.8 \times 10^{31} \text{ particles/m}^3)\tau \geq 3 \times 10^{20} \text{ s/m}^3$, which gives $\tau \geq 6 \times 10^{-12}$ s.

36. Because the quality factor for gamma rays is ≈ 1, we have

effective dose (rad) = effective dose (rem)/QF = (500 rem)/1 = 500 rad.

37. Because the quality factor for α-particle radiation is ≈ 20, we have

effective dose (rem) = effective dose (rad) $\times$ QF;

$\text{rad}_\alpha \times 20 = \text{rad}_X \times 1$;

$\text{rad}_X = (50 \text{ rad})(20) =$ 1000 rad.

38. Because the quality factor for slow neutrons is ≈ 3 and fast neutrons is ≈ 10, we have

effective dose (rem) = effective dose (rad) $\times$ QF;

$\text{rad}_{\text{slow}} \times 3 = \text{rad}_{\text{fast}} \times 10$;

$\text{rad}_{\text{slow}} = (50 \text{ rad})(10)/(3) =$ 167 rad.

39. The energy deposited is

$$E = (50\text{ rad})(1.00 \times 10^{-2}\text{ J/kg}\cdot\text{rad})(70\text{ kg}) = \boxed{35\text{ J.}}$$

40. If the counter counts 90% of the intercepted β particles, we have

$$n = (0.90)(0.20)(0.018 \times 10^{-6}\text{ Ci})(3.7 \times 10^{10}\text{ decays/s}\cdot\text{Ci}) = \boxed{120\text{ counts/s.}}$$

41. If the decay rate was constant, the time required would be

$$t_1 = (5000\text{ rad})/(1.0\text{ rad/min})(60\text{ min/h})(24\text{ h/day}) = 3.47\text{ days.}$$

This time is (3.47 days)/(14.3 days) = 0.243 half-lives.

The activity of the source at this time will be

$$(1.0\text{ rad/min})(\tfrac{1}{2})^{0.243} = 0.845\text{ rad/min.}$$

If we approximate the exponential decay as linear, and use the average activity, we get

$$t_2 = (5000\text{ rad})/\tfrac{1}{2}(1.0\text{ rad/min} + 0.845\text{ rad/min})(60\text{ min/h})(24\text{ h/day}) = 3.8\text{ days} \approx \boxed{4\text{ days.}}$$

42. If we start with the current definition of the roentgen, we get

$$1\text{ R} = (0.878 \times 10^{-2}\text{ J/kg})/(1.60 \times 10^{-19}\text{ J/eV})(1000\text{ g/kg})(35\text{ eV/pair})$$
$$= 1.57 \times 10^{12}\text{ pairs/g} \approx 1.6 \times 10^{12}\text{ pairs/g.}$$

43. Because each decay gives one gamma ray, the rate at which energy is emitted is

$$P = (2.0\text{ Ci})(3.7 \times 10^{10}\text{ decays/s}\cdot\text{Ci})(122 \times 10^{3}\text{ eV/decay})(1.60 \times 10^{-19}\text{ J/eV}) = 1.44 \times 10^{-9}\text{ J/s.}$$

If 50% of the energy is absorbed by the body, the dose rate is

$$\text{dose rate} = (0.50)(1.44 \times 10^{-9}\text{ J/s})(86{,}400\text{ s/day})/(1.00 \times 10^{-2}\text{ J/kg}\cdot\text{rad})(70\text{ kg}) = \boxed{8.9\text{ rad/day.}}$$

44. The decay constant is

$$\lambda = 0.693/T_{1/2} = 0.693/(5730\text{ yr}) = 1.209 \times 10^{-4}\text{ /yr} = 3.83 \times 10^{-12}\text{ s}^{-1}.$$

We find the number of ^{14}C atoms from

$$\text{Activity} = \lambda N;$$
$$(1.00 \times 10^{-6}\text{ Ci})(3.7 \times 10^{10}\text{ decays/s}\cdot\text{Ci}) = (3.83 \times 10^{-12}\text{ s}^{-1})N,\text{ which gives } N = 9.67 \times 10^{15}\text{ atoms.}$$

The mass is

$$m = [(9.67 \times 10^{15}\text{ atoms})/(6.02 \times 10^{23}\text{ atoms/mol})](14\text{ g/mol}) = 2.25 \times 10^{-7}\text{ g} = \boxed{0.225\ \mu\text{g.}}$$

45. (*a*) $\boxed{{}^{131}_{53}\text{I} \rightarrow {}^{131}_{54}\text{Xe} + {}^{0}_{-1}\text{e} + \bar{\nu} + \gamma.}$

(*b*) We find the number of half-lives from

$$N/N_0 = (\tfrac{1}{2})^n;$$
$$(0.10) = (\tfrac{1}{2})^n,\quad\text{or}\quad n\log 2 = \log 10,\text{ which gives } n = 3.32.$$

Thus the elapsed time is

$$\Delta t = nT_{1/2} = (3.32)(8.0\text{ days}) = \boxed{27\text{ days.}}$$

(*c*) We find the number of atoms from

$$\text{Activity} = \lambda N;$$
$$(1.00 \times 10^{-3}\text{ Ci})(3.7 \times 10^{10}\text{ decays/s}\cdot\text{Ci}) = [0.693/(8.0\text{ days})(86{,}400\text{ s/day})]N,$$

which gives $N = 3.69 \times 10^{13}$ atoms.

The mass is

$$m = [(3.69 \times 10^{13}\text{ atoms})/(6.02 \times 10^{23}\text{ atoms/mol})](131\text{ g/mol}) = \boxed{8.0 \times 10^{-9}\text{ g.}}$$

46. Because each decay gives one gamma ray, the rate at which energy is emitted is

$$P = (2000 \times 10^{-12}\ \text{Ci/L})(3.7 \times 10^{10}\ \text{decays/s}\cdot\text{Ci})(1.5\ \text{MeV/decay})(1.60 \times 10^{-13}\ \text{J/MeV})$$
$$= 1.78 \times 10^{-11}\ \text{J/s}\cdot\text{L}.$$

If 10% of the energy is absorbed by the body for half a year (12 h/day), the total absorbed energy rate is

$$\text{rate} = (0.10)(1.78 \times 10^{-11}\ \text{J/s}\cdot\text{L})(0.5\ \text{L})\tfrac{1}{2}(3.16 \times 10^{7}\ \text{s/yr}) = 1.4 \times 10^{-5}\ \text{J/yr}.$$

For beta particles and gamma rays, QF = 1.

(*a*) For an adult, the dose is

$$\text{dose} = (1.4 \times 10^{-5}\ \text{J/yr})(1)/(1.00 \times 10^{-2}\ \text{J/kg}\cdot\text{rad})(50\ \text{kg})$$
$$= 2.8 \times 10^{-5}\ \text{rem} = \boxed{0.028\ \text{mrem} \approx 0.006\%\ \text{of allowed dose.}}$$

(*b*) For a baby, the dose is

$$\text{dose} = (1.4 \times 10^{-5}\ \text{J/yr})(1)/(1.00 \times 10^{-2}\ \text{J/kg}\cdot\text{rad})(5\ \text{kg})$$
$$= 2.8 \times 10^{-4}\ \text{rem} = \boxed{0.28\ \text{mrem} \approx 0.06\%\ \text{of allowed dose.}}$$

47. (*a*) We find the daughter nucleus by balancing the mass and charge numbers:

$$Z(\text{X}) = Z(^{222}\text{Rn}) - Z(^{4}\text{He}) = 86 - 2 = 84;$$
$$A(\text{X}) = A(^{222}\text{Rn}) - A(^{4}\text{He}) = 222 - 4 = 218,\ \text{so the product nucleus is}\ \boxed{^{218}_{84}\text{Po}.}$$

(*b*) From Figure 30–10, we see that $^{218}_{84}\text{Po}$ is $\boxed{\text{radioactive,}}$ and decays by both α and β^- emission:

$$\boxed{^{218}_{84}\text{Po} \to {}^{214}_{82}\text{Pb} + {}^{4}_{2}\text{He}\,; \quad {}^{218}_{84}\text{Po} \to {}^{218}_{85}\text{Xe} + {}^{0}_{-1}\text{e} + \bar{\nu}\,.}$$

The half life for both decays is $\boxed{3.1\ \text{min.}}$

(*c*) The daughter nucleus from α decay is $\boxed{\text{chemically reactive.}}$

The daughter nucleus from β^- decay is $\boxed{\text{a noble gas.}}$

(*d*) The decay constant is

$$\lambda = 0.693/T_{1/2} = 0.693/(3.8\ \text{days})(86{,}400\ \text{s/day}) = 2.11 \times 10^{-6}\ \text{s}^{-1}.$$

The number of radon atoms in 1.0 ng is

$$N_0 = [(1.0 \times 10^{-9}\ \text{g})/(222\ \text{g/mol})](6.02 \times 10^{23}\ \text{atoms/mol}) = 2.71 \times 10^{12}\ \text{atoms}.$$

The initial activity is

$$\lambda N_0 = (2.11 \times 10^{-6}\ \text{s}^{-1})(2.71 \times 10^{12})/(3.7 \times 10^{10}\ \text{decays/s}\cdot\text{Ci}) = 1.5 \times 10^{-4}\ \text{Ci} = \boxed{150\ \mu\text{Ci.}}$$

After 1 month, the number of half-lives is

$$n = (30\ \text{days})/(3.8\ \text{days}) = 7.89.$$

The activity is

$$\lambda N = \lambda N_0\,(\tfrac{1}{2})^n = (150\ \mu\text{Ci})(\tfrac{1}{2})^{7.89} = \boxed{0.63\ \mu\text{Ci.}}$$

48. From the text we know that in a magnetic field of 1.000 T, resonance occurs when the photon has a frequency of 42.58 MHz. The wavelength is

$$\lambda = c/f = (3.00 \times 10^{8}\ \text{m/s})/(42.58 \times 10^{6}\ \text{Hz}) = \boxed{7.05\ \text{m, FM radio wave.}}$$

49. (*a*) We find the other nucleus by balancing the mass and charge numbers:

$$Z(\text{X}) = Z(^{9}\text{Be}) + Z(^{4}\text{He}) - Z(\text{n}) = 4 + 2 - 0 = 6;$$
$$A(\text{X}) = A(^{9}\text{Be}) + A(^{4}\text{He}) - A(\text{n}) = 9 + 4 - 1 = 12,\ \text{so the other nucleus is}\ \boxed{^{12}_{6}\text{C}.}$$

(*b*) For the reaction $^{9}_{4}\text{Be} + {}^{4}_{2}\text{He} \to {}^{12}_{6}\text{C} + {}^{1}_{0}\text{n}$, we determine the Q-value:

$$Q = [m(^{9}\text{Be}) + m(^{4}\text{He}) - m(^{12}\text{C}) - m(^{1}\text{n})]c^2$$
$$= [(9.012182\ \text{u}) + (4.002602\ \text{u}) - (12.000000\ \text{u}) - (1.008665\ \text{u})]c^2(931.5\ \text{MeV}/\text{u}c^2) = \boxed{+\,5.70\ \text{MeV.}}$$

50. We find the conversion from

$$\text{KE} = kT$$

by converting the units for k:

$$1.38 \times 10^{-23}\ \text{J/K} = (1.38 \times 10^{-23}\ \text{J/K})/(1.60 \times 10^{-16}\ \text{J/keV}) = \boxed{8.638 \times 10^{-8}\ \text{keV/K.}}$$

51. The kinetic energy is a function of the temperature:

$\text{KE} = \frac{1}{2}mv^2 = \frac{3}{2}kT.$

When we form the ratio for the molecules of the two isotopes at the same temperature, we get

$v_{235}/v_{238} = (m_{238}/m_{235})^{1/2} = \{[238\text{ u} + 6(19\text{ u})]/[235\text{ u} + 6(19\text{ u})]\}^{1/2} =$ **1.0043.**

52. (*a*) We find the number of fissions from

$n = (20\text{ kt})(5 \times 10^{12}\text{ J/kt})/(200\text{ MeV/fission})(1.60 \times 10^{-13}\text{ J/MeV}) = 3.13 \times 10^{24}$ fissions.

Each fission uses one uranium atom, so the required mass is

$M = [(3.13 \times 10^{24}\text{ atoms})/(6.02 \times 10^{23}\text{ atoms/mol})](235\text{ g/mol}) =$ **1.2 kg.**

(*b*) We find the mass transformation from

$m = E/c^2 = (20\text{ kt})(5 \times 10^{12}\text{ J/kt})/(3.00 \times 10^8\text{ m/s})^2 = 1.1 \times 10^{-3}\text{ kg} =$ **1.1 g.**

53. The effective dose in rem = effective dose (rad) × QF. For the two radiations, we get

dose (rem) = dose (rad) × QF = (25 mrad/yr)(1) + (3.0 mrad/yr)(10) = **55 mrem/yr.**

54. If we assume the oceans cover 70% of the Earth's surface and the average depth is 3 km, the mass of the ocean water is

$M = \rho V = (1000\text{ kg/m}^3)(0.70)4\pi(6.38 \times 10^6\text{ m})^2(3 \times 10^3\text{ m}) = 1 \times 10^{21}\text{ kg}.$

Each water molecule has two H atoms, so the number of deuterium atoms is

$N = [(1 \times 10^{21}\text{ kg})(10^3\text{ g/kg})/(18\text{ g/mol})](6.02 \times 10^{23}\text{ atoms/mol})(2)(0.015 \times 10^{-2}) = 1 \times 10^{43}$ atoms.

The mass of deuterium is

$m = (1 \times 10^{43}\text{ atoms})(2\text{ g/atom})/(6.02 \times 10^{23}\text{ atoms/mol}) = 3 \times 10^{19}\text{ g} =$ **3×10^{16} kg.**

If we assume equal contributions from the reactions in Eq. 31–5a and Eq. 31–5b, the average energy released by a deuterium nucleus is

(4.03 MeV + 3.27 MeV)/(2 + 2) = 1.8 MeV/nucleus.

Thus the total energy released is

$E = (1.8\text{ MeV/nucleus})(1 \times 10^{43}\text{ nuclei}) = 1.8 \times 10^{43}\text{ MeV} =$ **3×10^{30} J.**

55. Because the QF for gamma rays is 1, the dose in rem is the dose in rad. The allowed dose rate is

$(5.0\text{ rem/yr})/(50\text{ wk/yr})(40\text{ h/wk}) = 2.5 \times 10^{-3}\text{ rem/h} = 2.5 \times 10^{-3}\text{ rad/h}.$

If the dose rate falls off as the square of the distance, we form the ratio:

$(\text{dose rate})_2/(\text{dose rate})_1 = (r_1/r_2)^2;$

$(2.5 \times 10^{-3}\text{ rad/h})/(0.050\text{ rad/h}) = [(1.0\text{ m})/r_2]^2$, which gives $r_2 =$ **4.5 m.**

56. (*a*) **${}^{226}_{88}\text{Ra} \rightarrow {}^{222}_{86}\text{Rn} + {}^{4}_{2}\text{He}$.**

(*b*) If we ignore the KE of the daughter, the KE of the α particle is the Q-value:

$(\text{KE})_\alpha = Q = [m(^{226}\text{Ra}) - m(^{222}\text{Rn}) - m(^4\text{He})]c^2$

$= [(226.025402\text{ u}) - (222.017571\text{ u}) - (4.002602\text{ u})]c^2(931.5\text{ MeV/uc}^2) =$ **4.871 MeV.**

(*c*) From momentum conservation, the momenta of the α particle and the daughter will have equal magnitudes:

$p_{\text{Rn}} = p_\alpha = [2m_\alpha(\text{KE})_\alpha]^{1/2}$

$= [2(4\text{ u})(1.66 \times 10^{-27}\text{ kg/u})(4.87\text{ MeV})(1.6 \times 10^{-13}\text{ J/MeV})]^{1/2} =$ **1.02×10^{-19} kg·m/s.**

(*d*) The kinetic energy of the daughter is

$(\text{KE})_{\text{Rn}} = p_{\text{Rn}}^2/2m_{\text{Rn}}$

$= (1.02 \times 10^{-19}\text{ kg·m/s})^2/2(222\text{ u})(1.66 \times 10^{-27}\text{ kg/u}) = 1.4 \times 10^{-14}\text{ J} =$ **0.088 MeV.**

Because this is less than 2% of the Q-value, our approximation is valid.

57. (*a*) We find the number of fissions from

$P = E/t$;

4000 MW = (200 MeV)(1.60×10^{-19} J/eV)n/(3.16×10^7 s), which gives $n = 3.95 \times 10^{27}$ fissions.

Each fission uses one uranium atom, so the required mass is

m = [(3.95×10^{27} atoms)/(6.02×10^{23} atoms/mol)](235 g/mol) = $\boxed{1.54 \times 10^3 \text{ kg.}}$

(*b*) We find the activity from the number of Sr atoms produced:

Activity = λN_{Sr}

= [(0.693)/(29 yr)(3.16×10^7 s/yr)](0.06)(3.95×10^{27})/(3.7×10^{10} decays/s · Ci)

= $\boxed{4.8 \times 10^6 \text{ Ci.}}$

58. In the net reaction four protons produce 26.2 MeV. Thus the heat of combustion is

[(26.2 MeV)/(4 u)](1.6×10^{-13} J/MeV)/(1.66×10^{-27} kg/u) = $\boxed{6.26 \times 10^{14} \text{ J/kg.}}$

This is $\boxed{\approx 10^7\times \text{ the heat of combustion of coal.}}$

59. (*a*) The energy is radiated uniformly over a sphere, so the total energy rate is

P = (1400 W/m²)4π(1.50×10^{11} m)² = $\boxed{4.0 \times 10^{26} \text{ J/s}}$ = 2.5×10^{39} MeV/s.

(*b*) In the net reaction four protons produce 26.2 MeV. Thus the consumption of protons is

n = (2.5×10^{39} MeV/s)/[(26.2 MeV)/4 protons] = $\boxed{3.8 \times 10^{38} \text{ protons/s.}}$

(*c*) We find the time from

$t = N/n$

= [(2.0×10^{30} kg)/(1.66×10^{-27} kg/proton)]/(3.8×10^{38} protons/s) = 3.2×10^{18} s $\boxed{\approx 10^{11} \text{ yr.}}$

60. In the net reaction four protons produce two neutrinos. If we use the result from Problem 59 for the rate at which protons are consumed, for the rate at which neutrinos are produced we have

n_ν = (3.8×10^{38} protons/s)(2 neutrinos/4 protons)(3.16×10^7 s/yr) = 6.0×10^{45} neutrinos/yr.

The neutrinos are spread uniformly over a sphere centered at the Sun, so the number that pass through an area of 100 m² at the Earth is

N_ν = [(6.0×10^{45} neutrinos/yr)/4π(1.50×10^{11} m)²](100 m²) = $\boxed{2.1 \times 10^{24} \text{ neutrinos/yr.}}$

61. (*a*) The rate of decay is

Activity = (0.10×10^{-6} Ci)(3.7×10^{10} decays/s · Ci) = $\boxed{3.7 \times 10^3 \text{ decays/s.}}$

(*b*) Because the QF for gamma rays is 1, the dose in rem is the dose in rad:

dose rate = (3.7×10^3 decays/s)(1.4 MeV/decay)(3.16×10^7 s/yr)(1.6×10^{-13} J/MeV)/

(50 kg)(1.00×10^{-2} J/kg · rad) = $\boxed{5.2 \times 10^{-2} \text{ rem/yr} \approx 0.15 \text{ background.}}$

CHAPTER 32

Note: A useful expression for the energy of a photon in terms of its wavelength is
$E = hf = hc/\lambda = (6.63 \times 10^{-34}\ \text{J}\cdot\text{s})(3 \times 10^{8}\ \text{m/s})(10^{-9}\ \text{nm/m})/(1.60 \times 10^{-19}\ \text{J/eV})\lambda;$
$E = (1.24 \times 10^{3}\ \text{eV}\cdot\text{nm})/\lambda = (1.24 \times 10^{-12}\ \text{MeV}\cdot\text{m})/\lambda.$

1. The total energy of the proton is
$E = \text{KE} + m_0c^2 = 8.50\ \text{GeV} + 0.939\ \text{GeV} = \boxed{9.44\ \text{GeV}.}$

2. The total energy of the electron is
$E = \text{KE} + m_0c^2 = 40\ \text{GeV} + 0.511\ \text{MeV} = 40\ \text{GeV}.$
Thus the momentum is $p = E/c$.
We find the wavelength from
$E = hc/\lambda = h/p = hc/E$
$= (1.24 \times 10^{-12}\ \text{MeV}\cdot\text{m})/(40\ \text{GeV})(10^{3}\ \text{MeV/GeV}) = \boxed{3.1 \times 10^{-17}\ \text{m}.}$

3. We find the magnetic field from the cyclotron frequency:
$f = qB/2\pi m;$
$2.4 \times 10^{7}\ \text{Hz} = (1.60 \times 10^{-19}\ \text{C})B/2\pi(1.67 \times 10^{-27}\ \text{kg})$, which gives $B = \boxed{1.6\ \text{T}.}$

4. Very high-energy protons will have a speed $v \approx c$. Thus the time for one revolution is
$t = 2\pi r/v = 2\pi(1.0 \times 10^{3}\ \text{m})/(3.00 \times 10^{8}\ \text{m/s}) = 2.1 \times 10^{-5}\ \text{s} = \boxed{21\ \mu\text{s}.}$

5. The cyclotron frequency is
$f = qB/2\pi m;$
If we form the ratio for the two particles, we get
$f_2/f_1 = (q_2/q_1)(m_1/m_2);$
$f_2/(26\ \text{MHz}) = (2)(1/4)$, which gives $f_2 = \boxed{13\ \text{MHz}.}$

6. The size of a nucleon is
$d \approx 2(1.2 \times 10^{-15}\ \text{m}) = 2.4 \times 10^{-15}\ \text{m}.$
Because 30 MeV « m_0c^2, we find the momentum from
$\text{KE} = p^2/2m$, so the wavelength is
$\lambda = h/p = h/[2m(\text{KE})]^{1/2}.$
For the α particle we have
$\lambda_\alpha = (6.63 \times 10^{-34}\ \text{J}\cdot\text{s})/[2(4)(1.67 \times 10^{-27}\ \text{kg})(30\ \text{GeV})(1.60 \times 10^{-10}\ \text{J/GeV})]^{1/2}$
$= \boxed{2.6 \times 10^{-15}\ \text{m} \approx \text{size of nucleon}.}$
For the proton we have
$\lambda_p = (6.63 \times 10^{-34}\ \text{J}\cdot\text{s})/[2(1.67 \times 10^{-27}\ \text{kg})(30\ \text{GeV})(1.60 \times 10^{-10}\ \text{J/GeV})]^{1/2}$
$= \boxed{5.2 \times 10^{-15}\ \text{m} \approx 2(\text{size of nucleon}).}$
Thus the $\boxed{\alpha\ \text{particle is better.}}$

7. (*a*) The maximum kinetic energy is

$\text{KE} = \frac{1}{2}mv^2 = q^2B^2R^2/2m.$

If we form the ratio for the two particles, we get

$\text{KE}_\alpha/\text{KE}_p = (q_\alpha/q_p)^2(m_p/m_\alpha);$

$\text{KE}_\alpha/(8.7\text{ MeV}) = (2)^2(1/4) = 1$, which gives $\text{KE}_\alpha = \boxed{8.7\text{ MeV.}}$

We find the speed from

$\text{KE} = \frac{1}{2}mv^2;$

$(8.7\text{ MeV})(1.60\times10^{-13}\text{ J/MeV}) = \frac{1}{2}(4)(1.67\times10^{-27}\text{ kg})v^2$, which gives $v = \boxed{2.0\times10^7\text{ m/s.}}$

(*b*) For deuterons we get

$\text{KE}_d/\text{KE}_p = (q_d/q_p)^2(m_p/m_d);$

$\text{KE}_d/(8.7\text{ MeV}) = (1)^2(1/2) = 1/2$, which gives $\text{KE}_d = \boxed{4.3\text{ MeV.}}$

We find the speed from

$\text{KE} = \frac{1}{2}mv^2;$

$(4.3\text{ MeV})(1.60\times10^{-13}\text{ J/MeV}) = \frac{1}{2}(2)(1.67\times10^{-27}\text{ kg})v^2$, which gives $v = \boxed{2.0\times10^7\text{ m/s.}}$

Note that the α particle and the deuteron have the same q/m.

(*c*) The cyclotron frequency is

$f = qB/2\pi m;$

If we form the ratio for the two particles, we get

$f_d/f_\alpha = (q_d/q_\alpha)(m_\alpha/m_d) = (1/2)(4/2) = 1,$

so they require the same frequency. We find the frequency from

$f_d/f_p = (q_d/q_p)(m_p/m_d);$

$f_d/(26\text{ MHz}) = (1)(1/2)$, which gives $f_d = f_\alpha = \boxed{13\text{ MHz.}}$

8. The alternating frequency means the proton is accelerated by the voltage twice in each revolution. The number of revolutions is

$n = E/2eV = (25\text{ MeV})(10^3\text{ keV/MeV})/2(1\text{ e})(45\text{ kV}) = \boxed{2.8\times10^2\text{ rev.}}$

9. (*a*) We find the magnetic field from the maximum kinetic energy:

$\text{KE} = \frac{1}{2}mv^2 = q^2B^2R^2/2m;$

$(10\text{ MeV})(1.60\times10^{-13}\text{ J/MeV}) = (1.60\times10^{-19}\text{ C})^2B^2(1.0\text{ m})^2/2(2)(1.67\times10^{-27}\text{ kg}),$

which gives $B = \boxed{0.65\text{ T.}}$

(*b*) The cyclotron frequency is

$f = qB/2\pi m$

$= (1.60\times10^{-19}\text{ C})(0.65\text{ T})/2\pi(2)(1.67\times10^{-27}\text{ kg}) = 4.9\times10^6\text{ Hz} = \boxed{4.9\text{ MHz.}}$

(*c*) The deuteron is accelerated by the voltage twice in each revolution. The number of revolutions is

$n = E/2eV = (10\text{ MeV})(10^3\text{ keV/MeV})/2(1\text{ e})(25\text{ kV}) = \boxed{2.0\times10^2\text{ rev.}}$

(*d*) Because the time for each revolution is the same, we have

$t = nT = n/f = (2.0\times10^2)/(4.9\times10^6\text{ Hz}) = 4.1\times10^{-5}\text{ s} = \boxed{41\ \mu\text{s.}}$

(*e*) The radius of the path varies linearly from 0 to R, so the average radius is $\frac{1}{2}R$. The total distance traveled is

$d = n2\pi\frac{1}{2}R = (2.0\times10^2)\pi(1.0\text{ m}) = 6.3\times10^2\text{ m} = \boxed{0.63\text{ km.}}$

10. The total energy of the proton is

$E = \text{KE} + m_0c^2 = 900\text{ GeV} + 0.939\text{ GeV} = 900\text{ GeV}.$

Thus the momentum is $p = E/c$.

We find the wavelength from

$E = hc/\lambda = h/p = hc/E$

$= (1.24\times10^{-12}\text{ MeV}\cdot\text{m})/(900\text{ GeV})(10^3\text{ MeV/GeV}) = \boxed{1.4\times10^{-18}\text{ m.}}$

This is the minimum resolving distance, or the maximum resolving power.

11. The number of revolutions is
$n = \Delta E/2eV = (900 \text{ GeV} - 8.0 \text{ GeV})(10^3 \text{ MeV/GeV})/(1 \text{ e})(2.5 \text{ MV}) = 3.57 \times 10^5$ rev.
The total distance traveled is
$d = n2\pi R = (3.57 \times 10^5)2\pi(1.0 \text{ km}) =$ **2.2 × 10⁶ km.**
Very high-energy protons will have a speed $v \approx c$. Thus the time is
$t = d/v = (2.2 \times 10^9 \text{ m})/(3.00 \times 10^8 \text{ m/s}) =$ **7.5 s.**

12. Very high-energy protons will have a speed $v \approx c$. Thus the time for one revolution is
$T = 2\pi R/v = 2\pi(1.0 \times 10^3 \text{ m})/(3.00 \times 10^8 \text{ m/s}) = 2.09 \times 10^{-5}$ s.
The number of revolutions is
$n = t/T = (20 \text{ s})/(2.09 \times 10^{-5} \text{ s}) = 9.57 \times 10^5$ turns.
We find the energy provided on each turn from
$\Delta E/n = (900 \text{ GeV} - 150 \text{ GeV})/(9.57 \times 10^5 \text{ turns}) = 7.8 \text{ GeV/turn} =$ **0.78 MeV/turn.**

13. Very high-energy protons will have a speed $v \approx c$, and $E = pc$. The magnetic field provides the radial acceleration:
$qvB = mv^2/R$, or
$B = mv/qR = p/qR = E/qRc$
$= (900 \text{ GeV})(1.60 \times 10^{-10} \text{ J/GeV})/(1.60 \times 10^{-19} \text{ C})(1.0 \times 10^3 \text{ m})(3.00 \times 10^8 \text{ m/s}) =$ **3.0 T.**

14. In the relativistic limit $v \approx c$, and $E = pc$. The magnetic field provides the radial acceleration:
$qvB = mv^2/r$, or
$B = mv/qr = p/qr = E/qrc$.
Thus the energy is
$E = qBrc$.
If the energy is in eV, the charge is 1 e, so we have
$E(\text{eV}) = Brc$.

15.

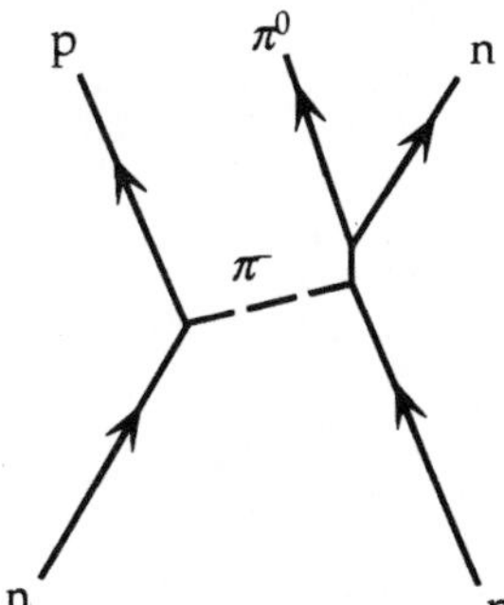

16.

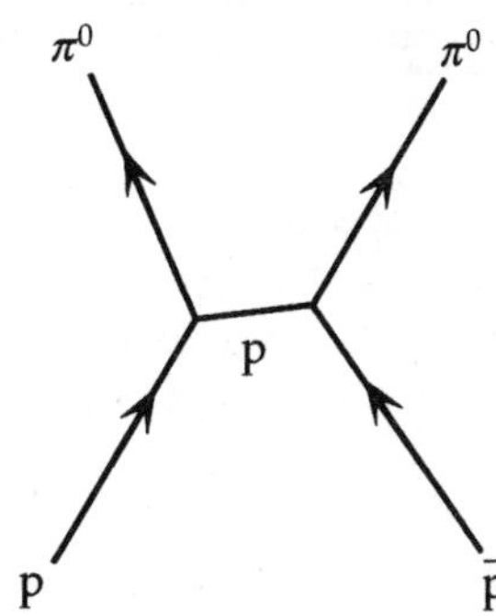

Note that the high-energy $p\bar{p}$ collisions usually produce many particles

17.

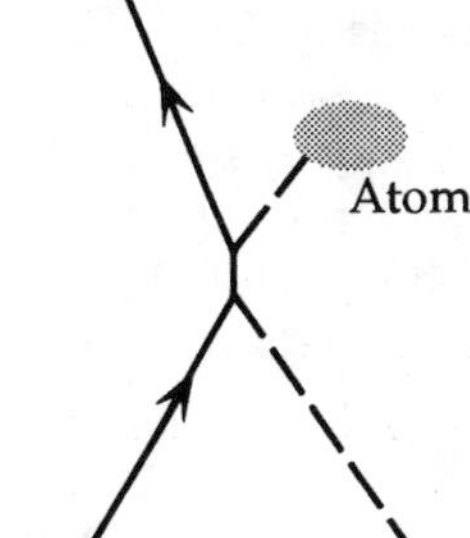

18. Because two protons are present before and after the process, and the total momentum is zero, the minimum kinetic energy will produce all three particles at rest. Thus the total initial kinetic energy must provide the rest mass of the π^0 meson:
$2(\text{KE}) = m_0c^2 = 135$ MeV, which gives KE = **67.5 MeV.**

19. For the reaction $\pi^+ \to \mu^+ + \nu_\mu$, we determine the Q-value:
$Q = [m(\pi^+) - m(\mu^+)]c^2$
$= [(139.6\text{ MeV}/c^2) - (105.7\text{ MeV}/c^2)]c^2 = 33.9\text{ MeV}.$
Thus **33.9 MeV** is released.

20. For the reaction $\Lambda^0 \to \text{n} + \pi^0$, we determine the Q-value:
$Q = [m(\Lambda^0) - m(\text{n}) - m(\pi^0)]c^2$
$= [(1115.7\text{ MeV}/c^2) - (939.6\text{ MeV}/c^2) - (135.0\text{ MeV}/c^2)]c^2 = 41.1\text{ MeV}.$
Thus **41.1 MeV** is released.

21. The minimum energy must provide the rest mass of the pair:
$E = 2m_0c^2 = 2(939.6\text{ MeV}) = 1879\text{ MeV} =$ **1.879 GeV.**

22. We estimate the range from
$mc^2 = hc/2\pi d;$
$(497.7\text{ MeV}) = (1.24 \times 10^{-12}\text{ MeV}\cdot\text{m})/2\pi d,$
which gives $d = 4.0 \times 10^{-16}\text{ m} =$ **0.40 fm.**

23. (*a*) For the decay $\Xi^0 \to \Sigma^+ + \pi^-$, the conservation laws are
Charge: $0 = +1 - 1;$
Energy (mass): $1314.9\text{ MeV} < 1189.4\text{ MeV} + 139.6\text{ MeV};$
Thus this decay is **forbidden, energy is not conserved.**

(*b*) For the decay $\Omega^- \to \Sigma^0 + \pi^- + \nu$, the conservation laws are
Charge: $-1 = 0 - 1;$
Energy (mass): $1672.5\text{ MeV} > 1314.9\text{ MeV} + 139.6\text{ MeV};$
Baryon number: $+1 = +1 + 0 + 0;$
Lepton number: $0 \neq 0 + 0 + 1.$
Thus this decay is **forbidden, lepton number is not conserved.**

(*c*) For the decay $\Sigma^0 \to \Lambda^0 + \gamma + \gamma$, the conservation laws are
Charge: $0 = 0 + 0 + 0;$
Energy (mass): $1314.9\text{ MeV} > 1115.7\text{ MeV} + 0 + 0;$
Baryon number: $+1 = +1 + 0 + 0;$
Lepton number: $-1 = -1 + 0 + 0.$
Thus this decay is **possible.**

24. If we use the average mass, we estimate the range from
$mc^2 = hc/2\pi d;$
$(85\text{ GeV})(10^3\text{ MeV/GeV}) = (1.24 \times 10^{-12}\text{ MeV}\cdot\text{m})/2\pi d,$
which gives $d =$ **2.3×10^{-18} m.**

25. The energy of the two photons must be the rest mass energy of the proton and antiproton:
$2m_0c^2 = 2hf = 2hc/\lambda;$
$2(938.3\text{ MeV}/c^2)c^2 = 2(1.24 \times 10^{-12}\text{ MeV}\cdot\text{m})/\lambda$, which gives $\lambda =$ **1.32×10^{-15} m.**

26. (*a*) For the reaction $e^- \rightarrow e^- + \gamma$, the isolated electron is at rest. For the photon, we have $E_\gamma = p_\gamma c$.
For energy conservation we have
$$m_0c^2 = [(p_ec)^2 + (m_0c^2)^2]^{1/2} + E_\gamma.$$
For momentum conservation we have
$$0 = p_e - p_\gamma.$$
When we eliminate p_e in the energy equation and rearrange, we have
$$m_0c^2 - E_\gamma = [E_\gamma^2 + (m_0c^2)^2]^{1/2}.$$
When we square both sides, we have
$$(m_0c^2)^2 + E_\gamma^2 - 2E_\gamma m_0c^2 = E_\gamma^2 + (m_0c^2)^2, \text{ which gives } E_\gamma = 0.$$
Thus no photon is emitted.

(*b*) For the photon exchange in Fig. 32–8, the photon exists for such a short time that the uncertainty principle allows energy to not be conserved during the exchange.

27. We assume the momentum of the electron-positron pair is zero. The two photons must have opposite and equal momenta; therefor, equal energies.
The energy of the two photons must be the total energy of the pair:
$$2(\text{KE} + m_0c^2) = 2hf = 2hc/\lambda;$$
$$2[(0.300\text{ MeV}) + (0.511\text{ MeV}/c^2)c^2] = 2(1.24\times10^{-12}\text{ MeV}\cdot\text{m})/\lambda,$$
which gives $\lambda = 1.53\times10^{-12}\text{ m} = \boxed{1.53\times10^{-3}\text{ nm.}}$

28. The total kinetic energy after the decay of the stationary π^+ is the Q-value:
$$Q = [m(\pi^+) - m(e^+) - m(\nu)]c^2$$
$$= [(139.6\text{ MeV}/c^2) - (0.511\text{ MeV}/c^2) - 0]c^2 = 139.1\text{ MeV}.$$
For momentum conservation we have
$$0 = p_e - p_\nu, \text{ or } (p_\nu c)^2 = (p_ec)^2 = E_e^2 - (m_ec^2)^2 = \text{KE}_e^2 + 2\text{KE}_e m_ec^2.$$
For energy conservation we have
$$Q = \text{KE}_e + p_\nu c = \text{KE}_e + p_ec, \text{ or } (p_ec)^2 = (Q - \text{KE}_e)^2 = Q^2 - 2\text{KE}_eQ + \text{KE}_e^2.$$
When we combine this with the result from momentum conservation, we get
$$\text{KE}_e = Q^2/2(Q + m_ec^2)$$
$$= (139.1\text{ MeV})^2/2(139.1\text{ MeV} + 0.511\text{ MeV}) = \boxed{69.3\text{ MeV.}}$$

29. The minimum initial kinetic energy of the neutron and proton must provide the rest mass of the K^+K^- pair:
$$\text{KE}_p + \text{KE}_n = 2m_Kc^2 = 2(493.7\text{ MeV}) = 987.4\text{ MeV}.$$
Because the neutron and proton have the same speed but different rest masses, they have slightly different kinetic energies:
$$\text{KE} = \left(\{1/[1 - (v/c)^2]\} - 1\right)m_0c^2 \propto m_0.$$
Thus we have
$$\text{KE}_n/(\text{KE}_p + \text{KE}_n) = m_{0n}/(m_{0p} + m_{0n});$$
$$\text{KE}_n/(987.4\text{ MeV}) = (939.6\text{ MeV})/(938.3\text{ MeV} + 939.6\text{ MeV}), \text{ which gives KE}_n = \boxed{494.0\text{ MeV;}}$$
$$\text{KE}_p/(\text{KE}_p + \text{KE}_n) = m_{0p}/(m_{0p} + m_{0n});$$
$$\text{KE}_p/(987.4\text{ MeV}) = (938.3\text{ MeV})/(938.3\text{ MeV} + 939.6\text{ MeV}), \text{ which gives KE}_p = \boxed{493.4\text{ MeV.}}$$
Note that we have ignored the small total momentum of the system.

30. The total kinetic energy after the decay of the stationary Ξ^- is the *Q*-value:

$$KE_\Lambda + KE_\pi = Q = [m(\Xi^-) - m(\Lambda^0) - m(\pi^-)]c^2$$
$$= [(1321.3\ \mathrm{MeV}/c^2) - (1115.7\ \mathrm{MeV}/c^2) - (139.6\ \mathrm{MeV}/c^2)]c^2 = 66.0\ \mathrm{MeV}.$$

For energy conservation we have

$$m_\Xi c^2 = E_\Lambda + E_\pi, \quad \text{or} \quad E_\pi = m_\Xi c^2 - E_\Lambda.$$

For momentum conservation we have

$$0 = p_\pi - p_\Lambda, \quad \text{or} \quad (p_\Lambda c)^2 = (p_\pi c)^2 = E_\Lambda{}^2 - (m_\Lambda c^2)^2 = E_\pi{}^2 - (m_\pi c^2)^2 .$$

When we combine this with the result from energy conservation, we get

$$E_\Lambda = (m_\Xi c^2 + m_\Lambda c^2 - m_\pi c^2)/2m_\Xi$$
$$= [(1321.3\ \mathrm{MeV})^2 + (1115.7\ \mathrm{MeV})^2 - (139.6\ \mathrm{MeV})^2]/2(1321.3\ \mathrm{MeV}) = 1124.3\ \mathrm{MeV}.$$

For the kinetic energies we have

$$KE_\Lambda = E_\Lambda - m_\Lambda c^2 = 1124.3\ \mathrm{MeV} - 1115.7\ \mathrm{MeV} = \boxed{8.6\ \mathrm{MeV};}$$
$$KE_\pi = Q - KE_\Lambda = 66.0\ \mathrm{MeV} - 8.6\ \mathrm{MeV} = \boxed{57.4\ \mathrm{MeV}.}$$

31. **No,** because the kinetic energy of the incoming proton is less than the rest mass energy of the π^+ meson, which is 139.6 MeV.
The reaction is $p + p \rightarrow p + n + \pi^+$. The minimum initial kinetic energy produces the three particles moving together with the same speed, which we can consider to be a single particle with rest mass

$$M_0 = m_{0p} + m_{0n} + m_{0\pi}.$$

For energy conservation we have

$$E_p + m_{0p}c^2 = E_M.$$

For momentum conservation we have

$$p_p + 0 = p_M, \quad \text{or} \quad (p_p c)^2 = (p_M c)^2 = E_p{}^2 - (m_{0p}c^2)^2 = E_M{}^2 - (M_0 c^2)^2 .$$

When we combine this with the result from energy conservation, we get

$$E_p = [(M_0c^2)^2 - 2(m_{0p}c^2)^2]/2m_{0p}c^2 = \{[(m_{0p} + m_{0n} + m_{0\pi})c^2]^2 - 2(m_{0p}c^2)^2\}/2m_{0p}c^2$$
$$= [(938.3\ \mathrm{MeV} + 939.6\ \mathrm{MeV} + 139.6\ \mathrm{MeV})^2 - 2(938.3\ \mathrm{MeV})^2]/2(938.3\ \mathrm{MeV}) = 1230.7\ \mathrm{MeV}.$$

For the kinetic energy we have

$$KE_p = E_p - m_{0p}c^2 = 1230.7\ \mathrm{MeV} - 938.3\ \mathrm{MeV} = \boxed{292.4\ \mathrm{MeV};}$$

32. For the reaction $p + p \rightarrow 3p + \bar{p}$ the *Q*-value is

$$Q = [2m_p - 4m_p]c^2 = -2m_pc^2.$$

The minimum initial kinetic energy produces the four particles moving together with the same speed, which we can consider to be a single particle with rest mass

$$M = 4m_p.$$

For energy conservation we have

$$E_p + m_{0p}c^2 = E_M.$$

For momentum conservation we have

$$p_p + 0 = p_M, \quad \text{or} \quad (p_pc)^2 = (p_Mc)^2 = E_p{}^2 - (m_pc^2)^2 = E_M{}^2 - (Mc^2)^2.$$

When we combine this with the result from energy conservation, we get

$$E_p = [(Mc^2)^2 - 2(m_pc^2)^2]/2m_pc^2 = \{[(4m_pc^2]^2 - 2(m_pc^2)^2\}/2m_pc^2 = 7m_pc^2.$$

For the threshold kinetic energy we have

$$KE_p = E_p - m_pc^2 = 7m_pc^2 - m_pc^2 = 6m_pc^2.$$

This is three times the magnitude of the *Q*-value.

33. We estimate the lifetime from

$$\Delta t = h/2\pi\,\Delta E = hc/2\pi c\,\Delta E = (1.24 \times 10^{-12}\ \mathrm{MeV \cdot m})/2\pi(3.00 \times 10^8\ \mathrm{m/s})(0.088\ \mathrm{MeV}) = \boxed{7.5 \times 10^{-21}\ \mathrm{s.}}$$

34. We estimate the lifetime from

$$\Delta t = h/2\pi\,\Delta E = hc/2\pi c\,\Delta E = (1.24 \times 10^{-12}\ \mathrm{MeV \cdot m})/2\pi(3.00 \times 10^8\ \mathrm{m/s})(0.277\ \mathrm{MeV}) = \boxed{2.38 \times 10^{-21}\ \mathrm{s.}}$$

35. We find the energy width from the lifetime given in Table 32–2.

(*a*) For η^0 we have

$$\Delta E = h/2\pi\,\Delta t = hc/2\pi c\,\Delta t$$
$$= (1.24 \times 10^{-12}\text{ MeV}\cdot\text{m})/2\pi(3.00 \times 10^8\text{ m/s})(5 \times 10^{-19}\text{ s}) = 1.3 \times 10^{-3}\text{ MeV} = \boxed{1.3\text{ keV.}}$$

(*b*) For Σ^0 we have

$$\Delta E = h/2\pi\,\Delta t = hc/2\pi c\,\Delta t$$
$$= (1.24 \times 10^{-12}\text{ MeV}\cdot\text{m})/2\pi(3.00 \times 10^8\text{ m/s})(7.4 \times 10^{-20}\text{ s}) = 8.9 \times 10^{-3}\text{ MeV} = \boxed{8.9\text{ keV.}}$$

36. From Fig. 32–12 we estimate the energy width as $\boxed{140\text{ MeV.}}$

We estimate the lifetime from

$$\Delta t = h/2\pi\,\Delta E = hc/2\pi c\,\Delta E = (1.24 \times 10^{-12}\text{ MeV}\cdot\text{m})/2\pi(3.00 \times 10^8\text{ m/s})(140\text{ MeV}) = \boxed{4.7 \times 10^{-24}\text{ s.}}$$

37. (*a*) For $B^- = b\bar{u}$ we have

Charge:	$-1 = -\frac{1}{3} - \frac{2}{3}$;
Spin:	$0 = +\frac{1}{2} - \frac{1}{2}$;
Baryon number:	$0 = +\frac{1}{3} - \frac{1}{3}$;
Strangeness:	$0 = 0 + 0$;
Charm:	$0 = 0 + 0$;
Bottomness:	$-1 = -1 + 0$;
Topness:	$0 = 0 + 0$.

(*b*) Because B^+ is the antiparticle of B^-, we have $\boxed{B^+ = \bar{b}u.}$

For B^0, to make the charge zero, we need the antiquark with the same properties as $\bar{u}$, but with a charge of $+\frac{1}{3}$. Thus we have $\boxed{B^0 = b\bar{d}.}$

Because $\bar{B}^0$ is the antiparticle of B^0, we have $\boxed{\bar{B}^0 = \bar{b}d.}$

38. (*a*) For the neutron we must have charge, strangeness, charm, bottomness, and topness = 0. For the baryon number to be 1, we need three quarks: $\boxed{n = ddu.}$

(*b*) For the antineutron we have $\boxed{\bar{n} = \bar{d}\bar{d}\bar{u}.}$

(*c*) For the Λ^0 we must have charge, charm, bottomness, and topness = 0. To get strangeness = – 1, we need an s quark. To get a baryon number of 1, we need three quarks. To get a charge = 0, we have $\boxed{\Lambda^0 = uds.}$

(*d*) For the $\bar{\Sigma}^0$ we must have charge, charm, bottomness, and topness = 0. To get strangeness = + 1, we need an $\bar{s}$ quark. To get a baryon number of – 1, we need three antiquarks. To get a charge = 0, we have $\boxed{\bar{\Sigma}^0 = \bar{u}\bar{d}\bar{s}.}$

39. (*a*) For uud we have charge +1, baryon number = + 1, while strangeness, charm, bottomness, and topness = 0. Thus we have $\boxed{p.}$

(*b*) For $\overline{uus} = \bar{u}\bar{u}\bar{s}$ we have charge = – 1, baryon number = – 1, strangeness = + 1, while charm, bottomness, and topness = 0. Thus we have $\boxed{\bar{\Sigma}^+.}$

(*c*) For $\bar{u}s$ we have charge = – 1, baryon number = 0, strangeness = – 1, while charm, bottomness, and topness = 0. Thus we have $\boxed{K^-.}$

(*d*) For $d\bar{u}$ we have charge = – 1, baryon number = 0, strangeness = 0, while charm, bottomness, and topness = 0. Thus we have $\boxed{\pi^-.}$

(*e*) For $\bar{c}s$ we have charge = – 1, baryon number = 0, strangeness = – 1, charm = – 1, while bottomness, and topness = 0. Thus we have $\boxed{D_S^-.}$

40. For the D^0 we have charge, baryon number, strangeness, bottomness, and topness = 0.
To get charm = + 1 we need the c quark. To get baryon number = 0, we need an antiquark.
Thus we have $\boxed{D^0 = c\bar{u}.}$

41. For the D_S^+ we have baryon number, bottomness, and topness = 0.
To get charm = + 1 we need the c quark. To get baryon number = 0, we need an antiquark.
To get strangeness = + 1 we need the $\bar{s}$ quark.
Thus we have $\boxed{D_S^+ = c\bar{s}.}$

42. (*a*) We estimate the energy from

$$\Delta x \approx h/2\pi\,\Delta p = hc/2\pi\,\Delta pc = hc/2\pi\,\Delta E;$$

$$10^{-30}\ \text{m} = (1.24\times10^{-12}\ \text{MeV}\cdot\text{m})/2\pi\,\Delta E\text{, which gives } \Delta E = 2\times10^{17}\ \text{MeV} = \boxed{2\times10^{14}\ \text{GeV.}}$$

(*b*) We find the temperature from

$$E = kT;$$

$$(2\times10^{14}\ \text{GeV})(1.60\times10^{-10}\ \text{J/GeV}) = (1.38\times10^{-23}\ \text{J/K})T\text{. which gives } T \approx \boxed{10^{27}\ \text{K.}}$$

43.

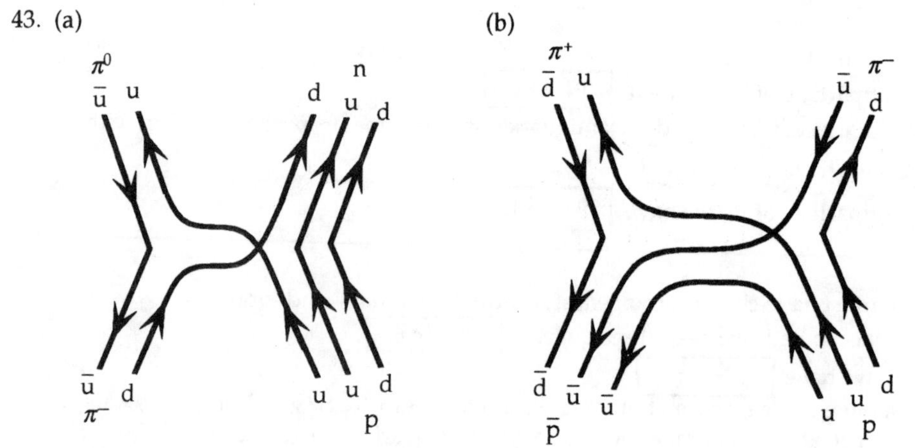

44.

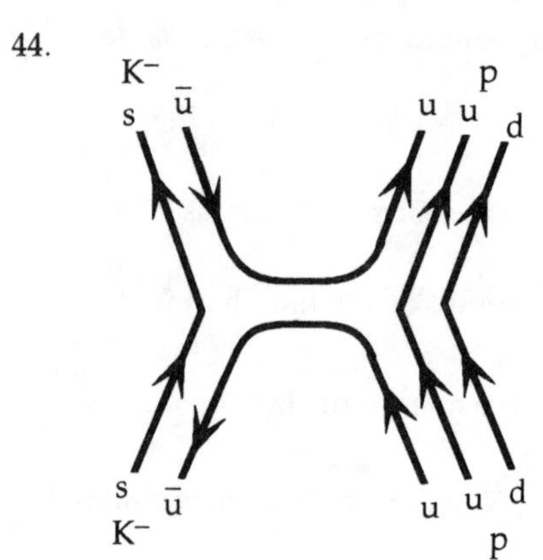

45. (*a*) At the high energy of 900 GeV the speed of the protons ≈ c. Thus the time for one proton to make a revolution is

$$T = 2\pi r/c = 2\pi(1.0 \times 10^3\ \text{m})/(3.00 \times 10^8\ \text{m/s}) = 2.09 \times 10^{-5}\ \text{s}.$$

The current carried by the beam of N protons is

$$I = Ne/T = (5 \times 10^{13})(1.60 \times 10^{-19}\ \text{C})/(2.09 \times 10^{-5}\ \text{s}) = \boxed{0.38\ \text{A.}}$$

(*b*) We find the speed of the car from

$$\text{KE} = \tfrac{1}{2}mv^2;$$

$$(5 \times 10^{13})(900\ \text{GeV})(1.60 \times 10^{-10}\ \text{J/GeV}) = \tfrac{1}{2}(1500\ \text{kg})v^2,$$ which gives $v = \boxed{98\ \text{m/s}}$ (350 km/h).

46. Because 7.0 TeV » m_pc^2, we have

$$\text{KE} = E = pc.$$

The magnetic field provides the radial acceleration:

$$vBq = mv^2/r, \text{ or } mv = p = qBr;$$

$$E = qBrc;$$

$$(7.0\ \text{TeV})(1.60 \times 10^{-7}\ \text{J/TeV}) = (1.60 \times 10^{-19}\ \text{C})B(4.25 \times 10^3\ \text{m})(3.00 \times 10^8\ \text{m/s}),$$

which gives $B = \boxed{5.5\ \text{T.}}$

47. (*a*) We assume the momentum of the electron-positron pair is zero. The released energy is

$$E = 2m_ec^2 = 2(0.511\ \text{MeV}) = \boxed{1.002\ \text{MeV.}}$$

(*b*) We assume the momentum of the proton-antiproton pair is zero. The released energy is

$$E = 2m_pc^2 = 2(938.3\ \text{MeV}) = \boxed{1876.6\ \text{MeV.}}$$

48. (*a*) For the reaction $\pi^-p \to K^+\Sigma^-$, the conservation laws are

Charge: $-1+1=+1-1;$ Spin: $0+\frac{1}{2}=0+\frac{1}{2};$

Baryon number: $0+1=0+1;$ Lepton number: $0+0=0+0;$

Strangeness: $0+0=+1-1.$

Thus the reaction is **possible, through the strong interaction.**

(*b*) For the reaction $\pi^+p \to K^+\Sigma^+$, the conservation laws are

Charge: $+1+1=+1+1;$ Spin: $0+\frac{1}{2}=0+\frac{1}{2};$

Baryon number : $0+1=0+1;$ Lepton number: $0+0=0+0;$

Strangeness: $0+0=+1-1.$

Thus the reaction is **possible, through the strong interaction.**

(*c*) For the reaction $\pi^-p \to \Lambda^0K^0\pi^0$, the conservation laws are

Charge: $-1+1=0+0+0;$ Spin: $0+\frac{1}{2}=\frac{1}{2}+0+0;$

Baryon number: $0+1=+1+0+0;$ Lepton number: $0+0=0+0;$

Strangeness: $0+0=-1+1+0.$

Thus the reaction is **possible, through the strong interaction.**

(*d*) For the reaction $\pi^+p \to \Sigma^0\pi^0$, the conservation laws are

Charge: $+1+1\neq 0+0;$

Thus the reaction is **forbidden, charge is not conserved.**

(*e*) For the reaction $\pi^-p \to pe^-\bar{\nu}_e$, the conservation laws are

Charge: $-1+1=+1-1+0;$ Spin: $0+\frac{1}{2}=\frac{1}{2}+\frac{1}{2}-\frac{1}{2};$

Baryon number: $0+1=+1+0+0;$ Lepton number: $0+0=0+1-1;$

Strangeness: $0+0=0+0+0.$

Thus the reaction is **possible, through the weak interaction.**

(*f*) For the reaction $\pi^-p \to K^0p\pi^0$, the conservation laws are

Charge: $-1+1\neq 0+1+0;$

Thus the reaction is **forbidden, charge is not conserved.**

(*g*) For the reaction $K^-p \to \Lambda^0\pi^0$, the conservation laws are

Charge: $-1+1=0+0;$ Spin: $0+\frac{1}{2}=\frac{1}{2}+0;$

Baryon number: $0+1=+1+0;$ Lepton number: $0+0=0+0;$

Strangeness: $-1+0=-1+0.$

Thus the reaction is **possible, through the strong interaction.**

(*h*) For the reaction $K^+n \to \Sigma^+\pi^0\gamma$, the conservation laws are

Charge: $+1+0=+1+0+0;$ Spin: $0+\frac{1}{2}=-\frac{1}{2}+0+1;$

Baryon number: $0+1=+1+0+0;$ Lepton number: $0+0=0+0.$

Strangeness: $+1+0\neq -1+0+0.$

Thus the reaction is **forbidden, strangeness is not conserved.**

(*i*) For the reaction $K^+ \to \pi^0\pi^0\pi^+$, the conservation laws are

Charge: $+1=0+0+1;$ Spin: $0=0+0+0;$

Baryon number: $0=0+0+0;$ Lepton number: $0=0+0+0;$

Strangeness: $+1=0+0+0.$

Thus the reaction is **possible, through the weak interaction.**

(*j*) For the reaction $\pi^+ \to e^+\nu_e$, the conservation laws are

Charge: $+1=+1+0;$ Spin: $0=\frac{1}{2}-\frac{1}{2};$

Baryon number: $0=0+0;$ Lepton number: $0=-1+1;$

Strangeness: $0=0+0.$

Thus the reaction is **possible, through the weak interaction.**

49. (*a*) For the decay $\Lambda^0 \to p\pi^-$, the *Q*-value is

$$Q = [m(\Lambda^0) - m(p) - m(\pi^-)]c^2$$
$$= [(1115.7\text{ MeV}/c^2) - (938.3\text{ MeV}/c^2) - (139.6\text{ MeV}/c^2)]c^2 = \boxed{37.8\text{ MeV.}}$$

(*b*) The total kinetic energy after the decay of the stationary Λ^0 is the *Q*-value:

$$\text{KE}_p + \text{KE}_\pi = Q = 37.8\text{ MeV}.$$

For energy conservation we have

$$m_\Lambda c^2 = E_p + E_\pi, \text{ or } E_\pi = m_\Lambda c^2 - E_p.$$

For momentum conservation we have

$$0 = p_\pi - p_p, \text{ or } (p_p c)^2 = (p_\pi c)^2 = E_p^2 - (m_p c^2)^2 = E_\pi^2 - (m_\pi c^2)^2.$$

When we combine this with the result from energy conservation, we get

$$E_p = (m_\Lambda c^2 + m_p c^2 - m_\pi c^2)/2m_\Lambda$$
$$= [(1115.7\text{ MeV})^2 + (938.3\text{ MeV})^2 - (139.6\text{ MeV})^2]/2(1115.7\text{ MeV}) = 943.7\text{ MeV}.$$

For the kinetic energies we have

$$\text{KE}_p = E_p - m_p c^2 = 943.7\text{ MeV} - 938.3\text{ MeV} = \boxed{5.4\text{ MeV;}}$$
$$\text{KE}_\pi = Q - \text{KE}_p = 37.8\text{ MeV} - 5.4\text{ MeV} = \boxed{32.4\text{ MeV.}}$$

50. We estimate the energy from

$$\Delta x \approx h/2\pi\,\Delta p = hc/2\pi\,\Delta pc = hc/2\pi\,\Delta E;$$
$$10^{-18}\text{ m} = (1.24 \times 10^{-12}\text{ MeV}\cdot\text{m})/2\pi\,\Delta E, \text{ which gives } \Delta E = 2 \times 10^5\text{ MeV} = 2 \times 10^2\text{ GeV}.$$

This is on the order of the 80 GeV rest energy of the W particles.

51. For energy conservation we have

$$m_\pi c^2 + m_p c^2 = E_{\pi^0} + E_n = E_{\pi^0} + \text{KE}_n + m_n c^2;$$
$$139.6\text{ MeV} + 938.3\text{ MeV} = E_{\pi^0} + 0.60\text{ MeV} + 939.6\text{ MeV, which gives } E_{\pi^0} = 137.7\text{ MeV}.$$

For momentum conservation we have

$$0 = p_{\pi^0} - p_n, \text{ or } (p_n c)^2 = (p_{\pi^0} c)^2 = E_n^2 - (m_n c^2)^2 = E_{\pi^0}^2 - (m_{\pi^0} c^2)^2;$$
$$(0.60\text{ MeV} + 939.6\text{ MeV})^2 - (939.6\text{ MeV})^2 = (137.7\text{ MeV})^2 - (m_{\pi^0} c^2)^2,$$

which gives $m_{\pi^0} = \boxed{133.5\text{ MeV}/c^2.}$

52. For the reaction $p + p \to p + p + \pi^0$, the *Q*-value is

$$Q = [2m_p - 2m_p - m_{\pi^0}]c^2 = -m_{\pi^0}c^2 = \boxed{-135.0\text{ MeV.}}$$

For the reaction $p + p \to p + n + \pi^+$, the *Q*-value is

$$Q = [2m_p - m_p - m_n - m_{\pi^+}]c^2 = 938.3\text{ MeV} - 939.6\text{ MeV} - 139.6\text{ MeV} = \boxed{-140.9\text{ MeV.}}$$

53. The electron will have maximum kinetic energy when the two neutrinos have the same momentum and move opposite to the direction of the electron. For a neutrino $E_\nu = p_\nu c$. For energy conservation we have

$$m_\mu c^2 = E_e + 2E_\nu, \text{ or } 2E_\nu = E_e - m_\mu c^2.$$

For momentum conservation we have

$$0 = p_e - 2p_\nu, \text{ or } 2(p_\nu c)^2 = 2E_\nu^2 = (p_e c)^2 = E_e^2 - (m_e c^2)^2.$$

When we combine this with the result from energy conservation, we get

$$E_e = (m_\mu c^2 - m_e c^2)/2m_\mu$$
$$= [(105.7\text{ MeV})^2 - (0.511\text{ MeV})^2]/2(105.7\text{ MeV}) = 52.85\text{ MeV}.$$

For the maximum kinetic energy we have

$$\text{KE}_e = E_e - m_e c^2 = 52.85\text{ MeV} - 0.511\text{ MeV} = \boxed{52.4\text{ MeV.}}$$

54. For the reaction $\pi^- p \rightarrow \Lambda^0 K^0$, the Q-value is

$$Q = [m(\pi^-) + m(p) - m(\Lambda^0) - m(K^0)]c^2$$
$$= [(139.6\ \text{MeV}/c^2) + (938.3\ \text{MeV}/c^2) - (1115.7\ \text{MeV}/c^2) - (497.7\ \text{MeV}/c^2)]c^2 = \boxed{-535.5\ \text{MeV.}}$$

The minimum initial kinetic energy produces the two particles moving together with the same speed, which we can consider to be a single particle with rest mass

$$M = m_\Lambda + m_K.$$

For energy conservation we have

$$E_\pi + m_{0p}c^2 = E_M.$$

For momentum conservation we have

$$p_\pi + 0 = p_M,\ \text{ or }\ (p_\pi c)^2 = (p_M c)^2 = E_\pi^2 - (m_\pi c^2)^2 = E_M^2 - (Mc^2)^2.$$

When we combine this with the result from energy conservation, we get

$$E_\pi = [(Mc^2)^2 - (m_\pi c^2)^2 - (m_p c^2)^2]/2m_p c^2 = [(m_\Lambda c^2 + m_K c^2)^2 - (m_\pi c^2)^2 - (m_p c^2)^2]/2m_\pi c^2$$
$$= [(1115.7\ \text{MeV} + 497.7\ \text{MeV})^2 - (938.3\ \text{MeV})^2 - (139.6\ \text{MeV})^2]/2(938.3\ \text{MeV}) = 907.6\ \text{MeV}.$$

For the threshold kinetic energy we have

$$\text{KE}_\pi = E_\pi - m_\pi c^2 = 907.6\ \text{MeV} - 139.6\ \text{MeV} = \boxed{768.0\ \text{MeV.}}$$

55. (*a*) (*b*) (*c*)

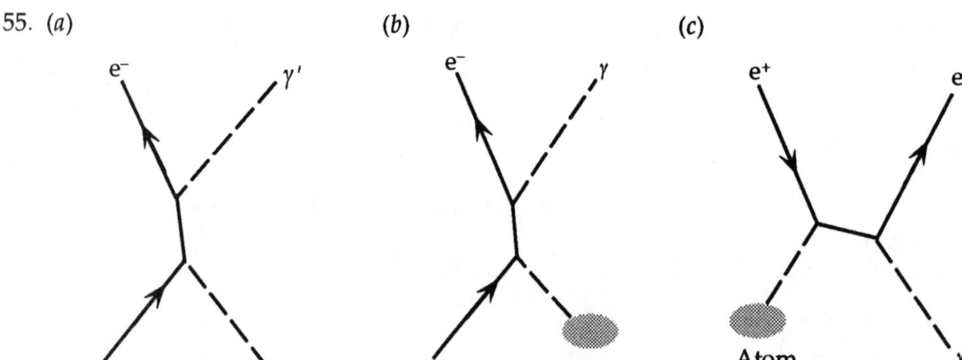

56. We consider the quarks and leptons as the fundamental fermions. A water molecule consists of two hydrogen atoms and one oxygen atom, or 2 + 8 = 10 protons, 2 + 8 = 10 electrons, and 8 neutrons. Each of the protons and neutrons is made up of three quarks, so the number of fundamental fermions is

$$N = (10 + 8)(3) + 10 = \boxed{64.}$$

CHAPTER 33

Note: A factor that appears in the analysis of energies is
$ke^2 = (9.00 \times 10^9 \text{ N}\cdot\text{m}^2/\text{C}^2)(1.60 \times 10^{-19} \text{ C})^2 = 2.30 \times 10^{-28} \text{ J}\cdot\text{m} = 1.44 \text{ MeV}\cdot\text{fm}.$

1. We find the distance from
$D = 1/\phi = [1/(0.00017°)(3600''/°)](3.26 \text{ ly/pc}) = \boxed{5.3 \text{ ly.}}$

2. We find the distance from
$D = 1/\phi = [1/(0.28'')](3.26 \text{ ly/pc}) = \boxed{12 \text{ ly.}}$

3. Using the definition of the parsec, we find the equivalent distance in m:
$D = d/\phi;$
$1 \text{ pc} = (1.50 \times 10^{11} \text{ m})/[(1.000'')/(3600''/°)(180°/\pi \text{ rad})] = 3.094 \times 10^{16} \text{ m}.$
From the definition of the light-year, we have
$1 \text{ pc} = (3.094 \times 10^{16} \text{ m})/(3.00 \times 10^8 \text{ m/s})(3.16 \times 10^7 \text{ s/yr}) = 3.26 \text{ ly}.$

4. (*a*) We find the parallax angle from
$D = 1/\phi;$
$24 \text{ pc} = 1/\phi$, which gives $\phi = \boxed{0.042''.}$
(*b*) We convert this to degrees:
$\phi = (0.042'')/(3600''/°) = \boxed{(1.2 \times 10^{-5})°.}$

5. We find the parallax angle from
$D = 1/\phi;$
$(42 \text{ ly})/(3.26 \text{ ly/pc}) = 1/\phi$, which gives $\phi = \boxed{0.078''.}$
The distance in parsecs is
$D = (42 \text{ ly})/(3.26 \text{ ly/pc}) = \boxed{13 \text{ pc.}}$

6. We find the distance in light-years:
$D = (25 \text{ pc})(3.26 \text{ ly/pc}) = 81.5 \text{ ly}.$
Thus the light takes $\boxed{82 \text{ yr}}$ to reach us.

7. The star farther away will subtend a smaller angle, so the parallax angle will be $\boxed{\text{less.}}$
From $D = 1/\phi$, we see that
$D_1/D_2 = 2 = \phi_2/\phi_1$, or $\boxed{\phi_1/\phi_2 = \tfrac{1}{2}.}$

8. (*a*) The apparent brightness of the Sun is $\boxed{\ell = 1.3 \times 10^3 \text{ W/m}^2.}$
(*b*) The absolute luminosity of the Sun is
$L = \ell A = (1.3 \times 10^3 \text{ W/m}^2)4\pi(1.5 \times 10^{11} \text{ m})^2 = \boxed{3.7 \times 10^{26} \text{ W.}}$

9. The apparent brightness is determined by the absolute luminosity and the distance:
$L = \ell A = \ell 4\pi D^2.$
If we form the ratio for the apparent brightness at the Earth and at Jupiter, we get
$\ell_E/\ell_J = (D_J/D_E)^2;$
$(1.3 \times 10^3 \text{ W/m}^2)/\ell_J = (5.2)^2$, which gives $\ell_J = \boxed{48 \text{ W/m}^2.}$

10. The diameter of our Galaxy is 100,000 ly so the angle subtended by our Galaxy is

$$\phi_{\text{Galaxy}} = d_1/D_1 = (100{,}000 \text{ ly})/(2 \times 10^6 \text{ ly}) = 0.05 \text{ rad} = \boxed{2.9°.}$$

The angle subtended by the Moon at the Earth is

$$\phi_{\text{Moon}} = d_2/D_2 = (3.48 \times 10^6 \text{ m})/(3.84 \times 10^8 \text{ m}) = 9.06 \times 10^{-3} \text{ rad} = 0.52°.$$

Thus we have

$$\boxed{\phi_{\text{Galaxy}} \approx 6\phi_{\text{Moon}}.}$$

11. If we assume negligible mass change, as a red giant, the density of the Sun will be

$$\rho = M/V = M/\tfrac{4}{3}\pi r^3;$$

$$= (2 \times 10^{30} \text{ kg})/\tfrac{4}{3}\pi(1.5 \times 10^{11} \text{ m})^3 = \boxed{1.4 \times 10^{-4} \text{ kg/m}^3.}$$

12. The angle subtended at the Earth by the Sun as a white dwarf will be

$$\phi = d/D = (3.48 \times 10^6 \text{ m})/(1.5 \times 10^{11} \text{ m}) = 2.3 \times 10^{-5} \text{ rad} = \boxed{4.8''.}$$

13. The density of the white dwarf is

$$\rho_{\text{dwarf}} = M/V = M/\tfrac{4}{3}\pi r^3;$$

$$= (2 \times 10^{30} \text{ kg})/\tfrac{4}{3}\pi(6.38 \times 10^6 \text{ m})^3 = \boxed{1.8 \times 10^9 \text{ kg/m}^3.}$$

The ratio of densities will be the ratio of masses:

$$\rho_{\text{dwarf}}/\rho_{\text{Earth}} = M_{\text{dwarf}}/M_{\text{Earth}} = (2 \times 10^{30} \text{ kg})/(5.98 \times 10^{24} \text{ kg}) = \boxed{3 \times 10^5.}$$

14. The density of the neutron star is

$$\rho_{\text{ns}} = M_{\text{ns}}/V_{\text{ns}} = M_{\text{ns}}/\tfrac{4}{3}\pi r_{\text{ns}}^3;$$

$$= (1.5)(2 \times 10^{30} \text{ kg})/\tfrac{4}{3}\pi(11 \times 10^3 \text{ m})^3 = \boxed{5.4 \times 10^{17} \text{ kg/m}^3.}$$

From the result for Problem 13 we see that

$$\boxed{\rho_{\text{ns}} = (3 \times 10^8)\rho_{\text{dwarf}}.}$$

For the density of nuclear matter we calculate the density of a proton:

$$\rho_{\text{p}} = m_{\text{p}}/\tfrac{4}{3}\pi r_{\text{p}}^3 = (1.67 \times 10^{-27} \text{ kg})/(1.2 \times 10^{-15} \text{ m})^3 = 2.3 \times 10^{17} \text{ kg/m}^3.$$

Thus we see that

$$\boxed{\rho_{\text{ns}} \approx \rho_{\text{nuclear matter}}.}$$

15. For the reaction ${}^4_2\text{He} + {}^4_2\text{He} \rightarrow {}^8_4\text{Be} + \gamma$, we determine the Q-value:

$$Q = [m({}^4\text{He}) + m({}^4\text{He}) - m({}^8\text{Be})]c^2$$

$$= [2(4.002602 \text{ u}) - (8.005305 \text{ u})]c^2(931.5 \text{ MeV}/uc^2) = -0.0941 \text{ MeV} = \boxed{-94.1 \text{ keV}.}$$

For the reaction ${}^4_2\text{He} + {}^8_4\text{Be} \rightarrow {}^{12}_6\text{C} + \gamma$, we determine the Q-value:

$$Q = [m({}^4\text{He}) + m({}^8\text{Be}) - m({}^{12}\text{C})]c^2$$

$$= [(4.002602 \text{ u}) + (8.005305 \text{ u}) - (12.000000 \text{ u})]c^2(931.5 \text{ MeV}/uc^2) = \boxed{7.365 \text{ MeV}.}$$

Note that the total Q-value is 7.27 MeV.

16. (*a*) For the reaction ${}^{12}_{6}\text{C} + {}^{12}_{6}\text{C} \rightarrow {}^{24}_{12}\text{Mg} + \gamma$, we determine the Q-value:

$$Q = [m(^{12}\text{C}) + m(^{12}\text{C}) - m(^{24}\text{Mg})]c^2$$
$$= [2(12.000000\ \text{u}) - (23.985042\ \text{u})]c2(931.5\ \text{MeV}/\text{u}c^2) = 13.93\ \text{MeV}.$$

Thus $\boxed{13.93\ \text{MeV is released.}}$

(*b*) We find the radius of the carbon nucleus from

$$r = (1.2\ \text{fm})A^{1/3} = (1.2\ \text{fm})(12)^{1/3} = 2.75\ \text{fm}.$$

If the two nuclei are just touching, the Coulomb potential energy must be the initial kinetic energies of the two nuclei:

$$2\text{KE} = \text{PE} = Z_C Z_C ke^2/2r = (6)(6)(1.44\ \text{MeV}\cdot\text{fm})/2(2.75\ \text{fm}) = 9.4\ \text{MeV}.$$

Thus each carbon nucleus has a kinetic energy of $\boxed{4.7\ \text{MeV.}}$

(*c*) We find the temperature from

$$\text{KE} = \tfrac{3}{2}kT;$$
$$(4.3\ \text{MeV})(1.60\times10^{-13}\ \text{J/MeV}) = \tfrac{3}{2}(1.38\times10^{-23}\ \text{J/K})T,\ \text{which gives}\ T = \boxed{3\times10^{10}\ \text{K.}}$$

17. (*a*) For the reaction ${}^{16}_{8}\text{O} + {}^{16}_{8}\text{O} \rightarrow {}^{28}_{14}\text{Si} + {}^{4}_{2}\text{He}$, we determine the Q-value:

$$Q = [m(^{16}\text{O}) + \text{m}(^{16}\text{O}) - \text{m}(^{28}\text{Si}) - \text{m}(^{4}\text{He})]c^2$$
$$= [2(15.994915\ \text{u}) - (27.976927\ \text{u}) - (4.002602\ \text{u})]c^2(931.5\ \text{MeV}/\text{u}c^2) = 9.595\ \text{MeV}.$$

Thus $\boxed{9.595\ \text{MeV is released.}}$

(*b*) We find the radius of the oxygen nucleus from

$$r = (1.2\ \text{fm})A^{1/3} = (1.2\ \text{fm})(16)^{1/3} = 3.02\ \text{fm}.$$

If the two nuclei are just touching, the Coulomb potential energy must be the initial kinetic energies of the two nuclei:

$$2\text{KE} = \text{PE} = Z_O Z_O ke^2/2r = (8)(8)(1.44\ \text{MeV}\cdot\text{fm})/2(3.02\ \text{fm}) = 15.25\ \text{MeV}.$$

Thus each oxygen nucleus has a kinetic energy of $\boxed{7.6\ \text{MeV.}}$

(*c*) We find the temperature from

$$\text{KE} = \tfrac{3}{2}kT;$$
$$(7.6\ \text{MeV})(1.60\times10^{-13}\ \text{J/MeV}) = \tfrac{3}{2}(1.38\times10^{-23}\ \text{J/K})T,\ \text{which gives}\ T = \boxed{6\times10^{10}\ \text{K.}}$$

18. From Wien's displacement law we can compare the temperatures of the two stars:

$$\lambda_1 T_1 = \lambda_2 T_2,\ \text{or}$$
$$T_2/T_1 = \lambda_1/\lambda_2 = (800\ \text{nm})/(400\ \text{nm}) = 2.$$

If r is the radius of a star, the radiating area is $A = 4\pi r^2$. From the Stefan-Boltzmann law we have

$$L = \ell 4\pi D^2 \propto AT^4 = 4\pi r^2 T^4.$$

If we form the ratio for the two stars, we get

$$(D_2/D_1)^2 = (T_2/T_1)^4 = (2)^4,\ \text{which gives}\ \boxed{D_2 = 4D_1.}$$

19. From Wien's displacement law we can compare the temperatures of the two stars:

$$\lambda_1 T_1 = \lambda_2 T_2,\ \text{or}$$
$$T_2/T_1 = \lambda_1/\lambda_2 = (500\ \text{nm})/(700\ \text{nm}) = 5/7.$$

If r is the radius of a star, the radiating area is $A = 4\pi r^2$. From the Stefan-Boltzmann law we have

$$L = \ell 4\pi D^2 \propto AT^4 = 4\pi r^2 T^4.$$

If we form the ratio for the two stars, we get

$$\ell_2/\ell_1 = (r_2/r_1)^2 (T_2/T_1)^4;$$
$$1/0.091 = (r_2/r_1)^2 (5/7)^4,\ \text{which gives}\ \boxed{r_2/r_1 = 6.5.}$$

20. (*a*) For the vertices of the triangle we choose the North pole and two points on a latitude line on opposite sides of the Earth, as shown on the diagram.

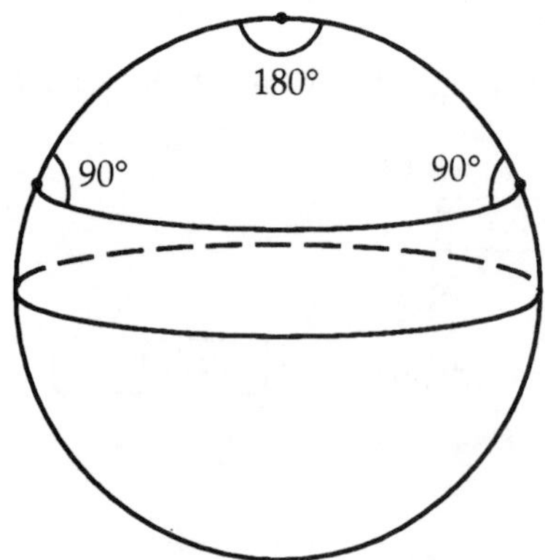

(*b*) We can get 180° only approximately. For the vertices of the triangle we choose the North pole and two points on a latitude line very close together, so the angle at the North pole is negligible, as shown on the diagram.

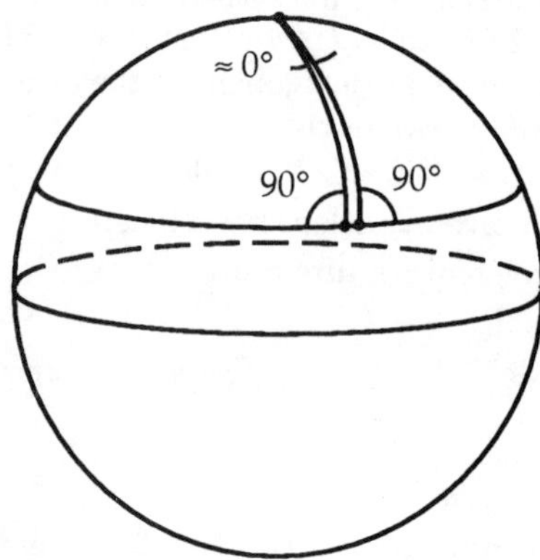

21. (*a*) The Schwarzschild radius for a star with mass equal to our Sun is

$R = 2GM/c^2$

$= 2(6.67 \times 10^{-11}\ \mathrm{N \cdot m^2/kg^2})(1.99 \times 10^{30}\ \mathrm{kg})/(3.00 \times 10^8\ \mathrm{m/s})^2 = 2.95 \times 10^3\ \mathrm{m} = 2.95\ \mathrm{km}.$

(*b*) The Schwarzschild radius for a star with mass equal to the Earth is

$R = 2GM/c^2$

$= 2(6.67 \times 10^{-11}\ \mathrm{N \cdot m^2/kg^2})(5.97 \times 10^{24}\ \mathrm{kg})/(3.00 \times 10^8\ \mathrm{m/s})^2 = 8.9 \times 10^{-3}\ \mathrm{m} = 8.9\ \mathrm{mm}.$

22. If we use the data for our galaxy, we have

$R = 2GM/c^2$

$= 2(6.6 \times 10^{-11}\ \mathrm{N \cdot m^2/kg^2})(3 \times 10^{41}\ \mathrm{kg})/(3.00 \times 10^8\ \mathrm{m/s})^2 = \boxed{4 \times 10^{14}\ \mathrm{m.}}$

23. If we consider a triangle with its three vertices on a great circle, such as one through the North and South poles as shown in the diagram, we see that the sum of the angles is $\boxed{540°.}$

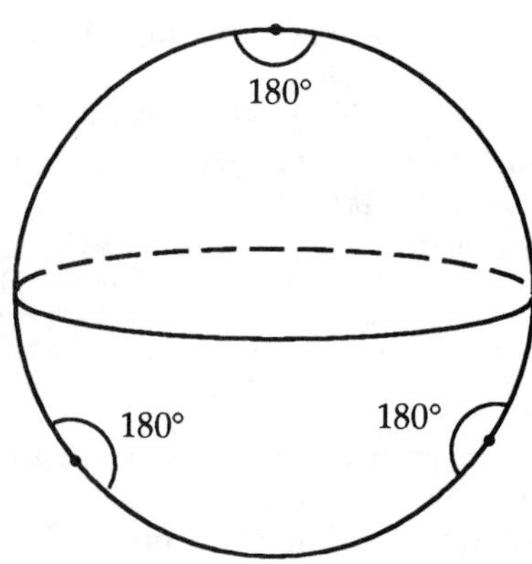

24. We find the distance from Hubble's law:

$v = Hd;$

$(0.01)(3.00 \times 10^8\ \mathrm{m/s}) = [(80 \times 10^3\ \mathrm{m/s/Mpc})/(10^6\ \mathrm{pc/Mpc})(3.26\ \mathrm{ly/pc})]d,$

which gives $d = \boxed{1.2 \times 10^8\ \mathrm{ly.}}$

25. We find the distance from Hubble's law:

$v = Hd;$

2.50×10^6 m/s = $[(80 \times 10^3$ m/s/Mpc)/(10^6 pc/Mpc)(3.26 ly/pc)]d,

which gives d = $\boxed{1.0 \times 10^8 \text{ ly.}}$

26. We estimate the speed of the galaxy from Hubble's law:

$v = Hd = [(80 \times 10^3$ m/s/Mpc)/(3.26 ly/pc)](10×10^3 Mly) = 2.5×10^8 m/s = $\boxed{0.8c.}$

27. We estimate the age of the universe from Hubble's law.

(*a*) $t = d/v = d/Hd = 1/H$

= [1/(50 km/s/Mpc)](3.26 ly/pc)(9.5×10^{-18} km/Mly) = 6.2×10^{17} s = $\boxed{2 \times 10^{10} \text{ yr.}}$

(*b*) We see that doubling the Hubble constant reduces the time by a factor of 2:

t = $\boxed{1 \times 10^{10} \text{ yr.}}$

28. (*a*) We find the receding speed of the galaxy from

$v = Hd = [(80 \times 10^3$ m/s/Mpc)/(3.26 ly/pc)](1.0 Mly) = 2.5×10^4 m/s = $(8 \times 10^{-5})c$.

For the Doppler shift of the wavelength we have

$\lambda'/\lambda = \{[1 + (v/c)]/[1 - (v/c)]\}^{1/2};$

λ'/(656 nm) = $[(1 + 8 \times 10^{-5})/(1 - 8 \times 10^{-5})]^{1/2}$, which gives λ' = $\boxed{656 \text{ nm.}}$

(*b*) We find the receding speed of the galaxy from

$v = Hd = [(80 \times 10^3$ m/s/Mpc)/(3.26 ly/pc)](1.0×10^2 Mly) = 2.5×10^6 m/s = $(8 \times 10^{-3})c$.

For the Doppler shift of the wavelength we have

$\lambda'/\lambda = \{[1 + (v/c)]/[1 - (v/c)]\}^{1/2};$

λ'/(656 nm) = $[(1 + 8 \times 10^{-3})/(1 - 8 \times 10^{-3})]^{1/2}$, which gives λ' = $\boxed{661 \text{ nm.}}$

(*c*) We find the receding speed of the galaxy from

$v = Hd = [(80 \times 10^3$ m/s/Mpc)/(3.26 ly/pc)](1.0×10^4 Mly) = 2.5×10^8 m/s = $0.833c$.

For the Doppler shift of the wavelength we have

$\lambda'/\lambda = \{[1 + (v/c)]/[1 - (v/c)]\}^{1/2};$

λ'/(656 nm) = $[(1 + 0.833)/(1 - 0.833)]^{1/2}$, which gives $\lambda' = 2.17 \times 10^3$ nm = $\boxed{2.17\ \mu\text{m.}}$

29. We find the speed from the Doppler shift:

$\lambda'/\lambda = \{[1 + (v/c)]/[1 - (v/c)]\}^{1/2};$

(610 nm)/(434 nm) = $\{[1 + (v/c)]/[1 - (v/c)]\}^{1/2}$, which gives v = $\boxed{0.328c.}$

We find the distance from Hubble's law:

$v = Hd;$

(0.328)(3.00×10^8 m/s) = $[(80 \times 10^3$ m/s/Mpc)/(10^6 pc/Mpc)(3.26 ly/pc)]d,

which gives d = $\boxed{4.0 \times 10^9 \text{ ly.}}$

30. The Doppler shift is

$\lambda'/\lambda = \{[1 + (v/c)]/[1 - (v/c)]\}^{1/2}.$

From the binomial expansion for small x, we have $(1 - x)^{-1} \approx 1 + x$. Thus when $v/c \ll 1$, we get

$\lambda'/\lambda = \{[1 + (v/c)]/[1 - (v/c)]\}^{1/2} \approx \{[1 + (v/c)]^2\}^{1/2} = 1 + (v/c).$ or $\lambda' \approx \lambda + (v/c)\lambda.$

Thus the fractional wavelength change is

$(\lambda' - \lambda)/\lambda = \Delta\lambda/\lambda \approx v/c.$

31. To an observer, the clocks in the reference frame of the moving source will run slow. If τ_0 is the period of the source in its frame, the period in the reference frame of the observer will be

$$\tau = \tau_0/[1 - (v^2/c^2)]^{1/2}.$$

The speed of the light is c in both frames. If the source emits a wavefront at $t = 0$ in the observer's frame, in the time τ it will travel $c\tau$ and the source will move $v\tau$. At this time the next wavefront is emitted, so the wavelength to the observer is

$$\lambda' = c\tau + v\tau = (c + v)\tau = (c + v)\tau_0/[1 - (v^2/c^2)]^{1/2}.$$

In the source frame we have $c = \lambda f_0 = \lambda/\tau_0$, so we get

$$\lambda' = (c + v)\lambda/c[1 - (v^2/c^2)]^{1/2}$$
$$= (c + v)\lambda/[c^2 - v^2]^{1/2}$$
$$= \lambda(c + v)/[(c + v)(c - v)]^{1/2} = \lambda[(c + v)/(c - v)]^{1/2} = \lambda\{[1 + (v/c)]/[1 - (v/c)]\}^{1/2}.$$

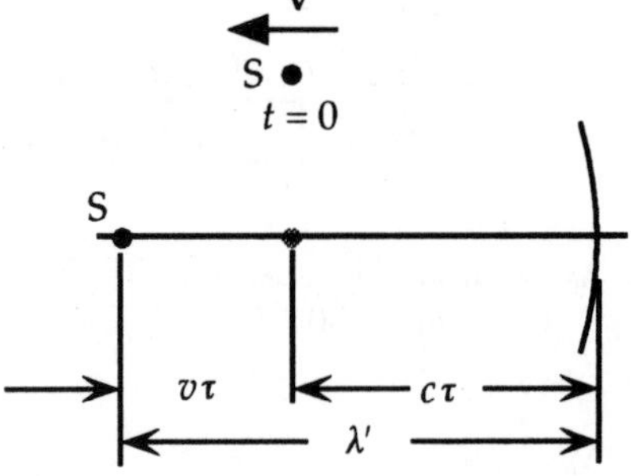

32. We find the peak wavelength from

$\lambda T = 2.90 \times 10^{-3}\ \text{m}\cdot\text{K};$

$\lambda(2.7\ \text{K}) = 2.90 \times 10^{-3}\ \text{m}\cdot\text{K}$, which gives $\lambda = 1.1 \times 10^{-3}\ \text{m} =$ **1.1 mm.**

33. We find the nucleon density from

$\rho_c = M/V = nm_n/V;$

$10^{-26}\ \text{kg/m}^3 = n(1.67 \times 10^{-27}\ \text{kg})/V$, which gives $n/V =$ **6 nucleons/m^3.**

34. If $d \propto 1/T$, when we form the ratio for two different times, we get

$d/d_0 = T_0/T$, where $T_0 = 2.7$ K is the temperature today.

(*a*) From Fig. 33–23 we estimate the temperature to be 3000 K, so we have

$d/d_0 = T_0/T = (2.7\ \text{K})/(3000\ \text{K}) \approx$ **10^{-3}.**

(*b*) From Fig. 33–23 we estimate the temperature to be 10^{10} K, so we have

$d/d_0 = T_0/T = (2.7\ \text{K})/(10^{10}\ \text{K}) \approx$ **10^{-10}.**

(*c*) From Fig. 33–23 we estimate the temperature to be 10^{13} K, so we have

$d/d_0 = T_0/T = (2.7\ \text{K})/(10^{13}\ \text{K}) \approx$ **10^{-13}.**

(*d*) From Fig. 33–23 we estimate the temperature to be 10^{27} K, so we have

$d/d_0 = T_0/T = (2.7\ \text{K})/(10^{27}\ \text{K}) \approx$ **10^{-27}.**

35. We find the equivalent temperature from

$E = Mc^2 = \frac{3}{2}kT$, or $T = \frac{2}{3}Mc^2/k.$

(*a*) For the kaon threshold we have

$T = \frac{2}{3}Mc^2/k = \frac{2}{3}(500\ \text{MeV})(1.60 \times 10^{-13}\ \text{J/MeV})/(1.380 \times 10^{-23}\ \text{J/K}) = 4 \times 10^{12}\ \text{K}.$

We estimate the time from Fig. 33–23: $t \approx$ **10^{-5} s.**

(*b*) For the Υ threshold we have

$T = \frac{2}{3}Mc^2/k = \frac{2}{3}(9500\ \text{MeV})(1.60 \times 10^{-13}\ \text{J/MeV})/(1.380 \times 10^{-23}\ \text{J/K}) = 7 \times 10^{13}\ \text{K}.$

We estimate the time from Fig. 33–23: $t \approx$ **10^{-7} s.**

(*c*) For the muon threshold we have

$T = \frac{2}{3}Mc^2/k = \frac{2}{3}(100\ \text{MeV})(1.60 \times 10^{-13}\ \text{J/MeV})/(1.380 \times 10^{-23}\ \text{J/K}) = 8 \times 10^{11}\ \text{K}.$

We estimate the time from Fig. 33–23: $t \approx$ **10^{-4} s.**

36. The absolute luminosity depends on the radius of the star and the temperature:
$L \propto AT^4 = 4\pi r^2 T^4$.
(*a*) For case A we have
temperature increases, luminosity is constant, and size decreases.
(*b*) For case B we have
temperature is constant, luminosity decreases, and size decreases.
(*c*) For case C we have
temperature decreases, luminosity increases, and size increases.

37. For the luminosity we have $L = \ell 4\pi d^2$. If we form the ratio with constant luminosity, we get
$(d_{\text{star}}/d_{\text{Sun}})^2 = \ell_{\text{Sun}}/\ell_{\text{star}}$;
$\{d_{\text{star}}/[(1.5 \times 10^{11}\text{ m})/(9.5 \times 10^{15}\text{ m/ly})]\}^2 = 10^{11}$, which gives $d_{\text{star}} =$ 5.0 ly.

38. For the conservation of angular momentum, with constant mass, we have
$I\omega = I_0\,\omega_0$;
$\frac{2}{5}Mr^2\omega = \frac{2}{5}Mr_0{}^2\omega_0$;
$(5\text{ km})^2\omega = (7 \times 10^5\text{ km})^2(1\text{ rev/mo})$,
which gives $\omega = 2 \times 10^{10}$ rev/mo $= (2 \times 10^{10}$ rev/mo$)/(30$ day/mo$)(24$ h/day$)(3600$ s/h$) =$ 8×10^3 rev/s.

39. We simplify the ratio of rotational kinetic energies by using the conservation of angular momentum:
$\text{KE}/\text{KE}_0 = \frac{1}{2}I\omega^2/\frac{1}{2}I_0\,\omega_0{}^2 = \omega/\omega_0 = (2 \times 10^{10}\text{ rev/mo})/(1\text{ rev/mo}) =$ 2×10^{10}.

40. The kinetic energy of the neutron star is
$\text{KE} = \frac{1}{2}I\omega^2 = \frac{1}{2}(\frac{2}{5}Mr^2)\omega^2$
$= \frac{1}{2}(\frac{2}{5})(1.5)(2.0 \times 10^{30}\text{ kg})(5.0 \times 10^3\text{ m})^2[(1.0\text{ rev/s})(2\pi\text{ rad/rev})]^2 = 5.92 \times 10^{38}\text{ J}.$
The power output is
$P = E/t = (10^{-9}\text{ day}^{-1})(5.92 \times 10^{38}\text{ J})/(86{,}400\text{ s/day}) =$ 6.9×10^{24} W.

41. If we use $R = 10 \times 10^{10}$ ly for the radius of the universe, the volume is
$V = \frac{4}{3}\pi R^3 = \frac{4}{3}\pi[(10 \times 10^{10}\text{ ly})(9.46 \times 10^{12}\text{ km/ly})]^3 = 3.5 \times 10^{72}\text{ km}^3.$
We find the increase in radius from Hubble's law:
$\Delta R = v\,\Delta t = HR\,\Delta t.$
Thus the rate at which the universe in expanding is
$\Delta V/\Delta t = 4\pi R^2\,\Delta R/\Delta t = 4\pi R^3 H = 3VH$
$= 3(3.5 \times 10^{72}\text{ km}^3)(80\text{ km/s/Mpc})/(3.26 \times 10^6\text{ ly/Mpc})(9.46 \times 10^{12}\text{ km/ly})$
$= 2.76 \times 10^{55}\text{ km}^3/\text{s}.$
The rate at which hydrogen atoms have to be created is given by
$\Delta m/\Delta t = m_{\text{H}}\,\Delta N/\Delta t = \rho\,\Delta V/\Delta t;$
$(1.67 \times 10^{-27}\text{ kg/atom})\,\Delta N/\Delta t = (10^{-27}\text{ kg/m}^3)(2.76 \times 10^{55}\text{ km}^3/\text{s})(10^3\text{ m/km})^3,$
which gives $\Delta N/\Delta t = 1.7 \times 10^{64}$ atoms/s.
If we find the rate per unit volume per year, we have
$(\Delta N/\Delta t)/V = (1.7 \times 10^{64}\text{ atoms/s})(3.16 \times 10^7\text{ s/yr})/(3.5 \times 10^{72}\text{ km}^3) = 0.15\text{ atoms/km}^3/\text{yr}.$
Thus hydrogen atoms would need to be created at the rate 1 H atom/km^3 every 7 years.

42. We use M for the mass of the universe and n for the number of nucleons.
(*a*) If nucleons provide 2% of the mass, we have
$$0.02M = nm_n, \text{ and } 0.98M = (10^9\, n)m_\nu.$$
When we combine these, we get
$$n(939 \times 10^6 \text{ eV})/0.02 = (10^9\, n)m_\nu/0.98, \text{ which gives } m_\nu = \boxed{46 \text{ eV.}}$$
(*b*) If nucleons provide 5% of the mass, we have
$$0.05M = nm_n, \text{ and } 0.95M = (10^9\, n)m_\nu.$$
When we combine these, we get
$$n(939 \times 10^6 \text{ eV})/0.05 = (10^9\, n)m_\nu/0.95, \text{ which gives } m_\nu = \boxed{18 \text{ eV.}}$$

43. We find the temperature of the stars from
$$\lambda T = 2.90 \times 10^{-3} \text{ m} \cdot \text{K};$$
$$(600 \times 10^{-9} \text{ m})T_1 = 2.90 \times 10^{-3} \text{ m} \cdot \text{K, which gives } T_1 = 4800 \text{ K};$$
$$(400 \times 10^{-9} \text{ m})T_2 = 2.90 \times 10^{-3} \text{ m} \cdot \text{K, which gives } T_2 = 7300 \text{ K}.$$
From the H–R diagram, we estimate the luminosities:
$$L_1 = 10^{26} \text{ W, and } L_2 = 2 \times 10^{27} \text{ W}.$$
From the Stefan-Boltzmann law, the absolute luminosity depends on the radius of the star and the temperature:
$$L \propto AT^4 = 4\pi r^2 T^4.$$
When we form the ratio, we get
$$L_2/L_1 = (r_2/r_1)^2(T_2/T_1)^4 = (r_2/r_1)^2(\lambda_1/\lambda_2)^4;$$
$$(2 \times 10^{27} \text{ W})/(10^{26} \text{ W}) = (r_2/r_1)^2(600 \text{ nm}/400 \text{ nm})^4, \text{ which gives } r_2/r_1 = \boxed{2.}$$

44. We find half the subtended angle from
$$\phi = 1/30 \text{ pc} = (1/30'')/(3600''/°) \approx (1 \times 10^{-9})°.$$
Thus the minimum subtended angle is $\boxed{\approx (2 \times 10^{-5})°.}$

45. The wavelengths for the same transition in hydrogen-like atoms are given by
$$1/\lambda = (\text{constant})Z^2.$$
Thus we have
$$\lambda'/\lambda = \lambda_H/\lambda_{He} = (Z_{He}/Z_H)^2 = (2/1)^2 = \{[1 + (v/c)]/[1 - (v/c)]\}^{1/2}, \text{ which gives } v = \boxed{0.88c.}$$

46. We find the temperature from
$$E = \tfrac{3}{2}kT, \text{ or } T = \tfrac{2}{3}Mc^2/k = \tfrac{2}{3}(1.8 \text{ TeV})(1.60 \times 10^{-7} \text{ J/TeV})/(1.38 \times 10^{-23} \text{ J/K}) = \boxed{1.4 \times 10^{16} \text{ K.}}$$
From Fig. 33–23 we see that this is the $\boxed{\text{hadron era.}}$

47. We find the speed of the gas clouds from the radial acceleration provided by the gravitational attraction:

$GMm/R^2 = mv^2/R$, or $v^2 = GM/R$,

where

$M = (2 \times 10^9)(2.0 \times 10^{30}\ \text{kg}) = 4.0 \times 10^{39}\ \text{kg}$, and

$R = (60\ \text{ly})(9.46 \times 10^{15}\ \text{m/ly}) = 5.7 \times 10^{17}\ \text{m}$.

To account for the gases moving in opposite directions on the two sides of the cloud, we have

$v/c = \pm (1/c)(GM/R)^{1/2}$

$= \pm (1/3.00 \times 10^8\ \text{m/s})[(6.67 \times 10^{-11}\ \text{N}\cdot\text{m}^2/\text{kg}^2)(4.0 \times 10^{39}\ \text{kg})/(5.7 \times 10^{17}\ \text{m})]^{1/2} = \pm 2.3 \times 10^{-3}$.

Because $v \ll c$, we use the result from Problem 30:

$\Delta\lambda/\lambda \approx v/c = \boxed{\pm 2.3 \times 10^{-3}.}$

This will be in addition to the shift from any overall receding of the galaxy.

48. Because Venus has a more negative value, **Venus is brighter.**

We write the logarithmic scale as

$m = k \log \ell$.

We find the value of k from

$m_2 - m_1 = k(\log \ell_2 - \log \ell_1) = k \log(\ell_2/\ell_1)$;

$+5 = k \log(1/100)$, which gives $k = -2.5$.

For Venus and Sirius, we have

$m_V - m_S = k \log(\ell_V/\ell_S)$;

$-4.4 - (-1.4) = -2.5 \log(\ell_V/\ell_S)$, which gives $\boxed{\ell_V/\ell_S = 16.}$

Answers to even-numbered problems

CHAPTER 1

2. (a) 86,900; (b) 7,100; (c) 0.66; (d) 876; (e) 0.000 086 2.
4. (a) 3 significant figures; (b) 4 significant figures; (c) 3 significant figures; (d) 1 significant figure; (e) 2 significant figures; (f) 4 significant figures; (g) 2, 3, or 4 significant figures, depending on the significance of the zeros.
6. 0.6%.
8. 1.5 m.
10. $(2.5 \pm 0.2) \times 10^9\ cm^2$.
12. (a) 1 Mvolt; (b) 1 μm; (c) 5 kdays; (d) 0.8 kbucks; (e) 8 npieces.
14. 5,000 kisses; 1,000,000 bucks/yr (millionaire).
16. (a) 1.5×10^{11} m; (b) 150 Gm.
18. 2.55 m.
20. 7.3%.
22. $3.80 \times 10^{13}\ m^2$; $A_{moon} = 7.42 \times 10^{-2} A_{Earth}$.
24. ≈ 100 days.
26. 2×10^9 beats.
28. (a) 10 days; (b) 80 days.
30. ≈ 3 h.
32. (a) 0.10 nm; (b) 1.0×10^5 fm; (c) 1.0×10^{10} Å; (d) 9.5×10^{25} Å.
34. (a) 1,000 drivers.
36. 1×10^{11} gal/yr.
38. 9 cm.
40. 4×10^5 t.
42. (a) $1.00 \times 10^9\ cm^3$; (b) $3.53 \times 10^4\ ft^3$; (c) $6.10 \times 10^7\ in^3$.
44. 150 m long, 25 m wide, and 15 m high.

CHAPTER 2

2. 0.60 h (36 min).
4. (a) 105 km/h; (b) 29 m/s; (c) 95 ft/s.
6. ≈ 300 m/s.
8. (a) 10.4 m/s; (b) + 3.5 m/s.
10. 4.43 h, 881 km/h.
12. 6.73 m/s.
14. 5.2 s
16. $-6.3\ m/s^2$, 0.64.
18. $v = at$, $x = \frac{1}{2}at^2$, $v^2 = 2ax$, $\bar{v} = \frac{1}{2}v$.
20. $-2.4\ m/s^2$.
22. $4.41\ m/s^2$, 2.61 s.
24. 33 m/s.
26. 40g.
30. She should decide to stop!
32. 28.6 g.
34. 60.0 m.
36. (a) 32 m; (b) 5.1 s.
38. 13 m.
40. (a) $v_0 = 4.85$ m/s; (b) $t = 0.990$ s.
44. (a) ± 12.8 m/s; (b) 0.735 s, 3.35 s; (c) on the way up and on the way down.
46. 1.8 m.
48. 9.1 m/s.
50. (a) 25 m/s; (b) 32 m; (c) 1.3 s; (d) 3.8 s.
52. (a) 0 to 17 s; (b) 28 s; (c) 38 s; (d) both directions.
54. (a) $2.5\ m/s^2$, $0.7\ m/s^2$; (b) 450 m.
56. (a) ≈ 1.7×10^3 m; (b) ≈ 5×10^2 m.
60. (a) $-1.5 \times 10^2\ m/s^2$; (b) the net should be loosened.
62. 1.3 m.
64. (b) $H_{50} = 9.8$ m; (c) $H_{100} = 39$ m.
66. 24 m/s.
68. (a) 52 min; (b) 31 min.
70. $\Delta v_{0down} = 0.8$ m/s; $\Delta v_{0up} = 0.9$ m/s.
72. 29.0 m.

CHAPTER 3

2. 20 blocks, 37° S of E.
4. 24.95, 41.10° below the x-axis.
6. (b) – 14.0, 19.9; (c) 24.3, 54.8° above – x-axis.
8. (a) $V_{1x} = -8.08$, $V_{1y} = 0$, $V_{2x} = 3.19$, $V_{2y} = 3.19$; (b) 5.84, $\theta = 33.1°$ above – x-axis.
10. (5.9, – 1.4, – 1.4), 6.2.
12. 97.2, 53.1° above + x-axis.
14. (a) 94.4, 31.3° below + x-axis; (b) 117, 72.1° above + x-axis; (c) 142, 11.7° below – x-axis.
16. (a) $1.90\ m/s^2$ down; (b) 18.8 s.
18. (a) $A_x = \pm 71.2$; (b) $B_x = -151.2$, $B_y = +55.0$, 161, 20° above – x-axis.
20. 44 m, 4.8 m.
22. 7.0 m/s.
24. 13 m/s.
26. 2.46 s.
28. 22 m.
30. (a) 9.39 m/s; (b) 0.80 m.
32. (a) 1.13 m; (b) 0.54°.
36. (a) 14.6 s; (b) 1.22 km; (c) 83.9 m/s, – 79.9 m/s; (d) 116 m/s; (e) 43.6° below the horizontal.
38. (a) 480 m; (b) – 8.41 m/s (down); (c) 97.5 m/s.
40. 10.5 m/s in the direction of the ship's motion, 6.5 m/s in the direction of the ship's motion.
42. $v_{SC} = 29$ m/s, $v_{SG} = 14$ m/s (down).

44. 0.00600 h = 21.6 s.
46. 8.13° west of south.
48. 1.9 m/s, 11° above the water.
50. 0.90 m/s.
52. 53°.
54. 0.366 m/s^2.
56. 15.8 s.
58. 6.2°.
60. $D_x = 50$ m, $D_y = -25$ m, $D_z = -10$ m, 27° from the x-axis toward the $-y$-axis, 10° below the horizontal.
62. ± 31.6°, ± 46.3.
64. 111 km/h, 37.0° N of E.
66. 0.88 s, 0.95 m.
68. 33 m/s.
70. 26.3 m/s, 0.714 s, 3.8 m beyond the net, good.
72. (*a*) $Dv/(v^2 - u^2)$; (*b*) $D/(v^2 - u^2)^{1/2}$.

CHAPTER 4

2. 116 kg.
4. 1.26×10^3 N.
6. (*a*) 196 N, 196 N; (*b*) 294 N, 98.0 N.
8. 153 N.
10. 1.29×10^3 N in the direction of the ball's motion.
12. 1.3×10^4 N.
14. 5.04×10^4 N, 4.47×10^4 N.
16. 2.5 m/s^2 (down).
18. (*a*) 7.4 m/s^2 (down); (*b*) 1176 N.
20. 3.2×10^3 N downward.
22. (*a*) 1.5×10^2 N; (*b*) 15.0 m/s.
24. Southwesterly direction.
28. (*b*) 62.2 N; (*c*) 204 N; (*d*) 100 N.
30. (*a*) 29 N, 58 N; (*b*) 34 N, 68 N.
32. (*a*) 3.2×10^2 N; (*b*) 0.98 m/s^2.
34. $\theta = 22°$.
36. 3.93 m.
38. 1.0×10^2 N, no force.
40. (*b*) Friction opposes motion down the plane; (*c*) friction opposes motion up the plane.
42. 2.3.
44. 4.1 m.
46. (*b*) 47 m; (*c*) 2.8×10^2 m.
48. Bonnie's.
50. 0.32.
52. (*a*) 3.67 m/s^2; (*b*) 8.17 m/s.
54. (*a*) 1.85 m/s^2, 5.81 m/s; (*b*) 0.82 m up the plane, 1.49 s.
56. 101 N, $\mu_k = 0.719$.
58. (*a*) 0.098 m/s^2; (*b*) 4.3 s.
60. $F_N = 4.9 \times 10^2$ N.
62. 2.2 m/s^2 up the plane.
64. 1.3×10^2 N.
66. -2.1×10^4 N, 1.1 m.
68. 1.5×10^3 N.
70. (*a*) 16 m/s; (*b*) 12 m/s.
72. 1.5×10^4 N.
74. (*a*) 75.0 kg; (*b*) 75.0 kg; (*c*) 75.0 kg; (*d*) 98.0 kg; (*e*) 52.0 kg.
76. Emerges with a speed of 7.3 m/s.
78. (*a*) 41 kg/s; (*b*) 1.6×10^2 N.
80. (*a*) $F = \frac{1}{2}Mg$; (*b*) $F_{T1} = F_{T2} = \frac{1}{2}Mg$, $F_{T3} = \frac{3}{2}Mg$, $F_{T4} = Mg$.

CHAPTER 5

2. (*a*) 1.52 m/s^2 toward the center; (*b*) 38.0 N toward the center.
4. 12 m/s.
6. 14 m/s, no effect.
8. (*a*) 3.14 N; (*b*) 9.02 N.
10. 27.6 m/s, 0.439 rev/s.
12. 9.2 m/s.
14. 11 rev/min.
16. $F_{T1} = 4\pi^2 f^2(m_1 r_1 + m_2 r_2)$, $F_{T2} = 4\pi^2 f^2 m_2 r_2$.
18. 0.20.
20. 0.23.
22. $\Sigma F_{tan} = 3.2 \times 10^3$ N, $\Sigma F_R = 1.8 \times 10^3$ N.
24. (*a*) 1.23 m/s; (*b*) 3.01 m/s.
26. 1.62 m/s^2.
28. 24.5 m/s^2.
30. $g_h = 0.91 g_{surface}$.
32. 2.0×10^7 m.
34. (*a*) 9.8 m/s^2; (*b*) 4.3 m/s^2.
36. 9.6×10^{17} N away from the Sun.
38. 2.0×10^{30} kg.
40. 3.14 m/s^2 upward.
42. 5.07×10^3 s (1.41 h), independent of the mass.
44. (*a*) 58 kg; (*b*) 58 kg; (*c*) 77 kg; (*d*) 39 kg; (*e*) 0.
46. (*a*) -0.309 (-30.9%); (*b*) $+0.309$ ($+30.9$%).
48. $M_{planet} = 4\pi^2 r^3/GT^2$.
50. (*b*) 4.8×10^3 kg/m^3.
52. 1.62×10^{11} m.
54. 5.97×10^{24} kg.
56. 3.4×10^{41} kg, 1.7×10^{11}.
58. $R_{Europa} = 6.71 \times 10^5$ km, $R_{Ganymede} = 1.07 \times 10^6$ km, $R_{Callisto} = 1.88 \times 10^6$ km.
60. (*b*) 3.02×10^4 m/s, 2.92×10^4 m/s.
62. 2.6×10^3 km.
64. Yes, $v_{top,min} = (gR)^{1/2}$.
66. -0.0337 m/s^2 (0.343% of g).

68. (*a*) Inside surface farthest from the center;
(*b*) 1.8×10^3 rev/day.
70. (*a*) $R_{min} = 3.0$ km; (*b*) $F_{Nbottom} = 5.5 \times 10^3$ N;
(*c*) $F_{Ntop} = 3.9 \times 10^3$ N.
72. (*a*) $\theta = \tan^{-1}(M_M R_E^2 / M_E D_M^2)$; (*b*) 7×10^{13} kg;
(*c*) $\theta \approx (1 \times 10^{-3})°$.
74. 5.07×10^3 s (1.41 h).
76. 29.2 m/s.
78. 5.2×10^{33} kg, $2.6 \times 10^3 \times$.

CHAPTER 6

2. (*a*) 1.1×10^3 J.
(*b*) 5.4×10^3 J.
4. 25 N.
6. 7.8 J.
8. $W_F = -W_{fr} = 1.7 \times 10^2$ J.
10. (*a*) 4.2×10^2 N; (*b*) -1.8×10^3 J; (*c*) -4.1×10^3 J;
(*d*) 5.9×10^3 J; (*e*) 0.
12. 5.0×10^3 J.
14. 0.084 J.
16. 2.4×10^{10} J.
18. (*a*) $\sqrt{2}$; (*b*) 4.
20. -4.67×10^5 J.
22. 44 m/s.
24. 2.25.
26. -1.1 N.
28. (*a*) 2.48×10^3 N; (*b*) 6.79×10^3 J; (*c*) 5.21×10^4 J;
(*d*) -4.53×10^4 J; (*e*) 7.86 m/s.
30. 71 J.
32. (*a*) 45.3 J; (*b*) 12.3 J; (*c*) 45.3 J.
34. (*a*) $\frac{1}{2}k(x^2 - x_0^2)$; (*b*) $\frac{1}{2}kx_0^2$, same.
36. 49.5 m/s.
38. 6.5 m/s.
40. $v_B = 24$ m/s, $v_C = 9.9$ m/s, $v_D = 19$ m/s.
42. (*a*) 1.0×10^2 N/m; (*b*) 22 m/s^2.
44. $h = 2.5r$.
46. $k = 12Mg/h$.
48. (*a*) 8.82×10^4 J; (*b*) 433 N.
50. 5.3×10^2 J.
52. 12 m/s.
54. 0.31.
56. 0.17 m.
58. 25.5 s.
60. (*b*) 0.134 hp.
62. 2.1×10^4 W = 28 hp.
64. 3.6×10^2 W.
66. 1.2×10^3 W.
68. $\theta_{max} = 24°$.
70. 9.0×10^2 W.
72. 3×10^7 N/m.
74. 8.0 m/s.
76. (*a*) $(2gL)^{1/2}$; (*b*) $(1.2gL)^{1/2}$.
78. (*a*) $v_{max} = [v_0^2 + (kx_0^2/m)]^{1/2}$;
(*b*) $x_{max} = [x_0^2 + (mv_0^2/k)]^{1/2}$.
80. (*a*) 2.6×10^5 J; (*b*) $v = 24$ m/s; (*c*) 1.2 m.
82. 20 km/h.
84. 4.4×10^4 N.
86. 3.8×10^2 W.

CHAPTER 7

2. -0.720 m/s (opposite to the direction of the package).
4. 4.8 m/s.
6. 1.6×10^4 kg.
8. 5.1×10^2 m/s.
10. 4.2×10^3 m/s.
12. (*a*) 4.70×10^3 m/s, 6.90×10^3 m/s; (*b*) 5.90×10^8 J.
14. 130 N, not large enough.
16. (*a*) 2.0 N · s; (*b*) 4.0×10^2 N.
18. (*a*) 4.6×10^2 kg · m/s (east);
(*b*) -4.6×10^2 kg · m/s (west);
(*c*) $+4.6 \times 10^2$ kg · m/s (east); (*d*) 6.1×10^2 N (east).
20. 69 m.
22. $v_1' = -1.00$ m/s (rebound), $v_2' = 2.00$ m/s.
24. $v_1' = 0.70$ m/s, $v_2' = 2.20$ m/s.
26. (*a*) $v_1' = 3.62$ m/s, $v_2' = 4.42$ m/s;
(*b*) -396 kg · m/s, $+396$ kg · m/s.
28. 0.28 m, 1.1 m.
30. 0.61 m.
32. KE(heavier) = 3000 J, KE(lighter) = 4500 J.
34. (*b*) $e = (h'/h)^{1/2}$.
36. (*a*) $v_1' = v_2' = 1.9$ m/s; (*b*) $v_1' = -1.6$ m/s, $v_2' = 7.9$ m/s;
(*c*) $v_1' = 0$, $v_2' = 5.2$ m/s; (*d*) $v_1' = 3.1$ m/s, $v_2' = 0$;
(*e*) $v_1' = -4.0$ m/s, $v_2' = 12$ m/s; result for (*c*) is reasonable, results for (*d*) and (*e*) are not reasonable.
38. 60°, 6.7 m/s.
40. (*a*) 30°; (*b*) $v_1' = v/\sqrt{3}$; (*c*) $\frac{2}{3}$.
42. (*a*) 3.47 m/s; (*b*) 11.9 m/s; (*c*) 3.29°.
44. *x*-axis, $v_1' = 3.7$ m/s, $v_2' = 2.0$ m/s.
46. 6.5×10^{-11} m from the carbon atom.
48. $3.83\ell_0$ from the outer edge of the small cube.
50. 1.2ℓ from the left, and 0.9ℓ from the back.
52. 12 kg.
54. 22 cm, 7.6 cm.
56. (*a*) 4.66×10^6 m; (*b*) elliptical path.
58. 24.8 cm.
60. $mv/(M + m)$ up, the balloon will also stop.

62. 8.16 m/s.
64. $v_x' = \frac{3}{2}v_0, v_y' = -v_0$.
66. (*a*) 0.194 m/s; (*b*) 700 N.
68. $m_2 = 5m/3$.
70. 0.14 m toward the initial position of the 75-kg person.
72. 2.
74. (*a*) $v_1' = -4.38$ m/s, $v_2' = 4.02$ m/s; (*b*) 1.96 m.
76. (*a*) 6.54 m; (*b*) 35 m.

CHAPTER 8

2. $R_{Sun} = 6.5 \times 10^5$ km.
4. 2.9×10^3 m.
6. 188 rad/s.
8. 1.9 rad/s^2.
10. -2.6×10^2 rad/s^2.
12. 3.3×10^3 revolutions.
14. (*a*) 1.99×10^{-7} rad/s; (*b*) 7.27×10^{-5} rad/s.
16. 3.6×10^4 rpm.
18. $\omega_1/\omega_2 = R_2/R_1$.
20. 0.56 rad/s^2.
22. (*a*) -84 rad/s^2; (*b*) 1.5×10^2 rev.
24. 38 m.
26. (*a*) 0.58 rad/s^2; (*b*) 12 s.
28. (*a*) $a_{tan} = \alpha r$, $a_R = \alpha^2 t^2 r$; (*b*) $\phi = \tan^{-1}(1/4\pi N)$.
30. (*a*) 38 m · N; (*b*) 33 m · N.
32. $\geq 6.5 \times 10^2$ m · N.
34. 1.89 kg · m^2.
36. (*a*) 6.1 kg · m^2; (*b*) 0.61 kg · m^2.
38. (*a*) 0.851 kg · m^2; (*b*) 0.0720 m · N.
40. (*a*) 1.99×10^{-3} kg · m^2; (*b*) $\tau_{applied} = 6.82 \times 10^{-2}$ m · N.
42. 5.9×10^2 m · N, 2.4×10^2 N.
44. (*a*) 7.5 m · N; (*b*) 3.0×10^2 N.
46. 2.25×10^3 kg · m^2, 8.8×10^3 m · N.
48. $a = (m_2 - m_1)g/(m_1 + m_2 + I/R_0^2)$,
    $a_0 = (m_2 - m_1)g/(m_1 + m_2)$, $a_0 > a$.
50. 94 J.
52. (*a*) 2.6×10^{29} J; (*b*) 2.7×10^{33} J, $KE_{total} = 2.7 \times 10^{33}$ J.
54. (*a*) 8.37 m/s, 41.8 rad/s; (*b*) 2.50; (*c*) none of the answers depends on the mass; the rotational speed depends on the radius.
56. 9.85 m/s.
58. (*a*) Angular momentum is conserved; (*b*) 1.6.
60. 0.77 kg · m^2, pulling her arms closer to her body.
62. (*a*) 14 kg · m^2/s; (*b*) -2.7 m · N.
64. $\omega/2$.
66. (*a*) 1.2 rad/s; (*b*) 2.0×10^3 J, 1.2×10^3 J, loss of 8.0×10^2 J, a decrease of 40%.
68. Angular speed of merry-go-round does not change.
70. $\omega_t = -0.30$ rad/s.
72. $L/2$, $L/2$.
74. 6.0 m/s.
76. (*a*) $\omega_R/\omega_F = R_F/R_R = N_F/N_R$; (*b*) 4.0; (*c*) 1.5.
78. (*b*) 2.2×10^3 rad/s; (*c*) 24 min.
80. (*a*) $\alpha = 3g/2L$; (*b*) $3g/2$.
82. (*b*) $\theta = 16°$; (*c*) 2.6 m.
84. $h_{min} = 2.7(R_0 - r)$.

CHAPTER 9

2. 1.7 N.
4. 1.7 m.
6. 6.52 kg.
8. $F_{N1} = 8.57 \times 10^3$ N, $F_{N2} = 6.13 \times 10^3$ N.
10. 3.0 m from the adult.
12. 3.9×10^3 N, 3.4×10^3 N.
14. 5.6×10^2 N.
16. 48 N, only the vertical components balance the weight.
18. 1.1 m from pivot on side of lighter boy.
20. $F_1 = -2.1 \times 10^3$ N (down), $F_2 = 3.0 \times 10^3$ N (up).
22. $F_1 = 5.8 \times 10^3$ N, $F_2 = 5.6 \times 10^3$ N.
24. (*a*) 2.1×10^2 N; (*b*) 2.0×10^3 N.
26. $F_T = 542$ N, $F_{hingex} = 409$ N, $F_{hingey} = -59$ N (down).
28. $\theta_{min} = \tan^{-1}(1/2\mu)$.
30. (*a*) 1.00 N; (*b*) 1.25 N.
32. (*a*) Tip over; (*b*) 50 cm.
34. 6.1 kg.
36. 2.5×10^2 N.
38. 2.4×10^2 N, 2.7×10^3 N.
40. $1.9w$.
42. Stable equilibrium, 2.5 m.
44. 1.91 cm.
46. 0.029 mm.
48. 0.12%, 0.15%.
50. 9.6×10^6 N/m^2.
52. -0.022%.
54. (*a*) 1.1×10^2 m · N; (*b*) wall; (*c*) all three.
56. 3.9×10^2 N, thicker strings, an increased tension.
58. (*a*) 4.4×10^{-5} m^2; (*b*) 2.7 mm.
60. 1.2 cm.
62. 12 m.
64. $M_C = 0.191$ kg, $M_D = 0.0544$ kg, $M_A = 0.245$ kg.
66. 2.0×10^3 N, No.
68. $\theta_{max} = 29°$.
70. 6, 2.0 m apart.
72. (*a*) 600 N; (*b*) $F_A = 0$, $F_B = 1200$ N; (*c*) $F_A = 150$ N, $F_B = 1050$ N; (*d*) $F_A = 750$ N, $F_B = 450$ N.

74. (*a*) $0.289mg$; (*b*) $0.577mg$; (*c*) horizontal at the lowest point, and 60° above the horizontal at the attachment.
76. (*a*) $F_{Gx} = 27$ N, $F_{Gy} = 147$ N; (*b*) 0.26.
78. (*a*) 6.5×10^3 m; (*b*) 6.4×10^3 m.
80. (*a*) $\mu_s < L/2h$; (*b*) $\mu_s > L/2h$.
82. Right end safe, left end not safe, 0.10 m.

CHAPTER 10

2. 80 kg.
4. 6.5×10^{-2} m^3.
6. 0.89.
8. (*a*) 9.8×10^7 N/m^2; (*b*) 1.8×10^5 N/m^2.
10. 1.1 m.
12. 7.1×10^3 kg.
14. 1.21×10^5 N/m^2, 3.2×10^7 N (down), 1.21×10^5 N/m^2.
16. 6.54×10^2 kg/m^3.
18. 4.0×10^5 N/m^2.
20. 1.057×10^3 kg/m^3, + 0.030 (3%).
22. 4.06×10^3 kg/m^3.
24. 3.0×10^3 kg.
26. Copper.
28. (*b*) Center of buoyancy is above the center of gravity.
30. 0.105 (10.5%).
32. 2.76 kg.
34. 47 cm/s.
36. 13 days.
38. 3.1×10^{-3} m^3/s.
40. 1.6×10^6 N.
44. (*b*) 0.24 m/s.
46. 1/7.7 = 0.13.
48. 4.2×10^{-3} Pa · s.
50. 0.71 (reduced by 29%).
52. (*a*) Laminar; (*b*) 3200, turbulent.
54. 1.9 m.
56. 9.1×10^{-3} N.
58. (*a*) $\gamma = F/4\pi r$; (*b*) 2.4 N/m.
62. (*a*) 5.7×10^{-4} N; (*b*) 0.32 N.
64. 0.051 atm.
66. 0.6 atm.
68. 954 kg/m^3.
70. 5.3×10^{18} kg.
72. 2.6 m.
74. 29 people.
76. 49 N, not float.
78. $d = D[v_0^2/(v_0^2 + 2gy)]^{1/4}$.
80. (*a*) 3.1 m/s; (*b*) 16 s.

CHAPTER 11

2. 1.3×10^2 N/m.
4. (*a*) 6.8×10^2 N/m; (*b*) $A = 2.5$ cm, 2.5 Hz.
6. (*a*) 9.5×10^{-2} N/m; (*b*) 2.2 Hz.
8. 3.8 Hz.
10. 0.52 kg.
12. 0.282 s.
14. (*a*) 0.489 s, 2.04 Hz; (*b*) 0.215 m; (*c*) 35.5 m/s^2; (*d*) (0.215 m) sin [(12.8 s^{-1})t]; (*e*) 2.86 J.
16. (*a*) $0.866\, x_0$; (*b*) $0.500\, x_0$.
18. (*a*) 0.35 m; (*b*) 0.875 Hz; (*c*) 1.14 s; (*d*) 0.74 J; (*e*) 0.68 J, 0.06 J.
20. (*a*) y = (0.10 m) cos [(11.4 s^{-1})t]; (*b*) 0.14 s; (*c*) 1.1 m/s; (*d*) 13 m/s^2, the release point.
22. $A_1 = 3.16A_2$.
26. 136 N/m, 20.3 m.
28. (a) 1.4 s; (*b*) 0.72 Hz.
30. (*a*) 1.4 s; (*b*) infinite (no swing).
32. $v_0 = \theta_0(gL)^{1/2}$.
34. 2.8 m/s.
36. 545 m, 188 m, 3.41 m, 2.78 m.
38. Less dense rod, 1.41.
40. 0.60 m.
42. (*a*) 1.9×10^3 km; (*b*) cannot be determined.
44. (*a*) 0.25; (*b*) 0.50.
48. 1.73.
50. (*c*) Kinetic energy.
52. 441 Hz.
54. n(0.40 Hz), n = 1, 2, 3, ... ; (2.5 s)/n, n = 1, 2, 3, ...
56. 70 Hz.
58. Decreased by 4.8%.
60. (*a*) 1.4 kg; (*b*) 0.36 kg; (*c*) 0.057 kg.
62. (*b*) 1.8% (increase); (*c*) will apply.
64. 25°.
66. (*a*) $\theta_{iM} = \sin^{-1}(v_1/v_2)$; (*b*) > 59°.
68. 2.9 cm, 1.4 m.
70. 11 cm.
72. (*a*) $1.22f$; (*b*) $0.71f$.
74. 260 Hz.
76. (*a*) 1.4×10^4 N/m; (*b*) 0.79 s.
78. 2.0 cm.
80. 1.3 Hz,
82. The block will oscillate with SHM, $k = 2\rho gA$, the density and the cross section.

CHAPTER 12

2. 1.6 km.
4. 0.0016 s, 0.024 s ($\approx 15\times t_{wire}$).
6. (*a*) 0.64 s; (*b*) 2.9 s.
8. 3.5%.
10. 63 dB.
12. 114 dB.
14. (*a*) 5×10^{-6} W; (*b*) 2×10^{7}.
16. (*a*) 123 dB, 115 dB; (*b*) almost twice as loud.
18. (*a*) 9; (*b*) 9.5 dB.
20. 4.
22. 8.17×10^{-5} m.
24. 150 Hz to 20,000 Hz.
26. 87 N.
28. 0.66 m.
30. 8.6 mm $< L <$ 8.6 m.
32. 7.2 cm.
34. – 2/6%.
36. (*a*) closed; (*b*) 88 Hz.
38. Open, 4.3 m.
40. 2.5 cm.
42. 0.50 Hz.
44. 28.5 kHz.
46. (*a*) 130.5 Hz, 133.5 Hz; (*b*) ± 2.3%.
48. (*a*) 343 Hz; (*b*) 1030 Hz, 1715 Hz.
52. 43,200 Hz.
54. 120 Hz.
56. 2091 Hz, 2087 Hz, not exactly the same, but close, 3554 Hz, 2875 Hz, 15950 Hz, 3750 Hz.
58. 0.205 m/s.
60. 5.8 m/s.
64. (*a*) 120; (*b*) 0.96°.
66. (*a*) 37°; (*b*) 1.7.
68. 55 m.
70. 410 km/h (257 mi/h).
72. 1, 0.444, 0.198, 0.0878, 0.0389.
74. 10 W.
76. 50 dB.
78. 12.3 m/s.
80. 2.29 kHz.
82. (*a*) 645 Hz; (*b*) 645 Hz; (*c*) 645 Hz; (*d*) 645 Hz; (*e*) 669 Hz; (*f*) 669 Hz.
84. (*a*) 7.8×10^{-3} m; (*b*) 0.13 s.
86. 55.39 kHz.

CHAPTER 13

2. 3.2×10^{22} atoms.
4. (*a*) 5°F; (*b*) – 26°C.
6. (*a*) 49.4°C; (*b*) 80.7°C.
8. – 40°F = – 40°C.
10. 2.0×10^{-6} m.
12. – 69°C.
14. – 6.4%.
16. (*a*) 74×10^{-6} $(C°)^{-1}$; (*b*) aluminum.
18. – 1.4×10^{2} °C.
20. (*b*) – 0.0057 (0.57%).
22. – 40 min.
24. 93°C.
26. (*a*) – 1.4×10^{8} N/m^2; (*b*) No; (*c*) – 1.4×10^{7} N/m^2, will fracture.
28. (*a*) 359 K; (*b*) 299 K; (*c*) 173 K; (*d*) 5773 K.
30. (*a*) 4273 K, 15×10^{6} K; (*b*) 6.4%, 0.0018%;
32. 1.43 kg/m^3.
34. (*a*) 14.8 m^3; (*b*) 1.81 atm.
36. 1.26×10^{3} atm.
38. 3.06 atm.
40. 1.4×.
42. 2.69×10^{25} molecules/m^3.
44. (*a*) 6×10^{22} mol; (*b*) 4×10^{46} molecules.
46. 20 molecules.
48. (*a*) 5.65×10^{-21} J; (*b*) 3.65×10^{3} J.
50. 819°C.
52. $\sqrt{2}$.
54. (*a*) 2.9×10^{2} m/s; (*b*) 12 m/s.
56. 1.00429.
58. Vapor.
60. 14°C.
62. 95°C.
64. 1.96 atm.
66. 3.0 kg.
68. 28%.
70. 14 h.
72. (*b*) 4×10^{-11} mol/s; (*c*) 0.6 s.
74. (*a*) Low; (*b*) 0.017%.
76. 0.18.
78. 9.4×10^{26} molecules, 1.6×10^{3} mol.
80. 4×10^{-17} N/m^2.
82. (*a*) $N_O = N_{He} = 4.8 \times 10^{23}$ molecules; (*b*) 1; (c) 2.8.
84. 1.61, 1.27.
88. 439 m/s.
90. 3.3×10^{-7} cm.
92. (*a*) Lower; (*b*) 0.36%.

CHAPTER 14

2. 3.14×10^6 J.
4. (*a*) 1.05×10^7 J/day; (*b*) 2.9 kWh/day; (*c*) \$0.29 /day.
6. 90 s.
8. 92 kcal.
10. 5.3×10^6 J.
12. 1.00 Btu/lb · F°.
14. 40.0°C.
16. 20.2°C.
18. 215°C.
20. 286 J/kg · C°.
22. 5.02×10^6 J.
24. 0.062 kg .
26. (*a*) 5.5 h; (*b*) 41.6 h.
28. 0.16 kg.
30. 1.7 g.
32. 1.6×10^4 W.
34. 0.75 mm.
36. (*a*) 13 W; (*b*) 3.4 W.
38. 20 h.
40. 295 K (22°C).
42. $\Delta Q/\Delta t = [(k_1A_1/L_1) + (k_2A_2/L_2)]\Delta T$.
44. (*b*) $\Delta Q/\Delta t = A(T_2 - T_1)/\Sigma(\ell_i/k_i)$.
46. 22%.
48. 69 m/s.
50. 2.8 C°.
52. (*a*) $C = mc$; (*b*) 4186 J/C°; (*c*) 2.09×10^5 J/C°.
54. (*a*) 66 W; (*b*) 1.0×10^4 W.
56. 23 W.
58. (*a*) 160 kW; (*b*) 2.4×10^8 J; (*c*) \$$6.8 \times 10^2$/month.
60. (*a*) 2.3 C°/s; (*b*) 84°C; (*c*) convection, conduction, evaporation.
62. 1.5×10^3 m/s.

CHAPTER 15

2. (*a*) 0; (*b*) 4.40×10^3 J.
6. (*a*) 0; (*b*) – 265 kJ.
8. (*a*) 1.3×10^2 J; (*b*) + 1.3×10^2 J.
10. (*b*) 2.70×10^3 J, 4.05×10^3 J; (*d*) 4.05×10^3 J.
12. (*a*) + 25 J; (*b*) + 63 J; (*c*) – 95 J; (*d*) – 120 J; (*e*) – 15 J.
14. 1.5×10^7 J (3.6×10^3 kcal).
16. 5,6 kg.
18. 14.3%.
20. 426°C.
22. 439°C.
24. 687°C.
26. 144 kg/s.
28. 5.7.
30. – 21°C.
32. 2.9.
34. – 0.145 kcal/K.
36. – 292 kcal/K.
38. + 0.0104 cal/K · s.
40. + 0.26 cal/K.
42. (*a*) 1/9; (*b*) 1/18.
44. (*a*) 5/16; (*b*) 1/64.
46. 46 m^2, would probably fit.
48. (*a*) 1.32×10^6 kWh; (*b*) 7.09×10^4 kW.
50. 4.2×10^{15} J.
52. Yes.
54. (*a*) 2.0×10^4 J/s; (*b*) 8.0×10^4 J/s; (*c*) 27 min.
56. 7.7%.
58. (*a*) 44.6°C; (*b*) + 0.58 J/K.
60. 24%.
62. – 8.3×10^5 J.
64. (*a*) 5.3 C°; (*b*) + 77 J/kg · K.

CHAPTER 16

2. 2.7 N.
4. 0.500 N.
6. 9.2 N.
8. 3.8×10^{14} electrons, 3.4×10^{-16} kg.
10. – 5.4×10^7 C.
12. 83.8 N away from the center of the triangle.
14. 2.96×10^5 N toward the center of the square.
16. 5.71×10^{13} C.
18. 2.1×10^{-10} m.
20. 50.0×10^{-6} C, 30.0×10^{-6} C; – 15.7×10^{-6} C, 95.7×10^{-6} C.
22. 5.6×10^{-16} N (west).
24. + 9.5×10^5 N/C (up).
26. 3.2×10^8 N/C toward the negative charge.
28. $Q = 6.2 \times 10^{-10}$ C.
30. $E_A = 5.8 \times 10^6$ N/C up,
    $E_B = 1.5 \times 10^7$ N/C 56° above the horizontal.
32. 4.82×10^4 N/C away from the opposite corner.
36. 3.42×10^5 km from the center of the Earth.
38. (*a*) 8.83×10^6 m/s.
40. (*a*) $\approx 3 \times 10^{-10}$ N; (*b*) $\approx 6 \times 10^{-10}$ N; (*c*) $\approx 10^{-4}$ N.
42. 3.2×10^{-10}.
44. 6.8×10^5 C, negative.
46. 8.1×10^5 N/C up.
48. $0.402Q_0$ 0.366ℓ from Q_0.
50. 3.6 m from the positive charge, and 1.6 m from the negative charge.
52. $(1.08 \times 10^7\ \text{N/C})/\{3 - \cos[(12.5\ \text{s}^{-1})t]\}^2$ up.
54. 5×10^{-9} C.
56. – 1.06×10^7 N/C (down).

CHAPTER 17

2. -2.40×10^{-17} J (done by the field), – 150 eV.
4. 2.16×10^3 V, plate *B*.
6. 7.04 V.
8. 3×10^{-5} m.
10. (*a*) 1.62×10^7 m/s; (*b*) 3.5×10^7 m/s.
12. 1.63×10^7 m/s.
14. 2.1 nC.
16. (*a*) 5.8×10^5 V; (*b*) 9.2×10^{-14} J.
18. 6.9×10^{-18} J.
20. 2.33×10^7 m/s.
22. (*a*) 27.2 V; (*b*) 13.6 eV; (*c*) – 13.6 eV; (*d*) 13.6 eV.
24. (*a*) 8.5×10^{-30} C · m; (*b*) zero.
26. (*a*) – 0.088 V; (*b*) 1%.
30. 2.6 μF.
32. 22.0 V.
34. 5.0×10^7 m^2.
36. 0.63 μF.
38. 0.24 m^2.
40. 4.32×10^{-9} F, 0.254 m^2.
42. 744 V, 1.86×10^{-3} C, 5.06×10^{-3} C.
44. 11 μF.
46. 2.
48. $1/2K$, $1/K$.
50. 1.17×10^5 m/s, 3.37×10^5 m/s.
52. $-4.8 \times 10^5 \text{ V/m} < E < 4.8 \times 10^5$ V/m.
54. (*a*) 1.1 MV; (*b*) 13 kg.
56. Yes, 1.7×10^{-12} V.
58. (*a*) 5.2 keV; (*b*) 42.8.
60. 6.3×10^2 V.
62. 0.15 μC.
64. $-3.5\,kQ/\ell$, $-5.2\,kQ/\ell$, $-6.8\,kQ/\ell$.
66. $\theta = 1.4°$.
68. (*a*) 9×10^{-16} m; (*b*) 1.1×10^8 m^2.
70. (*a*) 23 J; (*b*) 0.23 MW.

CHAPTER 18

2. 6.25×10^{18} electron/s.
4. 7.5×10^2 V.
6. (*a*) 4.95 A; (*b*) 6.11 A.
8. (*a*) 13 Ω; (*b*) 8.1×10^3 C.
10. 2.5×10^{-3} V.
12. 0.57 mm.
14. Yes, 4.6 mm.
16. 29 C°.
18. 363°C.
20. (*a*) 3.8×10^{-4} Ω; (*b*) 1.5×10^{-3} Ω; (*c*) 6.0×10^{-3} Ω.
22. $R_{N0} = 2.61$ kΩ, $R_{C0} = 2.09$ kΩ.
24. 3.2 W.
26. (*a*) 600 W; (*b*) 24 Ω; (*c*) 12 Ω.
28. 15 W.
30. $38.50.
32. 1.1 kWh.
34. 3.
36. 5.3 kW.
38. 0.111 kg/s.
40. 0.077 A.
42. (*a*) Infinite; (*b*) 1.9×10^2 Ω.
44. 636 V, 5.66 A.
46. 1.7 kW, 3.4 kW, zero.
48. (*a*) 0.0293 Ω; (*b*) 1.8×10^{-8} Ω · m; (*c*) 8.8×10^{28} m^{-3}.
50. 2.7 A/m^2 north.
52. 35 m/s.
54. 5.4×10^{-9} W.
56. 6.2 A.
58. 0.29 mm.
60. (*a*) $44; (*b*) 1.8×10^3 kg.
62. (*a*) – 19.5%; (*b*) percentage decrease in the power output would be less.
64. (*a*) 33.4 Hz; (*b*) 1.27 A; (*c*) (75.6 V) sin (210 s^{-1})t .
66. (*a*) 1.5 kW; (*b*) 12.5 A.
68. 2.
70. (*a*) 3.6 hp; (*b*) 2.4×10^2 km.
72. (*a*) 1.2 kW; (*b*) 100 W.
74. 1.4×10^{12} protons.

CHAPTER 19

2. (*a*) 360 Ω; (*b*) 8.9 Ω.
4. 261 Ω.
6. 720 Ω, 80 Ω, 360 Ω, 160 Ω.
8. (*a*) 13.8 V; (*b*) 34 Ω, 5.5 W.
10. 27 W.
12. 2.20 kΩ.
14. 3.36 mA, 2.11 mA, 1.26 mA, 0.84 mA, 0.42 mA, $V_{AB} = 1.85$ V.
16. (*a*) V_1 and V_3 decrease; V_2 increases; (*b*) I_1 and I_3 decrease; I_2 increases; (*c*) increase; (*d*) 17.0 V, 17.2 V.
18. (*a*) 8.41 V; (*b*) 8.49 V.
20. 0.4 Ω.
22. 0.053 Ω, 0.15 Ω.
24. 0.41 A.
26. – 26 V.
28. $I_1 = 0.68$ A, $I_2 = -0.40$ A.
30. $I_1 = 0.156$ A right, $I_2 = 0.333$ A left, $I_3 = 0.177$ A up.

32. $I_1 = 0.274$ A, $I_2 = 0.222$ A, $I_3 = 0.266$ A, $I_4 = 0.229$ A, $I_5 = 0.007$ A, $I = 0.496$ A.
34. $I_1 = 0.77$ A, $I_2 = 0.71$ A, $I_3 = 0.055$ A, 5.95 V.
36. 2.5 V.
38. 11 μF in parallel.
40. 0.0195 μF, 1.4 nF.
42. (*a*) $(C_1C_2 + C_1C_3 + C_2C_3)/(C_2 + C_3)$; (*b*) 188 μC.
44. (*a*) $V_1 = 5.0$ V, $V_2 = 4.0$ V; (*b*) 2.0 μC; (*c*) 9.0 V, 3.6 μC, 4.5 μC.
46. $\frac{1}{2}(K_1 + K_2)C_0$.
48. 5.13 pF.
50. 1.0 MW.
52. (*a*) 2.3 nF; (*b*) 14 μs.
54. (*a*) 8.0 V; (*b*) 16 V; (*c*) 8.0 V; (*d*) – 5.8 μC.
56. 100 μA.
58. (*a*) 2.9×10^{-5} Ω in parallel; (*b*) 35 kΩ in series.
60. 22 V, 17 V, 14% low, 14% low.
62. 0.82 mA, 4.6 V.
64. 8.9 V.
66. $R_2 = 222$ kΩ, 26 V.
68. $R_1 = 2$ Ω, $R_2 = 70$ Ω.
70. 3.7 V, 116 V.
72. (*a*) 1.0×10^{-3} J; (*b*) 2.4×10^{-4} J; (*c*) $Q_{\text{parallel}} = 45$ μC, $Q_{\text{series}} = 11$ μC.
74. Mercury cells would have a higher terminal voltage.
76. (*b*) 65.7 Ω.
78. (*b*) 1.836 V; (*c*) ± 0.90 mV; (*d*) no effect of the internal resistance.
80. 50.1 V, 1.25 W.
82. 52 V, – 28 V.
84. 13 W, 30 kΩ.
86. (*a*) 0.10 MV; (*b*) so it can be used for only short bursts.

CHAPTER 20

2. 3.4×10^{-4} N perpendicular to the wire and to **B**.
4. 0.753 T.
6. Clockwise vertical circle, 4.10×10^{-5} m.
8. (*a*) Down; (*b*) in; (*c*) right.
10. 7.6 T west.
12. (*a*) South pole; (*b*) 3.5×10^{2} A; (*c*) 5.22 N.
18. 3.0×10^{-7} m.
20. 83 A.
22. 26 A upward.
24. 1.3×10^{-3} *g*.
26. 4.3×10^{-5} T, 28° below the horizontal.
28. (*a*) $(2.0 \times 10^{-5}$ T/A$)(15$ A $- I)$ up; (*b*) $(2.0 \times 10^{-5}$ T/A$)(15$ A $+ I)$ down.
30. 14 A.
32. 12 T.
34. (*a*) 6.7×10^{3} A to the right; (*b*) unstable; (*c*) 6.7×10^{3} A to the left, stable for vertical displacements.
36. 91.9 A.
38. (*c*) Not uniform, $1/R$.
40. (*c*) At the surface of the wire, 6.0×10^{-3} T; (*d*) $r = 0.10r_0$, $r = 10r_0$.
42. 29 μA.
44. 1.07 T.
46. (*a*) 6.63×10^{-5} m · N; (*b*) south edge.
48. 0.820 T.
50. 1.57×10^{6} V/m.
52. 1.12 mm, 0.56 mm.
54. 2.5 m.
56. 2.7×10^{-5} T · m/A.
58. 2.6×10^{-6} N toward the wire.
60. 1.1×10^{-6} m/s west.
62. 0.17 N 68° above the horizontal toward the north.
64. 1.37×10^{3} A.
66. (*a*) $ILBt/m$; (*b*) $[(ILB/m) - \mu_k g]t$; (*c*) east.
68. 0.38 mm, 1.9 cm.
70. $\theta = 9.1°$.
72. (*a*) 1.5×10^{2} turns; (*b*) 5.2×10^{-3} T; (*c*) will not change.

CHAPTER 21

2. 0.15 V.
4. From right to left.
6. 24 mV.
8. (*a*) Counterclockwise; (*b*) clockwise; (*c*) zero; (*d*) counterclockwise.
10. (*a*) 0.17 V; (*b*) clockwise.
12. (*a*) Clockwise; (*b*) 43 mV; (*c*) 17 mA.
14. (*a*) 0.43 V; (*b*) 0.016 A; (*c*) 3.5×10^{-3} N.
16. (*a*) 0.21 A; (*b*) 2.8 mW.
18. 5.86 C.
20. 61 turns.
22. 0.925 rev/s.
24. 86 V.
26. $B_2/B_1 = 0.26$.
28. (*b*) 24.7 A; (*c*) 3.7 A.
30. 4.17×10^{5} turns.
32. 0.13.
34. (*a*) 6.0 V; (*b*) step-down.
36. (*a*) Step-down; (*b*) 3.3 A; (*c*) 0.33 A; (*d*) 3.6 Ω.
38. (*a*) 48 kV; (*b*) 0.056 (5.6%).
40. 33.4 kW.
42. 5.14 V, direction of the current.
44. 6.05 H.

46. 2.2 mH, 32 turns.
48. 2.
50. (*a*) $L_{series} = L_1 + L_2$; (*b*) $L_{parallel} = L_1L_2/(L_1 + L_2)$;
(*c*) $L_{series} = L_1 + L_2 + 2M$,
$L_{parallel} = (L_1L_2 - M^2)(L_1 + L_2 + 2M)$.
52. (*a*) 7.1×10^3 V; (*b*) 19.0 H.
54. 29 J.
56. 5.1×10^{15} J.
58. $(\Delta I/\Delta t)_0 = V/L$.
60. (*a*) 4; (*b*) $\frac{1}{2}$.
62. 69.2 Hz.
66. 50 mH.
68. (*a*) 6.4%; (*b*) 99%.
70. (*a*) 30 kΩ; (*b*) 99 kΩ.
72. 1.7 kW.
74. (*a*) 4.14 mA; (*b*) – 6.6°; (*c*) 0.49 W;
(*d*) $V_R = 119$ V, $V_C = 14$ V.
76. 34.5 mA.
78. 7.5 mW.
82. (*a*) 368 pF; (*b*) 26.8 μH.
84. (*a*) 21 mH; (*b*) 45 mA; (*b*) 2.2×10^{-5} J.
86. 61.
88. (*a*) Clockwise; (*b*) no induced current;
(*c*) counterclockwise.
90. $v = 0.76$ m/s.
94. (*a*) 41 kV; (*b*) 29.4 MW; (*c*) 0.78 MW; (*d*) 28.6 MW.
96. (*a*) 23 A; (*b*) 90 V; (*c*) 6.9×10^2 W; (*d*) 75%.
98. 9.76 nF.
100. 3.56×10^5 turns.
102. (*b*) 2.5 μH, 1.6×10^{-2} Ω.

CHAPTER 22

2. 2.3×10^{-9} A.
4. 5.00×10^{-8} T, zero.
8. 1.4×10^{-13} T.
10. 1.08 cm.
12. 2.00×10^{10} Hz.
14. 314 nm, ultraviolet.
16. 4.0×10^{16} m.
18. (*a*) 4.3 min; (*b*) 71 min.
20. Person at the radio hears the voice 0.14 s sooner.
22. 9.32×10^{-7} W/m^2.
24. 3.31×10^{-7} J/h.
26. 4.50×10^{-6} J.
28. 3.8×10^{26} W.
30. 302 pF.
32. $2.59 \text{ nH} \le L \le 3.89 \text{ nH}$.
34. 7.14×10^{-4} V/m, 1.35×10^{-9} W/m^2.
36. 2.979 m.
38. 5.56 m, 0.372 m.
40. (*a*) 1.28 s; (*b*) 4.3 min.
44. 469 V/m.
46. 3 m.
48. (*a*) 0.40 μW; (*b*) 0.012 V/m; (*c*) 0.012 V.
50. 0.36 mV.

CHAPTER 23

4. 76 cm.
6. 5.9×10^{-6} m^2.
8. (*b*) Reversed along the line of the original direction.
10. 13.5 cm.
12. 2.09 cm behind the surface, virtual, upright.
14. Concave, 5.7 cm.
16. (*a*) – 7 cm; (*c*) 1.0 mm.
18. Convex, – 28.0 cm.
22. – 8.8 m.
24. (*a*) Convex mirror; (*b*) 22 cm behind the surface;
(*c*) – 98 cm; (*d*) – 196 cm.
26. (*a*) 1.97×10^8 m/s; (*b*) 1.99×10^8 m/s.
28. 1.56.
30. 36°.
32. 41°.
34. 54.3° from the normal.
36. 81.0°.
40. 61.7°, Lucite.
42. 70.7 cm.
44. $n_{liquid} \ge 1.5$.
48. (*a*) 390 mm; (*b*) 39.0 cm.
50. Converging lens, + 39.8 cm, real.
52. (*a*) 72 cm behind the lens (virtual); (*b*) + 4.0.
54. (*a*) 81 mm; (*b*) 82 mm; (*c*) 87 mm; (*d*) 24 cm.
56. (*a*) Virtual; (*b*) converging lens; (*c*) 7.5 D.
58. (*a*) – 75.0 mm (beyond the lens);
(*b*) – 25.0 mm (beyond the lens).
60. (*a*) 152 mm (real, behind lens), – 0.279 cm (inverted);
(*b*) – 121 mm (virtual, in front of lens), + 0.222 cm
(upright).
62. 383 cm.
64. Real 21.3 cm beyond second lens, – 0.708 (inverted)
66. – 10.0 cm.
68. – 26.0 cm.
72. – 12.7 cm.
74. – 156 cm (in front of the lens), + 1.56 (upright).
76. (*a*) 3.0 m, 4.0 m, 7.0 m;
(*b*) toward you, away from you, toward you.
78. 53.7°.

82. Real, upright, real and upright.
84. $\geq 51.3°$.
88. + 20.0 cm (converging).
90. Chose different signs for the magnification, $d_o = 2f/3, 4f/3 = 13.3$ cm, 26.7 cm.

CHAPTER 24

2. 1.1 μm.
4. 440 nm, 6.82×10^{14} Hz.
6. 0.52 m, or 2.3 m away from the line perpendicular to the board midway between the openings.
8. 0.8 mm.
10. 613 nm.
12. 450 nm.
14. – 1.7%.
16. $\theta_{d1} = 72.2°$ from the normal, $\theta_{d2} = 65.2°$ from the normal.
18. 1.35°.
20. 2.4 m.
22. 8.46 cm.
24. 2.2 cm.
26. 6.49°.
28. 2.20 μm, no fourth order.
30. 2.07×10^3 lines/cm.
32. Two orders.
34. 0.84 m.
36. 529 nm, 775 nm.
38. Orange-red.
40. 169 nm.
42. 8.5 μm.
44. 113 nm, 225 nm.
46. 1.39.
48. 32 m.
50. 80.1 μm.
52. 1.00554.
54. 0.058.
56. 61.2°.
58. 45°.
60. 60°.
62. $\theta_p = 36.9°$, $\theta_c = 48.8°$, $\theta_p' = 53.1°$.
64. 0.5 cm.
66. 600 nm.
68. The 600 nm to 750 nm of the second order overlaps with 400 nm to 500 nm of the third order.
70. 19°, 41°, 78°, 102°, 139°, 161° above and below; 9.4°, 29°, 55°, 125°, 151°, 171° above and below.
72. 36.9°.
74. No film with $n <$ glass is possible, 2.00.
76. Central maximum at $\theta = 30°$, $\sin\theta = \sin 30° - (m\lambda/D)$, where $m = \pm 1, \pm 2, \ldots$.
78. (*a*) 81.5 nm; (*b*) 127 nm.
80. $\sin\theta_m = \sin\theta_i \pm (m\lambda/d)$, where $m = 0, 1, 2, \ldots$.
82. $d_1 = \frac{1}{2}(\lambda/n_1)$, $d_2 = \frac{1}{4}(\lambda/n_2)$.

CHAPTER 25

2. $f/2.3$.
4. $f/4$.
6. (1/6) s.
8. 6.87 m $\leq d_o \leq \infty$.
10. $f/4.0$.
12. + 1.2 D.
14. (*a*) Nearsighted; (*b*) 22 cm.
16. 0.3 cm.
18. (*a*) 2.0 cm; (*b*) 1.9 cm.
20. 8.3 cm.
22. 7.1 cm from the lens, 3.5×.
24. (*a*) – 49.4 cm; (*b*) 4.67×.
26. (*a*) 6.0×; (*b*) 1.9×.
28. – 29×, 83 cm.
30. 24 cm.
32. – 50×.
34. – 78×.
36. 6.91 mm, 1.24 m.
38. 450×.
40. (*a*) 744×; (*b*) 2.08 cm, 0.284 cm; (*c*) 0.289 cm.
42. (*a*) 0.85 cm; (*b*) 250×.
44. (*a*) converging; (*b*) 1.3 m.
46. 2.64×10^{-7} rad = $(1.51 \times 10^{-5})°$.
48. (*a*) 1.39; (*b*) 240 nm.
50. 347 nm, will be resolved by violet light, will not be resolved by red light.
52. 160×.
56. 0.245 nm.
58. Cannot be separately determined.
60. (*a*) Farsighted; (*b*) 31 cm; (*c*) 89 cm; (*d*) 15 cm.
64. – 0.56 D (upper part), + 1.7 D (lower part).
66. 1.0×.
68. 820 lines/mm, 98 lines/mm.
70. 34 cm.
74. 79.4 cm, 75.5 cm.

CHAPTER 26

2. 91.5 m.
4. 49.9 ly.
6. 0.773c.
8. 0.141c.
10. (*a*) 78.9 yr; (*b*) 24.6 yr; (*c*) 23.4 ly; (*d*) 0.950c.
12. 0.789c = 2.37×10^8 m/s.
14. 0.866c.
16. 6.8×10^{-8} %.
18. 1.6×10^{-13} J (1.0 MeV).
20. 3.56×10^{-28} kg.
22. 939 MeV/c^2.
24. 9.0×10^{13} J, 9.2×10^{10} kg.
26. (*a*) 13.9 GeV (2.23×10^{-9} J); (*b*) 7.91×10^{-18} kg · m/s.
28. 760 MeV (1.22×10^{-10} J), 7.55×10^{-19} kg · m/s.
30. 0.377c.
32. $1.05m_0$, 0.302c.
34. 3.34×10^{-27} kg, 0.866c.
36. (*a*) 7.6×10^{19} J; (*b*) – 4%.
38. 237.04832 u.
42. 234 MeV.
46. (*a*) 0.80c; (*b*) – 0.80c.
48. (*a*) 0.98c; (*b*) – 0.64c.
50. 3.8×10^{-5} s.
52. $\rho_0/[1-(v^2/c^2)]$
54. 3.5×10^{-8} kg.
56. (*a*) 1.17 s; (*b*) 2.68 s.
58. 0.78 MeV.
60. Electron.
62. 5.19×10^{-13} %.

CHAPTER 27

2. 4.76×10^6 m/s, 9.34 mm.
6. (*a*) 1.93×10^5 K; (*b*) 1.45 μm.
8. 5.4×10^{-20} J = 0.34 eV.
10. (*a*) 114 J; (*b*) 228 J; (*c*) 342 J; (*d*) 114n J; (*e*) – 456 J.
12. 1.77 eV < E < 3.10 eV.
14. 6.22×10^{-3} nm, insignificant diffraction.
16. 6.5×10^{14} Hz.
18. 0.70 eV, 5.0×10^5 m/s.
20. (*a*) 2.18 eV; (*b*) 0.92 V.
22. 4.11 eV.
24. 1.88 eV, 43.3 kcal/mol.
26. 1.52 MeV.
28. 6.6×10^{-16} m.
30. 1.71 MeV, 7.24×10^{-13} m.
32. (*a*) 7.12×10^{-13} m; (*b*) 2.43×10^{-12} m; (*c*) 4.86×10^{-12} m.
36. 3.2×10^{-32} m.
38. 29 V.
40. (*a*) 0.39 nm; (*b*) 0.12 nm; (*c*) 0.039 nm.
42. 1.84×10^3.
44. 3.3×10^{-38} m/s, no diffraction.
46. 150 V.
48. (*a*) Absorption, final state; (*b*) emission, initial state; (*c*) absorption, final state; $n = 1$ to $n' = 3$.
50. $n = 6$ to $n' = 3$.
54. (*a*) 486 nm; (*b*) 102 nm; (*c*) 434 nm.
56. 91.2 nm.
60. 4.35×10^3, -7.2×10^{-7} eV.
62. 1.20×10^{29} m, -4.23×10^{-97} J.
66. 1.1 mm.
68. 0.108 nm.
70. 4.66×10^{26} photons/s.
72. 5.3×10^{18} photons/s.
74. (*a*) 33 V; (*b*) 57 kV.
76. -1.36×10^{-8}
78. (*a*) 4.12×10^{16} rad/s, 6.55×10^{15} Hz; (*b*) 5.15×10^{15} rad/s, 8.19×10^{14} Hz.
80. 653 nm, 102 nm, 122 nm.
82. (*b*) 0.046 nm.
84. 0.039 eV, 0.15 nm.
86. 7 N.

CHAPTER 28

2. 1.1×10^{28} m.
4. 5.8×10^3 m/s.
6. 2.65×10^{-25} s.
8. 1.30×10^{-54} kg.
10. 21 MeV.
12. 7.13×10^{-10} m.
14. $\ell = 0, 1, 2, 3, 4, 5$.
16. 14 electrons.
20. $n \geq 5$, $m_\ell = -4, -3, -2, -1, 0, 1, 2, 3, 4$, $m_s = -\frac{1}{2}, +\frac{1}{2}$.
22. 3.66×10^{-34} kg · m²/s.
24. (*a*) $1s^2 2s^2 2p^6 3s^2 3p^6 3d^7 4s^2$; (*b*) $1s^2 2s^2 2p^6 3s^2 3p^6 3d^{10} 4s^2 4p^6$; (*c*) $1s^2 2s^2 2p^6 3s^2 3p^6 3d^{10} 4s^2 4p^6 5s^2$.
26. (*a*) Forbidden; (*b*) allowed; (*c*) forbidden; (*d*) forbidden; (*e*) allowed.
28. 3.4 eV.
32. 0.041 nm, 1 nm.
36. 0.23 nm.
38. Chromium.
40. 5.6×10^{-4} rad, 1.7×10^2 m.
42. (*a*) 6.6×10^{-8} eV; (*b*) 6.5×10^{-9}; (*c*) 7.9×10^{-7} nm.
44. (*a*) 2.5×10^{74}; (*b*) 5.0×10^{74}
46. 6.5×10^{-13} m, 91 keV.
48. 2869 $m^{-1/2}$.

CHAPTER 29

2. 0.70 eV.
4. 1 kcal/mol = 0.0435 eV/molecule, 102 kcal/mol.
8. 1.10×10^{-10} m.
10. (*a*) 1.52×10^{-2} eV, 0.082 mm;
    (*b*) 3.04×10^{-2} eV, 0.041 mm;
    (*c*) 4.56×10^{-2} eV, 0.027 mm.
12. (*a*) 2.68×10^{-47} kg · m^2; (*b*) 1.3×10^{-10} m.
14. (*b*) 4.5×10^{-48} kg · m^2; (*c*) 1.94×10^{-46} kg · m^2;
    (*d*) 1.3×10^{-45} kg · m^2; (*e*) 1.45×10^{-46} kg · m^2.
16. 0.34 nm.
18. 0.315 nm.
20. 1.94 eV.
22. $\lambda \leq 1.7\ \mu$m.
24. (*a*) $2N$; (*b*) $6N$; (*c*) $6N$; (*d*) $2N(2\ell + 1)$.
26. 5×10^6.
28. 1.91 eV.
30. 13 mA.
32. (*a*) 3.3 mA; (*b*) 6.7 mA.
34. (*a*) 9.4 mA (smooth); (*b*) 6.7 mA (rippled).
36. 4.0 kW.
38. 0.40 mA.
40. (*a*) 3.5×10^4 K; (*b*) 1.2×10^3 K.
42. 1.24 eV.
44. ± 1.4%.
46. 1.3×10^{-33} J, no.
48. 2.8×10^4 J/kg.

CHAPTER 30

2. 0.149 u.
4. 5.3%.
6. (*a*) 4.8×10^{-15} m; (*b*) 29.
8. 29 MeV.
10. 340 MeV.
12. 2.22 MeV.
14. 32.0 MeV, 5.33 MeV.
16. 12.4 MeV, 7.0 MeV.
18. (*b*) Stable.
20. 18.6 keV.
24. (*a*) Not possible; (*b*) not possible; (*c*) not possible.
26. 4.4 MeV, 0, $0 \leq E_\nu \leq 4.4$ MeV.
28. 6.12 MeV, 0.27 MeV.
30. 5.41 MeV.
34. 98.2%.
36. 4.91×10^{-18} s^{-1}.
38. 1.6×10^9 decays/s.
40. 5.60×10^{19} nuclei.
42. 7 α particles, 4 β^- particles.
44. (*a*) 3.8×10^{16} nuclei; (*b*) 2.5×10^{15} nuclei;
    (*c*) 5.7×10^{13} decays/s; (*d*) 2.5 min.
46. 2.25×10^{-10} g.
48. 2.9 min.
50. 53 min.
52. 1.1×10^9 yr.
54. 272 days, 2.7×10^{-17} kg.
56. 1.8×10^4 yr.
60. 0.25 MeV.
62. 1.9×10^4 yr.
64. (*a*) 7.2×10^{55}; (*b*) 1.2×10^{29} kg; (*c*) 3.2×10^{11} m/s^2.
66. 2.22 MeV.
68. 0.19 mg, 1.5 g.
70. Calcium, stored by the body in bones, 193 yr.
    ${}^{90}_{38}\mathrm{Sr} \rightarrow {}^{90}_{39}\mathrm{Y} + {}^{0}_{-1}\mathrm{e} + \bar{\nu}$, ${}^{90}_{39}\mathrm{Y}$ is radioactive;
    ${}^{90}_{39}\mathrm{Y} \rightarrow {}^{90}_{40}\mathrm{Zr} + {}^{0}_{-1}\mathrm{e} + \bar{\nu}$ ${}^{90}_{40}\mathrm{Zr}$ is stable.
74. (*a*) 0.002602 u, 2.424 MeV/c^2; (*b*) 0;
    (*c*) – 0.094909 u, – 88.41 MeV/c^2;
    (*d*) 0.043924 u, 40.92 MeV/c^2;
    (*e*) $\Delta \geq 0$ for $0 \leq Z \leq 8$ and $Z \geq 85$, $\Delta < 0$ for $9 \leq Z \leq 84$.
76. 0.083%.

CHAPTER 31

2. No threshold energy.
4. 17.35 MeV is released.
6. 5.700 MeV is released.
8. (*a*) Can occur; (*b*) 19.85 MeV.
10. + 4.734 MeV.
12. (*a*) ${}^{7}_{3}\mathrm{Li}$; (*b*) neutron is stripped from the deuteron;
    (*c*) + 5.025 MeV, exothermic.
14. (*a*) ${}^{31}_{15}\mathrm{P}(\mathrm{p}, \gamma){}^{32}_{16}\mathrm{S}$; (*b*) + 8.864 MeV.
18. 173.2 MeV.
20. 0.116 g.
22. 25.
24. 1.3 keV.
28. 6.1×10^{23} MeV/g, 4.9×10^{23} MeV/g,
    2.1×10^{24} MeV/g, 5.1×10^{23} MeV/g.
32. 4.1×10^3 kg/h.
34. (*b*) 4.69 MeV is released; (*c*) + 1.94 MeV, + 1.20 MeV,
    + 7.55 MeV, + 7.30 MeV, + 1.73 MeV, + 4.97 MeV;
    (*d*) greater repulsion.
36. 500 rad.
38. 167 rad.
40. 120 counts/s.
44. 0.225 μg.
46. (*a*) 0.028 mrem ≈ 0.006% of allowed dose;
    (*b*) 0.28 mrem ≈ 0.06% of allowed dose.

48. 7.05 m, FM radio wave.
50. 8.638×10^{-8} keV/K.
52. (*a*) 1.2 kg; (*b*) 1.1 g.
54. 3×10^{16} kg, 3×10^{30} J.
56. (*a*) ${}^{226}_{88}\text{Ra} \rightarrow {}^{222}_{86}\text{Rn} + {}^{4}_{2}\text{He}$; (*b*) 4.871 MeV;
(*c*) 1.02×10^{-19} kg · m/s; (*d*) 0.088 MeV.
58. 6.26×10^{14} J/kg, $\approx 10^7\times$ the heat of combustion of coal.
60. 2.1×10^{24} neutrinos/yr.

CHAPTER 32

2. 3.1×10^{-17} m.
4. 21 μs.
6. $\lambda_\alpha = 2.6 \times 10^{-15}$ m $\approx$ size of nucleon,
$\lambda_p = 5.2 \times 10^{-15}$ m $\approx$ 2(size of nucleon),
α particle is better.
8. 2.8×10^2 rev.
10. 1.4×10^{-18} m.
12. 0.78 MeV/turn.
18. 67.5 MeV.
20. 41.1 MeV.
22. 0.40 fm.
24. 2.3×10^{-18} m.
28. 69.3 MeV.
30. 8.6 MeV, 57.4 MeV.
34. 2.38×10^{-21} s.
36. 140 MeV, 4.7×10^{-24} s.
38. (*a*) n = d d u; (*b*) $\bar{\text{n}} = \bar{\text{d}}\bar{\text{d}}\bar{\text{u}}$; (*c*) Λ^0 = u d s; (*d*) $\bar{\Sigma}^0 = \bar{\text{u}}\bar{\text{d}}\bar{\text{s}}$.
40. $\text{D}^0 = \text{c}\bar{\text{u}}$.
42. (*a*) 2×10^{14} GeV; (*b*) 10^{27} K.
46. 5.5 T.
48. (*a*) Possible, strong interaction;
(*b*) possible, strong interaction;
(*c*) possible, strong interaction;
(*d*) forbidden, charge is not conserved;
(*e*) possible, weak interaction;
(*f*) forbidden, charge is not conserved;
(*g*) possible, strong interaction;
(*h*) forbidden, strangeness is not conserved;
(*i*) possible, weak interaction;
(*j*) possible, weak interaction.
52. – 135.0 MeV, – 140.9 MeV.
54. – 535.5 MeV, 768.0 MeV.
56. 64.

CHAPTER 33

2. 12 ly.
4. (*a*) 0.042″; (*b*) $(1.2 \times 10^{-5})°$.
6. 82 yr.
8. (*a*) 1.3×10^3 W/m^2; (*b*) 3.7×10^{26} W.
10. 2.9°, $\phi_{\text{Galaxy}} \approx 6\phi_{\text{Moon}}$.
12. 4.8″.
14. 5.4×10^{17} kg/m^3, $\rho_{\text{ns}} = (3 \times 10^8)\rho_{\text{dwarf}}$,
$\rho_{\text{ns}} \approx \rho_{\text{nuclear matter}}$.
16. (*a*) 13.93 MeV; (*b*) 4.7 MeV; (*c*) 3×10^{10} K.
18. $D_2 = 4D_1$.
22. 4×10^{14} m.
24. 1.2×10^8 ly.
26. 0.8*c*.
28. (*a*) 656 nm; (*b*) 661 nm; (*c*) 2.17 μm.
32. 1.1 mm.
34. (*a*) $\approx 10^{-3}$; (*b*) $\approx 10^{-10}$; (*c*) $\approx 10^{-13}$; (*d*) $\approx 10^{-27}$.
36. (*a*) temperature increases, luminosity is constant, and size decreases; (*b*) temperature is constant, luminosity decreases, and size decreases; (*c*) temperature decreases, luminosity increases, and size increases.
38. 8×10^3 rev/s.
40. 6.9×10^{24} W.
42. (*a*) 46 eV; (*b*) 18 eV.
44. $\approx (2 \times 10^{-5})°$.
46. 1.4×10^{16} K, hadron era.
48. Venus is brighter, $\ell_V/\ell_S = 16$.

COMPARISON OF PROBLEMS WITH 4th EDITION

CHAPTER 1

4th Ed.	Action	5th Ed.
	N	1
1	C	4
2	C	3
3	C	2
4		5
5	C	6
6		7
	N	8
	N	9
7		10
8		11
9	C	12
10	C	13
11		14
12		15
13		16
14	C	17
	N	18
	N	19
15		20
16		21
	N	22
	N	23
17		24
18		25
19		26
20		27
21	C	29
22		28
23		30
	N	31
24		32
25		33
26		41
27		34
28		35
29		36
30	C	37
31	C	38
32		39
	N	40
	N	42
	N	43
	N	44

CHAPTER 2

4th Ed.	Action	5th Ed.
1	C	1
2	X	
3	C	2
4	C	3
5	C	4
6	X	
	N	5
	N	6
7	C	7
8	C	8
9		11
10	C	10
11	C	9
12		12
13	C	13
14	C	14
15	C	15
16		16
17		17
18		18
19	C	19
20		20
21		21
22		22
23		23
24	C	24
25	C	25
26	C	26
27		27
28		28
29		29
30	C	30
31		31
32		32
33	C	33
34	C	34
35		35
36	C	36
37	C	37
38	C	38
39		39
40		40
41	C	41
42		72
43		42

4th Ed.	Action	5th Ed.
44		43
45	C	44
46		45
47		46
48		47
49		48
50		49
51		50
52		51
53		52
54		53
55		54
56		55
57		56
58		57
59		58
	N	59
60		60
61		61
62		62
63		63
64	X	
65		64
66		65
67		66
68		67
69		68
70	X	
71		69
72		70
	N	71
	N	73

X = deleted C = changed N = New

COMPARISON OF PROBLEMS WITH 4th EDITION

CHAPTER 3

4th Ed.	Action	5th Ed.
1	C	1
2	C	2
3		3
4	X	
5	C	4
6	C	6
7	C	7
8	C	5
9	C	8
10	C	9
11	C	10
12	X	
13	C	11
14	C	12
15	C	13
16	C	14
17	C	15
18	C	16
19	C	17
20	C	18
21		40
22	C	41
	N	42
23	C	43
24	C	44
25		45
	N	46
26		47
27		48
28		49
29		50
30		51
31		52
32		55
33		53
34		54
35		56
36		57
37	C	19
38		20
39		69
40	C	21
	N	22
	N	23
41	C	24

4th Ed.	Action	5th Ed.
42		25
43		26
44	X	
	N	27
45		28
	N	29
46		33
47		34
48		58
49		30
	N	31
	N	32
50		35
51	C	36
52	C	37
53	C	38
54		39
55		59
56		60
57		61
	N	62
58		63
59	C	64
60	C	65
61		66
62		67
63	C	68
64		70
65		72
66		71
67	X	
	N	73

CHAPTER 4

4th Ed.	Action	5th Ed.
1	C	4
2	C	1
3	C	2
4		3
5	C	5
6		6
7	C	7
8	C	8
9	C	9
10	C	10
11		11
12	C	12
13	C	13
14	C	14
15	C	15
	N	16
16	C	18
17	C	17
	N	19
18	C	20
19	C	21
20	C	22
21		23
22	C	24
	N	25
23		26
24	C	27
25	C	28
26	C	29
27	C	30
28	X	
29	C	31
30	C	32
31		33
	N	34
32	C	35
33	C	36
34		37
35	C	38
36	C	39
37		40
38	X	
39	X	
	N	41
40		42

4th Ed.	Action	5th Ed.
41	X	
42		43
43	C	44
44	C	45
45	C	46
46		47
	N	48
	N	49
47		50
48		51
49	C	52
50	C	53
51	C	54
52	C	55
53		56
	N	57
	N	58
54		59
	N	60
55		61
56		62
57		63
58		64
59		65
60		66
61		67
62		68
63		69
64		70
65	C	71
66	C	72
67		74
68		73
69		75
70	C	76
71		77
72		78
	N	79
	N	80
	N	81

X = deleted C = changed N = New

COMPARISON OF PROBLEMS WITH 4th EDITION

CHAPTER 5

4th Ed.	Action	5th Ed.
1	C	1
2	C	2
3		3
4	C	4
5		5
6	C	8
7	C	6
8	C	7
9	C	9
10		10
11	C	11
12	C	12
	N	13
	N	14
13		15
14		16
15		18
16		19
17	C	20
18		17
	N	21
19		22
20		23
21		24
22	X	
23	X	
24		25
25		26
26		27
27	C	28
28	C	29
	N	30
29		31
30		32
31		33
32		34
33	C	35
	N	36
34		37
35		38
36		39
37		40
38	C	41
39		42
40		43

4th Ed.	Action	5th Ed.
41		44
	N	45
42	C	46
43	C	47
44		48
45		49
46	C	50
47		51
48		52
49		53
50		54
51		55
52		56
53		57
54		58
55		59
56		60
	N	61
57		62
58		63
59		64
60		65
61		66
62		67
63	C	68
	N	69
64		70
65		71
66	C	72
67		73
68		74
69		75
70		76
	N	77
	N	78

CHAPTER 6

4th Ed.	Action	5th Ed.
1	C	1
2	C	2
3		3
4		4
5	C	5
6	C	6
7		7
	N	8
8	X	
9		9
10	C	10
11		11
12		12
13		13
14	C	14
15		15
16	C	16
17	X	
18		17
19		18
20		19
21		20
	N	21
22		22
23		23
24		24
25		25
26	C	26
27	C	27
28		28
29	C	29
30		30
31	C	31
32		32
33		33
34		34
35		35
36	C	36
37	C	37
38		38
39		39
40		40
41	C	41
	N	42
42	C	43

4th Ed.	Action	5th Ed.
43		44
44		45
45		46
46		47
47	C	48
48	C	49
49	C	50
50		51
51	C	52
52		53
53		54
54	C	55
55		56
56		57
57	C	58
58		59
59		60
60		61
61		62
62		63
63		64
64		65
65	C	66
66	C	67
67		68
	N	69
68		70
69	C	71
70		72
71		73
72		74
73		75
74		76
75		77
76		78
77		79
78		80
79	X	
80		81
81	C	82
	N	83
	N	84
	N	85
	N	86

X = deleted C = changed N = New

COMPARISON OF PROBLEMS WITH 4th EDITION

CHAPTER 7

4th Ed.	Action	5th Ed.
1	C	1
2	C	3
3	C	2
4	C	6
5	C	5
6	C	4
7		8
8	C	7
9		9
10	C	10
11	C	11
12	C	12
13	C	13
14		14
15		15
16	C	16
17		18
18		17
19	C	19
20		20
21		21
22		22
23	C	24
24		23
25	C	25
26		26
27	X	
28	X	
29		27
30		28
	N	29
31		37
32		39
33		40
34	X	
35		45
36	C	44
37		42
38		43
39	C	30
40	C	31
41	C	32
	N	33
42	C	38
43		34

4th Ed.	Action	5th Ed.
	N	35
44	X	
45		41
46		36
47		46
48		47
49		52
50		53
51		54
52		55
53		48
54	C	49
	N	50
55		51
56		56
57		57
58		58
59		59
60		60
61		61
62	C	62
63		63
64		64
	N	65
65		66
66		67
67		68
	N	69
	N	70
68		71
69		72
70	C	73
71	C	74
72		75
	N	76
	N	77

CHAPTER 8

4th Ed.	Action	5th Ed.
1		1
2		2
	N	3
3		4
4		5
5	C	6
6	C	7
7		8
	N	9
8		10
9	C	11
10		12
11		13
12		14
13		15
14	C	16
15	C	17
16		18
17		19
18		20
19	C	21
20		22
21		23
22		24
23		25
24		26
25		27
26		28
27		29
28	C	30
29		31
30	C	32
31		33
32	C	34
33		35
34	C	36
35		37
36		38
37	C	39
38	C	40
39		41
40		42
41		43
42		44
43		45
	N	46
	N	47

4th Ed.	Action	5th Ed.
44		48
45	C	49
46		50
47	C	51
48		52
49	C	53
50		54
51	X	
52		55
53	C	56
54	C	57
55	C	58
56		59
57		60
58		61
59	C	62
60		63
61		64
62		65
	N	66
	N	67
63		68
64		69
65		70
66		71
67	X	
68	X	
69	X	
70	X	
71	X	
72	X	
73		72
74		74
75		73
	N	75
	N	76
76		77
77	X	
78		78
79	C	79
80		80
81		81
82	C	82
83	X	
84		83
85		84
86		85

X = deleted C = changed N = New

COMPARISON OF PROBLEMS WITH 4th EDITION

CHAPTER 9

4th Ed.	Action	5th Ed.
1	C	1
2		2
3	C	3
4	C	4
5		6
6	C	5
7	C	7
8		8
9		9
10	C	10
11	C	11
12	C	12
13	C	13
14		14
15		15
	N	16
16		17
17		18
18		19
19		20
20		21
21		22
22		23
23		25
24		24
25	C	26
26	C	27
27		28
28	C	29
29	C	30
30		31
31		32
32		33
33		34
34		35
35	C	36
36	C	37
37	C	38
38		39
	N	40
39		41
40	X	
41		
42		42
43		43

4th Ed.	Action	5th Ed.
44		45
45		46
46		47
47		44
48	C	48
49		50
50		51
51		52
52		49
53		53
54		54
55		55
56		56
57	C	57
58		58
59	C	59
60		60
61		61
62		62
63		63
	N	64
64		65
65		66
66		67
67		68
68		69
69		70
70		71
71		72
72	C	74
73		73
74		75
75		76
76		77
77		78
78		79
79		80
	N	81
	N	82

CHAPTER 10

4th Ed.	Action	5th Ed.
1		1
2	C	2
3	C	3
	N	4
4		5
5		6
6	X	
7		7
8		8
9		9
10	C	10
11	C	11
12	X	
13		12
14		13
15	C	14
16		15
17		16
18		17
19	C	18
20	X	
21		19
22		20
23		21
24		22
25		23
26	C	24
27		25
28		26
29		27
30		28
31	C	29
32		30
33		31
34	C	32
35		33
36		34
37	C	35
38	C	36
39	C	37
40		38
41	C	39
42	C	40
	N	41
43		42
44		43
45		44
46		45

4th Ed.	Action	5th Ed.
47		46
48	X	
49	C	47
50	C	48
51		51
52	C	49
53		50
	N	52
	N	53
54		
55		54
56	X	
57	X	
58	X	
59	X	
60	X	
61	X	
62		55
63	C	56
64		57
65		58
66		59
67	X	
68		60
69	C	61
70	C	62
	N	63
71		64
72		65
	N	66
73		67
74		68
75	X	
76		69
77		70
78		71
79	C	72
80		73
81	C	74
82		75
83	C	76
	N	77
84		78
85		79
	N	80

X = deleted C = changed N = New

COMPARISON OF PROBLEMS WITH 4th EDITION

CHAPTER 11

4th Ed.	Action	5th Ed.
1	C	1
2	C	2
3	C	3
4	C	4
5		5
6	C	6
7		7
8	C	8
9		9
10		10
11		11
12		12
13	X	
14		13
15		14
16		15
17		16
18	C	17
19	C	18
20	C	19
21	C	20
22	C	21
23	C	22
24		23
25		24
	N	25
	N	26
26	X	
27		27
28	C	28
29	X	
	N	29
30		30
31		31
32		32
33		33
34	C	34
35		35
36		36
37	C	37
38		38
39		39
40	C	40
41		41
42		42

4th Ed.	Action	5th Ed.
43		43
44		47
45		48
46	C	49
47		44
48		45
49		46
50		50
51		51
52		52
53		53
54	C	54
55	C	55
56		56
57		57
58		58
59		59
60	C	60
61	C	61
62		62
63		63
64	C	64
65	C	65
66	C	66
67		67
68		68
	N	69
	N	70
69		71
70		72
71	C	73
72	C	74
73		75
74	C	76
75		77
76	C	78
77		79
78		80
79		81
80		82
81	X	

CHAPTER 12

4th Ed.	Action	5th Ed.
1	X	
2	C	1
3	C	2
4		3
	N	4
5	C	6
6		5
7		7
8		8
9		9
10		10
11		11
12		13
13		12
14	C	15
15		14
16	X	
17	C	20
18		18
19		21
20		22
21		16
22		17
23		19
24		24
25		23
26	C	25
27	C	26
28		27
29		29
30		28
	N	30
31		32
32	C	31
33		33
34		34
35		36
36		35
37		38
38	C	37
39		39
40		40
41		41
42	X	
43		42

4th Ed.	Action	5th Ed.
44		44
45		43
	N	45
46		47
47		46
48	X	
49		48
50		49
51		50
52	C	51
53		52
54		53
55		55
56		54
57		56
58		57
59		58
60		59
61	C	61
62		60
63		62
64	C	63
65		64
66		65
	N	66
	N	67
67		68
68		69
	N	70
69		71
70		72
71		73
72		74
73		75
74		76
75		77
76		83
77	C	78
78		79
79		81
80		80
81	C	82
	N	84
	N	85
	N	86

X = deleted C = changed N = New

COMPARISON OF PROBLEMS WITH 4th EDITION

CHAPTER 13

4th Ed.	Action	5th Ed.
1		1
	N	2
2		3
3	C	4
	N	5
4	C	6
5		7
6		8
7	C	9
	N	10
8		11
9		12
10		13
11		14
12	C	15
	N	16
13	C	17
14		18
15		19
16	C	20
17		21
18	C	22
19	C	23
20		24
21		25
22	C	26
23		27
24	C	28
25		29
26		30
27		31
28		32
29	C	33
30	C	34
31		35
32		36
33		37
34		38
35		39
36		40
37	C	41
38		42
39		43
40		44
41	C	45
	N	46
42		47
43		48
44	C	49
45	C	50
46	C	51
47		52
48		53
49		54
50	X	
51		55
52		56
53		57
54		58
55	C	59
56	C	60
57		61
58	C	62
59		63
60		64
61	C	65
62	C	66
63		67
64		68
65		69
66		70
67		71
68		72
69		73
70	C	74
71		75
72	C	76
73		77
	N	78
74		79
75		80
76		81
77	C	82
78		83
	N	84
79	C	85
80		86
81		87
82	C	88
83	C	89
84		90
85		91
86		92
	N	93

CHAPTER 14

4th Ed.	Action	5th Ed.
1	C	1
2		2
3		3
	N	4
4		5
5	C	6
6	C	7
7	C	8
8		10
9		9
10		11
11		12
12	C	13
13	C	14
14	C	15
15		19
16	C	16
17		17
18	C	18
19		21
20		20
21	X	
22		22
23		23
24		24
25		25
26	C	26
	N	27
27		28
28		29
29	C	30
	N	31
30		32
31	C	33
32		34
33	C	35
34		36
35		37
	N	38
	N	39
36		40
37		41
38	X	
39		42
40		43
41		44
42	C	45
43	C	46
44	C	47
45	C	48
46		49
47		50
48	C	51
49		52
50		53
51		54
52	C	55
53		56
54		57
55	C	58
56	C	59
57		60
58		61
	N	62

X = deleted C = changed N = New

COMPARISON OF PROBLEMS WITH 4th EDITION

CHAPTER 15

4th Ed.	Action	5th Ed.
1		5
2	C	1
3		3
4		4
5	C	2
6	C	6
7		7
8	C	8
9	C	9
	N	10
10	C	12
11		11
12		13
13		14
14		15
15		16
16	C	17
17		18
18	C	19
19	C	20
20	C	21
21	C	22
22		23
23		24
24	C	25
25	C	26
26	C	27
	N	28
27		29
28		30
29	C	31
30	C	32
31	C	33
32		34
33		35
34		36
35		37
36	C	38
37		39
38	C	40
39	C	41
40		42
41		43
42		44
43	X	
44		45
45	C	46
46		47
47	C	48
48		49
49		50
50		51
	N	52
51		53
	N	54
	N	55
52		56
53	C	57
54	C	58
55		59
56	C	60
57	C	61
58	C	62
59	C	63
60	C	64

CHAPTER 16

4th Ed.	Action	5th Ed.
1	C	1
2	C	2
3	C	3
4	C	4
5		5
6		6
7	C	7
8		8
9	C	9
10		10
11		11
12	C	12
13	C	13
14	C	14
	N	15
15		16
16		17
17	C	19
18	C	20
19	C	21
20		22
21		223
22		24
23	C	25
24		26
25		27
26	C	28
27		29
28		30
29		31
30		32
31	X	
32		33
33		34
34		35
35		36
36		37
37	C	38
38		39
39		40
40		41
41	C	42
42		43
43		44
44		45
45	C	46
46		47
47		18
48		48
49	C	49
50	C	50
51	C	51
52		
53		52
54	C	53
	N	54
	N	55
	N	56

X = deleted C = changed N = New

COMPARISON OF PROBLEMS WITH 4th EDITION

CHAPTER 17

4th Ed.	Action	5th Ed.
1	C	1
2		2
3	C	3
4	C	4
5	C	5
6		6
7	C	7
8		8
9	C	9
10	C	10
11	C	11
12		12
13	C	13
14	C	14
15	C	15
16		16
17	C	17
18		18
19	C	19
20	C	20
21	C	21
22		22
23		23
24		24
25		25
26		26
27		27
28		28
29		29
30	C	30
31	C	31
32	C	32
33		33
34	C	34
35	C	35
36	C	36
37	C	37
38	C	38
39	C	39
40	C	40
41	C	41
42		42
43	C	43
44	X	
	N	44

4th Ed.	Action	5th Ed.
45		45
46		46
47		47
48		48
49	C	49
50		50
51		51
52	C	52
53		53
54	C	54
55		55
56		56
57	C	57
58	C	58
59		59
60	C	60
61		61
62	C	62
63	C	63
64		64
65		65
66	C	66
67	X	
	N	67
	N	68
	N	69
	N	70
	N	71

CHAPTER 18

4th Ed.	Action	5th Ed.
1	C	1
2		2
3	C	3
4	C	4
5		5
6		6
7	C	7
8	C	8
9		9
10		10
11	C	11
12		12
13		13
14	C	14
	N	15
15		16
16		17
17		18
18		19
19		20
20		21
21		22
22		23
23	C	24
24		25
25	C	26
26	C	27
	N	28
27	C	29
28		30
	N	31
29	C	32
30		33
31	C	34
32		35
33		36
34	C	37
35	C	38
36		39
37	C	40
38	C	41
39		42
40		43
41		44
42	C	45

4th Ed.	Action	5th Ed.
43	C	46
44	C	47
45		48
46		49
47	C	50
48		51
49		52
50		53
51		54
52		55
53		56
54	C	57
55	C	58
56		59
57		60
58		61
59		62
60		63
61		64
62		65
63		66
64	C	67
65		68
66	C	69
67	C	70
68		71
	N	72
	N	73
	N	74
	N	75

X = deleted C = changed N = New

COMPARISON OF PROBLEMS WITH 4th EDITION

CHAPTER 19

4th Ed.	Action	5th Ed.
1	C	1
2		2
3		3
4		4
5		5
6	C	6
7	C	7
8	C	8
9		9
10		10
11		11
12		12
13		13
14	C	14
15	C	15
16	C	16
17		17
18	C	18
19		19
20		20
21		21
22		22
23		23
24		24
25		25
26		26
27		27
28		28
29		29
30	C	30
31	C	31
32		32
33	C	33
34		34
35		35
36		36
37	C	37
38	C	38
39	C	39
40		40
41	C	41
42		42
43	C	43
44		45
45		44
	N	46
	N	47
46		48
47		49
48		50
49	C	51
50		52
51	C	53
52		54
53	C	55
54		56
55		57
56	C	58
57		59
58		60
59		61
60		62
61		63
62	C	64
63		65
64		66
65	C	67
66		68
67		69
68	C	70
69		71
70		72
71	C	73
72	X	
73		74
74		75
75		76
76	C	77
77	C	78
78		79
79		80
80		81
81		82
82		83
	N	84
	N	85
	N	86

CHAPTER 20

4th Ed.	Action	5th Ed.
1	C	1
2		2
3		4
4	C	3
5	C	5
6	C	6
7		7
8		8
9	C	9
10	C	10
11	C	11
12	C	12
13	C	13
14		14
15		15
16		16
17	C	17
18		18
19	C	19
20		20
21	C	21
22		22
23		23
24	C	24
25		25
26	C	26
27		27
28	C	28
29		29
30	C	30
31		31
32		32
33		33
34	C	35
35		34
36	C	36
37		37
38		38
39		39
40		40
41		41
42		42
43	C	43
44	C	44
45		45
46		46
47		47
48		48
49		49
50	C	50
51		51
52	C	52
53	C	53
54		54
55		55
56		56
	N	57
	N	58
	N	59
57		60
58	C	61
59	C	62
60	C	63
61		64
62		65
63		66
64		67
65	C	68
66		69
67	C	70
68	C	71
69		72

X = deleted C = changed N = New

COMPARISON OF PROBLEMS WITH 4th EDITION

CHAPTER 21

4th Ed.	Action	5th Ed.	4th Ed.	Action	5th Ed.	4th Ed.	Action	5th Ed.
1		2	44	C	46		N	90
2	C	1	45	X		89		91
3		3	46		48	90		92
4		4	47	C	47		N	93
5	C	5	48		49	91	C	94
6	C	6	49		50	92	C	95
7		7	50		51	93		96
8		8	51		52	94	C	97
9		9	52		53	95	C	98
10	C	10	53	C	54	96	C	99
11		11	54	C	55	97	C	100
12	C	13	55		56	98		101
13	C	12	56	C	57	99		102
14	C	14	57		58		N	103
	N	15	58		59			
15	C	16	59		60			
16		17	60	C	61			
17	C	18	61	C	62			
18		19	62		63			
19	C	20	63		64			
20		21	64	C	65			
21	C	22	65	C	66			
22	C	23	66		67			
23	C	24	67		68			
24	C	25	68		69			
25		26	69		71			
26		27	70		70			
27		28	71		72			
	N	29	72	C	73			
28		30	73	C	74			
29		31	74	C	75			
30	C	32	75	C	76			
31		33	76		77			
32		34	77	C	78			
33		36	78	C	79			
34		35	79		80			
35	C	37	80		81			
36		38	81		82			
37		39	82	C	83			
38		40	83	C	84			
39		41	84		85			
40	C	42	85		86			
41	C	43	86		87			
42	C	44	87		88			
43	C	45	88		89			

X = deleted C = changed N = New

COMPARISON OF PROBLEMS WITH 4th EDITION

CHAPTER 22

4th Ed.	Action	5th Ed.
1	C	1
2	C	2
3	C	3
4	C	4
5		5
6		6
7	C	7
8	C	8
9		9
10		10
11		11
12	C	12
13		13
14	C	14
Ch 23–3		15
Ch 23–4		16
	N	17
	N	18
Ch 23–7		19
15		20
	N	21
16	C	22
17	C	23
18	C	24
19		25
20		26
21	C	27
22		28
23	C	29
24	C	30
25	C	31
26	C	32
27		33
28		34
29	C	35
30	C	36
31		37
32		38
33		39
34		40
35		41
36		42
37	C	43
38		44
39		45
	N	46
40		47
41	C	48
42		49
43	C	50
44	C	51

CHAPTER 23

4th Ed.	Action	5th Ed.
1		26
2		27
3		Ch 22–15
4		Ch 22–16
5		28
6		29
7		Ch 22–19
8	C	1
9		2
10		3
11		4
	N	5
12	C	6
13		7
14		8
15	C	9
16	C	10
17		11
18	C	12
19	C	13
20	C	14
21	C	15
22		16
	N	17
23	C	18
24		19
25		20
26		21
27	C	22
28		23
29	C	24
30	C	25
31	C	30
32	C	31
33	C	32
34	C	33
35		34
36	C	35
37		36
38		37
39		38
40		39
41		40
42	C	41
43	C	42
	N	43
44		44
45		45
46		46
47		47
48		48
49	C	49
50	C	50
51	C	51
52	C	52
53		53
54	C	54
55	C	55
56	C	56
57		57
58		58
59	C	59
60		60
61	C	61
62		62
63		63
64		64
65		65
66		66
67		67
68	C	68
69		69
70		70
71	C	71
72		72
73		73
74		74
75		75
76	X	
	N	76
	N	77
	N	78
77	C	79
	N	80
78	C	81
79		82
80		83
81	C	84
82	X	
83		85
84		86
85		87
86	X	
	N	88
87	C	89
88		91
89		90

X = deleted C = changed N = New

COMPARISON OF PROBLEMS WITH 4th EDITION

CHAPTER 24

4th Ed.	Action	5th Ed.
1		1
2	C	2
3	C	3
4		4
5	C	5
6	C	6
7	C	7
8	C	8
9		9
10		10
11	C	11
12	C	12
13		13
14	C	14
15		15
16	X	
17		16
18	C	17
19	C	18
20	C	19
21		20
22	C	21
23	C	22
24	C	23
25	C	24
26		25
27		26
28	C	27
29	C	28
30		29
31	C	30
32	C	31
33		32
34		33
35	C	34
36	C	35
37	C	36
38	C	37
39		38
40	C	39
41	C	40
42		41
43		42
44	C	43
45	C	44

4th Ed.	Action	5th Ed.
46		45
47		46
48		47
49	C	48
50	C	49
51	C	50
52		51
53		52
54		53
55		55
56		54
57		56
58		57
59		58
60		59
61		60
62		61
63		62
64		63
	N	64
65	C	65
66	C	66
67		67
68		68
69		69
70	C	70
71	C	71
72		72
73	C	73
74		74
75		75
76		76
77	C	77
78		78
79		79
80	X	
81		80
	N	81
	N	82
	N	83

CHAPTER 25

4th Ed.	Action	5th Ed.
1		1
2		2
3	C	3
4	C	4
5		5
6	C	6
7	C	7
8		8
9	C	9
10		10
11	C	11
12		12
	N	13
13	C	14
14	C	15
15		16
16	C	17
17		18
18	C	19
19		20
20	C	21
21	C	22
22	C	23
23		24
24	C	25
25	C	26
	N	27
26	C	28
27	C	29
28	C	30
29	C	31
30		32
31		33
32	C	34
33	C	35
34	C	36
35	C	37
36	C	38
37	C	39
38	C	40
39		41
40		42
41		43
42		44
	N	45

4th Ed.	Action	5th Ed.
43	C	46
44		47
45		48
46		49
47	C	50
48	C	51
49		52
50	C	53
51		54
52		55
53		56
54	C	57
55		58
56		59
	N	60
57	C	61
58		62
59		63
60		64
61	C	65
62	C	66
63		67
64		68
65	C	69
	N	70
66		71
67		72
	N	73
	N	74
	N	75

X = deleted   C = changed   N = New

COMPARISON OF PROBLEMS WITH 4th EDITION

CHAPTER 26

4th Ed.	Action	5th Ed.
1		1
2	C	2
3	C	3
4	C	4
5	C	5
6	C	6
7	C	7
8	C	8
9	C	9
10	C	10
11	C	11
12	C	12
13	C	13
14		14
15		15
16	C	16
17		17
18	C	18
19	C	19
20		20
21		21
22		22
	N	23
23		24
24		25
25	C	26
26		27
27	C	28
28	C	29
29	C	30
30	C	31
31		32
32		33
33		34
34		35
35	C	36
36	C	37
37		38
38		39
39		40
40	C	41
41	C	42
42		43
43		44
44		45
45		46
46	C	47
47		48
	N	49
	N	50
48	C	51
49		52
50		53
51	C	54
52		55
53	C	56
54	C	57
55		58
56		59
57		60
58		61
59		62
60	C	63
61		64
62		65
	N	66

CHAPTER 27

4th Ed.	Action	5th Ed.
1		1
2	C	2
3	C	3
4		4
5	C	5
6	C	6
7	C	7
8		8
9		9
10	C	10
	N	11
11		12
12		13
13		14
14		15
15	C	16
16	C	17
17	C	18
	N	19
18	C	20
19		21
20	C	22
21	C	23
22	C	24
23		25
24	C	26
25	C	27
26		28
27		29
28	C	30
29		31
30		32
31		33
	N	34
32		35
33	C	36
34	C	37
35	C	38
36		39
37		40
38		41
39		42
40		43
41		44
42		45
43		46
44	C	47
	N	48
45		49
46		50
47		51
48		52
49		53
50		54
	N	55
51		56
52		57
53		58
54		59
55		60
56		61
57		62
58		63
59		64
60		65
	N	66
61		67
62	C	68
63		69
	N	70
64		71
65	C	72
66	C	73
67	C	74
68		75
69	C	76
70		77
71		78
72		79
73		80
74		81
75	C	82
76		83
77		84
78		85
	N	86

X = deleted C = changed N = New

COMPARISON OF PROBLEMS WITH 4th EDITION

CHAPTER 28

4th Ed.	Action	5th Ed.
1		1
2		2
3		3
4		4
5		5
	N	6
6		7
7		8
8		9
9		10
10		11
11		12
12	X	
13		13
14		14
15		15
16		16
17		17
18		18
19		19
20		20
21		21
22		22
23		23
24		24
25		25
26		26
27		27
28		28
29		29
30		30
31		31
32		32
33		33
34		34
35		35
36		36
37		37
38		38
39		39
40		40
41		41
42		42
43		43
44		44
45		45
46		46
47		47
48		48
49		49
	N	50

CHAPTER 29

4th Ed.	Action	5th Ed.
1		1
2		2
3		3
4		4
5		5
6		6
7		7
8		8
9		9
10		10
11		11
12		12
13		13
14		14
15	C	15
16		16
17		17
18		18
19		19
20		20
21		21
22		22
23		23
24		24
25		25
26		26
27		27
28		28
29	C	29
30		30
	N	31
31	C	32
32		33
33	C	34
34	C	35
35	C	36
36	C	37
37	C	38
38		39
39		40
40	C	41
41		42
42		43
43	C	44
44		45
45		46
	N	47
	N	48

X = deleted C = changed N = New

COMPARISON OF PROBLEMS WITH 4th EDITION

CHAPTER 30

4th Ed.	Action	5th Ed.
1		1
2		2
3		3
4		4
	N	5
5	C	6
6		7
7		8
8		9
9		10
10		11
11		12
12		13
13		14
14		15
15		16
16		17
17		18
18		19
19		20
20		21
21		22
22		23
23		24
24		25
25		26
26		27
27		28
28		29
29		30
30		31
31		32
32		33
33		34
34		35
35		36
36		37
37	C	38
38		39
39	C	40
40		41
41		42
42		43
43		44
44	C	45

4th Ed.	Action	5th Ed.
45	C	46
46	C	47
47		48
48	C	49
49		50
50	C	51
51		52
52		53
53	C	54
54		55
55		56
56		57
57		58
58		59
	N	60
59		61
60		62
61		63
62		64
63		65
64		66
65		67
66		68
67		69
68		70
69		71
70		72
71		73
72		74
	N	75
	N	76

CHAPTER 31

4th Ed.	Action	5th Ed.
1		1
2		2
3		3
4		4
5		5
6		6
7		7
8		8
9		9
10		10
11		11
12		12
13		13
14		14
15		15
16		16
17		17
18		18
19		19
20		20
21		21
22		22
23		23
24		24
25		25
26		26
27		27
28		28
29		29
30		30
31		31
32		32
33		33
34		34
35		35
36		36
37		37
38		38
39		39
40	C	61
41		40
42		41
43		42
44		43
45		44

4th Ed.	Action	5th Ed.
46		45
47		46
48		47
49		48
	N	49
50		50
51		51
52		52
53		53
54		54
55		55
56	C	56
57		57
58		58
59		59
60		60

X = deleted C = changed N = New

COMPARISON OF PROBLEMS WITH 4th EDITION

CHAPTER 32

4th Ed.	Action	5th Ed.
1	C	1
2	C	2
3	C	3
4		4
5		5
6		6
7		7
8	C	8
9		9
10	C	10
11	C	11
	N	12
12	C	13
13		14
14		15
15		16
16		17
17		18
18		19
19		20
20		21
21		22
22		23
23		24
24		25
	N	26
25		27
26		28
27		29
28		30
29		31
30		32
31		33
32		34
33		35
34		36
35		37
36		38
37		39
38		40
39		41
40		42
41		43
42		44
	N	45
	N	46
43		47
44		48
45		49
46		50
47		51
48		52
49		53
50		54
	N	55
	N	56

CHAPTER 33

4th Ed.	Action	5th Ed.
1		1
2		2
	N	3
3		4
4		5
5		6
6		7
7		8
	N	9
8		10
9		11
10		12
11		13
12		14
13	X	
14		15
15		16
16		17
17		18
18		19
19		20
20		21
21		22
22		23
23		24
24		25
25		26
26		27
27		28
28		29
29		30
30		31
31		32
32		33
33		34
34		35
35		36
36		37
37		38
38		39
39		40
40	C	41
41		42
42		43
	N	44
	N	45
	N	46
	N	47
	N	48

X = deleted C = changed N = New